AF698828

Physics of Photonic Devices

Physics of Photonic Devices

Second Edition

SHUN LIEN CHUANG
Professor of Electrical and Computer Engineering
University of Illinois at Urbana-Champaign

WILEY

A JOHN WILEY & SONS, INC., PUBLICATION

Copyright © 2009 by John Wiley & Sons, Inc. All rights reserved

Published by John Wiley & Sons, Inc., Hoboken, New Jersey
Published simultaneously in Canada

No part of this publication may be reproduced, stored in a retrieval system, or transmitted in any form or by any means, electronic, mechanical, photocopying, recording, scanning, or otherwise, except as permitted under Sections 107 or 108 of the 1976 United States Copyright Act, without either the prior written permission of the Publisher, or authorization through payment of the appropriate per-copy fee to the Copyright Clearance Center, Inc., 222 Rosewood Drive, Danvers, MA 01923, (978) 750-8400, fax (978) 750-4470, or on the web at www.copyright.com. Requests to the Publisher for permission should be addressed to the Permissions Department, John Wiley & Sons, Inc., 111 River Street, Hoboken, NJ 07030, (201) 748-6011, fax (201) 748-6008, or online at http://www.wiley.com/go/permission.

Limit of Liability/Disclaimer of Warranty: While the publisher and author have used their best efforts in preparing this book, they make no representations or warranties with respect to the accuracy or completeness of the contents of this book and specifically disclaim any implied warranties of merchantability or fitness for a particular purpose. No warranty may be created or extended by sales representatives or written sales materials. The advice and strategies contained herein may not be suitable for your situation. You should consult with a professional where appropriate. Neither the publisher nor author shall be liable for any loss of profit or any other commercial damages, including but not limited to special, incidental, consequential, or other damages.

For general information on our other products and services or for technical support, please contact our Customer Care Department within the United States at (800) 762-2974, outside the United States at (317) 572-3993 or fax (317) 572-4002.

Wiley also publishes its books in variety of electronic formats. Some content that appears in print may not be available in electronic format. For more information about Wiley products, visit our web site at www.wiley.com.

Library of Congress Cataloging-in-Publication Data:

Chuang, Shun Lien.
Physics of photonic devices / Shun Lien Chuang.—2nd ed.
p. cm.
Includes bibliographical references and index.
ISBN 978-0-470-29319-5 (cloth)
1. Electrooptics. 2. Electrooptical devices. 3. Semiconductors. I. Title.
QC673.C482009
621.381'045—dc22 2008022814

Printed in Singapore

MWEP419832121225

To My Wife, Shu-Jung, with Love

Contents

Preface

Since the publication of the first edition, *Physics of Optoelectronic Devices*, by Wiley in 1995, significant advancements in the scientific field of optoelectronics, or photonics in general, have been made. The purpose of this new edition is to incorporate the new device concepts and to introduce novel photonic devices developed over the past years.

The new topics covered in this edition include a brief history on the invention of semiconductor lasers, the Lorentz dipole model and metal plasma, matrix optics, surface plasma waveguides, and optical ring resonators. Surface plasmonics and microring resonators have emerged as a new field of research for near-field imaging and biophotonics sensing applications. Therefore, we include them in the new edition.

On the generation of light, quantum dots have been researched in the past decade for applications to semiconductor lasers and nanophotonics applications. I include the theory of optical absorption in quantum dots and quantum wires, and their applications to semiconductor lasers. The sections on DFB lasers and VCSELs are revised with a more compact analysis with numerical examples. Novel microcavity and photonic crystal lasers, quantum-cascade lasers, and GaN blue-green lasers are exciting research subjects, and they are discussed within the context of advanced semiconductor lasers.

High-speed modulation of quantum-well and quantum-dot lasers, electrical and optical modulations, relative-intensity noise, and integrated electroabsorption modulator-laser (EML) play important roles in optical communications. They are presented in the section on the modulation of light.

Solar cells have played an important role in clean energy for the environment, and discussion of III–V *p-n* junction based solar cells has been added.

INTENDED AUDIENCE

The book is intended as a textbook for senior undergraduate and graduate students in the areas of optics and photonics. Chapters 5–8 and 13 on propagation of light and electrooptical modulators can be used independently for students in the optics and electromagnetics areas for undergraduate seniors. Chapters 2–4, 9–12, and 14–15 on semiconductor band structures, semiconductor lasers, electroabsorption modulators, photodetectors, and solar cells will be useful to graduate students as well as professionals in the photonics and optoelectronics community. This book will be useful for researchers and graduate students in physics, electrical engineering, mechanical engineering, and material science. The book covers strained quantum wells and quantum dots, semiconductor band structures, and optoelectronic device physics. Most of the required formulations are shown in great detail. Selective experimental results are included.

ACKNOWLEDGMENTS

I am indebted to many colleagues and students for their collaboration and technical discussions during the preparation of the manuscript. During the past years, I have had interactions with many colleagues. I am especially grateful to Professor Nick Holonyak, Jr., for his guidance, encouragement, and inspiration through coffee hours. I collaborate extensively with many colleagues and would like to thank them for their technical discussions, challenges, and contributions on many research ideas and projects: Professors Connie Chang-Hasnain, Paul D. Coleman, K. Y. Cheng, Russell Dupuis, Cun-Zheng Ning, S. H. Park, Hailing Wang, Weng Wang, Ming Wu, and Peidong Yang. I also benefited from technical discussions with Professors Yasuhiko Arakawa, Dieter Bimberg, Peter Blood, Yong-Hee Lee and Dr. Mitsuru Sugawara. I would like to thank my former and current students, postdocs, and visitors for their technical contributions to many publications in the field. Special thanks to Doyeol Ahn, Wei-chiao Fang, Matt Fisher, Alan Hsu, X. Jin, Tom Keating, Jungho Kim, Piotr Kondratko, Donghan Lee, Maytee Letteramb, Guobin Liu, Jeff Minch, and Jean-Francois Seurin for their recent contributions.

In preparation of the revised edition, I am indebted to Shu-Wei Chang for proofreading the entire manuscript. My group members, Guoen Chang, Adrian C. Y. Ni, T. R. Lin, Jian Li, Akira Matsudaira, Shin Mou, David Nielsen, and Adam Petschke, have contributed technically to the preparation of the figures and proofreading. I also thank Kelly C. Voyles for typing most of the chapters. I would like to express my gratitude to the students whose enthusiastic response and feedback help with my presentation of the concepts.

I am deeply indebted to my mother and brother for their unconditional support. I am grateful to my wife, Shu-Jung, for her love, dedication, and for inspiration from her poetic writing. Without her constant support, this work would have been impossible. I thank my children, Kendall, Kanglin, and Kangway, who have been a constant joy to our lives.

SHUN LIEN CHUANG

May 2008
Urbana-Champaign, Illinois

Preface to the First Edition

This textbook is intended for graduate students and advanced undergraduate students in electrical engineering, physics, and material science. It also provides an overview of the theoretical background for professional researchers in optoelectronic industries and research organizations. This book deals with the fundamental principles in semiconductor electronics, physics, and electromagnetics, and then systematically presents practical optoelectronic devices including semiconductor lasers, optical waveguides, directional couplers, optical modulators, and photodetectors. Both bulk and quantum-well semiconductor devices are discussed. Rigorous derivations are presented and the author attempts to make the theories self-contained.

Research on optoelectronic devices has been advancing rapidly. To keep up with the progress in optoelectronic devices, it is important to grasp the fundamental physical principles. Only through a solid understanding of the fundamental physics are we able to develop new concepts and design novel devices with superior performances. The physics of optoelectronic devices is a broad field with interesting applications based on electromagnetics, semiconductor physics, and quantum mechanics.

I have developed this book for a course on optoelectronic devices which I have taught at the University of Illinois at Urbana-Champaign during the past ten years. Many of our students are stimulated by the practical applications of quantum mechanics in semiconductor optoelectronic devices because many quantum phenomena can be observed directly using artificial materials such as quantum-well heterostructures with absorption or emission wavelengths determined by the quantized energy levels.

SCOPE

This book emphasizes the theory of semiconductor optoelectronic devices. Comparisons between theoretical and experimental results are also shown. The book starts with the fundamentals, including Maxwell's equations, the continuity equation, and the basic semiconductor equations of solid-state electronics. These equations are essential in learning semiconductor physics applied to optoelectronics. We then discuss the *propagation, generation, modulation, and detection of light*, which are the keys to understanding the physics behind the operation of optoelectronic devices. For example, knowledge of the generation and propagation of light is crucial for understanding how a semiconductor laser operates. The theory of gain coefficient of semiconductor lasers shows how light is amplified, and waveguide theory shows how light is confined to the waveguide in a laser cavity. An understanding of the modulation of light is useful in designing optical switches and modulators.

The absorption coefficient of bulk and quantum-well semiconductors demonstrates how light is detected and leads to a discussion on the operating principles of photodetectors.

FEATURES

- Important topics such as semiconductor heterojunctions and band structure calculations near the band edges for both bulk and quantum-well semiconductors are presented. Both Kane's model assuming parabolic bands and Luttinger-Kohn's model with valence-band mixing effects in quantum wells are presented.
- Optical dielectric waveguide theory is discussed and applied to semiconductor lasers, directional couplers, and electrooptic modulators.
- Basic optical transitions, absorption, and gain are discussed with the time-dependent perturbation theory. The general theory for gain and absorption is then applied to studying interband and intersubband transitions in bulk and quantum-well semiconductors.
- Important semiconductor lasers such as double-heterostructure, stripe-geometry gain-guided semiconductor lasers, quantum-well lasers, distributed feedback lasers, coupled laser arrays, and surface-emitting lasers are treated in great detail.
- High-speed modulation of semiconductor lasers using both linear and nonlinear gains is investigated systematically. The analytical theory for the laser spectral linewidth enhancement factor is derived.
- New subjects such as the theories on the band structures of strained semiconductors and strained quantum-well lasers are investigated.
- The electroabsorptions, in bulk (Franz-Keldysh effects) and quantum-well semiconductors (quantum confined Stark effects), are discussed systematically including exciton effects. Both the bound and continuum states of excitons using the hydrogen atom model are discussed.
- Intersubband transitions in quantum wells, in addition to conventional interband absorptions for far-infrared photodetector applications, are presented.

ACKNOWLEDGMENTS

After receiving a rigorous training in my Ph.D. work on electromagnetics at Massachusetts Institute of Technology, I became interested in semiconductor optoelectronics because of recent development in quantum-well devices with many applications of wave mechanics. I thank Professor J. A. Kong, my Ph.D. thesis adviser, and many of my professors for their inspirations and insight.

Because of the significant number of research results appearing in the literature, it is difficult to list all of the important contributions in the field. For a textbook, only the fundamental principles are emphasized. I thank those colleagues who granted me permission to reproduce their figures. I apologize to all of my colleagues whose important contributions have not been cited. I am grateful to many colleagues and friends in the field, especially D. A. B. Miller, W. H. Knox, M. C. Nuss, A. F. J. Levi, J. O'Gorman, D. S. Chemla, and the late S. Schmitt-Rink, with whom I had many stimulating discussions on quantum-well physics during and after my sabbatical leave at AT&T Bell Laboratories. I would also like to thank many of my students who provided valuable comments, especially C. S. Chang and W. Fang, who proofread the manuscript. I thank many of my research assistants, especially D. Ahn, C. Y. P. Chao, and S. P. Wu, for their interactions on research subjects related to this book. The support of my research on quantum-well optoelectronic devices by the Office of Naval Research during the past years is greatly appreciated. I am grateful to L. Beck for reading the whole manuscript and Kelly C. Voyles for typing many revisions of the manuscript in the past years. The constant support and encouragement of my wife, Shu-Jung, are deeply appreciated. Teaching and conducting research have been the stimulus for writing this book; it was an enjoyable learning experience.

SHUN LIEN CHUANG

Illinois, March 1995

1

Introduction

Semiconductor photonic devices such as laser diodes, light-emitting diodes, optical waveguides, directional couplers, electrooptic modulators, and photodetectors have important applications in lightwave technology systems. To understand the physics and the operational characteristics of these photonic devices, we have to understand the fundamental principles. In this chapter, we review some of the basic concepts of semiconductor electronics, provide a brief history of the invention of semiconductor lasers and light-emitting diodes, review the general field of optoelectronics, then present the overview of this book.

1.1 BASIC CONCEPTS OF SEMICONDUCTOR BAND AND BONDING DIAGRAMS

The basic idea is that for a semiconductor, such as GaAs or InP, many interesting optical properties occur near the band edges. For example, Table 1.1 shows part of the periodic table with many of the elements that are important for semiconductors [1], including group IV, III–V, and II–VI compounds. For a III–V compound semiconductor such as GaAs, the gallium (Ga) and arsenic (As) atoms form a zinc-blende structure, which consists of two interpenetrating face-centered-cubic lattices, one made of gallium atoms and the other made of arsenic atoms (Fig. 1.1). The Ga atom has an atomic number of 31, which has an [Ar] $3d^{10}4s^{2}4p^{1}$ configuration; that is, three valence electrons on the outermost shell (4*s* and 4*p* states). (Here [Ar] denotes the configuration of Ar, which has an atomic number of 18, and the 18 electrons are distributed as $1s^{2}2s^{2}2p^{6}3s^{2}3p^{6}$.)

The As atom has an atomic number of 33 with an [Ar] $3d^{10}4s^{2}4p^{3}$ configuration or five valence electrons in the outermost shell (4*s* and 4*p* states). For a simplified view, we show a planar bonding diagram [2, 3] in Fig. 1.2a, where each bond between two nearby atoms is indicated with two dots representing two *valence electrons*. These valence electrons are contributed by either Ga or As atoms. The bonding diagram shows that each atom such as Ga is connected to four nearby As atoms by four valence bonds or eight valence electrons. If we assume that none of the bonds is broken, then all of the electrons are in the valence band, and no free electrons are

---

Physics of Photonic Devices, Second Edition. By Shun Lien Chuang
Copyright © 2009 John Wiley & Sons, Inc.

Table 1.1 Part of the Periodic Table Containing Group II–VI Elements

Group II		Group III		Group IV		Group V		Group VI	
A	B	A	B	A	B	A	B	A	B
4 Be $1s^22s^2$		5 B $1s^22s^22p^1$		6 C $1s^22s^22p^2$		7 N $1s^22s^22p^3$		8 O $1s^22s^22p^4$	
12 Mg [Ne] $3s^2$		13 Al [Ne] $3s^23p^1$		14 Si [Ne] $3s^23p^2$		15 P [Ne] $3s^23p^3$		16 S [Ne] $3s^23p^4$	
20 Ca [Ar] $4s^2$			21 Sc [Ar] $3d^14s^2$		22 Ti [Ar] $3d^24s^2$		23 V [Ar] $3d^34s^2$		24 Cr [Ar] $3d^54s^1$
	30 Zn [Ar] $3d^{10}4s^2$	31 Ga [Ar] $3d^{10}4s^24p^1$		32 Ge [Ar] $3d^{10}4s^24p^2$		33 As [Ar] $3d^{10}4s^24p^3$		34 Se [Ar] $3d^{10}4s^24p^4$	
38 Sr [Kr] $5s^2$			39 Y [Kr] $4d^15s^2$		40 Zr [Kr] $4d^25s^2$		41Nb [Kr] $4d^45s^1$		42 Mo [Kr] $4d^55s^1$
	48 Cd [Kr]$4d^{10}5s^2$	49 In [Kr] $4d^{10}5s^25p^1$		50 Sn [Kr] $4d^{10}5s^25p^2$		51 Sb [Kr] $4d^{10}5s^25p^3$		52 Te [Kr] $4d^{10}5s^25p^4$	
56 Ba [Xe] $6s^2$									
	80 Hg [Xe]$4f^{14}5d^{10}6s^2$								

Note: [Ne] = $1s^22s^22p^6$; [Ar] = [Ne] $3s^23p^6$; [Kr] = [Ar] $3d^{10}4s^24p^6$; [Xe] = [Kr] $4d^{10}5s^25p^6$.

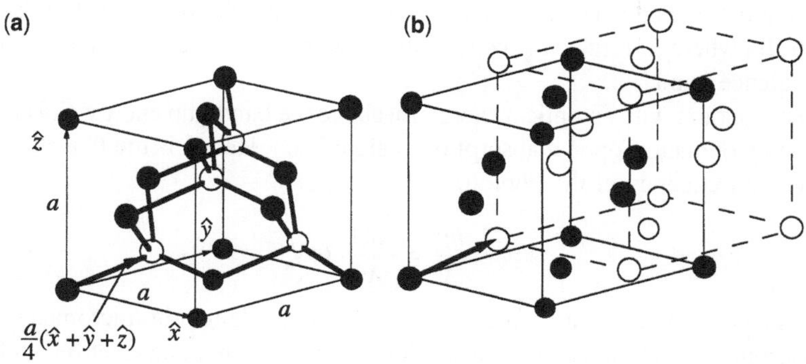

Figure 1.1 (a) A zinc-blende structure such as those of GaAs and InP semiconductors. (b) The zinc-blende structure consists of two interpenetrating face-centered-cubic lattices separated by a constant vector $(a/4)(\hat{x}+\hat{y}+\hat{z})$, where a is the lattice constant of the semiconductor.

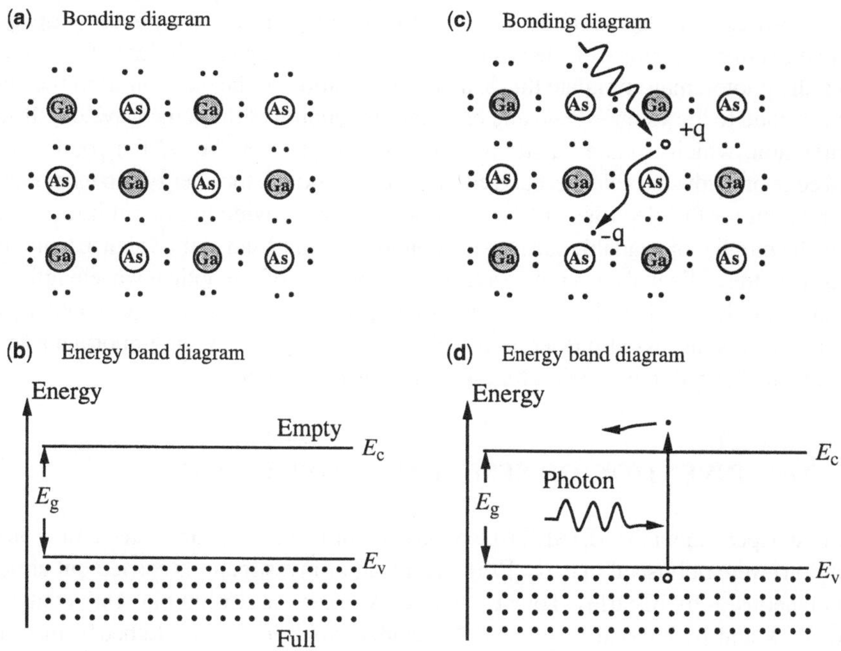

Figure 1.2 (a) A planar bonding diagram for a GaAs lattice. Each bond consists of two valence electrons shared by a gallium and an arsenic atom. (b) The energy band diagram in real space shows the valence-band edge E_v below which all states are occupied and the conduction-band edge E_c above which all states are empty. The separation $E_c - E_v$ is the band gap E_g. (c) A bonding diagram showing a broken bond due to the absorption of a photon with energy above the band gap. A free electron–hole pair is created. Note that the photogenerated electron is free to move around, and the hole is also free to hop around at different bonds between the Ga and As atoms. (d) The energy band diagram showing the energy levels of the electron and the hole.

in the conduction band. The energy band diagram as a function of position is shown in Fig. 1.2b, where E_c is the band edge of the conduction band and E_v is the band edge of the valence band.

When a photon with an optical energy $h\nu$ above the band-gap energy E_g is incident on the semiconductor, optical absorption is significant. Here h is the Planck constant and ν is the frequency of the photon,

$$h\nu = \frac{hc}{\lambda} = \frac{1.24}{\lambda} \quad (\text{eV}) \tag{1.1.1}$$

where c is the speed of light in free space, and λ is wavelength in micrometers (μm). The absorption of a photon may break a valence bond and create an electron–hole pair, shown in Fig. 1.2c, where an empty position in the bond is represented by a hole. The same concept in the energy band diagram is illustrated in Fig. 1.2d, where the free electron propagating in the crystal is represented by a dot in the conduction band. It is equivalent to acquiring an energy larger than the band gap of the semiconductor, and the kinetic energy of the electron is that amount above the conduction-band edge. The reverse process can also occur if an electron in the conduction band recombines with a hole in the valence band; this excess energy may emerge as a photon, and the process is called spontaneous emission. In the presence of a photon propagating in the semiconductor with electrons in the conduction band and holes in the valence band, the photon may stimulate the downward transition of the electron from the conduction band to the valence band and emit another photon of the same wavelength and polarization, which is called a stimulated emission process. Above the conduction-band edge or below the valence-band edge, we have to know the energy versus momentum relation for the electrons or holes. These relations provide important information about the number of available states in the conduction band and in the valence band. By measuring the optical absorption spectrum as a function of the optical wavelength, we can map out the number of states per energy interval. This concept of joint density of states, which is discussed further in the following chapters, plays an important role in the optical absorption and gain processes in semiconductors.

1.2 THE INVENTION OF SEMICONDUCTOR LASERS

The first papers about the MASER (microwave amplification by stimulated emission of radiation) were published in early 1951 as a result of investigations carried out almost simultaneously by Charles Townes and co-workers at Columbia University in New York and by Nikolai Basov and Alexander Prokhorov at the Lebedev Institute in Moscow [4]. The experimental demonstration of the maser was realized in 1954 using ammonia gas at 23,870 MHz, which was utilized as a frequency standard for some years. The concept of LASER (light amplification by stimulated emission of radiation) was proposed during 1958 to 1960 (called optical maser initially) first by Arthur Schawlow and Charles Townes [5] and by other groups [6]. There was a 30-year patent war that ended with the award of a few laser patents to Gordon Gould [6], who was the first person to use the word *laser*. The first experimental demonstration was realized in 1960 by Theodore Maiman [7] who designed a three-level ruby laser pumped by high-power flashes of intense light. It was then followed by the invention of helium–neon (HeNe) gas lasers by Ali Javan et al. [8] at Bell Laboratories in 1960.

At the Solid State Device Research Conference in July 1962, an MIT Lincoln Laboratory group and RCA Laboratories reported extremely high efficiency (85% to 100%) electroluminescence from GaAs diffused junction diodes. Semiconductor lasers were invented during September to October 1962 by four groups within 30 days [9–12] (see the review article by Dupuis in Ref. 13). They were led by Robert N. Hall of General Electric Research Development Center, Schenectady, New York; Nick Holonyak Jr. of General Electric, Syracuse, New York; Marshall I. Nathan of the IBM Research Laboratory, Yorktown Heights, New York; and Robert Rediker of the MIT Lincoln Laboratory, Lexington, Massachusetts. Among the four groups, only Holonyak's laser diodes and light-emitting diodes (LEDs) were created from single-crystal $GaAs_xP_{1-x}$ alloy material grown by vapor-phase transport and were the only devices emitting in the visible region. This work was also the beginning of band-gap engineering of ternary compound semiconductors beyond the binary compounds. The other three groups used zinc (Zn) diffused GaAs *p-n* junction emitting in the infrared. Figure 1.3 shows the evolution of visible-spectrum LEDs in terms of the electrical to optical power conversion (lumens/watt) since 1962 [14–16]. Over the past decades, remarkable progress in the performance of LEDs using compound semiconductors has been realized. Improvement of three orders of magnitude in power conversion efficiency has been achieved, hence the term *Alloy Road* (based on various compound semiconductors as opposed to single elements such as silicon

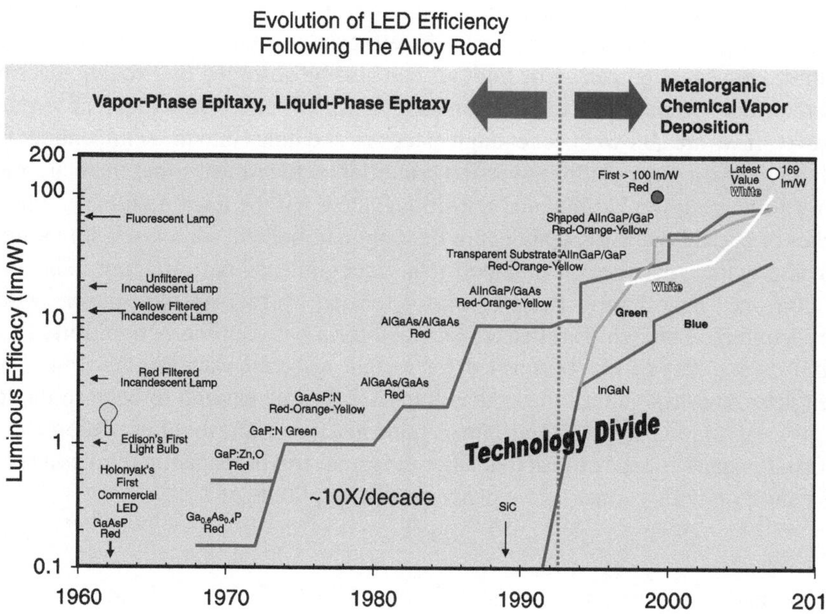

Figure 1.3 The evolution of the electrical to optical power conversion of visible LEDs or the *Alloy Road* based on compound semiconductors. The vertical axis labels the performance (Lumens/Watt) of LEDs starting from Holonyak's first commercial GaAsP LED. The performance of Edison's first light bulb is also shown. (Reprinted with permission from [16] © 2008 IEEE.)

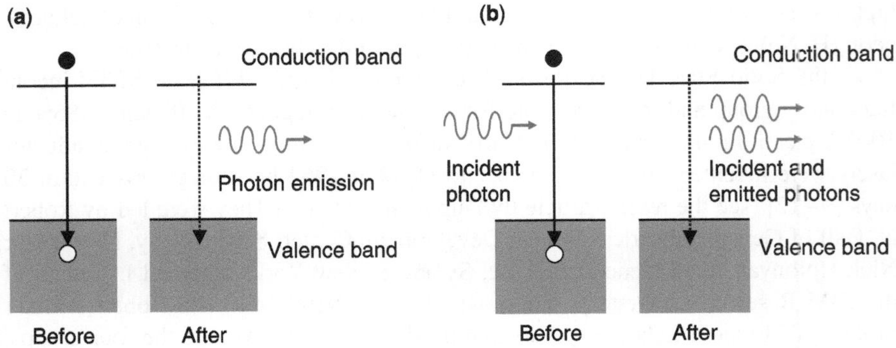

Figure 1.4 The electron occupation in the conduction and valence bands of a semiconductor. (a) Spontaneous emission occurs when an electron in the conduction band recombines with a hole in the valence band. (b) Stimulated emission occurs when an incident photon stimulates the recombination of an electron–hole pair and generates another photon of the same energy (and polarization).

and germanium) shown in Fig. 1.3 was coined after the presentation of LumiLeds [14–16]. More recently, the realization of the blue and green LEDs based on InGaN and their power conversion started to take off in the early 1990s.

An electron in the conduction band can recombine with a hole (or empty state) in the valence band and emit a photon close to the band-gap energy—a process called spontaneous emission or more commonly called radiative recombination (Fig. 1.4a). When a photon is propagating in the semiconductor with electrons in the conduction band and holes in the valence band, a stimulated emission process can cause the number of photons to increase. Figure 1.4b shows the electron occupation before and after the stimulated emission process. The photons can also be absorbed by exciting electrons from the valence band to the conduction band, a process called *stimulated* absorption. However, if we are able to inject enough electrons and holes into the semiconductor to reach the so-called population inversion condition, which means that there are more downward than upward stimulated transitions, there will be a net gain of the photon number or optical intensity. Gain is not the only requirement for a laser. It requires a resonator, which can be a one-, two-, or three-dimensional structure. The most common one is the Fabry–Perot resonator formed by two parallel mirrors with a cavity length L. The light is reflected back and forth between the two mirrors, thus a standing wave pattern can be formed for certain resonant wavelengths (Fig. 1.5a). When the round-trip gain of the optical intensity is large enough to balance the loss due to waveguide absorption and mirror transmission, a threshold condition can be reached. It means that the optical field after the round-trip propagation reaches a resonance condition with a constructive phase and an amplitude of 1,

$$r_1 r_2 e^{i2kL+(G-\alpha)L} = 1 \tag{1.2.1}$$

where r_1 and r_2 are the reflection coefficients of the optical fields from the two end facets, k is the propagation constant,

$$k = 2\pi n/\lambda = 2\pi \nu n/c, \tag{1.2.2}$$

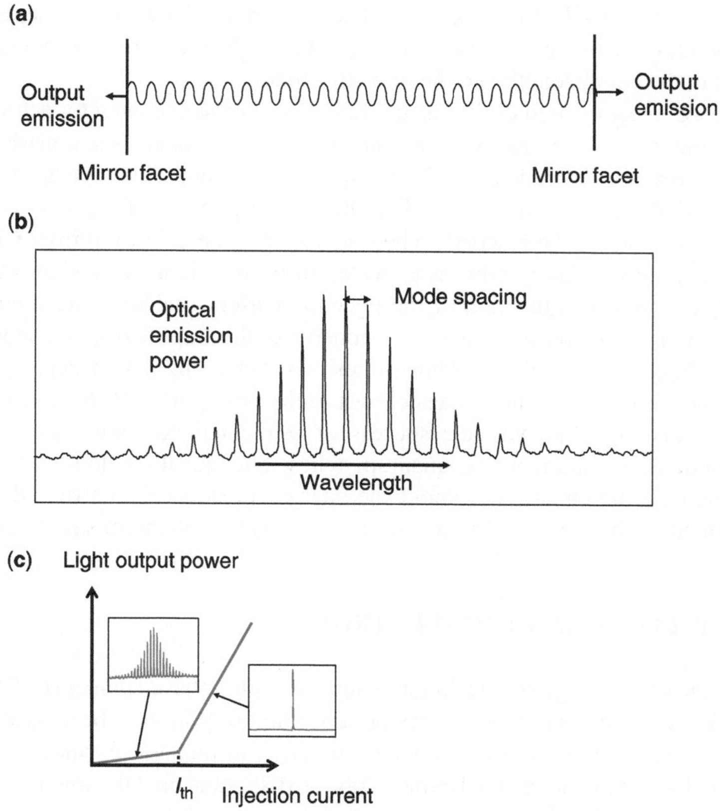

Figure 1.5 (a) Standing wave in a Fabry–Perot cavity. (b) Amplified spontaneous emission (ASE) spectrum from a Fabry–Perot cavity. (c) The light output power of a laser diode as a function of the injection current. The inset shows the ASE spectrum below threshold and the lasing spectrum above threshold. I_{th} indicates the threshold current.

and n is the refractive index of the semiconductor. G is the modal gain coefficient of the guided optical mode in the semiconductor waveguide, and α is the absorption coefficient. Equation (1.2.1) leads to the phase and magnitude conditions for lasing,

$$2kL = 2m\pi \tag{1.2.3}$$

$$G = \alpha + \frac{1}{2L}\ln\left(\frac{1}{R_1 R_2}\right) \tag{1.2.4}$$

where $R_1 = |r_1|^2$ and $R_2 = |r_2|^2$ are the power reflectivities. The phase condition (1.2.3) leads to the Fabry–Perot resonance spectrum

$$\nu_m = \frac{mc}{2nL} \qquad m = \text{integer}. \tag{1.2.5}$$

If we ignore the dispersion of the semiconductor (i.e., assuming that n is independent of the frequency), the mode spacing is given by $\Delta\nu = c/(2nL)$, which is inversely proportional to the cavity length and the refractive index.

When we inject enough electrons and holes into the semiconductor active region such that the gain of the propagation mode becomes significant, the emission of the photons gives the so-called amplified spontaneous emission spectrum (ASE), Fig. 1.5b. If the gain is large enough to balance the losses of the cavity, (1.2.4), the laser threshold can be reached. When we increase the gain by further increasing the injection current, laser action is expected to occur. Figure 1.5c shows the laser light output power as a function of the injection current. Below a threshold current value, the light output power is small and consists of the amplified spontaneous emission from Fabry–Perot modes. Above threshold, the lasing action occurs, and the optical power comes from the modes closest to the peak gain. The field of optoelectronics became possible with the success of semiconductor lasers using various heterojunctions, quantum wells, quantum wires, and quantum dots such that the optical gain can be enhanced without the energy spread of the carriers due to the improvement of the density of states in low-dimensional quantum structures.

1.3 THE FIELD OF OPTOELECTRONICS

Semiconductor crystal growth techniques such as liquid-phase epitaxy (LPE), vapor-phase epitaxy (VPE), and chemical-vapor deposition (CVD) have been used to grow wafers for device applications. The recent progress in modern crystal growth techniques [17] such as molecular beam epitaxy (MBE) and metal–organic chemical vapor deposition (MOCVD) has demonstrated that it is possible to grow semiconductors of different atomic compositions on top of another semiconductor substrate with monolayer precision. This opens up extremely exciting possibilities of the so-called band-gap engineering. For example, aluminum arsenide (AlAs) has a similar lattice constant to that of gallium arsenide (GaAs). One can grow a few atomic layers of AlAs on top of a GaAs substrate, then grow alternate layers of GaAs and AlAs. One can also grow a ternary compound such as $Al_xGa_{1-x}As$ (where the aluminum mole fraction x can be between 0 and 1) on a GaAs substrate and form a heterojunction, Fig. 1.6a. Interesting applications have been found using heterojunction structures. For example, when the wide band-gap $Al_xGa_{1-x}As$ is doped by donors, the free electrons from the ionized donors tend to fall to the conduction band of the GaAs region because of the lower potential energy on that side, as shown by the band diagram in Fig. 1.6b. (This band bending is investigated in Chapter 2.) An applied field in a direction parallel to the junction interface will create conduction current. Because these electrons conduct in a channel on the GaAs region, which is undoped, the amount of impurity scattering can be reduced. Therefore, the electron mobility can be enhanced. Based on this concept, the high-electron-mobility transistor (HEMT) has been realized.

For optoelectronic device applications, heterojunction structures [18] play important roles. For example, when semiconductor lasers were invented, they had to be

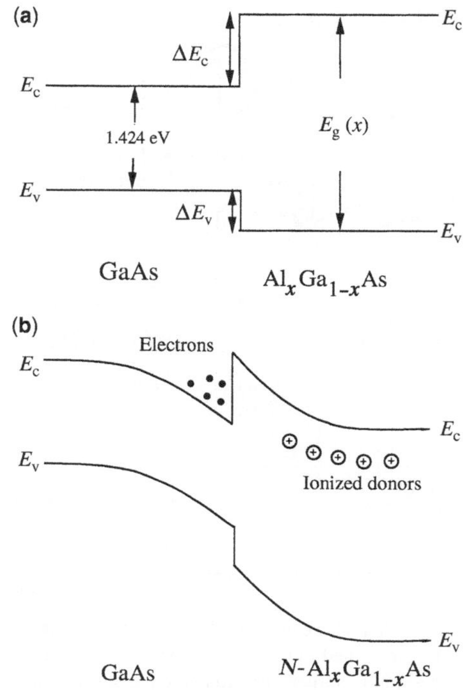

Figure 1.6 (a) A $GaAs/Al_xGa_{1-x}As$ heterojunction formed with different band gaps. The band edge discontinuities in the conduction band and the valence band are $\Delta E_c = 0.67\Delta E_g$ and $\Delta E_v = 0.33\Delta E_g$, where ΔE_g is the difference of the two band gaps. (b) With n-type doping in the wide gap $Al_xGa_{1-x}As$ region, the electrons ionized from the donors fall into the heterojunction surface layer on the GaAs side where the energy is smaller. An internal electric field pointing from the ionized (positive) donors in the $Al_xGa_{1-x}As$ region toward the electrons with negative charges creates band bending, which forms a triangular potential in the conduction band to confine the electrons.

cooled down to cryogenic temperature (77K), and the lasers could lase only in a pulsed mode. These lasers had large threshold current densities, which mean that a large amount of current has to be injected before the laser can start lasing. With the introduction of the heterojunction semiconductor lasers, the concept of carrier and photon confinements makes room temperature continuous wave (cw) operation possible, because the electrons and holes, once injected by the electrodes on both sides of the wide band-gap *P-N* regions (Fig. 1.7), will be confined in the central GaAs region, where the band gap is smaller, resulting in a smaller potential energy for the electrons in the conduction band as well as a smaller potential energy for holes in the valence band. We note that the energy for the holes is measured downward, which is opposite to that of the electrons. For the photons, it turns out that the optical refractive index of the narrow band-gap material (GaAs) is larger than that of the wide band-gap material ($Al_xGa_{1-x}As$). Therefore, the photons can be confined in the active region as well. This double confinement of both carriers and photons makes the stimulated emission process more efficient and leads to the room temperature operation of laser diodes.

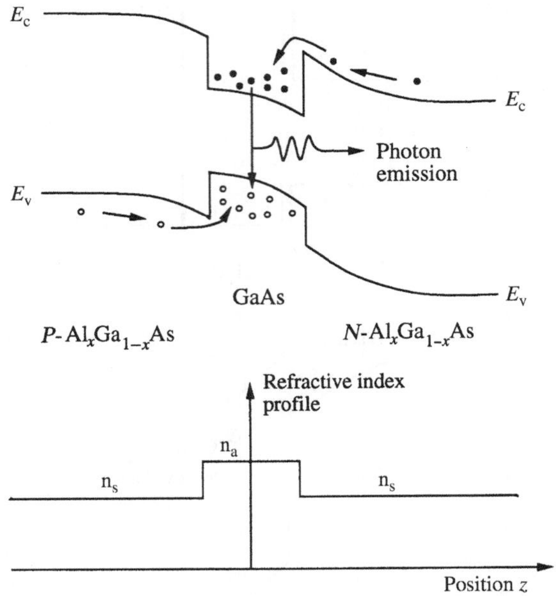

Figure 1.7 A double-heterojunction semiconductor laser structure, where the central GaAs region provides both the carrier confinement and optical confinement because of the conduction and valence band profiles and the refractive index profile. This double confinement enhances stimulated emissions and the optical modal gain.

The control of the mole fractions of different atoms also makes the band-gap engineering extremely exciting. For optical communication systems, it has been found that minimum attenuation [19] in the silica optical fibers occurs at 1.30 μm and 1.55 μm (Fig. 1.8a). The dispersion of light at 1.30 μm is actually zero (Fig. 1.8b). It is therefore natural to design sources such as light-emitting diodes and laser diodes, semiconductor modulators, and photodetectors operating at these desired wavelengths. In addition, many wavelengths, or the so-called optical channels for dense wavelength-division multiplexing (DWDM) applications, near 1550 nm with constant frequency spacing such as 50, 100, or 200 GHz can be used to take advantage of the broad 24 THz frequency bandwidth near the minimum attenuation. For example, by controlling the mole fraction of gallium and indium in an $In_{1-x}Ga_xAs$ material, a wide tunable range of band gap is possible because InAs has a 0.354 eV band gap and GaAs has a 1.424 eV band gap at room temperature. The lattice constant of the ternary alloy has a linear dependence on the mole fraction

$$a(A_xB_{1-x}C) = x\,a(AB) + (1-x)\,a(BC) \tag{1.3.1}$$

where $a(AB)$ is the lattice constant of the binary compound AB and $a(BC)$ is that of the compound BC. This linear interpolation formula works very well for the lattice constant, but not for the band gap. For the band-gap dependence, a quadratic dependence

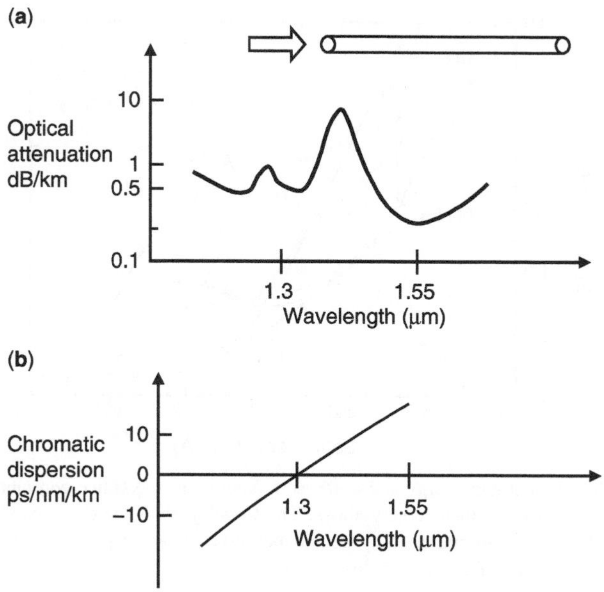

Figure 1.8 Simplified illustrations of (a) attenuation spectrum and (b) the dispersion parameter of light propagating in silica optical fiber.

on the mole fraction x is usually required (see Appendices C and D on pages 801–809 for some important material systems)

$$E_g(A_xB_{1-x}C) = x\,E_g(AB) + (1-x)\,E_g(BC) - bx(1-x) \qquad (1.3.2)$$

where b is called the bowing parameter because it causes a deviation of the ternary band-gap energy away from a linear interpolation of the two band-gap energies of the binary compounds. Figure 1.9 plots the band-gap energy at $T = 0$K as a function of the lattice constant for many binary and ternary compound semiconductors [20]. For example, GaAs has a band gap of 1.519 eV at low temperature and a lattice constant of 5.6533 Å, whereas InAs has a band gap of 0.417 eV and a lattice constant of 6.0584 Å, as indicated. A ternary $In_{1-x}Ga_xAs$ compound has the two end points at GaAs ($x = 0$) and InAs ($x = 1$) and its band gap has a slight downward bowing below a linear interpolation. At $x = 0.468$, the $In_{0.532}Ga_{0.468}As$ alloy has a lattice constant matched to that of the InP (5.8688 Å).

For $Al_xGa_{1-x}As$ ternary compounds with $0 \leq x < 0.4$, the following formula is commonly used at room temperature:

$$E_g(Al_xGa_{1-x}As) = 1.424 + 1.247x \text{ (eV)}. \qquad (1.3.3)$$

Most ternary compounds require a quadratic term because the bowing parameter is not zero. From the above formula, we can calculate the conduction and valence band edge discontinuities between a GaAs and an $Al_xGa_{1-x}As$ heterojunction using

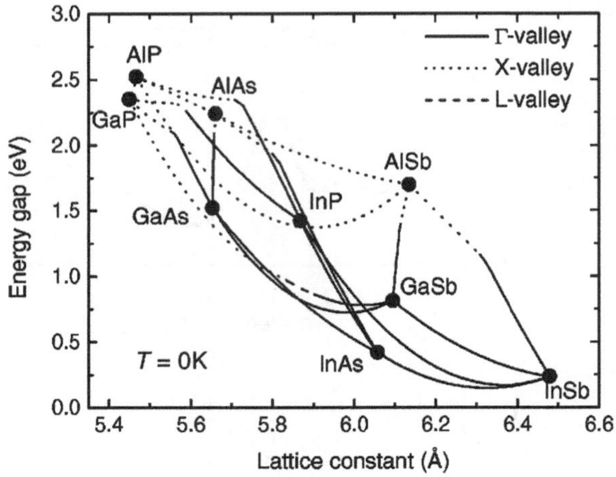

Figure 1.9 A plot of band-gap energy at $T = 0\text{K}$ as a function of the lattice constant for a few III–V compound semiconductors and their ternary alloys. The Γ-valley gives the direct band gap (solid), and the indirect X- and L-valleys are indicated by dotted and dashed lines, respectively. After [20] © 2001, American Institute of Physics [Reprint with permission].

$\Delta E_c = 0.67\Delta E_g$ and $\Delta E_v = 0.33\Delta E_g$ where $\Delta E_g = 1.247x$ (eV). When very thin layers of heterojunction structures are grown with a layer thickness thinner than the coherent length of the conduction band electrons, quantum size effects occur. These include the quantization of the subband energies with corresponding wave functions (Fig. 1.10a).

The success in the growth of quantum-well (QW) structures makes a study of the introductory quantum physics realizable in these man-made semiconductor materials [21, 22]. Because of the low-dimensional confinement of electrons and holes in the quantum structures, many electronic and optical properties differ significantly from those of the bulk materials. Many interesting quantum mechanical phenomena using quantum structures and their applications have been predicted and confirmed experimentally [22]. For a simple quantum-well potential, we have the particle in a box model. These quantized energy levels appear in the optical absorption and gain spectra with exciting applications to electroabsorption modulators, quantum-well lasers and photodetectors, because an enhanced absorption occurs when the optical energy is close to the difference between the conduction and hole subband levels, as shown in Fig. 1.10a. The density of states in the quasi-two-dimensional (2D) structure is also different from that of a bulk semiconductor. A significant discovery is that the quantum mechanical phenomena can be observed at room temperature using these structures. When an electric field is applied across the quantum-well region using a *p-n* junction diode structure, the potential energy profile is tilted and the energy levels of the quantized subbands are shifted (Fig. 1.10b). Therefore, the optical absorption spectrum can be changed by an electric field bias. This makes practical the applications of electroabsorption modulators using these quantum-well structures.

Experimental work on low threshold current quantum-well lasers [21–23] has been reported for different material systems, such as GaAs/AlGaAs, InGaAsP/InP, and InGaAs/InGaAsP. Advantages of the quantum-well lasers, such as a higher temperature

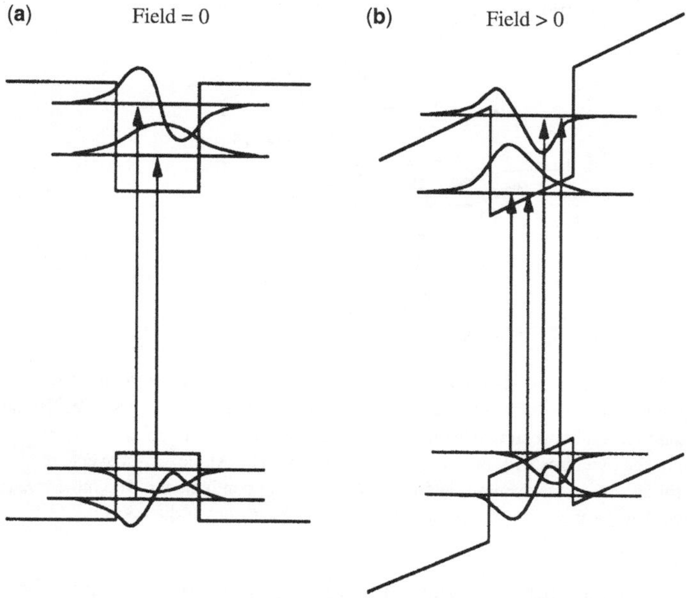

Figure 1.10 (a) A semiconductor quantum well without an applied electric field bias showing the quantized subbands and the corresponding wave functions. (b) With an applied electric field, the tilted potential shifts the quantized energy levels due to the field. The wave functions are skewed from the previous even or odd symmetric wave functions of the zero field case.

stability and an improved linewidth enhancement factor and wavelength tunability, have also been demonstrated. These devices are based on the band structure engineering concept using, for example, a separate-confinement-heterostructure (SCH) quantum-well structure to enhance the carrier and the optical confinements (Fig. 1.11).

The effect of uniaxial stress perpendicular to the junction on the threshold current of GaAs double-heterostructure lasers was studied experimentally in the 1970s. The idea of using strained quantum wells [22–24] by growing semiconductors with different lattice constants for tunable wavelength photodetectors and semiconductors has also been explored in the 1980s. For example, if a "thin" or quantum-well layer of semiconductor Q is grown on top of a thick substrate of semiconductor P, an in-plane strain due to the deformation of the lattice of semiconductor Q in order to match the in-plane lattice constant of the substrate P occurs (Fig. 1.12). If the growth axis is along the z direction, the in-plane (x–y plane) strain parallel to the quantum-well plane is defined as

$$\varepsilon_{\parallel} = \frac{a(P) - a(Q)}{a(Q)}. \tag{1.3.4}$$

If the in-plane strain is negative, that is, $a(Q)$ is larger than $a(P)$, the in-plane strain forces the lattice of the epitaxy layer Q in the x and y directions to match those of the

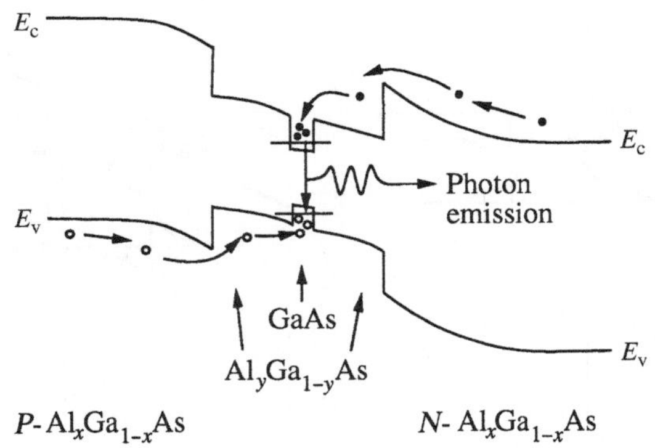

Figure 1.11 The energy band diagram of a separate-confinement heterostructure (SCH) quantum-well laser structure. The active GaAs layer, which has a dimension around 100 Å, provides the carrier confinement and is sandwiched between two $Al_yGa_{1-y}As$ layers, where the aluminum mole fraction y is smaller than those (x) of the outermost $Al_xGa_{1-x}As$ cladding regions. The $Al_yGa_{1-y}As$ layers are of the order submicrometers (or an optical wavelength) and provide the optical confinement. The mole fraction y can also be graded such that it varies with the position along the crystal growth direction.

substrate P, and the strain is called biaxially compressive. On the other hand, the strain is called biaxially tensile if $a(Q)$ is smaller than $a(P)$, because a tension exists along the x and y directions in order to "stretch" the lattice of the epilayer Q to match the lattice of the substrate P. For example, if we grow a thin $In_{1-x}Ga_xAs$ layer on top of InP substrate, we will have an in-plane strain of

$$\varepsilon_{\|} = \frac{a(\mathrm{InP}) - a(\mathrm{In}_{1-x}\mathrm{Ga}_x\mathrm{As})}{a(\mathrm{In}_{1-x}\mathrm{Ga}_x\mathrm{As})} \simeq \frac{a(\mathrm{InP}) - a(\mathrm{In}_{1-x}\mathrm{Ga}_x\mathrm{As})}{a(\mathrm{InP})} \tag{1.3.5}$$

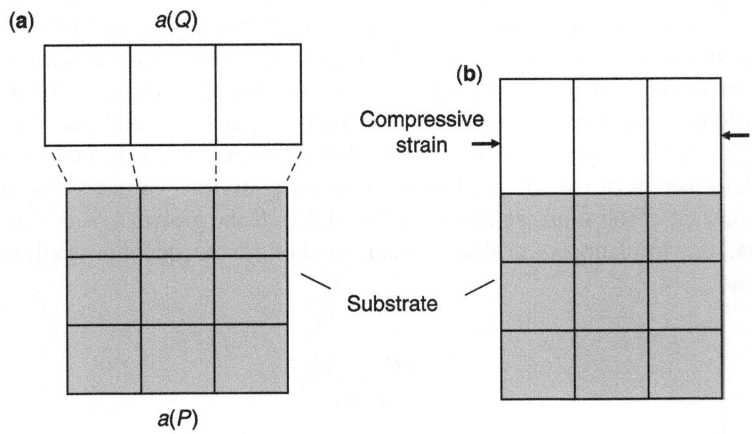

Figure 1.12 Growth of a semiconductor layer of lattice constant $a(Q)$ on top of a substrate with a lattice constant $a(P)$ before (a) and after (b) growth with a coherent strain deformation that forces the in-plane lattice constant of the layer Q equal to that of the substrate. Therefore, an in-plane strain given by (1.3.4) is induced.

At $x = 0.468$, we have zero strain (the so-called lattice-matched condition). When $x < 0.468$, we have an indium-rich compound and the in-plane strain is compressive in nature because the InGaAs lattice constant is larger than that of InP. The ternary InGaAs lattice will be deformed elastically in the x–y plane to match the lattice of the InP substrate, and an in-plane compressive strain exists with a corresponding tension along the growth (z) axis.

Strained-layer quantum-well lasers have been investigated for low threshold current operation, polarization switching, and bistability applications [23–28]. Important advantages using a strained-layer superlattice or quantum wells include the reduction of the threshold current density due to the lowering of the light-hole band relative to the heavy-hole band, the reduction of the in-plane heavy-hole effective mass, the elimination of intervalence band absorption, and the reduction of Auger recombination [25, 26]. Because of the selection rule for optical transitions, the polarization-dependent gains are also changed by the strain because the optical gain is mainly transverse magnetically (TM) polarized for the transition between the electron and the light-hole bands and transverse electrically (TE) polarized for the transition between the electron and the heavy-hole bands [27, 28]. Many of these details for valence subband electronic properties and polarization selection rules in quantum-well devices will be explained in Chapters 4, 9, and 10.

The next major breakthrough in crystal growth [29–31] was the realization of semiconductor quantum dots, which confine electrons and holes in a three-dimensional structure. The quantization of electron energy in all three directions makes the energy levels discrete instead of being continuous. Therefore, applications to devices such as quantum-dot semiconductor lasers, similar to gas lasers with a discrete density of states, become a reality. The reason is that the injected electrons and holes will occupy a discrete density of states; therefore, population inversion for laser action is easier to achieve at a lower carrier density.

A layer of material having a lattice constant different from that of the substrate may transform spontaneously to an array of three-dimensional islands after some critical thickness is deposited. The evolution of an initially two-dimensional growth into a three-dimensional corrugated growth front has been observed in many systems. Actually, Stranski and Krastanow (SK) initially proposed the possibility of island formation on an initially flat heteroepitaxial surface for the growth of *lattice matched* ionic crystals that had different charges [29]. It was then discovered that the growth of islands relaxed by misfit dislocations in strained heteroepitaxy is possible. The formation of coherent or defect-free islands as a result of SK growth of strained heterostructures becomes an important method of growing quantum dots. However, there remain challenges to grow uniform quantum dots due to the self-assembled or spontaneous formation process of the strained dots. We will discuss quantum-dot lasers in Chapters 9 and 10.

1.4 OVERVIEW OF THE BOOK

This book is divided into five parts: I, Fundamentals; II, Propagation of Light; III, Generation of Light; IV, Modulation of Light; and V, Detection of Light and Solar

Cells. We start with the fundamentals on semiconductor electronics, quantum mechanics, solid-state physics, and electromagnetics, with the emphasis on their applications to optoelectronic devices. In Fig. 1.13, we illustrate the important fundamental equations and their applications. In the presence of injection of electrons and holes using a current bias or an optical source, the semiconductor materials may change from being absorptive to gain media due to population inversion effect. This implies that the optical dielectric function is also changed. This change can be modeled with the knowledge of the electronic band structures, which require the solutions of the Schrödinger equation or the so-called effective-mass equation for the given bulk or quantum-well semiconductors. By solving Maxwell's equations, we obtain the optical field from the dielectric function of the semiconductors. The electronic band structure is also dependent on the static electric bias voltage, which determines the electron and hole current densities.

The semiconductor electronic equations governing the electron and hole carrier concentrations and their corresponding current densities have to be solved. The device operation characteristics such as the current–voltage relation in a *p-n* junction structure, the external quantum efficiency for the conversion of electric to optical power in a semiconductor laser, or the quantum efficiency for converting optical power to current in a photodetector have to be investigated using these semiconductor electronic equations. These fundamental equations are coupled to each other, and the most complete solution would require a self-consistent scheme, which would require

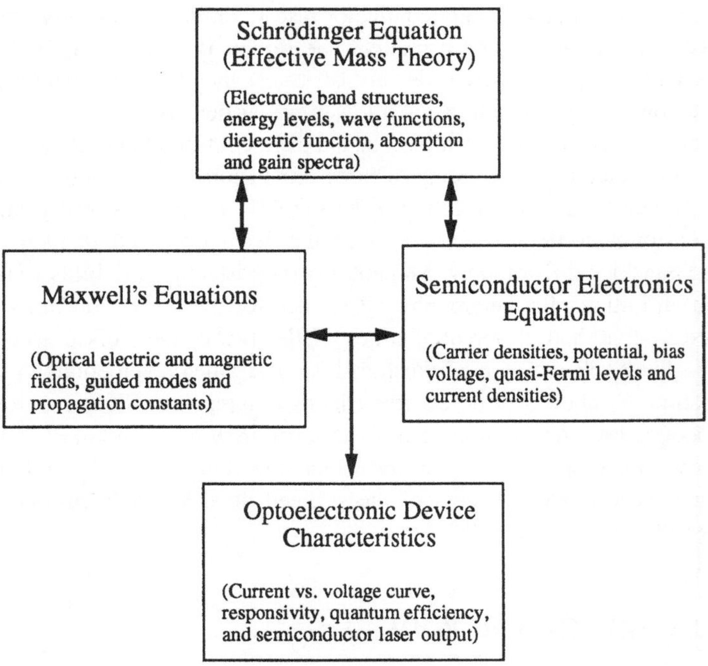

Figure 1.13 Fundamental equations and their applications to optoelectronic device characteristics.

heavy computations. Fortunately, with good understandings of most of the device physics, various approximation methods, such as the depletion approximation and perturbation theories for various device operation conditions, are possible. The validity of the models can be checked with full numerical solutions and confirmed with experimental observations. On the basic semiconductor electronics (Chapter 2), we investigate questions such as: How are the energy band diagrams drawn for heterojunctions such as *P-n*, *N-p*, *p-N*, or *n-P* junctions? (Here a capital letter such as *P* refers to a wide band-gap material doped *P* type and a small *p* refers to a smaller band-gap semiconductor doped *p* type.) We then review the basic quantum mechanics (Chapter 3) and theory of electronic band structures in semiconductors (Chapter 4).

In Part II, Propagation of Light, we review electromagnetics and light wave propagation (Chapter 5), anisotropic media and radiation (Chapter 6), optical waveguide theory (Chapter 7), and coupled-mode theory (Chapter 8). The optical waveguide plays an important role not only in guiding the light but also functions as a gain medium where the electrons and holes are confined in a double heterojunction laser structure, where the two end facets of the waveguide serve as reflection mirrors, Fig. 1.14a. Surface plasma waveguides, which have been intensively researched recently, are discussed. We then describe the coupled-mode theory and its applications to parallel waveguide directional couplers and optical ring resonators as add-drop filters. We also investigate the coupled waves that propagate in opposite directions in a distributed feedback structure. The distributed feedback (DFB) structure has important applications in a single-wavelength semiconductor laser for telecommunication applications, Fig. 1.14b.

In Part III, Generation of Light, we start with the fundamental optical processes in semiconductors (Chapter 9) and present the theories for optical gain and absorption in semiconductor bulk, quantum-well, and quantum-dot structures. We then discuss semiconductor lasers in Chapters 10 and 11. We cover quantum-well and quantum-dot Fabry–Perot lasers and distributed-feedback lasers. We also cover more advanced structures such as vertical-cavity surface-emitting lasers (VCSELs), photonic crystal lasers, and quantum-cascade lasers in the mid-infrared (3–12 μm) and terahertz (THz) wavelength ranges.

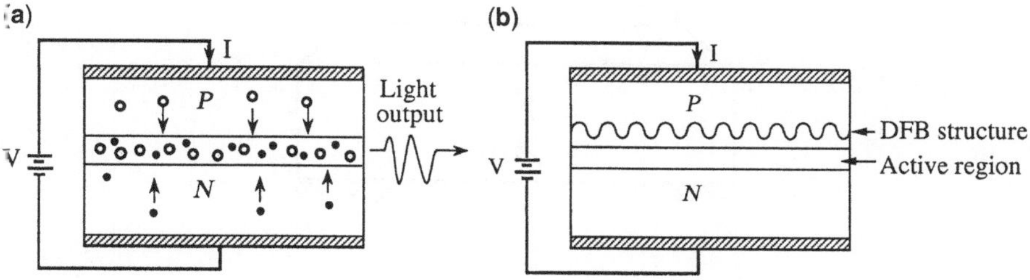

Figure 1.14 A cross section of (a) a double-heterojunction (DH) Fabry–Perot (FP) semiconductor laser and (b) a distributed-feedback (DFB) semiconductor laser which contains a grating structure for wavelength selectivity for single-wavelength operation.

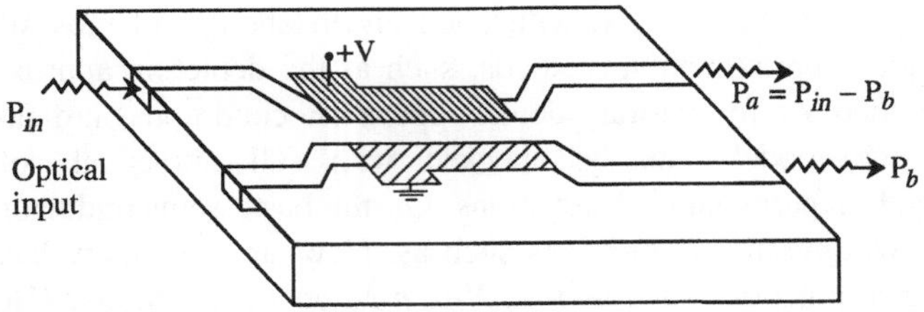

Figure 1.15 A directional-coupler modulator of which the output light power may be switched by an electric field bias.

In Part IV, Modulation of Light, we investigate how to modulate the optical intensity or phase. We include direct modulation of semiconductor lasers (Chapter 12), electrooptic and acoustooptic modulators (Chapter 13), and electroabsorption modulators (Chapter 14). An example of an electrically controlled directional coupler is shown in Fig. 1.15, where the coupling of two parallel waveguides is controlled by a pair of electrodes via the change of the refractive index through the electrooptical effects. The voltage bias can switch the output light from the same waveguide *a* to the other waveguide *b* when the input light is into the waveguide *a*.

In Part V, Detection of Light and Solar Cells, we discuss photodetectors and solar cells (Chapter 15). What are the different types of photodetectors and their operational characteristics? A simple example is a *p-n* junction photodiode as shown in Fig. 1.16. The absorption of photons and the conversion of optical energy to electric current will be investigated. Novel photodetectors using quantum wells will be presented. The physical processes in solar cells are strikingly similar to those of the *p-n* junction photodetectors, and they will be discussed.

Many of these novel optoelectronic devices and physics are still research issues under intensive investigation. The materials presented in this book emphasize the

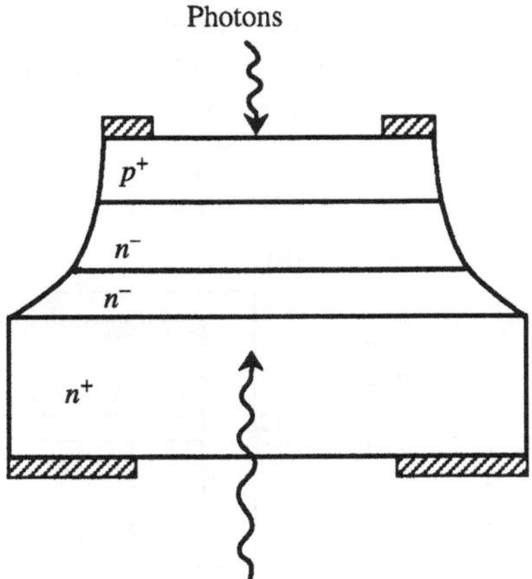

Figure 1.16 A *p-n* junction photodiode with optical illumination from the top or from the bottom.

fundamental principles and analytical skills on the essentials of the physics of optoelectronic devices. The book was written with the hope that the readers of this book will acquire fundamental analytical skills and knowledge in the physics of optoelectronic devices to analyze their research results, to understand more advanced materials in journal papers and research monographs, and to generate novel designs of optoelectronic devices. Many books that are highly recommended for further reading are listed in the bibliography.

PROBLEMS

1.1 (**a**) Calculate the band-gap wavelength λ_g for Si, GaAs, InAs, InP, and GaP at 300K. Use the band-gap energies in Table C.1 of Appendix C (page 802).

(**b**) Find the optical energy corresponding with the wavelength 1.30 μm and with the wavelength 1.55 μm.

1.2 (**a**) Find the gallium mole fraction x for $In_{1-x}Ga_xAs$ compound semiconductor such that its lattice constant matches that of InP.

(**b**) Find the aluminum mole fraction x for $Al_xIn_{1-x}As$ such that its lattice constant is the same as that of InP.

1.3 Calculate the band edge discontinuities ΔE_c and ΔE_v for $GaAs/Al_xGa_{1-x}As$ heterojunction if $x = 0.2$ and 0.3.

1.4 Plot the energy band-gap E_g versus the lattice constant for the two ternary compounds, $In_{1-x}Ga_xAs$ and $Al_xIn_{1-x}As$. Label the values for the binary compounds, InAs, GaAs, AlAs, and InP. The band-gap formulas at 300K are

$$E_g(In_{1-x}Ga_xAs) = 0.36 + 0.505x + 0.555x^2$$
$$E_g(Al_xIn_{1-x}As) = 0.36 + 2.35x + 0.24x^2.$$

Show on your plot the locations (and values) of the lattice match conditions for the InP substrate.

1.5 For telecommunication wavelength near 1.55 μm, find the optical wavelength spacing in nanometer (nm) for dense wavelength division multiplexing (DWDM) systems when the frequency spacing is 50 GHz.

REFERENCES

1. C. Kittel, *Introduction to Solid State Physics*, 8th ed., Wiley, New York, 2005.
2. B. G. Streetman and S. Banerjee, *Solid State Electronic Devices*, 6th ed., Prentice Hall, Upper Saddle River, NJ, 2005.
3. R. F. Pierret, *Solid State Semiconductor Fundamentals*, Prentice Hall, Upper Saddle River, NJ, 2007.
4. K. Iga, *Fundamentals of Laser Optics*, Plenum, New York, 1994.

5. A. Schawlow and C. H. Townes, "Infrared and optical masers," *Phys. Rev.* **112**, 1940–1949 (1958).
6. J. Hecht, "Winning the laser patent war," *Laser Focus World*, 49–51, December 1994.
7. T. H. Maiman, "Stimulated optical radiation in ruby," *Nature* **187**, 493–494 (1960).
8. A. Javan, W. R. Bennett, Jr., and D. R. Herriot, "Population inversion and continuous optical maser oscillation in a gas discharge containing a He-Ne mixture," *Phys. Rev. Lett.* **6**, 106–110 (1961).
9. R. N. Hall, G. E. Fenner, J. D. Kingsley, T. J. Soltys, and R. O. Carlson, "Coherent light emission from GaAs junctions," *Phys. Rev. Lett.* **9**, 366–368 (1962). [Received date: Sept. 24, 1962.]
10. M. I. Nathan, W. P. Dumke, G. Burns, F. H. Dill, Jr., and G. Lasher, "Simulated emission of radiation from GaAs p-n junctions," *Appl. Phys. Lett.* **1**, 62–64 (1962). [Received date: Oct. 6, 1962.]
11. N. Holonyak, Jr., and S. F. Bevacqua, "Coherent (visible) light emission from $Ga(As_{1-x}P_x)$ junctions," *Appl. Phys. Lett.* **1**, 82–83 (1962). [Received date: Oct. 17, 1962.]
12. T. M. Quist, R. H. Rediker, R. J. Keyes, W. E. Krag, B. Lax, A. L. McWhorter, and H. J. Zeiger, "Semiconductor maser of GaAs," *Appl. Phys. Lett.* **1**, 91–92 (1962). [Received date: Oct. 23, 1962.]
13. R. D. Dupuis, "The diode laser—the first 30 days, 40 years ago," *Optics and Photonics News*, April, 30–35 (2004).
14. M. G. Craford, N. Holonyak, Jr., and F. A. Kish, "In pursue of the ultimate lamp," *Sci. Am.* **284,** 63–67 (2001).
15. N. Holonyak, Jr., "From transistors to lasers and light-emitting diodes," *Mater. Res. Bull.* **30**, 509–517 (2005).
16. R. D. Dupuis and M. R. Krames, "History, development, and applications of high-brightness visible light-emitting diodes," *J. Lightwave Technol.* **26**, 1154–1171 (2008).
17. W. T. Tsang, Vol. Ed., *Lightwave Communications Technology*, vol. 22, Parts A–E, in R. K. Willardson and A. C. Beer, Eds., *Semiconductor and Semimetals*, Academic Press, New York, 1985.
18. H. C. Casey, Jr., and M. B. Panish, *Heterostructure Lasers*, Part A and B, Academic Press, Orlando, 1978.
19. T. Miya, Y. Terunuma, T. Hosaka, and T. Miyashita, "An ultimate low loss single mode fiber at 1.55 μm," *Electron. Lett.* **15**, 106–108 (1979).
20. I. Vurgaftman, J. R. Meyer, and L. R. Ram-Mohan, "Band parameters for III-V compound semiconductors and their alloys," *Appl. Phys. Rev.* **89**, 5815–5875 (2001).
21. N. Holonyak, Jr., R. M. Kolbas, R. D. Dupuis, and P. D. Dapkus, "Quantum-well heterostructure lasers," *IEEE J. Quantum Electron.* **QE-16**, 170–186 (1980).
22. D. S. Chemla and A. Pinczuk, Guest Editor, Special issue on Semiconductor Quantum Wells and Superlattices: Physics and Applications. *IEEE J. Quantum Electron.* **QE-22** (1986).
23. P. S. Zory, Jr., Ed., *Quantum Well Lasers*, Academic Press, San Diego, 1993.
24. T. P. Pearsall, Vol. Ed., *Strained Layer Superlattices: Physics*, vol. 32, 1990; and *Strained Layer Superlattices: Materials Science and Technology*, vol. 33, 1991, in R. K. Willardson and A. C. Beer, Eds., *Semiconductor and Semimetals*, Academic Press, New York.

25. A. R. Adams, "Band-structure engineering for low-threshold high-efficiency semiconductor lasers," *Electron. Lett.* **22**, 249–250 (1986).
26. E. Yablonovitch and E. O. Kane, "Reduction of lasing threshold current density by the lowering of valence band effective mass," *J. Lightwave Technol.* **4**, 504–506 (1986).
27. D. Ahn and S. L. Chuang, "Optical gain in a strained-layer quantum-well laser," *IEEE J. Quantum Electron.* **24**, 2400–2406 (1988).
28. C. S. Chang and S. L. Chuang, "Modeling of strained quantum-well lasers with spin-orbit coupling," *IEEE J. Sel. Top. Quantum Electron.* **1**, 218–229 (1995).
29. D. Bimberg, M. Grundmann, and N. N. Ledentsov, *Quantum Dot Heterostructures*, Wiley, New York, 2001.
30. D. Bimberg, M. Grundmann, N. N. Ledentsov, M. H. Mao, Ch. Ribbat, R. Sellin, V. M. Ustinov, A. E. Zhukov, Zh. I. Alferov, and J. A. Lott, "Novel infrared quantum dot lasers: theory and reality," *Phys. Stat. Sol.* (b) **224**, 787–796 (2001).
31. N. N. Ledendsov, "Long-wavelength quantum-dot lasers on GaAs substrate: from media to device concepts," *IEEE J. Quantum Electron.* **8**, 1015–1024 (2002).

BIBLIOGRAPHY

Historical

1. J. Orton, *The Story of Semiconductors*, Oxford University Press, Oxford, 2004.
2. C. H. Townes, *How the Laser Happened—Adventures of a Scientist*, Oxford University Press, Oxford, 1999.
3. T. Maiman, *The Laser Odyssey*, Laser Press, Blaine, WA, 2000.
4. N. Taylor, *Laser—The Inventor, the Nobel Laureate, and the Thirty-Year Patent War*, Simon & Schuster, New York, 2000.

Semiconductor Physics and Devices

5. M. Shur, *Physics of Semiconductor Devices*, Prentice Hall, Englewood Cliffs, NJ, 1990.
6. G. Bastard, *Wave Mechanics Applied to Semiconductor Heterostructures*, Wiley, New York, 1991.
7. C. Weisbuch and B. Vinter, *Quantum Semiconductor Structures*, Academic Press, New York, 1991.
8. M. Fox, *Optical Properties of Solids*, Oxford University Press, Oxford, 2002.
9. B. K. Ridley, *Quantum Processes in Semiconductors*, 4th ed., Oxford University Press, Oxford, 2000.
10. P. K. Basu, *Theory of Optical Processes in Semiconductors—Bulk and Microstructures*, Oxford University Press, Oxford, 1997.
11. C. F. Klingshirn, *Semiconductor Optics*, 3rd ed., Springer, Berlin, 2006.
12. N. W. Ashcroft and N. D. Mermin, *Solid State Physics*, Holt, Rinehart and Winston, New York, 1976.
13. S. M. Sze and K. K. Ng, *Physics of Semiconductor Devices*, Wiley, New York, 2007.
14. S. Wang, *Fundamentals of Semiconductor Theory and Device Physics*, Prentice Hall, Englewood Cliffs, NJ, 1989.

15. F. Bassani and G. P. Parravicini, *Electronic States and Optical Transitions in Solids*, Pergamon Press, Oxford, 1975.
16. K. Seeger, *Semiconductor Physics—An Introduction*, 9th ed., Springer, Berlin, 2004.
17. K. Hess, *Advanced Theory of Semiconductor Devices*, Wiley-IEEE Press, New York, 1999.
18. P. Y. Yu and M. Cardona, *Fundamental of Semiconductors: Physics and Materials Properties*, 2nd ed., Springer, New York, 1999.
19. K. W. Böer, *Survey of Semiconductor Physics-Electrons and Other Particles in Bulk Semiconductors*, Van Nostrand Reinhold, New York, 1990.

Quantum Electronics, Photonics, and Optoelectronics

20. A. Yariv and P. Yeh, *Photonics–Optical Electronics in Modern Communications*, 6th ed., Oxford University Press, New York, 2007.
21. A. Yariv, *Quantum Electronics*, 3rd ed., Wiley, New York, 2001.
22. B. E. A. Saleh and M. C. Teich, *Fundamentals of Photonics*, 2nd ed., Wiley, New York, 2007.
23. S. O. Kasap, *Optoelectronics and Photonics—Principles and Practices*, Prentice Hall, Upper Saddle River, NJ, 2001.
24. E. Rosencher and B. Vinter, *Optoelectronics*, Cambridge University Press, Cambridge, UK, 2002.
25. H. A. Haus, *Waves and Fields in Optoelectronics*, Prentice Hall, Englewood Cliffs, NJ, 1984.
26. P. Bhattacharya, *Semiconductor Optoelectronic Devices*, 2nd ed., Prentice Hall, Upper Saddle River, NJ, 1996.
27. J. Singh, *Semiconductor Optoelectronics: Physics and Technology*, McGraw-Hill, New York, 1995.
28. K. Iizuka, *Elements of Photonics*, Vol. I and II, Wiley, New York, 2002.
29. J. Wilson and J. F. B. Hawkes, *Optoelectronics: An Introduction*, 3rd ed., Prentice Hall, Englewood Cliffs, NJ, 1998.
30. C. R. Pollock, *Fundamentals of Optoelectronics*, Irwin, Chicago, 1995.
31. K. J. Ebeling, *Integrated Optoelectronics*, Springer-Verlag, Berlin, 1993.
32. V. V. Mitin, V. A. Kochelap, and M. A. Stroscio, *Quantum Heterostructures-Microelectronics and Optoelectronics*, Cambridge University Press, Cambridge, UK, 1999.
33. A. K. Ghatak and K. Thyagarajan, *Optical Electronics*, Cambridge University Press, Cambridge, UK, 1989.
34. J. T. Verdeyen, *Laser Electronics*, 3rd ed., Prentice Hall, Englewood Cliffs, NJ, 1994.
35. H. Haken, *Light, Vol. 1: Waves, Photons, Atoms*, North-Holland, Amsterdam, 1986.
36. R. Loudon, *The Quantum Theory of Light*, 3rd ed., Oxford University Press, Oxford, 2000.
37. H. Haug and S. W. Koch, *Quantum Theory of the Optical and Electronic Properties of Semiconductors*, 4th ed., World Scientific, Singapore, 2004.
38. W. T. Tsang, Vol. Ed., *Lightwave Communications Technology*, vol. 22, Part A to E, in *Semiconductor and Semimetals*, R. K. Willardson and A. C. Beer, Eds., Academic Press, New York, 1985.
39. R. Dingle, Vol. Ed., *Applications of Multiquantum Wells, Selective Doping, and Superlattices*, vol. 24 in *Semiconductor and Semimetals*, R. K. Willardson and A. C. Beer, Eds., Academic Press, New York, 1985.

40. M. M-K. Liu, *Principles and Applications of Optical Communications*, Irwin, Chicago, 1996.
41. P. Meystre and M. Sargent III, *Elements of Quantum Optics*, 3rd ed., Springer, Berlin, 1999.

Semiconductor Lasers and LEDs

42. E. Kapon, Ed., *Semiconductor Lasers I. Fundamentals, II. Materials and Structures*, Academic Press, San Diego, 1999.
43. L. A. Coldren and S. W. Corzine, *Diode Lasers and Photonic Integrated Circuits*, Wiley, New York, 1995.
44. S. Nakamura and G. Fasol, *The Blue Laser Diode—GaN Based Light Emitters and Lasers*, Springer, Berlin, 1997.
45. E. F. Schubert, *Light-Emitting Diodes*, 2nd ed., Cambridge University Press, Cambridge, UK, 2006.
46. M. Fukuda, *Reliability and Degradation of Semiconductor Lasers and LEDs*, Artech House, Boston, 1991.
47. O. Ueda, *Reliability and Degradation of III-V Optical Devices*, Artech House, Boston, 1996.
48. C. Wilmsen, H. Temkin, and L. A. Coldren, Eds., *Vertical-Cavity Surface-Emitting Lasers: Design, Fabrication, Characterization, and Applications*, Cambridge University Press, 1999.
49. H. Ghafouri-Shiraz, *Distributed Feedback Laser Diodes: Principles and Physical Modelling*, Wiley, New York, 1996.
50. G. Morthier and P. Vankwikelberge, *Handbook of Distributed-Feedback Laser Diodes*, Artech House, Boston, 1997.
51. G. H. B. Thompson, *Physics of Semiconductor Laser Devices*, Wiley, New York, 1980.
52. H. C. Casey, Jr., and M. B. Panish, *Heterostructure Lasers*, Part A and B, Academic Press, Orlando, 1978.
53. G. P. Agrawal and N. K. Dutta, *Long-Wavelength Semiconductor Lasers*, Van Nostrand Reinhold, New York, 1986.
54. P. S. Zory, Jr., Ed., *Quantum Well Lasers*, Academic Press, San Diego, 1993.
55. G. A. Evans and J. M. Hammer, Eds., *Surface Emitting Semiconductor Lasers and Arrays*, Academic Press, San Diego, 1993.
56. R. K. Willardson and A. C. Beer, Eds., *Lasers, Junctions, Transport*, vol. 14 in *Semiconductors and Semimetals*, Academic Press, New York, 1979.
57. W. W. Chow, S. W. Koch, and M. Sargent III, *Semiconductor-Laser Physics*, Springer, Berlin, 1994.
58. W. W. Chow and S. W. Koch, *Semiconductor Laser Fundamentals: Physics of the Gain Materials*, Springer, Berlin, 2006.
59. J. K. Butler, Ed., *Semiconductor Injection Lasers*, IEEE Press, New York, 1980.
60. J. J. Coleman, Ed., *Selected Papers on Semiconductor Diode Lasers*, SPIE Milestone Series, Vol. MS50, SPIE Optical Engineering Press, Bellingham, WA, 1992.

Optical Waveguides, Modulators, and Fiber Optical Communications

61. H. Nishihara, M. Haruna, and T. Suhara, *Optical Integrated Circuits*, McGraw-Hill, New York, 1989.

62. R. G. Hunsperger, *Integrated Optics: Theory and Technology*, 5th ed., Springer, Berlin, 2002.
63. T. Tamir, *Integrated Optics*, Springer-Verlag, Berlin, 1979.
64. T. Tamir, *Guided Wave Optoelectronics*, 2nd ed., Springer, Berlin, 1990.
65. D. Marcuse, *Theory of Dielectric Optical Waveguides*, Academic Press, New York, 1974.
66. A. W. Snyder and J. D. Love, *Optical Waveguide Theory*, Chapman and Hall, London, 1983.
67. A. B. Buckman, *Guided-Wave Photonics*, Saunders College Publishing, New York, 1992.
68. A. Yariv and P. Yeh, *Optical Waves in Crystals*, Wiley, New York, 1984.
69. S. V. Kartalopoulos, *Introduction to DWDM Technology: Data in a Rainbow*, Wiley-IEEE Press, New York, 2000.
70. D. K. Mynbaev and L. L. Scheiner, *Fiber-Optic Communications Technology*, Prentice Hall, Upper Saddle River, NJ, 2001.
71. J. C. Palais, *Fiber Optic Communications*, 5th ed., Prentice Hall, Upper Saddle River, NJ, 2005.
72. G. Keiser, *Optical Fiber Communications*, 3rd ed., McGraw-Hill, Boston, 2000.
73. G. Agrawal, *Fiber-Optic Communication Systems*, 3rd ed., Wiley, New York, 2002.
74. J. Hecht, *Understanding Fiber Optics*, Prentice Hall, Upper Saddle River, NJ, 2006.

Photodetectors

75. J. D. Vincent, *Fundamentals of Infrared Detector Operation and Testing*, Wiley, New York, 1990.
76. R. K. Willardson and A. C. Beer, Eds., *Infrared Detectors*, vol. 5, in *Semiconductors and Semimetals*, Academic Press, New York, 1970.
77. R. K. Willardson and A. C. Beer, Eds., *Infrared Detectors II*, vol. 12 in *Semiconductors and Semimetals*, Academic Press, New York, 1977.
78. R. K. Willardson and A. C. Beer, Eds., *Mercury Cadmium Telluride*, vol. 18 in *Semiconductors and Semimetals*, Academic Press, New York, 1981.
79. A. Rogalski, Ed., *Selected Papers on Semiconductor Infrared Detectors*, SPIE Milestone Series, Vol. MS66, SPIE Optical Engineering Press, Bellingham, WA, 1992.
80. M. O. Manasreh, Ed., *Semiconductor Quantum Wells and Superlattices for Long-Wavelength Infrared Detectors*, Artech House, Norwood, 1993.

Nonlinear Optics

81. N. Bloembergen, Nonlinear Optics, Addison-Wesley, Redwood City, 1992. [Originally published by W. A. Benjamin, Inc., 1965.]
82. R. W. Boyd, *Nonlinear Optics*, 2nd ed., Academic Press, San Diego, 2002.
83. A. C. Newell and J. V. Moloney, *Nonlinear Optics*, New ed., Westview Press, 2003.
84. Y. R. Shen, *The Principles of Nonlinear Optics*, reprint ed., Wiley, New York, 2002.
85. H. M. Gibbs, *Optical Bistability: Controlling Light with Light*, Academic Press, San Diego, 1985.
86. H. Haug, Ed., *Optical Nonlinearities and Instabilities in Semiconductors*, Academic Press, San Diego, 1988.

Part I
Fundamentals

2

Basic Semiconductor Electronics

In the study of semiconductor devices such as diodes and transistors, the characteristics of the devices are described by the voltage–current relations. The injection of electrons and holes by a voltage bias and their transport properties are studied. When optical injection or emission is involved, such as in laser diodes and photodetectors, we are interested in the optical field in the device as well as the light–matter interaction. In this case, we look for the light output versus the device bias current for a laser diode, or the change in the voltage–current relation due to the illumination of light in a photodetector. In general, it is useful to know the voltage, current, or quasi-static potentials and electric field in the electronic devices and the optical electric and magnetic fields in the optoelectronic devices. Thus, a full understanding of the basic equations for the modeling of these devices is very important. In this chapter, we will review the basic Maxwell's equations, semiconductor electronics equations, and boundary conditions. We also study the generation and recombination of carriers in semiconductors. The general theory for semiconductor heterojunctions and semiconductor/metal junctions is also investigated.

2.1 MAXWELL'S EQUATIONS AND BOUNDARY CONDITIONS

Maxwell's equations are the fundamental equations in electromagnetics. They were first established by James Clerk Maxwell in 1873 and were verified experimentally by Heinrich Hertz in 1888 [1–3]. Maxwell unified all knowledge of electricity and magnetism, added a displacement current density term $\partial \mathbf{D}/\partial t$ in Ampére's law, and predicted electromagnetic wave motion. He explained light propagation as an electromagnetic wave phenomenon. Heinrich Hertz demonstrated experimentally the electromagnetic wave phenomenon using a spark-gap generator as a transmitter and a loop of wire with a very small gap as a receiver. He then set off a spark in the transmitter and showed that a spark at the receiver was produced. By moving the relative position of the receiver from a metallic reflecting wall, standing wave effects were demonstrated. Thus electromagnetic wave theory was confirmed. For a historical account of the classic and quantum theory of light, see, for example, Ref. 3.

Physics of Photonic Devices, Second Edition. By Shun Lien Chuang
Copyright © 2009 John Wiley & Sons, Inc.

2.1.1 Maxwell's Equations in MKS Units

The well-known Maxwell's equations in MKS (meter, kilogram, and second) units are written as

$$\nabla \times \mathbf{E} = -\frac{\partial}{\partial t}\mathbf{B} \qquad \text{Faraday's law} \tag{2.1.1}$$

$$\nabla \times \mathbf{H} = \mathbf{J} + \frac{\partial \mathbf{D}}{\partial t} \qquad \text{Ampére's law} \tag{2.1.2}$$

$$\nabla \cdot \mathbf{D} = \rho \qquad \text{Gauss's law} \tag{2.1.3}$$

$$\nabla \cdot \mathbf{B} = 0 \qquad \text{Gauss's law} \tag{2.1.4}$$

where **E** is the electric field (V/m), **H** is the magnetic field (A/m), **D** is the electric displacement flux density (C/m^2), and **B** is the magnetic flux density (Vs/m^2 or $Webers/m^2$). The two source terms, the charge density $\rho(C/m^3)$ and the current density **J** (A/m^2), are related by the continuity equation

$$\nabla \cdot \mathbf{J} + \frac{\partial}{\partial t}\rho = 0 \tag{2.1.5}$$

where no net generation or recombination of electrons is assumed. In the study of electromagnetics, one usually assumes that the source terms ρ and **J** are given quantities. It is noted that (2.1.4) is derivable from (2.1.1) by taking the divergence of (2.1.1) and noting that $\nabla \cdot (\nabla \times \mathbf{E}) = 0$ for any vector **E**. Similarly, (2.1.3) is derivable from (2.1.2) using (2.1.5). Thus, we have only two independent vector equations (2.1.1) and (2.1.2), or six scalar equations as each vector has three components. However, there are **E**, **H**, **D**, and **B**, 12 scalar unknown components. Thus, we need six more scalar equations. These are the so-called constitutive relations that describe the properties of a medium. In isotropic media, they are given by

$$\mathbf{D} = \varepsilon \mathbf{E} \qquad \mathbf{B} = \mu \mathbf{H}. \tag{2.1.6}$$

In anisotropic media, they may be given by

$$\mathbf{D} = \boldsymbol{\varepsilon} \cdot \mathbf{E} \qquad \mathbf{B} = \boldsymbol{\mu} \cdot \mathbf{H} \tag{2.1.7}$$

where $\boldsymbol{\varepsilon}$ is the permittivity tensor and $\boldsymbol{\mu}$ is the permeability tensor:

$$\boldsymbol{\varepsilon} = \begin{bmatrix} \varepsilon_{xx} & \varepsilon_{xy} & \varepsilon_{xz} \\ \varepsilon_{yx} & \varepsilon_{yy} & \varepsilon_{yz} \\ \varepsilon_{zx} & \varepsilon_{zy} & \varepsilon_{zz} \end{bmatrix} \qquad \boldsymbol{\mu} = \begin{bmatrix} \mu_{xx} & \mu_{xy} & \mu_{xz} \\ \mu_{yx} & \mu_{yy} & \mu_{yz} \\ \mu_{zx} & \mu_{zy} & \mu_{zz} \end{bmatrix}. \tag{2.1.8}$$

For electromagnetic fields at optical frequencies, $\rho = 0$ and $\mathbf{J} = 0$.

2.1.2 Boundary Conditions

By applying the first two Maxwell's equations over a small rectangular surface with a width δ (dashed line in Fig. 2.1a) across the interface of a boundary and using Stokes' theorem between a line integral over a contour C and the surface S enclosed by the contour

$$\oint_C \mathbf{E} \cdot d\ell = \int_S \nabla \times \mathbf{E} \cdot \hat{\mathbf{n}} dS = -\frac{d}{dt}\int_S \mathbf{B} \cdot \hat{\mathbf{n}} dS \tag{2.1.9a}$$

$$\oint_C \mathbf{H} \cdot d\ell = \int_S \nabla \times \mathbf{H} \cdot \hat{\mathbf{n}} dS = \int_S \mathbf{J} \cdot \hat{\mathbf{n}} dS + \frac{d}{dt}\int_S \mathbf{D} \cdot \hat{\mathbf{n}} dS, \tag{2.1.9b}$$

the following boundary conditions can be derived by letting the width δ approach zero [1]:

$$\hat{\mathbf{n}} \times (\mathbf{E}_1 - \mathbf{E}_2) = 0 \tag{2.1.10}$$

$$\hat{\mathbf{n}} \times (\mathbf{H}_1 - \mathbf{H}_2) = \mathbf{J}_s, \tag{2.1.11}$$

where $\mathbf{J}_s (= \lim_{\mathbf{J} \to \infty, \delta \to 0} \mathbf{J}\delta)$ is the surface current density (A/m). Note that the unit normal vector $\hat{n}$ points from medium 2 to medium 1. Similarly, if we apply Gauss's laws (2.1.3) and (2.1.4) and integrate over a small volume (Fig. 2.1b) with a surface area A and a thickness δ and let δ approach zero, for example,

$$\oint_S \mathbf{D} \cdot \hat{\mathbf{n}} dS = \int_V \nabla \cdot \mathbf{D} dv = \int_V \rho \, dv = \rho \delta A,$$

we obtain the following boundary conditions:

$$\hat{\mathbf{n}} \cdot (\mathbf{D}_1 - \mathbf{D}_2) = \rho_s \tag{2.1.12}$$

$$\hat{\mathbf{n}} \cdot (\mathbf{B}_1 - \mathbf{B}_2) = 0, \tag{2.1.13}$$

where $\rho_s (= \lim_{\rho \to \infty, \delta \to 0} \rho\delta)$ is the surface charge density (C/m^2).

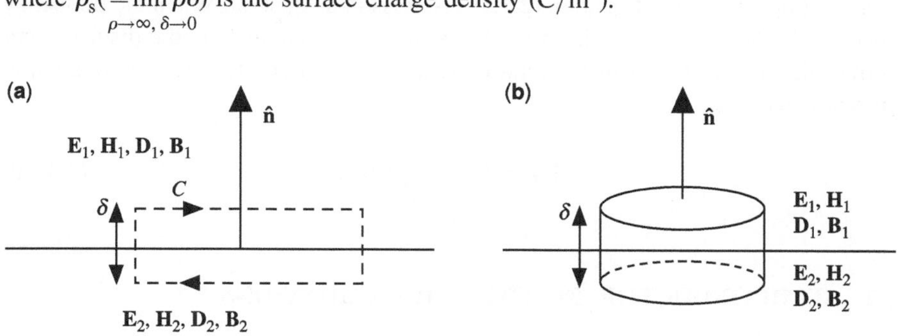

Figure 2.1 Geometry for deriving the boundary conditions across the interface of two media, region 1 and 2. (a) A rectangular surface is enclosed by the contour C (dashed line). (b) A small volume with a thickness δ. The unit normal vector points from region 2 toward region 1.

For an interface across two dielectric media, where no surface current or charge density can be supported, $\mathbf{J}_s = 0$, and $\rho_s = 0$, we have

$$\begin{aligned} \hat{\mathbf{n}} \times \mathbf{E}_1 &= \hat{\mathbf{n}} \times \mathbf{E}_2 \qquad & \hat{\mathbf{n}} \times \mathbf{H}_1 &= \hat{\mathbf{n}} \times \mathbf{H}_2 \\ \hat{\mathbf{n}} \cdot \mathbf{D}_1 &= \hat{\mathbf{n}} \cdot \mathbf{D}_2 \qquad & \hat{\mathbf{n}} \cdot \mathbf{B}_1 &= \hat{\mathbf{n}} \cdot \mathbf{B}_2. \end{aligned} \tag{2.1.14}$$

For an interface between a dielectric medium and a perfect conductor,

$$\begin{aligned} \hat{\mathbf{n}} \times \mathbf{E}_1 &= 0 \qquad & \hat{\mathbf{n}} \times \mathbf{H}_1 &= \mathbf{J}_s \\ \hat{\mathbf{n}} \cdot \mathbf{D}_1 &= \rho_s \qquad & \hat{\mathbf{n}} \cdot \mathbf{B}_1 &= 0 \end{aligned} \tag{2.1.15}$$

as the fields $\mathbf{E}_2$, $\mathbf{H}_2$, $\mathbf{D}_2$, and $\mathbf{B}_2$ inside the perfect conductor vanish. The surface charge density and the current density are supported by the perfect conductor surface.

2.1.3 Quasi-electrostatic Fields

For devices with a dc or low-frequency bias, because the time variation is very slow ($\partial/\partial t \simeq 0$), we usually have

$$\nabla \times \mathbf{E} = 0 \tag{2.1.16}$$

$$\nabla \cdot \mathbf{D} = \rho \tag{2.1.17}$$

and $\mathbf{H} \cong 0$, $\mathbf{B} \cong 0$ for the electronic devices for which no external magnetic fields are applied. In this case, the solution of the electric field can be put in the form of the gradient of an electrostatic potential ϕ

$$\mathbf{E} = -\nabla \phi \tag{2.1.18}$$

and

$$\nabla \cdot (\varepsilon \nabla \phi) = -\rho \tag{2.1.19}$$

in an isotropic medium. Equation (2.1.19) is Poisson's equation. When the frequency becomes higher (e.g., in a microwave transistor), one may include the displacement current density $\partial(\varepsilon\mathbf{E})/\partial t$ in the total current density in addition to the conduction current density $\mathbf{J}_{\text{con}}$

$$\mathbf{J}_{\text{tot}} = \mathbf{J}_{\text{con}} + \frac{\partial}{\partial t}(\varepsilon \mathbf{E}). \tag{2.1.20}$$

2.2 SEMICONDUCTOR ELECTRONICS EQUATIONS

In this section, we present the basic semiconductor electronics equations, which are very useful in the modeling of semiconductor devices [4–7]. These equations are actually based on the Maxwell's equations and the charge continuity equations.

2.2.1 Poisson's Equation

As shown in the previous section, Poisson's equation in the semiconductor is given by (2.1.19),

$$\nabla\cdot(\varepsilon\nabla\phi) = -\rho, \tag{2.2.1}$$

where ϕ is the electrostatic potential and ρ is the charge density given by

$$\rho = q\,(p - n + C_0) \tag{2.2.2}$$

$$C_0 = N_D^+ - N_A^-. \tag{2.2.3}$$

Here q (=1.6×10^{-19} C) is the magnitude of a unit charge, p is the hole concentration, n is the electron concentration, N_D^+ is the ionized donor concentration, and N_A^- is the ionized acceptor concentration.

2.2.2 Continuity Equations

From the Ampére's law

$$\nabla\times\mathbf{H} = \mathbf{J}_{\text{con}} + \frac{\partial}{\partial t}\mathbf{D} \tag{2.2.4}$$

where the conduction current density is

$$\mathbf{J}_{\text{con}} = \mathbf{J}_p + \mathbf{J}_n \tag{2.2.5}$$

and $\mathbf{J}_p$ and $\mathbf{J}_n$ are the hole and electron current densities, respectively, we have

$$0 = \nabla\cdot(\nabla\times\mathbf{H}) = \nabla\cdot\mathbf{J}_{\text{con}} + \frac{\partial}{\partial t}\nabla\cdot\mathbf{D} \tag{2.2.6}$$

or

$$\nabla\cdot\mathbf{J}_{\text{con}} + \frac{\partial}{\partial t}\rho = 0. \tag{2.2.7}$$

Assuming $C_0 = N_D^+ - N_A^-$ is independent of time, we have

$$\nabla\cdot(\mathbf{J}_p + \mathbf{J}_n) + q\frac{\partial}{\partial t}(p - n) = 0. \tag{2.2.8}$$

Thus, we may separate the above equation into two parts for electrons and holes,

$$\nabla\cdot\mathbf{J}_n - q\frac{\partial}{\partial t}n = +qR \tag{2.2.9}$$

$$\nabla\cdot\mathbf{J}_p + q\frac{\partial}{\partial t}p = -qR \tag{2.2.10}$$

where R is the net recombination rate ($cm^{-3}\ s^{-1}$) of electron–hole pairs. Sometimes it is convenient to write the generation rates (G_p and G_n) and recombination rates (R_p and R_n) explicitly

$$R = R_n - G_n \tag{2.2.11}$$

for electrons and

$$R = R_p - G_p \tag{2.2.12}$$

for holes. Thus, we have the current continuity equations for the carriers.

$$\frac{\partial n}{\partial t} = G_n - R_n + \frac{1}{q}\nabla \cdot \mathbf{J}_n \tag{2.2.13}$$

$$\frac{\partial p}{\partial t} = G_p - R_p - \frac{1}{q}\nabla \cdot \mathbf{J}_p. \tag{2.2.14}$$

2.2.3 Carrier Transport Equations

The carrier transport equations assuming Boltzmann distributions for carriers can be written as

$$\mathbf{J}_n = q\mu_n n\,\mathbf{E} + qD_n\nabla n \tag{2.2.15}$$

$$\mathbf{J}_p = q\mu_p p\,\mathbf{E} - qD_p\nabla p \tag{2.2.16}$$

where $\mathbf{E} = -\nabla\phi$ is the electric field, μ_n and μ_p are the electron and hole mobility, and D_n and D_p are the electron and hole diffusion coefficient, respectively, which follow the Einstein relation, $D_n/\mu_n = D_p/\mu_p = kT/q$. We may express the electric field in terms of the electrostatic potential in the carrier transport equation. We then have $\mathbf{J}_p$, $\mathbf{J}_n$, ϕ, p, n, or nine scalar components as unknowns. We also have (2.2.1), (2.2.13), (2.2.14), (2.2.15), (2.2.16), or nine scalar equations. We may also eliminate $\mathbf{J}_p$ and $\mathbf{J}_n$ using (2.1.15) and (2.2.16) and reduce the number of equations to three,

$$\frac{\partial n}{\partial t} = G_n - R_n + \frac{1}{q}\nabla \cdot [-q\mu_n n\nabla\phi + qD_n\nabla n] \tag{2.2.17}$$

$$\frac{\partial p}{\partial t} = G_p - R_p - \frac{1}{q}\nabla \cdot [-q\mu_p p\nabla\phi - qD_p\nabla p] \tag{2.2.18}$$

$$\nabla \cdot (\varepsilon\nabla\phi) = -q(p - n + C_0) \tag{2.2.19}$$

with three unknowns n, p, and ϕ. In principle, these three unknowns can be solved using the above three equations once we specify the boundary conditions for a given device geometry.

2.2.4 Auxiliary Relations

Often, it is convenient to introduce two auxiliary relations with two "imref" functions, $F_n(\mathbf{r})$ and $F_p(\mathbf{r})$, or the quasi-Fermi levels for the electrons and holes, respectively,

$$n(\mathbf{r}) = n_\mathrm{i} \exp\left[\frac{F_n(\mathbf{r}) - E_\mathrm{i}(\mathbf{r})}{k_\mathrm{B}T}\right] \tag{2.2.20}$$

$$p(\mathbf{r}) = n_\mathrm{i} \exp\left[\frac{E_\mathrm{i}(\mathbf{r}) - F_p(\mathbf{r})}{k_\mathrm{B}T}\right], \tag{2.2.21}$$

where the intrinsic carrier concentration n_i depends on the band edge concentration parameters N_c and N_v, the band gap and the temperature

$$n_\mathrm{i} = \sqrt{N_\mathrm{c}N_\mathrm{v}}e^{-E_\mathrm{g}/(2k_\mathrm{B}T)} \tag{2.2.22}$$

and the intrinsic energy level is

$$E_\mathrm{i}(\mathbf{r}) = -q\phi(\mathbf{r}) + E_\mathrm{ref.} \tag{2.2.23}$$

Here, E_ref is a reference constant energy. We may substitute (2.2.20) and (2.2.21) into (2.2.17)–(2.2.19), and use only three unknowns, $\phi(\mathbf{r})$, $\phi_n(\mathbf{r})$, and $\phi_p(\mathbf{r})$, which have the same units (volts)

$$F_n(\mathbf{r}) = -q\phi_n(\mathbf{r}) + E_\mathrm{ref} \tag{2.2.24}$$

$$F_p(\mathbf{r}) = -q\phi_p(\mathbf{r}) + E_\mathrm{ref.} \tag{2.2.25}$$

The two auxiliary relations are for nondegenerate semiconductors, for which the Maxwell–Boltzmann statistics are applicable. To take into account the effect of degeneracy or high carrier density case, one may modify (2.2.20) and (2.2.21) simply by using the Fermi–Dirac statistics together with the electron and hole density-of-state functions $\rho_\mathrm{e}(E)$ and $\rho_\mathrm{h}(E)$

$$n = \int_{-\infty}^{\infty} f_n(E)\rho_\mathrm{e}(E)\, dE \tag{2.2.26}$$

and

$$p = \int_{-\infty}^{\infty} f_p(E)\rho_\mathrm{h}(E)\, dE \tag{2.2.27}$$

where

$$f_n(E) = \frac{1}{1 + e^{(E-F_n)/k_\mathrm{B}T}} \tag{2.2.28}$$

is the Fermi–Dirac distribution for electrons and

$$f_p(E) = \frac{1}{1 + e^{(F_p - E)/k_B T}} \tag{2.2.29}$$

is the Fermi–Dirac distribution for the holes.

Density of States The density of states for electrons, $\rho_e(E)$, is derived as follows. The number of electrons per unit volume is given by

$$n = \frac{2}{V} \sum_{k_x} \sum_{k_y} \sum_{k_z} f_n(E) \tag{2.2.30}$$

where the factor of 2 takes care of both spins of the electrons. For the electron states above the conduction band, we may assume that the electrons are in a box with a volume $L_x L_y L_z$ with wave numbers satisfying

$$k_x = m\frac{2\pi}{L_x}, \quad k_y = n\frac{2\pi}{L_y}, \quad k_z = \ell\frac{2\pi}{L_z}, \qquad m, n, \ell = \text{integers.}$$

Thus, the number of available states in a small cube $dk_x dk_y dk_z$ ($= d^3\mathbf{k}$) in the $\mathbf{k}$-space is $d^3\mathbf{k}$ divided by the amount

$$\left(\frac{2\pi}{L_x}\right)\left(\frac{2\pi}{L_y}\right)\left(\frac{2\pi}{L_z}\right) = \frac{(2\pi)^3}{V} \tag{2.2.31}$$

for each state. In other words, we replace the sum over wavenumbers (or states) by integrations

$$\begin{aligned} \frac{2}{V} \sum_{k_x} \sum_{k_y} \sum_{k_z} &= \frac{2}{V} \int \frac{dk_x}{2\pi/L_x} \int \frac{dk_y}{2\pi/L_y} \int \frac{dk_z}{2\pi/L_z} \\ &= \int \frac{d^3\mathbf{k}}{4\pi^3} = \int \frac{4\pi k^2 dk}{4\pi^3} = \int \frac{k^2 dk}{\pi^2} \end{aligned} \tag{2.2.32}$$

for a parabolic band model with a spherical symmetry. For electrons in the conduction band, the energy band structure is

$$E = E_c + \frac{\hbar^2 k^2}{2m_e^*} \tag{2.2.33}$$

where E_c is the conduction-band edge; we obtain

$$\int \frac{k^2 dk}{\pi^2} = \int_{-\infty}^{\infty} dE \rho_e(E) \tag{2.2.34}$$

where $\rho_e(E)$ is called the density of states for the electrons in the conduction band

$$\rho_e(E) = \frac{1}{2\pi^2}\left(\frac{2m_e^*}{\hbar^2}\right)^{3/2}(E - E_c)^{1/2} \quad \text{for } E > E_c \tag{2.2.35}$$

and $\rho_e(E)$ is zero if $E < E_c$. A similar expression holds for the density of states of the holes in the valence band,

$$\rho_h(E) = \frac{1}{2\pi^2}\left(\frac{2m_h^*}{\hbar^2}\right)^{3/2}(E_v - E)^{1/2} \quad \text{for } E < E_v \tag{2.2.36}$$

and $\rho_h(E)$ is zero for $E > E_v$, where E_v is the valence-band edge. In the nondegenerate limit, it can be shown that (2.2.26) and (2.2.27) reduce to (2.2.20) and (2.2.21).

Using the Fermi–Dirac integral defined by

$$F_j(\eta) = \frac{1}{\Gamma(j+1)}\int_0^\infty \frac{x^j dx}{1 + e^{(x-\eta)}} \tag{2.2.37}$$

where Γ is the Gamma function, we may rewrite (2.2.26) and (2.2.27) as

$$n = N_c F_{1/2}\left(\frac{F_n - E_c}{k_B T}\right), \quad p = N_v F_{1/2}\left(\frac{E_v - F_p}{k_B T}\right) \tag{2.2.38}$$

where

$$N_c = 2\left(\frac{m_e^* k_B T}{2\pi\hbar^2}\right)^{3/2} = 2.51 \times 10^{19}\left(\frac{m_e^*}{m_0}\frac{T}{300}\right)^{3/2} \text{cm}^{-3} \tag{2.2.39a}$$

$$N_v = 2\left(\frac{m_h^* k_B T}{2\pi\hbar^2}\right)^{3/2} = 2.51 \times 10^{19}\left(\frac{m_h^*}{m_0}\frac{T}{300}\right)^{3/2} \text{cm}^{-3} \tag{2.2.39b}$$

and $\Gamma(3/2) = \sqrt{\pi}/2$ has been used.

Approximate Formula for the Fermi–Dirac Integral An approximate analytic form for the Fermi–Dirac integral $F_j(\eta)$ valid for $-\infty < \eta < \infty$ is given by [5, 8–10]

$$F_j(\eta) = \frac{1}{e^{-\eta} + C_j(\eta)} \tag{2.2.40a}$$

where, for $j = 1/2$, one can use [8] (or other expression in [9])

$$C_{1/2}(\eta) = \frac{3(\pi/2)^{1/2}}{\left[\eta + 2.13 + \left(|\eta - 2.13|^{12/5} + 9.6\right)^{5/12}\right]^{3/2}}. \tag{2.2.40b}$$

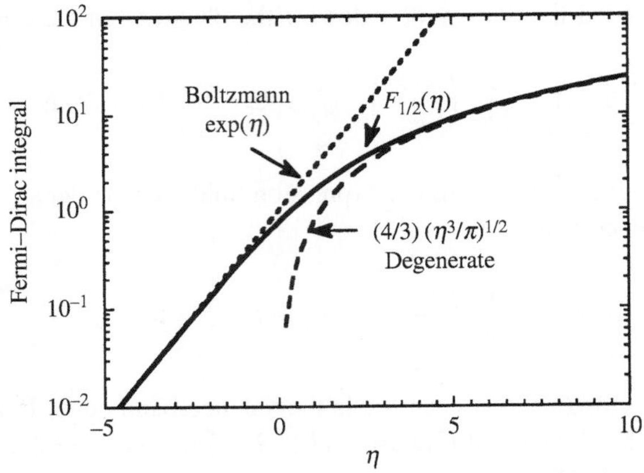

Figure 2.2 A plot of the Fermi–Dirac integral $F_{1/2}(\eta)$ versus η and its asymptotic limits for $\eta \ll -1$ and $\eta \gg 1$.

The maximum error of (2.2.40) is only 0.5%. We also see that

$$F_{1/2}(\eta) \sim e^{\eta} \qquad \text{for } \eta \ll -1$$

$$\sim \frac{4}{3}\left(\frac{\eta^3}{\pi}\right)^{1/2} \qquad \text{for } \eta \gg 1. \tag{2.2.41}$$

These asymptotic limits are shown as dashed lines and compared with the exact numerical integration in Fig. 2.2.

Determination of the Fermi Level E_F For a bulk semiconductor under thermal equilibrium, the Fermi level E_F $(= F_n = F_p)$ is determined by the charge neutrality condition

$$n_0 + N_A^- = p_0 + N_D^+ \tag{2.2.42}$$

where N_A^- is the ionized acceptor concentration and N_D^+ is the ionized donor concentration. We have

$$N_A^- = \frac{N_A}{1 + g_A \exp\left(\dfrac{E_A - E_F}{k_B T}\right)} \tag{2.2.43}$$

where g_A is the ground state degeneracy factor for acceptor levels. g_A equals 4 because in Si, Ge, and GaAs, each acceptor level can accept one hole of either spin and the acceptor level is doubly degenerate as a result of the two degenerate valence bands at $k = 0$, and

$$N_D^+ = N_D \left[1 - \frac{1}{1 + \frac{1}{g_D} \exp\left(\frac{E_D - E_F}{k_B T} \right)} \right] \tag{2.2.44}$$

where g_D is the ground state degeneracy of the donor impurity level and equals 2. The Fermi level is the only unknown in the charge neutrality condition and can be found numerically from (2.2.42).

The carrier densities at thermal equilibrium for nondegenerate carrier densities are

$$n_0 = n_i \exp\left[\frac{E_F - E_i}{k_B T} \right] = N_c \exp\left[\frac{E_F - E_c}{k_B T} \right] \tag{2.2.45a}$$

$$p_0 = n_i \exp\left[\frac{E_i - E_F}{k_B T} \right] = N_v \exp\left[\frac{E_v - E_F}{k_B T} \right]. \tag{2.2.45b}$$

(Note that at thermal equilibrium, $n_0 p_0 = n_i^2$.) For extrinsic semiconductors, we have two cases:

1. n-type, $N_D^+ - N_A^- \gg n_i$, therefore,

$$n_0 \simeq N_D^+ - N_A^- \quad \text{and} \quad p_0 = n_i^2/(N_D^+ - N_A^-). \tag{2.2.46a}$$

2. p-type, $N_A^- - N_D^+ \gg n_i$, therefore,

$$p_0 \simeq N_A^- - N_D^+ \quad \text{and} \quad n_0 = n_i^2/(N_A^- - N_D^+). \tag{2.2.46b}$$

The intrinsic energy level can be found from (2.2.45) and the definitions of N_c and N_v in (2.2.39a) and (2.2.39b)

$$E_i = \frac{E_c + E_v}{2} + \frac{3 k_B T}{4} \ln\left(\frac{m_h^*}{m_e^*} \right). \tag{2.2.47}$$

Thus far, we assume p is given by the heavy holes only, as the density of states of the heavy holes is usually much larger than that of the light holes. If we take into account the contribution due to light holes, the total hole concentration should be

$$p = p_{hh} + p_{\ell h} \tag{2.2.48a}$$

where

$$p_{\rm hh} = N_{\rm v}^{\rm hh} F_{1/2}\left(\frac{E_{\rm v} - E_{\rm F}}{k_{\rm B}T}\right), \quad N_{\rm v}^{\rm hh} = 2\left(\frac{m_{\rm hh}^{*} k_{\rm B}T}{2\pi\hbar^{2}}\right)^{3/2} \tag{2.2.48b}$$

$$p_{\ell\rm h} = N_{\rm v}^{\ell\rm h} F_{1/2}\left(\frac{E_{\rm v} - E_{\rm F}}{k_{\rm B}T}\right), \quad N_{\rm v}^{\ell\rm h} = 2\left(\frac{m_{\ell\rm h}^{*} k_{\rm B}T}{2\pi\hbar^{2}}\right)^{3/2}. \tag{2.2.48c}$$

Both terms have to be used in determining the Fermi level $E_{\rm F}$ using the charge neutrality condition.

An inversion formula is sometimes convenient if we know the carrier concentration n. Define

$$u = \frac{n}{N_{\rm C}} = F_{1/2}(\eta) \qquad \eta = \frac{E_{\rm F} - E_{\rm C}}{k_{\rm B}T} \tag{2.2.49}$$

Then, an approximate formula is [11]

$$\begin{aligned} \eta &\simeq \frac{\ln(u)}{1 - u^{2}} + \frac{v}{1 + (0.24 + 1.08v)^{-2}} \\ v &= \left(\frac{3\sqrt{\pi}\,u}{4}\right)^{2/3}. \end{aligned} \tag{2.2.50}$$

The maximum error [10, 11] of the above formula is 0.006. Further improvement by two orders of magnitude is possible [12].

2.2.5 Boundary Conditions

On the surface of ideal ohmic contacts [6]

$$\phi = \phi_0 + V_{\rm appl.} \tag{2.2.51a}$$

$$n = n_0. \tag{2.2.51b}$$

$$p = p_0. \tag{2.2.51c}$$

The quasi-Fermi potentials satisfy

$$\phi_n = \phi_p = V_{\rm appl.} \tag{2.2.52}$$

at the ohmic contacts. Here ϕ_0, n_0, and p_0 are the values of the corresponding variables for space-charge neutrality at equilibrium. On some interface with surface recombination, such as that along a Si—SiO_2 interface in a metal-oxide-semiconductors field-effect transistor (MOSFET), surface of a semiconductor laser, or back surface of a solar cell, one may have

$$\hat{\mathbf{n}} \cdot \mathbf{J}_n = -qR_s \qquad \hat{\mathbf{n}} \cdot \mathbf{J}_p = qR_s \tag{2.2.53}$$

where R_s is the surface recombination rate and the unit vector is pointing toward the surface from the semiconductor to account for carrier loss due to surface recombination. When interface charges ρ_s exist, such as those of the effective oxide charges on the Si—SiO_2 interface, one has

$$-\hat{\mathbf{n}} \cdot (\varepsilon_1 \nabla \phi_1 - \varepsilon_2 \nabla \phi_2) = \rho_s \tag{2.2.54}$$

where $\hat{\mathbf{n}}$ points from region 2 to region 1. On a semiconductor–insulator interface where no surface recombination exists, one has

$$\hat{\mathbf{n}} \cdot \mathbf{J}_n = \hat{\mathbf{n}} \cdot \mathbf{J}_p = 0. \tag{2.2.55}$$

Example We assume the donor concentration N_D is $1 \times 10^{17}\ \text{cm}^{-3}$ in a GaAs sample and these donors are ionized at room temperature. Because $N_D \gg n_i = 2.1 \times 10^6\,\text{cm}^{-3}$, the intrinsic carrier concentration, we have the electron concentration

$$n_0 = N_D = 1 \times 10^{17}\ \text{cm}^{-3}$$

and

$$p_0 = \frac{n_i^2}{N_D} = 4.41 \times 10^{-5}\ \text{cm}^{-3} \ll n_0.$$

We can also calculate the band edge concentration parameters using $m_e^* = 0.0665\ m_0$ and $m_h^* = 0.50\ m_0$ and $T = 300\text{K}$

$$N_c = 2\left(\frac{m_e^* k_B T}{2\pi\hbar^2}\right)^{3/2} = 2.51 \times 10^{19}\left(\frac{m_e^*}{m_0}\frac{T}{300}\right)^{3/2}\ \text{cm}^{-3} = 4.30 \times 10^{17}\ \text{cm}^{-3}$$

$$N_v = 2.51 \times 10^{19}\left(\frac{m_h^*}{m_0}\frac{T}{300}\right)^{3/2}\ \text{cm}^{-3} = 8.87 \times 10^{18}\ \text{cm}^{-3}.$$

The Fermi level can be obtained from the inversion formula (2.2.50) using

$$u = \frac{n_0}{N_c} = \frac{1 \times 10^{17}}{4.30 \times 10^{17}} = 0.2326$$

and we find

$$\eta \simeq -1.381$$

$$E_F - E_c = \eta\ k_B T = -35.7\ \text{meV}.$$

That is the Fermi level is 35.7 meV below the conduction band edge. If we use $\eta \simeq \ln u = -1.458$, we obtain $E_F - E_c = -37.7$ meV. The error is only 2 meV.

2.3 GENERATION AND RECOMBINATION IN SEMICONDUCTORS

In this section, we describe a phenomenological approach to the carrier generation and recombination processes in semiconductors. A quantum mechanical approach can also be taken using the time-dependent perturbation theory (i.e., Fermi's golden rule). The latter approach will be discussed when we study the optical absorption or gain in semiconductors in Chapter 9. In general, these generation/recombination processes can be classified into radiative and nonradiative. The radiative transitions involve the creation or annihilation of photons. The nonradiative transitions do not involve photons; they may involve the interaction with phonons or the exchange of energy and momentum with another electron or hole. The fundamental mechanisms can all be described using Fermi's golden rule with the energy and momentum conservation satisfied by these processes. The processes may also be discussed in terms of band-to-bound state transitions and band-to-band transitions.

2.3.1 Radiative Band-to-Band Generation–Recombination Processes

For radiative band-to-band (or interband) transitions [13–15] especially due to spontaneous emission of photons from semiconductors as shown in Fig. 2.3, the recombination rate of electrons and holes should be proportional to the electron and hole concentrations

$$R_n = R_p = Bnp \tag{2.3.1}$$

where B is called the bimolecular recombination coefficient or capture coefficient. The generation rate due to thermal or optical processes may be written as

$$G_n = G_p = e_r \tag{2.3.2}$$

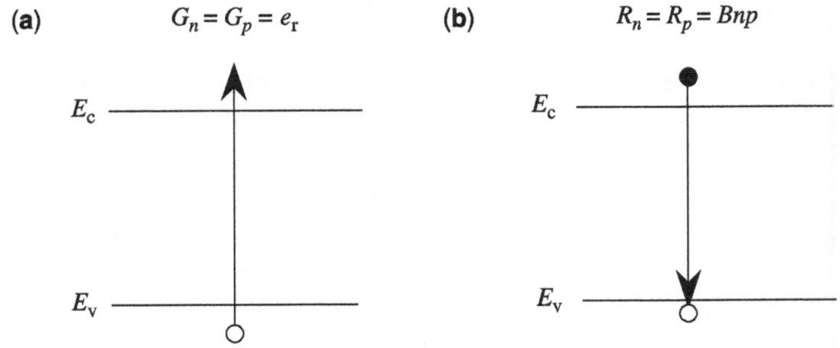

Figure 2.3 The energy band diagrams for (a) generation and (b) recombination of an electron–hole pair.

where e_r is an emission rate. The net recombination rate is

$$R = R_n - G_n = R_p - G_p = Bnp - e_r. \tag{2.3.3}$$

If there is no external perturbation such as electric or optical injection of carriers, the net recombination rate should vanish at thermal equilibrium

$$0 = Bn_0p_0 = e_r$$

where n_0 and p_0 are the electron and hole concentrations at thermal equilibrium $n_0p_0 = n_i^2$. The net recombination rate can be written as

$$R = B(np - n_0p_0). \tag{2.3.4}$$

If the carrier concentrations n and p deviate from their thermal equilibrium values by δn and δp, respectively,

$$n = n_0 + \delta n \qquad p = p_0 + \delta p \tag{2.3.5}$$

we find that the recombination rate under the condition of low-level injection, δn, $\delta p \ll (n_0 + p_0)$ is given by

$$R = B(n_0 + p_0 + \delta n)\delta n = \frac{\delta n}{\tau} \tag{2.3.6}$$

where $\delta n = \delta p$ because electrons and holes are created in pairs for interband transitions, and the radiative lifetime is

$$\tau \simeq \frac{1}{B(n_0 + p_0 + \delta n)}. \tag{2.3.7}$$

In the case of high-level injection such as a semiconductor laser with forward bias, $\delta n = \delta p \gg (n_0 + p_0)$, we have the radiative or spontaneous emission rate:

$$R_{sp} = B(n_0 + p_0 + \delta n)\delta n \simeq B(\delta n)^2 \simeq Bn^2. \tag{2.3.8}$$

Generation of electron–hole pairs by optical injection is commonly used in optical pumping or photoluminescence setups. When the semiconductor sample is illuminated with photons with energy larger than the band gap, electron–hole pairs can be generated due to the absorption of photons. Figure 2.4 shows the optical absorption spectrum of a few semiconductor materials including indirect band gap materials as well as direct band gap materials [16]. The open symbols were obtained from simultaneous measurements of reflectance and transmittance to extract the absorption coefficient for a given thickness of the sample. The solid symbols were obtained from spectral ellipsometry (SE) [16]. A typical absorption coefficient is above $1/\mu\text{m} = 10{,}000/\text{cm}$ for photon energies above the band gap. The exact lineshape of the absorption spectrum depends on the interband joint density of states and other

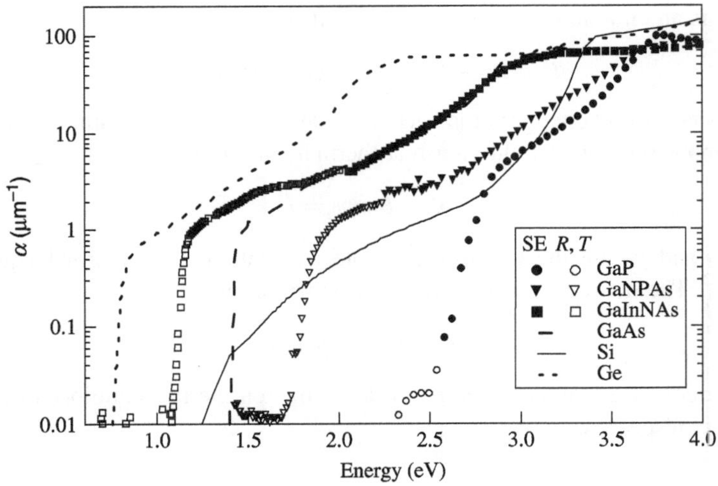

Figure 2.4 Optical absorption spectrum of a few semiconductors obtained from spectral ellipsometry (SE) and simultaneous measurements of reflectance and transmittance (R, T). GaNPAs has approximate composition $GaN_{0.04}P_{0.76}As_{0.2}$. GaInNAs has approximate composition $Ga_{0.95}In_{0.05}N_{0.01}As_{0.99}$. (Reprinted with permission from [16] © 2002 IOP Publishing Ltd.)

effects such as electron–hole Coulomb interaction, which will be discussed later in Chapters 9 and 14. The key idea is that the generation of electron–hole pairs depends on the incident optical power or number of photons injected per second and the absorption coefficient. This concept will be useful in studying photodetectors and solar cells.

2.3.2 Nonradiative Generation–Recombination Processes

Nonradiative recombination occurs mainly as a result of crystalline defects, impurities, and Auger processes [17–22]. The energy released through the recombination usually excites phonons and enhances the degradation of the devices. The crystalline defects include point defects (such as vacancies, interstitial atoms, and their complex), dislocations, and stacking faults. These defects change the perfect periodic structure nature of the crystal or its band diagram and generate either discrete trap states or continuum of states, where electrons or holes recombine most often nonradiatively or sometimes radiatively in a different photon wavelength from that of the band gap.

Hall–Shockley–Read (HSR) Recombination–Generation at Defects The simplest model is that net recombination rate is proportional to the excess carrier density beyond the thermal equilibrium value and the trap density [17]

$$R = \sigma v_{\text{th}} N_{\text{t}} \delta n \tag{2.3.9}$$

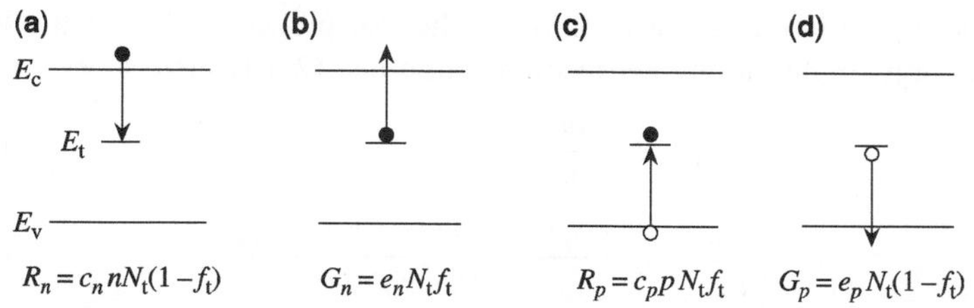

Figure 2.5 The energy band diagrams for the Hall–Shockley–Read generation–recombination processes: (a) electron capture, (b) electron emission, (c) hole capture, and (d) hole emission.

where σ is the capture cross section, v_{th} is the thermal velocity of the electrons, N_t is the trap (or defect) density for electron capture, and δn is the excessive electron concentration. More details analysis is basically from four processes [18–21] shown in Fig. 2.5. These processes may all be caused by the absorption or emission of phonons [20, 21].

1. **Electron Capture:** The recombination rate for the electrons is proportional to the density of electrons n, and the concentration of the traps N_t, multiplied by the probability that the trap is empty $(1 - f_t)$, where f_t is the occupation probability of the trap, $R_n = c_n n N_t(1 - f_t)$, where c_n is the capture coefficient for the electrons.
2. **Electron Emission:** The generation rate of the electrons due to this process is $G_n = e_n N_t f_t$, where e_n is the emission coefficient and $N_t f_t \equiv n_t$ is the density of the traps that are occupied by the electrons.
3. **Hole Capture:** $R_p = c_p p N_t f_t$.
4. **Hole Emission:** $G_p = e_p N_t(1 - f_t)$, where e_p is the emission coefficient for the holes.

   Based on the principle of detail balancing, there should be zero net generation–recombination of electrons and holes, respectively, at thermal equilibrium. We have $R_n - G_n = 0$, and $R_p - G_p = 0$. Thus, we obtain relations between c_n and e_n as well as c_p and e_p. The net recombination rate is

$$R_n - G_n = R_p - G_p = \frac{c_n c_p np - e_n e_p}{c_n n + e_n + c_p p + e_p} N_t$$
$$= \frac{np - n_i^2}{\tau_p(n + n_1) + \tau_n(p + p_1)} \quad (2.3.10)$$

where $\tau_p = 1/c_p N_t$ and $\tau_n = 1/c_n N_t$, which are the hole and electron lifetimes, respectively. We also have $n_1 = e_n/c_n$, $p_1 = e_p/c_p$, $n_1 p_1 = n_i^2$. We note that the above capture and emission processes can be due to optical illumination at the proper photon energy in addition to thermal processes. Under the low-level injection condition, the net recombination rate (2.3.10) is

$$R = \frac{\delta n}{\tau_0} \tag{2.3.11a}$$

$$\tau_0 = \frac{\tau_p(n + n_1) + \tau_n(p + p_1)}{(n_0 + p_0)}. \tag{2.3.11b}$$

Surface Recombination Surface of a semiconductor may have many defects such as dangling bonds and impurities generated from the ambient or the processing procedures. Therefore, nonradiative recombination occurs at a high rate near the surface. The net surface recombination rate is characterized by a surface recombination current density,

$$\hat{\mathbf{n}} \cdot \mathbf{J}_n = -qR_s \qquad R_s = S_n \delta n \tag{2.3.12a}$$

$$\hat{\mathbf{n}} \cdot \mathbf{J}_p = qR_s \qquad R_s = S_p \delta p \tag{2.3.12b}$$

where S_n is the surface recombination velocity (cm/s) of electrons and δn is the excess electron density. Similarly, S_p is the surface recombination velocity (cm/s) of holes and δp is the excess hole density. The unit vector points toward the surface from the semiconductor to account for the carrier loss by surface recombination at the surface states or dangling bonds or other types of defects, which trap carriers.

Auger Generation and Recombination Processes We consider band-to-band Auger-impact ionization processes [22], which account for nonradiative recombination in long-wavelength semiconductor lasers. (Band-to-bound Auger processes are possible only if the defect densities are significant.) We have the four possible processes, Fig. 2.6, shown in a simplified energy picture.

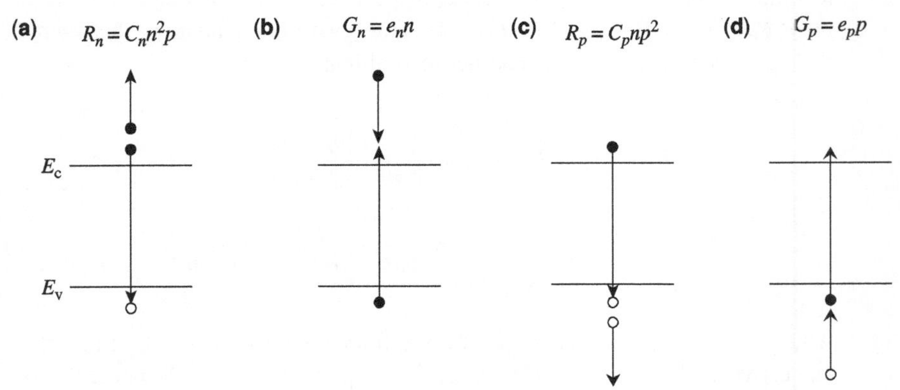

Figure 2.6 The energy band diagrams for the band-to-band Auger generation–recombination processes: (a) electron capture, (b) electron emission, (c) hole capture, and (d) hole emission.

1. **Electron Capture:** An electron in the conduction band recombines with a hole in the valence band and releases its energy to a nearby electron. This process destroys an electron–hole pair. The recombination rate is

$$R_n = C_n n^2 p. \tag{2.3.13}$$

An important Auger process, called CHCC process, is shown in Fig. 2.7a in the momentum space. An electron (1) in the conduction (C) band interacts with another electron (2) in the conduction (C) band via Coulomb interaction, resulting in the recombination of electron (2) with a heavy (H) hole in the valence band (2′). The emitted band-gap energy due to this electron–hole recombination is picked up by electron (1) being excited to a higher energy state (1′). In this process, the hole density in the heavy-hole band and the electron density in the conductor band should be used in (2.3.13).

2. **Electron Emission:** An incident (energetic) electron in the conduction band creates impact ionization by breaking a bond, thus causing an electron in the valence band jump to the conduction band. This process creates an electron–hole pair. The generation rate is

$$G_n = e_n n. \tag{2.3.14}$$

3. **Hole Capture:** An electron in the conduction band recombines with a hole in the valence band with the released energy taken up by a nearby hole. This process destroys an electron–hole pair. The recombination rate is

$$R_p = C_p n p^2. \tag{2.3.15}$$

Two important Auger processes related to hole captures, called CHSH and CHLH processes, are shown in Fig. 2.7b and Fig. 2.7c, respectively. In the CHSH process, a conduction (C) band electron recombines with a heavy hole (H), and the emitted band-gap energy is picked up by a nearby heavy

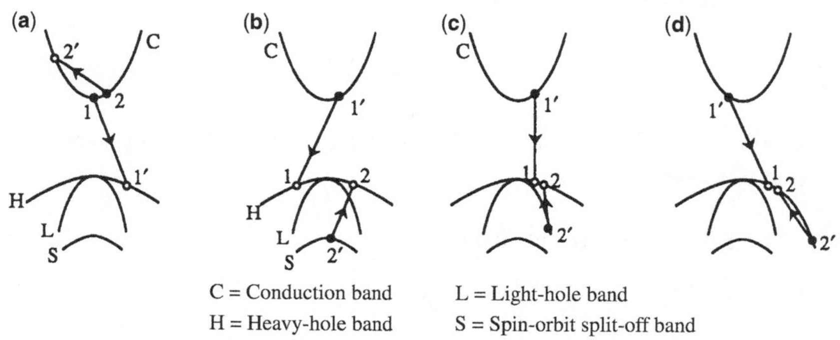

Figure 2.7 Band-to-band Auger recombination processes. (a) CHCC, (b) CHSH, (c) CHLH, and (d) CHHH processes.

(H) hole being excited to a spin-orbit split-off (S) hole band. In the CHLH process, a conduction (C) band electron recombines with a heavy hole (H), and the emitted band-gap energy is picked up by a nearby heavy (H) hole being excited to a light (L) hole band. In this case, C_p is the sum of CHLH and CHSH processes. The fourth process is the CHHH process, Fig. 2.7d, where an nearby heavy (H) hole is excited to the heavy (H) hole band when an electron (C) and heavy hole (H) recombine.

4. **Hole Emission:** An electron in the valence band jumps to the conduction band (or the breaking of a bond to create an electron–hole pair) due to the impact ionization of an energetic hole in the valence band. The generation rate is

$$G_p = e_p p. \tag{2.3.16}$$

At thermal equilibrium, no net generation/recombination exists. We thus have process a and its reverse process b balancing each other, as do processes c and d. Therefore,

$$R_n - G_n = 0 = C_n n_0^2 p_0 - e_n n_0$$

or

$$e_n = C_n n_0 p_0 = C_n n_\mathrm{i}^2. \tag{2.3.17}$$

Similarly, $e_p = C_p n_\mathrm{i}^2$. The total net Auger recombination rate is the sum of the net rates of electrons and holes as each process creates or destroys one electron–hole pair. Therefore,

$$R = R_n - G_n + R_p - G_p = (C_n n + C_p p)(np - n_\mathrm{i}^2). \tag{2.3.18}$$

In semiconductor lasers with a high-level of carrier injection, $n = p \gg n_\mathrm{i}$, we have simply the total Auger recombination rate

$$R_\mathrm{Aug} = Cn^3 \tag{2.3.19}$$

where C is due to the sum of all CHCC, CHLH, and CHSH Auger recombination processes.

2.3.3 The Intrinsic Quantum Efficiency

The total recombination rate when $n \gg (n_0 + p_0)$ is

$$R = An + Bn^2 + Cn^3 = \frac{n}{\tau(n)} \tag{2.3.20}$$

where the carrier lifetime is

$$\tau(n) = (A + Bn + Cn^2)^{-1}. \tag{2.3.21}$$

Often, we define the radiative and radiative lifetimes:

$$\tau_r = \frac{1}{Bn} \qquad \tau_{nr} = \frac{1}{A + Cn^2}. \tag{2.3.22}$$

The intrinsic quantum efficiency is the ratio of the radiative recombination rate to the total recombination rate

$$\eta_{in} = \frac{1/\tau_r}{1/\tau_r + 1/\tau_{nr}} = \frac{\tau_{nr}}{\tau_r + \tau_{nr}}, \tag{2.3.23}$$

which is an important parameter for LEDs. Under a forward bias of a LED, the current injection determines the carrier density; therefore, the radiative recombination due to the spontaneous emission of photons (Bn^2) and the nonradiative recombinations ($An + Cn^3$).

2.3.4 Recombination by the Stimulated Emission Process

In semiconductor lasers and LEDs, stimulated emission of photons by incident photons causes recombination of electron–hole pairs, as shown in Fig. 1.4b. This recombination rate is given by the photon density S multiplied by the rate of growth, or group velocity (v_g) times gain coefficient $g(n)$

$$R = v_g g(n) S. \tag{2.3.24}$$

The stimulated rate plays an important process in semiconductor lasers and will be used again in Chapter 10 for the analysis of semiconductor lasers to predict the light output power versus the injection current. It is also useful to predict the bandwidth of a directly modulated semiconductor laser when the injection current is modulated by an ac signal above threshold.

2.3.5 Impact Ionization Generation–Recombination Process [6]

This process is very much the same as the reverse Auger processes discussed above. However, the hot electron impact ionization processes usually depend on the incident current densities instead of the carrier concentrations. Microscopically, the processes are identical to the Auger-generation processes 2 and 4, which create an electron–hole pair due to an incident energetic electron or hole. These rates are usually given by

$$G_n = \alpha_n \frac{|J_n|}{q} \qquad G_p = \beta_p \frac{|J_p|}{q} \tag{2.3.25}$$

where α_n and β_p are the ionization coefficients for electrons and holes, respectively. α_n is the number of electron–hole pairs generated per unit distance due to an incident electron. β_p is the number of electron–hole pairs created per unit distance due to an incident hole.

The total net recombination rate is

$$R = -G_n - G_p = -\alpha_n \frac{|J_n|}{q} - \beta_p \frac{|J_p|}{q}. \tag{2.3.26}$$

Usually the ionization coefficients are related to the electric field E in the ionization region by the empirical formulas

$$\alpha_n(E) = \alpha_n^{\infty} \exp[-(E_{\mathrm{nc}}/E)^{\gamma}] \tag{2.3.27}$$

$$\beta_p(E) = \beta_p^{\infty} \exp[-(E_{\mathrm{pc}}/E)^{\gamma}] \tag{2.3.28}$$

where E_{nc} and E_{pc} are the critical fields and usually $1 \leq \gamma \leq 2$. These impact ionization coefficients will be used in the study of avalanche photodiodes in Chapter 15.

2.4 EXAMPLES AND APPLICATIONS TO OPTOELECTRONIC DEVICES

In this section, we consider a few simple examples to illustrate the optical generation and recombination processes and their effects on photodetectors. More details about photodetectors are discussed in Chapter 15. In the presence of optical injection, the continuity equations are

$$\frac{\partial n}{\partial t} = G_n - R_n + \frac{1}{q}\frac{\partial}{\partial x} J_n(x) \tag{2.4.1}$$

$$\frac{\partial p}{\partial t} = G_p - R_p - \frac{1}{q}\frac{\partial}{\partial x} J_p(x). \tag{2.4.2}$$

If a semiconductor is p-type doped with an acceptor concentration N_{A}, we know that

$$p_0 \simeq N_{\mathrm{A}} \quad \text{and} \quad n_0 \simeq n_{\mathrm{i}}^2/N_{\mathrm{A}}. \tag{2.4.3}$$

2.4.1 Uniform Optical Injection

Suppose the optical generation rate is uniform across the semiconductor, $G_n = G_n(t)$, which is independent of the position (Fig. 2.8) and the recombination rate is $R_n = \delta n/\tau_n$. We then have

$$\frac{\partial n}{\partial t} = G_n(t) - \frac{\delta n}{\tau_n}. \tag{2.4.4}$$

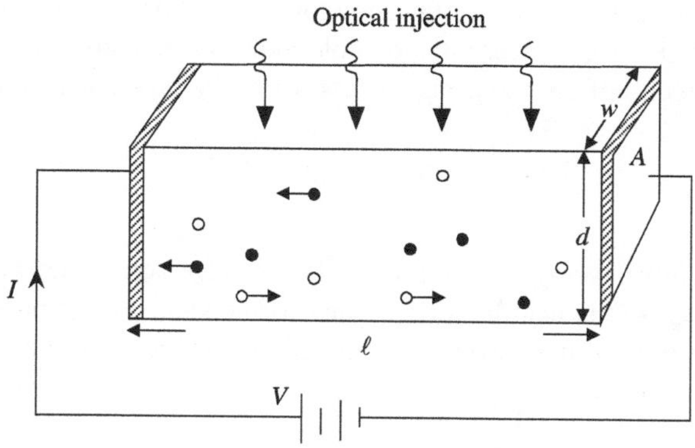

Figure 2.8 A homogeneous semiconductor under a uniform optical illumination from the side.

The new carrier densities are

$$p = p_0 + \delta p, \qquad n = n_0 + \delta n \tag{2.4.5}$$

where the excess carrier concentrations are equal $\delta p = \delta n$ for the interband optical generation process, as we assume that electrons and holes are created in pairs. Furthermore, for a constant light intensity, $G_n(t) = G_0$, we find the steady-state solution

$$\delta n = G_0 \tau_n. \tag{2.4.6}$$

Therefore, the amount of the excess carrier concentration is simply the optical generation rate G_0 multiplied by the carrier lifetime. In the presence of an electric field E along the x direction, the conduction current density is

$$J = J_n + J_p = q(\mu_n n + \mu_p p)E = \sigma E = \sigma \frac{V}{\ell} \tag{2.4.7}$$

with the conductivity

$$\sigma = q(\mu_n n + \mu_p p) \tag{2.4.8}$$

and V is the applied voltage and ℓ is the length of the semiconductor. At thermal equilibrium when there is no optical illumination

$$\sigma_0 = q(\mu_n n_0 + \mu_p p_0) \tag{2.4.9}$$

is the dark conductivity, and

$$\Delta\sigma = q(\mu_n \delta n + \mu_p \delta p) = q(\mu_n + \mu_p) G_0 \tau_n \tag{2.4.10}$$

is called the photoconductivity, which is proportional to the generation rate and the carrier lifetime. The photocurrent is

$$\Delta I = \Delta\sigma \frac{A}{\ell} V = q(\mu_n + \mu_p)\, G_0 \tau_n \frac{A}{\ell} V \tag{2.4.11}$$

and A is the cross-sectional area of the semiconductor. For most semiconductors, the electron mobility is much larger than the hole mobility. Because $v_d = \mu_n(V/\ell)$ is the average drift velocity of the electrons, we can write the photocurrent as

$$\Delta I \simeq q(G_0 A\ell)\frac{\tau_n}{\tau_t} \tag{2.4.12}$$

where $\tau_t = \ell/v_d$ is the average transit time of the electrons across the photoconductor of a length ℓ and $(G_0 A\ell)$ is just the total number of generated electron–hole pairs in a volume $A\ell$. The ratio of the carrier recombination lifetime τ_n to the transit time τ_t is the photoconductive gain.

Example A homogeneous germanium photoconductor (Fig. 2.8) is illuminated with an optical beam with the wavelength $\lambda = 1.55$ μm and an optical power $P = 1$ mW. Assume the optical beam is absorbed uniformly by the photoconductor and each photon creates one electron–hole pair, that is, the intrinsic quantum efficiency η_i is unity. The optical energy $h\nu$ is

$$h\nu = h\frac{c}{\lambda} = \frac{1.24}{\lambda} = 0.8 \text{ (eV)}$$

where λ is in micrometers in the above formula. The photon flux Φ_f is the number of photons injected per second

$$\Phi_f = \frac{P}{h\nu} = \frac{1\text{ mW}}{0.8 \times (1.6 \times 10^{-19})\text{ J}} = 7.81 \times 10^{15}\text{ s}^{-1}.$$

Because we assume all the photons are absorbed and the intrinsic quantum efficiency is unity, we have the generated rate per unit volume in $(wd\ell)$

$$G_0 = \eta_i\Phi_f/(wd\ell) = 7.81 \times 10^{15}/(wd\ell)\ (\text{s}^{-1}\text{ cm}^{-3}).$$

We use

$$\begin{array}{lll} \mu_n = 3900\text{ cm}^2\text{ V}^{-1}\text{s}^{-1} & \mu_p = 1800\text{ cm}^2\text{ V}^{-1}\text{ s}^{-1} & \tau_n = 10^{-3}\text{ s} \\ w = 10\ \mu\text{m} & d = 1\ \mu\text{m} & \ell = 1\text{ mm} \end{array}$$

The injected "primary" photocurrent is $q\eta_i\Phi_f = 1.6 \times 10^{-19} \times 7.81 \times 10^{15}\text{ A} = 1.25$ mA. The average electron transit time is

$$\tau_{tn} = \frac{\ell}{\mu_n \frac{V}{\ell}} = \frac{0.1\text{ cm}}{\left(3900\frac{\text{cm}^2}{\text{V}\cdot\text{s}}\right)\left(\frac{1\text{ V}}{0.1\text{ cm}}\right)} = 2.56 \times 10^{-6}\text{ s}$$

and the average transit time for holes is

$$\tau_{\mathrm{tp}} = \frac{\ell}{\mu_p \frac{V}{\ell}} = 5.56 \times 10^{-6}\,\mathrm{s}.$$

The photocurrent in the photoconductor is

$$\begin{aligned}\Delta I &= q(\mu_n + \mu_p) G_0 \tau_n \frac{wd}{\ell} V = (q\eta_{\mathrm{i}}\Phi_{\mathrm{f}})\left(\frac{\tau_n}{\tau_{\mathrm{tn}}} + \frac{\tau_n}{\tau_{\mathrm{tp}}}\right) \\ &= 1.25\,\mathrm{mA}\left(\frac{10^{-3}}{2.56\times10^{-6}} + \frac{10^{-3}}{5.56\times10^{-6}}\right) \\ &= 0.71\mathrm{A}\end{aligned}$$

where a photoconductive gain of around 570 occurs in this example.

2.4.2 Nonuniform Carrier Generation

If the carrier generation rate is not uniform, carrier diffusion will be important. For example, if the optical illumination on the semiconductor is limited to only a width S, instead of the length ℓ, Fig. 2.9a, the excess carriers generated in the illumination region will diffuse in both directions, assuming there is no external field applied in the x-direction [13]. Another example is in a stripe-geometry semiconductor laser [23] with an intrinsic region as the active layer where the carriers are injected

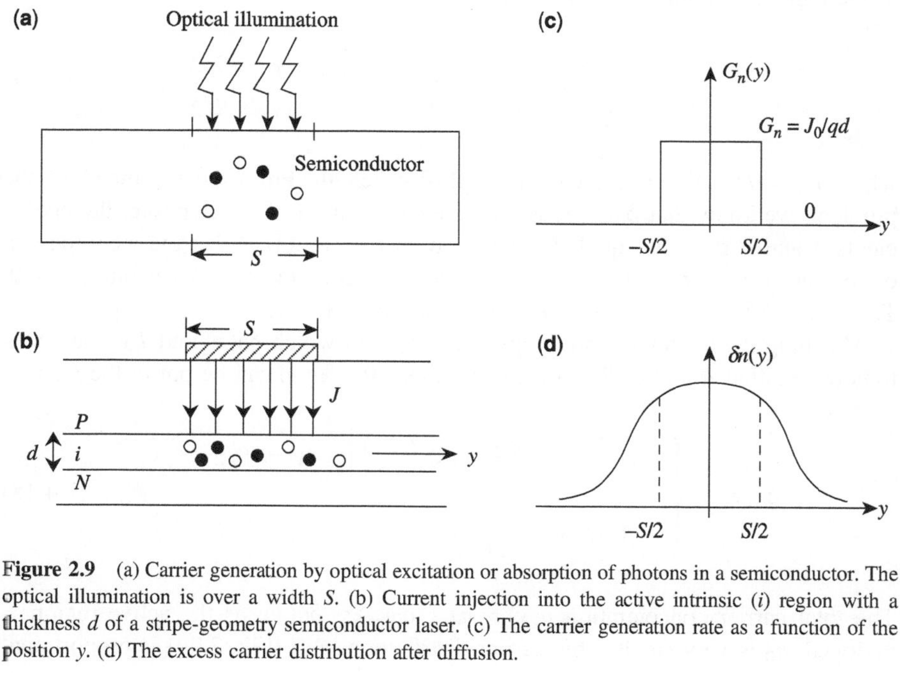

Figure 2.9 (a) Carrier generation by optical excitation or absorption of photons in a semiconductor. The optical illumination is over a width S. (b) Current injection into the active intrinsic (i) region with a thickness d of a stripe-geometry semiconductor laser. (c) The carrier generation rate as a function of the position y. (d) The excess carrier distribution after diffusion.

by a uniform current density within a stripe width S, Fig. 2.9b. We ignore the current spreading outside the region $|y| \leq S/2$.

The continuity equation for the electrons in the bulk semiconductor in Fig. 2.9a or in the intrinsic region in Fig. 2.9b is

$$\frac{\partial n}{\partial t} = G_n - \frac{\delta n}{\tau_n} + \frac{1}{q}\frac{\partial}{\partial y} J_n(y) \tag{2.4.13}$$

and the electron current density in the P-region is dominated by diffusion:

$$J_n(y) \simeq q D_n \frac{\partial n}{\partial y}. \tag{2.4.14}$$

At steady state, $\partial/\partial t = 0$, and

$$D_n \frac{\partial^2}{\partial y^2}\delta n - \frac{\delta n}{\tau_n} = -G_n(y). \tag{2.4.15}$$

Here the generation rate $G_n(y)$ is given by the optical intensity in Fig. 2.9a or, in the case of electric injection with a current density J_0 from the stripe S, Fig. 2.9c,

$$G_n(y) = \begin{cases} G_0 = \eta_i \dfrac{J_0}{qd}, & |y| \leq S/2 \\ 0, & |y| > S/2 \end{cases} \tag{2.4.16}$$

where $q = 1.6 \times 10^{-19}$ C, η_i is the injection efficiency, and d is the thickness of the active region. The solution to Eq. (2.4.13) is

$$\delta n(y) = \begin{cases} Ae^{-y/L_n} + Be^{+y/L_n} + G_0\tau_n & |y| \leq S/2 \\ Ce^{-(|y|-\frac{S}{2})/L_n} & |y| \geq S/2 \end{cases} \tag{2.4.17}$$

where $L_n = (D_n\tau_n)^{1/2}$ is the electron diffusion length. From the symmetry of the problem, we know that $\delta n(y)$ must be an even function of y. Therefore, the coefficients A and B must be equal. Another way to look at this is that the y-component of the electron current density $J_n(y)$ must be zero at the symmetry plane $y = 0$. Therefore, $d(\delta n)/dy = 0$ at $y = 0$, which also gives $A = B$.

Matching the boundary conditions at $y = S/2$, in which $\delta n(y)$ and $J_n(y)$ are continuous, we find A and C. The final expressions for $\delta n(y)$ can be put in the form

$$\delta n(y) = \begin{cases} G_0\tau_n\left[1 - e^{-S/2L_n}\cosh\left(\dfrac{y}{L_n}\right)\right] & |y| \leq S/2 \\ G_0\tau_n \sinh\left(\dfrac{S}{2L_n}\right)e^{-|y|/L_n} & |y| \geq S/2. \end{cases} \tag{2.4.18}$$

The total carrier concentration $n(y) = n_0 + \delta n(y) \simeq \delta n(y)$; as the active region is undoped, n_0 is very small. This carrier distribution $n(y)$ (Fig. 2.9d) is proportional to the spontaneous emission profile from the stripe-geometry semiconductor laser measured experimentally. In Chapter 10, we will discuss in more detail the stripe-geometry gain-guided semiconductor laser.

2.5 SEMICONDUCTOR *p-N* AND *n-P* HETEROJUNCTIONS

When two crystals of semiconductors with different energy gaps are combined, a heterojunction is formed [23–28]. The conductivity type of the smaller energy gap crystal is denoted by a lowercase n or p and that of the larger energy gap crystal is denoted by an uppercase N or P. Here we discuss the basic Anderson model for the heterojunctions [24, 25]. It has been pointed out that a more fundamental approach using the bulk and interface properties of the semiconductors should be used for the heterojunction model [26–28]. In Appendix A (page 787), we also discuss the model–solid theory for the band lineups of semiconductor heterojunctions including the strain effects, which are convenient for the estimation of band edge discontinuities.

Consider first a p-type narrow gap semiconductor such as GaAs in contact with an N-type wide band-gap semiconductor such as $Al_xGa_{1-x}As$ (Fig. 2.10). Let χ be the electron affinity, which is the energy required to take an electron from the conduction-band edge to the vacuum level, and let Φ be the work function, which is the energy difference between the vacuum level and the Fermi level. In each region, the Fermi level is determined by the charge neutrality condition. For example, in the p region, the electron density n and hole density p satisfy

$$n + N_a^- = p + N_d^+ \tag{2.5.1}$$

where n and p are related to the quasi-Fermi levels F_n and F_p, respectively, through (2.2.39) or (2.2.40). We may denote $N_a \equiv N_a^- - N_d^+$ as the "net" acceptor concentration. If $N_a \gg n_i$, we then have, $p = N_a$, which will determine the Fermi level F_p in the p region, Fig. 2.10a. A similar equation holds for the carrier densities, N and P, in the N region. Using $N_D \equiv N_D^+ - N_A^-$ as the "net" ionized donor concentration, we have, $N \simeq N_D \gg n_i$, which determines the bulk Fermi levels F_N in the N region before contact. When the two crystals are in contact, the Fermi levels will line up to be a constant across the junction under thermal equilibrium conditions without any voltage bias, Fig. 2.10b. Thus there will be redistributions of electrons and holes such that a built-in electric field exists to prevent any current flow in the crystal. To find the band bending, we use the depletion approximation.

2.5.1 Depletion Approximation for an Unbiased *p-N* Junction

The charge density is

$$\rho(x) = q(p - n + N_d^+ - N_a^-) \tag{2.5.2}$$

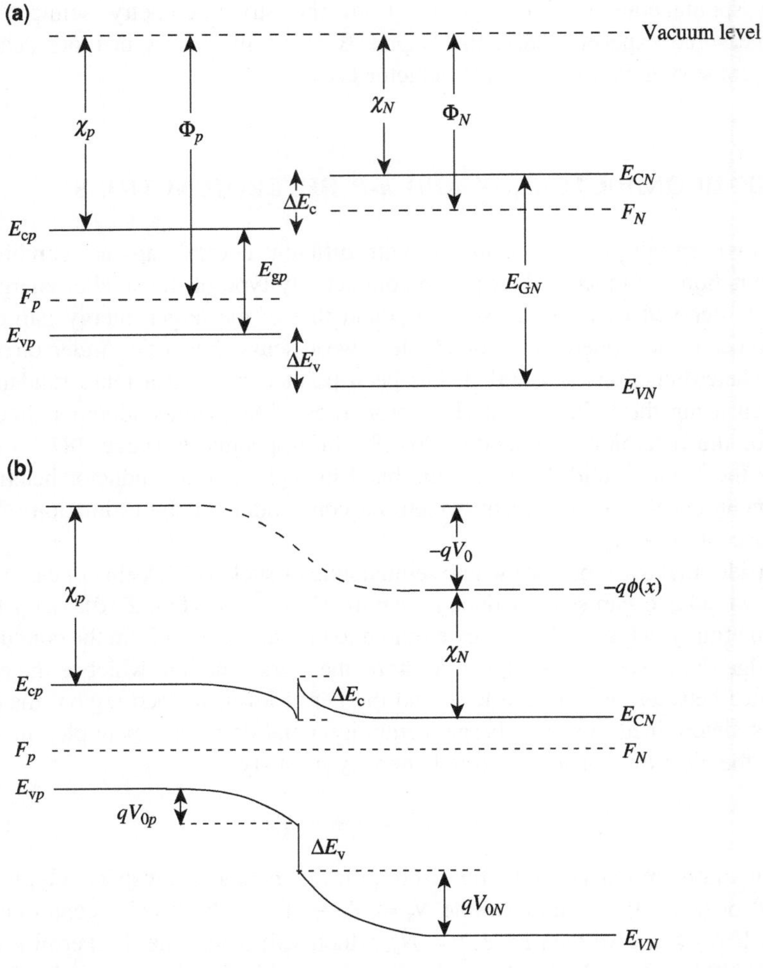

Figure 2.10 Energy band diagram for a *p-N* heterojunction (a) before contact and (b) after contact.

in the narrow gap (p) region and similar expressions with P and N in the wide gap N region. Because free carriers are depleted in the space charge region near the junction, we have, Fig. 2.11a and 2.11b,

$$\rho(x) = \begin{cases} -qN_a & -x_p < x < 0 \\ +qN_D & 0 < x < x_N \end{cases} \tag{2.5.3}$$

where N_a is the net acceptor concentration in the p-side and N_D is the net donor concentration on the N-side.

From Gauss's law, $\nabla \cdot (\varepsilon \mathbf{E}) = \rho$, we have

$$\frac{d}{dx}E(x) = \begin{cases} -q\, N_a/\varepsilon_p & -x_p < x < 0 \\ +q\, N_D/\varepsilon_N & 0 < x < x_N \end{cases} \tag{2.5.4}$$

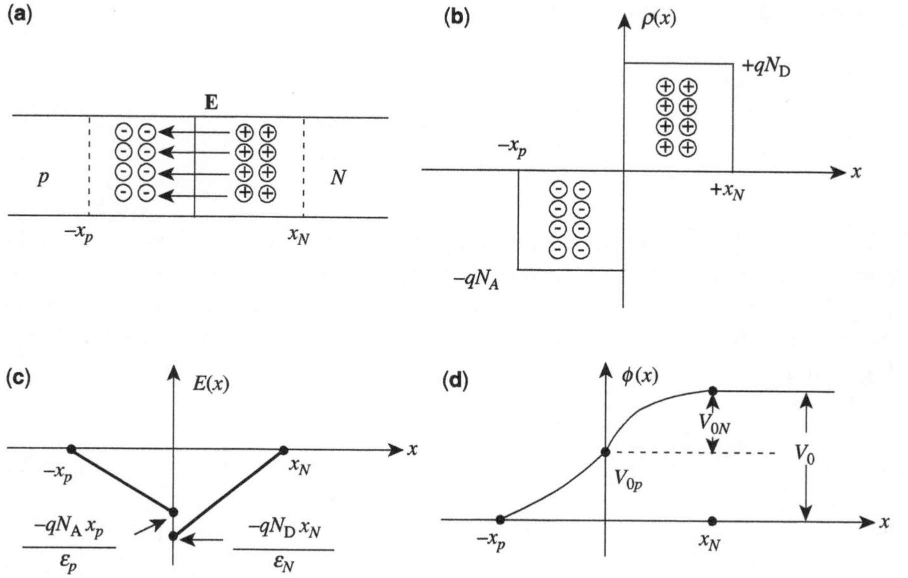

Figure 2.11 Illustrations of (a) a p-N junction geometry, (b) the charge distribution, (c) the electric field, and (d) the electrostatic potential based on the depletion approximation.

where ε_p and ε_N are the permittivity in the p and N regions, respectively. The above Gauss's law states that the slope of the $E(x)$ profile is given by the charge density divided by the permittivity. Thus the electric field is given by two linear functions

$$E(x) = \begin{cases} -q\,N_{\mathrm{a}}(x+x_p)/\varepsilon_p & -x_p < x < 0 \\ +q\,N_{\mathrm{D}}(x-x_N)/\varepsilon_N & 0 < x < x_N \end{cases} \tag{2.5.5}$$

in the depletion region and zero outside, Fig. 2.11c. The boundary condition states that the normal displacement vector $\mathbf{D} = \varepsilon\mathbf{E}$ is continuous at $x = 0$

$$\varepsilon_p\,E(0_-) = \varepsilon_N\,E(0_+) \tag{2.5.6}$$

or

$$N_{\mathrm{a}}x_p = N_{\mathrm{D}}x_N. \tag{2.5.7}$$

The electrostatic potential distribution $\phi(x)$ across the junction is related to the electric field by $E(x) = -d\phi(x)/dx$, which means that the slope of the potential profile is given by the negative of the electric field profile. If we choose the reference potential to be zero for $x < -x_p$, we have, Fig. 2.11d,

$$\phi(x) = \begin{cases} \dfrac{qN_{\mathrm{a}}}{2\varepsilon_p}(x+x_p)^2 & -x_p \le x \le 0 \\ \dfrac{qN_{\mathrm{a}}x_p^2}{2\varepsilon_p} + \dfrac{qN_{\mathrm{D}}}{2\varepsilon_N}(2xx_N - x^2) & 0 \le x \le x_N \\ V_0 & x_N \le x \end{cases} \tag{2.5.8}$$

where the contact potential V_0 is

$$\begin{aligned} V_0 &= V_{0p} + V_{0N} \\ V_{0p} &= \phi(0) = qN_a x_p^2/2\varepsilon_p \\ V_{0n} &= qN_D x_N^2/2\varepsilon_N. \end{aligned} \tag{2.5.9}$$

We see that V_0 is the total potential drop across the junction, whereas V_{0p} is the portion of the voltage drop on the p-side and V_{0N} on the N-side. The contact potential is evaluated using the bulk values of the Fermi levels F_p and F_N measured from the valence or conduction band edges E_{vp} and E_{CN}, respectively, before contact (see Fig. 2.10a),

$$\begin{aligned} V_0 &= (F_N - F_p)/q \\ &= [E_{gp} + \Delta E_c - (F_p - E_{vp}) - (E_{CN} - F_N)]/q \end{aligned} \tag{2.5.10}$$

and

$$\Delta E_c \simeq 0.67\,(E_{GN} - E_{gp}) \quad \text{for GaAs-Al}_x\text{Ga}_{1-x}\text{As system.} \tag{2.5.11}$$

For nondegenerate semiconductors,

$$E_{CN} - F_N \simeq -k_B T \ln (N/N_{CN}) \tag{2.5.12a}$$

$$F_p - E_{vp} \simeq -k_B T \ln (p/N_{vp}) \tag{2.5.12b}$$

where N_{CN} and N_{vp} are evaluated from (2.2.39a) and (2.2.39b) for the N and p regions, respectively. Because the electron concentration on the N-side is $N \simeq N_D$, and the hole concentration on the p-side is $p \simeq N_a$, the contact potential V_0 can be evaluated from the above equations when the doping concentrations N_a and N_D are known. Using the two conditions (2.5.7) and (2.5.9) and the total width of the depletion region x_W from

$$x_W = x_p + x_N \tag{2.5.13}$$

we find immediately that

$$x_p = \frac{N_D}{N_a + N_D} x_W \qquad x_N = \frac{N_a}{N_a + N_D} x_W. \tag{2.5.14}$$

Thus we may relate x_W to V_0 directly,

$$x_W = \left[\frac{2\varepsilon_p V_0}{qN_a N_D \left(N_D + \dfrac{\varepsilon_p}{\varepsilon_N} N_a \right)} \right]^{1/2} (N_a + N_D). \tag{2.5.15}$$

The band edge $E_v(x)$ from the *p*-side to the *N*-side is given by (choosing $E_v(-\infty) = 0$ as the reference potential energy)

$$E_v(x) = \begin{cases} -q\phi(x) & p\text{-side} \\ -\Delta E_v - q\phi(x) & N\text{-side} \end{cases} \tag{2.5.16}$$

where $\phi(x)$ is given in Eq. (2.5.8). The conduction band edge $E_c(x)$ is above $E_v(x)$ by an amount E_{gp} on the *p*-side and by an amount E_{GN} on the *N*-side, as shown in Fig. 2.9b

$$E_c(x) = \begin{cases} E_v(x) + E_{gp} & x < 0 \\ E_v(x) + E_{GN} & x > 0. \end{cases} \tag{2.5.17}$$

Example A p-GaAs/$N-$Al$_x$Ga$_{1-x}$As ($x = 0.3$) heterojunction is formed at thermal equilibrium without an external bias at room temperature. The doping concentration is $N_a = 1 \times 10^{18}\ \text{cm}^{-3}$ in the *p*-side and $N_D = 2 \times 10^{17}\ \text{cm}^{-3}$ in the *N*-side. Assume that the density-of-states hole effective mass for Al$_x$Ga$_{1-x}$As is

$$m_h^*(x) = (0.50 + 0.29x)m_0 \qquad 0 \le x \le 0.45$$

which accounts for both the heavy-hole and light-hole density of states. Other parameters are

$$\begin{aligned} m_e^* &= (0.0665 + 0.083x)m_0 & E_g(x) &= (1.424 + 1.247x)\ (\text{eV}) \\ \varepsilon(x) &= (13.1 - 3.0x)\varepsilon_0 & &(0 \le x \le 0.45) \end{aligned}$$

where x is the mole fraction of aluminum.

(a) We obtain for $x = 0.3$

p-GaAs: $m_e^* = 0.0665\ m_0$, $m_h^* = 0.50\ m_0$, $\varepsilon_p = 13.1\ \varepsilon_0$, $E_{gp} = 1.424$ eV.
N-Al$_{0.3}$Ga$_{0.7}$As: $m_e^* = 0.0914\ m_0$, $m_h^* = 0.587\ m_0$, $\varepsilon_N = 12.2\ \varepsilon_0$, $E_{gN} = 1.798$ eV.

The band edge discontinuities are

$$\begin{aligned} \Delta E_g &= 1.247 \times 0.3 = 0.374\ \text{eV} = 374\ \text{meV} \\ \Delta E_c &= 0.67\Delta E_g = 250.6\ \text{meV} \qquad \Delta E_v = 0.33\Delta E_g = 123.4\ \text{meV}. \end{aligned}$$

(b) We calculate the quasi-Fermi levels F_p and F_N for the bulk semiconductors for the given N_a and N_D separately.

p-GaAs region:

$$N_c = 2.51 \times 10^{19}\left(\frac{m_e^*}{m_0}\frac{T}{300}\right)^{3/2} \text{cm}^{-3} = 4.30 \times 10^{17}\,\text{cm}^{-3}$$

$$N_v = 2.51 \times 10^{19}\left(\frac{m_h^*}{m_0}\frac{T}{300}\right)^{3/2} \text{cm}^{-3} = 8.87 \times 10^{18}\,\text{cm}^{-3}$$

$$p = N_a = N_v F_{1/2}\left(\frac{E_{vp} - F_p}{k_B T}\right) = N_v \exp\left(\frac{E_{vp} - F_p}{k_B T}\right)$$

$$F_p - E_{vp} = -k_B T \ln\frac{N_a}{N_v} = 56.4\text{ meV}.$$

N-$Al_{0.3}Ga_{0.7}As$ region:

$$N_c = 2.51 \times 10^{19}(0.0914)^{3/2} = 6.94 \times 10^{17}\,\text{cm}^{-3}$$

$$N_v = 2.51 \times 10^{19}(0.587)^{3/2} = 1.13 \times 10^{19}\,\text{cm}^{-3}$$

$$N = N_D = N_c F_{1/2}\left(\frac{F_N - E_{cN}}{k_B T}\right) \simeq N_c \exp\left(\frac{F_N - E_{cN}}{k_B T}\right)$$

$$E_{cN} - F_N \simeq -k_B T \ln\frac{N_D}{N_c} = -25.85 \ln\frac{2 \times 10^{17}}{6.94 \times 10^{17}} = 32.2\text{ meV}.$$

(c) The contact potential is

$$\begin{aligned} V_0 &= [E_{gp} + \Delta E_c - (F_p - E_{vp}) - (E_{cN} - F_N)]/q \\ &= (1424 + 250.6 - 56.4 - 32.2)\text{ mV} = 1586\text{ mV}. \end{aligned}$$

(d) The depletion widths are

$$x_p = \left[\frac{2\varepsilon_p V_0}{qN_a N_D\left(N_D + \frac{\varepsilon_p}{\varepsilon_N}N_a\right)}\right]^{1/2} N_D$$

$$= \left[\frac{2 \times 13.1 \times 8.854 \times 10^{-14} \times 1.586}{1.6 \times 10^{-19} \times 10^{18} \times 2 \times 10^{7}\left(2 \times 10^{17} + \frac{13.1}{12.2} \times 10^{18}\right)}\right]^{1/2} \times 2 \times 10^{17}$$

$$= 0.019\ \mu\text{m}$$

$$x_N = \frac{x_p}{N_D} N_a = 0.095\ \mu\text{m}$$

$$x_W = x_p + x_N = 0.114\ \mu\text{m}.$$

The energy band diagram is plotted in Fig. 2.12.

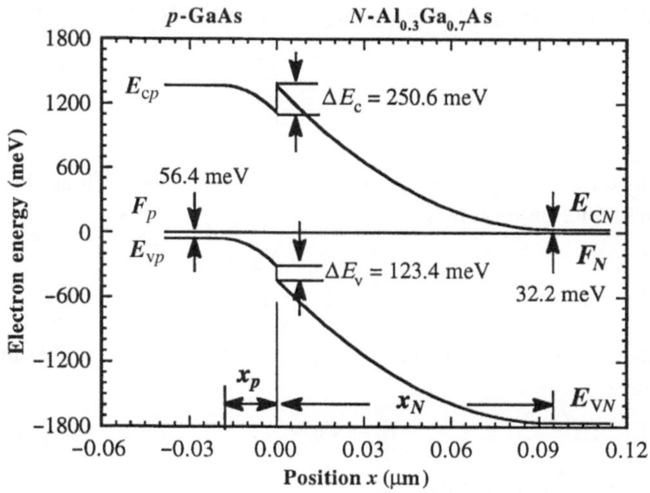

Figure 2.12 Band diagram of a p-GaAs/N-$Al_{0.3}Ga_{0.7}As$ heterojunction with $N_a = 1 \times 10^{18}\ cm^{-3}$ in the narrow gap p region and $N_D = 2 \times 10^{17}\ cm^{-3}$ in the wide gap N region.

2.5.2 Biased *p-N* Junction

With an applied voltage V across the diode, Fig. 2.13a, the potential barrier is reduced by qV if the voltage is a forward bias. Note that our convention for the polarity of the bias voltage V is that the positive electrode is connected to the p-side of the diode. Thus the depletion width x_W is reduced. Explicitly,

$$\phi(x) = \begin{cases} \dfrac{qN_a}{2\varepsilon_p}(x + x_p)^2 & -x_p \le x < 0 \\ \dfrac{qN_a x_p^2}{2\varepsilon_p} + \dfrac{qN_D}{2\varepsilon_N}(2xx_N - x^2) & 0 \le x \le x_N \\ V_0 - V & x_N \le x. \end{cases} \tag{2.5.18}$$

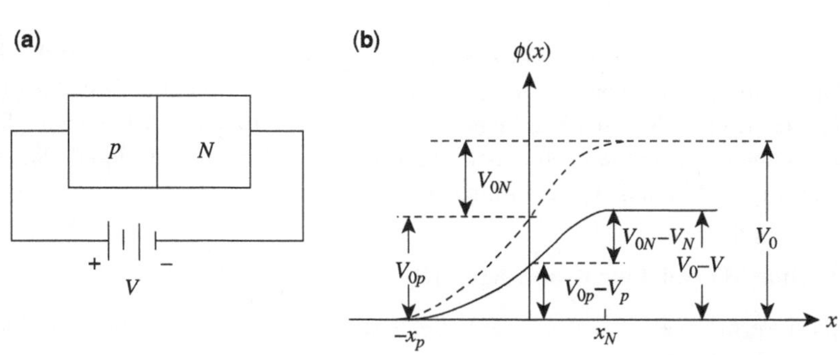

Figure 2.13 (a) A p-N junction with a bias V and (b) the corresponding electrostatic potential $\phi(x)$ (solid curve for $V > 0$ and dashed curve for $V = 0$).

The potential drop on the p-side of the depletion region is $\phi(0)$ shown in Fig. 2.13b,

$$V_{0p} - V_p = \frac{qN_a}{2\varepsilon_p} x_p^2 \tag{2.5.19}$$

which is reduced from V_{0p} by an amount V_p. Similarly, the voltage drop across the N-side of the depletion region is

$$V_{0N} - V_N = \frac{qN_D}{2\varepsilon_N} x_N^2 \tag{2.5.20}$$

which is reduced from V_{0N} by an amount V_N. The sum of V_p and V_N has to be equal to the bias voltage V

$$V_p + V_N = V. \tag{2.5.21}$$

Again the charge neutrality condition

$$N_a x_p = N_D x_N \tag{2.5.22}$$

and

$$x_W = x_p + x_N \tag{2.5.23}$$

together with

$$V_0 - V = \frac{qN_a}{2\varepsilon_p} x_p^2 + \frac{qN_D}{2\varepsilon_N} x_N^2 \tag{2.5.24}$$

lead to

$$x_W = \left[\frac{2\varepsilon_p (V_0 - V)}{qN_a N_D \left(N_D + \frac{\varepsilon_p}{\varepsilon_N} N_a \right)} \right]^{1/2} (N_a + N_D). \tag{2.5.25}$$

Thus the depletion region is reduced as x_W is proportional to $(V_0 - V)^{1/2}$. If the diode is reverse-biased, V is negative and x_W becomes wider. The potential function $\phi(x)$ in (2.5.18) determines the band bendings $E_v(x)$ and $E_c(x)$ through (2.5.16) and (2.5.17). The resultant band diagrams for a forward-biased and a reverse-biased diode are shown in Fig. 2.14a and Fig. 2.14b, respectively.

2.5.3 Quasi-Fermi Levels and Minority Carrier Injections

The quasi-Fermi levels $F_p(x)$ and $F_N(x)$ are separated by the bias voltage V multiplied by the magnitude of the electron charge $q = 1.6 \times 10^{-19}$ C

$$F_N - F_p = qV. \tag{2.5.26}$$

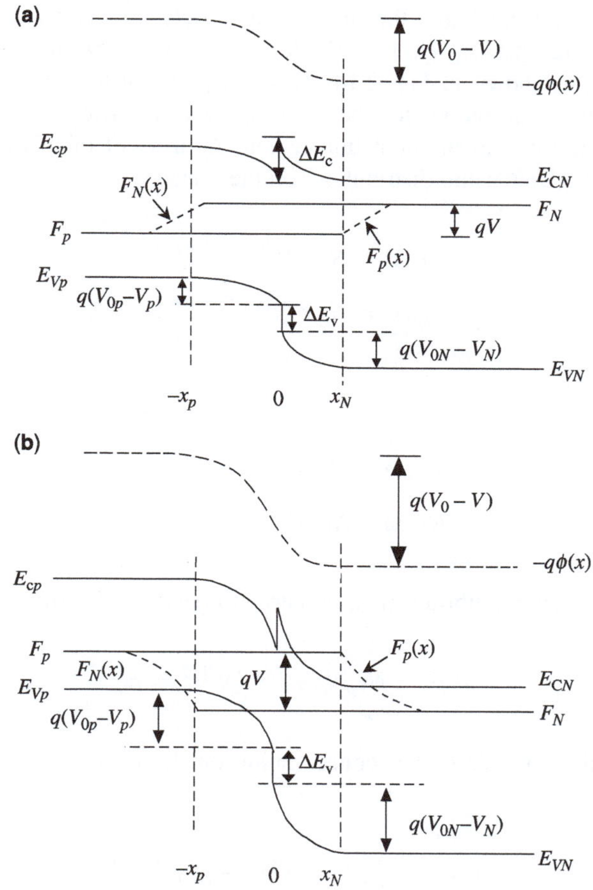

Figure 2.14 Energy band diagrams for a *p-N* heterojunction under (a) forward bias and (b) reverse bias based on the depletion approximation.

This fact can be seen clearly by comparing the band diagrams of the zero bias and forward bias cases in Fig. 2.10b and Fig. 2.14a, respectively. To find the variations of the quasi-Fermi levels $F_p(x)$ and $F_N(x)$ as functions of the position, we have to find the carrier concentrations everywhere in the diode. Assuming a low-level injection condition, which means that the amount of injected excess carriers due to the voltage bias is always much less than the majority carrier concentration in each region, the minority carrier concentration is much more strongly affected than is the majority carrier concentration in each region. As a result of this, there are two different quasi-Fermi levels $F_N(x)$ and $F_p(x)$ for the electron and the hole concentrations, respectively, near the junction region. Far away from the junction region, the quasi-Fermi level should approach its thermal equilibrium value in the bulk region, respectively, that is, $F_N(x \rightarrow +\infty) \rightarrow F_N$ (bulk value on the *N*-side) and $F_p(x \rightarrow -\infty) \rightarrow F_p$ (bulk value on the *p*-side), as the carrier concentrations are hardly affected away from the junction.

It is usually assumed that the quasi-Fermi levels stay as constants across the depletion region, that is, $F_N(x) = F_N$ for $-x_p \leq x < +\infty$, and $F_p(x) = F_p$ for $-\infty < x < x_N$, as shown in Fig. 2.14. This is equivalent to the statement that the carrier distribution in energy for the same species of carrier across the depletion region stays the same in the depletion region. As a result of this assumption, we find that, for the Boltzmann distribution, on the p-side

$$p_p(x) = N_{\mathrm{v}p} e^{(E_{\mathrm{v}}(x) - F_p(x))/k_{\mathrm{B}}T} \tag{2.5.27a}$$

$$n_p(x) = N_{\mathrm{c}p} e^{(F_N(x) - E_{\mathrm{c}}(x))/k_{\mathrm{B}}T}. \tag{2.5.27b}$$

On the N-side

$$P_N(x) = N_{VN} e^{(E_{\mathrm{v}}(x) - F_p(x))/k_{\mathrm{B}}T} \tag{2.5.28a}$$

$$N_N(x) = N_{CN} e^{(F_N(x) - E_{\mathrm{c}}(x))/k_{\mathrm{B}}T}. \tag{2.5.28b}$$

Because at thermal equilibrium (i.e., no current injection, $V = 0$), $F_p(x) = F_N(x)$

$$n_p p_p = N_{Vp} N_{\mathrm{c}p} e^{-(E_{\mathrm{c}} - E_{\mathrm{v}})/k_{\mathrm{B}}T} = n_{\mathrm{i}p}^2 \tag{2.5.29}$$

where $n_{\mathrm{i}p}$ is the intrinsic carrier concentration on the p-side. Thus, if $V \neq 0$, for $-x_p < x < 0$,

$$n_p p_p = n_{\mathrm{i}p}^2 e^{(F_N - F_p)/k_{\mathrm{B}}T} = n_{\mathrm{i}p}^2 e^{qV/k_{\mathrm{B}}T}. \tag{2.5.30}$$

At the edge of the depletion region, $x = -x_p$, $p_p(-x_p) \simeq p_{p0} \simeq N_{\mathrm{a}}$. Therefore,

$$n_p(-x_p) \simeq \frac{n_{\mathrm{i}p}^2}{N_{\mathrm{a}}} e^{qV/k_{\mathrm{B}}T} = n_{p0} e^{qV/k_{\mathrm{B}}T}. \tag{2.5.31}$$

Here the subscript "0" in p_{p0} and n_{p0} refers to their thermal equilibrium values. The minority carrier near the edge of the depletion region, $n_p(-x_p)$, differs by a factor $\exp(qV/k_{\mathrm{B}}T)$ from its thermal equilibrium value n_{p0} due to the carrier injection. Similarly, at the edge of the depletion region on the N-side, $x = x_N$, the minority carrier concentration is

$$P(x) = P_{N0} e^{qV/k_{\mathrm{B}}T} \tag{2.5.32}$$

where $P_{N0} = n_{\mathrm{i}N}^2/N_{\mathrm{D}}$, where $n_{\mathrm{i}N}$ is the intrinsic carrier concentration on the N-side.

If we compare the current density equations for the p-side of the diode, assuming that n_{p0} and p_{p0} are independent of x, $\partial n/\partial x = \partial \delta n/\partial x$, $\partial p/\partial x = \partial \delta p/\partial x$,

$$J_n = q\mu_n nE + qD_n \frac{\partial}{\partial x}\delta n \tag{2.5.33a}$$

$$J_p = q\mu_p pE - qD_p \frac{\partial}{\partial x}\delta p \tag{2.5.33b}$$

we see that in general, $\mu_n nE \ll \mu_p pE$, as $n \ll p$ on the p-side of the diode. If J_n and J_p are of the same order of magnitude and the charge neutrality condition, $\delta n(x) \simeq \delta p(x)$, applies in the quasi-neutral region (i.e., $x < -x_p$), where E is very small, we expect that the drift current is much less than the diffusion current for the electrons. Thus, the minority current density J_n is only dominated by the diffusion current density (this is not true for the majority current density as will be seen later)

$$J_n \simeq qD_n \frac{\partial}{\partial x}\delta n(x). \tag{2.5.34}$$

Using the charge continuity equation for the electrons

$$\frac{\partial}{\partial t}\delta n = -\frac{\delta n}{\tau_n} + \frac{1}{q}\frac{\partial}{\partial x}J_n \tag{2.5.35}$$

under steady-state conditions, we obtain the diffusion equation:

$$D_n \frac{\partial^2}{\partial x^2}\delta n(x) - \frac{\delta n(x)}{\tau_n} = 0. \tag{2.5.36}$$

The solution with the boundary condition, $\delta n(x = -\infty) = 0$, has the form

$$\delta n(x) = \delta n(-x_p)e^{(x+x_p)/L_n} \tag{2.5.37}$$

where

$$\begin{aligned}\delta n(-x_p) &= n_p(-x_p) - n_{p0}\\ &= n_{p0}(e^{qV/k_BT} - 1)\end{aligned} \tag{2.5.38}$$

and $L_n = \sqrt{D_n \tau_n}$ is the diffusion length for electrons in the p-region. The injection condition from (2.5.31) has been used. The total carrier concentrations on the p-side for $x < -x_p$ are

$$\begin{aligned}n_p(x) &= n_{p0} + \delta n(x)\\ &= n_{p0} + n_{p0}(e^{qV/k_BT} - 1)e^{(x+x_p)/L_n}\end{aligned} \tag{2.5.39}$$

and

$$\begin{aligned}p_p(x) &= p_{p0} + \delta p(x)\\ &= p_{p0} + n_{p0}(e^{qV/k_BT} - 1)e^{(x+x_p)/L_n}\end{aligned} \tag{2.5.40}$$

where $\delta p(x) = \delta n(x)$, $p_{p0} = N_a$, and $n_{p0} = n_{ip}^2/N_a$.

Similarly, in the quasi-neutral region on the N-side of the diode, $x > x_N$, the minority (hole) current density is approximately

$$J_P \simeq -qD_P \frac{\partial}{\partial x} \delta P(x) \tag{2.5.41}$$

and

$$\frac{\partial}{\partial t} \delta P = -\frac{\delta P}{\tau_P} - \frac{1}{q} \frac{\partial}{\partial x} J_P \tag{2.5.42}$$

which lead to

$$\delta P(x) = \delta P(x_N) e^{-(x-x_N)/L_p} \tag{2.5.43}$$

where

$$\begin{aligned} \delta P(x_N) &= P(x_N) - P_{N0} \\ &= P_{N0}(e^{qV/k_B T} - 1) \end{aligned} \tag{2.5.44}$$

and $L_P = \sqrt{D_P \tau_P}$ is the diffusion length of the holes on the N-side. The total carrier concentrations on the N-side of the diode, $x > x_N$, are

$$\begin{aligned} P_N(x) &= P_{N0} + \delta P(x) \\ &= P_{N0} + P_{N0}(e^{qV/k_B T} - 1) e^{-(x-x_N)/L_p} \end{aligned} \tag{2.5.45}$$

$$\begin{aligned} N_N(x) &= N_{N0} + \delta N(x) \\ &= N_{N0} + P_{N0}(e^{qV/k_B T} - 1) e^{-(x-x_N)/L_p}. \end{aligned} \tag{2.5.46}$$

Knowing the carrier concentration for $x < -x_p$ and $x > x_N$, the quasi-Fermi levels can be obtained from (2.5.27) and (2.5.28),

$$F_N(x) - E_{cp} = F_p - E_{cp} + k_B T \,\ell\text{n}[1 + (e^{qV/k_B T} - 1) e^{(x+x_p)/L_n}] \quad x \le -x_p \tag{2.5.47}$$

where $k_B T \,\ell\text{n}\, (n_{p0}/N_{cp}) = F_p - E_{cp}$ on the p-side has been used. Similarly,

$$E_{VN} - F_p(x) = E_{VN} - F_N + k_B T \,\ell\text{n}[1 + (e^{qV/k_B T} - 1) e^{-(x-x_N)/L_p}] \quad x > x_N \tag{2.5.48}$$

where $k_B T \,\ell\text{n}\, (P_{N0}/N_{VN}) = E_{VN} - F_N$ has been used. From (2.5.47) and (2.5.48), we have

$$F_N(-x_p) = F_p + qV = F_N \tag{2.5.49a}$$

$$F_p(x_N) = F_N - qV = F_p \tag{2.5.49b}$$

and $F_N\,(x \to -\infty) \to F_p$, $F_p\,(x \to +\infty) \to F_N$.

For the forward bias case, $V > 0$, when $\exp(qV/k_BT) \gg 1$, the above results in (2.5.47) and (2.5.48) can be further simplified and the quasi-Fermi levels are plotted in Fig. 2.14a. Similarly, for the reverse bias, $\exp(qV/k_BT) \ll 1$, the quasi-Fermi levels are plotted in Fig. 2.14b.

2.5.4 Current Densities and $I-V$ Characteristics

The current densities are obtained for the minority carriers first. On the p-side, $J_n = qD_n \frac{\partial}{\partial x}\delta n(x)$, and on the N-side, $J_P \simeq -qD_P \frac{\partial}{\partial x}\delta P(x)$. Assuming that there is no generation or recombination current in the space charge region, that is, J_n and J_P are constant over the space charge region, the total current density is thus the

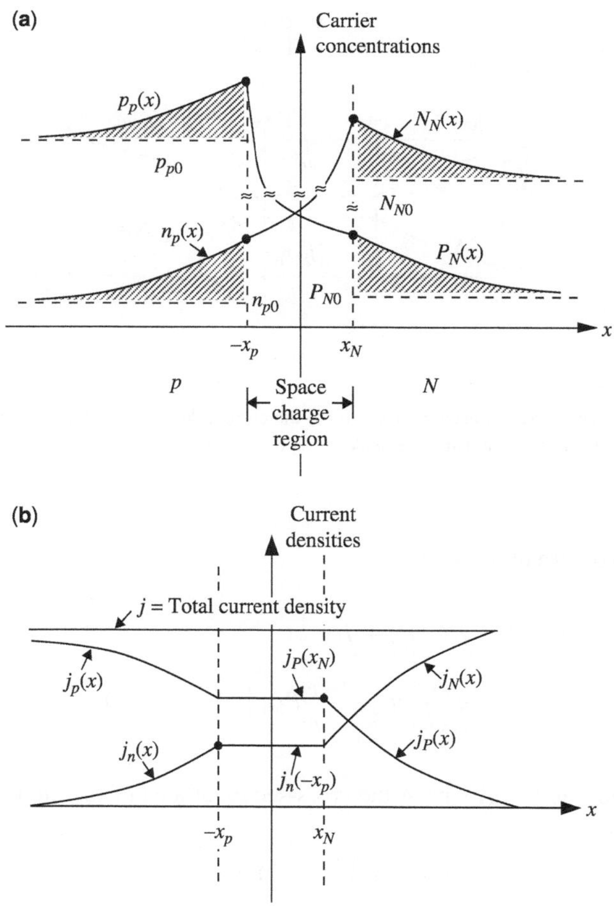

Figure 2.15 (a) The carrier concentrations and (b) the current densities as functions of position x in a forward biased p-N heterojunction diode using the depletion approximation.

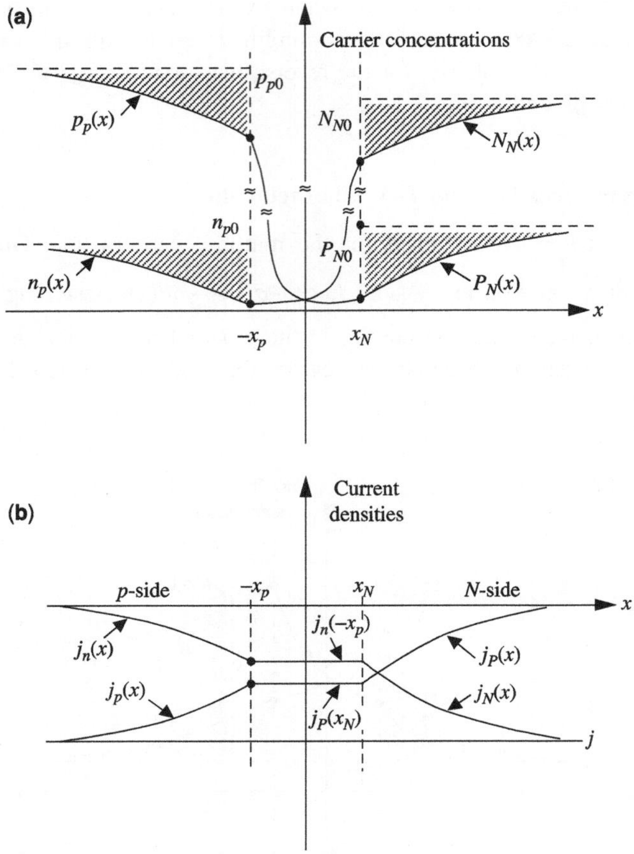

Figure 2.16 (a) The carrier concentrations and (b) the current densities as functions of x for a reverse biased p-N heterojunction diode using the depletion approximation.

sum of the two current densities

$$\begin{aligned} J &= J_n(-x_p) + J_P(x_N) \\ &= q\left(\frac{D_n}{L_n} n_{p0} + \frac{D_P}{L_P} P_{N0}\right)(e^{qV/k_B T} - 1). \end{aligned} \tag{2.5.50}$$

The total current is $I = jA$ with A the cross-sectional area of the diode.

$$I = I_0(e^{qV/k_B T} - 1) \tag{2.5.51a}$$

$$I_0 = q\left(\frac{D_n}{L_n} n_{p0} + \frac{D_P}{L_P} P_{N0}\right)A. \tag{2.5.51b}$$

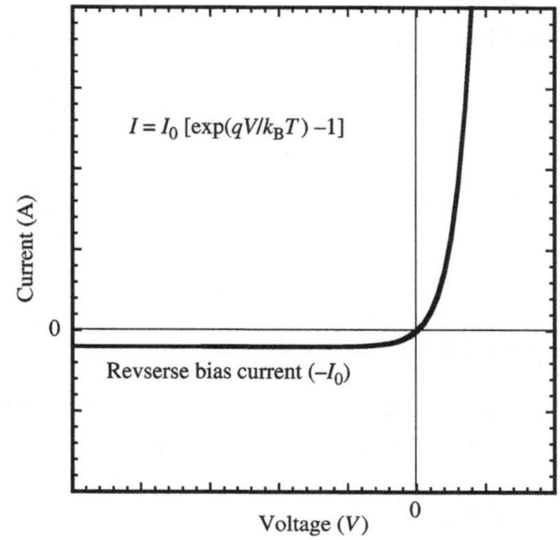

Figure 2.17 The I–V curve of a semiconductor diode.

Knowing the total current density J, which is a constant across the diode, the majority carrier current densities $J_p(x)$ and $J_N(x)$ can be obtained from

$$J_p(x) = J - J_n(x) \quad \text{on the } p\text{-side}, \quad x \leq -x_p \tag{2.5.52a}$$

$$J_N(x) = J - J_P(x) \quad \text{on the } N\text{-side}, \quad x \geq x_N. \tag{2.5.52b}$$

The complete current distributions are plotted in Fig. 2.15 for the forward bias case and in Fig. 2.16 for the reverse bias case. The well-known diode I–V curve is shown in Fig. 2.17. The current increases rapidly with the forward bias voltage, and saturates at a small negative current when the bias voltage is negative.

2.5.5 Semiconductor *n-P* Heterojunction

If the narrow gap semiconductor is doped *n*-type and the wide gap semiconductor is doped *P*-type, the procedure to find the band diagram is similar to that for the *p-N* heterojunction in the previous sections. We will show a numerical example as follows.

Example An *n*-GaAs/*P*-$Al_{0.3}Ga_{0.7}As$ heterojunction is formed at room temperature and a zero bias. The doping concentrations are $N_d = 4 \times 10^{16}\,\text{cm}^{-3}$ in the *n* region and $N_A = 2 \times 10^{17}\,\text{cm}^{-3}$ in the *P* region. The parameters have been given in the previous example on a *p-N* heterojunction. We have:

(a) *n*-side (GaAs) $N_c = 2.5 \times 10^{19}\left(\dfrac{m_e^*}{m_0}\dfrac{T}{300}\right)^{3/2} = 4.30 \times 10^{17}\,\text{cm}^{-3}$

$$E_{cn} - F_n = -k_BT \ln\frac{N_d}{N_c} = -25.85(-2.3749) = 61.4\ \text{meV}$$

$$P\text{-side }(\mathrm{Al_{0.3}Ga_{0.7}As})\ N_{\mathrm{V}} = 2.5 \times 10^{19}\left(\frac{m_{\mathrm{h}}^*}{m_0}\frac{T}{300}\right)^{3/2} = 1.13 \times 10^{19}\,\mathrm{cm}^{-3}$$

$$P = N_{\mathrm{A}} = N_{\mathrm{V}}\exp\left(\frac{E_{\mathrm{V}P} - F_P}{k_{\mathrm{B}}T}\right)$$

$$F_P - E_{VP} = -k_{\mathrm{B}}T\ln\left(\frac{N_{\mathrm{A}}}{N_{\mathrm{V}}}\right) = -25.85\left(\ln\frac{2\times 10^{17}}{1.13\times 10^{19}}\right)$$

$$= 104.3\,\mathrm{meV}.$$

(b) The contact potential is

$$V_0 = [E_{\mathrm{G}P} - \Delta E_{\mathrm{c}} - (E_{cn} - F_n) - (F_P - E_{\mathrm{V}P})]/q$$
$$= 1798 - 250.6 - 61.4 - 104.3 = 1381.7\ \mathrm{mV}.$$

(c) The depletion widths are

$$x_n = \left[\frac{2\varepsilon_n V_0}{qN_{\mathrm{d}}N_{\mathrm{A}}\left(N_{\mathrm{A}} + \dfrac{\varepsilon_n}{\varepsilon_p}N_{\mathrm{d}}\right)}\right]^{1/2} N_{\mathrm{A}}$$

$$= \left[\frac{2 \times 13.1 \times 8.854 \times 10^{-14} \times 1.3817}{1.6 \times 10^{-19} \times 4 \times 10^{16} \times 2 \times 10^{17}\left(2 \times 10^{17} + \dfrac{13.1}{12.2} \times 4 \times 10^{16}\right)}\right]^{1/2}$$

$$\times 2 \times 10^{17}\ (\mathrm{cm})$$
$$= 5.1\ \mu\mathrm{m}$$

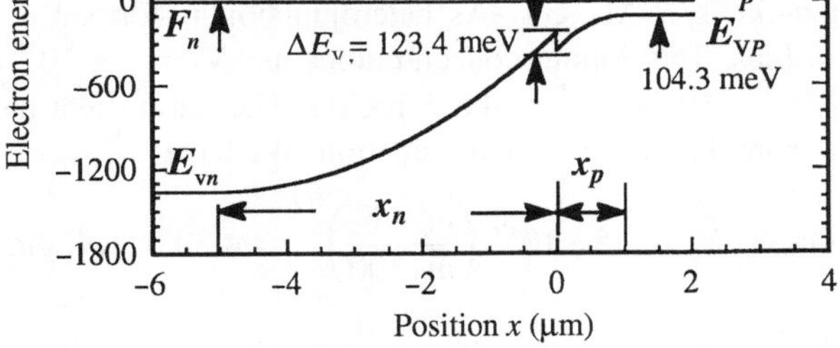

Figure 2.18 Band diagram of an unbiased n-GaAs/P-$\mathrm{Al_{0.3}Ga_{0.7}As}$ heterojunction with $N_{\mathrm{D}} = 4\times 10^{16}$ cm^{-3} in the n region and $N_{\mathrm{A}} = 2 \times 10^{17}\ \mathrm{cm}^{-3}$ in the P region.

$$x_p = x_n \frac{N_d}{N_A} = 1.0\ \mu\text{m}$$

$$x_W = x_n + x_p = 6.1\ \mu\text{m}.$$

(d) The energy band diagram of the n-GaAs/P-$Al_{0.3}Ga_{0.7}As$ junction is plotted in Fig. 2.18.

2.6 SEMICONDUCTOR *n-N* HETEROJUNCTIONS AND METAL–SEMICONDUCTOR JUNCTIONS

2.6.1 Semiconductor *n-N* Heterojunctions

For an n-N heterojunction, the theory is a bit more involved. With the development of high electron mobility transistor (HEMT), also called modulation doped field-effect transistor (MODFET), the n-N heterojunction theory is very useful for modeling these devices [29]. Depending on the model, the results may be different. A qualitative approach is shown in Fig. 2.19. When the two crystals with n-type (N_d) and N-type (N_D) dopings are brought in contact, the charge redistribution produces a space charge region. There will be electrons that spill over from the N region to the n region for the alignment of the Fermi levels. The charge distribution is

$$\rho(x) = \begin{cases} -q(n(x) - N_d), & x < 0 \\ qN_D, & 0 < x < x_N. \end{cases} \tag{2.6.1}$$

Note that far away from the junction, $n(x \to -\infty) \to N_d$

$$E(x) = \begin{cases} -q\int_{-\infty}^{x} [n(x') - N_d]dx'/\varepsilon_n & x < 0 \\ qN_D(x - x_N)/\varepsilon_N & 0 < x < x_N. \end{cases} \tag{2.6.2}$$

We obtain charge neutrality from the boundary condition for normal displacement vector component at $x = 0$, $n_s = N_D x_N$, where

$$n_s = \int_{-\infty}^{0} [n(x) - N_d]dx. \tag{2.6.3}$$

Judging from the charge distribution, the electric field, and the potential profile, one sees that the band diagram in Fig. 2.19 (Step 5) is very similar to that of the p-N heterojunction in Fig. 2.10b.

Using the fact that the contact potential is given by

$$V_0 = (F_N - F_n)/q = [(E_{cn} - F_n) + \Delta E_c - (E_{CN} - F_N)]/q, \tag{2.6.4}$$

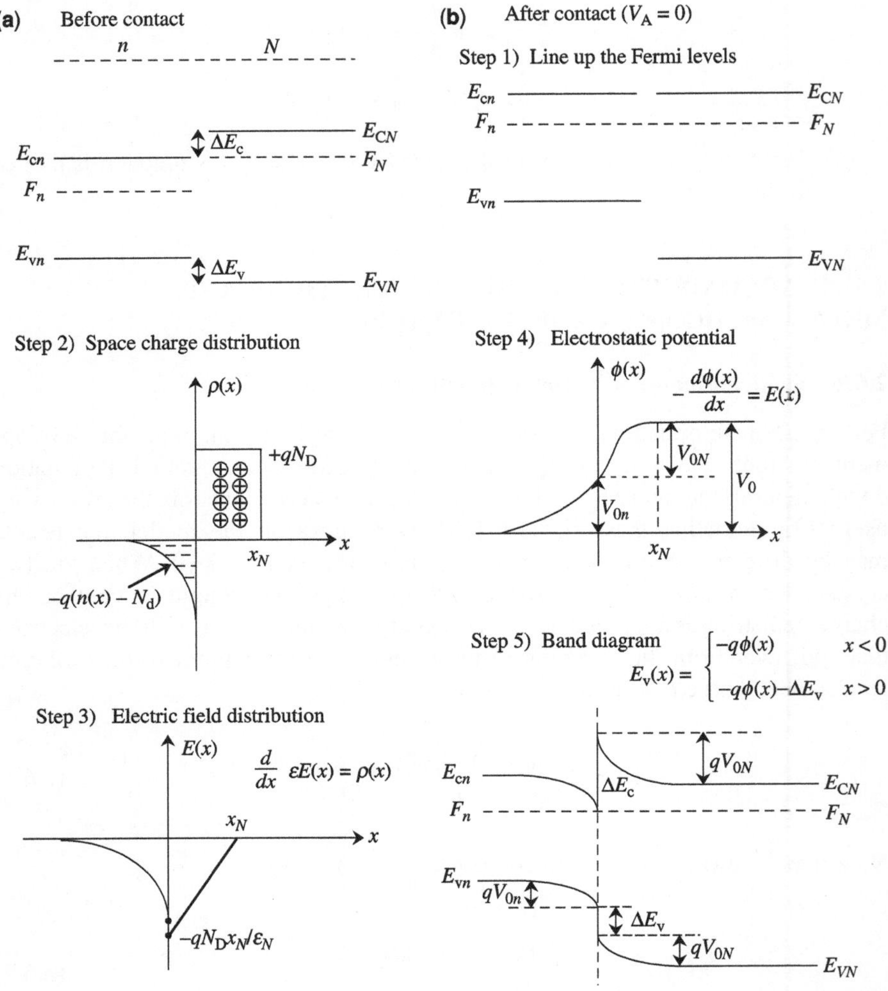

Figure 2.19 A step-by-step illustration for the energy band diagram of an n-N (isotope) heterojunction (a) before contact and (b) after contact showing five steps to obtain the steady-state energy band diagram.

define the partition:

$$V_0 = V_{0N} + V_{0n}. \tag{2.6.5}$$

Relating the carrier density $n(x)$ to the conduction band edge potential $\phi(x)$ and the electric field, the procedure is to set up the equations for the partitions of the contact potential, V_{0n} and V_{0N}, on the n-side and N-side, respectively. We skip the detailed equations and show an example in Fig. 2.20, the energy band diagrams for a P-n-N heterojunction. The n region is assumed to be wide enough such that

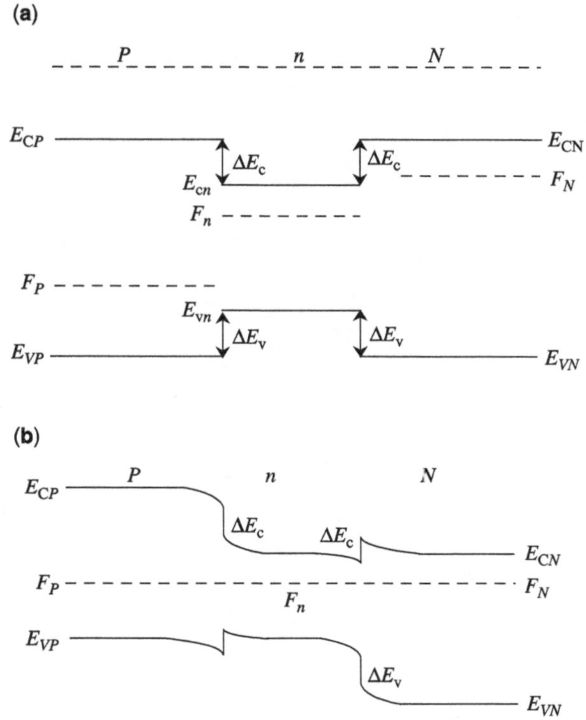

Figure 2.20 Energy band diagrams for a *P-n-N* heterojunction structure: (a) before contact and (b) after contact. Assume thermal equilibrium.

the two space charge regions near the two junctions do not merge together. If they do, the flat portion in the center *n*-type semiconductor will not exist, and the band diagram is further distorted.

2.6.2 Metal–Semiconductor Junction

Consider a metal with a work function Φ_m brought in contact with an *n*-type semiconductor with an electron affinity χ. For an ideal contact, the barrier height is given by (see Fig. 2.21)

$$\Phi_{ms} = \Phi_m - \chi \tag{2.6.6}$$

and a negative charge distribution builds up at the metal surface with a space charge region formed in the semiconductor region. The charge distribution in the semiconductor region is given by

$$\rho(x) = +q(N_D - n(x)). \tag{2.6.7}$$

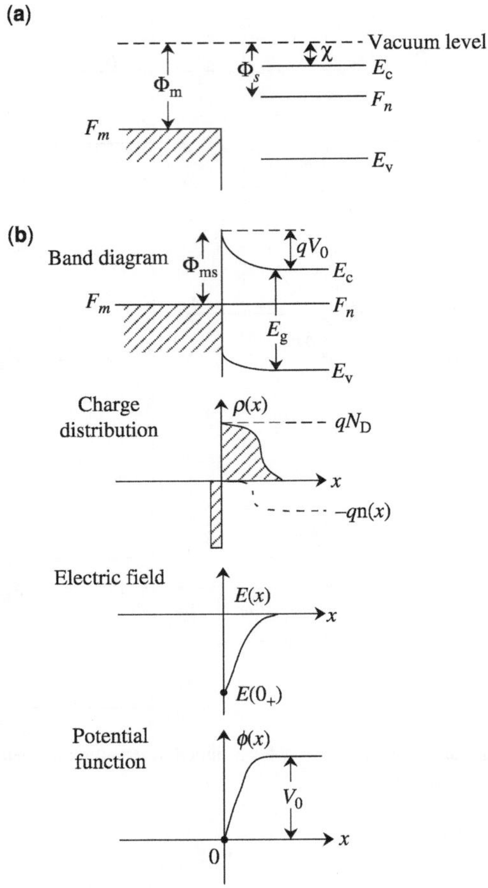

Figure 2.21 (a) The energy band diagram of a metal–semiconductor junction before contact. (b) The band diagram, charge distribution, electric field, and potential function after contact.

From Gauss's law, the electric field satisfies

$$\frac{dE(x)}{dx} = \frac{qN_D}{\varepsilon}[1 - e^{(E_{cn}-E_c(x))/k_BT}] \tag{2.6.8}$$

where $E_{cn} = E_c(\infty)$. We have

$$E_{cn} - E_c(x) = q[\phi(x) - V_0] \tag{2.6.9}$$

where $V_0 = \phi(x \to \infty)$, with the reference potential $\phi(0) = 0$, and

$$E(x) = -\frac{d\phi(x)}{dx}. \tag{2.6.10}$$

Multiplying (2.6.8) by $E(x)$ on both sides

$$d\frac{E^2(x)}{2} = \frac{qN_D}{\varepsilon}\left[-1 + e^{[\phi(x)-V_0]q/k_BT}\right]d\phi(x) \tag{2.6.11}$$

and integrating from 0_+ to $+\infty$, we obtain

$$E^2(0_+) = \frac{2qN_D}{\varepsilon}\left[V_0 - \frac{k_BT}{q} + \frac{k_BT}{q}e^{-qV_0/k_BT}\right]. \tag{2.6.12}$$

The contact potential is given by $V_0 = [\Phi_{ms} - (E_c - F_n)]/q$. The factor $\exp(-qV_0/k_BT)$ in the last expression in (2.6.12) is usually negligible for $V_0 > 0$.

When the metal semiconductor is forward biased with a voltage V and the barrier is reduced to $V_0 - V$, we have

$$|E(0_+)| = \sqrt{\frac{2qN_D}{\varepsilon}\left[V_0 - V - \frac{k_BT}{q}(1 - e^{-q(V_0-V)/k_BT})\right]}$$

$$\simeq \sqrt{\frac{2qN_D}{\varepsilon}\left(V_0 - V - \frac{k_BT}{q}\right)}. \tag{2.6.13}$$

If the electron concentration $n(x)$ is assumed to be depleted in a region of width x_n,

$$\frac{dE(x)}{dx} = \frac{qN_D}{\varepsilon} \quad 0 < x < x_n. \tag{2.6.14}$$

Then

$$E(x) = qN_D(x - x_n)/\varepsilon \tag{2.6.15}$$

and

$$E(0_+) = -qN_Dx_n/\varepsilon. \tag{2.6.16}$$

An iterative approach to include the effect of the electron is to use $E(0_+)$ from (2.6.13) and obtain

$$x_n \simeq \sqrt{\frac{2\varepsilon}{qN_D}\left(V_0 - V - \frac{k_BT}{q}\right)}. \tag{2.6.17}$$

PROBLEMS

2.1 Derive the boundary conditions using the integral form of the Maxwell's equations for a given charge density ρ_s and current density $\mathbf{J}_s$. Check the dimensions of all quantities appearing in the boundary conditions.

2.2 Consider an ideal case in which the electrons are distributed in a two-dimensional space with a surface carrier density n_s ($1/\text{cm}^2$).

(**a**) Derive a two-dimensional density of states and plot it versus the energy E.

(**b**) Find the relation between n_s and the Fermi level at a given temperature T.

(**c**) Repeat part (**b**) for $T = 0$.

2.3 If the electrons are distributed in a "quantum-wire" geometry, that is, an idealized one-dimensional space with a given "line" carrier density n_ℓ ($1/\text{cm}$),

(**a**) derive a one-dimensional density of states and plot it versus the energy E.

(**b**) Find the relation between n_ℓ and the Fermi level at a given temperature T.

(**c**) Repeat (**b**) for $T = 0$.

2.4 Use the inversion formula (2.2.50), and plot η versus u. Compare this curve with the curve in Fig. 2.2. What is their relation?

2.5 Derive Eq. (2.3.11b) for the average carrier lifetime τ_0.

2.6 Explain the Auger recombination processes in k space by considering CHCC, CHSH, CHLH, and CHHH processes [22].

2.7 Suppose the carrier generation rate in Fig. 2.9 is

$$G_n(y) = \begin{cases} G_0 \cos\left(\dfrac{\pi y}{S}\right) & |y| \leq S/2 \\ 0 & |y| > S/2. \end{cases}$$

(**a**) Find the excess carrier concentration $\delta n(y)$.

(**b**) Plot $G_n(y)$ and $\delta n(y)$ as a function of y for $S = 10\ \mu\text{m}$ and $L_n = 4\ \mu\text{m}$.

2.8 Plot the band diagram for a GaAs/$\text{Al}_{0.3}\text{Ga}_{0.7}\text{As}$ *p-N* junction. The GaAs region is doped with an acceptor concentration $N_a = 1 \times 10^{18}/\text{cm}^3$. The $\text{Al}_{0.3}\text{Ga}_{0.7}\text{As}$ region is doped with a donor concentration $N_D = 1 \times 10^{18}/\text{cm}^3$. Assume zero bias. $\Delta E_c = 0.67\Delta E_g$ and $\Delta E_v = 0.33\Delta E_g$, where $\Delta E_g = 1.247x$ (eV) is the band-gap difference between $\text{Al}_x\text{Ga}_{1-x}\text{As}$ and GaAs. Evaluate the Fermi levels, the contact potential V_0, and the depletion width.

2.9 Repeat Problem 2.8 for a reverse bias of -2 V.

REFERENCES

1. J. A. Kong, *Electromagnetic Wave Theory*, Wiley, New York, 2000.
2. J. D. Jackson, *Classical Electrodynamics*, 3rd ed., Wiley, New York, 1999.
3. M. Born and E. Wolf, *Principles of Optics*, Pergamon Press, Oxford, 1975.
4. S. M. Sze and K. K. Ng, *Physics of Semiconductor Devices*, Wiley, New York, 2007.
5. S. Selberherr, *Analysis and Simulation of Semiconductor Devices*, Springer-Verlag, New York, 1984.

6. W. L. Engl, H. K. Dirks, and B. Meinerzhagen, "Device modeling," *Proc. IEEE* **71**, 10–33 (1983).

7. A. H. Marshak and C. M. Van Vliet, "Electrical current and carrier density in degenerate materials with nonuniform band structure," *Proc. IEEE* **72**, 148–164 (1984).

8. X. Aymerich-Humett, F. Serra-Mestres, and J. Millan, "An analytical approximation for the Fermi-Dirac integral $F_{3/2}(x)$," *Solid-State Electron.* **24**, 981–982 (1981).

9. D. Bednarczyk and J. Bednarczyk, "The approximation of the Fermi-Dirac integral $F_{1/2}(\eta)$," *Phys. Lett.* **64A**, 409–410 (1978).

10. J. S. Blakemore, "Approximations for Fermi-Dirac integrals, especially the function $F_{1/2}(\eta)$ used to describe electron density in a semiconductor," *Solid-State Electron.* **25**, 1067–1076 (1982).

11. N. G. Nilsson, "An accurate approximation of the generalized Einstein relation for degenerate semiconductors," *Phys. Stat. Sol. (a)* **19**, K75–K78 (1973).

12. T. Y. Chang and A. Izabelle, "Full range analytic approximations for Fermi energy and Fermi-Dirac integral $F_{-1/2}$ in terms of $F_{1/2}$," *J. Appl. Phys.* **65**, 2162–2164 (1989).

13. R. B. Adler, A. C. Smith, and R. L. Longini, *Introduction to Semiconductor Physics, Semiconductor Electronics Education Committee*, Vol. 1, Wiley, New York, 1964.

14. R. F. Pierret, *Semiconductor Fundamentals*, in R. F. Pierret and G. W. Neudeck, Eds., *Modular Series on Solid State Devices*, Vol. 1, Addison-Wesley, Reading, MA, 1983.

15. S. Wang, *Fundamentals of Semiconductor Theory and Device Physics*, Prentice Hall, Englewood Cliffs, NJ, 1989.

16. J. F. Geisz and D. J. Friedman, "III-N-V semiconductors for solar photovoltaic applications," *Semicond. Sci. Technol.* **17**, 769–777 (2002).

17. M. Fukuda, *Reliability and Degradation of Semiconductor Lasers and LEDs*, Artech House, Norwood, MA, 1991.

18. W. Shockley and W. T. Read, Jr., "Statistics of the recombinations of holes and electrons," *Phys. Rev.* **87**, 835–842 (1952).

19. R. N. Hall, "Electron-hole recombination in germanium," *Phys. Rev.* **87**, 387 (1952).

20. C. T. Sah, "The equivalent circuit model in solid-state electronics-Part I: The single energy level defect centers," *Proc. IEEE* **55**, 654–671 (1967). "Part II: The multiple energy level impurity centers," *Proc. IEEE* **55**, 672–684 (1967). "Part III: Conduction and displacement currents," *Solid-State Electron.* **13**, 1547–1575 (1970). "Equivalent circuit models in semiconductor transport for thermal, optical, Auger-impact, and tunnelling recombination-generation-trapping processes," *Phys. Stat. Sol. (a)* **7**, 541–559 (1971).

21. C. T. Sah, *Fundamentals of Solid-State Electronics*, World Scientific, Singapore, 1991.

22. A. Sugimura, "Band-to-band Auger effect in long wavelength multinary III-V alloy semiconductor lasers," *IEEE J. Quantum Electron.* **QE-18**, 352–363 (1982).

23. H. C. Casey, Jr. and M. B. Panish, *Heterostructure Lasers, Part A: Fundamental Principles, Part B: Materials and Operating Characteristics*, Academic Press, Orlando, 1978.

24. R. L. Anderson, "Germanium-gallium arsenide heterojunctions," *IBM J.* **July**, 283–287 (1960).

25. R. L. Anderson, "Experiments on Ge-GaAs heterojunctions," *Solid-State Electron.* **5**, 341–351 (1962).

26. W. R. Frensley and H. Kroemer, "Prediction of semiconductor heterojunction discontinuities from bulk band structures," *J. Vac. Sci. Technol.* **13**, 810–815 (1976).
27. W. A. Harrison, "Elementary theory of heterojunctions," *J. Vac. Sci. Technol.* **14**, 1016–1021 (1977).
28. H. Kroemer, "Heterostructure devices: A device physicist looks at interfaces," *Surf. Sci.* **132**, 543–576 (1983).
29. R. F. Pierret, "Extension of the approximate two-dimensional electron gas formulation," *IEEE Trans. Electron Devices* **ED-32**, 1279–1287 (1985).

3

Basic Quantum Mechanics

In this chapter, we present some basic quantum mechanics with the aim that these results will be useful in later chapters to understand the physics of optoelectronic processes and devices. In quantum mechanics, there are three important potentials, which have exact analytical solutions and are presented in most undergraduate texts [1, 2] on modern physics or introductory quantum mechanics. These three potentials are

1. **A Square Well Potential.** This simplest case is usually taught in undergraduate modern physics and will be presented in Section 3.2. Here, we take into account the fact that the effective masses inside and outside the quantum well are different for a semiconductor quantum well. The solutions for the bound states are simple cosine and exponential functions inside and outside the quantum well, respectively.
2. **A Harmonic Oscillator.** The solutions are Hermite–Gaussian functions [1–3] and will be discussed in Section 3.3. A similar situation exists for the electric field of an optical waveguide with a parabolic permittivity or gain profile, as the solutions to the wave equation with a parabolic dependence on the position are also the Hermite–Gaussian functions.
3. **A Hydrogen Atom.** The solutions for both the bound states ($E < 0$) and the unbound (continuum) states ($E > 0$) will be discussed. We use the hydrogen atom model to describe an exciton, which is formed by an electron–hole pair. The optical absorption process involving the Coulomb interaction between the electron and hole is called the excitonic absorption. The excitonic absorption spectrum is strongly dependent on the energies and wave functions of the exciton bound and continuum states. In bulk semiconductors, the electron–hole interaction is a Coulomb potential in a three-dimensional space. In quantum-well semiconductors, electrons and holes are confined to a quasi-two-dimensional structure; thus, their binding energy is different from the three-dimensional results. We summarize in Section 3.4 the main results for the hydrogen model in three-dimensional and two-dimensional space for both the bound and unbound states [4–8]. These results will be

Physics of Photonic Devices, Second Edition. By Shun Lien Chuang
Copyright © 2009 John Wiley & Sons, Inc.

used in Chapter 14 where we discuss excitonic absorption and quantum-confined Stark effects (QCSE).

Because most potentials except the above three do not have exact, analytical solutions, we have to use perturbation methods to find their eigenfunctions and eigenenergies. We present a conventional time-independent perturbation theory in Section 3.5. In Section 3.6, we present the time-dependent perturbation theory and derive Fermi's golden rule, which is very important for the understanding of the emission and absorption of photons. In Appendix 3A, a slightly modified perturbation theory, called the Löwdin's method [9], is shown. The Löwdin's method is applied to derive the semiconductor band structures such as the Luttinger–Kohn Hamiltonian in Chapter 4.

3.1 SCHRÖDINGER EQUATION

The Schrödinger equation in a nonrelativistic quantum mechanical description of a single particle is

$$H\Psi(\mathbf{r}, t) = i\hbar \frac{\partial}{\partial t}\Psi(\mathbf{r}, t) \tag{3.1.1}$$

where the Hamiltonian H is

$$H = -\frac{\hbar^2}{2m}\nabla^2 + V(\mathbf{r}, t) \tag{3.1.2}$$

where the potential energy function $V(\mathbf{r}, t)$ is real, $\hbar$ is the Planck constant h divided by 2π, and m is the mass of the particle. For a free particle in an unbounded space, $V(\mathbf{r}, t) = 0$, the solution is simply a plane wave

$$\Psi(\mathbf{r}, t) = \frac{1}{\sqrt{V}} e^{i\mathbf{k}\cdot\mathbf{r} - iEt/\hbar} \tag{3.1.3}$$

where the energy is

$$E = \frac{\hbar^2 k^2}{2m} \tag{3.1.4}$$

and V is the volume of the space. The probability density is defined as

$$\rho(\mathbf{r}, t) = \Psi^*(\mathbf{r}, t)\Psi(\mathbf{r}, t) \tag{3.1.5}$$

and the probability current density as

$$\mathbf{j}(\mathbf{r}, t) = \frac{\hbar}{2mi}[\Psi^* \nabla \Psi - \Psi \nabla \Psi^*] \tag{3.1.6}$$

Here $\Psi^*(\mathbf{r},t)\Psi(\mathbf{r},t)d^3\mathbf{r}$ is the probability of finding the particle in a volume $d^3\mathbf{r}$ near the position $\mathbf{r}$ at time t. The wave function is normalized such that $\int \Psi^*(\mathbf{r}, t)\Psi(\mathbf{r}, t)d^3\mathbf{r} = 1$, that is, the probability of finding the particle in the whole space is unity. By substitution, it is straightforward to show that

$$\nabla \cdot \mathbf{j} + \frac{\partial}{\partial t}\rho = 0 \tag{3.1.7}$$

which is the continuity equation or the conservation of probability density. It is analogous to the charge continuity equation in electromagnetics.

The expectation value of any physical quantity defined by an operator O is given by

$$\langle O \rangle = \int_V \Psi^*(\mathbf{r}, t) O \Psi(\mathbf{r}, t) d^3\mathbf{r} = \langle \Psi | O | \Psi \rangle \tag{3.1.8}$$

where the volume of integration is over the whole space. The ket and bra vectors, $|\Psi\rangle$ and $\langle\Psi|$, are used as short-hand notation. In the above real-space representation, that is, $\psi \equiv \psi(\mathbf{r}, t)$, the correspondence for the operators is

$$\text{Position: } \mathbf{r}_{\text{op}} = \mathbf{r} \qquad \text{Momentum: } \mathbf{p}_{\text{op}} = \frac{\hbar}{i}\nabla. \tag{3.1.9}$$

If $V(\mathbf{r}, t)$ is independent of t, the solution $\Psi(\mathbf{r}, t)$ can always be obtained using the separation of variables

$$\Psi(\mathbf{r}, t) = \Psi(\mathbf{r})e^{-iEt/\hbar} \tag{3.1.10}$$

and

$$\left[-\frac{\hbar^2}{2m}\nabla^2 + V(\mathbf{r})\right]\Psi(\mathbf{r}) = E\Psi(\mathbf{r}) \tag{3.1.11}$$

which is the so-called time-independent Schrödinger equation. The solution may be in terms of quantized energy levels E_n with corresponding wave functions $\psi_n(\mathbf{r})$ or a continuous spectrum E with corresponding wave functions $\psi_E(\mathbf{r})$. In general, any solution of the Schrödinger equation may be constructed from the superposition of these stationary solutions

$$\Psi(\mathbf{r}, t) = \sum_n a_n \psi_n(\mathbf{r})e^{-iE_n t/\hbar} + \int a_E \psi_E(\mathbf{r})e^{-iEt/\hbar}\, dE \tag{3.1.12}$$

where $|a_n|^2$ gives the probability that the particle will be in the nth stationary (bound) state $\psi_n(\mathbf{r}, t)$ with an energy E_n, and a_E satisfies the normalization condition for the continuum states. In studying a time-dependent potential problem, very often a perturbation approach is used if the time-dependent perturbing potential is small compared with the unperturbed Hamiltonian, and the above expansion in terms of the stationary states $\psi_n(\mathbf{r})$ or $\psi_E(\mathbf{r})$, which are solutions to the unperturbed problem, is very useful. In this case, a_n (and a_E) will be functions of time as the perturbation is time-dependent, and $|a_n(t)|^2$ will give the time-dependent probability that the particle is in state n of the unperturbed problem.

If we write the Fourier transforms of $\psi(\mathbf{r})$ and $V(\mathbf{r})$ in the single-particle Schrödinger equation (3.1.11),

$$\tilde{\psi}(\mathbf{k}) = \int \psi(\mathbf{r}) e^{-i\mathbf{k}\cdot\mathbf{r}} d^3\mathbf{r} \tag{3.1.13}$$

$$\tilde{V}(\mathbf{k}) = \int V(\mathbf{r}) e^{-i\mathbf{k}\cdot\mathbf{r}} d^3\mathbf{r} \tag{3.1.14}$$

we then obtain the momentum-space representation of the Schrödinger equation

$$\frac{\hbar^2 k^2}{2m}\tilde{\psi}(\mathbf{k}) + \int \frac{d^3\mathbf{k}'}{(2\pi)3}\tilde{V}(\mathbf{k}-\mathbf{k}')\tilde{\psi}(\mathbf{k}') = E\tilde{\psi}(\mathbf{k}) \tag{3.1.15}$$

which becomes an integral equation. This integral equation will be discussed in Chapter 14.

3.2 THE SQUARE WELL

We consider a square (or rectangular) quantum well with a barrier height V_0. In the one-dimensional case, the time-independent Schrödinger equation is

$$\left[-\frac{\hbar^2}{2m}\frac{d^2}{dz^2} + V(z)\right]\phi(z) = E\phi(z) \tag{3.2.1}$$

3.2.1 Infinite Barrier Model

First we assume V_0 is infinitely high; thus, the wave function vanishes at the boundaries, Fig. 3.1a. The solution to the infinite barrier model satisfying the boundary conditions $\phi_n(0) = 0$ and $\phi_n(L) = 0$ is

$$\phi_n(z) = \sqrt{\frac{2}{L}}\sin\left(\frac{n\pi}{L}z\right), \qquad n = 1, 2, 3, \ldots, \tag{3.2.2}$$

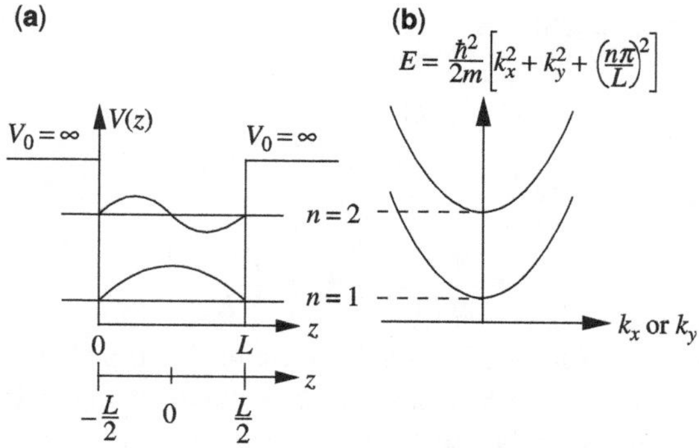

Figure 3.1 (a) A quantum well with a width L and an infinite barrier $V_0 = \infty$. (b) The energy dispersion in the k_x or k_y space using Eq. (3.2.7).

with corresponding energy and wave number

$$E_n = \frac{\hbar^2}{2m}\left(\frac{n\pi}{L}\right)^2, \qquad k_z = \frac{n\pi}{L} \tag{3.2.3}$$

that is, the energy E_n and the wave number k_z in the z-direction are quantized. The wave function has been normalized $\int_0^L |\phi_n(Z)|^2 dz = 1$.

If the origin of the coordinate $z = 0$ is chosen at the center of the well, $V(+z) = V(-z)$, the solution can always be put in terms of even or odd functions by parity consideration, which can be obtained from (3.2.2) by replacing z by $(z + 0.5L)$. The parity consideration is useful because the symmetry properties of the system are employed; thus, the wave functions have associated symmetry properties. In general, if $V(z) = V(-z)$, for any solution $\phi(z)$ of the Schrödinger equation (3.2.1), $\phi(-z)$ will also be a solution with the same eigenenergy. If we form linear combinations of $\phi(z)$ and $\phi(-z)$, they will also be solutions. Specifically,

$$\phi_e(z) = \frac{1}{\sqrt{2}}[\phi(z) + \phi(-z)] \tag{3.2.4a}$$

$$\phi_o(z) = \frac{1}{\sqrt{2}}[\phi(z) - \phi(-z)] \tag{3.2.4b}$$

where $\phi_e(z)$ is an even function and $\phi_o(z)$ is an odd function.

The complete solution for the potential well $V(z)$ in a three-dimensional space is solved from

$$\left[-\frac{\hbar^2}{2m}\nabla^2 + V(z)\right]\Psi(x, y, z) = E\Psi(x, y, z). \tag{3.2.5}$$

The normalized wave function is

$$\Psi(x, y, z) = \frac{e^{ik_x x + ik_y y}}{\sqrt{A}} \phi_n(z) \tag{3.2.6}$$

with a corresponding energy

$$E = \frac{\hbar^2}{2m}\left[k_x^2 + k_y^2 + \left(\frac{n\pi}{L}\right)^2\right] \tag{3.2.7}$$

where $\phi_n(z)$ is the same as in (3.2.2), and A is the cross-sectional area in the x–y plane. This energy dispersion diagram E versus k_x or k_y is plotted in Fig. 3.1b for $n = 1$ and 2. In semiconductors, m should be replaced by the effective mass m^* of the carriers.

Example For a GaAs/Al$_x$Ga$_{1-x}$As quantum-well structure, suppose the GaAs well width is 100 Å and the aluminum mole fraction x in the barrier region is large such that an infinite barrier model is applicable. We have the effective mass of the electrons $m_e^* = 0.0665\, m_0$ in the well region. The quantized electron subband energies are

$$E_n = \frac{\hbar^2}{2m_e^*}\left(\frac{n\pi}{L}\right)^2 = n^2 E_1 = E_1, 4E_1, 9E_1, \ldots$$

where

$$E_1 = \frac{\hbar^2}{2m_e^*}\left(\frac{\pi}{L}\right)^2 = \frac{1}{(m_e^*/m_0)L^2} \times 37.6 \text{ eV} = 56.5 \text{ meV}.$$

Therefore, we find the subband energies are

$$E_n = 56.5, 226, 508.5, \ldots (\text{meV}).$$

Two-dimensional (2D) Density of States The electron concentration in a quantum well can be calculated using

$$n = \frac{2}{V}\sum_n \sum_{k_x} \sum_{k_y} f(E) \tag{3.2.8}$$

where the summation is over all the occupied subbands. Because the electron energy is quantized in the k_z quantum number, and the dependence on x and y are still plane-wave like, we convert the summation over k_x and k_y into integrations

$$\frac{2}{V}\sum_{k_x}\sum_{k_y} = \frac{2}{V}\int \frac{dk_x}{\left(\frac{2\pi}{L_x}\right)} \int \frac{dk_y}{\left(\frac{2\pi}{L_y}\right)} = \frac{2}{L_z}\int \frac{dk_x dk_y}{(2\pi)^2}$$

$$= \frac{2}{L_z}\int_0^\infty \frac{2\pi k_t dk_t}{(2\pi)^2} = \frac{m^*}{\pi\hbar^2 L_z}\int_{E_n}^\infty dE \tag{3.2.9}$$

where we have used the polar coordinates,

$$dk_x dk_y = k_t d\phi dk_t, \quad k_t^2 = k_x^2 + k_y^2, \quad dE = \hbar^2 k_t dk_t / m^*.$$

We can then write the 2D density of states as

$$\rho_{2D}(E) = \frac{m^*}{\pi\hbar^2 L_z} \sum_n H(E - E_n) \tag{3.2.10}$$

where $H(x)$ is a Heaviside step function $H(x) = 1$ for $x > 0$ and $H(x) = 0$ for $x < 0$. The electron density n is then given by

$$n = \int dE \rho_{2D}(E) f(E). \tag{3.2.11}$$

The density of states $\rho_{2D}(E)$ is plotted in Fig. 3.2 as the steplike function with each step occurs wherever there is a new subband energy level E_1, $E_2 = 4E_1$, $E_3 = 9E_1$, and so forth. It is interesting to compare the two-dimensional density of states with the three-dimensional density of states (2.2.36)

$$\rho_{3D}(E) = \frac{1}{2\pi^2} \left(\frac{2m^*}{\hbar^2}\right)^{3/2} \sqrt{E}. \tag{3.2.12}$$

At $E = E_n$ from (3.2.3)

$$\begin{aligned} \rho_{3D}(E) &= \frac{1}{2\pi^2} \left(\frac{2m^*}{\hbar^2}\right)^{3/2} \left(\frac{\hbar^2 n^2 \pi^2}{2m^* L_z^2}\right) \\ &= n \left(\frac{m^*}{\pi\hbar^2 L_z}\right) \end{aligned} \tag{3.2.13}$$

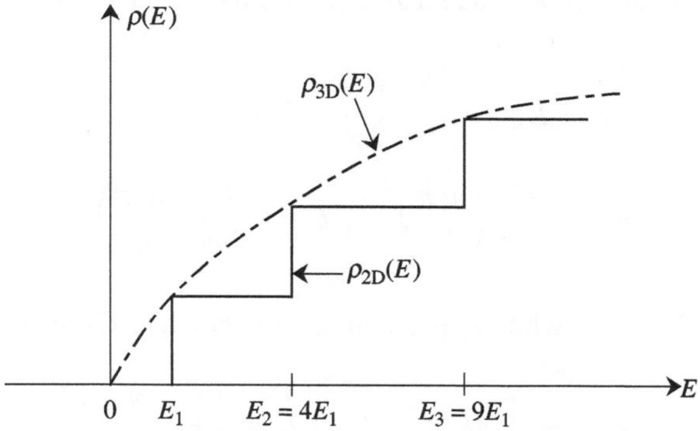

Figure 3.2 The electron density of states $\rho_{2D}(E)$ (solid line) for a two-dimensional quantum-well structure is compared with the three-dimensional density of states $\rho_{3D}(E)$ (dashed curve).

which is the same as $\rho_{2D}(E = E_n)$. The three-dimensional density of states $\rho_{3D}(E)$ is shown as the dashed curve in Fig. 3.2 with a smooth $\sqrt{E}$ behavior above the conduction band edge, whereas the two-dimensional $\rho_{2D}(E)$ shows a sharp step-like behavior above each subband edge.

Example As a numerical example, we calculate the electron density of states for a bulk GaAs semiconductor at an energy 0.1 eV above the conduction band edge

$$\rho_{3D}(E - E_c = 0.1 \text{ eV}) = \frac{1}{2\pi^2}\left(\frac{2m_e^*}{\hbar^2}\right)^{3/2}(E - E_c)^{1/2} = 3.69 \times 10^{19}(\text{cm}^{-3}\text{eV}^{-1}).$$

For a quantum well with $L_z = 100$ Å, we estimate the density of states of the first step (with $E_1 = 56.5$ meV)

$$\rho_{2D}(E - E_1) = \frac{m_e^*}{\pi\hbar^2 L_z} = 2.78 \times 10^{19}(\text{cm}^{-3}\text{eV}^{-1}).$$

We can also estimate the carrier concentration with an energy spread of $k_B T = 0.026$ eV at 300K and obtain

$$n \sim 2.78 \times 10^{19} \times 0.026 = 7.2 \times 10^{17} \text{ cm}^{-3}.$$

One-dimensional (1D) Density of States Similar to (3.2.10), the electron concentration in a 1D quantum wire along the z direction can be obtained from

$$n = \frac{2}{V}\sum_{n_x, n_y}\sum_{k_z} f(E) \tag{3.2.14}$$

where the electron energy E is quantized in both the x and y directions

$$E = E_{n_x} + E_{n_y} + \frac{\hbar^2 k_z^2}{2m^*}$$

$$E_{n_x} = \frac{\hbar^2}{2m^*}\left(\frac{n_x \pi}{L_x}\right)^2, \quad E_{n_y} = \frac{\hbar^2}{2m^*}\left(\frac{n_y \pi}{L_y}\right)^2,$$

and n_x, $n_y = 1, 2, 3, \ldots$, while k_z is a continuous variable. Using

$$\frac{2}{V}\sum_{n_x, n_y}\sum_{k_z} = \frac{2}{L_x L_y}\sum_{n_x, n_y}\int_{-\infty}^{\infty}\frac{dk_z}{2\pi} = \frac{1}{\pi L_x L_y}\sqrt{\frac{2m^*}{\hbar^2}}\sum_{n_x, n_y}\int_{E_{n_x}+E_{n_y}}^{\infty}\frac{dE}{\sqrt{E - E_{n_x} - E_{n_y}}}$$

the 1D density of states $\rho_{1D}(E)$ is

$$\rho_{1D}(E) = \frac{1}{\pi L_x L_y}\sqrt{\frac{2m^*}{\hbar^2}}\sum_{n_x, n_y}\frac{1}{\sqrt{E - E_{n_x} - E_{n_y}}}, \quad E > E_{n_x} + E_{n_y}. \tag{3.2.15}$$

3.2.2 Finite Barrier Model

For a finite barrier quantum well as shown in Fig. 3.3, we have

$$V(z) = \begin{cases} V_0, & |z| = L/2 \\ 0, & |z| < L/2. \end{cases} \tag{3.2.16}$$

We solve the Schrödinger equation

$$\left[-\frac{\hbar^2}{2}\frac{d}{dz}\frac{1}{m^*}\frac{d}{dz} + V(z)\right]\phi(z) = E\phi(z) \tag{3.2.17}$$

where m^* equals m_w in the well and m_b in the barrier region. We consider the bound-state solutions, which have energies in the range between 0 and V_0.

Even solutions For the even wave functions, we have a solution of the form

$$\phi(z) = \begin{cases} C_1 e^{-\alpha(|z|-L/2)} & |z| = \dfrac{L}{2} \\ C_2 \cos k z & |z| < \dfrac{L}{2} \end{cases} \tag{3.2.18}$$

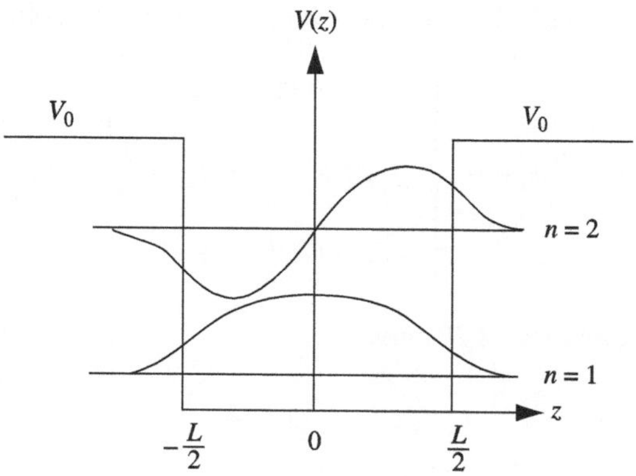

Figure 3.3 A quantum well with a width L and a finite barrier height V_0. The energy levels for $n = 1$ and $n = 2$ and their corresponding wave functions are shown.

where

$$k = \sqrt{2m_{\mathrm{w}}E}/\hbar \tag{3.2.19a}$$

$$\alpha = \sqrt{2m_{\mathrm{b}}(V_0 - E)}/\hbar\,. \tag{3.2.19b}$$

Using boundary conditions in which the wave function ϕ and its first derivative divided by the effective mass $(1/m^*)(d\phi/dz)$ are continuous at the interface between the barrier and the well, that is,

$$\phi\left(\frac{L^+}{2}\right) = \phi\left(\frac{L^-}{2}\right) \tag{3.2.20a}$$

and

$$\frac{1}{m_{\mathrm{b}}}\frac{d}{dz}\phi\left(\frac{L^+}{2}\right) = \frac{1}{m_{\mathrm{w}}}\frac{d}{dz}\phi\left(\frac{L^-}{2}\right) \tag{3.2.20b}$$

we obtain

$$\begin{aligned} C_1 &= C_2 \cos k\,\frac{L}{2} \\ \frac{\alpha}{m_{\mathrm{b}}}C_1 &= \frac{k}{m_{\mathrm{w}}}C_2 \sin k\,\frac{L}{2}. \end{aligned} \tag{3.2.21}$$

Eliminating C_1 and C_2, we obtain the eigenequation or the quantization condition

$$\alpha = \frac{m_{\mathrm{b}}k}{m_{\mathrm{w}}}\tan k\frac{L}{2}. \tag{3.2.22}$$

The eigenenergy E can be found from the above equation by substituting k and α from (3.2.19a) and (3.2.19b) into (3.2.22).

Odd Solutions Similarly for odd wave functions, we have solutions of the form

$$\phi(z) = \begin{cases} C_1 e^{-\alpha(z-L2)} & z > \dfrac{L}{2} \\ C_2 \sin kz & |z| = \dfrac{L}{2} \\ -C_1 e^{+\alpha(z+L2)} & z < -\dfrac{L}{2}. \end{cases} \tag{3.2.23}$$

The boundary conditions (3.2.20) give

$$\begin{aligned} C_1 &= C_2 \sin k\frac{L}{2} \\ -\frac{\alpha C_1}{m_{\mathrm{b}}} &= \frac{k}{m_{\mathrm{w}}}C_2 \cos k\frac{L}{2}. \end{aligned} \tag{3.2.24}$$

Thus, the eigenequation is given by

$$\alpha = -\frac{m_b k}{m_w} \cot k\frac{L}{2} \tag{3.2.25}$$

which determines the eigenenergy E for the odd wave function again by using k and α in (3.2.19a) and (3.2.19b). In general, the solutions for the quantized eigenenergies can be obtained by finding $(\alpha L/2)$ and $(kL/2)$ directly from a graphical approach because from (3.2.19a) and (3.2.19b):

$$\left(k\frac{L}{2}\right)^2 + \frac{m_w}{m_b}\left(\alpha\frac{L}{2}\right)^2 = \frac{2m_w V_0}{\hbar^2}\left(\frac{L}{2}\right)^2 \tag{3.2.26}$$

and

$$\alpha\frac{L}{2} = \frac{m_b}{m_w} k\frac{L}{2} \tan k\frac{L}{2} \quad \text{even solution} \tag{3.2.27a}$$

$$\alpha\frac{L}{2} = -\frac{m_b}{m_w} k\frac{L}{2} \cot k\frac{L}{2} \quad \text{odd solution.} \tag{3.2.27b}$$

The above equations can be solved by plotting them on the $(\alpha L/2)$ versus $(kL/2)$ plane as shown in Fig. 3.4. If $m_w = m_b$, Eq. (3.2.26) is a circle with a radius $\sqrt{2m_w V_0}\,(L/2\hbar)$. If $m_w \neq m_b$, we define $\alpha' \equiv \alpha\sqrt{m_w/m_b}$, and rewrite

$$\alpha'\frac{L}{2} = \sqrt{\frac{m_b}{m_w}}\, k\frac{L}{2} \tan k\frac{L}{2} \tag{3.2.28a}$$

$$\alpha'\frac{L}{2} = -\sqrt{\frac{m_b}{m_w}}\, k\frac{L}{2} \cot k\frac{L}{2}. \tag{3.2.28b}$$

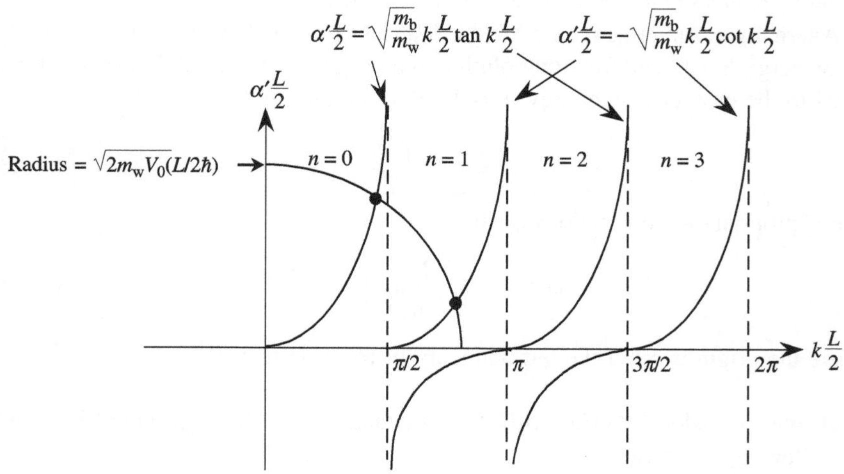

Figure 3.4 A graphical solution for the decaying constant α and wave number k of a finite quantum well. Here we define $\alpha' = \alpha\sqrt{m_w/m_b}$ to account for the different effective masses in the well and in the barrier. The label $n = 0$ refers to the ground (or the first bound) state. $n = 1$ refers to the first excited (or the second bound) state, etc.

On the $\alpha' L/2$ versus $kL/2$ plane, Eq. (3.2.26) is still a circle with the same radius $\sqrt{2m_w V_0}(L/2\hbar)$. The solutions to α' and k are obtained from the intersection points between the circle (3.2.26) and either the tangent curve (3.2.28a) or the cotangent curve (3.2.28b), as shown in Fig. 3.4. If the radius of the circle is in the range

$$(N-1)\frac{\pi}{2} \leq \sqrt{2m_w V_0}\left(\frac{L}{2\hbar}\right) < N\frac{\pi}{2} \tag{3.2.29}$$

there are N solutions. After solving the eigenequations for α and k, we obtain the quantized energy E.

The bound state solution ($E < V_0$) is obtained by using the normalization condition $\int_{-\infty}^{\infty} |\phi(z)|^2 dz = 1$ to determine the last unknown coefficient:

$$C_2 = \frac{\sqrt{2}}{\sqrt{L + \dfrac{2}{\alpha}\left(\cos^2 k\dfrac{L}{2} + \dfrac{m_b}{m_w}\sin^2 k\dfrac{L}{2}\right)}}. \tag{3.2.30}$$

If $m_w = m_b$, we have $C_2 = \sqrt{2/(L+2/\alpha)}$. The length $L + (2/\alpha)$ is the well width plus twice the penetration depth $(1/\alpha)$ on each side of the well. The normalization factor is very similar to that of the infinite well (3.2.2) except that we have an effective well width $L_e = L + (2/\alpha)$. For odd solutions in (3.2.23) we find

$$C_2 = \frac{\sqrt{2}}{\sqrt{L + \dfrac{2}{\alpha}\left(\sin^2 k\dfrac{L}{2} + \dfrac{m_b}{m_w}\cos^2 k\dfrac{L}{2}\right)}}. \tag{3.2.31}$$

Again C_2 reduces to the expression $\sqrt{2/(L+2/\alpha)}$ when $m_b = m_w$.

Alternatively, an effective well width L_{eff} can be defined by using the ground state eigenenergy E obtained from the solution of the eigenequation (3.2.28a) and setting it equal to the ground state energy ($n = 1$) of an infinite well

$$E = \frac{\hbar^2}{2m_w}\left(\frac{\pi}{L_{\text{eff}}}\right)^2. \tag{3.2.32}$$

The appropriate wave functions are then

$$\phi_n(z) = \sqrt{\frac{2}{L_{\text{eff}}}}\sin\left(\frac{n\pi}{L_{\text{eff}}}z\right) \tag{3.2.33}$$

where the origin is set at the left boundary of the infinite well.

Example Consider a GaAs/Al$_x$Ga$_{1-x}$As quantum well shown in Fig. 3.5. Assume the following parameters:

$m_e^* = (0.0665 + 0.0835x)\, m_0$ $\quad$ $m_{hh}^* = 0.34 + 0.42x)\, m_0$

$m_{\ell h}^* = (0.094 + 0.043x)\, m_0$ $\quad$ $E_g(x) = 1.424 + 1.247x\,(\text{eV})$

$(0 \leq x < 0.45$, room temperature).

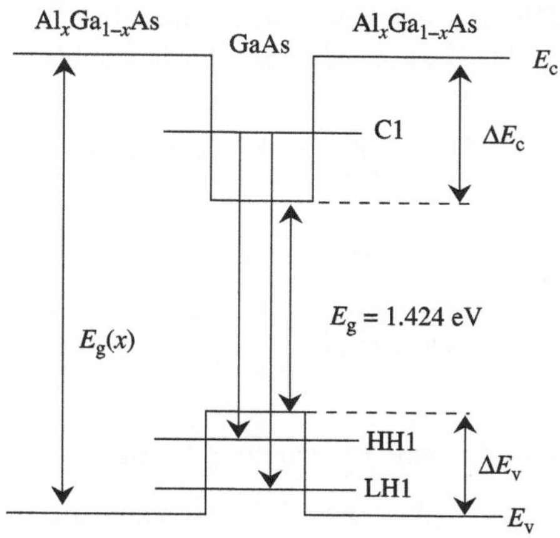

Figure 3.5 A GaAs/Al_xGa_{1-x} As quantum-well structure showing the band gaps and the interband transition energies.

$\Delta E_g(x) = 1.247x$ (eV), $\Delta E_c = 0.67\Delta E_g$, and $\Delta E_v = 0.33\Delta E_g$, where m_0 is the free electron mass.

(a) Consider the aluminum mole fraction $x = 0.3$ in the barrier regions ($x = 0$ in the well region) and the well width $L_w = 100$ Å. How many bound states are there in the conduction band? How many bound heavy-hole and light-hole subbands are there?

(b) Find the lowest bound state energies for the conduction subband (C1), the heavy-hole (HH1) and the light-hole (LH1) subbands for a 100 Å GaAs quantum well in part (a). What are the C1-HH1 and C1-LH1 transition energies?

(c) Assume that we define an effective well width L_{eff} using an infinite barrier model such that its ground state energy is the same as the energy of the first conduction subband E_{C1} in (b). What is L_{eff}? If we repeat the same procedure for the HH1 and the LH1 subbands, what are L_{eff}?

Solutions We tabulate the physical parameters as follows.

	m_e^*	m_{hh}^*	$m_{\ell h}^*$	E_g
Well	0.0665 m_0	0.34 m_0	0.094 m_0	1.424 eV
Barrier ($x = 0.3$)	0.0916 m_0	0.466 m_0	0.107 m_0	1.798 eV

$$L_w = 100\,\text{Å},\ \Delta E_c = 0.67,\ \Delta E_g = 0.2506\ \text{eV},\ \Delta E_v = 0.1235\ \text{eV}.$$

(a) The number of bound states N is determined by (3.2.29) where the barrier height V_0 is ΔE_c for electrons and ΔE_v for holes, respectively.

For electrons: $\sqrt{2m_e^*\Delta E_c}\left(\dfrac{L}{2\hbar}\right) = 3.30 < N\dfrac{\pi}{2}$ $\quad N = 3$ bound states

For heavy holes: $\sqrt{2m_{hh}^*\Delta E_v}\left(\dfrac{L}{2\hbar}\right) = 5.25 < N\dfrac{\pi}{2}$ $\quad N = 4$ bound states

For light holes: $\sqrt{2m_{\ell h}^*\Delta E_v}\left(\dfrac{L}{2\hbar}\right) = 2.76 < N\dfrac{\pi}{2}$ $\quad N = 2$ bound states.

(b) The eigenenergy E is found by searching for the root in $\alpha = \dfrac{m_b^*}{m_w^*}k\tan\left(\dfrac{kL_w}{2}\right)$, where k and α are given by (3.2.19a) and (3.2.19b), respectively. For electrons, $m_w^* = 0.0665\,m_0$, $m_b^* = 0.0916\,m_0$, and $V_0 = 250.6$ meV, the first subband energy is

$$E_{C1} = 30.7 \text{ meV}.$$

Similarly, $E_{HH1} = 7.4$ meV and $E_{LH1} = 20.6$ meV. The transition energies are

$$E_{C1\text{-}HH1} = E_g + E_{C1} + E_{HH1} = 1462 \text{ meV}$$
$$E_{C1\text{-}LH1} = E_g + E_{C1} + E_{LH1} = 1475 \text{ meV}.$$

(c) The energy for an infinite barrier model is $E_1 = \pi^2\hbar^2/(2m^*L_{eff}^2)$

$$\text{For C1,} \quad L_{eff} = \sqrt{\frac{\pi^2\hbar^2}{2m_e^*E_{C1}}} = 136\,\text{Å}.$$

$$\text{For HH1,} \quad L_{eff} = \sqrt{\frac{\pi^2\hbar^2}{2m_{hh}^*E_{HH1}}} = 122\,\text{Å}.$$

$$\text{For LH1,} \quad L_{eff} = \sqrt{\frac{\pi^2\hbar^2}{2m_{\ell h}^*E_{LH1}}} = 139.6\,\text{Å}.$$

3.3 THE HARMONIC OSCILLATOR

If an electron is in a parabolic potential of the form $V(z) = \frac{K}{2}z^2$, as shown in Fig. 3.6, the time-independent Schrödinger equation in one dimension is

$$\left(-\frac{\hbar^2}{2m}\frac{d^2}{dz^2} + \frac{K}{2}z^2\right)\phi(z) = E\phi(z). \tag{3.3.1}$$

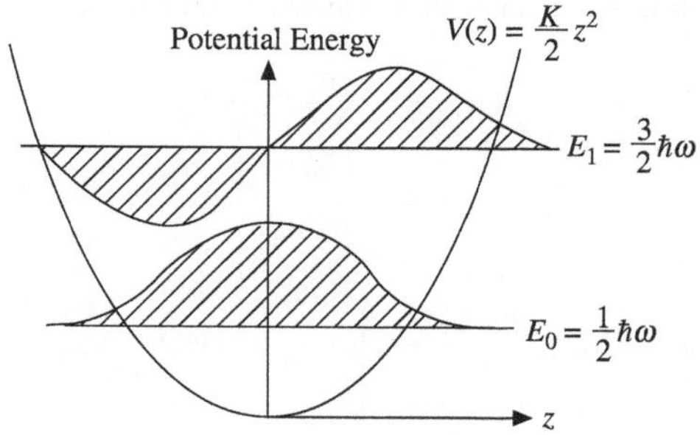

Figure 3.6 A parabolic quantum well with its quantized energy levels and wave functions.

If we define

$$\omega = \sqrt{\frac{K}{m}} \tag{3.3.2}$$

and change the variable from z to ξ

$$\xi = \alpha z \tag{3.3.3}$$

$$\alpha = \sqrt{\frac{m\omega}{\hbar}} \tag{3.3.4}$$

Equation (3.3.1) becomes

$$\left[\frac{d^2}{d\xi^2} + \left(\frac{2E}{\hbar\omega} - \xi^2\right)\right]\phi(\xi) = 0. \tag{3.3.5}$$

The solutions are the Hermite–Gaussian functions

$$\phi_n(\xi) = \left(\frac{\alpha}{\sqrt{\pi}2\pi n!}\right)^{1/2} H_n(\xi)e^{-\xi^2/2} \tag{3.3.6}$$

where $H_n(\xi)$ are the Hermite polynomials satisfying the differential equation

$$\left(\frac{d^2}{d\xi^2} - 2\xi\frac{d}{d\xi} + 2n\right)H_n(\xi) = 0 \tag{3.3.7}$$

and n is related to the energy E by

$$E = \left(n + \frac{1}{2}\right)\hbar\omega \quad n = 0, 1, 2, 3, \ldots \tag{3.3.8}$$

The Hermite polynomials can also be obtained from the generating function

$$e^{-t^2+2t\xi} = \sum_{n=0}^{\infty} \frac{H_n(\xi)}{n!} t^n \tag{3.3.9}$$

or

$$H_n(\xi) = \left[\frac{d^n}{dt^n} e^{-t^2+2t\xi}\right]_{t=0} = (-1)^n e^{\xi^2} \frac{d^n}{d\xi^n} e^{-\xi^2}. \tag{3.3.10}$$

The first few Hermite polynomials are

$$\begin{aligned} &H_0(\xi) = 1 && H_1(\xi) = 2\xi \\ &H_2(\xi) = -2 + 4\xi^2 && H_3(\xi) = -12\xi + 8\xi^3 \\ &H_4(\xi) = 12 - 48\xi^2 + 16\xi^4. \end{aligned} \tag{3.3.11}$$

Another elegant way [3] to find the solutions for the harmonic oscillator is to use the matrix approach by defining the annihilation operator

$$a = \sqrt{\frac{m\omega}{2\hbar}} z + i \frac{p}{\sqrt{2m\hbar\omega}} \tag{3.3.12}$$

and the creation operator

$$a^+ = \sqrt{\frac{m\omega}{2\hbar}} z - i \frac{p}{\sqrt{2m\hbar\omega}}. \tag{3.3.13}$$

Note the relation

$$zp - pz = i\hbar \tag{3.3.14}$$

which can be proved using $p = (\hbar/i)(\partial/\partial z)$, and

$$zp - pz = z \frac{\hbar}{i} \frac{\partial}{\partial z} \psi - \frac{\hbar}{i} \frac{\partial}{\partial z} (z\psi) = i\hbar\psi$$

for any function $\psi(z)$. Therefore, we find

$$\begin{aligned} a^+ a &= \frac{m\omega}{2\hbar} z^2 + \frac{p^2}{2m\hbar\omega} + \frac{i}{2\hbar}(zp - pz) \\ &= \frac{1}{\hbar\omega}\left(\frac{p^2}{2m} + \frac{m\omega^2}{2} z^2\right) - \frac{1}{2}. \end{aligned} \tag{3.3.15}$$

The Hamiltonian (3.3.1) can be rewritten in terms of a^+a

$$\begin{aligned} H &= \frac{p^2}{2m} + \frac{1}{2}m\omega^2 z^2 \\ &= \hbar\omega\left(a^+a + \frac{1}{2}\right). \end{aligned} \tag{3.3.16}$$

Because the last term $\hbar\omega/2$ is a constant, the problem of solving the Schrödinger equation

$$H\phi_n = E_n\phi_n \tag{3.3.17}$$

is the same as finding the eigenvector and the eigenvalue of the operator a^+a. Defining

$$N \equiv a^+a \tag{3.3.18}$$

and noting that the commutator [,]

$$[a,a^+] \equiv aa^+ - a^+a = \frac{i}{2h}\{(pz - zp) - (zp - pz)\} = 1 \tag{3.3.19}$$

from (3.3.12) and (3.3.13), we find

$$Na = a^+aa = (aa^+ - 1)a = aN - a \tag{3.3.20}$$

$$Na^+ = a^+aa^+ = a^+(a^+a + 1) = a^+(N + 1). \tag{3.3.21}$$

For any eigenstate of H or N, say $\phi_n \equiv |n\rangle$, we have

$$\langle n|N|n\rangle = \langle n|a^+a|n\rangle = \langle\psi|\psi\rangle \geq 0 \tag{3.3.22}$$

as the wave function $|\psi\rangle = a|n\rangle$ has a norm $\langle\psi|\psi\rangle$ that is always nonnegative. Let n be the eigenvalue of the operator N with the corresponding eigenvector $|n\rangle$, $N|n\rangle = n|n\rangle$. We have

$$Na|n\rangle = (aN - a)|n\rangle = (n - 1)a|n\rangle. \tag{3.3.23}$$

Thus $a|n\rangle$ is an eigenstate of the operator N with eigenvalue $n - 1$. Assume that the eigenfunctions are normalized for all n

$$\langle n|n\rangle = 1. \tag{3.3.24}$$

Then

$$n = \langle n|a^+a|n\rangle = ((\langle n|a^+)(a|n\rangle)). \tag{3.3.25}$$

Because $(a|n\rangle)$ is an eigenstate of the operator N with eigenvalue $n-1$ from (3.3.23) and its magnitude squared is n from (3.3.25), we have

$$a|n\rangle = \sqrt{n}|n-1\rangle \tag{3.3.26}$$

where the normalization condition $\langle n-1|n-1\rangle = 1$ has been used. Repeating the above procedure, we have

$$a^2|n\rangle = \sqrt{n}a|n-1\rangle = \sqrt{n}\sqrt{n-1}|n-2\rangle. \tag{3.3.27}$$

One finds the lowest state will be $|0\rangle$ with its eigenvalue $n = 0$ because of the nonnegative condition (3.3.22). Similarly, using (3.3.21), we find

$$a^+|n\rangle = \sqrt{n+1}|n+1\rangle. \tag{3.3.28}$$

In general, the nth eigenstate can be obtained from the ground state $|0\rangle$ by the creation operator

$$|n\rangle = \frac{(a^+)^n}{\sqrt{n!}}|0\rangle \tag{3.3.29}$$

with

$$E_n = \left(n+\frac{1}{2}\right)\hbar\omega, \quad n = 0, 1, 2, 3, \ldots \tag{3.3.30}$$

The ground state wave function can also be obtained using the fact that

$$a|0\rangle = 0.$$

Therefore,

$$\begin{aligned} 0 = \langle z|a|0\rangle &= \left\langle z\left|\sqrt{\frac{m\omega}{2\hbar}}z + \sqrt{\frac{\hbar}{2m\omega}}\frac{d}{dz}\right|0\right\rangle \\ &= \sqrt{\frac{\hbar}{2m\omega}}\left(\frac{d}{dz}+\frac{m\omega}{\hbar}z\right)\phi_0(z). \end{aligned} \tag{3.3.31}$$

The solution to the above first-order differential equation is

$$\phi_0(z) = \left(\frac{m\omega}{\pi\hbar}\right)^{1/4} e^{-m\omega z^2/(2\hbar)} \tag{3.3.32}$$

which has been normalized. Knowing that $\phi_0(z) = \langle z|0\rangle$, all the other eigenfunctions can be created sequentially using the creation operator in (3.3.28).

3.4 THE HYDROGEN ATOM AND EXCITON IN 2D AND 3D

In this section, we summarize the major results of the energies and wave functions for the hydrogen atom model with both bound ($E < 0$) and continuum ($E > 0$) state solutions [4–8]. The hydrogen atom is a two-particle system for the positive nucleus (with a mass m_1) at a position $\mathbf{r}_1$ and an electron (with mass m_2) at a position $\mathbf{r}_2$. The two-particle system is described by a wave function $\psi(\mathbf{r}_1, \mathbf{r}_2, t)$, which is a solution from the Hamiltonian

$$H = \frac{\mathbf{p}_1^2}{2m_1} + \frac{\mathbf{p}_2^2}{2m_2} + V(\mathbf{r}_1 - \mathbf{r}_2) \tag{3.4.1}$$

where $\mathbf{p}_1 = (\hbar/i)\nabla_1$ and $\mathbf{p}_2 = (\hbar/i)\nabla_2$, and ∇_i refers to the gradient operator with respect to $\mathbf{r}_i$ ($i = 1, 2$). A general solution is to transform from $\mathbf{r}_1$ and $\mathbf{r}_2$ coordinates to the center-of-mass coordinate $\mathbf{R}$,

$$\mathbf{R} = \frac{m_1\mathbf{r}_1 + m_2\mathbf{r}_2}{m_1 + m_2} \tag{3.4.2a}$$

and the difference coordinate $\mathbf{r}$,

$$\mathbf{r} = \mathbf{r}_1 - \mathbf{r}_2. \tag{3.4.2b}$$

The complete solution is of the form

$$\psi(\mathbf{r}_1, \mathbf{r}_2) = \frac{e^{i\mathbf{k}\cdot\mathbf{r}}}{\sqrt{V}}\,\psi(\mathbf{r}) \tag{3.4.3}$$

where $\psi(\mathbf{r})$ satisfies

$$\left[-\frac{\hbar^2}{2m_r}\nabla_r^2 - \frac{e^2}{4\pi\varepsilon r}\right]\psi(\mathbf{r}) = E\psi(\mathbf{r}) \tag{3.4.4}$$

where m_r is the reduced effective mass, $1/m_r = 1/m_1 + 1/m_2$.

3.4.1 3D Solutions

In three-dimensional space, the eigenfunctions can be expressed as

$$\Psi(\mathbf{r}) = \begin{cases} R_{nl}(r)\,Y_{lm}(\theta, \phi): & \text{bound-states } (E < 0) \\ R_{El}(r)\,Y_{lm}(\theta, \phi): & \text{continuum-states } (E < 0) \end{cases} \tag{3.4.5}$$

where the radial functions $R_{nl}(r)$ and $R_{El}(r)$ can be found in Refs. 4–8 and 10. The first few spherical harmonics $Y_{lm}(\theta, \phi)$ are given as follows:

$l = 0$ (s orbit)

$$Y_{00} = \frac{1}{\sqrt{4\pi}} \tag{3.4.6}$$

$l = 1$ (p orbits)

$$Y_{10}(\theta, \phi) = \sqrt{\frac{3}{4\pi}} \cos\theta = \sqrt{\frac{3}{4\pi}} \frac{z}{r} \equiv |Z\rangle \tag{3.4.7a}$$

$$\begin{aligned} Y_{1\pm1}(\theta, \phi) &= \mp\sqrt{\frac{3}{8\pi}} \sin\theta e^{\pm i\phi} = \mp\sqrt{\frac{3}{8\pi}} \frac{x \pm iy}{r} \\ &\equiv \mp\frac{1}{\sqrt{2}} |X \pm iY\rangle. \end{aligned} \tag{3.4.7b}$$

For bound state solutions, the energy levels are quantized as

$$E_n = -\frac{1}{n^2} R_y \qquad (n = 1, 2, 3, \ldots) \tag{3.4.8}$$

The magnitude, R_y/n^2, is called the binding energy, where the Rydberg energy for the hydrogen atom is

$$R_y = \frac{m_r e^4}{2(4\pi\varepsilon)^2 \hbar^2} = \frac{\hbar^2}{2m_r} \frac{1}{a_0^2} \tag{3.4.9}$$

and the Bohr radius a_0 is

$$a_0 = \frac{4\pi\varepsilon\hbar^2}{e^2 m_r}. \tag{3.4.10}$$

The wavefunction at the origin is

$$|\psi_{n00}(\mathbf{r} = 0)|^2 = \frac{1}{\pi a_0^3 n^3}. \tag{3.4.11}$$

For continuum state solutions, the energy E is a continuous variable and the wave function at the origin is given by

$$|\psi_{E00}(\mathbf{r} = 0)|^2 = \frac{1}{R_y a_0^3 4\pi} \left[\frac{e^{\pi/(ka_0)}}{\sinh\left(\dfrac{\pi}{ka_0}\right)} \right] \tag{3.4.12}$$

where $E = \hbar^2 k^2 / 2m_r$. The expression in the above square brackets is called the Sommerfeld (or Coulomb) enhancement factor for a 3D hydrogen atom.

3.4.2 2D Solutions

The solutions for the two-dimensional hydrogen atom problem are given by

$$\Psi(\mathbf{r}) = \begin{cases} R_{nm}(r)\dfrac{e^{im\phi}}{\sqrt{2\pi}}, & \text{bound-states } (E < 0) \\ R_{Em}(r)\dfrac{e^{im\phi}}{\sqrt{2\pi}}, & \text{continuum-states } (E > 0). \end{cases} \tag{3.4.13}$$

The eigenenergies for the bound states are quantized

$$E_n = -\frac{R_y}{\left(n - \dfrac{1}{2}\right)^2} \quad (n = 1, 2, 3, \ldots). \tag{3.4.14}$$

We have

$$E_1 = -4R_y,\ E_2 = -\frac{4}{9}R_y,\ E_3 = -\frac{4}{25}R_y, \ldots \tag{3.4.15}$$

It is noted that the binding energy for the 1s states $|E_1|$ is four times that in the three-dimensional case. This enhancement of the binding energy will be very useful in understanding the excitonic effects in semiconductor quantum wells and the observation of the excitonic optical absorption spectra, which will be investigated in Chapter 14. The wavefunction at the origin is

$$|\psi_{n0}(\mathbf{r} = 0)|^2 = \frac{1}{\pi a_0^3\left(n - \dfrac{1}{2}\right)^3} \quad (n = 1, 2, 3, \ldots). \tag{3.4.16}$$

For continuum states, the energy E is a continuous variable, and the wave function at the origin is

$$|\psi_{E0}(\mathbf{r} = 0)|^2 = \frac{1}{R_y a_0^3 4\pi}\left[\frac{e^{\pi/(ka_0)}}{\cosh\left(\dfrac{\pi}{ka_0}\right)}\right] \tag{3.4.17}$$

where $E = \hbar^2 k^2/2m_r$. The expression inside the square bracket is the Sommerfeld enhancement factor.

3.5 TIME-INDEPENDENT PERTURBATION THEORY

3.5.1 Perturbation Method

In most practical physical systems, the Schrödinger equations do not have exact or analytical solutions. It is always convenient to find the solutions using the perturbation method [11] when the problem of interest can be separated into two parts: one part consists of an "unperturbed" Hamiltonian with known solutions, $H^{(0)}$; the other part consists of a small perturbing potential, H',

$$H = H^{(0)} + H'. \tag{3.5.1}$$

The unperturbed wave functions $\phi_n^{(0)}$ and eigenvalues $E_n^{(0)}$ are assumed known

$$H^{(0)}\phi_n^{(0)} = E_n^{(0)}\phi_n^{(0)}. \tag{3.5.2}$$

In order to find the solutions for the total Hamiltonian

$$H\psi = E\psi \tag{3.5.3}$$

it is convenient to introduce a perturbation parameter λ (set $\lambda = 1$ later)

$$H = H^{(0)} + \lambda H'. \tag{3.5.4}$$

We look for the solutions of the form

$$E = E^{(0)} + \lambda E^{(1)} + \lambda^2 E^{(2)} + \cdots \tag{3.5.5a}$$

$$\psi = \psi^{(0)} + \lambda\,\psi^{(1)} + \lambda^2\psi^{(2)} + \cdots \tag{3.5.5b}$$

Substituting the above expressions for H, E, and ψ into the Schrödinger equation, we find, to each order in λ,

$$\text{Zeroth order} \quad H^{(0)}\psi^{(0)} = E^{(0)}\psi^{(0)} \tag{3.5.6}$$

$$\text{First order} \quad H^{(0)}\psi^{(1)} + H'\,\psi^{(0)} = E^{(0)}\psi^{(1)} + E^{(1)}\psi^{(0)} \tag{3.5.7}$$

$$\text{Second order} \quad H^{(0)}\psi^{(2)} + H'\psi^{(1)} = E^{(0)}\psi^{(2)} + E^{(1)}\psi^{(1)} + E^{(2)}\psi^{(0)}. \tag{3.5.8}$$

Zeroth-Order Solutions Thus, it is clearly seen that the zeroth-order solutions are the unperturbed solutions

$$\psi_n^{(0)} = \phi_n^{(0)} \tag{3.5.9a}$$

$$E_n^{(0)} = E_n^{(0)}. \tag{3.5.9b}$$

First-Order Solutions The first-order wave function $\psi^{(1)}$ may be expanded in terms of a linear combination of the unperturbed solutions

$$\psi_n^{(1)} = \sum_m a_{mn}^{(1)}\phi_m^{(0)}. \tag{3.5.10}$$

Thus

$$(H^{(0)} - E_n^{(0)})\psi_n^{(1)} = E^{(1)}\phi_n^{(0)} - H'\phi_n^{(0)}. \tag{3.5.11}$$

Multiplying the above equation by $\phi_m^{(0)^*}$ and integrating over space, using

$$\langle \phi_m^{(0)} | \phi_n^{(0)} \rangle = \delta_{mn} \tag{3.5.12}$$

we obtain

$$E_n^{(1)} = H'_{nn} \tag{3.5.13a}$$

and

$$a_{mn}^{(1)} = \frac{H'_{mn}}{E_n^{(0)} - E_m^{(0)}} \quad \text{if } m \neq n \tag{3.5.13b}$$

where

$$H'_{mn} = \int \phi_m^{(0)^*} H' \phi_n^{(0)} d^3\mathbf{r}. \tag{3.5.14}$$

To the first order in perturbation, we need to normalize the wave function

$$\int \left(\phi_n^{(0)} + \sum_m a_{mn}^{(1)} \phi_m^{(0)} \right)^* \left(\phi_n^{(0)} + \sum_m a_{mn}^{(1)} \phi_m^{(0)} \right) d^3r = 1$$

or

$$1 + a_{nn}^{(1)^*} + a_{nn}^{(1)} + \sum_m a_{mn}^{(1)} a_{mn}^{(1)^*} = 1. \tag{3.5.15}$$

Thus we have $a_{nn}^{(1)} = 0$ to the first order in perturbation. The result is therefore

$$\psi_n = \phi_n^{(0)} + \sum_{m \neq n} \frac{H'_{mn}}{E_n^{(0)} - E_m^{(0)}} \phi_m^{(0)} \tag{3.5.16a}$$

$$E_n = E_n^{(0)} + H'_{nn}. \tag{3.5.16b}$$

Second-Order Solutions Similarly, the second-order wave function $\psi^{(2)}$ can be expanded in terms of the zero-order solutions

$$\psi_n^{(2)} = \sum_m a_{mn}^{(2)} \phi_m^{(0)} \tag{3.5.17}$$

we find

$$E_n^{(2)} = \sum_{m \neq n} a_{mn}^{(1)} H'_{nm} = \sum_{m \neq n} \frac{H'_{mn} H'_{nm}}{E_n^{(0)} - E_m^{(0)}} \tag{3.5.18a}$$

and

$$a_{mn}^{(2)} = \sum_{k \neq n} \frac{H'_{mk} H'_{kn}}{\left(E_n^{(0)} - E_m^{(0)}\right)\left(E_n^{(0)} - E_k^{(0)}\right)} - \frac{H'_{mn} H'_{nn}}{\left(E_n^{(0)} - E_m^{(0)}\right)^2}, \quad m \neq n. \tag{3.5.18b}$$

To the second-order correction, the wave function may be normalized

$$\int\left(\phi_n^{(0)}+\sum_{m\neq n}a_{mn}^{(1)}\phi_m^{(0)}+\sum_m a_{mn}^{(2)}\phi_m^{(0)}\right)^*\left(\phi_n^{(0)}+\sum_{m\neq n}a_{mn}^{(1)}\phi_m^{(0)}+\sum_m a_{mn}^{(2)}\phi_m^{(0)}\right)d^3\mathbf{r}=1. \tag{3.5.19}$$

Neglecting terms of third and higher orders, we obtain

$$a_{nn}^{(2)}=-\frac{1}{2}\sum_{m\neq n}\left|a_{mn}^{(1)}\right|^2. \tag{3.5.20}$$

The normalized wave function ψ_n and its eigenenergy E_n, to the second order in perturbation, are given by

$$\psi_n=\phi_n^{(0)}+\sum_{m\neq n}\frac{H'_{mn}}{E_n^{(0)}-E_m^{(0)}}\phi_m^{(0)}+\sum_{m\neq n}\left\{\left[\sum_{k\neq n}\frac{H'_{mk}H'_{kn}}{\left(E_n^{(0)}-E_m^{(0)}\right)\left(E_n^{(0)}-E_k^{(0)}\right)}\right.\right.$$
$$\left.\left.-\frac{H'_{mn}H'_{nn}}{\left(E_n^{(0)}-E_m^{(0)}\right)^2}\right]\phi_m^{(0)}-\frac{\left|H'_{mn}\right|^2}{2\left(E_n^{(0)}-E_m^{(0)}\right)^2}\phi_n^{(0)}\right\} \tag{3.5.21a}$$

$$E_n=E_n^{(0)}+H'_{nn}+\sum_{m\neq n}\frac{\left|H'_{nm}\right|^2}{E_n^{(0)}-E_m^{(0)}}. \tag{3.5.21b}$$

Example

(a) When an infinite quantum well has an applied electric field F in the z-direction (Fig. 3.7), the Hamiltonian can be written as

$$H=H_0+eFz \tag{3.5.22}$$

where H_0 describes an electron in the infinite quantum well *without* the electric field. Treating the term eFz as a perturbation, show that the energy shift due to the applied electric field, $E-E_n^{(0)}$, can be written as

$$\Delta E=E-E_n^{(0)}=C_n\frac{\pi^2}{2}\frac{(eFL)^2}{E_1^{(0)}}\left(=C_n\frac{m^*e^2F^2L^4}{\hbar^2}\right) \tag{3.5.23}$$

where $E_1^{(0)}$ and $E_n^{(0)}$ are the energies of the ground state and the nth state of the infinite quantum well without an applied electric field. Find a general expression for C_n. All energies are measured from the center of the well.

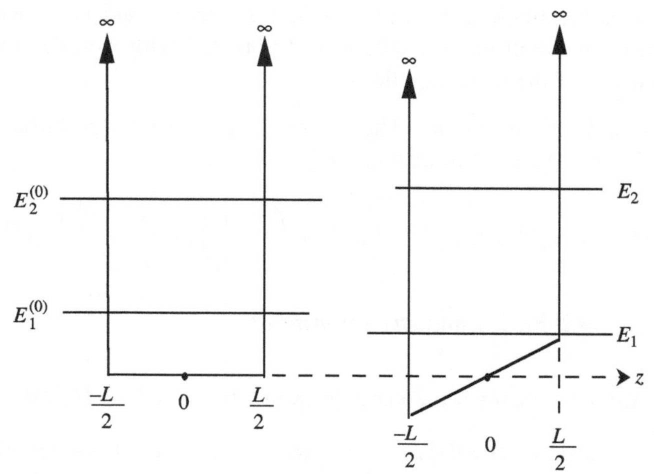

Figure 3.7 An infinite quantum well without an applied field (left diagram) and with an applied field (right diagram) where the potential due to the applied field eFz is treated as a perturbation.

(b) Evaluate C_n for $n = 1$, 2, and 3 numerically by keeping the first two nonzero terms in the summation over the index m. Show that C_1 is negative whereas C_2 and C_3 are positive.

(c) Compare numerically C_n, $n = 1$, 2, 3, with those in part (b), where C_n is given by

$$C_n = (n^2\pi^2 - 15)/(24\,n^4\pi^4). \tag{3.5.24}$$

Solutions

(a) We treat $H' = eFz$ as a perturbation from H_0. We write down the unperturbed wave functions and eigenvalues from the infinite barrier model.

Zeroth-order Solution

$$\begin{aligned} H_0 u_n^{(0)}(z) &= \frac{-\hbar^2}{2m^*}\frac{d^2}{dz^2}\,u_n^{(0)}(z) = E_n^{(0)}u_n^{(0)}(z) \\ u_n^{(0)}(z) &= \sqrt{\frac{2}{L}}\sin\left[\frac{n\pi}{L}\left(z+\frac{L}{2}\right)\right], \quad E_n^{(0)} = n^2\frac{\hbar^2\pi^2}{2m^*L^2}. \end{aligned} \tag{3.5.25}$$

First-order Perturbation

$$H'_{nn} = \langle u_n^{(0)}|eFz|u_n^{(0)}\rangle = eF\int_{-L/2}^{L/2} u_n^{(0)}(z)z\,u_n^{(0)}(z)\,dz = 0. \tag{3.5.26}$$

Therefore, the first-order correction in the energy vanishes because of the symmetry of the original quantum-well potential with well-defined even and odd parities of the wave functions.

Second-order Perturbation: The energy to second-order perturbation is given by (3.5.21b). We need to evaluate H'_{nm}, $n \neq m$

$$H'_{nm} = \langle u_n^{(0)} | eFz | u_m^{(0)} \rangle = \left\langle u_n^{(0)} \middle| eF\left(z + \frac{L}{2}\right) \middle| u_m^{(0)} \right\rangle - eF\frac{L}{2}\langle u_n^{(0)} | u_m^{(0)} \rangle$$

$$= 2eFL \int_0^1 t \sin(n\pi t)\ \sin(m\pi t)\, dt$$

where we have changed the variable from z to t, $t = [z + (L/2)]/L$. Write

$$2\sin(n\pi t)\sin(m\pi t) = \cos[(n-m)\pi t] - \cos[(n+m)\pi t]$$

and use

$$\int_0^1 t\cos(M\pi t)\, dt = \begin{cases} \dfrac{(-1)^M - 1}{(M\pi)^2} & \text{for } M \neq 0 \\ \dfrac{1}{2} & \text{for } M = 0. \end{cases}$$

We obtain

$$\begin{aligned} H'_{nm} &= \frac{eFL}{\pi^2}\left[\frac{(-1)^{n-m} - 1}{(n-m)^2} - \frac{(-1)^{n+m} - 1}{(n+m)^2}\right] \\ &= \frac{eFL}{\pi^2}[(-1)^{n-m} - 1]\frac{4nm}{(n^2 - m^2)^2} \quad (n \neq m). \end{aligned} \tag{3.5.27}$$

The perturbation to second order gives

$$\begin{aligned} E &= E_n^{(0)} + H'_{nn} + \sum_{m \neq n} \frac{|H'_{nm}|^2}{E_n^{(0)} - E_m^{(0)}} \\ &= E_n^{(0)} + C_n \frac{\pi^2}{2}\frac{(eFL)^2}{E_1^{(0)}} \end{aligned} \tag{3.5.28}$$

where we use $E_n^{(0)} - E_m^{(0)} = (n^2 - m^2)E_1^{(0)}$ and

$$C_n = \frac{32n^2}{\pi^6}\sum_{m \neq n} \frac{[(-1)^{n-m} - 1]^2 m^2}{(n^2 - m^2)^5}. \tag{3.5.29}$$

The above results show that the energy shifts depend quadratically on the applied electric field, which is an important phenomenon when we study the quantum-confined Stark effects in Chapter 14.

(b) If we keep only the first two nonvanishing terms in the above summation for C_n, we obtain

$$C_1 = \frac{32}{\pi^6} \times \left[\frac{4 \times 2^2}{(1 - 2^2)^5} + \frac{4 \times 4^2}{(1^2 - 4^2)^5}\right] = -2.194 \times 10^{-3}$$

$$C_2 = \frac{32 \times 2^2}{\pi^6} \times \left[\frac{4 \times 1^2}{(4 - 1)^5} + \frac{4 \times 3^2}{(4 - 9)^5}\right] = 6.578 \times 10^{-4}$$

$$C_3 = \frac{32 \times 3^2}{\pi^6} \times \left[\frac{4 \times 2^2}{(9 - 4)^5} + \frac{4 \times 4^2}{(9 - 16)^5}\right] = 3.931 \times 10^{-4}.$$

We see that E_1 decreases with increasing field F as $C_1 < 0$ and both E_2 and E_3 increase slightly with increasing field F.

(c) The above C_n in part (a) can be summed up to an analytical expression [12–14]

$$C_n = \frac{n^2\pi^2 - 15}{24n^4\pi^5} \quad \text{or} \quad C_1 = -2.194 \times 10^{-3}, \quad C_2 = 6.544 \times 10^{-4}, \text{ and}$$

$$C_3 = 3.899 \times 10^{-4}$$

which are close to those in (b). These results using the second-order perturbation theory agree very well with those obtained from variational methods [14–17].

3.5.2 Matrix Formulation

Alternatively, the eigenvalue problem

$$H\psi = \left(H^{(0)} + H'\right)\psi = E\psi \tag{3.5.30}$$

can be solved directly by letting

$$\psi = \sum_m a_m \phi_m^{(0)} \tag{3.5.31}$$

where $\{\phi_m^{(0)}\}$ are the eigenfunctions of the unperturbed Hamiltonian,

$$H_0 \phi_m^{(0)} = E_m^{(0)} \phi_m^{(0)}. \tag{3.5.32}$$

The second subscript n in a is dropped for convenience. Here $\phi_m^{(0)}, m = 1, 2, \ldots, N$, may also be degenerate wave functions. A direct substitution of (3.5.31) into (3.5.30), and taking the inner product with respect to $\phi_k^{(0)}, k = 1, \ldots, N$, gives

$$\sum_m (H_{km} - E\delta_{km})a_m = 0 \tag{3.5.33}$$

if $\{\phi_m^{(0)}\}$ form an orthonormal set. If $\{\phi_m^{(0)},\ m = 1, 2, \ldots, N\}$ are not orthonormal to each other,

$$\langle \phi_k^{(0)} | \phi_m^{(0)} \rangle = N_{km} \tag{3.5.34}$$

then, the above eigenequation becomes

$$\sum_m (H_{km} - EN_{km}) a_m = 0. \tag{3.5.35}$$

The eigenequations (3.5.33) or (3.5.35) can be solved by letting

$$\det|H_{km} - E\delta_{km}| = 0 \tag{3.5.36}$$

for (3.5.33), or

$$\det|H_{km} - EN_{km}| = 0 \tag{3.5.37}$$

for (3.5.35). The above procedure is equivalent to diagonalizing the matrix H_{km} in (3.5.33) or simultaneously diagonalizing H_{km} and N_{km} in (3.5.35) (this is possible because N is also Hermitian). Finally, one finds the eigenvalues for E and the eigenvector a_m. The wave function is obtained from (3.5.31), which should be normalized.

3.6 TIME-DEPENDENT PERTURBATION THEORY

Consider the Schrödinger equation

$$H\psi(\mathbf{r}, t) = -\frac{\hbar}{i}\frac{\partial}{\partial t}\psi(\mathbf{r}, t) \tag{3.6.1}$$

where the Hamiltonian H consists of an unperturbed part H_0, which is time-independent, and a small perturbation $H'(\mathbf{r}, t)$, which depends on time

$$H = H_0 + H'(\mathbf{r}, t). \tag{3.6.2}$$

The solution to the unperturbed part is assumed known

$$H_0\phi_n(\mathbf{r}, t) = -\frac{\hbar}{i}\frac{\partial}{\partial t}\phi_n(\mathbf{r}, t) \tag{3.6.3}$$

$$\phi_n(\mathbf{r}, t) = \phi_n(\mathbf{r}) e^{-iE_n t/\hbar}. \tag{3.6.4}$$

The time-dependent perturbation is assumed to have the form

$$H'(\mathbf{r}, t) = \begin{cases} H'(\mathbf{r})e^{-i\omega t} + H'^{+}(\mathbf{r})e^{+i\omega t} & t \geq 0 \\ 0, & t < 0. \end{cases} \tag{3.6.5}$$

To find the solution $\psi(\mathbf{r}, t)$ to the time-dependent Schrödinger equation, we expand the wave function in terms of the unperturbed solutions

$$\psi(\mathbf{r}, t) = \sum_n a_n(t)\phi_n(\mathbf{r})e^{-iE_nt/\hbar} \tag{3.6.6}$$

where $|a_n(t)|^2$ gives the probability that the electron is in the state n at time t. Substituting the expansion for ψ into the Schrödinger equation and using (3.6.3), we obtain

$$\sum_n \frac{da_n(t)}{dt}\phi_n(\mathbf{r})e^{-iE_nt/\hbar} = -\frac{i}{\hbar}\sum_n H'(\mathbf{r}, t)a_n(t)\phi_n(\mathbf{r})e^{-iE_nt/\hbar}. \tag{3.6.7}$$

Taking the inner product with the wave function $\phi_m^*(\mathbf{r})$, and using the orthonormal property, $\int d^3\mathbf{r}\phi_m^*(\mathbf{r})\phi_n(\mathbf{r}) = \delta_{mn}$, we find

$$\frac{da_m(t)}{dt} = -\frac{i}{\hbar}\sum_n a_n(t)H'_{mn}(t)e^{i\omega_{mn}t} \tag{3.6.8}$$

where

$$\begin{aligned} H'_{mn}(t) &= \langle m|H'(\mathbf{r}, t)|n\rangle \\ &= \int \phi_m^*(\mathbf{r})H'(\mathbf{r}, t)\phi_n(\mathbf{r})d^3\mathbf{r} \\ &= H'_{mn}e^{-i\omega t} + H'^{+}_{mn}e^{i\omega t} \qquad (3.6.9) \\ \omega_{mn} &= (E_m - E_n)/\hbar \qquad (3.6.10) \end{aligned}$$

and the matrix element is defined as

$$H'_{mn} = \int \phi_m^*(\mathbf{r})H'(\mathbf{r})\phi_n(\mathbf{r})d^3\mathbf{r}. \tag{3.6.11}$$

Introducing the perturbation parameter λ (set $\lambda = 1$ later)

$$H = H_0 + \lambda H'(\mathbf{r}, t) \tag{3.6.12}$$

and letting

$$a_n(t) = a_n^{(0)} + \lambda a_n^{(1)}(t) + \lambda^2 a_n^{(2)}(t) + \cdots \tag{3.6.13}$$

we obtain

$$\frac{da_m^{(0)}}{dt} = 0 \tag{3.6.14a}$$

$$\frac{da_m^{(1)}}{dt} = -\frac{i}{\hbar}\sum_n a_n^{(0)}(t)H'_{mn}(t)\, e^{i\omega_{mn}t} \tag{3.6.14b}$$

$$\frac{da_m^{(2)}}{dt} = -\frac{i}{\hbar}\sum_n a_n^{(1)}(t)H'_{mn}(t)\, e^{i\omega_{mn}t}. \tag{3.6.14c}$$

Thus, the zeroth-order solutions for (3.6.14a) are constant. Let the electron be at the state i initially

$$a_i^{(0)}(t=0) = 1; \quad a_m^{(0)}(t) = 0, \quad m \neq i. \tag{3.6.15}$$

We have the zeroth-order solution

$$\begin{aligned} a_i^{(0)}(t) &= 1; \\ a_m^{(0)}(t) &= 0, \quad m \neq i. \end{aligned} \tag{3.6.16}$$

Therefore, the electron stays at the state i in the absence of any perturbation. The first-order solution is obtained from (3.6.14b)

$$\begin{aligned} \frac{da_m^{(1)}}{dt} &= -\frac{i}{\hbar}H'_{mi}(t)e^{i\omega_{mi}t} \\ &= -\frac{i}{\hbar}[H'_{mi}e^{i(\omega_{mi}-\omega)t} + H'^{+}_{mi}e^{i(\omega_{mi}+\omega)t}]. \end{aligned} \tag{3.6.17}$$

Suppose we are interested in a final state $m = f$; the above equation can be solved directly by integration

$$a_f^{(1)}(t) = -\frac{1}{\hbar}\left[H'_{fi}\frac{e^{i(\omega_{fi}-\omega)t} - 1}{\omega_{fi} - \omega} + H'^{+}_{fi}\frac{e^{i(\omega_{fi}+\omega)t} - 1}{\omega_{fi} + \omega}\right]. \tag{3.6.18}$$

If we consider the photon energy to be near resonance, either $\omega \sim \omega_{fi}$ or $\omega \sim -\omega_{fi}$, we find the dominant terms:

$$\left|a_f^{(1)}(t)\right|^2 = \frac{4\left|H'_{fi}\right|^2}{\hbar^2}\frac{\sin^2\frac{t}{2}(\omega_{fi}-\omega)}{(\omega_{fi}-\omega)^2} + \frac{4\left|H'^{+}_{fi}\right|^2}{\hbar^2}\frac{\sin^2\frac{t}{2}(\omega_{fi}+\omega)}{(\omega_{fi}+\omega)^2} \tag{3.6.19}$$

where the cross-term has been dropped because it is small compared with either of the above two terms. When the interaction time is long enough, using

$$\frac{\sin^2\left(\frac{xt}{2}\right)}{x^2} \longrightarrow \frac{\pi t}{2}\delta(x) \tag{3.6.20}$$

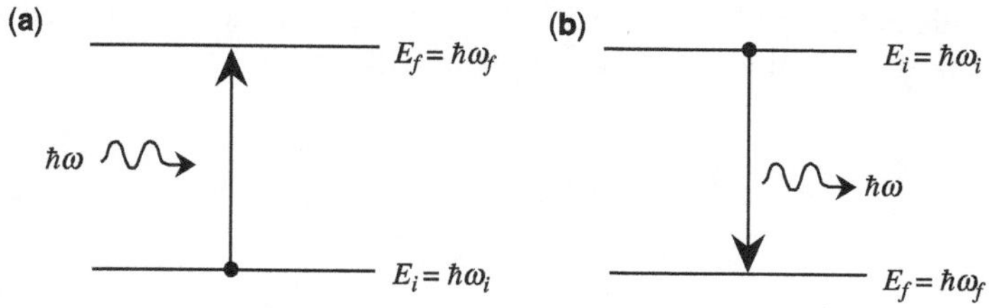

Figure 3.8 Diagrams showing the two processes in Fermi's golden rule for electron–photon interaction: (a) stimulated absorption and (b) stimulated emission of a photon in a two-level system.

we find

$$\left|a_f^{(1)}(t)\right|^2 = \frac{2\pi t}{\hbar^2}\left|H'_{fi}\right|^2 \delta(\omega_{fi} - \omega) + \frac{2\pi t}{\hbar^2}\left|H'^{+}_{fi}\right|^2 \delta(\omega_{fi} + \omega). \tag{3.6.21}$$

The transition rate is given by

$$\begin{aligned} W_{if} &= \frac{d}{dt}\left|a_f^{(1)}(t)\right|^2 \\ &= \frac{2\pi}{\hbar}\left|H'_{fi}\right|^2 \delta(E_f - E_i - \hbar\omega) + \frac{2\pi}{\hbar}\left|H'^{+}_{fi}\right|^2 \delta(E_f - E_i + \hbar\omega) \end{aligned} \tag{3.6.22}$$

where the property $\delta(\hbar\omega) = \delta(\omega)/\hbar$ has been used. The first term corresponds with the absorption of a photon by an electron, as $E_f = E_i + \hbar\omega$, whereas the second corresponds with the emission of a photon, as $E_f = E_i - \hbar\omega$. These processes are illustrated in Fig. 3.8 for a two-level system.

In summary, we have Fermi's golden rule: For a time-harmonic perturbation, turned on at $t = 0$,

$$H'(\mathbf{r}, t) = H'(\mathbf{r})e^{-i\omega t} + H'^{+}(\mathbf{r})e^{+i\omega t}, \quad t \geq 0. \tag{3.6.23}$$

The transition rate of an electron from an initial state i with energy E_i to a final state f with energy E_f is given by

$$W_{if} = \frac{2\pi}{\hbar}\left|H'_{fi}\right|^2 \delta(E_f - E_i - \hbar\omega) + \frac{2\pi}{\hbar}\left|H'^{+}_{fi}\right|^2 \delta(E_f - E_i + \hbar\omega). \tag{3.6.24}$$

APPENDIX 3A: LÖWDIN'S RENORMALIZATION METHOD

A useful method on the perturbation theory is Löwdin's perturbation method [9]. It may be necessary to divide the eigenfunctions and energies into two classes, A and B. Suppose we are mainly interested in the states in class A, and we look for a perturbation formula for class A with the influence from the other states denoted as class B.

We start with the eigenequation (3.5.33) assuming that the unperturbed states are orthonormalized. Equation (3.5.33) may be rewritten as

$$(E - H_{mm})a_m = \sum_{n \neq m}^{A} H_{mn} a_n + \sum_{\alpha \neq m}^{B} H_{m\alpha} a_\alpha \tag{3A.1}$$

or

$$a_m = \sum_{n \neq m}^{A} \frac{H_{mn}}{E - H_{mm}} a_n + \sum_{\alpha \neq m}^{B} \frac{H_{m\alpha}}{E - H_{mm}} a_\alpha \tag{3A.2}$$

where the first sum on the right-hand side is over the states in class A only, whereas the second sum is over the states in class B. Because we are interested in the coefficients a_m for m in class A, we may eliminate those in class B by an iteration procedure and obtain

$$a_m = \sum_{n}^{A} \frac{U_{mn}^A - H_{mn}\delta_{mn}}{E - H_{mm}} a_n \tag{3A.3}$$

and

$$U_{mn}^A = H_{mn} + \sum_{\alpha \neq m}^{B} \frac{H_{m\alpha} H_{\alpha n}}{E - H_{\alpha\alpha}} + \sum_{\substack{\alpha,\beta \neq m,n \\ \alpha \neq \beta}}^{B} \frac{H_{m\alpha} H_{\alpha\beta} H_{\beta n}}{(E - H_{\alpha\alpha})(E - H_{\beta\beta})} + \cdots \tag{3A.4}$$

Or, equivalently, we solve the eigenvalue problems for a_n, $(n \in A)$

$$\sum_{n}^{A} \left(U_{mn}^A - E\delta_{mn}\right) a_n = 0, \quad m \in A \tag{3A.5}$$

and

$$a_\gamma = \sum_{n}^{A} \frac{U_{\gamma n}^A - H_{\gamma n}\delta_{\gamma n}}{E - H_{\gamma\gamma}} a_n, \quad \gamma \in B. \tag{3A.6}$$

When the coefficients a_n belonging to class A are determined from the eigenequation (3A.5), the coefficients a_γ in class B can be found from (3A.6). A necessary condition for the expansion of (3A.4) to be convergent is

$$|H_{m\alpha}| \ll |E - H_{\alpha\alpha}|, \quad m \in A, \ \alpha \in B. \tag{3A.7}$$

In practice, to the second order in perturbation, we may replace E by E_A in (3A.4), where E_A is an average energy of states in class A, and truncate the series (3A.4) at the second-order term. For example, if the class A consists of only a single nondegenerate state n, then the class B consists of the rest. Equation (3A.5) gives only one equation

$$E = U^A_{nn}$$
$$= H_{nn} + \sum_{\alpha \neq n} \frac{H_{n\alpha} H_{\alpha n}}{E - H_{\alpha\alpha}} + \sum_{\substack{\alpha,\beta \neq n \\ \alpha \neq \beta}} \frac{H_{n\alpha} H_{\alpha\beta} H_{\beta n}}{(E - H_{\alpha\alpha})(E - H_{\beta\beta})} + \cdots \tag{3A.8}$$

If we separate H into $H^{(0)}$ and a perturbation H'

$$H = H^{(0)} + H' \tag{3A.9}$$

we obtain, to the second order in H',

$$E = E^{(0)}_n + H'_{nn} + \sum_{\alpha \neq n} \frac{H'_{n\alpha} H'_{\alpha n}}{E^{(0)}_n - E^{(0)}_\alpha} \tag{3A.10}$$

where $E^{(0)}_n = H_{nn}$ has been used.

If, however, the states in class A are degenerate, the diagonal elements are exactly or almost the same

$$H_{nn} \cong E_A \tag{3A.11}$$

with differences of the first and higher orders. We have

$$U^A_{mn} = H_{mn} + \sum_{\alpha}^{B} \frac{H'_{m\alpha} H'_{\alpha n}}{E_A - H_{\alpha\alpha}}. \tag{3A.12}$$

Solving the eigenequation using U^A_{mn} above by letting

$$\det\left| U^A_{mn} - E\delta_{mn} \right| = 0, \tag{3A.13}$$

the eigenvalues for E and eigenvectors for a_n are thus obtained, and the wave functions are given by

$$\psi = \sum_n a_n \phi^{(0)}_n \tag{3A.14}$$

where a_n for $n \in$ class B are obtained from (3A.6). The above wave function can be normalized directly by dividing the above expression by $\left\langle \sum_n a_n \phi^{(0)}_n \middle| \sum_n a_n \phi^{(0)}_n \right\rangle$, or

$$\psi = \frac{\sum_n a_n \phi^{(0)}_n}{\sum_n |a_n|^2}. \tag{3A.15}$$

PROBLEMS

3.1 Consider an $In_{0.53}Ga_{0.47}As/InP$ quantum well. Assume the following parameters at 300K:

$In_{0.53}Ga_{0.47}As$ region	InP region
$m_e^* = 0.041\, m_0$	$m_e^* = 0.077\, m_0$
$m_{hh}^* = 0.50\, m_0$	$m_{hh}^* = 0.60\, m_0$
$m_{\ell h}^* = 0.0503\, m_0$	$m_{\ell h}^* = 0.12\, m_0$
$E_g = 0.75\, eV$	$E_g = 1.344\, eV$

where m_0 is the free electron mass.
The band edge discontinuity is $\Delta E_c = 0.40 \Delta E_g$, and $\Delta E_v = 0.60 \Delta E_g$.

(**a**) Consider a well width $L_w = 100$ Å. How many bound states are there in the conduction band? How many bound heavy-hole and light-hole subbands are there?

(**b**) Find the lowest bound state energies for the conduction subband (C1), the heavy-hole (HH1) and the light-hole (LH1) subbands in part (**a**). What are the C1-HH1 and C1-LH1 transition energies?

(**c**) Assume that we define an effective well width L_{eff} using an infinite barrier model such that its ground state energy is the same as the energy of the first conduction subband E_{C1} in (**b**). What is L_{eff}? If we repeat the same procedure for the HH1 subband, what is L_{eff}?

3.2 In the infinite barrier model, the dispersion relation for the electron subband is given by Eq. (3.2.7).

(**a**) Plot the dispersion relations E versus $k_t = \sqrt{k_x^2 + k_y^2}$ for $n = 1$, 2, and 3 on the same chart. Plot the corresponding wave functions $\phi_1(z)$, $\phi_2(z)$, and $\phi_3(z)$ with $k_t = 0$.

(**b**) Derive the electron density of states $\rho_e(E)$ versus the energy E and plot $\rho_e(E)$ versus E.

(**c**) Assume that there are n_s electrons per unit area in the quantum wells at $T = 0$. Find an expression for the Fermi-level position in terms of n_s.

3.3 Calculate the Fermi level for the electrons in a $GaAs/Al_{0.3}Ga_{0.7}As$ quantum well described in the example in Section 3.2 with a surface electron concentration $n_s = 1 \times 10^{12}\, cm^{-2}$ at $T = 0K$. How many subbands are occupied by electrons?

3.4 (**a**) Find an analytical expression relating the surface electron concentration n_s $(1/cm^2)$ to the Fermi level at a finite temperature T in a quantum well based on the infinite barrier model.

(**b**) What is the carrier concentration at the ith subband?

3.5 If a finite barrier model is used, what is the electron density of states function $\rho_e(E)$ in Problem 3.2. Plot $\rho_e(E)$ versus the energy E.

3.6 Consider a graded quantum-well structure with a parabolic band edge profile along the growth (z) axis in real space, $E_c(z) = Kz^2/2$.

(**a**) What are the general dispersion relations of the conduction subbands: $E = E_n(k_x, k_y)$?

(**b**) What is the density of states of the electrons in the conduction band $\rho_e(E)$? Plot $\rho_e(E)$ versus the energy E.

3.7 Show that the wave function in (3.3.32) is the solution of the differential equation (3.3.31) for the ground state of the harmonic oscillator problem.

3.8 An exciton consisting of an electron with an effective mass m_e^* and a hole with an effective mass m_h^* can be described using a hydrogen model.

(**a**) Calculate the Rydberg energy R_y for a GaAs semiconductor. Assume that $\varepsilon = 12.5\varepsilon_0$; the other parameters are given in the numerical example in Section 3.2.

(**b**) Calculate the $1s$ bound state energy for GaAs if the electron–hole pair is restricted in a pure two-dimensional space.

3.9 (**a**) Plot the continuum state wave function at the origin $|\psi_{E00}\,(r = 0)|^2$ versus the energy E for the three-dimensional hydrogen model.

(**b**) Plot the continuum state wave function at the origin $|\psi_{E0}\,(r = 0)|^2$ versus the energy E for the two-dimensional hydrogen model.

3.10 From the perturbation results in the example of Section 3.5, calculate the energy shift at an applied field of 100 kV/cm using the effective well widths $L_{\text{eff}} =$

(**a**) 136 Å for electrons in C1 subband,

(**b**) 122 Å for holes in HH1 subband, and

(**c**) 139.6 Å for holes in LH1 subband in a GaAs/$Al_{0.3}Ga_{0.7}As$ quantum well.

3.11 Derive Fermi's golden rule and discuss the regime of validity for applying this rule.

3.12 Discuss the differences between Löwdin's perturbation method in Appendix 3A and the conventional perturbation method in Section 3.5.

REFERENCES

1. A. Goswami, *Quantum Mechanics*, Wm. C. Brown Publishers, Dubuque, Iowa, 1992.
2. R. P. Feynman, R. B. Leighton, and M. Sands, *The Feynman Lectures on Physics, Quantum Mechanics*, Vol. III, Addison-Wesley, Reading, MA, 1965.
3. R. P. Feynman, *Statistical Mechanics, A Set of Lectures*, Addison-Wesley, Redwood City, CA, 1972, Chapter 6.
4. For both bound and continuum state solutions in the three-dimensional case see L. D. Landau and E. M. Lifshitz, *Quantum Mechanics*, 3rd ed., Pergamon Press,

Oxford, 1977, p. 117; and H. A. Bethe and E. E. Salpeter, *Quantum Mechanics of One- and Two-Electron Atoms*, Springer-Verlag, Berlin, 1957.

5. For an n-dimensional space, $n \geq 2$, see M. Bander and C. Itzykson, "Group theory and the hydrogen atom (I) and (II)," *Rev. Mod. Phys.* **38**, 330–345 (1966) and **38**, 346–358 (1966).
6. C. Y. P. Chao and S. L. Chuang, "Analytical and numerical solutions for a two-dimensional exciton in momentum space," *Phys. Rev. B* **43**, 6530–6543 (1991).
7. E. Menzbacher, *Quantum Mechanics*, 2nd ed., Wiley, New York, 1970.
8. H. Haug and S. W. Koch, *Quantum Theory of the Optical and Electronic Properties of Semiconductors*, World Scientific, Singapore, 1990.
9. P. Löwdin, "A note on the quantum-mechanical perturbation theory," *J. Chem. Phys.* **19**, 1396–1401 (1951).
10. S. L. Chuang, *Physics of Optoelectronic Devices*, 1st ed., Wiley, New York, 1995, Table 3.1.
11. A. Yariv, *Quantum Electronics*, 3rd ed., Wiley, New York, 1989.
12. A. Lukes, G. A. Ringwood, and B. Suprapto, "A particle in a box in the presence of an electric field and applications to disordered systems," *Physica* **84A**, 421–434 (1976).
13. F. M. Fernandez and E. A. Castro, "Hypervirial-perturbational treatment of a particle in a box in the presence of an electric field," *Physica* **111A**, 334–342 (1982).
14. M. Matsuura and T. Kamizato, "Subbands and excitons in a quantum well in an electric field," *Phys. Rev. B* **33**, 8385–8389 (1986).
15. G. Bastard, E. E. Mendez, L. L. Chang, and L. Esaki, "Variational calculations on a quantum well in an electric field," *Phys. Rev. B* **28**, 3241–3245 (1983).
16. D. Ahn and S. L. Chuang, "Variational calculations of subbands in a quantum well with uniform electric field: Gram-Schmidt orthogonalization approach," *Appl. Phys. Lett.* **49**, 1450–1452 (1986).
17. S. Nojima, "Electric field dependence of the exciton binding energy in $GaAs/Al_xGa_{1-x}As$ quantum wells," *Phys. Rev. B* **37**, 9087–9088 (1988).

4

Theory of Electronic Band Structures in Semiconductors

To understand the optical properties of semiconductors, such as absorption or gain due to electronic transitions in the presence of an incident optical wave, we have to know the electronic band structure including the energy band and the corresponding wave function. Knowing the initial and final states of the electrons, we may calculate the optical absorption using Fermi's golden rule derived in Section 3.6. In this chapter, we discuss the theory of the band structures. For optical devices, most semiconductors have direct band gaps, and many physical phenomena near the band edges are of great interest. We will focus on the conduction and valence band structures near the band edges, where the $\mathbf{k} \cdot \mathbf{p}$ method is extremely useful.

4.1 THE BLOCH THEOREM AND THE $\mathbf{k} \cdot \mathbf{p}$ METHOD FOR SIMPLE BANDS

For a periodic potential, the electronic band structure and the wave function can be derived from the Hamiltonian, which satisfies the symmetry of the semiconductor crystals. The general theory follows the Bloch theorem [1], which will be discussed in this section. Numerical methods to find the band structures and the wave functions include the tight binding, the pseudopotential, the orthogonalized plane wave, the augmented plane wave, the Green's function, and the cellular methods. Many texts on solid-state physics have detailed discussions on these methods [1, 2]. Our interest here is near the band edges of direct band-gap semiconductors, where the wave vector $\mathbf{k}$ deviates by a small amount from a vector $\mathbf{k}_0$ where a local minimum or maximum occurs. The $\mathbf{k} \cdot \mathbf{p}$ method was introduced by Bardeen [3] and Seitz [4]. The method has also been applied by many researchers to semiconductors [5–10]. Here we will discuss Kane's model [7, 8], which takes into account the spin–orbit interaction, and Luttinger–Kohn's models [9] for degenerate bands. These models are very popular in studying bulk and quantum-well semiconductors and have been used during the past three decades. They are much easier to apply than most other numerical methods. For a general discussion, see Refs. 11–14.

Physics of Photonic Devices, Second Edition. By Shun Lien Chuang
Copyright © 2009 John Wiley & Sons, Inc.

For an electron in a periodic potential

$$V(\mathbf{r}) = V(\mathbf{r} + \mathbf{R}) \tag{4.1.1}$$

where $\mathbf{R} = n_1\,\mathbf{a}_1 + n_2\,\mathbf{a}_2 + n_3\,\mathbf{a}_3$, and $\mathbf{a}_1, \mathbf{a}_2, \mathbf{a}_3$ are the lattice vectors, and n_1, n_2, and n_3 are integers, the electron wave function satisfies the Schrödinger equation

$$H\Psi(\mathbf{r}) = \left[\frac{\hbar^2}{2m_0}\nabla^2 + V(\mathbf{r})\right]\Psi(\mathbf{r}) = E(\mathbf{k})\Psi(\mathbf{r}). \tag{4.1.2}$$

The Hamiltonian is invariant under translation by the lattice vectors, $\mathbf{r} \rightarrow \mathbf{r} + \mathbf{R}$. If $\psi(\mathbf{r})$ describes an electron moving in the crystal, $\psi(\mathbf{r} + \mathbf{R})$ will also be a solution to (4.1.2). Thus, $\psi(\mathbf{r} + \mathbf{R})$ will differ from $\psi(\mathbf{r})$ at most by a constant. This constant must have a unity magnitude; otherwise, the wave function may grow to infinity if we repeat the translation $\mathbf{R}$ indefinitely. The general solution of the above equation is given by

$$\psi_{n\mathbf{k}}(\mathbf{r}) = e^{i\mathbf{k}\cdot\mathbf{r}}u_{n\mathbf{k}}(\mathbf{r}) \tag{4.1.3a}$$

where

$$u_{n\mathbf{k}}(\mathbf{r} + \mathbf{R}) = u_{n\mathbf{k}}(\mathbf{r}) \tag{4.1.3b}$$

is a periodic function. This result is the Bloch theorem. The wave function $\psi_{n\mathbf{k}}(\mathbf{r})$ is usually called the Bloch function. The energy is given by

$$E = E_n(\mathbf{k}).$$

Here n refers to the band and $\mathbf{k}$ the wave vector of the electron. A full description of the band structure requires numerical methods [1, 2]. An example of the GaAs band structure calculated [15] by the pseudopotential method is shown in Fig. 4.1a, which represents the general energy bands along different $\mathbf{k}$ directions. Our interest will focus near the direct band gap (Γ valley) between the valence-band edges and the conduction-band edge as shown in Fig. 4.1b, where the energy dispersions for a small $\mathbf{k}$ vector will be considered.

The $\mathbf{k}\cdot\mathbf{p}$ method is a very useful technique to analyze the band structure near a particular point $\mathbf{k}_0$, especially when it is near an extremum of the band structure. Here we consider that the extremum occurs at the zone center where $\mathbf{k}_0 = 0$. This is a very useful case for III–V direct band-gap semiconductors.

Consider the general Schrödinger equation for an electron wave function $\psi_{n\mathbf{k}}(\mathbf{r})$ in the nth band with a wave vector $\mathbf{k}$,

$$\left\{\frac{p^2}{2m_0} + V(\mathbf{r})\right\}\psi_{n\mathbf{k}}(\mathbf{r}) = E_n(\mathbf{k})\psi_{n\mathbf{k}}(\mathbf{r}). \tag{4.1.4}$$

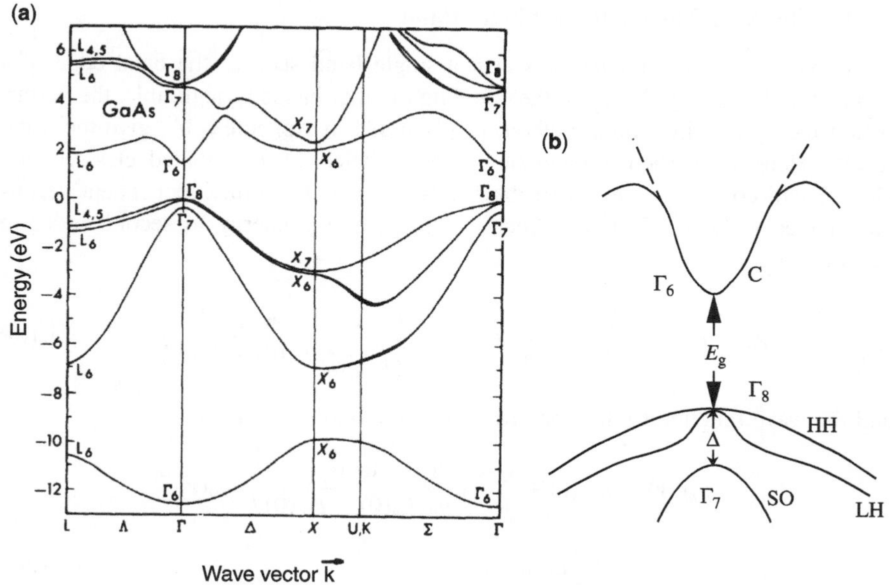

Figure 4.1 (a) GaAs band structure calculated by the pseudopotential method. (Reprinted with permission from [15] © 1976 American Physical Society.) (b) The band structure near the band edges of the direct band gap showing the conduction (C), heavy-hole (HH), light-hole (LH), and spin–orbit (SO) split-off bands.

When written in terms of $u_{n\mathbf{k}}(\mathbf{r})$, it becomes

$$\left\{\frac{p^2}{2m_0}+\frac{\hbar}{m_0}\mathbf{k}\cdot\mathbf{p}+V(\mathbf{r})\right\}u_{n\mathbf{k}}(\mathbf{r})=\left[E_n(\mathbf{k})-\frac{\hbar^2k^2}{2m_0}\right]u_{n\mathbf{k}}(\mathbf{r}). \tag{4.1.5}$$

The above equation can be expanded near a particular point $\mathbf{k}_0$ of interest in the band structure. When $\mathbf{k}_0 = 0$, the above equation is expanded near $E_n(0)$,

$$\left\{H_0+\frac{\hbar^2}{m_0}\mathbf{k}\cdot\mathbf{p}\right\}u_{n\mathbf{k}}(\mathbf{r})=\left[E_n(\mathbf{k})-\frac{\hbar^2k^2}{2m_0}\right]u_{n\mathbf{k}}(\mathbf{r}) \tag{4.1.6}$$

where

$$H_0=\frac{p^2}{2m_0}+V(\mathbf{r}) \tag{4.1.7a}$$

$$H_0u_{n0}(\mathbf{r})=E_n(0)u_{n0}(\mathbf{r}). \tag{4.1.7b}$$

4.1.1 The k • p Theory for a Single Band

If the band structure of interest is near a single band such as the band edge of a conduction band, Fig. 4.2a, and the coupling to other bands is negligible, the perturbation theory and Löwdin's method, as discussed in Appendix 3A, give the same results. Here the particular band of interest, labeled as n, is called class A, and class B will consist of the rest of the bands, $n' \neq n$. The time-independent perturbation theory, Eq. (3.5.21b) in Section 3.5, gives the energy to second order in perturbation:

$$E_n(\mathbf{k}) = E_n(0) + \frac{\hbar^2 k^2}{2m_0} + \frac{\hbar}{m_0}\mathbf{k}\cdot\mathbf{p}_{nn} + \frac{\hbar^2}{m_0^2}\sum_{n'\neq n}\frac{|\mathbf{k}\cdot\mathbf{p}_{nn'}|^2}{E_n(0) - E_{n'}(0)} \tag{4.1.8}$$

and the wave function to the first order in perturbation, (3.5.16a):

$$\begin{aligned} u_{n\mathbf{k}}(\mathbf{r}) &= u_{n0}(\mathbf{r}) + \sum_{n'\neq n}\left(\frac{\hbar}{m_0}\frac{\mathbf{k}\cdot\mathbf{p}_{n'n}}{E_n(0) - E_{n'}(0)}\right)u_{n'0}(\mathbf{r}) \\ &\equiv \sum_{n'} a_{n'}u_{n'0}(\mathbf{r}) \end{aligned} \tag{4.1.9a}$$

$$\psi_{n\mathbf{k}}(\mathbf{r}) = e^{i\mathbf{k}\cdot\mathbf{r}}u_{n\mathbf{k}}(\mathbf{r}) \tag{4.1.9b}$$

where the momentum matrix elements are defined as

$$\mathbf{p}_{nn'} = \int_{\text{unit cell}} u_{n0}^*(\mathbf{r})\,\mathbf{p}u_{n'0}(\mathbf{r})\,d^3\mathbf{r} \tag{4.1.10}$$

and $u_{n\mathbf{k}}(\mathbf{r})$'s are normalized as

$$\int_{\text{unit cell}} u_{n0}^*(\mathbf{r})\,u_{n'0}(\mathbf{r})\,d^3\mathbf{r} = \delta_{nn'}. \tag{4.1.11}$$

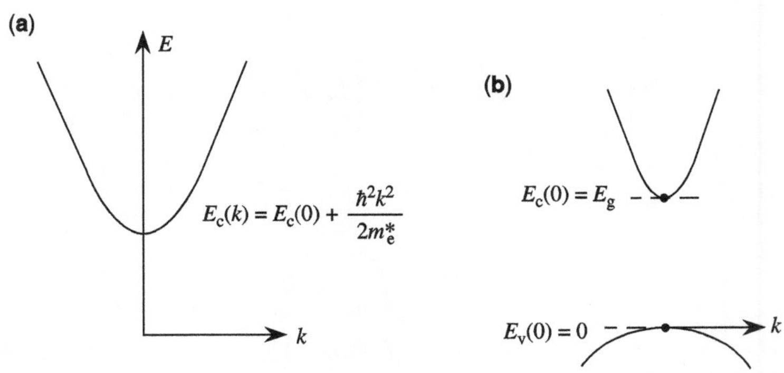

Figure 4.2 (a) A single-band model in the **k • p** theory. (b) The two-band model in the **k • p** theory.

If $\mathbf{k}_0$ is at an extremum of $E_n(\mathbf{k})$, $E_n(\mathbf{k})$ must depend quadratically on $\mathbf{k}$ near $\mathbf{k}_0$ and $\mathbf{p}_{nn} = 0$. That is why we need to go to second-order perturbation theory for the energy correction and only the first-order correction is needed for the wave function. Because we set $\mathbf{k}_0$ to 0, we have

$$E_n(\mathbf{k}) - E_n(0) = \sum_{\alpha,\beta} D^{\alpha\beta} k_\alpha k_\beta = \frac{\hbar^2}{2} \sum_{\alpha,\beta} \left(\frac{1}{m^*}\right)_{\alpha\beta} k_\alpha k_\beta \tag{4.1.12}$$

$$D^{\alpha\beta} = \frac{\hbar^2}{2m_0} \delta_{\alpha\beta} + \frac{\hbar^2}{2m_0^2} \sum_{n' \neq n} \frac{p^\alpha_{nn'} p^\beta_{n'n} + p^\beta_{nn'} p^\alpha_{n'n}}{E_n(0) - E_{n'}(0)} = \frac{\hbar^2}{2} \left(\frac{1}{m^*}\right)_{\alpha\beta} \tag{4.1.13}$$

where $\alpha, \beta = x, y$, and z. It should be noted that the $D^{\alpha\beta}$ matrix in the quadratic form has been defined to be symmetric. The matrix $D^{\alpha\beta}$ is the inverse effective mass in matrix form multiplied by $\hbar^2/2$.

4.1.2 The k • p Theory for Two-Band (or Nondegenerate Multibands) Model

If only two (or multi-) strongly interacting nondegenerate bands are considered, we call them class A, as shown in Fig. 4.2b. To solve (4.1.6), we assume

$$u_{n\mathbf{k}}(\mathbf{r}) = \sum_{n'} a_{n'}(\mathbf{k})\, u_{n'0}(\mathbf{r}). \tag{4.1.14}$$

Substituting the above expression into (4.1.6) for $u_{n\mathbf{k}}(\mathbf{r})$, and multiplying by $u^*_{n0}(\mathbf{r})$ and integrating over a unit cell, we have

$$\sum_{n'} \left\{ \left[E_n(0) + \frac{\hbar^2 k^2}{2m_0} \right] \delta_{nn'} + \frac{\hbar}{m_0} \mathbf{k} \cdot \mathbf{p}_{nn'} \right\} a_{n'} = E_n(\mathbf{k}) a_n \tag{4.1.15}$$

where the orthogonality relation $\int u^*_{n0} u_{n'0} d^3\mathbf{r} = \delta_{nn'}$ has been used. For two coupled bands, labeled by n and n', the above equation can be solved from the determinantal equation

$$\begin{vmatrix} E_n(0) + \dfrac{\hbar^2 k^2}{2m_0} - E & \dfrac{\hbar}{m_0} \mathbf{k} \cdot \mathbf{p}_{nn'} \\ \dfrac{\hbar}{m_0} \mathbf{k} \cdot \mathbf{p}_{n'n} & E_{n'}(0) + \dfrac{\hbar^2 k^2}{2m_0} - E \end{vmatrix} = 0. \tag{4.1.16}$$

The standard procedure is to find the eigenvalue E with the corresponding eigenvector. The two eigenvalues for (4.1.16) can also be compared with those obtained from a direct perturbation theory (see Problem 4.3).

4.2 KANE'S MODEL FOR BAND STRUCTURE: THE k • p METHOD WITH THE SPIN–ORBIT INTERACTION

In Kane's model for direct band semiconductors, the spin–orbit interaction is taken into account [7, 8]. Four bands—the conduction, the heavy-hole, the light-hole, and the spin–orbit split-off bands—are considered, which have double degeneracy with their spin counterparts, Fig. 4.3a.

4.2.1 The Schrödinger Equation for the Function $u_{n\mathbf{k}}(\mathbf{r})$

Consider the Hamiltonian near $\mathbf{k}_0 = \mathbf{0}$,

$$H = H_0 + \frac{\hbar}{4m_0^2c^2}\sigma \cdot \nabla V \times \mathbf{p} \tag{4.2.1a}$$

$$H_0 = \frac{p^2}{2m_0} + V(\mathbf{r}) \tag{4.2.1b}$$

where the second term in (4.2.1a) accounts for the spin–orbit interaction and σ is the Pauli spin matrix with components

$$\sigma_x = \begin{bmatrix} 0 & 1 \\ 1 & 0 \end{bmatrix} \quad \sigma_y = \begin{bmatrix} 0 & -i \\ i & 0 \end{bmatrix} \quad \sigma_z = \begin{bmatrix} 1 & 0 \\ 0 & -1 \end{bmatrix}, \tag{4.2.2}$$

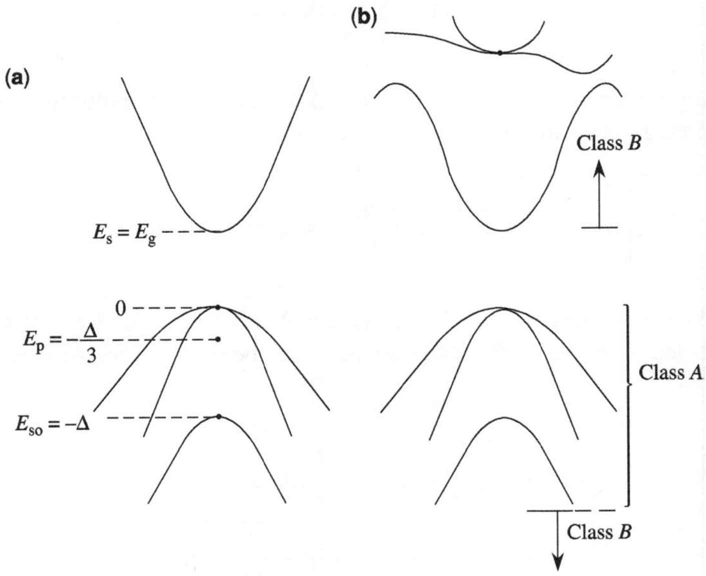

Figure 4.3 (a) The **k • p** method in Kane's model: only a conduction band, a heavy-hole, a light-hole, and a spin–orbit split-off band with double degeneracy are considered. All other higher and lower bands are discarded. (b) Luttinger–Kohn's model: the heavy-hole, light-hole, and spin split-off bands in double degeneracy are of interest and are called class A. All other bands are denoted as class B. The effects of bands in class B on those in class A are taken into account in this model.

which, when operating on the spins,

$$\uparrow \equiv \begin{bmatrix} 1 \\ 0 \end{bmatrix} \qquad \downarrow \equiv \begin{bmatrix} 0 \\ 1 \end{bmatrix} \tag{4.2.3}$$

give, by simple matrix product of (4.2.2) and (4.2.3),

$$\begin{array}{lll} \sigma_x \uparrow = \downarrow & \sigma_y \uparrow = i\downarrow & \sigma_z \uparrow = \uparrow \\ \sigma_x \downarrow = \uparrow & \sigma_y \downarrow = -i\uparrow & \sigma_z \downarrow = -\downarrow . \end{array} \tag{4.2.4}$$

From the original Schrödinger equation for the Bloch function,

$$\left\{ \frac{p^2}{2m_0} + V(\mathbf{r}) + \frac{\hbar}{4m_0^2c^2}[\nabla V \times \mathbf{p}] \cdot \boldsymbol{\sigma} \right\} \psi_{n\mathbf{k}}(\mathbf{r}) = E_n(\mathbf{k})\psi_{n\mathbf{k}}(\mathbf{r}) \tag{4.2.5}$$

the Schrödinger equation for the cell periodic function $u_{n\mathbf{k}}(\mathbf{r})$ is obtained

$$\left\{ \frac{p^2}{2m_0} + V(\mathbf{r}) + \frac{\hbar}{m_0}\mathbf{k} \cdot \mathbf{p} + \frac{\hbar}{4m_0^2c^2}[\nabla V \times \mathbf{p}] \cdot \boldsymbol{\sigma} + \frac{\hbar^2}{4m_0^2c^2} \nabla V \times \mathbf{k} \cdot \boldsymbol{\sigma} \right\} u_{n\mathbf{k}}(\mathbf{r})$$
$$= E' u_{n\mathbf{k}}(\mathbf{r}) \tag{4.2.6}$$

where $E' = E_n(\mathbf{k}) - \dfrac{\hbar^2 k^2}{2m_0}$. The last term on the left-hand side is a $\mathbf{k}$-dependent spin–orbit interaction, which is small compared with the other terms because the crystal momentum $\hbar\mathbf{k}$ is very small compared with the atomic momentum $\mathbf{p}$ in the far interior of the atom where most of the spin–orbit interaction occurs. Thus, only the first four terms on the left-hand side are considered

$$Hu_{n\mathbf{k}}(\mathbf{r}) \simeq \left(H_0 + \frac{\hbar}{m_0}\mathbf{k} \cdot \mathbf{p} + \frac{\hbar}{4m_0^2c^2} \nabla V \times \mathbf{p} \cdot \sigma \right) u_{n\mathbf{k}}(\mathbf{r}) = E' u_{n\mathbf{k}}(\mathbf{r}). \tag{4.2.7}$$

4.2.2 Basis Functions and the Hamiltonian Matrix

We look for the eigenvalue E' with corresponding eigenfunction

$$u_{n\mathbf{k}}(\mathbf{r}) = \sum_{n'} a_{n'} \, u_{n'0}(\mathbf{r}).$$

The band edge functions $u_{n0}(\mathbf{r})$ are

Conduction band: $|S\uparrow\rangle$, $|S\downarrow\rangle$ with corresponding eigenenergy E_s

Valence band: $|X\uparrow\rangle$, $|Y\uparrow\rangle$, $|Z\uparrow\rangle$, $|X\downarrow\rangle$, $|Y\downarrow\rangle$, $|Z\downarrow\rangle$ with eigenenergy E_p

where the wave functions in each band are degenerate with respect to H_0. $H_0|S\uparrow\rangle = E_s|S\uparrow\rangle$, $H_0|S\downarrow\rangle = E_s|S\downarrow\rangle$, $H_0|X\uparrow\rangle = E_p|X\uparrow\rangle$, $H_0|Y\uparrow\rangle = E_p|Y\uparrow\rangle$, $H_0|Z\uparrow\rangle = E_p|Z\uparrow\rangle$, and so forth. It is convenient to choose the basis functions

$$|iS\downarrow\rangle,\ \left|\frac{X-iY}{\sqrt{2}}\uparrow\right\rangle,\ |Z\downarrow\rangle,\ \left|-\frac{X+iY}{\sqrt{2}}\uparrow\right\rangle$$

and

$$|iS\uparrow\rangle,\ \left|-\frac{X+iY}{\sqrt{2}}\downarrow\right\rangle,\ |Z\uparrow\rangle,\ \left|\frac{X-iY}{\sqrt{2}}\downarrow\right\rangle$$

where the conduction-band wave function is the s-state wave function and the valence-band basis functions are taken from the the p-state wave functions of a hydrogen atom model, which are the spherical harmonics $Y_{10} = |Z\rangle$, and $Y_{1\pm1} = \mp\frac{1}{\sqrt{2}}(X \pm iY)$. The first few spherical harmonics $Y_{\ell m}(\theta, \varphi)$ are

$$\ell = 0\ (s \text{ orbit}): \qquad Y_{00} = \frac{1}{\sqrt{4\pi}}$$

$$\ell = 1\ (p \text{ orbits}): \quad Y_{10}(\theta, \varphi) = \sqrt{\frac{3}{4\pi}}\cos\theta = \sqrt{\frac{3}{4\pi}}\frac{z}{r} \equiv |Z\rangle$$

$$Y_{1\pm1}(\theta, \varphi) = \mp\sqrt{\frac{3}{8\pi}}\sin\theta\, e^{\pm i\varphi} = \mp\sqrt{\frac{3}{8\pi}}\,\frac{x \pm iy}{r} = \mp\frac{1}{\sqrt{2}}|X \pm iY\rangle.$$

The reason for this choice is that the electron wave functions are p-like near the top of the valence band and s-like near the bottom of the conduction band, which are contributed by the outermost *valence electrons*, say $4s^2$ and $4p^1$ electrons of the valence-III (Ga) atoms and $4s^2$ and $4p^3$ electrons of the valence-V (As) atoms when they form the GaAs zinc-blende crystal. The first four basis functions are respectively degenerate with the last four basis functions. The 8×8 interaction matrix becomes

$$\begin{bmatrix} \mathbf{H} & 0 \\ 0 & \mathbf{H} \end{bmatrix}$$

where, assuming $\mathbf{k} = k\hat{z}$,

$$\mathbf{H} = \begin{bmatrix} E_s & 0 & kP & 0 \\ 0 & E_p - \frac{\Delta}{3} & \sqrt{2}\Delta/3 & 0 \\ kP & \sqrt{2}\Delta/3 & E_p & 0 \\ 0 & 0 & 0 & E_p + \frac{\Delta}{3} \end{bmatrix} \tag{4.2.8}$$

and the Kane's parameter P and the spin–orbit split-off energy Δ are defined as

$$P \equiv -i\frac{\hbar}{m_0}\langle S|p_z|Z\rangle \tag{4.2.9}$$

$$\Delta \equiv \frac{3\hbar i}{4m_0^2c^2}\left\langle X\left|\frac{\partial V}{\partial x}p_y - \frac{\partial V}{\partial y}p_x\right|Y\right\rangle. \tag{4.2.10}$$

4.2.3 Solutions for the Eigenvalues and the Eigenfunctions of the Hamiltonian Matrix

Define the reference energy to be $E_p = -\Delta/3$, and define $E_s = E_g$. The Hamiltonian in (4.2.8) becomes

$$\mathbf{H} = \begin{bmatrix} E_g & 0 & kP & 0 \\ 0 & -\dfrac{2\Delta}{3} & \sqrt{2}\Delta/3 & 0 \\ kP & \sqrt{2}\Delta/3 & \dfrac{-\Delta}{3} & 0 \\ 0 & 0 & 0 & 0 \end{bmatrix}. \tag{4.2.11}$$

The determinantal equation $\det|\mathbf{H} - E'\mathbf{I}| = 0$ gives four eigenvalues for E'. We can see that the last band in (4.2.11) is decoupled from the first three bands.

(1) $E' = 0$ (i.e., band-edge energy is zero), (4.2.12a)

and

$$\text{(2)}\quad E'(E' - E_g)(E' + \Delta) - k^2P^2\left(E' + \frac{2}{3}\Delta\right) = 0. \tag{4.2.12b}$$

The second equation gives three roots. Because k^2 is very small, we expect that the roots of Eq. (4.2.12b) will be very close to $E' = E_g$, $E' = 0$, and $E' = -\Delta$, the three band edges.

(i) Let $E' = E_g + \varepsilon(k^2)$ where $\varepsilon \ll \Delta$ and E_g. We find from (4.2.12b)

$$\varepsilon = \frac{k^2P^2\left(E_g + \dfrac{2\Delta}{3}\right)}{E_g(E_g + \Delta)}. \tag{4.2.13}$$

(ii) Let $E' = 0 + \varepsilon(k^2)$. Equation (4.2.12b) gives

$$\varepsilon = -\frac{2k^2P^2}{3E_g}. \tag{4.2.14}$$

(iii) Let $E' = -\Delta + \varepsilon(k^2)$. We find

$$\varepsilon = -\frac{k^2P^2}{3(E_g + \Delta)}. \tag{4.2.15}$$

Because $E' = E_n(k) - \dfrac{\hbar^2 k^2}{2m_0}$, we obtain four eigenvalues from (4.2.12a) and (4.2.13)–(4.2.15). They are, starting from the highest energy level,

$$\text{(4.2.13)} \quad n = c \qquad E_c(k) = E_g + \frac{\hbar^2 k^2}{2m_0} + \frac{k^2P^2\left(E_g + \dfrac{2\Delta}{3}\right)}{E_g(E_g + \Delta)} \tag{4.2.16a}$$

$$\text{(4.2.12a)} \quad n = \text{hh} \qquad E_{\text{hh}}(k) = \frac{\hbar^2 k^2}{2m_0} \tag{4.2.16b}$$

$$\text{(4.2.14)} \quad n = \ell\text{h} \qquad E_{\ell\text{h}}(k) = \frac{\hbar^2 k^2}{2m_0} - \frac{2\,k^2P^2}{3E_g} \tag{4.2.16c}$$

$$\text{(4.2.15)} \quad n = \text{so} \qquad E_{\text{so}}(k) = -\Delta + \frac{\hbar^2 k^2}{2m_0} - \frac{k^2P^2}{3(E_g + \Delta)}. \tag{4.2.16d}$$

These results are not complete because the effects of higher bands have not been included, and they will be considered when we discuss the Luttinger–Kohn model. Note that the above result gives an incorrect effective mass for the heavy-hole band.

The eigenfunctions can be obtained from (4.2.11) for the first 4×4 matrix,

$$\phi_{\text{hh},\alpha} = \left| -\left(\frac{X + iY}{\sqrt{2}}\right)\uparrow \right\rangle \qquad \text{hh band} \tag{4.2.17a}$$

$$\phi_{n,\alpha} = a_n|iS\downarrow\rangle + b_n\left|\frac{X - iY}{\sqrt{2}}\uparrow\right\rangle + c_n|Z\downarrow\rangle \qquad n = c,\ \ell\text{h, so} \tag{4.2.17b}$$

and the second 4×4 matrix

$$\phi_{\text{hh},\beta} = \left|\frac{X - iY}{\sqrt{2}}\downarrow\right\rangle \qquad \text{hh band} \tag{4.2.18a}$$

$$\phi_{n,\beta} = a_n|iS\uparrow\rangle + b_n\left|-\frac{(X + iY)}{\sqrt{2}}\downarrow\right\rangle + c_n|Z\uparrow\rangle \qquad n = c,\ \ell\text{h, so}. \tag{4.2.18b}$$

The eigenvectors column $[a_n, b_n, c_n]$ are obtained by substituting each eigenvalue into the eigenequation

$$\begin{bmatrix} E_g - E'_n & 0 & kP \\ 0 & \frac{-2\Delta}{3} - E'_n & \sqrt{2}\frac{\Delta}{3} \\ kP & \sqrt{2}\frac{\Delta}{3} & -\frac{\Delta}{3} - E'_n \end{bmatrix} \begin{bmatrix} a_n \\ b_n \\ c_n \end{bmatrix} = 0 \tag{4.2.19}$$

and then normalizing such that $(a_n^2 + b_n^2 + c_n^2)^{1/2} = 1$.

The results in the limit $k^2 \to 0$ give

$$\begin{aligned} n &= c & a &\simeq 1, b \simeq 0, c \simeq 0 \\ n &= \ell h & a &\simeq 0, b = \frac{1}{\sqrt{3}}, c = \sqrt{\frac{2}{3}} \\ n &= so & a &\simeq 0, b = \sqrt{\frac{2}{3}}, c = -\sqrt{\frac{1}{3}}. \end{aligned} \tag{4.2.20}$$

4.2.4 Summary of the Eigenenergies and Corresponding Band-Edge Basis Functions

We summarized the results below and in Fig. 4.4. The commonly used parabolic band models for the energy dispersions are also redefined in the parentheses.

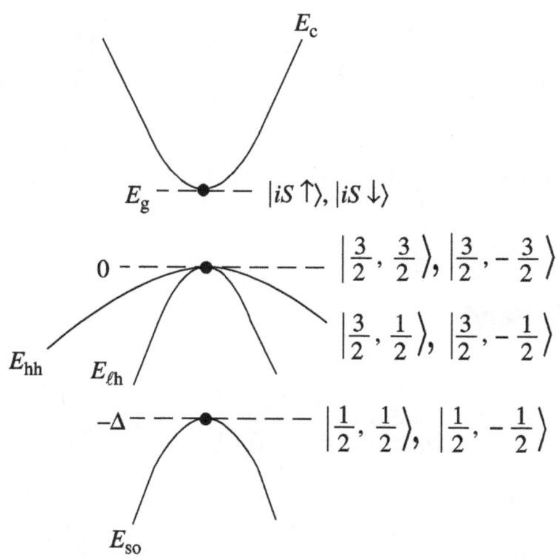

Figure 4.4 The band-edge energies E_g, 0, 0, and $-\Delta$ for the conduction, heavy-hole, light-hole, and spin split-off bands with their corresponding band-edge Bloch functions. Note that the dispersion relation E-k for the heavy-hole band E_{hh} should curve down and should follow the result of the Luttinger–Kohn model.

Conduction Band

$$E_c(k) = E_g + \frac{\hbar^2 k^2}{2m_0} + \frac{k^2P^2}{3}\frac{(3E_g + 2\Delta)}{E_g(E_g + \Delta)}\left(\equiv E_g + \frac{\hbar^2 k^2}{2m_e^*}\right)$$
$$\phi_{c\alpha} = |iS\downarrow\rangle$$
$$\phi_{c\beta} = |iS\uparrow\rangle. \tag{4.2.21}$$

Valence Band

Heavy Hole

$$E_{hh}(k) = \frac{\hbar^2 k^2}{2m_0}\left(\text{should be} - \frac{\hbar^2 k^2}{2m_{hh}^*}\right)$$
$$\phi_{hh,\alpha} = \frac{-1}{\sqrt{2}}|(X + iY)\uparrow\rangle \equiv \left|\frac{3}{2}, \frac{3}{2}\right\rangle \tag{4.2.22}$$
$$\phi_{hh,\beta} = \frac{1}{\sqrt{2}}|(X - iY)\downarrow\rangle \equiv \left|\frac{3}{2}, -\frac{3}{2}\right\rangle.$$

Light Hole

$$E_{\ell h}(k) = \frac{\hbar^2 k^2}{2m_0} - \frac{2k^2P^2}{3E_g}\left(\equiv -\frac{\hbar^2 k^2}{2m_{\ell h}^*}\right)$$
$$\phi_{\ell h,\alpha} = \frac{1}{\sqrt{6}}|(X - iY)\uparrow\rangle + \sqrt{\frac{2}{3}}|Z\downarrow\rangle \equiv \left|\frac{3}{2}, -\frac{1}{2}\right\rangle \tag{4.2.23}$$
$$\phi_{\ell h,\beta} = -\frac{1}{\sqrt{6}}|(X + iY)\downarrow\rangle + \sqrt{\frac{2}{3}}|Z\uparrow\rangle \equiv \left|\frac{3}{2}, \frac{1}{2}\right\rangle.$$

Spin-orbit Split-off Band

$$E_{so}(k) = -\Delta + \frac{\hbar^2 k^2}{2m_0} - \frac{k^2P^2}{3(E_g + \Delta)}\left(\equiv -\Delta - \frac{\hbar^2 k^2}{2m_{so}^*}\right)$$
$$\phi_{so,\alpha} = \frac{1}{\sqrt{3}}|(X - iY)\uparrow\rangle - \frac{1}{\sqrt{3}}|Z\downarrow\rangle \equiv \left|\frac{1}{2}, \frac{-1}{2}\right\rangle \tag{4.2.24}$$
$$\phi_{so,\beta} = \frac{1}{\sqrt{3}}|(X + iY)\downarrow\rangle + \frac{1}{\sqrt{3}}|Z\uparrow\rangle \equiv \left|\frac{1}{2}, \frac{1}{2}\right\rangle.$$

The Kane's parameter P can also be related to the effective mass of the electron m_e^* using

$$E_c(k) - E_g = \frac{\hbar^2 k^2}{2m_0} + \frac{k^2 P^2\left(E_g + \frac{2}{3}\Delta\right)}{E_g(E_g + \Delta)} = \frac{\hbar^2 k^2}{2m_e^*} \tag{4.2.25}$$

$$P^2 = \left(1 - \frac{m_e^*}{m_0}\right) \frac{\hbar^2 E_g(E_g + \Delta)}{2m_e^*\left(E_g + \frac{2\Delta}{3}\right)}. \tag{4.2.26}$$

Sometimes, the $\hbar^2 k^2/2m_0$ term is ignored, as $m_e^* \simeq 0.067m_0 \ll m_0$ for GaAs; therefore, the term m_e^*/m_0 in (4.2.26) is ignored.

We note that these wave functions in (4.2.21)–(4.2.24) are eigenvectors of the Hamiltonian

$$H = H_0 + \frac{\hbar^2 k^2}{2m_0} + \frac{\hbar}{m_0}\mathbf{k} \cdot \mathbf{p} + \frac{\hbar}{4m_0^2 c^2}\sigma \cdot \nabla V \times \mathbf{p} \tag{4.2.27}$$

with eigenenergies $E = E_g, 0, 0, -\Delta$ as $k \longrightarrow 0$ for the conduction, heavy-hole, light-hole, and spin split-off bands, respectively.

4.2.5 General Coordinate Direction

If k is not along the z-direction,

$$\mathbf{k} = k \sin\theta \cos\varphi \hat{x} + k \sin\theta \sin\varphi \hat{y} + k \cos\theta\, \hat{z} \tag{4.2.28}$$

the following transformations can be used to find the basis functions in the general coordinate system

$$\begin{bmatrix} \uparrow' \\ \downarrow' \end{bmatrix} = \begin{pmatrix} e^{-i\phi/2}\cos\frac{\theta}{2} & e^{i\phi/2}\sin\frac{\theta}{2} \\ -e^{-i\phi/2}\sin\frac{\theta}{2} & e^{i\phi/2}\cos\frac{\theta}{2} \end{pmatrix} \begin{bmatrix} \uparrow \\ \downarrow \end{bmatrix} \tag{4.2.29}$$

$$\begin{bmatrix} X' \\ Y' \\ Z' \end{bmatrix} = \begin{bmatrix} \cos\theta \cos\phi & \cos\theta \cos\phi & -\sin\theta \\ -\sin\phi & \cos\phi & 0 \\ \sin\theta\cos\phi & \sin\theta\cos\phi & \cos\theta \end{bmatrix} \begin{bmatrix} X \\ Y \\ Z \end{bmatrix}. \tag{4.2.30}$$

The spherically symmetrical function maintains its symmetry, $S(r') = S(r)$ as $r' = r$ because the length scale is preserved in a unitary transformation. The above transformation will be useful in Chapter 9 when we discuss optical matrix elements for quantum wells.

4.3 LUTTINGER–KOHN MODEL: THE k · p METHOD FOR DEGENERATE BANDS

Suppose we are mainly interested in the six valence bands (the heavy-hole, light-hole, and spin split-off bands, all doubly degenerate) and ignore the coupling to the two degenerate conduction bands with both spins. It is convenient to use Löwdin's perturbation method in Appendix 3A and treat the six valence bands as in class A and put the rest of the bands in class B, Fig. 4.3b. Note that Luttinger–Kohn's model can also be generalized to include both conduction bands in class A, especially for narrow band-gap semiconductors [9–13].

4.3.1 The Hamiltonian and the Basis Functions

Write the total Hamiltonian in (4.2.6) for $u_{\mathbf{k}}(\mathbf{r})$ (dropping the band index n for convenience)

$$Hu_{\mathbf{k}}(\mathbf{r}) = E(\mathbf{k})\, u_{\mathbf{k}}(\mathbf{r}) \tag{4.3.1a}$$

$$H = H_0 + \frac{\hbar^2 k^2}{2m_0} + \frac{\hbar}{4m_0^2c^2}\nabla V \times \mathbf{p}\cdot\boldsymbol{\sigma} + H' \tag{4.3.1b}$$

where

$$H_0 = \frac{p^2}{2m_0} + V(\mathbf{r}) \tag{4.3.1c}$$

$$H' = \frac{\hbar}{m_0}\mathbf{k}\cdot\boldsymbol{\Pi} \tag{4.3.1d}$$

where

$$\boldsymbol{\Pi} = \mathbf{p} + \frac{\hbar}{4m_0c^2}\boldsymbol{\sigma}\times\nabla V. \tag{4.3.1e}$$

Note again that the last term in (4.2.6) is much smaller than the fourth term because $\hbar k \ll p = |\langle u_k|\mathbf{p}|u_k\rangle| \simeq \hbar/a$, because the electron velocity in the atomic orbit is much larger than the velocity of the wave packet with the wave vectors in the vicinity of $\mathbf{k}_0(=\mathbf{0})$ [11].

We expand the function

$$u_{\mathbf{k}}(\mathbf{r}) = \sum_{j'}^{A} a_{j'}(\mathbf{k})u_{j'0}(\mathbf{r}) + \sum_{\gamma}^{B} a_{\gamma}(\mathbf{k})u_{\gamma 0}(\mathbf{r}) \tag{4.3.2}$$

where j' is in class A and γ is in class B. Or specifically, we have in class A

$$
\begin{aligned}
u_{10}(\mathbf{r}) &= \left|\frac{3}{2}, \frac{3}{2}\right\rangle = \frac{-1}{\sqrt{2}}|(X+iY)\uparrow\rangle \\
u_{20}(\mathbf{r}) &= \left|\frac{3}{2}, \frac{1}{2}\right\rangle = \frac{-1}{\sqrt{6}}|(X+iY)\downarrow\rangle + \sqrt{\frac{2}{3}}|Z\uparrow\rangle \\
u_{30}(\mathbf{r}) &= \left|\frac{3}{2}, \frac{-1}{2}\right\rangle = \frac{1}{\sqrt{6}}|(X-iY)\uparrow\rangle + \sqrt{\frac{2}{3}}|Z\downarrow\rangle \\
u_{40}(\mathbf{r}) &= \left|\frac{3}{2}, \frac{-3}{2}\right\rangle = \frac{1}{\sqrt{2}}|(X-iY)\downarrow\rangle \\
u_{50}(\mathbf{r}) &= \left|\frac{1}{2}, \frac{1}{2}\right\rangle = \frac{1}{\sqrt{3}}|(X+iY)\downarrow\rangle + \frac{1}{\sqrt{3}}|Z\uparrow\rangle \\
u_{60}(\mathbf{r}) &= \left|\frac{1}{2}, \frac{-1}{2}\right\rangle = \frac{1}{\sqrt{3}}|(X-iY)\uparrow\rangle - \frac{1}{\sqrt{3}}|Z\downarrow\rangle
\end{aligned}
\tag{4.3.3}
$$

from (4.2.22)–(4.2.24). At $\mathbf{k} = 0$, the above band-edge functions satisfy

$$
H(\mathbf{k}=\mathbf{0})\, u_{j0}(\mathbf{r}) = E_j(0)\, u_{j0}(\mathbf{r}) \tag{4.3.4}
$$

where

$$
E_j(0) = E_p + \Delta/3 = 0 \qquad \text{for } j = 1, 2, 3, 4, \tag{4.3.5a}
$$

$$
E_j(0) = E_p - 2\Delta/3 = -\Delta \qquad \text{for } j = 5, 6, \tag{4.3.5b}
$$

as we set $E_p = -\Delta/3$. The above band-edge energies and basis functions are also shown in Fig. 4.4.

4.3.2 Solution of the Hamiltonian Using Löwdin's Perturbation Method

The Löwdin's method is that we only need to solve the eigenequation

$$
\sum_{j'}^{A} \left(U_{jj'}^{A} - E\delta_{jj'}\right) a_{j'}(\mathbf{k}) = 0 \tag{4.3.6}
$$

instead of

$$
\sum_{j'}^{A} \left(H_{jj'} - E\delta_{jj'}\right) a_{j'}(\mathbf{k}) = 0 \tag{4.3.7}
$$

where

$$U_{jj'}^{A} = H_{jj'} + \sum_{\gamma \neq j, j'}^{B} \frac{H_{j\gamma} H_{\gamma j'}}{E_0 - E_\gamma} = H_{jj'} + \sum_{\gamma \neq j, j'}^{B} \frac{H'_{j\gamma} H'_{\gamma j'}}{E_0 - E_\gamma} \tag{4.3.8a}$$

$$H_{jj'} = \langle u_{j0} | H | u_{j'0} \rangle = \left[E_j(0) + \frac{\hbar^2 k^2}{2m_0} \right] \delta_{jj'} \qquad (j, j' \in A) \tag{4.3.8b}$$

$$H'_{j\gamma} = \langle u_{j0} | \frac{\hbar}{m_0} \mathbf{k} \cdot \Pi | u_{\gamma 0} \rangle \cong \sum_{\alpha} \frac{\hbar k_\alpha}{m_0} p_{j\gamma}^{\alpha} \qquad (j \in A, \gamma \notin A) \tag{4.3.8c}$$

where we note that $\Pi_{jj'} = 0$ for $j, j' \in A$, and $\Pi_{j\gamma}^{\alpha} \simeq p_{j\gamma}^{\alpha}$ for $j \in A$ and $\gamma \notin A$. Because $\gamma \neq j$, adding the unperturbed part to the perturbed part in $H'_{j\gamma}$ does not affect the results (i.e., $H_{j\gamma} = H'_{j\gamma}$). We thus obtain

$$U_{jj'}^{A} = \left[E_j(0) + \frac{\hbar^2 k^2}{2m_0} \right] \delta_{jj'} + \frac{\hbar^2}{m_0^2} \sum_{\gamma \neq j, j'}^{B} \sum_{\alpha, \beta} \frac{k_\alpha k_\beta p_{j\gamma}^{\alpha} p_{\gamma j'}^{\beta}}{E_0 - E_\gamma}. \tag{4.3.9}$$

Let $U_{jj'}^{A} \equiv D_{jj'}$. We obtain the matrix of the form $D_{jj'}$

$$D_{jj'} = E_j(0) \delta_{jj'} + \sum_{\alpha, \beta} D_{jj'}^{\alpha\beta} k_\alpha k_\beta \tag{4.3.10}$$

where the expression of $D_{jj'}^{\alpha\beta}$ is defined as

$$D_{jj'}^{\alpha\beta} = \frac{\hbar^2}{2m_0} \left\{ \delta_{jj'} \delta_{\alpha\beta} + \sum_{\gamma}^{B} \frac{p_{j\gamma}^{\alpha} p_{\gamma j'}^{\beta} + p_{j\gamma}^{\beta} p_{\gamma j'}^{\alpha}}{m_0 (E_0 - E_\gamma)} \right\} \tag{4.3.11}$$

which is similar to (4.1.13), the single band case (where $j = j' =$ the single band index n). Here we have generalized (4.1.13) to include the degenerate bands.

Explicit Expression for the Luttinger–Kohn Hamiltonian Matrix $D_{jj'}$ To write out the matrix elements $D_{jj'}$ in (4.3.10) explicitly, we define

$$A_0 = \frac{\hbar^2}{2m_0} + \frac{\hbar^2}{m_0^2} \sum_{\gamma}^{B} \frac{p_{x\gamma}^{x} p_{\gamma x}^{x}}{E_0 - E_\gamma} \tag{4.3.12a}$$

$$B_0 = \frac{\hbar^2}{2m_0} + \frac{\hbar^2}{m_0^2} \sum_{\gamma}^{B} \frac{p_{x\gamma}^{y} p_{\gamma x}^{y}}{E_0 - E_\gamma} \tag{4.3.12b}$$

$$C_0 = \frac{\hbar^2}{m_0^2} \sum_{\gamma}^{B} \frac{p_{x\gamma}^{x} p_{\gamma y}^{y} + p_{x\gamma}^{y} p_{\gamma y}^{x}}{E_0 - E_\gamma} \tag{4.3.12c}$$

and define the band structure (or Luttinger–Kohn) parameters γ_1, γ_2, and γ_3 as

$$
\begin{aligned}
-\frac{\hbar^2}{2m_0}\gamma_1 &= \frac{1}{3}(A_0 + 2B_0) \\
-\frac{\hbar^2}{2m_0}\gamma_2 &= \frac{1}{6}(A_0 - B_0) \\
-\frac{\hbar^2}{2m_0}\gamma_3 &= \frac{C_0}{6}.
\end{aligned}
\tag{4.3.13}
$$

We obtain the Luttinger–Kohn Hamiltonian $\mathbf{U}^A \equiv \mathbf{D}$, denoted as $\mathbf{H}^{LK}$, in the basis functions given by (4.3.3)

$$
\mathbf{H}^{LK} = -\begin{bmatrix}
P+Q & -S & R & 0 & -S/\sqrt{2} & \sqrt{2}R \\
-S^+ & P-Q & 0 & R & -\sqrt{2}Q & \sqrt{3/2}S \\
R^+ & 0 & P-Q & S & \sqrt{3/2}S^+ & \sqrt{2}Q \\
0 & R^+ & S^+ & P+Q & -\sqrt{2}R^+ & -S^+/\sqrt{2} \\
-S^+/\sqrt{2} & -\sqrt{2}Q^+ & \sqrt{3/2}S & -\sqrt{2}R & P+\Delta & 0 \\
\sqrt{2}R^+ & \sqrt{3/2}S^+ & \sqrt{2}Q^+ & -S/\sqrt{2} & 0 & P+\Delta
\end{bmatrix}
\tag{4.3.14}
$$

$$
\begin{aligned}
P &= \frac{\hbar^2\gamma_1}{2m_0}\left(k_x^2 + k_y^2 + k_z^2\right) \qquad Q = \frac{\hbar^2\gamma_2}{2m_0}\left(k_x^2 + k_y^2 - 2k_z^2\right) \\
R &= \frac{\hbar^2}{2m_0}\left[-\sqrt{3}\,\gamma_2\left(k_x^2 - k_y^2\right) + i\,2\sqrt{3}\,\gamma_3\,k_xk_y\right] \\
S &= \frac{\hbar^2\gamma_3}{m_0}\sqrt{3}(k_x - i\,k_y)k_z
\end{aligned}
\tag{4.3.15}
$$

where the superscript "+" means Hermitian conjugate.

4.3.3 Summary

In summary, for the valence hole subbands, we only have to solve for the eigenvalue equation

$$
\sum_{j'=1}^{6} H_{jj'}^{LK}\, a_{j'}(\mathbf{k}) = E a_j(\mathbf{k})
\tag{4.3.16}
$$

or for the eigenvalue E and the corresponding eigenvector column $[a_1, a_2, \ldots, a_6]$, where the matrix elements $H_{jj'}^{LK} = E_j(0)\delta_{jj'} + \sum_{\alpha,\beta} D_{jj'}^{\alpha\beta} k_\alpha k_\beta$ are given in (4.3.14).

The wave function $\psi_{n\mathbf{k}}(\mathbf{r})$ satisfying

$$\left[\frac{p^2}{2m_0} + V(\mathbf{r}) + \frac{\hbar^2}{4m_0^2c^2}\nabla V \times \mathbf{p}\cdot\sigma\right] \psi_{n\mathbf{k}}(\mathbf{r}) = E_{n\mathbf{k}}\ \psi_{n\mathbf{k}}(\mathbf{r}) \tag{4.3.17}$$

is then given by

$$\psi_{n\mathbf{k}}(\mathbf{r}) = e^{i\mathbf{k}\cdot\mathbf{r}} u_{n\mathbf{k}}(\mathbf{r}) \qquad u_{n\mathbf{k}}(\mathbf{r}) = \sum_{j=1}^{6} a_j(\mathbf{k})\ u_{j0}(\mathbf{r}) \tag{4.3.18}$$

and

$$E_{n\mathbf{k}} = E.$$

4.4 THE EFFECTIVE MASS THEORY FOR A SINGLE BAND AND DEGENERATE BANDS

In this section, we summarize the effective mass theory [9] for both a single band and degenerate bands in semiconductors.

4.4.1 The Effective Mass Theory for a Single Band

The most important conclusion of the effective mass theory (EMT) for a single band is as follows. If the energy dispersion relation for a single band n near $\mathbf{k}_0$ (assuming at the origin 0) is given by

$$E_n(\mathbf{k}) = E_n(0) + \sum_{\alpha,\beta} \frac{\hbar^2}{2}\left(\frac{1}{m^*}\right)_{\alpha\beta} k_\alpha k_\beta \tag{4.4.1}$$

for the Hamiltonian H_0 with a periodic potential $V(\mathbf{r})$

$$H_0 = \frac{p^2}{2m_0} + V(\mathbf{r}) \tag{4.4.2}$$

$$H_0\psi_{n\mathbf{k}}(\mathbf{r}) = E_n(\mathbf{k})\psi_{n\mathbf{k}}(\mathbf{r}), \tag{4.4.3}$$

then the solution for the Schrödinger equation with a perturbation $U(\mathbf{r})$ such as an impurity potential or a quantum-well potential

$$[H_0 + U(\mathbf{r})]\ \psi(\mathbf{r}) = E\psi(\mathbf{r}) \tag{4.4.4}$$

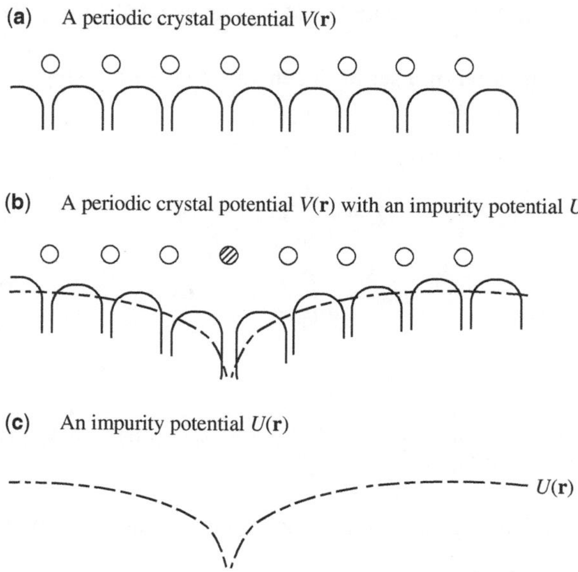

Figure 4.5 Illustrations of (a) the periodic potential $V(\mathbf{r})$, (b) the sum of the periodic potential $V(\mathbf{r})$ and the impurity potential $U(\mathbf{r})$, and (c) only the impurity potential $U(\mathbf{r})$ for the effective mass theory.

is obtainable by solving

$$\left[\sum_{\alpha,\beta}\frac{\hbar^2}{2}\left(\frac{1}{m^*}\right)_{\alpha\beta}\left(-i\frac{\partial}{\partial x_\alpha}\right)\left(-i\frac{\partial}{\partial x_\beta}\right)+U(\mathbf{r})\right]F(\mathbf{r})=(E-E_n(0))F(\mathbf{r}) \qquad (4.4.5)$$

for the envelope function $F(\mathbf{r})$ and the energy E. The wave function is approximated by

$$\psi(\mathbf{r})=F(\mathbf{r})u_{n\mathbf{k}_0}(\mathbf{r}). \qquad (4.4.6)$$

The most important result is that the periodic potential $V(\mathbf{r})$ determines the energy bands and the effective masses, $(1/m^*)_{\alpha\beta}$, and the effective mass Eq. (4.4.5) contains only the extra perturbation potential $U(\mathbf{r})$, as the effective masses already take into account the periodic potential, Fig. 4.5. The perturbation potential can also be a quantum-well potential, as in a semiconductor heterostructure such as GaAs/$\mathrm{Al}_x\mathrm{Ga}_{1-x}\mathrm{As}$ quantum wells.

4.4.2 The Effective Mass Theory for Degenerate Bands

In Section 4.3, we discuss the $\mathbf{k}\cdot\mathbf{p}$ method for degenerate bands such as the heavy-hole, light-hole, and the spin split-off bands of semiconductors. The effective mass theory for a perturbation potential $U(\mathbf{r})$ for degenerate bands is stated as follows [9, 11]:

If the dispersion relation of a set of degenerate bands satisfying

$$H\psi_{n\mathbf{k}}(\mathbf{r}) = E(\mathbf{k})\psi_{n\mathbf{k}}(\mathbf{r}) \tag{4.4.7a}$$

$$H = \frac{p^2}{2m_0} + V(\mathbf{r}) + H_{so} \tag{4.4.7b}$$

$$H_{so} = \frac{\hbar}{4m_0^2c^2}\nabla V \times \mathbf{p} \cdot \sigma \tag{4.4.7c}$$

is given by

$$\sum_{j'=1}^{6} H_{jj'}^{LK} a_{j'}(\mathbf{k}) \equiv \sum_{j'=1}^{6}\left[E_j(0)\delta_{jj'} + \sum_{\alpha,\beta} D_{jj'}^{\alpha\beta} k_\alpha k_\beta\right] a_{j'}(\mathbf{k}) = E(\mathbf{k})a_j(\mathbf{k}) \tag{4.4.8}$$

then the solution $\psi(\mathbf{r})$ for the semiconductors in the presence of a perturbation $U(\mathbf{r})$,

$$[H + U(\mathbf{r})]\psi(\mathbf{r}) = E\psi(\mathbf{r}) \tag{4.4.9}$$

is given by

$$\psi(\mathbf{r}) = \sum_{j=1}^{6} F_j(\mathbf{r})\, u_{j0}(\mathbf{r}) \tag{4.4.10}$$

where $F_j(\mathbf{r})$ satisfies

$$\sum_{j'=1}^{6}\left[E_j(0)\,\delta_{jj'} + \sum_{\alpha,\beta} D_{jj'}^{\alpha\beta}\left(-i\frac{\partial}{\partial x_\alpha}\right)\left(-i\frac{\partial}{\partial x_\beta}\right) + U(\mathbf{r})\,\delta_{jj'}\right]F_{j'}(\mathbf{r}) = EF_j(\mathbf{r}). \tag{4.4.11}$$

4.5 STRAIN EFFECTS ON BAND STRUCTURES

Strained-layer superlattices [16, 17] have been of great interest since the early 1980s. It has been demonstrated that it is possible to vary important material properties: lattice constant, band gap, and perpendicular transport using ternary strained-layer superlattices. Applications of the strained-layer superlattices or quantum wells to long-wavelength photodetectors [17] and semiconductor lasers [18] have been proposed and later demonstrated. For example, strained quantum-well lasers have been shown to exhibit superior performance compared with that for conventional diode lasers [19, 20] in many aspects, which will be discussed later in Chapter 10. Detailed discussions on the semiconductor growth and the physics of strained-layer quantum wells can be found in Ref. 20.

When a crystal is under a uniform deformation, it may preserve the periodic property such that the Bloch theorem may still be applicable. The modulating part of the Bloch function remains periodic with a period equal to that of the new elementary cell because the elementary cell is also deformed. In this section, we derive the Pikus–Bir Hamiltonian [21, 22] for strained semiconductors and discuss their band structures.

4.5.1 The Pikus–Bir Hamiltonian for a Strained Semiconductor

Suppose near the band extremum $\mathbf{k}_0 = \mathbf{0}$ of a semiconductor, we have

$$[H_0 + V_0(\mathbf{r})]\psi_{n\mathbf{k}_0}(\mathbf{r}) = E_n(\mathbf{k}_0)\psi_{n\mathbf{k}_0}(\mathbf{r}) \tag{4.5.1}$$

with the Bloch function $\psi_{n\mathbf{k}_0}(\mathbf{r}) = e^{i\mathbf{k}_0\cdot\mathbf{r}}u_{n\mathbf{k}_0}(\mathbf{r})$, where $V_0(\mathbf{r})$ is a periodic potential in the undeformed crystal. Here we present a simple picture for the strain analysis [23–25]. As shown in Fig. 4.6, the unit vectors $\hat{x}$, $\hat{y}$, (and $\hat{z}$) (for simplicity, assuming they are basis vectors, too) in the undeformed crystal are related to $\hat{x}'$, $\hat{y}'$, (and $\hat{z}'$) in the uniformly deformed crystal by

$$\begin{aligned}
\mathbf{x}' &= (1+\varepsilon_{xx})\hat{x} + \varepsilon_{xy}\hat{y} + \varepsilon_{xz}\hat{z} \\
\mathbf{y}' &= \varepsilon_{yx}\hat{x} + (1+\varepsilon_{yy})\hat{y} + \varepsilon_{yz}\hat{z} \\
\mathbf{z}' &= \varepsilon_{zx}\hat{x} + \varepsilon_{zy}\hat{y} + (1+\varepsilon_{zz})\hat{z}.
\end{aligned} \tag{4.5.2}$$

Obviously $\mathbf{x}'$, $\mathbf{y}'$, and $\mathbf{z}'$ are not unit vectors anymore. We assume a homogeneous strain and $\varepsilon_{ij} = \varepsilon_{ji}$. We can define six strain components as

$$\begin{aligned}
&e_1 = \varepsilon_{xx} \qquad e_2 = \varepsilon_{yy} \qquad e_3 = \varepsilon_{zz} \\
&e_4 = \mathbf{x}'\cdot\mathbf{y}' = \varepsilon_{xy} + \varepsilon_{yx} \\
&e_5 = \mathbf{y}'\cdot\mathbf{z}' = \varepsilon_{yz} + \varepsilon_{zy} \\
&e_6 = \mathbf{z}'\cdot\mathbf{x}' = \varepsilon_{zx} + \varepsilon_{xz}
\end{aligned} \tag{4.5.3}$$

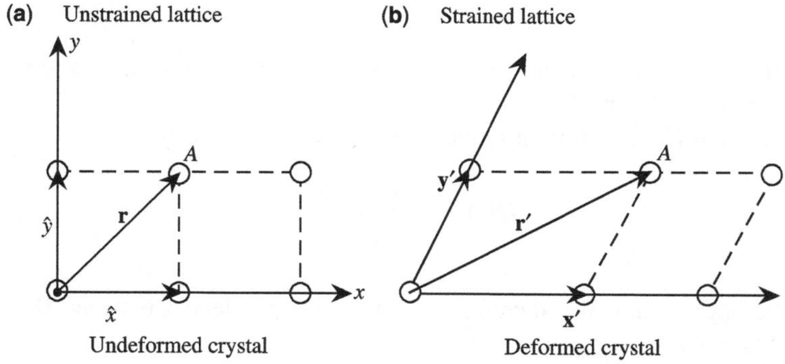

Figure 4.6 (a) Position vector $\mathbf{r}$ for atom A in an unstrained lattice. (b) Position vector $\mathbf{r}'$ for atom A in a strained lattice.

keeping only the linear terms in strain. To label a position A (or atom A) in the undeformed crystal, we have

$$\mathbf{r} = x\hat{x} + y\hat{y} + z\hat{z} \equiv (x, y, z). \tag{4.5.4a}$$

The same atom in the deformed crystal can be labeled either as

$$\mathbf{r}' = x\mathbf{x}' + y\mathbf{y}' + z\mathbf{z}' \tag{4.5.4b}$$

using the new basis vectors $\mathbf{x}'$, $\mathbf{y}'$, and $\mathbf{z}'$, or labeled as

$$\mathbf{r}' = x'\hat{x} + y'\hat{y} + z'\hat{z} \equiv (x', y', z') \tag{4.5.4c}$$

in the original bases in the undeformed crystal. An example is shown in Fig. 4.6. We have $\mathbf{r} = \hat{x} + \hat{y} \equiv (1, 1, 0)$ in the undeformed crystal and $\mathbf{r}' = \mathbf{x}' + \mathbf{y}'$ in the deformed crystal. We can also see that the change of the volume in the linear strain regime becomes

$$\frac{V + \delta V}{V} = \mathbf{x}' \cdot \mathbf{y}' \times \mathbf{z}' = 1 + (\varepsilon_{xx} + \varepsilon_{yy} + \varepsilon_{zz}). \tag{4.5.5}$$

The quantity $\varepsilon_{xx} + \varepsilon_{yy} + \varepsilon_{zz}$ is the trace of the matrix $\overline{\overline{\varepsilon}}$, or $\mathrm{Tr}(\overline{\overline{\varepsilon}})$, which is exactly the fractional change of the volume $\delta V/V$ of the crystal under uniform deformation

$$\delta V/V = \varepsilon_{xx} + \varepsilon_{yy} + \varepsilon_{zz}. \tag{4.5.6}$$

Major Results for the Pikus–Bir Hamiltonian for a Strained Semiconductor Using the above relations (4.5.2)–(4.5.6) between the deformed coordinates and the undeformed coordinates, the Hamiltonian for a strained semiconductor can be derived. Here we summarize the major conclusions. In the unstrained semiconductor, the full 6×6 Luttinger–Kohn Hamiltonian is given by (4.3.10)

$$D_{jj'} = H^{\mathrm{LK}}_{jj'} = E_j(0)\,\delta_{jj'} + \sum_{\alpha,\beta} D^{\alpha\beta}_{jj'}\,k_\alpha k_\beta \tag{4.5.7}$$

and its full matrix form is shown in (4.3.14) and (4.3.15) explicitly in terms of the expressions P, Q, R, and S.

The strained Hamiltonian introduces extra terms denoted by

$$(H_\varepsilon)_{jj'} = \sum_{\alpha,\beta} \hat{D}^{\alpha\beta}_{jj'}\,\varepsilon_{\alpha\beta} \tag{4.5.8}$$

due to the linear strain. We, therefore, use the correspondences between $D^{\alpha\beta}_{jj'}$ in $H^{\mathrm{LK}}_{jj'}$ and $\hat{D}^{\alpha\beta}_{jj'}$ in $(H_\varepsilon)_{jj'}$, we find

$$k_\alpha k_\beta \longrightarrow \varepsilon_{\alpha\beta}. \tag{4.5.9}$$

Conduction Band (Isotropic Case) We replace the prefactor with the inverse effective mass by the conduction deformation potential

$$\frac{\hbar^2}{2m_e^*} \longleftrightarrow a_c \tag{4.5.10}$$

and replace the wave vector components by strain components using (4.5.9). We obtain the conduction band edge dispersion

$$E(k) = E_c(0) + \frac{\hbar^2}{2m_e^*}(k_x^2 + k_y^2 + k_z^2) + a_c(\varepsilon_{xx} + \varepsilon_{yy} + \varepsilon_{zz}). \tag{4.5.11}$$

We note that the inverse effective mass tensor is diagonal ($\alpha = \beta$) in the principal axis system.

For Valence Band

$$\begin{aligned} \frac{\hbar^2\gamma_1}{2m_0} &\longleftrightarrow D_v^d \equiv -a_v \\ \frac{\hbar^2\gamma_2}{2m_0} &\longleftrightarrow \frac{D_u}{3} \equiv -\frac{b}{2} \\ \frac{\hbar^2\gamma_3}{2m_0} &\longleftrightarrow \frac{D_u'}{3} \equiv -\frac{d}{2\sqrt{3}}. \end{aligned} \tag{4.5.12}$$

Therefore, P_k, Q_k, R_k, and S_k can be added to their corresponding strain counterparts P_ε, Q_ε, R_ε, and S_ε, of which the explicit expressions are shown in (4.5.13) in Section 4.5.2. The energy parameters a_c, a_v, b, and d are called the deformation potentials for the valence bands and they are tabulated in Appendix C for a few important semiconductors. The strained Hamiltonian has been used extensively in the study of the strain effects [26–38] on the band structures of semiconductors.

4.5.2 Band Structures Without the Spin–Orbit Split-Off Band Coupling

Next we illustrate how the strain modifies the valence-band structures, including the band-edge energies and the effective masses, which are among the most important parameters characterizing any semiconductor materials. For most III–V semiconductors, the split-off bands are several hundred millielectron volts below the heavy-hole and light-hole bands. Because the energy range of interest is only several tens of millielectron volts, it is usual to assume that the split-off bands can be ignored. The band structures of the heavy-hole and light-hole bands are approximately described by the 4×4 Hamiltonian [37, 38]

$$\mathbf{H} = -\begin{bmatrix} P+Q & -S & R & 0 \\ -S^+ & P-Q & 0 & R \\ R^+ & 0 & P-Q & S \\ 0 & R^+ & S^+ & P-Q \end{bmatrix} \begin{matrix} \left|\frac{3}{2},\frac{3}{2}\right\rangle \\ \left|\frac{3}{2},\frac{1}{2}\right\rangle \\ \left|\frac{3}{2},-\frac{1}{2}\right\rangle \\ \left|\frac{3}{2},-\frac{3}{2}\right\rangle \end{matrix} \tag{4.5.13a}$$

$$\begin{aligned} P &= P_k + P_\varepsilon \quad Q = Q_k + Q_\varepsilon \\ R &= R_k + R_\varepsilon \quad S = S_k + S_\varepsilon \end{aligned} \tag{4.5.13b}$$

$$\begin{aligned} P_k &= \frac{\hbar^2\gamma_1}{2m_0}\left(k_x^2+k_y^2+k_z^2\right), \qquad Q_k = \frac{\hbar^2\gamma_2}{2m_0}\left(k_x^2+k_y^2-2k_z^2\right) \\ R_k &= \frac{\hbar^2}{2m_0}\left[-\sqrt{3}\,\gamma_2\left(k_x^2-k_y^2\right)+i\,2\sqrt{3}\,\gamma_3\,k_xk_y\right] \\ S_k &= \frac{\hbar^2\gamma_3}{m_0}\sqrt{3}(k_x-i\,k_y)k_z \end{aligned} \tag{4.5.13c}$$

$$\begin{aligned} P_\varepsilon &= -a_v(\varepsilon_{xx}+\varepsilon_{yy}+\varepsilon_{zz}) \qquad Q_\varepsilon = -\frac{b}{2}(\varepsilon_{xx}+\varepsilon_{yy}-2\varepsilon_{zz}) \\ R_\varepsilon &= \frac{\sqrt{3}b}{2}(\varepsilon_{xx}-\varepsilon_{yy})-id\varepsilon_{xy} \qquad S_\varepsilon = -d(\varepsilon_{xz}-i\varepsilon_{yz}). \end{aligned} \tag{4.5.13d}$$

The above Hamiltonian H is written for an arbitrary strain. For simplicity, we restrict ourselves to the special case of a biaxial strain, namely,

$$\begin{aligned} \varepsilon_{xx} &= \varepsilon_{yy} \neq \varepsilon_{zz} \\ \varepsilon_{xy} &= \varepsilon_{yz} = \varepsilon_{zx} = 0. \end{aligned} \tag{4.5.14}$$

Thus

$$R_\varepsilon = S_\varepsilon = 0$$

which essentially covers two of the most important strained systems: (i) a strained-layer semiconductor pseudomorphically grown on a (001)-oriented substrate and (ii) a bulk semiconductor under an external uniaxial stress along the z-direction. For the case of the lattice-mismatched strain, we obtain

$$\varepsilon_{xx} = \varepsilon_{yy} = \frac{a_0 - a}{a} \tag{4.5.15a}$$

$$\varepsilon_{zz} = -\frac{2C_{12}}{C_{11}}\varepsilon_{xx} \tag{4.5.15b}$$

where a_0 and a are the lattice constants of the substrate and the layer material, Fig. 4.7, and C_{11} and C_{12} are the elastic stiffness constants. Equations (4.5.15a) and (4.5.15b) can be derived by using the fact that in the plane of the heterojunction, the layered

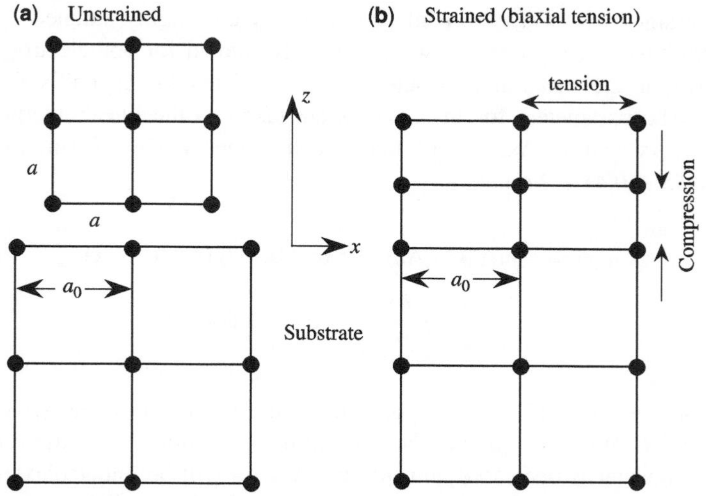

Figure 4.7 An epilayer material with a lattice constant a to be grown on a substrate with a lattice constant a_0: (a) unstrained and (b) strained.

material is strained such that the lattice constant along the plane of the layer is equal to a_0. Therefore, $\varepsilon_{xx} = \varepsilon_{yy} = (a_0 - a)/a$. Because the stress tensor is related to strain by the elastic stiffness tensor with elements C_{ij},

$$\begin{bmatrix} \tau_{xx} \\ \tau_{yy} \\ \tau_{zz} \\ \tau_{xy} \\ \tau_{yz} \\ \tau_{zx} \end{bmatrix} = \begin{bmatrix} C_{11} & C_{12} & C_{12} & 0 & 0 & 0 \\ C_{12} & C_{11} & C_{12} & 0 & 0 & 0 \\ C_{12} & C_{12} & C_{11} & 0 & 0 & 0 \\ 0 & 0 & 0 & C_{44} & 0 & 0 \\ 0 & 0 & 0 & 0 & C_{44} & 0 \\ 0 & 0 & 0 & 0 & 0 & C_{44} \end{bmatrix} \begin{bmatrix} \varepsilon_{xx} \\ \varepsilon_{yy} \\ \varepsilon_{zz} \\ 0 \\ 0 \\ 0 \end{bmatrix}. \tag{4.5.16}$$

We find $\tau_{xy} = \tau_{zy} = \tau_{zx} = 0$. There should also be no stress in the z-direction:

$$0 = \tau_{zz} = C_{12}(\varepsilon_{xx} + \varepsilon_{yy}) + C_{11}\varepsilon_{zz}.$$

Therefore, $\varepsilon_{zz} = -\dfrac{2C_{12}}{C_{11}}\varepsilon_{xx}$. For the case of an external uniaxial stress T along the z axis, $\tau_{zz} = T$ and $\tau_{xx} = \tau_{yy} = 0$, we find

$$\begin{aligned} \varepsilon_{xx} = \varepsilon_{yy} &= \frac{-C_{12}}{C_{11}^2 + C_{11}C_{12} - 2C_{12}^2}T \\ \varepsilon_{zz} &= \frac{C_{11} + C_{22}}{C_{11}^2 + C_{11}C_{12} - 2C_{12}^2}T. \end{aligned} \tag{4.5.17}$$

The results and conclusions presented here can be generalized to other crystal orientations or stress directions. As an example, we discuss one of the most important systems: strained $In_{1-x}Ga_xAs$ on InP. High-quality and highly strained samples of this system have already been grown and widely studied for optoelectronics applications. All of the material parameters used are listed in Appendix C on page 802–804. The parameters for $In_{1-x}Ga_xAs$ are taken as the linear interpolation of those of InAs and GaAs, except that for the energy gap, $E_g(In_{1-x}Ga_xAs) = 0.324 + 0.7x + 0.4x^2$. We have

$$a(x) = a(\mathrm{In}_{1-x}\mathrm{Ga}_x\mathrm{As}) = 5.6533x + 6.0584(1 - x) \tag{4.5.18a}$$

$$\varepsilon_{xx} = \varepsilon_{yy} = \frac{a_0 - a(x)}{a(x)} \qquad a_0 = 5.8688\ \text{Å}. \tag{4.5.18b}$$

At $x = 0.468 \simeq 0.47$, $a(0.468) = a_0$ and the strain is zero. $In_{0.53}Ga_{0.47}As$ is lattice matched to InP. When $x > 0.468$, the gallium mole fraction is increased; therefore, the lattice constant is decreased and the $In_{1-x}Ga_xAs$ will be under biaxial tension, Fig. 4.7. Here biaxial tension means that the lattice in the parallel (x–y) plane will experience a tensile strain with a simultaneous compressive strain along the growth (z) direction. On the other hand, if $x < 0.468$, we have $a(x) > a_0$, and we will have the case of biaxial compression.

At the zone center, $\mathbf{k} = 0$, we have only P_ε and Q_ε appearing in the diagonal terms of the matrix (4.5.13) nonvanishing. Therefore, we obtain the band-edge energies of the heavy-hole and light-hole bands

$$E_{\mathrm{HH}}(\mathbf{k} = 0) = -P_\varepsilon - Q_\varepsilon = a_v(\varepsilon_{xx} + \varepsilon_{yy} + \varepsilon_{zz}) + \frac{b}{2}(\varepsilon_{xx} + \varepsilon_{yy} - 2\varepsilon_{zz}) \tag{4.5.19a}$$

$$E_{\mathrm{LH}}(\mathbf{k} = 0) = -P_\varepsilon + Q_\varepsilon = a_v(\varepsilon_{xx} + \varepsilon_{yy} + \varepsilon_{zz}) - \frac{b}{2}(\varepsilon_{xx} + \varepsilon_{yy} - 2\varepsilon_{zz}). \tag{4.5.19b}$$

On the other hand, the conduction band edge energy of the electron is given by

$$E_c(\mathbf{k} = 0) = E_g + a_c(\varepsilon_{xx} + \varepsilon_{yy} + \varepsilon_{zz}). \tag{4.5.20}$$

Note that both the conduction and valence band energies are defined to be positive for the upward direction of the energy. The net energy transitions will be

$$\begin{aligned} E_{\text{C-HH}}(\mathbf{k} = 0) &= E_g + a_c(\varepsilon_{xx} + \varepsilon_{yy} + \varepsilon_{zz}) + P_\varepsilon + Q_\varepsilon \\ &= E_g + a(\varepsilon_{xx} + \varepsilon_{yy} + \varepsilon_{zz}) - \frac{b}{2}(\varepsilon_{xx} + \varepsilon_{yy} - 2\varepsilon_{zz}) \end{aligned} \tag{4.5.21a}$$

for the conduction to heavy-hole band, and

$$\begin{aligned} E_{\text{C-LH}}(\mathbf{k} = 0) &= E_g + a_c(\varepsilon_{xx} + \varepsilon_{yy} + \varepsilon_{zz}) + P_\varepsilon - Q_\varepsilon \\ &= E_g + a(\varepsilon_{xx} + \varepsilon_{yy} + \varepsilon_{zz}) + \frac{b}{2}(\varepsilon_{xx} + \varepsilon_{yy} - 2\varepsilon_{zz}) \end{aligned} \tag{4.5.21b}$$

for the conduction to the light-hole band, where E_g is the band gap of the unstrained semiconductor, and

$$a = a_c - a_v \tag{4.5.22}$$

is the hydrostatic deformation potential. Sometimes the hydrostatic and shear deformation energies, δE_{hy} and δE_{sh}, are defined, respectively, as

$$\delta E_{hy} = -a(\varepsilon_{xx} + \varepsilon_{yy} + \varepsilon_{zz}) \tag{4.5.23}$$

and

$$\frac{1}{2}\delta E_{sh} = Q_\varepsilon = -\frac{b}{2}(\varepsilon_{xx} + \varepsilon_{yy} - 2\varepsilon_{zz}). \tag{4.5.24}$$

The effective band gaps are given by

$$E_{C\text{-}HH} = E_g - \delta E_{hy} + \frac{1}{2}\delta E_{sh} \tag{4.5.25a}$$

$$E_{C\text{-}LH} = E_g - \delta E_{hy} - \frac{1}{2}\delta E_{sh}. \tag{4.5.25b}$$

For the Hamiltonian in Eq. (4.5.13), the valence-band structure of a bulk semiconductor is determined by the algebraic equation

$$\det\left[H_{ij}(k) - \delta_{ij}E\right] = 0 \tag{4.5.26}$$

where k is now interpreted as a real vector, and the envelope functions are simply taken as plane waves. For the 4×4 Hamiltonian, the solutions of Eq. (4.5.26) are simply

$$E_{HH}(\mathbf{k}) = -P_\varepsilon - P_k - \mathrm{sgn}(Q_\varepsilon)\sqrt{(Q_\varepsilon + Q_k)^2 + |R_k|^2 + |S_k|^2} \tag{4.5.27a}$$

$$E_{LH}(\mathbf{k}) = -P_\varepsilon - P_k + \mathrm{sgn}(Q_\varepsilon)\sqrt{(Q_\varepsilon + Q_k)^2 + |R_k|^2 + |S_k|^2} \tag{4.5.27b}$$

for the heavy holes and light holes, respectively. Each of the solutions is doubly degenerate. Note that it is important to include the sign factor $\mathrm{sgn}(Q_\varepsilon)$ (= +1 for $Q_\varepsilon > 0$ and = −1 for $Q_\varepsilon < 0$) in front of the square root, because Q_ε can be either negative (compressive strain) or positive (tensile strain) whereas the square root is conventionally taken as positive. As **k** approaches zero, the band-edge energies of the heavy hole and light hole in (4.5.19) should be recovered. Note that for the unstrained case,

$$E(\mathbf{k}) = -P_k \pm \sqrt{|Q_k|^2 + |R_k|^2 + |S_k|^2}$$
$$= Ak^2 \pm \sqrt{B^2 k^4 + C^2\left(k_x^2 k_y^2 + k_y^2 k_z^2 + k_z^2 k_x^2\right)} \tag{4.5.27c}$$

gives the heavy-hole and light-hole dispersion relations.

The dispersion relations for the heavy-hole band $E_{\mathrm{HH}}(\mathbf{k})$ and the light-hole band $E_{\mathrm{LH}}(\mathbf{k})$ versus the crystal growth direction k_z and the parallel direction k_x can be obtained analytically from (4.5.27a) and (4.5.27b), respectively. Along the parallel plane, for example, the k_x direction ($k_y = k_z = 0$), we have for $Q_\varepsilon < 0$ (compression) and k_x is finite

$$E_{\mathrm{HH}}(k_x) = -P_\varepsilon - \frac{\hbar^2\gamma_1}{2m_0}k_x^2 + \left[\left(Q_\varepsilon + \frac{\hbar^2\gamma_2}{2m_0}k_x^2\right)^2 + 3\left(\frac{\hbar^2\gamma_2}{2m_0}\right)^2 k_x^4\right]^{1/2} \tag{4.5.28a}$$

from (4.5.27a) and

$$E_{\mathrm{LH}}(k_x) = -P_\varepsilon - \frac{\hbar^2\gamma_1}{2m_0}k_x^2 - \left[\left(Q_\varepsilon + \frac{\hbar^2\gamma_2}{2m_0}k_x^2\right)^2 + 3\left(\frac{\hbar^2\gamma_2}{2m_0}\right)^2 k_x^4\right]^{1/2} \tag{4.5.28b}$$

for the light hole from (4.5.27b).

In the case of tension ($Q_\varepsilon > 0$), we have

$$E_{\mathrm{HH}}(k_x) = -P_\varepsilon - \frac{\hbar^2\gamma_1}{2m_0}k_x^2 - \left[\left(Q_\varepsilon + \frac{\hbar^2\gamma_2}{2m_0}k_x^2\right)^2 + 3\left(\frac{\hbar^2\gamma_2}{2m_0}\right)^2 k_x^4\right]^{1/2} \tag{4.5.29a}$$

$$E_{\mathrm{LH}}(k_x) = -P_\varepsilon - \frac{\hbar^2\gamma_1}{2m_0}k_x^2 + \left[\left(Q_\varepsilon + \frac{\hbar^2\gamma_2}{2m_0}k_x^2\right)^2 + 3\left(\frac{\hbar^2\gamma_2}{2m_0}\right)^2 k_x^4\right]^{1/2}. \tag{4.5.29b}$$

Again, at $k_x = 0$, $E_{\mathrm{HH}}(0) + P_\varepsilon = -Q_\varepsilon < 0$, and $E_{\mathrm{LH}}(0) + P_\varepsilon = +Q_\varepsilon > 0$, which agree with (4.5.19). Along the k_z direction, we obtain for both compression and tension,

$$E_{\mathrm{HH}}(k_z) = -P_\varepsilon - \frac{\hbar^2}{2m_0}(\gamma_1 - 2\gamma_2)k_z^2 - Q_\varepsilon \tag{4.5.30a}$$

$$E_{\mathrm{LH}}(k_z) = -P_\varepsilon - \frac{\hbar^2}{2m_0}(\gamma_1 - 2\gamma_2)k_z^2 + Q_\varepsilon. \tag{4.5.30b}$$

The results of the valence-band structure for E_{HH} and E_{LH} versus k_x and k_z for both compression and tension are shown in Fig. 4.8, for $Ga_xIn_{1-x}As$ grown on InP substrate. We can see clearly that the heavy-hole band has a lighter effective mass than does the light-hole band in the k_x direction near $k_x = 0$ for the compression case ($Q_\varepsilon < 0$). The heavy hole still keeps its feature of a heavy effective mass along the k_z direction, and it is above the light-hole band at $k = 0$ for the compression case. On the other hand, the light-hole band is above the heavy-hole band in the case of tension ($x > 0.468$).

For a finite and fixed strain, the small-k expansion of the above dispersion relation can be written as

$$E_{HH}(\mathbf{k}) \approx -P_\varepsilon - Q_\varepsilon - \left(\frac{\hbar^2}{2m_0}\right)[(\gamma_1 + \gamma_2)k_t^2 + (\gamma_1 - 2\gamma_2)k_z^2] \qquad (k \to 0) \qquad (4.5.31a)$$

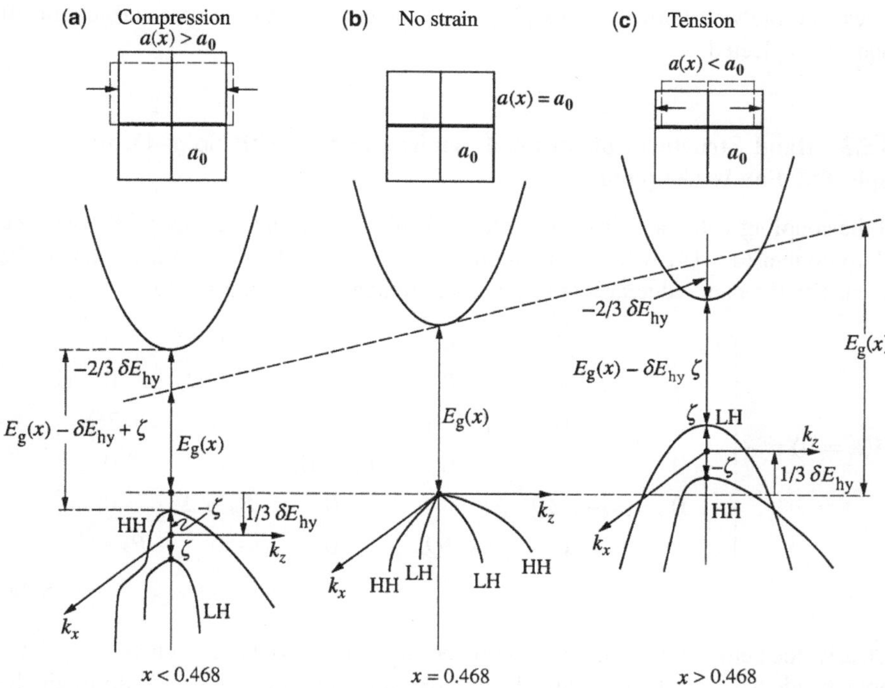

Figure 4.8 The energy-band structure in the momentum space for a bulk $Ga_xIn_{1-x}As$ material under (a) biaxial compression, (b) lattice-matched condition, and (c) biaxial tension for different Ga mole fractions x. The heavy-hole band is above the light-hole band, and its effective mass in the transverse plane (the k_x or k_y direction) is lighter than that of the light-hole band in the compressive strain case in (a). The light-hole band shifts above the heavy-hole band in the case of tension in (c). (Reprinted with permission from [37] © 1991 American Physical Society.)

$$E_{\text{LH}}(\mathbf{k}) \approx -P_\varepsilon + Q_\varepsilon - \left(\frac{\hbar^2}{2m_0}\right)[(\gamma_1 - \gamma_2)k_t^2 + (\gamma_1 + 2\gamma_2)k_z^2] \qquad (k \to 0) \tag{4.5.31b}$$

where the transverse wave vector has a magnitude $k_t = \sqrt{k_x^2 + k_y^2}$. From (4.5.31), we immediately obtain the band-edge energies

$$E_{\text{HH}}(0) \approx -P_\varepsilon - Q_\varepsilon \tag{4.5.32a}$$

$$E_{\text{LH}}(0) \approx -P_\varepsilon + Q_\varepsilon \tag{4.5.32b}$$

and the effective masses parallel ($\|$ or t) or perpendicular ($\perp$or z) to the $x-y$ plane

$$\begin{aligned} \frac{m_{\text{hh}}^z}{m_0} &= \frac{1}{\gamma_1 - 2\gamma_2} & \frac{m_{\text{hh}}^t}{m_0} &= \frac{1}{\gamma_1 + \gamma_2} \\ \frac{m_{\ell\text{h}}^z}{m_0} &= \frac{1}{\gamma_1 + 2\gamma_2} & \frac{m_{\ell\text{h}}^t}{m_0} &= \frac{1}{\gamma_1 - \gamma_2}. \end{aligned} \tag{4.5.33}$$

These are the well-known results [26] when the coupling to the spin–orbit split-off band is neglected.

4.5.3 Band Structures of Strained Semiconductors with Spin–Orbit Split-Off Bands Coupling

If the coupling with the spin–orbit split-off bands is taken into account [38], the 6×6 Hamiltonian (4.3.14) with the additional strain terms defined in (4.5.13) has to be used. For the band edge at $\mathbf{k} = 0$, the Hamiltonian $\mathbf{H}$ is simplified to

$$\mathbf{H}(\mathbf{k}=0) = -\begin{bmatrix} P_\varepsilon + Q_\varepsilon & 0 & 0 & 0 & 0 & 0 \\ 0 & P_\varepsilon - Q_\varepsilon & 0 & 0 & -\sqrt{2}Q_\varepsilon & 0 \\ 0 & 0 & P_\varepsilon - Q_\varepsilon & 0 & 0 & \sqrt{2}Q_\varepsilon \\ 0 & 0 & 0 & P_\varepsilon + Q_\varepsilon & 0 & 0 \\ 0 & -\sqrt{2}Q_\varepsilon & 0 & 0 & P_\varepsilon + \Delta & 0 \\ 0 & 0 & \sqrt{2}Q_\varepsilon & 0 & 0 & P_\varepsilon + \Delta \end{bmatrix}. \tag{4.5.34}$$

Clearly, the heavy-hole bands are decoupled from the rest of bands, whereas the light-hole bands $\left(\left|\frac{3}{2}, \pm\frac{1}{2}\right\rangle\right)$ are coupled with the split-off bands $\left(\left|\frac{1}{2}, \pm\frac{1}{2}\right\rangle\right)$through the strain-dependent off-diagonal terms. This coupling would be totally unaccounted for in the 4×4 approximation. For $\text{In}_{1-x}\text{Ga}_x\text{As}$ on InP, the transition energies from the heavy-hole and the light-hole bands to the conduction band with and without the *SO* coupling are shown [38] in Fig. 4.9. The comparison evidently demonstrates how important it is to include the spin–orbit split-off bands, because the error in

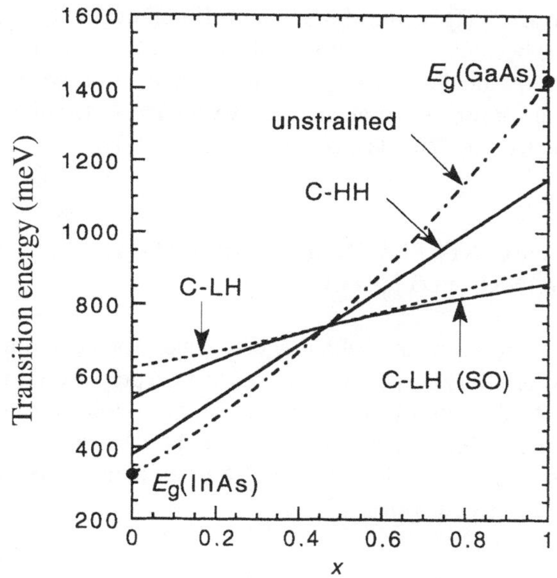

Figure 4.9 The energy band gap of a bulk $In_{1-x}Ga_xAs$ versus the Ga mole fraction x. The dotted-dashed curve: unstrained $In_{1-x}Ga_xAs$; the solid curves: transition energies from the conduction band (C) to the heavy-hole (HH) and light-hole (LH) bands for a bulk $In_{1-x}Ga_xAs$ pseudomorphically grown on InP; the dashed curve: the conduction to light-hole transition energy calculated without the spin–orbit (SO) split-off band coupling. (Reprinted with permission from [38] © 1992 American Physical Society.)

the light-hole energies could be as large as several tens of millielectron volts. The error is comparable with the heavy-hole and light-hole energy splits and is certainly too large to be ignored. As a consequence of the coupling, the eigenvectors corresponding to the energy $E(0)$ (= $E_{LH}(0)$ or $E_{SO}(0)$), determined by

$$\begin{bmatrix} -P_\varepsilon + Q_\varepsilon & \pm\sqrt{2}Q_\varepsilon \\ \pm\sqrt{2}Q_\varepsilon & -P_\varepsilon - \Delta \end{bmatrix} \begin{bmatrix} F_{3/2,\pm 1/2} \\ F_{1/2,\pm 1/2} \end{bmatrix} = E(0) \begin{bmatrix} F_{3/2,\pm 1/2} \\ F_{1/2,\pm 1/2} \end{bmatrix} \tag{4.5.35a}$$

$$\left|F_{3/2,\pm 1/2}\right|^2 + \left|F_{1/2,\pm 1/2}\right|^2 = 1 \tag{4.5.35b}$$

are not a *pure* light-hole state or a pure split-off state, but an admixture of the light-hole state and the split-off state. The band-edge energies can be readily solved from (4.5.34) or (4.5.35a),

$$E_{HH}(0) = -P_\varepsilon - Q_\varepsilon \tag{4.5.36a}$$

$$E_{LH}(0) = -P_\varepsilon + \frac{1}{2}\left(Q_\varepsilon - \Delta + \sqrt{\Delta^2 + 2\Delta Q_\varepsilon + 9Q_\varepsilon^2}\right) \tag{4.5.36b}$$

$$E_{SO}(0) = -P_\varepsilon + \frac{1}{2}\left(Q_\varepsilon - \Delta - \sqrt{\Delta^2 + 2\Delta Q_\varepsilon + 9Q_\varepsilon^2}\right). \tag{4.5.36c}$$

If the split-off bands are included in the 6×6 Hamiltonian, the E-k relation determined by (4.5.26) becomes a sixth-order polynomial of E, which apparently can be decomposed into two identical cubic polynomials because of the symmetry property of the Hamiltonian. However, an attempt to expand and factor directly the determinantal equation is tedious. The details are given in Ref. 38.

4.6 ELECTRONIC STATES IN AN ARBITRARY ONE-DIMENSIONAL POTENTIAL

In this section, we show how the Schrödinger equation for a one-dimensional potential profile with an arbitrary shape can be solved using a propagation-matrix approach, which is similar to that used in electromagnetic wave reflection or guidance in a multilayered medium [39, 40]. An arbitrary profile, $V(z)$, can always be approximated by a piecewise step profile as shown in Fig. 4.10 as long as the original potential profile does not have singularities.

4.6.1 Derivation of the Propagation Matrix Equation and Its Solution for the Eigenvalues

For a Schrödinger equation of the form

$$-\frac{\hbar^2}{2}\frac{d}{dz}\frac{1}{m(z)}\frac{d}{dz}\psi(z) + V(z)\psi(z) = E\psi(z) \tag{4.6.1}$$

we find that in the region ℓ, $z_{\ell-1} \leq z \leq z_\ell$,

$$-\frac{\hbar^2}{2m_\ell}\frac{d^2}{dz^2}\psi_\ell(z) + V_\ell\psi_\ell(z) = E\psi_\ell(z) \tag{4.6.2}$$

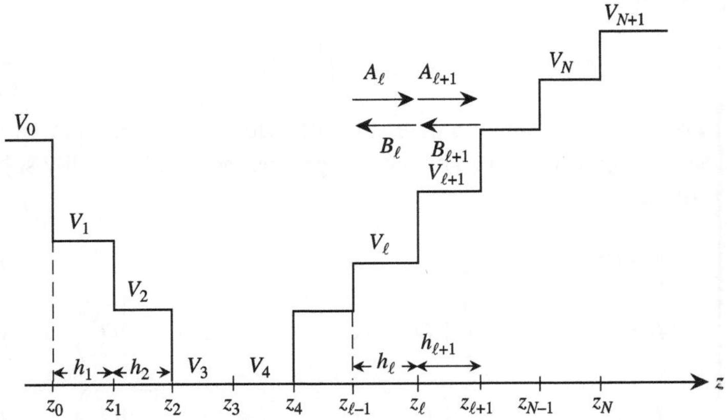

Figure 4.10 An arbitrary potential profile is subdivided into a stepwise potential.

where $V(z) = V_\ell$ and $m(z) = m_\ell$ in region ℓ. The solution can be written in the form

$$\psi_\ell(z) = A_\ell e^{ik_\ell(z-z_\ell)} + B_\ell e^{-ik_\ell(z-z_\ell)} \quad \text{for } z_{\ell-1} \le z \le z_\ell \tag{4.6.3}$$

where the wavenumber in region ℓ is

$$k_\ell = \sqrt{\frac{2m_\ell}{\hbar^2}(E - V_\ell)}. \tag{4.6.4}$$

Matching the boundary conditions in which $\psi(z)$ and $(1/m(z))d\psi(z)/dz$ are continuous at $z = z_\ell$, we find

$$A_\ell + B_\ell = A_{\ell+1}e^{ik_{\ell+1}(z_\ell - z_{\ell+1})} + B_{\ell+1}e^{-ik_{\ell+1}(z_\ell - z_{\ell+1})} \tag{4.6.5a}$$

$$\frac{k_\ell}{m_\ell}(A_\ell - B_\ell) = \frac{k_{\ell+1}}{m_{\ell+1}}[A_{\ell+1}e^{ik_{\ell+1}(z_\ell - z_{\ell+1})} - B_{\ell+1}e^{-ik_{\ell+1}(z_\ell - z_{\ell+1})}]. \tag{4.6.5b}$$

Define

$$P_{\ell(\ell+1)} = \frac{m_\ell k_{\ell+1}}{m_{\ell+1}k_\ell}, \tag{4.6.6}$$

and $z_{\ell+1} - z_\ell = h_{\ell+1}$ is the thickness of the region $\ell + 1$.

We can express $A_{\ell+1}$ and $B_{\ell+1}$ in terms of A_ℓ and B_ℓ in a matrix form

$$\begin{bmatrix} A_{\ell+1} \\ B_{\ell+1} \end{bmatrix} = \mathbf{F}_{(\ell+1)\ell}\begin{bmatrix} A_\ell \\ B_\ell \end{bmatrix} \tag{4.6.7a}$$

where we have defined a forward-propagation matrix

$$\mathbf{F}_{(\ell+1)\ell} = \frac{1}{2}\begin{bmatrix} (1 + P_{(\ell+1)\ell})e^{ik_{\ell+1}h_{\ell+1}} & (1 - P_{(\ell+1)\ell})e^{ik_{\ell+1}h_{\ell+1}} \\ (1 - P_{(\ell+1)\ell})e^{-ik_{\ell+1}h_{\ell+1}} & (1 + P_{(\ell+1)\ell})e^{-ik_{\ell+1}h_{\ell+1}} \end{bmatrix}. \tag{4.6.7b}$$

The relation can propagate from layer to layer

$$\begin{bmatrix} A_{N+1} \\ B_{N+1} \end{bmatrix} = \mathbf{F}_{(N+1)N}\,\mathbf{F}_{N(N-1)} \cdots \mathbf{F}_{10}\begin{bmatrix} A_0 \\ B_0 \end{bmatrix}. \tag{4.6.8}$$

For bound state solutions, we have $E < V_0$ and V_{N+1}. Therefore,

$$k_0 = i\alpha_0 \qquad \alpha_0 = \sqrt{\frac{2m_0}{\hbar^2}(V_0 - E)} \tag{4.6.9a}$$

$$k_{N+1} = i\alpha_{N+1} \quad \alpha_{N+1} = \sqrt{\frac{2m_{N+1}}{\hbar^2}(V_{N+1} - E)}. \tag{4.6.9b}$$

The solutions in region 0 and region $N + 1$ must be decaying solutions of the forms

$$\psi_0(z) = B_0\, e^{+\alpha_0(z-z_0)} \tag{4.6.10a}$$

and

$$\psi_{N+1}(z) = A_{N+1} e^{-\alpha_{N+1}(z - z_{N+1})} \tag{4.6.10b}$$

that is, $A_0 = 0$ and $B_{N+1} = 0$. Note the last position z_{N+1} is introduced for the use of the propagation matrix and it is arbitrary.

We write the product of the matrices

$$\mathbf{F}_{(N+1)N} \cdots \mathbf{F}_{10} \equiv \begin{bmatrix} f_{11} & f_{12} \\ f_{21} & f_{22} \end{bmatrix} \tag{4.6.11}$$

and, therefore,

$$\begin{bmatrix} A_{N+1} \\ 0 \end{bmatrix} = \begin{bmatrix} f_{11} & f_{12} \\ f_{21} & f_{22} \end{bmatrix} \begin{bmatrix} 0 \\ B_0 \end{bmatrix}. \tag{4.6.12}$$

For nontrivial solutions, we must have the eigenequation satisfied,

$$f_{22}(E) = 0 \tag{4.6.13}$$

as B_0 cannot be zero (otherwise, all field amplitudes are zero). Solving the eigenequation (4.6.13), we obtain the eigenvalues E. In practice, $f_{22}(E)$ is a complex function and the eigenvalues, E, are real. Therefore, a convenient method to find the eigenvalues is to search for the minima of $|f_{22}(E)|$ (or $\log_{10} |f_{22}(E)|$) in the range of energy given by the lowest and highest energies of the potential profile $V(z)$.

4.6.2 Self-Consistent Solution for a Modulation-Doped Quantum Well

Consider a quantum-well structure with a built-in potential

$$V_{\text{bi}}^{\text{e}}(z) = \begin{cases} \Delta E_{\text{c}} & |z| \geq L_{\text{w}}/2 \\ 0 & |z| < L_{\text{w}}/2 \end{cases} \tag{4.6.14a}$$

$$V_{\text{bi}}^{\text{h}}(z) = \begin{cases} -\Delta E_{\text{v}} & |z| \geq L_{\text{w}}/2 \\ 0 & |z| < L_{\text{w}}/2 \end{cases} \tag{4.6.14b}$$

and a doping profile $N_{\text{D}}(z)$ and $N_{\text{A}}(z)$. The Schrödinger equation

$$\left[\frac{-\hbar^2}{2m_{\text{e}}^*}\frac{d^2}{dz^2} + V_{\text{e}}(z)\right] f(z) = E_{\text{e}}(0)\, f(z) \tag{4.6.15a}$$

$$\left[\frac{+\hbar^2}{2m_{\text{h}}^*}\frac{d^2}{dz^2} + V_{\text{h}}(z)\right] g(z) = E_{\text{h}}(0)\, g(z) \tag{4.6.15b}$$

where

$$V_e(z) = V_{bi}^e(z) + |e|Fz + V_H(z) \tag{4.6.16a}$$

$$V_h(z) = V_{bi}^h(z) + |e|Fz + V_H(z). \tag{4.6.16b}$$

And the Hartree potential energy $V_H(z)$ is related to the electrostatic potential $\phi(z)$

$$V_H(z) = -|e|\phi(z). \tag{4.6.17}$$

Note that we define $V_{bi}^h(z)$ as in (4.6.14b) such that all energies are measured upward. Therefore, the same expressions for the field-induced potential and the Hartree potential are used in (4.6.16a) and (4.6.16b) without any sign change. Here F is the externally applied field (= 0 here), $\phi(z)$ is the electrostatic potential and it satisfies Gauss's law or Poisson's equation

$$\nabla \cdot (\varepsilon \mathbf{E}) = \rho(z). \tag{4.6.18}$$

For a one-dimensional problem, we have the electric field

$$\mathbf{E} = \hat{z}\, E(z) \quad \text{and} \quad E(z) = -\frac{\partial \phi}{\partial z}. \tag{4.6.19}$$

Therefore,

$$\frac{\partial}{\partial z}[\varepsilon E(z)] = \rho(z) \tag{4.6.20}$$

or

$$\frac{\partial}{\partial z}\left(\varepsilon \frac{\partial \phi}{\partial z}\right) = -\rho(z). \tag{4.6.21}$$

The charge distribution is given by

$$\rho(z) = |e|[p(z) - n(z) + N_D^+(z) - N_A^-(z)] \tag{4.6.22}$$

where $N_D^+(z)$ and $N_A^-(z)$ are the ionized donor and acceptor concentrations, respectively. The electron and hole concentrations, $n(z)$ and $p(z)$, are related to the wave functions of the nth conduction subband and the mth valence subband by

$$n(z) = \sum_n |f_n(z)|^2 N_{ns} \tag{4.6.23a}$$

$$p(z) = \sum_m |g_m(z)|^2 P_{ms} \tag{4.6.23b}$$

where the sums over n and m are only over the (lowest few) occupied subbands. Here the surface electron concentration in the nth conduction subband is

$$N_{ns} = \frac{k_B T m_e^*}{\pi \hbar^2} \ln\left[1 + e^{(F_c - E_{en}(0))/k_B T}\right]. \tag{4.6.24a}$$

The surface hole concentration in the mth valence subband is

$$P_{ms} = \frac{k_B T\, m_h^*}{\pi \hbar^2} \ln\left[1 + e^{(E_{hm}(0) - F_v)/k_B T}\right] \tag{4.6.24b}$$

where $F_c = F_v = E_F$ in a modulation-doped sample without any external injection of carriers. The parabolic energy dispersion relations

$$E_{en}(k_t) = E_{en}(0) + \frac{\hbar^2 k_t^2}{2m_e^*} \tag{4.6.25a}$$

$$E_{hm}(k_t) = E_{hm}(0) - \frac{\hbar^2 k_t^2}{2m_h^*} \tag{4.6.25b}$$

have been used in the summation over the two-dimensional wave vector $\mathbf{k}$ in the $x-y$ plane. The Fermi level is obtained from the charge neutrality condition:

$$\int_{-L/2}^{L/2} \rho(z) = 0 \tag{4.6.26a}$$

or

$$N_D^+ L_D + \sum_m P_{ms} = N_A^- L_A + \sum_n N_{ns}. \tag{4.6.26b}$$

4.6.3 *N*-type Modulation-Doped Quantum Well

For an n-type modulation-doped quantum well, Fig. 4.11, we have $N_A = 0$ and $p(z) \simeq 0$. We only have to solve for the electron wave function $f(z)$ and the electrostatic potential $\phi(z)$ self-consistently. The electric field $E(z)$ is obtained by integration

$$E(z) = \frac{1}{\varepsilon} \int_{-L/2}^{z} \rho(z')dz' + E(-L/2). \tag{4.6.27}$$

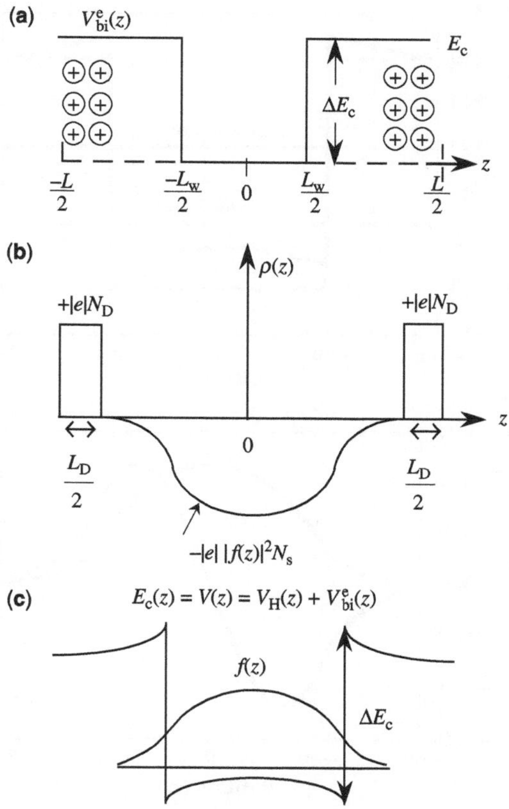

Figure 4.11 (a) The built-in potential, (b) the charge density, and (c) the screened electron potential energy profile of an n-type modulation doped quantum well.

For a symmetrically doped quantum well without any external bias, we have $E(-L/2) = 0$ by a symmetry consideration. Because $V_H(z) = -|e|\phi(z)$, we have

$$\frac{\partial V_H(z)}{\partial z} = +|e|E(z) \tag{4.6.28a}$$

or

$$V_H(z) = |e| \int_{-L/2}^{z} E(z')dz' + V_H(0). \tag{4.6.28b}$$

Here $V_H(0)$ can be chosen to be zero as it is only a reference potential energy.

Example We consider an n-type modulation-doped quantum well with doping widths $L_D/2 = 10$ Å at both ends (see Fig. 4.11), and $N_D = 4 \times 10^{18}\ \text{cm}^{-3}$ in a GaAs/$Al_{0.3}Ga_{0.7}As$ quantum-well structure. The surface carrier concentration in a

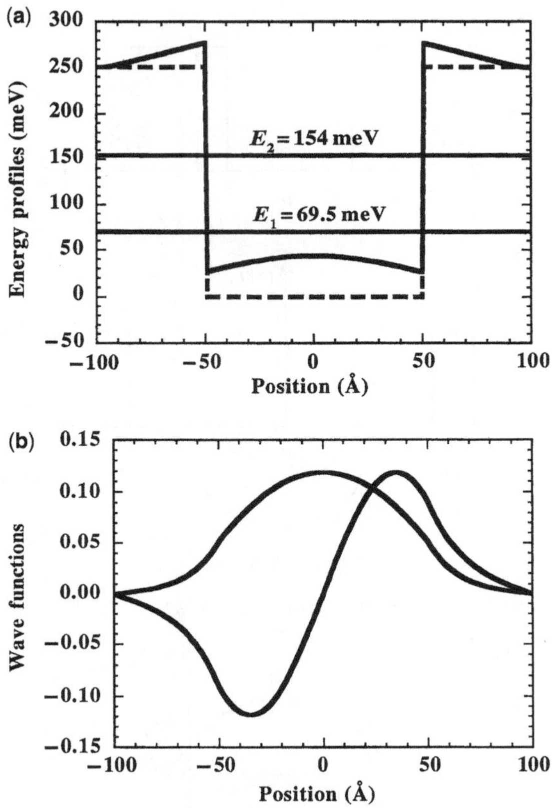

Figure 4.12 (a) Self-consistent potential profile (solid curve) $V_e(z)$ and the built-in potential profile (dashed) $V_{bi}^e(z)$ for a $GaAs/Al_{0.3}Ga_{0.7}As$ quantum well with modulation dopings within 10 Å at both ends of the profile $-100\text{ Å} \leq z \leq -90\text{ Å}$ and $90\text{ Å} \leq z \leq 100\text{ Å}$. The two solid horizontal lines represent the energy levels E_1 and E_2 of the self-consistent potential. (b) The corresponding wave functions $f_1(z)$ and $f_2(z)$ of the self-consistent potential in (a).

period is $N_D L_D = 8 \times 10^{11}\text{ cm}^{-2}$. The conduction band edge discontinuity ΔE_c is 251 meV and the built-in potential profile $V_{bi}^e(z)$ is shown as the dashed line in Fig. 4.12a. The final conduction band potential profile $V_e(z) = V_{bi}^e(z) + V_H(z)$ after solving the Schrödinger equation and the Poisson's equation self-consistently is shown as the solid curves together with the corresponding eigenenergies of the lowest two states, E_1 and E_2, plotted as two horizontal solid lines. The wave functions $f_1(z)$ and $f_2(z)$ for the self-consistent potential $V_e(z)$ are plotted in Fig. 4.12b. We can see that the effects of the charge distribution with positively ionized donors at the two ends and negative electrons in the wells create a net electric field pointing inward toward the center of the quantum well. Therefore, the band bending is curved upward as the slope of the potential energy profile gives the electric field and its direction. More examples of the modulation-doped potentials with an externally applied electric field F and their applications to intersubband photodetectors and resonant tunneling diodes can be found in Refs. 41–43.

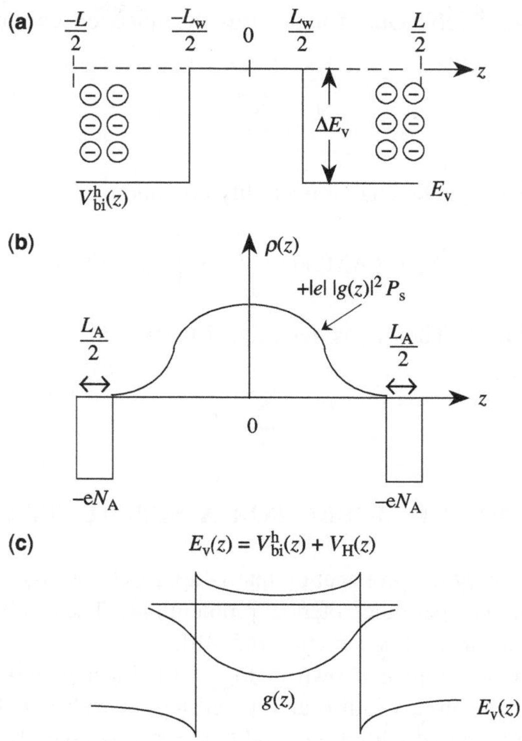

Figure 4.13 Same as Fig. 4.11 except for p-type.

4.6.4 *P*-type Modulation-Doped Quantum Well

For the hole distribution, Fig. 4.13, it is also possible to take into consideration both the heavy-hole and the light-hole dispersion relations

$$p(z) = \sum_{m=\ell h,\, hh} |g_m(z)|^2 P_{ms} \tag{4.6.29a}$$

$$P_{ms} = \frac{k_B T}{\pi \hbar^2} \left[m^*_{hh} \sum_{m=hh} \ln\left(1 + e^{(E_{hm}-E_F)/k_BT}\right) + m^*_{\ell h} \sum_{m=\ell h} \ln\left(1 + e^{(E_{hm}-E_F)/k_BT}\right) \right]. \tag{4.6.29b}$$

4.6.5 Populations by Both Electrons and Holes

In a laser structure, both electrons and holes are injected by an external electric or optical pumping. Two quasi-Fermi levels, F_c and F_v, can be used to describe

$n(z)$ and $p(z)$, respectively. One usually uses the injection current density to find N_s, then

$$N_s = \sum_n N_{ns} \tag{4.6.30}$$

is used to determine F_c. The charge neutrality condition

$$P_s + \int N_D^+(z)dz = N_s + \int N_A^-(z)dz \tag{4.6.31}$$

is used to determine P_s. Then F_v is determined from

$$P_s = \sum_m P_{ms}. \tag{4.6.32}$$

4.7 KRONIG–PENNEY MODEL FOR A SUPERLATTICE

In this section, we apply the propagation matrix approach discussed in Section 4.6 to study a one-dimensional periodic potential problem [44]. The results are the Kronig–Penney model for a superlattice structure [45–48].

The superlattice structure is shown in Fig. 4.14. Each period (or cell) consists of one barrier region with a width b and a well region with a width w. The period is $L = b + w$. The potential within a period $0 < z < L$ is given by

$$V(z) = \begin{cases} V_0, & 0 < z < b \\ 0, & b < z < b + w = L \end{cases} \tag{4.7.1}$$

and $V(z + nL) = V(z)$ for any integer n.

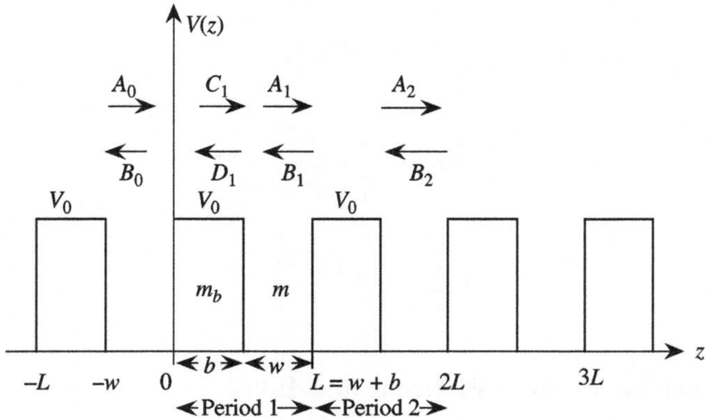

Figure 4.14 A periodic potential (Kronig–Penney model) consists of one barrier region with a thickness b and a quantum-well region with a width w. The period is $L = w + b$.

4.7.1 Derivation of the Propagation Matrix

The wave function in the period n is given by the plane wave solution of the wave equation

$$\psi_n(z) = \begin{cases} A_n e^{ik(z-nL)} + B_n e^{-ik(z-nL)} & nL - w \le z \le nL \\ C_n e^{ik_b[z-b-(n-1)L]} + D_n e^{-ik_b[z-b-(n-1)L]} & (n-1)L \le z \le (n-1)L + b \end{cases} \tag{4.7.2}$$

$$k = \sqrt{\frac{2m}{\hbar^2}E} \qquad k_b = \sqrt{\frac{2m_b}{\hbar^2}(E - V_0)}. \tag{4.7.3}$$

More specifically, we write

$$\psi(z) = \begin{cases} A_0\, e^{ikz} + B_0\, e^{-ikz} & -w \le z \le 0 \\ C_1\, e^{ik_b(z-b)} + D_1\, e^{-ik_b(z-b)} & 0 \le z \le b \\ A_1\, e^{ik(z-L)} + B_1\, e^{-ik(z-b)} & b \le z \le L. \end{cases} \tag{4.7.4}$$

Using the boundary conditions in which $\psi(z)$ and $(1/m(z))(\partial\psi/\partial z)$ are continuous at $z = 0$, we find the matrix equation for the coefficients of the wave functions in region 0 and the barrier b

$$\begin{bmatrix} C_1 \\ D_1 \end{bmatrix} = \mathbf{F}_{b0} \begin{bmatrix} A_0 \\ B_0 \end{bmatrix} \tag{4.7.5a}$$

where

$$\mathbf{F}_{b0} = \frac{1}{2}\begin{bmatrix} (1+P_{b0})\, e^{ik_b b} & (1-P_{b0})\, e^{ik_b b} \\ (1-P_{b0})\, e^{-ik_b b} & (1+P_{b0})\, e^{-ik_b b} \end{bmatrix} \tag{4.7.5b}$$

$$P_{b0} = \frac{m_b k}{m k_b}. \tag{4.7.5c}$$

Similarly, the boundary conditions at $z = b$ give the matrix equation for region 1 and the barrier b

$$\begin{bmatrix} A_1 \\ B_1 \end{bmatrix} = \mathbf{F}_{1b} \begin{bmatrix} C_1 \\ D_1 \end{bmatrix} \tag{4.7.6a}$$

$$\mathbf{F}_{1b} = \frac{1}{2}\begin{bmatrix} (1+P_{1b})e^{ikw} & (1-P_{1b})e^{ikw} \\ (1-P_{1b})e^{-ikw} & (1+P_{1b})e^{-ikw} \end{bmatrix} \tag{4.7.6b}$$

where

$$P_{1b} = \frac{m k_b}{m_b k}. \tag{4.7.6c}$$

Define

$$P = P_{1b} = m k_b/(m_b k) = 1/P_{b0}. \tag{4.7.7}$$

We have the transition matrix for one period consisting of one well and one barrier

$$\begin{bmatrix} A_1 \\ B_1 \end{bmatrix} = \mathbf{T} \begin{bmatrix} A_0 \\ B_0 \end{bmatrix} \tag{4.7.8a}$$

where

$$\mathbf{T} = \mathbf{F}_{1b}\mathbf{F}_{b0} = \begin{bmatrix} t_{11} & t_{12} \\ t_{21} & t_{22} \end{bmatrix} \tag{4.7.8b}$$

and the matrix elements are given by

$$\begin{aligned}
t_{11} &= e^{ikw}\left[\cos k_b b + \frac{i}{2}\left(P + \frac{1}{P}\right)\sin k_b b\right] \\
t_{12} &= e^{ikw}\frac{i}{2}\left(P - \frac{1}{P}\right)\sin k_b b \\
t_{21} &= e^{-ikw}\left(-\frac{i}{2}\right)\left(P - \frac{1}{P}\right)\sin k_b b \\
t_{22} &= e^{-ikw}\left[\cos k_b b - \frac{i}{2}\left(P + \frac{1}{P}\right)\sin k_b b\right].
\end{aligned} \tag{4.7.8c}$$

Note that the determinant of the $\mathbf{T}$ matrix is unity

$$\det|\mathbf{T}| = 1. \tag{4.7.9}$$

If we continue the relation (4.7.8a) to the nth period, we find

$$\begin{bmatrix} A_n \\ B_n \end{bmatrix} = \mathbf{T}^n \begin{bmatrix} A_0 \\ B_0 \end{bmatrix}. \tag{4.7.10}$$

4.7.2 Solutions for the Eigenvalues and Eigenvectors

The eigenvalues and eigenvectors of the 2×2 matrix $\mathbf{T}$ are solutions of the determinantal equation

$$\mathbf{T}\begin{bmatrix} A_0 \\ B_0 \end{bmatrix} = t\begin{bmatrix} A_0 \\ B_0 \end{bmatrix} \tag{4.7.11}$$

$$\det\begin{vmatrix} t_{11} - t & t_{12} \\ t_{21} & t_{22} - t \end{vmatrix} = 0. \tag{4.7.12}$$

We obtain a second-order polynomial equation for the eigenvalue t

$$t^2 - (t_{11} + t_{22})t + \det\mathbf{T} = 0. \tag{4.7.13}$$

Because det $|\mathbf{T}| = 1$, we obtain two roots.

$$t_{\pm} = \left(\frac{t_{11} + t_{22}}{2}\right) \pm \sqrt{\left(\frac{t_{11} + t_{22}}{2}\right)^2 - 1}. \tag{4.7.14}$$

If $|(t_{11} + t_{22})/2| > 1$, t_+ and t_- are real and either $\lim_{n\to\infty} |t^n_{\pm}| \to \infty$ or $\lim_{n\to-\infty} |t^n_{\pm}| \to \infty$. In that case, the condition in which the wave function (i.e., $|A_n|$ and $|B_n|$) must remain finite as $n \to \pm\infty$ is violated. Hence, the eigenvalue E must satisfy the condition that

$$\left|\frac{t_{11} + t_{22}}{2}\right| = \left|\cos kw \cos k_b b - \frac{1}{2}\left(P + \frac{1}{P}\right) \sin kw \sin k_b b\right| \leq 1. \tag{4.7.15}$$

The two eigenvalues can be written in the forms

$$t_+ = e^{iqL} \quad \text{and} \quad t_- = e^{-iqL} \tag{4.7.16}$$

as $|t_{\pm}| = 1$ and $t_+ t_- = 1$. That is, the determinantal equation can be written as

$$(t - t_+)(t - t_-) = 0 \tag{4.7.17a}$$

or

$$t_+ + t_- = t_{11} + t_{22} \tag{4.7.17b}$$

and

$$t_+ t_- = \det|\mathbf{T}| = 1. \tag{4.7.17c}$$

We conclude that the eigenequation is

$$\cos qL = \cos kw \cos k_b b - \frac{1}{2}\left(P + \frac{1}{P}\right) \sin kw \sin k_b b \tag{4.7.18}$$

with the two eigenvectors

$$\begin{bmatrix} A_0^+ \\ B_0^+ \end{bmatrix} \quad \text{and} \quad \begin{bmatrix} A_0^- \\ B_0^- \end{bmatrix}$$

corresponding to eigenvalues, e^{iqL} and e^{-iqL}, respectively. The eigenfunctions can be obtained from (i) the ratio A_0^+/B_0^+ following the eigenequation (4.7.11) and the normalization condition $\int_0^L |\psi(z)|^2 dz = 1$,

$$\frac{A_0^+}{B_0^+} = \frac{t_{12}}{e^{iqL} - t_{11}} \left(= \frac{e^{iqL} - t_{22}}{t_{21}}\right) \tag{4.7.19}$$

and (ii) the ratio A_0^-/B_0^- for $\exp(-iqL)$ with $q \to -q$ and the normalization condition.

For bound state solutions, $E < V_0$, k_b is purely imaginary

$$k_b = i\alpha_b, \qquad \alpha_b = \sqrt{\frac{2m_b}{\hbar^2}(V_0 - E)}. \tag{4.7.20}$$

The determinantal equation is given by

$$\cos qL = \cos kw \cosh \alpha_b b + \frac{1}{2}\left(\eta - \frac{1}{\eta}\right) \sin kw \sinh \alpha_b b \tag{4.7.21}$$

where η is defined in the following equation when P is purely imaginary

$$P = \frac{mk_b}{m_b k} = i\frac{m\alpha_b}{m_b k} = i\eta. \tag{4.7.22}$$

In summary, the Kronig–Penney model for a superlattice can be obtained from solving the determinantal equation

$$\cos qL = f(E) \tag{4.7.23a}$$

where the eigenequation is defined as

$$f(E) = \begin{cases} \cos kw \cosh \alpha_b b + \dfrac{1}{2}\left(\eta - \dfrac{1}{\eta}\right) \sin kw \sinh \alpha_b b & 0 < E < V_0 \\ \cos kw \cos k_b b - \dfrac{1}{2}\left(P + \dfrac{1}{P}\right) \sin kw \sin k_b b & V_0 < E. \end{cases} \tag{4.7.23b}$$

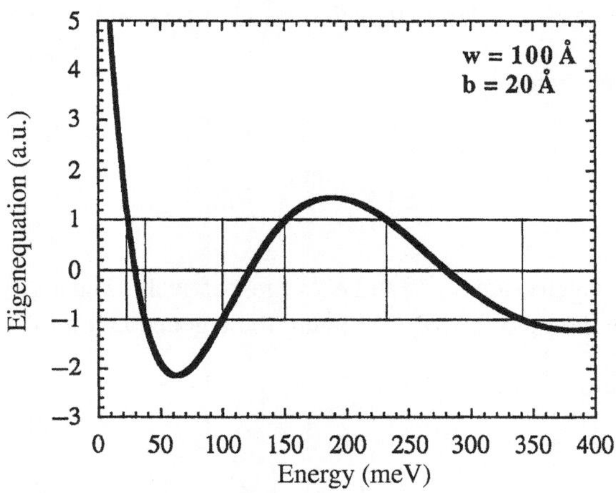

Figure 4.15 A plot of the eigenequation $f(E)$ ($=\cos qL$) from Eq. (4.7.23b) versus the energy E. Only the miniband ranges such that $-1 \leq f(E) \leq 1$ are acceptable solutions. The plot is for the conduction minibands of a $GaAs/Al_{0.3}Ga_{0.7}As$ superlattice with a well width $w = 100$ Å and a barrier width $b = 20$ Å. The barrier height is 250.6 meV.

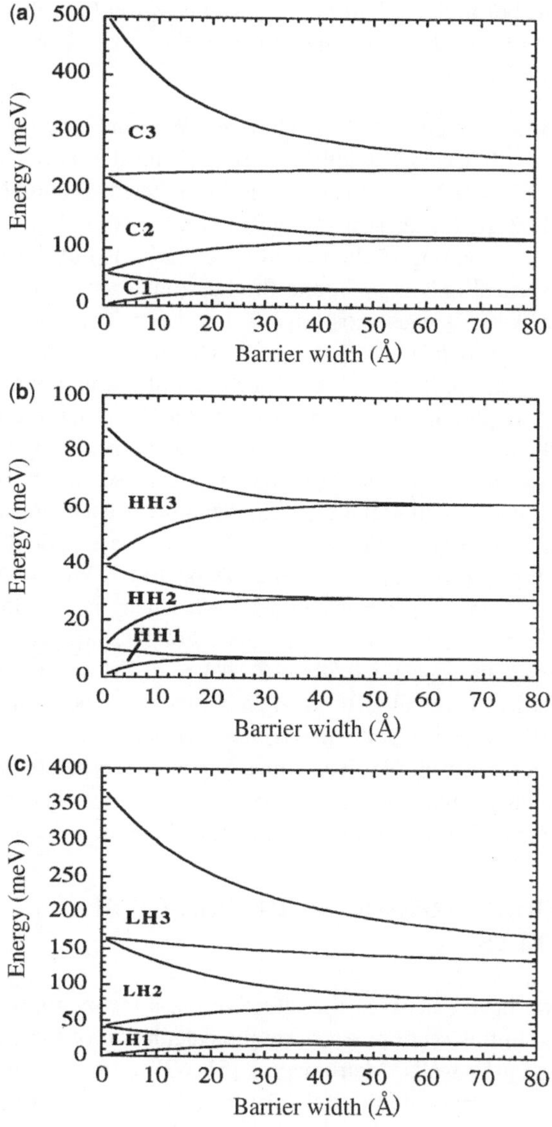

Figure 4.16 The minibands of a $GaAs/Al_{0.3}Ga_{0.7}As$ superlattice with a well width $w = 100$ Å are plotted versus the barrier width b. We show (a) the conduction minibands, (b) the heavy-hole minibands, and (c) the light-hole minibands.

A method to find the solution E is to plot $f(E)$ versus $E \geq 0$. The region of E such that the condition $|f(E)| \leq 1$ is satisfied will be acceptable. The quantum number q is obtained from

$$qL = \cos^{-1} f(E) \tag{4.7.24}$$

where $f(E)$ is real. For a given E in the acceptable region (or called miniband), the solution q is used to find the eigenfunction from the ratio A_0^+/B_0^+ (or A_0^-/B_0^-) and the normalization condition of the wave function.

Example of a GaAs/Al$_{0.3}$Ga$_{0.7}$As Superlattice We consider a GaAs/Al$_x$Ga$_{1-x}$As superlattice with $x = 0.3$, a well width $w = 100$ Å, and the barrier width $b = 20$ Å. The barrier height $V_0 = 0.2506$ eV is obtained from the conduction band discontinuity $V_0 = \Delta E_c = 0.67\ \Delta E_g(x)$, $\Delta E_g(x) = 1.247x$ (eV). We see in Fig. 4.15 that there are two minibands corresponding to the bound states ($E < V_0$) of the superlattice with $|f(E)| \leq 1$. The vertical axis is $f(E)$, and the horizontal axis is the electron energy E in millielectron volts. Because cosh $\alpha_b b \geq 1$ at $kw = N\pi$, where N is an integer, we obtain $f(E) = \cosh \alpha_b b \geq 1$ at $k = N\pi/w$. Therefore, $k = N\pi/w$ always occurs either in the forbidden miniband gap or at the edge of the miniband.

In Fig. 4.16(a), we plot the energy spectrum for the electron minibands for a superlattice with a well width $w = 100$ Å and the barrier width b varying from 1 Å to 80 Å. We see that the width of the miniband energy decreases as the barrier width is increased because the coupling among wells becomes weaker for a thicker barrier. The heavy-hole and light-hole minibands for the same GaAs/Al$_{0.3}$Ga$_{0.7}$As superlattice are shown in Fig. 4.16(b) and Fig. 4.16(c), respectively. We use $\Delta E_v = 0.33\ \Delta E_g$, and $m_{hh}^z = m_0/(\gamma_1 - 2\gamma_2)$, $m_{\ell h}^z = m_0/(\gamma_1 + 2\gamma_2)$, where $\gamma_1 = 6.85$, and $\gamma_2 = 2.10$ for GaAs, $\gamma_1 = 3.45$, and $\gamma_2 = 0.68$ for AlAs. For Al$_x$Ga$_{1-x}$As, we use $\gamma_1(x) = 3.45x + 6.85(1 - x)$ and $\gamma_2(x) = 0.68x + 2.10(1 - x)$. More discussions on semiconductor superlattices using the Kronig–Penny model can be found in Refs. 44–50. The above formulation can be extended to any periodic potential with each period consisting of an arbitrary profile. The procedure is to discretize each period with the periodic boundary condition at the two ends and solve the problem using the finite difference method or the propagation matrix method.

4.8 BAND STRUCTURES OF SEMICONDUCTOR QUANTUM WELLS

In Section 4.4, we have discussed the effective mass theory for a single band and degenerate bands. In this section, we study the calculation of the band structures of semiconductor quantum wells. Many papers [51–67] on the recent development of the effective mass theory or the $\mathbf{k} \cdot \mathbf{p}$ theory for quantum-well structures have been published. Theoretical methods such as the tight-binding [68] and the bond-orbital models [69, 70] have also been introduced. Here we focus on the $\mathbf{k} \cdot \mathbf{p}$ method of the Luttinger–Kohn Hamiltonian together with the strain terms of Pikus and Bir, as discussed in Sections 4.4 and 4.5.

4.8.1 Conduction Band

The effective mass theory for the conduction band is obtained from the dispersion relation

$$E(k) = \frac{\hbar^2 k^2}{2m^*} \tag{4.8.1}$$

where the effective mass of the electron in the conduction band is $m^* = m_b^*$ in the barrier region and $m^* = m_w^*$ in the quantum well. In the presence of the quantum-well potential,

$$V(z) = \begin{cases} V_0(=\Delta E_c) & |z| > L_w/2 \\ 0 & |z| \leq L_w/2 \end{cases} \tag{4.8.2}$$

where the energies are all measured from the conduction-band edge. The effective mass Eq. (4.4.5) for a single band is

$$\left[-\frac{\hbar^2}{2}\frac{\partial}{\partial z}\frac{1}{m(z)}\frac{\partial}{\partial z} + \frac{\hbar^2}{2m(z)}\nabla_t^2 + V(z)\right]\psi(r) = E\psi(r) \tag{4.8.3}$$

where $\dfrac{1}{m(z)}\dfrac{\partial}{\partial z}$ appears inside $\partial/\partial z$ to ensure that the probability current density

$$j_z(z) \sim \frac{1}{m(z)}\left[\psi^*\frac{\partial}{\partial z}\psi - \psi\frac{\partial\psi^*}{\partial z}\right]$$

is continuous at the heterojunction.

In general, the wave function $\psi(\mathbf{r})$ can be written in the form

$$\psi(\mathbf{r}) = \frac{e^{i\mathbf{k}_t\cdot\mathbf{r}}}{\sqrt{A}}\psi(z) \tag{4.8.4}$$

and

$$\frac{-\hbar^2}{2}\frac{\partial}{\partial z}\left[\frac{1}{m(z)}\frac{\partial}{\partial z}\psi(z)\right] + V(z)\,\psi(z) = \left(E(k_t) - \frac{\hbar^2 k_t^2}{2m(z)}\right)\psi(z). \tag{4.8.5}$$

The eigenvalue and the eigenfunction are obtained from the above equation, following the procedures discussed in Chapter 3. Here, we ignore the k_t dependence of $\psi(z)$. Equation (4.8.5) is usually solved at $k_t = 0$ for the nth subband energy $E_n(0)$ with a wavefunction $\psi(z) = f_n(z)$. Then we have $E_n(k_t) = E_n(0) + \dfrac{\hbar^2 k_t^2}{2m_w^*}$.

4.8.2 Valence Band

Band-Edge Energy For a given quantum-well potential,

$$V_h(z) = \begin{cases} 0 & |z| \leq L_w/2 \\ -\Delta E_v & |z| > L_w/2 \end{cases} \tag{4.8.6}$$

let us find the band-edge energy at $\mathbf{k}_t = 0$ first. The Luttinger–Kohn Hamiltonian (4.3.14) or (4.5.13) is diagonal for $k_x = k_y = 0$

$$E_{\mathrm{HH}}(k_z) = -\frac{\hbar^2}{2m_0}(\gamma_1 - 2\gamma_2)k_z^2 \tag{4.8.7a}$$

$$E_{\mathrm{LH}}(k_z) = -\frac{\hbar^2}{2m_0}(\gamma_1 + 2\gamma_2)k_z^2. \tag{4.8.7b}$$

Define

$$m_{\mathrm{hh}}^z \equiv \frac{m_0}{\gamma_1 - 2\gamma_2} \tag{4.8.8a}$$

$$m_{\ell\mathrm{h}}^z \equiv \frac{m_0}{\gamma_1 + 2\gamma_2}. \tag{4.8.8b}$$

Because the parameters γ_1 and γ_2 in the well are different from those in the barrier regions, we solve

$$\left[+\frac{\hbar^2}{2}\frac{\partial}{\partial z}\frac{1}{m_{(m)}^z}\frac{\partial}{\partial z} + V_{\mathrm{h}}(z)\right] g_{(m)}(z) = E g_{(m)}(z) \tag{4.8.9}$$

where $(m) \equiv (hhm)$ or (ℓhm). Therefore, the band-edge energies for the hhm or ℓhm subbands can all be found from the equation identical to that for the parabolic band model. We only have to use the appropriate effective masses (4.8.8a) and (4.8.8b) in the corresponding regions.

Example As an example, we consider a GaAs/$Al_{0.3}Ga_{0.7}As$ quantum well as shown in Fig. 4.17a. The energies for the electron subbands, Fig. 4.17b, and for the heavy-hole and light-hole subbands, Fig. 4.17c, are plotted versus the well width L_w. We can see that the magnitude of all energy levels decrease with an increasing well width.

Valence Subbands Dispersion Relations The effective mass equation for four degenerate valence bands (two heavy-hole and two light-hole bands) follows Eq. (4.4.11) for a quantum-well potential $V_{\mathrm{h}}(z)$ given by (4.8.6)

$$\left[\mathbf{H}^{\mathrm{LK}}\left(k_x,\ k_y,\ k_z = -i\frac{\partial}{\partial z}\right) + V_{\mathrm{h}}(z)\mathbf{I}\right] \cdot \begin{bmatrix} F_1 \\ F_2 \\ F_3 \\ F_4 \end{bmatrix} = E \begin{bmatrix} F_1 \\ F_2 \\ F_3 \\ F_4 \end{bmatrix} \tag{4.8.10}$$

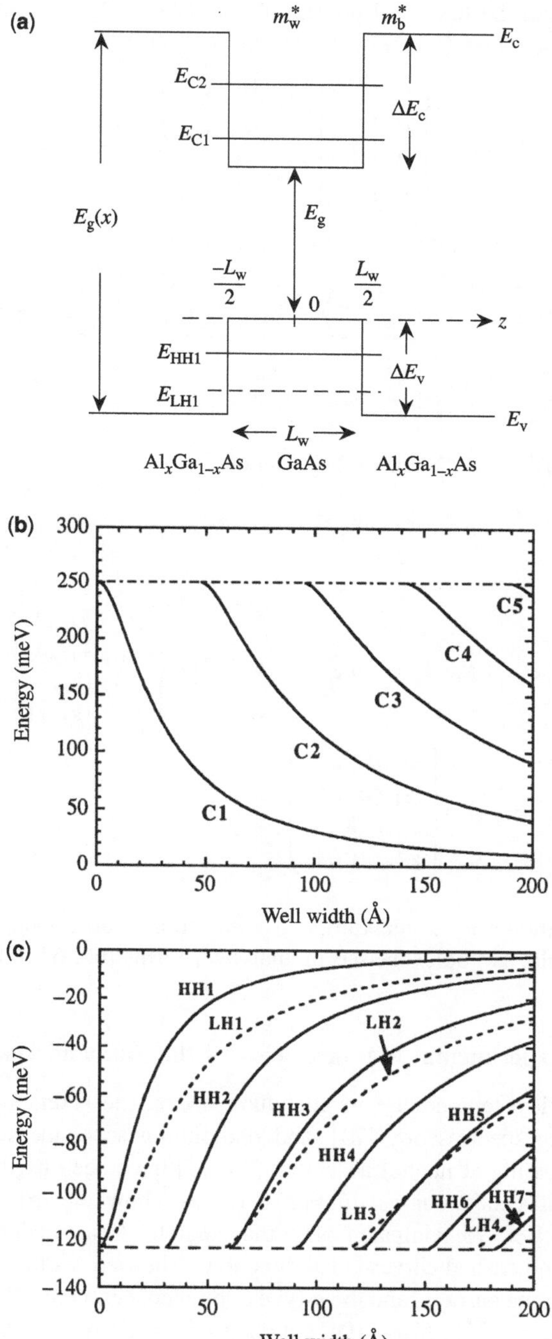

Figure 4.17 (a) Quantum-well profiles for the conduction and valence bands of a GaAs/$Al_xGa_{1-x}As$ system. (b) Conduction subband energies, E_{C1}, E_{C2}, ..., and (c) valence subband energies E_{HH1}, E_{HH2}, ..., and E_{LH1}, E_{LH2} ... versus the well width L_w.

where $\mathbf{H}^{\mathrm{LK}}$ is from the first 4×4 portion of (4.3.14) or (4.5.13), and the envelope functions F_1, F_2, F_3, and F_4 can be written in the vector form

$$\mathbf{F}_{\mathbf{k}}(\mathbf{r}) = \begin{bmatrix} F_1 \\ F_2 \\ F_3 \\ F_4 \end{bmatrix} = \begin{bmatrix} g_{3/2}(k_x, k_y, z) \\ g_{1/2}(k_x, k_y, z) \\ g_{-1/2}(k_x, k_y, z) \\ g_{-3/2}(k_x, k_y, z) \end{bmatrix} \frac{e^{ik_x + ik_y y}}{\sqrt{A}}. \tag{4.8.11}$$

The wavefunction in component form is expressed as

$$\begin{aligned} \psi_{\mathbf{k}}(\mathbf{r}) &= F_1\left|\frac{3}{2}, \frac{3}{2}\right\rangle + F_2\left|\frac{3}{2}, \frac{1}{2}\right\rangle + F_3\left|\frac{3}{2}, -\frac{1}{2}\right\rangle + F_4\left|\frac{3}{2}, -\frac{3}{2}\right\rangle \\ &= \frac{e^{ik_x x + ik_y y}}{\sqrt{A}} \sum_{\nu} g_\nu(k_x, k_y, z)\left|\frac{3}{2}, \nu\right\rangle \end{aligned} \tag{4.8.12}$$

where $\nu = 3/2,\ 1/2,\ -1/2$, and $-3/2$. Denote

$$\mathbf{k}_t \equiv \hat{x}k_x + \hat{y}k_y. \tag{4.8.13}$$

We write

$$\begin{aligned} \left[\mathbf{H}^{\mathrm{LK}}\left(\mathbf{k}_t,\ k_z = -i\frac{\partial}{\partial z}\right) + V_{\mathrm{h}}(z)\mathbf{I}\right] \cdot \begin{bmatrix} g_{3/2}(\mathbf{k}_t, z) \\ g_{1/2}(\mathbf{k}_t, z) \\ g_{-1/2}(\mathbf{k}_t, z) \\ g_{-3/2}(\mathbf{k}_t, z) \end{bmatrix} \\ = E(\mathbf{k}_t) \begin{bmatrix} g_{3/2}(\mathbf{k}_t, z) \\ g_{1/2}(\mathbf{k}_t, z) \\ g_{-1/2}(\mathbf{k}_t, z) \\ g_{-3/2}(\mathbf{k}_t, z) \end{bmatrix}. \end{aligned} \tag{4.8.14}$$

Theoretically, a model for a quantum well with infinite barriers has been used, and many of the results can be expressed in analytical forms [60, 63, 64, 71].

4.8.3 Direct Experimental Measurements of the Subband Dispersions

Experimentally, low-temperature magnetoluminescent measurement [72], resonant magnetotunneling spectroscopy [73], and photoluminescent measurements of hot electrons recombining at neutral acceptors [74, 75] have been used to map out the hole subband dispersion curves. In Fig. 4.18, we show the experimental results (data points) of Ref. 75 compared with the theoretical calculations (solid curves [100] direction and dashed curves [110] direction). The well width L_{w}, the aluminum mole fraction x in the barriers, and the p-type doping concentration N_{A} in the wells are (a) $L_{\mathrm{w}} = 33$ Å, $x = 0.325$, $N_{\mathrm{A}} = 10^{18}\ \mathrm{cm}^{-3}$; (b) $L_{\mathrm{w}} = 51$ Å, $x = 0.315$, $N_{\mathrm{A}} = 10^{18}\ \mathrm{cm}^{-3}$, (c) $L_{\mathrm{w}} = 75$ Å, $x = 0.32$, $N_{\mathrm{A}} = 2 \times 10^{18}\ \mathrm{cm}^{-3}$; and (d) $L_{\mathrm{w}} = 98$ Å, $x = 0.38$, $N_{\mathrm{A}} = 3 \times 10^{17}\ \mathrm{cm}^{-3}$. Each well is Be-doped in the central 10 Å except for the 98 Å wells (doped in the center 35 Å) at the above-indicated concentration N_{A}. It is

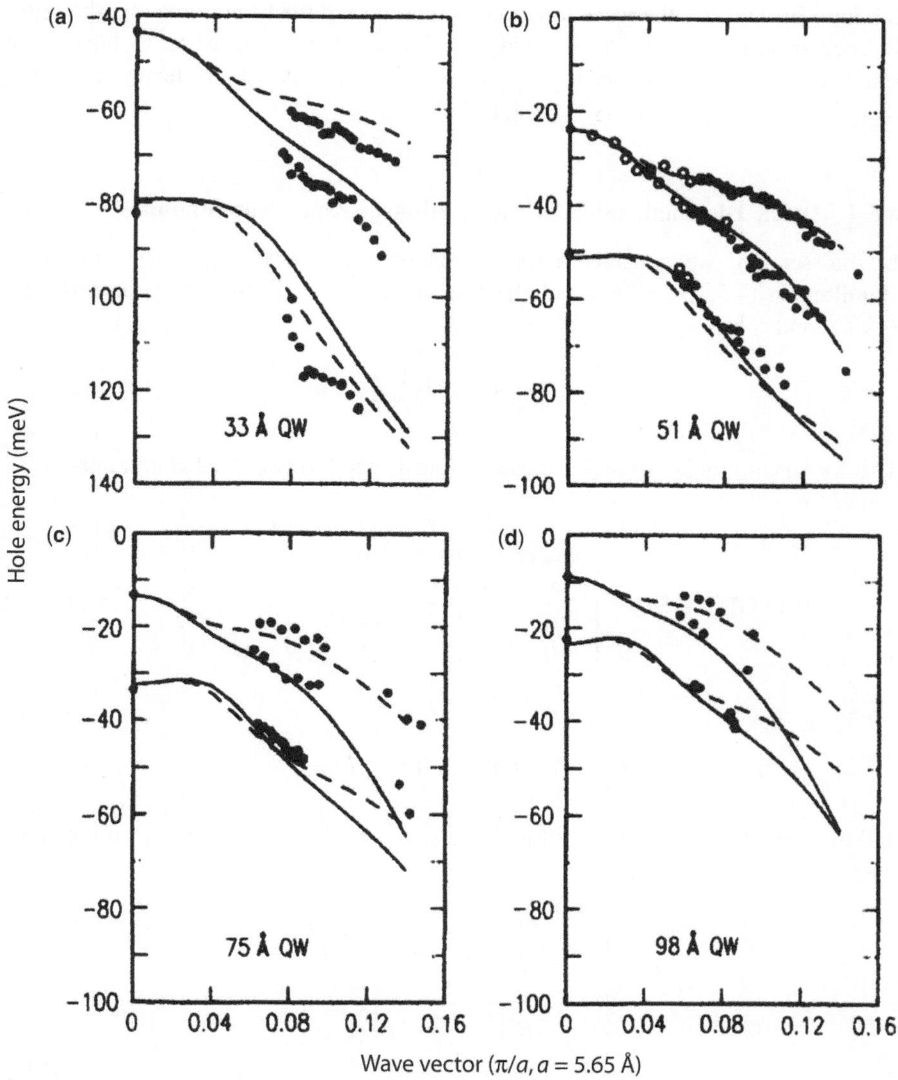

Figure 4.18 Experimental data (dots) and theoretical calculations (solid curves: wave vector along [100] direction; dashed curves: wave vector along [110] direction) for the hole subband energies of GaAs/$Al_xGa_{1-x}As$ quantum wells for (a) 33 Å quantum wells with $x = 0.325$ in the barriers, (b) 51 Å quantum wells with $x = 0.315$ in the barriers, (c) 75 Å quantum wells with $x = 0.32$, and (d) 98 Å quantum wells with $x = 0.38$. The horizontal axis is the wave vector normalized by π/a, where $a = 5.65$ Å is the lattice constant of GaAs. (Reprinted with permission from [75] © 1994 Elsevier: Surface Science.)

noted that these nonparabolic subband dispersion curves exhibit the valence band mixing behavior in quantum wells.

We can see that the coupling between the heavy-hole HH1 and the light LH1 subbands leads to warping (or nonparabolicity) of the band structure. These valence band mixing effects cause the wave function to exhibit both the heavy-hole and the light-hole characteristics. Here the labeling of HH and LH is according to their wave functions at $k_x = k_y = 0$, as $|\frac{3}{2}, \pm\frac{3}{2}\rangle$ are assigned to that of the heavy hole, and $|\frac{3}{2}, \pm\frac{1}{2}\rangle$ are assigned to that of the light hole.

4.8.4 Block Diagonalization of the Luttinger–Kohn Hamiltonian

In this section, we consider a transformation that block diagonalizes the 4×4 Hamiltonian (4.5.13), which includes the strain effects. Define the phases θ_R and θ_S of R and S by

$$R = |R|e^{i\theta_R} \qquad S = |S|e^{i\theta_S}. \tag{4.8.15}$$

The 4×4 Hamiltonian in (4.5.13) can be transformed to two 2×2 Hamiltonians H^{U} and H^{L}

$$\mathbf{H} = \mathbf{U}\mathbf{H}^{\mathrm{LK}}\mathbf{U}^{+} = -\begin{bmatrix} P+Q & \tilde{R} & 0 & 0 \\ \tilde{R}^{+} & P-Q & 0 & 0 \\ 0 & 0 & P-Q & \tilde{R} \\ 0 & 0 & \tilde{R}^{+} & P+Q \end{bmatrix} \begin{matrix} |1\rangle \\ |2\rangle \\ |3\rangle \\ |4\rangle \end{matrix} \tag{4.8.16}$$

where

$$\tilde{R} = |R| - i|S| \qquad \tilde{R}^{+} = |R| + i|S| \tag{4.8.17}$$

and the transformation between the old bases $\left|\frac{3}{2}, \nu\right\rangle \left(\nu = \pm\frac{3}{2}, \pm\frac{1}{2}\right)$ and the new bases is

$$\begin{aligned} |1\rangle &= \alpha\left|\frac{3}{2}, \frac{3}{2}\right\rangle - \alpha^{*}\left|\frac{3}{2}, -\frac{3}{2}\right\rangle \\ |2\rangle &= -\beta^{*}\left|\frac{3}{2}, \frac{1}{2}\right\rangle + \beta\left|\frac{3}{2}, -\frac{1}{2}\right\rangle \\ |3\rangle &= \beta^{*}\left|\frac{3}{2}, \frac{1}{2}\right\rangle + \beta\left|\frac{3}{2}, -\frac{1}{2}\right\rangle \\ |4\rangle &= \alpha\left|\frac{3}{2}, \frac{3}{2}\right\rangle + \alpha^{*}\left|\frac{3}{2}, -\frac{3}{2}\right\rangle \end{aligned} \tag{4.8.18}$$

where

$$\alpha = \frac{1}{\sqrt{2}} e^{i\left(\frac{\theta_S+\theta_R}{2} + \frac{\pi}{4}\right)} \tag{4.8.19a}$$

$$\beta = \frac{1}{\sqrt{2}} e^{i\left(\frac{\theta_S-\theta_R}{2} + \frac{\pi}{4}\right)} \tag{4.8.19b}$$

and the transformation matrix is

$$\mathbf{U} = \begin{bmatrix} \alpha^* & 0 & 0 & -\alpha \\ 0 & -\beta & \beta^* & 0 \\ 0 & \beta & \beta^* & 0 \\ \alpha^* & 0 & 0 & \alpha \end{bmatrix}. \tag{4.8.20}$$

For special cases such as an external stress $T \parallel [100]$, $[001]$, $[110]$, $\varepsilon_{xz} = \varepsilon_{yz} = 0$, θ_S is independent of k_z. Therefore, the Hamiltonian can be applied to the quantum-well problem directly; $k_z \to -i\partial/\partial z$ becomes an operator in (4.8.20), which is difficult to treat in general. When $\varepsilon_{yz} = \varepsilon_{xz} = 0$, k_z is canceled in (4.8.20) and both θ_R and θ_S are independent of k_z.

For the case in which an external stress T is applied along the [110] direction or the case in which strain is caused by the lattice-mismatch accommodated elastic strain

$$\varepsilon_{xx} = \varepsilon_{yy} \neq \varepsilon_{zz}, \quad \varepsilon_{xy} = \varepsilon_{xz} = \varepsilon_{yz} = 0 \tag{4.8.21}$$

we have

$$\theta_S = -\phi \tag{4.8.22}$$

where $\phi = \tan^{-1}(k_y/k_x)$.

4.8.5 Axial Approximation for the Luttinger–Kohn Hamiltonian

If we approximate in the R_k term

$$\begin{aligned} R_k &= -\frac{\hbar^2\sqrt{3}}{2m_0}\left[\frac{\gamma_2+\gamma_3}{2}(k_x - ik_y)^2 + \frac{\gamma_2-\gamma_3}{2}(k_x + ik_y)^2\right] \\ &\simeq -\frac{\hbar^2\sqrt{3}}{2m_0}\bar{\gamma}(k_x - ik_y)^2 \end{aligned} \tag{4.8.23}$$

where $\bar{\gamma} = (\gamma_2 + \gamma_3)/2$ and using

$$k_x - ik_y = k_t e^{-i\phi} \tag{4.8.24}$$

we find that the energy subband dispersion relation is independent of the angle ϕ because

$$\tilde{R} = |R| - i|S| = \frac{\hbar^2\sqrt{3}}{2m_0}\bar{\gamma}\, k_t^2 - i\frac{\hbar^2\gamma_3}{2m_0}2\sqrt{3}\, k_t k_z \tag{4.8.25}$$

and the Hamiltonian depends only on the magnitude of the vector $\mathbf{k}_t$. This is called the axial approximation. Note that in this approximation, we assume $\gamma_2 \simeq \gamma_3$ in the

R_k term only, whereas we still use γ_2 and γ_3 in the other terms. We obtain

$$\theta_R = \pi - 2\phi \tag{4.8.26a}$$

and

$$\alpha = \frac{1}{\sqrt{2}} e^{i\left(\frac{3\pi}{4} - \frac{3\phi}{2}\right)} \tag{4.8.26b}$$

$$\beta = \frac{1}{\sqrt{2}} e^{i\left(\frac{\phi}{2} - \frac{\pi}{4}\right)}. \tag{4.8.26c}$$

4.8.6 Numerical Approach for the Solutions of the Upper 2×2 Hamiltonian

Let us look at the upper 2×2 Hamiltonian in (4.8.16). The wave function for the hole subbands can be written generally as

$$\begin{aligned}\psi^U(\mathbf{k}_t, \mathbf{r}) &= \frac{e^{i\mathbf{k}_t \cdot \mathbf{r}_t}}{\sqrt{A}} \left[g^{(1)}(k_t, z)|1\rangle + g^{(2)}(k_t, z)|2\rangle\right] \\ &= \begin{bmatrix} g^{(1)}(k_t, z) \\ g^{(2)}(k_t, z) \end{bmatrix} \frac{e^{i\mathbf{k}_t \cdot \mathbf{r}_t}}{\sqrt{A}} \equiv \begin{bmatrix} F_1 \\ F_2 \end{bmatrix}.\end{aligned} \tag{4.8.27}$$

This wave function satisfies the Hamiltonian equation

$$-\begin{bmatrix} P + Q - V_h(z) & \tilde{R} \\ \tilde{R}^+ & P - Q - V_h(z) \end{bmatrix} \begin{bmatrix} g^{(1)}(k_t, z) \\ g^{(2)}(k_t, z) \end{bmatrix} = E(k_t) \begin{bmatrix} g^{(1)}(k_t, z) \\ g^{(2)}(k_t, z) \end{bmatrix} \tag{4.8.28}$$

where P, Q, and $\tilde{R}$ are all differential operators and can be obtained from (4.8.25) with k_z replaced by $-i\partial/\partial z$. These wave functions $g^{(1)}$ and $g^{(2)}$ depend on the magnitude of the wave vector k_t and position z and are independent of the direction of the wave vector (or the angle ϕ)

$$\begin{aligned} P &= P_\varepsilon + \frac{\hbar^2 \gamma_1}{2m_0}\left(k_t^2 - \frac{\partial^2}{\partial z^2}\right) \qquad Q = Q_\varepsilon + \frac{\hbar^2 \gamma_2}{2m_0}\left(k_t^2 + 2\frac{\partial^2}{\partial z^2}\right) \\ \tilde{R} &= \frac{\hbar^2 \sqrt{3}}{2m_0} \bar{\gamma} k_t^2 - \frac{\hbar^2 \gamma_3}{m_0} \sqrt{3}\, k_t \frac{\partial}{\partial z} \\ \tilde{R}^+ &= \frac{\hbar^2 \sqrt{3}}{2m_0} \bar{\gamma} k_t^2 + \frac{\hbar^2 \gamma_3}{m_0} \sqrt{3}\, k_t \frac{\partial}{\partial z}. \end{aligned} \tag{4.8.29}$$

The built-in quantum-well potential for the holes $V_h(z)$ has been incorporated into the diagonal terms of the Hamiltonian. Because $V_h(z)$ is a stepwise potential, boundary conditions between the well and the barrier interface have to be used properly.

The basic idea is that the envelope function and the probability current density across the heterojunction should be continuous. Therefore, to ensure the Hermitian property of H, we have to write all operators of the form

$$A(z)\frac{\partial^2}{\partial z^2} \quad \text{as} \quad \frac{\partial}{\partial z}A(z)\frac{\partial}{\partial z} \tag{4.8.30a}$$

and

$$B(z)\frac{\partial}{\partial z} \quad \text{as} \quad \frac{1}{2}\left[B(z)\frac{\partial}{\partial z}+\frac{\partial}{\partial z}B(z)\right]. \tag{4.8.30b}$$

Equation (4.8.28) can be solved using a propagation-matrix method [37], which is very efficient, or using a finite-difference method [76]. The solution to the Schrödinger equation in matrix form (4.8.28) will be a set of subband energies

$$E(\mathbf{k}_t) = E_m^{\mathrm{U}}(\mathbf{k}_t)$$

where the subband index m refers to HH1, HH2, ... and LH1, LH2, ... subbands. The superscript "U" refers to the upper Hamiltonian.

4.8.7 Solutions for the Lower 2×2 Hamiltonian

Similar procedures hold for the lower 2×2 Hamiltonian of (4.8.16). The wave function is

$$\begin{aligned}\psi^L(k_t, \mathbf{r}) &= \frac{e^{i\mathbf{k}_t\cdot\mathbf{r}_t}}{\sqrt{A}}\left[g^{(3)}(k_t, z)|3\rangle + g^{(4)}(k_t, z)|4\rangle\right]\\ &= \begin{bmatrix} g^{(3)}(k_t, z)\\ g^{(4)}(k_t, z)\end{bmatrix}\frac{e^{i\mathbf{k}_t\cdot\mathbf{r}_t}}{\sqrt{A}} = \begin{bmatrix} F_3\\ F_4\end{bmatrix}\end{aligned} \tag{4.8.31}$$

which satisfies the Hamiltonian equation

$$-\begin{bmatrix} P-Q-V_{\mathrm{h}}(z) & \tilde{R}\\ \tilde{R}^{+} & P+Q-V_{\mathrm{h}}(z)\end{bmatrix}\begin{bmatrix} g^{(3)}(k_t, z)\\ g^{(4)}(k_t, z)\end{bmatrix} = E^{\mathrm{L}}(k_t)\begin{bmatrix} g^{(3)}(k_t, z)\\ g^{(4)}(k_t, z)\end{bmatrix}. \tag{4.8.32}$$

The solution will be a set of dispersion curves labeled as m in $E_m^{\mathrm{L}}(k_t)$. For a symmetrical potential, we find that $E_m^{\mathrm{L}}(k_t)$ of the lower Hamiltonian is degenerate with that of the upper Hamiltonian $E_m^{\mathrm{U}}(k_t)$ for each subband m. The wave functions $g^{(3)}(k_t, z)$ and $g^{(4)}(k_t, z)$ can also be related to $g^{(2)}(k_t, z)$ and $g^{(1)}(k_t, z)$, respectively, if we change $z \to -z$.

Example The valence subband structures for a GaAs/$Al_{0.3}Ga_{0.7}As$ quantum well with $L_{\mathrm{w}} = 100$ Å and $L_{\mathrm{w}} = 50$ Å are plotted in Fig. 4.19a and Fig. 4.19b,

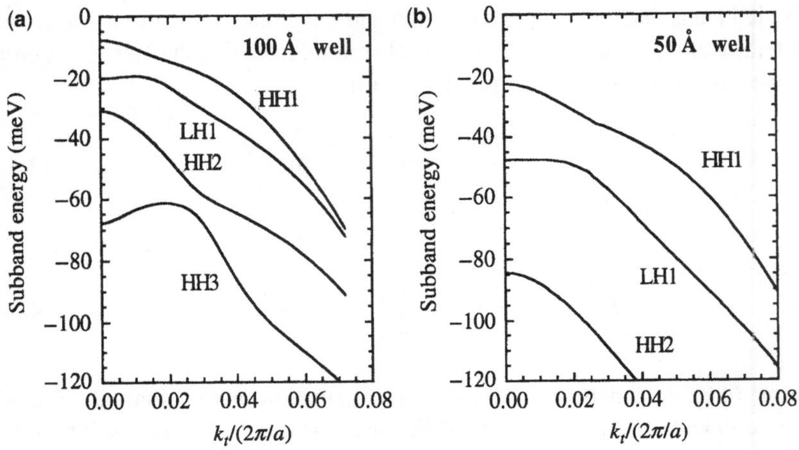

Figure 4.19 The valence subbands for (a) a 100 Å and (b) a 50 Å $GaAs/Al_{0.3}Ga_{0.7}As$ quantum well for k_t in the plane of the quantum well under the axial approximation. The vertical axis is the hole subband energy and the horizontal axis is the in-plane wave vector k_t normalized by $(2\pi/a)$, where $a = 5.6533$ Å is the lattice constant of GaAs. These results in (a) and (b) can be compared with the experimental data for similar GaAs quantum-well structures in Fig. 4.18d and Fig. 4.18b, respectively.

respectively. The horizontal wave vector k_t is normalized by $2\pi/a$, where $a = 5.6533$ Å is the lattice constant of GaAs. These subband dispersions can also be compared with the experimental data for similar GaAs quantum well structures with p-type dopings in Fig. 4.18d and Fig. 4.18b, respectively. The axial approximation gives very good results for a small k_t, and the results are exactly the same as those using the original 4×4 Hamiltonian at $k_t = 0$. This approximation is attractive because the ϕ dependence in the k_x–k_y plane can be taken into account analytically in the basis functions, and the valence subband energies are independent of ϕ. These wave functions can be applied to study optical absorption and gain in quantum-well structures.

4.9 BAND STRUCTURES OF STRAINED SEMICONDUCTOR QUANTUM WELLS

For a strained quantum well, the conduction-band edge is given by

$$E_c(k=0) = a_c \mathrm{Tr}(\bar{\bar{\varepsilon}}) = a_c(\varepsilon_{xx} + \varepsilon_{yy} + \varepsilon_{zz}) \tag{4.9.1}$$

where $\varepsilon_{xx} = \varepsilon_{yy} = \dfrac{a_0 - a}{a}$, and $\varepsilon_{zz} = -2\dfrac{C_{12}}{C_{11}}\varepsilon_{xx}$. Here a_0 is the lattice constant of the substrate, and a is the lattice constant of the quantum well. The energy is measured from the conduction-band edge of an unstrained quantum well, as shown in Fig. 4.20. We assume that the barriers are lattice matched to the substrate; therefore, their strains

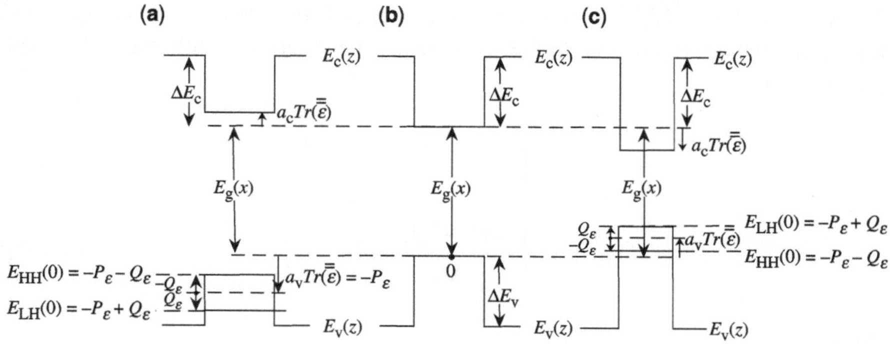

Figure 4.20 Band-edge profile for (a) a compressive strained, (b) an unstrained, and (c) a tensile strained quantum well.

are zero. For an unstrained quantum well, we have the potential energy for the electron given by

$$E_c^0(z) = \begin{cases} 0 & |z| \leq L_w/2 \\ \Delta E_c & |z| > L_w/2. \end{cases} \tag{4.9.2}$$

For example, $E_g\ (In_{1-x}Ga_xAs) = 0.324 + 0.7x + 0.4x^2$ for an unstrained $In_{1-x}Ga_xAs$ compound, and $E_g(InP) = 1.344$ eV at room temperature

$$\begin{aligned} \Delta E_g &= 1.344 - E_g(In_{1-x}Ga_xAs) \\ a(x) &= xa(GaAs) + (1-x)a(InAs). \end{aligned} \tag{4.9.3}$$

From previous experimental data, we take

$$\begin{aligned} \Delta E_c &= 0.4\Delta E_g(x) \\ \Delta E_v &= 0.6\Delta E_g(x). \end{aligned} \tag{4.9.4}$$

At $x = 0.468$, $In_{1-x}Ga_xAs$ is lattice matched to the InP substrate. Strain exists when $x \neq 0.468$.

4.9.1 Subband Energies in a Strained Quantum Well

The conduction band edge energy for a strained quantum well is given by

$$E_c(z) = \begin{cases} a_c(\varepsilon_{xx} + \varepsilon_{yy} + \varepsilon_{zz}) & |z| \leq L_w/2 \\ \Delta E_c & |z| > L_w/2. \end{cases} \tag{4.9.5}$$

Similarly, the valence band edge energy for the unstrained quantum well is defined as

$$V_h(z) = E_v^0(z) = \begin{cases} 0 & |z| \leq L_w/2 \\ -\Delta E_v & |z| > L_w/2. \end{cases} \tag{4.9.6}$$

The valence band edge energies for the strained quantum well are obtained from $E_{\mathrm{HH}}(0)$ and $E_{\mathrm{LH}}(0)$ in (4.5.36) ignoring the coupling with the spin–orbit split-off band

$$E_{\mathrm{HH}}(z) = \begin{cases} -P_\varepsilon - Q_\varepsilon & |z| \leq L_{\mathrm{w}}/2 \\ -\Delta E_{\mathrm{v}} & |z| > L_{\mathrm{w}}/2 \end{cases} \tag{4.9.7a}$$

$$E_{\mathrm{LH}}(z) = \begin{cases} -P_\varepsilon + Q_\varepsilon & |z| \leq L_{\mathrm{w}}/2 \\ -\Delta E_{\mathrm{v}} & |z| > L_{\mathrm{w}}/2 \end{cases} \tag{4.9.7b}$$

where

$$P_\varepsilon = -a_{\mathrm{v}}(\varepsilon_{xx} + \varepsilon_{yy} + \varepsilon_{zz}) \tag{4.9.8a}$$

$$Q_\varepsilon = \frac{-b}{2}(\varepsilon_{xx} + \varepsilon_{yy} - 2\varepsilon_{zz}). \tag{4.9.8b}$$

The subband edge energies can all be obtained from a simple quantum-well model using the band-edge effective masses

$$m_{\mathrm{e}}^* = \begin{cases} m_{\mathrm{w}}^* & |z| \leq L_{\mathrm{w}}/2 \\ m_{\mathrm{b}}^* & |z| > L_{\mathrm{w}}/2 \end{cases} \tag{4.9.9a}$$

$$m_{\mathrm{hh}}^z = \begin{cases} m_0/(\gamma_{1\mathrm{w}} - 2\gamma_{2\mathrm{w}}) & |z| \leq L_{\mathrm{w}}/2 \\ m_0/(\gamma_{1\mathrm{b}} - 2\gamma_{2\mathrm{b}}) & |z| > L_{\mathrm{w}}/2 \end{cases} \tag{4.9.9b}$$

$$m_{\ell\mathrm{h}}^z = \begin{cases} m_0/(\gamma_{1\mathrm{w}} + 2\gamma_{2\mathrm{w}}) & |z| \leq L_{\mathrm{w}}/2 \\ m_0/(\gamma_{1\mathrm{b}} + 2\gamma_{2\mathrm{b}}) & |z| > L_{\mathrm{w}}/2 \end{cases} \tag{4.9.9c}$$

where the subscripts "w" and "b" in the effective mass parameters designate the well and the barrier region, respectively.

If the spin–orbit split-off band coupling effect is included, we find that the band-edge energies for heavy hole $E_{\mathrm{HH}}(z)$ and m_{hh}^z are not affected, but

$$E_{\mathrm{LH}}(z) = \begin{cases} -P_\varepsilon + \frac{1}{2}\left(Q_\varepsilon - \Delta + \sqrt{\Delta^2 + 2\Delta Q_\varepsilon + 9Q_\varepsilon^2}\right) & |z| \leq L_{\mathrm{w}}/2 \\ -\Delta E_{\mathrm{v}} & |z| > L_{\mathrm{w}}/2 \end{cases} \tag{4.9.10}$$

and

$$m_{\ell\mathrm{h}}^z = \begin{cases} m_0/(\gamma_{1\mathrm{w}} + 2f_+\gamma_{2\mathrm{w}}) & |z| \leq L_{\mathrm{w}}/2 \\ m_0/(\gamma_{1\mathrm{b}} + 2\gamma_{2\mathrm{b}}) & |z| > L_{\mathrm{w}}/2 \end{cases} \tag{4.9.11}$$

where f_+ is defined in Ref. 38.

4.9.2 Valence Subband Energy Dispersions in a Strained Quantum Well

Using **H** from (4.5.13), we can solve the valence subband energies and eigenfunctions using (4.8.14)

$$\left[\mathbf{H}\left(k_z = -i\frac{\partial}{\partial z}\right) + V_h(z)\,\mathbf{I}\right] \cdot \begin{bmatrix} g_{3/2}(\mathbf{k}_t, z) \\ g_{1/2}(\mathbf{k}_t, z) \\ g_{-1/2}(\mathbf{k}_t, z) \\ g_{-3/2}(\mathbf{k}_t, z) \end{bmatrix} = E(\mathbf{k}_t) \begin{bmatrix} g_{3/2}(\mathbf{k}_t, z) \\ g_{1/2}(\mathbf{k}_t, z) \\ g_{-1/2}(\mathbf{k}_t, z) \\ g_{-3/2}(\mathbf{k}_t, z) \end{bmatrix} \quad (4.9.12)$$

where $V_h(z)$ is given by the unstrained (built-in) potential $E_v^0(z)$ from (4.9.6), and the band edge shifts due to $-P_\varepsilon$ and $\pm Q_\varepsilon$ have all been included in **H** for $|z| \le L_w/2$, and both P_ε and Q_ε are zero for $|z| > L_w/2$.

Example: An $In_{1-x}Ga_xAs/In_{1-x}Ga_xAs_yP_{1-y}$ Quantum Well Again, a convenient way to find the valence subband energies and eigenfunctions is to use the block-diagonalized Hamiltonians (4.8.28) and (4.8.32), which simplify the numerical calculations by a significant amount. As a numerical example [37], we plot in Fig. 4.21 the energy dispersion curves for the valence subbands of an $In_{1-x}Ga_xAs$

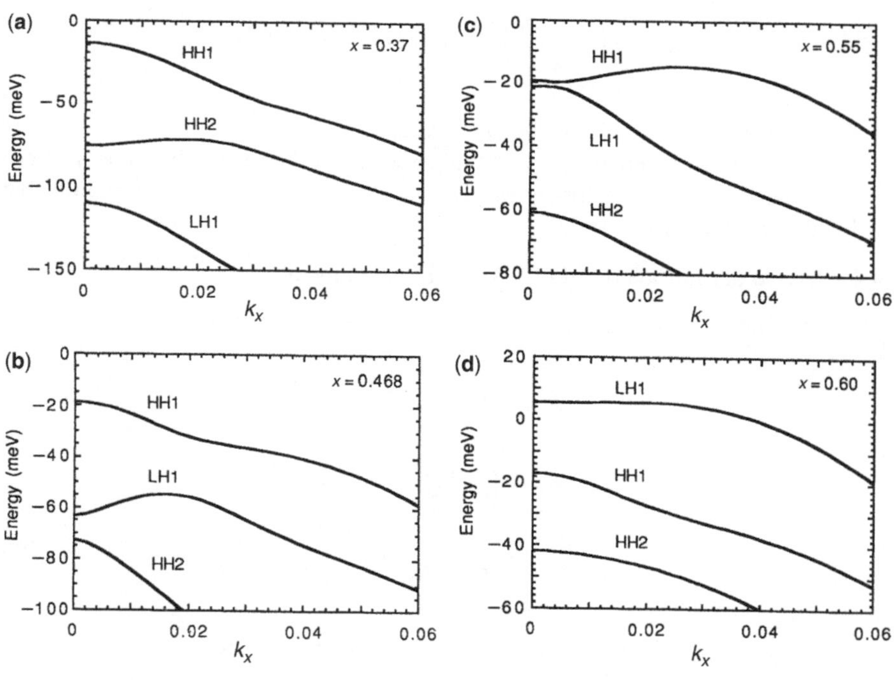

Figure 4.21 The valence subband dispersions of an $In_{1-x}Ga_xAs$ grown on $In_{1-x}Ga_xAs_yP_{1-y}$ quaternary barriers (with a band gap of 1.3 μm) lattice matched to InP substrate. (a) $x = 0.37$ (biaxial compression), (b) $x = 0.468$ (unstrained), (c) $x = 0.55$ (small tension), and (d) $x = 0.60$ (large tension). Here the wave vector along [100] direction k_x is normalized by $2\pi/a$, where $a = 5.6533$ Å is the lattice constant of GaAs. (Reprinted with permission from [37] © 1991 American Physical Society.)

quantum well grown on an $In_{1-x}Ga_xAs_yP_{1-y}$ substrate with a band-gap wavelength 1.3 μm lattice matched to an InP substrate. The gallium mode fraction x is varied: (a) $x = 0.37$ (biaxial compression), (b) $x = 0.468$ (unstrained), (c) $x = 0.55$ (small tension), and (d) $x = 0.60$ (large tension). These band structures will be useful to investigate the strain effects on the effective mass, the density of states, optical gain, and absorption in quantum wells. We will discuss these effects and their applications further in Chapters 9 and 10.

PROBLEMS

4.1 Derive Eqs. (4.1.6) and (4.1.7) for the wave function $u_{n\mathbf{k}}(\mathbf{r})$ in the $\mathbf{k} \cdot \mathbf{p}$ theory using (4.1.4) for $\psi_{n\mathbf{k}}(\mathbf{r})$.

4.2 Derive Eq. (4.1.8) using the perturbation theory.

4.3 Consider the two-band model, one is the conduction band ($n = c$) and the other is the heavy hole in the valence band ($n' = \mathrm{v}$). We have $\mathbf{k}_0 = \mathbf{0}$ for a direct band-gap material, $E_n = E_c$, and $E_{n'} = E_v$. Show that Eq. (4.1.16) gives two eigenvalues

$$E = \frac{1}{2}\left(E_g + \frac{\hbar^2 k^2}{m_0}\right) \pm \frac{1}{2}\left[E_g^2 + 4\frac{\hbar^2}{m_0^2}|\mathbf{k} \cdot \mathbf{p}_{cv}|^2\right]^{1/2}$$

where $E_v = 0$, $E_c = E_g =$ the energy gap.

Check if the energies using the perturbation theory are the same as those obtained by the Taylor-series expansion of the expressions assuming that the $(\mathbf{k} \cdot \mathbf{p}_{cv})$ term is small.

4.4 Derive the full matrix in (4.2.8) for Kane's model.

4.5 **(a)** Show that the eigenvector for (4.2.19) can be written as

$$a_n = kP\left(E_n' + \frac{2\Delta}{3}\right) \Big/ N$$

$$b_n = \left(\sqrt{2}\frac{\Delta}{3}\right)(E_n' - E_g) \Big/ N$$

$$c_n = (E_n' - E_g)\left(E_n' + \frac{2\Delta}{3}\right) \Big/ N$$

where N is obtained from the normalization condition

$$a_n^2 + b_n^2 + c_n^2 = 1.$$

(b) Show that in the limit $k = 0$, the above eigenvectors lead to Eqs. (4.2.21)–(4.2.24).

4.6 Derive Eq. (4.2.26) for Kane's parameter P.

4.7 Tabulate all the band-edge energies and their corresponding wave functions for the conduction, heavy-hole, light-hole, and spin–orbit split-off bands.

4.8 Derive the Hamiltonian Eq. (4.3.1a)–(4.3.1e) for $u_{\mathbf{k}}(\mathbf{r})$ using (4.2.5).

4.9 Using the definition for $D_{jj'}^{\alpha\beta}$ in (4.3.11)–(4.3.13), show that the matrix representation of the Luttinger–Kohn Hamiltonian H^{LK} is given by Eqs. (4.3.14) and (4.3.15).

4.10 Summarize the effects of the strain on the conduction-band edge and the valence-band edges from the results in Section 4.5.

4.11 Calculate the strains (ε_{xx}, ε_{yy}, ε_{zz}), P_ε, Q_ε, and the band-edge shifts for $\text{In}_{1-x}\text{Ga}_x\text{As}$ quantum-well layers grown in InP substrate with **(a)** $x = 0.37$ and **(b)** $x = 0.57$.

4.12 Calculate the band-edge effective masses using (4.5.33) for Problem 4.11. Compare the values of the structure of part **(a)** with those in part **(b)**.

4.13 Show that the eigenvalues for the matrix $H(\mathbf{k} = 0)$ in (4.5.34) are given by (4.5.36a)–(4.5.36c).

4.14 Calculate the band-edge energies with spin–orbit coupling effects using (4.5.36a)–(4.5.36c) and compare with those using (4.5.19) without the spin–orbit coupling effects for the $\text{In}_{1-x}\text{Ga}_x\text{As/InP}$ system with (a) $x = 0.37$, (b) $x = 0.47$, and (c) $x = 0.57$.

4.15 Show that the eigenvalues of the matrix (4.5.13) are given by Eqs. (4.5.27a) and (4.5.27b).

4.16 Derive the propagation matrix (4.6.7) from the boundary conditions.

4.17 Discuss in the infinite barrier model (i.e., $V_0 = \infty$ and $V_{N+1} = \infty$), **(a)** how to modify the boundary conditions in the propagation matrix method and **(b)** how to find the eigenenergy E and its corresponding wave function.

4.18 Summarize the procedures for solving **(a)** an n-type and **(b)** a p-type modulation-doped quantum well self-consistently.

4.19 Discuss qualitatively how Fig. 4.13 should be modified if the donor profile is within 20 Å at the center of the quantum well, that is, $N_\text{D}(z) = N_\text{D}$ for $|z| \leq 20$ Å and $N_\text{D}(z) = 0$, otherwise.

4.20 Derive Eqs. (4.7.23a) and (4.7.23b) in the Kronig–Penney model.

4.21 **(a)** Discuss the procedure for determining the number of bands that can be considered as "bound states" of the superlattice from Fig. 4.15.

(b) What happens if the barrier height V_0 falls within a miniband of Fig. 4.15?

4.22 Compare the miniband energies of the $GaAs/Al_{0.3}Ga_{0.7}As$ superlattice in Fig. 4.16 with those calculated for a single isolated quantum well with the same well width $w = 100$ Å from Chapter 3.

4.23 **(a)** Using the band-edge discontinuity rules discussed in Appendix A at the end of the text, design an $In_{1-x}Ga_xAs/InP$ heterojunction such that $\Delta E_c = 0.3$ eV. Assume that strain only occurs in the $In_{1-x}Ga_xAs$ layer because the substrate InP is thick.

(b) Calculate ΔE_v, $E_g(In_{1-x}Ga_xAs)$, P_ε, Q_ε, and so forth, for part **(a)**.

4.24 Comment on the results from the subband energies in Figs. 4.17b and 4.17c.

4.25 Check the subband energies at the band edges ($k_t = 0$) in Fig. 4.18. Comment on the discrepancies if there are any.

4.26 Derive the block-diagonalized Hamiltonian (4.8.16) using the axial approximation and the basis transformation (4.8.18).

4.27 **(a)** Find the eigenvalues and the corresponding eigenvectors for the 2×2 Hamiltonian $\mathbf{H}^U$ in (4.8.16) in the plane-wave representation.

$$\mathbf{H}^{\mathrm{U}}\begin{bmatrix} g^{(1)} \\ g^{(2)} \end{bmatrix} = E\begin{bmatrix} g^{(1)} \\ g^{(2)} \end{bmatrix}, \qquad \mathbf{H}^{\mathrm{U}} = -\begin{bmatrix} P+Q & \tilde{R} \\ \tilde{R}^+ & P-Q \end{bmatrix}$$

(b) Repeat **(a)** for

$$\mathbf{H}^{\mathrm{L}} = -\begin{bmatrix} P-Q & \tilde{R} \\ \tilde{R}^+ & P+Q \end{bmatrix}.$$

4.28 **(a)** Express the wave function $\Psi^U(\mathbf{k}_t, \mathbf{r})$ in (4.8.27), which is in the bases $|1\rangle$ and $|2\rangle$, in the original basis $|\frac{3}{2}, \frac{3}{2}\rangle$, $|\frac{3}{2}, \frac{1}{2}\rangle$, $|\frac{3}{2}, -\frac{1}{2}\rangle$ and $|\frac{3}{2}, -\frac{3}{2}\rangle$.

(b) Express the bases $|\frac{3}{2}, \nu\rangle$ ($\nu = 3/2,\ 1/2,\ -1/2,\ -3/2$) in terms of the bases $|1\rangle$, $|2\rangle$, $|3\rangle$, and $|4\rangle$.

4.29 Label all the energies in Fig. 4.20a–c for $In_{1-x}Ga_xAs/InP$ quantum wells with **(a)** $x = 0.37$, **(b)** $x = 0.47$, and **(c)** $x = 0.57$.

4.30 Calculate the first quantized subband energies for the electron, heavy-hole, and light-hole subbands in Problem 4.29.

REFERENCES

1. N. W. Ashcroft and N. D. Mermin, *Solid State Physics*, Holt, Rinehart and Winston, New York, 1976.
2. F. Bassani and G. Pastori Parravicini, *Electronic States and Optical Transitions in Solids*, Pergamon Press, Oxford, 1975.

3. J. Bardeen, "An improved calculation of the energies of metallic Li and Na," *J. Chem. Phys.* **6**, 367–371 (1938).

4. F. Seitz, *The Modern Theory of Solids*, McGraw-Hill, New York, 1940, p. 352.

5. W. Shockley, "Energy band structures in semiconductors," *Phys. Rev.* **78**, 173–174 (1950).

6. G. Dresselhaus, A. F. Kip, and C. Kittel, "Cyclotron resonance of electrons and holes in silicon and germanium crystals," *Phys. Rev.* **98**, 368–384 (1955).

7. E. O. Kane, "Band structure of indium antimonide," *J. Phys. Chem. Solids* **1**, 249–261 (1957).

8. E. O. Kane, "The **k·p** method," Chapter 3, Vol. 1 in *Semiconductors and Semimetals*, edited by R. K. Willardson and A. C. Beer, Academic Press, New York, 1966.

9. J. M. Luttinger and W. Kohn, "Motion of electrons and holes in perturbed periodic fields," *Phys. Rev.* **97**, 869–883 (1955).

10. J. M. Luttinger, "Quantum theory of cyclotron resonance in semiconductors: General theory," *Phys. Rev.* **102**, 1030–1041 (1956).

11. I. M. Tsidilkovski, *Band Structure of Semiconductors*, Pergamon Press, Oxford, 1982.

12. B. R. Nag, *Electron Transport in Compound Semiconductors*, Springer-Verlag, Berlin, 1980.

13. C. R. Pidgeon, "Free carrier optical properties of semiconductors," Chapter 5, Vol. 2 in *Handbook on Semiconductors*, edited by M. Balkanski, North-Holland, Amsterdam, New York, 1980.

14. For some recent discussions on the effective-mass theory, see M. G. Burt, "The justification for applying the effective-mass approximation to microstructures," *J. Phys. Condens. Matter* **4**, 6651–6690 (1992); M. G. Burt, "An exact formulation of the envelope function method for the determination of electronic states in semiconductor microstructures," *Semicond. Sci. Technol.* **3**, 739–753 (1988); M. G. Burt, "The evaluation of the matrix element for interband optical transitions in quantum wells using envelope functions," *J. Phys. Condens. Matter* **5**, 4091–4098 (1993).

15. J. R. Chelikowsky and M. L. Cohen, "Nonlocal pseudopotential calculations for the electronic structure of eleven diamond and zinc-blende semiconductors," *Phys. Rev. B* **14**, 556–582 (1976).

16. G. C. Osbourn, "$In_xGa_{1-x}As$-$In_yGa_{1-y}As$ strained-layer superlattices: A proposal for useful, new electronic materials," *Phys. Rev. B* **27**, 5126–5128 (1983).

17. G. C. Osbourn, "InAsSb strained-layer superlattices for long wavelength detector applications," *J. Vac. Sci. Technol.* **B2**, 176–178 (1984).

18. E. Yablonovitch, "Band structure engineering of semiconductor lasers for optical communications," *J. Lightwave Technol.* **6**, 1292–1299 (1988).

19. T. P. Pearsall, Vol. Ed., "Strained-layer superlattices: Physics," Vol. 32 in R. K. Willardson and A. C. Beer, Eds., *Semiconductors and Semimetals*, Academic Press, New York, 1990.

20. P. S. Zory, *Quantum Well Lasers*, Academic Press, New York, 1993.

21. G. E. Pikus and G. L. Bir, "Effects of deformation on the hole energy spectrum of germanium and silicon," *Sov. Phys. Solid State* **1**, 1502–1517 (1960).

22. G. L. Bir and G. E. Pikus, *Symmetry and Strain-Induced Effects in Semiconductors*, Wiley, New York, 1974.

23. R. P. Feynman, R. B. Leighton, and M. Sands, *The Feynman Lectures on Physics*, Chapter 39, Addison-Wesley, Reading, Massachusetts, 1964.
24. J. F. Nye, *Physical Properties of Crystals*, Oxford University Press, Oxford, 1985.
25. K. Seeger, *Semiconductor Physics*, Springer-Verlag, Berlin, 1982.
26. J. C. Hensel and G. Feher, "Cyclotron resonance experiments in uniaxially stressed silicon: Valence band inverse mass parameters and deformation potentials," *Phys. Rev.* **129**, 1041–1062 (1963).
27. H. Hasegawa, "Theory of cyclotron resonance in strained silicon crystals," *Phys. Rev.* **129**, 1029–1040 (1963).
28. A. Gavini and M. Cardona, "Modulated piezoreflectance in semiconductors," *Phys. Rev. B* **1**, 672–682 (1970).
29. L. D. Laude, F. H. Pollak, and M. Cardona, "Effects of uniaxial stress on the indirect exciton spectrum of silicon," *Phys. Rev. B* **3**, 2623–2636 (1971).
30. K. Suzuki and J. C. Hensel, "Quantum resonances in the valence bands of germanium. I. Theoretical considerations," *Phys. Rev. B* **9**, 4184–4218 (1974).
31. J. C. Hensel and K. Suzuki, "Quantum resonances in the valence bands of germanium. II. Cyclotron resonances in uniaxially stressed crystals," *Phys. Rev. B* **9**, 4219–4257 (1974).
32. M. Chandrasekhar and F. H. Pollak, "Effects of uniaxial stress on the electroreflectance spectrum of Ge and GaAs," *Phys. Rev. B* **15**, 2127–2144 (1977).
33. D. E. Aspnes and M. Cardona, "Strain dependence of effective masses in tetrahedral semiconductors," *Phys. Rev. B* **17**, 726–740 (1978).
34. H. Mathieu, P. Merle, E. I. Ameziane, B. Archilla, J. Camassel, and G. Poiblaud, "Deformation potentials of the direct and indirect absorption edges of GaP," *Phys. Rev. B* **19**, 2209–2223 (1979).
35. D. L. Smith and C. Mailhiot, "Theory of semiconductor superlattice electronic structure," *Rev. Mod. Phys.* **62**, 173–234 (1990).
36. D. L. Smith and C. Mailhiot, "Strained-layer semiconductor superlattices," *Crit. Rev. Solid State Mater. Sci.* **16**, 131–160 (1990).
37. S. L. Chuang, "Efficient band-structure calculations of strained quantum wells using a two-by-two Hamiltonian," *Phys. Rev. B* **43**, 9649–9661 (1991). (Note that a_v in this paper should be taken as $-a_v$ to compare with data in the literature.)
38. C. Y. P. Chao and S. L. Chuang, "Spin-orbit-coupling effects on the valence-band structure of strained semiconductor quantum wells," *Phys. Rev. B* **46**, 4110–4122 (1992).
39. J. A. Kong, *Electromagnetic Wave Theory*, Wiley, New York, 1990.
40. P. Yeh, *Optical Waves in Layered Media*, Wiley, New York, 1988.
41. E. J. Roan and S. L. Chuang, "Linear and nonlinear intersubband electroabsorptions in a modulation-doped quantum well," *J. Appl. Phys.* **69**, 3249–3260 (1991).
42. K. T. Kim, S. S. Lee, and S. L. Chuang, "Inter-miniband optical absorption in a modulation-doped $Al_xGa_{1-x}As/GaAs$ superlattice," *J. Appl. Phys.* **69**, 6617–6624 (1991).
43. F. Stern and S. Das Sarma, "Electron energy levels in GaAs-$Ga_{1-x}Al_xAs$ heterojunctions," *Phys. Rev. B* **30**, 840–848 (1984).
44. E. Merzbacher, *Quantum Mechanics*, 2nd ed., Wiley, New York, 1970.
45. C. Kittel, *Introduction to Solid State Physics*, 8th ed., Wiley, New York, 2005.

46. G. Bastard, "Superlattice band structure in the envelope-function approximation," *Phys. Rev. B* **24**, 5693–5697 (1981).

47. G. Bastard and J. A. Brum, "Electronic states in semiconductor heterostructures," *IEEE J. Quantum Electronics* **QE-22**, 1625–1644 (1986).

48. G. Bastard, *Wave Mechanics Applied to Semiconductor Heterostructures*, Halsted Press, Wiley, New York, 1988.

49. K. H. Yoo, L. R. Ram-Mohan, and D. F. Nelson, "Effect of nonparabolicity in GaAs/$Ga_{1-x}Al_xAs$ semiconductor quantum wells," *Phys. Rev. B* **39**, 12808–12813 (1989).

50. D. Mukherji and B. R. Nag, "Band structure of semiconductor superlattices," *Phys. Rev. B* **12**, 4338–4345 (1975).

51. C. Wetzel, B. K. Meyer, and P. Omling, "Electron effective mass in direct-band-gap $GaAs_{1-x}P_x$ alloys," *Phys. Rev. B* **47**, 15588–15592 (1993).

52. C. Hermann and C. Weisbuch, "**k·p** perturbation theory in III-V compounds and alloys: a reexamination," *Phys. Rev. B* **15**, 823–833 (1977).

53. W. Zawadzki, P. Pfeffer, and H. Sigg, "Five-level **k·p** model for conduction electrons in GaAs. Description of cyclotron resonance experiments," *Solid State Commun.* **53**, 777–782 (1985).

54. D. A. Broido and L. J. Sham, "Effective masses of holes at GaAs-AlGaAs heterojunctions," *Phys. Rev. B* **31**, 888–892 (1985); and **31**, 6831 (1985).

55. A. Twardowski and C. Hermann, "Variational calculation of polarization of quantum-well photoluminescence," *Phys. Rev. B* **35**, 8144–8153 (1987).

56. R. Eppenga, M. F. H. Schuurmans, and S. Colak, "New **k·p** theory for GaAs/$Ga_{1-x}Al_x$ As-type quantum wells," *Phys. Rev. B* **36**, 1554–1564 (1987).

57. M. Altarelli, "Band structure, impurities and excitons in superlattices," pp. 12–37, in *Proceedings of Les Houches Winterschool Semiconductor Superlattices and Heterojunctions*, edited by G. Allan, G. Bastard, N. Boccara, M. Lannoo, and M. Voos, Springer-Verlag, Berlin, 1986.

58. D. L. Smith and C. Mailhiot, "**k·p** theory of semiconductor superlattice electronic structure. I. Formal results," *Phys. Rev. B* **33**, 8345–8359 (1986).

59. C. Mailhiot and D. L. Smith, "**k·p** theory of semiconductor superlattice electronic structure. II. Application to $Ga_{1-x}In_xAs$-$Al_{1-y}In_yAs$ [100] superlattices," *Phys. Rev. B* **33**, 8360–8372 (1986).

60. L. C. Andreani, A. Pasquarello, and F. Bassani, "Hole subbands in strained GaAs-$Ga_{1-x}Al_xAs$ quantum wells: Exact solution of the effective-mass equation," *Phys. Rev. B* **36**, 5887–5894 (1987).

61. L. R. Ram-Mohan, K. H. Yoo, and R. L. Aggarwal, "Transfer-matrix algorithm for the calculation of the band structure of semiconductor superlattices," *Phys. Rev. B* **38**, 6151–6159 (1988).

62. G. Y. Wu, T. C. McGill, C. Mailhiot, and D. L. Smith, "**k·p** theory of semiconductor superlattice electronic structure in an applied magnetic field," *Phys. Rev. B* **39**, 6060–6070 (1989).

63. B. K. Ridley, "The in-plane effective mass in strained-layer quantum wells," *J. Appl. Phys.* **68**, 4667–4673 (1990).

64. I. Suemune, "Band-edge hole mass in strained-quantum-well structures," *Phys. Rev. B* **43**, 14099–14106 (1991).

65. M. Silver, W. Batty, A. Ghiti, and E. P. O'Reilly, "Strain-induced valence-subband splitting in III-V semiconductors," *Phys. Rev. B* **46**, 6781–6788 (1992).

66. T. Manku and A. Nathan, "Valence energy-band structure for strained group-IV semiconductors," *J. Appl. Phys.* **73**, 1205–1213 (1993).

67. D. Munzar, "Heavy hole-light hole splitting in GaAs/AlAs superlattices," *Phys. Stat. Sol.* (b) **175**, 395–401 (1993).

68. J. N. Schulman and Y. C. Chang, "Band mixing in semiconductor superlattices," *Phys. Rev. B* **31**, 2056–2068 (1985).

69. Y. C. Chang, "Bond-orbital models for superlattices," *Phys. Rev. B* **37**, 8215–8222 (1988).

70. M. P. Houng, Y. C. Chang, and W. I. Wang, "Orientation dependence of valence-subband structures in GaAs-$Ga_{1-x}Al_xAs$ quantum-well structures," *J. Appl. Phys.* **64**, 4609–4613 (1988).

71. S. S. Nedorezov, "Space quantization in semiconductor films," *Soviet Phys. Solid State* **12**, 1814–1819 (1971).

72. E. D. Jones, S. K. Lyo, I. J. Fritz, J. F. Klem, J. E. Schirber, C. P. Tigges, and T. J. Drummond, "Determination of energy-band dispersion curves in strained-layer structures," *Appl. Phys. Lett.* **54**, 2227–2229 (1989).

73. R. K. Hayden, L. Eaves, M. Henini, T. Takamasu, N. Miura, and U. Ekenberg, "Investigating the cubic anisotropy of the confined hole subbands of an AlAs/GaAs/AlAs quantum well using resonant magnetotunneling spectroscopy," *Appl. Phys. Lett.* **61**, 84–86 (1992).

74. J. A. Kash, M. Zachau, and M. A. Tischer, "Anisotropic valence bands in quantum wells: Quantitative comparison of theory and experiment," *Phys. Rev. Lett.* **69**, 2260–2263 (1992).

75. J. A. Kash, M. Zachau, M. A. Tischlet, and U. Ekenberg, "Optical measurements of warped valence bands in quantum wells," *Surf. Sci.* **305**, 251–255 (1994).

76. D. Ahn, S. L. Chuang, and Y. C. Chang, "Valence-band mixing effects on the gain and the refractive index change of quantum-well lasers," *J. Appl. Phys.* **64**, 4056–4064 (1988).

Part II
Propagation of Light

5

Electromagnetics and Light Propagation

In this chapter, we discuss a few important results using electromagnetic theory. First, the time-harmonic fields and the duality principle, which will be useful in studying the propagation of light in waveguides and laser cavities, will be presented in Section 5.1. The duality principle allows us to obtain the solutions for plane wave reflection and guidance for the transverse magnetic (TM) polarization of light once we obtain the solutions for the transverse electric (TE) polarization. We then present Poynting's theorem and the reciprocity theorem in Section 5.2, followed by plane wave solutions in homogenous media in Section 5.3. We then consider the simplest case, isotropic media, in Section 5.4. The complex permittivity functions for resonant dielectric media and conducting media are presented in Section 5.5. Plane wave reflection from a planar surface and a multilayered medium is then investigated in Sections 5.6 to 5.8, which include matrix optics, a convenient tool to investigate forward and backward propagation plane waves. An important case of plane wave reflection from a distributed Bragg reflector (DBR) is presented in Section 5.9. DBR structures provide extremely high reflectivity and are used in vertical-cavity surface-emitting lasers (VCSELs) as well as microcavity lasers. In Appendix 5A, we discuss the Kramers–Kronig relations, which relate the real part and imaginary part of a permittivity function to each other. These relations are useful in optical materials because if the absorption coefficients of the semiconductors are measured, the real parts of the refractive indices are calculated based on the Kramers–Kronig relation or vice versa.

5.1 TIME-HARMONIC FIELDS AND DUALITY PRINCIPLE

5.1.1 Time-Harmonic Fields

Very often, the excitation sources described by real functions $\mathbf{J}(\mathbf{r}, t)$ and $\rho(\mathbf{r}, t)$ depend on time sinusoidally. We may write the source terms as

$$\mathbf{J}(\mathbf{r}, t) = \mathrm{Re}\,[\mathbf{J}(\mathbf{r}, \omega)\, e^{-i\omega t}] \tag{5.1.1}$$

$$\rho(\mathbf{r}, t) = \mathrm{Re}\,[\rho(\mathbf{r}, \omega)\, e^{-i\omega t}] \tag{5.1.2}$$

Physics of Photonic Devices, Second Edition. By Shun Lien Chuang
Copyright © 2009 John Wiley & Sons, Inc.

where $\mathbf{J}(\mathbf{r}, \omega)$ and $\rho(\mathbf{r}, \omega)$ are called phasors, which are complex quantities. Similar expressions apply for the fields **E**, **H**, **B**, and **D**, where Re[ ] means the real part of the function in the bracket. For example, an electric field in the time domain of the form

$$\mathbf{E}(\mathbf{r}, t) = \hat{x}E_x\cot(\omega t - \phi_x) + \hat{y}E_y\cos(\omega t - \phi_y) + \hat{z}E_z\cos(\omega t - \phi_z) \tag{5.1.3a}$$

can be expressed in terms of the phasor $\mathbf{E}(\mathbf{r}, \omega)$ or simply $\mathbf{E}(\mathbf{r})$

$$\mathbf{E}(\mathbf{r}, t) = \mathrm{Re}[\mathbf{E}(\mathbf{r}, \omega)e^{-i\omega t}] \tag{5.1.3b}$$

$$\mathbf{E}(\mathbf{r}, \omega) = \hat{x}E_x e^{i\phi_x} + \hat{y}E_y e^{i\phi_y} + \hat{z}E_z e^{i\phi_z}. \tag{5.1.3c}$$

We then obtain Maxwell's equations in the frequency domain, noting $\partial/\partial t \to -i\omega$,

$$\nabla \times \mathbf{E}(\mathbf{r}) = i\omega\mathbf{B}(\mathbf{r}) \tag{5.1.4a}$$
$$\nabla \times \mathbf{H}(\mathbf{r}) = \mathbf{J}(\mathbf{r}) - i\omega\mathbf{D}(\mathbf{r}) \tag{5.1.4b}$$
$$\nabla \cdot \mathbf{D}(\mathbf{r}) = \rho(\mathbf{r}) \tag{5.1.4c}$$
$$\nabla \cdot \mathbf{B}(\mathbf{r}) = 0 \tag{5.1.4d}$$

where all quantities **E**, **H**, **B**, **D**, ρ, and **J** are in phasor representations and their dependencies on the angular frequency ω are implicitly assumed.

The relation of the displacement vector $\mathbf{D}(\mathbf{r}, t)$ to the electric field $\mathbf{E}(\mathbf{r}, t)$ is

$$\mathbf{D}(\mathbf{r}, t) = \varepsilon_0\mathbf{E}(\mathbf{r}, t) + \mathbf{P}(\mathbf{r}, t) = \varepsilon_0\mathbf{E}(\mathbf{r}, t) + \varepsilon_0 \int_{-\infty}^{t} \chi(\mathbf{r}, t - t')\mathbf{E}(\mathbf{r}, t')dt' \tag{5.1.5}$$

or

$$\mathbf{D}(\mathbf{r}, t) = \int_{-\infty}^{t} \varepsilon(\mathbf{r}, t - t')\mathbf{E}(\mathbf{r}, t')dt' \tag{5.1.6}$$

where **P** is the polarization density ($\mathrm{C/m^2}$) and χ the susceptibility. In the frequency domain, we have

$$\mathbf{D}(\mathbf{r}, \omega) = \varepsilon(\mathbf{r}, \omega)\mathbf{E}(\mathbf{r}, \omega) \tag{5.1.7}$$
$$\varepsilon(\mathbf{r}, \omega) = \varepsilon_0(1 + \chi(\mathbf{r}, \omega)). \tag{5.1.8}$$

Outside of the source region, we find

$$\nabla \times \mathbf{E} = i\omega\mu\mathbf{H} \tag{5.1.9a}$$
$$\nabla \times \mathbf{H} = -i\omega\varepsilon\mathbf{E} \tag{5.1.9b}$$
$$\nabla \cdot \varepsilon\mathbf{E} = 0 \tag{5.1.9c}$$
$$\nabla \cdot \mu\mathbf{H} = 0. \tag{5.1.9d}$$

5.1.2 Duality Principle in Electromagnetics

There is certain symmetry among Maxwell's equations [1]. If we make the following changes,

$$\mathbf{E} \to \mathbf{H} \qquad \mathbf{H} \to -\mathbf{E} \qquad \mu \to \varepsilon \qquad \varepsilon \to \mu, \tag{5.1.10}$$

Maxwell's equations stay unchanged. Thus for a given solution ($\mathbf{E}, \mathbf{H}$) in a medium (μ, ε), we have a corresponding solution ($\mathbf{H}, -\mathbf{E}$) in a medium (ε, μ). This is the duality principle [1]. Because the boundary conditions in which the tangential components of the electric and magnetic fields are continuous across the dielectric interfaces and stay the same in both media, we have a dual situation such that the simultaneous replacements in (5.1.10) give the solutions to Maxwell's equations as well. If including both electric and magnetic sources, we write

$$\begin{aligned} \nabla \times \mathbf{E} &= i\omega\mu\mathbf{H} - \mathbf{M} \\ \nabla \times \mathbf{H} &= -i\omega\varepsilon\mathbf{E} + \mathbf{J} \\ \nabla \cdot \varepsilon\mathbf{E} &= \rho_{\mathrm{e}} \\ \nabla \cdot \mu\mathbf{H} &= \rho_{\mathrm{M}}. \end{aligned} \tag{5.1.11}$$

Once we make the following identification,

$$\begin{matrix} \mathbf{E} \to \mathbf{H} & \mathbf{H} \to -\mathbf{E} & \mathbf{J} \to \mathbf{M} & \mathbf{M} \to -\mathbf{J} \\ \mu \to \varepsilon & \varepsilon \to \mu & \rho_{\mathrm{e}} \to \rho_{\mathrm{M}} & \rho_{\mathrm{M}} \to -\rho_{\mathrm{e}} \end{matrix} \tag{5.1.12}$$

we find that Maxwell's equations (5.1.11) are invariant. As an application, if we know the solutions ($\mathbf{E}, \mathbf{H}$) to the original Maxwell's equations in a medium (μ, ε) due to the electric sources ρ_{e} and $\mathbf{J}$, we can obtain the "dual" set of solutions due to the magnetic sources ρ_{M} and $\mathbf{M}$, in a medium (ε, μ) with "dual" boundary conditions (e.g., tangential magnetic fields are continuous vs. tangential electric fields continuous in the original problem with electric sources). Applications of (5.1.10) will be discussed in Chapters 6 and 7, where we discuss plane wave reflection and the optical waveguide theory for both the TE and TM polarizations. The idea is that once we obtain the field solutions for the TE polarization in a dielectric waveguide from Maxwell's equations, we can use the duality principle to write down the TM solutions simply by inspection.

5.2 POYNTING'S THEOREM AND RECIPROCITY RELATIONS

5.2.1 Poynting's Theorem

The power conservation is a very useful law [1–3]. From Maxwell's equations in the time domain,

$$\nabla \times \mathbf{E} = -\frac{\partial}{\partial t}\mathbf{B} \tag{5.2.1}$$

$$\nabla \times \mathbf{H} = \mathbf{J} + \frac{\partial}{\partial t}\mathbf{D}. \tag{5.2.2}$$

Dot-multiplying (5.2.1) by **H** and (5.2.2) by **E**, and taking the difference, we obtain

$$\nabla \cdot (\mathbf{E} \times \mathbf{H}) = -\mathbf{H} \cdot \frac{\partial}{\partial t}\mathbf{B} - \mathbf{E} \cdot \frac{\partial \mathbf{D}}{\partial t} - \mathbf{E} \cdot \mathbf{J} \tag{5.2.3}$$

where $\nabla \cdot (\mathbf{E} \times \mathbf{H}) = \mathbf{H} \cdot \nabla \times \mathbf{E} - \mathbf{E} \cdot \nabla \times \mathbf{H}$ has been used. Define the Poynting vector as

$$\mathbf{S} = \mathbf{E} \times \mathbf{H} \tag{5.2.4}$$

which gives the energy flux density (W/m^2). For an isotropic medium, $\mathbf{D} = \varepsilon\mathbf{E}$ and $\mathbf{B} = \mu\mathbf{H}$, the electric and magnetic energy densities are

$$\begin{aligned} w_e &= \frac{\varepsilon}{2}\mathbf{E} \cdot \mathbf{E} \\ w_m &= \frac{\mu}{2}\mathbf{H} \cdot \mathbf{H}. \end{aligned} \tag{5.2.5}$$

Therefore, Poynting's theorem in the time domain is simply

$$\nabla \cdot \mathbf{S} = -\frac{\partial}{\partial t}(w_e + w_m) - \mathbf{E} \cdot \mathbf{J}. \tag{5.2.6}$$

The theorem simply states the power flow, or the divergence of the power density **S**, is given by the decreasing rate of the total electric and magnetic energy density plus the power generated by the current source, $-\mathbf{E} \cdot \mathbf{J}$. If integrating over a volume V enclosed by a surface S, we obtain Poynting's theorem in the form

$$\iint_S \mathbf{E} \times \mathbf{H} \cdot d\mathbf{S} = -\frac{\partial}{\partial t}\iiint_V (w_e + w_m)\, dV - \iiint_V \mathbf{E} \cdot \mathbf{J}\, dV \tag{5.2.7}$$

that is, the power flow out of the surface S equals the decreasing rate of the stored electric and magnetic energies plus the power supplied by the source, $-\iiint \mathbf{E} \cdot \mathbf{J}\, dV$.

A complex Poynting's theorem can also be derived from Maxwell's equations in frequency domain

$$\nabla \cdot (\mathbf{E} \times \mathbf{H}^*) = -i\omega\left(\frac{1}{2}\mathbf{E} \cdot \mathbf{D}^* - \frac{1}{2}\mathbf{B} \cdot \mathbf{H}^*\right) - \frac{1}{2}\mathbf{E} \cdot \mathbf{J}^*. \tag{5.2.8}$$

If $\mathbf{J} = \mathbf{J}_d + \mathbf{J}_f$ where $\mathbf{J}_d$ accounts for dissipation (e.g., $\mathbf{J}_d = \sigma\mathbf{E}$ in a conductor), we then have

$$\nabla \cdot (\mathbf{E} \times \mathbf{H}^*) + i\frac{\omega}{2}(\mathbf{E} \cdot \mathbf{D}^* - \mathbf{B} \cdot \mathbf{H}^*) + \frac{1}{2}\mathbf{E} \cdot \mathbf{J}_d^* = \frac{-1}{2}\mathbf{E} \cdot \mathbf{J}_f^*. \tag{5.2.9}$$

where the right-hand side is the time-averaged power supplied by the source $\mathbf{J}_f$ and the terms on the left-hand side are the time-averaged power flux, the difference in the electric and magnetic stored energy density, and the time-average dissipated power, respectively.

5.2.2 Reciprocity Relations

Consider two sources $\mathbf{J}^{(1)}$ and $\mathbf{J}^{(2)}$ producing two sets of fields in the same medium described by $\mathbf{D} = \varepsilon\mathbf{E}$ and $\mathbf{B} = \mu\mathbf{H}$,

$$\nabla \times \mathbf{E}^{(1)} = -\frac{\partial}{\partial t}\mathbf{B}^{(1)} \tag{5.2.10a}$$

$$\nabla \times \mathbf{H}^{(1)} = \mathbf{J}^{(1)} + \frac{\partial}{\partial t}\mathbf{D}^{(1)} \tag{5.2.10b}$$

$$\nabla \times \mathbf{E}^{(2)} = -\frac{\partial}{\partial t}\mathbf{B}^{(2)} \tag{5.2.11a}$$

$$\nabla \times \mathbf{H}^{(2)} = \mathbf{J}^{(2)} + \frac{\partial}{\partial t}\mathbf{D}^{(2)}. \tag{5.2.11b}$$

If we take $\mathbf{H}^{(2)} \cdot$ (5.2.10a) $-$ $\mathbf{E}^{(1)} \cdot$ (5.2.11b) and subtract from $\mathbf{H}^{(1)} \cdot$ (5.2.11a) $-\mathbf{E}^{(2)} \cdot$ (5.2.10b), we find

$$\nabla \cdot (\mathbf{E}^{(1)} \times \mathbf{H}^{(2)} - \mathbf{E}^{(2)} \times \mathbf{H}^{(1)}) = -\mathbf{E}^{(1)} \cdot \mathbf{J}^{(2)} + \mathbf{E}^{(2)} \cdot \mathbf{J}^{(1)}. \tag{5.2.12}$$

Integrating over an infinite volume and using the divergence theorem, we obtain

$$\oint_S (\mathbf{E}^{(1)} \times \mathbf{H}^{(2)} - \mathbf{E}^{(2)} \times \mathbf{H}^{(1)}) \cdot d\mathbf{S} = \int_V \mathbf{E}^{(2)} \cdot \mathbf{J}^{(1)} dV - \int_V \mathbf{E}^{(1)} \cdot \mathbf{J}^{(2)} dV. \tag{5.2.13}$$

Using the property in which the surface integral on the left-hand side goes to zero as the radius of the surface goes to infinity, we obtain

$$\int_{V_1} \mathbf{E}^{(2)} \cdot \mathbf{J}^{(1)} dV = \int_{V_2} \mathbf{E}^{(1)} \cdot \mathbf{J}^{(2)} dV. \tag{5.2.14}$$

Because the two sources $\mathbf{J}_1$ and $\mathbf{J}_2$ are distributed over finite regions, such as those of two dipoles, the volume integrals are only over the regions of the two sources V_1 and V_2, respectively. The reciprocity relation (5.2.14) states that the electric field $\mathbf{E}^{(2)}$ (generated by $\mathbf{J}^{(2)}$) evaluated at the other source $\mathbf{J}^{(1)}$ is equal to $\mathbf{E}^{(1)}$ (generated by $\mathbf{J}^{(1)}$) evaluated at $\mathbf{J}^{(2)}$. On the other hand, if we choose a *finite* surface S enclosing a volume V outside of the source regions, then we have

$$\oint \mathbf{E}^{(1)} \times \mathbf{H}^{(2)} \cdot d\mathbf{S} = \oint \mathbf{E}^{(2)} \times \mathbf{H}^{(1)} \cdot d\mathbf{S} \tag{5.2.15}$$

which is another form of reciprocity relation [4, 5].

5.3 PLANE WAVE SOLUTIONS FOR MAXWELL'S EQUATIONS IN HOMOGENEOUS MEDIA

We shall investigate solutions to Maxwell's equations in regions where $\mathbf{J} = \rho = 0$. Maxwell's equations for a harmonic field become

$$\nabla \times \mathbf{E} = i\omega\mathbf{B} \tag{5.3.1a}$$

$$\nabla \times \mathbf{H} = -i\omega\mathbf{D} \tag{5.3.1b}$$

$$\nabla \cdot \mathbf{B} = 0 \tag{5.3.1c}$$

$$\nabla \cdot \mathbf{D} = 0. \tag{5.3.1d}$$

We also assume that the media are homogeneous; therefore, we have plane wave solutions of the form $\exp(i\mathbf{k} \cdot \mathbf{r})$. Because all complex field vectors **E**, **H**, **B**, and **D** have the same spatial dependence, $\exp(i\mathbf{k} \cdot \mathbf{r})$, we obtain the basic relations for plane waves propagating in source-free homogeneous media, using $\nabla e^{i\mathbf{k}\cdot\mathbf{r}} \rightarrow i\mathbf{k}\, e^{i\mathbf{k}\cdot\mathbf{r}}$

$$\mathbf{k} \times \mathbf{E} = \omega\mathbf{B} \tag{5.3.2a}$$

$$\mathbf{k} \times \mathbf{H} = -\omega\mathbf{D} \tag{5.3.2b}$$

$$\mathbf{k} \cdot \mathbf{B} = 0 \tag{5.3.2c}$$

$$\mathbf{k} \cdot \mathbf{D} = 0. \tag{5.3.2d}$$

We see that the wave vector **k** is always perpendicular to **B** and **D** from (5.3.2c) and (5.3.2d). But we cannot say that **k** is always perpendicular to **H** or **E** unless the media are isotropic. In isotropic media, $\mathbf{D} = \varepsilon\mathbf{E}$ and $\mathbf{B} = \mu\mathbf{H}$. Therefore, **D** and **B** are parallel to **E** and **H**, respectively. Otherwise, the media are called anisotropic. Thus, the time-averaged Poynting vector, which is given by $(1/2)\mathrm{Re}(\mathbf{E} \times \mathbf{H}^*)$, does not necessarily point in the direction of **k** in an anisotropic medium. In other words, the direction of power flow of a plane wave may not always be in the direction of the wave vector **k**.

In Section 5.4 and in Chapter 6, we discuss how characteristic polarizations and propagation constants in a given medium are obtained. A characteristic mode or normal mode is defined as a wave with a polarization, which does not change while propagating in the homogeneous medium. This polarization is called the characteristic polarization of the medium.

5.4 LIGHT PROPAGATION IN ISOTROPIC MEDIA

In isotropic media, the constitutive relations are simply

$$\mathbf{D} = \varepsilon\mathbf{E} \tag{5.4.1}$$

$$\mathbf{B} = \mu\mathbf{H}. \tag{5.4.2}$$

Thus, **D** is always parallel to **E**, and **B** is parallel to **H**. Any polarization direction of **E** can be a characteristic polarization of the isotropic medium. Equations (5.3.2a)–(5.3.2d) become

$$\mathbf{k} \times \mathbf{E} = \omega\mu\mathbf{H} \tag{5.4.3a}$$

$$\mathbf{k} \times \mathbf{H} = -\omega\varepsilon\mathbf{E} \tag{5.4.3b}$$

$$\mathbf{k} \cdot \mathbf{H} = 0 \tag{5.4.3c}$$

$$\mathbf{k} \cdot \mathbf{E} = 0. \tag{5.4.3d}$$

These are the equations satisfied by plane waves in isotropic media. One sees clearly that

(a) **k** is perpendicular to both **E** and **H**.
(b) $\mathbf{k} \times \mathbf{E}$ points in the direction of **H**, and $\mathbf{k} \times \mathbf{H}$ points in the direction of $-\mathbf{E}$.

Thus, the three vectors **E**, **H**, and **k** are perpendicular to each other and they form a right-hand coordinate system.

Let us take the cross-product of (5.4.3a) by **k** from the left-hand side,

$$\mathbf{k} \times (\mathbf{k} \times \mathbf{E}) = \omega\mu\mathbf{k} \times \mathbf{H}. \tag{5.4.4}$$

The left-hand side can be simplified using the vector identity

$$\begin{aligned}\mathbf{k} \times (\mathbf{k} \times \mathbf{E}) &= (\mathbf{k} \cdot \mathbf{E})\mathbf{k} - k^2\mathbf{E} \\ &= -k^2\mathbf{E}\end{aligned} \tag{5.4.5}$$

where we have made use of (5.4.3d). We obtain

$$\begin{aligned}-k^2\mathbf{E} &= \omega\mu\mathbf{k} \times \mathbf{H} \\ &= \omega\mu(-\omega\varepsilon\mathbf{E})\end{aligned} \tag{5.4.6}$$

using (5.4.3b). The result is simply

$$(k^2 - \omega^2\mu\varepsilon)\mathbf{E} = 0. \tag{5.4.7}$$

The solutions to the above equation are either a trivial solution $\mathbf{E} = 0$ (no field, which is not of interest), or

$$k^2 = \omega^2\mu\varepsilon, \tag{5.4.8}$$

which is the dispersion relation in an isotropic medium. We obtain the propagation constant $k = \omega\sqrt{\mu\varepsilon}$. If we define the refractive index $n = \sqrt{\mu\varepsilon/\mu_0\varepsilon_0}$, where $\mu_0 = 4\pi \times 10^{-7}\,\mathrm{H/m}$, and $\varepsilon_0 = 8.854 \times 10^{-12}\,\mathrm{F/m} \simeq 10^{-9}/(36\pi)\,\mathrm{F/m}$, we have

$$k = \frac{\omega}{c}n$$

and $c = 1/\sqrt{\mu_0 \varepsilon_0} = 3 \times 10^8$ m/s is the speed of light in free space. Let us define the unit vectors $\hat{e}$, $\hat{h}$, and $\hat{k}$ such that

$$\mathbf{E} = \hat{e}E \qquad \mathbf{H} = \hat{h}H \qquad \mathbf{k} = \hat{k}k. \tag{5.4.9}$$

Substituting (5.4.9) into (5.4.3a), we obtain

$$\hat{k}k \times \hat{e}E = \omega\mu\hat{h}H. \tag{5.4.10}$$

Or simply, $\hat{k} \times \hat{e}$ is in the direction of $\hat{h}$. Because $\hat{k}$, $\hat{e}$, and $\hat{h}$ are all unit vectors and $\hat{k}$ is perpendicular to $\hat{e}$, (5.4.10) leads to

$$\hat{k} \times \hat{e} = \hat{h} \qquad kE = \omega\mu H. \tag{5.4.11}$$

Using $k = \omega\sqrt{\mu\varepsilon}$, we have

$$E = \eta H \qquad \eta = \sqrt{\mu/\varepsilon} \tag{5.4.12}$$

where η is called the characteristic impedance of the medium. We conclude, as shown in Fig. 5.1, that

(a) The three unit vectors $\hat{e}$, $\hat{h}$, and $\hat{k}$ construct a right-hand rectangular coordinate system.
(b) The magnitude of the electric field is equal to the magnitude of the magnetic field multiplied by the characteristic impedance of the medium η.
(c) The time-averaged Poynting's power density (W/m^2) is

$$\mathbf{S} = \frac{1}{2}\text{Re}\,\mathbf{E} \times \mathbf{H}^* = \frac{|E|^2}{2\eta}\hat{k} = \frac{\eta}{2}|H|^2\hat{k}. \tag{5.4.13}$$

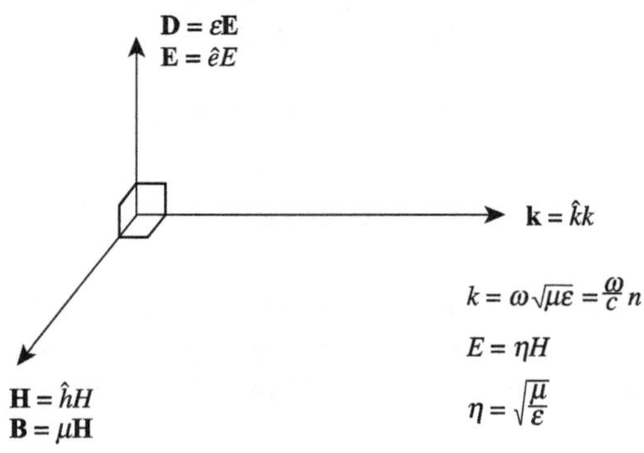

Figure 5.1 In isotropic media, $\mathbf{D} = \varepsilon\mathbf{E}$, $\mathbf{B} = \mu\mathbf{H}$. The three vectors **E**, **H**, and **k**, or **D**, **B**, and **k**, form a right-handed rectangular coordinate system.

5.5 WAVE PROPAGATION IN LOSSY MEDIA: LORENTZ OSCILLATOR MODEL AND METAL PLASMA

5.5.1 Propagation Constant and Optical Refractive Index in Semiconductors

In most semiconductors, $\mu = \mu_0$, and the permittivity function is complex,

$$\varepsilon(\omega) = \varepsilon'(\omega) + i\varepsilon''(\omega) \tag{5.5.1}$$

where the real and imaginary parts of the permittivity function satisfy the Kramers–Kronig relations (Appendix 5A). The complex propagation constant $k = k' + ik''$ can be written in terms of the complex refractive index $\bar{n}(\omega) = n(\omega) + i\kappa(\omega)$

$$\begin{aligned} k(\omega) &= \omega\sqrt{\mu_0\varepsilon_0}\left[\varepsilon_r'(\omega) + i\varepsilon_r''(\omega)\right]^{1/2} \\ &= \frac{\omega}{c}(n + i\kappa) \end{aligned} \tag{5.5.2}$$

where we have used the real and the imaginary parts of the relative permittivity

$$\varepsilon_r'(\omega) = \varepsilon'(\omega)/\varepsilon_0, \qquad \varepsilon_r''(\omega) = \varepsilon''(\omega)/\varepsilon_0. \tag{5.5.3}$$

The real and the imaginary parts of the refractive index $n(\omega)$ and $\kappa(\omega)$ can be expressed as

$$n^2 = \frac{1}{2}\left[\varepsilon_r'(\omega) + \sqrt{\varepsilon_r'^2(\omega) + \varepsilon_r''^2(\omega)}\right] \tag{5.5.4a}$$

$$\kappa^2 = \frac{1}{2}\left[-\varepsilon_r'(\omega) + \sqrt{\varepsilon_r'^2(\omega) + \varepsilon_r''^2(\omega)}\right]. \tag{5.5.4b}$$

Experimental data for n and κ of GaAs and InP semiconductors as a function of optical energy and wavelength are tabulated and plotted in Appendix B at the end of the book. The imaginary part of the complex propagation constant is half of the intensity absorption coefficient α.

$$k(\omega) = \frac{\omega}{c}n + i\frac{\alpha}{2} = k' + ik'' \tag{5.5.5}$$

where

$$k' = \frac{\omega}{c}n = \frac{2\pi}{\lambda}n \tag{5.5.6a}$$

and

$$\alpha(\omega) = 2k'' = 2\frac{\omega}{c}\kappa(\omega) = \frac{4\pi}{\lambda}\kappa(\omega) \tag{5.5.6b}$$

where λ is the wavelength in free space. For a plane wave propagating in the $+z$-direction with a polarization along $\hat{x}$, the electric field behaves as follows:

$$\mathbf{E} = \hat{x}\, E_0\, e^{ikz} = \hat{x}\, E_0 \exp\left(i\frac{2\pi}{\lambda}nz - \frac{\alpha}{2}z\right) \tag{5.5.7a}$$

$$\mathbf{H} = \hat{y}\, \frac{E_0}{\eta}\, e^{ikz} = \hat{y}\, \frac{\bar{n}E_0}{\eta_0} \exp\left(i\frac{2\pi}{\lambda}nz - \frac{\alpha}{2}z\right), \tag{5.5.7b}$$

where $\eta_0 = \sqrt{\mu_0/\varepsilon_0} = 120\pi$ is the characteristic impedance of the free space. The time-averaged Poynting vector for the optical power density is

$$\mathbf{S} = \frac{1}{2}\mathrm{Re}[\mathbf{E} \times \mathbf{H}^*] = \hat{z}\, \frac{n}{2\eta_0}\left|E_0\right|^2 e^{-\alpha z} \tag{5.5.8}$$

which decays exponentially as the wave propagates farther along the z-direction with a decay constant determined by the absorption coefficient.

5.5.2 Lorentz Dipole Model

In a resonant dielectric medium such as a collection of resonant atoms in the presence of a driving electromagnetic field $E(t)$, we can model each resonant atom as a classic harmonic oscillator, Fig. 5.2. The equation is motion is determined by the reduced mass for a free electron with mass m_0 bound to a nucleus with mass M [6–8]

$$\frac{1}{m_r} = \frac{1}{m_0} + \frac{1}{M}. \tag{5.5.9}$$

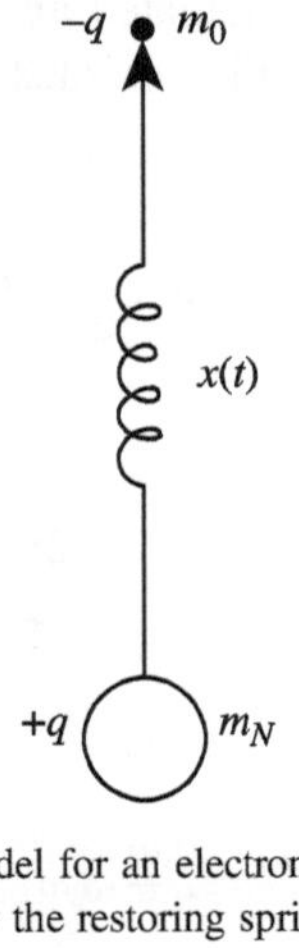

Figure 5.2 A classic dipole oscillator model for an electron bound to a positive nucleus. The resonant frequency $\omega_0 = \sqrt{k/m_r}$ is characterized by the restoring spring constant k and the reduced mass m_r.

Usually, $M \gg m_0$, therefore, $m_r \approx m_0$. The equation of motion for the displacement of the electron with a unit charge $-q$ from its equilibrium position is governed by

$$m_0 \frac{d^2x}{\partial t^2} + m_0 \gamma \frac{dx}{dt} + kx = -qE(t) \tag{5.5.10}$$

where γ is a damping factor and k is the spring constant. For a time harmonic optical electromagnetic field, $E(t) = E_0 e^{-i\omega t}$, the dipole moment of the electron $-q$ and the positive nucleus $+q$ is given by

$$p = -qx(t) = \frac{q^2 E(t)}{m_0(-\omega^2 + i\gamma\omega + \omega_0^2)} \tag{5.5.11}$$

where a resonant frequency $\omega_0 = \sqrt{k/m_0}$ has been defined. The polarization density P due to N atoms or electron–nucleus pairs per unit volume ($1/\text{cm}^3$) is

$$P = Np = \frac{Nq^2 E(t)}{m_0(-\omega^2 - i\gamma\omega + \omega_0^2)}. \tag{5.5.12}$$

The displacement vector D is given by

$$\begin{aligned} D &= \varepsilon_0 E + P_b + P \\ &= \varepsilon_0 \left[1 + \chi_b + \frac{Nq^2}{m_0 \varepsilon_0 (\omega_0^2 - \omega^2 - i\gamma\omega)} \right] E \end{aligned} \tag{5.5.13}$$

where we have included the background contribution to the polarization density $P_b = \varepsilon_0 \chi_b E$. We obtain the permittivity function for the resonant dielectric medium or an atomic gas:

$$\varepsilon(\omega) = \varepsilon_0 [1 + \chi_b + \chi(\omega)] \tag{5.5.14a}$$

$$\chi(\omega) = \frac{Nq^2}{m_0 \varepsilon_0 (\omega_0^2 - \omega^2 - i\gamma\omega)} = \frac{\omega_p^2}{(\omega_0^2 - \omega^2 - i\gamma\omega)} \tag{5.5.14b}$$

where

$$\omega_p = \sqrt{\frac{Nq^2}{m_0 \varepsilon_0}} \tag{5.5.15}$$

is the plasma frequency. Separating the real and imaginary parts, $\chi = \chi' + i\chi''$

$$\chi'(\omega) = \frac{\omega_p^2 (\omega_0^2 - \omega^2)}{(\omega_0^2 - \omega^2)^2 + (\gamma\omega)^2} \tag{5.5.16a}$$

$$\chi''(\omega) = \frac{\omega_p^2 \gamma\omega}{(\omega_0^2 - \omega^2)^2 + (\gamma\omega)^2}. \tag{5.5.16b}$$

The relative permittivity or dielectric function is given by

$$\begin{aligned}\varepsilon_r(\omega) &= \frac{\varepsilon(\omega)}{\varepsilon_0} = \varepsilon_r'(\omega) + i\varepsilon_r''(\omega) \\ &= 1 + \chi_b + \chi'(\omega) + i\chi''(\omega) \qquad (5.5.17a)\\ \varepsilon_r'(\omega) &= 1 + \chi_b + \chi'(\omega) \qquad (5.5.17b)\\ \varepsilon_r''(\omega) &= \chi''(\omega). \qquad (5.5.17c)\end{aligned}$$

We denote the static (dc) value of the permittivity at zero frequency,

$$\chi_s = \chi(0) = \frac{\omega_p^2}{\omega_0^2} \qquad (5.5.18a)$$

$$\varepsilon_s = \varepsilon_r(0) = 1 + \chi_b + \chi_0. \qquad (5.5.18b)$$

The other limit as ω approaches infinity is $\chi(\infty) = 0$, that is, the medium polarization simply cannot respond to the fast varying driving electromagnetic field, and the dielectric constant in the high-frequency limit is given by the background value:

$$\varepsilon_\infty = \varepsilon_r(\infty) = 1 + \chi_b. \qquad (5.5.19)$$

Figure 5.3 plots the real and imaginary parts of the susceptibility χ. The imaginary part peaks at the resonant frequency ω_0 as expected, with a full-width at half maximum of γ. The plasma resonance is also given as

$$Q = \frac{\omega_0}{(\Delta\omega)_{\text{FWHM}}} = \frac{\omega_0}{\gamma} \qquad \chi''_{\text{peak}}(\omega_0) = \chi_0 Q. \qquad (5.5.20)$$

The real part has a zero crossing at the resonant frequency and is positive for frequency below resonance and is negative for frequency above resonance. The complex propagation constant is

$$\begin{aligned}k &= \omega\sqrt{\mu_0\varepsilon} = \omega\sqrt{\mu_0[\varepsilon_b + \varepsilon_0\chi(\omega)]} \\ &\approx \omega\sqrt{\mu_0\varepsilon_b}\left\{1 + \frac{\varepsilon_0}{2\varepsilon_b}[\chi' + i\chi'']\right\} \\ &= k_0 n_b + k_0\Delta n + i\frac{\alpha}{2} \qquad (5.5.21)\end{aligned}$$

where $n_b = \sqrt{\varepsilon_b/\varepsilon_0}$ is the background dielectric constant.

Here the approximation is valid only if the second term is much smaller than the first background permittivity. Otherwise, (5.5.4a) and (5.5.4b) have to be used. We have then,

$$\Delta n = \frac{1}{2n_b}\chi' = \text{ the change in refractive index} \qquad (5.5.22a)$$

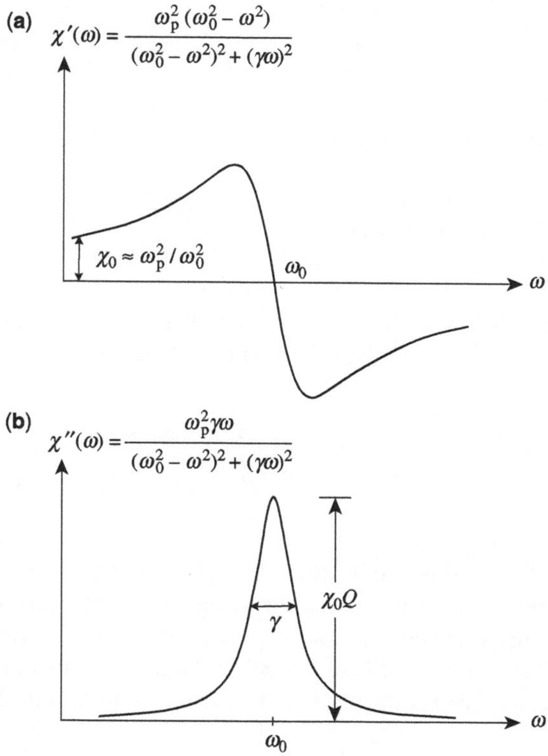

Figure 5.3 Plots for (a) the real part $\chi'(\omega)$ and (b) the imaginary part $\chi''(\omega)$ of the susceptibility function $\chi(\omega)$ for a resonant dielectric medium with a single resonant frequency ω_0 and a damping factor γ, which determines the full width at half maximum (FWHM) of the imaginary part.

and the "intensity" absorption coefficient $\alpha = 2\mathrm{Im}(k)$ is

$$\alpha = \frac{k_0}{n_b}\chi''. \tag{5.5.22b}$$

The total refractive index is, therefore,

$$n(\omega) = n_b + \Delta n(\omega) = n_b + \frac{1}{2n_b}\chi'(\omega). \tag{5.5.22c}$$

5.5.3 Conducting Media

The conductivity of a metal is described by the relation between the current density and the electric field: $\mathbf{J} = \sigma\mathbf{E}$

$$\begin{aligned}\nabla \times \mathbf{H} &= -i\omega\varepsilon\mathbf{E} + \mathbf{J} \\ &= -i\omega\varepsilon\mathbf{E} + \sigma\mathbf{E} \\ &= -i\omega\left(\varepsilon + i\frac{\sigma}{\omega}\right)\mathbf{E}.\end{aligned} \tag{5.5.23}$$

Therefore, we define the complex permittivity as

$$\varepsilon(\omega) = \varepsilon + i\frac{\sigma}{\omega}. \tag{5.5.24}$$

The complex propagation constant is

$$k = \omega\sqrt{\mu_0\left(\varepsilon + i\frac{\sigma}{\omega}\right)} \approx \sqrt{\frac{\omega\mu_0\sigma}{2}}(1+i) \tag{5.5.25}$$

when the frequency is low such that $\sigma/\omega\varepsilon \gg 1$. The distance when the electromagnetic field decays by e^{-1} is call the skin depth, δ, which is

$$\delta = \sqrt{\frac{2}{\omega\mu_0\sigma}}. \tag{5.5.26}$$

Drude Model for ac Conductivity [6] At high frequency, especially near optical frequency, the conductivity is not a constant. A simple model, called the Drude model for the free electrons, has been very useful. The model for free carrier conductivity is similar to that of the Lorentz oscillator model for bound electrons except that there is no restoring force, as if we set the resonant frequency ω_0 to zero in (5.5.10)–(5.5.14)

$$m_0\frac{d\mathbf{v}}{dt} + m_0\gamma\mathbf{v} = -q\mathbf{E}. \tag{5.5.27}$$

Another way of understanding the physical process is that the momentum of the electron changes due to the scattering process and the above equation is written as

$$\frac{d\mathbf{p}}{dt} = -\frac{\mathbf{p}}{\tau} - q\mathbf{E} \tag{5.5.28}$$

where $\tau\,(= 1/\gamma)$ is the momentum lifetime, the average time the electron changes its momentum due to scattering processes in the conductor. In the presence of a time-harmonic electromagnetic field, the solution for the velocity can be obtained, and the current density due to the flow of N electrons per unit volume is given by

$$\mathbf{J} = -Nq\mathbf{v} = \frac{Nq^2}{m_0(\gamma - i\omega)}\mathbf{E} = \sigma(\omega)\mathbf{E} \tag{5.5.29}$$

where we obtain the ac conductivity:

$$\sigma(\omega) = \frac{Nq^2}{m_0(\gamma - i\omega)} = \frac{Nq^2\tau}{m_0(1 - i\omega\tau)} = \frac{\sigma_0}{1 - i\omega\tau}. \tag{5.5.30}$$

The dc conductivity σ_0 is given by

$$\sigma_0 = \frac{Nq^2\tau}{m_0} \tag{5.5.31a}$$

$$\varepsilon(\omega) = \varepsilon_\infty + i\frac{\sigma(\omega)}{\omega} = \varepsilon_\infty - \frac{\omega_p^2\varepsilon_0}{\omega^2 + i\omega\gamma} \tag{5.5.31b}$$

$$\omega_p^2 = \frac{Nq^2}{m_0\varepsilon_0}. \tag{5.5.31c}$$

The background ε_∞ is used when we include the contributions due to other polarization contributions. Otherwise, it is given by the free space value. We have the complex permittivity for metals:

$$\begin{aligned}\varepsilon(\omega) &= \varepsilon_\infty - \frac{\sigma_0\tau}{1+\omega^2\tau^2} + i\frac{\sigma_0}{\omega(1+\omega^2\tau^2)} \\ &= \varepsilon_\infty - \varepsilon_0\frac{\omega_p^2\tau^2}{1+\omega^2\tau^2} + i\varepsilon_0\frac{\omega_p^2\tau}{\omega(1+\omega^2\tau^2)}.\end{aligned} \tag{5.5.32}$$

Typical values [9] for gold are $\varepsilon_\infty = 9.84\,\varepsilon_0$, $\omega_p = 11.27\times 10^{15}$ (1/s), and $\gamma = 1/\tau = 2.0\times 10^{14}$ (1/s). For silver, $\varepsilon_\infty = 4.785\ \varepsilon_0$, $\omega_p = 14.385\times 10^{15}$ (1/s), and $\gamma = 7.95\times 10^{13}$ (1/s). Figure 5.4a and Fig. 5.4b show the plots of the real and imaginary parts of silver (Ag) and gold (Au) over the range from visible to near infrared calculated using various models (with more fitting terms than the above simple expressions) and compare with experimental data [10–16]. For optical wavelengths, the real part of the permittivity is negative. Metal plasma has been used recently for surface plasma waveguides and near-field enhancement applications [10–17].

Doped Semiconductors The complex permittivity of a doped semiconductor is very similar to that of the metal at near optical frequency, except that the effective mass of the electrons for the *n*-type doped semiconductors or the holes for the *p*-type doped semiconductors have to be used. In addition, the background permittivity at optical frequency ε_{op} to account for polarization due to other processes beyond the free carrier contributions has to be added [6]

$$\begin{aligned}\varepsilon(\omega) &= \varepsilon_{op} - \frac{Nq^2}{m^*(\omega^2 + i\gamma\omega)} \\ &= \varepsilon_{op} - \frac{\omega_p^2\varepsilon_0}{\omega^2 + i\gamma\omega}\end{aligned} \tag{5.5.33}$$

$$\omega_p^2 = \frac{Nq^2}{m^*\varepsilon_0}. \tag{5.5.34}$$

When both electrons and holes are present, the free carrier plasma contributions due to both are added.

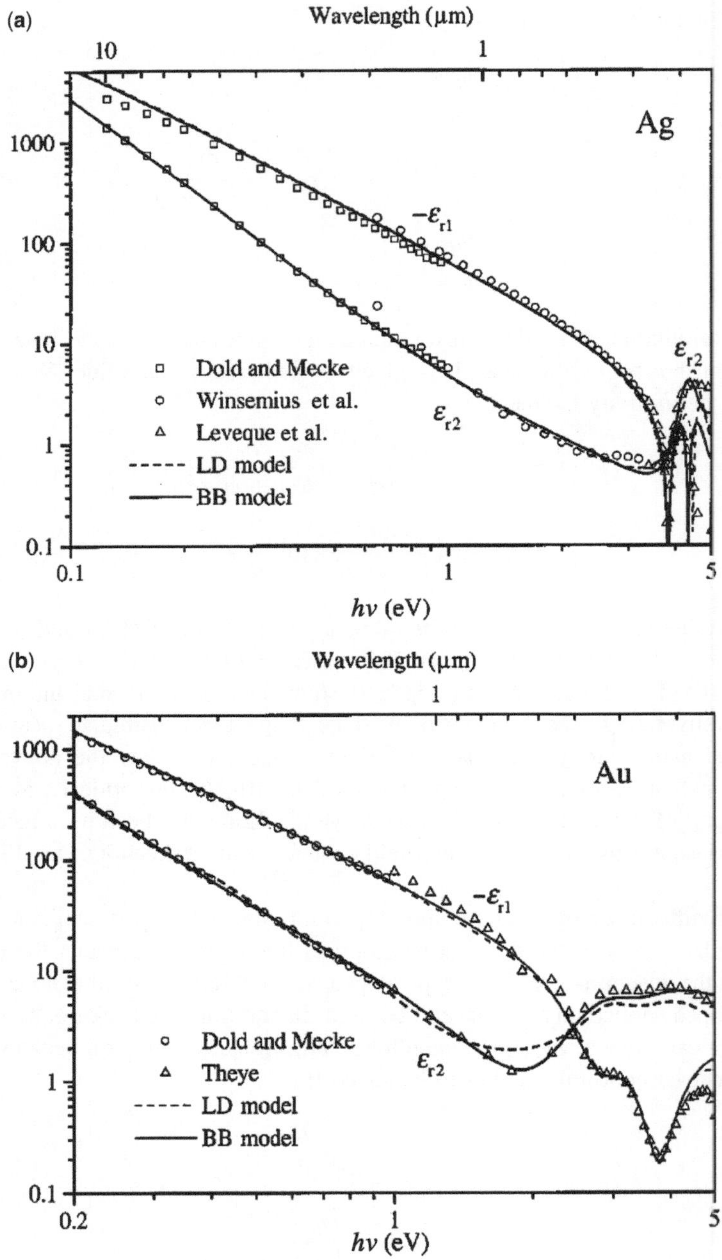

Figure 5.4 (a) Real $\varepsilon'(\omega)$ and imaginary $\varepsilon''(\omega)$ parts of the optical dielectric function of Ag: solid curves, values that we calculated using the Brendel–Bormann (BB) model [11]; dashed curves, the Lorentz–Drude (LD) model. Also shown are the selected experimental data points from Dold and Mecke [12], Winsemius et al. [13], and Leveque et al. [14]. (b) Real and imaginary parts of the optical dielectric function of Au: solid curves, values that we calculated using the BB model; dashed curves, the LD model. Also shown are selected experimental data points from Dold and Mecke [12] and Thèye [15]. (Reprinted with permissions from [10] © 1998 Optical Society of America.)

5.6 PLANE WAVE REFLECTION FROM A SURFACE

In this section, we first discuss plane wave reflection from a single planar interface between two dielectric media [1, 3]. We then present a propagation-matrix approach to plane wave reflection from a multiple layered medium.

5.6.1 TE Polarization

Consider a plane wave incident on a planar boundary as shown in Fig. 5.5a with the electric field given by

$$\mathbf{E}_\mathrm{i} = \hat{y}E_y = \hat{y}E_0 e^{-ik_{1x}x + ik_{1z}z} \tag{5.6.1}$$

and the magnetic field given by

$$\begin{aligned}\mathbf{H}_\mathrm{i} &= \frac{1}{i\omega\mu_1}\nabla\times\mathbf{E}_\mathrm{i}\\ &= \frac{-1}{\omega\mu_1}(k_{1z}\hat{x} + k_{1x}\hat{z})E_0 e^{-ik_{1x}x + ik_{1z}z}\end{aligned} \tag{5.6.2}$$

where the components of the wave vector satisfy the dispersion relation

$$k_{1x}^2 + k_{1z}^2 = \omega^2\mu_1\varepsilon_1 = k_0^2 n_1^2. \tag{5.6.3}$$

Here $k_0 = \omega\sqrt{\mu_0\varepsilon_0} = \omega/c$ is the wave number in free space, and $n_1 = \sqrt{\mu_1\varepsilon_1/\mu_0\varepsilon_0}$ is the refractive index of the material. Usually, $\mu_1 = \mu_0$ for dielectric materials.

The reflected fields are given by

$$\mathbf{E}_\mathrm{r} = \hat{y}rE_0 e^{ik_{1x}^r x + ik_{1z}^r z} \tag{5.6.4a}$$

$$\mathbf{H}_\mathrm{r} = \frac{1}{\omega\mu_1}(-k_{1z}^r\hat{x} + k_{1x}^r\hat{z})r\,E_0 e^{ik_{1x}^r x + ik_{1z}z} \tag{5.6.4b}$$

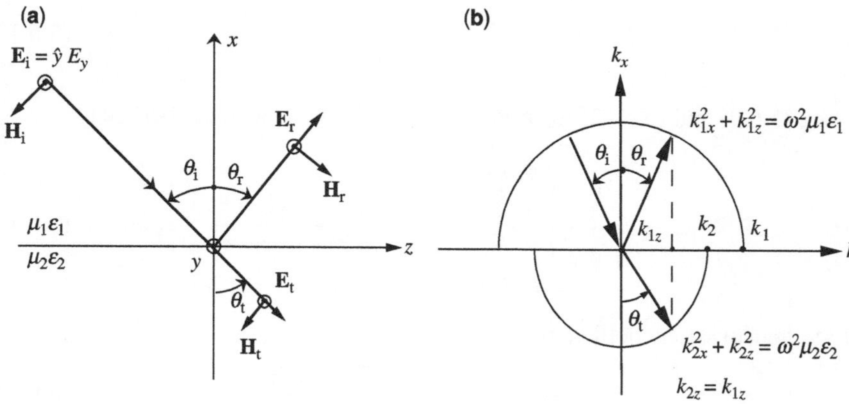

Figure 5.5 (a) Plane wave reflection from a planar surface. (b) The phase-matching diagram for plane wave reflection and transmission.

The transmitted fields are

$$\mathbf{E}_{\mathrm{t}} = \hat{y} t E_0 e^{ik_{2x}x + ik_{2z}z} \tag{5.6.5a}$$

$$\mathbf{H}_{\mathrm{t}} = \frac{-1}{\omega\mu_2}(-k_{2z}\hat{x} + k_{2x}\hat{z}) t E_0 e^{ik_{2x}x + ik_{2z}z} \tag{5.6.5b}$$

where

$$k_{2x}^2 + k_{2z}^2 = \omega^2 \mu_2 \varepsilon_2 = k_0^2 n_2^2 \tag{5.6.6}$$

and $n_2 = \sqrt{\mu_2\varepsilon_2/\mu_0\varepsilon_0}$. Matching the boundary conditions in which the tangential electric field (E_y) is continuous at $x = 0$,

$$E_0 e^{ik_{1z}z} + r\, E_0 e^{ik_{1z}^r z} = t E_0 e^{ik_{2z}z} \tag{5.6.7}$$

for all z, we obtain

$$k_{1z} = k_{1z}^r = k_{2z} \tag{5.6.8}$$

$$1 + r = t. \tag{5.6.9}$$

Equation (5.6.8) is Snell's law or the phase-matching condition, Fig. 5.5b,

$$k_1 \sin\theta_{\mathrm{i}} = k_1 \sin\theta_{\mathrm{r}} = k_2 \sin\theta_{\mathrm{t}} \tag{5.6.10}$$

or

$$\begin{aligned} \theta_{\mathrm{i}} &= \theta_{\mathrm{r}} \\ n_1 \sin\theta_{\mathrm{i}} &= n_2 \sin\theta_{\mathrm{t}}, \end{aligned} \tag{5.6.11}$$

where $k_i = \omega\sqrt{\mu_i\varepsilon_i} = k_0 n_i$, $i = 1, 2$. The other boundary condition in which the tangential magnetic field (H_z) is continuous leads to

$$1 - r = \frac{\mu_1\, k_{2x}}{\mu_2\, k_{1x}} t \tag{5.6.12}$$

Solving (5.6.9) and (5.6.12) for r and t, we obtain the reflection coefficient

$$r = \frac{1 - \left(\dfrac{\mu_1\, k_{2x}}{\mu_2\, k_{1x}}\right)}{1 + \left(\dfrac{\mu_1\, k_{2x}}{\mu_2\, k_{1x}}\right)} = \frac{n_1\cos\theta_{\mathrm{i}} - n_2\cos\theta_{\mathrm{t}}}{n_1\cos\theta_{\mathrm{i}} + n_2\cos\theta_{\mathrm{t}}} \tag{5.6.13}$$

and the transmission coefficient

$$t = \frac{2}{1 + \left(\dfrac{\mu_1\, k_{2x}}{\mu_2\, k_{1x}}\right)} = \frac{2 n_1\cos\theta_{\mathrm{i}}}{n_1\cos\theta_{\mathrm{i}} + n_2\cos\theta_{\mathrm{t}}} \tag{5.6.14}$$

for the electric field. The fraction of the power reflected from its boundary is given by the reflectivity R

$$R = \left.\frac{\frac{1}{2}\mathrm{Re}(\mathbf{E}_\mathrm{r} \times \mathbf{H}_\mathrm{r}^*)_x}{-\frac{1}{2}\mathrm{Re}(\mathbf{E}_\mathrm{i} \times \mathbf{H}_\mathrm{i}^*)_x}\right|_{x=0} = |r|^2. \tag{5.6.15}$$

The transmissivity is given by

$$T = \left.\frac{-\frac{1}{2}\mathrm{Re}(\mathbf{E}_\mathrm{t} \times \mathbf{H}_\mathrm{t}^*)_x}{-\frac{1}{2}\mathrm{Re}(\mathbf{E}_\mathrm{i} \times \mathbf{H}_\mathrm{i}^*)_x}\right|_{x=0} = \frac{\mu_1}{k_{1x}}\mathrm{Re}\left(\frac{k_{2x}}{\mu_2}\right)|t|^2. \tag{5.6.16}$$

Power conservation requires $R + T = 1$, which can also be checked by substituting (5.6.13) and (5.6.14) into (5.6.15) and (5.6.16).

Total Internal Reflection We note that when $k_1 > k_2$, the total internal reflection occurs at an angle of incidence larger than the critical angle θ_c, given by the condition when $\theta_\mathrm{t} = 90°$

$$k_1 \sin\theta_\mathrm{c} = k_2 \tag{5.6.17}$$

or $\theta_\mathrm{c} = \sin^{-1}\left(\frac{n_2}{n_1}\right)$. When $\theta_\mathrm{i} > \theta_\mathrm{c}$, $k_{1z} = k_1 \sin\theta_\mathrm{i} > k_2$, we have

$$k_{2x}^2 = k_2^2 - k_{2z}^2 = k_2^2 - k_{1z}^2 < 0. \tag{5.6.18}$$

Thus, k_{2x} is purely imaginary, $k_{2x} = i\alpha_2$, $\alpha_2 > 0$, and the field is decaying along the $-x$-direction,

$$E_y = tE_0 e^{\alpha_2 x + ik_{1z}z}. \tag{5.6.19}$$

The reflection coefficient becomes

$$r = \frac{1 - i\dfrac{\mu_1\alpha_2}{\mu_2 k_{1x}}}{1 + i\dfrac{\mu_1\alpha_2}{\mu_2 k_{1x}}} = e^{-i2\phi_{12}} \tag{5.6.20a}$$

where

$$\phi_{12} = \tan^{-1}\left(\frac{\mu_1\alpha_2}{\mu_2 k_{1x}}\right). \tag{5.6.20b}$$

It is clear that if $\theta_\mathrm{i} > \theta_\mathrm{c}$, total internal reflection occurs; the transmissivity is zero since k_{2x} is purely imaginary, and the reflectivity is unity, $|r|^2 = \left|e^{-i2\phi_{12}}\right| = 1$. However, the reflected field experiences an optical phase shift of an amount $-2\phi_{12}$. This angle is called the Goos–Hänchen phase shift.

Special Case for Normal Incidence At normal incidence, $\theta_i = 0°$, $k_{2x} = k_2 = k_0 n_2$ and $k_{1x} = k_1 = k_0 n_1$. We find the reflection and transmission coefficients for the field are, respectively,

$$r_{12} = \frac{n_1 - n_2}{n_1 + n_2} \qquad t_{12} = \frac{2n_1}{n_1 + n_2} \tag{5.6.21}$$

At an optical energy below the band gap of most semiconductors, the absorption is usually small or negligible. Above the band gap, the optical absorption is important. When a plane wave is normally incident from the air to a semiconductor surface, the reflectivity of the power is

$$R = \left| \frac{n_0 - \bar{n}}{n_0 + \bar{n}} \right|^2 = \frac{(n-1)^2 + \kappa^2}{(n+1)^2 + \kappa^2} \tag{5.6.22}$$

which takes into account the absorption effect when the refractive index $\bar{n}$ is complex and n_0 of the air is 1.

Numerical Example For InP material, the dispersive effects of the real and the imaginary parts $n(\omega)$ and $\kappa(\omega)$ are given in Appendix B at the end of the text as a function of the photon energy $\hbar\omega$ or the free-space wavelength λ. At an energy $\hbar\omega = 2.0\,\text{eV}$ ($\lambda = 0.62\ \mu\text{m}$), which is close to that of a HeNe laser wavelength, we have $n = 3.549$, and $\kappa = 0.317$. The absorption coefficient is

$$\alpha = \frac{4\pi}{\lambda}\kappa = \frac{4\pi}{0.62\ \mu\text{m}} \times 0.317 = 6.43 \times 10^4\,\text{cm}^{-1}.$$

The reflectivity for the reflected power from the semiconductor is

$$R = \frac{(3.549 - 1)^2 + 0.317^2}{(3.549 - 1)^2 + 0.317^2} = 0.317.$$

5.6.2 TM Polarization

The result of TM polarization can be obtained by the duality principle using the exchange of the physical quantities in Section 5.6.1. The results are

$$\mathbf{H}_i = \hat{y}\, H_y = \hat{y}\, H_0 e^{-ik_{1x}x + ik_{1z}z} \tag{5.6.23a}$$

$$\mathbf{H}_r = \hat{y}\, r_{TM}\, H_0 e^{+ik_{1x}x + ik_{1z}z} \tag{5.6.23b}$$

$$\mathbf{H}_t = \hat{y} t_{TM}\, H_0 e^{-ik_{2x}x + ik_{2z}z} \tag{5.6.23c}$$

where $k_{2z} = k_{1z}$ and

$$r_{\mathrm{TM}} = \frac{1 - \left(\dfrac{\varepsilon_1 k_{2x}}{\varepsilon_2 k_{1x}}\right)}{1 + \left(\dfrac{\varepsilon_1 k_{2x}}{\varepsilon_2 k_{1x}}\right)} = \frac{n_2 \cos\theta_{\mathrm{i}} - n_1 \cos\theta_{\mathrm{t}}}{n_2 \cos\theta_{\mathrm{i}} + n_1 \cos\theta_{\mathrm{t}}} \tag{5.6.24a}$$

$$t_{\mathrm{TM}} = \frac{2}{1 + \left(\dfrac{\varepsilon_1 k_{2x}}{\varepsilon_2 k_{1x}}\right)} = \frac{2n_2 \cos\theta_{\mathrm{i}}}{n_2 \cos\theta_{\mathrm{i}} + n_1 \cos\theta_{\mathrm{t}}}. \tag{5.6.24b}$$

Brewster Angle For two dielectric media, the reflection coefficient for the TM polarized light vanishes when $\varepsilon_2 k_{1x} = \varepsilon_1 k_{2x}$, which means

$$n_2 \cos\theta_{\mathrm{i}} = n_1 \cos\theta_{\mathrm{t}}. \tag{5.6.25a}$$

Snell's law requires that

$$n_1 \sin\theta_{\mathrm{i}} = n_2 \sin\theta_{\mathrm{t}} \tag{5.6.25b}$$

The above two equations are satisfied if $\theta_{\mathrm{i}} + \theta_{\mathrm{t}} = 90°$. This angle of incidence θ_{i} at which the reflection vanishes is called the Brewster angle θ_{B}

$$\theta_{\mathrm{B}} = \tan^{-1}(n_2/n_1). \tag{5.6.26}$$

Unlike the critical angle, $\theta_{\mathrm{c}} = \sin^{-1}(n_2/n_1)$, which requires $n_2 < n_1$, the Brewster angle exists for wave transmission in either direction.

For total internal reflection, $k_{2x} = i\alpha_2$, and $r_{\mathrm{TM}} = e^{-i2\phi_{12}^{\mathrm{TM}}}$ where

$$\phi_{12}^{\mathrm{TM}} = \tan^{-1}\left(\frac{\varepsilon_1 \alpha_2}{\varepsilon_2 k_{1x}}\right). \tag{5.6.27}$$

The magnetic field experiences a Goos–Hänchen phase shift of an amount $-2\phi_{12}^{\mathrm{TM}}$ when the angle of incidence is larger than the critical angle θ_{c}.

Example Consider InP ($n = 3.16$) and air interface. We plot the reflectivity of power for TE and TM polarizations for plane wave reflection (a) from air to InP surface, and (b) from InP toward the air interface in Fig. 5.6a and Fig. 5.6b, respectively. For case (a), there is no total internal reflection. The reflectivity of the TE polarization is always larger than that of the TM polarization. The Brewster angle occurs at 72.4°. For case (b), total internal reflection occurs at 18.4°, beyond which both TE and TM polarized lights have a unity reflectivity. The Brewster angle occurs at 17.6°.

5.6.3 Concept of Impedance for Plane Wave Propagation

In free space, the ratio of the transverse electric and magnetic fields for a plane wave is simply $\eta_0 = \sqrt{\mu_0/\varepsilon_0} = 120\pi\Omega$. For a plane wave incident obliquely onto a surface

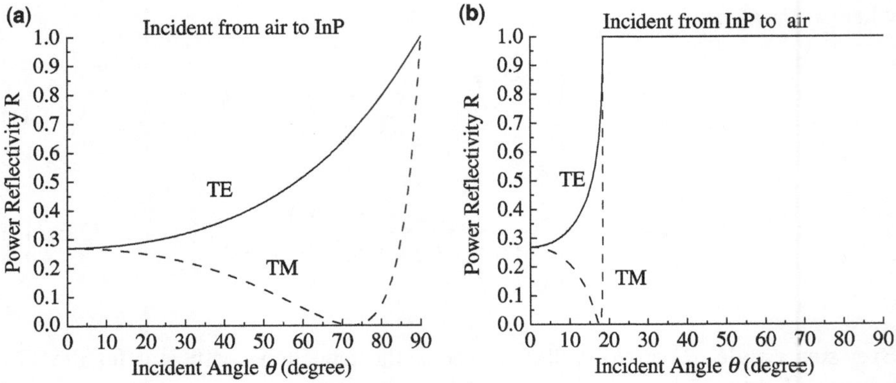

Figure 5.6 (a) The power reflectivities of TE and TM polarized light from air ($n_0 = 1$) to a semiconductor (InP, $n = 3.16$) interface are plotted versus the angle of incidence. The reflectivity of the TM polarized light is generally smaller than that of the TE polarized light and vanishes at the Brewster angle, $\theta_B = \tan^{-1}(n/n_0) = 72.4°$. (b) When the wave is incident from InP semiconductor into the air interface, total internal reflection occurs at $\theta_c = \sin^{-1}(n_0/n) = 18.4°$, and Brewster angle occurs at $\theta_B = \tan^{-1}(n_0/n) = 17.6°$.

at $x = 0$ between medium 1 and 2, we define an impedance for the TE polarized light with $\mathbf{E}_i = \hat{y}E_y$ as

$$\eta_1 = \frac{E_y}{-H_z} = \frac{\omega\mu_1}{k_{1x}} = \sqrt{\frac{\mu_1}{\varepsilon_1}}\frac{1}{\cos\theta_1} = \frac{\eta_0}{n_1\cos\theta_1}. \tag{5.6.28}$$

Using (5.6.1) and (5.6.2), we can define η_2 for the transmitted wave similarly.

The reflection coefficient for the TE polarization is

$$r_{12} = \frac{1-(\eta_1/\eta_2)}{1+(\eta_1/\eta_2)} = \frac{\eta_2-\eta_1}{\eta_2+\eta_1} \tag{5.6.29}$$

$$t_{12} = \frac{2\eta_2}{\eta_2+\eta_1}. \tag{5.6.30}$$

For lossless case and the region of no total internal reflection (i.e., η_1 and η_2 are real),

$$r_{12}^2 + t_{12}t_{21} = 1. \tag{5.6.31}$$

5.7 MATRIX OPTICS

For plane wave reflection from surfaces or a multilayered medium, the matrix approach for forward and backward propagation of plane waves becomes a useful tool. Here we present a simple formulation for the TE polarized wave or E_y component. For TM polarization, we can use the duality principle to obtain similar results for the transverse magnetic field H_y.

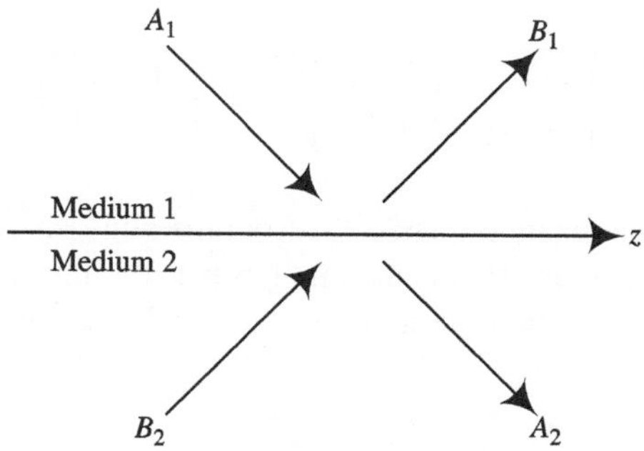

Figure 5.7 Schematics for the incoming and outgoing field amplitudes at a dielectric interface between medium 1 and medium 2 for the construction of the transition matrix.

Transmission at an Interface The incident and reflected electric field E_y in medium 1 can be written as, Fig. 5.7,

$$E_y = (A_1 e^{-ik_{1x}x} + B_1 e^{ik_{1x}x})e^{ik_{1z}z}. \tag{5.7.1}$$

We write similar expressions for E_y in medium 2 with both upward and downward components

$$E_y = (A_2 e^{-ik_{2x}x} + B_2 e^{ik_{2x}x})e^{ik_{2z}z} \tag{5.7.2}$$

where $k_{2z} = k_{1z}$ due to phase matching. Matching the boundary conditions that tangential components E_y and $H_z = (1/i\omega\mu)(\partial E_y/\partial x)$ are continuous at $x = 0$, we obtain two equations relating the four field amplitudes

$$A_1 + B_1 = A_2 + B_2 \tag{5.7.3a}$$

$$\frac{k_{1x}}{\omega\mu_1}(A_1 - B_1) = \frac{k_{2x}}{\omega\mu_2}(A_2 - B_2). \tag{5.7.3b}$$

We can recognize that the plane wave impedance

$$\eta_i = \frac{\omega\mu_i}{k_{ix}}, \quad i = 1, 2 \tag{5.7.4}$$

for TE polarization. Note the definition

$$P_{12} = \frac{\mu_1 k_{2x}}{\mu_2 k_{1x}} = \frac{\eta_1}{\eta_2}. \tag{5.7.5}$$

The field amplitudes are related by a 2×2 matrix

$$\begin{bmatrix} A_1 \\ B_1 \end{bmatrix} = \mathbf{M} \begin{bmatrix} A_2 \\ B_2 \end{bmatrix} \tag{5.7.6}$$

where

$$\mathbf{M} = \frac{1}{2}\begin{bmatrix} 1+P_{12} & 1-P_{12} \\ 1-P_{12} & 1+P_{12} \end{bmatrix} = \frac{1}{t_{12}}\begin{bmatrix} 1 & r_{12} \\ r_{12} & 1 \end{bmatrix} = \mathbf{T}_{12}, \tag{5.7.7}$$

which defines a 2×2 transition matrix $\mathbf{T}_{12}$ from medium 1 into medium 2 interface.

Alternatively, the relations for the incoming wave amplitudes A_1 and B_2 toward the interface and the outgoing wave amplitudes A_2 and B_1 can be written in terms of a scattering matrix **S**.

$$\begin{bmatrix} B_1 \\ A_2 \end{bmatrix} = \mathbf{S}\begin{bmatrix} A_1 \\ B_2 \end{bmatrix} \tag{5.7.8a}$$

$$\mathbf{S} = \begin{bmatrix} r_{12} & t_{21} \\ t_{12} & r_{21} \end{bmatrix}. \tag{5.7.8b}$$

Propagation over a Uniform Layer of Dielectric Medium If we consider the plane wave at $x = 0$ as

$$E_y = (Ae^{-ik_x x} + Be^{ik_x x})e^{ik_z z} \tag{5.7.9}$$

when the wave is expressed at a uniform layer of a dielectric medium with a thickness h, Fig. 5.8, we have (at $x = -h$)

$$E_y = [A'e^{-ik_x(x+h)} + B'e^{ik_x(x+h)}]e^{ik_z z}, \tag{5.7.10}$$

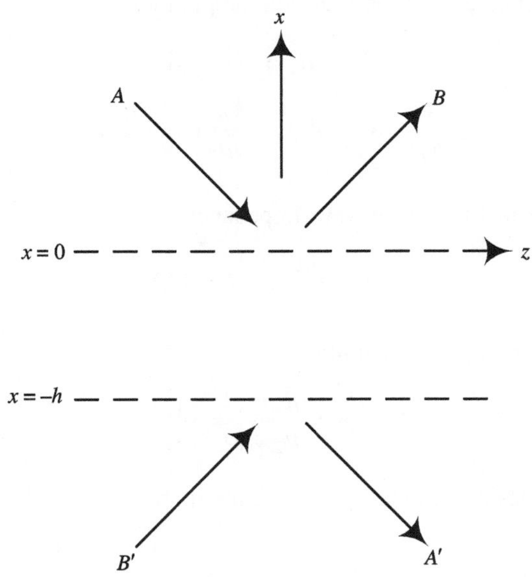

Figure 5.8 Propagation of plane waves over a uniform dielectric layer of thickness h.

we can see that the field amplitudes are related by a propagation matrix **P** as below:

$$\begin{bmatrix} A \\ B \end{bmatrix} = \begin{bmatrix} e^{-ik_x h} & 0 \\ 0 & e^{ik_x h} \end{bmatrix} \begin{bmatrix} A' \\ B' \end{bmatrix}. \tag{5.7.11}$$

Combination of an Interface Transition Plus a Uniform Layer Propagation

$$\begin{aligned} \mathbf{M} = \mathbf{T}_{12}\mathbf{P}_2 &= \frac{1}{2}\begin{bmatrix} 1+P_{12} & 1-P_{12} \\ 1-P_{12} & 1+P_{12} \end{bmatrix} \begin{bmatrix} e^{-ik_{2x}h} & 0 \\ 0 & e^{ik_{2x}h} \end{bmatrix} \\ &= \frac{1}{2}\begin{bmatrix} (1+P_{12})e^{-ik_{2x}h} & (1-P_{12})e^{ik_{2x}h} \\ (1-P_{12})e^{-ik_{2x}h} & (1+P_{12})e^{ik_{2x}h} \end{bmatrix}. \end{aligned} \tag{5.7.12}$$

Power Normalized Formulation Because the impedance η_i differs from medium 1 to medium 2, the scattering matrix in (5.7.8b) is not symmetric, namely $t_{12} \neq t_{21}$. Very often, the field amplitude can be redefined to account for the difference in impedance

$$E_y = a_1\sqrt{2\eta_1}\, e^{-ik_{1x}x+ik_{1z}z} \tag{5.7.13a}$$

such that the incident power along $-x$-direction is

$$P_1 = |a_1|^2. \tag{5.7.13b}$$

That is, we redefine

$$\begin{aligned} A_1 &= \sqrt{2\eta_1}\, a_1, \qquad & B_1 &= \sqrt{2\eta_1}\, b_1 \\ A_2 &= \sqrt{2\eta_2}\, a_2, \qquad & B_2 &= \sqrt{2\eta_2}\, b_2. \end{aligned} \tag{5.7.14}$$

Then,

$$\begin{bmatrix} b_1 \\ a_2 \end{bmatrix} = \mathbf{S} \begin{bmatrix} a_1 \\ b_2 \end{bmatrix} \tag{5.7.15}$$

$$\mathbf{S} = \begin{bmatrix} r_{12} & t \\ t & r_{21} \end{bmatrix} \tag{5.7.16}$$

$$t = 2\frac{\sqrt{\eta_1\eta_2}}{\eta_1+\eta_2} \tag{5.7.17}$$

$$r_{12} = \frac{\eta_2-\eta_1}{\eta_2+\eta_1}. \tag{5.7.18}$$

We can see the matrix has the property

$$S_{12} = S_{21} \tag{5.7.19}$$

which follows the reciprocity relation. Power conservation requires

$$|a_1|^2 - |b_1|^2 = |a_2|^2 - |b_2|^2 \tag{5.7.20}$$

which means $|a_1|^2 + |b_2|^2 = |a_2|^2 + |b_1|^2$,

$$\mathbf{S}^{+}\mathbf{S} = \mathbf{I} \tag{5.7.21}$$

or **S** is unitary. Here **I** is a unity matrix.

$$\begin{aligned} S_{11}^* S_{11} + S_{21}^* S_{21} &= 1 \\ S_{11}^* S_{12} + S_{21}^* S_{22} &= 0 \\ S_{12}^* S_{11} + S_{22}^* S_{21} &= 0 \\ S_{12}^* S_{12} + S_{22}^* S_{22} &= 1. \end{aligned} \tag{5.7.22}$$

The **M**-matrix becomes

$$\begin{bmatrix} a_1 \\ b_1 \end{bmatrix} = \mathbf{M} \begin{bmatrix} a_2 \\ b_2 \end{bmatrix} \tag{5.7.23a}$$

$$\mathbf{M} = \frac{1}{t} \begin{bmatrix} 1 & r_{12} \\ r_{12} & 1 \end{bmatrix} \tag{5.7.23b}$$

$$t = 2\frac{\sqrt{\eta_1 \eta_2}}{\eta_1 + \eta_2}. \tag{5.7.23c}$$

5.8 PROPAGATION MATRIX APPROACH FOR PLANE WAVE REFLECTION FROM A MULTILAYERED MEDIUM

If the medium is inhomogeneous along the x-direction $\varepsilon = \varepsilon(x)$ and $\mu = \mu(x)$, we can discretize the permittivity and permeability and use the propagation matrix method. We use

$$\begin{aligned} \mu_\ell &= \mu(x_\ell) \\ \varepsilon_\ell &= \varepsilon(x_\ell) \text{ for } -d_{\ell-1} \geq x \geq -d_\ell \end{aligned}$$

as shown in Fig. 5.9. For a TE polarized incident wave

$$\mathbf{E}_{\mathrm{i}} = \hat{y} E_0 e^{-ik_{0x}x + ik_{0z}z} \tag{5.8.1}$$

we have the reflected wave

$$\mathbf{E}_{\mathrm{r}} = \hat{y} r E_0 e^{ik_{0x}x + ik_{0z}z}. \tag{5.8.2}$$

In the ℓth layer, $x_{\ell-1} \geq x \geq x_\ell$, the electric field is given by $\mathbf{E}_\ell = \hat{y} E_y^\ell$ where

$$\begin{aligned} E_y^\ell &= \left[A_\ell e^{-ik_{\ell x}(x+d_\ell)} + B_\ell e^{ik_{\ell x}(x+d_\ell)}\right] e^{ik_{\ell z}z} \\ H_z^\ell &= \frac{1}{i\omega\mu_\ell}\frac{\partial}{\partial x} E_y^\ell \\ &= \frac{-k_{\ell x}}{\omega\mu_\ell}[A_\ell e^{-ik_{\ell x}(x+d_\ell)} - B_\ell e^{ik_{\ell x}(x+d_\ell)}] e^{ik_{\ell z}z} \end{aligned} \tag{5.8.3}$$

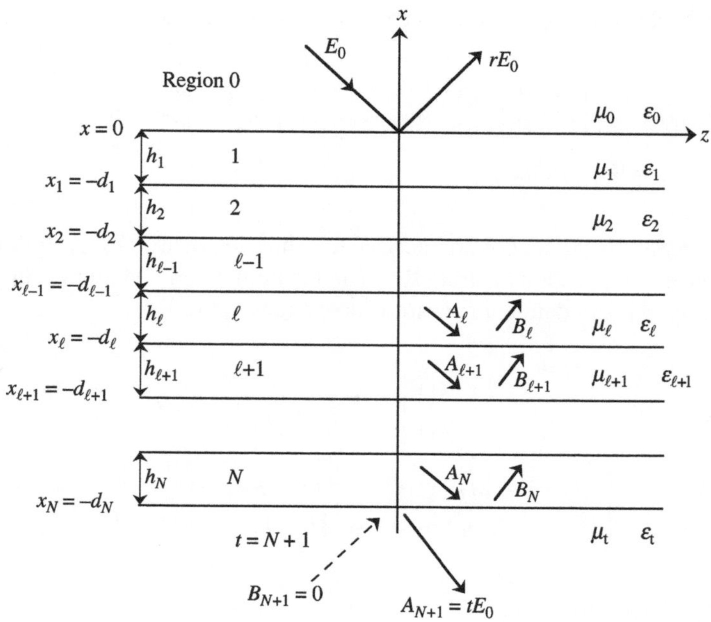

Figure 5.9 A plane wave with TE polarization $\mathbf{E} = \hat{y}$, $E_y = \hat{y}$, $E_0\, e^{-ik_{0x}x+ik_{0z}z}$ is incident on a multilayered medium.

and $k_{\ell x} = \sqrt{\omega^2 \mu_\ell \varepsilon_\ell - k_{0z}^2}$, $k_{\ell z} = k_{0z}$ for all ℓ. The boundary conditions in which E_y and H_z are continuous at $x = -d_\ell$ lead to

$$A_\ell + B_\ell = A_{\ell+1} e^{-ik_{(\ell+1)x}(-d_\ell + d_{\ell+1})} + B_{\ell+1} e^{ik_{(\ell+1)x}(-d_\ell + d_{\ell+1})}$$

$$A_\ell - B_\ell = \frac{\mu_\ell k_{(\ell+1)x}}{\mu_{\ell+1} k_{\ell x}} \left[A_{\ell+1} e^{-ik_{(\ell+1)x}(-d_\ell + d_{\ell+1})} - B_{\ell+1} e^{ik_{(\ell+1)x}(-d_\ell + d_{\ell+1})} \right]. \tag{5.8.4}$$

Again, similar to the propagation method in a one-dimensional potential in the Schrödinger equation, we define

$$P_{\ell(\ell+1)} = \frac{\mu_\ell k_{(\ell+1)x}}{\mu_{\ell+1} k_{\ell x}} \tag{5.8.5}$$

and the thickness of the region $\ell + 1$

$$h_{\ell+1} = d_{\ell+1} - d_\ell. \tag{5.8.6}$$

We have

$$\begin{bmatrix} A_\ell \\ B_\ell \end{bmatrix} = \mathbf{B}_{\ell(\ell+1)} \begin{bmatrix} A_{\ell+1} \\ B_{\ell+1} \end{bmatrix} \tag{5.8.7a}$$

where the backward-propagation matrix $\mathbf{B}_{\ell(\ell+1)}$ is defined as

$$\mathbf{B}_{\ell(\ell+1)} = \frac{1}{2}\begin{bmatrix} (1+P_{\ell(\ell+1)})e^{-ik_{(\ell+1)x}h_{\ell+1}} & (1-P_{\ell(\ell+1)})e^{ik_{(\ell+1)x}h_{\ell+1}} \\ (1-P_{\ell(\ell+1)})e^{-ik_{(\ell+1)x}h_{\ell+1}} & (1+P_{\ell(\ell+1)})e^{ik_{(\ell+1)x}h_{\ell+1}} \end{bmatrix}$$
$$= \mathbf{T}_{\ell(\ell+1)}\mathbf{P}_{(\ell+1)} \tag{5.8.7b}$$

which is simply the product of an interface transmission matrix $\mathbf{T}_{\ell(\ell+1)}$ followed by a uniform layer propagation matrix, $\mathbf{P}_{\ell+1}$, in medium $\ell+1$, as shown in (5.7.12). Alternatively, we can define a forward-propagation matrix $\mathbf{F}_{(\ell+1)\ell}$

$$\begin{bmatrix} A_{\ell+1} \\ B_{\ell+1} \end{bmatrix} = \mathbf{F}_{(\ell+1)\ell}\begin{bmatrix} A_\ell \\ B_\ell \end{bmatrix} \tag{5.8.8a}$$

$$\mathbf{F}_{(\ell+1)\ell} = \frac{1}{2}\begin{bmatrix} (1+P_{(\ell+1)\ell})e^{ik_{(\ell+1)x}h_{\ell+1}} & (1-P_{(\ell+1)\ell})e^{ik_{(\ell+1)x}h_{\ell+1}} \\ (1-P_{(\ell+1)\ell})e^{-ik_{(\ell+1)x}h_{\ell+1}} & (1+P_{(\ell+1)\ell})e^{-ik_{(\ell+1)x}h_{\ell+1}} \end{bmatrix}. \tag{5.8.8b}$$

Note that $P_{(\ell+1)\ell} = 1/P_{\ell(\ell+1)}$.

The amplitudes of the incident and reflected waves are related to those amplitudes in the transmitted region $N+1$ by

$$\begin{bmatrix} E_0 \\ rE_0 \end{bmatrix} = \mathbf{B}_{01}\mathbf{B}_{12}\mathbf{B}_{23}\cdots\mathbf{B}_{N(N+1)}\begin{bmatrix} A_{N+1} \\ B_{N+1} \end{bmatrix}$$
$$= \begin{bmatrix} b_{11} & b_{12} \\ b_{21} & b_{22} \end{bmatrix}\begin{bmatrix} tE_0 \\ 0 \end{bmatrix} \tag{5.8.9}$$

where $B_{N+1} = 0$ as there is no incident wave from the bottom region in Fig. 5.9. We obtain the transmission coefficient of the multilayered medium

$$t = \frac{1}{b_{11}} \tag{5.8.10}$$

and the reflection coefficient of the multilayered medium

$$r = \frac{b_{21}}{b_{11}}. \tag{5.8.11}$$

The electric field in the transmitted region is given by

$$E_y = tE_0 e^{-ik_{(N+1)x}(x-d_{N+1})} e^{ik_{(N+1)z}z}. \tag{5.8.12}$$

Because we use d_N as the bottom boundary, the last region $N+1$ can be chosen such that $h_{N+1} = d_{N+1} - d_N = 0$ for convenience. It means that the field will be measured with $d_{N+1} = d_N$.

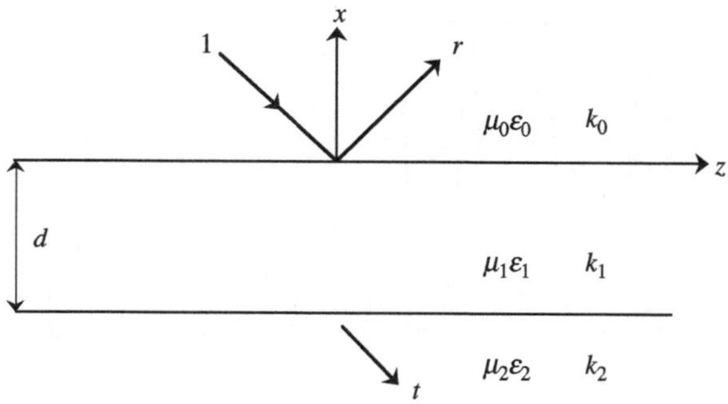

Figure 5.10 A plane wave reflection from a dielectric slab with a thickness d.

As an example, we consider a dielectric slab with a thickness d as shown in Fig. 5.10. The backward-propagation matrices $\mathbf{B}_{01}$ and $\mathbf{B}_{12}$ are

$$\mathbf{B}_{01} = \frac{1}{2}\begin{bmatrix} (1+P_{01})e^{-ik_{1x}d} & (1-P_{01})e^{ik_{1x}d} \\ (1-P_{01})e^{-ik_{1x}d} & (1+P_{01})e^{ik_{1x}d} \end{bmatrix} \tag{5.8.13}$$

$$\mathbf{B}_{12} = \frac{1}{2}\begin{bmatrix} (1+P_{12}) & (1-P_{12}) \\ (1-P_{12}) & (1+P_{12}) \end{bmatrix}. \tag{5.8.14}$$

We calculate the matrix product $\mathbf{B}_{01}\mathbf{B}_{12}$, then the transmission coefficient,

$$t = \frac{1}{b_{11}} = \frac{4e^{ik_{1x}d}}{(1+P_{01})(1+P_{12})[1+r_{01}r_{12}e^{i2k_{1x}d}]} = \frac{t_{01}t_{12}e^{ik_{1x}d}}{1+r_{01}r_{12}e^{i2k_{1x}d}} \tag{5.8.15}$$

and the reflection coefficient

$$r = \frac{b_{21}}{b_{11}} = \frac{r_{01} + r_{12}\ e^{i2k_{1x}d}}{1 + r_{01}r_{12}\ e^{i2k_{1x}d}} \tag{5.8.16}$$

where

$$r_{\ell(\ell+1)} = \frac{1 - P_{\ell(\ell+1)}}{1 + P_{\ell(\ell+1)}} \tag{5.8.17}$$

is the reflection coefficient of a planar surface between medium ℓ and medium $\ell + 1$. Note that $r_{\ell(\ell+1)} = -r_{(\ell+1)\ell}$ as $P_{\ell(\ell+1)} = 1/P_{(\ell+1)\ell}$.

For TM polarization, we use the duality principle again: exchange $\mathbf{E} \rightarrow \mathbf{H}$, $\mathbf{H} \rightarrow -\mathbf{E}$, $\mu_\ell \rightarrow \varepsilon_\ell$, and $\varepsilon_\ell \rightarrow \mu_\ell$.

Example In Fig. 5.11, we plot the power reflectivity of a slab of InP semiconductor assuming a refractive index of $n = 3.16$ near the wavelength of 1.55 μm. We can see the multiple Fabry–Perot interferences with maxima and minima of the reflectivity determined by $2k_1d = 2m\pi$ or $2k_1d = (2m+1)\pi$, respectively. In other words, the reflectivity vanishes at $r_{01} + r_{12}\exp(i2k_1d) = 0$, which gives the conditions for the wavelengths for minimum or maximum reflectivity.

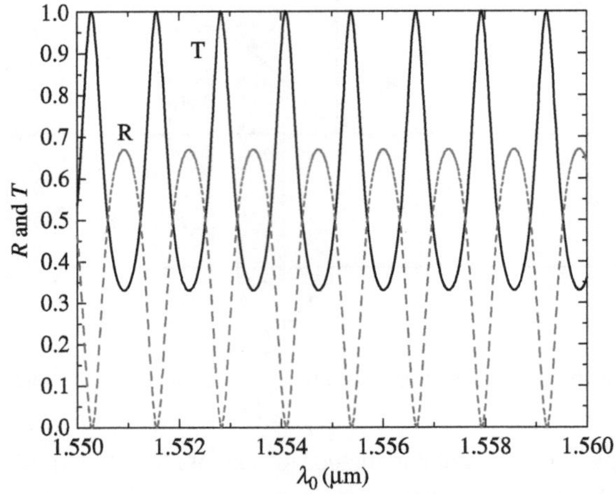

Figure 5.11 Power reflectivity (R) and transmissivity (T) of a slab of InP semiconductor with a refractive index $n = 3.16$ near 1.55 μm wavelength. The dispersion of the refractive index is ignored.

5.9 WAVE PROPAGATION IN PERIODIC MEDIA

5.9.1 Dispersion Diagram and Stop Bands

When the dielectric medium is periodic, the concept of dispersion relation similar to that of band structure of semiconductor crystals exists. There are "stop bands" similar to those of forbidden band gaps in semiconductors. For a simple one-dimensional periodic structure with a period Λ,

$$\varepsilon_r(x) = \varepsilon(x + \Lambda) = \sum_{p=-\infty}^{+\infty} \varepsilon_p e^{ip\frac{2\pi}{\Lambda}x} \tag{5.9.1}$$

where ε_p is the pth component of the Fourier series expansion. The wave solution (e.g., TE polarization) $\mathbf{E} = \hat{y}\, E_y(x)$ has the form

$$E_y(x) = e^{ikx} u_k(x) = \sum_n E_n e^{ikx + in\frac{2\pi}{\Lambda}x} \tag{5.9.2}$$

where $u_k(x)$ is a periodic function, $u_k(x + \Lambda) = u_k(x)$. The wave equation

$$\left[\frac{\partial^2}{\partial x^2} + \frac{\omega^2}{c^2}\varepsilon_r(x)\right] E_y(x) = 0. \tag{5.9.3}$$

Let $g = 2\pi/\Lambda$ as the grating wave number. The wave equation can be written in terms of the Fourier series expansion

$$-\sum_n (k + ng)^2 E_n e^{i(k+ng)x} + \frac{\omega^2}{c^2} \sum_p \sum_n \varepsilon_p E_n e^{ikx + i(n+p)gx} = 0. \tag{5.9.4}$$

Multiplying the above by $e^{-i(k+mg)x}$ and integrating x over a period Λ, we retain the mth component of E_m

$$-(k+mg)^2 E_m + \frac{\omega^2}{c^2} \sum_{p=-\infty}^{\infty} \varepsilon_p E_{m-p} = 0. \tag{5.9.5}$$

The above equations for different m become a set of linearly coupled equations, which require numerical solution of the determinantal equation with truncation by keeping a finite number of m. If we separate the E_m term in the summation from the rest of $p \neq 0$

$$E_m = \frac{\omega^2}{c^2} \sum_{p \neq 0} \frac{\varepsilon_p}{(k+mg)^2 - \dfrac{\omega^2}{c^2} n_0^2} E_{m-p} \tag{5.9.6}$$

where $n_0^2 = \varepsilon_0$ or the dc term is used. If E_0 is the dominant term (i.e., $E_0\, e^{ikx}$), the rest are

$$E_m \simeq \frac{\omega^2}{c^2} \frac{\varepsilon_m}{(k+mg)^2 - \dfrac{\omega^2}{c^2} n_0^2} E_0, \quad m \neq 0. \tag{5.9.7}$$

On the other hand, strong coupling between various E_m components occurs near

$$k + mg = \pm \frac{\omega}{c} n_0, \quad m = 0,\ \pm 1,\ \pm 2,\ \ldots \tag{5.9.8}$$

The set of curves on the ω-k diagram indicate various Fourier series components or "Floquet" modes with wave vector given by

$$(k+mg)\hat{x} = \pm \frac{\omega}{c} n_0 \hat{x}, \tag{5.9.9}$$

Fig. 5.12a. For example, the dispersion curve

$$k = \frac{\omega}{c} n_0 \tag{5.9.10}$$

for the $E_0\, e^{ikx}$ mode will be coupled to another mode

$$k - g = -\frac{\omega}{c} n_0 \tag{5.9.11}$$

or the $E_{-1} e^{i(k-g)x}$ mode with ω-k relation given above, Fig. 5.12b. The resonance occurs near the intercept of two lines at the "Bragg" wave number

$$k_B = \frac{g}{2} = \frac{\omega_B}{c} n_0 = \frac{\pi}{\Lambda}. \tag{5.9.12}$$

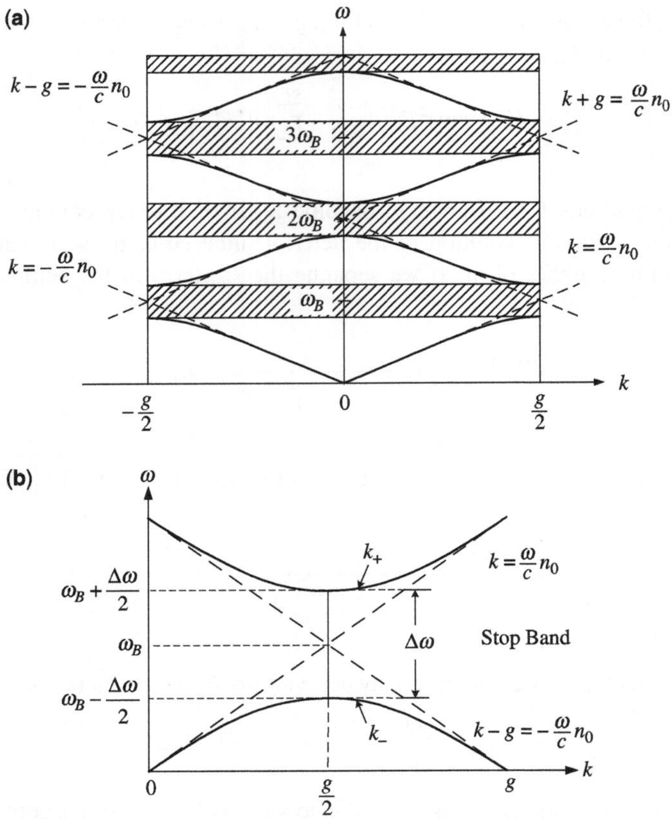

Figure 5.12 (a) The dispersion diagrams of the Floquet modes, where the intercepts of positive and negative traveling waves form stop bands. (b) An expansion near the stop band formed between the 0th order and −1 order Floquet mode.

We can write the dominant terms near the Bragg wavelength by keeping the $m=0$ and $m=-1$ terms in the coupled equations

$$\left(\frac{\omega^2}{c^2}n_0^2-k^2\right)E_0+\frac{\omega^2}{c^2}\varepsilon_1 E_{-1}=0 \tag{5.9.13a}$$

$$\frac{\omega^2}{c^2}\varepsilon_{-1}E_0+\left[\frac{\omega^2}{c^2}n_0^2-(k-g)^2\right]E_{-1}=0. \tag{5.9.13b}$$

For nontrivial solutions, the determinant vanishes

$$\left[\left(\frac{\omega}{c}n_0\right)^2-k^2\right]\left[\left(\frac{\omega}{c}n_0\right)^2-(k-g)^2\right]-k_1^4=0 \tag{5.9.14}$$

where we have defined $k_1^2 = \frac{\omega^2}{c^2}\sqrt{\varepsilon_1\varepsilon_{-1}}$. To understand the stop band behavior near the Bragg wavelength, we further approximate

$$\left(\frac{\omega}{c}n_0\right)^2 - k^2 = \left(\frac{\omega}{c}n_0 + k\right)\left(\frac{\omega}{c}n_0 - k\right)$$

$$\simeq g\left(\frac{\omega}{c}n_0 - k\right) \tag{5.9.15a}$$

$$\left(\frac{\omega}{c}n_0\right)^2 - (k-g)^2 \simeq g\left[\frac{\omega}{c}n_0 + (k-g)\right]. \tag{5.9.15b}$$

The determinantal equation is reduced to

$$\left(\frac{\omega}{c}n_0 - k\right)\left[\frac{\omega}{c}n_0 + (k-g)\right] - \frac{k_1^4}{g^2} = 0 \tag{5.9.16}$$

which allows for the solution for k

$$k = \frac{g}{2} \pm \sqrt{\left(\frac{\omega}{c}n_0 - \frac{g}{2}\right)^2 - \frac{k_1^4}{g^2}}. \tag{5.9.17}$$

The stop band occurs near, Fig. 5.12b,

$$\left|\frac{\omega}{c}n_0 - \frac{\omega_B}{c}n_0\right| \leq \frac{k_1^2}{g} \simeq \frac{\omega_B^2}{c^2}\frac{\sqrt{\varepsilon_1\varepsilon_{-1}}}{g} \simeq \frac{\omega_B}{c}\frac{\sqrt{\varepsilon_1\varepsilon_{-1}}}{2n_0} \tag{5.9.18}$$

or

$$|\omega - \omega_B| \leq \frac{\omega_B\sqrt{\varepsilon_1\varepsilon_{-1}}}{2n_0^2} \equiv \frac{\Delta\omega}{2}. \tag{5.9.19}$$

5.9.2 Plane Wave Reflection from a Distributed-Bragg Reflector

Distributed-Bragg reflectors (DBRs) play an important role in optical filters and mirrors, which are used in vertical-cavity surface-emitting lasers (VCSELs) and microcavity lasers [18]. In this section, we apply the transition matrix or propagation matrix approach, which allows for numerical calculation of the DBR structure. A typical design requires a structure consisting of $(2N + 1)$ layers of alternating high and low refractive indices n_1 and n_2, with the layer thicknesses h_1 and h_2 close to a quarter of the resonant (Bragg) wavelength in each layer, Fig. 5.13. For example, if we choose the Bragg resonant wavelength as λ_B,

$$h_1 = \frac{\lambda_B}{4n_1}, \quad h_2 = \frac{\lambda_B}{4n_2}.$$

In general, we can define the phase factors

$$\phi_1 = k_1h_1 = \frac{2\pi}{\lambda}n_1h_1, \quad \phi_2 = k_2h_2 = \frac{2\pi}{\lambda}n_2k_2 \tag{5.9.20}$$

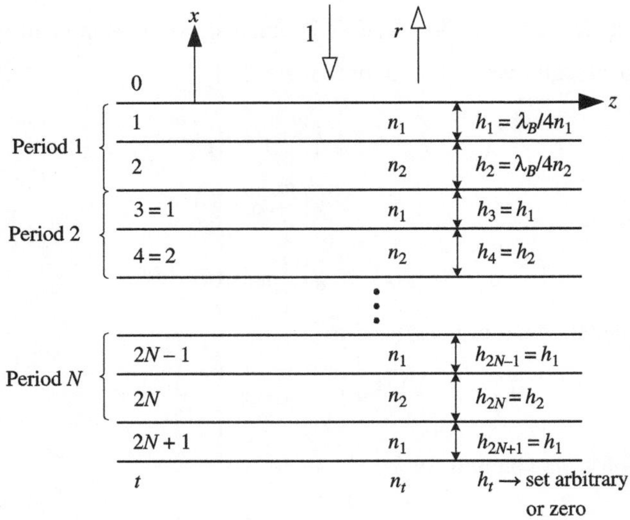

Figure 5.13 A distributed-Bragg reflector (DBR) structure consisting of $(2N + 1)$ layers of alternating quarter-wavelength high and low index materials.

for arbitrary thicknesses h_1 and h_2. The propagate matrix gives the relation

$$
\begin{aligned}
\begin{bmatrix} A_0 \\ B_0 \end{bmatrix} &= \mathbf{B}_{01}(\mathbf{B}_{12}\mathbf{B}_{23})(\mathbf{B}_{34}\mathbf{B}_{45})\cdots(\mathbf{B}_{(2N-1)2N}\mathbf{B}_{2N(2N+1)})\mathbf{B}_{(2N+1)t}\begin{bmatrix} tA_0 \\ 0 \end{bmatrix} \\
&= \mathbf{B}_{01}\mathbf{M}^N\mathbf{B}_{(2N+1)t}\begin{bmatrix} tA_0 \\ 0 \end{bmatrix}
\end{aligned} \tag{5.9.21}
$$

where $\mathbf{M}$ is the product of two matrices for one period of n_1 and n_2 layers

$$
\mathbf{M} = \mathbf{B}_{12}\mathbf{B}_{21} = \begin{bmatrix} A & B \\ C & D \end{bmatrix} \tag{5.9.22}
$$

and the matrix elements are

$$
\begin{aligned}
A &= \frac{(n_1 + n_2)^2}{4n_1 n_2} e^{-i(\phi_1+\phi_2)} - \frac{(n_1 - n_2)^2}{4n_1 n_2} e^{-i(\phi_1-\phi_2)} \\
B &= \left(\frac{n_2^2 - n_1^2}{4n_1 n_2}\right) e^{i\phi_1}(e^{-i\phi_2} - e^{i\phi_2}) \\
C &= \left(\frac{n_2^2 - n_1^2}{4n_1 n_2}\right) e^{-i\phi_1}(e^{i\phi_2} - e^{-i\phi_2}) = B^* \\
D &= \frac{(n_1 + n_2)^2}{4n_1 n_2} e^{i(\phi_1+\phi_2)} - \frac{(n_1 - n_2)^2}{4n_1 n_2} e^{i(\phi_1-\phi_2)} = A^*.
\end{aligned} \tag{5.9.23}
$$

A, B, C, and D satisfy the property

$$AD - BC = 1, \tag{5.9.24}$$

that is, the matrix M is unimodular. In addition, the above structure gives the property that $A = D^*$ and $B = C^*$. Therefore,

$$|A|^2 - |C|^2 = 1 \tag{5.9.25}$$

The Nth power of the unimodular matrix can be written in terms of the Chebychev polynomials $U_N(x)$

$$\begin{aligned} \mathbf{M}^N &= \begin{bmatrix} A & B \\ C & D \end{bmatrix}^N = \begin{bmatrix} m_{11} & m_{12} \\ m_{21} & m_{22} \end{bmatrix} \\ &= \begin{bmatrix} AU_{N-1}(x) - U_{N-2}(x) & BU_{N-1}(x) \\ CU_{N-1}(x) & DU_{N-1}(x) - U_{N-2}(x) \end{bmatrix} \end{aligned} \tag{5.9.26}$$

where $x = (A + D)/2$, and the Chebychev polynomials satisfy the recurrence relation,

$$U_{N+1}(x) = 2xU_N(x) - U_{N-1}(x), \tag{5.9.27}$$

with $U_0(x) = 1$, $U_1(x) = 2x$, $U_2(x) = 4x^2 - 1$, etc., and

$$U_i(x) = 0 \quad \text{if } i < 0. \tag{5.9.28}$$

Another form of writing $U_N(x)$ is

$$U_N(x) = \frac{\sin(N+1)\theta}{\sin\theta} \tag{5.9.29}$$

where

$$x = \cos\theta = (A + D)/2. \tag{5.9.30}$$

When $|(A + D)/2| > 1$, the solution for θ becomes complex

$$\theta = m\pi + ia \tag{5.9.31}$$

$$U_N(x) = \pm\frac{\sinh(N+1)\alpha}{\sinh\alpha} \tag{5.9.32}$$

where $\pm$ signs refer to

$$\begin{aligned}\sin(m\pi + i\alpha) &= \cos m\pi i \sinh\alpha \\ &= (-1)^m i \sinh\alpha = \pm i \sinh\alpha \\ \sin[(N+1)(m\pi + i\alpha)] &= (-1)^{m(N+1)} i \sinh[(N+1)\alpha] \\ &= \pm i \sinh[(N+1)\alpha]\end{aligned} \tag{5.9.33}$$

which depends on $m =$ even or odd, and $(N+1)m$ is even or odd. Note that

$$\mathbf{B}_{01} = \frac{1}{2}\begin{bmatrix} \left(1 + \dfrac{n_1}{n_0}\right)e^{-i\phi_1} & \left(1 - \dfrac{n_1}{n_0}\right)e^{i\phi_1} \\ \left(1 - \dfrac{n_1}{n_0}\right)e^{-i\phi_1} & \left(1 + \dfrac{n_1}{n_0}\right)e^{i\phi_1} \end{bmatrix} \tag{5.9.34}$$

is just a propagation matrix for a uniform layer. $\mathbf{B}_{(2N+1)t}$ is

$$\mathbf{B}_{(2N+1)t} = \frac{1}{2}\begin{bmatrix} 1 + \dfrac{n_t}{n_1} & 1 - \dfrac{n_t}{n_1} \\ 1 - \dfrac{n_t}{n_1} & 1 + \dfrac{n_t}{n_1} \end{bmatrix}. \tag{5.9.35}$$

If we consider $n_0 = n_1$ and $n_t = n_1$, which implies matched index layers for region 0 and t to n_1, we have $\mathbf{B}_{(2N+1)t} =$ a unity matrix. Therefore,

$$\begin{aligned}\begin{bmatrix} 1 \\ r \end{bmatrix} &= \begin{bmatrix} e^{-i\phi_1} & 0 \\ 0 & e^{i\phi_1} \end{bmatrix}\begin{bmatrix} m_{11} & m_{12} \\ m_{21} & m_{22} \end{bmatrix}\begin{bmatrix} t \\ 0 \end{bmatrix} \\ &= \begin{bmatrix} m_{11}e^{-i\phi_1} & m_{12}e^{-i\phi_1} \\ m_{21}e^{i\phi_1} & m_{22}e^{i\phi_1} \end{bmatrix}\begin{bmatrix} t \\ 0 \end{bmatrix}.\end{aligned} \tag{5.9.36}$$

The reflection coefficient for the $(2N+1)$ layer structure (when $n_0 = n_1$ and $n_t = n_1$) is, therefore,

$$r_N = \frac{m_{21}}{m_{11}}e^{i2\phi_1} = \frac{CU_{N-1}(x)}{AU_{N-1}(x) - U_{N-2}(x)}e^{i2\phi_1} \tag{5.9.37}$$

and

$$t_N = \frac{1}{m_{11}}e^{i\phi_1}. \tag{5.9.38}$$

The magnitude of $|r_N|^2$ is

$$|r_N|^2 = \left|\frac{CU_{N-1}(x)}{AU_{N-1}(x) - U_{N-2}(x)}\right|^2. \tag{5.9.39}$$

Using the fact that $|A|^2 = |C|^2 + 1$ and some algebraic manipulation, we can write $|r_N|^2$ as

$$|r_N|^2 = \frac{|C|^2}{|C|^2 + \left|\dfrac{\sin\theta}{\sin N\theta}\right|^2} \tag{5.9.40}$$

or inside the stop band $\theta = m\pi + i\alpha$.

$$|r_N|^2 = \frac{|C|^2}{|C|^2 + \left|\dfrac{\sinh\alpha}{\sinh N\alpha}\right|^2}. \tag{5.9.41}$$

It is noted that when $N = 1$,

$$|r_1|^2 = \frac{|C|^2}{|C|^2 + 1}. \tag{5.9.42}$$

Therefore,

$$|C|^2 = \frac{|r_1|^2}{1 - |r_1|^2} = \frac{R_1}{1 - R_1}. \tag{5.9.43}$$

The magnitude of $|C|^2$ is related to the power reflectivity of a single period, $R_1 = |r_1|^2$. For $(2N + 1)$ layers, the power reflectivity becomes

$$|r_N|^2 = \frac{R_1}{R_1 + (1 - R_1)\left(\dfrac{\sinh\alpha}{\sinh(N\alpha)}\right)^2}. \tag{5.9.44}$$

As N approaches a large number, $(\sinh\alpha/\sinh(N\alpha))^2 \to \exp(-2(N-1)\alpha) \to$ zero. Therefore $|r_N|^2 \to$ unity in the stop band. Figure 5.14a plots the power reflectivity spectrum $|r_N|^2$ as a function of frequency ω near the stop band. In Fig. 5.14b, we plot the phase θ_N of $r_N = |r_N|e^{i\theta_N}$ from (5.9.37).

Special Case Right at Resonance $\lambda = \lambda_B$ At exact resonance wavelength, $\phi_1 = k_1 h_1 = \pi/2$ and $\phi_2 = k_2 h_2 = \pi/2$. The matrix $\mathbf{M}$ can be simplified as

$$\begin{aligned}\mathbf{M} &= \mathbf{B}_{12}\mathbf{B}_{21}\\ &= \frac{-1}{2}\begin{bmatrix} \dfrac{n_1}{n_2} + \dfrac{n_2}{n_1} & \dfrac{n_1}{n_2} - \dfrac{n_2}{n_1} \\ \dfrac{n_1}{n_2} - \dfrac{n_2}{n_1} & \dfrac{n_1}{n_2} + \dfrac{n_2}{n_1} \end{bmatrix}.\end{aligned} \tag{5.9.45}$$

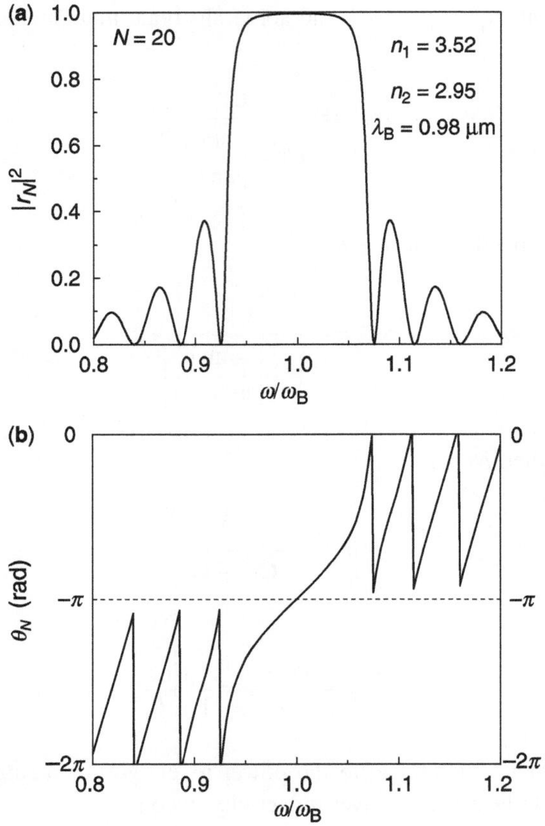

Figure 5.14 (a) The power reflectivity of a DBR structure $|r_N|^2$ is plotted as a function of wavelength near the stop band wavelength. (b) The phase of the reflection coefficient r_N. The parameters are $n_1 = 3.52$, $n_2 = 2.95$, $\lambda_B = 0.98\ \mu\text{m}$, $h_1 = \lambda_B/(4n_1) = 69.60$ nm, and $h_2 = \lambda_B/(4n_2) = 83.05$ nm.

$\mathbf{M}^N$ can be evaluated using the simple identity

$$\frac{(-1)^N}{2^N}\begin{bmatrix} a+b & a-b \\ a-b & a+b \end{bmatrix}^N = \frac{(-1)^N}{2}\begin{bmatrix} a^N+b^N & a^N-b^N \\ a^N-b^N & a^N+b^N \end{bmatrix} \tag{5.9.46}$$

which can be proved easily by induction, that is, it is true when $N = 1$. Assume the above form is valid at $n = N$ (as it is true at $N = 1$), then prove that it is true when $n = N + 1$. We then find

$$\begin{aligned}\begin{bmatrix} 1 \\ r \end{bmatrix} &= \mathbf{B}_{01}\mathbf{M}^N\mathbf{B}_{(2N+1)t}\begin{bmatrix} t \\ 0 \end{bmatrix} \\ &= \begin{bmatrix} \alpha_{11} & \alpha_{12} \\ \alpha_{21} & \alpha_{22} \end{bmatrix}\begin{bmatrix} t \\ 0 \end{bmatrix}\end{aligned} \tag{5.9.47}$$

where the matrix elements are

$$
\begin{aligned}
\alpha_{11} &= \frac{(-1)^N}{2}(-i)\left(a^N \frac{n_1}{n_0} + b^N \frac{n_t}{n_1}\right) \\
\alpha_{12} &= \frac{(-1)^N}{2}(-i)\left(a^N \frac{n_1}{n_0} - b^N \frac{n_t}{n_1}\right) \\
\alpha_{21} &= \frac{(-1)^N}{2}(-i)\left(-a^N \frac{n_1}{n_0} + b^N \frac{n_t}{n_1}\right) \\
\alpha_{22} &= \frac{(-1)^N}{2} i\left(a^N \frac{n_1}{n_0} + b^N \frac{n_t}{n_1}\right).
\end{aligned}
\tag{5.9.48}
$$

We obtain the reflection coefficient at resonance wavelength

$$
\begin{aligned}
r_N &= \frac{b^N\left(\dfrac{n_t}{n_1}\right) - a^N\left(\dfrac{n_1}{n_0}\right)}{b^N\left(\dfrac{n_t}{n_1}\right) + a^N\left(\dfrac{n_1}{n_0}\right)} \\
&= \frac{\left(\dfrac{n_2}{n_1}\right)^{2N} - \left(\dfrac{n_1^2}{n_0 n_t}\right)}{\left(\dfrac{n_2}{n_1}\right)^{2N} + \left(\dfrac{n_1^2}{n_0 n_t}\right)}.
\end{aligned}
\tag{5.9.49}
$$

The above expressions for the reflection of the electric field at resonance Bragg wavelength λ_B has many important applications. Notice that as $N \to \infty$, we have

$$
r_N \to \begin{cases} +1 & \text{if } n_2 > n_1 \\ -1 & \text{if } n_2 < n_1. \end{cases} \tag{5.9.50}
$$

The DBR layers work as a perfect magnetic or electric conductor depending on $n_2 > n_1$ or $n_2 < n_1$. In addition, the reflection coefficient vanishes for the index matched structure, that is,

$$
r_N = 0 \quad \text{if} \quad \left(\frac{n_2}{n_1}\right)^{2N} = \frac{n_1^2}{n_0 n_t} \tag{5.9.51}
$$

which is the condition for perfect transmission. In Fig. 5.15, we plot $|r_N|^2$ as a function of the number of periods N for n_1 (GaAs) = 3.52 and n_2 (AlAs) = 2.95 at $\lambda_B =$ 0.98 μm [18] assuming $n_0 = n_t = n_1$ (GaAs) = 3.52

$$
|r_N|^2 = \left|\frac{(n_2/n_1)^{2N} - 1}{(n_2/n_1)^{2N} + 1}\right|^2. \tag{5.9.52}
$$

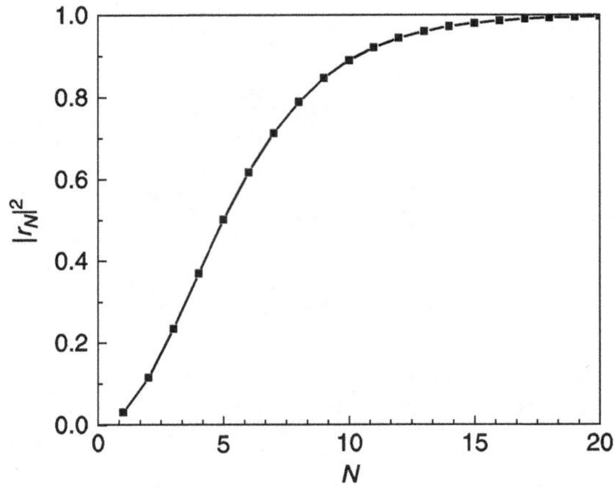

Figure 5.15 The power reflectivity at resonant wavelength as a function of the number of periods, N, which gives a total of $2N + 1$ quarter-wavelength layers. The parameters are $n_1 = 3.52$, $n_2 = 2.95$, and the resonant wavelength is 0.98 μm.

We can see the power reflectivity $|r_N|^2$ approaches unity quickly as N is above 10 for $n_2/n_1 = 2.95/3.52 = 0.838$

$$|r_N|^2 \simeq \begin{cases} 1 - 4\left(\dfrac{n_2}{n_1}\right)^{2N} & \text{if } n_1 > n_2 \text{ and } N \to \infty \\ 1 - 4\left(\dfrac{n_1}{n_2}\right)^{2N} & \text{if } n_1 < n_2 \text{ and } N \to \infty. \end{cases} \tag{5.9.53}$$

APPENDIX 5A: KRAMERS–KRONIG RELATIONS

The induced electric polarization $\mathbf{P}$ in a material is due to the response of the medium to an electric field $\mathbf{E}$ (ignoring the spatial dependence on $\mathbf{r}$):

$$\mathbf{P}(t) = \varepsilon_0 \int_{-\infty}^{t} \chi(t-\tau)\mathbf{E}(\tau)d\tau$$

$$= \varepsilon_0 \int_{0}^{\infty} \chi(\tau)\mathbf{E}(t-\tau)d\tau. \tag{5A.1}$$

The integration over τ is from $-\infty$ to t in the first expression as the response $\mathbf{P}(t)$ at time t comes from the excitation field before t. In other words, the function $\chi(\tau)$ has the property that

$$\chi(\tau) = 0 \quad \text{for } \tau < 0, \tag{5A.2}$$

that is, the system is causal. Because

$$\mathbf{D}(t) = \varepsilon_0\mathbf{E}(t) + \mathbf{P}(t), \tag{5A.3}$$

we find in the frequency domain

$$\mathbf{D}(\omega) = \varepsilon(\omega)\mathbf{E}(\omega) \tag{5A.4}$$

by taking the Fourier transform of (5A.3) where

$$\varepsilon(\omega) = \varepsilon_0 \left[1 + \int_0^\infty \chi(\tau) e^{i\omega\tau} d\tau \right]. \tag{5A.5}$$

Because $\chi(\tau)$ is a real function (it is in the time domain), we see that

$$\varepsilon(-\omega) = \varepsilon^*(\omega). \tag{5A.6}$$

If we write $\varepsilon(\omega)$ in terms of its real and imaginary parts,

$$\varepsilon(\omega) = \varepsilon'(\omega) + i\varepsilon''(\omega) \tag{5A.7}$$

we find that

$$\varepsilon'(-\omega) = \varepsilon'(\omega) \tag{5A.8}$$

$$\varepsilon''(-\omega) = -\varepsilon''(\omega) \tag{5A.9}$$

that is, the real part of $\varepsilon(\omega)$ is an even function, and the imaginary part is an odd function of ω. Another physical property of $\varepsilon(\omega)$ is that, in the high-frequency limit, $\varepsilon(\omega)$ tends to ε_0, because the polarization processes $\mathbf{P}(t)$, which are responsible for χ, cannot occur when the field changes sufficiently rapidly [4]

$$\varepsilon(\infty) \to \varepsilon_0. \tag{5A.10}$$

Define an integral I in the complex ω' plane,

$$I = \frac{1}{2\pi i} \oint_C \frac{\varepsilon(\omega') - \varepsilon_0}{\omega' - \omega} d\omega' \tag{5A.11}$$

where the closed contour C is shown in Fig. 5.16. It is in the upper ω'-plane. The function $\varepsilon(\omega)$ is analytic in the upper-half plane, as when $\omega' = \omega_R + i\omega_I$, the integrand in (5A.5) includes an exponentially decreasing factor $e^{-\omega_I \tau}$ when $\omega_I > 0$. Because the function $\chi(\tau)$ is finite (for a physical process) throughout $0 < \tau < \infty$, the integral in (5A.5) converges.

Because the function $\varepsilon(\omega') - \varepsilon_0$ is analytic in the upper ω'-plane, the closed contour does not have any poles; therefore, the integral I vanishes using Cauchy's theorem. By breaking the contour into three parts, (i) on the real axis, $-R < \omega' < \omega - \delta$, $\omega + \delta < \omega' < R$, (ii) along the infinitesimally small semicircle near ω with a radius δ, and (iii) along the big semicircle with a large radius R, we find that the contribution due to part (iii) is zero as $R \to \infty$. Therefore, as $\delta \to 0$, the integration

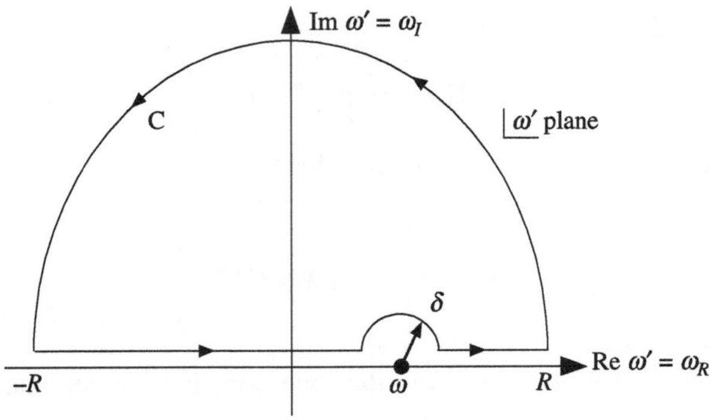

Figure 5.16 The integration contour on the upper-half ω' plane for the derivation of the Kramers–Kronig relations.

along part (i) is the principle value (denoted by P) of the integration along the real axis from $-\infty$ to $+\infty$, and we obtain

$$\frac{1}{2\pi i}P\int_{-\infty}^{\infty}\frac{\varepsilon(\omega')-\varepsilon_0}{\omega'-\omega}d\omega'+\frac{1}{2\pi i}\int_{\delta}\frac{\varepsilon(\omega')-\varepsilon_0}{\omega'-\omega}d\omega'=0. \tag{5A.12}$$

The second integral equals $-\frac{1}{2}[\varepsilon(\omega)-\varepsilon_0]$, which can be evaluated simply by a change of variable from ω' to θ, $\omega'-\omega=\delta e^{i\theta}$ at a constant radius δ. We find

$$\varepsilon(\omega)-\varepsilon_0=\frac{1}{\pi i}P\int_{-\infty}^{\infty}\frac{\varepsilon(\omega')-\varepsilon_0}{\omega'-\omega}d\omega'. \tag{5A.13}$$

We separate (5A.13) into the real and imaginary parts and find

$$\varepsilon'(\omega)-\varepsilon_0=\frac{1}{\pi}P\int_{-\infty}^{\infty}\frac{\varepsilon''(\omega')}{\omega'-\omega}d\omega' \tag{5A.14}$$

$$\varepsilon''(\omega)=-\frac{1}{\pi}P\int_{-\infty}^{\infty}\frac{\varepsilon'(\omega')-\varepsilon_0}{\omega'-\omega}d\omega'. \tag{5A.15}$$

The above results are the Kramers–Kronig relations, which relate the real and the imaginary parts of $\varepsilon(\omega)$ to each other. If we make use of the even property of the real part $\varepsilon'(\omega)$ and the odd property of the imaginary part $\varepsilon''(\omega)$, we obtain alternatively

$$\varepsilon'(\omega)-\varepsilon_0=\frac{2}{\pi}P\int_{0}^{\infty}\frac{\omega'\varepsilon''(\omega')}{\omega'^2-\omega^2}d\omega' \tag{5A.16}$$

$$\varepsilon''(\omega)=-\frac{2\omega}{\pi}P\int_{0}^{\infty}\frac{\varepsilon'(\omega')-\varepsilon_0}{\omega'^2-\omega^2}d\omega'. \tag{5A.17}$$

Similar relations between the real part and imaginary part of the complex refractive index, $\bar{n}(\omega) = n(\omega) + i\kappa(\omega)$ exist

$$n(\omega) - 1 = P \int_{-\infty}^{\infty} \frac{d\omega'}{\pi} \frac{\kappa(\omega)}{\omega' - \omega} \tag{5A.18}$$

$$\kappa(\omega) = -P \int_{-\infty}^{\infty} \frac{d\omega'}{\pi} \frac{n(\omega') - 1}{\omega' - \omega}. \tag{5A.19}$$

PROBLEMS

5.1 Check the duality principle using (5.1.12).

5.2 For a laser light with a power 1 mW propagating in a GaAs semiconductor ($n = 3.5$) waveguide with a cross section 10 μm × 1 μm, find (**a**) the power density if we assume the intensity is uniform with the waveguide cross section, (**b**) the electric field strength and the magnetic field strength assuming it is a uniform plane wave.

5.3 Derive (5.5.4a) and (5.5.4b).

5.4 Plot the real and imaginary parts of the relative permittivity using (5.5.32) for gold and check with Fig. 5.4.

5.5 (**a**) A plane wave is reflected between the free space and a bulk GaAs semiconductor with a refractive index assumed to be $n = 3.5$. Calculate the reflection and the transmission coefficients of the field, r and t, at normal incidence.

(**b**) Repeat part (**a**) if the wave is incident from the GaAs region onto the GaAs/air interface at normal incidence.

(**c**) For oblique incidence in part (**b**), find the critical angle and the Brewster angle.

5.6 For a plane wave incident from an InP region ($n = 3.16$ and $\lambda = 1.55$ μm) and reflected with an angle of incidence $\theta_i = 15°$, calculate (**a**) the reflection and the transmission coefficients, r and t, of the optical field for both TE and TM polarizations, and (**b**) the reflectivity R and the transmissivity T for both polarizations.

5.7 Calculate the Goos–Hänchen phase shifts for an angle of incidence $\theta_i = 45°$ between an InP/air interface for both TE and TM polarizations ($n = 3.16$ and $\lambda = 1.55$ μm).

5.8 (**a**) Calculate the reflection and transmission coefficients, r and t, for a plane wave normally incident on a slab of GaAs sample with a thickness $d = 10$ μm and a refractive index $n = 3.5$ at a wavelength $\lambda = 1$ μm.

(**b**) Find the power reflectivity and transmissivity in part (**a**).

5.9 (**a**) Derive the **M** matrix for a slab structure of refractive index n_2 sandwiched by medium n_1 on both sides. $\mathbf{M} = \mathbf{T}_{12}\mathbf{P}_2\mathbf{T}_{21}$.

(**b**) Using the result in (**a**), derive the reflection and transmission coefficient through the slab.

5.10 Derive the S matrix in (5.7.15)–(5.7.18).

5.11 Simplify (5.9.23) for a DBR structure at resonant wavelength and show that the results agree with (5.9.45).

5.12 Prove (5.9.26) by induction. **Hint**: Check $N = 1$ is true. Assume (5.9.26) is true for N. Then show it is true for $N + 1$.

5.13 Plot $|r_N|^2$ for a DBR structure with $N = 8$ as a function of ω near Bragg frequency (or Bragg wavelength $\lambda_B = 1.55\ \mu$m, $\omega_B = 2\pi c/\lambda_B$). Assume $n_1 = 3.5$, $n_2 = 3.2$, $h_1 = \lambda_B/4n_1$, and $h_2 = \lambda_B/4n_2$. Also, $n_0 = n_t = n_1$.

5.14 Derive (5.9.40) from (5.9.39).

5.15 Prove (5.9.46) by induction.

5.16 Plot $|r_N|^2$ as a function of N for n_1 (GaAs) = 3.52, n_2 ($Al_{0.2}$ $Ga_{0.8}$ As) = 3.39 at a resonant wavelength of $\lambda_B = 0.98\ \mu$m.

REFERENCES

1. J. A. Kong, *Electromagnetic Wave Theory*, Wiley, New York, 2000.
2. J. D. Jackson, *Classical Electrodynamics*, 3rd ed., Wiley, New York, 1999.
3. L. C. Shen and J. A. Kong, *Applied Electromagnetism*, Brooks/Cole Engineering Division, Monterey, CA, 1983.
4. L. D. Landau, E. M. Lifshitz, and L. P. Pitaevskii, *Electrodynamics of Continuous Media*, 2nd ed., Pergamon Press, New York, 1984.
5. H. A. Haus, *Waves and Fields in Optoelectronics*, Prentice Hall, Englewood Cliffs, NJ, 1984.
6. M. Fox, *Optical Properties of Solids*, Oxford University Press, Oxford, 2002.
7. A. Yariv and P. Yeh, *Photonics—Optical Electronics in Modern Communications*, 6th ed., Oxford University Press, New York, 2007.
8. B. E. A. Saleh and M. C. Teich, *Fundamentals of Photonics*, 2nd ed., Wiley, New York, 2007.
9. E. D. Palik, *Handbook of Optical Constants*, Academic Press, Orlando, 1985.
10. A. D. Rakić, A. B. Djurišić, J. M. Elazar, and M. L. Majewski, "Optical properties of metallic films for vertical-cavity optoelectronic devices," *Appl. Optics*. **37**, 5271–5283, (1998).
11. R. Brendel and D. Bormann, "An infrared dielectric function model for amorphous solids," *J. Appl. Phys.* **71**, 1–6 (1992).
12. B. Dold and R. Mecke, "Optische eigenschaften von edelmetallen, übergangsmetallen und deren legierungen im infrarot (1. Teil)," *Optik* **22**, 435–446 (1965).

13. P. Winsemius, H. P. Langkeek, and F. F. van Kampen, "Structure dependence of the optical properties of Cu, Ag and Au," *Physica* **79B**, 529–546 (1975).

14. G. Leveque, C. G. Olson, and D. W. Lynch, "Reflectance spectra and dielectric functions of Ag in the region of interband transitions," *Phys. Rev. B* **27**, 4654–4660 (1983).

15. M. L. Thèye, "Investigation of the optical properties of Au by means of thin semitransparent films," *Phys. Rev. B* **2**, 3060–3078 (1970).

16. P. B. Johnson and R. W. Christy, "Optical constants of the Nobel Metals," *Phys. Rev. B* **6**, 4370–4379 (1972).

17. C. Oubre and P. Nordlander, "Optical properties of metallodielectric nanostructures calculated using the finite difference time domain method," *J. Phys. Chem. B* **108**, 17740–17747 (2004).

18. L. A. Coldren and E. R. Hegblom, "Fundamental issues in VCSEL design," Chapter 2 in *Vertical-Cavity Surface-Emitting Lasers*, C. W. Wilmsen, H. Temkin, and L. A. Coldren, eds., Cambridge University Press, Cambridge, UK, 1999.

6

Light Propagation in Anisotropic Media and Radiation

The propagation of light is an extremely interesting topic because many common phenomena such as refraction of light, polarization properties of light, and scattering of light can be observed every day. The Rayleigh scattering of light by water molecules has been used to explain why the sky is blue in the daytime and red in the evening. In Section 6.1, we discuss some basic properties of the propagation of electromagnetic waves in uniaxial media [1, 2]. We then present light propagation in gyrotropic media and the magnetooptic effects [3, 4] in Section 6.2. The phenomenon of Faraday rotation and its application to optical isolators are discussed. In Section 6.3, we present the general solutions to Maxwell's equations for a given current density $\mathbf{J}$ and a charge density ρ, which satisfy the continuity equation. We discuss the gauge transformation including the Lorentz gauge and the Coulomb gauge; the latter will be used in the Hamiltonian to account for the interaction between the electrons and the photon field in semiconductors in Chapter 9. The radiation of the electromagnetic field is presented in Section 6.4 with the aim that the far-field pattern from a diode laser and laser arrays [5] will be derived once we know the laser mode on the facet of the laser cavity.

6.1 LIGHT PROPAGATION IN UNIAXIAL MEDIA

Consider the case of a uniaxial medium [1, 2] described by

$$\mathbf{D} = \boldsymbol{\varepsilon} \cdot \mathbf{E} \tag{6.1.1a}$$

$$\boldsymbol{\varepsilon} = \begin{bmatrix} \varepsilon & 0 & 0 \\ 0 & \varepsilon & 0 \\ 0 & 0 & \varepsilon_z \end{bmatrix} \tag{6.1.1b}$$

$$\mathbf{B} = \mu \mathbf{H} \tag{6.1.2}$$

Physics of Photonic Devices, Second Edition. By Shun Lien Chuang
Copyright © 2009 John Wiley & Sons, Inc.

in the principal coordinate system, which means that the coordinate system has been chosen such that the permittivity matrix $\boldsymbol{\varepsilon}$ has been diagonalized. Two of the three diagonal values are equal in a uniaxial medium. The medium is positive uniaxial if $\varepsilon_z > \varepsilon$ and negative uniaxial if $\varepsilon_z < \varepsilon$. The z axis is called the optical axis.

6.1.1 Field Solutions

Because the medium described by (6.1.1) and (6.1.2) is invariant under any rotation around the z axis, we can always choose the plane containing the wave vector $\mathbf{k}$ and the z axis to be the x–z plane, that is,

$$\mathbf{k} = \hat{x}k_x + \hat{z}k_z \quad k_x = k\sin\theta,\ k_z = k\cos\theta$$

without loss of generality, as shown in Fig. 6.1. In Table 6.1, we summarize the forms of vectors and dyadics and their corresponding matrix representations.

Using (5.3.2a) and (5.3.2b) for plane wave solutions to Maxwell's equations, we have

$$\mathbf{k}\times(\mathbf{k}\times\mathbf{E}) = \mathbf{k}\times\omega\mathbf{B} \tag{6.1.3a}$$

$$= \omega\mu\mathbf{k}\times\mathbf{H} \tag{6.1.3b}$$

$$= -\omega^2\mu\boldsymbol{\varepsilon}\cdot\mathbf{E} \tag{6.1.3c}$$

and the matrix representation (see Table 6.1) for the cross-product, $\mathbf{k}\times$, we have

$$\begin{bmatrix} 0 & -k_z & 0 \\ k_z & 0 & -k_x \\ 0 & k_x & 0 \end{bmatrix}^2 \begin{bmatrix} E_x \\ E_y \\ E_z \end{bmatrix} = -\omega^2\mu \begin{bmatrix} \varepsilon & 0 & 0 \\ 0 & \varepsilon & 0 \\ 0 & 0 & \varepsilon_z \end{bmatrix} \begin{bmatrix} E_x \\ E_y \\ E_z \end{bmatrix}. \tag{6.1.4}$$

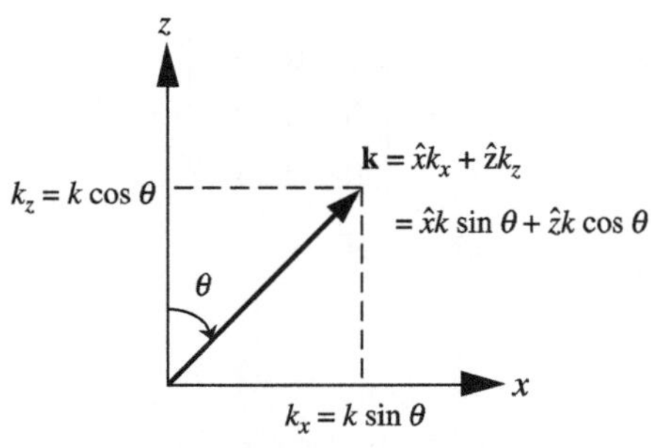

Figure 6.1 The wave vector $\mathbf{k}$ in the x–z plane.

Table 6.1 Representations of Vectors, Dyadics, and Their Corresponding Matrix Forms

Vectors and Dyadics	Matrices
$\mathbf{a} = a_x\hat{x} + a_y\hat{y} + a_z\hat{z}$	$\mathbf{a} = \begin{bmatrix} a_x \\ a_y \\ a_z \end{bmatrix}$
$\mathbf{a}\cdot\mathbf{b} = a_xb_x + a_yb_y + a_zb_z$	$\mathbf{a}^t\mathbf{b}$
$(\mathbf{a}\cdot\mathbf{b})(\mathbf{c}\cdot\mathbf{d}) = \mathbf{a}\cdot(\mathbf{bc})\cdot\mathbf{d}$	$(\mathbf{a}^t\mathbf{b})(\mathbf{c}^t\mathbf{d}) = \mathbf{a}^t(\mathbf{bc}^t)\mathbf{d}$
where $\mathbf{bc} = b_xc_x\hat{x}\hat{x} + b_xc_y\hat{x}\hat{y} + b_xc_z\hat{x}\hat{z} + b_yc_x\hat{y}\hat{x} + b_yc_y\hat{y}\hat{y} + b_yc_z\hat{y}\hat{z} + b_zc_x\hat{z}\hat{x} + b_zc_y\hat{z}\hat{y} + b_zc_z\hat{z}\hat{z}$	$\mathbf{bc}^t = \begin{bmatrix} b_x \\ b_y \\ b_z \end{bmatrix}[c_xc_yc_z] = \begin{bmatrix} b_xc_x & b_xc_y & b_xc_z \\ b_yc_x & b_yc_y & b_yc_z \\ b_zc_x & b_zc_y & b_zc_z \end{bmatrix}$
$\mathbf{k}\times\mathbf{E} = \begin{vmatrix} \hat{x} & \hat{y} & \hat{z} \\ k_x & k_y & k_z \\ E_x & E_y & E_z \end{vmatrix} = \hat{x}(k_yE_z - k_zE_y) + \hat{y}(k_zE_x - k_xE_z) + \hat{z}(k_xE_y - k_yE_x)$	$\mathbf{k}\times\mathbf{E} = \begin{bmatrix} 0 & -k_z & k_y \\ k_z & 0 & -k_x \\ -k_y & k_x & 0 \end{bmatrix}\begin{bmatrix} E_x \\ E_y \\ E_z \end{bmatrix} = \begin{bmatrix} (k_yE_z - k_zE_y) \\ (k_zE_x - k_xE_z) \\ (k_xE_y - k_yE_x) \end{bmatrix}$
$\hat{z}\times$	$\hat{z}\times \equiv \begin{bmatrix} 0 & -1 & 0 \\ 1 & 0 & 0 \\ 0 & 0 & 0 \end{bmatrix}$
$(k_x\hat{x} + k_z\hat{z})\times$	$(k_x\hat{x} + k_z\hat{z})\times \equiv \begin{bmatrix} 0 & -k_z & 0 \\ k_z & 0 & -k_x \\ 0 & k_x & 0 \end{bmatrix}$

The left-hand side of (6.1.4) is

$$\begin{bmatrix} -k_z^2 & 0 & k_xk_z \\ 0 & -k_x^2 - k_z^2 & 0 \\ k_xk_z & 0 & -k_x^2 \end{bmatrix}\begin{bmatrix} E_x \\ E_y \\ E_z \end{bmatrix}.$$

The matrix equation (6.1.4) can be written as

$$\begin{bmatrix} k_z^2 - \omega^2\mu\varepsilon & 0 & -k_xk_z \\ 0 & k_x^2 + k_z^2 - \omega^2\mu\varepsilon & 0 \\ -k_xk_z & 0 & k_x^2 - \omega^2\mu\varepsilon_z \end{bmatrix}\begin{bmatrix} E_x \\ E_y \\ E_z \end{bmatrix} = 0. \tag{6.1.5}$$

To have nontrivial solutions for the electric field, we require that the determinant of the matrix in the above equation be zero. That is,

$$\left(k_x^2 + k_z^2 - \omega^2 \mu\varepsilon\right)\left[\left(k_z^2 - \omega^2 \mu\varepsilon\right)\left(k_x^2 - \omega^2 \mu\varepsilon_z\right) - k_x^2 k_z^2\right] = 0. \tag{6.1.6}$$

There are two possible solutions.

Solution 1 Ordinary Waves

$$k_x^2 + k_z^2 - \omega^2 \mu\varepsilon = 0. \tag{6.1.7}$$

If (6.1.7) is true, we find from (6.1.5), which contains three algebraic equations, that

$$E_y \text{ can be arbitrary except zero (for nontrivial solution)} \tag{6.1.8a}$$

and

$$E_x = E_z = 0. \tag{6.1.8b}$$

That is, if the electric field is polarized only in the y direction, the wave vector must satisfy the dispersion relation given by (6.1.7). The solutions of the fields are, therefore,

$$\mathbf{E} = \hat{y} E_0 e^{i\mathbf{k}\cdot\mathbf{r}} \tag{6.1.9a}$$

$$\mathbf{H} = \frac{1}{\omega\mu}\mathbf{k} \times \mathbf{E} = \frac{1}{\omega\mu}(-\hat{x}k_z + \hat{z}k_x)E_0 e^{i\mathbf{k}\cdot\mathbf{r}} \tag{6.1.9b}$$

$$\mathbf{B} = \mu\mathbf{H} = \frac{1}{\omega}(-\hat{x}k_z + \hat{z}k_x)E_0 e^{i\mathbf{k}\cdot\mathbf{r}} \tag{6.1.9c}$$

$$\mathbf{D} = \boldsymbol{\varepsilon} \cdot \mathbf{E} = \hat{y}\varepsilon E_0 e^{i\mathbf{k}\cdot\mathbf{r}}. \tag{6.1.9d}$$

We can see that for an electric field polarized in the y direction, it will propagate with a wave number $k = \omega\sqrt{\mu\varepsilon}$ and its polarization remains y-polarized. Therefore, this polarization (6.1.9a) is a characteristic polarization of the uniaxial medium.

Solution 2 Extraordinary Waves

$$\frac{k_x^2}{\omega^2\mu\varepsilon_z} + \frac{k_z^2}{\omega^2\mu\varepsilon} = 1 \tag{6.1.10}$$

which is obtained by setting the square bracket in (6.1.6) to zero. If (6.1.10) is true, we find immediately from (6.1.5) that

$$E_y = 0 \text{ (because } k_x^2 + k_z^2 - \omega^2 \mu\varepsilon \neq 0 \quad \text{if (6.1.10) is true)} \tag{6.1.11}$$

and

$$\left(k_z^2 - \omega^2 \mu \varepsilon\right) E_x - k_x k_z E_z = 0. \tag{6.1.12a}$$

Using (6.1.10) and (6.1.12a), we obtain

$$\left(-\frac{\varepsilon}{\varepsilon_z} k_x^2\right) E_x - k_x k_z E_z = 0 \tag{6.1.12b}$$

or simply

$$k_x \varepsilon E_x + k_z \varepsilon_z E_z = 0 \tag{6.1.12c}$$

which is the same as

$$\mathbf{k} \cdot \mathbf{D} = 0. \tag{6.1.12d}$$

The complete solutions of the fields obeying (6.1.11) and (6.1.12c) are given by

$$\mathbf{E} = \hat{x} E_x + \hat{z} E_z = \left(\hat{x} - \hat{z} \frac{k_x \varepsilon}{k_z \varepsilon_z}\right) E_{x0} e^{i\mathbf{k}\cdot\mathbf{r}} \tag{6.1.13a}$$

$$\mathbf{H} = \frac{1}{\omega\mu} \mathbf{k} \times \mathbf{E} = \frac{1}{\omega\mu} \begin{vmatrix} \hat{x} & \hat{y} & \hat{z} \\ k_x & 0 & k_z \\ E_x & 0 & E_z \end{vmatrix} = \hat{y} \frac{\omega\varepsilon}{k_z} E_{x0} e^{i\mathbf{k}\cdot\mathbf{r}} \tag{6.1.13b}$$

$$\mathbf{B} = \mu \mathbf{H} = \hat{y} \frac{\omega\mu\varepsilon}{k_z} E_{x0} e^{i\mathbf{k}\cdot\mathbf{r}} \tag{6.1.13c}$$

$$\begin{aligned} \mathbf{D} &= \boldsymbol{\varepsilon} \cdot \mathbf{E} = \hat{x} \varepsilon E_x + \hat{z} \varepsilon_z E_z \\ &= \left(\hat{x} - \hat{z} \frac{k_x}{k_z}\right) \varepsilon E_{x0} e^{i\mathbf{k}\cdot\mathbf{r}}. \end{aligned} \tag{6.1.13d}$$

The polarization of the electric field given by Eq. (6.1.13a) is another characteristic polarization of the uniaxial medium because it is a characteristic mode of the medium from the fact that the fields (6.1.13a)–(6.1.13d) satisfy all of Maxwell's equations. This wave propagates with a wave vector determined by Eq. (6.1.10). We note again that

$$\mathbf{k} \cdot \mathbf{D} = 0 \tag{6.1.14a}$$

$$\mathbf{k} \cdot \mathbf{B} = 0 \tag{6.1.14b}$$

$$\mathbf{k} \cdot \mathbf{H} = 0 \text{ as } (\mathbf{B} // \mathbf{H}) \tag{6.1.14c}$$

$$\mathbf{k} \cdot \mathbf{E} \neq 0! \tag{6.1.14d}$$

Thus the wave vector **k** is not perpendicular to the electric field **E**. The complex Poynting vector, $\mathbf{E} \times \mathbf{H}^*$, is therefore not in the direction of **k**. The solutions of the fields (6.1.13a)–(6.1.13d) are called extraordinary waves. The dispersion relation of the extraordinary waves is given by Eq. (6.1.10). On the other hand, the solutions of the fields (6.1.9a)–(6.1.9d), for which the wave vector **k** is perpendicular to **E**, **H**,

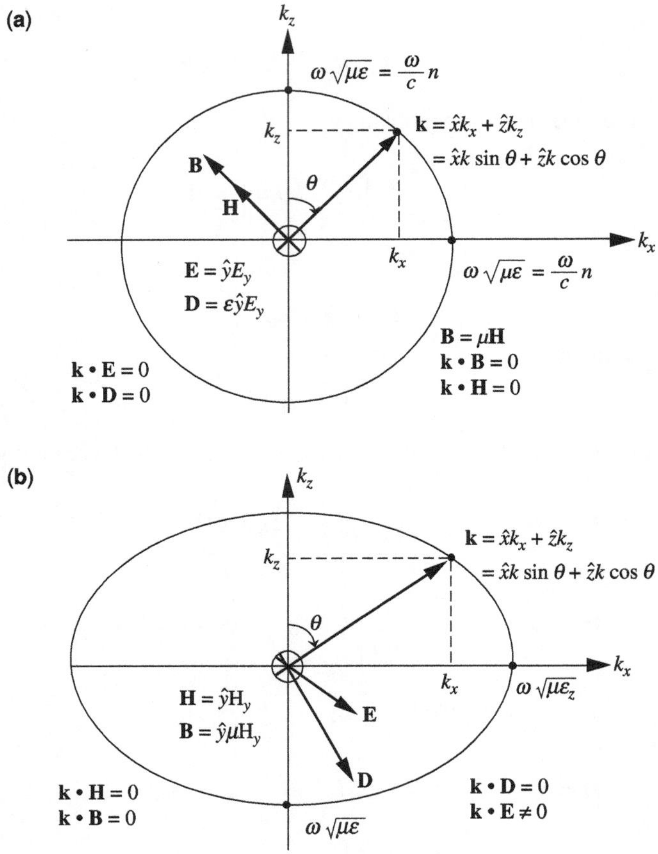

Figure 6.2 A summary of the **k** vector and the field vectors for (a) ordinary waves and (b) extraordinary waves.

D, and **B**, are called ordinary waves. The complex Poynting vector points in the direction of **k**. The ordinary waves behave the same as those in isotropic media with a permittivity ε. A summary of results of the ordinary and extraordinary waves is shown in Fig. 6.2.

6.1.2 *k* Surfaces

We may plot the dispersion relations such as (6.1.7) and (6.1.10) in the **k**-space. For the ordinary waves in case (a), Eq. (6.1.7) is a circle with a radius $\omega\sqrt{\mu\varepsilon}$ in the k_x–k_z plane. For the extraordinary waves in case (b), Eq. (6.1.10) is an ellipse in the k_x–k_z plane. We summarize the results as follows.

(a) Ordinary wave surface:

$$k_x^2 + k_z^2 = \omega^2\mu\varepsilon \quad (\mathbf{E} \text{ must be polarized in the } y \text{ direction}). \tag{6.1.15}$$

In other words, **E** must be polarized in the direction perpendicular to the plane containing both the optical axis and the wave vector **k** for ordinary waves. The wave number $k = \omega\sqrt{\mu\varepsilon} = \omega n_o/c$ and the refractive index $n_o = \sqrt{\mu\varepsilon/\mu_0\varepsilon_0}$.

(b) Extraordinary wave surface:

$$\frac{k_x^2}{\omega^2\mu\varepsilon_z} + \frac{k_z^2}{\omega^2\mu\varepsilon} = 1 \quad (\mathbf{E} \text{ must lie on the } x\text{–}z \text{ plane}). \tag{6.1.16}$$

Here $\omega\sqrt{\mu\varepsilon_z} = \omega n_e/c$. The extraordinary wave surface has two principal axes given by $\omega n_o/c$ and $\omega n_e/c$. The electric field **E** must be polarized in the plane containing both the optical axis and the wave vector **k**. The results are shown in Fig. 6.2 and Fig. 6.3.

Remember we have made use of the symmetry of the media with respect to the z axis (the optical axis). The plane containing the optical axis and the wave vector **k** is chosen to be the x–z plane. If we allow the wave vector **k** to rotate around the z axis, we would expect that:

(a) The ordinary waves will have the electric field perpendicular to the wave vector **k** and the optical axis, and the dispersion relation is a spherical surface in the **k**-space

$$k_x^2 + k_y^2 + k_z^2 = \omega^2\mu\varepsilon. \tag{6.1.17}$$

(b) The extraordinary waves will have the electric field polarized in the plane containing the optical axis and the wave vector **k**. The dispersion relation is given

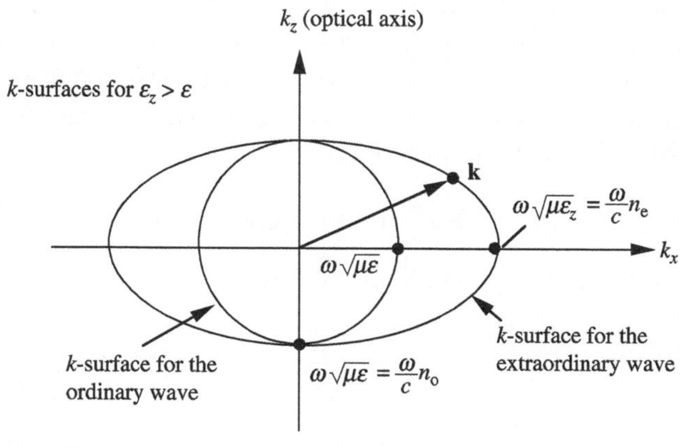

Figure 6.3 A plot of the k surfaces for the ordinary wave and the extraordinary wave of a uniaxial medium.

by an ellipsoid in the **k**-space

$$\frac{k_x^2}{\omega^2 \mu \varepsilon_z} + \frac{k_y^2}{\omega^2 \mu \varepsilon_z} + \frac{k_z^2}{\omega^2 \mu \varepsilon} = 1, \qquad (6.1.18)$$

which is a result of rotating the ellipse of (6.1.16) around the optical axis.

Special Cases

(1) If the wave propagates perpendicularly to the optical axis, for example, $\mathbf{k} = \hat{x}k$, $k_y = k_z = 0$, we have from (6.1.9a) and (6.1.13a) either

$$\mathbf{E} = \hat{y}E_0 e^{i\mathbf{k}\cdot\mathbf{r}} = \hat{y}E_0 e^{ik_o x} \quad k_o = \omega\sqrt{\mu\varepsilon} \quad \text{(ordinary waves)} \qquad (6.1.19)$$

or

$$\mathbf{E} = \hat{z}E_0 e^{i\mathbf{k}\cdot\mathbf{r}} = \hat{z}E_0 e^{ik_e x} \quad k_e = \omega\sqrt{\mu\varepsilon_z} \quad \text{(extraordinary waves).} \qquad (6.1.20)$$

Note that when $k_z = 0$, one finds $E_x = 0$ from (6.1.12c), and (6.1.13a) should reduce to (6.1.20).

(2) If the wave propagates parallel to the optical axis, that is, $\mathbf{k} = \hat{z}k$, $k_x = k_y = 0$, we have from (6.1.9a) and (6.1.13a) again

$$\mathbf{E} = \hat{y}E_0 e^{ik_o z} \quad \text{(for } \mathbf{E} \text{ polarized in the } y \text{ direction)} \qquad (6.1.21)$$

and

$$\mathbf{E} = \hat{x}E_0 e^{ik_o z} \quad \text{(for } \mathbf{E} \text{ polarized in the } x\text{–}z \text{ plane).} \qquad (6.1.22)$$

Both electric fields propagate with the same wave number $k_o = \omega\sqrt{\mu\varepsilon}$ and both are perpendicular to the direction of propagation! The ordinary wave and the extraordinary wave become degenerate (both are ordinary waves now).

(3) In general, the wave propagates in a direction with an angle θ to the optical axis. We have the dispersion relation

$$k^2 = k^2 \sin^2\theta + k^2 \cos^2\theta = \omega^2 \mu\varepsilon \qquad (6.1.23)$$

for the ordinary waves. The phase velocity

$$v_p = \frac{\omega}{k} = \frac{1}{\sqrt{\mu\varepsilon}} \qquad (6.1.24)$$

is independent of the angle θ for ordinary waves (remember that the polarization of the electric field is perpendicular to **k** and the optical axis). For the extraordinary waves, for which **E** is on the $x-z$ plane, we have

$$\frac{k^2}{\omega^2}\left(\frac{\sin^2\theta}{\mu\varepsilon_z}+\frac{\cos^2\theta}{\mu\varepsilon}\right)=1. \tag{6.1.25}$$

The phase velocity is

$$v_{\mathrm{p}}=\frac{\omega}{k}=\left(\frac{\sin^2\theta}{\mu\varepsilon_z}+\frac{\cos^2\theta}{\mu\varepsilon}\right)^{1/2}=\frac{1}{\sqrt{\mu\varepsilon}}\left(\frac{\varepsilon}{\varepsilon_z}\sin^2\theta+\cos^2\theta\right)^{1/2} \tag{6.1.26}$$

which depends on the angle of propagation.

6.1.3 Index Ellipsoid

For a given permittivity matrix $\boldsymbol{\varepsilon}$, we define an impermeability matrix **K** as its inverse

$$\mathbf{K}=\boldsymbol{\varepsilon}^{-1}. \tag{6.1.27}$$

An index ellipsoid is defined as

$$\varepsilon_0\sum_{i,j=1}^{3}K_{ij}x_ix_j=1 \tag{6.1.28}$$

where K_{ij} are the elements of the matrix **K**, and $x_1=x$, $x_2=y$, $x_3=z$. In the principal axis system such that $\boldsymbol{\varepsilon}$ is diagonalized

$$\boldsymbol{\varepsilon}=\varepsilon_0\begin{bmatrix} n_x^2 & 0 & 0 \\ 0 & n_y^2 & 0 \\ 0 & 0 & n_z^2 \end{bmatrix}, \tag{6.1.29}$$

the index ellipsoid Eq. (6.1.28) becomes

$$\frac{x^2}{n_x^2}+\frac{y^2}{n_y^2}+\frac{z^2}{n_z^2}=1, \tag{6.1.30}$$

which is plotted in Fig. 6.4. If $n_x=n_y=n_{\mathrm{o}}$ (ordinary refractive index) and $n_z=n_{\mathrm{e}}$ (extraordinary refractive index) in a uniaxial medium as described by (6.1.1), we have

$$\frac{x^2}{n_{\mathrm{o}}^2}+\frac{y^2}{n_{\mathrm{o}}^2}+\frac{z^2}{n_{\mathrm{e}}^2}=1. \tag{6.1.31}$$

It should be noted that the index ellipsoid is directly defined from the inverse of the permittivity function, and it contains information about the refractive indices of the characteristic polarizations. On the other hand, the k surfaces depend on the polarization and the direction of wave propagation. For a uniaxial medium, there are two k-surfaces: one for the ordinary wave and the other for the extraordinary wave, with only one index ellipsoid to define the medium.

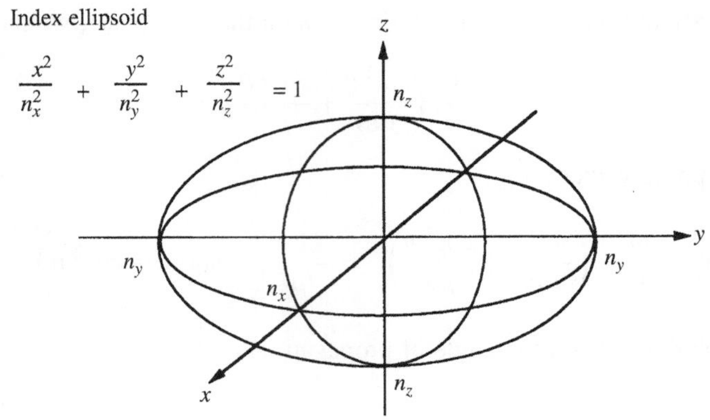

Figure 6.4 An index ellipsoid in the principal axis system for which the permittivity matrix is diagonalized.

6.1.4 Applications

Quarter-Wave Plate (QWP) Based on the special case in (6.1.19) and (6.1.20), we assume a plane wave incident from the free space to a slab of uniaxial medium with

$$\boldsymbol{\varepsilon} = \begin{bmatrix} \varepsilon & 0 & 0 \\ 0 & \varepsilon & 0 \\ 0 & 0 & \varepsilon_z \end{bmatrix}. \tag{6.1.32}$$

Consider an incident electric field from free space given by

$$\mathbf{E} = \left(\hat{y}\frac{1}{\sqrt{2}} + \hat{z}\frac{1}{\sqrt{2}}\right)E_0 e^{ikx}, \tag{6.1.33}$$

which is linearly polarized and $k = \omega\sqrt{\mu_0\varepsilon_0}$ is the propagation constant in free space, Fig. 6.5. We also assume that the permeability of the uniaxial medium μ is the same as that of the free space μ_0. The electric field makes an angle of 45° with the y and z axes. After passing through the uniaxial medium, the electric field will be (ignoring the reflections at the boundaries)

$$\mathbf{E} = \hat{y}\frac{E_0}{\sqrt{2}}e^{ik_o d} + \hat{z}\frac{E_0}{\sqrt{2}}e^{ik_e d} \tag{6.1.34}$$

because the y component of the electric field will be an ordinary wave and it propagates with a wavenumber $k_o = \omega\sqrt{\mu\varepsilon}$, and the z component of the electric field will be an extraordinary wave with a wavenumber $k_e = \omega\sqrt{\mu\varepsilon_z}$. If we choose the thickness d such that the phase difference between the two components of the electric field to be 90° or multiplied by an odd integer (assuming $\varepsilon_z > \varepsilon$),

$$(k_e - k_o)d = \frac{\pi}{2}, \frac{3\pi}{2}, \ldots, \frac{(2n+1)\pi}{2}, \text{ etc.} \quad (n \text{ an integer}), \tag{6.1.35}$$

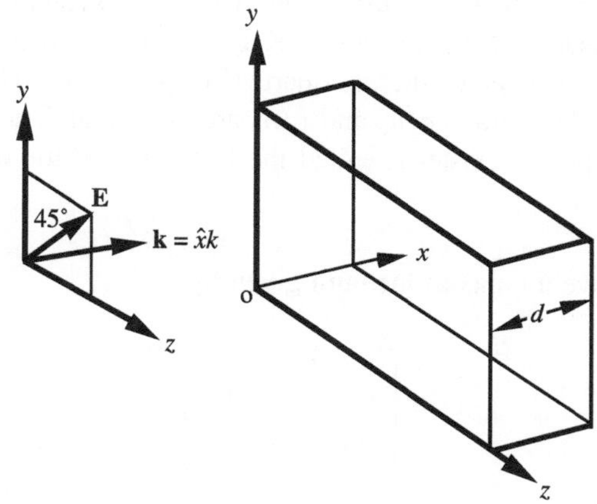

Figure 6.5 A plane wave **E** incident on a uniaxial medium with a thickness d. The electric field vector makes an angle of 45° with the two principal (y and z) axes.

we obtain

$$E = \left[\hat{y} + \hat{z}e^{i(k_e - k_o)d}\right]\frac{E_0}{\sqrt{2}}e^{ik_o d}$$

$$= (\hat{y} \pm i\hat{z})\frac{E_0}{\sqrt{2}}e^{ik_o d} \quad \text{for } (k_e - k_o)d = \begin{cases} \frac{\pi}{2}, & \frac{5\pi}{2}, \quad \frac{9\pi}{2}, \ldots \\ \frac{3\pi}{2}, & \frac{7\pi}{2}, \quad \frac{11\pi}{2}, \ldots \end{cases} \tag{6.1.36}$$

which becomes circularly polarized at the output end $x = d$. Using

$$k_e = \frac{2\pi}{\lambda_e} = \frac{\omega}{c}n_e \tag{6.1.37a}$$

$$k_o = \frac{2\pi}{\lambda_0} = \frac{\omega}{c}n_o \tag{6.1.37b}$$

and defining

$$k_e - k_o = \frac{2\pi}{\lambda_d} \quad \text{or} \quad \lambda_d = \frac{\lambda}{|n_e - n_o|} \tag{6.1.37c}$$

where λ is the wavelength in free space ($\omega/c = 2\pi/\lambda$), we have

$$d = \frac{\lambda_d}{4}, \frac{3\lambda_d}{4}, \frac{5\lambda_d}{4}, \ldots \tag{6.1.38}$$

Thus the plate, which can transform an incident linearly polarized wave into a circularly polarized wave, is called a quarter-wave plate. Note that the value λ_d is neither the wavelength in the free space nor the wavelength in the uniaxial medium (it corresponds with 2π divided by the difference in the two wave numbers k_e and k_O). In many crystals such as lithium niobate ($LiNbO_3$) or KDP (KH_2PO_4), the refractive indices have the property $n_e > n_O$. Some crystals such as quartz (SiO_2), $n_e < n_O$. The velocities of the two characteristic polarizations c/n_e and c/n_O are not equal. The axis along which the polarization propagates faster is called the fast axis and the other axis is called the slow axis.

Polaroid If we have a uniaxial medium given by

$$\boldsymbol{\varepsilon} = \begin{bmatrix} \varepsilon & 0 & 0 \\ 0 & \varepsilon & 0 \\ 0 & 0 & \varepsilon_z + i\dfrac{\sigma_z}{\omega} \end{bmatrix} \tag{6.1.39}$$

we see that for an incident wave propagating in the x direction, if the electric field is polarized in the y direction, it will propagate through with a propagation constant $k_O = \omega\sqrt{\mu\varepsilon}$, which is real. However, if $\mathbf{E}$ is polarized in the z direction, it will propagate with a complex propagation constant

$$k_e = \omega\sqrt{\mu\left(\varepsilon_z + i\frac{\sigma_z}{\omega}\right)} \cong \omega\sqrt{i\mu\frac{\sigma_z}{\omega}} \quad \text{for } \frac{\sigma_z}{\omega} \gg \varepsilon_z \tag{6.1.40}$$

and the wave will be attenuated significantly in the medium. Thus, if the thickness of the plate is large enough, an incident field with an arbitrary polarization will have its z component attenuated when passing through the plate. The transmitted field will be essentially polarized in the y direction only, which is linearly polarized.

Example An experimental setup for an absorber of laser light uses a polaroid, a quarter-wave plate, and a mirror as shown in Fig. 6.6. For an incident light with

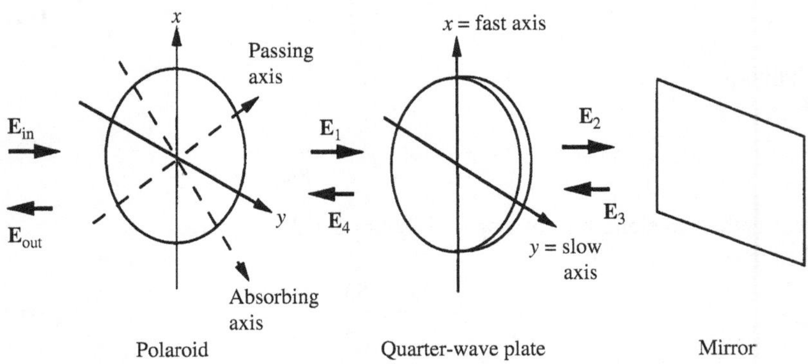

Figure 6.6 An experimental setup with a polaroid, a quarter-wave plate (QWP), and a mirror for complete absorption of the light incident from the left side.

an optical field $\mathbf{E}_{in}$ randomly polarized and propagating along the z direction, the transmitted field $\mathbf{E}_1$ after passing through the polaroid is linearly polarized along the direction of the passing axis, which makes 45° with the x and y axes

$$\mathbf{E}_1 = \left(\frac{\hat{x}+\hat{y}}{\sqrt{2}}\right) E_0 e^{ikz}.$$

The wave $\mathbf{E}_1$ also makes 45° with the fast and slow axes of the quarter-wave plate. The transmitted field at $z = d$ is

$$\begin{aligned}
\mathbf{E}_2 &= \hat{x}\frac{E_0}{\sqrt{2}}\exp\left(i\frac{\omega}{c}n_f d\right) + \hat{y}\frac{E_0}{\sqrt{2}}\exp\left(i\frac{\omega}{c}n_s d\right) \\
&= \left(\hat{x} + \hat{y}e^{i\pi/2}\right)\frac{E_0}{\sqrt{2}}\exp\left(i\frac{\omega}{c}n_f d\right)
\end{aligned}$$

where $\frac{\omega}{c}(n_s - n_f)d = \pi/2$ for a quarter-wave plate. The reflected wave from the mirror is

$$\mathbf{E}_3 = -\left(\hat{x} + \hat{y}e^{i\pi/2}\right)\frac{E_0}{\sqrt{2}}\exp\left(i\frac{\omega}{c}n_f d\right)\exp(-ik_o z)$$

propagating in the $-z$ direction. Upon impinging on the quarter-wave plate at $z = d$, the x component of the electric field propagates with a propagation constant $\omega n_f/c$, and the y component propagates with a propagation constant $\omega n_s/c$ again with an additional phase difference of $\pi/2$

$$\mathbf{E}_4 = -\left(\hat{x} + \hat{y}e^{i\pi}\right)\frac{E_0}{\sqrt{2}}\exp\left(i\frac{\omega}{c}n_f 2d\right)\exp(-ik_o z).$$

Therefore, $\mathbf{E}_4$ is linearly polarized in the direction $(\hat{x} - \hat{y})$, which is along the absorbing axis of the polaroid. The final output light $\mathbf{E}_{out} = 0$ because $\mathbf{E}_4$ is absorbed by the polaroid.

This setup actually works as a light absorber that can be used to absorb laser light to avoid stray light reflections in the laboratory.

6.2 WAVE PROPAGATION IN GYROTROPIC MEDIA: MAGNETOOPTIC EFFECTS

In this section, we present a general formulation to find the electromagnetic fields with the characteristic polarizations of gyrotropic media [1, 3, 4]. The magnetooptic effects are then investigated.

The constitutive relations for gyrotropic media are given by [1]

$$\mathbf{D} = \boldsymbol{\varepsilon} \cdot \mathbf{E} \tag{6.2.1a}$$

$$\boldsymbol{\varepsilon} = \begin{bmatrix} \varepsilon & i\varepsilon_{\mathrm{g}} & 0 \\ -i\varepsilon_{\mathrm{g}} & \varepsilon & 0 \\ 0 & 0 & \varepsilon_z \end{bmatrix} \tag{6.2.1b}$$

$$\mathbf{B} = \mu \mathbf{H}. \tag{6.2.2}$$

One example is a plasma with an externally applied dc magnetic field B_0 in the $\hat{z}$ direction, of which the parameters are:

$$\varepsilon = \varepsilon_0 \left[1 - \frac{\omega_{\mathrm{p}}^2}{\omega^2 - \omega_{\mathrm{c}}^2} \right] \tag{6.2.3a}$$

$$\varepsilon_{\mathrm{g}} = \varepsilon_0 \left[\frac{-\omega_{\mathrm{p}}^2 \omega_{\mathrm{c}}}{\omega(\omega^2 - \omega_{\mathrm{c}}^2)} \right] \tag{6.2.3b}$$

$$\varepsilon_z = \varepsilon_0 \left[1 - \frac{\omega_{\mathrm{p}}^2}{\omega^2} \right] \tag{6.2.3c}$$

where

$$\omega_{\mathrm{c}} = qB_0/m \tag{6.2.4}$$

is the cyclotron frequency, and

$$\omega_{\mathrm{p}} = \sqrt{\frac{Nq^2}{m\varepsilon_0}} \tag{6.2.5}$$

is the plasma frequency for a carrier density N. Here $m = m_0$ is the free electron mass. For the carrier effects in semiconductors, the effective mass of the carriers (electrons and holes) should be used.

Note the geometry of the medium considered is rotation-invariant around the z axis. Thus, we may consider the propagation wave vector $\mathbf{k}$ to be in the x–z plane without loss of generality

$$\mathbf{k} = \hat{x} k_x + \hat{z} k_z. \tag{6.2.6}$$

We repeat the Maxwell equations for plane-wave solutions here:

$$\mathbf{k} \times \mathbf{E} = \omega \mathbf{B} \tag{6.2.7a}$$

$$\mathbf{k} \times \mathbf{H} = -\omega \mathbf{D} \tag{6.2.7b}$$

$$\mathbf{k} \cdot \mathbf{B} = 0 \tag{6.2.7c}$$

$$\mathbf{k} \cdot \mathbf{D} = 0. \tag{6.2.7d}$$

Using (6.2.7a), (6.2.7b), and (6.2.2), we obtain

$$\mathbf{k} \times (\mathbf{k} \times \mathbf{E}) = \omega\mu\mathbf{k} \times \mathbf{H} = -\omega^2\mu\boldsymbol{\varepsilon} \cdot \mathbf{E}.$$

Or equivalently, the above vector equation can be written in a matrix representation following Table 6.1:

$$\begin{bmatrix} 0 & -k_z & 0 \\ k_z & 0 & -k_x \\ 0 & k_x & 0 \end{bmatrix}^2 \begin{bmatrix} E_x \\ E_y \\ E_z \end{bmatrix} = -\omega^2\mu \begin{bmatrix} \varepsilon & i\varepsilon_g & 0 \\ -i\varepsilon_g & \varepsilon & 0 \\ 0 & 0 & \varepsilon_z \end{bmatrix} \begin{bmatrix} E_x \\ E_y \\ E_z \end{bmatrix}. \tag{6.2.8}$$

We carry out the square of the matrix on the left-hand side and move to the right to obtain:

$$\begin{bmatrix} k_z^2 - \omega^2\mu\varepsilon & -i\omega^2\mu\varepsilon_g & -k_xk_z \\ i\omega^2\mu\varepsilon_g & k_x^2 + k_z^2 - \omega^2\mu\varepsilon & 0 \\ -k_xk_z & 0 & k_x^2 - \omega^2\mu\varepsilon_z \end{bmatrix} \begin{bmatrix} E_x \\ E_y \\ E_z \end{bmatrix} = 0. \tag{6.2.9}$$

The dispersion relation is obtained by setting the determinant of the matrix to zero for nontrivial solutions for the electric field. We obtain after some algebra

$$(k_x^2 + k_z^2 - \omega^2\mu\varepsilon)(-\omega^2\mu\varepsilon k_x^2 - \omega^2\mu\varepsilon_z k_z^2 + \omega^4\mu^2\varepsilon\varepsilon_z) - \omega^4\mu^2\varepsilon_g^2(k_x^2 - \omega^2\mu\varepsilon_z) = 0. \tag{6.2.10}$$

Rewrite the wave vector $\mathbf{k}$ in terms of the angle θ with respect to the z axis,

$$\mathbf{k} = \hat{x}k_x + \hat{z}k_z = \hat{x}k\sin\theta + \hat{z}k\cos\theta$$

and define the constants

$$K = \omega\sqrt{\mu\varepsilon} \tag{6.2.11a}$$

$$K_z = \omega\sqrt{\mu\varepsilon_z} \tag{6.2.11b}$$

$$K_g = \omega\sqrt{\mu\varepsilon_g}. \tag{6.2.11c}$$

Equation (6.2.10) reduces to

$$\begin{aligned} &k^4(K^2\sin^2\theta + K_z^2\cos^2\theta) - k^2[(K^4 - K_g^4)\sin^2\theta + K^2K_z^2\cos^2\theta + K^2K_z^2] \\ &\quad + (K^4 - K_g^4)K_z^2 = 0 \end{aligned} \tag{6.2.12}$$

which has the form

$$k^4A - k^2B + C = 0.$$

The solution k^2 is easily obtained from

$$k^2 = \frac{B \pm \sqrt{B^2 - 4AC}}{2A} \tag{6.2.13}$$

where

$$A = K^2 \sin^2\theta + K_z^2 \cos^2\theta$$

$$B = \left(K^4 - K_g^4\right)\sin^2\theta + K^2K_z^2(1+\cos^2\theta)$$

$$C = \left(K^4 - K_g^4\right)K_z^2.$$

Using (6.2.9) again, we have

$$\begin{bmatrix} k^2\cos^2\theta - K^2 & -iK_g^2 & -k^2\sin\theta\cos\theta \\ iK_g^2 & k^2 - K^2 & 0 \\ -k^2\sin\theta\cos\theta & 0 & -k^2\sin\theta - K_z^2 \end{bmatrix}\begin{bmatrix} E_x \\ E_y \\ E_z \end{bmatrix} = 0 \tag{6.2.14}$$

where k^2 is given by the two possible roots in (6.2.13). Because the above determinant of the matrix is zero, Eq. (6.2.14) contains three algebraic equations, which are linearly dependent. Using the second and the third equations in (6.2.14), we find

$$\frac{E_x}{E_y} = \frac{k^2 - K^2}{-iK_g^2} \tag{6.2.15a}$$

$$\frac{E_x}{E_z} = \frac{k^2\sin^2\theta - K_z^2}{k^2\sin\theta\cos\theta} \tag{6.2.15b}$$

where two possible values of k^2 are given by the roots in (6.2.13).

Special Cases

Case (1): $\theta = 0$. The wave is propagating parallel to the magnetic field. Equation (6.2.14) becomes

$$\begin{bmatrix} k^2 - K^2 & -iK_g^2 & 0 \\ iK_g^2 & k^2 - K^2 & 0 \\ 0 & 0 & -K_z^2 \end{bmatrix}\begin{bmatrix} E_x \\ E_y \\ E_z \end{bmatrix} = 0. \tag{6.2.16}$$

We have

$$E_z = 0, \quad \text{as } K_z^2 \neq 0 \tag{6.2.17}$$

and

$$(k^2 - K^2)^2 - K_g^4 = 0. \tag{6.2.18}$$

We obtain two roots for k^2,

$$k_\pm^2 = K^2 \pm K_g^2 = \omega^2\mu(\varepsilon \pm \varepsilon_g). \tag{6.2.19}$$

Substituting the above roots back into (6.2.16), we find

$$\frac{E_y}{E_x} = \frac{k^2 - K^2}{iK_g^2} = \frac{\pm K_g^2}{iK_g^2} = \mp i \quad \text{for } k^2 = \omega^2 \mu\varepsilon(\varepsilon \pm \varepsilon_g). \tag{6.2.20}$$

Combining (6.2.17) and (6.2.20), we see that if the wave is propagating parallel to the magnetic field, the wave will be circularly polarized. Let us assume that the wave is propagating in the $+\hat{z}$ direction; we have either

(a)
$$\frac{E_y}{E_x} = -i, \ k = \omega\sqrt{\mu(\varepsilon + \varepsilon_g)} \tag{6.2.21}$$

the wave is left-hand circularly polarized (LHCP), or

(b)
$$\frac{E_y}{E_x} = +i, \ k = \omega\sqrt{\mu(\varepsilon - \varepsilon_g)} \tag{6.2.22}$$

the wave is right-hand circularly polarized (RHCP).

Case (2): $\theta = \frac{\pi}{2}$. The wave is propagating perpendicularly to the magnetic field. Equation (6.2.14) becomes

$$\begin{bmatrix} -K^2 & -iK_g^2 & 0 \\ iK_g^2 & k^2 - K^2 & 0 \\ 0 & 0 & k^2 - K_z^2 \end{bmatrix} \begin{bmatrix} E_x \\ E_y \\ E_z \end{bmatrix} = 0. \tag{6.2.23}$$

Possible nontrivial solutions for the electric field will be

(a)
$$k^2 = K_z^2, \ E_z \neq 0. \tag{6.2.24}$$

If $k^2 = K_z^2$ is true, we find that $E_x = 0$, $E_y = 0$ from the first two equations in (6.2.23). Therefore, we obtain that the characteristic polarization is linear polarization in the z direction, $\mathbf{E} = \hat{z}E_z$ and the propagation constant is $k = K_z = \omega\sqrt{\mu\varepsilon_z}$.

(b) $-K^2(k^2 - K^2) - K_g^4 = 0$, which leads to

$$k^2 = K^2 - \frac{K_g^4}{K^2}. \tag{6.2.25}$$

If (6.2.25) is true, then $k^2 \neq K_z^2$, $E_z = 0$, and

$$E_x = -i\frac{K_g^4}{K^2}E_y. \tag{6.2.26}$$

The wave is generally elliptically polarized in the $x-y$ plane with a propagation constant $k = \sqrt{\left(K^4 - K_g^4\right)/K^2}$. The wave vector is always on the

$x-y$ plane as $\theta = \frac{\pi}{2}$. Furthermore, if $\varepsilon_g = 0$, the medium becomes uniaxial. Equations (6.2.24) to (6.2.26) show that either the electric field is polarized in the z direction and $k^2 = K_z^2$ or polarized in the y direction [as $E_x = 0$, $E_z = 0$ from (6.2.26)] and $k^2 = K^2$. Both waves are linearly polarized propagating with different velocities. This birefringence is called the Cotton–Mouton effect.

Faraday Rotation Let us consider a slab of gyrotropic medium with a dc magnetic field applied in the $+\hat{z}$ direction and the wave propagated parallel to the dc magnetic field. This is the special case (1) $\theta = 0$ discussed before, and the two characteristic polarizations are left- and right-hand circularly polarized with corresponding propagation constants given by Eqs. (6.2.21) and (6.2.22).

Consider an incident plane wave as shown in Fig. 6.7 with

$$\mathbf{E} = \hat{x}E_0 e^{ikz}. \tag{6.2.27}$$

Upon striking the interface at $z = 0$, the wave will break up into two circularly polarized waves:

$$E = (\hat{x} - i\hat{y})\frac{E_0}{2}e^{ik_+ z} + (\hat{x} + i\hat{y})\frac{E_0}{2}e^{ik_- z}. \tag{6.2.28}$$

These two circularly polarized waves propagate with two different wave numbers, k_+ and k_-. As discussed before, $k_\pm = \omega\sqrt{\mu(\varepsilon \pm \varepsilon_g)}$. At $z = d$, we have

$$\begin{aligned}\mathbf{E} &= (\hat{x} - i\hat{y})\frac{E_0}{2}e^{ik_+ d} + (\hat{x} + i\hat{y})\frac{E_0}{2}e^{ik_- d} \\ &= \hat{x}E_0\left(\frac{e^{ik_+ d} + e^{ik_- d}}{2}\right) + \hat{y}E_0\left(\frac{-ie^{ik_+ d} + ie^{ik_- d}}{2}\right).\end{aligned} \tag{6.2.29}$$

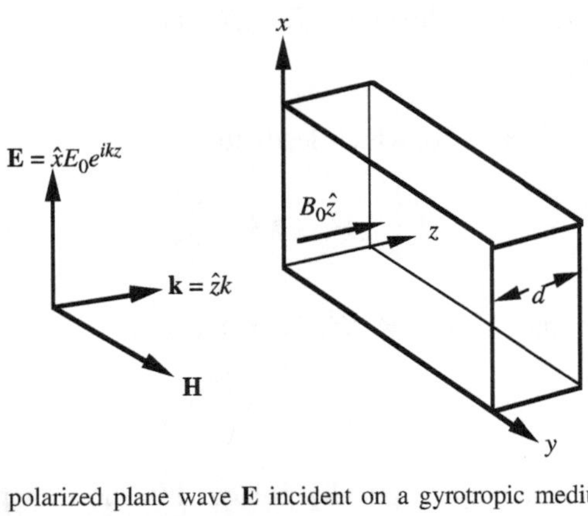

Figure 6.7 A linearly polarized plane wave **E** incident on a gyrotropic medium experiences Faraday rotation after passing through the medium at $z = d$.

Thus, we have the ratio of the $\hat{y}$ component to the $\hat{x}$ component of the electric field

$$\begin{aligned} \frac{E_y}{E_x} &= -i\frac{e^{ik_+d} - e^{ik_-d}}{e^{ik_+d} + e^{ik_-d}} \\ &= \tan\left[\frac{(k_+ - k_-)d}{2}\right] \end{aligned} \tag{6.2.30}$$

which is a real number. Thus the electric field at $z = d$ is again linearly polarized making an angle

$$\theta_{\mathrm{F}} = \frac{(k_+ - k_-)}{2}d \tag{6.2.31}$$

with the $\hat{x}$ axis. We conclude that the incident linearly polarized wave (in the $\hat{x}$ direction) is rotated by an angle θ_{F} at $x = d$, which is called the Faraday rotation.

Because the Faraday angle θ_{F} depends on the difference between $k_+ = \omega\sqrt{\mu(\varepsilon + \varepsilon_{\mathrm{g}})}$ and $k_- = \omega\sqrt{\mu(\varepsilon - \varepsilon_{\mathrm{g}})}$, the carrier density n and the effective mass of the electrons can be measured from the magnetooptic effects using the Faraday rotation as discussed above. Another setup is called the Voigt configuration for which the propagation direction is perpendicular to the direction of the applied dc magnetic field $\hat{z}B_0$. The incident wave is chosen to be linearly polarized at an angle of 45° with respect to the static magnetic field. The transmitted wave does not experience a rotation; it becomes elliptically polarized, however. The phase angle or the amount of ellipticity is determined by the difference of the propagation constants of the two characteristic polarizations, which is related to the plasma frequency ω_{p} and the cyclotron frequency ω_{c}. For more discussions on the magnetooptic effects and their measurements in semiconductors, see Ref. 3.

Optical Isolator An important optical component is an optical isolator, which allows light to transmit in only one direction. Figure 6.8 shows the configuration using two polarizers making 45° with each other and a Faraday rotator. Incident light propagating along the z direction becomes linearly polarized along the x direction after passing through polarizer 1 with x as the passing axis. The x-polarized light is rotated 45° by the Faraday rotator and becomes aligned with the passing axis with polarizer 2. Therefore, light is transmitted in the $+z$ direction. On the other hand, if the light is incident from the right side and propagates along the $-z$ direction, it becomes polarized along 45° with respect to the x axis after passing through polarizer 2. The light polarization is further rotated by 45° after passing through the Faraday rotator and becomes polarized along the y direction, which aligns with the absorbing axis of polarizer 1. Therefore, this light is absorbed. Thin film materials such as yttrium iron garnet (YIG) and terbium gallium garnet (TGG) have been used in a permanent magnetic field to make compact optical isolators.

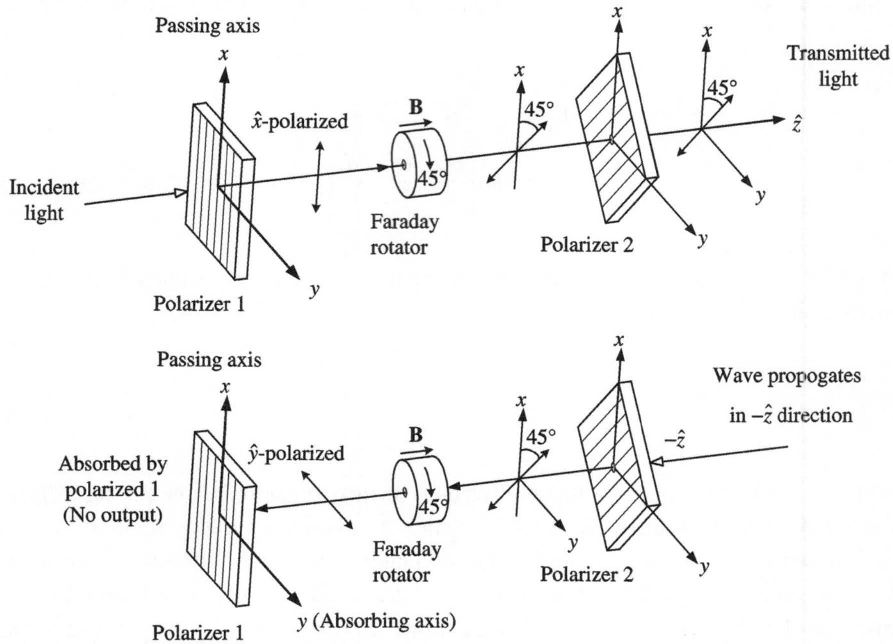

Figure 6.8 An optical isolator consists of a Faraday rotator between two polarizers (1 and 2), which make 45° with respect to each other. The light propagating along the $+z$ direction is transmitted, whereas the light propagating along the $-z$ direction is absorbed by polarizer 1.

6.3 GENERAL SOLUTIONS TO MAXWELL'S EQUATIONS AND GAUGE TRANSFORMATIONS

In this section, we study the general solutions [1] to Maxwell's equations,

$$\nabla \times \mathbf{E} = -\frac{\partial}{\partial t}\mathbf{B} \tag{6.3.1}$$

$$\nabla \times \mathbf{H} = \mathbf{J} + \frac{\partial}{\partial t}\mathbf{D} \tag{6.3.2}$$

$$\nabla \cdot \mathbf{D} = \rho \tag{6.3.3}$$

$$\nabla \cdot \mathbf{B} = 0 \tag{6.3.4}$$

where the source terms ρ and $\mathbf{J}$ satisfy the continuity equation

$$\nabla \cdot \mathbf{J} + \frac{\partial}{\partial t}\rho = 0. \tag{6.3.5}$$

Because

$$\nabla \cdot (\nabla \times \mathbf{A}) \equiv 0 \tag{6.3.6}$$

for any vector **A**, we can write **B** in the form

$$\mathbf{B} = \nabla \times \mathbf{A}. \tag{6.3.7}$$

However, this equation does not uniquely specify **A**. If we add $\nabla\xi$ to **A**, where ξ is an arbitrary function, we find that

$$\mathbf{A}' = \mathbf{A} + \nabla\xi \tag{6.3.8}$$

still satisfies $\nabla \times \mathbf{A}' = \mathbf{B}$. From the fundamental theorem of vector analysis, to uniquely define a vector **A**, we have to specify both its curl and its divergence.

Substituting (6.3.7) into (6.3.1), we obtain

$$\nabla \times \left(\mathbf{E} + \frac{\partial}{\partial t}\mathbf{A}\right) = 0. \tag{6.3.9}$$

Because $\nabla \times (-\nabla\phi) \equiv 0$ for any function ϕ, we may write

$$\mathbf{E} = -\frac{\partial}{\partial t}\mathbf{A} - \nabla\phi. \tag{6.3.10}$$

Assume an isotropic homogeneous medium,

$$\mathbf{D} = \varepsilon\mathbf{E} \tag{6.3.11}$$

$$\mathbf{B} = \mu\mathbf{H}. \tag{6.3.12}$$

Substituting the expressions for **B** and **E** from (6.3.7) and (6.3.10) into (6.3.2), we find

$$\left(\nabla^2\mathbf{A} - \mu\varepsilon\frac{\partial^2}{\partial t^2}\mathbf{A}\right) - \nabla\left(\nabla\cdot\mathbf{A} + \mu\varepsilon\frac{\partial\phi}{\partial t}\right) = -\mu\mathbf{J} \tag{6.3.13}$$

where the vector identity

$$\nabla \times \nabla \times \mathbf{A} = \nabla(\nabla\cdot\mathbf{A}) - \nabla^2\mathbf{A} \tag{6.3.14}$$

has been used. Gauss's law for the electric field (6.3.3) gives

$$\nabla^2\phi + \frac{\partial}{\partial t}(\nabla\cdot\mathbf{A}) = -\frac{\rho}{\varepsilon}. \tag{6.3.15}$$

Thus, generally speaking, we should find the solutions for the scalar potential ϕ and the vector potential **A** from (6.3.13) and (6.3.15) in terms of the sources ρ and **J**. We still have to choose a gauge to uniquely specify the potentials ϕ and **A**. For any ϕ and **A** satisfying (6.3.7) and (6.3.10), the following gauge transformations

$$\mathbf{A}' = \mathbf{A} + \nabla\xi \tag{6.3.16}$$

$$\phi' = \phi - \frac{\partial}{\partial t}\xi \tag{6.3.17}$$

also satisfy (6.3.7) and (6.3.10) where ξ is an arbitrary function. Thus, we have to specify $\nabla \cdot \mathbf{A}$. There are two common ways to specify $\nabla \cdot \mathbf{A}$, called the Lorentz gauge and the Coulomb gauge.

(1) Lorentz Gauge. In the Lorentz gauge, we choose

$$\nabla \cdot \mathbf{A} = -\mu\varepsilon \frac{\partial \phi}{\partial t}. \tag{6.3.18}$$

We then have

$$\nabla^2 \mathbf{A} - \mu\varepsilon \frac{\partial^2}{\partial t^2} \mathbf{A} = -\mu \mathbf{J} \tag{6.3.19}$$

$$\nabla^2 \phi - \mu\varepsilon \frac{\partial^2}{\partial t^2} \phi = -\frac{\rho}{\varepsilon}. \tag{6.3.20}$$

That is, the vector and scalar potentials satisfy the wave equations with sources $\mathbf{J}$ and ρ, respectively. The solutions can be written generally as

$$\mathbf{A}(\mathbf{r}, t) = \frac{\mu}{4\pi} \int \frac{d^3\mathbf{r}' \mathbf{J}\left(\mathbf{r}', t - \frac{|\mathbf{r} - \mathbf{r}'|}{c}\right)}{|\mathbf{r} - \mathbf{r}'|} \tag{6.3.21}$$

$$\phi(\mathbf{r}, t) = \frac{1}{4\pi\varepsilon} \int \frac{d^3\mathbf{r}' \rho\left(\mathbf{r}', t - \frac{|\mathbf{r} - \mathbf{r}'|}{c}\right)}{|\mathbf{r} - \mathbf{r}'|} \tag{6.3.22}$$

where the time delay due to propagation, $|\mathbf{r} - \mathbf{r}'|/c$, appears explicitly in the arguments of the source terms.

(2) Coulomb Gauge [2]. In the Coulomb gauge, we choose

$$\nabla \cdot \mathbf{A} = 0. \tag{6.3.23}$$

Therefore,

$$\nabla^2 \phi = -\frac{\rho}{\varepsilon} \tag{6.3.24}$$

which is Poisson's equation. The solution is

$$\phi(\mathbf{r}, t) = \frac{1}{4\pi\varepsilon} \int \frac{d^3\mathbf{r}' \rho(\mathbf{r}', t)}{|\mathbf{r} - \mathbf{r}'|}. \tag{6.3.25}$$

The solution for $\mathbf{A}$ has to be found from (6.3.13) and (6.3.23). For an optical field, $\rho = 0$, we also find $\phi = 0$ and $E = -\frac{\partial}{\partial t}\mathbf{A}$.

6.4 RADIATION AND THE FAR-FIELD PATTERN

6.4.1 General Expressions for Radiation Fields

We write (6.3.19) and (6.3.20) in the Lorentz gauge ($\nabla \cdot \mathbf{A} = -\mu\varepsilon\, \partial\phi/\partial t$) in the frequency domain

$$\nabla^2 \mathbf{A}(\mathbf{r}) + \omega^2 \mu\varepsilon \mathbf{A}(\mathbf{r}) = -\mu \mathbf{J}(\mathbf{r}) \tag{6.4.1}$$

$$\nabla^2 \phi(\mathbf{r}) + \omega^2 \mu\varepsilon \phi(\mathbf{r}) = -\frac{\rho(r)}{\varepsilon}. \tag{6.4.2}$$

We find

$$\mathbf{A}(\mathbf{r}) = \frac{\mu}{4\pi} \int \frac{e^{ik|\mathbf{r}-\mathbf{r}'|}\mathbf{J}(\mathbf{r}')}{|\mathbf{r}-\mathbf{r}'|} d^3\mathbf{r}' \tag{6.4.3}$$

and

$$\phi(\mathbf{r}) = \frac{1}{4\pi\varepsilon} \int \frac{e^{ik|\mathbf{r}-\mathbf{r}'|}\rho(\mathbf{r}')}{|\mathbf{r}-\mathbf{r}'|} d^3\mathbf{r}' \tag{6.4.4}$$

where $k = \omega\sqrt{\mu\varepsilon}$. From the charge continuity equation, we have

$$\rho(\mathbf{r}) = \frac{1}{i\omega} \nabla \cdot \mathbf{J}(\mathbf{r}). \tag{6.4.5}$$

The electric field in the time domain is given by

$$\mathbf{E} = -\frac{\partial}{\partial t}\mathbf{A} - \nabla\phi \tag{6.4.6}$$

and in the frequency domain is

$$\begin{aligned} \mathbf{E}(\mathbf{r}) &= i\omega \mathbf{A}(\mathbf{r}) - \nabla\phi(\mathbf{r}) \\ &= i\omega\mu \int \overline{\overline{\mathbf{G}}}(\mathbf{r}, \mathbf{r}') \cdot \mathbf{J}(\mathbf{r}') d^3\mathbf{r}' \end{aligned} \tag{6.4.7}$$

where a dyadic Green's function has been defined

$$\overline{\overline{\mathbf{G}}}(\mathbf{r}, \mathbf{r}') \equiv \left[\overline{\overline{\mathbf{I}}} + \frac{1}{k^2}\nabla\nabla\right] \frac{e^{ik|\mathbf{r}-\mathbf{r}'|}}{4\pi|\mathbf{r}-\mathbf{r}'|} \tag{6.4.8}$$

and $\overline{\overline{\mathbf{I}}}$ is a unity matrix.

$$\overline{\overline{\mathbf{I}}} = \begin{bmatrix} 1 & 0 & 0 \\ 0 & 1 & 0 \\ 0 & 0 & 1 \end{bmatrix} = \hat{x}\hat{x} + \hat{y}\hat{y} + \hat{z}\hat{z} = \hat{r}\hat{r} + \hat{\theta}\hat{\theta} + \hat{\phi}\hat{\phi}.$$

In deriving Eqs. (6.4.7)–(6.4.8), we have used

$$\begin{aligned}\int g(\mathbf{r}-\mathbf{r}')\,\nabla'\cdot\mathbf{J}(\mathbf{r}')\,d^3\mathbf{r}' &= \int \nabla'\cdot(g\mathbf{J})\,d^3\mathbf{r}' - \int \nabla' g(\mathbf{r}-\mathbf{r}')\cdot\mathbf{J}(\mathbf{r}')\,d^3\mathbf{r}' \\ &= \nabla\cdot\int g(\mathbf{r}-\mathbf{r}')\mathbf{J}(\mathbf{r}')\,d^3\mathbf{r}'.\end{aligned}$$

The term $\int \nabla'\cdot(g\mathbf{J})d^3\mathbf{r}' = \oint g\mathbf{J}\cdot dS' = 0$ because the surface integral at infinity should vanish assuming there is no source at infinity and $g(\mathbf{r}-\mathbf{r}') \to 0$ as $\mathbf{r}'$ approaches infinity.

The magnetic field is given by

$$\begin{aligned}\mathbf{H} &= \frac{1}{i\omega\mu}\nabla\times\mathbf{E}(\mathbf{r}) \\ &= \nabla\times\int \overline{\overline{\mathbf{G}}}(\mathbf{r},\mathbf{r}')\cdot\mathbf{J}(\mathbf{r}')\,d^3\mathbf{r}'.\end{aligned} \tag{6.4.9}$$

Suppose we consider an equivalent magnetic current source $\mathbf{M}$ and an equivalent magnetic charge density ρ_M,

$$\begin{aligned}&\nabla\times\mathbf{E} = i\omega\mu\mathbf{H} - \mathbf{M} \qquad \nabla\times\mathbf{H} = -i\omega\varepsilon\mathbf{E} \\ &\nabla\cdot\varepsilon\mathbf{E} = 0 \qquad \nabla\cdot\mu\mathbf{H} = \rho_M \qquad (6.4.10) \\ &\nabla\cdot\mathbf{M} - i\omega\rho_M = 0. \qquad (6.4.11)\end{aligned}$$

We can find equivalently from the duality principle $\mathbf{H}(\mathbf{r})$ and $\mathbf{E}(\mathbf{r})$ in terms of $\mathbf{M}(\mathbf{r})$

$$\mathbf{H}(\mathbf{r}) = i\omega\varepsilon\int \overline{\overline{\mathbf{G}}}(\mathbf{r},\mathbf{r}')\cdot\mathbf{M}(\mathbf{r}')\,d^3\mathbf{r}' \tag{6.4.12}$$

$$\mathbf{E}(\mathbf{r}) = -\nabla\times\int \overline{\overline{\mathbf{G}}}(\mathbf{r},\mathbf{r}')\cdot\mathbf{M}(\mathbf{r}')\,d^3\mathbf{r}'. \tag{6.4.13}$$

Thus, in the presence of both electric and magnetic sources, $\mathbf{J}$ and $\mathbf{M}$, we use the superposition principle and add the solutions (6.4.7) and (6.4.13) for the electric field, and similarly add Eqs. (6.4.9) and (6.4.12) for the magnetic field [1].

6.4.2 Far-Field Approximation

In many cases of radiation of the fields, for example, if the source distribution is over a finite region and the observation point $\mathbf{r}$ is far away from the source, we may approximate

$$|\mathbf{r}-\mathbf{r}'| \simeq r - \hat{r}\cdot\mathbf{r}' \tag{6.4.14}$$

where $\hat{r}$ is a unit vector along the $\mathbf{r}$ direction as shown in Fig. 6.9. The distance $\hat{r} \cdot \mathbf{r}'$ is the projection of $\mathbf{r}'$ on $\hat{r}$ and the distance $|\mathbf{r} - \mathbf{r}'|$ is approximately the difference between the length r and $\hat{r} \cdot \mathbf{r}'$. With this approximation, we further approximate

$$\frac{e^{ik|\mathbf{r}-\mathbf{r}'|}}{4\pi|\mathbf{r}-\mathbf{r}'|} \simeq \frac{e^{ikr}}{4\pi r} e^{-i\mathbf{k}\cdot\mathbf{r}'} \tag{6.4.15}$$

where $\mathbf{k} = k\hat{r}$ is along the direction of observation. We only keep the term $\hat{r} \cdot \mathbf{r}'$ in the phase factor because any small variation in $\mathbf{r}'$ comparable with wavelength causes a big change in the phase of $e^{-i\mathbf{k}\cdot\mathbf{r}'}$ but a negligible change in the denominator of (6.4.15), as $\mathbf{r}$ is very large in the far-field region. We consider radiation into the free space, $k = k_0 = \omega\sqrt{\mu_0\varepsilon_0}$. Therefore,

$$\overline{\overline{\mathbf{G}}}(\mathbf{r}, \mathbf{r}') \simeq \left[\overline{\overline{\mathbf{I}}} - \hat{r}\hat{r}\right] \frac{e^{ikr}}{4\pi r} e^{-i\mathbf{k}\cdot\mathbf{r}'} \tag{6.4.16a}$$

$$\nabla \times \overline{\overline{\mathbf{G}}}(\mathbf{r}, \mathbf{r}') \simeq ik\hat{r} \times \overline{\overline{\mathbf{G}}}(\mathbf{r}, \mathbf{r}'). \tag{6.4.16b}$$

As an example, for a semiconductor laser as shown in Fig. 6.10, if the aperture field at the output facet of the laser is given by $\mathbf{E}_A(\mathbf{r}')$, we may use an *equivalent source* by simply taking [1]

$$\mathbf{J}(\mathbf{r}') = 0, \tag{6.4.17}$$

and the magnetic current source is a current sheet with

$$\mathbf{M}_S(\mathbf{r}') = -2\hat{n} \times \mathbf{E}_A(\mathbf{r}') = -2\hat{z} \times \mathbf{E}_A(\mathbf{r}') \tag{6.4.18}$$

where $\hat{n}$ is a unit normal to the aperture surface. Using

$$\hat{r} \times \left(\overline{\overline{I}} - \hat{r}\hat{r}\right) \cdot \mathbf{M} = \hat{r} \times \left(\hat{\theta} M_\theta + \hat{\phi} M_\phi\right) = \hat{r} \times \mathbf{M},$$

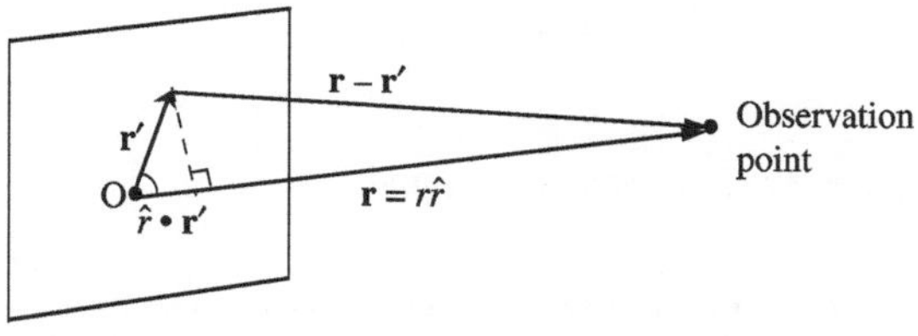

Figure 6.9 Far-field approximation $|\mathbf{r} - \mathbf{r}'| \simeq r - \hat{r} \cdot \mathbf{r}'$, where $\hat{r}$ is a unit vector along the direction of $\mathbf{r}$. The length $|\mathbf{r} - \mathbf{r}'|$ is approximately r subtracted by the distance of the projection of $\mathbf{r}'$ on $\hat{r}$.

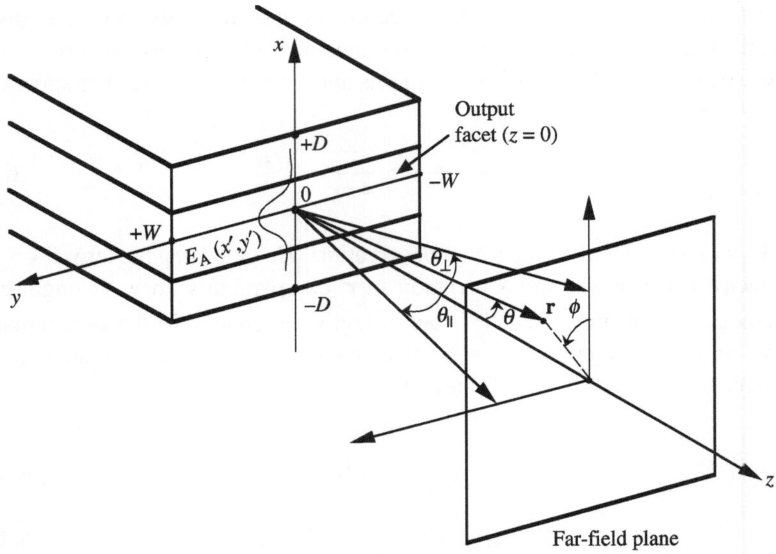

Figure 6.10 The coordinates for calculating the far-field pattern at position $\mathbf{r} = (r, \theta, \phi)$ from an aperture field $E_A(x', y')$ at the laser output facet plane $z = 0$. The angle $\theta_\perp$ is measured from the z axis along the x direction, and $\theta_\parallel$ is measured from the z axis toward the y direction.

we find $\mathbf{J}(\mathbf{r}') = 0$, (6.4.13) in the far field becomes a Fourier transform of the source field:

$$\mathbf{E}(\mathbf{r}) = \frac{e^{ikr}}{4\pi r}\left\{-ik\hat{r} \times \int d^3\mathbf{r}' e^{i\mathbf{k}\cdot\mathbf{r}'}\mathbf{M}(\mathbf{r}')\right\}. \tag{6.4.19}$$

We then integrate over the surface of the aperture to find the radiation field,

$$\mathbf{E}(\mathbf{r}) = \frac{-ike^{ikr}}{4\pi r}\hat{r} \times \iint_{\text{Aperture}} dx'dy' e^{-ik\hat{r}\cdot\mathbf{r}'}\mathbf{M}_S(\mathbf{r}'). \tag{6.4.20}$$

The radiated power density is

$$P = \frac{1}{2\eta}|\mathbf{E}(\mathbf{r})|^2 \tag{6.4.21}$$

where $\eta = \sqrt{\mu_0/\varepsilon_0} = 377\Omega$ in free space, and

$$\hat{r} = \sin\theta\cos\phi\,\hat{x} + \sin\theta\sin\phi\,\hat{y} + \cos\theta\,\hat{z} \tag{6.4.22}$$

is a unit vector along the direction of observation $\mathbf{r}$, which is the same as the direction of wave propagation in the far field $\mathbf{k}$. If we consider the aperture field to be due to a

TE mode in the active region of the semiconductor laser

$$\mathbf{E}_{\mathrm{A}} = \hat{y} E_{\mathrm{A}}(x, y), \tag{6.4.23}$$

then the magnetic current sheet is

$$\mathbf{M}_{\mathrm{S}} = -2\hat{n} \times \mathbf{E}_{\mathrm{A}} = 2\hat{x} E_{\mathrm{A}}(x, y). \tag{6.4.24}$$

Consider two cases.

(1) $\theta_\perp$ measurement. If the far-field pattern is measured along the x direction in the far-field plane, $\phi = 0$ (or $\phi = \pi$), and $\theta = \theta_\perp$, we obtain

$$\hat{r} = \sin\theta_\perp \hat{x} + \cos\theta_\perp \hat{z} \tag{6.4.25}$$

$$\mathbf{r}' = x'\hat{x} + y'\hat{y} \quad \text{as } z' = 0 \text{ on the aperture} \tag{6.4.26}$$

$$\mathbf{E}(\mathbf{r}) = \frac{-ike^{ikr}}{4\pi r} \hat{r} \times 2\hat{x} \int_{-D}^{D} dx' \int_{-W}^{W} dy' e^{-ik\sin\theta_\perp x'} E_{\mathrm{A}}(x', y'). \tag{6.4.27}$$

The optical power intensity pattern is

$$P(\theta_\perp) \propto \cos^2\theta_\perp \left| \int_{-D}^{D} dx' \int_{-W}^{W} dy' e^{-ik\sin\theta_\perp x'} E_{\mathrm{A}}(x', y') \right|^2 \tag{6.4.28}$$

where the extra factor $\cos^2\theta_\perp$ is necessary due to the polarization of the aperture field.

(2) $\theta_\parallel$ measurement. If the far-field pattern is measured along the y direction in the far-field plane, $\phi = \pi/2$, and $\theta = \theta_\parallel$, we obtain

$$\hat{r} = \sin\theta_\parallel \hat{y} + \cos\theta_\parallel \hat{z} \tag{6.4.29}$$

$$\mathbf{E}(\mathbf{r}) = \frac{-ike^{ikr}}{2\pi r} (\cos\theta_\perp \hat{y} - \sin\theta_\perp \hat{z}) \int_{-D}^{D} dx' \int_{-W}^{W} dy' e^{-ik\sin\theta_\perp y'} E_{\mathrm{A}}(x', y') \tag{6.4.30}$$

and the radiation power intensity pattern is

$$P(\theta_\perp) \propto \left| \int_{-D}^{D} dx' \int_{-W}^{W} dy' e^{-ik\sin\theta_\perp y'} E_{\mathrm{A}}(x', y') \right|^2 . \tag{6.4.31}$$

The above results for the far-field patterns have useful applications for studying the beam divergence for a single-element semiconductor laser and coupled laser arrays. A simple scalar formulation can be found in Ref. 5, where numerical results for various diode lasers are shown.

PROBLEMS

6.1 **(a)** Find the time-averaged Poynting's vector for the ordinary wave and the extraordinary wave using the electric and magnetic fields derived in Section 6.1. Is the Poynting's vector pointing in the direction of k in each case?

(b) Show that the Poynting's vector always points in a direction perpendicular to the k surface at the point (k_x, k_z) for the ordinary and the extraordinary waves.

6.2 Consider a $LiNbO_3$ crystal with the ordinary and extraordinary refractive indices, $n_o = 2.297$ and $n_e = 2.208$. Write down the dispersion relations for the ordinary waves and the extraordinary waves.

6.3 For a $LiNbO_3$ crystal with the permittivity matrix given by

$$\boldsymbol{\varepsilon} = \varepsilon_0 \begin{pmatrix} n_o^2 & 0 & 0 \\ 0 & n_o^2 & 0 \\ 0 & 0 & n_e^2 \end{pmatrix} \qquad \text{where } n_o = 2.297 \text{ and } n_e = 2.209,$$

(a) Find the two characteristic polarizations with corresponding propagation constants for a wave propagating in the y direction.

(b) If a plane wave propagates in $+z$ direction, what are the possible characteristic polarizations?

6.4 In Problem 6.3, if a plane wave propagates in a direction $\mathbf{k} = k(\hat{x}\sin\theta + \hat{z}\cos\theta)$ on the x–z plane, find the explicit expressions of the fields **E**, **H**, **D**, and **B** for **(a)** the ordinary wave and **(b)** the extraordinary wave.

6.5 **(a)** Calculate the thickness $\lambda_d/4$ for a quarter-wave-plate for a $LiNbO_3$ assuming $n_o = 2.297$, $n_e = 2.208$ at a wavelength $\lambda = 0.633\ \mu m$.

(b) Repeat part **(a)** for a KDP crystal with $n_o = 1.5074$, $n_e = 1.4669$ at the same wavelength.

6.6 If we replace the mirror by a polaroid in Fig. 6.6 with the passing axis perpendicular to that of the entrance polaroid, find the optical transmitted field passing through this new polaroid.

6.7 A quartz (SiO_2) crystal is a uniaxial medium described by two refractive indices, $n_o = 1.544$ and $n_e = 1.553$, at a free space wavelength of $0.589\ \mu m$. Draw a diagram when answering the following questions. You may assume that all transmission coefficients at the crystal surfaces are unity (i.e., ignore reflections) in this problem.

(a) Calculate the possible thicknesses of a quarter-wave plate.

(b) If the incident electric field onto a quartz quarter-wave plate is left-hand circularly polarized, what is the polarization of the transmitted field after passing through the quarter-wave plate (QWP) (Label your own slow and fast axes and show your derivations.) What about an incident right-hand circularly polarized wave? Draw a figure with labeled coordinates for the incident wave, the QWP, and the transmitted wave, and their polarization directions.

(c) A half-wave plate (HWP) has the property that the two components along the slow and fast axes of an incident linearly polarized wave have a phase difference of 180° (or π radians) after passing through the plate. Find the possible *thickness(es)* of a half-wave plate at a free-space wavelength of 0.589 μm using a quartz crystal. If a linearly polarized wave has its electric field vector at a 45° angle with respect to the slow and the fast axes of the half-wave plate, what is the polarization of the transmitted wave? Draw a figure to illustrate your answer.

6.8 Two polaroid analyzers are arranged with their passing axes perpendicular to each other and a quarter-wave-plate is inserted between them. Initially, the fast axis is assumed to be aligned with the passing axis of the entrance polaroid. Find the optical transmission power intensity for an incident randomly polarized light as a function of the rotation angle θ between the fast axis and the passing axis of the entrance polaroid. Plot the transmission power versus θ.

6.9 Derive the two eigenvalues for the propagation constants in (6.2.21) and (6.2.22) and their corresponding polarizations (LHCP and RHCP) of the optical electric fields.

6.10 Examine Faraday rotation when the optical field propagates along the $-z$ direction while the applied dc magnetic field is along the $+z$ direction. Show that the angle of rotation remains the same as (6.2.31), not negative of that value. The medium is nonreciprocal.

6.11 **(a)** Take the Fourier transforms of $\mathbf{A}(\mathbf{r}, t)$ and $\phi(\mathbf{r}, t)$ with respect to time in (6.3.21) and (6.3.22) and express $\mathbf{A}(\mathbf{r}, \omega)$ and $\phi(\mathbf{r}, \omega)$ in terms of $\mathbf{J}(\mathbf{r}, \omega)$ and $\rho(\mathbf{r}, \omega)$.

(b) Take the Fourier transforms of $\mathbf{A}(\mathbf{r}, t)$ and $\phi(\mathbf{r}, t)$ with respect to both the spatial and time variables r and t and express $\mathbf{A}(\mathbf{k}, \omega)$ and $\phi(\mathbf{k}, \omega)$ in terms of $\mathbf{J}(\mathbf{k}, \omega)$ and $\rho(\mathbf{k}, \omega)$.

6.12 Find the far-field radiation patterns $P(\theta_\perp)$ and $P(\theta_\parallel)$ in Fig. 6.10 assuming a uniform aperture field $E_A(x, y) = E_0$ for $|x| \le D$ and $|y| \le W$, and $E_A(x, y) = 0$ otherwise.

6.13 If the aperture field is a Gaussian function in both the x- and y-directions, $E_A(x, y) \approx E_D \exp\left[-\left(\frac{x}{x_0}\right)^2 - \left(\frac{y}{y_0}\right)^2\right]$ in Fig. 6.10, find the far-field radiation patterns $P(\theta_\perp)$ and $P(\theta_\parallel)$.

REFERENCES

1. J. A. Kong, *Electromagnetic Wave Theory*, EMW Publishing, Cambridge, MA, 2000.
2. M. Born and E. Wolf, *Principles of Optics: Electromagnetic Theory of Propagation, Interference and Diffraction of Light*, 7th ed., Cambridge University Press, Cambridge, UK, 1999.
3. K. C. Yeh and C. H. Liu, *Theory of Ionospheric Waves*, Academic Press, New York, 1972.
4. K. Seeger, *Semiconductor Physics—An Introduction*, 9th ed., Springer, Berlin, 2004.
5. H. C. Casey, Jr., and M. B. Panish, *Heterostructure Lasers, Part A: Fundamental Principles*, Academic Press, Orlando, 1978.

7

Optical Waveguide Theory

This chapter covers topics on optical dielectric waveguide theory that will be useful for studying heterojunction semiconductor lasers. We will present the fundamental results in Section 7.1 for transverse electric (TE) and transverse magnetic (TM) modes in a symmetric dielectric waveguide including the propagation constant, the optical field pattern, the optical confinement factor, and the cutoff conditions. We then discuss an asymmetric dielectric waveguide in Section 7.2. A ray optics approach is shown in Section 7.3 that explains the waveguide behavior. More practical structures, such as rectangular dielectric guides used in index-guidance diode lasers, are then discussed using an approximate Marcatili's method in Section 7.4 and the effective index method in Section 7.5. The results for a laser cavity with gain and loss in the media are derived to obtain the lasing conditions of a Fabry–Perot semiconductor laser cavity in Section 7.6. We then investigate surface plasmon waveguides in Section 7.7. Reference books on general waveguide theory can be found in Refs. 1–4 if the readers are interested in more extensive treatment of the subject.

7.1 SYMMETRIC DIELECTRIC SLAB WAVEGUIDES

The dielectric slab waveguide theory is very important because it provides almost all of the basic principles for general dielectric waveguides and the guidelines for more complicated cross sections such as the optical fibers. Let us consider a slab waveguide as shown in Fig. 7.1 with the width w much larger than the thickness d, and the field dependence on y is negligible ($\partial/\partial y \equiv 0$). From the wave equation

$$(\nabla^2 + \omega^2 \mu \varepsilon)\mathbf{E} = 0 \tag{7.1.1}$$

we shall find the solutions for the fields everywhere.

We assume that the waveguide is symmetric; that is, the permittivity and the permeability are ε and μ, respectively, for $|x| \geq d/2$, and ε_1 and μ_1 for $|x| < d/2$. The origin has been chosen to be at the center of the guide because the waveguide is symmetric; therefore, we have even mode and odd mode solutions. We separate the solutions of the fields into two classes, TE polarization and TM polarization.

Physics of Photonic Devices, Second Edition. By Shun Lien Chuang
Copyright © 2009 John Wiley & Sons, Inc.

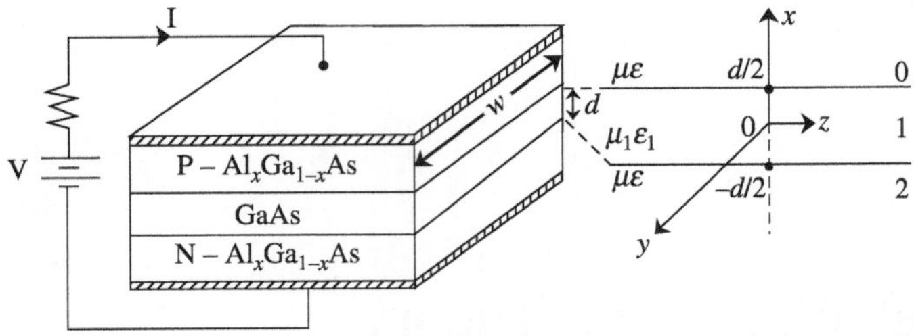

Figure 7.1 A simplified heterojunction diode laser structure for the waveguide analysis.

Here, a waveguide mode or a normal mode is defined as the wave solution to Maxwell's equations with all of the boundary conditions satisfied; the transverse spatial profile and its polarization remain unchanged while propagating down the waveguide.

7.1.1 Derivations of Electric Fields and Guidance Conditions for TE Polarization

For TE polarization, the electric field has only an E_y component: $\mathbf{E} = \hat{y}E_y$

$$\left(\frac{\partial^2}{\partial x^2} + \frac{\partial^2}{\partial z^2} + \omega^2\mu\varepsilon\right)E_y = 0 \tag{7.1.2a}$$

where

$$\mathbf{H} = \frac{1}{i\omega\mu_{\rm i}}\left(-\hat{x}\, ik_z E_y + \hat{z}\frac{\partial}{\partial x}E_y\right) \tag{7.1.2b}$$

($\mu_{\rm i} = \mu_1$ inside and $\mu_{\rm i} = \mu$ outside the guide).

TE Even Modes We obtain the guided mode solutions by examining wave equation (7.1.2), which has solutions of the form $[\exp(ik_z z),\ \exp(-ik_z z)] \times [\exp(ik_x x),\ \exp(-ik_x x),\ \text{or}\ \cos k_x x,\ \sin k_x x]$. Because the guide is translationally invariant along the z direction, we choose $\exp(ik_z z)$ for a guided mode propagating in the $+z$ direction. We then choose the standing-wave solution $\cos k_x x$ or $\sin k_x x$ inside the waveguide and $\exp(-\alpha x)$ or $\exp(+\alpha x)$ outside the waveguide because the wave is a guided mode solution.

With the above observations, the electric field for the even modes can be written in the form

$$E_y = e^{ik_z z}\begin{cases} C_0 e^{-\alpha(|x|-d/2)} & |x| \geq d/2 \\ C_1 \cos k_x x & |x| \leq d/2 \end{cases} \tag{7.1.3}$$

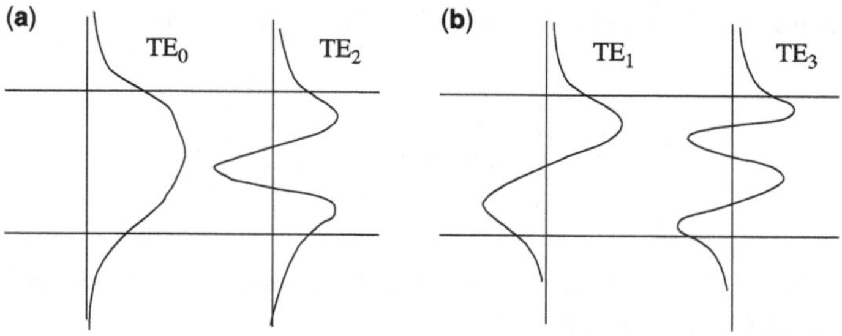

Figure 7.2 (a) The electric field profiles of the TE even modes. (b) The electric field profiles of the TE odd modes.

as shown in Fig. 7.2a, where k_x, k_z, and α satisfy

$$k_x^2 + k_z^2 = \omega^2 \mu_1 \varepsilon_1 \tag{7.1.4a}$$

$$-\alpha^2 + k_z^2 = \omega^2 \mu \varepsilon \tag{7.1.4b}$$

which are obtained by substituting (7.1.3) into the wave equation (7.1.2). Matching the boundary conditions in which E_y and H_z are continuous at $x = d/2$ and $x = -d/2$, we obtain

$$C_0 = C_1 \cos\left(k_x \frac{d}{2}\right) \tag{7.1.5a}$$

$$\frac{\alpha}{\mu} C_0 = C_1 \frac{k_x}{\mu_1} \sin\left(k_x \frac{d}{2}\right). \tag{7.1.5b}$$

Note that the permeability is μ_1 for $|x| \leq d/2$. Eliminating C_0 and C_1, from (7.1.5a) and (7.1.5b), we find the transcendental equation or the guidance condition

$$\alpha = \frac{\mu}{\mu_1} k_x \tan\left(k_x \frac{d}{2}\right). \tag{7.1.6}$$

The above guidance condition also shows that in order for E_y of the form (7.1.3) to be a guided mode solution, the standing-wave pattern along the transverse x direction or the amount of oscillations (determined by $k_x d$) has to match the decay constant α outside the guide. An interesting limit is to consider that the decay rate α is infinitely large, E_y will be zero outside the guide, and Eq. (7.1.6) gives $k_x d/2 = \pi/2, 3\pi/2, \ldots$ or $k_x d = m\pi$, which is the same as the guidance condition for a metallic waveguide, of which the tangential electric field vanishes at the surfaces of the perfectly conducting waveguide.

TE Odd Modes The electric field can be written in the form

$$E_y = e^{ik_z z}\begin{cases} C_0 e^{-\alpha(x-d/2)} & x \geq d/2 \\ C_1 \sin k_x x & |x| \leq d/2 \\ -C_0 e^{\alpha(x+d/2)} & x \leq -d/2 \end{cases} \tag{7.1.7}$$

as shown in Fig. 7.2b. Matching the tangential field components E_y and H_z at $x = d/2$ and eliminating C_0 and C_1, we obtain

$$\alpha = -\frac{\mu}{\mu_1} k_x \cot\left(k_x \frac{d}{2}\right). \tag{7.1.8}$$

7.1.2 Graphical Solution for the Guidance Conditions

To find the propagation constant k_z of the guided mode, we have to solve for k_x, k_z, and α from Eqs. (7.1.4a), (7.1.4b), (7.1.6), and (7.1.8). Define $X = k_x \frac{d}{2}$ and $Y = \alpha \frac{d}{2}$. We rewrite the eigenequations in the following forms

$$Y = \begin{cases} \dfrac{\mu}{\mu_1} X \tan X, & \text{TE even modes} \quad (7.1.9a) \\ -\dfrac{\mu}{\mu_1} X \cot X, & \text{TE odd modes} \quad (7.1.9b) \end{cases}$$

and subtract (7.1.4b) from (7.1.4a) to eliminate k_z, and multiply by $(d/2)^2$

$$X^2 + Y^2 = R^2 \tag{7.1.10}$$

where

$$R = \omega\sqrt{\mu_1\varepsilon_1 - \mu\varepsilon}\left(\frac{d}{2}\right) = \left(k_0 \frac{d}{2}\right)\sqrt{n_1^2 - n^2}. \tag{7.1.11}$$

We have two unknowns k_x and α. The solutions for α and k_x can be obtained from a graphical approach by looking at the $\left(\alpha\frac{d}{2}\right)$ versus $\left(k_x\frac{d}{2}\right)$ plane [4, 5]. Equation (7.1.10) appears as a circle with a radius R in Fig. 7.3, and the two transcendental equations (7.1.9a) and (7.1.9b) are also plotted on the same plane. Their interceptions give the solutions for $\left(\alpha\frac{d}{2}\right)$ and $\left(k_x\frac{d}{2}\right)$. Here the refractive index inside the guide $n_1 = \sqrt{\mu_1\varepsilon_1/(\mu_0\varepsilon_0)}$ and outside $n = \sqrt{\mu\varepsilon/(\mu_0\varepsilon_0)}$ have been used and $k_0 = \omega\sqrt{\mu_0\varepsilon_0} = \omega/c$ is the free-space wave number. Given the waveguide dimension d, we find the propagation constant k_z from either (7.1.4a) or (7.1.4b) after k_x or α is determined.

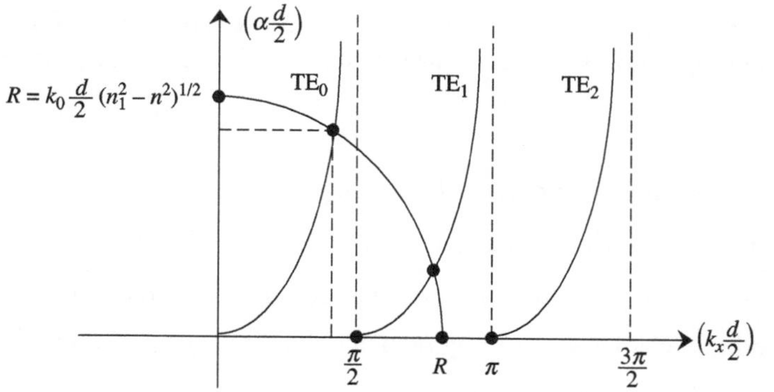

Figure 7.3 A graphical solution for the eigenequations to determine $\alpha d/2$ and $k_x d/2$ for the even modes, TE_0, TE_2, and so forth, and the odd modes, TE_1, TE_3, and so forth.

7.1.3 Cutoff Condition

It is clear from the graph that the cutoff condition occurs at $R = m\dfrac{\pi}{2}$ (and $X = 0$) for the TE_m mode, that is,

$$k_0 \frac{d}{2}\sqrt{n_1^2 - n^2} = m\frac{\pi}{2}, \quad m = 0, 1, 2, \ldots \tag{7.1.12}$$

Thus the TE_0 mode has no cutoff frequency. (Note: here we assume that the refractive indices are independent of the frequency. In semiconductors, the frequency-dependent refractive indices should be used.) For

$$m\frac{\pi}{2} < k_0 \frac{d}{2}\sqrt{n_1^2 - n^2} < (m+1)\frac{\pi}{2} \tag{7.1.13}$$

there are $(m + 1)$ guided TE modes in the dielectric slab.

For a single mode operation, the condition $k_0 \dfrac{d}{2}\sqrt{n_1^2 - n^2} < \dfrac{\pi}{2}$ is required. This puts a limit on ω, d, and $n_1^2 - n^2$. For example, given the wavelength λ_0 in free space and the waveguide dimension d,

$$n_1^2 - n^2 < \frac{(\pi/2)^2}{\left(k_0 \dfrac{d}{2}\right)^2} = \left(\frac{\lambda_0}{2d}\right)^2. \tag{7.1.14}$$

If n_1 is almost equal to n such as the case in an $Al_xGa_{1-x}As/GaAs/Al_xGa_{1-x}As$ waveguide, we have

$$\Delta n < \frac{1}{8n_1}\left(\frac{\lambda_0}{d}\right)^2 \tag{7.1.15}$$

using $n_1^2 - n^2 \simeq 2n_1\Delta n$. For example, at $n_1 \simeq 3.6$, we have $\Delta n < 0.035$ at $\lambda_0 = d$. Single mode operation is achieved with a small difference in the refractive index $\Delta n < 0.035$.

7.1.4 Low-Frequency and High-Frequency Limits

1. In the low-frequency limit, the waveguide modes can be cut off. At cutoff, $k_0 \frac{d}{2}\sqrt{n_1^2 - n^2} = m\frac{\pi}{2}$, and $\alpha = 0$, as can be seen from the graphical solution. (The cutoff condition can be achieved by reducing the frequency, or d, or $n_1^2 - n^2$.) We find $k_x \frac{d}{2} = m\frac{\pi}{2}$, and from (7.1.4b)

$$k_z = \omega\sqrt{\mu\varepsilon} = \omega n/c \tag{7.1.16}$$

because $\alpha = 0$. Therefore, the propagation constant of the guided mode k_z approaches the propagation constant $\omega n/c$ outside the waveguide. This is expected because the decaying constant $\alpha = 0$ at cutoff, which means that the waveguide mode does not decay outside the guide. Almost all of the mode power propagates outside the guide; therefore, the velocity of the mode will equal the speed of light outside the guide, Fig. 7.4b.

2. In the high-frequency limit, $R \to \infty$. The graphical solution in Fig. 7.3 shows that $\left(k_x \frac{d}{2}\right) \to \frac{(m+1)\pi}{2}$ and $\left(\alpha \frac{d}{2}\right) \to \infty$ for the TE$_m$ mode. Because m is

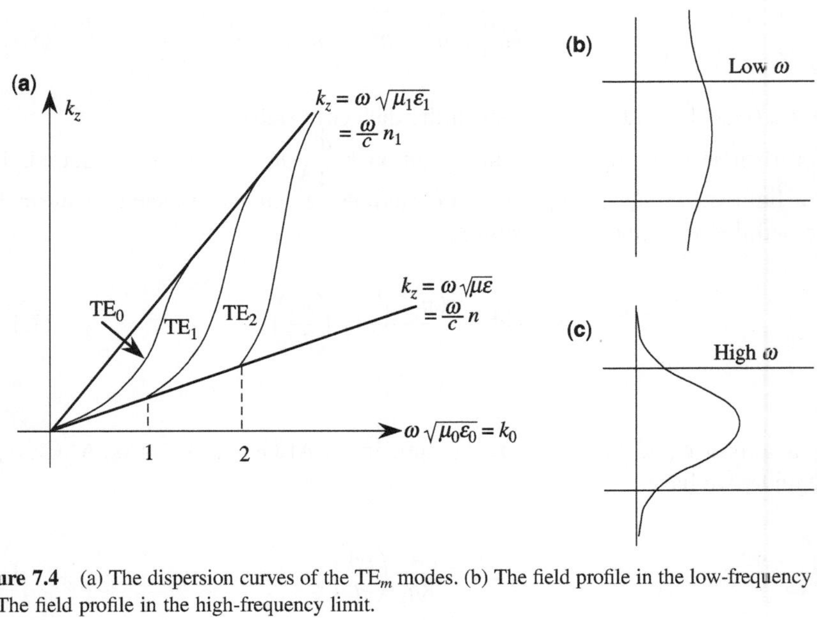

Figure 7.4 (a) The dispersion curves of the TE$_m$ modes. (b) The field profile in the low-frequency limit. (c) The field profile in the high-frequency limit.

a fixed mode number, we find that $\left(\alpha \frac{d}{2}\right) \to R \to \infty$, as can also be seen from the graphical solution. Therefore,

$$k_z^2 = \omega^2 \mu_1 \varepsilon_1 - k_x^2 \longrightarrow \omega^2 \mu_1 \varepsilon_1 \tag{7.1.17}$$

or $k_z = \omega\sqrt{\mu_1 \varepsilon_1} = \omega n_1/c$. The propagation constant k_z approaches that in the waveguide because the mode decays rapidly outside the guide ($\alpha \to \infty$). Most of the power is guided inside the waveguide, Fig. 7.4c.

7.1.5 Propagation Constant k_z and the Effective Index n_{eff}

From Eqs. (7.1.16) and (7.1.17), we see that the propagation constant k_z starts from $\omega\sqrt{\mu\varepsilon}$ at cutoff and increases to $\omega\sqrt{\mu_1\varepsilon_1}$ as the frequency goes to infinity. The dispersion curves k_z versus $\omega\sqrt{\mu_0\varepsilon_0}$ are plotted in Fig. 7.4a. An effective index for the guided mode is defined as

$$n_{\text{eff}} = \frac{k_z}{k_0}. \tag{7.1.18}$$

Again we have the lower bound $\omega\sqrt{\mu\varepsilon}/k_0 = n$ and the upper bound $\omega\sqrt{\mu_1\varepsilon_1}/k_0 = n_1$ at the low- and high-frequency limits, and the dispersion curves in Fig. 7.4a map to a set of effective index curves as shown in Fig. 7.5a. Very often a normalized propagation parameter

$$\tilde{\beta} = \frac{k_z^2 - \omega^2\mu\varepsilon}{\omega^2\mu_1\varepsilon_1 - \omega^2\mu\varepsilon} = \frac{n_{\text{eff}}^2 - n^2}{n_1^2 - n^2} \tag{7.1.19}$$

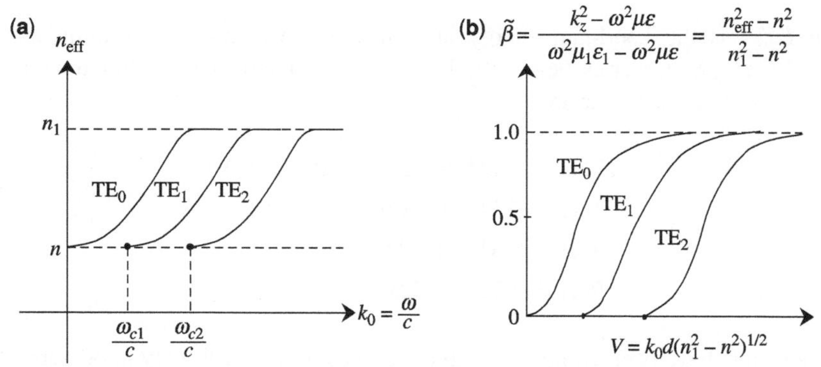

Figure 7.5 (a) The effective index $n_{\text{eff}} = k_z/k_0$ of the TE_m modes versus the free-space wave number k_0. (b) The normalized propagation parameter $\tilde{\beta}$ versus the normalized parameter V number for the TE_m modes.

is plotted versus the V number or normalized frequency

$$V = k_0 d(n_1^2 - n^2)^{1/2} \tag{7.1.20}$$

as shown in Fig. 7.5b, showing a set of universal curves for the lowest three TE modes.

The above discussions assume that n and n_1 are independent of the frequency. In semiconductors, the dispersion effect of the refractive index near the band edge can be significant. It is important to know the refractive index at the operation wavelength $n(\lambda)$. The effective index $n_{\text{eff}}(\lambda)$ still falls between $n(\lambda) < n_{\text{eff}}(\lambda) < n_1(\lambda)$ when the material dispersion is taken into account.

7.1.6 Refractive Index of $Al_xGa_{1-x}As$ System

The refractive index of this system and of many III–V materials [6–18] below the direct band gap can be approximately given by the form

$$n(\omega) \simeq \sqrt{\varepsilon'(\omega)/\varepsilon_0} \tag{7.1.21}$$

where $\varepsilon'(\omega)$ is the real part of the permittivity function, and the imaginary part $\varepsilon''(\omega)$ is negligible for optical energy below the band gap [12, 13, 15]:

$$\begin{aligned} \varepsilon'(\omega)/\varepsilon_0 &\simeq A(x)\left\{ f(y) + \frac{1}{2}\left[\frac{E_g(x)}{E_g(x)+\Delta(x)}\right]^{3/2} f(y_{so}) \right\} + B(x) \\ f(y) &= \frac{1}{y^2}\left[2 - (1+y)^{1/2} - (1-y)^{1/2}\right] \end{aligned} \tag{7.1.22a}$$

$$\begin{aligned} y &= \hbar\omega/E_g(x) \\ y_{so} &= \hbar\omega/[E_g(x) + \Delta(x)] \end{aligned} \tag{7.1.22b}$$

Here $E_g(x)$ is the band-gap energy and $\Delta(x)$ is the spin–orbit splitting energy. For $Al_xGa_{1-x}As$ ternaries below the band gap, the parameters as functions of the aluminum mole fraction x are

$$\begin{aligned} E_g(x) &= 1.424 + 1.266x + 0.26x^2 \text{ (eV)} \\ \Delta(x) &= 0.34 - 0.5x \text{ (eV)} \\ A(x) &= 6.64 + 16.92x \\ B(x) &= 9.20 - 9.22x. \end{aligned} \tag{7.1.23}$$

These results have been compared with experimental data with very good agreement as shown in Fig. 7.6a [11] for the real part $\varepsilon_1(\omega) = \varepsilon'(\omega)/\varepsilon_0$ below the band gap. The refractive index $n(\omega)$ below the band gap is plotted in Fig. 7.6b. In Fig. 7.6c, we show the refractive index above and below the band gap for $Al_xGa_{1-x}As$ with various

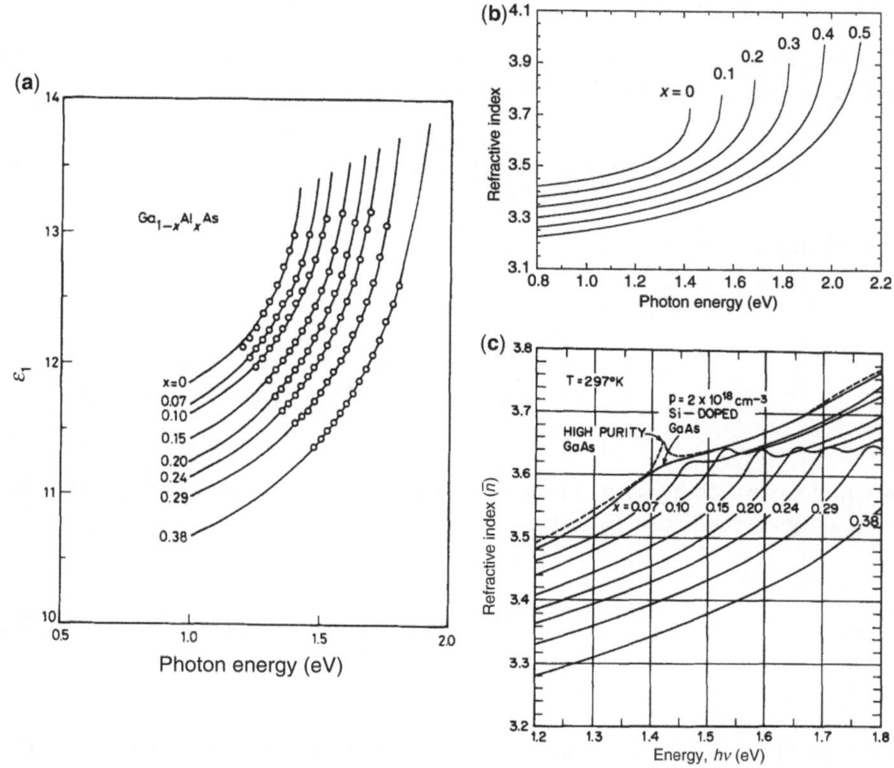

Figure 7.6 (a) The real part of the relative permittivity function $\varepsilon_1(\omega) = \varepsilon'(\omega)/\varepsilon_0$ versus the photon energy $\hbar\omega$ (eV) for $Ga_{1-x}Al_xAs$ alloy. Solid lines are calculated using (7.1.22) and (7.1.23). The circles are experimental data. After [11] © 1983 American Institute of Physics. (Reprint with permission.) (b) The theoretical refractive index $n(\omega) \simeq \sqrt{\varepsilon'(\omega)/\varepsilon_0}$ below the band gap calculated using (7.1.21)–(7.1.23). (c) Experimental data for the refractive index of $Al_xGa_{1-x}As$ as a function of photon energy $h\nu$. The top dashed curve is the refractive index of a high-purity GaAs, and the top solid curve is that of a silicon-doped p-type GaAs with $p = 2 \times 10^{18}\ \text{cm}^{-3}$. After [14] © 1974 American Institute of Physics. (Reprint with permission.)

aluminum mole fraction x. For comparison, the refractive index of a high-purity GaAs is shown as the top dashed curve and that of a p-type doped GaAs is shown as the top solid curve. We can see the features near the band gap of the $Al_xGa_{1-x}As$ alloys. More data and theoretical fits on GaAs, $Al_xGa_{1-x}As$, $In_{1-x}Ga_xAs_yP_{1-y}$, InP, and other material systems are documented in Refs. 9–19. It is noted that the refractive index is generally dispersive near or above the band gap with a fast variation near various interband transition energies. It also depends slightly on the doping concentration of the semiconductors, the temperature, and the strain. Above the band gap, both the real and the imaginary parts of the permittivity function are important, and Eqs. (5.5.4a) and (5.5.4b) in Chapter 5 should be used to relate $\varepsilon'(\omega)$ and $\varepsilon''(\omega)$ to the complex refractive index $n + i\kappa$.

7.1.7 Normalization Constant for the Optical Mode

The undetermined coefficient C_1 is determined by the normalization condition in which the total guided power is assumed to be unity. That is

$$P = \frac{1}{2}\mathrm{Re}\int_{-\infty}^{\infty} (\mathbf{E} \times \mathbf{H}^*) \cdot \hat{z}\, dx = 1. \tag{7.1.24}$$

For TE modes we have

$$\begin{aligned} H_x &= -\frac{k_z}{\omega\mu} E_y \\ P &= \frac{-1}{2}\int E_y H_x^* dx = \frac{1}{2}\int_{-\infty}^{\infty} \frac{k_z}{\omega\mu} |E_y|^2 dx \end{aligned} \tag{7.1.25}$$

where the permeability is μ_1 inside the guide and μ outside the guide. We find the normalization coefficient C_1 to be

$$C_1 = \left[\frac{4\omega\mu}{k_z\left(d + \dfrac{2}{\alpha}\right)}\right]^{1/2} \quad (\text{for } \mu = \mu_1) \tag{7.1.26}$$

for both TE even and odd modes. The above coefficient can be derived using the expressions for E_y and the eigenequations (7.1.9a) and (7.1.9b) for the even and odd modes, respectively.

7.1.8 Optical Confinement Factor Γ

An important quantity, called the optical confinement factor, is defined as the fraction of power guided in the waveguide.

$$\Gamma = \frac{\dfrac{1}{2}\displaystyle\int_{\text{inside}} \mathrm{Re}(\mathbf{E} \times \mathbf{H}^*) \cdot \hat{z}\, dx}{\dfrac{1}{2}\displaystyle\int_{\text{total}} \mathrm{Re}(\mathbf{E} \times \mathbf{H}^*) \cdot \hat{z}\, dx}. \tag{7.1.27}$$

Using the expression in (7.1.3), we find

$$\Gamma = \left[1 + \left(\frac{\mu_1}{\mu}\right)\left(\frac{2}{\alpha d}\right)\frac{\cos^2(k_x d/2)}{(1 + \sin k_x d/k_x d)}\right]^{-1} \quad \text{Even modes} \tag{7.1.28a}$$

$$\Gamma = \left[1 + \left(\frac{\mu_1}{\mu}\right)\left(\frac{2}{\alpha d}\right)\frac{\sin^2(k_x d/2)}{(1 - \sin k_x d/k_x d)}\right]^{-1} \quad \text{Odd modes.} \tag{7.1.28b}$$

It is obvious that $\Gamma < 1$ for dielectric waveguides. As the waveguide dimension becomes smaller $d \to 0$, or $k_x d \to 0$, we find that for the fundamental TE_0 mode

$$\alpha\frac{d}{2} = \frac{\mu}{\mu_1} X \tan X \simeq \frac{\mu}{\mu_1} X^2 \ll X \ll 1$$

and

$$\Gamma \longrightarrow 2\frac{\mu}{\mu_1}\left(\alpha\frac{d}{2}\right) \longrightarrow 2\left(\frac{\mu}{\mu_1}\right)^2\left(k_x\frac{d}{2}\right)^2 \longrightarrow 2\left(\frac{\mu}{\mu_1}\right)^2\left(k_0\frac{d}{2}\right)^2(n_1^2 - n^2), \quad (7.1.29)$$

which gives an estimation of the confinement factor for a single mode in the weakly guidance limit.

7.1.9 TM Modes $\mathbf{H} = \hat{y}H_y$

For TM polarization, we may obtain the results by the duality principle: replacing the field solutions **E** and **H** of the TE modes by **H** and $-\mathbf{E}$, respectively, μ by ε, and ε by μ as discussed in Section 5.1. We obtain

$$E = \frac{1}{i\omega\varepsilon_i}\left(\hat{x}\, ik_z H_y - \hat{z}\frac{\partial}{\partial x}H_y\right), (\varepsilon_i = \varepsilon_1 \text{ inside, } \varepsilon_i = \varepsilon \text{ outside the guide}). \quad (7.1.30)$$

The field expressions for H_y has the same forms as (7.1.3) and (7.1.7) for the even and odd modes, respectively, except that the coefficients C_0 and C_1 have to be found by matching the boundary conditions and the normalization condition of the guided power along the z direction.

The eigenequations from matching the boundary conditions are

$$\alpha\frac{d}{2} = \frac{\varepsilon}{\varepsilon_1}\left(k_x\frac{d}{2}\right)\tan\left(k_x\frac{d}{2}\right) \quad \text{TM even modes} \quad (7.1.31a)$$

$$\alpha\frac{d}{2} = -\frac{\varepsilon}{\varepsilon_1}\left(k_x\frac{d}{2}\right)\cot\left(k_x\frac{d}{2}\right) \quad \text{TM odd modes.} \quad (7.1.31b)$$

These two guidance conditions can also be obtained from duality principle by changing μ and μ_i in TE mode to ε and ε_1 for TM modes. The optical confinement factors for the TM modes can be similarly obtained from (7.1.28a) and (7.1.28b).

For dielectric materials, $\mu = \mu_1 = \mu_0$, we can replace all factors containing μ/μ_1 by 1 in the guidance conditions for TE modes and $\varepsilon/\varepsilon_1$ by n^2/n_1^2 in the expressions for the eigenequations and the optical confinement factors for both TE and TM polarizations. In Fig. 7.7, we plot the optical confinement factor Γ for the TE (solid curves) and TM (dashed curves) modes. All optical confinement factors start from zero at cutoff and approach unity at the high-frequency limit.

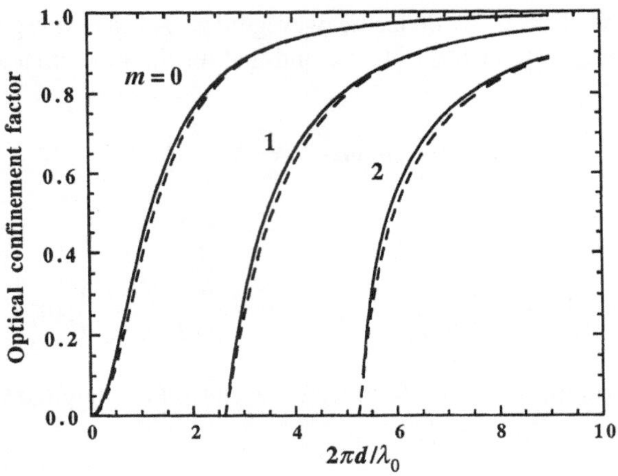

Figure 7.7 Optical confinement factors for the TE_m (solid curves) and the TM_m (dashed curves) modes.

7.2 ASYMMETRIC DIELECTRIC SLAB WAVEGUIDES

7.2.1 TE Polarization, $E = \hat{y}E_y$

If the dielectric constant in the substrate ε_2 is different from that of ε in the top medium, or $\mu_2 \neq \mu$, the structure becomes asymmetric (Fig. 7.8). In this case, the field solutions cannot be either even or odd modes. The general solution can be put in the form

$$E_y = e^{ik_z z} \begin{cases} C_0 \, e^{-\alpha x} & x \geq 0 \\ C_1 \cos(k_{1x} x + \phi) & -d \leq x \leq 0 \, . \\ C_2 \, e^{\alpha_2 (x+d)} & x \leq -d \end{cases} \tag{7.2.1}$$

Again, the transverse mode profile is decays away from the guide with different decay constants α and α_2 in region 0 and region 2, respectively. The

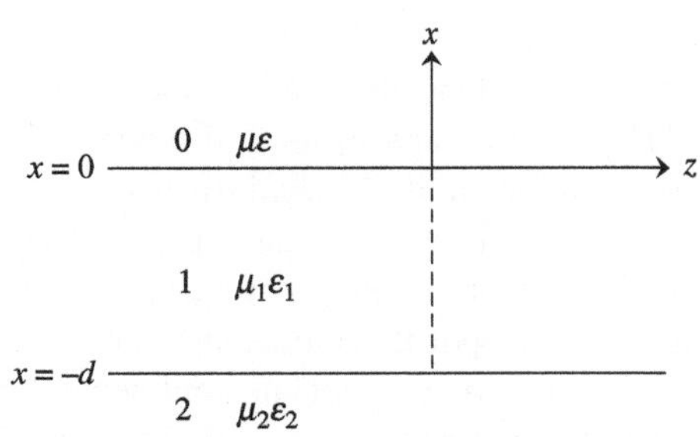

Figure 7.8 An asymmetric dielectric slab waveguide.

standing-wave solution inside the guide is written as $C_1 \cos(k_{1x}x + \phi)$, which is equivalent to $A \cos k_{1x}x + B \sin k_{1x}x$. We choose this form because it is easier to match the boundary conditions later and the phase angle ϕ is also related to the Goos–Hänchen phase shift (discussed in Section 5.6) when total internal reflection occurs. Because E_y satisfies the wave equation in each region, we obtain

$$-\alpha^2 + k_z^2 = \omega^2 \mu\varepsilon \tag{7.2.2a}$$

$$k_{1x}^2 + k_z^2 = \omega^2 \mu_1 \varepsilon_1 \tag{7.2.2b}$$

$$-\alpha_2^2 + k_z^2 = \omega^2 \mu_2 \varepsilon_2. \tag{7.2.2c}$$

Matching the boundary conditions in which E_y and $H_z = (1/i\omega\mu)\,\partial E_y/\partial x$ are continuous at $x = 0$ and $x = -d$, we obtain four equations relating C_0, C_1, C_2, and ϕ. We then eliminate C_0, C_1, and C_2 and obtain

$$\alpha = \frac{\mu}{\mu_1} k_{1x} \tan\phi \tag{7.2.3a}$$

$$\alpha_2 = -\frac{\mu_2}{\mu_1} k_{1x} \tan(-k_{1x}d + \phi). \tag{7.2.3b}$$

Eliminating ϕ, we obtain the eigenequation or the guidance condition for the TE_m mode:

$$k_{1x}d = \tan^{-1}\left(\frac{\mu_1 \alpha}{\mu k_{1x}}\right) + \tan^{-1}\left(\frac{\mu_1 \alpha_2}{\mu_2 k_{1x}}\right) + m\pi \quad (m = 0, 1, 2, \ldots). \tag{7.2.4}$$

If $\mu = \mu_1 = \mu_2$, the above equation can be simplified to

$$\tan k_{1x}d = \frac{(\alpha + \alpha_2)k_{1x}}{k_{1x}^2 - \alpha\alpha_2}. \tag{7.2.5}$$

After solving for α, α_2, k_{1x}, and k_z from (7.2.2) and (7.2.4), the complete electric field can be obtained from the normalization condition for the guided power along the z direction. We obtain

$$E_y = C_1\, e^{ik_z z} \begin{cases} \cos\phi\, e^{-\alpha x} & x \geq 0 \\ \cos(k_{1x}x + \phi) & -d \leq x \leq 0 \\ \cos(-k_{1x}d + \phi) e^{\alpha_2(x+d)} & x \leq -d \end{cases} \tag{7.2.6}$$

where $C_1 = \left[\dfrac{4\omega\mu}{k_z\left(d + \dfrac{1}{\alpha} + \dfrac{1}{\alpha_2}\right)}\right]^{1/2}$ if $\mu = \mu_1 = \mu_2$; otherwise, it can be cast in a more complicated analytical expression.

Cutoff Condition For the wave to be guided, ε_1 has to be larger than both ε_2 and ε. Let us assume that $\varepsilon_2 > \varepsilon$ $(\mu = \mu_1 = \mu_2)$. When reducing the frequency until $k_z = \omega\sqrt{\mu_2\varepsilon_2} = k_0 n_2$, the decay constant in region 2, α_2, will vanish before α does. Thus, at cutoff

$$k_{1x}d = \tan^{-1}\left(\frac{\mu_1\alpha}{\mu k_{1x}}\right) + m\pi \tag{7.2.7}$$

for the TE$_m$ mode at $\alpha_2 = 0$, and $k_z = k_0 n_2$. We also have

$$\alpha = \sqrt{k_z^2 - \omega^2\mu\varepsilon} = k_0\sqrt{n_2^2 - n^2} \tag{7.2.8a}$$

$$k_{1x} = k_0\sqrt{n_1^2 - n_2^2} \tag{7.2.8b}$$

at cutoff. The cutoff frequency is determined from

$$k_0 d\sqrt{n_1^2 - n_2^2} = \tan^{-1}\left(\frac{\mu_1\sqrt{n_2^2 - n^2}}{\mu\sqrt{n_1^2 - n_2^2}}\right) + m\pi \tag{7.2.9}$$

where $k_0 = \omega\sqrt{\mu_0\varepsilon_0} = \omega/c$. The dispersion curves for the TE$_m$ modes are plotted in Fig. 7.9. They start from the cutoff condition $k_z = k_0 n_2$ and approach the upper limit $k_z \rightarrow k_0 n_1$ when the frequency is increased. Note that if $n_1 \gtrsim n_2 \gg n$, then

$$\alpha/k_{1x} = \sqrt{n_2^2 - n^2}\Big/\sqrt{n_1^2 - n_2^2} \gg 1.$$

We obtain

$$k_0 d\sqrt{n_1^2 - n_2^2} = \left(m + \frac{1}{2}\right)\pi \tag{7.2.10}$$

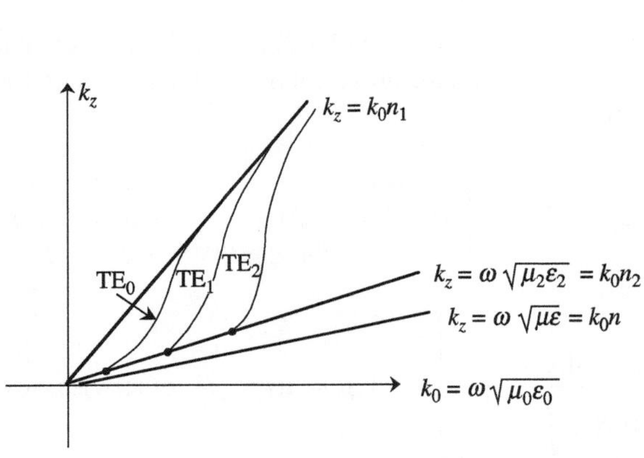

Figure 7.9 The dispersion relations for an asymmetric dielectric waveguide.

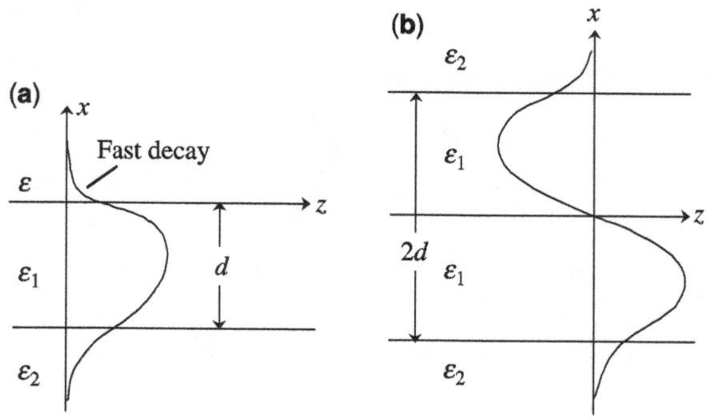

Figure 7.10 (a) An asymmetric waveguide with $\varepsilon_1 \gtrsim \varepsilon_2 \gg \varepsilon$ and a thickness d. (b) A symmetric waveguide with a thickness $2d$. The cutoff condition for the TE_m mode in (a) is equivalent to that of the $TE_{(2m+1)}$ mode in (b).

which is equivalent to the cutoff condition of the $TE_{(2m+1)}$ mode in a symmetric waveguide with a $2d$ thickness. This can be easily understood from a comparison of the two electric field profiles in Fig. 7.10a and Fig. 7.10b. Because the decaying constant α in region ε decays very fast in Fig. 7.10a, its field profile looks very much like half of the field profile of Fig. 7.10b.

7.2.2 TM Polarization $H = \hat{y}H_y$

We can easily obtain the solutions for the TM polarization using the duality principle. The filed has the same form as (7.2.6), except that the constant C_1 is different from that in (7.2.6) and can be derived from the normalization condition or duality principle, noting that $\varepsilon \neq \varepsilon_1 \neq \varepsilon_2$. The guidance condition is obtained from (7.2.4) after replacing μ_i by ε_i. We find

$$k_{1x}d = \tan^{-1}\left(\frac{\varepsilon_1 \alpha}{\varepsilon k_{1x}}\right) + \tan^{-1}\left(\frac{\varepsilon_1 \alpha_2}{\varepsilon_2 k_{1x}}\right) + m\pi \tag{7.2.11}$$

for the TM_m mode.

7.3 RAY OPTICS APPROACH TO WAVEGUIDE PROBLEMS

An efficient method to find the eigenequation for the dielectric slab waveguide problem is to use the ray optics picture. We know from Section 5.3 that when a plane wave is incident on a planar dielectric boundary with an angle of incidence θ larger than the critical angle, the reflection coefficient r_{12} has a Goos–Hänchen phase shift $-2\phi_{12}$.

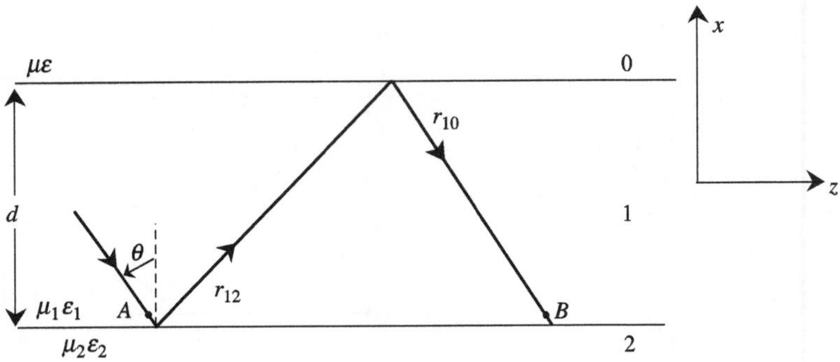

Figure 7.11 A ray optics picture for a guided mode in a slab waveguide.

For a mode to be guided, the total field after reflecting back and forth (from point A to B) by the two boundaries should repeat itself, Fig. 7.11,

$$(r_{12}\ e^{ik_{1x}d})r_{10}\ e^{ik_{1x}d} = 1 \tag{7.3.1}$$

where the phase $2k_{1x}d$ accounts for the round-trip optical path delay along the x direction.

The above condition is also called the transverse resonance condition. In terms of the Goos–Hänchen phase shifts, we obtain

$$e^{-i2\phi_{12}-i2\phi_{10}+i2k_{1x}d} = 1 \tag{7.3.2}$$

where for TE modes

$$\phi_{12} = \tan^{-1}\left(\frac{\mu_1\alpha_2}{\mu_2 k_{1x}}\right) \tag{7.3.3a}$$

$$\phi_{10} = \tan^{-1}\left(\frac{\mu_1\alpha}{\mu k_{1x}}\right). \tag{7.3.3b}$$

Therefore, we find

$$2k_{1x}d - 2\phi_{12} - 2\phi_{10} = 2m\pi. \tag{7.3.4}$$

The total phase delay (noting that the Goos–Hänchen phase shifts are negative $-2\phi_{12} - 2\phi_{10}$, or advance phase shifts) should be an even multiple of π. Equation (7.3.4) gives the guidance condition for the TE_m mode:

$$k_{1x}d = \tan^{-1}\left(\frac{\mu_1\alpha_2}{\mu_2 k_{1x}}\right) + \tan^{-1}\left(\frac{\mu_1\alpha}{\mu k_{1x}}\right) + m\pi. \tag{7.3.5}$$

If $\mu_2 = \mu$ and $\varepsilon_2 = \varepsilon$ (i.e., the slab waveguide is symmetric), we have $\phi_{12} = \phi_{10}$ and

$$k_{1x}d = m\pi + 2\phi_{10} \tag{7.3.6}$$

which leads to

$$\tan\left(k_{1x}\frac{d}{2}\right) = \begin{cases} \tan\phi_{10} = \dfrac{\mu_1\alpha}{\mu k_{1x}} & \text{if } m \text{ is even} \\ -\tan\phi_{10} = -\dfrac{\mu k_{1x}}{\mu_1\alpha} & \text{if } m \text{ is odd} \end{cases} \tag{7.3.7}$$

or

$$\alpha = \begin{cases} \dfrac{\mu}{\mu_1}k_{1x}\tan\left(k_{1x}\dfrac{d}{2}\right) & m = \text{even} \quad (7.3.8a) \\ -\dfrac{\mu}{\mu_1}k_{1x}\cot\left(k_{1x}\dfrac{d}{2}\right) & m = \text{odd} \quad (7.3.8b) \end{cases}$$

which are the same as (7.1.9a) and (7.1.9b), as expected.

7.4 RECTANGULAR DIELECTRIC WAVEGUIDES

The rectangular dielectric waveguide theory is very useful because most waveguides have finite dimensions in both the x and y directions. There are two possible classes of modes. The analyses are approximate here, assuming $w > d$, as shown in Fig. 7.12. The method we use here is sometimes called Marcatili's method for rectangular dielectric waveguides [20, 21]. These modes are

1. HE_{pq} modes or $E^y_{(p+1)(q+1)}$ modes: H_x and E_y are the dominant components.
2. EH_{pq} modes or $E^x_{(p+1)(q+1)}$ modes: E_x and H_y are the dominant components.

In general, we have $\dfrac{\partial}{\partial z} = ik_z$, and Maxwell's equations in the following forms will be used as the starting equations in the later analysis for both HE_{pq}

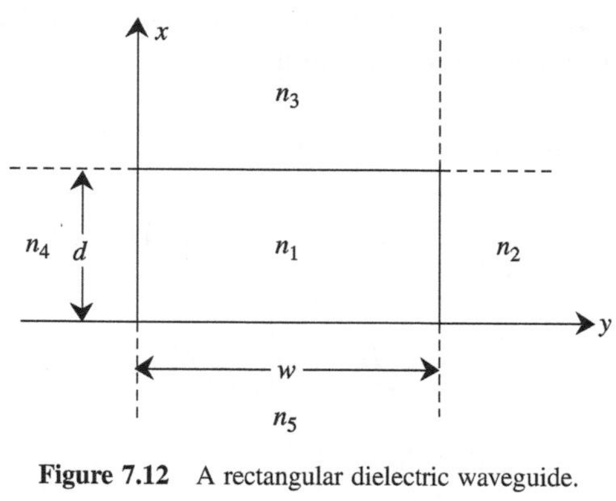

Figure 7.12 A rectangular dielectric waveguide.

modes and EH_{pq} modes:

$$\begin{aligned}\mathbf{H} &= \frac{1}{i\omega\mu}\nabla\times\mathbf{E}\\ &= \frac{1}{i\omega\mu}\left[\hat{x}\left(\frac{\partial}{\partial y}E_z - ik_zE_y\right) + \hat{y}\left(ik_zE_x - \frac{\partial}{\partial x}E_z\right) + \hat{z}\left(\frac{\partial}{\partial x}E_y - \frac{\partial}{\partial y}E_x\right)\right]\end{aligned} \tag{7.4.1}$$

$$\begin{aligned}\mathbf{E} &= \frac{1}{-i\omega\varepsilon}\nabla\times\mathbf{H}\\ &= \frac{1}{-i\omega\varepsilon}\left[\hat{x}\left(\frac{\partial}{\partial y}H_z - ik_zH_y\right) + \hat{y}\left(ik_zH_x - \frac{\partial}{\partial x}H_z\right) + \hat{z}\left(\frac{\partial}{\partial x}H_y - \frac{\partial}{\partial y}H_x\right)\right]\end{aligned} \tag{7.4.2}$$

$$\frac{\partial}{\partial x}E_x + \frac{\partial}{\partial y}E_y + ik_zE_z = 0 \tag{7.4.3}$$

$$\frac{\partial}{\partial x}H_x + \frac{\partial}{\partial y}H_y + ik_zH_z = 0. \tag{7.4.4}$$

Here we have assumed that the permittivity is a constant in each region.

7.4.1 HE_{pq} Modes (or $E^y_{(p+1)(q+1)}$ Modes)

We start with the HE_{pq} modes. The large components are H_x and E_y. The other components E_x, E_z, H_y, and H_z are assumed to be small. In principle, we can assume $H_y \simeq 0$, and express E_x, E_y, E_z, and H_z in terms of H_x using Maxwell's equations. These are called the HE_{pq} modes or E^y_{pq} modes. Alternatively, we can also assume $E_x \simeq 0$, and express E_z, H_x, H_y, and H_z in terms of E_y. Because these are all approximate solutions, they have to be compared with the exact solutions [22, 23]. We find that both approximations give essentially the same transcendental equations with minor differences. We simply call these modes HE_{pq} modes, where p labels the number of nodes for a wave function in the x direction and q for the y direction. The solution procedure is

(i) Solve the wave equation for H_x everywhere

$$\left(\frac{\partial^2}{\partial x^2} + \frac{\partial^2}{\partial y^2} - k_z^2 + \omega^2\mu\varepsilon\right)H_x = 0. \tag{7.4.5}$$

(ii)

$$H_z = \frac{1}{-ik_z}\frac{\partial}{\partial x}H_x. \tag{7.4.6}$$

(iii)
$$E_z = \frac{1}{i\omega\varepsilon}\frac{\partial}{\partial y}H_x. \tag{7.4.7}$$

(iv) Because $H_y = 0$, we obtain
$$E_x = \frac{-1}{i\omega\varepsilon}\frac{\partial}{\partial y}H_z = \frac{-1}{\omega\varepsilon k_z}\frac{\partial}{\partial x}\frac{\partial}{\partial y}H_x. \tag{7.4.8}$$

(v)
$$\begin{aligned} E_y &= \frac{1}{-i\omega\varepsilon}\left(ik_z H_x - \frac{\partial}{\partial x}H_z\right) = \frac{1}{\omega\varepsilon k_z}\left(-k_z^2 + \frac{\partial^2}{\partial x^2}\right)H_x \\ &= \frac{1}{\omega\varepsilon k_z}\left(-\omega^2\mu\varepsilon - \frac{\partial^2}{\partial y^2}\right)H_x. \end{aligned} \tag{7.4.9}$$

Once H_x is found, all other components (except that $H_y = 0$) can be expressed in terms of H_x using Eqs. (7.4.6)–(7.4.9).

The expression for H_x can be written approximately as

$$H_x = e^{ik_z z}\begin{cases} C_1 \cos(k_x x + \phi_x)\cos(k_y y + \phi_y) & \text{Region 1} \\ C_2 \cos(k_x x + \phi_x)\, e^{-\alpha_2(y-w)} & \text{Region 2} \\ C_3\, e^{-\alpha_3(x-d)}\cos(k_y y + \phi_y) & \text{Region 3} \\ C_4 \cos(k_x x + \phi_x)\, e^{\alpha_4 y} & \text{Region 4} \\ C_5\, e^{\alpha_5 x}\cos(k_y y + \phi_y) & \text{Region 5} \end{cases} \tag{7.4.10}$$

where $k_x^2 + k_y^2 + k_z^2 = \omega^2\mu_1\varepsilon_1$, $k_x^2 - \alpha_2^2 + k_z^2 = \omega^2\mu_2\varepsilon_2$, and so forth, which are obtained by substituting the field expressions (7.4.10) into the wave equation (7.4.5) using the material parameters μ_i and ε_i for region i. The choice of the standing-wave solutions $\cos(k_x + \phi_x)$ or decaying solutions is made simply by inspecting Fig. 7.12 and keeping in mind that we are looking for the guided mode solution. The same x dependence is used for regions 1 and 2 because both regions share the same boundary at $y = w$, and the phase matching condition or Snell's law requires the same dependence on the spatial variable x. Similar rules apply to the field expressions in other regions.

Boundary Conditions To derive the guidance conditions and the relations between the field coefficients $C_1 \ldots C_5$, we match the boundary conditions.

(i) At $x = 0$ and $x = d$, H_z and E_y are continuous. The other boundary condition for E_z continuous is ignored because E_z is small. We obtain relations between the coefficients C_1, C_3, C_5, and the phase ϕ_x.

Eliminating C_1, C_3, and C_5 and then eliminating ϕ_x, we find

$$k_x d = p\pi + \tan^{-1}\left[\frac{\varepsilon_3\left(\omega^2\mu_1\varepsilon_1 - k_y^2\right)\alpha_3}{\varepsilon_1\left(\omega^2\mu_3\varepsilon_3 - k_y^2\right)k_x}\right] + \tan^{-1}\left[\frac{\varepsilon_5\left(\omega^2\mu_1\varepsilon_1 - k_y^2\right)\alpha_5}{\varepsilon_1\left(\omega^2\mu_5\varepsilon_5 - k_y^2\right)k_x}\right]. \tag{7.4.11a}$$

The above formula can be further simplified for $k_y \ll k_1$, k_3, and k_5:

$$k_x d = p\pi + \tan^{-1}\left[\frac{\mu_1\alpha_3}{\mu_3 k_x}\right] + \tan^{-1}\left[\frac{\mu_1\alpha_5}{\mu_5 k_x}\right]. \tag{7.4.11b}$$

(ii) At $y = 0$ and $y = w$, match H_x and E_z. (The other conditions that H_z and E_x are continuous give identical results as H_x and E_z.) We obtain the relations between C_1, C_2, C_4, as well as ϕ_y. By eliminating these coefficients and ϕ_y, we find

$$k_y w = q\pi + \tan^{-1}\left[\frac{\varepsilon_1\alpha_2}{\varepsilon_2 k_y}\right] + \tan^{-1}\left[\frac{\varepsilon_1\alpha_4}{\varepsilon_4 k_y}\right]. \tag{7.4.12}$$

The two guidance conditions can also be seen from the ray optics approach. An intuitive way to understand the approximate boundary conditions is that for (i) at $x = 0$ and $x = d$, the optical power density of interest is its x component bouncing back and forth between the two interfaces at $x = 0$ and d. We need therefore E_y, H_z and H_y, E_z. However, $H_y \simeq 0$. We only need E_y and H_z and drop E_z because it is small. For boundary conditions (ii) at $y = 0$ and w, we use only H_z and E_x, or H_x and E_z for the y component of the optical power density. Both pairs give the same results because H_y is not of concern.

7.4.2 EH_{pq} Modes (or $E^x_{(p+1)(q+1)}$ Modes)

For the EH_{pq} mode, the dominant field components are E_x and H_y. We can assume that $H_x \simeq 0$ and H_y have exactly the same form as in (7.4.10). Expressing all other components in terms of H_y only, we then match the boundary conditions. We use

$$H_z = \frac{1}{-ik_z}\frac{\partial}{\partial y}H_y, \quad E_z = \frac{1}{-i\omega\varepsilon}\frac{\partial}{\partial x}H_y, \quad E_y = \frac{1}{\omega\varepsilon k_z}\frac{\partial}{\partial y}\frac{\partial}{\partial x}H_y$$
$$E_x = \frac{1}{\omega\varepsilon k_z}\left(\omega^2\mu\varepsilon + \frac{\partial^2}{\partial x^2}\right)H_y, \tag{7.4.13}$$

which are derivable from Eqs. (7.4.1)–(7.4.4). The following guidance conditions are found after matching the boundary conditions,

$$k_x d = p\pi + \tan^{-1}\left(\frac{\varepsilon_1 \alpha_3}{\varepsilon_3 k_x}\right) + \tan^{-1}\left(\frac{\varepsilon_1 \alpha_5}{\varepsilon_5 k_x}\right) \tag{7.4.14}$$

$$k_y w = q\pi + \tan^{-1}\left[\frac{\varepsilon_2(\omega^2\mu_1\varepsilon_1 - k_x^2)\alpha_2}{\varepsilon_1(\omega^2\mu_2\varepsilon_2 - k_x^2)k_y}\right] + \tan^{-1}\left[\frac{\varepsilon_4(\omega^2\mu_1\varepsilon_1 - k_x^2)\alpha_4}{\varepsilon_1(\omega^2\mu_4\varepsilon_4 - k_x^2)k_y}\right] \tag{7.4.15}$$

where (7.4.22) can be further simplified for $k_x \ll k_1$, k_2, and k_4:

$$k_y w = q\pi + \tan^{-1}\left(\frac{\mu_1 \alpha_2}{\mu_2 k_y}\right) + \tan^{-1}\left(\frac{\mu_1 \alpha_4}{\mu_4 k_y}\right). \tag{7.4.16}$$

Alternatively, we may assume that $E_y \simeq 0$ and E_x is of the form (7.4.10), and then express all the other components in terms of the E_x component using Maxwell's equations. These results will give almost the same eigenequations and mode patterns.

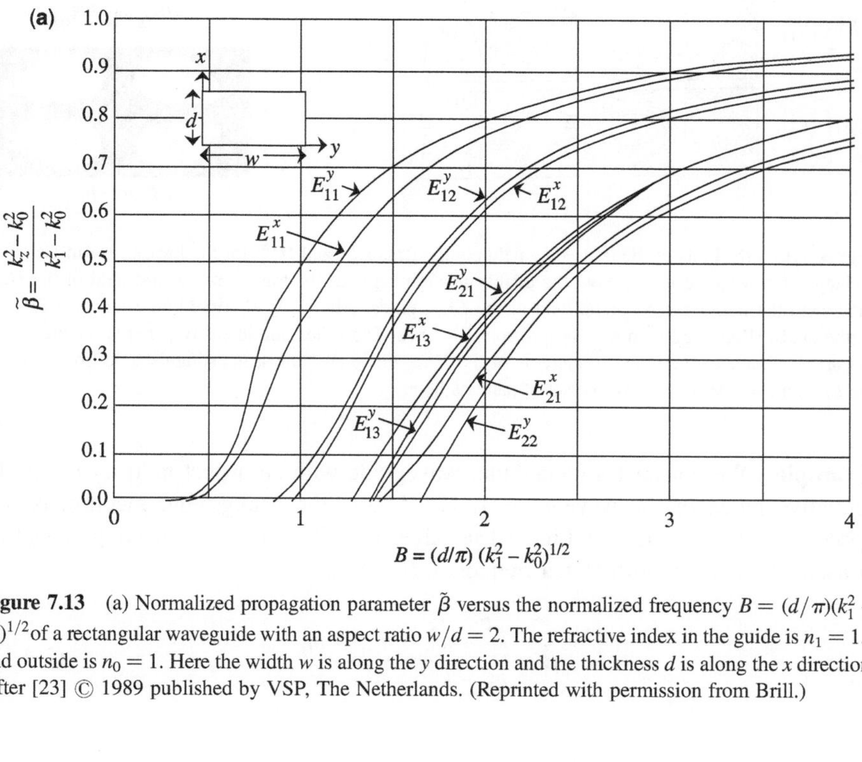

Figure 7.13 (a) Normalized propagation parameter $\tilde{\beta}$ versus the normalized frequency $B = (d/\pi)(k_1^2 - k_0^2)^{1/2}$ of a rectangular waveguide with an aspect ratio $w/d = 2$. The refractive index in the guide is $n_1 = 1.5$ and outside is $n_0 = 1$. Here the width w is along the y direction and the thickness d is along the x direction. After [23] © 1989 published by VSP, The Netherlands. (Reprinted with permission from Brill.)

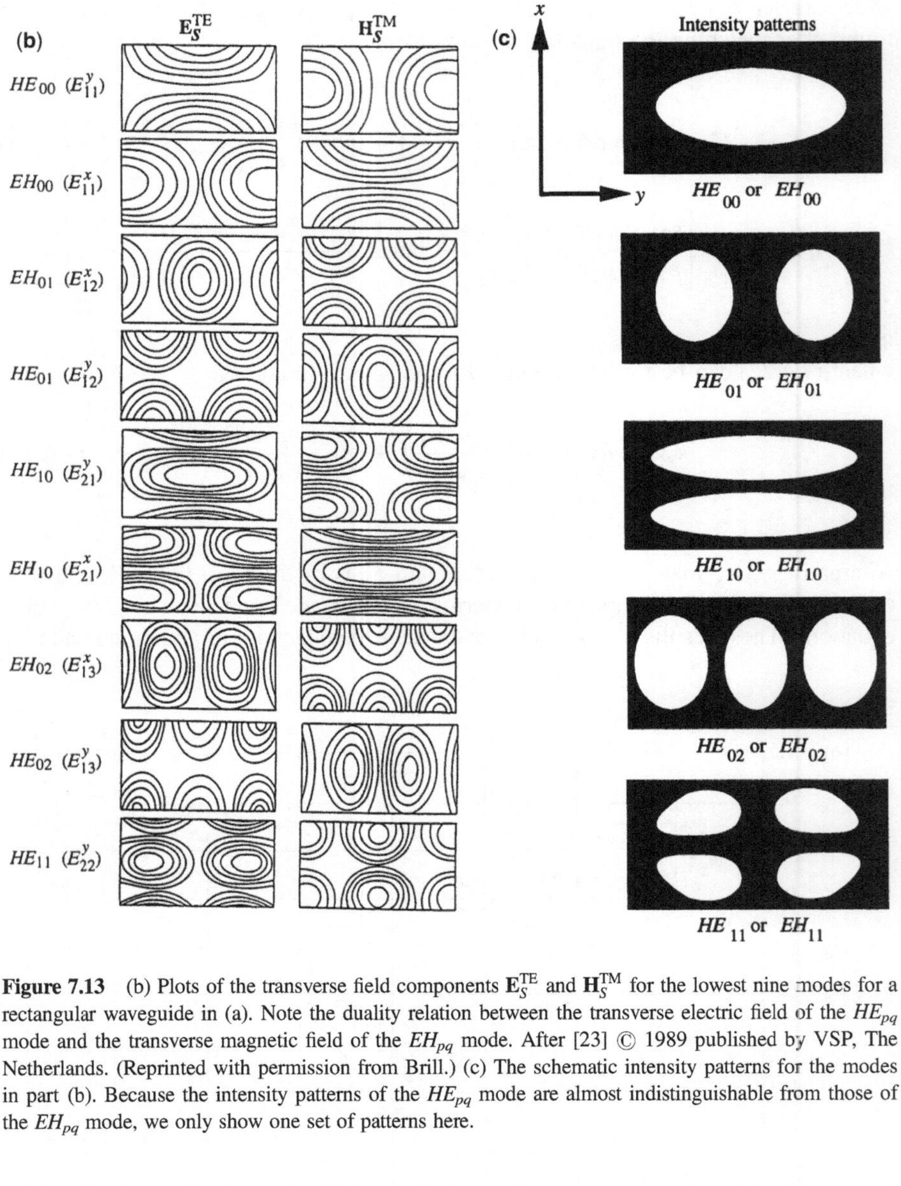

Figure 7.13 (b) Plots of the transverse field components $\mathbf{E}_S^{TE}$ and $\mathbf{H}_S^{TM}$ for the lowest nine modes for a rectangular waveguide in (a). Note the duality relation between the transverse electric field of the HE_{pq} mode and the transverse magnetic field of the EH_{pq} mode. After [23] © 1989 published by VSP, The Netherlands. (Reprinted with permission from Brill.) (c) The schematic intensity patterns for the modes in part (b). Because the intensity patterns of the HE_{pq} mode are almost indistinguishable from those of the EH_{pq} mode, we only show one set of patterns here.

Example We consider a rectangular waveguide with an aspect ratio $w/d = 2$, the refractive index of the waveguide n_1 is 1.5, and the background medium is free space $n_0 = 1$. We plot on Fig. 7.13a, after Ref. 23, the normalized propagation constant $\tilde{\beta}$ versus a normalized frequency $B = V/\pi$,

$$\tilde{\beta} = \frac{k_z^2 - k_0^2}{k_1^2 - k_0^2} = \frac{n_{\text{eff}}^2 - n_0^2}{n_1^2 - n_0^2} \tag{7.4.17}$$

and

$$B = V/\pi = (d/\pi)(k_1^2 - k_0^2)^{1/2} \tag{7.4.18}$$

where the V number is defined the same as that of a slab waveguide with a thickness d,

$$V = k_0 d(n_1^2 - n_0^2)^{1/2}. \tag{7.4.19}$$

The corresponding transverse electric and magnetic field patterns inside the dielectric waveguide are also plotted in Fig. 7.13b. Note the dual relations between the transverse electric field $\mathbf{E}_S^{\mathrm{TE}}$ of the HE_{pq} mode and the transverse magnetic field $\mathbf{H}_S^{\mathrm{TM}}$ of the EH_{pq} mode. We can see that the field patterns provide the information on the intensity variations of the E_x, E_y, H_x, and H_y components of the waveguide modes, Fig. 7.13c. The intensity pattern of the HE_{pq} mode is almost indistinguishable from that of the EH_{pq} mode. Therefore, only one set of intensity patterns is shown. These mode patterns should be interesting to compare with those of the metallic waveguides [24]. These results in Fig. 7.13a have been compared with those of Goell [22] with good agreement.

The original labels of the modes in Ref. 23 are slightly different. In Fig. 7.13a and Fig. 7.13b, we use our convention for the $HE_{pq}(= E^y_{(p+1)(q+1)})$ modes and the EH_{pq} $(= E^x_{(p+1)(q+1)})$ modes, and the coordinate system is shown in Fig. 7.12, where the vertical axis is the x axis and the horizontal axis is the y axis. This coordinate system is chosen such that when the width $w \rightarrow \infty$, the results here approach those of the slab waveguides discussed in Sections 7.1–7.3 using the same coordinate system. Our convention is the same as those for Marcatili [20] and Goell [22] using the $E^x_{(p+1)(q+1)}$ and $E^y_{(p+1)(q+1)}$ modes except that our aspect ratio is $w/d > 1$ where w is the width along the y direction instead of the x direction.

7.5 THE EFFECTIVE INDEX METHOD

The effective index method is a very useful technique to find the propagation constant of a dielectric waveguide. As an example, assume that we are interested in the solutions for a rectangular dielectric waveguide as in Section 7.4. Consider first the $E^y_{(p+1)(q+1)}$ modes for which E_y is the dominant component (Fig. 7.14a).

Step 1. We solve first a slab waveguide problem with refractive index n_1 inside and n_3, n_5 outside the waveguide, Fig. 7.14b. The eigenequation for the E_y component will be that of the TE modes of a slab guide discussed in Section 7.1 because E_y is parallel to the boundaries

$$k_{1x} d = \tan^{-1}\left(\frac{\mu_1 \alpha_3}{\mu_3 k_{1x}}\right) + \tan^{-1}\left(\frac{\mu_1 \alpha_5}{\mu_5 k_{1x}}\right) + p\pi \tag{7.5.1}$$

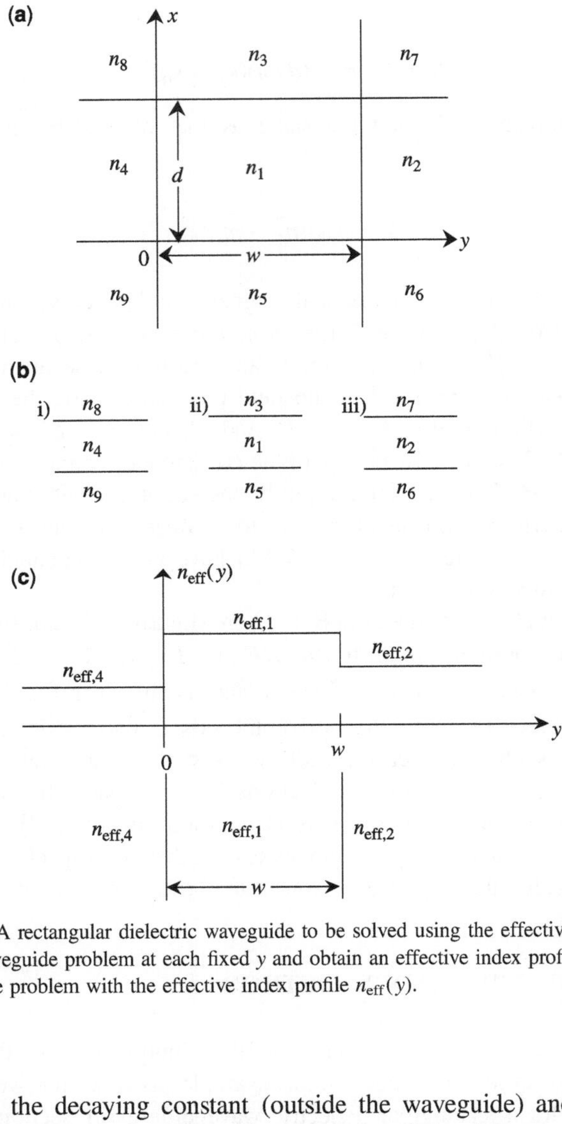

Figure 7.14 (a) A rectangular dielectric waveguide to be solved using the effective index method. (b) Solve the slab waveguide problem at each fixed y and obtain an effective index profile $n_{eff}(y)$. (c) Solve the slab waveguide problem with the effective index profile $n_{eff}(y)$.

where α_i is the decaying constant (outside the waveguide) and μ_i is the permeability in the ith region. From the solutions of the slab waveguide problem, we obtain the propagation constant, therefore, the effective index $n_{eff,1}$. For $y > w$ or $y < 0$, similar equations hold, if $n_2 > n_6, n_7$, and $n_4 > n_8, n_9$. Otherwise, approximations have to be made by assuming $n_{eff,2} \simeq n_2$ and $n_{eff,4} \simeq n_4$ when the fields in regions 6, 7, 8, and 9 are negligible.

Step 2. We then solve the slab waveguide problem as shown in Fig. 7.14c, and obtain

$$k_{1y}w = \tan^{-1}\left(\frac{\varepsilon_{eff,1}\,\alpha_2}{\varepsilon_{eff,2}\,k_{1y}}\right) + \tan^{-1}\left(\frac{\varepsilon_{eff,1}\,\alpha_4}{\varepsilon_{eff,4}\,k_{1y}}\right) + q\pi \tag{7.5.2}$$

where $\varepsilon_{\text{eff,i}} = n^2_{\text{eff,i}}\,\varepsilon_0$. Note that the E_y component is perpendicular to the slab boundaries; therefore, we should use the guidance condition for the TM modes of the slab waveguide problem. In step 1, for each y, we find a modal distribution as a function of x. This field distribution will also vary as y changes. Thus we call this function $F(x, y)$. In step 2, the solution for the modal distribution will be a function of only y, called $G(y)$. Thus the total electric field E_y can be written as

$$E_y(x, y) \simeq F(x, y)\,G(y). \tag{7.5.3}$$

It is interesting to compare the two eigenequations (7.5.1) and (7.5.2) obtained here with those derived in Section 7.4, Eqs. (7.4.11b) and (7.4.12). The difference is that the *effective permittivities* appear in (7.5.2) in the effective index method whereas the material permittivities are used in (7.4.12) in the previous approximate Marcatili's method. When comparing the propagation constants with the numerical approach from the full-wave analysis, the results of both Marcatili's method in Section 7.4 and the effective index method agree very well with those of the numerical method, generally speaking. The effective index method usually has a better agreement with the numerical method especially near cutoff.

The effective index method has also been applied [25, 26] to the analysis of diode lasers with lateral variations in thickness of the active region or in the complex permittivity. The resultant analytical formulas in many cases are extremely helpful in understanding the semiconductor laser operation characteristics.

7.6 WAVE GUIDANCE IN A LOSSY OR GAIN MEDIUM

When the permittivity is complex,

$$\varepsilon = \varepsilon' + i\varepsilon'', \tag{7.6.1}$$

that is, the medium is lossy ($\varepsilon'' > 0$) or it has gain ($\varepsilon'' < 0$), and the propagation constant becomes

$$\begin{aligned} k &= \omega\sqrt{\mu(\varepsilon' + i\varepsilon'')} = \frac{\omega}{c}(n + i\kappa) \\ &= k' + ik''. \end{aligned} \tag{7.6.2}$$

For a waveguide that may contain gain in the active region and loss in the passive region, the propagation constant is generally complex. A typical analysis is based on the perturbation theory or a variational method. Consider a dielectric slab waveguide that has complex permittivities everywhere, as shown in Fig. 7.15

$$\varepsilon(x) = \begin{cases} \varepsilon_1' + i\varepsilon_1'' & \text{inside} \\ \varepsilon' + i\varepsilon'' & \text{outside.} \end{cases}$$

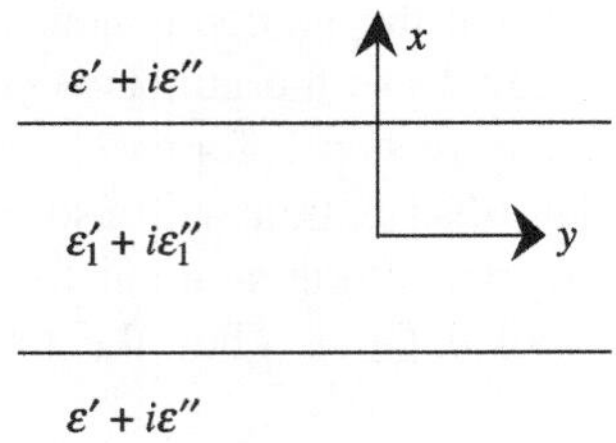

Figure 7.15 A lossy (gain) waveguide structure.

Inside the guide,

$$k_1 = \omega\sqrt{\mu(\varepsilon_1' + i\varepsilon_1'')} \simeq \omega\sqrt{\mu\varepsilon'}\left(1 + i\frac{\varepsilon_1''}{2\varepsilon_1'}\right)$$
$$= k_1' - i\frac{g}{2} \tag{7.6.3}$$

where $k_1' = \omega\sqrt{\mu\varepsilon_1'} = k_0 n_1$, and $g = -k_1'\dfrac{\varepsilon_1''}{\varepsilon_1'}$ is the gain coefficient, if $\varepsilon_1'' < 0$.

Outside the waveguide, we have

$$k = \omega\sqrt{\mu(\varepsilon' + i\varepsilon'')} \cong \omega\sqrt{\mu\varepsilon'}\left(1 + i\frac{\varepsilon''}{2\varepsilon'}\right)$$
$$= k' + i\frac{\alpha}{2} \tag{7.6.4}$$

where $\alpha = k'\dfrac{\varepsilon''}{\varepsilon'}$ is the decaying constant of the absorption coefficient for the optical intensity. Suppose we try to solve the TE mode

$$\mathbf{E} = \hat{y}\, E_y = \hat{y}\, \phi(x)e^{ik_z z} \tag{7.6.5}$$

$$\left[\frac{d^2}{dx^2} - k_z^2 + \omega^2\mu\varepsilon(x)\right]\phi(x) = 0. \tag{7.6.6}$$

A simple method is to use the variational approach

$$k_z^2 = \frac{\displaystyle\int_{-\infty}^{\infty} \phi^*(x)\left[\frac{d^2}{dx^2} + \omega^2\mu\varepsilon(x)\right]\phi(x)\, dx}{\displaystyle\int_{-\infty}^{\infty} \phi^*(x)\phi(x)\, dx}. \tag{7.6.7}$$

When we write the permittivity as the sum of a lossless part $\varepsilon^{(0)}(x)$ and a part due to gain or loss $\Delta\varepsilon(x)$, we obtain

$$\varepsilon^{(0)}(x) = \begin{cases} \varepsilon_1' & \text{inside} \\ \varepsilon' & \text{outside} \end{cases} \qquad \Delta\varepsilon(x) = \begin{cases} i\varepsilon_1'' & \text{inside} \\ i\varepsilon'' & \text{outside} \end{cases} \tag{7.6.8}$$

where $\Delta\varepsilon(x)$ is considered a perturbation. The unperturbed solution for $\varepsilon^{(0)}(x)$ has been discussed previously, and the solution $\phi^{(0)}(x)$ is just that of a lossless slab waveguide. Note that

$$\left[\frac{d^2}{dx^2} - k_z^{(0)^2} + \omega^2 \mu \varepsilon^{(0)}(x)\right] \phi^{(0)}(x) = 0. \tag{7.6.9}$$

If we approximate the trial function $\phi(x)$ in the variational expression (7.6.7) by the unperturbed solution $\phi^{(0)}(x)$, we obtain

$$k_z^2 = k_z^{(0)^2} + i\omega^2 \mu \varepsilon_1'' \, \Gamma + i\omega^2 \mu \varepsilon'' \, (1 - \Gamma) \tag{7.6.10}$$

where

$$\Gamma = \frac{\int_{\text{inside}} \left|\phi^{(0)}(x)\right|^2 dx}{\int_{-\infty}^{\infty} \left|\phi^{(0)}(x)\right|^2 dx} \tag{7.6.11}$$

is the optical confinement factor.

When ε_1' is close to ε', the propagation constant can be written in the form

$$k_z \simeq k_z^{(0)} - i\Gamma \frac{g}{2} + i(1 - \Gamma)\frac{\alpha}{2} \tag{7.6.12}$$

which is obtained by taking the square roots of both sides of (7.6.10) and using the binomial expansion assuming that the imaginary part is small compared with the real part, and $k_z^{(0)}$ is close to k' and k_1'. When applying (7.6.12) to a laser cavity as shown in Fig. 7.16, we have the wave bouncing back and forth in a round-trip of distance $2L$

$$e^{ik_z 2L} r^2 = 1 \tag{7.6.13}$$

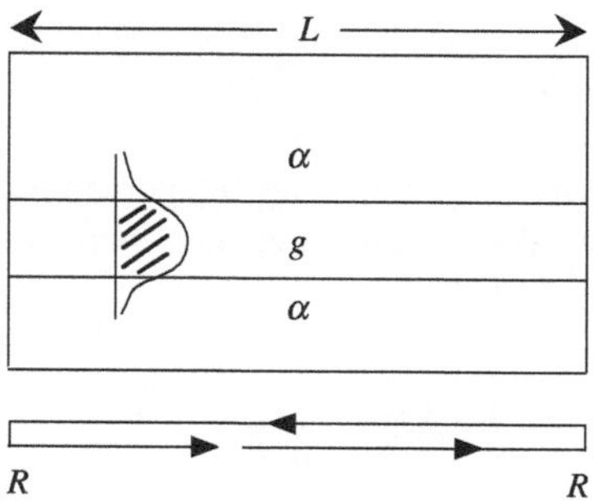

Figure 7.16 A semiconductor laser cavity with a net gain g in the active region and an absorption coefficient α outside the active region. The cavity length is L and the reflectivity is R at both ends.

where r is the reflection coefficient of the electric field. The reflectivity R is related to r by $R = |r|^2$. Ignoring the small imaginary part of r, which is a good approximation formost cases, we have

$$e^{i2k_z^{(0)}L}\, e^{\Gamma g L - (1-\Gamma)\alpha L} R = 1. \tag{7.6.14}$$

Thus

$$2k_z^{(0)}L = 2m\pi \quad m = \text{an integer} \tag{7.6.15}$$

which determines the longitudinal mode spectrum of the semiconductor laser, and

$$\Gamma g = (1-\Gamma)\alpha + \frac{1}{L}\ln\frac{1}{R} \tag{7.6.16}$$

which determines the threshold gain condition. Here Γg is the modal gain of the dielectric waveguide, and $k_z^{(0)} \simeq \frac{\omega}{c} n_e$, where n_e is the effective index. Equation (7.6.15) gives

$$\frac{\omega}{c} n_e L = m\pi. \tag{7.6.17}$$

The frequency spectrum is

$$f_m = m\frac{c}{2n_e L}. \tag{7.6.18}$$

If we ignore the frequency dependence of n_e, the Fabry–Perot frequency spacing is

$$\Delta f = \frac{c}{2n_e L} \equiv \text{FSR} \tag{7.6.19}$$

which is called the free spectral range (FSR) of the Fabry–Perot spectrum. Alternatively, we write in terms of the free-space wavelength λ:

$$L = m\frac{\lambda}{2n_e} \tag{7.6.20}$$

which means that the cavity length L is an integer multiple of half-wavelength in the semiconductor cavity. The spacing between two nearby Fabry–Perot modes ($\Delta m = 1$) is

$$\Delta\lambda = \frac{\lambda^2}{2n_g L} \tag{7.6.21a}$$

$$n_g = n_e - \lambda\frac{dn_e}{d\lambda} \tag{7.6.21b}$$

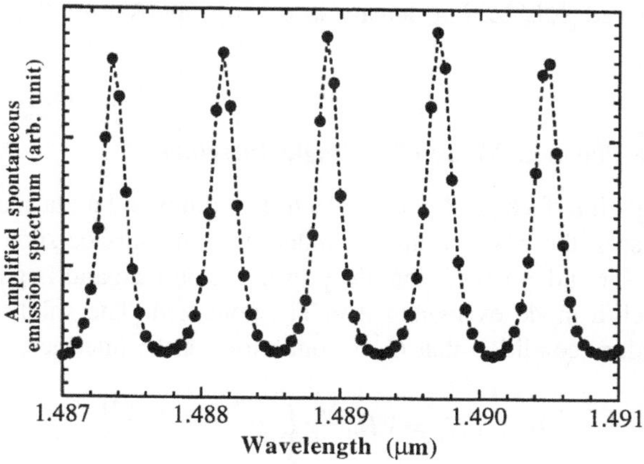

Figure 7.17 Amplified spontaneous emission spectrum of a 320-μm semiconductor laser cavity at room temperature below threshold. The injection current is $I = 8$ mA. (Above the laser threshold current, $I_{th} = 13.5$ mA, lasing starts.)

where the dispersion effect ($dn_e/d\lambda \neq 0$) is taken into account and n_g is the group index of the guided mode, which accounts for the dispersion of the effective index $n_e(\lambda)$.

In Fig. 7.17, we show the amplified spontaneous emission spectrum of a InGaAsP/InGaAsP strained quantum-well laser at 300K. The cavity length L is 370 μm. The threshold current is $I_{th} = 13.5$ mA at this temperature. Using $n_e \simeq 3.395$ and [10, 16] $dn_e/d\lambda \simeq -0.264\ (\mu\text{m})^{-1}$ near $\lambda = 1.49\ \mu\text{m}$, we obtain the wavelength spacing $\Delta\lambda \simeq 7.9$ Å, which agrees with the spacing shown in the figure.

7.7 SURFACE PLASMON WAVEGUIDES

In recent years, surface plasmon waveguides, or plasmonics, has become an intensive subject of research. The main idea is to develop waveguide structures of small dimensions of subwavelength scales at optical frequencies. From Section 5.5, we understand that metals behave like plasma at optical frequencies. If the optical frequency ω is less than the plasma frequency ω_p, the permittivity

$$\varepsilon_p(\omega) = \varepsilon_0\left(1 - \omega_p^2/\omega^2\right) \tag{7.7.1}$$

becomes negative, and the wave is strongly attenuated in the metal. If $\omega > \omega_p$, the permittivity $\varepsilon(\omega)$ becomes real. At $\omega = \omega_p$, $\varepsilon(\omega) = 0$, and the propagation constant $k = \omega\sqrt{\mu\varepsilon(\omega)}$ vanishes (i.e., the electromagnetic wave does not propagate any mode). The medium response gives

$$\begin{aligned} 0 = \mathbf{D} = \varepsilon_p(\omega)\mathbf{E} &= \varepsilon_0\mathbf{E} + \mathbf{P} \\ \mathbf{P} &= -\varepsilon_0\mathbf{E}, \end{aligned} \tag{7.7.2}$$

that is, the induced polarization density is exactly opposite of that of the electric field, $-\varepsilon_0\mathbf{E}$.

7.7.1 Surface Plasmon Mode of a Single Interface

Consider a single interface between a dielectric medium and a plasma medium [27], Fig. 7.18. We seek the possible guided mode near the surface such that the mode decays in both $+x$ and $-x$ direction away from the surface and is guided along the z direction. Such a mode exists and it is TM polarized. The solution is given by, using the boundary condition that H_y is continuous at the interface,

$$\mathbf{H} = \hat{y} H_y = \hat{y} H_0 e^{ik_z z} \begin{cases} e^{-\alpha_1 x}, & x \geq 0 \\ e^{\alpha_2 x}, & x \leq 0 \end{cases} \tag{7.7.3}$$

where the wave equation in each region gives

$$-\alpha_1^2 + k_z^2 = \omega^2 \mu_0 \varepsilon_1 \tag{7.7.4a}$$

$$-\alpha_2^2 + k_z^2 = \omega^2 \mu_0 \varepsilon_p(\omega). \tag{7.7.4b}$$

The electric field is obtained from the Maxwell's equations, $\mathbf{E} = (\nabla \times \mathbf{H})/(-i\omega\varepsilon)$

$$\mathbf{E} = \begin{cases} \dfrac{1}{i\omega\varepsilon_1}(\alpha_1 \hat{z} + \hat{x} i k_z) H_0 e^{-\alpha_1 x + i k_z z}, & x \geq 0 \\ \dfrac{1}{i\omega\varepsilon_p}(-\alpha_2 \hat{z} + \hat{x} i k_z) H_0 e^{\alpha_2 x + i k_z z} & x \leq 0. \end{cases} \tag{7.7.5}$$

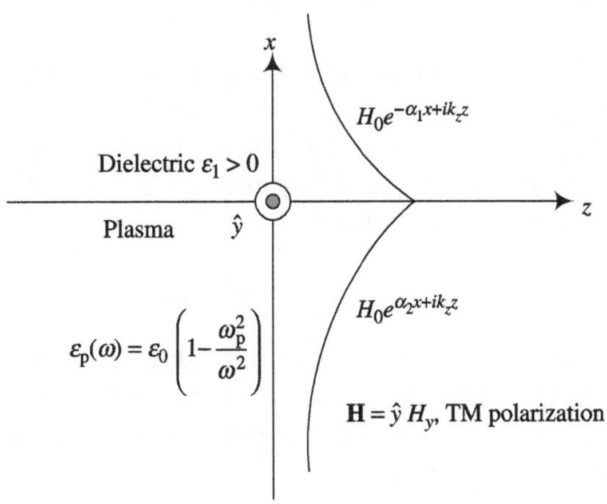

Figure 7.18 The magnetic field profile for the surface plasmon mode in a single dielectric–plasma interface. It is TM polarized and propagates along the z direction.

The tangential **E** or E_z component is continuous. We have

$$\frac{\alpha_1}{\varepsilon_1} = -\frac{\alpha_2}{\varepsilon_p}. \qquad (7.7.6)$$

Therefore, we obtain the solutions for α_1, α_2, and k_z in terms of ε_1 and ε_p:

$$\alpha_1 = \omega\sqrt{\frac{-\mu_0\varepsilon_1^2}{\varepsilon_1 + \varepsilon_p}} \qquad (7.7.7a)$$

$$\alpha_2 = \omega\sqrt{\frac{-\mu_0\varepsilon_p^2}{\varepsilon_1 + \varepsilon_p}} \qquad (7.7.7b)$$

$$k_z = \omega\sqrt{\frac{\mu_0\varepsilon_1\varepsilon_p}{\varepsilon_1 + \varepsilon_p}}. \qquad (7.7.7c)$$

Real solutions for α_1, α_2, and k_z exist for $\varepsilon_p < -\varepsilon_1 < 0$. The Poynting' vector or power density flow along the z direction is given by

$$\mathbf{P} = \frac{1}{2}\mathrm{Re}(\mathbf{E} \times \mathbf{H}^*)$$
$$= \hat{z}\begin{cases} \dfrac{k_z}{2\omega\varepsilon_1}|H_0|^2 e^{-2\alpha_1 x}, & x \geq 0 \\ \dfrac{k_z}{2\omega\varepsilon_p}|H_0|^2 e^{2\alpha_2 x}, & x \leq 0. \end{cases} \qquad (7.7.8)$$

7.7.2 Surface Plasmon Modes in a Metallic Slab

In a thin metallic slab surrounded by a symmetric dielectric ($\varepsilon_1 > 0$) on both sides, we have even and odd modes, similar to those of a symmetric dielectric waveguide [28]. By the same token, we can have a thin air (or dielectric) layer sandwiched between two metals and we look for the guided mode in such a waveguide.

Below we consider a thin metallic slab, and the optical frequency ω is below the plasma frequency of the metal such that $\varepsilon_p < 0$.

TM Even Modes (Fig. 7.19a)

$$\mathbf{H} = \hat{y}\, e^{ik_z z}\begin{cases} C_0\, e^{-\alpha_1\left(x-\frac{d}{2}\right)}, & x \geq \frac{d}{2} \\ C_1 \cosh \alpha_2 x, & |x| \leq \frac{d}{2} \\ C_0\, e^{\alpha_1\left(x+\frac{d}{2}\right)}, & x \leq -\frac{d}{2} \end{cases} \qquad (7.7.9)$$

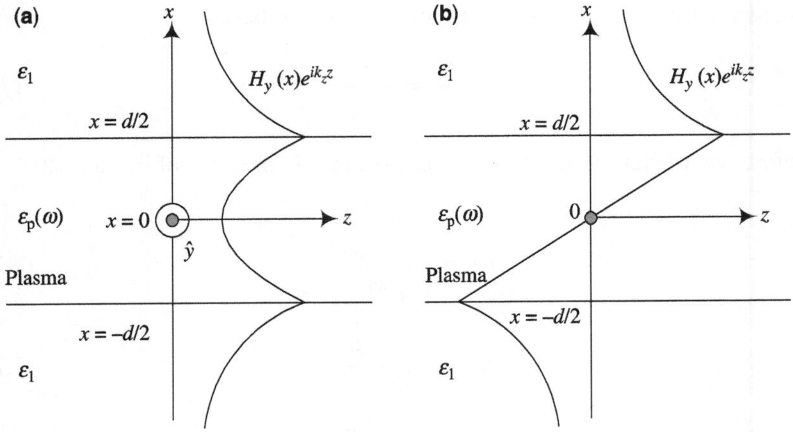

Figure 7.19 The magnetic field profile of a guided (a) even mode and (b) odd mode in a plasma waveguide using a thin plasma layer surrounded by dielectrics on top and below. The modes are TM polarized and propagate along the z direction.

where the wave equation in each region gives

$$-\alpha_1^2 + k_z^2 = \omega^2\mu\varepsilon_1 \tag{7.7.10a}$$

$$-\alpha_2^2 + k_z^2 = \omega^2\mu\varepsilon_p. \tag{7.7.10b}$$

We then use the boundary conditions in which tangential electric field, $E_z = (-1/i\omega\varepsilon)\,(\partial H_y/\partial x)$, and magnetic field H_y are continuous at $x = \pm d/2$. We find

$$C_0 = C_1 \cosh \alpha_2 \frac{d}{2} \tag{7.7.11a}$$

$$-\frac{\alpha_1}{\varepsilon_1} C_0 = C_1 \frac{\alpha_2}{\varepsilon_p} \sinh \alpha_2 \frac{d}{2}. \tag{7.7.11b}$$

Taking the ratio of the above two equations to eliminate C_0 and C_1, we find the guidance condition for the TM even mode

$$\alpha_1 = -\frac{\varepsilon_1}{\varepsilon_p} \alpha_2 \tanh \alpha_2 \frac{d}{2} \tag{7.7.12}$$

which should be solved together with

$$\alpha_2^2 - \alpha_1^2 = \omega^2\mu_0(\varepsilon_1 - \varepsilon_p). \tag{7.7.13}$$

TM Odd Modes (Fig. 7.19b) The procedure is similar to that of the even modes. We summarize the results below:

$$\mathbf{H} = \hat{y}\, e^{ik_z z} C_1 \begin{cases} \sinh \alpha_2 \frac{d}{2} e^{-\alpha_1\left(x - \frac{d}{2}\right)}, & x \geq \frac{d}{2} \\ \sinh \alpha_2 x, & |x| \leq \frac{d}{2} \\ -\sinh \alpha_2 \frac{d}{2} e^{\alpha_1\left(x + \frac{d}{2}\right)}, & x \leq -\frac{d}{2} \end{cases} \tag{7.7.14}$$

$$\alpha_1 = -\frac{\varepsilon_1}{\varepsilon_p} \alpha_2 \coth \alpha_2 \frac{d}{2}. \tag{7.7.15}$$

Solution Procedure If we define

$$X = \alpha_2 \frac{d}{2} \quad \text{and} \quad Y = \alpha_1 \frac{d}{2} \tag{7.7.16}$$

we can write (7.7.12), (7.7.13), and (7.7.15) as

$$Y = \begin{cases} -\frac{\varepsilon_1}{\varepsilon_p} X \tanh X & \text{even modes} \quad (7.7.17a) \\ -\frac{\varepsilon_1}{\varepsilon_p} X \coth X & \text{odd modes} \quad (7.7.17b) \end{cases}$$

$$X^2 - Y^2 = R^2 \tag{7.7.18}$$

where

$$R = \omega \sqrt{\mu_0 (\varepsilon_1 - \varepsilon_p)} \frac{d}{2}. \tag{7.7.19}$$

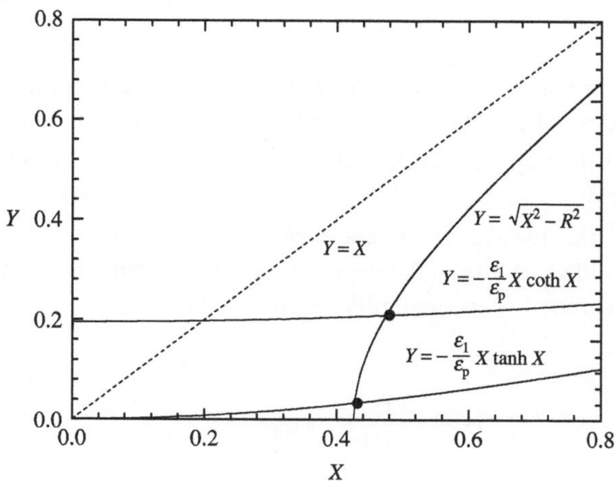

Figure 7.20 A graphic method to find the solutions for the even and odd eigenmodes of a lossless plasma waveguide. The intercepts between the curve, $Y = \sqrt{X^2 - R^2}$, and the two eigenequations or so-called guidance conditions give the solutions for (X, Y) for the even and odd modes. We assume $\varepsilon_p = -14.8\varepsilon_0$, $\varepsilon_1 = 2.89\varepsilon_0$, d = 0.02 μm, and the wavelength is 0.62 μm.

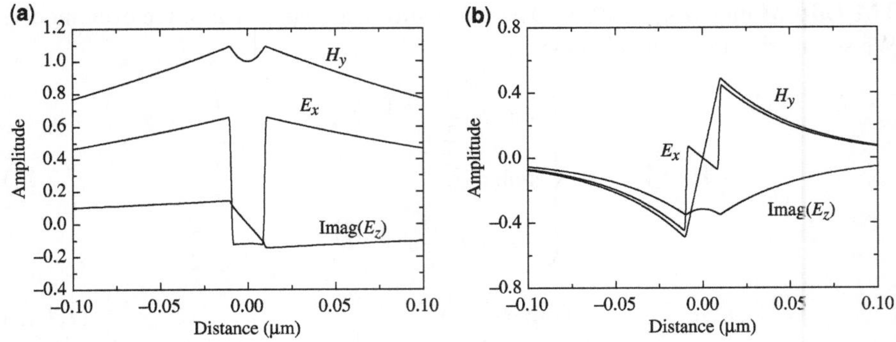

Figure 7.21 The field profiles for (a) the even mode and (b) the odd mode of a silver layer waveguide. The thin silver has a thickness of 0.02 μm. We have assumed that the loss is negligible for the plots. The field amplitudes are normalized to the amplitudes in vacuum.

The intercepts between (7.7.18) with (7.7.17a) or (7.7.17b) give the solutions for $\alpha_1 d/2$ and $\alpha_2 d/2$ of the even or odd modes, respectively. Figure 7.20 shows a graphical method by plotting the three equations on the $X - Y$ plane for a thin silver planar layer as a plasmon waveguide (with the approximation that the imaginary part of the permittivity is negligible, and the real part is $\varepsilon_p = -14.8\varepsilon_0$ for silver at a wavelength of 0.62 μm [28]) surrounded by a dielectric material with $\varepsilon_1 = 2.89\varepsilon_0$ on both sides. Figure 7.21a shows the field profiles of H_y, E_x, and imaginary part of E_z (as E_z is purely imaginary) for the even mode. Figure 7.21b shows the field profiles for the odd mode. The even or odd symmetry of the wave functions are clear.

PROBLEMS

7.1 Consider a $Ga_xIn_{1-x}As$ waveguide with a thickness d confined between two InP regions, where the gallium mole fraction x is 0.47, as shown in Fig. 7.22.

(a) If the free-space wavelength λ_0 is 1.65 μm, what is the photon energy (in eV)?

(b) Find the maximum thickness d_0 (in μm) so that only the TE_0 mode is guided at the wavelength λ_0 given in part **(a)**. Assume that the waveguide loss is negligible at this wavelength.

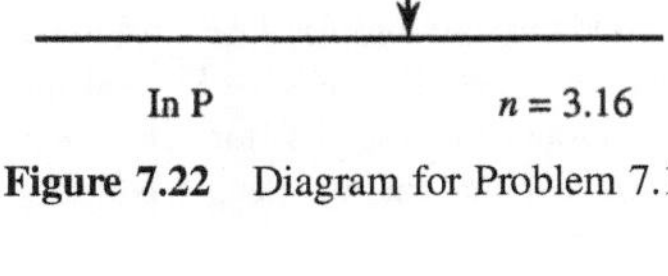

Figure 7.22 Diagram for Problem 7.1.

(c) If the thickness d of the waveguide is 0.3 μm, find the following parameters approximately (in 1/μm) for the TE_0 mode using the graphical approach in the text, (i) the decay constant α in the InP regions, (ii) k_x, and (iii) k_z in the waveguide.

(d) Calculate the optical confinement factor for the TE_0 mode in part **(c)**.

7.2 Repeat Problems 7.1(**b**) and 7.1(**c**) for the TM polarization. Are the answers the same as those for the TE polarization?

7.3 Calculate the Goos–Hänchen phase shifts for the TE_0 and TM_0 guided modes in Problem 7.1(**c**).

7.4 Show that for a symmetric dielectric waveguide with the V-number

$$V = k_0 d\sqrt{n_1^2 - n^2}$$

approaching zero, the propagation constant k_z of the TE_0 mode can be written as

$$k_z = k_0 n + C_0 V^4.$$

Find the expression for C_0. (Hint: Start with the guidance condition and the dispersion relation for $V \rightarrow 0$.)

7.5 Consider a symmetric slab waveguide with a thickness d (Fig. 7.23a) and a symmetric quantum well with a well width d (Fig. 7.23b). Consider the guided TE even modes for the optical waveguide and the general even bound states for the heterojunction quantum-well structure. Summarize in a table an exact one-to-one correspondence of the wave functions, material parameters, eigenequations (guidance conditions), propagation constants, and decay constants for the two physical problems above.

7.6 Consider a symmetric optical slab waveguide with a refractive index of 3.5 inside the guide and a refractive index of 3.2 outside the guide; the waveguide dimension is 2 μm, and assume the magnetic permeability is the same everywhere.

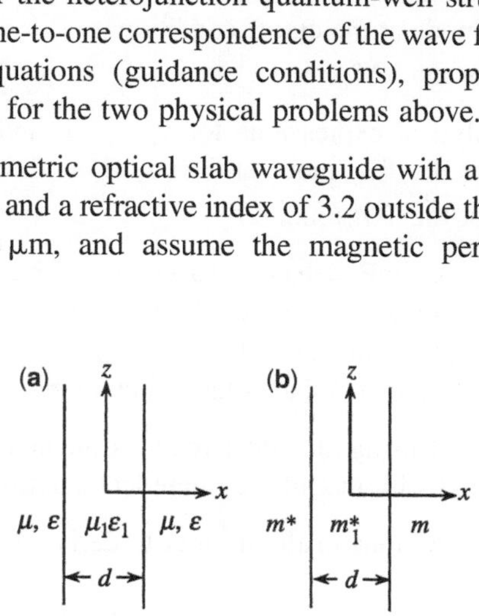

Figure 7.23 (a), (b) Diagram for Problem 7.5.

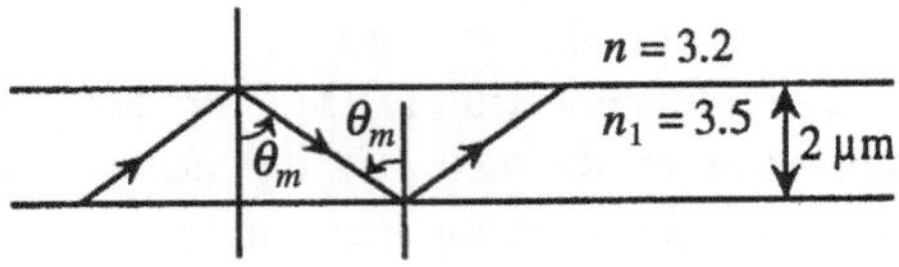

Figure 7.24 Diagram for Problem 7.6.

(a) For a light with a wavelength of 1.5 μm in free space, find the number of guided TE modes in this waveguide.

(b) Using the ray optics approach, find the angle θ_m of the guided TE_m mode bouncing back and forth between the two boundaries of the waveguide, Fig. 7.24.

(c) [Independent of parts (**a**) and (**b**).] Find the cutoff wavelength λ_m of the TE_m mode. Evaluate the decay constant α outside the waveguide and the propagation constant k_z at the cutoff condition.

7.7 **(a)** Plot the refractive index n versus the optical energy $\hbar\omega$ using Eqs. (7.1.21) to (7.1.23) for $x = 0, 0.1, 0.2$, and 0.3.

(b) Design a $GaAs/Al_xGa_{1-x}As$ waveguide using the above results such that only the fundamental TE_0 mode can be guided at a wavelength λ_0 near the GaAs band-gap wavelength at room temperature. Ignore the absorption effects for simplicity. Specify the possible parameters such as the aluminum mole fraction x and the waveguide thickness d.

7.8 **(a)** Derive the normalization constant C_1 for the electric field of the TE mode in an asymmetric dielectric slab waveguide.

(b) Repeat (**a**) for the magnetic field of the TM mode.

7.9 **(a)** Find the analytical expressions for k_x, k_z, and α in a symmetric waveguide for the TE_m mode at the cutoff condition.

(b) Repeat (**a**) for the TM_m mode.

7.10 **(a)** Find the analytical expressions for k_x, k_z, α, and α_2 in an asymmetric waveguide for the TE_m mode at the cutoff condition.

(b) Repeat (**a**) for the TM_m mode.

7.11 For an $In_{0.53}Ga_{0.47}As/InP$ dielectric waveguide with a cross section (2 μm × 1 μm), find the propagation constant of the lowest HE_{00} (or E^y_{11}) mode. You may use the graph in Fig. 7.13 assuming $n_1 = 3.56$ and $n = 3.16$ at a wavelength $\lambda_0 = 1.65$ μm. What is the effective index of this mode?

7.12 Repeat Problem 7.11 using the effective index method. Check the accuracy of the result compared with that of the numerical solution from Fig. 7.13.

7.13 List all of the assumptions made in order to derive the threshold condition

$$\Gamma g = (1 - \Gamma)\,\alpha + \frac{1}{L}\ln\frac{1}{R}.$$

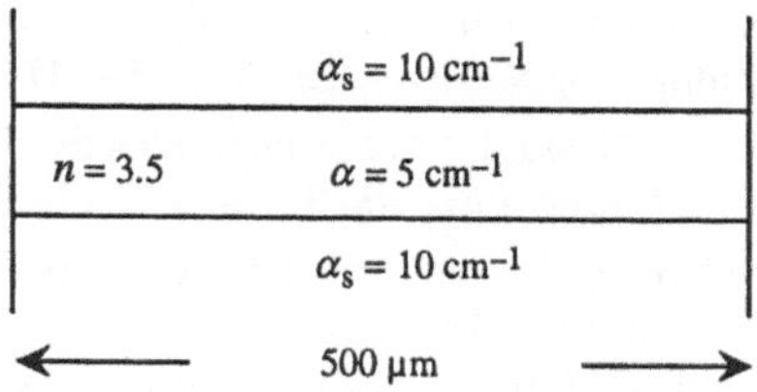

Figure 7.25 Diagram for Problem 7.15.

7.14 Show that if the reflectivities from the two end facets of a Fabry–Perot cavity are different, $R_1 \neq R_2$, the threshold condition is given by

$$\Gamma g = (1 - \Gamma)\,\alpha + \frac{1}{2L}\ln\left(\frac{1}{R_1 R_2}\right).$$

7.15 Consider a laser structure shown in Fig. 7.25 with losses described by the absorption coefficient $\alpha_s = 10\,\text{cm}^{-1}$ in the substrates and $\alpha = 5\,\text{cm}^{-1}$ in the active region. The reflectivity R is 0.3 at both ends, and the waveguide length is 500 μm. Assume, for simplicity, that the propagation constant of the guided mode is approximately $\frac{\omega}{c}n$, where $n = 3.5$.

(a) What is the threshold gain g_{th} of the structure if the optical confinement factor of the guided mode is 0.9?

(b) What is the longitudinal mode spectrum of this laser structure? Calculate the frequency spacing Δf and the wavelength spacing $\Delta\lambda$ of two nearby longitudinal modes at $\lambda_0 \simeq 0.8$ μm and ignore the dispersion effect for simplicity.

7.16 **(a)** Find the expression for the guided power density for the even mode of the plasma waveguide in each region of the waveguide. Comment on the signs of the power densities.

(b) Integrate over the transverse (x) direction and find the expressions for the power guided inside and outside the waveguide.

REFERENCES

1. D. Marcuse, *Theory of Dielectric Optical Waveguides*, Academic Press, New York, 1974.
2. A. W. Snyder and J. D. Love, *Optical Waveguide Theory*, Chapman and Hall, London, 1983.
3. A. Yariv and P. Yeh, *Photonics–Optical Electronics in Modern Communications*, 6th ed., Oxford University Press, New York, 2007.
4. H. C. Casey, Jr., and M. B. Panish, *Heterostructure Lasers, Part A: Fundamental Principles*, Academic Press, Orlando, 1978.
5. J. A. Kong, *Electromagnetic Wave Theory*, EMW, Cambridge, MA, 2000.
6. M. A. Afromowitz, "Refractive index of $Ga_{1-x}Al_xAs$," *Solid State Commun.* **15**, 59–63 (1974).

7. J. P. van der Ziel, M. Ilegems, and R. M. Mikulyak, "Optical birefringence of thin GaAs-AlAs multilayer films," *Appl. Phys. Lett.* **28**, 735–737 (1976).
8. J. P. van der Ziel and A. C. Gossard, "Absorption, refractive index, and birefringence of AlAs-GaAs monolayers," *J. Appl. Phys.* **48**, 3018–3023 (1977).
9. J. S. Blakemore, "Semiconducting and other major properties of gallium arsenide," *J. Appl. Phys.* **53**, R123–R181 (1982).
10. S. Adachi, "Refractive indices of III–V compounds: key properties of InGaAsP relevant to device designa," *J. Appl. Phys.* **53**, 5863–5869 (1982).
11. S. Adachi and K. Oe, "Internal strain and photoelastic effects in $Ga_{1-x}Al_xAs/GaAs$ and $In_{1-x}Ga_xAs_yP_{1-y}$ /InP crystals," *J. Appl. Phys.* **54**, 6620–6627 (1983).
12. S. Adachi, "GaAs, AlAs, and $Ga_{1-x}Al_xAs$: material parameters for use in research and device applications," *J. Appl. Phys.* **58**, R1–R29 (1985).
13. S. Adachi, "Optical properties of $Al_xGa_{1-x}As$ alloys," *Phys. Rev. B* **38**, 12345–12352 (1988).
14. H. C. Casey, Jr., D. D. Sell, and M. B. Panish, "Refractive index of $Al_xGa_{1-x}As$ between 1.2 and 1.8 eV," *Appl. Phys. Lett.* **24**, 63–65 (1974).
15. M. Amiotti and G. Landgren, "Ellipsometric determination of thickness and refractive index at 1.3, 1.55 and 1.7 μm for $In_{(1-x)}Ga_xAs_yP_{(1-y)}$ films on InP," *J. Appl. Phys.* **73**, 2965–2971 (1993).
16. S. Adachi, *Physical Properties of III–V Semiconductor Compounds*, Wiley, New York, 1992.
17. S. Adachi, *Properties of Indium Phosphide*, INSPEC, The Institute of Electrical Engineers, London and New York, 1991.
18. K. H. Hellwege, Ed., *Landolt-Börnstein Numerical Data and Functional Relationships in Science and Technology*, New Series, Group III **17a**, Springer-Verlag, Berlin, 1982; Groups III–V **22a**, Springer-Verlag, Berlin, 1986.
19. O. Madelung, Ed., *Semiconductors, Group IV Elements and III–V Compounds*, in *Data in Science and Technology*, P. Poerschke, Ed., Springer-Verlag, Berlin, 1991.
20. E. A. J. Marcatili, "Dielectric rectangular waveguide and directional coupler for integrated optics," *Bell Syst. Tech. J.* **48**, 2071–2102 (1969).
21. E. A. J. Marcatili, "Slab-coupled waveguides," *Bell Syst. Tech. J.* **53**, 645–674 (1974).
22. J. E. Goell, "A circular-harmonic computer analysis of rectangular dielectric waveguide," *Bell Syst. Tech. J.* **48**, 2133–2160 (1969).
23. W. C. Chew, "Analysis of optical and millimeter wave dielectric waveguide," *J. Electromagnetic Waves Appl.* **3**, 359–377 (1989).
24. C. S. Lee, S. W. Lee, and S. L. Chuang, "Plot of modal field distribution in rectangular and circular waveguides," *IEEE Trans. Microwave Theory Tech.* **MTT-33**, 271–274 (1985).
25. W. Streifer, R. D. Burnham, and D. R. Scifres, "Analysis of diode lasers with lateral spatial variations in thickness," *Appl. Phys. Lett.* **37**, 121–123 (1980).
26. J. Buus, "The effective index method and its application to semiconductor lasers," *IEEE J. Quantum Electron.* **QE-18**, 1083–1089 (1982).
27. C. Kittel, Solid State Physics, Chapter 14, 8th ed., Wiley, New York, 2005.
28. G. Winter, S. Wedge, and W. L. Barnes, "Can lasing at visible wavelengths be achieved using the low-loss long-range surface plasmon-polariton mode?" *New J. Phys.* **8**, 125 (2006).

8

Coupled-Mode Theory

In this chapter, we discuss in Section 8.1 how a light can be coupled into and out of a waveguide, then how it can be coupled between two parallel waveguides or coupled within the same waveguide with a distributed feedback (DFB) coupling. We present the coupled-mode theory for wave propagation in parallel waveguides in Section 8.2 An analogy of this coupling between two propagation modes in dielectric waveguides is a coupled double-pendulum system or a coupled two-resonator (inductor–capacitor or LC) system. The coupling of two resonators leads to energy transfer between the resonators at a frequency determined by half of the beat frequency of the two system modes: the in-phase mode and the out-of-phase mode. For coupling of propagation modes between the two parallel waveguides, a beat length can also be defined, which is twice the minimum distance required for the exchange of the guided power between the waveguides. Many interesting physical systems exhibit the phenomena of the coupling of modes, which can be understood from a simple model of two coupled pendulums. We apply the coupled-mode theory to optical switches (Section 8.3) and optical ring resonators with applications as optical add and drop filters (Section 8.4). We then investigate in Section 8.5 the counter-propagation waves coupled in a distributed feedback structure under the same framework of coupled-mode theory.

8.1 WAVEGUIDE COUPLERS

We discuss transverse couplers, prism couplers, and grating couplers and point out the significance of the phase-matching condition in this section [1–3].

8.1.1 Transverse Couplers

Direct Focusing As shown in Fig. 8.1a, a lens is used to focus and shape the beam profile so that it is matched to the modal distribution in the waveguide. The coupling efficiency depends on the overlap integral between the field profile of the incident wave and the modal profile of the guided mode in the thin film. This configuration can achieve nearly 100% efficiency. However, it may not be suitable for general integrated optics applications because of the nonplanar configuration.

Physics of Photonic Devices, Second Edition. By Shun Lien Chuang
Copyright © 2009 John Wiley & Sons, Inc.

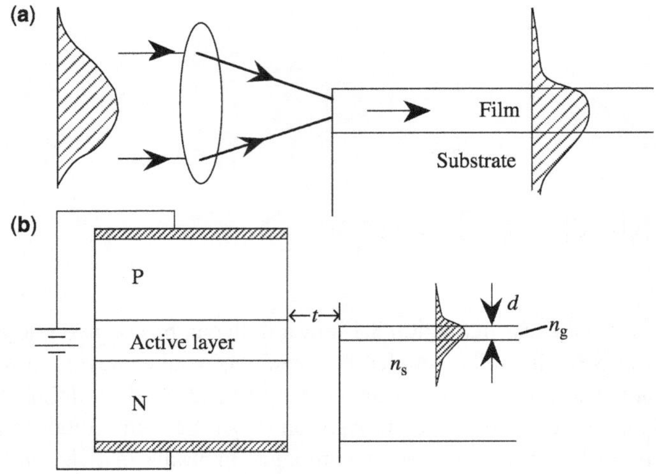

Figure 8.1 (a) Direct focusing. (b) End-butt coupling.

End-Butt Coupling As shown in Fig. 8.1b, the efficiency of an end-butt coupling can be controlled by the gap width t, the waveguide dimension d, and refractive indices of the waveguide and the substrate, n_g and n_s, respectively.

8.1.2 Prism Couplers

For a mode of a waveguide with a refractive index n_1 inside (and n_0 outside) the guide, the propagation constant β along the z direction has to satisfy the conditions, $k_0 n_0 < \beta < k_0 n_1$ and $k_0 n_s < \beta < k_0 n_1$, where $k_0 = 2\pi/\lambda_0$ is the wave number and λ_0 is the wavelength in free space. For an incident plane wave from region 0, its horizontal wavenumber is $k_z = k_0 n_0 \sin\theta_i < k_0 n_0$ for any real incident angle θ_i. Thus, the wave cannot be phase matched to the guided mode because $k_0 n_0 < \beta$, Fig. 8.2b. The

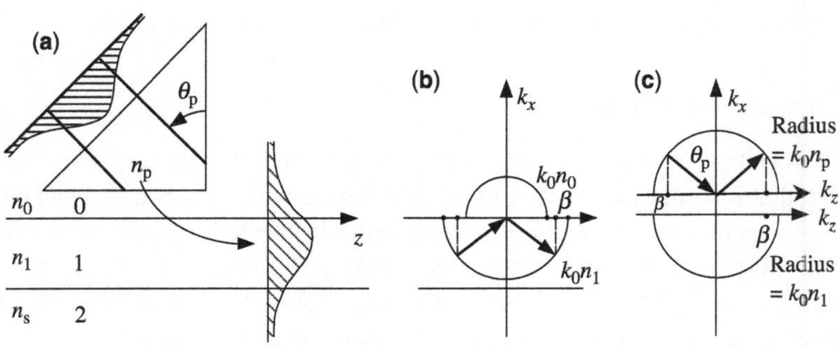

Figure 8.2 (a) A prism coupler. (b) The phase-matching diagram for a guided mode with a propagation constant $\beta > k_0 n_0$. (c) The phase-matching diagram for an incident wave from the prism region and coupling to the guided mode with $k_0 n_p \sin\theta_p = \beta$.

excitation of the guided mode is very weak. However, using a prism above the waveguide with a small gap (assuming $n_p > n_0$ and $n_p > n_s$), Fig. 8.2a, the phase-matching condition can be satisfied: $k_0 n_p \sin\theta_p = \beta$, Fig. 8.2c. This angle of incidence θ_p will be larger than the critical angle for the interface between the prism and region 0,

$$\theta_c = \sin^{-1}\left(\frac{n_0}{n_p}\right) \tag{8.1.1}$$

because $\beta > k_0 n_0$. Therefore, the transmitted wave in the gap region is an evanescent type and can be resonantly coupled to the guided mode. In general, for a particular TE_m mode with a propagation constant β_m, we can choose θ_p such that

$$k_0 n_p \sin\theta_p = \beta_m. \tag{8.1.2}$$

Other important parameters are the incident beam profile, coupling length, and gap dimension.

8.1.3 Grating Couplers

Consider first a plane wave incident on a periodic grating that is assumed to be a perfect conductor with the surface profile described by $x = h(z)$ with a period Λ, as shown in Fig. 8.3,

$$\begin{aligned} \mathbf{E}_i &= \hat{y}\, E_0\, e^{-ik_{x0}x + ik_{z0}z} \\ k_{x0} &= k\cos\theta_i \qquad k_{z0} = k\sin\theta_i. \end{aligned} \tag{8.1.3}$$

The reflected wave is $\mathbf{E}_R = \hat{y}E_R(x, z)$. Because the boundary condition requires that the tangential electric field vanish at the conductor surface, the reflected wave

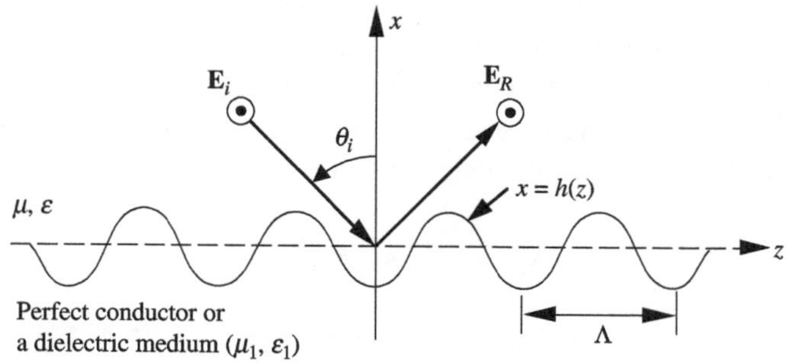

Figure 8.3 A plane wave $\mathbf{E}_i$ with the polarization along the y direction is incident on a grating surface with a profile $x = h(z)$ and a period Λ.

on the surface of the grating should satisfy

$$\begin{aligned} E_R(x=h(z),\ z) &= -E_0\, e^{-ik_{x0}h(z)+ik_{z0}z} \\ &= E_0\, e^{ik_{z0}z} \sum_{m=-\infty}^{\infty} A_m\, e^{i\frac{2\pi}{\Lambda}mz} \end{aligned} \tag{8.1.4a}$$

$$A_m = \frac{1}{\Lambda}\int_0^{\Lambda} dz[-\, e^{-ik_{x0}h(z)}]e^{-i\frac{2\pi}{\Lambda}mz} \tag{8.1.4b}$$

where we have made use of the Fourier series expansion because $e^{-ik_{x0}h(z)}$ is a periodic function with a period Λ. In the region above the grating, $E_R(x, z)$ satisfies the wave equation and should have solutions of the form $\exp(ik_x x + ik_z z)$ as long as $k_x^2 + k_z^2 = \omega^2\mu\varepsilon$. Therefore, we see that a general solution for $E_R(x, z)$ will be

$$E_R(x, z) = E_0 \sum_{m=-\infty}^{\infty} R_m\, e^{ik_{xm}x+ik_{zm}z} \tag{8.1.5}$$

where the horizontal wave numbers are

$$k_{zm} = k_{z0} + m\frac{2\pi}{\Lambda} \tag{8.1.6}$$

and their corresponding vertical components can be obtained from the dispersion relation

$$k_{xm} = \sqrt{\omega^2\mu\varepsilon - k_{zm}^2}. \tag{8.1.7}$$

The horizontal wave number k_{zm} for a particular m differs from its nearby value by $2\pi/\Lambda$. We can generalize the above discussions to the case of a penetrable dielectric grating. A phase-matching diagram is shown in Fig. 8.4. These plane waves for the reflected and transmitted fields are called the space harmonics. It can be seen that for a particular m such that $k_{zm} > \omega\sqrt{\mu\varepsilon}$, the x-component k_{xm} becomes purely imaginary; and we choose the imaginary part of k_{xm} to be positive such that it decays away from the surface. In the transmission region with a refractive index $n_1 = \sqrt{\mu_1\varepsilon_1/(\mu_0\varepsilon_0)}$, the transmitted wave vectors $\mathbf{k}_m^t$ will have the same horizontal wave numbers k_{zm} as those of the reflected waves (8.1.6),

$$\mathbf{k}_m^t = \hat{x}\, k_{xm}^t + \hat{z}\, k_{zm}^t \tag{8.1.8a}$$

$$k_{zm}^t = k_{zm} = k_{z0} + m\frac{2\pi}{\Lambda} \tag{8.1.8b}$$

$$k_{xm}^t = \sqrt{\omega^2\mu_1\varepsilon_1 - k_{zm}^2}. \tag{8.1.8c}$$

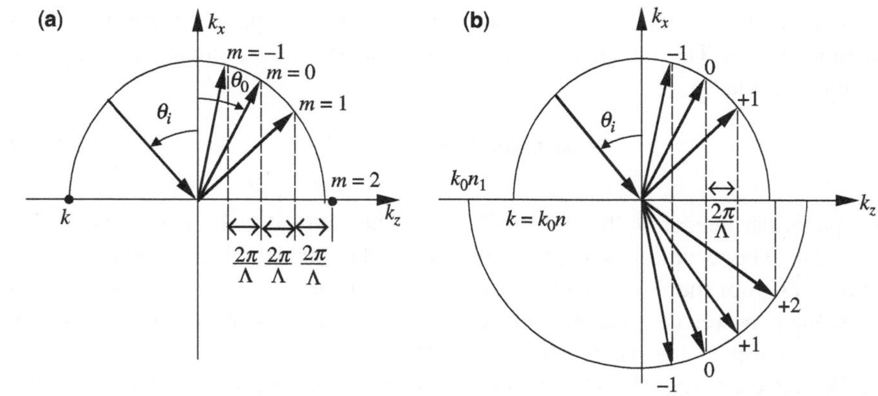

Figure 8.4 Phase-matching diagram for a plane wave (a) reflection and (b) both reflection and transmission from a grating structure with a period Λ.

For a general theory to find the reflection and transmission coefficients of the space harmonics, see Refs. 4 and 5.

Phase-Matching Condition for Grating Couplers A dielectric grating can therefore be used as a waveguide input or output coupler as shown in Fig. 8.5.

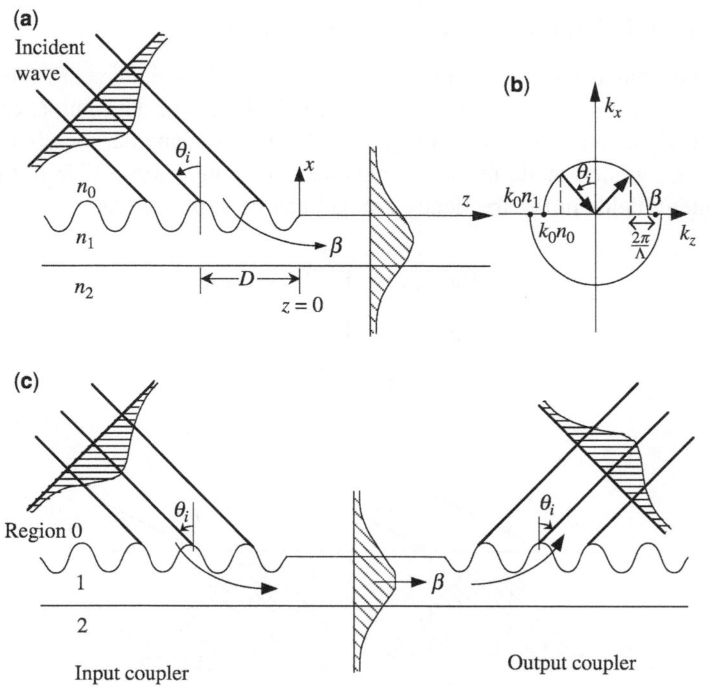

Figure 8.5 (a) A grating coupler. (b) Phase-matching diagram for $\beta = k_0 n_0 \sin\theta_i + \frac{2\pi}{\Lambda}$. (c) Grating input and output couplers.

Suppose we have a dielectric slab waveguide with a propagation constant β for the fundamental TE_0 mode. If the incident angle θ_i and the grating period Λ are chosen such that

$$k_0 n_0 \sin\theta_i + m\frac{2\pi}{\Lambda} = \beta \tag{8.1.9}$$

for a particular integer m, the incident wave will be coupled to the guided TE_0 mode, Fig. 8.5b. The coupling strength depends on the incident beam profile and its distance D from the edge of the grating as shown in Fig. 8.5a. If D is too long, the incident power coupled into the waveguide with the grating section can leak back the power to the incident region. If D is too short, half of the beam profile will be reflected from the planar interface in the section $z > 0$, and the coupling into the guided mode in the region $z > 0$ will be small. We can use the grating as an output coupler based on the reciprocity concept. The guided wave in the center waveguide section in Fig. 8.5c can radiate back into region 0 in the right grating section while propagating along the waveguide. The exit angle can be determined using (8.1.9).

8.2 COUPLED OPTICAL WAVEGUIDES

8.2.1 General Formulation of Coupled-Mode Theory

The coupled-mode theory [6–32] is also very useful to understand the coupling mechanisms in parallel waveguides as shown in Fig. 8.6. The guidance is along the z direction. Waveguides a and b can be two rectangular or slab dielectric guides. For a guided mode in the $+z$ direction in a single waveguide a, the electric and magnetic fields in the frequency domain can be expressed as

$$\mathbf{E}(x, y, z) = \mathbf{E}^{(a)}(x, y)\, a(z) \tag{8.2.1a}$$

$$\mathbf{H}(x, y, z) = \mathbf{H}^{(a)}(x, y)\, a(z) \tag{8.2.1b}$$

$$a(z) = a_0\, e^{i\beta_a z} \tag{8.2.1c}$$

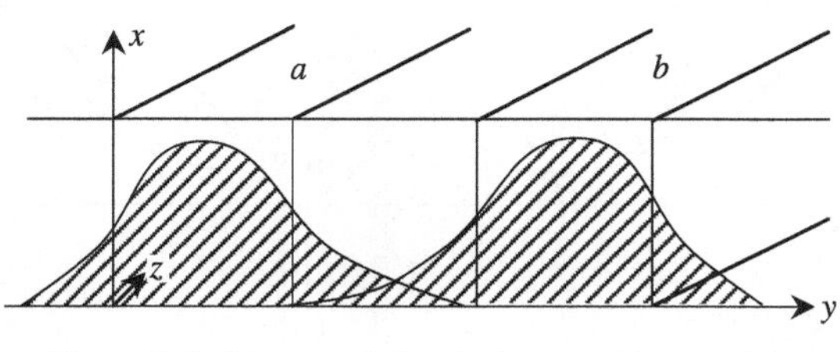

Figure 8.6 Two coupled optical waveguides a and b.

where $\mathbf{E}^{(a)}(x, y)$ and $\mathbf{H}^{(a)}(x, y)$ are the modal distributions in the $x - y$ plane, which satisfy the normalization condition

$$\frac{1}{2}\text{Re}\int_{-\infty}^{\infty}\int_{-\infty}^{\infty}\mathbf{E}^{(a)}(x, y) \times \mathbf{H}^{(a)*}(x, y)\cdot \hat{z}\, dxdy = 1. \tag{8.2.2}$$

Note that the time conversion $\exp(-i\omega t)$ in Maxwell's equations has been adopted. We also see that

$$\frac{d}{dz}a(z) = i\beta_a\, a(z). \tag{8.2.3}$$

The total guided power for the fields in (8.2.1) and (8.2.2) is

$$P = \frac{1}{2}\text{Re}\int_{-\infty}^{\infty}\int_{-\infty}^{\infty}\mathbf{E}(x, y, z) \times \mathbf{H}^{*}(x, y, z) \cdot \hat{z}\, dxdy = |a(z)|^2. \tag{8.2.4}$$

If the mode is guided in the $-z$ direction, then $a(z) = a_0 \exp(-i\beta_a z)$

$$\frac{d}{dz}a(z) = -i\beta_a\, a(z) \tag{8.2.5}$$

and the Poynting vector $\mathbf{E} \times \mathbf{H}^*$ will be in the $-z$ direction. If we define the power to be the $+z$ component of the guided power, we obtain

$$P = -|a(z)|^2 \tag{8.2.6}$$

for the guided power in the $-z$ direction for (8.2.5).

Consider two parallel waveguides, a and b, for which the total field solutions can be written as linear combinations of the individual waveguide modes

$$\begin{aligned}\mathbf{E}(x, y, z) &= a(z)\,\mathbf{E}^{(a)}(x, y) + b(z)\,\mathbf{E}^{(b)}(x, y)\\ \mathbf{H}(x, y, z) &= a(z)\,\mathbf{H}^{(a)}(x, y) + b(z)\,\mathbf{H}^{(b)}(x, y).\end{aligned} \tag{8.2.7}$$

The amplitudes $a(z)$ and $b(z)$ satisfy

$$\begin{aligned}\frac{d}{dz}a &= i\beta_a\, a + iK_{ab}\, b\\ \frac{d}{dz}b &= iK_{ba}\, a + i\beta_b\, b\end{aligned} \tag{8.2.8}$$

where K_{ab} and K_{ba} are the coupling coefficients. The expressions for K_{ab} and K_{ba} are discussed in Appendix 8A. The total guided power is

$$P = \frac{1}{2}\mathrm{Re}\int_{-\infty}^{\infty}\int_{-\infty}^{\infty} \mathbf{E}(x, y, z) \times \mathbf{H}^{*}(x, y, z) \cdot \hat{z}\, dxdy$$
$$= s_a|a(z)|^2 + s_b|b(z)|^2 + \mathrm{Re}[a(z)b^{*}(z)C_{ba} + b(z)a^{*}(z)C_{ab}] \tag{8.2.9}$$

where the cross-overlap integrals

$$C_{pq} = \frac{1}{2}\int_{-\infty}^{\infty}\int_{-\infty}^{\infty} \mathbf{E}^{(q)}(x, y) \times \mathbf{H}^{(p)*}(x, y) \cdot \hat{z}\, dxdy \tag{8.2.10}$$

and s_a, $s_b = +1$ for propagation in the $+z$ direction and -1 for propagation in the $-z$ direction.

If we assume that the coupling of the two waveguide modes is very weak so that the cross-overlap integrals C_{ab} and C_{ba} are negligible, the total power is then

$$P = s_a|a(z)|^2 + s_b|b(z)|^2. \tag{8.2.11}$$

If the system is lossless, power conservation requires

$$\frac{dP}{dz} = 0. \tag{8.2.12}$$

Substituting (8.2.11) into (8.2.12) and using the coupled-mode equations, we find

$$K_{ab} = K_{ba}^{*} \quad \text{if } s_a s_b > 0 \text{ (co-directional coupling)} \tag{8.2.13a}$$
$$K_{ab} = -K_{ba}^{*} \quad \text{if } s_a s_b < 0 \text{ (contra-directional coupling).} \tag{8.2.13b}$$

The coupled-mode equations can be expressed in a matrix form

$$\frac{d}{dz}\begin{bmatrix} a \\ b \end{bmatrix} = i\mathbf{M}\begin{bmatrix} a \\ b \end{bmatrix} \tag{8.2.14a}$$

where

$$\mathbf{M} = \begin{bmatrix} \beta_a & K_{ab} \\ K_{ba} & \beta_b \end{bmatrix}. \tag{8.2.14b}$$

8.2.2 Eigenstate Solutions

The solution can be obtained by substituting

$$\begin{bmatrix} a(z) \\ b(z) \end{bmatrix} = \begin{bmatrix} A \\ B \end{bmatrix} e^{i\beta z} \tag{8.2.15}$$

into (8.2.14). We find

$$(\mathbf{M} - \beta \mathbf{I}) \begin{bmatrix} A \\ B \end{bmatrix} = 0$$

where I is an identity matrix, or equivalently,

$$\begin{bmatrix} \beta_a - \beta & K_{ab} \\ K_{ba} & \beta_b - \beta \end{bmatrix} \begin{bmatrix} A \\ B \end{bmatrix} = 0. \tag{8.2.16}$$

For nontrivial solutions to (8.2.16), the determinant must vanish:

$$\det|\mathbf{M} - \beta \mathbf{I}| = (\beta_a - \beta)(\beta_b - \beta) - K_{ab}K_{ba} = 0. \tag{8.2.17}$$

There are two eigenvalues for β:

$$\beta = \frac{\beta_a + \beta_b}{2} \pm q \tag{8.2.18}$$

where

$$q = \sqrt{\Delta^2 + K_{ab}K_{ba}} \tag{8.2.19}$$

$$\Delta \equiv \frac{\beta_a - \beta_b}{2}. \tag{8.2.20}$$

For co-directional coupling, $K_{ba} = K_{ab}^*$, q is real for a lossless system. As an example, we assume for co-directional coupling that

$$\beta_a \equiv \beta_a(\omega) \qquad \beta_b \equiv \beta_b(\omega).$$

The $\beta_a(\omega)$ and $\beta_b(\omega)$ versus ω curves are shown as dashed lines in Fig. 8.7a. Both curves have positive slopes and intercept each other at (ω_0, β_0). The eigensolutions β_+ and β_- are the solid curves. At $\omega = \omega_0$, $\beta_a = \beta_b$, $q = |K_{ab}|(\equiv K)$, the splitting $(\beta_+ - \beta_-)$ is $2K$. Because

$$\frac{A}{B} = \frac{K_{ab}}{\beta - \beta_a} = \frac{\beta - \beta_b}{K_{ba}}, \tag{8.2.21}$$

we find if $K_{ab} > 0$,

$$\text{(1) for } \beta = \beta_+ \quad \frac{A^+}{B^+} = \frac{K_{ab}}{-\Delta + q} = \frac{\Delta + q}{K_{ba}} > 0 \tag{8.2.22}$$

and

$$\text{(2) for } \beta = \beta_- \quad \frac{A^-}{B^-} = \frac{K_{ab}}{-\Delta - q} = \frac{\Delta - q}{K_{ba}} < 0. \tag{8.2.23}$$

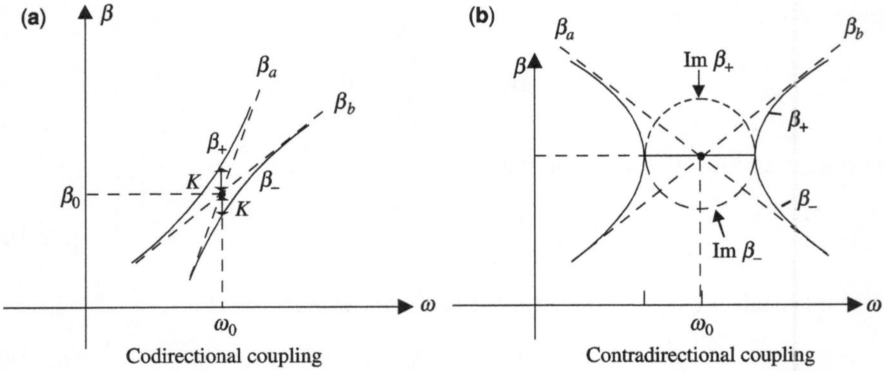

Figure 8.7 The dispersion curves β_+ and β_- for (a) co-directional coupling and (b) contra-directional coupling.

Furthermore, if the two guides have the same propagation constants, $\beta_a = \beta_b$, we find (for real K_{ab} and co-directional coupling)

$$\frac{A^+}{B^+} = 1 \qquad \text{for } \beta = \beta_+ \tag{8.2.24}$$

$$\frac{A^-}{B^-} = -1 \qquad \text{for } \beta = \beta_- \tag{8.2.25}$$

We thus have two possible system modes, the β_+ (in-phase) mode and the β_- (out-of-phase) mode, which are plotted in Fig. 8.8.

For contra-directional coupling, $q = \sqrt{\Delta^2 - |K_{ab}|^2}$, q can be purely imaginary if

$$\left|\frac{\beta_a - \beta_b}{2}\right| < |K_{ab}|.$$

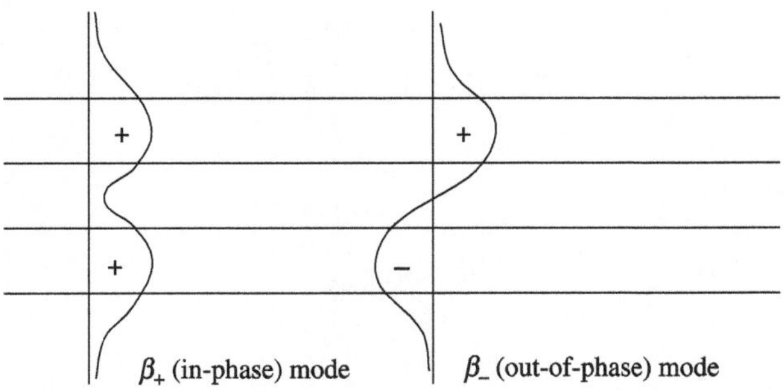

Figure 8.8 A simple sketch of the two system modes: the β_+ (in-phase) mode and the β_- (out-of-phase) mode.

The propagation constants β_+ and β_- are complex in the region, Fig. 8.7b,

$$\left|\frac{\beta_a - \beta_b}{2}\right| < |K_{ab}|.$$

The imaginary parts $\mathrm{Im}\,\beta_+$ and $\mathrm{Im}\,\beta_-$ are shown as the dashed semicircles. We also see that a stop band exists with its bandwidth $\Delta\omega$ determined by $\left|\frac{\beta_a^{(\omega)} - \beta_b^{(\omega)}}{2}\right| = |K_{ab}|$.

8.2.3 General Solution of the Coupled Waveguides

After finding the eigenvectors $\mathbf{v}_1$ and $\mathbf{v}_2$, where

$$\mathbf{v}_1 = \begin{bmatrix} K_{ab} \\ q - \Delta \end{bmatrix} \quad \text{or} \quad \begin{bmatrix} q + \Delta \\ K_{ba} \end{bmatrix} \quad \text{for } \beta_+ = \frac{\beta_a + \beta_b}{2} + q \tag{8.2.26a}$$

$$\mathbf{v}_2 = \begin{bmatrix} K_{ab} \\ -q - \Delta \end{bmatrix} \quad \text{or} \quad \begin{bmatrix} -q + \Delta \\ K_{ba} \end{bmatrix} \quad \text{for } \beta_- = \frac{\beta_a + \beta_b}{2} - q, \tag{8.2.26b}$$

the solution for $a(z)$ and $b(z)$ given the initial condition $a(0)$ and $b(0)$ can be obtained in the same way as (8.2.19). It can also be derived using the eigenmatrix $\mathbf{V}$ where its columns consist of the two eigenvectors of the matrix $\mathbf{M}$

$$\mathbf{V} = \begin{bmatrix} \mathbf{v}_1 & \vdots & \mathbf{v}_2 \end{bmatrix}. \tag{8.2.27}$$

The general solution is

$$\begin{bmatrix} a(z) \\ b(z) \end{bmatrix} = \mathbf{V} \begin{bmatrix} e^{i\beta_+ z} & 0 \\ 0 & e^{i\beta_- z} \end{bmatrix} \mathbf{V}^{-1} \begin{bmatrix} a(0) \\ b(0) \end{bmatrix}. \tag{8.2.28}$$

The derivation is straightforward. The general solution is a linear combination of the eigensolutions

$$\begin{aligned} \begin{bmatrix} a(z) \\ b(z) \end{bmatrix} &= c_1 \begin{bmatrix} K_{ab} \\ q - \Delta \end{bmatrix} e^{i\beta_+ z} + c_2 \begin{bmatrix} K_{ab} \\ -q - \Delta \end{bmatrix} e^{i\beta_- z} \\ &= \mathbf{V} \begin{bmatrix} e^{i\beta_+ z} & 0 \\ 0 & e^{i\beta_- z} \end{bmatrix} \begin{bmatrix} c_1 \\ c_2 \end{bmatrix}. \end{aligned} \tag{8.2.29}$$

Therefore, at $z = 0$, we find

$$\begin{bmatrix} c_1 \\ c_2 \end{bmatrix} = \mathbf{V}^{-1} \begin{bmatrix} a(0) \\ b(0) \end{bmatrix}. \tag{8.2.30}$$

Substituting (8.2.30) into (8.2.29), we obtain the solution (8.2.28). The solutions can also be expressed in terms of a matrix $\mathbf{S}(z)$ and the initial conditions:

$$\begin{bmatrix} a(z) \\ b(z) \end{bmatrix} = \mathbf{S}(z) \begin{bmatrix} a(0) \\ b(0) \end{bmatrix} \tag{8.2.31}$$

where

$$\begin{aligned} \mathbf{S} &= \mathbf{V} \begin{bmatrix} e^{i\beta_+ z} & 0 \\ 0 & e^{i\beta_- z} \end{bmatrix} \mathbf{V}^{-1} \\ &= \begin{bmatrix} \cos qz + i\dfrac{\Delta}{q}\sin qz & i\dfrac{K_{ab}}{q}\sin qz \\ i\dfrac{K_{ba}}{q}\sin qz & \cos qz - i\dfrac{\Delta}{q}\sin qz \end{bmatrix} e^{i\phi z} \end{aligned} \tag{8.2.32}$$

where

$$\phi = \frac{\beta_a + \beta_b}{2}.$$

If at $z = 0$, the optical power is incident only in waveguide 1, ($a(0) = 1$, $b(0) = 0$), we find

$$|b(z)|^2 = \left|\frac{K_{ba}}{q}\right|^2 \sin^2 qz. \tag{8.2.33}$$

Therefore, at $qz = \pi/2, 3\pi/2, \ldots, (2n+1)\pi/2$ ($n =$ an integer), the power transfer from guide a to guide b is maximum. Because

$$\left|\frac{K_{ba}}{q}\right|^2 = \frac{|K_{ba}|^2}{\Delta^2 + |K_{ba}|^2} < 1 \tag{8.2.34}$$

for $\beta_a \neq \beta_b$, the power transfer is never complete, Fig. 8.9a. The minimum distance, $L_B = \pi/q$, at which the output power reappears in the same guide as the input, is called a beat length. If $\beta_a = \beta_b$, we have $q = |K_{ab}|$, and the solutions are

$$a(z) = \cos Kz\, e^{i\beta z} \qquad b(z) = i\sin Kz\, e^{i\beta z} \tag{8.2.35}$$

where $K = K_{ab} = K_{ba}$, $\beta = \beta_a = \beta_b$ have been used. Complete power transfer occurs for synchronous coupling, Fig. 8.9b. For $K\ell = (2n+1)\pi/2$, complete power transfer occurs. It is called a cross state $\otimes$. For $K\ell = n\pi$, there is no power transfer from guide a to guide b. It is called a parallel state $\ominus$.

The above theory has been improved [18–32] taking into consideration effects such as the modes in separate guides are not orthogonal to each other, and additional

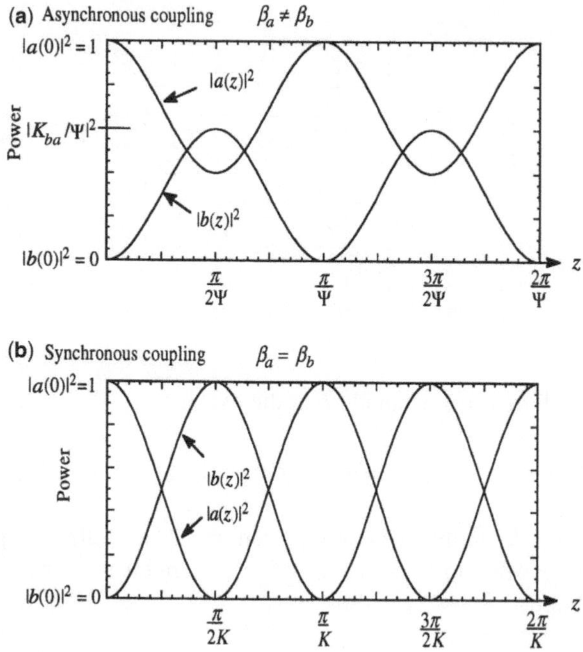

Figure 8.9 Guided powers $|a(z)|^2$ and $|b(z)|^2$ versus the coupling distance z: (a) asynchronous coupling $\beta_a \neq \beta_b$. (b) synchronous coupling $\beta_a = \beta_b$.

terms may affect the parameters used in the coupled-mode theory, although the form of the coupled-mode equations remains the same. Experimental results on directional couplers and three coupled waveguides [24, 33–38], extension to anisotropic waveguides [39–41], and optical tunable filters [42–45] have been reported.

8.3 APPLICATIONS OF OPTICAL WAVEGUIDE COUPLERS

In this section, we discuss some applications of the coupled waveguide theory, which include optical waveguide switch and the $\Delta\beta$ coupler.

8.3.1 Optical Waveguide Switch

Consider a directional coupler with an incident light into waveguide a as shown in Fig. 8.10. Assume $a(0) = 1$. The output power from waveguide b is

$$P_b = |b(\ell)|^2 \simeq \frac{K^2}{\Delta^2 + K^2} \sin^2(q\ell). \tag{8.3.1}$$

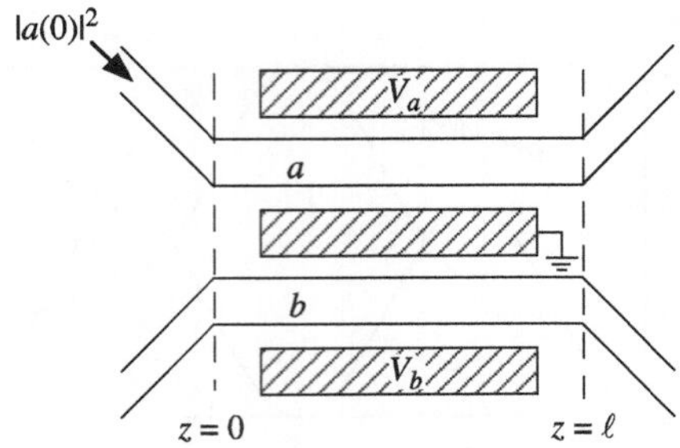

Figure 8.10 An optical waveguide switch.

No power is transferred to waveguide b at the exit if

$$q\ell = n\pi, \quad n = 1, 2, 3, \ldots$$

We can plot $\Delta\ell$ and $K\ell$ in the plane to represent this state, called the parallel state ⊖ because all input power into guide a at $z = \ell$ exits from the same waveguide. In other words, the equations

$$(\Delta\ell)^2 + (K\ell)^2 = (n\pi)^2, \quad n = 1, 2, 3, \ldots \tag{8.3.2}$$

represent a set of circles on the $K\ell$ versus $\Delta\ell$ plane, as shown in the switching diagram, Fig. 8.11. On the other hand, complete power transfer can only occur if $\Delta^2 = 0$, and $K\ell = (2m + 1)\pi/2$, where $m = 0, 1, 2, \ldots$. It is called a cross state ⊗ when the total power is transferred from one waveguide to the other. These cross states are represented by only discrete points labeled as ⊗ on the vertical ($K\ell$) axis.

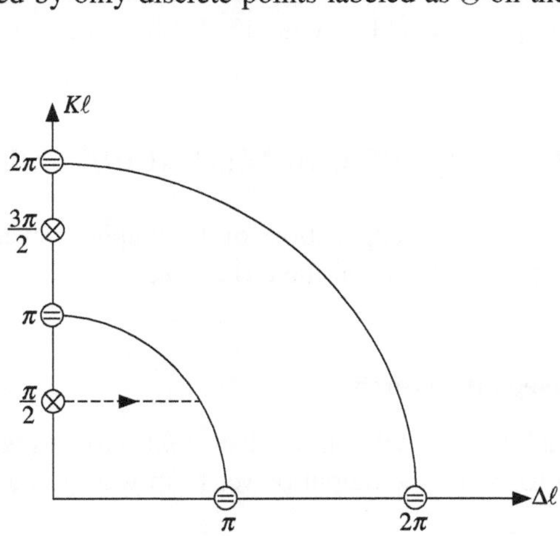

Figure 8.11 A switching diagram for an optical waveguide switch. The parallel states ⊖ are represented by the solid curves and the cross states ⊗ are discrete points on the $K\ell$ axis. The horizontal dashed line shows a switching from a cross state ⊗ toward the parallel state ⊖.

Using the electrooptical effects, which will be discussed further in Chapter 13, the refractive indices in the waveguides can be changed by an applied voltage bias. The refractive index difference between two waveguides, or the mismatch factor $\Delta = (\beta_a - \beta_b)/2$, can be tuned by the voltage. Therefore, we can switch the directional coupler from a cross state at $K\ell = \pi/2$ (and $\Delta = 0$) to a parallel state shown as the horizontal dashed line in Fig. 8.11.

8.3.2 The $\Delta\beta$ Coupler

It is difficult to achieve switching in the design of a single-section waveguide coupler because the coupling coefficient is a function of the spacing and the material parameters; these parameters are harder to adjust in the fabrication processes. It is generally easier to control coupling by an applied voltage to change $\Delta\beta = \beta_a - \beta_b$.

A two-section waveguide coupler has been proposed [46] to achieve switching with more flexibility, as shown in Fig. 8.12. The amplitudes $a(z)$ and $b(z)$ for a coupled waveguide configuration as shown in the first half section of Fig. 8.12 are related to the initial conditions by

$$\begin{bmatrix} a(z) \\ b(z) \end{bmatrix} = \mathbf{S}(z; \beta_a, \beta_b) \begin{bmatrix} a(0) \\ b(0) \end{bmatrix} \tag{8.3.3}$$

where $\mathbf{S}(z; \beta_a, \beta_b)$ is given in (8.2.32) and we keep the order β_a and β_b in the arguments of the S matrix

$$\Delta = \frac{\beta_a - \beta_b}{2}. \tag{8.3.4}$$

The output mode amplitudes at the exit, $z = \ell$, are obtained by the product of two matrices because of the cascade connection

$$\begin{bmatrix} a(\ell) \\ b(\ell) \end{bmatrix} = \mathbf{S}\left(\frac{\ell}{2}; \beta_b, \beta_a\right) \mathbf{S}\left(\frac{\ell}{2}; \beta_a, \beta_b\right) \begin{bmatrix} a(0) \\ b(0) \end{bmatrix}. \tag{8.3.5}$$

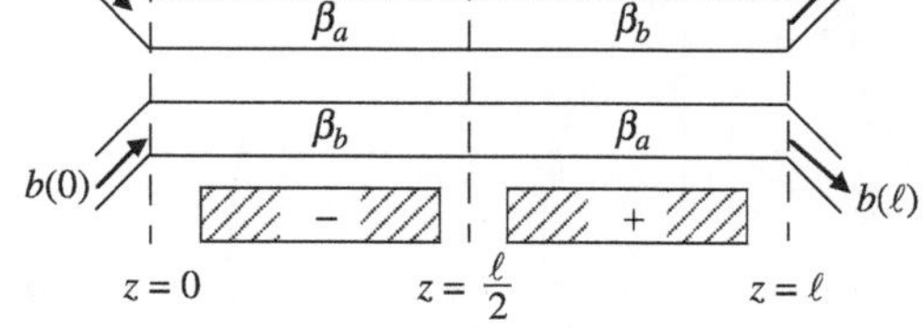

Figure 8.12 A $\Delta\beta$ coupler with two sections where the propagation constants β_a and β_b exchange positions because of the change in the applied voltages.

If $K_{12} \simeq K_{21} \equiv K$, $a(0) = 1$, and $b(0) = 0$, we find

$$a(\ell) = \left[\cos^2 \frac{q\ell}{2} + \frac{\Delta^2 - K^2}{q^2} \sin^2 q \frac{\ell}{2}\right] e^{i\phi\ell} \tag{8.3.6a}$$

$$b(\ell) = i\frac{2K}{q} \sin\frac{q\ell}{2}\left(\cos\frac{q\ell}{2} + i\frac{\Delta}{q}\sin\frac{q\ell}{2}\right) e^{i\phi\ell}. \tag{8.3.6b}$$

The full power transfer (cross state) occurs when $a(\ell) = 0$, or

$$\cot^2 \frac{q\ell}{2} = \frac{K^2 - \Delta^2}{K^2 + \Delta^2}. \tag{8.3.7}$$

For example, if $\Delta = 0$, we find $q = K$, and $\cot^2 \dfrac{K\ell}{2} = 1$. Therefore, $K\ell = \pi/2, 3\pi/2, 5\pi/2$, and so forth. No power transfer (parallel state) occurs if $|b(\ell)|^2 = 0$. This can happen if either

$$\text{Case 1:} \quad \sin\frac{q\ell}{2} = 0 \tag{8.3.8}$$

or

$$\text{Case 2:} \quad \cos^2\frac{q\ell}{2} + \frac{\Delta^2}{q^2}\sin^2\frac{q\ell}{2} = 0. \tag{8.3.9}$$

For case 1, $q\ell = 2m\pi$, or

$$(\Delta\ell)^2 + (K\ell)^2 = (2m\pi)^2, \tag{8.3.10}$$

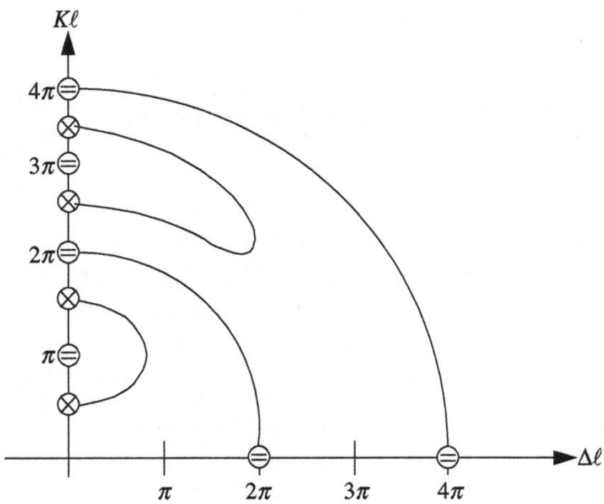

Figure 8.13 Switching diagram for a $\Delta\beta$ coupler.

which are circles in the $K\ell$ versus $\Delta\ell$ plane for $m =$ a nonzero integer. For case 2, it is possibly satisfied only when $\Delta = 0$, and $\cos(K\ell/2) = 0$, that is, $K\ell = (2m+1)\pi$, $m =$ an integer. A plot of the switching diagram is shown in Fig. 8.13. The switching diagram shows the degree of freedom to switch from a parallel state $\ominus$ to a cross state $\otimes$ or vice versa.

The theory for coupled waveguides has also been extended to gain media [47, 48], coupled fiber Fabry–Perot resonators [49, 50], laser amplifiers [51, 52], and coupled laser arrays [53–55]. In Chapter 11, we will discuss coupled semiconductor laser arrays, and the coupled-mode theory will be applied to investigate the super-mode behaviors.

8.4 OPTICAL RING RESONATORS AND ADD-DROP FILTERS

Optical ring resonators [56–76] have been under intensive research recently because of their applications to a variety of devices such as optical add-drop filters, modulation, switching, and dispersion compensation devices. In addition, when gain is added into the resonators, wavelength selective amplification, oscillation, and lasing become possible [59, 62, 63].

8.4.1 Formulation of the Waveguide–Ring Resonator System

The simplest design is to couple the light between a straight waveguide and a ring resonator. We consider a dielectric waveguide coupled to a circular ring resonator as shown in Fig. 8.14. The coupling is assumed to occur near the proximity of the ring edge and the straight waveguide. If the coupling is considered lossless, the input and output waves can be described simply by the solutions from the coupled-mode equations in the previous section for a two-port input and output structure

$$\begin{bmatrix} b_1 \\ b_2 \end{bmatrix} = \begin{bmatrix} t & i\kappa \\ i\kappa^* & t^* \end{bmatrix} \begin{bmatrix} a_1 \\ a_2 \end{bmatrix} \tag{8.4.1}$$

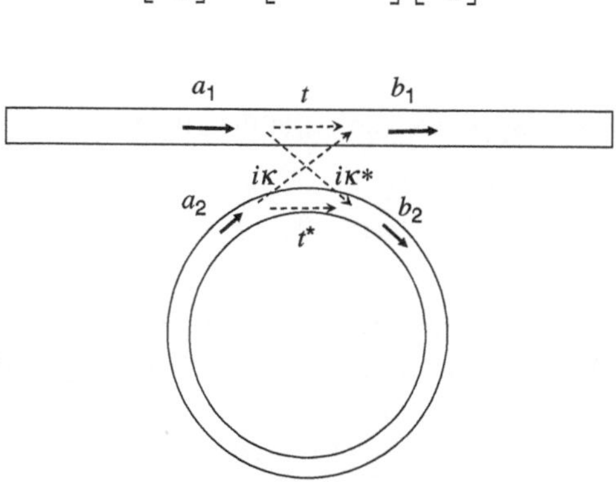

Figure 8.14 A straight waveguide (single bus) coupled to a ring resonator.

where, see Eq. (8.2.32),

$$tt^* + \kappa\kappa^* = 1 \tag{8.4.2}$$

and the common propagation common phase factor $e^{i\phi z}$ has been ignored. In addition, we have the circulation condition of the traveling wave in the ring resonator

$$a_2 = b_2\, ae^{i\theta}, \tag{8.4.3}$$

$$a = e^{-\alpha L/2}, \qquad \theta = \frac{\omega}{c} n_{\text{eff}} L \tag{8.4.4}$$

where a is the amplitude after wave attenuation over a ring distance of L, and $\alpha/2$ is the field attenuation coefficient. The angle θ is the phase delay due to the propagation around the ring of distance L with an effective index n_{eff}. Solving the above equations for b_1 in terms of a_1, we obtain

$$\frac{b_1}{a_1} = \frac{t - ae^{i\theta}}{1 - ate^{i\theta}} \tag{8.4.5}$$

or

$$T = \left|\frac{b_1}{a_1}\right|^2 = \frac{|t|^2 + |a|^2 - 2a|t|\cos(\theta - \theta_t)}{1 + |at|^2 - 2a|t|\cos(\theta - \theta_t)} \tag{8.4.6}$$

where $t = |t|\exp(i\theta_t)$ has been used. The power transmission spectrum has dips at resonance frequencies determined by:

$$\theta - \theta_t = 2m\pi = \frac{2\pi f}{c} n_{\text{eff}} L - \theta_t = \frac{2\pi f}{\text{FSR}} - \theta_t \tag{8.4.7}$$

where FSR is the free spectral range or the frequency spacing between two nearby modes in the frequency spectrum and is determined by the total length of the ring resonator and the effective index of its guided mode

$$\text{FSR} = \frac{c}{n_{\text{eff}} L}. \tag{8.4.8}$$

Near resonance $f_m = mc/n_{\text{eff}}L$, we can write the angle as

$$\theta - \theta_t = 2m\pi + 2\pi(f - f_m)\frac{n_{\text{eff}} L}{c}. \tag{8.4.9}$$

Figure 8.15 shows a plot of the transmission spectrum as a function of the angle for a loss parameter $a = 0.8$. At resonance, the transmission magnitude is given by

$$T = \frac{(|t| - a)^2}{(1 - a|t|)^2}. \tag{8.4.10}$$

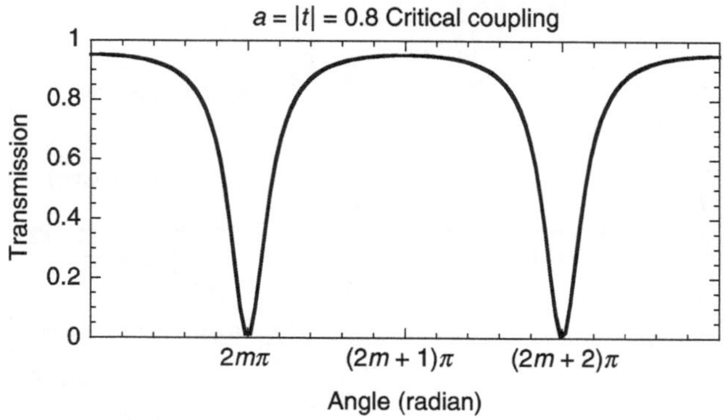

Figure 8.15 Transmission spectrum of a single bus coupled to a ring resonator.

Figure 8.16 shows a plot of the transmission T at resonance as a function of the loss parameter a. It vanishes at $a = |t|$ and becomes unity at $a = 1$. We have a few conditions for the transmission spectrum shown in Fig. 8.17:

1. Critical coupling occurs if $a = |t|$. Under critical coupling, the transmission T vanishes at resonance frequencies. Because $b_1 = ta_1 + i\kappa a_2$, it implies that completely destructive interference between the directly transmitted field ta_1 and the internal field in the ring coupled into the output waveguide $i\kappa a_2$; that is, they cancel each other.
2. Undercoupling occurs if $a < |t|$. The transmission spectrum does not vanish at resonance. A smaller amplitude a means a larger attenuation loss in the resonator propagation, hence, the spectral dip appears broader than that of critical coupling.

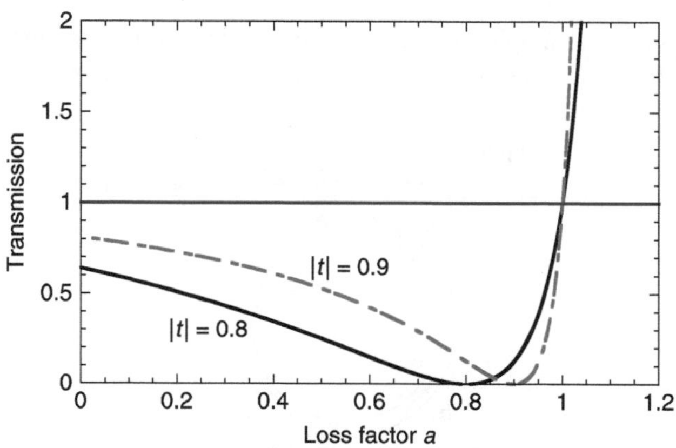

Figure 8.16 The transmission power as a function of the ring loss factor a for two different values of $|t|$ in the coupling matrix.

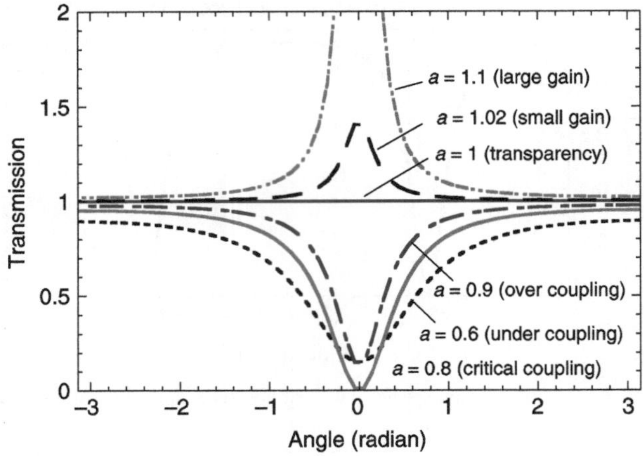

Figure 8.17 The transmission spectrum near resonance for various values of the loss factor a from undercoupling $a < |t| = 0.8$, critical coupling $a = |t|$, overcoupling $a = 0.9$ ($>|t|$), transparency $a = 1$, and gain $a > 1$.

3. Overcoupling occurs if $a > |t|$. The transmission spectrum shows a sharper dip when a increases, yet it does not vanish at resonance. Further increase of a increases the transmission at resonance.
4. At $a = 1$, meaning no attenuation in the resonator at all, transparency occurs. The transmission spectrum becomes unity for all angles θ.
5. When $a > 1$, meaning that gain is introduced into the resonator so the amplitude a is larger than unity, the transmission spectrum is larger than unity and peaks at resonant frequency. Lasing action occurs when $a = 1/|t|$ because T approaches infinity.

8.4.2 Optical Add-Drop Filter

Figure 8.18 shows a simple configuration consisting of two straight waveguides coupled by a ring resonator, a so-called double-bus ring resonator. The fundamental

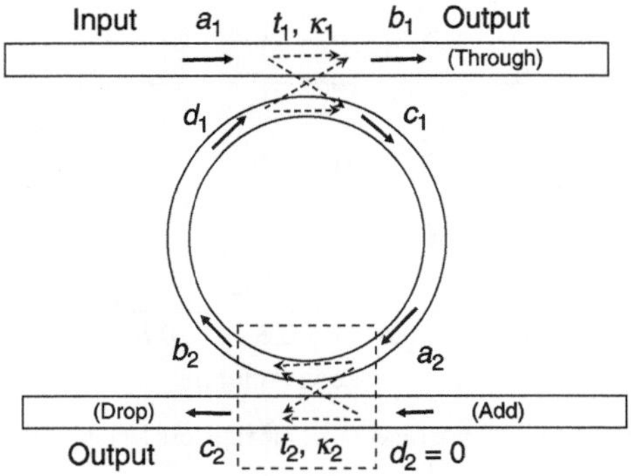

Figure 8.18 A double-bus ring resonator as an optical add-drop filter.

equations governing the configuration are a straightforward generalization of the formulation in Section 8.4.1. Using (8.4.1) with

$$\begin{bmatrix} b_1 \\ c_1 \end{bmatrix} = \begin{bmatrix} t_1 & i\kappa_1 \\ i\kappa_1^* & t_1^* \end{bmatrix} \begin{bmatrix} a_1 \\ d_1 \end{bmatrix} \tag{8.4.11a}$$

$$a_2 = c_1\, a^{1/2}\, e^{i\theta/2}, \quad d_1 = b_2\, a^{1/2}\, e^{i\theta/2} \tag{8.4.11b}$$

$$\begin{bmatrix} c_2 \\ b_2 \end{bmatrix} = \begin{bmatrix} t_2 & i\kappa_2 \\ i\kappa_2^* & t_2^* \end{bmatrix} \begin{bmatrix} d_2 \\ a_2 \end{bmatrix} \tag{8.4.11c}$$

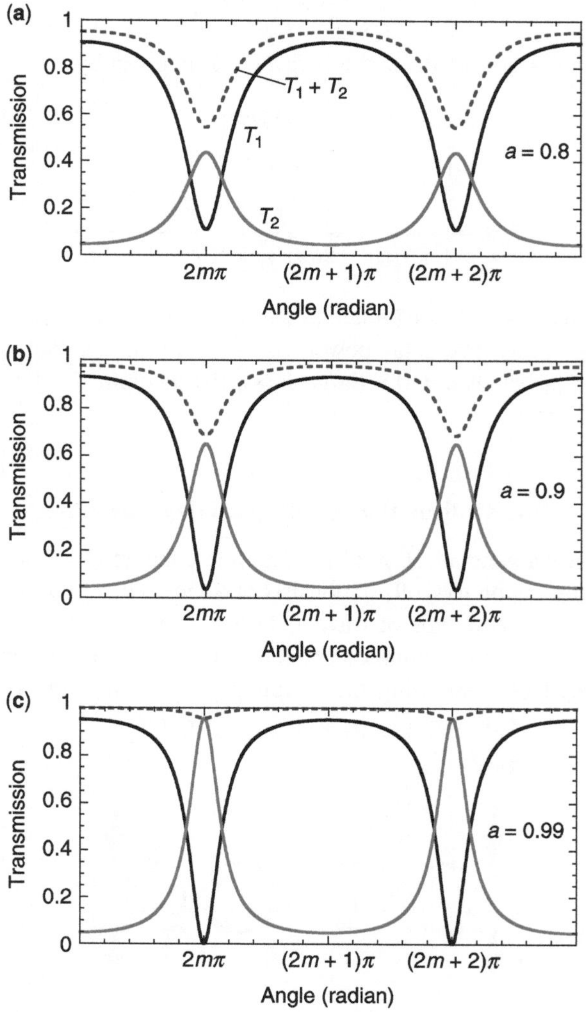

Figure 8.19 The power spectrum of the through port (T_1) and the drop port (T_2) for three loss factors (a) $a = 0.8$, (b) $a = 0.9$, and (c) $a = 0.99$. The dashed line is the sum of T_1 and T_2, showing the largest loss occurs at resonances due to attenuation loss in the ring.

and $d_2 = 0$ (because there is no incident wave from the d_2 port), we obtain

$$\frac{b_1}{a_1} = \frac{t_1 - at_2^* e^{i\theta}}{1 - at_2^* t_1^* e^{i\theta}}. \tag{8.4.12}$$

The results for the transmission in the b_1 port are simply replacing a by $t_2^* a$ in Section 8.4.1 because of the extra transmission factor in the additional coupling junction between the ring and the second waveguide

$$T_1 = \left|\frac{b_1}{a_1}\right|^2 = \frac{|t_1|^2 + |at_2|^2 - 2a|t_1 t_2|\cos(\theta - \theta_{t_1} - \theta_{t_2})}{1 + |at_1 t_2|^2 - 2a|t_1 t_2|\cos(\theta - \theta_{t_1} - \theta_{t_2})}. \tag{8.4.13}$$

The output in the c_2 port of the second waveguide is given by

$$\begin{aligned} T_2 &= \left|\frac{c_2}{a_1}\right|^2 = \left|\frac{-\kappa_2 \kappa_1^* a^{1/2} e^{i\theta/2}}{1 - at_2^* t_1^* e^{i\theta}}\right|^2 \\ &= \frac{(1 - |t_1|^2)(1 - |t_2|^2)a}{1 + |at_1 t_2|^2 - 2a|t_1 t_2|\cos(\theta - \theta_{t_1} - \theta_{t_2})}. \end{aligned} \tag{8.4.14}$$

The transmissions T_1 and T_2 are plotted as a function of the angle $(\theta - \theta_{t_1} - \theta_{t_2})$ near resonances, Fig. 8.19. The total power $T = T_1 + T_2$ of the two output ports is also plotted for (a) small $a = 0.8$ (large loss), (b) medium $a = 0.9$, and (c) large $a = 0.99$ (small loss).

8.4.3 Coupled-Ring Optical Waveguide (CROW) Structure

Figure 8.20 shows a generalization of the multiple coupled ring resonators [57] in a double-bus configuration [68, 69]. It consists of a sequence of coupled high-Q resonators, which form a new type of waveguide behavior coupling the light from one resonator to another. The transmission and reflection of such a sequence of resonators can be engineered by controlling the coupling factor κ as well as the resonance

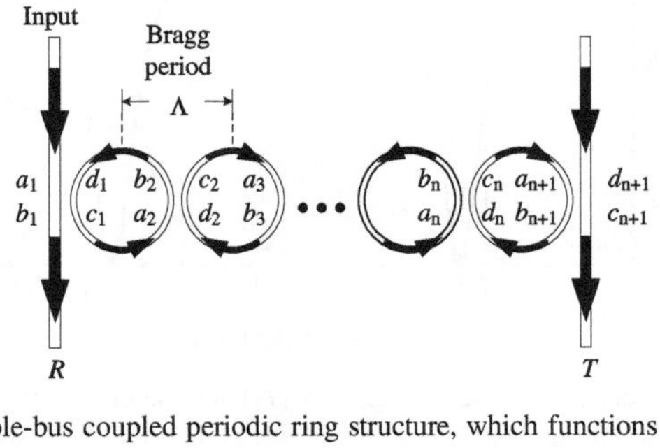

Figure 8.20 A double-bus coupled periodic ring structure, which functions as a broad band filter.

frequency. A photonic band gap type of dispersion forms when we cascade a periodic ring structure

$$\begin{bmatrix} b_n \\ c_n \end{bmatrix} = \begin{bmatrix} t & i\kappa \\ i\kappa & t \end{bmatrix} \begin{bmatrix} a_n \\ d_n \end{bmatrix} \tag{8.4.15}$$

which can be rewritten as

$$\begin{bmatrix} c_n \\ d_n \end{bmatrix} = \frac{1}{i\kappa} \begin{bmatrix} -1 & t \\ -t & 1 \end{bmatrix} \begin{bmatrix} a_n \\ b_n \end{bmatrix}. \tag{8.4.16}$$

Using the relations for half a loop that

$$a_{n+1} = c_n \, e^{i\theta/2} \sqrt{a} \tag{8.4.17a}$$

$$d_n = b_{n+1} \, e^{i\theta/2} \sqrt{a} \tag{8.4.17b}$$

where one loop of phase delay is θ and amplitude attenuation is a, the recurrence relation is

$$\begin{aligned} \begin{bmatrix} a_{n+1} \\ b_{n+1} \end{bmatrix} &= \begin{bmatrix} \sqrt{a}\, e^{i\theta/2} & 0 \\ 0 & \dfrac{1}{\sqrt{a}}\, e^{-i\theta/2} \end{bmatrix} \begin{bmatrix} c_n \\ d_n \end{bmatrix} \\ &= \begin{bmatrix} A & B \\ C & D \end{bmatrix} \begin{bmatrix} a_n \\ b_n \end{bmatrix} \end{aligned} \tag{8.4.18}$$

where

$$\begin{aligned} A &= -\frac{\sqrt{a}}{i\kappa}\, e^{i\theta/2}, \qquad & B &= \frac{t\sqrt{a}}{i\kappa}\, e^{i\theta/2} \\ C &= -\frac{t}{i\kappa\sqrt{a}}\, e^{-i\theta/2}, \qquad & D &= \frac{1}{i\kappa\sqrt{a}}\, e^{-i\theta/2}. \end{aligned} \tag{8.4.19}$$

Note $AD - BC = 1$ even in the presence of loss, $a \neq 1$. Cascading N times, we have

$$\begin{bmatrix} a_{n+1} \\ b_{n+1} \end{bmatrix} = \mathbf{M}^N \begin{bmatrix} a_1 \\ b_1 \end{bmatrix} \tag{8.4.20}$$

where [77]

$$\begin{aligned} \mathbf{M}^N &= \begin{bmatrix} A & B \\ C & D \end{bmatrix}^N \equiv \begin{bmatrix} m_{11} & m_{12} \\ m_{21} & m_{22} \end{bmatrix} \\ &= \begin{bmatrix} AU_{N-1}(x) - U_{N-2}(x) & BU_{N-1}(x) \\ CU_{N-1}(x) & DU_{N-1}(x) - U_{N-2}(x) \end{bmatrix} \end{aligned} \tag{8.4.21}$$

where $U(x)$ is the Chebyshev polynomial and $x = (A + D)/2$ (see page 215). The output at the double-bus CROW structure is

$$\begin{aligned}\begin{bmatrix} c_{n+1} \\ d_{n+1} \end{bmatrix} &= \frac{1}{i\kappa}\begin{bmatrix} -1 & t \\ -t & 1 \end{bmatrix}\begin{bmatrix} a_{n+1} \\ b_{n+1} \end{bmatrix} \\ &= \frac{1}{i\kappa}\begin{bmatrix} -m_{11} + tm_{21} & -m_{12} + tm_{22} \\ -tm_{11} + m_{21} & -tm_{12} + m_{22} \end{bmatrix}\begin{bmatrix} a_1 \\ b_1 \end{bmatrix}\end{aligned} \tag{8.4.22}$$

where $d_{n+1} = 0$ because there is not input from the d_{n+1} port. The reflection and transmission coefficient is

$$r_{DB} = \frac{b_1}{a_1} = \frac{tm_{11} - m_{21}}{m_{22} - tm_{12}} = \frac{t[AU_{N-1}(x) - U_{N-2}(x)] - CU_{N-1}(x)}{DU_{N-1}(x) - U_{N-2}(x) - tBU_{N-1}(x)} \tag{8.4.23}$$

$$t_{DB} = \frac{c_{n+1}}{a_1} = \frac{i\kappa}{m_{22} - tm_{12}} = \frac{i\kappa}{DU_{N-1}(x) - U_{N-2}(x) - tBU_{N-1}(x)}. \tag{8.4.24}$$

The zeros of the reflection give the conditions for the reflection minima

$$tA - C = tU_{N-2}(x)/U_{N-1}(x). \tag{8.4.25}$$

The poles give the oscillation conditions

$$D - tB = U_{N-2}(x)/U_{N-1}(x). \tag{8.4.26}$$

When $N = 1$, we have

$$tA - C = 0 \tag{8.4.27}$$

which has the solution $\theta = 2m\pi$. The oscillator condition is given by

$$D = tB \quad \text{or} \quad \theta = 2m\pi \quad \text{and} \quad t = 1/\sqrt{a}. \tag{8.4.28}$$

Periodic Ring Resonators For lossless structure $a = 1$, the Bloch theorem for a periodic structure allows for

$$\begin{bmatrix} a_{n+1} \\ b_{n+1} \end{bmatrix} = e^{i\beta\Lambda}\begin{bmatrix} a_n \\ b_n \end{bmatrix} \tag{8.4.29}$$

where $\Lambda = L_b$ is the Bragg period. We expect

$$\begin{bmatrix} A & B \\ C & D \end{bmatrix}\begin{bmatrix} a_n \\ b_n \end{bmatrix} = q\begin{bmatrix} a_n \\ b_n \end{bmatrix} \tag{8.4.30}$$

where the eigenvalue equation

$$(A-q)(D-q)-BC=0 \tag{8.4.31}$$

leads to two possible eigenvalues

$$q_{\pm}=\frac{A+D}{2}\pm\sqrt{\left(\frac{A+D}{2}\right)^2-1} \tag{8.4.32}$$

where

$$\begin{aligned} q_+ + q_- &= A+D \\ q_+ q_- &= AD-BC=1. \end{aligned} \tag{8.4.33}$$

Therefore, $q_{\pm}=e^{\pm i\beta\Lambda}$ are the two possible eigenvalues, which satisfy

$$\frac{1}{2}(q_+ + q_-)=\cos\beta\Lambda=\frac{A+D}{2}=-\frac{\sin(\theta/2)}{\kappa}. \tag{8.4.34}$$

The dispersion relation between the propagation constant β and the frequency ω or $\theta=\frac{\omega}{c}n_{\text{eff}}L$ is given by

$$\cos\beta\Lambda=-\frac{\sin(\theta/2)}{\kappa} \tag{8.4.35}$$

which has real solutions for $\beta\Lambda$ that are only

$$|\cos\beta\Lambda|=\left|\frac{\sin(\theta/2)}{\kappa}\right|\leq 1. \tag{8.4.36}$$

The centers of the pass bands occur near

$$\theta=2m\pi,\quad m=0,\ \pm1,\ \pm2,\ldots \tag{8.4.37}$$

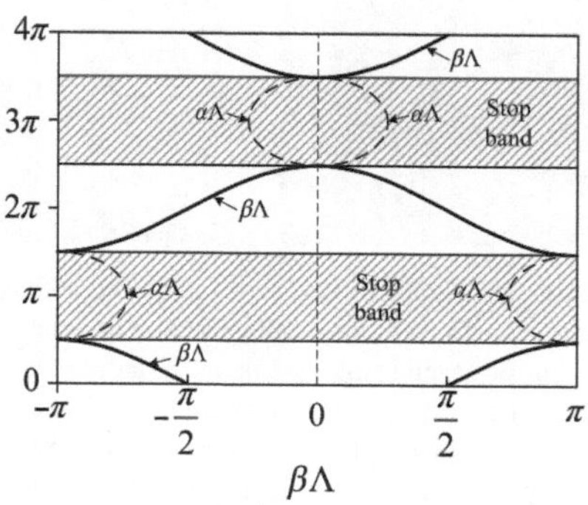

Figure 8.21 The dispersion diagram for the double-bus periodic ring structure.

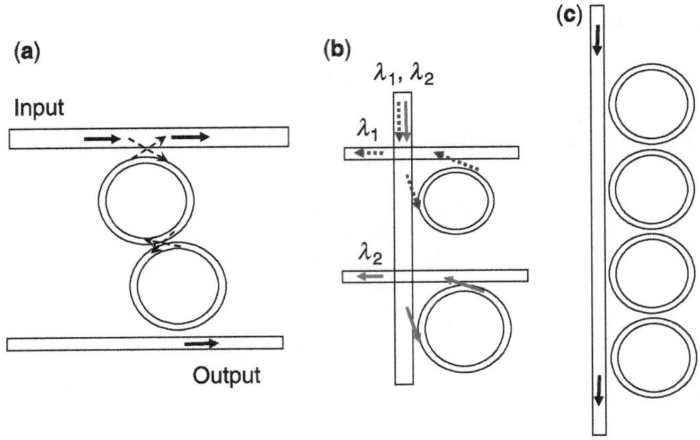

Figure 8.22 (a) A double-bus double-ring architecture. (b) A two-wavelength drop filter. (c) A single-bus periodic ring structure as a broad band filter or slow light device.

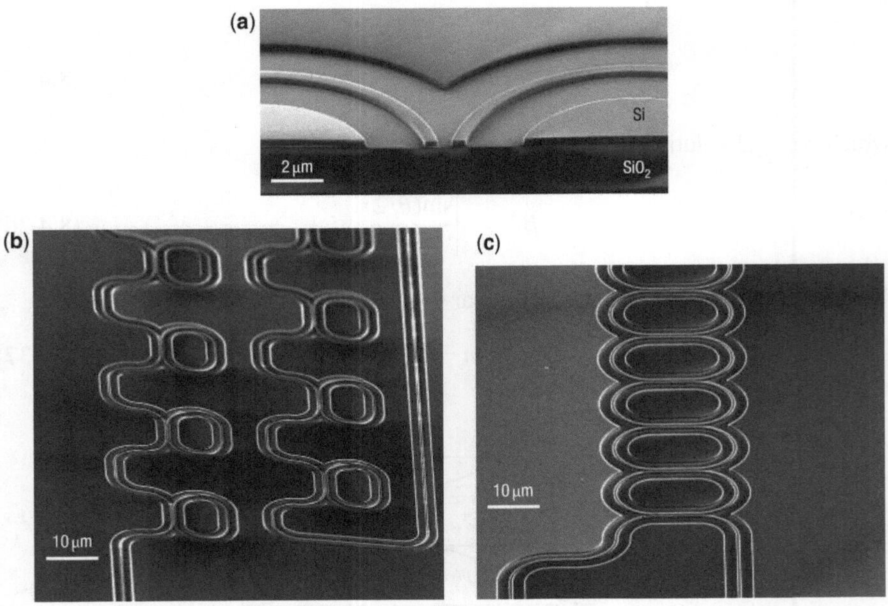

Figure 8.23 Scanning electron micrograph of resonantly enhanced optical delay lines based on photonic-wire waveguides. (a) Detailed view of a cleaved coupling section of a bent waveguide bus and a resonator. (b) Optical delay line consisting of several ring resonators in an all-pass filter (APF) configuration in a cascade. Each ring is side coupled to a common waveguide bus. The gap between the ring resonator and the bus is about 200 nm. (c) Optical delay line consisting of several ring resonators in a CROW configuration. Each ring is coupled to its nearest neighbor with a gap of about 200 nm. (Adapted by permission from [73] © 2007 Macmillan Publishers Ltd: Nature Publishing Group.)

On the other hand, stop band occurs when

$$\left|\frac{\sin(\theta/2)}{\kappa}\right| > 1. \tag{8.4.38}$$

The centers of stop bands occur near

$$\theta = (2m+1)\pi \tag{8.4.39}$$

when

$$\theta = \pi, 5\pi, 9\pi, \ldots, (4n+1)\pi, \ \sin(\theta/2) = +1$$
$$\beta\Lambda = \pi \pm i\alpha\Lambda \tag{8.4.40}$$

$$\cos(\pi \pm i\alpha\Lambda) = -\frac{\sin\theta/2}{\kappa} \quad \text{or} \quad \cosh\alpha\Lambda = \frac{\sin\theta/2}{\kappa}. \tag{8.4.41}$$

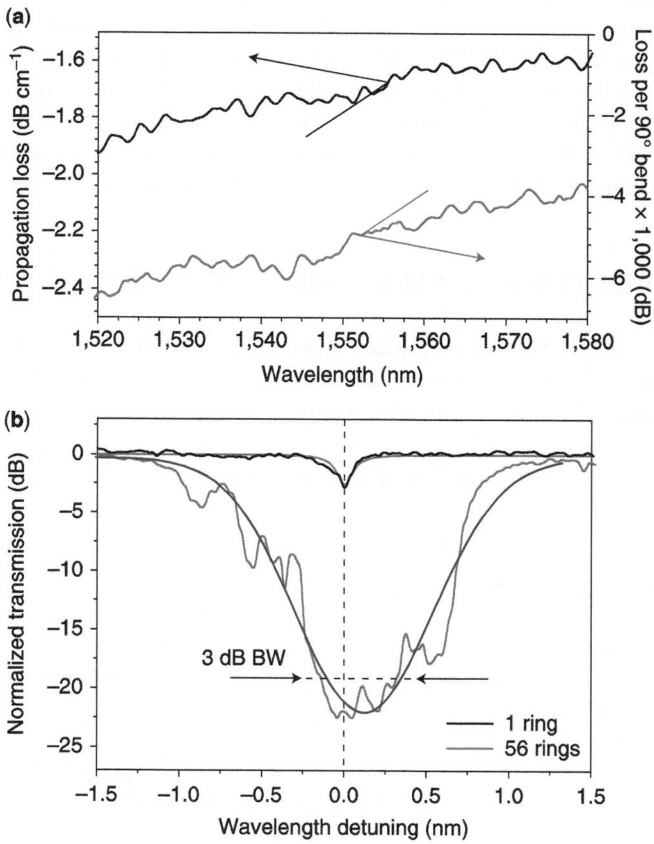

Figure 8.24 (a) Propagation loss (dark curve) and bending loss (light curve) measured in photonic wire waveguide. (b) The measured (light curves) and calculated (dark and smooth curves) transmission spectrum for a single ring resonator (top curves) and for 56 rings (bottom curves). The broader transmission bandwidth using 56 rings is clearly shown. (Adapted by permission from [73] © 2007 Macmillan Publishers Ltd: Nature Publishing Group.)

Near $\theta = 3\pi, 7\pi, 11\pi, \ldots (4n+3)\pi,\ \sin(\theta/2) = -1$

$$\beta\Delta = 2m\pi \pm i\alpha\Lambda \tag{8.4.42}$$

$$\cosh \alpha\Lambda = -\frac{\sin(\theta/2)}{\kappa}. \tag{8.4.43}$$

A schematic plot of the dispersion relation between $\theta(=\omega n_{\text{eff}}L/c)$ versus $\beta\Lambda$ is shown in Fig. 8.21. The solid curves are the $\beta\Lambda$ or the real part of the propagation constant, and the dashed curves are for the imaginary parts $\alpha\Lambda$ in the stop bands when $\beta\Lambda = 2m\pi \pm i\alpha\Lambda$.

Figure 8.22a shows a double-ring double-bus configuration with the output as shown. Figure 8.22b shows an optical filter design for an input with two wavelengths and the drop of each wavelength at two separate horizontal straight waveguides. Figure 8.22c shows a single-bus periodic ring design for engineering the dispersion or group delay. Such a structure has been demonstrated on silicon-on-insulator (SOI) submicrometer photonic wire waveguides shown in the scanning electron microscopy (SEM) images in Fig. 8.23 [73]. The design uses coupled resonators to a curved waveguide bus for slow light or optical delay line applications. Figure 8.24a and Fig. 8.24b show the measured losses and the transmission spectra of 1 ring and 56 rings [73]. We can see a broader bandwidth when 56 rings are used. Delay of more than 10 bits at 20 Gb/s is reported using these structures.

8.5 DISTRIBUTED FEEDBACK (DFB) STRUCTURES

Consider a two-dimensional corrugated waveguide or a DFB structure [78–86] as shown in Fig. 8.25. The permittivity function $\varepsilon(x, z)$ can be written as an unperturbed slab waveguide $\varepsilon^{(0)}(x)$ and a perturbed periodic part $\Delta\varepsilon(x, z)$

$$\varepsilon(x, z) = \varepsilon^{(0)}(x) + \Delta\varepsilon(x, z). \tag{8.5.1}$$

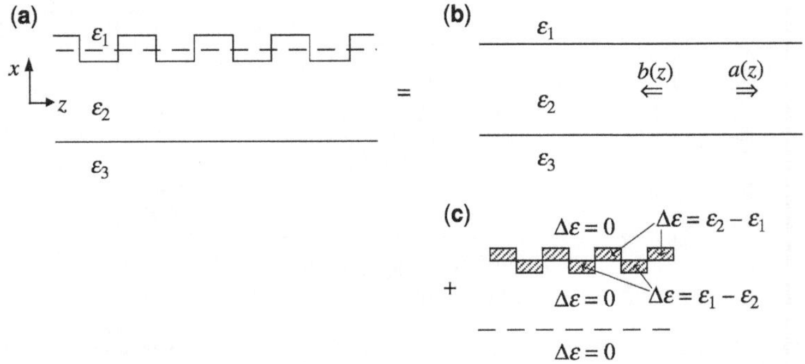

Figure 8.25 (a) A corrugated waveguide with a permittivity function $\varepsilon(x, z)$ can be written as the sum of (b) an unperturbed part $\varepsilon^{(0)}(\mathbf{r})$, which describes a uniform slab waveguide, and (c) a perturbed part $\Delta\varepsilon = \varepsilon_2 - \varepsilon_1$ or $\Delta\varepsilon = \varepsilon_1 - \varepsilon_2$ in the corrugation regions.

For the unperturbed slab waveguide $\varepsilon^{(0)}(x)$, assume that the fundamental TE_0 mode with a forward (or backward) propagation constant β_0 (or $-\beta_0$) is of interest. The amplitude for the forward propagating mode is $A_0 e^{i\beta_0 z}$ and backward propagating mode is $B_0 e^{-i\beta_0 z}$. Because $\varepsilon(x, z)$ is periodic in z, we may write the Fourier series expansion

$$\Delta\varepsilon(x, z) = \varepsilon_0 \sum_{p=-\infty}^{\infty} \Delta\varepsilon_p(x) e^{ip\frac{2\pi}{\Lambda}z} \tag{8.5.2}$$

where Λ is the period. For a lossless structure, $\Delta\varepsilon(x, z)$ is real; therefore,

$$\Delta\varepsilon_p^*(x) = \Delta\varepsilon_{-p}(x). \tag{8.5.3}$$

8.5.1 Derivations of the Coupled-Mode Equations

From Maxwell's equations, we write the wave equation for TE polarization

$$\mathbf{E} = \hat{y}\, E_y(x, z), \tag{8.5.4}$$

$$\left[\frac{\partial^2}{\partial x^2} + \frac{\partial^2}{\partial z^2} + \omega^2 \mu \varepsilon^{(0)}(x)\right] E_y = -\omega^2 \mu \Delta\varepsilon(x, z)\, E_y \tag{8.5.5}$$

for a time-harmonic field with an $e^{-i\omega t}$ dependence. We expand the solution E_y in terms of the unperturbed fundamental TE_0 waveguide mode traveling in $+z$ and $-z$ directions

$$E_y(x, z) = \left[A_0(z)\, e^{i\beta_0 z} + B_0(z)\, e^{-i\beta_0 z}\right] E_y^{(0)}(x) \tag{8.5.6}$$

where we assume $A_0(z)$ and $B_0(z)$ are slowly varying functions of z due to the distributed-feedback effect. $E_y^{(0)}(x)$ is the transverse modal distribution of the TE_0 mode satisfying

$$\left[\frac{\partial^2}{\partial x^2} + \frac{\partial^2}{\partial z^2} + \omega^2 \mu \varepsilon^{(0)}(x)\right] E_y^{(0)}(x)\, e^{\pm i\beta_0 z} = 0 \tag{8.5.7}$$

which has been solved in the waveguide problem in Chapter 7. We substitute the expression (8.5.6) into the wave equation (8.5.5) and ignore the second-order derivatives of $A_0(z)$ and $B_0(z)$ because they are slowly varying functions. We find

$$\begin{aligned}&\left(2i\beta_0 \frac{\partial A_0}{\partial z} e^{i\beta_0 z} - 2i\beta_0 \frac{\partial B_0}{\partial z} e^{-i\beta_0 z}\right) E_y^{(0)}(x) \\ &\quad = -\omega^2 \mu \varepsilon_0 \sum_p \Delta\varepsilon_p(x) e^{ip\frac{2\pi}{\Lambda}z} E_y^{(0)}(x) \left[A_0(z) e^{i\beta_0 z} + B_0(z) e^{-i\beta_0 z}\right]\end{aligned} \tag{8.5.8}$$

If we multiply both sides of (8.5.8) by $-H_x^{(0)*}(x) = (\beta_0/2\omega\mu)E_y^{(0)*}(x)$, which is the $-x$ component of the magnetic field of the guided TE_0 mode, and make use of the normalization condition of the guided power along the z direction,

$$\frac{\beta_0}{2\omega\mu}\int_{-\infty}^{\infty}\left|E_y^{(0)}(x)\right|^2 dx = 1, \tag{8.5.9}$$

we can eliminate the $E_y^{(0)}(x)$ on the left-hand side of (8.5.8). We then focus on the coupling of the forward propagating mode $e^{i\beta_0 z}$ to the backward propagating mode $e^{-i\beta_0 z + ip\frac{2\pi}{\Lambda}z}$ so that $-\beta_0 + (2\pi/\Lambda)p$ is close to β_0, for a particular $p = \ell$. Assuming the couplings to the other modes are negligible, we then have

$$\frac{dA_0(z)}{dz} = iK_{ab}B_0(z)\, e^{-i2\left(\beta_0 - \frac{\ell\pi}{\Lambda}\right)z} \tag{8.5.10}$$

where

$$K_{ab} = \frac{\omega}{4}\varepsilon_0 \int_{-\infty}^{+\infty} \Delta\varepsilon_\ell(x)\left|E_y^{(0)}(x)\right|^2 dx. \tag{8.5.11}$$

A phase-matching diagram is shown in Fig. 8.26, where we consider the first-order grating $\ell = \pm 1$ such that $\beta_0 = -\beta_0 + 2\pi/\Lambda$ and $A_0(z)$ is coupled to $B_0(z)$ by (8.5.10). Equation (8.5.11) defines the coupling coefficient for TE_0 mode in a corrugated waveguide. Extensive research work has been done on both TE and TM mode coupling coefficients in distributed feedback structures and applications to DFB lasers [80–86]. If we focus on the backward propagating mode $e^{-i\beta_0 z}$ on the left-hand side with phase matching to the term $e^{+i\beta_0 z - i(\ell\pi/\Lambda z)}$ on the right-hand side

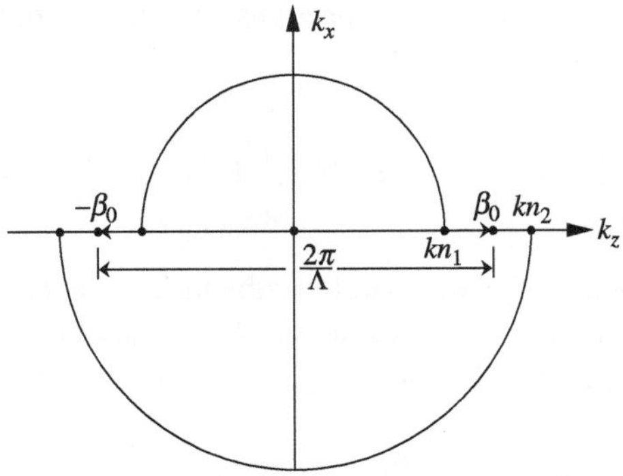

Figure 8.26 Phase-matching diagram for a forward propagating mode with a propagation constant β_0 coupling to a backward propagating mode with a propagation constant β_{-0} $(= -\beta_0)$ in the negative z direction. $\beta_0 = -\beta_0 + (2\pi/\Lambda)$ is the phase-matching condition for the z components of the propagation wave vectors. The difference $\Delta\beta = \beta - (\pi/\Lambda)$ is the detuning parameter.

of (8.5.8), we obtain the second coupled-mode equation

$$\frac{dB_0(z)}{dz} = iK_{ba}A_0(z)\,e^{i2\left(\beta_0-\frac{\ell\pi}{\Lambda}\right)z} \tag{8.5.12}$$

where

$$K_{ba} = -\frac{\omega}{4}\varepsilon_0 \int_{-\infty}^{+\infty} \Delta\varepsilon_{-\ell}(x)\left|E_y^{(0)}(x)\right|^2 dx. \tag{8.5.13}$$

We also see that the condition for contra-directional coupling in a lossless coupler

$$K_{ba^*} = -K_{ab}^* \tag{8.5.14}$$

is satisfied because $\Delta\varepsilon_\ell^* = \Delta\varepsilon_{-\ell}$. As shown in Fig. 8.26, the forward propagation amplitude $A_0(z)$ is coupled to the backward propagation amplitude $B_0(z)$ by (8.5.12). Therefore, the phase-matching condition requires $-\beta_0 \simeq \beta_0 - (2\pi/\Lambda)$. Define

$$\Delta\beta = \beta_0 - \beta_B, \quad \beta_B = \frac{\pi}{\Lambda} \tag{8.5.15}$$

$$\begin{aligned} A(z) &= A_0(z)\,e^{i\Delta\beta z} \\ B(z) &= B_0(z)\,e^{-i\Delta\beta z}. \end{aligned} \tag{8.5.16}$$

We obtain the well-known coupled-mode equations for DFB structures

$$\frac{d}{dz}\begin{bmatrix} A(z) \\ B(z) \end{bmatrix} = i\begin{bmatrix} \Delta\beta & K_{ab} \\ K_{ba} & -\Delta\beta \end{bmatrix}\begin{bmatrix} A(z) \\ B(z) \end{bmatrix}. \tag{8.5.17}$$

The above coupled-mode equation in matrix form for the two new unknowns $A(z)$ and $B(z)$ is very useful because the 2×2 matrix contains constant elements, which are independent of the variable z. Equations (8.5.10) and (8.5.12) contain z-dependent coefficients, and their solutions are less obvious. The transformation (8.5.16) helps simplify the algebraic equations.

Example: Index Grating If the refractive index in the waveguide is given by

$$n = n_0 + \Delta n(z) \quad \text{for } |x| \leq d/2 \text{ (inside the guide)} \tag{8.5.18}$$

where n_0 is the background index of the guide, we find that

$$n^2 \simeq n_0^2 + 2n_0\,\Delta n(z) \tag{8.5.19}$$

where n_0^2 enters the unperturbed waveguide permittivity function $\varepsilon^{(0)}(x)$, and we have ignore the second-order terms $(\Delta n)^2$. We find

$$\Delta\varepsilon(x, z) = 2n_0\Delta n(z)\varepsilon_0 \quad \text{for } |x| \leq d/2 \text{ (inside the guide).} \tag{8.5.20}$$

Because $\Delta n(z)$ is periodic, it can be decomposed into the Fourier series

$$\Delta n(z) = \sum_{p=-\infty}^{\infty} \Delta n_p \, e^{ip\frac{2\pi}{\Lambda}z}. \tag{8.5.21}$$

Therefore,

$$\begin{aligned}
&\Delta\varepsilon_p = 2n_0\,\Delta n_p \quad \text{for } |x| \le d/2, \quad \text{and} \quad \Delta\varepsilon_p = 0 \quad \text{for } |x| > d/2 \\
&K_{ab} = \frac{\omega\varepsilon_0}{4} \int_{|x| \le \frac{d}{2}} \Delta\varepsilon_p(x) \left| E_y^{(0)}(x) \right|^2 dx \\
&\quad = \frac{\omega\varepsilon_0}{4} (2n_0 \Delta n_{+1}) \frac{2\omega\mu}{\beta} \Gamma \\
&\quad \simeq \Gamma k_0 \Delta n_{+1}
\end{aligned} \tag{8.5.22}$$

where $k_0 = \omega\sqrt{\mu_0\varepsilon_0}$, and $\beta_0 = k_0 n_0$ have been used, and Γ is exactly the optical confinement factor because we assume the index grating is dependent on z inside the guide. Similarly,

$$K_{ba} \simeq -\Gamma k_0 \Delta n_{-1} \tag{8.5.23}$$

and we have kept $p = +1$ or -1 for the first-order grating.

For an index phase grating with a magnitude Δn and phase φ,

$$\begin{aligned}
\Delta n(z) &= \Delta n \cos\left(\frac{2\pi}{\Lambda} z + \varphi\right) \\
&= \Delta n_{+1}\, e^{i\frac{2\pi}{\Lambda}z} + \Delta n_{-1}\, e^{-i\frac{2\pi}{\Lambda}z} \\
\Delta n_{+1} &= \frac{\Delta n}{2}\, e^{i\varphi}, \qquad \Delta n_{-1} = \frac{\Delta n}{2}\, e^{-i\varphi}.
\end{aligned} \tag{8.5.24}$$

The coupling coefficients are

$$K_{ab} = \Gamma k_0 \frac{\Delta n}{2}\, e^{i\varphi} \tag{8.5.25a}$$

$$K_{ba} = -\Gamma k_0 \frac{\Delta n}{2}\, e^{-i\varphi} = -K_{ab}^{*} \tag{8.5.25a}$$

In other words, the coupling coefficients can be put in very simple forms and they satisfy the relation $K_{ba} = -K_{ab}^{*}$ because the index grating is lossless.

8.5.2 Eigenstate Solutions of the Coupled-Mode Equations

We follow a similar procedure as before for coupling in two parallel waveguides by assuming the eigensolution of the form

$$\begin{bmatrix} A(z) \\ B(z) \end{bmatrix} = \begin{bmatrix} A \\ B \end{bmatrix} e^{iqz}. \tag{8.5.26}$$

The eigenequations become

$$\begin{bmatrix} \Delta\beta - q & K_{ab} \\ K_{ba} & -\Delta\beta - q \end{bmatrix} \begin{bmatrix} A \\ B \end{bmatrix} = 0. \tag{8.5.27}$$

Solving the eigenequation

$$(\Delta\beta - q)(-\Delta\beta - q) - K_{ba}K_{ab} = 0 \tag{8.5.28}$$

for nontrivial solutions, we obtain the two eigenvalues

$$q_{\pm} = \pm\sqrt{(\Delta\beta)^2 + K_{ab}K_{ba}} = \pm q. \tag{8.5.29}$$

In the original expression for the electric field (8.5.6), we keep only the forward and backward propagating TE_0 mode

$$\begin{aligned} E_y(x, z) &= [A_0(z)\, e^{i\beta_0 z} + B_0(z)\, e^{-i\beta_0 z}]E_y^{(0)}(x) \\ &= [A(z)\, e^{i\beta_B z} + B(z)\, e^{-i\beta_B z}]E_y^{(0)}(x). \end{aligned} \tag{8.5.30}$$

We can see that the propagation constant for the forward propagating mode is dominated by $A(z)\, e^{i\beta_B z}$ or $\beta_B \pm q$, and the propagation constant for the backward propagating mode $B(z)\, e^{-i\beta_B z}$ is dominated by $-\beta_B \mp q$. In other words, we are looking at the coupling between the two counter-propagating TE_0 modes

$$\begin{aligned} \beta_a &= \beta_0(\omega) \\ \beta_b &= -\beta_0(\omega) + \frac{2\pi}{\Lambda} \end{aligned} \tag{8.5.31}$$

where $\beta_0(\omega) = \omega n_{\text{eff}}/c$ is the propagation constant of the TE_0 mode. The two system propagation constants are

$$\begin{aligned} \beta_{\pm} &= \frac{\beta_a + \beta_b}{2} \pm \sqrt{\left(\frac{\beta_a - \beta_b}{2}\right)^2 + K_{ab}K_{ba}} \\ &= \frac{\pi}{\Lambda} \pm q. \end{aligned} \tag{8.5.32}$$

As ω is far away from the Bragg frequency at $\beta_0(\omega_0) = \frac{\omega_0}{c} n_{\text{eff}} = \frac{\pi}{\Lambda}$, we have

$$\beta_\pm \longrightarrow \begin{cases} \beta_a = \beta_0(\omega) \\ \beta_b = -\beta_0(\omega) + \dfrac{2\pi}{\Lambda}. \end{cases} \tag{8.5.33}$$

For a lossless system, $K_{ba} = -K_{ab}^*$. We have $K_{ab}K_{ba} = -|K|^2$. Then

$$\beta_\pm = \beta_B \pm \sqrt{(\Delta\beta)^2 - |K|^2}. \tag{8.5.34}$$

For $|\Delta\beta| = \left|\beta(\omega) - \frac{\pi}{\Lambda}\right| < |K|$, we have

$$\beta_\pm = \frac{\pi}{\Lambda} \pm i\sqrt{|K|^2 - |\Delta\beta|^2} \tag{8.5.35a}$$

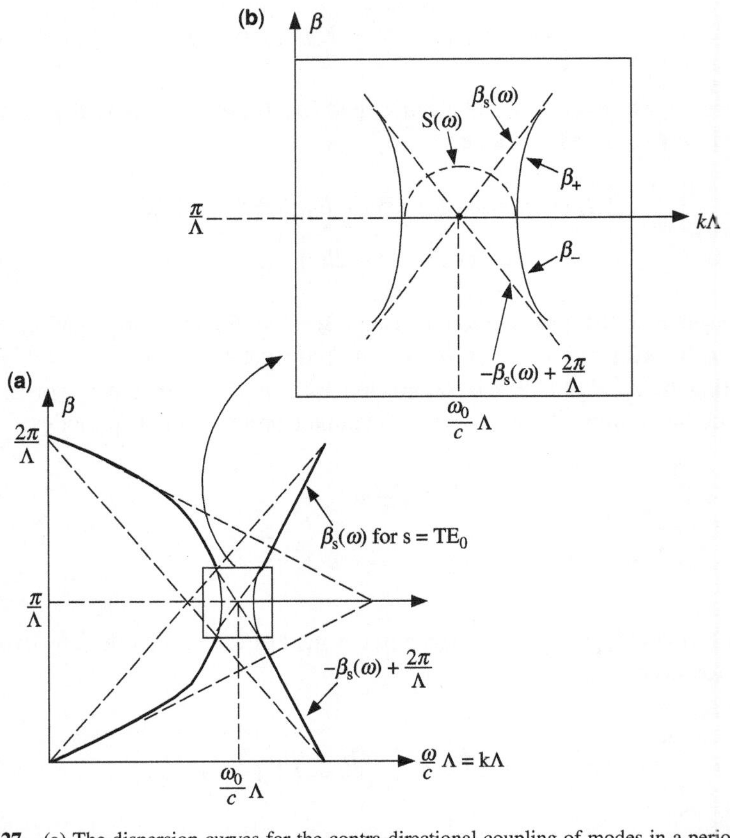

Figure 8.27 (a) The dispersion curves for the contra-directional coupling of modes in a periodic waveguide structure. (b) The enlarged portion of the intersection region.

which gives

$$\mathrm{Re}(\beta_{\pm}) = \frac{\pi}{\Lambda} \tag{8.5.35b}$$

$$\mathrm{Im}(\beta_{\pm}) = \pm\sqrt{|K|^2 - |\Delta\beta|^2}. \tag{8.5.35c}$$

In Fig. 8.27, we plot the dispersion relations for $\ell = 1$. We see that the stop band occurs at the vicinity of the frequency ω_0.

For the two eigenvalues $\pm q$, their corresponding eigenvectors are determined by

$$(1) \qquad q_+ = q, \quad \frac{A^+}{B^+} = \frac{K_{ab}}{-\Delta\beta + q} = \frac{\Delta\beta + q}{K_{ba}} \tag{8.5.36a}$$

and

$$(2) \qquad q_- = -q, \quad \frac{A^-}{B^-} = \frac{K_{ab}}{-\Delta\beta - q} = \frac{\Delta\beta - q}{K_{ba}}. \tag{8.5.36b}$$

The general solution can be written as

$$\begin{bmatrix} A(z) \\ B(z) \end{bmatrix} = c_1 \begin{bmatrix} K_{ab} \\ q - \Delta\beta \end{bmatrix} e^{iqz} + c_2 \begin{bmatrix} K_{ab} \\ -q - \Delta\beta \end{bmatrix} e^{-iqz}. \tag{8.5.37}$$

8.5.3 Reflection and Transmission of a DFB Structure

For an incident wave $A(0)$ from the left side of the distributed feedback structure at $z = 0$, as shown in Fig. 8.28, we may find the general solution using the fact that at $z = L$, $B(L) = 0$ because there is no incident wave from the right-hand side. Therefore,

$$K_{ab}(c_1 + c_2) = A(0) \tag{8.5.38}$$

and

$$(-\Delta\beta + q)\, e^{iqL} c_1 - (\Delta\beta + q)\, e^{-iqL} c_2 = B(L) = 0. \tag{8.5.39}$$

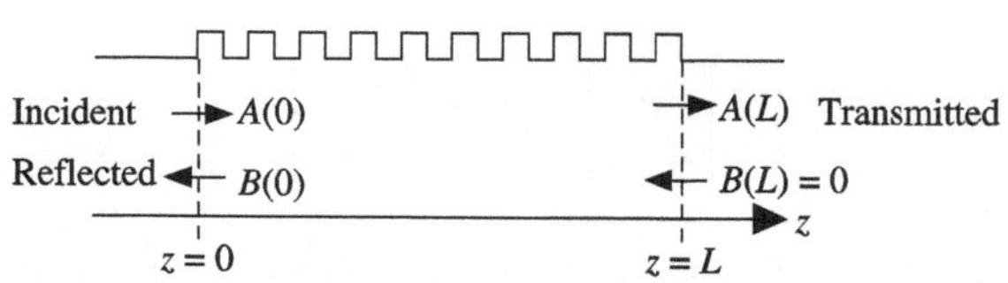

Figure 8.28 A periodic corrugated waveguide structure with an incident wave at $z = 0$. $A(0)$ is the field amplitude of the incident wave, and $B(0)$ is the field amplitude of the reflected wave. $A(L)$ is the field amplitude of the transmitted wave, and $B(L) = 0$ because we assume no wave is incident from the right-hand side.

Solving for c_1 and c_2 in terms of $A(0)$ and substituting into (8.5.37), we obtain the reflection coefficient $B(0)/A(0)$ and the transmission coefficient $A(L)/A(0)$.

Matrix Method We can solve generally the wave propagation problem for the DFB laser structure using a matrix formulation, similar to what we have done for the parallel coupled waveguides. We write the general solution (8.5.37) in a matrix form

$$\begin{bmatrix} A(z) \\ B(z) \end{bmatrix} = \mathbf{V} \begin{bmatrix} e^{iqz} & 0 \\ 0 & e^{-iqz} \end{bmatrix} \begin{bmatrix} c_1 \\ c_2 \end{bmatrix} \tag{8.5.40}$$

where the matrix $\mathbf{V}$ contains the two eigenvectors in its two columns

$$\mathbf{V} = \left[\begin{array}{c|c} K_{ab} & K_{ab} \\ q - \Delta\beta & -q - \Delta\beta \end{array} \right]. \tag{8.5.41}$$

Solving c_1 and c_2 in terms of the inverse $\mathbf{V}^{-1}$, $A(0)$ and $B(0)$, we have

$$\begin{bmatrix} A(z) \\ B(z) \end{bmatrix} = \mathbf{F}(z) \begin{bmatrix} A(0) \\ B(0) \end{bmatrix} \tag{8.5.42}$$

where the "forward" propagation matrix is

$$\begin{aligned} \mathbf{F}(z) &= \mathbf{V} \begin{bmatrix} e^{iqz} & 0 \\ 0 & e^{-iqz} \end{bmatrix} \mathbf{V}^{-1} \\ &= \begin{bmatrix} \cos qz + i\dfrac{\Delta\beta}{q}\sin qz & i\dfrac{K_{ab}}{q}\sin qz \\ i\dfrac{K_{ba}}{q}\sin qz & \cos qz - i\dfrac{\Delta\beta}{q}\sin qz \end{bmatrix} \end{aligned} \tag{8.5.43}$$

which is almost identical to (8.2.32) of the coupled parallel waveguides, except for a phase factor $\exp(i\phi z)$ therein. Given an incident wave $A(0)$ at $z = 0$, and no incident wave from the other end at $z = L$, $B(L) = 0$, we have

$$\begin{bmatrix} A(L) \\ B(L) \end{bmatrix} = \mathbf{F}(L) \begin{bmatrix} A(0) \\ B(0) \end{bmatrix} \tag{8.5.44}$$

which can be solved easily for the reflection coefficient of the field

$$\Gamma(0) = \frac{B(0)}{A(0)} = -\frac{F_{21}(L)}{F_{22}(L)} = \frac{-iK_{ba}\sin qL}{q\cos qL - i\Delta\beta\sin qL} \tag{8.5.45}$$

and the transmission coefficient

$$t(L) = \frac{A(L)}{A(0)} = F_{11} - \frac{F_{12}F_{21}}{F_{22}} = \frac{1}{F_{22}(z = L)} = \frac{q}{q\cos qL - i\Delta\beta\sin qL}. \tag{8.5.46}$$

Within the stop band, $q = iS$, we use $\cos qL = \cosh SL$ and $\sin qL = i\sinh SL$. If we look for $A(z)$ and $B(z)$ at an arbitrary position z, we use (8.5.42) and (8.5.45),

$$\begin{aligned} A(z) &= F_{11}(z)\,A(0) + F_{12}(z)\,B(0) \\ &= \left[F_{11}(z) - F_{12}(z)\frac{F_{21}(z = L)}{F_{22}(z = L)}\right]A(0) \\ &= \frac{S\cosh S(z - L) + i\Delta\beta\sinh S(z - L)}{S\cosh SL - i\Delta\beta\sinh SL}A(0) \end{aligned} \tag{8.5.47}$$

$$B(z) = \frac{iK_{ba}\sinh S(z - L)}{S\cosh SL - i\Delta\beta\sinh SL}A(0). \tag{8.5.48}$$

The reflection coefficient at $z = 0$, using $K_{ab} = K$, $K_{ba} = -K^*$,

$$\Gamma(0) = \frac{B(0)}{A(0)} = \frac{iK^*\sinh SL}{S\cosh SL - i\Delta\beta\sinh SL}. \tag{8.5.49}$$

The transmission coefficient is

$$t = \frac{A(L)}{A(0)} = \frac{iS}{\Delta\beta\sinh SL + iS\cosh SL}. \tag{8.5.50}$$

At $\Delta\beta = 0$, $S = |K|$

$$|\Gamma(0)| = \tanh|KL|, \tag{8.5.51}$$

which approaches 1 if $|KL|$ is large enough. Also, when $|\Delta\beta| > |K|$, S becomes purely imaginary and $|\Gamma(0)|^2$ vanishes when $\sin^2\sqrt{(\Delta\beta L)^2 - |KL|^2} = 0$, or $|\Delta\beta/K| = \sqrt{1 + (n\pi/|K|L)^2}$. A plot of the reflectivity $|\Gamma(0)|^2$ versus $\Delta\beta/|K|$ for three different values of $|K|L$ (i.e., three lengths of DFB reflectors) is shown in Fig. 8.29. We see that if $|K|L$ increases, the reflectivity for small $\Delta\beta\ell$ increases. When $|S| \to 0$, $|\Delta\beta| \to |K|$, and $|\Gamma(0)|^2 \to |KL|^2/(|KL|^2 + 1)$.

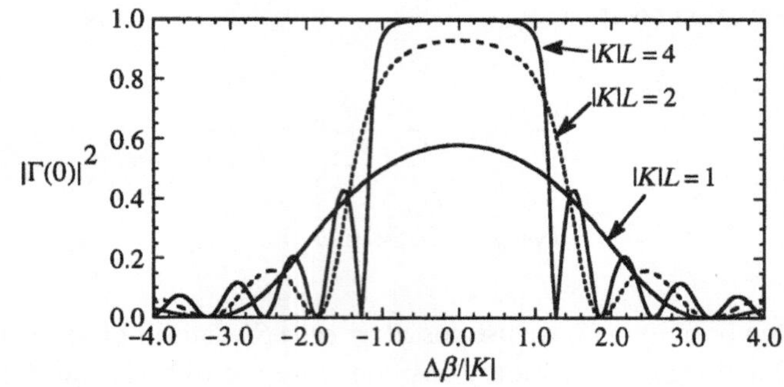

Figure 8.29 A plot of the reflectivity $|\Gamma(0)|^2$ versus the ratio $\Delta\beta/|K|$ for three different values of $|K|L$. For a large value of $|K|L$, the bandwidth for $\Delta\beta$ is about $2|K|$.

APPENDIX 8A: COUPLING COEFFICIENTS FOR PARALLEL WAVEGUIDES

The coupling coefficients between two parallel waveguides p and q can be written as [27–29]

$$K_{pq} = \frac{\omega}{4}\iint \Delta\varepsilon^{(q)}(x, y)[\mathbf{E}_t^{(p)} \cdot \mathbf{E}_t^{(q)} - \mathbf{E}_z^{(p)} \cdot \mathbf{E}_z^{(q)}]\, dxdy \tag{8A.1}$$

$$\Delta\varepsilon^{(q)}(x, y) = \varepsilon(x, y) - \varepsilon^{(q)}(x, y) \tag{8A.2}$$

where $\varepsilon(x, y)$ is the permittivity distribution for the two waveguides, $\varepsilon^{(a)}(x, y)$ means only waveguide a exists, and $\varepsilon^{(b)}(x, y)$ means only waveguide b exists, Fig. 8.30.

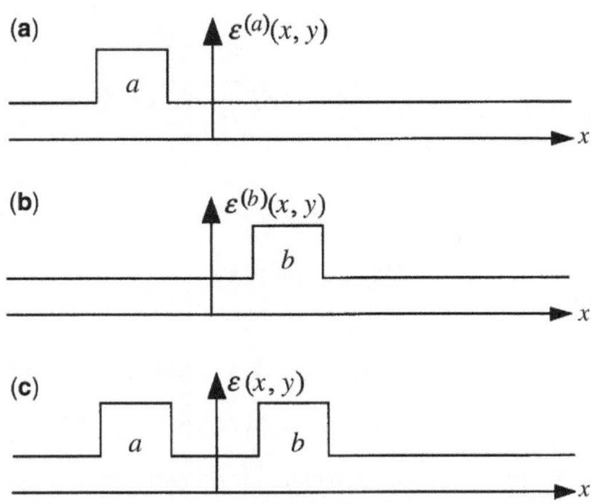

Figure 8.30 Three different permittivity functions of interest for applications in deriving the improved coupled-mode theory using the reciprocity theorem. (a) Only waveguide a exists in the whole space. (b) Only waveguide b exists in the whole space. (c) Both waveguides a and b exist.

The difference $\Delta\varepsilon^{(q)}(x, y)$ between the two permittivity functions indicate that $\Delta\varepsilon^{(q)}(x, y)$ is zero in most part of space, therefore, the integral in (8A.1) is over a finite region. In addition, for TE modes in a slab waveguide, E_z is zero. Equation (8A.1) is the product of the TE_0 mode $E_y^{(a)}(x)$ of guide a and $E_y^{(b)}(x)$ of guide b integrated over the layer thickness of guide a or guide b for K_{ab} and K_{ba}, respectively.

APPENDIX 8B: IMPROVED COUPLED-MODE THEORY

A detailed formulation of the improved coupled-mode theory is shown in Refs. 18–32. The major results are that we have the same form in the coupled-mode equations

$$\frac{d}{dz}a(z) = i\gamma_a\, a(z) + ik_{ab}\, b(z) \tag{8B.1a}$$

$$\frac{d}{dz}b(z) = ik_{ba}\, a(z) + i\gamma_b\, b(z) \tag{8B.1b}$$

where the coefficients γ_a, γ_b, k_{ab}, and k_{ba} are all defined in [27, 28]. The transverse fields have been assumed to be

$$\mathbf{E}_t = a(z)\,\mathbf{E}_t^{(a)}(x, y) + b(z)\,\mathbf{E}_t^{(b)}(x, y) \tag{8B.2a}$$

$$\mathbf{H}_t = a(z)\,\mathbf{H}_t^{(a)}(x, y) + b(z)\,\mathbf{H}_t^{(b)}(x, y) \tag{8B.2b}$$

and as all the transverse field components are all real for a lossless system, we find γ_a, γ_b, k_{ab}, and k_{ba} to be real. Power conservation leads to

$$\begin{aligned}
0 &= \frac{d}{dz}P(z) = \frac{d}{dz}\frac{1}{2}\text{Re}\int_{-\infty}^{\infty}\int_{-\infty}^{\infty}\mathbf{E}_t \times \mathbf{H}_t^* \cdot \hat{z}\,dxdy \\
&= \frac{d}{dz}\text{Re}[aa^* + ab^*C_{ba} + ba^*C_{ab} + bb^*] \\
&= \frac{d}{dz}[aa^* + (ab^* + ba^*)\,C + bb^*] \\
&= ab^*i(\gamma_a C - \gamma_b C + k_{ba} - k_{ab}) + ba^*i(\gamma_b C - \gamma_a C + k_{ab} - k_{ba}).
\end{aligned} \tag{8B.3}$$

Here the overlap integrals C_{ab} and C_{ba} have been defined in (8.2.10). $C = (C_{ab} + C_{ba})/2$. Because both a and b are arbitrary, we conclude that the coefficients in front of ab^* and ba^* must vanish. Therefore,

$$k_{ab} - k_{ba} = (\gamma_a - \gamma_b)C. \tag{8B.4}$$

What is important is that when the overlap coefficient C is not small, the total guided power should be given by

$$P(z) = |a|^2 + |b|^2 + (ab^* + ba^*)C \tag{8B.5}$$

where the last two terms due to the cross powers between $\mathbf{E}^{(a)}$ and $\mathbf{H}^{(b)}$ or $\mathbf{E}^{(b)}$ and $\mathbf{H}^{(a)}$ are not negligible anymore. The two coupling coefficients, κ_{ab} and κ_{ba}, are not equal to each other when γ_a is not equal to γ_b or C is not negligible.

PROBLEMS

8.1 A TE light ($\lambda = 0.8\ \mu$m) is normally incident on a prism from air as shown in Fig. 8.31. The waveguide width d is chosen such that the light can be resonantly coupled into the waveguide along the z direction (i.e., the phase-matching condition is satisfied).

(a) Find the propagation constant k_z of the incident light.

(b) Find the Goos–Hänchen phase shifts for the guided mode for both the air/waveguide and substrate/waveguide interfaces.

(c) From the ray optics picture, determine the width d from the propagation constant k_z obtained in (**a**).

8.2 Consider an InP/$In_{0.53}Ga_{0.47}As$/InP dielectric waveguide with a dimension $d = 0.3\ \mu$m. The wavelength in free space λ_0 is 1.65 μm. The refractive indices are n_1 ($In_{0.53}Ga_{0.47}As$) = 3.56 and $n = 3.16$. The propagation constant for the TE_0 mode is found to be $\beta = 12.7\ (\mu\text{m})^{-1}$.

(a) If a grating coupler is designed using the same materials, Fig. 8.32, so that the first-order space harmonic mode ($m = 1$) is coupled to the TE_0 mode at $\theta_i = 45^\circ$, find the grating period Λ.

(b) Repeat part (**a**) for $m = 2$. Draw a phase-matching diagram to show all of the propagating (nonevanescent) space harmonics (also called Floquet modes) in the InP region for this case and show their reflection angles measured from the x axis.

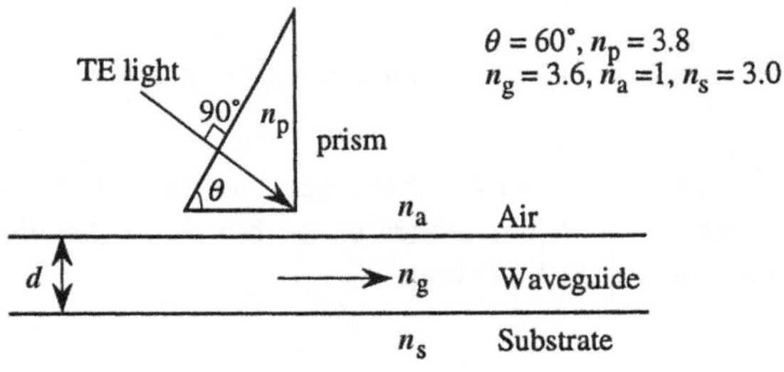

Figure 8.31 Diagram for Problem 8.1.

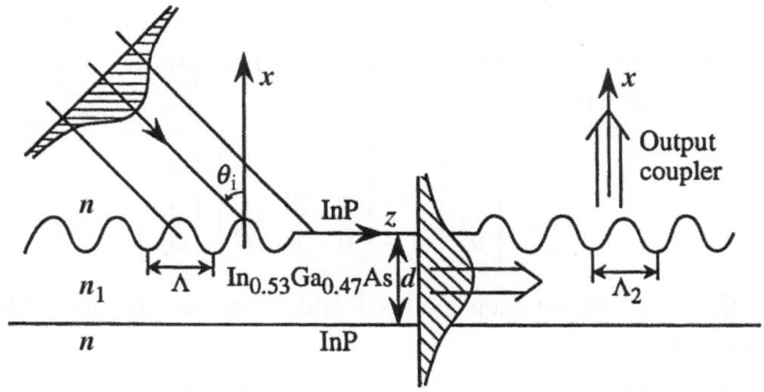

Figure 8.32 Diagram for Problem 8.2.

(c) Consider a grating output coupler shown in Fig. 8.32. If the output coupler is designed so that there is one and only one Floquet mode propagating in the x direction (this configuration also has applications for surface-emitting diode lasers), what should the period Λ_2 be?

8.3 (a) Describe how a $\Delta\beta$ coupler works. What are its advantages compared with a single-section directional coupler.

(b) Show the switching diagram ($K\ell$ vs. $\Delta\ell$) with detailed illustration.

(c) Consider an input with $a(0) = 1$, $b(0) = 0$, $\Delta\ell = \pi/2$, and $K\ell = \pi$, where ℓ is the total length of the $\Delta\beta$ coupler. Find the output powers, $|a(\ell)|^2$ and $|b(\ell)|^2$.

8.4 Consider a three-parallel-waveguide coupler shown in Fig. 8.33. Using the coupled-mode theory, the electric field of the three-guide coupler can be written as

$$\mathbf{E}(x, z) = \hat{y}[a_1(z)E_y^{(1)}(x) + a_2(z)E_y^{(2)}(x) + a_3(z)E_y^{(3)}(x)]$$

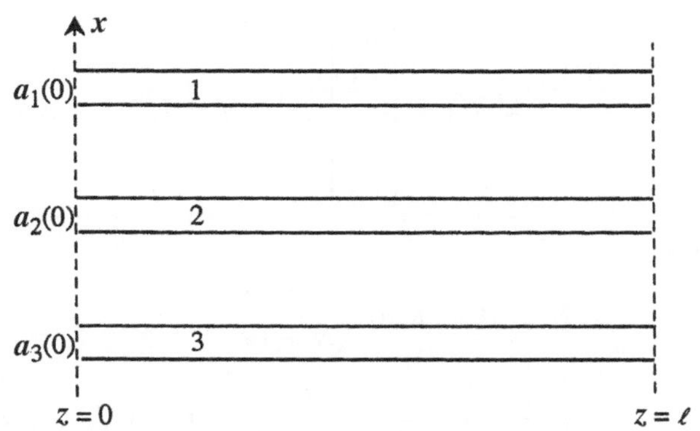

Figure 8.33 Diagram for Problem 8.4.

where $E_y^{(j)}(x)$ describes the x dependence of the TE_0 mode of waveguide j ($j = 1, 2, 3$) in the absence of the other two guides. The coupled-mode equation can be written as

$$\frac{d}{dz}\begin{bmatrix} a_1 \\ a_2 \\ a_3 \end{bmatrix} = i\begin{bmatrix} \beta & \kappa & 0 \\ \kappa & \beta & \kappa \\ 0 & \kappa & \beta \end{bmatrix}\begin{bmatrix} a_1 \\ a_2 \\ a_3 \end{bmatrix}$$

where the coupling between guides 1 and 3 has been neglected because they are far apart.

(a) Find the three eigenvalues and the corresponding eigenvectors of the three-waveguide coupler.

(b) Sketch approximately the mode profiles of the three eigenmodes as a function of x.

(c) Given the initial values $a_1(0)$, $a_2(0)$, and $a_3(0)$, find $a_1(z)$, $a_2(z)$, and $a_3(z)$ in terms of their initial values. Put your results in a matrix form.

8.5 The coupled-mode equations for N identical equally spaced waveguides, assuming only adjacent waveguide coupling, can be written as

$$\frac{d}{dz}\mathbf{A}(z) = i\mathbf{M}\mathbf{A}(z)$$

where

$$\mathbf{A}(z) = \begin{bmatrix} a_1 \\ a_2 \\ \vdots \\ a_N \end{bmatrix}, \qquad \mathbf{M} = \begin{bmatrix} \beta & K & & \mathbf{0} & \\ K & \beta & K & & \\ & \ddots & \ddots & \ddots & \\ & & K & \beta & K \\ & \mathbf{0} & & K & \beta \end{bmatrix}.$$

(a) Show that the general solutions for the normalized eigenvectors and their eigenvalues for an arbitrary N are

$$\mathbf{A}^{(\ell)}(z) = \begin{bmatrix} a_1^\ell \\ a_2^\ell \\ \vdots \\ a_N^\ell \end{bmatrix} e^{i\beta_\ell z} \quad \ell = 1, 2, \ldots, N$$

where

$$\beta_\ell = \beta + 2K\cos\left(\frac{\ell\pi}{N+1}\right)$$

$$a_k^\ell = \sqrt{\frac{2}{N+1}}\sin\left(\frac{k\ell\pi}{N+1}\right) \qquad k = 1, 2, \ldots, N.$$

The order of β_ℓ is chosen such that $\beta_1 > \beta_2 > \beta_3 \cdots > \beta_N$.

Hint:

$$\det\begin{bmatrix} 2\cos\theta & 1 & & \mathbf{0} & \\ 1 & 2\cos\theta & 1 & & \\ & \ddots & \ddots & \ddots & \\ & & 1 & 2\cos\theta & 1 \\ & \mathbf{0} & & 1 & 2\cos\theta \end{bmatrix} = \frac{\sin[(N+1)\theta]}{\sin\theta}$$

where N is the dimension of the matrix.

(b) Find the eigenvalues and their corresponding eigenvectors for $N = 4$, and plot the transverse field profile $E(x) = \sum_{i=1}^{4} a_i E_i(x)$ for all four eigenvalues.

8.6 **(a)** Sketch qualitatively the dispersion relation β_a of the TE_0 mode versus ω for a single slab waveguide in Fig. 8.34a with a fixed refractive index n_a at two different thicknesses ($d_a < d_b$). Explain your sketch of β versus ω curves using the graphical solution for $\alpha(d/2)$versus $k_x(d/2)$.

(b) Consider the two coupled slab waveguides, Fig. 8.34b. With $n_a > n_b$, derive a condition between d_a and d_b such that the dispersion curves $\beta_a(\omega)$ and $\beta_b(\omega)$ intercept at a frequency ω_0.

(c) Describe how the above result can be used in the design of a direction coupler as an optical wavelength filter.

8.7 Plot the transmission spectrum (8.4.6) for a ring resonator with (a) $a = 0.6$, $|t| = 0.6$, and (b) $a = 0.9$, $|t| = 0.9$.

8.8 Prove the matrix property for $\mathbf{M}^N$ in (8.4.21) by induction.

8.9 Plot the dispersion relation (8.4.35) between $\beta\Lambda$ and θ. Check the stop bands and pass bands.

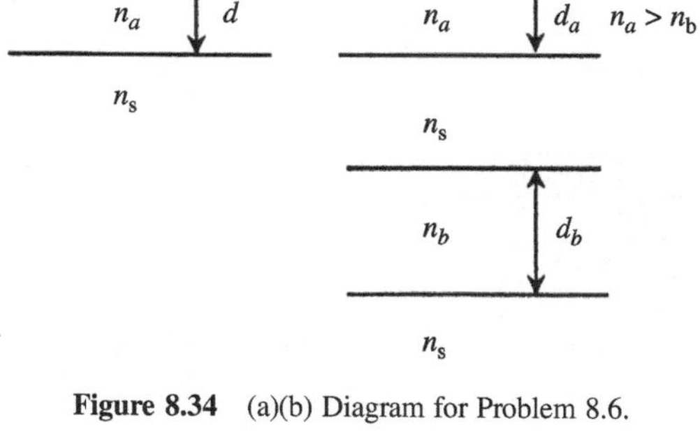

Figure 8.34 (a)(b) Diagram for Problem 8.6.

8.10 If we have index grating $\Delta n(z) = \Delta n \cos\left(\frac{2\pi}{\Lambda}z\right)$, where $\Delta n = 0.2$, calculate the coupling coefficients K_{ab} and K_{ba} for $\lambda_0 = 1.55\ \mu\text{m}$. Assume the optical confinement factor of the guided TE_0 mode is 0.5. Calculate $|K_{ab}L|$ for a cavity length $L = 500\ \mu\text{m}$ of a DFB laser.

8.11 Derive the propagation matrix **F** as shown in (8.5.43). When the wavelength is within the stop band, find the expression for **F** if $q = iS = i\sqrt{|K|^2 - (\Delta\beta)^2}$.

8.12 Consider a distributed feedback structure as shown in Fig. 8.35a, where

$$\Delta\varepsilon(x, z) = \begin{cases} \varepsilon - \varepsilon_1 & \text{if } -h < x < 0 \quad \text{and} \quad 0 < z < \dfrac{\Lambda}{2} \\ \varepsilon_1 - \varepsilon & \text{if } 0 < x < h \quad \text{and} \quad \dfrac{\Lambda}{2} < z < \Lambda \end{cases}$$

and $\Delta\varepsilon(x, z)$ is periodic in z, Fig. 8.35b.

(a) Find the Fourier series expansion of $\Delta\varepsilon(x, z)$, that is, $\Delta\varepsilon_p(x)$, where

$$\Delta\varepsilon(x, z) = \varepsilon_0 \sum_{p=-\infty}^{\infty} \Delta\varepsilon_p(x) \exp\left(i\frac{2\pi}{\Lambda}pz\right)$$

(b) Evaluate the coupling coefficient K_{ab} and K_{ba} for the TE_0 mode. Assume that the first orders $p = \pm 1$ provide the coupling.

(c) Let $K \equiv K_{ab} = -K_{ba}^*$. Consider an incident wave from $z = 0$ with an amplitude $A(0)$. Derive the general expressions for $B(0)$ and $A(\ell)$.

(d) Plot the reflectivity $|B(0)/A(0)|^2$ versus $|\Delta\beta/K|$ for $K\ell = 4$.

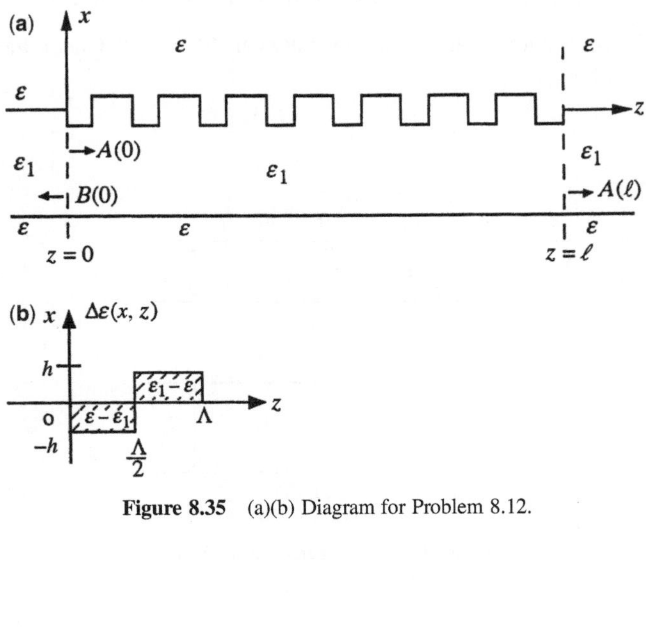

Figure 8.35 (a)(b) Diagram for Problem 8.12.

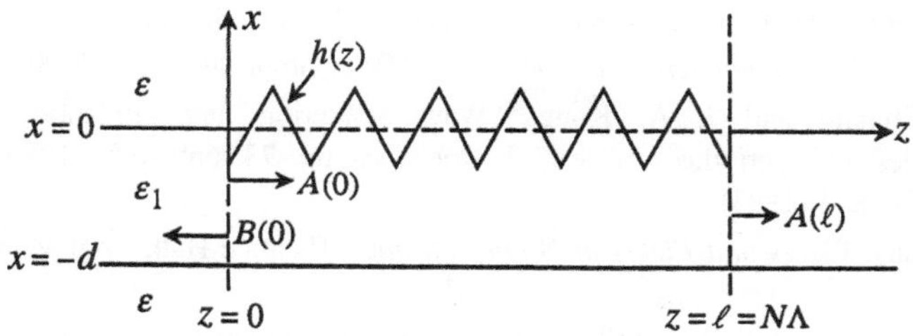

Figure 8.36 Diagram for Problem 8.13.

8.13 Consider a distributed feedback structure as shown in Fig. 8.36 where the height function is

$$h(z) = \begin{cases} -h_0 + h_0(z - n\Lambda)\dfrac{4}{\Lambda} & n\Lambda \le z \le n\Lambda + \dfrac{\Lambda}{2} \\ h_0 - h_0\left(z - n\Lambda - \dfrac{\Lambda}{2}\right)\dfrac{4}{\Lambda} & n\Lambda + \dfrac{\Lambda}{2} \le z \le (n+1)\Lambda \\ 0 & n = 0, 1, 2, \ldots, N-1 \quad z > \ell,\ z < 0. \end{cases}$$

Let the unperturbed $\varepsilon^{(0)}(x)$ be $\varepsilon^{(0)} = \begin{cases} \varepsilon_1 & -d < x < 0 \\ \varepsilon & \text{otherwise.} \end{cases}$

Repeat Problem 8.12(a)–(d) for this sawtooth grating structure.

8.14 **(a)** Write down the expression for the coupling coefficient K for two identical parallel waveguides. Consider only the TE_0 mode. Discuss qualitatively all possible ways and trade-offs to increase the coupling coefficient.

(b) Repeat **(a)** for the coupling coefficient K for a distributed feedback structure.

(c) Compare the coupling coefficients in **(a)** and **(b)**. What are their similarities and differences?

8.15 **(a)** Calculate the coupling coefficient (8.5.25) for an index phase grating with $\Delta n = 0.1$, zero phase ($\varphi = 0$), and the confinement factor is 1 for a semiconductor background index $n_0 = 3.16$ at 1.55-μm wavelength.

(b) Plot the reflectivity of the power $|\Gamma(0)|^2$ in (8.5.45) as a function of detuning $\Delta\beta$ for a cavity length $L = 10$ μm. You should allow the argument qL to be complex to cover the detuning range both inside and outside the stop band.

REFERENCES

1. R. G. Hunsperger, *Integrated Optics: Theory and Technology*, Springer-Verlag, Berlin, 1984.
2. T. Tamir, *Integrated Optics*, Springer-Verlag, Berlin, 1979.

3. T. Tamir, *Guided Wave Optoelectronics*, Springer-Verlag, Berlin, 1990.
4. J. A. Kong, *Electromagnetic Wave Theory*, EMW, Cambridge, MA, 2000.
5. S. L. Chuang and J. A. Kong, "Wave scattering and guidance by dielectric waveguides with periodic surfaces," *J. Opt. Soc. Am.* **73**, 669–679 (1983); *Addendum* **73**, 1823–1824 (1983).
6. H. A. Haus, *Waves and Fields in Optoelectronics*, Prentice Hall, Englewood Cliffs, NJ, 1984.
7. S. L. Chuang and B. Do, "Electron states in two coupled quantum wells—a strong coupling of modes approach," *J. Appl. Phys.* **62**, 1290–1297 (1987).
8. A. Yariv and P. Yeh, *Photonics—Optical Electronics in Modern Communications*, 6th Edition, Oxford University Press, New York, USA, 2007.
9. A. Yariv, "Coupled-mode theory for guided-wave optics," *IEEE J. Quantum Electron.* **QE-9**, 919–933 (1973).
10. H. F. Taylor and A. Yariv, "Guided wave optics," *Proc. IEEE* **62**, 1044–1060 (1974).
11. H. Nishihara, M. Haruna, and T. Suhara, *Optical Integrated Circuits*, McGraw-Hill, New York, 1989.
12. E. Marom, O. G. Ramer, and S. Ruschin, "Relation between normal-mode and coupled-mode analyses of parallel waveguides," *IEEE J. Quantum Electron.* **QE-20**, 1311–1319 (1984).
13. J. E. Watson, M. A. Milbrodt, and T. C. Rice, "A polarization-independent 1×16 guided-wave optical switch integrated on lithium niobate," *J. Lightwave Technol.* **LT-4**, 1717–1721 (1986).
14. L. A. Molter-Orr and H. A. Haus, "$N \times N$ coupled waveguide switch," *Opt. Lett.* **9**, 466–467 (1984).
15. N. Kishi, E. Yamashita, and K. Atsuki, "Modal and coupling field analysis of optical fibers with circularly distributed multiple cores and a central core," *J. Lightwave Technol.* **LT-4**, 991–996 (1986).
16. D. B. Mortimore and J. W. Arkwright, "Theory and fabrication of wavelength flattened $1 \times N$ single-mode couplers," *Appl. Opt.* **29**, 1814–1818 (1990).
17. Y. Suematsu and K. Kishino, "Coupling coefficient in strongly coupled dielectric waveguides," *Radio Science* **12**, 587–592 (1977).
18. A. Hardy and W. Streifer, "Coupled mode theory of parallel waveguides," *J. Lightwave Technol.* **LT-3**, 1135–1146 (1985).
19. A. Hardy and W. Streifer, "Coupled mode solutions of multiwaveguide systems," *IEEE J. Quantum Electron.* **QE-22**, 528–534 (1986).
20. J. R. Qian and W. P. Huang, "Coupled-mode theory for LP modes," *J. Lightwave Technol.* **LT-4**, 619–625 (1986).
21. J. R. Qian and W. P. Huang, "LP modes and ideal modes on optical fibers," *J. Lightwave Technol.* **LT-4**, 626–630 (1986).
22. D. Marcuse, "Directional couplers made of nonidentical asymmetric slabs. Part I: synchronous couplers," *J. Lightwave Technol.* **LT-5**, 113–118 (1987).
23. E. A. J. Marcatili, "Improved coupled-mode equations for dielectric guides," *IEEE J. Quantum Electron.* **QE-22**, 988–993 (1986).
24. E. A. J. Marcatili, L. L. Buhl, and R. C. Alferness, "Experimental verification of the improved coupled-mode equations," *Appl. Phys. Lett.* **49**, 1692–1693 (1986).

25. W. Streifer, M. Osiński, and A. Hardy, "Reformation of the coupled-mode theory of multiwaveguide systems," *J. Lightwave Technol.* **LT-5**, 1–4 (1987).

26. H. A. Haus, W. P. Huang, S. Kawakami, and N. A. Whitaker, "Coupled-mode theory of optical waveguides," *J. Lightwave Technol.* **LT-5**, 16–23 (1987).

27. S. L. Chuang, "A coupled-mode formulation by reciprocity and a variational principle," *J. Lightwave Technol.* **LT-5**, 5–15 (1987).

28. S. L. Chuang, "A coupled-mode theory for multiwaveguide systems satisfying the reciprocity theorem and power conservation," *J. Lightwave Technol.* **LT-5**, 174–183 (1987).

29. S. L. Chuang, "Application of the strongly coupled-mode theory to integrated optical devices," *IEEE J. Quantum Electron.* **QE-23**, 499–509 (1987).

30. A. Ankiewicz, A. W. Snyder, and X. H. Zheng, "Coupling between parallel optical fiber cores–critical examination," *J. Lightwave Technol.* **LT-4**, 1317–1323 (1986).

31. A. W. Snyder, A. Ankiewicz, and A. Altintas, "Fundamental error of recent coupled mode formulations," *Electron. Lett.* **23**, 1097–1098 (1987).

32. H. C. Chang, "Coupled-mode equations for dielectric waveguides based on projection and partition modal amplitudes," *IEEE J. Quantum Electron.* **QE-23**, 1929–1937 (1987).

33. J. P. Donnelly, N. L. DeMeo, Jr., and G. A. Ferrante, "Three-guide optical couplers in GaAs," *J. Lightwave Technol.* **LT-1**, 417–424 (1983).

34. K. L. Chen and S. Wang, "Cross-talk problems in optical directional couplers," *Appl. Phys. Lett.* **44**, 166–168 (1984).

35. J. P. Donnelly, N. L. DeMeo, Jr., G. A. Ferrante, K. B. Nichols, and F. J. O'Donnell, "Optical guided-wave gallium arsenide monolithic interferometer," *Appl. Phys. Lett.* **45**, 360–362 (1984).

36. K. L. Chen and S. Wang, "The cross talk in three-waveguide optical directional couplers," *IEEE J. Quantum Electron.* **QE-22**, 1039–1041 (1986).

37. L. McCaughan and S. K. Korotky, "Three-electrode Ti:$LiNbO_3$ Optical switch," *J. Lightwave Technol.* **LT-4**, 1324–1327 (1986).

38. J. P. Donnelly, "Limitations on power-transfer efficiency in three-guide optical couplers," *IEEE J. Quantum Electron.* **QE-22**, 610–616 (1986).

39. D. Marcuse, "Coupled-mode theory for anisotropic optical waveguides," *Bell Syst. Tech. J.* **54**, 985–995 (1975).

40. A. Hardy, W. Streifer, and M. Osiński, "Coupled-mode equations for multimode waveguide systems in isotropic or anisotropic media," *Opt. Lett.* **11**, 742–744 (1986).

41. L. Tsang and S. L. Chuang, "Coupled GaAs multiple-quantum-well channel waveguides including quadratic electrooptic effect," *J. Lightwave Technol.* and *IEEE J. Quantum Electron.* **6**, 832–836 (1988).

42. R. C. Alferness and R. V. Schmidt, "Tunable optical waveguide directional coupler filter," *Appl. Phys. Lett.* **33**, 161–163 (1978).

43. R. C. Alferness and J. J. Veselka, "Tunable Ti:$LiNbO_3$ waveguide filter for long-wavelength ($\lambda = 1.3$–$1.6\ \mu m$) multiplexing/demultiplexing," *Tech. Dig. Conf. Lasers Electro-optics* (Anaheim, CA), 230–231 (1984).

44. B. Broberg, B. S. Lindgren, M. G. Öberg, and H. Jiang, "A novel integrated opticswavelength filter in InGaAsP-InP," *J. Lightwave Technol.* **LT-4**, 196–203 (1986).

45. B. Broberg and S. Lindgren, "Refractive index of $In_{1-x}Ga_xAs_yP_{1-y}$ layers and InP in the transparent wavelength region," *J. Appl. Phys.* **55**, 3376–3381 (1984).

46. H. Kogelnik and R. V. Schmidt, "Switched directional couplers with alternating $\Delta\beta$," *IEEE J. Quantum Electron.* **QE-12**, 396–401 (1976).

47. C. J. Setterlind and L. Thylen, "Directional coupler switches with optical gain," *IEEE J. Quantum Electron.* **QE-22**, 595–602 (1986).

48. G. H. B. Thompson, "Analysis of optical directional couplers that include gain or loss and their application to semiconductor slab dielectric guides," *J. Lightwave Technol.* **LT-4**, 1678–1693 (1986).

49. P. Urquhart, "Transversely coupled fiber Fabry-Perot resonator: theory," *Appl. Opt.* **26**, 456–463 (1987).

50. M. Brierley and P. Urquhart, "Transversely coupled fiber Fabry-Perot resonators: performance characteristics," *Appl. Opt.* **26**, 4841–4845 (1987).

51. M. J. Adams, "Theory of twin-guide Fabry-Perot laser amplifiers," *IEE Proc.* **136**, 287–292 (1989).

52. M. J. Adams, D. A. H. Mace, J. Singh, and M. A. Fisher, "Optical switching in the twin-guide Fabry-Perot laser amplifier," *IEEE J. Quantum Electron.* **26**, 1764–1771 (1990).

53. W. J. Fader and G. E. Palma, "Normal modes of N coupled lasers," *Opt. Lett.* **10**, 381–383 (1985).

54. E. Kapon, J. Katz, and A. Yariv, "Supermode analysis of phase-locked arrays of semiconductor lasers," *Opt. Lett.* **10**, 125–127 (1984); "Erratum," *Opt. Lett.* **9**, 318 (1984).

55. A. Hardy and W. Streifer, "Analysis of phased-array diode lasers," *Opt. Lett.* **10**, 335–337 (1985).

56. B. E. Little, S. T. Chu, H. A. Haus, J. S. Foresi, and J.-P. Laine, "Microring resonator channel dropping filter," *J. Lightwave Technol.* **15**, 998–1005 (1997).

57. A. Yariv, Y. Xu, R. K. Lee, and A. Scherer, "Coupled-resonator optical waveguide: a proposal and analysis," *Opt. Lett.* **24**, 711–713 (1999).

58. M. Cai and K. Vahala, "Highly efficient optical power transfer to whispering- gallery-modes by use of a symmetrical dual-coupling configuration," *Opt. Lett.* **25**, 260–262 (2000).

59. A. Yariv, "Universal relations for coupling of optical power between microresonators and dielectric waveguides," *Electron. Lett.* **361**, 321–322 (2000).

60. M. Cai, O. Painter, K. J. Vahala, and P. C. Sercel, "Fiber-coupled microsphere laser," *Opt. Lett.* **25**, 1430–1432 (2000).

61. Y. Xu, Y. Li, R. K. Lee, and A. Yariv, "Scattering-theory analysis of waveguide-resonator coupling," *Phys. Rev.* **E 62**, 7389–7404 (2000).

62. J. M. Choi, R. K. Lee, and A. Yariv, "Control of critical coupling in a ring resonator-fiber configuration: application to wavelength-selective switching, modulation, amplification, and oscillation," *Opt. Lett.* **26**, 12361–238 (2001).

63. A. Yariv, "Critical coupling and its control in optical waveguide-ring resonator systems," *IEEE Photon. Technol. Lett.* **4**, 483–485 (2002).

64. K. J. Vahala, "Optical microcavities," *Nature* **424**, 839–846 (2003).

65. C. J. Kaalund, "Critically coupled ring resonator for add-drop filtering," *Opt. Comm.* **237**, 357–362 (2004).

66. V. M. Menon, W. Tong, and S. R. Forrest, "Control of quality factor and critical coupling in microring resonators through integration of a semiconductor optical amplifier," *IEEE Photon Technol Lett.* **16**, 1343–1345 (2004).

67. W. M. J. Green, R. K. Lee, G. A. DeRose, A. Scherer, and A. Yariv, "Hybrid InGaAsP-InP Mach-Zehnder racetrack resonator for thermooptic switching and coupling control," *Opt. Express* **13**, 1651–1659 (2005).

68. Y. M. Landobasa, S. Darmawan, and M. K. Chin, "Matrix analysis of 2-D micro-resonator lattice optical filters," *IEEE J. Quantum Electron.* **41**, 1410–1418 (2005).

69. S. Darmawan and M. K. Chin, "Critical coupling, oscillation, reflection, and transmission in optical waveguide ring resonator systems," *J. Opt. Soc. Am. B.* **23**, 834–841 (2006).

70. M. C. Lee and M. C. Wu, "Tunable coupling regimes of silicon microdisk resonators using MEMS actuators," *Optics Express* **14**, 4703–4712 (2006).

71. S. I. Bozhevolnyi, V. S. Volkov, E. Devaux, J.-Y. Laluet, and T. W. Ebbesen, "Channel plasmon sub-wavelength waveguide components including interferometers and ring resonators," *Nature* **440**, 508–511 (2006).

72. M. Hossein-Zadeh and K. J. Vahala, "Free ultra-high-Q microtoroid: a tool for designing photonic devices," *Opt. Express* **15**, 166–175 (2007).

73. F. Xia, L. Sekaric, and Y. A. Vlasov, "Ultracompact optical buffers on a silicon chip," *Nature Photonics* **1**, 65–71 (2007).

74. E. Parra and J. Lowell, "Toward applications of slow light technology," *Opt. Photon. News* **18**, 41–45 (2007).

75. D. X. Xu, A. Densmore, P. Waldron, J. Lapointe, E. Post, A. Delâge, S. Janz, P. Cheben, J. H. Schmid, and B. Lamontagne, "High bandwidth SOI photonic wire ring resonators using MMI couplers," *Opt. Express* **15**, 3149–3155 (2007).

76. Q. Xu, P. Dong, and M. Lipson, "Breaking the delay-bandwidth limit in a photonic structure," *Nature* **3**, 406–410 (2007).

77. A. A. Tovar and L. W. Casperson, "Generalized Sylvester theorems for periodic applications in matrix optics," *J. Opt. Soc. Am. A.* **12**, 578–590 (1995).

78. H. Kogelnik, "Coupled wave theory for thick hologram gratings," *Bell Syst. Tech. J.* **48**, 2909–2947 (1969).

79. W. S. C. Chang, "Periodic structures and their application in integrated optics," *IEEE Trans. Microwave Theory Tech.* **MTT-21**, 775–785 (1973).

80. W. Streifer, D. R. Scifres, and R. D. Burnham, "Coupling coefficients for distributed feedback single- and double-heterostructure diode lasers," *IEEE J. Quantum Electron.* **QE-11**, 867–873 (1975).

81. W. Streifer, D. R. Scifres, and R. D. Burnham, "TM-mode coupling coefficients in guided-wave distributed feedback lasers," *IEEE J. Quantum Electron.* **QE-12**, 74–78 (1976).

82. P. H. H. Hsieh and C. G. Fonstad, "Coupling coefficient calculations for lead-tin telluride distributed feedback lasers," *IEEE J. Quantum Electron.* **QE-13**, 17–23 (1977).

83. Y. Yamamoto, T. Kamiya, and H. Yanai, "Improved coupled mode analysis of corrugated waveguides and lasers," *IEEE J. Quantum Electron.* **QE-14**, 245–258 (1978).

84. Y. Yamamoto, T. Kamiya, and H. 114Yanai, "Improved coupled mode analysis of corrugated waveguides and lasers-II: TM mode," *IEEE J. Quantum Electron.* **QE-14**, 620–624 (1978).

85. S. H. Kim and C. G. Fonstad, "Tunable narrow-band thin-film waveguide grating filters," *IEEE J. Quantum Electron.* **QE-15**, 1405–1408 (1979).

86. E. Kapon, A. Hardy, and A. Katzir, "TE and TM coupling coefficients in distributed feedback double-heterostructure PbSnTe lasers," *IEEE J. Quantum Electron.* **QE-17**, 391–397 (1981).

Part III
Generation of Light

9

Optical Processes in Semiconductors

The quantum theory of light has been an intriguing subject since 1900 when Planck [1] proposed the idea of quantization of electromagnetic energy. In this theory, the light or electromagnetic wave with an oscillating frequency ω is quantized into indivisible packets of energy $\hbar\omega$. In 1905, Einstein [2] proposed that the photoelectric effect could be explained using a corpuscle concept of the electromagnetic radiation. With the invention of lasers, the quantum theory of light also plays an important role in our understanding of many optical and electronic processes in materials such as semiconductors. Rigorous in-depth treatments of the quantum theory of light can be found in Refs. 3 and 4.

In this chapter, we study optical transitions, absorptions, and gains in semiconductors. In Section 9.1, we start with Fermi's golden rule using the time-dependent perturbation theory derived in Section 3.6. We discuss in Section 9.2 the Einstein A and B coefficients for spontaneous and stimulated emissions and show how they can be calculated quantum mechanically. We then discuss in Sections 9.3–9.6 the optical processes in semiconductor bulk, quantum-well (QW), quantum-wire (QWR), and quantum-dot (QD) structures. The bulk optical matrix elements are derived based on Kane's model for the semiconductor conduction- and valence-band structures. Intersubband transitions in quantum wells are presented in Section 9.7. For semiconductor quantum wells, valence-band mixing effects play an important role in the optical gain of a quantum-well laser. We show in Section 9.8 how these heavy-hole and light-hole band mixing effects can be taken into account in the modeling of interband optical gain spectrum in quantum-well structures.

9.1 OPTICAL TRANSITIONS USING FERMI'S GOLDEN RULE

Consider a semiconductor illuminated by light. The interaction between the photons and the electrons in the semiconductor can be described by the Hamiltonian [5]

$$H = \frac{1}{2m_0}(\mathbf{p} - e\mathbf{A})^2 + V(\mathbf{r}) \tag{9.1.1}$$

Physics of Photonic Devices, Second Edition. By Shun Lien Chuang
Copyright © 2009 John Wiley & Sons, Inc.

where m_0 is the free electron mass, $e = -|e|$ for electrons, $\mathbf{A}$ is the vector potential accounting for the presence of the electromagnetic field, and $V(\mathbf{r})$ is the periodic crystal potential. The above Hamiltonian can be understood because taking the partial derivatives of H with respect to the x, y, and z components of the momentum variable $\mathbf{p}$ and the position vector $\mathbf{r}$ will lead to the classic equation of motion described by the Lorentz force equation [5] in the presence of an electric and a magnetic field (see Problem 9.1).

9.1.1 The Electron–Photon Interaction Hamiltonian

The Hamiltonian can be expanded into

$$\begin{aligned} H &= \frac{p^2}{2m_0} + V(\mathbf{r}) - \frac{e}{2m_0}(\mathbf{p}\cdot\mathbf{A} + \mathbf{A}\cdot\mathbf{p}) + \frac{e^2A^2}{2m_0} \\ &\simeq H_0 + H' \end{aligned} \tag{9.1.2}$$

where H_0 is the unperturbed Hamiltonian and H' is considered as a perturbation due to light

$$H_0 = \frac{p^2}{2m_0} + V(\mathbf{r}) \tag{9.1.3}$$

$$H' \simeq -\frac{e}{m_0}\mathbf{A}\cdot\mathbf{p}, \tag{9.1.4}$$

and the Coulomb gauge (discussed in Section 6.3)

$$\nabla\cdot\mathbf{A} = 0 \tag{9.1.5}$$

has been used such that $\mathbf{p}\cdot\mathbf{A} = \mathbf{A}\cdot\mathbf{p}$, noting that $\mathbf{p} = (\hbar/i)\nabla$. The last term $e^2A^2/2m_0$ is much smaller than the terms linear in $\mathbf{A}$ because $|e\mathbf{A}| \ll |\mathbf{p}|$ for most practical optical field intensities. This can be checked using $p \simeq \hbar k \simeq \hbar\pi/a$, and $a \sim 5$ Å is the lattice constant. Assuming the vector potential for the optical electric field of the form

$$\begin{aligned} \mathbf{A} &= \hat{e}A_0\cos(\boldsymbol{k}_{\mathrm{op}}\cdot\mathbf{r} - \omega t) \\ &= \hat{e}\frac{A_0}{2}e^{i\mathbf{k}_{\mathrm{op}}\cdot\mathbf{r}}e^{-i\omega t} + \hat{e}\frac{A_0}{2}e^{-i\mathbf{k}_{\mathrm{op}}\cdot\mathbf{r}}e^{+i\omega t} \end{aligned} \tag{9.1.6}$$

where $\mathbf{k}_{\mathrm{op}}$ is the wave vector, ω is the optical angular frequency, and $\hat{e}$ is a unit vector in the direction of the optical electric field, we have

$$\mathbf{E}(\mathbf{r}, t) = -\frac{\partial\mathbf{A}}{dt} = -\hat{e}\omega A_0\sin(\mathbf{k}_{\mathrm{op}}\cdot\mathbf{r} - \omega t) \tag{9.1.7}$$

$$\mathbf{H}(\mathbf{r}, t) = \frac{1}{\mu}\nabla\times\mathbf{A} = -\frac{1}{\mu}\mathbf{k}_{\mathrm{op}}\times\hat{e}A_0\sin(\mathbf{k}_{\mathrm{op}}\cdot\mathbf{r} - \omega t) \tag{9.1.8}$$

where we have used the fact that the scalar potential φ vanishes ($\rho = 0$) for the optical field, and $\mu = \mu_0$, the permeability of the free space. The Poynting vector for the power intensity (W/cm^2) is given by

$$\mathbf{P}(\mathbf{r}, t) = \mathbf{E}(\mathbf{r}, t) \times \mathbf{H}(\mathbf{r}, t) = \hat{k}k_{op}\frac{\omega A_0^2}{\mu}\sin^2(\mathbf{k}_{op} \cdot \mathbf{r} - \omega t) \tag{9.1.9}$$

which is pointing along the direction of wave propagation $\mathbf{k}_{op}$. The time average of the Poynting flux is simply

$$P = |\langle \mathbf{P}(\mathbf{r}, t) \rangle| = \frac{\omega A_0^2}{2\mu}k_{op} = \frac{n_r c \varepsilon_0 \omega^2 A_0^2}{2} \tag{9.1.10}$$

noting that the time average of the $\sin^2(\bullet)$ function is $\frac{1}{2}$. Here n_r is the refractive index of the material, and c is the speed of light in free space. The interaction Hamiltonian $H'(\mathbf{r}, t)$ can be written as

$$\begin{aligned} H'(\mathbf{r}, t) &= -\frac{e}{m_0}\mathbf{A}(\mathbf{r}, t) \cdot \mathbf{p} \\ &= H'(\mathbf{r})e^{-i\omega t} + H'^{+}(\mathbf{r})e^{+i\omega t} \end{aligned} \tag{9.1.11}$$

where

$$H'(\mathbf{r}) = -\frac{eA_0 e^{i\mathbf{k}_{op}\cdot\mathbf{r}}}{2m_0}\hat{e} \cdot \mathbf{p}. \tag{9.1.12}$$

The superscript "+" means the Hermitian adjoint operator.

9.1.2 Transition Rate due to Electron–Photon Interaction

The transition rate for the absorption of a photon (Fig. 9.1a) assuming an electron is initially at state a is given by Fermi's golden rule and has been derived in Section 3.6 using the time-dependent perturbation theory

$$W_{abs} = \frac{2\pi}{\hbar}|\langle b|H'(r)|a\rangle|^2 \delta(E_b - E_a - \hbar\omega) \tag{9.1.13}$$

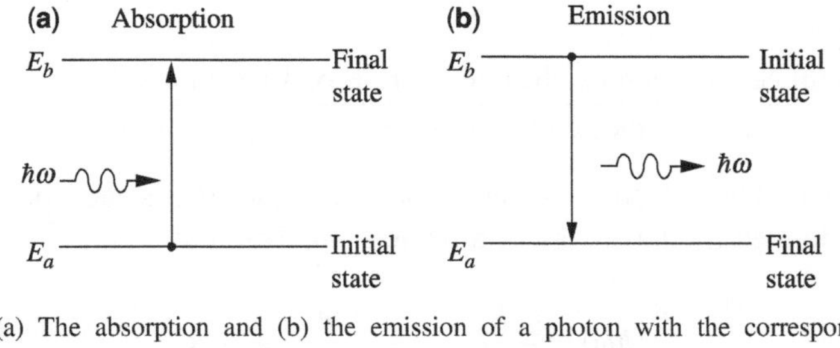

Figure 9.1 (a) The absorption and (b) the emission of a photon with the corresponding electron transitions.

where $E_b > E_a$ has been assumed. The total upward transition rate per unit volume (s^{-1} cm^{-3}) in the crystal taking into account the probability that state a is occupied and state b is empty is

$$R_{a\to b} = \frac{2}{V}\sum_{\mathbf{k}_a}\sum_{\mathbf{k}_b}\frac{2\pi}{\hbar}\left|H'_{ba}\right|^2 \delta(E_b - E_a - \hbar\omega) f_a(1 - f_b) \tag{9.1.14}$$

where we sum over the initial and final states and assume that the Fermi–Dirac distribution f_a is the probability that the state a is occupied. A similar expression holds for f_b with E_a replaced by E_b, and $(1 - f_b)$ is the probability that the state b is empty. The prefactor 2 takes into account the sum over spins, and the matrix element H'_{ba} is given by

$$H'_{ba} \equiv \langle b|H'(\mathbf{r})|a\rangle = \int \Psi_b^*(\mathbf{r})H'(\mathbf{r})\Psi_a(\mathbf{r})d^3\mathbf{r}. \tag{9.1.15}$$

Similarly, the transition rate for the emission of a photon (Fig. 9.1b) if an electron is initially at state b is

$$W_{\text{ems}} = \frac{2\pi}{\hbar}\left|\langle a|H'^{+}(r)|b\rangle\right|^2 \delta(E_a - E_b + \hbar\omega). \tag{9.1.16}$$

The downward transition rate per unit volume (s^{-1} cm^{-3}) is

$$R_{b\to a} = \frac{2}{V}\sum_{\mathbf{k}_a}\sum_{\mathbf{k}_b}\frac{2\pi}{\hbar}\left|H'^{+}_{ab}\right|^2 \delta(E_a - E_b + \hbar\omega) f_b(1 - f_a). \tag{9.1.17}$$

Using the even property of the delta function, $\delta(-x) = \delta(x)$, and $|H'_{ba}| = |H'^{+}_{ab}|$, the net upward transition rate per unit volume can be written as

$$\begin{aligned} R &= R_{a\to b} - R_{b\to a} \\ &= \frac{2}{V}\sum_{\mathbf{k}_a}\sum_{\mathbf{k}_b}\frac{2\pi}{\hbar}\left|H'_{ba}\right|^2 \delta(E_b - E_a - \hbar\omega)(f_a - f_b). \end{aligned} \tag{9.1.18}$$

9.1.3 Optical Absorption Coefficient

The absorption coefficient α (1/cm) in the crystal is the fraction of photons absorbed per unit distance:

$$\alpha = \frac{\text{Number of photons absorbed per second per unit volume}}{\text{Number of injected photons per second per unit area}}. \tag{9.1.19}$$

The injected number of photons per second per unit area is the optical intensity P (W/cm^2) divided by the energy of a photon $\hbar\omega$; therefore,

$$\alpha(\hbar\omega) = \frac{R}{P/\hbar\omega} = \frac{\hbar\omega}{(n_r c\varepsilon_0\omega^2 A_0^2/2)}R. \tag{9.1.20}$$

Using the dipole (long wavelength) approximation that $\mathbf{A}(\mathbf{r}) = \mathbf{A}e^{i\mathbf{k}_{op}\cdot\mathbf{r}} \simeq \mathbf{A}$, we find that the matrix elements can be written in terms of the momentum matrix element

$$H'_{ba} = -\frac{e}{m_0}\mathbf{A}\cdot\langle b|\mathbf{p}|a\rangle = -\frac{eA_0}{2m_0}\hat{e}\cdot\mathbf{p}_{ba}. \tag{9.1.21}$$

The absorption coefficient (9.1.20) becomes

$$\alpha(\hbar\omega) = C_0\frac{2}{V}\sum_{\mathbf{k}_a}\sum_{\mathbf{k}_b}|\hat{e}\cdot\mathbf{p}_{ba}|^2\delta(E_b-E_a-\hbar\omega)(f_a-f_b) \tag{9.1.22a}$$

$$C_0 = \frac{\pi e^2}{n_r c\varepsilon_0 m_0^2\omega}. \tag{9.1.22b}$$

We can see that the factors containing A_0^2 are canceled because the linear optical absorption coefficient is independent of the optical intensity.

The Hamiltonian can also be written in terms of the optical dipole moment

$$H'_{ba} = -\langle b|e\mathbf{r}\cdot\mathbf{E}|a\rangle = -\mu_{ba}\cdot\mathbf{E} \tag{9.1.23}$$

$$\mu_{ba} = e\langle b|\mathbf{r}|a\rangle = e\,\mathbf{r}_{ba}. \tag{9.1.24}$$

The above form can be derived using first the property:

$$\mathbf{p} = m_0\frac{d}{dt}\mathbf{r} = \frac{m_0}{i\hbar}(\mathbf{r}H_0 - H_0\mathbf{r}) \tag{9.1.25}$$

which can be proved first by writing out the x component of (9.1.25) and checking both sides that they are identical, followed by proving the y and z component by the same token (see Problem 9.3)

$$\begin{aligned} H'_{ba} &= \left\langle b\left|\frac{-e}{m_0}\mathbf{A}(\mathbf{r})\cdot\mathbf{p}\right|a\right\rangle \simeq \frac{-e}{i\hbar}\mathbf{A}\cdot\langle b|\mathbf{r}H_0 - H_0\mathbf{r}|a\rangle \\ &= -\frac{e(E_a-E_b)}{i\hbar}\mathbf{A}\cdot\langle b|\mathbf{r}|a\rangle \simeq -\boldsymbol{\mu}_{ba}\cdot\mathbf{E} \end{aligned} \tag{9.1.26}$$

where we have used $H_0|a\rangle = E_a|a\rangle$, $\langle b|H_0 = \langle b|E_b$, $E_b - E_a \simeq \hbar\omega$, and $\mathbf{E} = i\omega\mathbf{A}$ for the first term in $\mathbf{A}(\mathbf{r}, t)$ with $\exp(-i\omega t)$ dependence.

In terms of the dipole moment, we write the absorption coefficient as

$$\alpha(\hbar\omega) = \frac{\pi\omega}{n_r c\varepsilon_0}\frac{2}{V}\sum_{\mathbf{k}_a}\sum_{\mathbf{k}_b}|\hat{e}\cdot\boldsymbol{\mu}_{ba}|^2\delta(E_b - E_a - \hbar\omega)(f_a - f_b). \tag{9.1.27}$$

When the scattering relaxation is included, the delta function may be replaced by a Lorentzian function with a full width at half maximum (FWHM) or linewidth 2γ

$$\delta(E_b - E_a - \hbar\omega) \longrightarrow L(E_b - E_a - \hbar\omega) = \frac{\gamma/\pi}{(E_b - E_a - \hbar\omega)^2 + \gamma^2} \tag{9.1.28}$$

where a factor π has been included such that the area under the Lorentzian and delta function is properly normalized

$$\int_{-\infty}^{+\infty} L(E_b - E_a - \hbar\omega)\, d(\hbar\omega) = 1. \tag{9.1.29}$$

9.1.4 Real and Imaginary Parts of the Permittivity Function

The absorption coefficient α is related to the imaginary part of the permittivity function by

$$\alpha = 2\ \text{Im}\ \omega\sqrt{\mu(\varepsilon_1 + i\varepsilon_2)} \simeq 2\ \text{Im}\left[\frac{\omega}{c}\, n_r\left(1 + i\frac{\varepsilon_2}{2\varepsilon_1}\right)\right] = \frac{\omega}{n_r c}\frac{\varepsilon_2}{\varepsilon_0} \tag{9.1.30}$$

assuming that $|\varepsilon_2| \ll \varepsilon_1$. Here a factor of 2 accounts for the fact that α refers to the absorption coefficient of the optical intensity, not the electric field. We obtain the imaginary part of the permittivity function

$$\begin{aligned} \varepsilon_2(\omega) &= \frac{n_r c\, \varepsilon_0}{\omega}\alpha(\hbar\omega) \\ &= \frac{\pi e^2}{m_0^2\omega^2}\frac{2}{V}\sum_{\mathbf{k}_a}\sum_{\mathbf{k}_b}|\hat{e}\cdot\mathbf{p}_{ba}|^2\delta(E_b - E_a - \hbar\omega)(f_a - f_b). \end{aligned} \tag{9.1.31}$$

In semiconductors, the conduction- and valence-band structures determine the energy-momentum relations $E_b = E(\mathbf{k}_b)$ and $E_a = E(\mathbf{k}_a)$, respectively. An important part of calculating the absorption coefficient is to find the band structures and the wave functions. The optical matrix element $\mathbf{p}_{ba}$ is calculated from the wave functions based on the parabolic band model. We would usually start from bulk semiconductors for which the band structures are known. Then we use the $\mathbf{k}\cdot\mathbf{p}$ perturbation method and the effective mass theory near the band edges, as discussed in Chapter 4, to study the optical processes near the band edges of bulk and quantum-well semiconductor structures. Using the Kramers–Kronig relation derived in Appendix 5A, we obtain the real part of the permittivity function from the imaginary part:

$$\varepsilon_1(\omega) = \varepsilon_0 + \frac{2e^2\hbar^2}{m_0^2}\frac{2}{V}\sum_{\mathbf{k}_a}\sum_{\mathbf{k}_b}|\hat{e}\cdot\mathbf{p}_{ba}|^2\frac{(f_a - f_b)}{(E_b - E_a)[(E_b - E_a)^2 - (\hbar\omega)^2]}. \tag{9.1.32}$$

If instead, Eq. (9.1.27) with the dipole moment matrix is used, similar expressions for $\varepsilon_1(\omega)$ and $\varepsilon_2(\omega)$ can be derived. In practice, linewidth broadening has to be considered to compare the theoretical absorption spectrum with experimental data. We replace the delta function in (9.1.31) by a Lorentzian function using (9.1.28).

9.2 SPONTANEOUS AND STIMULATED EMISSIONS

In Section 9.1, we presented a quantum-mechanical derivation for optical absorption of semiconductors in the presence of a monochromatic electromagnetic field using Fermi's golden rule. Here we discuss the spontaneous and stimulated emissions in semiconductors and derive the relations between the spontaneous emission, stimulated emission, and absorption spectra [4, 6–8].

First we consider a *discrete* two-level system in the presence of an electromagnetic field with a broad spectrum (Fig. 9.2a). The total transition rate per unit volume (s^{-1} cm^{-3}) is given by

$$R_{12} = \frac{1}{V}\sum_{k} \frac{2\pi}{\hbar} \left|H'_{12}\right|^2 \delta(E_2 - E_1 - \hbar\omega_k) 2n_{ph} \tag{9.2.1}$$

where H'_{12} is the matrix element of the interaction Hamiltonian due to the electromagnetic field, $\hbar\omega_k$ is the energy of a photon with a wave vector $\mathbf{k}$,

$$n_{ph} = \frac{1}{e^{\hbar\omega_k/k_B T} - 1} \tag{9.2.2}$$

is the number of photons per state following the Bose–Einstein statistics for identical particles (photons), and a factor of 2 accounts for two polarization states for each $\mathbf{k}$ vector. The photon field can be described by a plane wave for simplicity as

$$e^{i\mathbf{k}\cdot\mathbf{r}} = e^{ik_x x + ik_y y + ik_z z}. \tag{9.2.3}$$

9.2.1 Density of States for Photons

To define the density of states for the photon field, we use the periodic boundary conditions that the wave function should be periodic in the x, y, and z directions with a

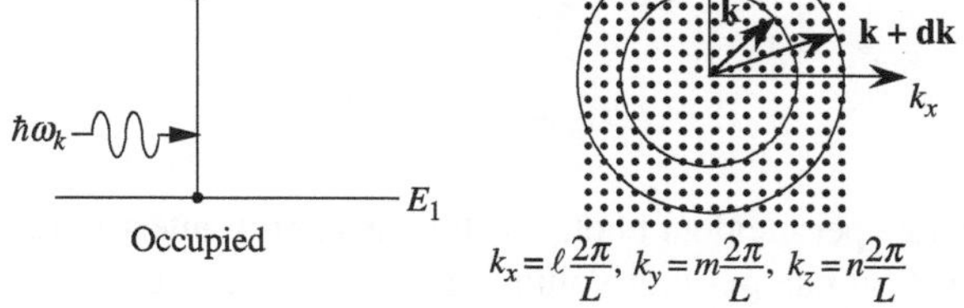

Figure 9.2 (a) A photon incident on a discrete two-level system where level 1 is occupied and level 2 is empty. (b) The k-space diagram for the density of photon states. A dot represents one state with two possible polarizations.

period L. Therefore,

$$k_x = \ell\frac{2\pi}{L},\ k_y = m\frac{2\pi}{L},\ \text{and}\ k_z = n\frac{2\pi}{L}. \tag{9.2.4}$$

The volume of a state in the k-space is therefore $(2\pi/L)^3$, Fig. 9.2b. Using the dispersion relation for the photon

$$\omega_k = kc/n_r \tag{9.2.5}$$

where c/n_r is the speed of light in the medium with a refractive index n_r, we can change the sum over the $\mathbf{k}$ vector to an integral.

Let us look at the integral using the number of states with a differential volume in the k-space $d^3k/(2\pi/L)^3 = k^2\,dk\,d\Omega/(2\pi/L)^3$, $d\Omega$ is the differential solid angle,

$$N(E_{21}) = \frac{2}{V}\sum_{\mathbf{k}} \delta(E_2 - E_1 - E_k) = 2\int \frac{k^2 dk\, d\Omega}{(2\pi)^3}\delta(E_2 - E_1 - E_k) \tag{9.2.6}$$

where V is the volume of the space,

$$E_k = \hbar\omega_k = \frac{\hbar kc}{n_r} \tag{9.2.7}$$

is the photon energy, and the integration over the solid angle is 4π. We find

$$N(E_{21}) = \frac{8\pi n_r^3\, E_{21}^2}{h^3 c^3} = \frac{n_r^3\, E_{21}^2}{\pi^2 \hbar^3 c^3}, \tag{9.2.8}$$

which is the number of states with photon energy E_{21} per unit volume per energy interval, $\text{cm}^{-3}(\text{eV})^{-1}$, and $E_{21} = E_2 - E_1$ is the energy spacing between the two levels.

9.2.2 Stimulated and Spontaneous Emissions: Einstein's A and B Coefficients

Define

$$B_{12} = \frac{2\pi}{\hbar}\left|H'_{12}\right|^2 \tag{9.2.9}$$

as the transition rate per incident photon within an energy interval, (eV/s). We obtain the upward transition rate per unit volume ($\text{s}^{-1}\ \text{cm}^{-3}$) for a broad spectrum or incoherent light (black body radiation)

$$R_{12} = B_{12}S(E_{21}) \tag{9.2.10}$$

where

$$S(E_{21}) = N(E_{21})n_{\text{ph}} \tag{9.2.11}$$

is the number of photons per unit volume per energy interval with a dimension of $\text{cm}^{-3}(\text{eV})^{-1}$, and

$$n_{\text{ph}} = \frac{1}{e^{E_{21}/k_B T} - 1} \tag{9.2.12}$$

is the average number of photons per state at an optical energy E_{21}.

If we take into account the occupation probabilities of level 1 and level 2 using Fermi–Dirac distributions, f_1 and f_2, respectively, the expression for R_{12} is slightly modified, Fig. 9.3,

$$\begin{aligned} R_{12}(= r_{12}(E)dE) &= \frac{1}{V}\sum_{k} B_{12}\,\delta(E_2 - E_1 - \hbar\omega_k)2n_{\text{ph}} \cdot f_1(1 - f_2) \\ &= B_{12} f_1(1 - f_2)\,S(E_{21}) \end{aligned} \tag{9.2.13}$$

where $r_{12}(E)dE$ means that the upward transition rate per unit volume has been integrated for a light with a spectral width dE near $E = E_{21}$. $r_{12}(E)$ is the number of $1 \to 2$ transitions per second per unit volume per energy interval ($\text{s}^{-1}\,\text{cm}^{-3}\,\text{eV}^{-1}$).

Similarly, a stimulated emission rate per unit volume can be given

$$R_{21}^{\text{stim}} = r_{21}^{\text{stim}}(E)dE = B_{21} f_2(1 - f_1)S(E_{21}). \tag{9.2.14}$$

The spontaneous emission rate per unit volume is independent of the photon density and is given by

$$R_{21}^{\text{spon}} = r_{21}^{\text{spon}}(E)dE = A_{21} f_2(1 - f_1). \tag{9.2.15}$$

At thermal equilibrium, there is only one Fermi level; therefore $F_1 = F_2$. We balance the total downward transition rate with the upward transition rate

$$R_{12} = R_{21}^{\text{stim}} + R_{21}^{\text{spon}} \tag{9.2.16}$$

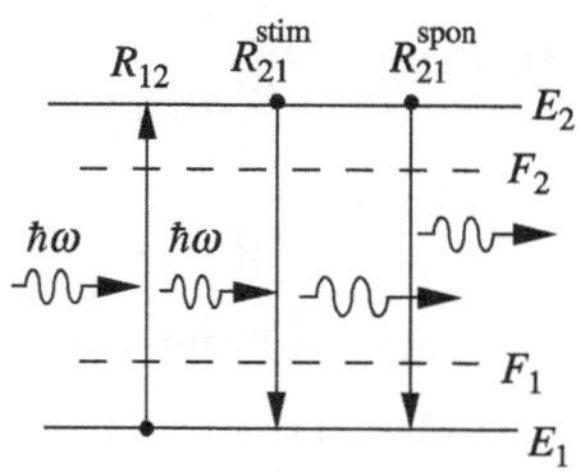

Figure 9.3 Schematic diagram for the stimulated absorption rate R_{12}, stimulated emission rate R_{21}^{stim}, and the spontaneous emission rate R_{21}^{spon} in the presence of two levels with quasi-Fermi levels, F_1 and F_2, respectively.

or

$$B_{12}f_1(1-f_2)S(E_{21}) = B_{21}f_2(1-f_1)S(E_{21}) + A_{21}f_2(1-f_1). \tag{9.2.17}$$

We obtain

$$\frac{A_{21}}{B_{12}e^{E_{21}/k_BT} - B_{21}} = N(E_{21})\frac{1}{e^{E_{21}/k_BT} - 1}. \tag{9.2.18}$$

By comparing the left- and right-hand sides of (9.2.18), we find

$$B_{12} = B_{21} \tag{9.2.19}$$

and

$$\frac{A_{21}}{B_{21}} = N(E_{21}) = \frac{n_r^3 E_{21}^2}{\pi^2 \hbar^3 c^3}. \tag{9.2.20}$$

The ratio of the stimulated and spontaneous emission rate is, therefore,

$$\frac{R_{21}^{\text{stim}}}{R_{21}^{\text{spon}}} = \frac{B_{21}S(E_{21})f_2(1-f_1)}{A_{21}f_2(1-f_1)} = n_{\text{ph}} \tag{9.2.21}$$

which is the number of photons per state, (9.2.12).

9.2.3 Derivation of the Optical Gain and Spontaneous Emission Spectrum

The net absorption rate per unit volume within a spectral width dE is

$$\begin{aligned} R_{\text{net}}^{\text{abs}} &= r_{\text{net}}^{\text{abs}}(E_{21})dE \\ &= B_{12}f_1(1-f_2)S(E_{21}) - B_{21}f_2(1-f_1)S(E_{21}) \\ &= B_{12}(f_1-f_2)S(E_{21}). \end{aligned} \tag{9.2.22}$$

Therefore, the absorption spectrum within a spectral width dE can be written as

$$\begin{aligned} \alpha(E_{21})dE &= \frac{r_{\text{net}}^{\text{abs}}(E_{21})dE}{P(E_{21})\frac{c}{n_r}} \\ &= \frac{n_r}{c}B_{12}(f_1-f_2). \end{aligned} \tag{9.2.23}$$

The ratio of the spontaneous emission spectrum and the absorption spectrum is

$$\begin{aligned} \frac{r_{21}^{\text{spon}}(E_{21})}{\alpha(E_{21})} &= \frac{A_{21}f_2(1-f_1)}{\frac{n_r}{c}B_{12}\,(f_1-f_2)} \\ &= \frac{c}{n_r}\cdot N(E_{21})\frac{1}{e^{(E_{21}-\Delta F)/k_BT} - 1} \end{aligned} \tag{9.2.24}$$

or

$$r_{21}^{\text{spon}}(E_{21}) = \left(\frac{n_r^2 E_{21}^2}{\pi^2 \hbar^3 c^2}\right) \frac{1}{e^{(E_{21}-\Delta F)/k_B T} - 1} \alpha(E_{21}) \tag{9.2.25}$$

and $\Delta F = F_2 - F_1$ is the quasi-Fermi level separation.

If we consider the net upward transition for a monochromatic light (a single photon with $E = \hbar\omega$) instead of a light spectrum, we use $r_{\text{net}}^{\text{abs}}(E)$, that is, the net upward transition rate per unit volume per energy interval from (9.2.22) with $n_{\text{ph}} = 1$,

$$r_{\text{net}}^{\text{abs}}(E)dE = B_{12}(f_1 - f_2)N(E). \tag{9.2.26a}$$

We change $1/dE$ to a delta function for a pair of discrete states E_2 and E_1,

$$r_{\text{net}}^{\text{abs}}(E) = N(E)B_{12}\delta(E_2 - E_1 - E)(f_1 - f_2). \tag{9.2.26b}$$

This has to be summed over all initial states 1 (using $\mathbf{k}_a$ as the quantum number) and final states 2 (using $\mathbf{k}_b$) for all the electron wave vectors,

$$r_{\text{net}}^{\text{abs}}(E = \hbar\omega) = 2N(E)\sum_{\mathbf{k}_a}\sum_{\mathbf{k}_b} B_{ab}\delta(E_b - E_a - \hbar\omega)(f_a - f_b). \tag{9.2.27}$$

Similarly from (9.2.23)

$$\alpha(\hbar\omega) = \frac{n_r}{c} 2\sum_{\mathbf{k}_a}\sum_{\mathbf{k}_b} B_{ab}\delta(E_b - E_a - \hbar\omega)(f_a - f_b) \tag{9.2.28}$$

where

$$B_{ab} = B_{ba} = \frac{2\pi}{\hbar}\left|\langle b| - \frac{eA_0}{2m_0}\hat{e}\cdot\mathbf{p}|a\rangle\right|^2 \tag{9.2.29}$$

for each incident photon with an energy $\hbar\omega$. Therefore, we have the time-averaged power density P from the Poynting vector in Eq. (9.1.10), which is the optical intensity of a photon in a volume V propagating at a velocity c/n_r

$$P = \hbar\omega\frac{c}{n_r}\frac{1}{V} = \frac{n_r c\,\varepsilon_0\,\omega^2 A_0^2}{2}. \tag{9.2.30}$$

Therefore, we obtain

$$A_0^2 = \frac{2\hbar}{n_r^2 \varepsilon_0 \omega}\frac{1}{V} \tag{9.2.31}$$

$$B_{ba} = \frac{2\pi}{\hbar}\frac{e^2 A_0^2}{4m_0^2}|\hat{e}\cdot\mathbf{p}_{ba}|^2 = \frac{\pi e^2}{n_r^2 \varepsilon_0\, m_0^2 \omega}|\hat{e}\cdot\mathbf{p}_{ba}|^2\frac{1}{V}, \tag{9.2.32}$$

which is proportional to the optical matrix element. By substituting $B_{ab} = B_{ba}$ in (9.2.32) into (9.2.28), we obtain the identical result as (9.1.22) for the absorption coefficient. Similarly, the spontaneous emission rate per unit volume per unit energy interval ($s^{-1}\ cm^{-3}\ eV^{-1}$) is given by

$$r^{\text{spon}}(\hbar\omega) = \left(\frac{n_r^2\omega^2}{\pi^2\hbar c^2}\right) C_0 \frac{2}{V} \sum_{\mathbf{k}_a} \sum_{\mathbf{k}_b} |\hat{e}\cdot\mathbf{p}_{ba}|^2 \delta(E_b - E_a - \hbar\omega) f_b(1 - f_a) \quad (9.2.33)$$

and the net stimulated emission rate per unit volume per energy interval is given by

$$r^{\text{stim}}(\hbar\omega) = \left(\frac{n_r^2\omega^2}{\pi^2\hbar c^2}\right) C_0 \frac{2}{V} \sum_{\mathbf{k}_a} \sum_{\mathbf{k}_b} |\hat{e}\cdot\mathbf{p}_{ba}|^2 \delta(E_b - E_a - \hbar\omega)(f_b - f_a). \quad (9.2.34)$$

When $\alpha(\hbar\omega)$ becomes negative, we have gain in the medium:

$$g(\hbar\omega) = -\alpha(\hbar\omega). \quad (9.2.35)$$

We can also write from (9.2.25)

$$g(\hbar\omega) = \left(\frac{\pi^2\hbar c^2}{n_r^2\omega^2}\right)\left(1 - e^{[\hbar\omega-(F_2-F_1)]/k_BT}\right) r^{\text{spon}}(\hbar\omega) \quad (9.2.36)$$

$$r^{\text{stim}}(\hbar\omega) = \left(\frac{n_r^2\omega^2}{\pi^2\hbar c^2}\right) g(\hbar\omega) \quad (9.2.37)$$

$$r^{\text{stim}}(\hbar\omega) = \left(1 - e^{[\hbar\omega-(F_2-F_1)]/k_BT}\right) r^{\text{spon}}(\hbar\omega). \quad (9.2.38)$$

The spontaneous emission spectrum $r^{\text{spon}}(\hbar\omega)$ and the gain spectrum $g(\hbar\omega)$ are plotted in Fig. 9.4 [9–14]. The emission spectrum is always a positive quantity, whereas the gain changes a sign to become absorption when the optical energy is larger than the quasi-Fermi level separation [8] $F_2 - F_1$. Similar expressions as Eqs. (9.2.33) and (9.2.36) have been compared with experimental data with general good agreement [9–13]. Note that the spontaneous emission power intensity per unit volume per energy interval is $\hbar\omega r^{\text{spon}}(\hbar\omega)$. Some effects of the optical matrix elements [14] due to doping effects or conduction band non-parabolicity [9] have been discussed. The

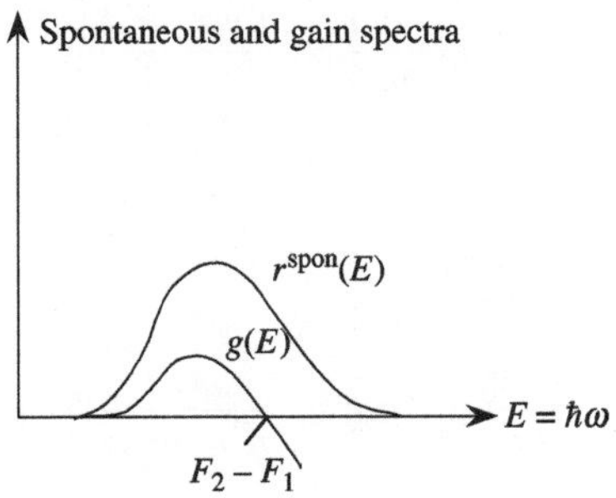

Figure 9.4 The spontaneous emission spectrum $r^{\text{spon}}(\hbar\omega)$ and the gain spectrum $g(\hbar\omega)$ are plotted. The gain changes to absorption (negative) when the optical energy ($\hbar\omega$) is larger than the quasi-Fermi level separation, $F_2 - F_1$.

crossing point $F_2 - F_1$ in the gain spectrum serves as a good way to estimate the carrier density in a laser diode or a light-emitting diode in addition to matching both the gain and the spontaneous emission spectra. More details on the gain spectrum and the population inversion condition, $\hbar\omega < F_2 - F_1$, will be discussed in Sections 9.3 and 9.4.

Figure 9.5a shows the measured spontaneous emission spectrum as a function of the injection current from the side (window light) of a semiconductor laser near 1.55 μm wavelength [12]. Using the fundamental relation [9–14] between the gain and the spontaneous emission spectrum (9.2.36), we can extract the gain from each curve in Fig. 9.5a, and the resultant gain spectrum is shown as the dashed curve at each bias current. In this approach, the quasi-Fermi level separation is taken as an input, $\Delta F = F_2 - F_1$, which can be measured independently (see Section 10.1). Another approach, called the Hakki–Paoli method [7, 15], is to directly measure the amplified spontaneous emission spectrum from the end facet of the laser at each bias and then extract the optical gain at each wavelength from the peaks and valleys of the Fabry–Perot spectrum. The results of the net modal gain, $\Gamma g(\hbar\omega) - \alpha_i$, are shown as the solid curves in Fig. 9.5b [10–13]. We can see the agreement between the two methods is very good. The net modal gain includes the

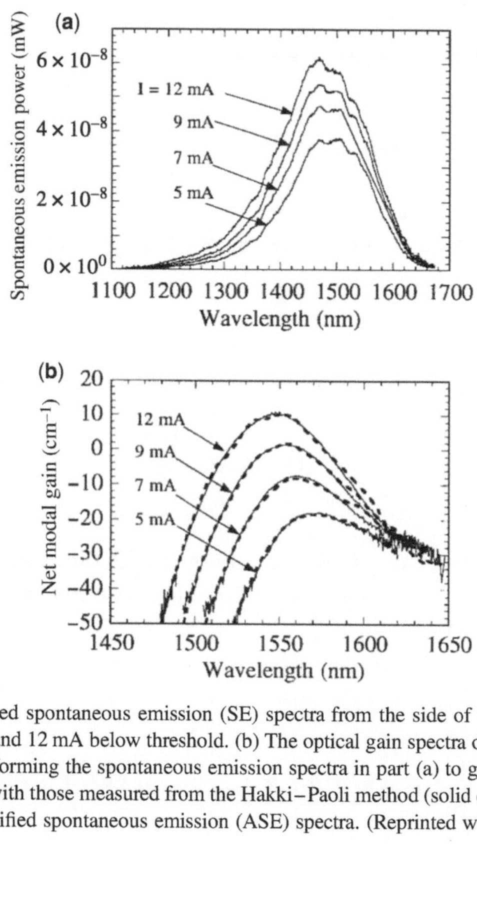

Figure 9.5 (a) Measured spontaneous emission (SE) spectra from the side of a semiconductor laser at bias currents of 5, 7, 9, and 12 mA below threshold. (b) The optical gain spectra obtained using the fundamental relation by transforming the spontaneous emission spectra in part (a) to gain (dashed curves). The gain spectra agree well with those measured from the Hakki–Paoli method (solid curves) extracted directly from the measured amplified spontaneous emission (ASE) spectra. (Reprinted with permission from [12] © 1999 IEEE.)

optical confinement factor and is shifted downward by the intrinsic absorption. By drawing a horizontal line at $-\alpha_i$, we can also check the quasi-Fermi level separation, $\Delta F = F_2 - F_1$, from the gain spectrum measured from the Hakki–Paoli method. More details are discussed in Section 10.1.

9.3 INTERBAND ABSORPTION AND GAIN OF BULK SEMICONDUCTORS

For interband transitions between the valence band and conduction band of a semiconductor, we have to evaluate the optical matrix element

$$H'_{ba} = \left\langle b \left| \frac{-eA(\mathbf{r})}{m_0} \cdot \mathbf{p} \right| a \right\rangle. \tag{9.3.1}$$

9.3.1 Evaluation of the Interband Optical Matrix Element and the k-Selection Rule

The vector potential for the optical field is

$$\mathbf{A}(\mathbf{r}) = \mathbf{A}\ e^{i\mathbf{k}_{\mathrm{op}} \cdot \mathbf{r}} = \frac{\hat{e}A_0}{2}\ e^{i\mathbf{k}_{\mathrm{op}} \cdot \mathbf{r}}. \tag{9.3.2}$$

The Bloch functions for electrons in the valence band E_a and the conduction band E_b are, respectively,

$$\psi_a(\mathbf{r}) = u_{\mathrm{v}}(\mathbf{r}) \frac{e^{i\mathbf{k}_{\mathrm{v}} \cdot \mathbf{r}}}{\sqrt{V}} \tag{9.3.3a}$$

$$\psi_b(\mathbf{r}) = u_{\mathrm{c}}(\mathbf{r}) \frac{e^{i\mathbf{k}_{\mathrm{c}} \cdot \mathbf{r}}}{\sqrt{V}} \tag{9.3.3b}$$

where $u_{\mathrm{v}}(\mathbf{r})$ and $u_{\mathrm{c}}(\mathbf{r})$ are the periodic parts of the Bloch functions, and the remainders are the envelope functions (plane waves) for a free electron. The momentum matrix element is derived from

$$\begin{aligned} H'_{ba} &= -\frac{eA_0}{2m_0} \hat{e} \cdot \int \psi_b^* e^{i\mathbf{k}_{\mathrm{op}} \cdot \mathbf{r}} \mathbf{p} \psi_a d^3\mathbf{r} \\ &= -\frac{eA_0}{2m_0} \hat{e} \cdot \int u_{\mathrm{c}}^*(\mathbf{r}) e^{-i\mathbf{k}_{\mathrm{c}} \cdot \mathbf{r}} e^{i\mathbf{k}_{\mathrm{op}} \cdot \mathbf{r}} \left[\left(\frac{\hbar}{i} \nabla u_{\mathrm{v}}(\mathbf{r}) \right) e^{i\mathbf{k}_{\mathrm{v}} \cdot \mathbf{r}} + \hbar \mathbf{k}_{\mathrm{v}} u_{\mathrm{v}}(\mathbf{r}) e^{i\mathbf{k}_{\mathrm{v}} \cdot \mathbf{r}} \right] \frac{d^3\mathbf{r}}{V} \end{aligned}$$

$$\simeq -\frac{eA_0}{2m_0}\hat{e}\cdot \int_{\Omega} u_c^*(\mathbf{r})\frac{\hbar}{i}\nabla u_v(\mathbf{r})\frac{d^3\mathbf{r}}{\Omega}\int_{V} e^{i(-\mathbf{k}_c+\mathbf{k}_{op}+\mathbf{k}_v)\cdot\mathbf{r}}\frac{d^3\mathbf{r}}{V}$$

$$= -\frac{eA_0}{2m_0}\hat{e}\cdot \mathbf{p}_{cv}\delta_{\mathbf{k}_c,\mathbf{k}_v+\mathbf{k}_{op}} \tag{9.3.4}$$

$$\mathbf{p}_{cv} = \int_{\Omega} u_c^*(\mathbf{r})\frac{\hbar}{i}\nabla u_v(\mathbf{r})\frac{d^3\mathbf{r}}{\Omega} \tag{9.3.5}$$

where we noted that $[u_c^*(\mathbf{r})\frac{\hbar}{i}\nabla u_v(\mathbf{r})]$ and $[u_c^*(\mathbf{r})\ u_v(\mathbf{r})]$ are periodic functions with the period of a unit cell, whereas the envelope functions are slowly varying functions over a unit cell. Therefore, the integral over $d^3\mathbf{r}$ can be separated into the product of two integrals, one over the unit cell Ω for the periodic part, and the other over the slowly varying part. In another words, we use the approximation:

$$\int_{V}\left[u_c^*(\mathbf{r})\frac{\hbar}{i}\nabla u_v(\mathbf{r})\right]F(\mathbf{r})d^3\mathbf{r} \simeq \int_{V} F(\mathbf{r})d^3\mathbf{r}\int_{\Omega} u_c^*(\mathbf{r})\frac{\hbar}{i}\nabla u_v(\mathbf{r})\frac{d^3\mathbf{r}}{\Omega} \tag{9.3.6}$$

where $F(\mathbf{r})$ is slowly varying over a unit cell, and we have used the periodic property of the Bloch periodic functions

$$u_c^*(\mathbf{r})\frac{\hbar}{i}\nabla u_v(\mathbf{r}) = \sum_{\mathbf{G}} \mathbf{C}_{\mathbf{G}}\ e^{i\mathbf{G}\cdot\mathbf{r}} \tag{9.3.7}$$

where the vectors **G**'s are the reciprocal lattice vectors. Because $F(\mathbf{r})$ is slowly varying over a unit cell, we may approximate $F(\mathbf{r}+\mathbf{R}) \simeq F(\mathbf{R})$ and put it outside of the integral over a unit cell. Here the **R**s are the lattice vectors, and $\exp[i\,\mathbf{G}\cdot\mathbf{R}] = 1$. Ω is the volume of a unit cell. Note that the orthogonal property

$$\int_{\Omega} u_c^* u_v d^3\mathbf{r} = 0, \qquad \frac{1}{\Omega}\int_{\Omega} u_c^* u_c d^3\mathbf{r} = \frac{1}{\Omega}\int_{\Omega} u_v^* u_v d^3\mathbf{r} = 1. \tag{9.3.8}$$

From the matrix element (9.3.4), we see that the momentum conservation

$$\hbar\mathbf{k}_c = \hbar\mathbf{k}_v + \hbar\mathbf{k}_{op} \tag{9.3.9}$$

is obeyed. The electron at the final state has its crystal momentum $\hbar\mathbf{k}_c$ equal to its initial momentum $\hbar\mathbf{k}_v$ plus the photon momentum $\hbar\mathbf{k}_{op}$. Because $k_{op} \sim 2\pi/\lambda_0$, and the magnitudes k_c, k_v are of the order $2\pi/a_0$, where a_0 is the lattice constant of the semiconductors, which is typically of the order 5.5 Å and is much smaller than λ_0, we may ignore $\mathbf{k}_{op}$ and obtain

$$H'_{ba} \simeq -\frac{eA_0}{2m_0}\hat{e}\cdot \mathbf{p}_{cv}\delta_{\mathbf{k}_c,\mathbf{k}_v} \tag{9.3.10}$$

which is the **k**-selection rule or *vertical transition* in the interband optical transitions of the band structure. Notice that the interband momentum matrix element, $\mathbf{p}_{cv}$, depends only on the periodic parts (u_c and u_v) of the Bloch functions and is derived from the original optical momentum matrix element $\mathbf{p}_{ba}$, which, on the other hand, depends on the full wave functions (i.e., including the envelope functions).

9.3.2 Optical Absorption Spectrum

Using the **k**-selection rule in the matrix element (9.3.10), we find that the absorption coefficient (9.1.22) for a bulk semiconductor is

$$\alpha(\hbar\omega) = C_0 \frac{2}{V} \sum_{\mathbf{k}} |\hat{e} \cdot \mathbf{p}_{cv}|^2 \delta(E_c - E_v - \hbar\omega)(f_v - f_c) \tag{9.3.11}$$

where the Fermi–Dirac distributions for the electrons in the valence band and in the conduction band are, respectively,

$$f_v(\mathbf{k}) = \frac{1}{1 + e^{(E_v(\mathbf{k}) - F_v)/k_B T}}, \quad f_c(\mathbf{k}) = \frac{1}{1 + e^{(E_c(\mathbf{k}) - F_c)/k_B T}} \tag{9.3.12}$$

and F_v, F_c are the quasi-Fermi levels, which are dependent on the level of carrier injection and doping concentration. We use $\mathbf{k}$ to represent both $\mathbf{k}_c$ and $\mathbf{k}_v$. In the case of thermal equilibrium, $F_v = F_c = E_F$. We assume that the semiconductor is undoped, the valence band is completely occupied, and the conduction band is empty, $f_v = 1$ and $f_c = 0$. We further assume that the matrix element $|\hat{e} \cdot \mathbf{p}_{cv}|^2$ is independent of $\mathbf{k}$ and denote the absorption spectrum at thermal equilibrium

$$\alpha_0(\hbar\omega) = C_0 |\hat{e} \cdot \mathbf{p}_{cv}|^2 \int \frac{2d^3\mathbf{k}}{(2\pi)^3} \delta\left(E_g + \frac{\hbar^2 k^2}{2m_r^*} - \hbar\omega\right) \tag{9.3.13}$$

where we have used the reduced effective mass m_r^*,

$$E_c = E_g + \frac{\hbar^2 k^2}{2m_e^*} \quad E_v = -\frac{\hbar^2 k^2}{2m_h^*}$$
$$\frac{1}{m_r^*} = \frac{1}{m_e^*} + \frac{1}{m_h^*}. \tag{9.3.14}$$

Here all energies are measured from *the top of the valence band*. Therefore, both E_c and F_c contain the band-gap energy E_g. Let

$$X = E_g + E - \hbar\omega \quad E = \frac{\hbar^2 k^2}{2m_r^*}. \tag{9.3.15}$$

We find, by a change of variables, the integration can be carried out with the contribution at $X = 0$, and $E = \hbar\omega - E_g$

$$\alpha_0(\hbar\omega) = C_0|\hat{e}\cdot \mathbf{p}_{cv}|^2 \rho_r(\hbar\omega - E_g) \tag{9.3.16a}$$

$$\rho_r(\hbar\omega - E_g) = \frac{1}{2\pi^2}\left(\frac{2m_r^*}{\hbar^2}\right)^{3/2}(\hbar\omega - E_g)^{1/2}. \tag{9.3.16b}$$

Therefore, the absorption coefficient depends on the momentum matrix element and the joint (or reduced) density of states. The absorption spectrum is plotted in Fig. 9.6. We can see the dominant square-root behavior of the optical energy above the band gap, $(\hbar\omega - E_g)^{1/2}$, representing the joint density of states. Below the band-gap energy E_g, the absorption does not occur because the photons see a forbidden band gap.

9.3.3 Optical Gain Spectrum

Under current injection or optical pumping, there will be both electrons and holes in semiconductors. Let us assume that a quasi-equilibrium state has been reached such that we have two quasi-Fermi levels, F_c and F_v, for the electrons and holes, respectively. Carrying out the integration as (9.3.13)–(9.3.16), we obtain

$$\alpha(\hbar\omega) = \alpha_0(\hbar\omega)[f_v(k_0) - f_c(k_0)] \tag{9.3.17a}$$

where

$$k_0 = \sqrt{\frac{2m_r^*}{\hbar^2}(\hbar\omega - E_g)}. \tag{9.3.17b}$$

We note that $\alpha(\hbar\omega)$ becomes negative if $f_v(k) - f_c(k) < 0$, which leads to

$$e^{(E_c - F_c)/k_BT} < e^{(E_v - F_v)/k_BT}$$

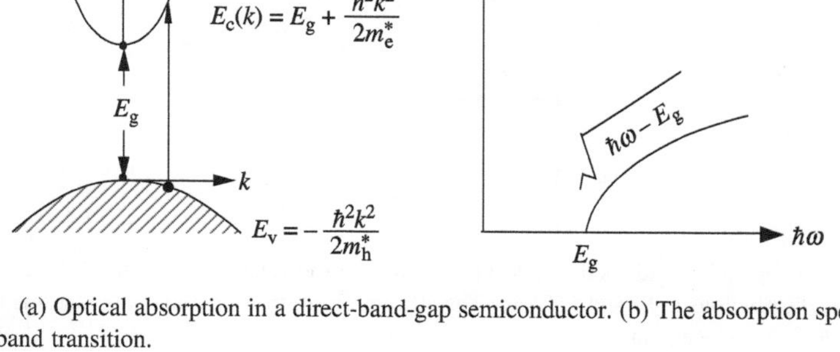

Figure 9.6 (a) Optical absorption in a direct-band-gap semiconductor. (b) The absorption spectrum due to the interband transition.

or

$$F_c - F_v > E_c - E_v = \hbar\omega. \tag{9.3.18}$$

The population inversion condition is satisfied if (9.3.18) holds and the absorption becomes negative, or equivalently, there is gain in the medium. The above condition (9.3.18) is called the Bernard–Duraffourg inversion condition [16]. Plots of $\alpha(\hbar\omega)$ and $(f_v - f_c)$ are illustrated [17] in Fig. 9.7. We can see that gain exists in the optical spectral region $E_g < \hbar\omega < F_c - F_v$. The gain spectrum is determined by the absorption spectrum $\alpha_0(\hbar\omega)$ of the semiconductors without optical excitations multiplied by the Fermi–Dirac inversion factor, $(f_v - f_c)$, which accounts for the population inversion probability. This factor, $f_v - f_c$, has a lower limit -1 at small optical energies and has a zero crossing at the quasi-Fermi level separation energy, $F_c - F_v$, then approaches $+1$ at large energies, as shown in Fig. 9.7. The increasing absorption coefficient with increasing energy $\hbar\omega$ and the decreasing magnitude of $|f_v - f_c|$ (dashed line) give a peak gain, which depends on the temperature and the effective masses of the electrons and holes near the band edges.

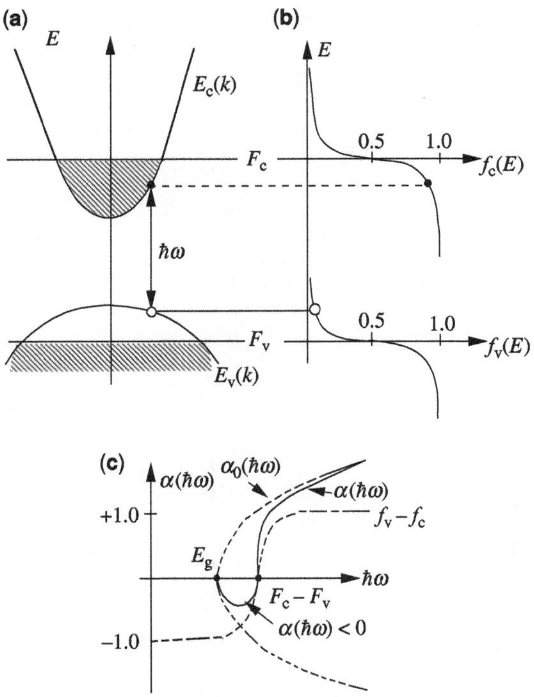

Figure 9.7 (a) Optical transitions between conduction band and valence band in the *E*-*k* diagram. (b) The Fermi–Dirac distributions $f_v(E)$ and $f_c(E)$. (c) The absorption spectrum $\alpha(\hbar\omega)$. Note that it is negative for $E_g < \hbar\omega < F_c - F_v$, where gain exists.

9.4 INTERBAND ABSORPTION AND GAIN IN A QUANTUM WELL

In this section, we consider the interband absorption and gain in a quantum well ignoring the excitonic effects due to the Coulomb interaction between electrons and holes. The excitonic effects will be considered in Chapter 14. The interband absorption or gain can be calculated from (9.1.22) analytically if scattering relaxation is not considered.

9.4.1 Interband Optical Matrix Element of a Quantum Well

Within a two-band model, the Bloch wave functions can be described by

$$\psi_a(\mathbf{r}) = u_v(\mathbf{r}) \frac{e^{i\mathbf{k}_t \cdot \rho}}{\sqrt{A}} g_m(z) \tag{9.4.1}$$

for a hole wave function in the heavy-hole or a light-hole subband m, and

$$\psi_b(\mathbf{r}) = u_c(\mathbf{r}) \frac{e^{i\mathbf{k}'_t \cdot \rho}}{\sqrt{A}} \phi_n(z) \tag{9.4.2}$$

for an electron in the conduction subband n. The momentum matrix $\mathbf{p}_{ba}$ is given by

$$\begin{aligned} \mathbf{p}_{ba} &= \langle \psi_b | \mathbf{p} | \psi_a \rangle \\ &\simeq \langle u_c | \mathbf{p} | u_v \rangle \; \delta_{\mathbf{k}_t, \mathbf{k}'_t} \; I_{hm}^{en} \end{aligned} \tag{9.4.3a}$$

where

$$I_{hm}^{en} = \int_{-\infty}^{+\infty} dz \; \phi_n^*(z) g_m(z) \tag{9.4.3b}$$

is the overlap integral of the conduction- and valence-band envelope functions. The **k**-selection rule in the plane of the quantum well is still satisfied. The polarization-dependent matrix element $\mathbf{p}_{cv} = \langle u_c | \mathbf{p} | u_v \rangle$ is discussed in Section 9.5. Here we have to take into account the quantizations of the electron and hole energies E_a and E_b

$$E_a = E_{hm} - \frac{\hbar^2 k_t^2}{2m_h^*} \tag{9.4.4a}$$

$$E_b = E_g + E_{en} + \frac{\hbar^2 k_t^2}{2m_e^*}. \tag{9.4.4b}$$

Note that $E_{hm} < 0$ and

$$E_b - E_a = E_{hm}^{en} + E_t, \quad E_t = \frac{\hbar^2 k_t^2}{2m_r^*} \tag{9.4.5a}$$

where

$$E_{hm}^{en} = E_g + E_{en} - E_{hm} \tag{9.4.5b}$$

is the band edge transition energy ($k_t = 0$). The summations over the quantum numbers $\mathbf{k}_a$ and $\mathbf{k}_b$ become summations over ($\mathbf{k}_t'$, m) and ($\mathbf{k}_t$, n). Because $\mathbf{k}_t = \mathbf{k}_t'$ in the matrix element (9.4.3a), we find

$$\alpha(\hbar\omega) = C_0 \sum_{n,m} |I_{hm}^{en}|^2 \frac{2}{V} \sum_{\mathbf{k}_t} |\hat{e} \cdot \mathbf{p}_{cv}|^2 \delta(E_{hm}^{en} + E_t - \hbar\omega)(f_v^m - f_c^n). \tag{9.4.6}$$

9.4.2 Joint Density of States and Optical Absorption Spectrum

We use the two-dimensional joint (or reduced) density of states ρ_r^{2D},

$$\frac{2}{V}\sum_{\mathbf{k}_t} = \frac{2A}{V}\int \frac{d^2\mathbf{k}_t}{(2\pi)^2} = \frac{1}{\pi L_z}\int_0^\infty k_t\, dk_t = \int_0^\infty dE_t \rho_r^{2D} \tag{9.4.7a}$$

$$\rho_r^{2D} = \frac{m_r^*}{\pi\hbar^2 L_z} \tag{9.4.7b}$$

where A is the area of the cross section, $AL_z = V$, L_z is an effective period of the quantum wells, and V is a volume of a period. The delta function gives the contribution at $E_{hm}^{en} + E_t = \hbar\omega$, and the absorption edges occur at $\hbar\omega = E_{hm}^{en}$. For an unpumped semiconductor, $f_v^m = 1$ and $f_c^n = 0$, we have the absorption spectrum at thermal equilibrium $\alpha_0(\hbar\omega)$

$$\alpha_0(\hbar\omega) = C_0 \sum_{n,m} |I_{hm}^{en}|^2 |\hat{e} \cdot \mathbf{p}_{cv}|^2 \rho_r^{2D} H(\hbar\omega - E_{hm}^{en}). \tag{9.4.8}$$

Because the integration of the delta function gives the step function, shown as H or the Heaviside step function, $H(x) = 1$ for $x > 0$, and 0 for $x < 0$. For a symmetric quantum well, we find $I_{hm}^{en} = \delta_{nm}$ using an infinite well model, and the absorption spectrum is

$$\alpha_0(\hbar\omega) = C_0 |\hat{e} \cdot \mathbf{p}_{cv}|^2 \begin{cases} \dfrac{m_r^*}{\pi\hbar^2 L_z} & \text{for } E_{h1}^{e1} < \hbar\omega < E_{h2}^{e2} \\ 2\dfrac{m_r^*}{\pi\hbar^2 L_z} & \text{for } E_{h2}^{e2} < \hbar\omega < E_{h3}^{e3} \\ 3\dfrac{m_r^*}{\pi\hbar^2 L_z} & \text{for } E_{h3}^{e3} < \hbar\omega < E_{h4}^{e4} \\ \text{etc.} \end{cases} \tag{9.4.9}$$

which has a spectrum given by the joint density of states of a two-dimensional structure. A plot of the above function is shown in Fig. 9.8.

With carrier injection, the Fermi–Dirac population inversion factor, $f_v^m - f_c^n$, has to be included, and we obtain the absorption spectrum in the presence of carrier injection

$$\alpha(\hbar\omega) = \alpha_0(\hbar\omega)[f_v^m(E_t = \hbar\omega - E_{hm}^{en}) - f_c^n(E_t = \hbar\omega - E_{hm}^{en})]. \tag{9.4.10}$$

The above gain process can also be understood from Fig. 9.9 and the following analysis for carrier populations in quantum wells.

9.4.3 Determination of the Quasi-Fermi Levels

For a given injected electron density n, which is usually determined by the injection current and the background doping, Fig. 9.9a, the quasi-Fermi level F_c for the electrons can be determined using

$$n = \sum_{\substack{n= \\ \text{occupied} \\ \text{subbands}}} N_n = \sum_n \int_0^\infty \rho_e^{2D}(E) f_c^n(E) dE \tag{9.4.11a}$$

$$N_n = n_c \ln\left(1 + e^{(F_c - E_{en})/k_B T}\right) \tag{9.4.11b}$$

$$n_c = \frac{m_e^* k_B T}{\pi \hbar^2 L_z} \tag{9.4.11c}$$

where N_n is the electron density in the nth conduction subband. The carrier concentration is just the area below the function $\rho_e^{2D}(E) f_c(E)$ in the top figure of Fig. 9.9b. Similarly, the quasi-Fermi level for the holes, F_v, can be determined from

$$P = N + N_A^- - N_D^+ = \sum_m P_m$$

$$= \sum_m \frac{2}{V} \sum_{\mathbf{k}_t} [1 - f_v^m(k_t)] = \sum_m \int_0^\infty dE_t \rho_h(E_t) f_h(E_t) \tag{9.4.12a}$$

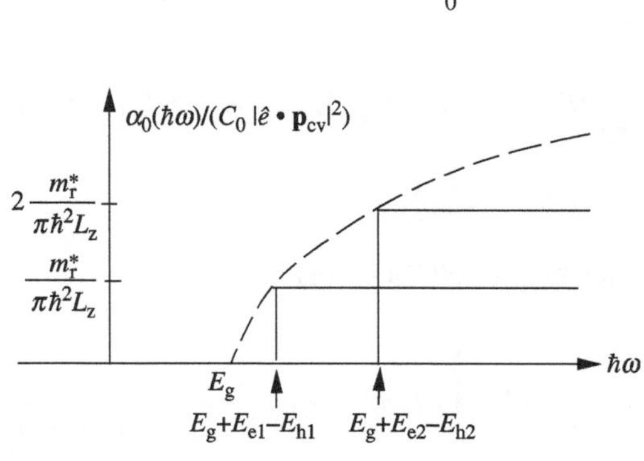

Figure 9.8 The stepwise absorption spectrum for a quantum-well structure.

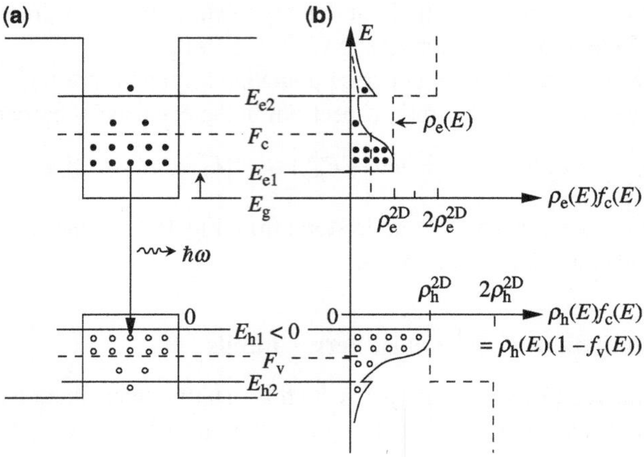

Figure 9.9 (a) Population inversion in a quantum well such that $F_c - F_v > \hbar\omega > E_g + E_{e1} - E_{h1}$. Here F_c is measured from the valence-band edge where the energy level is chosen to be zero. (b) The products of the density of states and the occupation probability for electrons in the conduction band $\rho_e(E)f_c(E)$ and holes in the valence band $\rho_h(E)f_h(E) = \rho_h(E)(1 - f_v(E))$ are plotted versus the energy E in the vertical scale.

$$P_m = n_v \ln\left[1 + e^{(E_{hm} - F_v)/k_B T}\right] \tag{9.4.12b}$$

$$n_v = \frac{m_h^* k_B T}{\pi\hbar^2 L_z}. \tag{9.4.12c}$$

Again, the hole concentration is just the area below the function $\rho_h(E)f_h(E)$ in the bottom figure of Fig. 9.9b.

9.4.4 Summary of the Gain Spectrum

We can write the gain spectrum from (9.4.9) and (9.4.11). Summarizing the expressions for the gain spectrum:

1. Zero linewidth

$$g(\hbar\omega) = C_0 \sum_{n,m} \left|I_{hm}^{en}\right|^2 |\hat{e} \cdot \mathbf{p}_{cv}|^2 \left[f_c^n\left(E_t = \hbar\omega - E_{hm}^{en}\right) - f_v^m\left(E_t = \hbar\omega - E_{hm}^{en}\right)\right]$$
$$\times \rho_r^{2D} H\left(\hbar\omega - E_{hm}^{en}\right). \tag{9.4.13}$$

2. Linewidth broadening (FWHM $= 2\gamma$)

$$g(\hbar\omega) = C_0 \sum_{n,m} \left|I_{hm}^{en}\right|^2 \int_0^\infty dE_t \rho_r^{2D} |\hat{e} \cdot \mathbf{p}_{cv}|^2 \frac{\gamma/\pi}{[E_{hm}^{en} + E_t - \hbar\omega]^2 + \gamma^2}$$
$$\times [f_c^n(E_t) - f_v^m(E_t)] \tag{9.4.14}$$

where

$$C_0 = \frac{\pi e^2}{n_r c \varepsilon_0 m_0^2 \omega} \qquad \rho_r^{2D} = \frac{m_r^*}{\pi \hbar^2 L_z} \tag{9.4.15}$$

$$I_{hm}^{en} = \int_{-\infty}^{\infty} dz \phi_n(z) g_m(z) \tag{9.4.16}$$

$$f_c^n(E_t) = \frac{1}{1 + \exp[(E_g + E_{en} + \frac{m_r^*}{m_e^*} E_t - F_c)/k_B T]} \tag{9.4.17a}$$

$$f_v^m(E_t) = \frac{1}{1 + \exp[(E_{hm} - \frac{m_r^*}{m_h^*} E_t - F_v)/k_B T]}. \tag{9.4.17b}$$

The proper momentum matrix elements $|\hat{e} \cdot \mathbf{p}_{cv}|^2$ for bulk and quantum wells are discussed in Section 9.5. It is important to note that the momentum matrix element is isotropic for bulk cubic semiconductors and it is polarization-dependent for

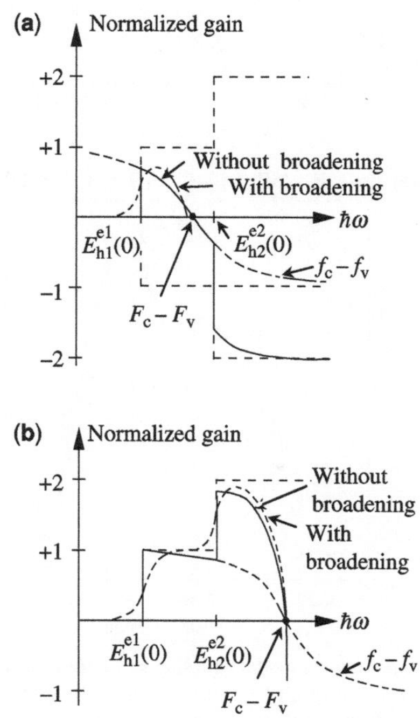

Figure 9.10 The gain spectra with (dashed curves) and without (solid) scattering broadenings for a quantum well with transition involving (a) a single electron and hole subband pair and (b) two electron and two hole subbands. The gain is normalized by the peak coefficient of the first step given by the coefficient in Eq. (9.4.13).

quantum well semiconductors. Theory and experiments on polarization-dependent gain in quantum well structures based on Kane's model [18] have been investigated [19–25]. It is noted that the momentum matrix $|\hat{e} \cdot \mathbf{p}_{cv}|^2$ should be used with care [21] because various missing factors exist in the literature.

9.4.5 Theoretical Gain Spectrum and Comparison with Experiments

The gain spectra using (9.4.13) for a zero linewidth and using (9.4.14) for a finite linewidth are plotted in Fig. 9.10a and Fig. 9.10b for transitions between a pair of

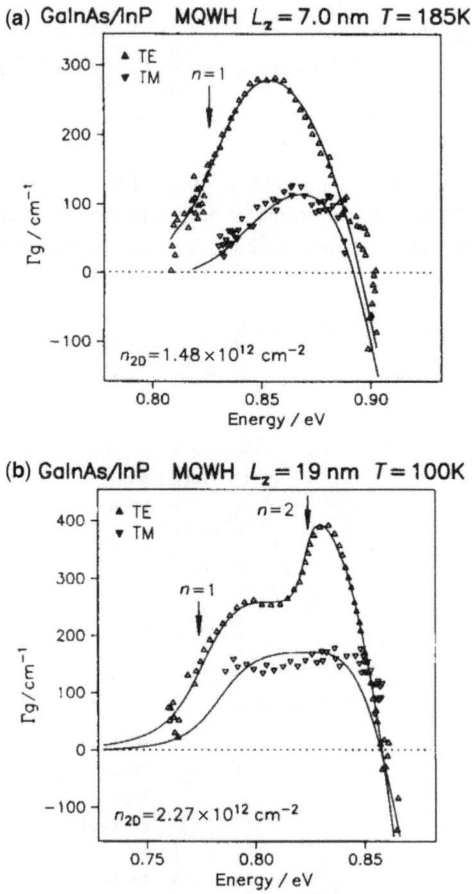

Figure 9.11 Theoretical and experimental modal gain spectra Γg for both the TE and TM polarizations of an $In_{0.53}Ga_{0.47}As/InP$ multiple-quantum-well laser (a) with a well width 70 Å at $T = 185$ K. The sheet carrier density is estimated to be $1.48 \times 10^{12}\,cm^{-2}$; (b) with a well width 190 Å at $T = 100$ K and the sheet carrier density $2.27 \times 10^{12} cm^{-2}$. The second electron and hole subbands are occupied, and they contribute to the second peak at the high-energy end. (Reprinted with permission from Ref. [24], © 1989 IEEE.)

electron and hole subbands and between two pairs of electron and hole subbands. Experimental data for quantum-well lasers have been shown to compare very well with the above theory based on the parabolic band model [24, 25], as can be seen from Fig. 9.11a and Fig. 9.11b. Improvement by the valence-band mixing will be discussed in Section 9.8. The materials are $In_{0.53}Ga_{0.47}As$ quantum wells lattice-matched to InP barriers and substrate. We can see at a small carrier concentration, in Fig. 9.11a, that only $n = 1$ electron and hole subbands contributed to the gain. The gain for TE polarization is larger than that of the TM polarization because most of the holes occupy the heavy-hole subband. At a higher carrier concentration in a quantum well with a larger well width in Fig. 9.11b, the second electron and hole subbands are occupied and features like double steps appear in the TE gain spectrum.

9.5 INTERBAND MOMENTUM MATRIX ELEMENTS OF BULK AND QUANTUM-WELL SEMICONDUCTORS

In a bulk semiconductor, the optical matrix element is usually isotropic. However, in a quantum-well or superlattice structure, the optical matrix element will depend on the polarization of the optical electromagnetic field [19]. In Kane's model [18] for the semiconductor band structures near the band edges, as discussed in Section 4.2, the electron wave vector is originally assumed to be in the z direction. The wave functions at the band edges have also been obtained as $|\frac{3}{2}, \pm\frac{3}{2}\rangle$ for heavy holes, $|\frac{3}{2}, \pm\frac{1}{2}\rangle$ for light holes, and $|\frac{1}{2}, \pm\frac{1}{2}\rangle$ for spin split-off holes. If the electron wave vector $\mathbf{k}$ has a general direction specified by (k, θ, ϕ) in spherical coordinates,

$$\mathbf{k} = k \sin\theta \cos\phi\, \hat{x} + k \sin\theta \sin\phi\, \hat{y} + k \cos\theta\, \hat{z}, \qquad (9.5.1)$$

the band-edge wave functions need be expressed in the general coordinate system, and they are listed in Appendix 9A. The band-edge momentum matrix elements will be averaged over the solid angle for the bulk semiconductors. For quantum wells, the angular average is taken over the x–y plane of quantum wells if the growth axis is the z axis.

9.5.1 Momentum Matrix Element of a Bulk Semiconductor

For a bulk semiconductor, we take the average of the momentum matrix element with respect to the solid angle $d\Omega$. For example, the momentum matrix element for TE polarization ($\hat{e} = \hat{x}$) due to the transitions from the conduction band with spin up $\langle iS\uparrow'|$ to both heavy-hole bands, $|\frac{3}{2}, \frac{3}{2}\rangle'$ and $|\frac{3}{2}, -\frac{3}{2}\rangle'$, is $|\langle iS\uparrow'|p_x|\frac{3}{2}, \frac{1}{2}\rangle'|^2 + |\langle iS\uparrow'|p_x|\frac{3}{2}, -\frac{1}{2}\rangle'|^2$ (one of the two transitions is zero). Here the superscript prime ($'$) means that the new z axis is set along the $\mathbf{k}$ direction. The average over the solid angle is as follows,

$$
\begin{aligned}
|\hat{e}\cdot \mathbf{p}_{\mathrm{cv}}|^2 &\equiv \left\langle |\hat{e}\cdot \mathbf{M}_{c-\mathrm{hh}}|^2 \right\rangle = \frac{1}{4\pi}\int_{\theta=0}^{\pi}\int_{\phi=0}^{2\pi} |\hat{x}\cdot \mathbf{M}_{c-\mathrm{hh}}|^2 \sin\theta\, d\theta\, d\phi \\
&= \frac{1}{4\pi}\int_0^{\pi}\sin\theta\, d\theta \int_0^{2\pi} d\phi(\cos^2\theta\,\cos^2\phi + \sin^2\phi)\frac{P_x^2}{2} \\
&= \frac{1}{3}P_x^2 \equiv M_{\mathrm{b}}^2
\end{aligned}
\tag{9.5.2}
$$

where

$$
M_{\mathrm{b}}^2 = \frac{1}{3}P_x^2 = \frac{m_0^2}{3\hbar^2}P^2 = \left(\frac{m_0}{m_{\mathrm{e}}^*} - 1\right)\frac{m_0 E_{\mathrm{g}}(E_{\mathrm{g}}+\Delta)}{6\left(E_{\mathrm{g}} + \frac{2}{3}\Delta\right)}
\tag{9.5.3}
$$

is the bulk momentum matrix element (squared). The above expression (9.5.3) from Kane's model may underestimate the momentum matrix element. In practice, an energy parameter E_{p} for the matrix element is defined,

$$
E_{\mathrm{p}} = \frac{2m_0}{\hbar^2}P^2 \tag{9.5.4a}
$$

$$
M_{\mathrm{b}}^2 = \frac{m_0}{6}E_{\mathrm{p}} \tag{9.5.4b}
$$

where E_{p} is taken from experimental data and is tabulated in Table C.2 in Appendix C at the end of the book. Notice that the above result (9.5.2) is the same for the other spin in the conduction band. If we take $\hat{e} = \hat{y}$ or $\hat{z}$, and average over the solid angle, we still obtain M_{b}^2 because the bulk crystal is isotropic. We could also take the spin-down component, $\left|\langle iS\downarrow'|p_x|\frac{3}{2},\frac{1}{2}\rangle'\right|^2 + \left|\langle iS\downarrow'|p_x|\frac{3}{2},-\frac{1}{2}\rangle'\right|^2$, and average over the solid angle and obtain M_{b}^2 as expected. Therefore, a factor of 2 is usually used in the interband transitions when we calculate the optical absorption or gain.

9.5.2 Momentum Matrix Elements of Quantum Wells

Next, we consider quantum-well structures. The optical matrix elements will become polarization dependent. The theory and experimental data on the polarization-dependent gain have been discussed in Refs. 19–23 based on the parabolic band model. Improvement of the optical matrix element using the valence-band mixing model in quantum wells is discussed in Section 9.8.

TE Polarization Let $\hat{e} = \hat{x}$; that is, the optical field is polarized along the x (or y) direction. The optical dipole matrix element is averaged over the azimuthal angle ϕ in the plane of quantum wells. We obtain for $\langle iS \uparrow'|$,

$$
\begin{aligned}
|\hat{e} \cdot \mathbf{p}_{cv}|^2 \equiv \left\langle |\hat{e} \cdot \mathbf{M}_{c-\mathrm{hh}}|^2 \right\rangle &= \frac{1}{2\pi} \int_0^{2\pi} d\phi |\hat{x} \cdot \mathbf{M}_{c-\mathrm{hh}}|^2 \\
&= \frac{1}{2\pi} \int_0^{2\pi} d\phi (\cos^2\theta \cos^2\phi + \sin^2\phi) \frac{P_x^2}{2} \\
&= \frac{3}{4}(1 + \cos^2\theta) M_b^2
\end{aligned} \tag{9.5.5}
$$

and the same matrix element is obtained for the other spin $\langle iS \downarrow'|$in the conduction band. Similarly, for the sum of transitions of an electron with spin $\downarrow'$ from the conduction band to both light-hole bands, we find

$$
\begin{aligned}
\langle |\hat{e} \cdot \mathbf{M}_{c-\ell\mathrm{h}}|^2 \rangle &= \frac{1}{2\pi} \int_0^{2\pi} d\phi \left(\left| \left\langle iS \downarrow' |p_x| \frac{3}{2}, \frac{1}{2} \right\rangle' \right|^2 + \left| \left\langle iS \downarrow' |p_x| \frac{3}{2}, -\frac{1}{2} \right\rangle' \right|^2 \right) \\
&= \left[\frac{2}{3} \sin^2\theta \langle \cos^2\phi \rangle + \frac{1}{6} \cos^2\theta \langle \cos^2\phi \rangle + \frac{1}{6} \langle \sin^2\phi \rangle \right] P_x^2 \\
&= \left[\sin^2\theta + \frac{1}{4} (\cos^2\theta + 1) \right] M_b^2 \\
&= \left[\frac{5}{4} - \frac{3}{4} \cos^2\theta \right] M_b^2 .
\end{aligned} \tag{9.5.6}
$$

The above result is the same for both spins $\langle iS \downarrow'|$ and $\langle iS \uparrow'|$. Note that a sum rule exists,

$$
\left\langle |\hat{e} \cdot \mathbf{M}_{c-\mathrm{hh}}|^2 \right\rangle + \left\langle |\hat{e} \cdot \mathbf{M}_{c-\ell\mathrm{h}}|^2 \right\rangle = 2M_b^2, \tag{9.5.7}
$$

which is independent of the angle θ.

TM Polarization Let $\hat{e} = \hat{z}$

$$
\langle |\hat{e} \cdot \mathbf{M}_{c-\mathrm{hh}}|^2 \rangle = \frac{1}{2\pi} \int_0^{2\pi} d\phi |\hat{z} \cdot \mathbf{M}_{c-\mathrm{hh}}|^2 = \frac{3}{2} \sin^2\theta M_b^2 \tag{9.5.8}
$$

$$\begin{aligned}\langle|\hat{e}\cdot\mathbf{M}_{c-\ell h}|^2\rangle &= \frac{1}{2\pi}\int_0^{2\pi} d\phi\left(\left|\left\langle iS\downarrow'|p_z|\frac{3}{2},\frac{1}{2}\right\rangle'\right|^2+\left|\left\langle iS\downarrow'|p_z|\frac{3}{2},-\frac{1}{2}\right\rangle'\right|^2\right)\\ &= \left[\frac{1}{6}\sin^2\theta+\frac{2}{3}\cos^2\theta\right]P_z^2\\ &= \frac{1+3\cos^2\theta}{2}M_b^2. \end{aligned} \tag{9.5.9}$$

The sum of (9.5.8) and (9.5.9) for $\hat{e}=\hat{z}$ is still $2M_b^2$. The above results (9.5.8) and (9.5.9) are the same for both spins in the conduction band. The angular factor $\cos^2\theta$ can be related to the electron or hole wave vectors by

$$\cos^2\theta = \frac{k_z^2}{k_x^2+k_y^2+k_z^2} = \frac{E_{en}}{E_{en}+\dfrac{\hbar^2k_t^2}{2m_e^*}} \tag{9.5.10}$$

for the electron wave vector $\mathbf{k}=\hat{\rho}\,k_t+\hat{z}\,k_{zn}$, with $E_{en}=\hbar^2k_{zn}^2/2m_e^*$ or

$$\cos^2\theta = \frac{|E_{hm}|}{|E_{hm}|+\dfrac{\hbar^2k_t^2}{2m_h^*}} \tag{9.5.11}$$

for the hole wave vector $\mathbf{k}=\hat{\rho}\,k_t+\hat{z}\,k_{zm}$, with $E_{hm}=-\dfrac{\hbar k_{zm}^2}{2m_h^*}$. Note that the hole energy E_{hm} is defined to be negative so all of the energies are measured upward. An alternative approximation is

$$\cos^2\theta_{nm} = \frac{E_{en}+|E_{hm}|}{E_{en}+|E_{hm}|+\dfrac{\hbar^2k_t^2}{2m_r^*}} \tag{9.5.12}$$

where the reduced effective mass m_r defined by

$$\frac{1}{m_r^*}=\frac{1}{m_e^*}+\frac{1}{m_h^*} \tag{9.5.13}$$

has been used. The results of the momentum matrix elements are summarized in Table 9.1. It is noted at the subband edges, where $k_t=0$, that

Table 9.1 Summary of the Momentum Matrix Elements in Parabolic Band Model ($|\hat{e}\cdot \mathbf{p}_{cv}|^2 = |\hat{e}\cdot \mathbf{M}|^2$)

Bulk

$$|\hat{x}\cdot \mathbf{p}_{cv}|^2 = |\hat{y}\cdot \mathbf{p}_{cv}|^2 = |\hat{z}\cdot \mathbf{p}_{cv}|^2 = M_b^2 = \frac{m_0}{6}E_p$$

Quantum Well

TE Polarization ($\hat{e} = \hat{x}$ or $\hat{y}$)	TM Polarization ($\hat{e} = \hat{z}$)
$\left\langle \|\hat{e}\cdot \mathbf{M}_{c-\mathrm{hh}}\|^2 \right\rangle = \frac{3}{4}(1+\cos^2\theta)M_b^2$	$\left\langle \|\hat{e}\cdot \mathbf{M}_{c-\mathrm{hh}}\|^2 \right\rangle = \frac{3}{2}\sin^2\theta M_b^2$
$\left\langle \|\hat{e}\cdot \mathbf{M}_{c-\ell\mathrm{h}}\|^2 \right\rangle = \left[\frac{5}{4} - \frac{3}{4}\cos^2\theta\right]M_b^2$	$\left\langle \|\hat{e}\cdot \mathbf{M}_{c-\ell\mathrm{h}}\|^2 \right\rangle = \frac{1}{2}(1+3\cos^2\theta)M_b^2$

Conservation Rule

$$\left\langle |\hat{x}\cdot \mathbf{M}_{c-h}|^2 \right\rangle + \left\langle |\hat{y}\cdot \mathbf{M}_{c-h}|^2 \right\rangle + \left\langle |\hat{z}\cdot \mathbf{M}_{c-h}|^2 \right\rangle = 3M_b^2, \quad (h = \mathrm{hh} \text{ or } \ell\mathrm{h}).$$

$$\left\langle |\hat{e}\cdot \mathbf{M}_{c-\mathrm{hh}}|^2 \right\rangle + \left\langle |\hat{e}\cdot \mathbf{M}_{c-\ell\mathrm{h}}|^2 \right\rangle = 2M_b^2$$

(a) for TE polarization, the matrix elements are

$$\left\langle |\hat{x}\cdot \mathbf{M}_{c-\mathrm{hh}}|^2 \right\rangle_{k_t=0} = \frac{3}{2}M_b^2 \tag{9.5.14a}$$

$$\left\langle |\hat{x}\cdot \mathbf{M}_{c-\ell\mathrm{h}}|^2 \right\rangle_{k_t=0} = \frac{1}{2}M_b^2 \tag{9.5.14b}$$

and (b) for TM polarization, the matrix elements are

$$\left\langle |\hat{z}\cdot \mathbf{M}_{c-\mathrm{hh}}|^2 \right\rangle_{k_t=0} = 0 \tag{9.5.15a}$$

$$\left\langle |\hat{z}\cdot \mathbf{M}_{c-\ell\mathrm{h}}|^2 \right\rangle_{k_t=0} = 2M_b^2. \tag{9.5.15b}$$

These matrix elements are useful to study the excitonic absorptions in quantum wells and they will be used in Chapter 14 where we discuss electroabsorption modulators using quantum-confined Stark effects.

9.6 QUANTUM DOTS AND QUANTUM WIRES

The theory developed in Sections 9.3 and 9.4 can be easily extended to quantum dots [26–54] and quantum wires [55–57]. Since the early theories of quantum-dot lasers [26–28] and the predictions of low threshold current density, low chirp, and high differential gain, progress on the experiments has been impressive, leading to the first electrical injection quantum dot lasers [29, 30]. It is remarkable that even though quantum dots suffer from inhomogeneous broadening, which results in the broadening of the gain spectrum and the reduction of the gain peak, quantum-dot lasers have demonstrated low threshold current density and high temperature stability [30, 32, 34]. In this section, we discuss the elementary model for quantum dots and quantum wires. We will discuss quantum-dot and quantum-wire lasers in Section 10.5.

9.6.1 Quantum Dots

We first assume that the dots are uniformly distributed in order to have a simple picture of the dot density and density of states, Fig. 9.12. We have an average of one dot per volume of $V = L_x L_y L_z$, whereas each dot takes a size $a \times b \times c$ along the x, y, and z directions. We then have the 3D and 2D (areal) dot densities for ease of counting the dots and carrier density:

$$N_{\text{dot}}^{3\text{D}} = \frac{1}{V} = \frac{1}{L_x L_y L_z} \qquad N_{\text{dot}}^{2\text{D}} = \frac{1}{L_x L_y}. \tag{9.6.1}$$

The *fill factor* is defined as

$$F^{3\text{D}} = \frac{abc}{L_x L_y L_z} \qquad F^{2\text{D}} = \frac{ab}{L_x L_y}. \tag{9.6.2}$$

For electrons in the conduction subband, labeled by $(mn\ell)$, we have the wave function

$$\psi_c(x, y, z) = \frac{\sqrt{8}}{\sqrt{abc}} \sin\left(\frac{m\pi}{a} x\right) \sin\left(\frac{n\pi}{b} y\right) \sin\left(\frac{\ell\pi}{c} z\right) u_c(\mathbf{r}) \tag{9.6.3}$$

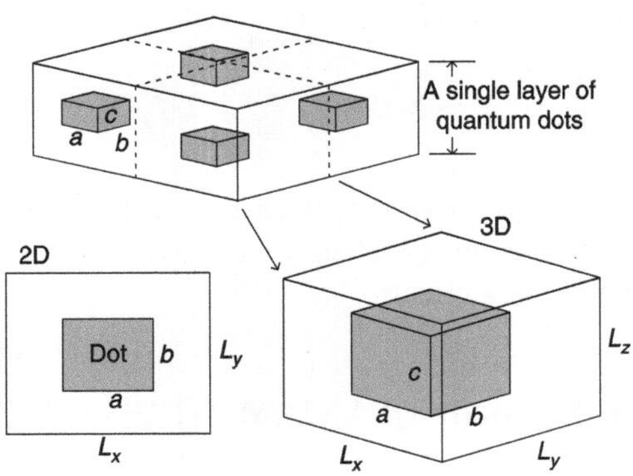

Figure 9.12 Schematics of a layer of quantum dots distributed uniformly in a simplified model for counting the number of the quantum dots and the density of states. A three-dimensional (3D) and two-dimensional (2D) *fill factor* can be defined.

in a quantum dot, and $u_c(\mathbf{r})$ is the Bloch periodic part of the conduction-band wave function. We have assumed an infinite barrier potential for the quantum dot. The corresponding conduction-band energy is

$$E_c^{mn\ell} = E_{c0} + \frac{\hbar^2}{2m_e^*}\left[\left(\frac{m\pi}{a}\right)^2 + \left(\frac{n\pi}{b}\right)^2 + \left(\frac{\ell\pi}{c}\right)^2\right] \tag{9.6.4}$$

where E_{c0} is the conduction-band edge ($= E_g$ for an unstrained semiconductor in a simplified model). The electron density n is related to the dot density and how the states of each dot are occupied:

$$\begin{aligned} n &= \frac{2}{V}\sum_{m,n,\ell} f_c(E) = N_{\text{dot}}^{3D}\, 2\sum_{m,n,\ell} f_c(E_c^{mn\ell}) \\ &= 2\frac{N_{\text{dot}}^{2D}}{L_z}\sum_{m,n,\ell}\frac{1}{1 + e^{(E_c^{mn\ell} - F_c)/k_B T}} \end{aligned} \tag{9.6.5}$$

where F_c is the quasi-Fermi level for the conduction band. Similarly, the valence-band wave function and corresponding energy is

$$\psi_v(x, y, z) = \frac{\sqrt{8}}{\sqrt{abc}}\sin\left(\frac{m'\pi}{a}x\right)\sin\left(\frac{n'\pi}{b}y\right)\sin\left(\frac{\ell'\pi}{c}z\right)u_v(\mathbf{r}) \tag{9.6.6}$$

$$E_v^{m'n'\ell'} = E_{v0} - \frac{\hbar^2}{2m_h^*}\left[\left(\frac{m'\pi}{a}\right)^2 + \left(\frac{n'\pi}{b}\right)^2 + \left(\frac{\ell'\pi}{c}\right)^2\right], \tag{9.6.7}$$

which is measured from the valence-band edge, E_{v0} ($= 0$ for unstrained sample). A similar equation as (9.6.5) holds for the hole density with the quasi-Fermi level in the valence band.

The interband momentum matrix element is

$$\langle\psi_c|\hat{e}\cdot\mathbf{p}|\psi_v\rangle \simeq \langle u_c|\hat{e}\cdot\mathbf{p}|u_v\rangle\,\delta_{mm'}\delta_{nn'}\delta_{\ell\ell'} \tag{9.6.8}$$

which can be derived using $\mathbf{p} = (\hbar/i)\nabla$ and the fact that $\langle u_c|u_v\rangle = 0$ and $\langle u_c|\hat{e}\cdot\mathbf{p}|u_v\rangle = \hat{e}\cdot\mathbf{p}_{cv}$, which is related to the bulk momentum matrix element or E_p matrix parameter. The interband absorption spectrum $\alpha(\hbar\omega)$ is given by

$$\begin{aligned} \alpha(\hbar\omega) &= C_0\frac{2}{V}\sum_{mn\ell}\sum_{m'n'\ell'}|\langle\psi_c|\hat{e}\cdot\mathbf{p}|\psi_v\rangle|^2\delta(E_c^{mn\ell} - E_v^{m'n'\ell'} - \hbar\omega)(f_v - f_c) \\ &= C_0\frac{2N_{\text{dot}}^{2D}}{L_z}\sum_{m,n,\ell}|\hat{e}\cdot\mathbf{p}_{cv}|^2\delta(E_c^{mn\ell} - E_v^{mn\ell} - \hbar\omega)(f_v - f_c). \end{aligned} \tag{9.6.9}$$

The interband transition energies occur at

$$E_{\mathrm{cv}}^{(mn\ell)} = E_{\mathrm{c}}^{(mn\ell)} - E_{\mathrm{v}}^{(mn\ell)} = E_{\mathrm{g}} + \frac{\hbar^2}{2m_{\mathrm{r}}^*}\left[\left(\frac{m\pi}{a}\right)^2 + \left(\frac{n\pi}{b}\right)^2 + \left(\frac{\ell\pi}{c}\right)^2\right] \tag{9.6.10}$$

where $N_{\mathrm{dot}}^{2\mathrm{D}}$ is the number of quantum dots per unit area (#/cm^2) and L_z is the thickness of the layer containing the dots. When there is no carrier injection, $f_{\mathrm{v}} = 1$ and $f_{\mathrm{c}} = 0$, the absorption spectrum is

$$\alpha_0(\hbar\omega) = C_0 \frac{2N_{\mathrm{dot}}^{2\mathrm{D}}}{L_z} \sum_{m,n,\ell} |\hat{e} \cdot \mathbf{p}_{\mathrm{cv}}|^2 L\left(E_{\mathrm{cv}}^{(mn\ell)} - \hbar\omega\right) \tag{9.6.11a}$$

$$L(E - \hbar\omega) = \frac{\gamma/\pi}{(E - \hbar\omega)^2 + \gamma^2} \tag{9.6.11b}$$

where we have replaced the delta function by a Lorentzian lineshape function with a full width at half maximum 2γ to account for homogeneous broadening caused by carrier scattering processes. In the presence of carrier injection, we insert f_{v} and f_{c},

$$\alpha(\hbar\omega) = C_0 \frac{2N_{\mathrm{dot}}^{2\mathrm{D}}}{L_z} \sum_{m,n,\ell} |\hat{e} \cdot \mathbf{p}_{\mathrm{cv}}|^2 L\left(E_{\mathrm{cv}}^{(mn\ell)} - \hbar\omega\right) (f_{\mathrm{v}} - f_{\mathrm{c}}). \tag{9.6.12}$$

Inhomogeneous Broadening When inhomogeneous broadening due to the size variation of quantum dots is taken into account, the above formula is replaced by

$$n = 2\frac{N_{\mathrm{dot}}^{2\mathrm{D}}}{L_z} \sum_{m,n,\ell} \int_0^\infty dE\, G(E) f_{\mathrm{c}}(E) \tag{9.6.13}$$

where $G(E)$ is a Gaussian function given by

$$G(E) = \frac{1}{\sqrt{2\pi}\sigma_{\mathrm{c}}} e^{-(E - E_{\mathrm{c}}^{mn\ell})^2/2\sigma_{\mathrm{c}}^2} \tag{9.6.14}$$

which has a mean value centered at $E_{\mathrm{c}}^{mn\ell}$ and a full width at half maximum (FWHM) equal to $2\sqrt{2\ln 2}\sigma_{\mathrm{c}} \simeq 2.35\sigma_{\mathrm{c}}$. A similar expression holds between the hole concentration p and f_{v}, which determines the quasi-Fermi level F_{v}. If the effect of inhomogeneous broadening is taken into account, the absorption/gain expression requires an integral

$$\alpha(\hbar\omega) = C_0 \sum_{m,n,\ell} \int_0^\infty dE |\hat{e} \cdot \mathbf{p}_{\mathrm{cv}}|^2 D(E) L(E - \hbar\omega)(f_{\mathrm{v}} - f_{\mathrm{c}}) \tag{9.6.15}$$

where

$$D(E) = \frac{2N_{\text{dot}}^{\text{2D}}}{L_z} \frac{1}{\sqrt{2\pi}\sigma} e^{-(E - E_{\text{cv}}^{mn\ell})^2/2\sigma^2} \tag{9.6.16a}$$

$$\sigma^2 = \sigma_{\text{c}}^2 + \sigma_{\text{v}}^2 \tag{9.6.16b}$$

which accounts for inhomogeneous broadening effects on the electron and hole energies.

Example: Absorption and Gain of a Quantum-Dot Structure Figure 9.13a shows the absorption spectrum of a uniform quantum dot sample at various injection carrier densities ($n = p$) when we consider only the homogenous broadening. The parameters are $a = b = 200$ Å, $c = 50$ Å, $N_{\text{dot}}^{\text{2D}} = 5 \times 10^{10}$ cm^{-2}, $L_z = 50$ Å $= c$, and $2\gamma = 30$ meV. The other parameters are $m_{\text{e}}^* = 0.023m_0$, $m_{\text{h}}^* = 0.40m_0$,

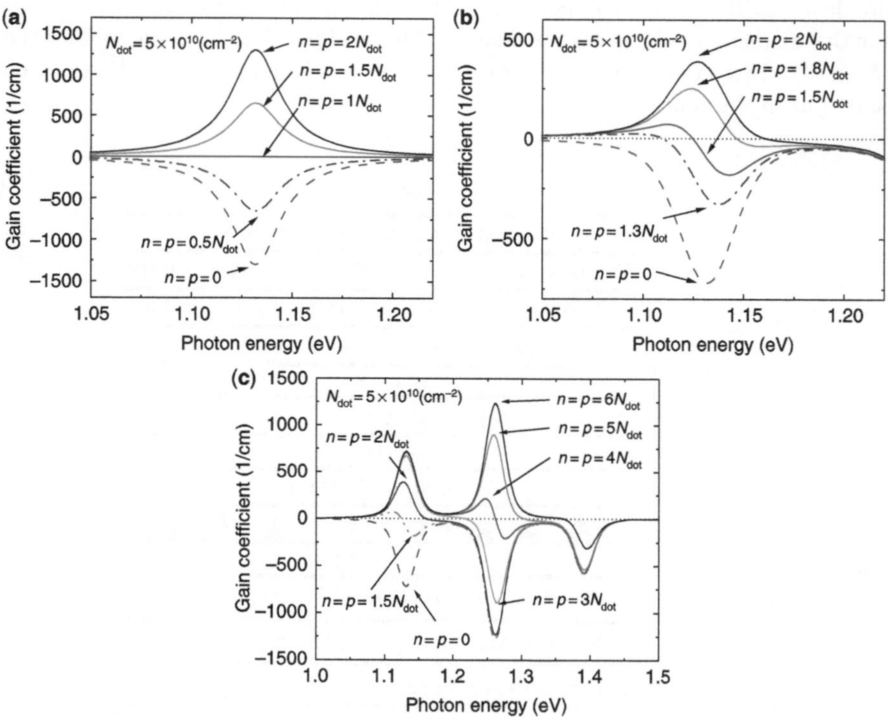

Figure 9.13 Optical absorption and gain spectra of a quantum-dot structure for (a) only homogeneous broadening with a FWHM of $2\gamma = 30$ meV; (b) and (c) both homogenous and inhomogeneous broadenings with increasing carrier density from low (b) to high (c) values. We assume the following parameters: $N_{\text{dot}}^{\text{2D}} = 5 \times 10^{10}$ cm^{-2}, $L_x = L_y = 200$ Å, $L_z = 50$ Å, and the full width at half maximum of the homogeneous broadening is $2\gamma = 30$ meV for (a). An inhomogeneous broadening width of 50 meV is used in (b) and (c).

$E_g = 0.354$ eV, $E_p = 22.2$ eV, $n_r = 3.51$. Figure 9.13b shows the plot of the optical gain spectrum at various injection carrier densities when we include the inhomogeneous broadening with a linewidth (FWHM) (i.e., $2\sqrt{2\ln 2}\sigma_c) = 50$ meV. We have $\sigma_c = 21.2$ meV, $\sigma_v = 1.2$ meV, and $\sigma = 21.25$ meV. We can see that the quasi-Fermi level separation (or the transparency energy between the gain and absorption) has a blue shift due to band filling as the carrier density is increased. This blue shift is similar to that in a quantum well or bulk semiconductor. Further increase of carrier density is shown in Fig. 9.13c.

Figure 9.14 [53] shows an example of the layered structure grown by molecular epitaxy from NanoSemiconductor GmbH in Germany [42, 52]. There are 10 layers of quantum dots, and the dot density is about $5.0 \times 10^{10}\,\text{cm}^{-2}$ per dot layer. The sample has a p-type doping of $5.0 \times 10^{17}\,\text{cm}^{-3}$ in the GaAs barrier of each period. By modeling the dots as microdisks, the electron and hole subband energies can be calculated, and then the optical gain spectrum at various carrier density. Figure 9.15a shows the energy band profile and subband energy levels [53]. Figure 9.15b shows the calculated optical gain spectrum and comparison with the measured data, with the 2D electron density per dot layer as the input parameter at each bias current. The comparison allows for the extraction of the carrier density. The intrinsic loss appears to be small ($5\,\text{cm}^{-1}$). The ionized acceptor concentration is taken as $3.0 \times 10^{11}\,\text{cm}^{-2}$ per dot layer in the model.

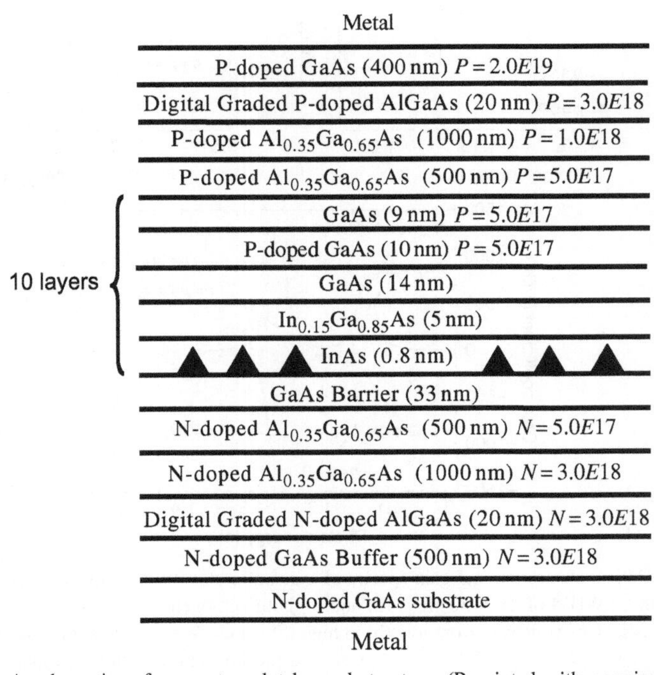

Figure 9.14 A schematics of a quantum-dot layered structure. (Reprinted with permission from [53], © 2006 IEEE.)

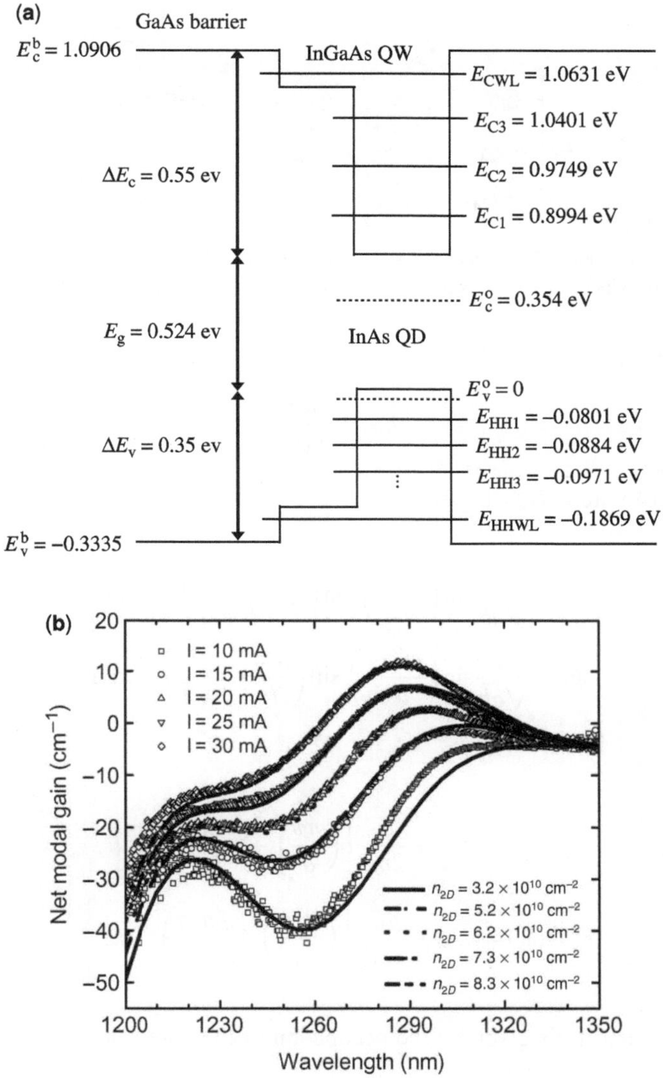

Figure 9.15 (a) Calculated energy levels of a quantum dot modeled as a disk. (b) The measured (solid) and calculated (dashed) optical gain spectra at various (surface) electron density per quantum-dot layer are plotted. (Reprinted with permission from [53] © 2006 IEEE.)

9.6.2 Quantum Wires

We consider a quantum wire [55–58] of a cross section $a \times b$, Fig. 9.16. The areal density of the wires in the $x-y$ cross section is

$$N_{wr} = \frac{1}{L_x L_y}. \tag{9.6.17}$$

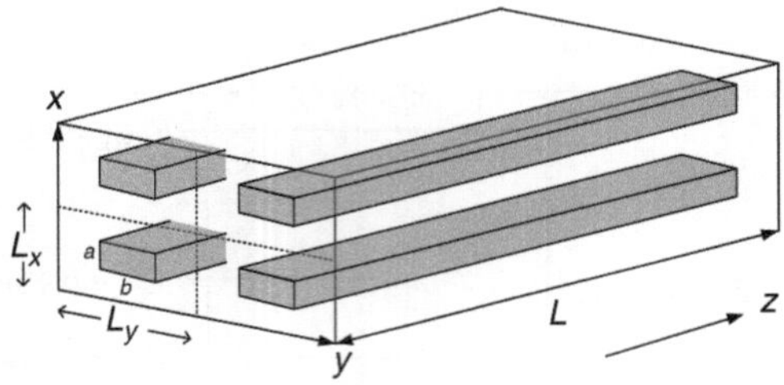

Figure 9.16 Schematics of a quantum-wire structure. The wire density in the $x-y$ plane is defined as $1/(L_xL_y)$ and the wire cross section is $a \times b$. The z axis is assumed to be the axis of the wire.

The fill factor is

$$F_{\mathrm{wr}} = \frac{ab}{L_xL_y}. \tag{9.6.18}$$

The length of wire is $L \gg a, b$.

$$\psi_{\mathrm{c}} = \frac{2}{\sqrt{ab}} \sin\left(\frac{m\pi}{a}x\right) \sin\left(\frac{n\pi}{b}y\right) \frac{1}{\sqrt{L}} e^{ik_z z} u_{\mathrm{c}}(\mathbf{r}) \tag{9.6.19a}$$

$$\psi_{\mathrm{v}} = \frac{2}{\sqrt{ab}} \sin\left(\frac{m'\pi}{a}x\right) \sin\left(\frac{n'\pi}{b}y\right) \frac{1}{\sqrt{L}} e^{ik'_z z} u_{\mathrm{v}}(\mathbf{r}) \tag{9.6.19b}$$

where

$$E_{\mathrm{c}}^{mn}(k_z) = E_{\mathrm{c}0} + \frac{\hbar^2}{2m_{\mathrm{e}}^*}\left[\left(\frac{m\pi}{a}\right)^2 + \left(\frac{n\pi}{b}\right)^2 + k_z^2\right] \tag{9.6.20a}$$

$$E_{\mathrm{v}}^{m'n'}(k'_z) = E_{\mathrm{v}0} - \frac{\hbar^2}{2m_{\mathrm{h}}^*}\left[\left(\frac{m'\pi}{a}\right)^2 + \left(\frac{n'\pi}{b}\right)^2 + k_z'^2\right]. \tag{9.6.20b}$$

The electron density is given by the occupation of the wire states:

$$n = \frac{2}{V}\sum_{m,n,k_z} f_c(E) = N_{\mathrm{wr}}\frac{2}{L}\sum_{m,n}\int_{-\infty}^{\infty}\frac{dk_z}{\left(\frac{2\pi}{L}\right)} f_c(E)$$

$$= N_{\mathrm{wr}}\int_0^{\infty} dE \rho_{\mathrm{e}}^{\mathrm{1D}}(E)\, f_c(E) \tag{9.6.21a}$$

$$\rho_{\mathrm{e}}^{\mathrm{1D}}(E) = \sum_{m,n}\frac{1}{\pi}\sqrt{\frac{2m_{\mathrm{e}}^*}{\hbar^2}}\frac{1}{\sqrt{E - E_{\mathrm{c}}^{mn}}} \quad \text{for } E > E_{\mathrm{c}}^{mn}. \tag{9.6.21b}$$

Similar expression holds for the hole density in the valence band. At zero temperature, the Fermi function is 1 for energy below the quasi-Fermi level, F_c, and (9.6.21) can be integrated analytically:

$$n = N_{wr} \frac{1}{\pi} \sqrt{\frac{2m_e^*}{\hbar^2}} \sum_{m,n} \sqrt{F_c - E_c^{mn}}. \tag{9.6.22}$$

The interband matrix element is

$$\langle \psi_c | \hat{e} \cdot \mathbf{p} | \psi_v \rangle = \langle u_c | \hat{e} \cdot \mathbf{p} | u_v \rangle \delta_{mm'} \delta_{nn'} \delta_{k_z k_z'}. \tag{9.6.23}$$

The absorption coefficient for a wire density of N_{wr} (number of wires per cross-sectional area perpendicular to the wire direction) is then

$$\alpha(\hbar\omega) = C_0 N_{wr} \frac{2}{L} \sum_{m,n} \int_{-\infty}^{\infty} \frac{dk_z}{\left(\frac{2\pi}{L}\right)} |\hat{e} \cdot \mathbf{p}_{cv}|^2 \; \delta\left(E_c^{mn}(k_z) - E_v^{mn}(k_z)\right)(f_v - f_c)$$

$$= C_0 N_{wr} \sum_{m,n} |\hat{e} \cdot \mathbf{p}_{cv}|^2 \rho_r^{1D}(\hbar\omega - E_{cv}^{(mn)})(f_v - f_c) \tag{9.6.24a}$$

where the one-dimensional joint density of states is

$$\rho_r^{1D}\left(\hbar\omega - E_{cv}^{mn}\right) = \frac{1}{\pi} \sqrt{\frac{2m_r^*}{\hbar^2}} \frac{1}{\sqrt{\hbar\omega - E_{cv}^{mn}}} \tag{9.6.24b}$$

$$E_{cv}^{mn} = E_g + \frac{\hbar^2}{2m_r^*} \left[\left(\frac{m\pi}{a}\right)^2 + \left(\frac{n\pi}{b}\right)^2 \right]. \tag{9.6.24c}$$

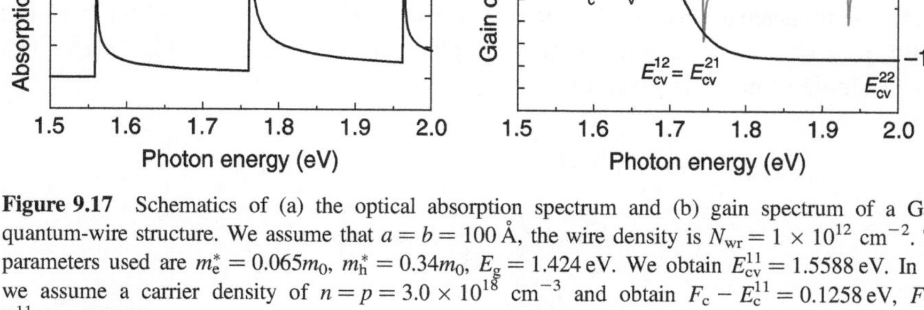

Figure 9.17 Schematics of (a) the optical absorption spectrum and (b) gain spectrum of a GaAs quantum-wire structure. We assume that $a = b = 100$ Å, the wire density is $N_{wr} = 1 \times 10^{12}$ cm^{-2}. The parameters used are $m_e^* = 0.065 m_0$, $m_h^* = 0.34 m_0$, $E_g = 1.424$ eV. We obtain $E_{cv}^{11} = 1.5588$ eV. In (b), we assume a carrier density of $n = p = 3.0 \times 10^{18}$ cm^{-3} and obtain $F_c - E_c^{11} = 0.1258$ eV, $F_v - E_v^{11} = -0.0253$ ev.

The above simplified model neglects the homogeneous and inhomogeneous broadenings of quantum wire states in the optical transitions. If all of those broadenings are taken into account, the 1D density of states will be broadened and the absorption or gain spectrum will be broadened as well. In Fig. 9.17, we plot a schematic diagram for (a) the optical absorption and (b) gain spectrum of a layer of GaAs quantum wires using (9.6.24), with the linewidth broadenings ignored. The major features are the 1D density of states, which behaves like a $1/\sqrt{\hbar\omega - E_{\mathrm{cv}}^{mn}}$ above each subband edge, and the presence of gain when population inversion is achieved.

9.7 INTERSUBBAND ABSORPTION

In an n-doped quantum-well structure, intersubband absorption is of interest because of its applications to far-infrared photodetectors [59–61], for example. In this section, we study the infrared absorption [62] due to intersubband transitions in the conduction band of a quantum-well structure with modulation doping. We assume that the doping is not large enough so that screening or many-body effects due to the electron–electron Coulomb interaction [63] can be ignored. The far-infrared photodetector applications using intersubband transitions will be discussed in Section 15.5.

9.7.1 Intersubband Dipole Moment

Consider the intersubband transition between the ground state

$$\psi_a(\mathbf{r}) = u_{\mathrm{c}}(\mathbf{r})\frac{e^{i\mathbf{k}_{\mathrm{t}}\cdot\rho}}{\sqrt{A}}\phi_1(z) \tag{9.7.1}$$

and the first excited state

$$\psi_b(\mathbf{r}) = u_{\mathrm{c}'}(\mathbf{r})\frac{e^{i\mathbf{k}'_{\mathrm{t}}\cdot\rho}}{\sqrt{A}}\phi_2(z) \tag{9.7.2}$$

where the transverse wave vectors $\mathbf{k}_{\mathrm{t}} = k_x\hat{x} + k_y\hat{y}$, $\mathbf{k}'_{\mathrm{t}} = k'_x\hat{x} + k'_y\hat{y}$ and the position vector $\boldsymbol{\rho} = x\hat{x} + y\hat{y}$ in the quantum-well plane have been used, Fig. 9.18. The optical dipole moment is given by

$$\begin{aligned}\boldsymbol{\mu}_{ba} &= \langle\psi_b|e\mathbf{r}|\psi_a\rangle \\ &\simeq \langle u_{\mathrm{c}}|u_{\mathrm{c}'}\rangle\left\langle\frac{e^{i\mathbf{k}'_{\mathrm{t}}\cdot\rho}}{\sqrt{A}}\phi_2|e\mathbf{r}|\frac{e^{i\mathbf{k}_{\mathrm{t}}\cdot\rho}}{\sqrt{A}}\phi_1\right\rangle \\ &\simeq \delta_{\mathbf{k}_{\mathrm{t}},\mathbf{k}'_{\mathrm{t}}}\langle\phi_2|ez|\phi_1\rangle\hat{z}\end{aligned} \tag{9.7.3}$$

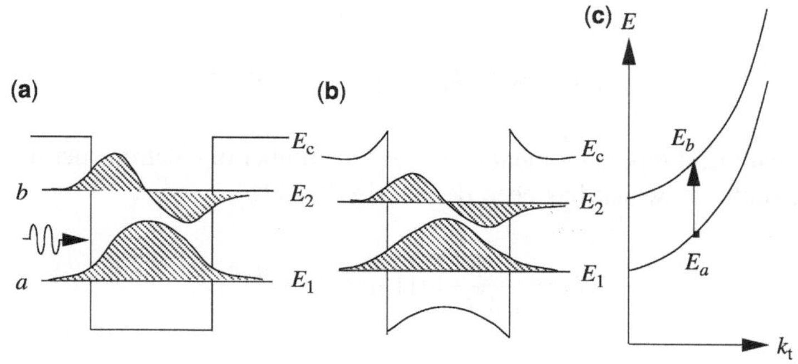

Figure 9.18 (a) A simple quantum well with a small doping concentration. (b) A modulation-doped quantum well with a significant amount of screening due to a large doping concentration. (c) The subband energy diagram in the $\mathbf{k}_t$ space. A direct vertical transition occurs because of the **k**-selection rule in the plane of quantum wells.

which has only a z component, where we have used the orthonormal conditions

$$\langle \phi_2 | \phi_1 \rangle = 0, \quad \text{and } \langle u_c | u_{c'} \rangle \simeq 1.$$

The dipole moment is z-polarized because the x and y components of the intersubband dipole moment in (9.7.3) are zero. Therefore, we obtain nonzero absorption coefficient $\alpha(\hbar\omega)$ only for $\hat{e} = \hat{z}$ (TM polarization).

9.7.2 Intersubband Absorption Spectrum

The energies of the initial state and the final states are, respectively,

$$E_a = E_1 + \frac{\hbar^2 k_t^2}{2m_e^*} \qquad E_b = E_2 + \frac{\hbar^2 k_t^2}{2m_e^*} \tag{9.7.4}$$

as shown in Fig. 9.18c

$$\begin{aligned} \alpha(\hbar\omega) &= \left(\frac{\omega}{n_r c \varepsilon_0}\right) \frac{2}{V} \sum_{\mathbf{k}_t} \sum_{\mathbf{k}'_t} \frac{|\hat{e} \cdot \boldsymbol{\mu}_{ba}|^2 \gamma}{(E_b - E_a - \hbar\omega)^2 + \gamma^2} (f_a - f_b) \\ &= \left(\frac{\omega}{n_r c \varepsilon_0}\right) \frac{|\mu_{21}|^2 \gamma}{(E_2 - E_1 - \hbar\omega)^2 + \gamma^2} \frac{2}{V} \sum_{\mathbf{k}_t} (f_a - f_b) \\ &= \left(\frac{\omega}{n_r c \varepsilon_0}\right) \frac{|\mu_{21}|^2 \gamma}{(E_2 - E_1 - \hbar\omega)^2 + \gamma^2} (N_1 - N_2) \end{aligned} \tag{9.7.5}$$

where

$$\mu_{21} = \langle \phi_2 | ez | \phi_1 \rangle = \int \phi_2^*(z)\, ez\, \phi_1(z)\, dz \tag{9.7.6}$$

is the intersubband dipole moment and N_i is the number of electrons per unit volume in the ith subband, which has been derived in (9.4.11b)

$$N_i = \frac{m_e^* k_B T}{\pi \hbar^2 L_z} \ln\left[1 + e^{(E_F - E_i)/k_B T}\right]. \tag{9.7.7}$$

In the low-temperature limit such that $(E_F - E_i) \gg k_B T$, we obtain

$$N_i = \frac{m_e^*}{\pi \hbar^2 L_z}(E_F - E_i) \tag{9.7.8}$$

and for occupations of the first two levels,

$$\alpha(\hbar\omega) = \left(\frac{\omega}{n_r c \varepsilon_0}\right) \frac{|\mu_{21}|^2 \gamma}{(E_2 - E_1 - \hbar\omega)^2 + \gamma^2} \left(\frac{m_e^*}{\pi \hbar^2 L_z}\right)(E_2 - E_1). \tag{9.7.9}$$

An integrated absorbance is given by

$$A = \int_0^\infty \alpha(\hbar\omega)\, d(\hbar\omega) \simeq \left(\frac{\omega_{21}}{n_r c \varepsilon_0}\right) |\mu_{21}|^2 \pi (N_1 - N_2) \tag{9.7.10}$$

where we have replaced the integration limits $(0, \infty)$ by $(-\infty, \infty)$ for the Lorentzian function in (9.7.5), and $\hbar\omega_{21} = E_2 - E_1$ has been used.

Example Let us estimate the peak absorption coefficient in an n-type doped GaAs quantum well using an infinite barrier model with a well width $L_z = 100$ Å. The first two subband energies and wave functions are ($m_e^* = 0.0665\, m_0$)

$$E_1 = \frac{\hbar^2}{2 m_e^*}\left(\frac{\pi}{L_z}\right)^2 = 56.5 \text{ meV}, \qquad \phi_1(z) = \sqrt{\frac{2}{L_z}} \sin\left(\frac{\pi}{L_z} z\right)$$

$$E_1 = 4E_1 = 226 \text{ meV}, \qquad \phi_2(z) = \sqrt{\frac{2}{L_z}} \sin\left(\frac{\pi}{L_z} z\right).$$

The intersubband dipole moment is

$$\mu_{21} = e \int_0^{L_z} \phi_2(z) z\, \phi_1(z) dz = -\frac{16}{9\pi^2} e L_z$$

$$= -18\, \text{eÅ} = -2.88 \times 10^{-18}\ C \cdot m.$$

If the carrier concentration N is $1 \times 10^{18}\,\text{cm}^{-3}$, we can assume that only the first subband is occupied and check if the second subband population N_2 is indeed small. We calculate

$$E_F - E_1 = k_B T[\exp(NL_z/N_s) - 1] = 78\,\text{meV}$$

where

$$N_s = \frac{m_e^* k_B T}{\pi \hbar^2} = 7.19 \times 10^{11}\,\text{cm}^{-2}.$$

We can check that $N_2 = (N_s/L_z)\ln\{1 + \exp[(E_F - E_2)/k_B T]\} = 2.4 \times 10^{15}\,\text{cm}^{-3} \ll N$. If N_2 is not negligible, we have to use $N_1 + N_2 = N$ to determine the Fermi level E_F assuming that the first two subbands are occupied.

The peak absorption coefficient occurs at $\hbar\omega \approx E_2 - E_1 = 170$ meV. The peak wavelength is $\lambda \approx 1.24/0.170 = 7.3$ μm. Assume that the linewidth is $2\gamma =$ 30 meV and the refractive index $n_r = 3.3$. We find

$$\alpha_{\text{peak}} = \frac{\omega}{n_r c \varepsilon_0} \frac{|\mu_{21}|^2}{\gamma} (N_1 - N_2) \approx 1.015 \times 10^4\,\text{cm}^{-1}.$$

In principle, the peak absorption coefficient increases with the doping concentration N. However, if the concentration is too high, the screening, band bending, and many-body effects will be important as they affect the energy levels, therefore, the peak absorption wavelength. The occupation of the second level N_2 will also decrease the absorption.

Example We consider GaAs/$Al_{0.3}Ga_{0.7}As$ superlattice structures as shown in Fig. 9.19 with a well width W and a barrier width b. The absorption spectrum is

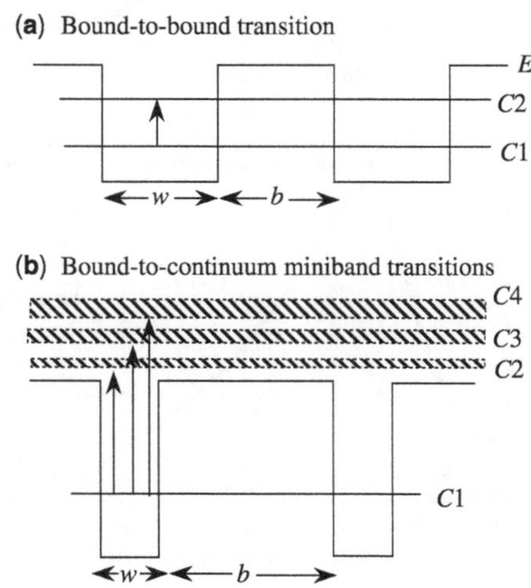

Figure 9.19 (a) Intersubband bound-to-bound transition in a superlattice. (b) Intersubband bound-to-continuum transitions in a superlattice.

shown in Fig. 9.20a for a structure with $W = b = 100$ Å and in Fig. 9.20b for $W = 100$ Å and $b = 20$ Å. As the barrier width b is reduced from 100 Å to 20 Å, the interminiband width is broadened because of the strong coupling of well states. We assume a surface concentration of $N_s = 2 \times 10^{11}\ \mathrm{cm}^{-2}$, $T = 77\mathrm{K}$, and a linewidth $2\gamma = 15$ meV.

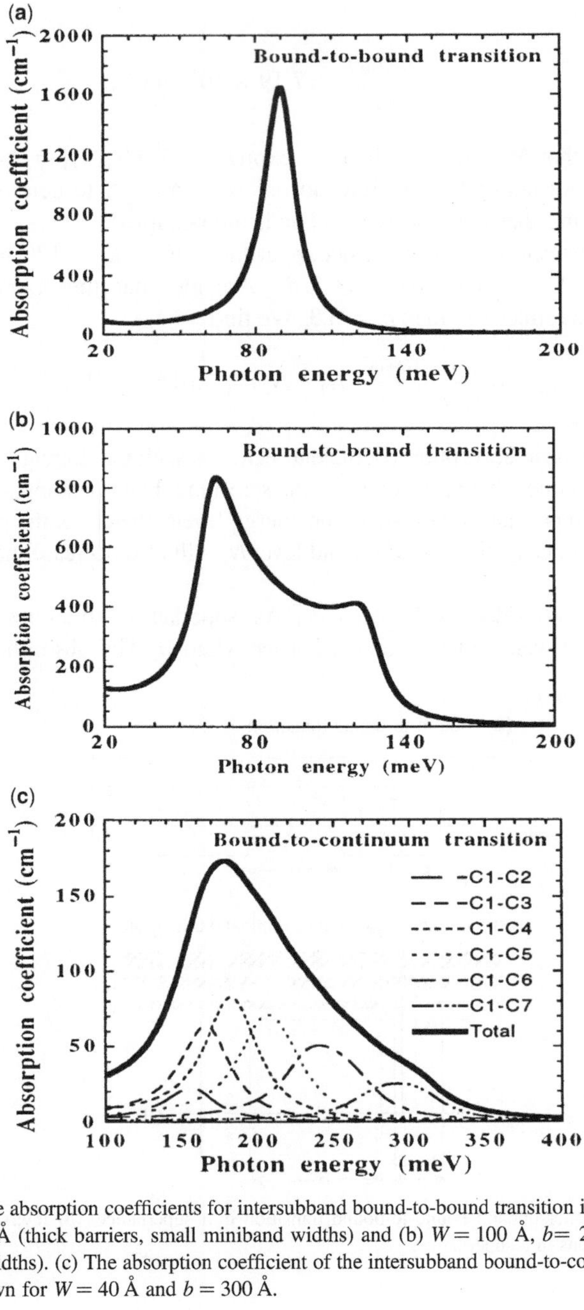

Figure 9.20 The absorption coefficients for intersubband bound-to-bound transition in a superlattice for (a) $W = b = 100$ Å (thick barriers, small miniband widths) and (b) $W = 100$ Å, $b = 20$ Å (thin barriers, large miniband widths). (c) The absorption coefficient of the intersubband bound-to-continuum miniband transitions is shown for $W = 40$ Å and $b = 300$ Å.

If we reduce the well width W to 40 Å, the intersubband transitions will occur from the bound to the continuum minibands. The resultant absorption spectrum is a broader spectrum with a longer energy tail at the high-energy side, as shown in Fig. 9.20c. Here we use the same $N_s = 2 \times 10^{11}\ \text{cm}^{-2}$ and a linewidth $2\gamma = 40$ meV. The broad spectrum is contributed by a larger γ as well as the transitions to multiple minibands.

9.7.3 Experimental Results

In Fig. 9.21, we show the experimental data for five different samples by Levine et al. [60]. The material parameters are listed as follows:

Sample	Well Width (Å)	Barrier Width (Å)	x	$N_D\,(10^{18}/\text{cm}^3)$
A	40	500	0.26	1
B	40	500	0.25	1.6
C	60	500	0.15	0.5
E	50	500	0.26	0.42
F	50	50	0.30	0.42
		500	0.26	

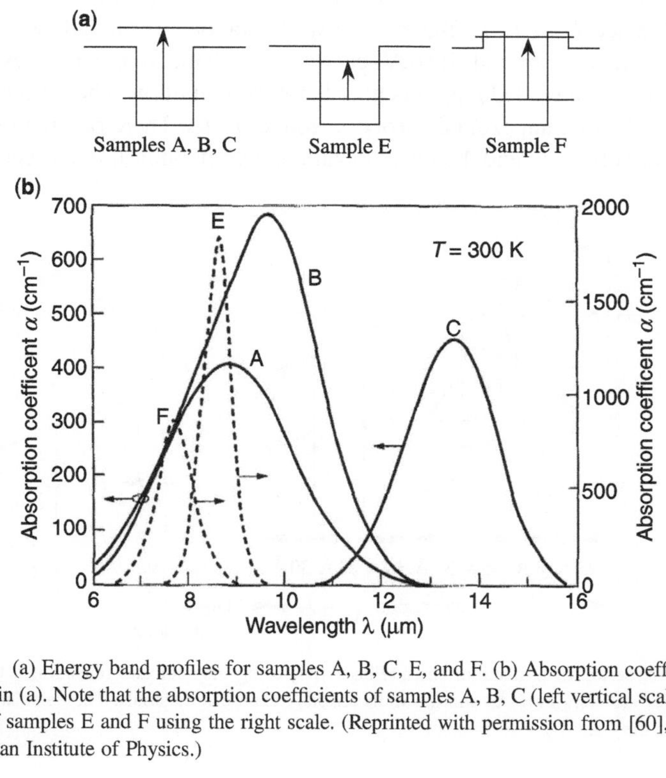

Figure 9.21 (a) Energy band profiles for samples A, B, C, E, and F. (b) Absorption coefficients for the five samples in (a). Note that the absorption coefficients of samples A, B, C (left vertical scale) are smaller than those of samples E and F using the right scale. (Reprinted with permission from [60], copyright © 1993 American Institute of Physics.)

Samples A, B, and C were designed for the bound-to-continuum transition. Sample E was designed for bound-to-bound state transition and F for bound-to-quasibound state transition. The bound-to-bound transition (sample) E has a large peak absorption spectrum with a narrow linewidth. Sample F has a narrower spectrum than those of the bound-to-continuum transitions in samples A, B, and C. These samples result in different responsivities because the carrier collection by the electrodes also play an important role. For example, the responsivity of sample F is larger than that of sample E because the electrons arriving at the final (quasi-bound) state can easily be collected compared with those in the final bound state of sample E.

9.7.4 Intersubband Quantum Cascade Laser

Intersubband lasers using a structure as shown in Fig. 9.22 as a period were demonstrated [64] in 1994. The laser structure was grown with the $Al_{0.48}In_{0.52}As/Ga_{0.47}In_{0.53}As$ heterojunctions lattice matched to InP substrate by molecular beam epitaxy (MBE). There are 25 periods of the active undoped regions, with graded regions consisting of an AlInAs/GaInAs superlattice with a constant period shorter than the electron thermal de Broglie wavelength. Electrons are injected through the 45 Å AlInAs barrier into the E_3 energy level of the active region as shown in Fig. 9.22. The reduced spatial overlap of the E_3 and E_2 wave functions and the strong coupling to the E_1 level in an adjacent well ensure a population inversion between these states. The laser operates between the E_3 and E_2 levels with an energy difference $E_3 - E_2 \approx 295$ meV or an operation wavelength $\approx 4.2\ \mu$m. The applied bias field is around 100 kV/cm, and the band diagram is as shown in Fig. 9.22, where the graded region is near flat-band condition. The estimated tunneling time into the first trapezoidal barrier is around 0.2 ps. Therefore, the filling of the state E_3 is very efficient, and the intersubband optical-phonon-limited relaxation time

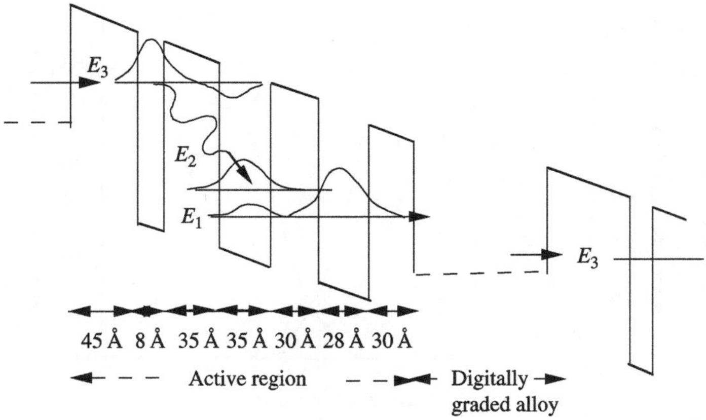

Figure 9.22 A period of a quantum cascade laser [64] using intersubband transition between E_3 and E_2. The barriers are $Al_{0.48}In_{0.52}As$ and the wells are $In_{0.47}Ga_{0.53}As$ materials. The calculated values are $E_3 - E_2 = 295$ meV and $E_2 - E_1 = 30$ meV.

τ_{32} is estimated to be around 4.3 ps at this bias field with a reduced overlap integral. Because τ_{32} is relatively long, the population inversion can be achieved because the lower state empties with a fast relaxation time (around 0.6 ps) due to the adjacent 28 Å GaInAs well with E_1 state. The relaxation between the E_2 and E_1 states is very fast because the energy separation is equal or close to an optical phonon energy so that scattering occurs with essentially zero momentum transfer. Finally, the tunneling out of the last 30 Å barrier is extremely fast (smaller than 0.5 ps). The laser emits photons at a wavelength of 4.2 μm with a peak power above 8 mW in pulsed operation. The operation temperature was as high as 88K in 1994. Since then, significant progress above room temperature operation has been made [65]. We will discuss type I quantum-cascade lasers in more detail in Section 11.4.

9.8 GAIN SPECTRUM IN A QUANTUM-WELL LASER WITH VALENCE-BAND MIXING EFFECTS

In Section 9.4, we presented a simplified model for the interband gain of a quantum-well laser based on the parabolic band structures. In quantum wells, valence-band mixing effects between the heavy-hole and light-hole subbands are important, and the subband structures can be highly non-parabolic, especially for strained quantum wells as discussed in Sections 4.8 and 4.9. In this section, we present a general theory for the gain spectrum in quantum wells taking into account the valence-band mixing effects [66–70]. The theory is applicable to both strained and unstrained quantum-well lasers.

9.8.1 General Formulation for the Gain Spectrum with Valence Band Mixing

Our starting equation follows (9.1.22) for the absorption spectrum with the delta function replaced by a Lorentzian function (9.1.28) to account for scattering broadenings,

$$\alpha(\hbar\omega) = C_0 \frac{1}{V} \sum_{\eta,\sigma} \sum_{\mathbf{k}_a} \sum_{\mathbf{k}_b} |\hat{e} \cdot \mathbf{p}_{ba}|^2 \frac{\gamma/\pi}{(E_b - E_a - \hbar\omega)^2 + \gamma^2} (f_a - f_b) \tag{9.8.1a}$$

$$C_0 = \frac{\pi e^2}{n_r c \varepsilon_0 m_0^2 \omega} \tag{9.8.1b}$$

where the quantum numbers are $|a\rangle = |\mathbf{k}_a, \sigma\rangle$ for initial states in the valence band and $|b\rangle = |\mathbf{k}_b, \eta\rangle$ for final states in the conduction band. The summations over the spins of the initial (hole) states σ and the final (electron) states η have been written out explicitly instead of a simple factor of 2. Here $\eta = \uparrow$, or $\downarrow$ for the conduction band, and σ is more complicated. For the formulation using the block-diagonalized Hamiltonian (see Section 4.8), $\sigma = U$ for the upper Hamiltonian with an eigenfunction in the mth subband denoted by

$$\Psi_m^U(\mathbf{k}_t, \mathbf{r}) = \frac{e^{i\mathbf{k}_t \cdot \rho}}{\sqrt{A}} \left[g_m^{(1)}(k_t, z)|1\rangle + g_m^{(2)}(k_t, z)|2\rangle \right] \tag{9.8.2}$$

and $\sigma = L$ for the lower Hamiltonian with an eigenfunction in the mth subband denoted by

$$\Psi_m^L(\mathbf{k}_t, \mathbf{r}) = \frac{e^{i\mathbf{k}_t \cdot \rho}}{\sqrt{A}} \left[g_m^{(3)}(k_t, z)|3\rangle + g_m^{(4)}(k_t, z)|4\rangle \right]. \tag{9.8.3}$$

The nth conduction subband wave function is denoted by

$$\Psi_n^{c\eta}(\mathbf{k}_t', \mathbf{r}) = \frac{e^{i\mathbf{k}_t' \cdot \rho}}{\sqrt{A}} \phi_n(z)|S\eta\rangle \qquad \eta = \uparrow \text{ or } \downarrow. \tag{9.8.4}$$

Here $|S\eta\rangle(= u_c(\mathbf{r}))$ denotes the spherically symmetric wave function of the Bloch periodic part of the electron state. We have to label the quantum number $\mathbf{k}_a$ by $(\mathbf{k}_t, m)$ for the mth subband, where m refers to the quantum number for the z-dependent wave functions. We label $\mathbf{k}_b$ as $(\mathbf{k}_t', n)$ for the nth conduction subband.

The matrix element contains a $\mathbf{k}_t$-selection rule when we take the inner product, $\mathbf{k}_t = \mathbf{k}_t'$. Therefore, we can write

$$\alpha(\hbar\omega) = C_0 \frac{1}{V} \sum_{\sigma,\eta} \sum_{n,m} \sum_{\mathbf{k}_t} |\hat{e} \cdot \mathbf{p}_{nm}^{\eta\sigma}|^2 \frac{\gamma/\pi}{\left[E_{hm}^{en}(k_t) - \hbar\omega\right]^2 + \gamma^2} \left(f_v^{\sigma m} - f_c^n\right) \tag{9.8.5}$$

where

$$E_{hm}^{en}(k_t) = E_{en}(k_t) - E_m^{\sigma}(k_t) + E_g \tag{9.8.6}$$

is the energy difference between the nth electron subband and the mth hole subband at $\mathbf{k}_t$. Here

$$E_{en}(k_t) = E_{en}(0) + \frac{\hbar^2 k_t^2}{2m_e^*} \tag{9.8.7}$$

is taken to be parabolic in the conduction subband. The non-parabolic valence subband structures are described by $E_m^{\sigma}(k_t)$ obtained from the upper ($\sigma = U$) or the lower ($\sigma = L$) Hamiltonian.

9.8.2 Evaluation of the Momentum Matrix Elements

The momentum matrix element is defined as

$$\left|\hat{e}\cdot\mathbf{p}_{nm}^{\eta\sigma}\right|^2 = \left|\left\langle\Psi_n^{c\eta}\middle|\hat{e}\cdot\mathbf{p}\middle|\Psi_m^{\sigma}\right\rangle\right|^2 \tag{9.8.8}$$

with the factor $\delta_{\mathbf{k}_t,\mathbf{k}'_t}$ dropped. Here we list the definitions of the old and new basis functions and their transformation relations, which have been discussed in Chapter 4.

Old Bases

$$\begin{aligned}
\left|\frac{3}{2},\frac{3}{2}\right\rangle &= \frac{-1}{\sqrt{2}}|(X+iY)\uparrow\rangle\\
\left|\frac{3}{2},\frac{1}{2}\right\rangle &= \frac{-1}{\sqrt{6}}|(X+iY)\downarrow\rangle + \frac{2}{\sqrt{6}}|Z\uparrow\rangle\\
\left|\frac{3}{2},\frac{-1}{2}\right\rangle &= \frac{1}{\sqrt{6}}|(X-iY)\uparrow\rangle + \frac{2}{\sqrt{6}}|Z\downarrow\rangle\\
\left|\frac{3}{2},\frac{-3}{2}\right\rangle &= \frac{1}{\sqrt{2}}|(X-iY)\downarrow\rangle.
\end{aligned} \tag{9.8.9a}$$

New Bases

$$\begin{aligned}
|1\rangle &= \alpha\left|\frac{3}{2},\frac{3}{2}\right\rangle - \alpha^*\left|\frac{3}{2},\frac{-3}{2}\right\rangle\\
|2\rangle &= -\beta^*\left|\frac{3}{2},\frac{1}{2}\right\rangle + \beta\left|\frac{3}{2},\frac{-1}{2}\right\rangle\\
|3\rangle &= \beta^*\left|\frac{3}{2},\frac{1}{2}\right\rangle + \beta\left|\frac{3}{2},\frac{-1}{2}\right\rangle\\
|4\rangle &= \alpha\left|\frac{3}{2},\frac{3}{2}\right\rangle + \alpha^*\left|\frac{3}{2},\frac{-3}{2}\right\rangle\\
\alpha &= \frac{1}{\sqrt{2}}e^{i\left(\frac{3\pi}{4}-\frac{3\phi}{2}\right)}, \quad \beta = \frac{1}{\sqrt{2}}e^{i\left(\frac{-\pi}{4}+\frac{\phi}{2}\right)}.
\end{aligned} \tag{9.8.9b}$$

The matrix element $\left|\hat{e}\cdot\mathbf{p}_{nm}^{\eta\sigma}\right|^2$ can be found as shown in the following example. For TE polarization ($\hat{e}=\hat{x}$ or $\hat{y}$) and $\eta=\uparrow$, $\sigma=U$, we obtain

$$\begin{aligned}
\left|\hat{x}\cdot\mathbf{p}_{nm}^{\uparrow U}\right|^2 &= \left|\left\langle\Psi_n^{c\uparrow}\middle|p_x\middle|\Psi_m^{U}\right\rangle\right|^2\\
&= \delta_{\mathbf{k}_t,\mathbf{k}'_t}\left|\left\langle\phi_n\middle|g_m^{(1)}\right\rangle\langle S\uparrow|p_x|1\rangle + \left\langle\phi_n\middle|g_m^{(2)}\right\rangle\langle S\uparrow|p_x|2\rangle\right|^2
\end{aligned}$$

$$= \delta_{\mathbf{k}_t,\mathbf{k}_t'} |\langle S|p_x|X\rangle|^2 \left| \langle \phi_n | g_m^{(1)} \rangle \left(\frac{-\alpha}{\sqrt{2}} \right) + \langle \phi_n | g_m^{(2)} \rangle \frac{\beta}{\sqrt{6}} \right|^2$$
$$= \delta_{\mathbf{k}_t,\mathbf{k}_t'} |\langle S|p_x|X\rangle|^2 \frac{1}{4} \left\{ \langle \phi_n | g_m^{(1)} \rangle^2 \right.$$
$$\left. + \frac{1}{3} \langle \phi_n | g_m^{(2)} \rangle^2 + \frac{2}{\sqrt{3}} \cos 2\phi \, \langle \phi_n | g_m^{(1)} \rangle \langle \phi_n | g_m^{(2)} \rangle \right\} \tag{9.8.10}$$

where we have used the following relations.

Spin ↑

$$\begin{aligned}
\langle S\uparrow|p_x|1\rangle &= \frac{-\alpha}{\sqrt{2}} \langle S|p_x|X\rangle & \langle S\uparrow|p_z|1\rangle &= 0 \\
\langle S\uparrow|p_x|2\rangle &= \frac{\beta}{\sqrt{6}} \langle S|p_x|X\rangle & \langle S\uparrow|p_z|2\rangle &= -\sqrt{\frac{2}{3}}\beta^* \langle S|p_z|Z\rangle \\
\langle S\uparrow|p_x|3\rangle &= \frac{\beta}{\sqrt{6}} \langle S|p_x|X\rangle & \langle S\uparrow|p_z|3\rangle &= \sqrt{\frac{2}{3}}\beta^* \langle S|p_z|Z\rangle \\
\langle S\uparrow|p_x|4\rangle &= \frac{-\alpha}{\sqrt{2}} \langle S|p_x|X\rangle & \langle S\uparrow|p_z|4\rangle &= 0.
\end{aligned} \tag{9.8.11a}$$

Spin ↓

$$\begin{aligned}
\langle S\downarrow|p_x|1\rangle &= \frac{-\alpha^*}{\sqrt{2}} \langle S|p_x|X\rangle & \langle S\downarrow|p_z|1\rangle &= 0 \\
\langle S\downarrow|p_x|2\rangle &= \frac{\beta^*}{\sqrt{6}} \langle S|p_x|X\rangle & \langle S\downarrow|p_z|2\rangle &= \beta\sqrt{\frac{2}{3}} \langle S|p_z|Z\rangle \\
\langle S\downarrow|p_x|3\rangle &= \frac{-\beta^*}{\sqrt{6}} \langle S|p_x|X\rangle & \langle S\downarrow|p_z|3\rangle &= \beta\sqrt{\frac{2}{3}} \langle S|p_z|Z\rangle \\
\langle S\downarrow|p_x|4\rangle &= \frac{\alpha^*}{\sqrt{2}} \langle S|p_x|X\rangle & \langle S\downarrow|p_z|4\rangle &= 0.
\end{aligned} \tag{9.8.11b}$$

Using the axial approximation, the band structures are isotropic in the $k_x - k_y$ plane; we, therefore, write

$$\frac{1}{V}\sum_{k_t} = \frac{1}{L_z}\int_0^\infty \frac{k_t dk_t}{2\pi}\int_0^{2\pi}\frac{d\phi}{2\pi}. \tag{9.8.12}$$

The term containing the $\cos 2\phi$ factor in (9.8.10) will not contribute to $\alpha(\hbar\omega)$ because its integration over ϕ vanishes. Let us pull out the factor $|\langle S|p_x|X\rangle|^2$ in the matrix element, where

$$|\langle S|p_x|X\rangle|^2 = 3M_b^2 = \frac{m_0^2}{\hbar^2}P^2 = \frac{m_0}{2}E_p \tag{9.8.13}$$

is defined in terms of the Kane parameter P or an energy parameter E_p. The expression for $\alpha(\hbar\omega)$ can be cast in the form

$$\alpha(\hbar\omega) = C_0\frac{2}{L_z}\sum_{\sigma}\sum_{n,m}\int_0^{\infty}\frac{k_t dk_t}{2\pi}M_{nm}^{\sigma}(k_t)\frac{\gamma/\pi}{\left[E_{hm}^{en}(k_t)-\hbar\omega\right]^2+\gamma^2}\left(f_v^{\sigma m}-f_c^n\right) \tag{9.8.14}$$

Here a factor of 2 for sum over η is explicitly shown, as we find the final expression for the magnitude squared $|\hat{e}\cdot\mathbf{p}_{nm}^{\eta\sigma}|^2$ is independent of η.

(a) TE polarization: $\hat{e}=\hat{x}$

$$\begin{aligned}M_{nm}^{\sigma}(k_t) &= \frac{3}{4}\left[\langle\phi_n|g_m^{(1)}\rangle^2+\frac{1}{3}\langle\phi_n|g_m^{(2)}\rangle^2\right]M_b^2 \qquad \sigma=U\\ &= \frac{3}{4}\left[\frac{1}{3}\langle\phi_n|g_m^{(3)}\rangle^2+\langle\phi_n|g_m^{(2)}\rangle^2\right]M_b^2 \qquad \sigma=L.\end{aligned} \tag{9.8.15}$$

(b) TM polarization: $\hat{e}=\hat{z}$

$$\begin{aligned}M_{nm}^{\sigma}(k_t) &= \langle\phi_n|g_m^{(2)}\rangle^2 M_b^2, \qquad \sigma=U\\ &= \langle\phi_n|g_m^{(3)}\rangle^2 M_b^2, \qquad \sigma=L.\end{aligned} \tag{9.8.16}$$

9.8.3 Final Expression for the Gain Spectrum and Numerical Examples

The optical gain is obtained from the negative of the absorption function $g(\hbar\omega) = -\alpha(\hbar\omega)$. Therefore, we have

$$g(\omega) = C_0\cdot\frac{2}{L_z}\sum_{\sigma}\sum_{n,m}\int_0^{\infty}\frac{k_t\,dk_t}{2\pi}M_{nm}^{\sigma}(k_t)\cdot\frac{\gamma/\pi}{\left[E_{hm}^{en}(k_t)-\hbar\omega\right]^2+\gamma^2}\left(f_c^n-f_v^{\sigma m}\right) \tag{9.8.17}$$

Notice that the form (9.8.17) or (9.8.14) is similar to (9.4.8) using the parabolic band model. For a symmetric quantum well without an applied electric field, we find that the two valence bands with $\sigma=U$ and $\sigma=L$ are degenerate. The matrix elements turn out to be the same for $\sigma=U$ and L. Therefore, the sum over σ can be replaced by a factor 2, and only $g_m^{(1)}$ and $g_m^{(2)}$ have to be calculated with the corresponding energy dispersion relation $E_m^U(k_t)$.

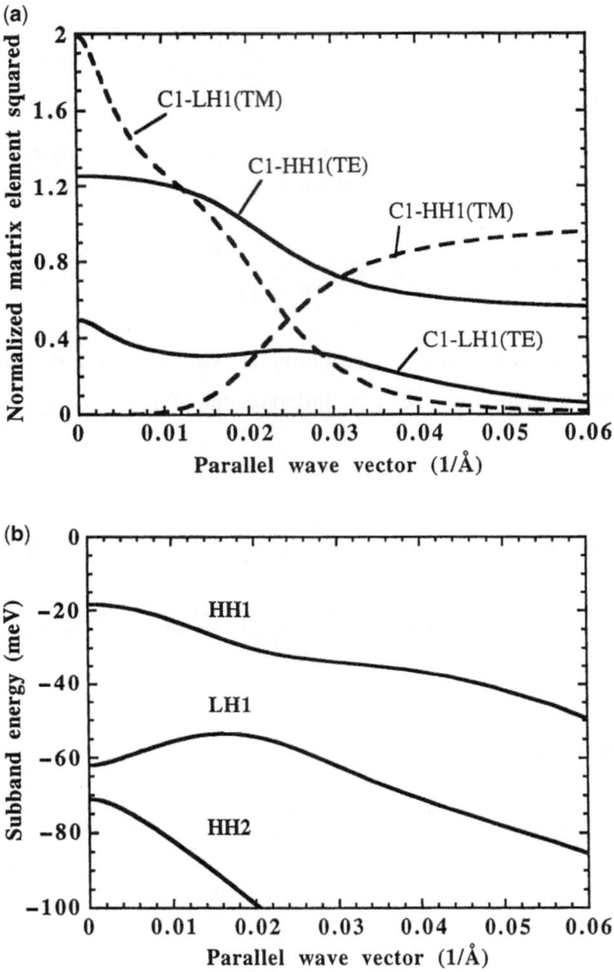

Figure 9.23 (a) The average of the square of the momentum matrix normalized by the bulk matrix element for TE and TM polarizations for the transition between the first conduction subband (C1) and the first heavy-hole subband (HH1) or the first light-hole subband (LH1). (b) The valence subband structures of the $In_{0.53}Ga_{0.47}As/In_{1-x}Ga_xAs_{1-y}P_y$ quantum well ($L_w = 60$ Å) lattice-matched to an InP substrate.

In Fig. 9.23a, we plot the matrix elements $2M(k_t)/M_b^2$ using Eqs. (9.8.15) and (9.8.16) for the TE polarization and the TM polarization for the first conduction band to the first heavy-hole subband (C1-HH1) and to the first light-hole subband (C1-LH1) transitions in a $In_{0.53}Ga_{0.47}As/In_{1-x}Ga_xAs_{1-y}P_y$ quantum well lattice-matched to InP substrate with a well width of 60 Å. The factor of 2 here accounts for sum over σ of the hole spins in the valence band (not in the conduction band). The conduction band spin degeneracy of a factor of 2 is always included in the density of states. The barrier $In_{1-x}Ga_xAs_{1-y}P_y$ has a band-gap wavelength of 1.3 μm and its parameters can be obtained from Table C.2 in Appendix C. The

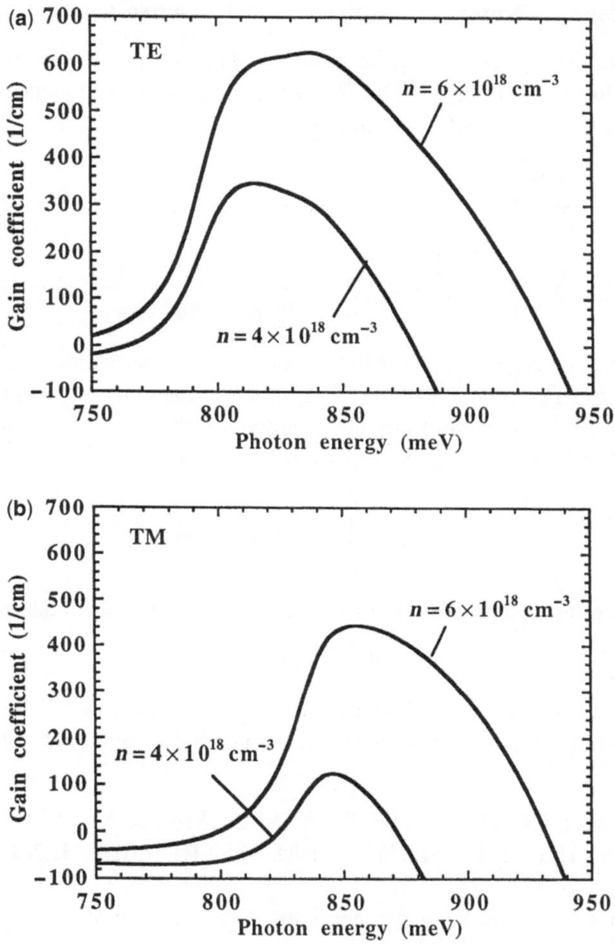

Figure 9.24 The linear gain versus the photon energy for (a) TE and (b) TM polarization at two different carrier concentrations.

valence subband structures are plotted in Fig. 9.23b with HH1 as the top valence subband. The non-parabolic behavior is clearly seen.

The gain spectra for both TE and TM polarizations are plotted in Fig. 9.24a and Fig. 9.24b for two carrier concentrations n. The surface carrier concentration is $n_s = nL_z$, and the gain coefficient in a period of a total width $L_{\mathrm{T}} = L_z + L_{\mathrm{b}}$ is proportional to $1/L_{\mathrm{T}}$ instead of $1/L_z$ in (9.8.17). Here L_{b} is the thickness of the barrier. We can see that for a given carrier density, the TE gain coefficient is larger than that of the TM gain coefficient because the holes will occupy the lowest subband, which is heavy hole in nature, and the conduction-to-heavy hole dipole matrix is mostly TE polarized as can be seen from Fig. 9.23a. Theoretical gain model taking into account the valence-band mixing and strain effects has compared very well with experimental data measured by the Hakki–Paoli method [10–13].

9.8.4 Spontaneous Emission Spectrum and Radiative Current Density

Comparing (9.2.33) with (9.8.17), the spontaneous emission rate per unit volume per unit energy interval ($s^{-1}\,cm^{-3}\,eV^{-1}$) taking into account valence-band-mixing effects in quantum wells can be written as

$$r^{\text{spon}}(\hbar\omega) = \left(\frac{n_r^2\omega^2}{\pi^2\hbar c^2}\right) C_0 \frac{2}{L_z} \times \sum_{\sigma}\sum_{n,m}\int_0^{\infty} \frac{k_t\,dk_t}{2\pi} M_{nm}^{\sigma}(k_t) \frac{\gamma/\pi}{\left[E_{hm}^{en}(k_t)-\hbar\omega\right]^2+\gamma^2} f_c^n(1-f_v^{\sigma m}) \quad (9.8.18)$$

which is similar to $g(\omega)$ except for the prefactor in the parenthesis and the Fermi–Dirac occupation factors $f_c^n\ (1-f_v^{\sigma m})$. The total spontaneous emission rate per unit volume ($s^{-1}\,cm^{-3}$) is to sum up the average of three polarizations $(2\times\text{TE}+\text{TM})/3$ and then integrate over the emission spectrum to obtain

$$R_{\text{spon}} = \int_0^{\infty} r^{\text{spon}}(\hbar\omega)\, d(\hbar\omega). \quad (9.8.19)$$

A radiative current density (A/cm^2) for a semiconductor laser is defined as

$$J_{\text{rad}} = qd\,R_{\text{sp}} \quad (9.8.20)$$

where $q = 1.6\times 10^{-19}$ C is a unit charge, and d is the thickness of the active region where an injection current provides the electrons and holes for radiative recombination.

APPENDIX 9A: COORDINATE TRANSFORMATION OF THE BASIS FUNCTIONS AND THE MOMENTUM MATRIX ELEMENTS

If the electron wave vector $\mathbf{k}$ has a general direction specified by (k, θ, ϕ) in spherical coordinates,

$$\mathbf{k} = k\sin\theta\cos\phi\,\hat{x} + k\sin\theta\sin\phi\,\hat{y} + k\cos\theta\,\hat{z}, \quad (9.A.1)$$

we find the band-edge wave functions are as follows, using the coordinate transformations for the spatial variables, X, Y, and Z, in (4.2.30) and keeping the spins in their primed coordinates.

Conduction Band

$$|iS\downarrow'\rangle \quad \text{and} \quad |iS\uparrow'\rangle. \quad (9.A.2)$$

Heavy-Hole Band

$$\left|\frac{3}{2},\frac{3}{2}\right\rangle' = \frac{-1}{\sqrt{2}}|(X'+iY')\uparrow'\rangle$$

$$= \frac{-1}{\sqrt{2}}|(\cos\theta\cos\phi - i\sin\phi)X + (\cos\theta\sin\phi + i\cos\phi)Y - \sin\theta Z\rangle|\uparrow'\rangle$$

$$\left|\frac{3}{2},-\frac{3}{2}\right\rangle' = \frac{1}{\sqrt{2}}|(X'-iY')\downarrow'\rangle$$
$$= \frac{1}{\sqrt{2}}|(\cos\theta\cos\phi + i\sin\phi)X + (\cos\theta\sin\phi - i\cos\phi)Y - \sin\theta Z\rangle|\downarrow'\rangle. \tag{9.A.3}$$

Light-Hole Band

$$\left|\frac{3}{2},\frac{1}{2}\right\rangle' = \frac{-1}{\sqrt{6}}|(X'+iY')\downarrow'\rangle + \sqrt{\frac{2}{3}}|Z'\uparrow'\rangle$$
$$= \frac{-1}{\sqrt{6}}|(\cos\theta\cos\phi - i\sin\phi)X + (\cos\theta\sin\phi + i\cos\phi)Y - \sin\theta Z\rangle|\downarrow'\rangle$$
$$+ \sqrt{\frac{2}{3}}|\sin\theta\cos\phi\, X + \sin\theta\sin\phi\, Y + \cos\theta\, Z\rangle|\uparrow'\rangle$$
$$\left|\frac{3}{2},-\frac{1}{2}\right\rangle' = \frac{1}{\sqrt{6}}|(X'-iY')\uparrow'\rangle + \sqrt{\frac{2}{3}}|Z'\downarrow'\rangle$$
$$= \frac{1}{\sqrt{6}}|(\cos\theta\cos\phi + i\sin\phi)X + (\cos\theta\sin\phi - i\cos\phi)Y - \sin\theta Z\rangle|\uparrow'\rangle$$
$$+ \sqrt{\frac{2}{3}}|\sin\theta\cos\phi\, X + \sin\theta\sin\phi\, Y + \cos\theta\, Z\rangle|\downarrow'\rangle. \tag{9.A.4}$$

Spin–Orbit Split-Off Band

$$\left|\frac{1}{2},\frac{1}{2}\right\rangle' = \frac{1}{\sqrt{3}}|(X'+iY')\downarrow'\rangle + \frac{1}{\sqrt{3}}|Z'\uparrow'\rangle$$
$$= \frac{1}{\sqrt{3}}|(\cos\theta\cos\phi - i\sin\phi)X + (\cos\theta\sin\phi + i\cos\phi)Y - \sin\theta Z\rangle|\downarrow'\rangle$$
$$+ \frac{1}{\sqrt{3}}|\sin\theta\cos\phi\, X + \sin\theta\sin\phi\, Y + \cos\theta\, Z\rangle|\uparrow'\rangle$$
$$\left|\frac{1}{2},-\frac{1}{2}\right\rangle' = \frac{1}{\sqrt{3}}|(X'-iY')\uparrow'\rangle - \frac{1}{\sqrt{3}}|Z'\downarrow'\rangle$$
$$= \frac{1}{\sqrt{3}}|(\cos\theta\cos\phi + i\sin\phi)X + (\cos\theta\sin\phi - i\cos\phi)Y - \sin\theta Z\rangle|\uparrow'\rangle$$
$$- \frac{1}{\sqrt{3}}|\sin\theta\cos\phi\, X + \sin\theta\sin\phi\, Y + \cos\theta\, Z\rangle|\downarrow'\rangle. \tag{9.A.5}$$

We keep the primes in $|\downarrow'\rangle$ and $|\uparrow'\rangle$ for spins in the new coordinate system for ease of calculations of the dipole matrix elements, as will be seen later.

Consider a quantum-well structure with the growth axis along the z direction. In general [19], the z axis for the p-state functions does not have to coincide with the z axis for the quantum-well structures. However, when averaged over the angle in the plane of the quantum well, it is found that the same matrix elements for both polarizations give the same results as those obtained assuming that both z axes coincide with each other.

To calculate the optical momentum matrix element

$$\mathbf{M} = \langle u_c|\mathbf{p}|u_v\rangle = \mathbf{p}_{cv}$$
$$= \hat{x}\left\langle u_c\left|\frac{\hbar}{i}\frac{\partial}{\partial x}\right|u_v\right\rangle + \hat{y}\left\langle u_c\left|\frac{\hbar}{i}\frac{\partial}{\partial y}\right|u_v\right\rangle + \hat{z}\left\langle u_c\left|\frac{\hbar}{i}\frac{\partial}{\partial z}\right|u_v\right\rangle \tag{9.A.6}$$

with polarization dependence, we evaluate $\hat{e}\cdot\mathbf{M}$ for $\hat{e} = \hat{x}$ or $\hat{y}$ for TE polarization, and $\hat{e} = \hat{z}$ for TM polarization. Here u_c is $|iS\uparrow'\rangle$ or $|iS\downarrow'\rangle$ and u_v can be u_{hh} or $u_{\ell h}$.

Define the Kane parameter P and a momentum matrix parameter P_x as

$$\begin{aligned} P &= \frac{\hbar}{m_0}\langle iS|p_z|Z\rangle = \frac{\hbar}{m_0}P_x \\ P_x &= \langle iS|p_x|X\rangle = \langle iS|p_y|Y\rangle = \langle iS|p_z|Z\rangle. \end{aligned} \tag{9.A.7}$$

Then for conduction to heavy-hole transitions, we obtain $\mathbf{M}_{c\text{-hh}}$ as

$$\begin{aligned}
\left\langle iS\uparrow'|\mathbf{p}|\frac{3}{2},\frac{3}{2}\right\rangle' &= -[(\cos\theta\cos\phi - i\sin\phi)\,\hat{x} \\
&\quad + (\cos\theta\sin\phi + i\cos\phi)\,\hat{y} - \sin\theta\,\hat{z}]P_x/\sqrt{2} \\
\left\langle iS\downarrow'|\mathbf{p}|\frac{3}{2},-\frac{3}{2}\right\rangle' &= [(\cos\theta\cos\phi + i\sin\phi)\,\hat{x} \\
&\quad + (\cos\theta\sin\phi - i\cos\phi)\,\hat{y} - \sin\theta\,\hat{z}]P_x/\sqrt{2} \\
\left\langle iS\uparrow'|\mathbf{p}|\frac{3}{2},-\frac{3}{2}\right\rangle' &= 0 \\
\left\langle iS\downarrow'|\mathbf{p}|\frac{3}{2},\frac{3}{2}\right\rangle' &= 0.
\end{aligned} \tag{9.A.8}$$

For conduction to light-hole transitions, $\mathbf{M}_{c\text{-}\ell h}$ are

$$\begin{aligned}
\left\langle iS\uparrow'|\mathbf{p}|\frac{3}{2},\frac{1}{2}\right\rangle' &= \sqrt{\frac{2}{3}}(\sin\theta\cos\phi\,\hat{x} + \sin\theta\sin\phi\,\hat{y} + \cos\theta\,\hat{z})P_x \\
\left\langle iS\downarrow'|\mathbf{p}|\frac{3}{2},-\frac{1}{2}\right\rangle' &= \sqrt{\frac{2}{3}}(\sin\theta\cos\phi\,\hat{x} + \sin\theta\sin\phi\,\hat{y} + \cos\theta\,\hat{z})P_x \\
\left\langle iS\uparrow'|\mathbf{p}|\frac{3}{2},-\frac{1}{2}\right\rangle' &= \frac{1}{\sqrt{6}}[(\cos\theta\cos\phi + i\sin\phi)\,\hat{x} \\
&\quad + (\cos\theta\sin\phi - i\cos\phi)\,\hat{y} - \sin\theta\,\hat{z}]P_x
\end{aligned}$$

$$\left\langle iS \downarrow' |\mathbf{p}| \frac{3}{2}, \frac{1}{2} \right\rangle' = \frac{-1}{\sqrt{6}} [(\cos\theta\cos\phi - i\sin\phi)\,\hat{x} + (\cos\theta\sin\phi + i\cos\phi)\,\hat{y} - \sin\theta\,\hat{z}] P_x. \qquad (9.A.9)$$

PROBLEMS

9.1 In classic physics, the equation of motion for a charge e in the presence of an electric field $\mathbf{E}$ and a magnetic flux density $\mathbf{B}$ is described by the Lorenz force equation

$$\frac{d}{dt}\mathbf{p} = \mathbf{F} = e(\mathbf{E} + \mathbf{v} \times \mathbf{B})$$

where $\mathbf{p} = m\mathbf{v}$. The magnetic flux density $\mathbf{B}$ and the electric field in classic electromagnetics are expressed in terms of the vector and scalar potentials (see Chapter 6)

$$\mathbf{B} = \nabla \times \mathbf{A} \quad \text{and} \quad \mathbf{E} = -\nabla\phi - \frac{\partial \mathbf{A}}{\partial t}.$$

Show that using the Hamiltonian in the presence of the electromagnetic field

$$H = \frac{1}{2m}(\mathbf{p} - e\mathbf{A})^2 + e\phi$$

and the equations for the Hamiltonian

$$\frac{d}{dt}r_i = \frac{\partial H}{\partial p_i} \qquad \frac{d}{dt}p_i = -\frac{\partial H}{\partial r_i}$$

lead to the classic Lorenz force equation. If we choose Coulomb gauge, we have $\nabla \cdot \mathbf{A} = 0$, and ϕ is zero because $\rho = 0$ for the optical field. The periodic potential $V(\mathbf{r})$ is the background static potential that is not coupled to the vector potential $\mathbf{A}$ for the optical field and is added to the Hamiltonian for completeness. Therefore,

$$H = \frac{1}{2m}(\mathbf{p} - e\mathbf{A})^2 + V(\mathbf{r}).$$

9.2 **(a)** For a semiconductor laser operating at 0.8 μm with an optical output power of 10 mW coming from a 20 μm × 1 μm cross-sectional area, estimate the electric field strength E_0 using $P = \omega A_0^2 k_{op}/2\mu = E_0^2\, n_r/2\mu_0 c$, where $n_r = 3.4$. Calculate k_{op} and A_0.

(b) Repeat (a) for 1.55-μm laser wavelength for the same power. Assume the other parameters are unchanged.

9.3 To prove (9.1.25), we start by proving the x component of the equation:

$$m_0 \frac{d}{dt} x = \frac{m_0}{i\hbar}(xH_0 - H_0 x).$$

The procedure is to apply the right-hand side of the above equation on the wave function Ψ satisfying, $H_0\Psi = E\Psi$. Show that

$$\frac{m_0}{i\hbar}(xH_0 - H_0 x)\Psi = \frac{\hbar}{i}\frac{\partial}{\partial x}\Psi = p_x \Psi.$$

Then use $p_x = m_0 dx/dt$.

9.4 Derive the relation (9.2.36) between the spontaneous emission rate per unit volume per unit energy interval r^{spon} $(\hbar\omega)$ and the gain coefficient $g(\hbar\omega)$. Check the units on both sides of (9.2.36).

9.5 Derive the Bernard–Duraffourg population inversion condition.

9.6 **(a)** Find the quasi-Fermi levels F_c and F_v for the bulk structure at zero temperature in terms of the injection carrier densities n and p, respectively.

(b) Derive an expression for the gain spectrum $g(\hbar\omega)$ at $T = 0\text{K}$.

9.7 Use the material parameters in Appendix C (page 803) for GaAs materials, except that the band gap at zero temperature is assumed to be 1.522 eV. Neglect the light-hole band for simplicity.

(a) Find the numerical values for the quasi-Fermi levels for the electrons and holes in a bulk semiconductor in the presence of carrier injection at *zero temperature* assuming $n = p = 5 \times 10^{18}\,\text{cm}^{-3}$.

(b) Calculate and plot the gain spectrum for part (a).

9.8 Repeat Problem 9.7 for a quantum-well structure with an effective well width 120 Å. Assume an infinite well model and a constant momentum matrix for simplicity. Label the major transition energies in your plots.

9.9 The $E-k$ relations of a semiconductor material with a non-parabolicity conduction band can be described as

$$\text{Conduction band:} \quad E_c(k) = E_g + \frac{\hbar^2 k_0^2}{2m_e^*}\left[\left(\frac{k}{k_0}\right)^2 + \beta\left(\frac{k}{k_0}\right)^4\right]$$

$$\text{Valence band:} \quad E_v(k) = -\frac{\hbar^2 k_0^2}{2m_h^*}\left(\frac{k}{k_0}\right)^2$$

where k_0 and β are positive material parameters describing the non-parabolicity.

(a) Derive an expression for the interband absorption spectrum $\alpha(\hbar\omega)$ for a bulk structure at zero temperature. Assume a constant momentum matrix.

(b) Plot the absorption spectrum $\alpha(\hbar\omega)$ versus the normalized energy $(\hbar\omega - E_g)/(\hbar^2 k_0^2/2m_r^*)$ for (i) $\beta = 0$ and (ii) $\beta = 0.1\ m_e^*/m_r^*$ where m_r^* is the reduced effective mass.

9.10 Consider a GaAs quantum well with an effective well width $L = 100$ Å (assuming an infinite well model). The electron and hole energies are

$$E_{en}(k) = E_{en}(0) + \frac{\hbar^2 k_t^2}{2m_e^*}$$

$$E_{hm}(k) = E_{hm}(0) - \frac{\hbar^2 k_t^2}{2m_h^*}$$

where $k_t^2 = k_x^2 + k_y^2$, $m_e^* = 0.0665\ m_0$, $m_h^* = 0.34\ m_0$. The other parameters are $E_g = 1.424$ eV; $m_0 = 9.1 \times 10^{-31}$ kg; $\hbar = 1.0546 \times 10^{-34}\ \mathrm{J \cdot s} = 0.6582 \times 10^{-15}$ eV s; $q = 1.60218 \times 10^{-19}$ C; $n_r = 3.68$; and $\varepsilon_0 = 8.854 \times 10^{-12}$ F/m.

(a) Calculate the subband energies, $E_{en}(0)$ and $E_{hm}(0)$, for the first two electron subbands and hole subbands, respectively.

(b) Using the normalized wave functions and the subband energies of the first two electron and hole subbands, calculate the absorption spectrum due to the interband transitions within the infinite well model. Assume the matrix element $|\hat{e} \cdot \mathbf{p}_{cv}|^2$ is a constant. (The overlap integrals between the electron wave functions $f_n(z)$ and the hole wave functions $g_m(z)$ are assumed to satisfy the relation $\int f_n(z)\, g_m(z)\, dz = 1$ if $n = m$, and 0 if $n \neq m$.)

(c) Assume that $|\hat{e} \cdot \mathbf{p}_{cv}|^2 \simeq m_0 E_p/6$, and $E_p = 25.7$ eV for GaAs. Calculate the magnitudes of the absorption coefficients (1/cm) for part (b) at (i) $\hbar\omega = 1.43$ eV, (ii) $\hbar\omega = 1.55$ eV, and (iii) $\hbar\omega = 1.75$ eV.

9.11 **(a)** Estimate the constants $C_0 = \pi e^2/(n_r c \varepsilon_0 m_0^2 \omega)$, the momentum matrix element $|\hat{e} \cdot \mathbf{p}_{cv}|^2 = 1.5 M_b^2 = 1.5(m_0 E_p/6)$, the reduced density of states $\rho_r^{2D} = m_r^*/(\pi \hbar^2 L_z)$, and the absorption coefficient (TE polarization) $\alpha_0(\hbar\omega)$ for a GaAs quantum well with an effective well width $L_z = 100$ Å near the absorption edge.

(b) Repeat (a) for an $In_{0.53}Ga_{0.47}As$ quantum well with $L_z = 100$ Å near the absorption edge. $E_g(In_{0.53}Ga_{0.47}As) = 0.75$ eV.

9.12 **(a)** Plot the optical matrix elements for both TE and TM polarizations (see Table 9.1) versus the transverse wave number k_t for a GaAs quantum well with an effective well width $L_z = 100$ Å.

(b) Compare the optical matrix elements in (a) with those of another GaAs quantum well with a smaller effective well width $L_z = 40$ Å.

9.13 If a Gaussian lineshape function is used in the expressions for optical absorptions, what is the complete expression for the Gaussian function to replace the delta function $\delta(E_c - E_v - \hbar\omega)$?

9.14 Assume an infinite barrier height model for a quantum well structure with an effective well width $L_w = 100$ Å. The well is doped p-type with a doping concentration of $N_A = 2.0 \times 10^{18}/\text{cm}^3$ and $T = 300\text{K}$. Consider *only one conduction and one heavy-hole subband.* Assume $m_e^* = 0.067m_0$, and $m_{hh}^* = 0.34m_0$. (a) Calculate the quantized electron and hole subband energies. (b) If we know the injected electron density, $n = 1.0 \times 10^{18}\ \text{cm}^{-3}$, calculate the quasi-Fermi levels for electrons and holes, respectively. Hint: Use charge neutrality condition for holes. (c) Calculate and plot the optical gain spectrum (no linewidth broadening). (d) Compare the above gain spectrum and peak value with those of an undoped quantum well ($N_A = 0$) at the same electron concentration $n = p$. Comment on the effects of p-type doping.

9.15 Consider a semiconductor laser structure with the active region consisting of one quantum well layer *and* one quantum dot layer. You can assume that the well width and the quantum dot height are both 80 Å so we can compare the gain coefficient per well width (or per quantum-dot layer thickness). Assume that the ground state transition energy is the same for both the well and the dots and that the active regions are GaAs for both cases (for simplicity because the dots are usually InAs).

(a) What is the absorption coefficient α_{QW} of the QW layer at the ground state transition energy using a step-like density of states (ignoring the linewidth broadening for simplicity)?

(b) What is the required quantum dot density (#/area) such that the absorption coefficient α_{QD} of the QD layer *at the ground state transition energy* is the same magnitude as that in part (a) if we assume that a Lorentzian lineshape with a full-width at half maximum (FWHM) of 30 meV is used? You can assume that the optical matrix elements and the prefactors such as refractive index (or C_0 in the text) are the same for both QW and QD.

9.16 Compare the interband and the intersubband absorption coefficients in a semiconductor quantum-well structure. Discuss the optical matrix elements, the density of states, and the absorption spectra.

9.17 **(a)** To calculate the intersubband absorption in the n-type modulation-doped $GaAs/Al_{0.3}Ga_{0.7}As$ quantum well with a well width of 100 Å, we use the following procedures:

(i) The ground-state energy E_1 for the first conduction subband for a finite barrier model can be calculated numerically. We can then define an effective well width L_{eff} such that

$$E_1 = \frac{\hbar^2}{2m_e^*}\left(\frac{\pi}{L_{\text{eff}}}\right)^2$$

Find L_{eff} for this GaAs/$\text{Al}_{0.3}\text{Ga}_{0.7}$As quantum-well structure.

(ii) Using the infinite well model with an effective well width L_{eff} from part (i), calculate the second subband energy E_2 and the intersubband dipole moment

$$M_{21} = |e| \ \phi_1(z) \ z\phi_2(z)dz$$

where the integration is over the width L_{eff}.

(iii) Calculate the absorption coefficient α for a doping density $N_D = 3 \times 10^{17}\,\text{cm}^{-3}$ assuming that only the first subband is occupied. Assume that the refractive $n_r = 3.6$, $T = 77$K, and the linewidth is 30 meV.

(b) Repeat part (a) for an $\text{In}_{0.53}\text{Ga}_{0.47}$As/InP quantum-well structure with a well width of 100 Å and the same doping density. Use the parameters such as band gaps and the electron effective masses from the Table C.3 (page 804) on the Physical Properties of InGaAs alloys in Appendix C. Assume $\Delta E_c = 40\% \Delta E_g$. Compare the results with the GaAs/$\text{Al}_{0.3}\text{Ga}_{0.7}$As system and comment on the differences.

REFERENCES

1. M. Planck, *Verh. dt. Phys. Ges.* **2**, 202 and 237 (1900).
2. A. Einstein, *Ann. d. Phys.* **17**, 132 (1905).
3. W. Heitler, *The Quantum Theory of Radiation*, Clarendon Press, Oxford, UK, 1954.
4. R. Loudon, *The Quantum Theory of Light*, 3rd ed., Clarendon, Oxford, UK, 2000.
5. H. Haken, *Light, vol. 1: Waves, Photons, Atoms*, p. 204, Chapter 7, North-Holland, Amsterdam, 1986.
6. G. H. B. Thompson, *Physics of Semiconductor Laser Devices*, Chapter 2, Wiley, New York, 1980.
7. H. C. Casey, Jr., and M. B. Panish, *Heterostructure Lasers, Part A: Fundamental Principles*, Chapter 3, Academic Press, Orlando, 1978.
8. G. Lasher and F. Stern, "Spontaneous and stimulated recombination radiation in semiconductors," *Phys. Rev.* **133**, A553–A563 (1964).
9. S. L. Chuang, J. O'Gorman, and A. F. J. Levi, "Amplified spontaneous emission and carrier pinning in laser diodes," *IEEE J. Quantum Electron.* **29**, 1631–1639 (1993).
10. C. S. Chang, S. L. Chuang, J. Minch, Y. K. Chen, and T. Tanbun-Ek, "Amplified spontaneous emission spectroscopy in strained quantum-well lasers," *IEEE J. Selected Topics Quantum Electron.*, Special issue on Applied Optical Diagnostics of Semiconductors **1**, 1100–1107 (1995).

11. W. Fang, M. Hattendorf, S. L. Chuang, J. Minch, C. S. Chang, C. G. Bethea, and Y. K. Chen, "Analysis of temperature sensitivity in semiconductor lasers using gain and spontaneous emission measurements," *Appl. Phys. Lett.* **70**, 796–798 (1997).
12. J. Minch, S. H. Park, and S. L. Chuang, "Theory and experiment of InGaAsP and InGaAlAs long-wavelength strained quantum-well lasers," *IEEE J. Quantum Electron.* **35**, 771–782 (1999).
13. T. Keating, S. H. Park, J. Minch, X. Jin, S. L. Chuang, and T. Tanbun-Ek, "Optical gain measurements based on fundamental properties and comparison with many-body theory," *J. Appl. Phys.* **86**, 2945–2952 (1999).
14. H. C. Casey, Jr., and F. Stern, "Concentration-dependent absorption and spontaneous emission of heavily doped GaAs," *J. Appl. Phys.* **47**, 631–643 (1976).
15. B. W. Hakki and T. L. Paoli, "Gain spectra in GaAs double heterostructure injection lasers," *J. Appl. Phys.* **46**, 1299–1306 (1975).
16. M. G. A. Bernard and G. Duraffourg, "Laser conditions in semiconductors," *Phys. Stat. Sol.* **1**, 699–703 (1961).
17. A. Yariv, *Quantum Electronics*, 3rd ed., Wiley, New York, 1989.
18. E. O. Kane, "Band structure of indium antimonide," *J. Phys. Chem. Solids* **1**, 249–261 (1957).
19. M. Yamanishi and I. Suemune, "Comment on polarization dependent momentum matrix elements in quantum well lasers," *Jpn. J. Appl. Phys.* **23**, L35–L36 (1984).
20. M. Asada, A. Kameyama, and Y. Suematsu, "Gain and intervalence band absorption in quantum-well lasers," *IEEE J. Quantum Electron.* **QE-20**, 745–753 (1984).
21. R. H. Yan, S. W. Corzine, L. A. Coldren, and I. Suemune, "Corrections to the expression for gain in GaAs," *IEEE J. Quantum Electron.* **26**, 213–216 (1990).
22. H. Kobayashi, H. Iwamura, T. Saku, and K. Otsuka, "Polarization-dependent gain-current relationship in GaAs-AlGaAs MQW laser diodes," *Electron. Lett.* **19**, 166–168 (1983).
23. M. Yamada, S. Ogita, M. Yamagishi, K. Tabata, N. Nakaya, M. Asada, and Y. Suematsu, "Polarization-dependent gain in GaAs/AlGaAs multi-quantum-well lasers: Theory and experiment," *Appl. Phys. Lett.* **45**, 324–325 (1984).
24. E. Zielinski, F. Keppler, S. Hausser, M. H. Pilkuhn, R. Sauer, and W. T. Tsang, "Optical gain and loss processes in GaInAs/InP MQW laser structures," *IEEE J. Quantum Electron.* **25**, 1407–1416 (1989).
25. E. Zielinski, H. Schweizer, S. Hausser, R. Stuber, M. H. Pilkuhn, and G. Weimann, "Systematics of laser operation in GaAs/AlGaAs multiquantum well heterostructures," *IEEE J. Quantum Electron.* **QE-23**, 969–976 (1987).
26. Y. Arakawa and H. Sakaki, "Multidimensional quantum well laser and temperature dependence of its threshold current," *Appl. Phys. Lett.*, **40**, 939–941 (1982).
27. Y. Arakawa and A. Yariv, "Theory of gain, modulation response, and spectral linewidth in AlGaAs quantum well lasers," *IEEE J. Quantum Electron.* **QE-21**, 1666–1674 (1985).
28. M. Asada, Y. Miyamoto, and Y. Suamatsu, "Gain and threshold of three-dimensional quantum-box lasers," *IEEE J. Quantum Electron.* **QE-22**, 1915–1921 (1986).
29. D. Bimberg, N. Kirstaedter, N. N. Ledenstov, Zh. I. Alferov, P. S. Kop'ev, and V. M. Ustinov, "InGaAs-GaAs quantum dot lasers," *IEEE J. Sel. Top. Quantum Electron.* **3**, 196–205 (1997).

30. N. Kirstaedter, N. N. Ledentsov, M. Grundmann, D. Bimberg, V. M. Ustinov, S. S. Rumimov, M. V. Maximov, P. S. Kop'ev, Zh. I. Alferov, U. Richter, P. Werner, U. Gösele, and J. Heydenreich, "Low threshold, large to injection laser emission from (InGa)As quantum dots," *Electron. Lett.* **30**, 1416–1417 (1994).

31. M. Grundmann, O. Stier, and D. Bimberg, "InAs/GaAs pyramidal quantum dots: strain distribution, optical phonons, and electronic structure," *Phys. Rev. B* **52**, 11969–11981 (1995).

32. H. Shoji, K. Mukai, N. Ohtsuka, M. Sugawara, T. Uchida, and H. Ishikawa, "Lasing at three-dimensionally quantum-confined sublevel of self-organized In/sub 0.5/Ga/sub 0.5/As quantum dots by current injection," *IEEE Photon. Technol. Lett*, **7**, 1385–1387 (1995).

33. M. Sugawara, "Theory of spontaneous-emission lifetime of Wannier excitons in mesoscopic semiconductor quantum disks," *Phys. Rev. B* **51**, 10743–10754 (1995).

34. N. Kirstaedter, O. G. Schmidt, N. N. Ledentsov, D. Bimberg, V. M. Ustinov, A. Y. Egorov, A. E. Zhukov, M. V. Maximov, P. S. Kopev, and Z. I. Alferov, "Gain and differential gain of single layer InAs/GaAs quantum dot injection lasers," *Appl. Phys. Lett.* **69**, 1226–1228 (1996).

35. M. Grundmann, N. N. Ledentsov, O. Stier, J. Bohrer, D. Bimberg, V. M. Ustinov, P. S. Kop'ev, and Zh. I. Alferov, "Nature of optical transitions in self-organized InAs/GaAs quantum dots," *Phys. Rev. B* **53**, R10511 (1996).

36. N. N. Ledentsov, V. A. Shchukin, M. Grundmann, N. Kirstaedter, J. Böhrer, O. Schmidt, D. Bimberg, V. M. Ustinov, A. Yu. Egorov, A. E. Zhukov, P. S. Kop'ev, S. V. Zaitsev, N. Yu. Gordeev, Zh. I. Alferov, A. I. Borovkov, A. O. Kosogov, S. S. Ruvimov, P. Werner, U. Gösele, and J. Heydenreich, "Direct formation of vertically coupled quantum dots in Stranski-Krastanow growth," *Phys. Rev. B* **54**, 8743–8750 (1996).

37. M. Grundmann and D. Bimberg, "Formation of quantum dots in twofold cleaved edge overgrowth," *Phys. Rev. B* **55**, 4054–4056 (1997).

38. M. Grundmann and D. Bimberg, "Theory of random population for quantum dots," *Phys. Rev. B* **55**, 9740–9745 (1997).

39. L. V. Asryan and R. A. Suris, "Inhomogeneous line broadening and the threshold current density of a semiconductor quantum dot laser," *Semicond. Sci. Technol.*, **11**, 554–567 (1996).

40. L. V. Sryan, and R. A. Suris, "Charge neutrality violation in quantum-dot lasers," *IEEE J. Sel. Topics Quantum Electron.* **3**, 148–157 (1997).

41. L. V. Asryan and R. A. Suris, "Temperature dependence of the threshold current density of a quantum dot laser," *IEEE J. Quantum Electron.*, **34**, 841–850 (1998).

42. V. M. Ustinov, N. A. Maleev, A. E. Zhukov, A. R. Kovsh, A. Yu. Egorov, A. V. Lunev, B. V. Volovik, I. L. Krestnikov, Yu. G. Musikhin, N. A. Bert, P. S. Kop'ev, Zh. I. Alferov, N. N. Ledentsov, and D. Bimberg, "InAs/InGaAs quantum dot structures on GaAs substrates emitting at 1.3 μm," *Appl. Phys. Lett.*, **74**, 2815–2817 (1999).

43. M. Sugawara, K. Mukai, and Y. Nakata, "Light emission spectra of columnar-shaped self-assembled InGaAs/GaAs quantum-dot lasers: Effect of homogeneous broadening of the optical gain on lasing characteristics," *Appl. Phys. Lett.*, **74**, 561–1563 (1999).

44. A. Sakamoto and M. Sugawara, "Theoretical calculation of lasing spectra of quantum-dot lasers: effect of homogeneous broadening of optical gain," *IEEE Photon. Technol. Lett.* **12**, 107–109 (2000).

45. A. Stintz, G. T. Liu, H. Li, L. F. Lester, and K. J. Malloy, "Low-threshold current density 1.3 μm InAs quantum-dot lasers with the dots-in-a-well (DWELL) structure," *IEEE Photon. Technol. Lett.* **12**, 591–593 (2000).
46. D. Bimberg, M. Grundmann, and N. N. Ledentsov, *Quantum Dot Heterostructures*, Wiley, New York, 2001.
47. D. Bimberg, M. Grundmann, N. N. Ledentsov, M. H. Mao, Ch. Ribbat, R. Sellin, V. M. Ustinov, A. E. Zhukov, Zh. I. Alferov, and J. A. Lott, "Novel infrared quantum dot lasers: theory and reality," *Phys. Stat. Sol.* (b) **224**, 787 (2001).
48. N. N. Ledendsov, "Long-wavelength quantum-dot lasers on GaAs substrate: from media to device concepts," *IEEE J. Quantum Electron.* **8**, 1015 (2002).
49. M. Sugawara, H. Ebe, N. Hatori, M. Ishida, Y. Arakawa, T. Akiyama, K. Otsubo, and Y. Nakata, "Theory of optical signal amplification and processing by quantum-dot semiconductor optical amplifires," *Phys. Rev. B* **69**, 235332-1 to 235332-39 (2004).
50. K. Mukai, Y. Nakata, K. Ohtsubo, M. Sugawara, N. Yokoyama, and H. Ishikawa, "1.3-μm CW lasing of InGaAs-GaAs quantum dots at room temperature with a threshold current of 8 mA," *IEEE Photon. Technol. Lett.* **11**, 1205–1207 (1999).
51. A. Fiore, U. Oesterle, R. P. Stanley, R. Houdré, F. Lelarge, M. Ilegems, P. Borri, W. Langbein, D. Birkedal, J. M. Hvam, M. Cantoni, and F. Bobard, "Structural and electrooptical characteristics of quantum dots emitting at 1.3 μm on gallium arsenide," *IEEE J. Quantum Electron.* **37**, 1050–1058 (2001).
52. S. S. Mikhrin, A. R. Kovsh, I. L. Krestnikov, A. V. Kozhukhov, D. A. Livshits, N. N. Ledentsov, Y. M. Shernyakov, I. I. Novikov, M. V. Maximov, V. M. Ustinov, and Z. I. Alferov, "High power temperature insensitive 1.3 μm InAs/InGaAs/GaAs quantum dot lasers," *Semicond. Sci. Technol.*, **20**, 340–342 (2005).
53. J. Kim and S. L. Chuang, "Theoretical and experimental study of optical gain, refractive index change, and linewidth enhancement factor of p-doped quantum-dot lasers" *IEEE J. Quantum Electron.*, **42**, 942–952 (2006).
54. J. Kim and S. L. Chuang, "Small signal cross-gain modulation of quantum-dot semiconductor optical amplifiers," *IEEE Photon. Technol. Lett.*, **18**, 2538–2540 (2006).
55. E. Kapon, D. M. Hwang, and R. Bhat, "Stimulated emission in semiconductor quantum wire heterostructures," *Phys. Rev. Lett.* **63**, 430–433 (1989).
56. A. Chavez-Pirson, H. Ando, H. Saito, and H. Kanbe, "Polarization properties of a vertical cavity surface emitting laser using a fractional layer superlattice gain medium," *Appl. Phys. Lett.* **62**, 3082–3084 (1993).
57. H. Ando, S. Nojima, and H. Kanbe, "Band-edge optical-absorption spectra of gaas quantum wires calculated by multiband effective-mass theory," *J. Appl. Phys.* **74**, 6383–6390 (1993).
58. T. Kojima, S. Tanaka, H. Yasumoto, S. Tamura, and S. Arai, "Evaluation of optical gain properties of GaInAsP/InP compressively strained quantum-wire lasers," *J. Appl. Phys.* **38**, 6327–6334 (1999).
59. L. C. West and S. J. Eglash, "First observation of an extremely large dipole infrared transition within the conduction band of a GaAs quantum well," *Appl. Phys. Lett.* **46**, 1156–1158 (1983).
60. For a review, see B. F. Levine, "Quantum-well infrared photodetectors," *J. Appl. Phys.* **74**, R1–R81 (1993).

61. M. O. Manasreh, Ed., *Semiconductor Quantum Wells and Superlattices for Long-Wavelength Infrared Detectors*, Artech House, Norwood, 1993.

62. D. Ahn and S. L. Chuang, "Calculation of linear and nonlinear intersubband optical absorption in a quantum well model with an applied electric field," *IEEE J. Quantum Electron.* **QE-23**, 2196–2204 (1987).

63. S. L. Chuang, M. S. C. Luo, S. Schmitt-Rink, and A. Pinczuk, "Many-body effects on intersubband transitions in semiconductor quantum-well structures," *Phys. Rev. B* **46**, 1897–1899 (1992).

64. J. Faist, F. Capasso, D. L. Sivco, C. Sirtori, A. L. Hutchinoson, and A. Y. Cho, "Quantum cascade laser," *Science* **264**, 553–556 (1994).

65. F. Capasso, R. Paiella, R. Martini, R. Colombelli, C. Gmachl, T. L. Myers, M. S. Taubman, R. M. Williams, C. G. Bethea, K. Unterrainer, H. Y. Hwang, D. L. Sivco, A. Y. Cho, A. M. Sergent, H. C. Liu, and E. A. Whittaker, "Quantum cascade lasers: Ultrahigh-speed operation, optical wireless communication, narrow line-width, and far-infrared emission," *IEEE J. Quantum Electron.* **38**, 511–532 (2002).

66. D. Ahn, S. L. Chuang, and Y. C. Chang, "Valence band mixing effects on the gain and the refractive index change of the quantum well laser," *J. Appl. Phys.* **64**, 4056–4064 (1988).

67. D. Ahn and S. L. Chuang, "Optical gain in a strained-layer quantum-well laser," *IEEE J. Quantum Electron.* **24**, 2400–2407 (1988).

68. D. Ahn and S. L. Chuang, "Optical gain and gain suppression of quantum-well lasers with valence band mixing," *IEEE J. Quantum Electron.* **26**, 13–24 (1990).

69. D. Ahn and S. L. Chuang, "The theory of strained-layer quantum-well lasers with bandgap renormalization," *IEEE J. Quantum Electron.*, **30**, 350–365 (1994).

70. C. S. Chang and S. L. Chuang, "Modeling of strained quantum-well lasers with spin-orbit coupling," *IEEE J. Selected Topics Quantum Electron.*, Special issue on Semiconductor Lasers **1**, 218–229 (1995).

10

Fundamentals of Semiconductor Lasers

Semiconductor lasers are important optoelectronic devices for optical communication systems. For a historical account of the inventions and development of semiconductor lasers, see Refs. 1–3. The first semiconductor lasers were fabricated in 1962 using homojunctions [4–7]. These lasers had high threshold current density (e.g., 19,000 A/cm^2) and operated at cryogenic temperatures. The estimated current density [2] for room temperature operation would be 50,000 A/cm^2. The concept of heterojunction semiconductor lasers was realized in 1969–1970, with a low threshold current density (1600 A/cm^2) operating at room temperature [8–11]. These double-heterostructure diode lasers provide both carrier and optical confinements, which improve the efficiency for stimulated emission, and make applications to optical communication systems practical as transmitters.

In the 1970s, the concept of quantum-well structures for semiconductor lasers was proposed and realized experimentally [12–15]. The first single and multiple quantum-well lasers were grown successfully by liquid phase epitaxy (LPE) [13, 14] by Holonyak's group at the University of Illinois in 1977, although the metal-organic chemical vapor deposition (MOCVD) [15] and molecular-beam epitaxy (MBE) [16] are commonly used in recent years. The threshold current density was reduced [16] to about 500 A/cm^2. The reduction of the confinement region for the electron–hole pairs to a tiny volume in the quantum-well region improves the laser performance significantly. The quantum-well structures provide a wavelength tunability by varying the well width and the barrier height. These devices are also excellent practical examples for textbook problems on basic quantum mechanics (i.e., particle in a box model).

In the late 1980s, strained quantum-well lasers were proposed to improve the laser performance [17, 18]. Superior diode lasers using strained quantum wells reduced [19] the threshold current density to 65 A/cm^2 at room temperature. Further reduction in threshold current density using strain effects has been possible [20]. An interesting note is that the threshold current density of semiconductor lasers was improved by approximately one order of magnitude smaller per decade since their invention in 1962. The improvements resulted from the development of novel ideas and a realizable technology such as LPE, MBE, or MOCVD [21].

Physics of Photonic Devices, Second Edition. By Shun Lien Chuang
Copyright © 2009 John Wiley & Sons, Inc.

In the 1990s, research on strained quantum-well lasers continued with remarkable progress toward high-performance diode lasers. Quantum-wire semiconductor lasers have been realized. The potential improvement using quantum-wire and quantum-dot semiconductor lasers has been under investigation both theoretically and experimentally [22–25]. Novel structures such as surface-emitting lasers [26–28] and microdisk lasers [29, 30] have also been realized. The microdisk lasers using a small disk with a radius R of the order 10 μm and a thickness of 0.6 μm show interesting resonator physics with whispering-gallery mode lasing characteristics. Many of these structures also make research on the cavity quantum electrodynamics practical [31].

In this chapter, we study the basic semiconductor lasers and their physical principles. Various important designs and their models will be discussed. Because the research on semiconductor lasers and technology is so rapid, these physical principles and theoretical models will provide a good tool to design diode lasers with the desired characteristics or to set up a foundation to design novel laser structures with superior performance.

10.1 DOUBLE-HETEROJUNCTION SEMICONDUCTOR LASERS

10.1.1 Energy Band Diagram and Carrier Injection

Let us first consider a P-InP/p-$\text{In}_{1-x}\text{Ga}_x\text{As}_{1-y}\text{P}_y$/$N$-InP double-heterojunction [32, 33] structure. Before the formation of the heterojunctions, the energy band diagram is shown schematically in Fig. 10.1a, where the energy levels are all measured from the vacuum level. At thermal equilibrium, Fig. 10.1b, the Fermi level is a constant across the three regions. The band diagram still contains the features of the bulk properties in the regions away from the heterojunctions, assuming that the central region is wide enough. The built-in potentials are determined by the Fermi levels

$$-qV_{\text{D}}^{P-p} = E_{\text{F}P} - E_{\text{f}p} = (E_{\text{F}P} - E_{\text{V}P}) - \Delta E_{\text{V}} - (E_{\text{f}p} - E_{\text{v}p}) \qquad (10.1.1)$$

$$-qV_{\text{D}}^{p-N} = E_{\text{f}p} - E_{\text{F}N} = (E_{\text{f}p} - E_{\text{v}p}) + \Delta E_{\text{V}} - (E_{\text{F}N} - E_{\text{V}N}) \qquad (10.1.2)$$

where

$$\Delta E_{\text{V}} = E_{\text{v}p} - E_{\text{V}P} = E_{\text{v}p} - E_{\text{V}N}. \qquad (10.1.3)$$

With an applied voltage V, the injection current is I, and the current density is

$$J = \frac{I}{wL} \qquad (10.1.4)$$

where w is the width of the active region in the lateral direction and L is the cavity length of the diode laser. The carrier concentration in the active region is determined by the rate equation

$$\frac{dn(t)}{dt} = \eta_{\text{i}} \frac{J(t)}{qd} - R(n) - \text{v}_{\text{g}} g(n) S(t) \qquad (10.1.5)$$

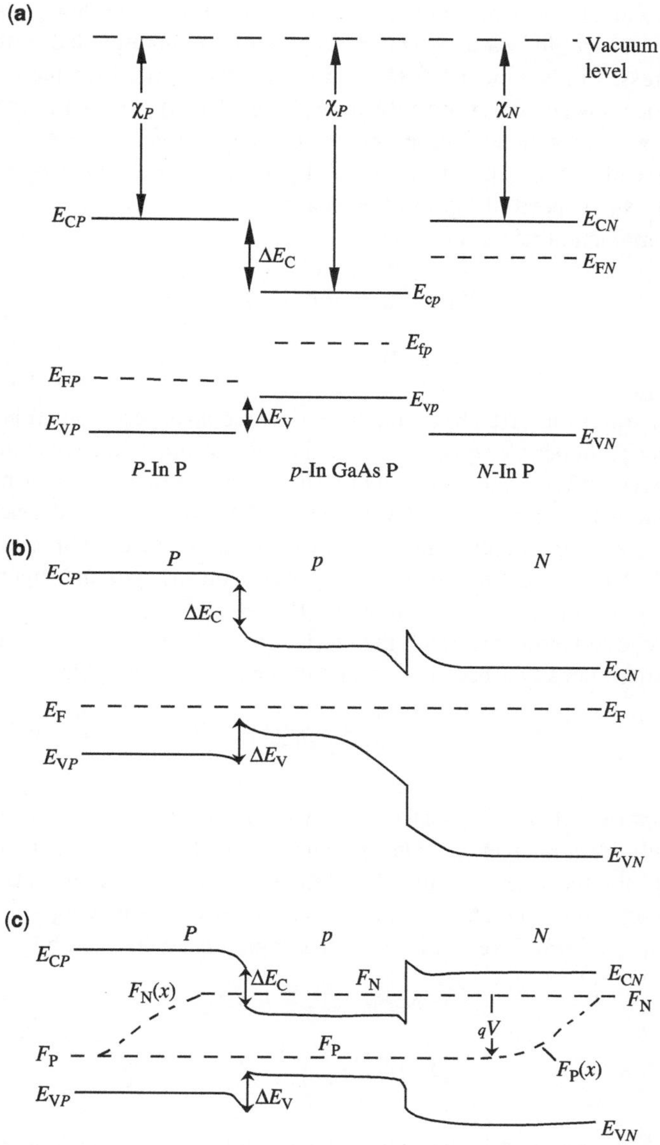

Figure 10.1 Energy band diagram for a P-InP/p-InGaAsP/N-InP double-heterojunction structure (a) before contact, (b) after contact at thermal equilibrium ($V = 0$), and (c) under a forward bias ($V > 0$).

where the first term on the right-hand side accounts for carrier injection into the active region with a thickness d. The unit charge q is 1.6×10^{-19} C. The collection efficiency of carriers into the active region is η_i, which is the percentage of electrons and holes transporting into the active region after surviving the carrier losses due to recombination outside the active regions or carrier over flow into the other side of the p-n junction. The second term, $R(n)$, accounts for the carrier recombinations due to both radiative (spontaneous emission) and nonradiative processes. We have ignored the carrier diffusion, which is important for gain-guided structures and will be addressed in Section 10.2. The last term is the carrier loss due to stimulated emission process when the laser cavity has a photon density $S(t)$, which is the number of photons per unit volume. These photons encounter a gain of $g(n)$ $(1/\text{cm})$ while propagating with the group velocity v_g of light, $v_g = c/n_g$, where n_g is the group index, and c is the speed of light in free space.

The recombination rate is given by

$$R(n) = An + Bn^2 + Cn^3 = \frac{n}{\tau} \tag{10.1.6}$$

$$\tau(n) = (A + Bn + Cn^2)^{-1} \tag{10.1.7}$$

where the first term in (10.1.6) is due to the nonradiative recombination of carriers with defects or monocarrier type recombination, Bn^2 is due to the spontaneous radiative recombination by electron–hole pairs (Bnp), and Cn^3 accounts for nonradiative Auger recombination processes. We have assumed that $n \simeq p$ in the active region (i.e., the doping concentration in the active region is small). Otherwise, we use Bnp instead of Bn^2, and Cn^2p or Cnp^2 depending on the type of Auger processes. B may depend on the carrier concentration, $B \simeq B_0 - B_1 n$.

At steady state below threshold, the carrier concentration n is determined by the injection current density J because the photon density is negligible

$$\eta_i \frac{J}{qd} = \frac{n}{\tau}. \tag{10.1.8}$$

Often, the carrier lifetime τ is taken as an input parameter, and we determine n approximately by $n = \eta_i J\tau/qd$. On the other hand, if we estimate the coefficients A, B, and C for the laser structure from experimental data or theoretical models, we can extract more accurate J-n relation using (10.1.6). Knowing n in the active region, the quasi-Fermi level F_n in the active region is determined by

$$n = N_c F_{1/2}((F_n - E_c)/k_B T) \tag{10.1.9a}$$

$$N_c = 2\left(\frac{m_e^* k_B T}{2\pi\hbar^2}\right)^{3/2} \tag{10.1.9b}$$

and the hole concentration in the active region is determined by the charge neutrality condition

$$n + N_{A^-} = p + N_{D^+} \tag{10.1.10}$$

where N_{A^-} and N_{D^+} are the ionized acceptor and donor concentrations, respectively. The quasi-Fermi level in the active region for the holes is then determined by

$$p = N_v F_{1/2}\big((E_v - F_p)/k_B T\big) \tag{10.1.11a}$$

$$N_v = 2\left(\frac{k_B T}{2\pi\hbar^2}\right)^{3/2}\left(m_{hh}^{*\,3/2} + m_{\ell h}^{*\,3/2}\right). \tag{10.1.11b}$$

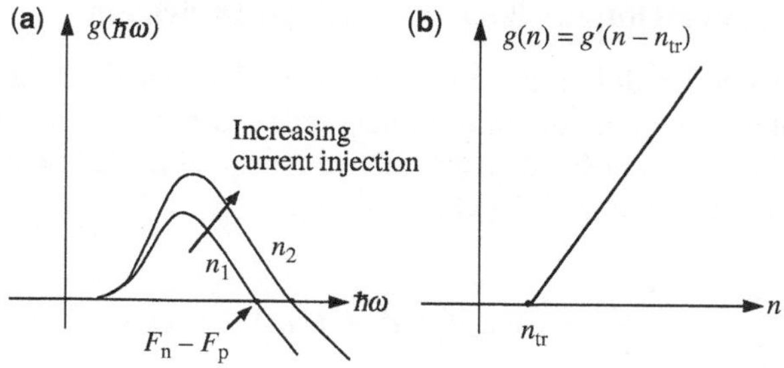

Figure 10.2 (a) The gain spectrum for different carrier densities, $n_2 > n_1$; (b) the peak gain versus the carrier density n.

From the electron and hole concentrations, n and p, or the quasi-Fermi levels, F_n and F_p, we can calculate the gain directly using the band structure. Typically, a linear relation for the peak gain coefficient versus the carrier density such as

$$g(n) = g'(n - n_{tr}) \tag{10.1.12}$$

for a bulk semiconductor laser is obtained over a region of carrier concentration, where g' is the differential gain, dg/dn, and n_{tr} is a transparency concentration when the gain is zero, Fig. 10.2a and 10.2b. For quantum-well lasers, an empirical formula using a logarithmic dependence for the gain versus carrier density or gain versus injection current density is usually used. This will be discussed further in Sections 10.3 and 10.4.

10.1.2 Threshold Condition

The threshold gain is determined by (7.6.16), that is, when the modal gain equals the sum of the intrinsic loss and the mirror transmission loss

$$\Gamma g_{th} = \alpha_i + \alpha_m \tag{10.1.13}$$

$$\alpha_m = \frac{1}{2L}\ln\frac{1}{R_1 R_2} \tag{10.1.14}$$

where Γ is the optical confinement factor, and α_i is the intrinsic loss due to mostly free carrier absorption of the waveguide materials and scattering loss caused by the waveguide wall roughness. The mirror loss α_m accounts for the transmissions (outputs) at the two end mirrors (facets). With an increase in the injection current density J, the gain increases due to the increase of the carrier concentration n. When the threshold condition is reached, the carrier concentration n is pinned at the threshold value n_{th} because the gain is pinned at the threshold value: $g_{th} = (\alpha_i + \alpha_m)/\Gamma$; therefore, $n_{th} = n_{tr} + (\alpha_i + \alpha_m)/(\Gamma g')$.

$$\eta_i \frac{J_{th}}{qd} = R(n_{th}) = An_{th} + Bn_{th}^2 + Cn_{th}^3. \tag{10.1.15}$$

10.1.3 Light Power Output: Back-of-Envelope Derivation

Below the threshold condition, the output light consists mainly of spontaneous and amplified spontaneous emission, and its magnitude is governed by Bn^2. A further increase in the injection current density J above threshold leads to the light emission of photons (S) through the stimulated emission process

$$\eta_i \frac{J}{qd} = (An_{th} + Bn_{th}^2 + Cn_{th}^3) + v_g g_{th} S \tag{10.1.16}$$

because the carrier density is pinned at n_{th}, which is caused by gain pinning at g_{th}. Therefore, the photon density is simply taken from the difference of (10.1.15) and (10.1.16):

$$S = \eta_i \frac{J - J_{th}}{qd v_g g_{th}}. \tag{10.1.17}$$

The photon lifetime τ_p is related to the threshold gain

$$\frac{1}{\tau_p} = v_g(\alpha_i + \alpha_m) = v_g \Gamma g_{th}, \tag{10.1.18}$$

which accounts for the loss rate of the photons in the laser cavity due to absorptions and transmissions. The photons have a mean lifetime as they disappear from the cavity by either absorption or escape via transmission through the mirrors. We obtain

$$\frac{S}{\tau_p} = \eta_i \frac{J - J_{th}}{qd} \Gamma. \tag{10.1.19}$$

The light output intensity P_{out} versus the injection current (L–I curve) is determined by

$$\begin{aligned} P_{out} &= \begin{pmatrix} \text{Energy of} \\ \text{a photon} \end{pmatrix} \begin{pmatrix} \text{Photon} \\ \text{density} \end{pmatrix} \begin{pmatrix} \text{Effective volume of} \\ \text{the optical mode} \end{pmatrix} \begin{pmatrix} \text{Escape rate} \\ \text{of photons} \end{pmatrix} \\ &= \hbar\omega\, S \left(wLd_{op}\right) \left(v_g \alpha_m\right) \end{aligned} \tag{10.1.20}$$

where $v_g\, \alpha_m$ is the escape rate of photons. Here w is the width and d_{op} is the effective thickness of the optical mode ($d_{op} = d/\Gamma$), and $I = wLJ$. We obtain a simple relation for the light output power versus the injection current (i.e., the well-known L–I curve):

$$P_{out} = \eta_i \frac{\hbar\omega}{q} \frac{\alpha_m}{\alpha_i + \alpha_m} (I - I_{th}). \tag{10.1.21}$$

The external differential quantum efficiency η_e is defined as

$$\eta_e = \frac{dP_{out}/dI}{\hbar\omega/q} = \eta_i \frac{\alpha_m}{\alpha_m + \alpha_i}$$
$$= \eta_i \frac{\ln(1/R)}{\alpha_i L + \ln(1/R)}. \quad (10.1.22a)$$

The inverse of the external differential quantum efficiency is

$$\eta_e^{-1} = \eta_i^{-1}\left[1 + \frac{\alpha_i L}{\ln(1/R)}\right]. \quad (10.1.22b)$$

A plot of η_e^{-1} versus the cavity length L shows a linear relationship with the intercept η_i^{-1} at $L = 0$, as shown in Fig. 10.3.

In real devices, leakage current exists. The effect of leakage current I_L can be taken into account in

$$I = I_A + I_L = JwL + I_L \quad (10.1.23)$$

where I_A is the current injected into the active region. Therefore, I_{th} is modified to be

$$I_{th} = J_{th}\, wL + I_L$$
$$= \frac{qn_{th}(wLd)}{\eta_i \tau_e(n_{th})} + I_L. \quad (10.1.24)$$

The leakage current I_L may also be increased with an increase in I. Very often, the threshold current density as a function of temperature is described by an empirical

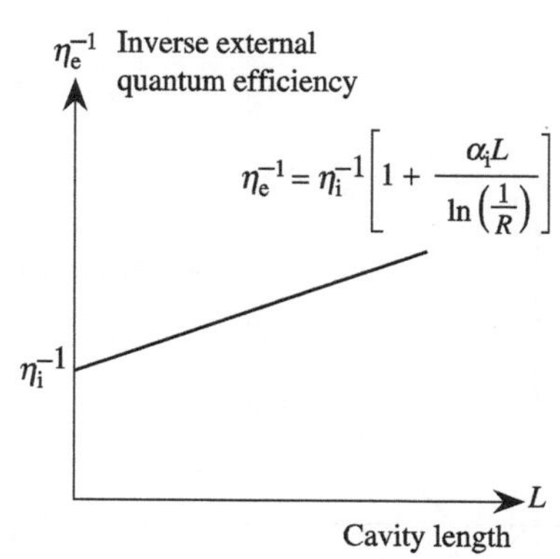

Figure 10.3 A plot of the inverse external quantum efficiency $1/\eta_e$ versus the laser cavity length L. The intercept with the vertical axis gives the inverse intrinsic quantum efficiency $1/\eta_i$.

formula using a characteristic temperature T_0 over a temperature range near a chosen temperature T_a.

$$I_{\text{th}}(T) = Const. \times e^{T/T_0} = I_{\text{th}}(T_a)e^{(T-T_a)/T_0}. \tag{10.1.25}$$

A large characteristic temperature is desirable since it implies that the threshold current does not change much with temperature. The output optical power P_{out} versus the injection current I relation can be modified to be

$$P_{\text{out}} = \eta_{\text{i}} \frac{\hbar\omega}{q} \frac{\alpha_{\text{m}}}{\alpha_{\text{m}} + \alpha_{\text{i}}} (I - I_{\text{th}} - \Delta I_{\text{L}}) \tag{10.1.26}$$

where ΔI_{L} accounts for any additional increase in the leakage current [34–36] due to the increase of I. A typical output power versus injection current relation is shown in Fig. 10.4. Below the threshold injection current, the output light intensity is negligibly small. Above threshold, the output power is linearly increasing until saturation effects occur.

There are three possible reasons for saturation [33].

(i) The leakage current increases with the injection current I (i.e., $\Delta I_{\text{L}} \neq 0$).

(ii) The threshold current I_{th} may also depend on the injection current I due to junction heating. The raise in temperature reduces the recombination lifetime τ_{e}. For example, the Auger recombination rate increases when the temperature is raised.

(iii) The intrinsic absorption α_{i} increases with an increase in the injection current I.

The output power versus the injection current becomes a nonlinear dependence when saturation occurs.

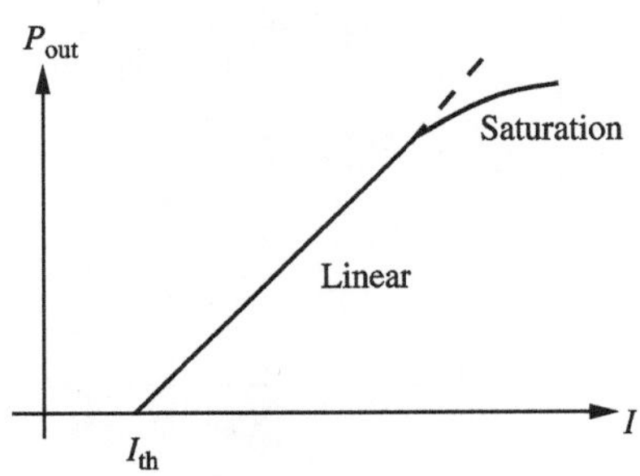

Figure 10.4 A typical diode laser output power versus injection current density relation (L–I curve).

10.1.4 Light-Emitting Diode versus Laser Diode: The Role of Spontaneous Emission versus Amplified Spontaneous Emission

In Chapter 9, we discuss the spontaneous emission $r^{\text{spon}}(\hbar\omega)$ from a forward biased *p-n* junction diode. The typical spontaneous emission spectrum is commonly used in light-emitting diodes (LEDs). The band gap of semiconductors is chosen such that LEDs in the visible region and the infrared region can be designed with a high quantum efficiency. For a simple planar structure such as shown [37–39] in Fig. 10.5, the light emission from the top surface of the cladding layer due to a spontaneously emitted photon will be limited by a cone angle of $2\theta_c$, where θ_c is the critical angle between the *p*-type semiconductor and the air. Light with an angle of emission outside of this cone angle will be reflected back to the semiconductor region. Therefore, the geometric design such as a hemispherical lens or parabolic lens above the *p-n* junction diode has been used to optimize the light emission. Other important considerations for visible LEDs include the responsivity of human eyes. Infrared LEDs for applications in optical communication are designed such that efficient coupling into optical fibers can be achieved, as shown in Fig. 10.6, with a surface emitter configuration [40]. The double-heterojunction AlGaAs/GaAs structure also provides the carrier confinement to increase the efficiency [41].

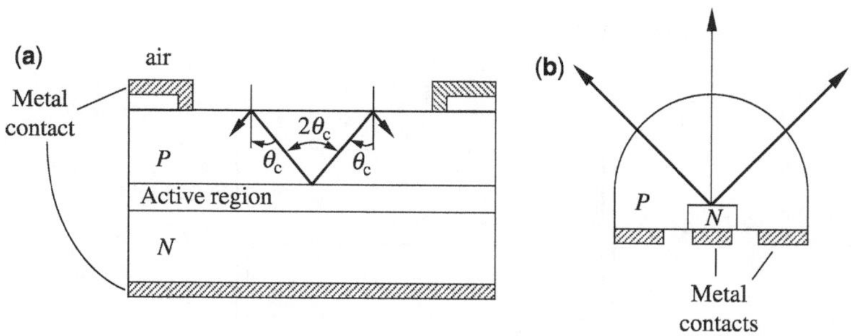

Figure 10.5 (a) Schematic diagram of a planar LED device to show the effect of total internal refraction. (b) The semiconductor hemisphere geometry for LED operation.

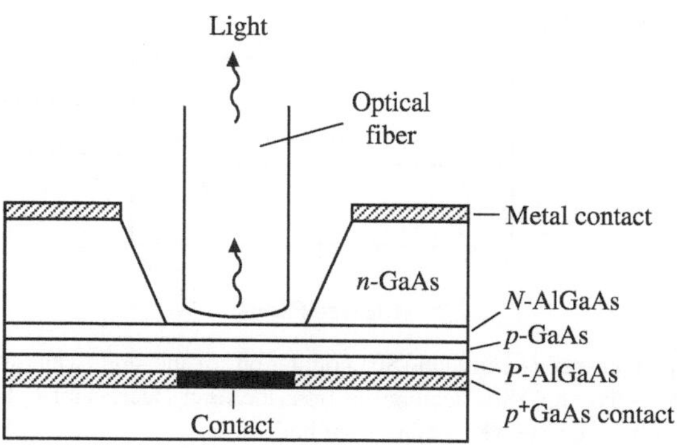

Figure 10.6 AlGaAs surface-emitting LED with an attached fiber to couple light output to an optical fiber.

For a laser diode, the cavity design and the amplified spontaneous emission will be important for the formation of laser modes. To understand the physics of the laser diode versus light-emitting diode operation, we consider an LED that is designed to have the same Fabry–Perot structure as an laser diode (*LD*) except that its two ends are antireflection coated. If we measure the spontaneous emission power spectrum from the top or bottom of the diode, called the window light L_W, it will be a broad spectrum determined by the spontaneous emission rate per unit volume $r^{\text{spon}}(\hbar\omega)$, Fig. 10.7a [42, 43],

$$L_W(\hbar\omega) = \hbar\omega \, r^{\text{spon}}(\hbar\omega) \, wdL \tag{10.1.27}$$

where w is the width, d is the thickness, and L is the cavity length of the index-guided structure. The facet light spectrum $L_F(\hbar\omega)$, taken from a facet of the LED, is the amplified spontaneous emission spectrum [42–45]

$$\begin{aligned} L_F(\hbar\omega) &= \hbar\omega \int_{z=0}^{L} r^{\text{spon}}(\hbar\omega) e^{G_n(\hbar\omega)z} \, dz \, wd \\ &= \hbar\omega \, r^{\text{spon}}(\hbar\omega) \left[\frac{e^{G_n(\hbar\omega)L} - 1}{G_n(\hbar\omega)} \right] wd \end{aligned} \tag{10.1.28}$$

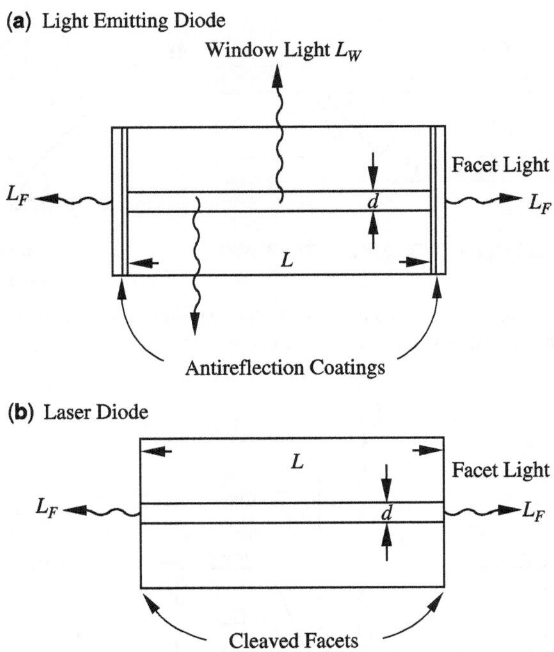

Figure 10.7 (a) The spontaneous emission (SE) coming out of the side (window light L_W) versus the amplified spontaneous emission (ASE) coming out from the facet (facet light L_F) in an LED structure. (b) A laser diode with the output power coming from both facets.

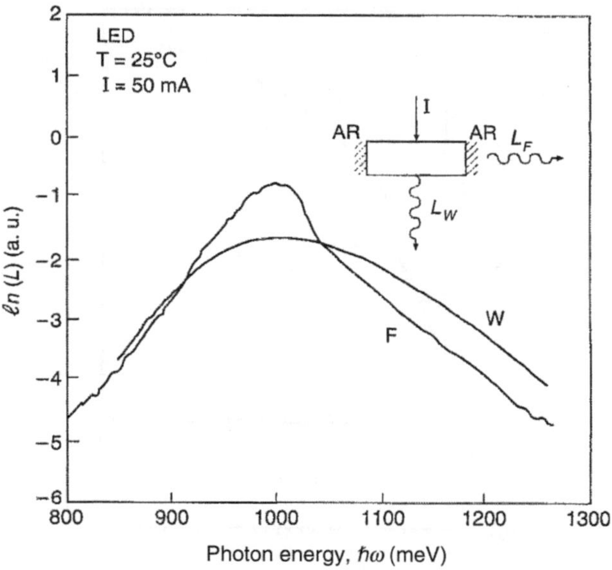

Figure 10.8 Measured spontaneous emission spectrum obtained by collecting photons through a window (W) in the substrate L_W and the amplified spontaneous emission spectrum obtained by collecting photons from the facet (F) L_F of an LED. (Reprinted with permission from [42] © 1993 American Institute of Physics.)

where $G_n(\hbar\omega) = \Gamma g - \alpha_i$ is the net modal gain (including all the losses due to absorptions and scatterings in the device). Note that at the transparency wavelength, $G_n(\hbar\omega) = 0$ and $L_F(\hbar\omega) = L_W(\hbar\omega)$ when no amplification by the gain action exists. With enough carrier injection such that $G_n > 0$, we see that the photon density in the spectral region where $g(\hbar\omega)$ is positive will experience amplification whereas that outside the positive gain region will experience absorption. Because the gain spectrum is narrower than that of the spontaneous emission spectrum [46], the facet light will be narrower than that of the window light. These results [42] are shown in Fig. 10.8. As a matter of fact, a comparison of these two spectra has been used to extract the gain spectrum of a laser diode structure. By measuring the two spectra $L_F(\hbar\omega)$ and $L_W(\hbar\omega)$ at the same current injection level, we can obtain the gain spectrum. Furthermore, if we take the logarithmic function of the ratio of the two spectra, ln $[L_F(\hbar\omega)/L_W(\hbar\omega)]$, it will be close to that of the gain spectrum because it is proportional to $\ln[(e^{G_nL} - 1)/G_n] \sim (\Gamma g_n - \alpha_i)L$ if the overall gain ($G_nL \gg 1$) is large enough [43]. By fitting the gain spectrum with a theoretical gain model, we can extract the carrier density n at a given injection current I [43]. The carrier density versus the injection current I is a monotonically increasing function of I, as shown in Fig. 10.9a at 25°C and Fig. 10.9b at 55°C. For laser diodes, the carrier density is pinned at the threshold value when the injection current is increased beyond threshold.

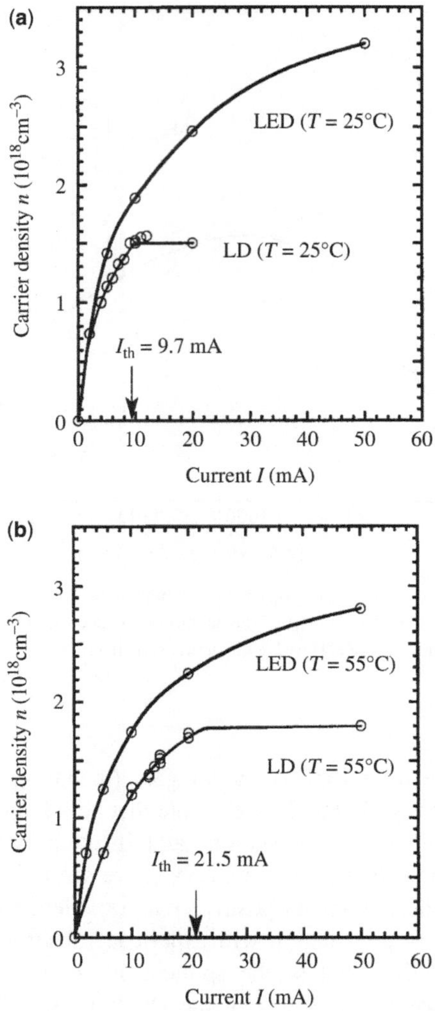

Figure 10.9 Carrier density versus injection current plotted for both LED and LD at (a) $T = 25°C$ and (b) $T = 55°C$. The solid curves are just visual guides. (Reprinted with permission from [43] © 1993 IEEE.)

10.1.5 Amplified Spontaneous Emission and Optical Gain Measurements

When the laser diode is biased with a forward voltage, current injection provides electron–hole pairs into the active region. With enough electron and hole densities, electron–hole recombination (the radiative process) can provide spontaneous emission. The spontaneously emitted photons propagating in the population inverted medium cause stimulated emission. It means that the absorption will be reduced due to the presence of gain caused by the amplified spontaneous emission process. As the injection current increases, gain increases. When the gain balances the background absorption loss, we have the "material" transparent condition at which the current density is called the transparent current density. Further increase of the current increases gain until it reaches the threshold condition, at which the gain is equal to the background absorption plus the mirror transmission loss (or laser output) of the cavity.

Consider an optical field $E_{sp}(\lambda)$ due to the spontaneous emission of photons by electron–hole radiative recombinations in a Fabry–Perot cavity. When the optical field propagates toward the minor facets with optical field reflection coefficients r_1 and r_2 on each facet, we can write down the multiple reflections of the optical field as

$$E_{ASE}(\lambda) = E_{sp}(\lambda)\left[1 + r_1 r_2 e^{i2kL} + \left(r_1 r_2 e^{i2kL}\right)^2 + \cdots\right]$$
$$= \frac{E_{sp}(\lambda)}{1 - r_1 r_2 e^{i2kL}}. \quad (10.1.29)$$

The complex propagation constant has a real and imaginary part

$$k = k' - i\frac{G_n}{2} = \frac{2\pi}{\lambda} n_e - i\frac{G_n}{2} \quad (10.1.30a)$$

$$G_n = \Gamma g - \alpha_i = \text{The } net \text{ modal gain.} \quad (10.1.30b)$$

The measured ASE power spectrum is proportional to $|E_{ASE}(\lambda)|^2$

$$I(\lambda) = |E_{ASE}(\lambda)|^2 = \frac{|E_{sp}(\lambda)|^2}{|1 - r_1 r_2 e^{i2kL}|^2} = \frac{|E_{sp}(\lambda)|^2}{(1-A)^2 + 4A\sin^2(k'L)} \quad (10.1.31)$$

where the amplitude A is

$$A = \sqrt{R_1 R_2}\, e^{G_n L}, \quad R_1 = |r_1|^2, \quad R_2 = |r_2|^2. \quad (10.1.32)$$

The ASE spectrum has maxima at

$$k'L = m\pi, \quad I_{max} = \frac{|E_{sp}(\lambda_{max})|^2}{(1-A)^2} \quad (10.1.33a)$$

and minima at

$$k'L = \left(m + \frac{1}{2}\right)\pi, \quad I_{min}(\lambda) = \frac{|E_{sp}(\lambda_{min})|^2}{(1+A)^2}. \quad (10.1.33b)$$

If we take the ratio between two *nearby* peak and valley of the ASE spectrum,

$$\frac{I_{max}}{I_{min}} = \frac{(1+A)^2}{(1-A)^2} \quad (10.1.34a)$$

we obtain

$$A = \frac{\sqrt{I_{max}/I_{min}} - 1}{\sqrt{I_{max}/I_{min}} + 1}. \tag{10.1.34b}$$

We solve for the net modal gain from (10.1.32)

$$G_n = \frac{1}{L}\ln\frac{\sqrt{I_{max}/I_{min}} - 1}{\sqrt{I_{max}/I_{min}} + 1} + \frac{1}{2L}\ln\frac{1}{R_1R_2}. \tag{10.1.35}$$

The above gain is the *net modal gain* of the optical modes

$$\Gamma g(\lambda) - \alpha_i = Q_r + \alpha_m \tag{10.1.36a}$$

$$Q_r = \frac{1}{L}\ln\frac{\sqrt{I_{max}/I_{min}} - 1}{\sqrt{I_{max}/I_{min}} + 1}. \tag{10.1.36b}$$

The procedure to extract the optical gain in the Hakki-Paoli method is as follows:

1. Measure the ASE spectrum $I(\lambda)$ as a function of wavelength. For example, we consider a multiple quantum-well sample as shown in Fig. 10.10 [47]. The measured ASE spectrum at room temperature is shown in Fig. 10.11a. We can see clear Fabry–Perot interference spectrum in expanded scale in Fig. 10.11b.
2. Find the magnitudes $I_{max}(\lambda)$ and $I_{min}(\lambda)$ of two nearby maximum and minimum near the mth longitudinal mode λ_m, and take the ratio.
3. The net optical gain $G_n(\lambda)$ spectrum is given by Eq. (10.1.36), where the mirror loss is usually calculated from the mirror reflectivity and the measured cavity length L. Alternatively, the cavity length can also be extracted from the spacing of the two nearby peaks if we know the effective index n_e.

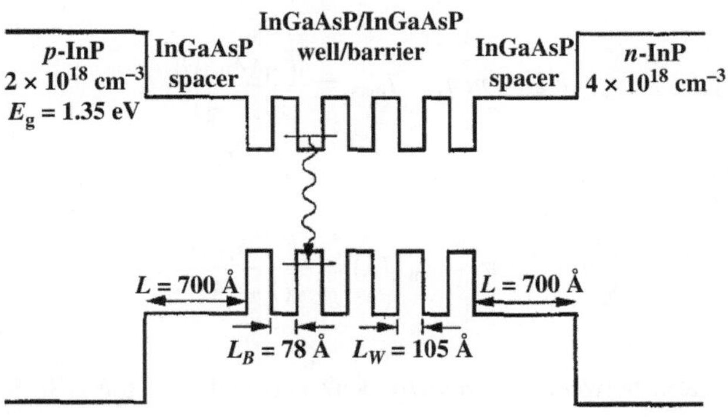

Figure 10.10 The band-edge profiles of conduction and valence bands for the InGaAsP strained QW laser structure. The PL wavelength of the slightly tensile-strained InGaAsP well region is 1.485 μm. The lattice-matched InP barrier has a band-gap wavelength of 1.28 μm. (Reprinted with permission from [47] © 1995 IEEE.)

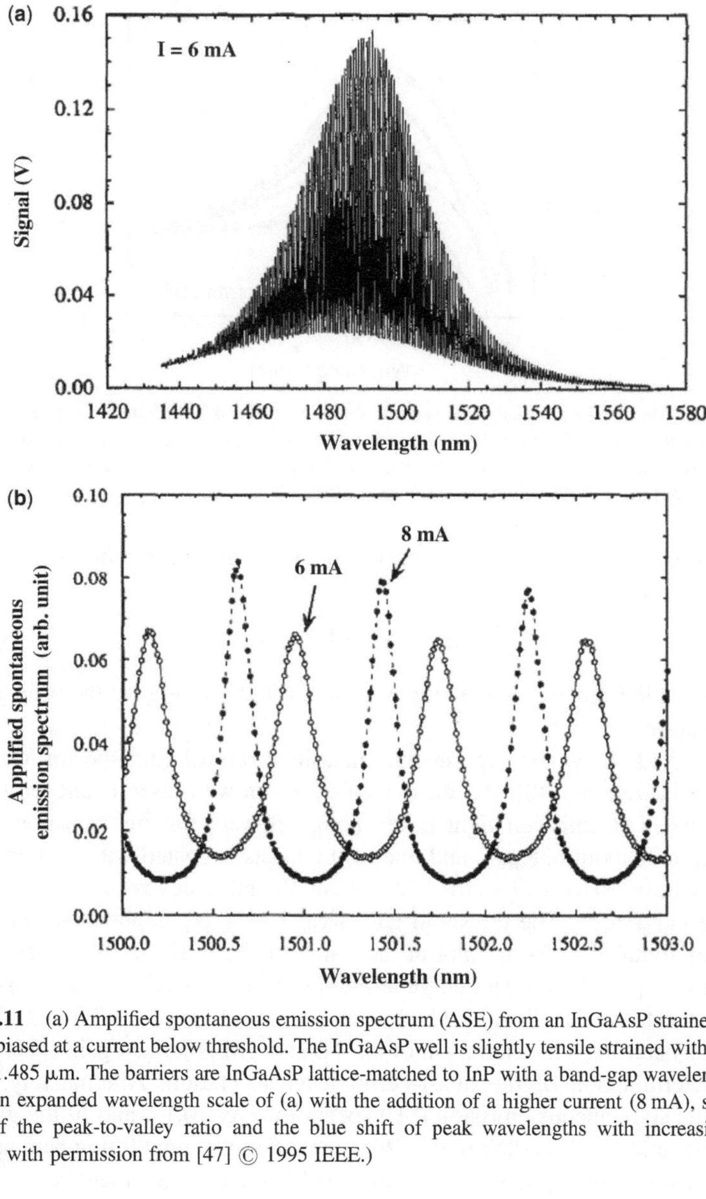

Figure 10.11 (a) Amplified spontaneous emission spectrum (ASE) from an InGaAsP strained quantum-well laser biased at a current below threshold. The InGaAsP well is slightly tensile strained with a PL wavelength of 1.485 μm. The barriers are InGaAsP lattice-matched to InP with a band-gap wavelength of 1.28 μm. (b) An expanded wavelength scale of (a) with the addition of a higher current (8 mA), showing the increase of the peak-to-valley ratio and the blue shift of peak wavelengths with increasing current. (Reprinted with permission from [47] © 1995 IEEE.)

4. Below the band edge (long wavelength side), the material gain is negligible and $\Gamma g - \alpha_i$ approaches the intrinsic absorption $-\alpha_i$, Fig. 10.12 [48]. As we increase current to the threshold condition, $I_{max}/I_{min} \rightarrow \infty$, and the peak gain occurs at the threshold gain value

$$\Gamma g_{th} - \alpha_i = \alpha_m. \tag{10.1.37}$$

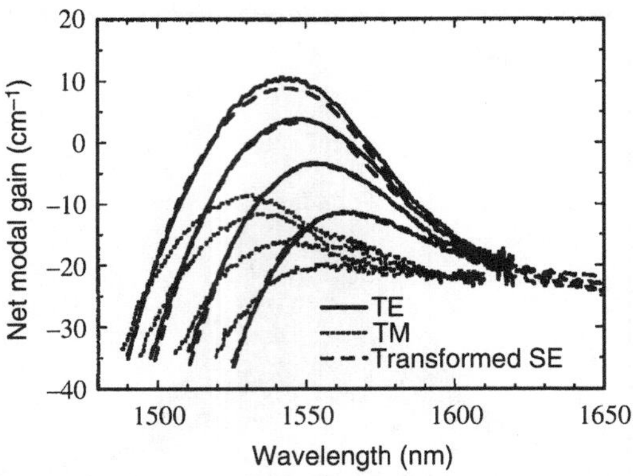

Figure 10.12 Measured net modal gain spectrum for TE (solid) and TM (dotted) polarizations based on the Hakki–Paoli method. The TE-polarized gain spectrum (dashed) obtained by the transformation from the spontaneous emission (SE) spectrum for Fig. 10.13 is also shown for comparison. (Reprinted with permission from [48] © 1999 American Institute of Physics.)

The top curve on Fig. 10.12 shows the threshold condition. In other words, the data from the expression

$$Q_r + \alpha_m = \Gamma g - \alpha_i \tag{10.1.38}$$

covers the range from $-\alpha_i$ at the long wavelength limit to $+\alpha_m$ at the peak gain near threshold current.

Figure 10.12 shows the experimental measured optical gain spectra for both TE and TM polarizations [48]. An unstrained quantum-well laser is chosen such that both TE and TM emission light have enough powers for our measurement. The cavity length is about 638 μm, and one of the facets is coated with high reflectivity (0.85). The mirror loss is 11.4 cm^{-1}. We obtain the intrinsic loss $\alpha_i = 22$ to 24 cm^{-1} below the band edge. The threshold gain $\Gamma g_{th} = \alpha_i + \alpha_m$ is about 36 cm^{-1}. If we draw a horizontal line on the plot at the value of the intrinsic loss level, $-\alpha_i$, we find the intercept with the gain curve at the quasi-Fermi level separation $\Delta F = F_n - F_p$ on the short wavelength side, which is also the intercept of the TE and TM polarized gain. Therefore, a measurement of the TE and TM polarized gain spectra allows for the determination of the quasi-Fermi level separation. Alternatively, we can measure the spontaneous emission (SE) spectrum, Fig. 10.13, and fit the gain spectrum by the fundamental relation (9.2.36) using ΔF as the input fitting parameter. The result is shown as the dashed curves in Fig. 10.12, and the agreement between the SE transformation method and the Hakki–Paoli method looks very good. The third method to measure the quasi-Fermi level separation is to apply a small signal ac modulated optical beam with a wavelength near the quasi-Fermi level separation energy, and measure the small ac signal of induced voltage of the semiconductor test laser by tuning its applied injection current. When the induced voltage changes a sign because of the crossing between gain and absorption, we obtain the transparency energy, which is exactly ΔF. The results of the above three methods for measuring the quasi-Fermi level separation are shown in Fig. 10.14a and the

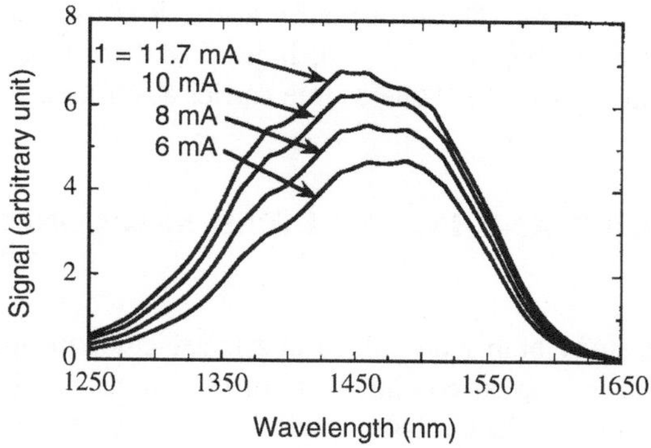

Figure 10.13 The measured spontaneous emission (SE) spectra from the side of the semiconductor laser using an optical fiber. The laser is temperature controlled and biased at a current from 6 mA to threshold. (Reprinted with permission from [48] © 1999 American Institute of Physics.)

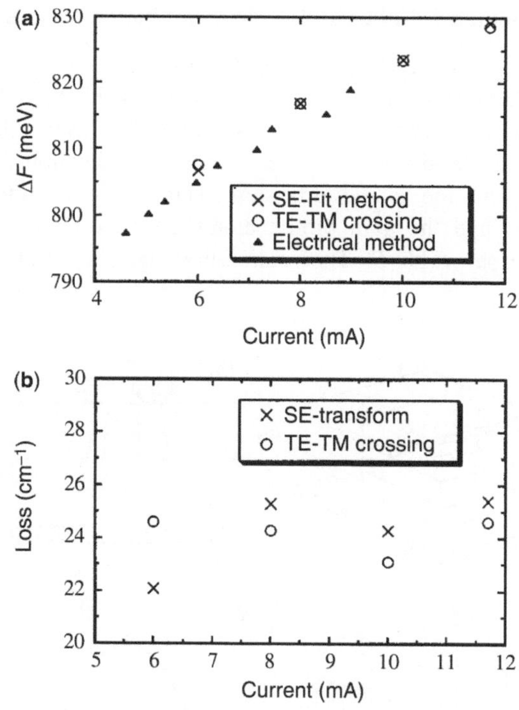

Figure 10.14 (a) The quasi-Fermi level separation and (b) the intrinsic absorption loss of a test laser. The experimental data were obtained from the intercepts of TE and TM polarized gain spectra (circles) and from the spontaneous emission transformation method (crosses). In (a), the results based on an electrical method (triangles) are also shown, which are obtained by measuring the induced ac voltage in the presence of an external pump laser with an ac modulation, and its wavelength is tuned across the quasi-Fermi level separation of the test laser. (Reprinted with permission from [48] © 1999 American Institute of Physics.)

agreement seems very good. The measured intrinsic loss as a function of the bias current is also shown in Fig. 10.14b [48], indicating there is a slight dependence on the carrier density, most likely due to free carrier absorption.

10.2 GAIN-GUIDED AND INDEX-GUIDED SEMICONDUCTOR LASERS

In this section, we study both gain-guided and index-guided semiconductor lasers. The difference is in the lateral confinements of the optical mode and the carriers. The index-guided semiconductor lasers have been demonstrated to have superior performance than that of the gain-guided geometry. However, the gain-guided structures are still used because of their ease of fabrication. They are investigated here because many interesting physical processes including carrier injection and diffusion, optical gain, and spatial carrier profiles can be extracted from these structures.

10.2.1 Stripe-Geometry Gain-Guided Semiconductor Lasers

A stripe-geometry gain-guided semiconductor laser [34, 38, 49–57] is shown in Fig. 10.15, where the metal contact is the stripe with a width S defined by processing procedure. The injected current J spreads along the lateral (i.e., y) direction. Therefore, the carrier density n distributes along the y direction determined by the current spreading and the lateral diffusion. The effective width of the carrier density distribution and the gain profile along the lateral direction are therefore wider than the stripe width S. This gain inhomogeneity along the y direction

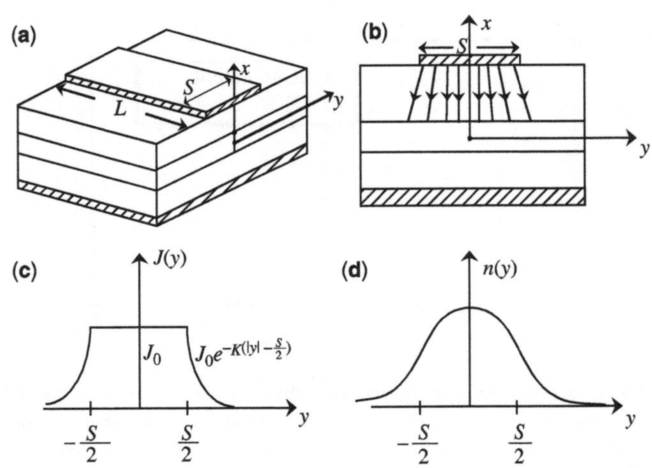

Figure 10.15 (a) A stripe-geometry gain-guided semiconductor laser; (b) cross section; (c) distribution of the current density $J(y)$; and (d) the carrier density profile $n(y)$.

provides the guiding mechanism for the optical mode confinement in the lateral direction. An associated effect is the antiguidance due to the reduction in the refractive index variation induced by the gain profile. In this section, we study an approximate analysis method to model the gain-guided stripe-geometry semiconductor laser.

Analysis of the Optical Modal Profile and Complex Propagation Constant
We consider the TE mode with index guidance in the x direction, perpendicular to the p-n junction plane, and with gain guidance along the y direction, which is transverse to the propagation direction. The optical electric field is approximated by

$$\mathbf{E} \simeq \hat{y}\, E_y(x, y, z) \simeq \hat{y}\, F(x)\, G(y)\, e^{ik_z z} \tag{10.2.1}$$

under scalar approximation, where $E_y(x, y, z)$ satisfies the wave equation

$$\nabla^2 E_y + \omega^2 \mu\varepsilon(x, y)\, E_y = 0 \tag{10.2.2}$$

where a parabolic permittivity profile in the active layer is assumed because of the carrier diffusion along the y direction and its resultant gain profile

$$\varepsilon(x, y) = \begin{cases} \varepsilon_1 & x > d/2 \\ \varepsilon_2(0) - a^2 y^2 & |x| \leq d/2 \\ \varepsilon_1 & x < -d/2. \end{cases} \tag{10.2.3}$$

Our goal is to find the electric field profile $F(x)G(y)$ and the complex propagation constant k_z for the inhomogeneous permittivity profile (10.2.3). Based on a perturbation method, the optical field is given by the Hermite–Gaussian functions

$$E_y(x, y, z) = E_y^{(0)}(x)\, H_n\left((k_0 a)^{1/2} \left(\frac{\Gamma}{\varepsilon_0}\right)^{1/4} y \right) e^{-k_0 a \left(\frac{\Gamma}{\varepsilon_0}\right)^{1/2} y^2/2}\, e^{ik_z z} \tag{10.2.4}$$

where Γ is the optical confinement factor, $E_y^{(0)}(x)$ is the unperturbed electric field profile, and H_n is a Hermite polynomial, see (3.3.11), which is a typical solution of wave equations with a parabolic spatial profile. The complex propagation constant of the gain-guided laser, k_z, is

$$k_z^2 = \beta_z^2 - \left(\omega^2 \mu a^2 \Gamma\right)^{1/2} (2n + 1) \tag{10.2.5}$$

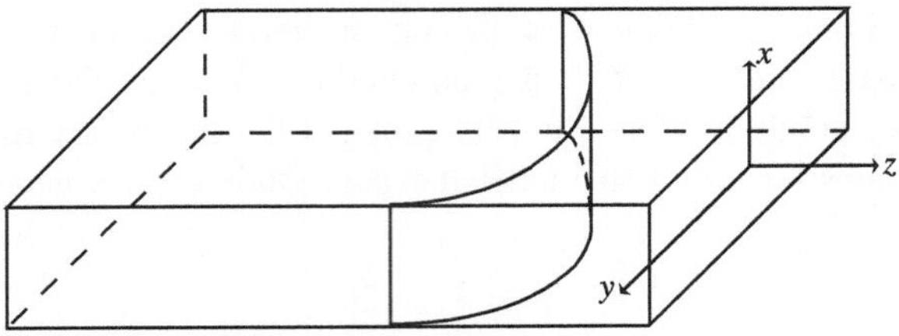

Figure 10.16 The constant phase front of a gain-guided mode is a cylindrical parabolic surface [51].

for the nth lateral mode. Using binomial expansion in (10.2.5), we have for $n = 0$

$$k_z \simeq \beta_z - \frac{k_0 a}{2\beta_z}\left(\frac{\Gamma}{\varepsilon_0}\right)^{1/2}. \tag{10.2.6}$$

Because $a = a_r + i\, a_i$ is a complex number, the constant phase front is determined by

$$k_z^r z - k_0 a_i \left(\frac{\Gamma}{\varepsilon_0}\right)^{1/2} \frac{y^2}{2} = \text{constant} \tag{10.2.7}$$

which is a cylindrical parabolic surface (Fig. 10.16), where k_z^r is the real part of k_z, which is close to β_z.

Current Spreading and Carrier Density Profile If we ignore the current spreading outside of the stripe region so that $J(y) = 0$ for $|y| \geq S/2$, the carrier density profile $n(y)$ $(=\delta n)$ is given by (2.4.17) in Chapter 2. Because the spontaneous emission spectrum depends on the local carrier density $n(y)$, a measurement of the emission profile as a function of y will determine the density profile. Therefore, the carrier density profile can be mapped out from the spontaneous emission intensity profile, and the results agree very well [50, 54] with the theoretical curve calculated using (2.4.17), as shown in Fig. 10.17.

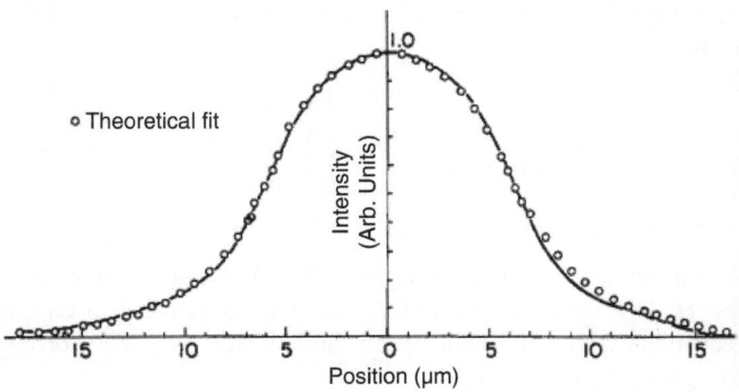

Figure 10.17 Theoretical plot of the carrier density profile $n(y)$ in a gain-guided semiconductor laser measured by the spontaneous emission intensity. The theoretical model assumes a stripe width $S = 11.7$ μm and a diffusion length $L_n = 3.6$ μm. (Reprinted with permission from [54] © 1977 IEEE.)

10.2.2 Index-Guided Semiconductor Lasers

In the gain-guided semiconductor lasers [33, 58, 59], the carrier spreading along the lateral direction degrades the laser performance, such as a high threshold current density and a low differential quantum efficiency. To improve the laser performance, index-guided structures have been used such that a better optical confinement, therefore an improved stimulated emission process, can occur in the active region. One example is a ridge waveguide laser with a weakly index-guided structure, as shown in Fig. 10.18a. Another example is an etched-mesa buried heterostructure with improved optical and carrier confinements in the active region, as shown in Fig. 10.18b.

For a ridge waveguide laser, an effective index profile $n_{\text{eff}}(y)$ along the lateral y direction can be obtained using the effective index method by solving the slab waveguide problem at a fixed cross section defined by a constant y. The guidance condition and the optical field pattern are then obtained from the waveguide theory for the index profile $n_{\text{eff}}(y)$, following the procedures described [60] in Section 7.5. The effective index method can also be applied to a rectangular dielectric waveguide structure as the active region of the etched-mesa buried-heterostructure laser. Once we find the optical electric field for the fundamental TE mode (or higher-order TE modes) from (7.5.3),

$$\mathbf{E} = \hat{y}\, E_y(x, y) \simeq \hat{y}\, F(x, y)\, G(y). \tag{10.2.8}$$

We obtain the optical confinement factor using the definition

$$\Gamma = \frac{\int\int_{\text{active region}} \frac{1}{2}\text{Re}(\mathbf{E} \times \mathbf{H}^*) \cdot \hat{z}\, dxdy}{\int_{-\infty}^{\infty}\int_{-\infty}^{\infty} \frac{1}{2}\text{Re}(\mathbf{E} \times \mathbf{H}^*) \cdot \hat{z}\, dxdy}. \tag{10.2.9}$$

Analytical expressions for the slab waveguide geometry have been derived in Chapter 7. Here we may use

$$\frac{1}{2}\text{Re}(\mathbf{E} \times \mathbf{H}^*) \cdot \hat{z} \simeq \frac{1}{2}\text{Re}(-\,E_y H_x^*) = \frac{k_z |E_y|^2}{2\omega\mu}. \tag{10.2.10}$$

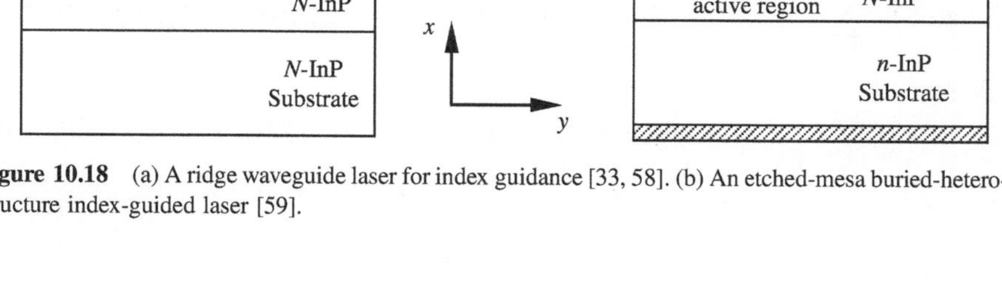

Figure 10.18 (a) A ridge waveguide laser for index guidance [33, 58]. (b) An etched-mesa buried-heterostructure index-guided laser [59].

Because $k_z/\omega\mu$ is a constant for a given mode, we can simply use

$$\Gamma = \frac{\int\int_{\text{active}} |E_y(x, y)|^2 dxdy}{\int_{-\infty}^{\infty}\int_{-\infty}^{\infty} |E_y(x, y)|^2 dxdy} \simeq \Gamma_x \Gamma_y \tag{10.2.11}$$

which is approximated by the product of the two optical confinement factors along the x direction (with a slab geometry in the effective index method) and along the y direction when the field is approximately given by

$$E_y \simeq F(x)\, G(y). \tag{10.2.12}$$

It is a good approximation for a strongly index-guided structure, and $F(x, y) \simeq F(x)$ if the geometry along the y direction is uniform such that the first part of the wave function $F(x, y)$ in the effective index method described in Section 7.5 can be assumed to be independent of y. Index-guided semiconductor lasers have been shown to exhibit excellent performance including the fundamental mode operation, low threshold, high quantum efficiency, and low temperature sensitivity [33].

10.3 QUANTUM-WELL LASERS

Quantum-well (QW) structures [12–16, 61], as shown in Fig. 10.19, have been used as the active layer of semiconductor laser diodes with reduced threshold current densities compared with those for conventional double-heterostructure (DH) semiconductor diode lasers. Research on quantum-well physics and semiconductor lasers has been of great interest recently. For a brief history, see Ref. 3. Various designs such as single quantum well (SQW), multiple quantum well (MQW), and graded-index separate-confinement heterostructures (GRINSCH) have been used for semiconductor lasers [21]. As we have seen in Chapter 9, quantum-well structures show quantized subbands and step-like densities of states. The density of states for a quasi-two-dimensional structure has been used to reduce threshold current density and improve temperature stability. Energy quantization provides another degree of freedom to tune the lasing wavelength by varying the well width and the barrier height. Scaling laws for quantum-well lasers and quantum-wire lasers show significant reduction of threshold current in reduced dimensions [24].

10.3.1 A Simplified Gain Model

The simplest model we will consider is the gain spectrum based on (9.4.18) for a finite temperature, assuming a zero scattering linewidth [62, 63]

$$g(\hbar\omega) = \sum_{n,m} g_{\max}[f_c^n(E_t = \hbar\omega - E_{hm}^{en}) - f_v^m(E_t = \hbar\omega - E_{hm}^{en})]H(\hbar\omega - E_{hm}^{en}) \tag{10.3.1a}$$

where

$$g_{\max} = C_0 |\hat{e}\cdot\mathbf{M}|^2 |I_{hm}^{en}|^2 \rho_r^{2D} \simeq C_0 |\hat{e}\cdot\mathbf{M}|^2 \rho_r^{2D} \delta_{nm} \tag{10.3.1b}$$

$$C_0 = \frac{\pi e^2}{n_r c \varepsilon_0 m_0^2 \omega} \qquad \rho_r^{2D} = \frac{m_r^*}{\pi \hbar^2 L_z} \tag{10.3.1c}$$

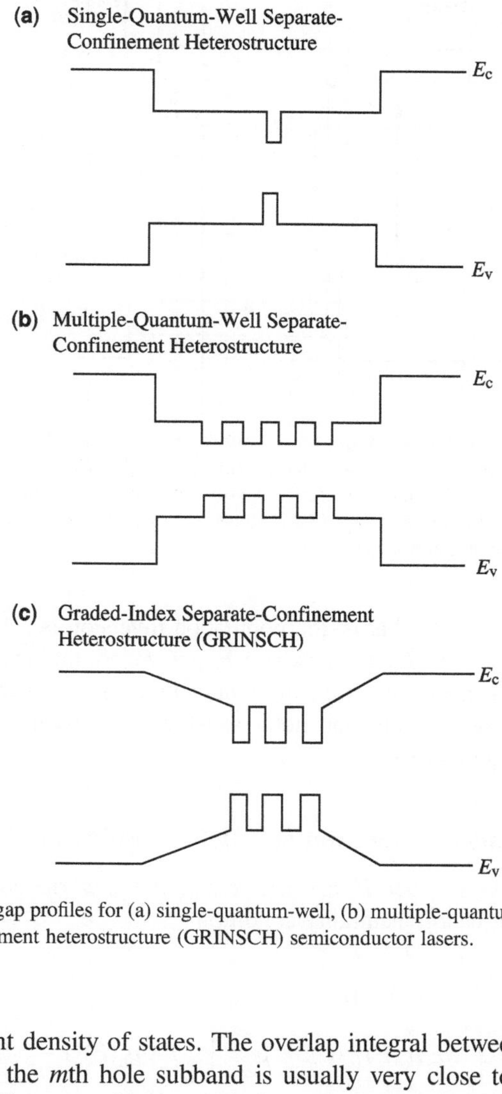

Figure 10.19 Band-gap profiles for (a) single-quantum-well, (b) multiple-quantum-well, and (c) graded-index separate-confinement heterostructure (GRINSCH) semiconductor lasers.

is the reduced joint density of states. The overlap integral between the nth conduction subband and the mth hole subband is usually very close to unity for $n = m$, and it vanishes if $n \neq m$ because of the even–odd parity consideration. The polarization-dependent momentum matrix element is listed in Table 9.1 for conduction to heavy-hole and light-hole band transitions. The occupation factors for the electrons in the nth conduction subband and the electrons in the mth hole subband are

$$f_{\mathrm{c}}^{n}(E_{\mathrm{t}}=\hbar\omega-E_{hm}^{en})=\frac{1}{1+e^{\left[E_{en}+(m_{\mathrm{r}}^{*}/m_{\mathrm{e}}^{*})(\hbar\omega-E_{hm}^{en})-F_{\mathrm{c}}\right]/k_{\mathrm{B}}T}} \tag{10.3.2a}$$

$$f_{\mathrm{v}}^{m}(E_{\mathrm{t}}=\hbar\omega-E_{hm}^{en})=\frac{1}{1+e^{\left[E_{hm}-(m_{\mathrm{r}}^{*}/m_{\mathrm{h}}^{*})(\hbar\omega-E_{hm}^{en})-F_{\mathrm{v}}\right]/k_{\mathrm{B}}T}}. \tag{10.3.2b}$$

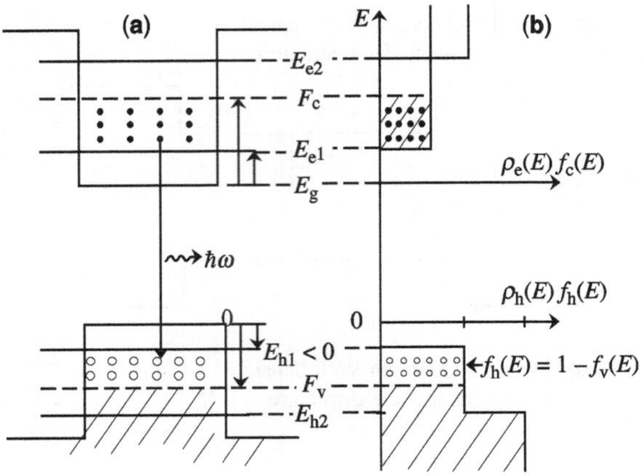

Figure 10.20 (a) Population inversion in a quantum-well structure, where $E_g + F_c - F_v > \hbar\omega > E_{e1} - E_{h1} + E_g$. (b) The product of the conduction band density of states $\rho_e(E)$ and the Fermi–Dirac occupation probability $f_c(E)$ for the calculation of the electron density n is plotted versus the energy E in the vertical scale. Similarly, $\rho_h(E) f_h(E) = \rho_h(E)\,[1 - f_v(E)]$ is plotted versus E for the energy $(E < 0)$ in the valence band. Assume that the temperature T is 0K.

Gain occurs when $f_c^n > f_v^m$, that is, the population inversion is achieved, Fig. 10.20. It also leads to $E_g + F_c - F_v > \hbar\omega$, where F_c and F_v are the quasi-Fermi levels for electrons and holes, measured from the conduction and valence band edges, respectively. Only those electrons and holes satisfying the k-selection rule contribute significantly to the gain process.

10.3.2 Determination of Electron and Hole quasi-Fermi Levels

The quasi-Fermi levels F_c and F_v are determined by the carrier concentrations n and p, which satisfy the charge neutrality condition

$$n + N_A^- = p + N_D^+ \tag{10.3.3}$$

$$n = \int_0^\infty dE\, \rho_e(E) f_c(E), \quad p = \int_{-\infty}^\infty dE\, \rho_h(E)[1 - f_v(E)]$$

$$\rho_e(E) = \frac{m_e^*}{\pi\hbar^2 L_z} \sum_{n=1}^{\infty} H(E - E_{en})$$

$$\rho_h(E) = \frac{m_h^*}{\pi\hbar^2 Lz} \sum_{m=1}^{\infty} H(E_{hm} - E) \tag{10.3.4}$$

where $H(x)$ is the Heaviside step function, $H(x) = 1$ if $x > 0$, and $H(x) = 0$ if $x < 0$. In Fig. 10.21a and Fig. 10.21b, we plot the products $\rho_e(E) f_c(E)$ and $\rho_h(E)\,[1 - f_v(E)]$ versus the energy E for $T = $ 0K and 300K, respectively. The areas below these functions give the carrier concentrations n and p.

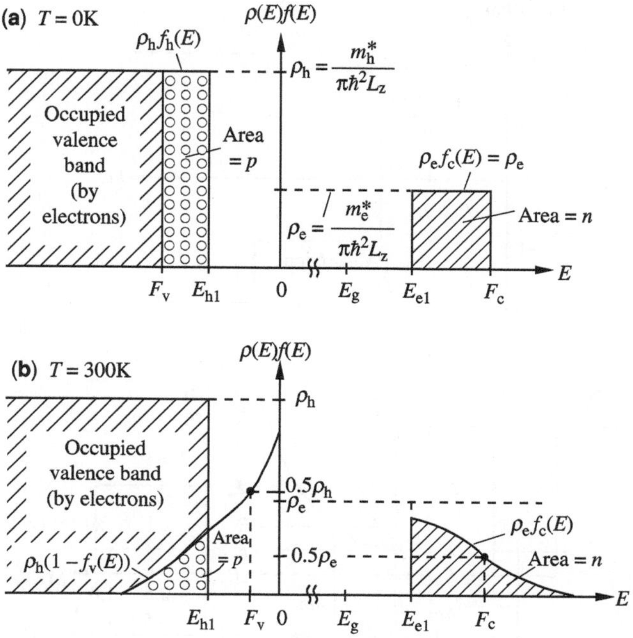

Figure 10.21 The functions of the products $\rho_e(E)\, f_c(E)$ and $\rho_h(E)\, f_h(E) = \rho_h(E)\, [1 - f_v(E)]$ versus the electron energy E for the calculations of the carrier concentrations n and p, respectively, for (a) $T = 0\text{K}$ and (b) $T = 300\text{K}$. The area under $\rho_e(E)\, f_c(E)$ is the electron concentration, and the area under $\rho_h(E)$ $f_h(E)$ is the hole concentration.

10.3.3 Zero-Temperature Gain Spectrum

To understand the gain spectrum, let us consider the gain at $T = 0\text{K}$. Remember that the model here is oversimplified because it does not include the effects of scatterings or inhomogeneous broadenings such as those due to well-width fluctuations and the electron–hole Coulomb interactions. Our model here is to show the simplest picture for population inversion for the optical gain process. At $T = 0\text{K}$, the Fermi–Dirac distributions are step functions. For the electrons in the conduction subband, we have

$$f_c(E) = \begin{cases} 1 & E < F_c \\ 0 & E > F_c. \end{cases} \tag{10.3.5}$$

The electron concentration is simply the area under the function $\rho_e(E)\, f_c(E)$, which is plotted versus E in Fig. 10.22d

$$n = \frac{m_e^*}{\pi\hbar^2 L_z} \sum_{\substack{n=\text{occupied}\\ \text{subbands}}} (F_c - E_{en}). \tag{10.3.6}$$

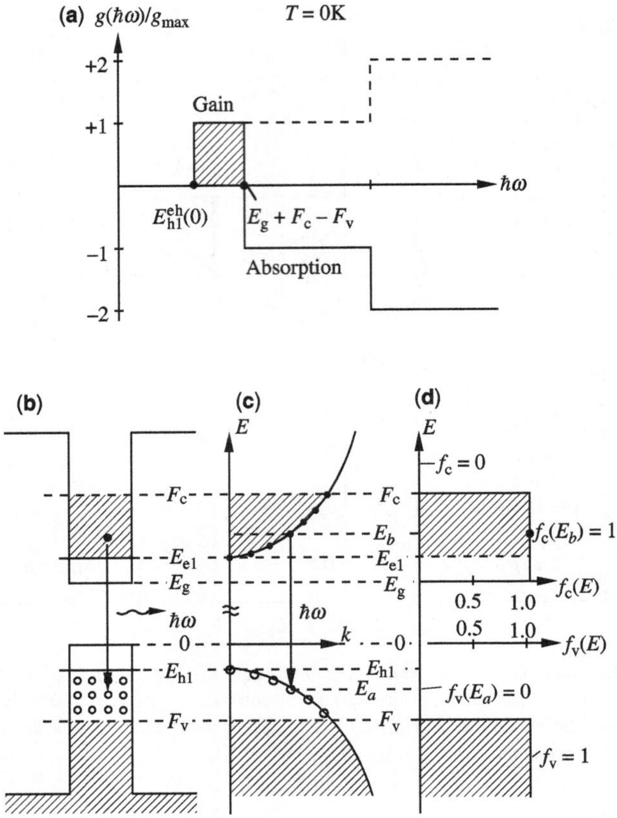

Figure 10.22 An illustration of optical gain of a quantum well at $T = 0$K. (a) Normalized gain spectrum, (b) the energy profile in real space, (c) the energy dispersion in k space, and (d) the Fermi–Dirac function versus the energy for the electrons in the conduction band and in the valence band.

For a single (ground state) subband case,

$$n = \frac{m_e^*}{\pi\hbar^2 L_z}(F_c - E_{e1}) \tag{10.3.7}$$

the location of the quasi-Fermi level is linearly proportional to the carrier concentration at zero temperature. Similar results hold for holes

$$p = \frac{m_h^*}{\pi\hbar^2 L_z}(E_{h1} - F_v). \tag{10.3.8}$$

For unstrained quantum wells, the topmost subband is a heavy-hole subband. The hole effective mass in the plane of the quantum well m_h^* is usually smaller than its value in the bulk semiconductors because of its coupling to the light-hole subbands. However, m_h^* is larger than m_e^* for most unstrained quantum wells. We find that for $n = p$,

$$m_h^*(E_{h1} - F_v) = m_e^*(F_c - E_{e1}) \tag{10.3.9}$$

which means that the areas under the function $\rho_e(E)\,f_c(E)$, and $\rho_h(E)\,f_v(E)$ should be equal. Because $m_h^* > m_e^*$, we find $E_{h1} - F_v < F_c - E_{e1}$. That is, the quasi-Fermi level F_v is closer to the hole subband edge than the separation of F_n and E_{c1}. The transparency optical energy occurs at $\hbar\omega = F_c - F_v$, where $f_c - f_v = 0$. As a matter of fact, $f_c = 1$ and $f_v = 0$ for $\hbar\omega < E_g + F_c - F_v$, and $f_c = 0$, $f_v = 1$ for $\hbar\omega > E_g + F_c - F_v$. The gain spectrum is a simple rectangular window, as shown in Fig. 10.22a

$$g(\hbar\omega) = \begin{cases} g_{\max} & E_{h1}^{e1} < \hbar\omega < E_g + F_c - F_v \\ -g_{\max} \sum\limits_{n=m} H\left(\hbar\omega - E_{hm}^{en}\right) & \text{otherwise.} \end{cases} \tag{10.3.10}$$

For the optical energy between the e1–h1 transition band edge and the quasi-Fermi level separation $E_g + F_c - F_v$, the photons experience gain. Otherwise, they experience absorption with a spectrum similar to that of an unpumped quantum well with a stepwise shape. The only modification is near the band edge. The occupations of states by electrons and holes in the quantum well are shown in Fig. 10.22b in real space and in Fig. 10.22c in k-space showing the k-selection rule. The Fermi–Dirac distributions for electrons in the conduction band $f_c(E)$ and for electrons in the valence band $f_v(E)$ are shown in Fig. 10.22d, where $E_b(k) - E_a(k) = \hbar\omega$ has to be satisfied from the k-selection rule and the energy conservation condition, which leads to the population inversion factor $f_c(E_b) - f_v(E_a)$. We see that for $E_g + E_{e1} - E_{h1} < \hbar\omega < E_g + F_c - F_v$, $f_c(E_b) = 1$ and $f_v(E_a) = 0$, therefore, gain occurs. Otherwise, $f_c = 0$ and $f_v = 1$, and absorption occurs as shown in Fig. 10.22a.

10.3.4 Finite-Temperature Gain Spectrum

At a finite temperature, the gain spectrum looks like that shown in Fig. 10.23a. The Fermi–Dirac distribution will deviate from a sharp step function and is equal to $1/2$ at the quasi-Fermi level. The electron concentration is the area below the function $\rho_e(E)\,f_c(E)$ as shown in Fig. 10.21b, which can be obtained analytically (9.4.11)

$$n = \sum_{n=1}^{\infty} n_c \ln\left[1 + e^{(F_c - E_{en})/k_B T}\right] \tag{103.11a}$$

where the prefactor

$$n_c = \frac{m_e^* k_B T}{\pi\hbar^2 L_z} \tag{10.3.11b}$$

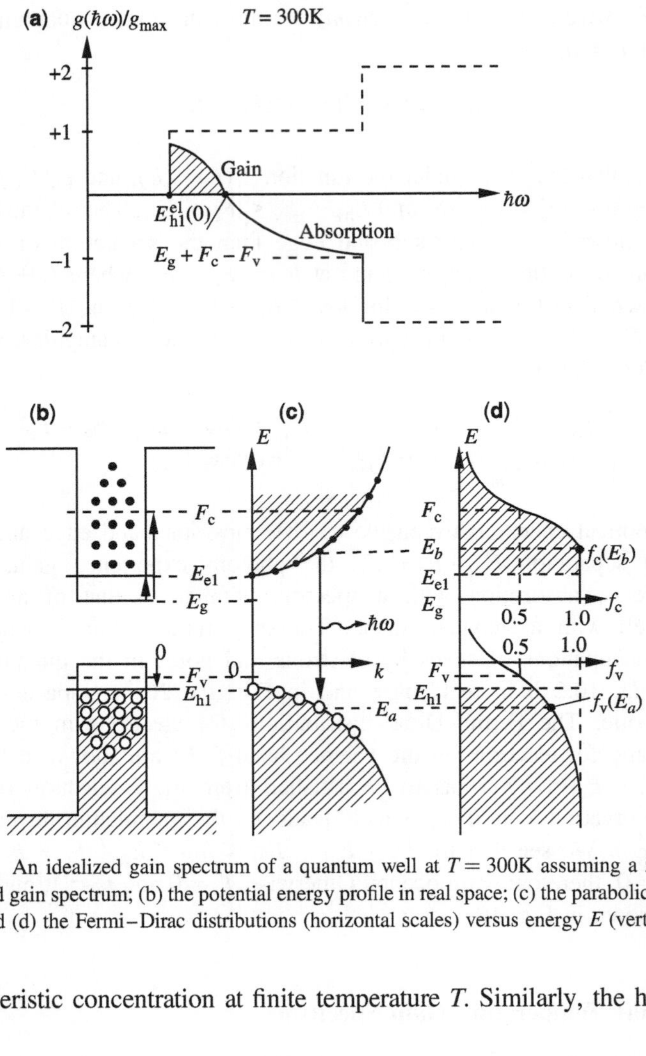

Figure 10.23 An idealized gain spectrum of a quantum well at $T = 300\text{K}$ assuming a zero linewidth. (a) Normalized gain spectrum; (b) the potential energy profile in real space; (c) the parabolic band structure in k space; and (d) the Fermi–Dirac distributions (horizontal scales) versus energy E (vertical scale).

is a characteristic concentration at finite temperature T. Similarly, the hole concentration is

$$p = \sum_{m=1}^{\infty} n_v \ln\left[1 + e^{(E_{hm} - F_v)/k_B T}\right] \tag{10.3.12a}$$

where

$$n_v = \frac{m_h^* k_B T}{\pi \hbar^2 L_z}, \tag{10.3.12b}$$

which can account for heavy-hole and light-hole subbands where m accounts for all hole subbands. The gain spectrum is given by Eq. (10.3.1a) and is plotted in Fig. 10.23a for illustration. It starts with a peak value at the transition edge $\hbar\omega = E^{e1}_{h1}(0) = E_g + E_{e1} - E_{h1}$, where $E_t = 0$ and decreases to zero at $E_g + F_c - F_v$, then becomes absorption at higher optical energies. The sharp rise in gain near the band edge compared with the soft increase in a bulk (3D) semiconductor is due to the step-like density of states in 2D versus the slow increase of the square-root ($\sqrt{E}$) density of states in 3D. For a single subband pair, $n = \mathrm{c1}$ and $m = \mathrm{h1}$, we have the gain spectrum

$$g(\hbar\omega) = g_{\max}\left[f_c\left(E_t = \hbar\omega - E_{\mathrm{h1}}^{\mathrm{e1}}(0)\right) - f_v\left(E_t = \hbar\omega - E_{\mathrm{h1}}^{\mathrm{e1}}(0)\right)\right] \qquad (10.3.13)$$

where $g_{\max}$ is given in (10.3.1b). The peak gain occurs at $E_t = 0$, or $\hbar\omega = E_{\mathrm{h1}}^{\mathrm{e1}}(0)$

$$g_p = g_{\max}\left[f_c\left(\hbar\omega = E_{\mathrm{h1}}^{\mathrm{e1}}\right) - f_v\left(\hbar\omega = E_{\mathrm{h1}}^{\mathrm{e1}}\right)\right] \qquad (10.3.14a)$$

where

$$f_c\left(\hbar\omega = E_{\mathrm{h1}}^{\mathrm{e1}}\right) = \frac{1}{1 + e^{(E_{\mathrm{e1}} - F_c)/k_B T}} \qquad (10.3.14b)$$

$$f_v\left(\hbar\omega = E_{\mathrm{h1}}^{\mathrm{e1}}\right) = \frac{1}{1 + e^{(E_{\mathrm{h1}} - F_v)/k_B T}}. \qquad (10.3.14c)$$

10.3.5 Peak Gain Coefficient versus the Carrier Density

As shown in Fig. 10.24, the gain increases when we increase the carrier density. If only one conduction and one valence subband are occupied, we obtain

$$f_c(E_t = 0) \simeq 1 - e^{-n/n_c} \qquad f_v(E_t = 0) \simeq e^{-p/n_v}. \qquad (10.3.15)$$

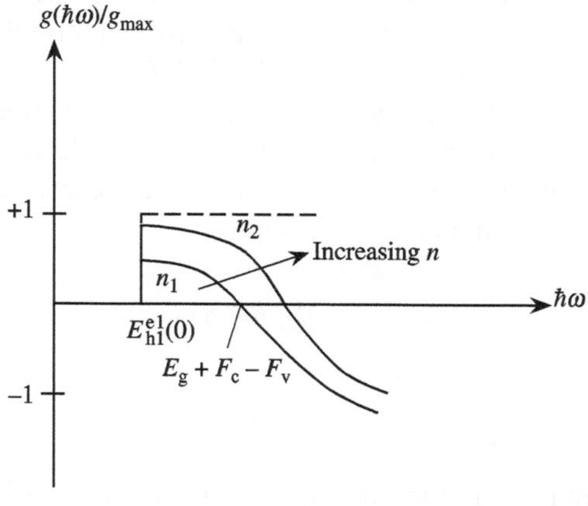

Figure 10.24 Normalized gain spectrum at two different injection levels ($n_2 > n_1$).

The above expressions are approximately true if we redefine [62]

$$n_c \simeq \frac{m_e^* k_B T}{\pi \hbar^2 L_z} \sum_{n=1}^{\infty} e^{(E_{e1} - E_{en})/k_B T} \tag{10.3.16a}$$

and a similar equation for holes

$$n_v \simeq \frac{m_h^* k_B T}{\pi \hbar^2 L_z} \sum_{m=1}^{\infty} e^{(E_{hm} - E_{h1})/k_B T}. \tag{10.3.16b}$$

Therefore, a simplified formula for the peak gain g_p as a function of the electron concentration n is given by

$$g_p = g_{max}\left[1 - e^{-n/n_c} - e^{-p/n_v}\right]. \tag{10.3.17}$$

Because $n_v/n_c = m_h^*/m_e^* = R$ is the ratio of the hole and electron effective mass for a single subband occupation, we have

$$g_p = g_{max}\left[1 - e^{-n/n_c} - e^{-n/(Rn_c)}\right] \tag{10.3.18}$$

which can be plotted as a function of n. The transparency carrier density occurs at $g_p = 0$,

$$e^{-n_{tr}/n_c} + e^{-n_{tr}/(Rn_c)} = 1. \tag{10.3.19}$$

If $R = 1$, we find $n_{tr} = n_c \ln 2$. We can also plot $f_c \sim 1 - e^{-n/n_c}$ and $f_v \sim e^{-n/n_v}$ versus the carrier concentration n; the intersection point gives the transparency density n_{tr}, which corresponds with $g_p = 0$ as shown in Fig. 10.25. The peak gain g_p versus the carrier density n is plotted in Fig. 10.26. The analytical formula for the differential gain, $dg_p(n)/dn$, can be derived from (10.3.18) [62, 64].

We can also calculate the peak gain directly versus the current density J, taking into account the nonradiative Auger recombination, the leakage current, and the radiative recombination. The process is first assuming a carrier concentration n; we then calculate the peak gain and

$$J_{rad} = qL_z R_{sp}(n) \tag{10.3.20a}$$

$$J_{Aug} = qL_z R_{Aug}(n) \tag{10.3.20b}$$

and the leakage current J_{leak} from the expressions in Section 10.1; then, $J = J_{rad} + J_{Aug} + J_{leak}$. Here the spontaneous emission rate per unit volume $R_{sp}(n)$

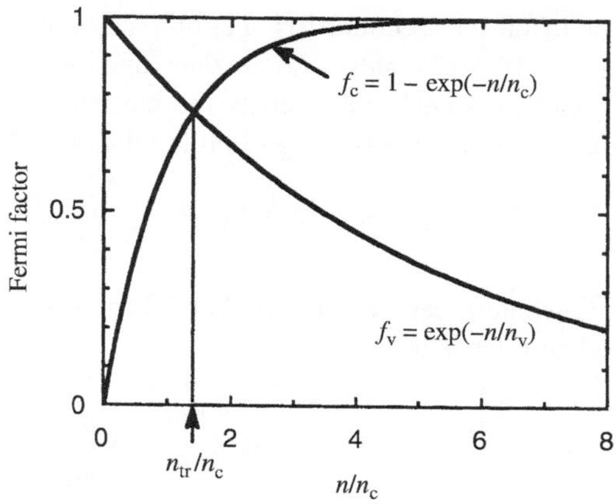

Figure 10.25 A plot of the occupation probability of electrons in the conduction band $f_c \simeq 1 - e^{-n/n_c}$ and that of electrons in the valence band $f_v \simeq e^{-n/n_v}$ (or the probability of holes in the valence band $f_h = 1 - f_v$). The interception point where $f_c = f_v$ gives the transparent carrier density n_{tr}.

(s^{-1} cm^{-3}) can be calculated using the formulation with the valence-band mixings discussed in Chapter 9 or simply using the simplified model [65, 66]

$$R_{sp}(n) \simeq Bn^2. \tag{10.3.21a}$$

Similarly,

$$R_{Aug}(n) \simeq Cn^3 \tag{10.3.21b}$$

which can also be calculated directly using the interaction Hamiltonians for Auger processes (i.e., carrier interactions via the Coulomb potential). The finite scattering

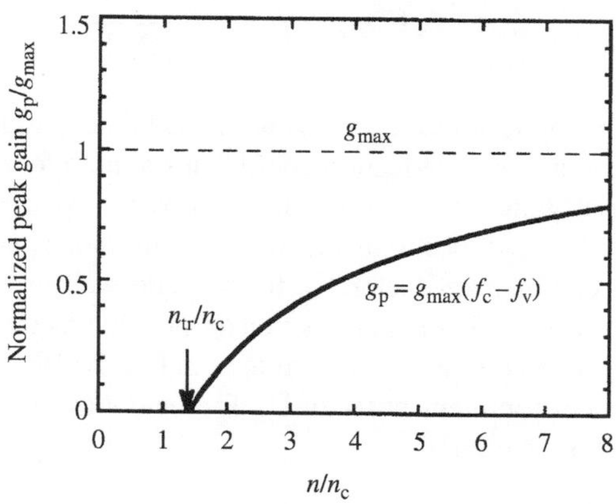

Figure 10.26 Peak gain coefficient g_p normalized by the maximum gain g_{max} is plotted as a function of the carrier density n (normalized by the characteristic concentration n_c).

linewidth can also be taken into account [67–71] by convolving the zero linewidth gain spectrum $g(\hbar\omega)$ in (10.3.13) with a Lorentzian function $L(\hbar\omega)$ as in (9.4.14). The procedure will lead to a peak gain versus the current density relation, $g_p = g_p(J)$. A common empirical formula is a logarithmic relation [72–76]

$$g_p(J) = g_0\left[1 + \ln\frac{J}{J_0}\right]. \tag{10.3.22}$$

We see that if $J \propto n^{\beta}$, where β can be between 2 and 3 for a certain range of current density, the above relation leads to

$$g_p = g_0\left[1 + \beta\ln\frac{n}{n_0}\right] = g_0\beta\left[\frac{1}{\beta} + \ln\frac{n}{n_0}\right], \tag{10.3.23}$$

which is also a logarithmic relation.

10.3.6 Scaling Laws for MQW Lasers

For a quantum-well laser lasing from only the first quantized electron and hole subbands, we use the empirical logarithmic formula for the peak gain–current density relation (10.3.23) for the rest of the discussions in this chapter

$$g_w = g_0\left[\ln\left(\frac{J_w}{J_0}\right) + 1\right] \tag{10.3.24}$$

where J_w and g_w are the injected current density and the peak gain coefficient of a single quantum-well structure. Such a relation is plotted in Fig. 10.27a. The transparency current density occurs at $J_{tr} = J_0 e^{-1}$.

For an MQW structure with n_w quantum wells and a cavity length L, the required modal gain G_{th} at threshold condition is

$$G_{th} = n_w\Gamma_w g_w = \alpha + \frac{1}{2L}\ln\frac{1}{R_1R_2} \tag{10.3.25}$$

where α is the internal optical loss, Γ_w is the optical confinement factor per well, and R_1 and R_2 are the optical power reflection coefficients at both facets. We should note that g_w is a "material gain" for the quantum-well region assuming that all of the electron–photon interactions occur in the well region and $g_w \propto 1/L_z$ in the gain model. The optical confinement factor Γ_w for a single well takes into account the optical modal distribution, and only that fraction of the photons inside the well experiences gain. For a separate-confinement structure as in Fig. 10.19a and Fig. 10.19b, we can calculate the optical confinement factor Γ_{op} for the optical waveguide part, ignoring the well region first, then use

$$\Gamma_w = \Gamma_{op}\frac{L_z}{W_{mode}} \tag{10.3.26}$$

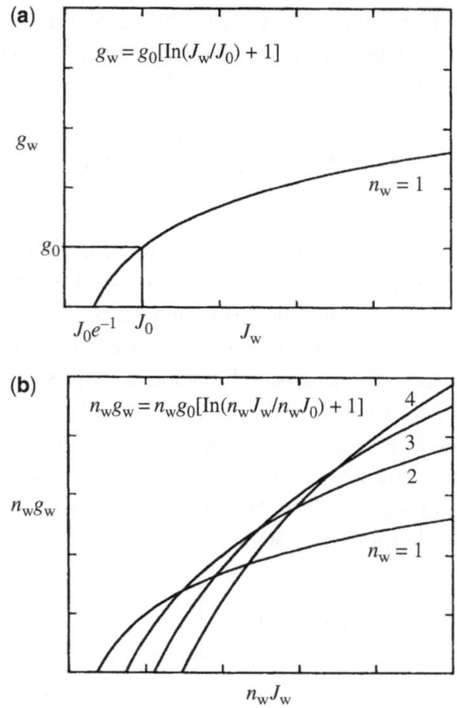

Figure 10.27 (a) Gain (g_w) versus injection current density (J_w) relation for a single-quantum-well structure; (b) gain (n_wg_w) versus injection current density (n_wJ_w) relation for a multiple-quantum-well structure with n_w wells.

where Γ_{op} takes into account the refractive index difference of the large optical waveguide for optical confinement and L_z/W_{mode} accounts for the carrier confinement and stimulated emission. The modal gain for a single quantum well should be

$$\begin{aligned} G &= \Gamma_w g_w \\ &= \Gamma_w g_{max}\left[f_c\left(\hbar\omega = E_{h1}^{e1}(0)\right) - f_v\left(\hbar\omega = E_{h1}^{e1}(0)\right)\right] \end{aligned} \tag{10.3.27}$$

where, using g_{max} from (10.3.1b),

$$\begin{aligned} \Gamma_w g_{max} &= \frac{L_z}{W_{mode}} g_{max}\Gamma_{op} \\ &= C_0|\hat{e}\cdot\mathbf{M}_{ch}|^2 \frac{m_r^*}{\pi\hbar^2}\frac{\Gamma_{op}}{W_{mode}}, \end{aligned} \tag{10.3.28}$$

which is also the maximum achievable gain for a single quantum well. For n_w quantum wells, the modal gain is approximately $n_w\Gamma_w g_w$. The threshold current density J_{th} for the MQW structure is then

$$\eta J_{th} = n_w J_w \tag{10.3.29}$$

where η is the injection quantum efficiency or the percentage of the injection current that gets captured by the quantum wells. Therefore, we can also plot the material gain ($n_w g_w$) versus the injection current density ($n_w J_w$) using (10.3.24) [72–78]

$$n_w g_w = n_w g_0 \left[\ln \left(\frac{n_w J_w}{n_w J_0} \right) + 1 \right] \qquad (10.3.30)$$

assuming the well-to-well coupling can be ignored for simplicity. This relation is plotted in Fig. 10.27b for $n_w = 1$, 2, 3, and 4. For $n_w = 1$, as shown in Fig. 10.27a, the intersection with the horizontal axis is $J_w = J_0 e^{-1}$ and $g_w = 0$. When $J_w = J_0$, we have $g_w = g_0$. For $n_w = 2$, we find that $2g_w = 0$ occurs at $(2J_w) = 2J_0 e^{-1}$; therefore, the horizontal intersection is shifted to the right by a factor of 2. At $2J_w = 2J_0$, we obtain the vertical axis $(2g_w) = (2g_0)$, which is twice that for $n_w = 1$. Therefore, the gain–current density curve for $n_w = 2$ starts at twice the transparency current density of that for $n_w = 1$ and increases by a factor of 2 faster than that for $n_w = 1$. We obtain the threshold current density by [72–76] substituting (10.3.24) into (10.3.30)

$$\begin{aligned} J_{th} &= \left(\frac{n_w J_0}{\eta} \right) \exp \left[\left(\frac{g_w}{g_0} \right) - 1 \right] \\ &= \left(\frac{n_w J_0}{\eta} \right) \exp \left[\left(\frac{1}{n_w \Gamma_w g_0} \right) \left(\alpha + \frac{1}{2L} \ln \frac{1}{R_1 R_2} \right) - 1 \right]. \end{aligned} \qquad (10.3.31)$$

If we take the logarithmic function of J_{th}, we see

$$\ln J_{th} = \ln \left(\frac{n_w J_0}{\eta} \right) + \frac{\alpha}{n_w \Gamma_w g_0} + \frac{L_{opt}}{L} - 1 \qquad (10.3.32)$$

where we define an optimum cavity length

$$L_{opt} \equiv \frac{1}{2} \frac{1}{(n_w \Gamma_w g_0)} \ln \left(\frac{1}{R_1 R_2} \right). \qquad (10.3.33)$$

Therefore, ln J_{th} varies linearly with $1/L$. Such a dependence has been plotted for various numbers of quantum wells as shown in Fig. 10.28a.

For real device measurements, the threshold current I_{th} ($= J_{th} wL$) is measured, where w is the width of the stripe. We have

$$I_{th} = \left(\frac{wL n_w J_0}{\eta} \right) \exp \left[\frac{1}{n_w \Gamma_w g_0} \left(\alpha + \frac{1}{2L} \ln \frac{1}{R_1 R_2} \right) - 1 \right]. \qquad (10.3.34)$$

The above expression for I_{th} contains a prefactor increasing linearly with L and an argument in the exponential function that decreases with L, resulting in a minimum of the threshold current occurring at $L = L_{opt}$ as shown in Fig. 10.28b.

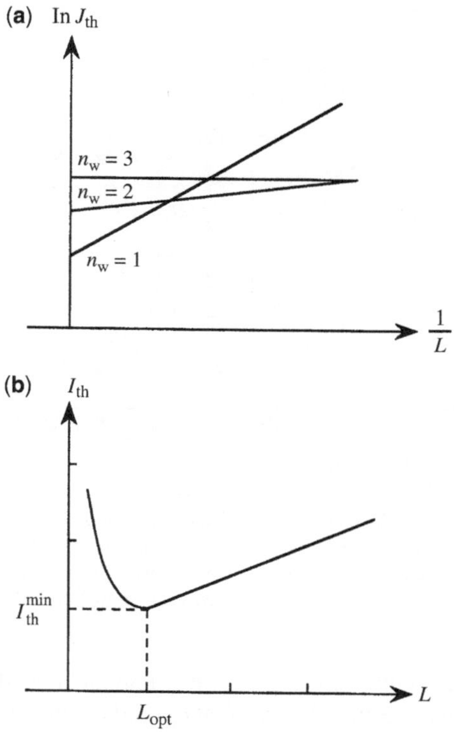

Figure 10.28 (a) Theoretical curves for the logarithm of the threshold current density, ln(J_{th}) using (10.3.32), versus the inverse cavity length for lasers with 1, 2, and 3 quantum wells. (b) A plot of the threshold current I_{th} versus the cavity length L for $n_w = 1$ using (10.3.34).

The minimum threshold current, I_{th}^{min}, is then given by

$$I_{th}^{min} = \frac{1}{2}\left(\frac{wJ_0}{\eta\Gamma_w g_0}\right)\ln\left(\frac{1}{R_1R_2}\right)\exp\left[\frac{\alpha}{n_w\Gamma_w g_0}\right]. \tag{10.3.35}$$

If we assume that the loss coefficient α and other parameters such as Γ_w, J_0, η, and g_0 are independent of the number of wells, we find an optimum number of wells, n_{opt}, such that I_{th} in (10.3.34) is minimized, $(\partial I_{th}/\partial n_w = 0)$

$$n_{opt} = \frac{1}{\Gamma_w g_0}\left[\alpha + \frac{1}{2L}\ln\left(\frac{1}{R_1R_2}\right)\right]. \tag{10.3.36}$$

Comparison with Experimental Data [73] The above simple relations between the threshold current density J_{th} and the cavity length L or the number of quantum wells n_w have been demonstrated experimentally. In Fig. 10.29, we show the experimental data [73] for (a) broad area contact and (b) ridge waveguide GaAs/$Al_{0.22}Ga_{0.78}As$ separate confinement heterostructure (SCH) quantum-well lasers.

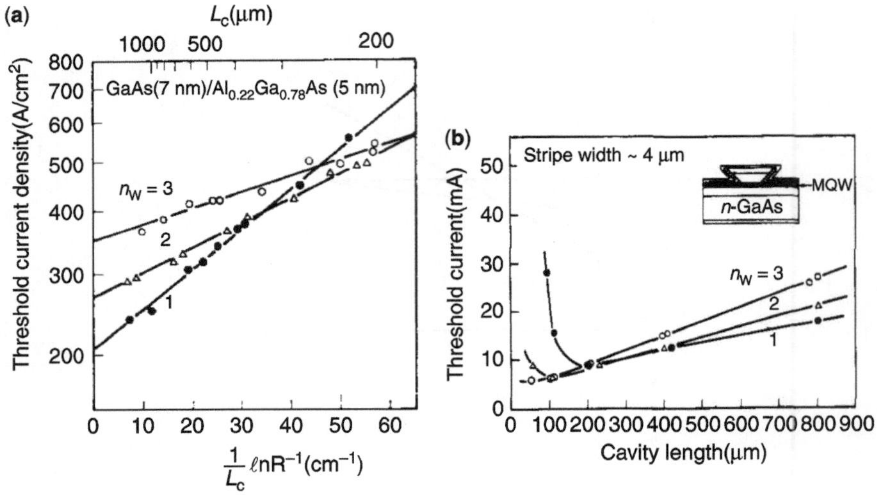

Figure 10.29 Experimental results showing (a) the linear behavior of the $\ln(J_{\text{th}})$ versus the inverse cavity length $1/L$ following (10.3.32) for lasers with 1, 2, and 3 quantum wells; and (b) the threshold current I_{th} as a function of the cavity length L. A minimum value of the threshold current exists at an optimum cavity length L_{opt} based on (10.3.34) and (10.3.45). (Reprinted with permission from [73] © 1988 IEEE.)

The undoped GaAs quantum wells are 70 Å in width and the $Al_{0.22}Ga_{0.78}As$ barriers are 50 Å wide. The linear relation between $\ln(J_{\text{th}})$ and the inverse cavity length $1/L$ is clearly shown in Fig. 10.29a for three numbers of quantum wells, $n_{\text{w}} = 1$, 2, and 3. The threshold current I_{th} is also shown to have a minimum at an optimum cavity length in Fig. 10.29b, in agreement with the analytical relations (10.3.34) and (10.3.35). The stripe widths are 4.0 μm, 4.5 μm, and 3.5 μm for $n_{\text{w}} = 1$, 2, and 3, respectively in Fig. 10.29b.

10.4 STRAINED QUANTUM-WELL LASERS

In recent years, strain effects in semiconductors have been proposed and explored intensively for optoelectronic device applications [79, 80]. The tunability of strained materials makes them attractive for designing semiconductor lasers, electrooptic modulators, and photodetectors operating at a desired wavelength. The concept of band structure engineering [17, 18, 81] is further explored in strained semiconductors beyond the previous applications using quantum wells by controlling the well width and the barrier height. Using strained material systems such as $In_{1-x}Ga_xAs/InP$, $In_{1-x}Ga_xAs/In_{1-x}Ga_xAs_{1-y}P_y$, and $In_{1-x}Ga_xAs/GaAs$ quantum wells, a wide tunable range of wavelength is possible. Furthermore, the band structure including the band gap, the effective masses, and therefore the density of states can be engineered. It has been proposed that by using a compressively strained quantum-well structure, semiconductor lasers with a lower threshold current density can be achieved because of the reduced in-plane heavy-hole effective mass [17, 18, 81]. A reduced threshold carrier density also reduces the non-radiative

Auger recombination, one of the limiting factors for long-wavelength (1.55 μm) semiconductor lasers.

Experimentally, high-performance strained quantum-well lasers [82, 83], such as low-threshold current density [19, 20, 84–89], high-power output [90–93], and high-temperature operation [94–97], have been demonstrated with remarkable results. Theoretical models using both simplified parabolic band structures and realistic band structures of strained quantum wells have also been used to investigate the optical gains in these semiconductor lasers [98–105].

To understand the strained quantum-well systems, we will use the $In_{1-x}Ga_xAs$/InP materials as an illustration. The ternary $In_{1-x}Ga_xAs$ alloy has a lattice constant $a(x)$, which is a linear interpolation of the lattice constants of GaAs and InAs, a(GaAs), and a(InAs)

$$\begin{aligned} a(x) &= x\,a(\mathrm{GaAs}) + (1-x)\,a(\mathrm{InAs}) \\ &= 5.6533x + 6.0584(1-x). \end{aligned} \tag{10.4.1}$$

The lattice constant of the InP substrate a_0 is 5.8688 Å. When a thin $In_{1-x}Ga_xAs$ layer is grown on the InP substrate, at $x \simeq 0.468 \simeq 0.47$, the lattice constant of $In_{0.53}Ga_{0.47}As$ is the same as that of InP. It is the lattice-matched condition.

If $a(x) > a_0$ (i.e., $x < 0.47$), an elastic strain occurs that will be compressive in the plane of the layer resulting in a tensile strain in the direction perpendicular to the interface. We call this case biaxial compression or just compressive strain. Similarly, if $a(x) < a_0$(InP) (i.e., $x > 0.47$), the strain in the plane of the layer is tensile and there is a compressive strain in the perpendicular direction. We usually call this case biaxial tension, or tensile strain. Theoretical work [106, 107] shows that beyond a critical layer thickness, large densities of misfit dislocations are formed to accommodate the strain. Therefore, the strained layer dimension is always kept below the critical thickness for optoelectronic material applications.

10.4.1 The Influence of the Effective Mass on Gain and Transparency Carrier Density

A band structure calculation shows that the effective mass of the heavy-hole band along the parallel plane of a quantum well has a reduction in its effective mass due to its coupling to the light hold band. Therefore, the density of states of the heavy-hole subband is reduced. This helps to reduce the transparency carrier density required for population inversion as shown in Fig. 10.30a and Fig. 10.30b. The Bernard–Duraffourg population inversion condition [108] requires the separation of the quasi-Fermi levels $F_c - F_v + E_g > \hbar\omega > E_g + E_{e1} - E_{h1}$ ($\equiv E_g^{\mathrm{eff}}$).

For conventional semiconductors, the heavy-hole effective mass is always several times that of the electron in the conduction band. The quasi-Fermi level of the hole is usually above the valence-band edge instead of being below the band edge, and the conduction band degenerates easily with electron population. Because a large separation of the quasi-Fermi levels $F_c - F_v + E_g > E_g^{\mathrm{eff}}$ (the effective band gap of the quantum well) is required, an ideal situation will be a symmetric band structure configuration such that $m_h^* = m_e^*$ as shown in Fig. 10.30b. The population inversion condition

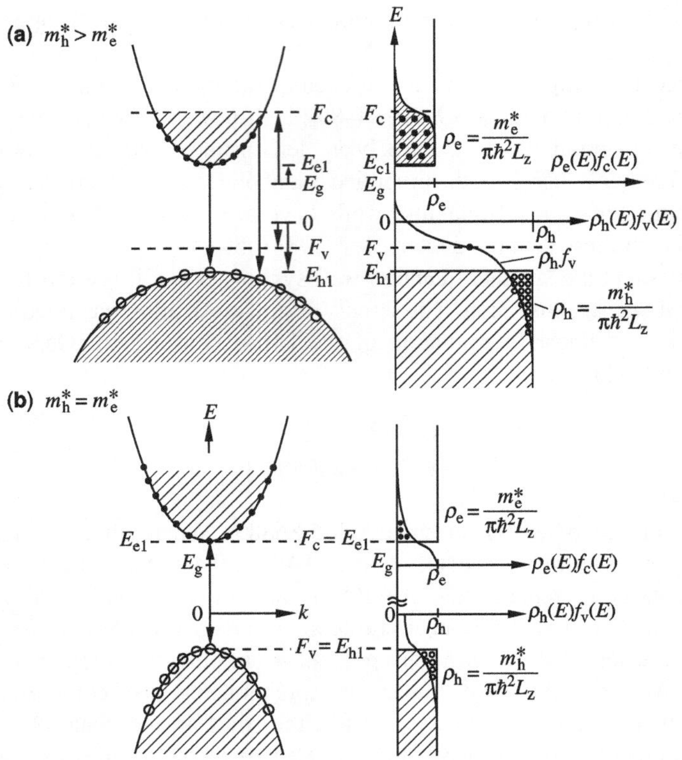

Figure 10.30 Population inversion in a quantum-well structure with (a) $m_h^* > m_e^*$ and (b) $m_h^* = m_e^*$.

$F_c - F_v + E_g = E_g^{eff}$ can then be satisfied easily at a smaller carrier density as shown. If we take the simplified model for the peak gain [62, 64] from the previous section,

$$g_p(n) = g_{max}(f_c - f_v) \simeq g_{max}(1 - e^{-n/n_c} - e^{-n/Rn_c}) \tag{10.4.2}$$

where $R = m_h^*/m_e^*$ is the ratio of the heavy hole and the conduction-band electron effective mass. We plot in Fig. 10.31a $f_c = 1 - e^{-n/n_c}$ and $f_v = e^{-n/Rn_c}$ versus the carrier concentration n for two values of R. We see that the intersection point $f_c = f_v$ determines the transparency density n_{tr}, at which the peak gain is zero. This transparency density n_{tr} is reduced when the mass rate is reduced from $R = 5$ to $R = 1$. The peak gain is also plotted in Fig. 10.31b for $R = 1$ and $R = 5$. We note that while the transparency density n_{tr} is smaller at $R = 1$, the maximum achievable gain also saturates at a smaller value when n increases because g_{max} is proportional to the joint density of states, which is proportional to the reduced effective mass $m_r^* = m_e^* m_h^*/(m_e^* + m_h^*) = m_e^* R/(R+1)$. If $m_h^* = m_e^*$, we have $m_r^* = m_e^*/2$. If $m_h^* = 5m_e^*$, we have $m_r^* = 5m_e^*/6$.

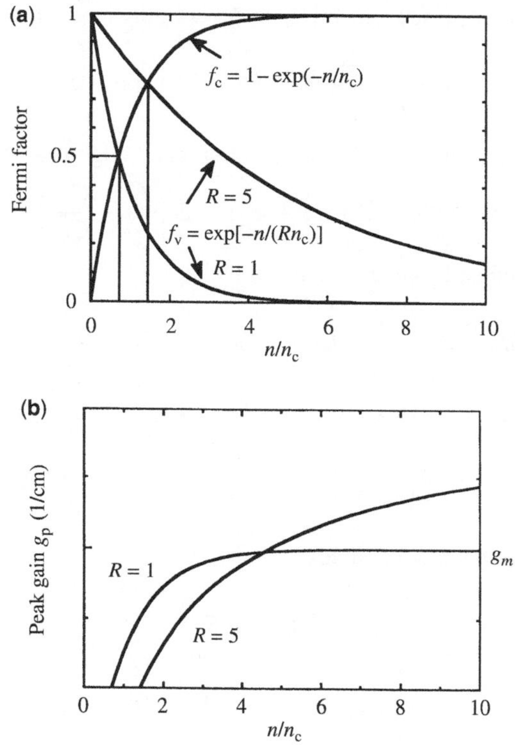

Figure 10.31 (a) Occupation probabilities $f_c = 1 - e^{-n/n_c}$ and $f_v = e^{-n/(Rn_c)}$ are plotted versus the normalized carrier concentration n/n_c for two heavy-hole-to-electron effective mass ratios $R\ (= m_h^*/m_e^*) = 5$ and $R = 1$. (b) Peak gain coefficient normalized by the maximum gain g_{max} versus the normalized carrier density n/n_c for $R = 5$ and $R = 1$.

10.4.2 Strain Effects on the Band-Edge Energies

For $In_{1-x}Ga_xAs$ quantum-well layers grown on InP substrate, we have the in-plane strain [109, 110]

$$\varepsilon = \varepsilon_{xx} = \varepsilon_{yy} = \frac{a_0 - a(x)}{a_0} \tag{10.4.3}$$

where a_0 is the lattice constant of InP. For a compressive strain, $a(x) > a_0$, therefore, the in-plane strain ε is negative. On the other hand, the in-plane strain ε is positive for a tensile strain because $a(x) < a_0$. The strain in the perpendicular direction is

$$\varepsilon_\perp = \varepsilon_{zz} = -2\frac{\sigma}{1-\sigma}\varepsilon = -2\frac{C_{12}}{C_{11}}\varepsilon \tag{10.4.4}$$

where σ is Poisson's ratio and C_{11}, C_{12} are the elastic stiffness constants. For most of the III–V compound semiconductors, $\sigma \simeq 1/3$ or $C_{12} \simeq 0.5\ C_{11}$.

The band gap of an unstrained $In_{1-x}Ga_xAs$ at 300K is given by

$$E_g(x) = 0.324 + 0.7x + 0.4x^2 \text{ (eV)}. \tag{10.4.5}$$

For a strained $In_{1-x}Ga_xAs$ layer, the conduction band edge is shifted by

$$\delta E_c = a_c\left(\varepsilon_{xx} + \varepsilon_{yy} + \varepsilon_{zz}\right) = 2a_c\left(1 - \frac{C_{12}}{C_{11}}\right)\varepsilon \tag{10.4.6}$$

and the valence subbands are shifted by

$$\begin{aligned}
\delta E_{HH} &= -P_\varepsilon - Q_\varepsilon, \qquad \delta E_{LH} = -P_\varepsilon + Q_\varepsilon \\
P_\varepsilon &= -a_v(\varepsilon_{xx} + \varepsilon_{yy} + \varepsilon_{zz}) = -2a_v\left(1 - \frac{C_{12}}{C_{11}}\right)\varepsilon \\
Q_\varepsilon &= -\frac{b}{2}(\varepsilon_{xx} + \varepsilon_{yy} - 2\varepsilon_{zz}) = -b\left(1 + 2\frac{C_{12}}{C_{11}}\right)\varepsilon.
\end{aligned} \tag{10.4.7}$$

Therefore, we have the new band edges

$$\begin{aligned}
E_c &= E_g(x) + \delta E_c \\
E_{HH} &= \delta E_{HH} = -P_\varepsilon - Q_\varepsilon \\
E_{LH} &= \delta E_{LH} = -P_\varepsilon + Q_\varepsilon.
\end{aligned} \tag{10.4.8}$$

The effective band gaps are

$$\begin{aligned}
E_{C-HH} &= E_g(x) + \delta E_c - \delta E_{HH} = E_g(x) + \delta E_c + P_\varepsilon + Q_\varepsilon \\
&= E_g(x) + 2a\left(1 - \frac{C_{12}}{C_{11}}\right)\varepsilon - b\left(1 + 2\frac{C_{12}}{C_{11}}\right)\varepsilon
\end{aligned} \tag{10.4.9a}$$

$$\begin{aligned}
E_{C-LH} &= E_g(x) + \delta E_c - \delta E_{LH} = E_g(x) + \delta E_c + P_\varepsilon - Q_\varepsilon \\
&= E_g(x) + 2a\left(1 - \frac{C_{12}}{C_{11}}\right)\varepsilon + b\left(1 + 2\frac{C_{12}}{C_{11}}\right)\varepsilon
\end{aligned} \tag{10.4.9b}$$

where the hydrostatic deformation potential a(eV) has been defined.

$$a = a_c - a_v.$$

As can be seen from Table 10.1, we have $a_c < 0$ and $a_v > 0$, using our definitions in this section. Therefore, we have the energy shift for compressive strain (with the notation for the trace of the strain matrix $\mathrm{Tr}(\bar{\bar{\varepsilon}}) = \left(\varepsilon_{xx} + \varepsilon_{yy} + \varepsilon_{zz}\right)$,

$$\delta E_c = a_c \mathrm{Tr}(\bar{\bar{\varepsilon}}) > 0, \quad a_v \mathrm{Tr}(\bar{\bar{\varepsilon}}) < 0, \quad b\left(1 + 2\frac{C_{12}}{C_{11}}\right)\varepsilon > 0,$$

Table 10.1 Physical Quantities of a Strained Semiconductor [109, 110]

Physical Quantity	Compressive		Tension
In-plane strain $\varepsilon = \varepsilon_{xx} = \varepsilon_{yy} = \frac{a_0 - a(x)}{a_0}$	Negative		Positive
$\varepsilon_{\perp} = -2\frac{C_{11}}{C_{12}}\varepsilon$	Positive		Negative
Conduction-band deformation potential a_c(eV)		Negative	
Valence-band deformation potential a_v(eV)		Positive	
$a = a_c - a_v$(eV)		Negative	
Shear deformation potential b(eV)		Negative	
$\delta E_c = a_c(\varepsilon_{xx} + \varepsilon_{yy} + \varepsilon_{zz})$	Positive		Negative
$P_\varepsilon = -a_v(\varepsilon_{xx} + \varepsilon_{yy} + \varepsilon_{zz})$	Positive		Negative
$Q_\varepsilon = -\frac{b}{2}(\varepsilon_{xx} + \varepsilon_{yy} - 2\varepsilon_{zz})$	Negative		Positive
Heavy-hole band-edge shift (eV)	$\delta E_{HH} = -P_\varepsilon - Q_\varepsilon$		
Light-hole band-edge shift (eV)	$\delta E_{LH} = -P_\varepsilon + Q_\varepsilon$		
Effective band gaps (eV)	$E_{C-HH} = E_g(x) + \delta E_c + P_\varepsilon + Q_\varepsilon$		
	$E_{C-LH} = E_g(x) + \delta E_c + P_\varepsilon - Q_\varepsilon$		

that is, the conduction-band edge is shifted upward by an amount δE_c, the valence-band edge is shifted downward by the magnitude of $a_v \mathrm{Tr}(\overline{\overline{\varepsilon}})$, then split upward by $b\left(1 + 2\frac{C_{12}}{C_{11}}\right)\varepsilon$ for the heavy hole and downward by $b\left(1 + 2\frac{C_{12}}{C_{11}}\right)\varepsilon$ for the light hole, as shown in Fig. 10.32. For tensile strain, the in-plane strain ε is positive because $a(x)$ is smaller than the lattice constant of the substrate a_0. Therefore, the directions of the band-edge shifts are opposite to those of the compressive strain.

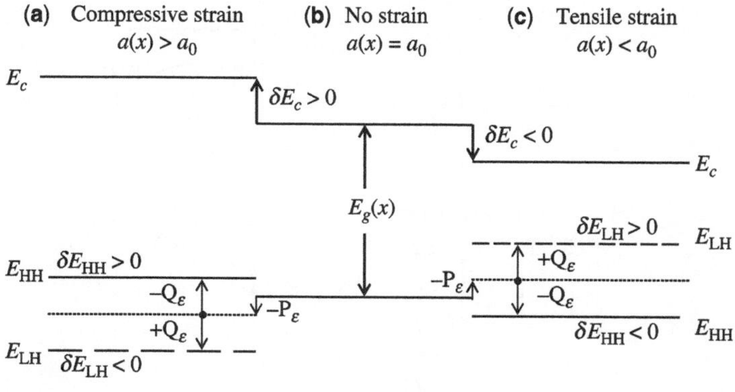

Figure 10.32 Conduction and valence band edges for semiconductors under (a) a compressive strain, (b) no strain, and (c) a tensile strain. The function $E_g(x)$ is the band gap of the unstrained $In_{1-x}Ga_xAs$ alloy. For example, $E_g(x) = 0.324 + 0.7x + 0.4x^2$ eV at 300K. The band edge shifts are $\delta E_c = 2a_c\left(1 - \frac{C_{12}}{C_{11}}\right)\varepsilon$, $\delta E_{HH} = -P_\varepsilon - Q_\varepsilon$, $\delta E_{LH} = -P_\varepsilon + Q_\varepsilon$, where $P_\varepsilon = -2a_v\left(1 - \frac{C_{12}}{C_{11}}\right)\varepsilon$, $Q_\varepsilon = -b\left(1 + 2\frac{C_{12}}{C_{11}}\right)\varepsilon$, and the in-plane strain $\varepsilon = [a_0 - a(x)]/a_0$. $a(x)$ is the lattice constant of the unstrained $In_{1-x}Ga_xAs$ alloy, and a_0 is that of the substrate, which is also the in-plane lattice constant of the strained $In_{1-x}Ga_xAs$ lattice.

10.4.3 Band Structure of a Strained Quantum Well

For a quantum-well structure such as an $In_{1-x}Ga_xAs$ layer sandwiched between InP barriers, the band structures are shown in Fig. 10.33 for (a) a compressive strain ($x < 0.468$), (b) no strain ($x = 0.468$), and (c) a tensile strain ($x > 0.468$). The left-hand side shows the quantum-well band structures in real space versus position along the growth (z) direction. The right-hand side shows the quantized subband dispersions in momentum space along the parallel (k_x) direction in the plane of the layer. These dispersion curves show the modification of the effective masses or the densities of states due to both the quantization and strain effects. More precise valence band structures can be found in Section 4.9.

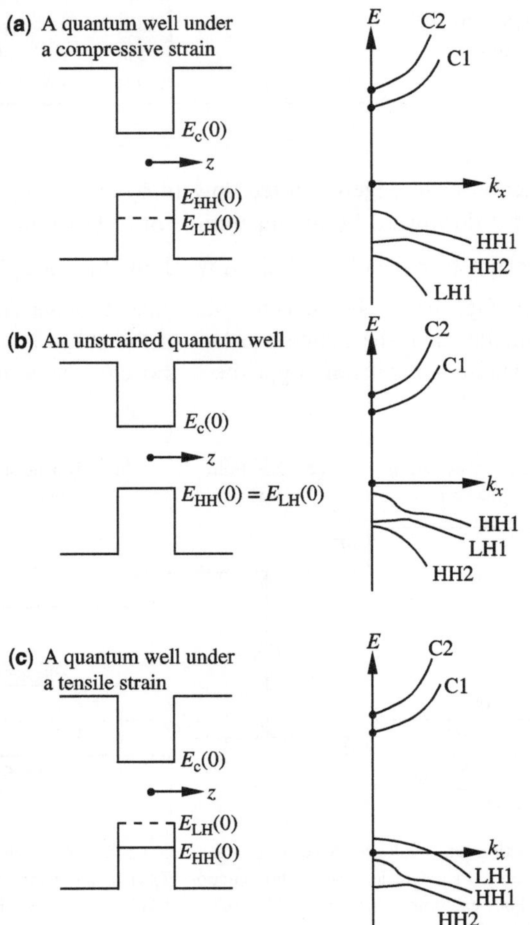

Figure 10.33 Band edge profiles in real space along the growth (z) direction and the quantized subband dispersions in k space along the k_x direction (perpendicular to the growth direction) for a quantum well with (a) a compressive strain, (b) no strain, and (c) a tensile strain.

Example: Gain Spectrum of a Strained Quantum Well Using the formulation for the gain spectrum in a strained quantum well, we plot in Fig. 10.34 the gain spectrum for light with a TE ($\hat{x}$ or $\hat{y}$) or TM ($\hat{z}$) polarization for the three cases in Fig. 10.33. We consider an $In_{1-x}Ga_xAs$ quantum-well system with

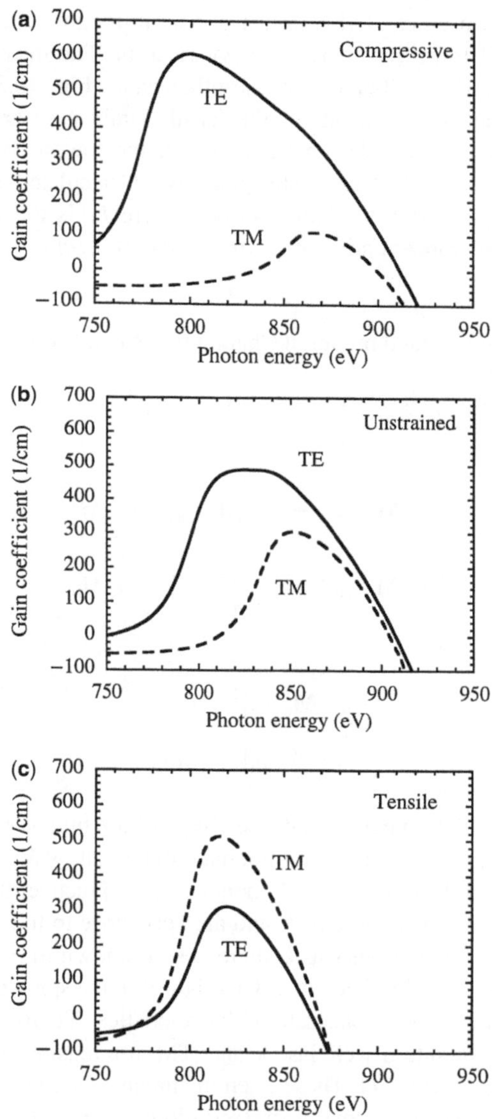

Figure 10.34 Gain spectra for both TE and TM polarizations of (a) a compressive strain ($x = 0.41$, well width = 45 Å), (b) an unstrained ($x = 0.47$, $L_z = 60$ Å), and (c) a tensile strain ($x = 0.53$, and $L_z = 115$ Å) $In_{1-x}Ga_xAs/In_{1-x}Ga_xAs_{1-y}P_y$ (barrier band-gap wavelength $\lambda_b = 1.3$ μm) quantum-well laser. These material gain coefficients were calculated for the same period $L_T = 200$ Å and surface carrier concentration $n_s = nL_z = 3\times10^{12}\ \mathrm{cm}^{-2}$ in all three structures.

$In_{1-x}Ga_xAs_{1-y}P_y$ barriers (band-gap wavelength $\lambda_b = 1.3$ μm) lattice-matched to InP substrate. Electronic and optical properties of similar quantum-well structures have been shown in Fig. 4.21a–d, Fig. 9.23, and Fig. 9.24. As shown in Fig. 10.34a, compressive strain (quantum well gallium mole fraction $x = 0.41$, well width = 45 Å), and Fig. 10.34b, lattice-matched case ($x = 0.47$, $L_z = 60$ Å), the gain of the TE polarization is always larger than that of the TM polarization, because the top valence subband for these two cases is always heavy hole in nature, and the optical matrix element for C1-HH1 transition prefers TE polarization. The injected holes populate mostly the ground (HH1) subband. On the other hand, Fig. 10.34c shows that the TM polarization may be dominant for the tensile strain case ($x = 0.53$, and $L_z =$ 115 Å). The well widths are chosen such that all the lowest band edge transition wavelengths are close to 1.55 μm. The gains were calculated for the same total width $L_T = L_z + L_b = 200$ Å in a unit period, where L_b is the barrier width. The same surface carrier concentration $n_s = nL_z = 3 \times 10^{12}/\text{cm}^2$ is used in all three structures.

If we recall the parabolic band model in Chapter 9, the band edge ($k_t = 0$) momentum matrix elements are

TE Polarization

$$|\hat{x} \cdot \mathbf{M}_{c-hh}|^2 = |\hat{y} \cdot \mathbf{M}_{c-hh}|^2 = \frac{3}{2} M_b^2 \tag{10.4.10a}$$

$$|\hat{x} \cdot \mathbf{M}_{c-\ell h}|^2 = |\hat{y} \cdot \mathbf{M}_{c-\ell h}|^2 = \frac{1}{2} M_b^2. \tag{10.4.10b}$$

TM Polarization

$$|\hat{z} \cdot \mathbf{M}_{c-hh}|^2 = 0 \tag{10.4.10c}$$

$$|\hat{z} \cdot \mathbf{M}_{c-\ell h}|^2 = 2M_b^2 \tag{10.4.10d}$$

where M_b is the optical momentum matrix element of the bulk semiconductor, which is related to the energy parameter E_p for the matrix element by $M_b^2 = m_0 E_p/6$, and E_p is usually tabulated in databooks (see Appendix C). In practice, the matrix elements depend on the transverse wave vector k_t and are very close to the above values at the band edge where $k_t = 0$, then vary as k_t increases, as shown in Fig. 10.35, using the valence-band mixing model in Section 9.8 and ignoring the spin–orbit split-off band. Laser actions with the above polarization characteristics, TE for compressive strain and lattice-matched quantum-well lasers and TM for tensile strain quantum-well lasers, have been reported [111]. These phenomena also agree with those in conventional externally stressed or thermally stressed diode lasers [112, 113].

In Fig. 10.36, we plot the optical modal gain versus the surface carrier concentration $n_s = nL_z$, where L_z is the well width, for three cases: (a) compressive, (b) lattice matched, and (c) tensile strain. For long-wavelength semiconductors, the well width L_z has to be adjusted such that lasing action at 1.55 μm can be achieved.

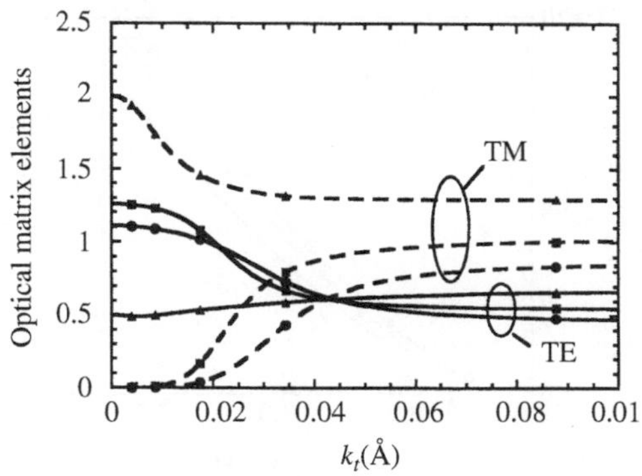

Figure 10.35 Normalized momentum matrix element $2|M_{nm}(k_t)|^2/M_b^2$ versus the transverse k_t vector using (9.8.15) and (9.8.16) in Chapter 9 for compressive (dots), unstrained (solid squares), and tensile (solid triangles) strained quantum-well lasers. Here M_{nm} refers to $n = \text{C1}$, and $m =$ the top valence subband (HH1 for compressive and unstrained and LH1 for tensile case).

The design usually requires L_z (compressive) $< L_z$ (lattice-matched) $< L_z$ (tensile). As discussed in Section 10.3, it is the modal gain Γg that is important in determining the threshold condition, and

$$\Gamma g \propto \frac{L_z}{W_{\text{mode}}} g \tag{10.4.11}$$

where L_z will cancel with the factor $(1/L_z)$ in g. Therefore, it is more useful to compare the modal gain Γg for three different strains if the well width L_z varies. We see that the compressive strain quantum-well laser has the smallest transparency carrier density at which $g = 0$. However, it also saturates faster than the other two

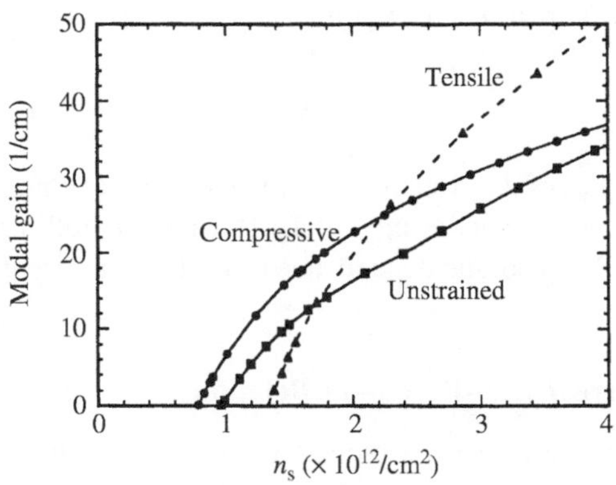

Figure 10.36 Modal gain coefficient versus the surface carrier concentration for compressive strain (dots), unstrained (squares), and tensile strain (triangles) $In_{1-x}Ga_xAs/In_{1-x}Ga_xAs_{1-y}P_y$ quantum-well lasers with the band-edge transition wavelengths near 1.55 μm. The solid curves with dots and squares are TE polarization, and the dashed curve with triangles is TM polarization.

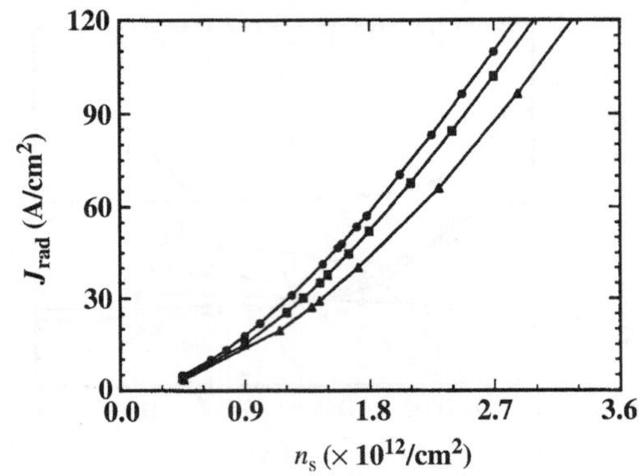

Figure 10.37 Radiative current density $J_{rad}(A/cm^2)$ versus the surface carrier concentration $n_s(1/cm^2)$ for compressive strain (dots), unstrained (squares), and tensile strain (triangles) quantum-well lasers.

cases. On the other hand, the tensile strain case (TM polarization) has a larger transparency current density, yet it increases faster implying a larger differential gain.

We can also plot the radiative current density

$$\begin{aligned} J_{rad} &= qL_z\, R_{sp}(n) \\ &= qL_z \int_0^\infty r^{spon}(\hbar\omega)d(\hbar\omega) \end{aligned} \tag{10.4.12}$$

where the well width factor L_z is canceled because $r^{spon}(\hbar\omega) \propto 1/L_z$ from (9.8.18), and $J_{rad}(n_s)$ is a two-dimension current density with a dimension (A/cm^2). A common approximation is

$$R_{sp}(n) = Bn^2 = B'n_s^2 \tag{10.4.13}$$

where $n_s = nL_z$ and $B' = B/L_z^2$. Therefore

$$J_{rad} = qB_s n_s^2 \tag{10.4.14}$$

with $B_s = B/L_z$. A plot of J_{rad} versus n_s for three different strain conditions is shown in Fig. 10.37 for comparison using (10.4.12) with the full valence-band mixing model. We see that the quadratic dependence (10.4.14) is a very good approximation.

10.4.4 Gain–Current Density (*G–J*) Relation

The peak optical gain can also be plotted versus the radiative current density J_{rad} directly for three strains, as shown in Fig. 10.38. We see that the compressive strain case has the lowest transparency current density and its gain saturates faster as the current density is increased. In real devices, the injected current density has other losses such as nonradiative Auger recombination and intervalence band

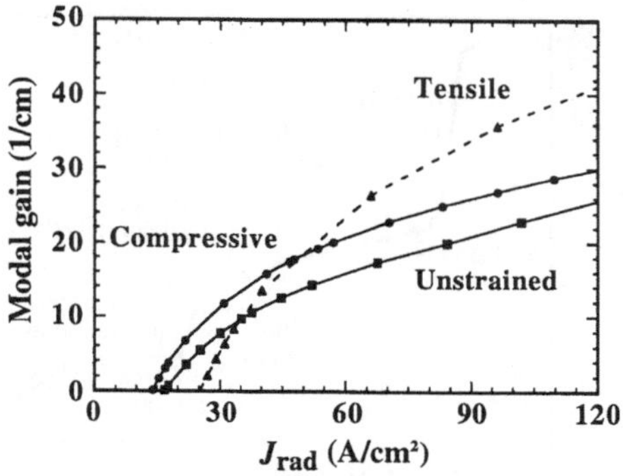

Figure 10.38 Modal gain $G = \Gamma g$ versus the radiative current density $J_{\rm rad}({\rm A/cm^2})$ for compressive strain (dots), unstrained (squares), and tensile strain (triangles) quantum-well lasers. The solid curves are TE polarization, and the dashed curve is TM polarization.

absorptions, which have to be added to the horizontal axis to obtain the true G–J relation. If these loss current densities are less sensitive to strains, we would expect the threshold gain $G_{\rm th} = \Gamma g_{\rm th}$ to determine the threshold current density by the intersection of the G–J curve with a horizontal line $G = G_{\rm th}$. We see that for a small $G_{\rm th}$, the compressive strain laser will have the smallest threshold current density. On the other hand, if $G_{\rm th}$ is large, the tensile strain laser may have the smallest threshold current density. These G–J curves provide good design rules for strained quantum-well lasers.

An empirical G–J relation using the logarithmic dependence

$$G = n_{\rm w}\Gamma_{\rm w} g_{\rm w} = n_{\rm w}\Gamma_{\rm w} g_0 \left[\ln\left(\frac{n_{\rm w}J_{\rm w}}{n_{\rm w}J_0}\right) + 1\right], \tag{10.4.15}$$

as discussed in Section 10.3, has also been used to study strained quantum-well lasers with very good agreement with experimental data [75, 76, 114]. The major equations are the same as those discussed in Section 10.3 and will not be repeated here. The same analysis using (10.3.29)–(10.3.36) has been applied to study strained quantum-well lasers.

Advanced issues such as Auger recombination rates [115–118], the temperature sensitivity [119], and high-speed carrier transport and capture in strained quantum wells [120, 121] are under intensive investigation. Many-body effects [122, 123] due to carrier–carrier Coulomb interactions on the gain spectrum have been investigated. Many of these require more theoretical and experimental work to fully understand the physics of strained quantum-well lasers.

10.5 STRAINED QUANTUM-DOT LASERS

As pointed out by Alferov [124], the history of semiconductor lasers can be seen from the evolution of the reduction of the threshold current density, Fig. 10.39. The

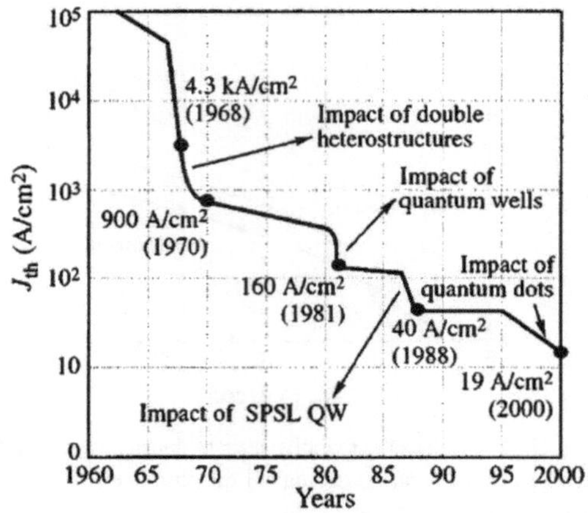

Figure 10.39 Evolution of the threshold current of semiconductor lasers. (Reprinted with permission from [124] © 2000 IEEE.)

realization of semiconductor lasers using homojunction to double-heterojunctions, quantum wells, followed by strained quantum wells, and then quantum dots (QDs), indicate a reduction of threshold current density. Figure 10.40 shows the density of states of a bulk semiconductor (3D), a quantum well (2D), a quantum wire (1D), and quantum dots (0D) [125–127]. The discrete nature of quantum dots leads to many advantages of quantum-dot lasers: they are expected to have low threshold current density, a large characteristic temperature, high differential gain, low chirp, and a high modulation speed. Figure 10.41 shows the predicted threshold current density as a function of temperature [124], indicating a temperature-stable threshold current for quantum dot lasers [124–126].

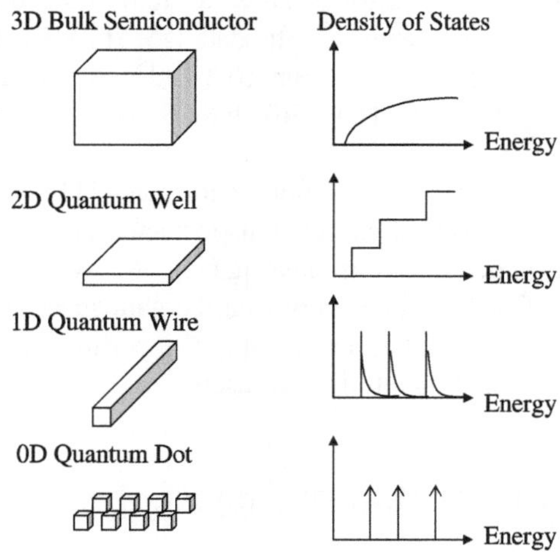

Figure 10.40 Density of states for 3D bulk, 2D quantum well, 1D quantum wire, and 0D quantum dot structures.

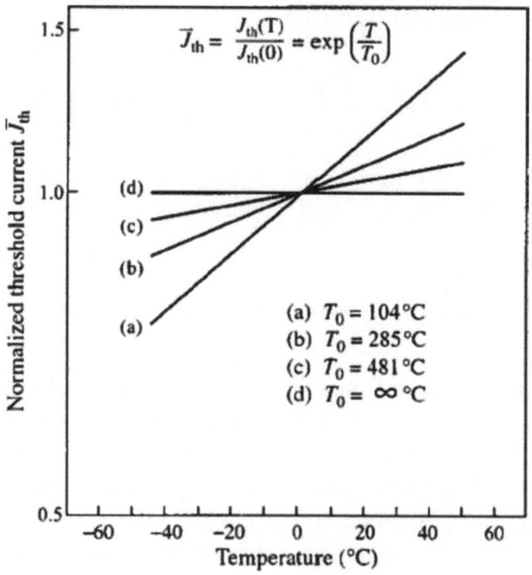

Figure 10.41 Normalized temperature dependence of the threshold current for various double-heterojunction-structure (DHS) lasers. (a) Bulk; (b) with quantum wells (QWs); (c) with quantum wires (QWRs); (d) with quantum dots (QDs). (Reprinted with permission from [124] © 2000 IEEE.)

10.5.1 Quantum Dot Structures

One breakthrough in crystal growth is the realization of semiconductor quantum dots [124–139], which confine electrons and holes in a three-dimensional structure. The quantization of electron energy in all three directions makes the energy levels discrete instead of continuous. Therefore, applications to devices such as quantum-dot semiconductor lasers similar to gas lasers with a discrete density of states become a reality. A layer of material that has a lattice constant different from that of the substrate may transform spontaneously to an array of three-dimensional islands after the height of the deposited material has reached some critical thickness. The evolution of an initially two-dimensional growth into a three-dimensional corrugated growth front has been observed in many systems. Actually, Stranski and Krastanow (SK) [128] initially proposed the possibility of island formation on an initially flat heteroepitaxial surface for the growth of *lattice-matched* ionic crystals that have different charges. It was then discovered that the growth of islands relaxed by misfit dislocations in strained heteroepitaxy is possible. The formation of coherent or defect-free islands as a result of the SK growth of strained heterostructures becomes an important method of growing quantum dots.

In the SK growth mode, a strained layer forms up to a thickness of a few atomic monolayers. The increasing strain energy of the layer relaxes partially by the spontaneous formation of islands on the top of this wetting layer. This spontaneous formation during the epitaxial growth of two highly lattice-mismatched semiconductors leads to dense arrays of dots. For example, InAs islands grown on GaAs surface

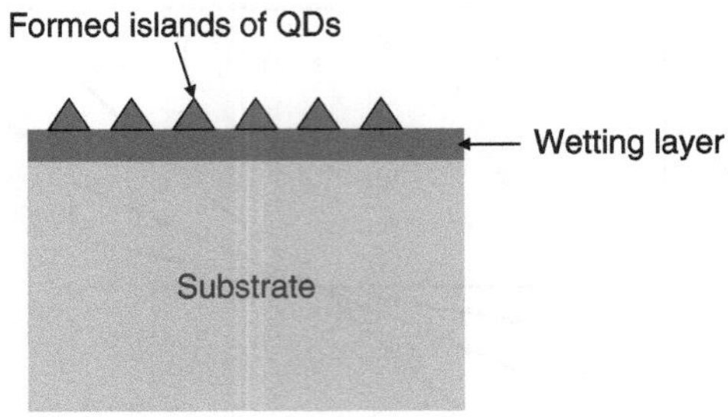

Figure 10.42 The formation of InAs islands during crystal growth.

has been studied with efficient photoluminescence from InAs islands covered by GaAs. Research has identified that 3D islands are formed in appropriate growth conditions caused by the strain-induced renormalization of the surface energy of the QD facets [136], Fig. 10.42. Repulsive interaction of the islands via the substrate allows for lateral ordering. For a strained epilayer with small interface energy, the initial growth may occur layer by layer, but a thicker layer has larger strain energy and can lower its energy by forming isolated islands in which strain is relaxed. Thus, the SK growth mode occurs. It was believed initially that islands formed in the SK growth are dislocated. However, further experiments have demonstrated the formation of three-dimensional coherently strained islands. If strained InGaAs islands are covered with a thin GaAs layer, InGaAs islands in a second sheet can be formed on top of the dots in the first sheet, resulting in 3D ordered arrays of QDs, where vertical coupling of dots is also possible [131]. The size and the shape of the InAs islands can be changed by using InGaAs or InGaAlAs and by changing the deposition mode.

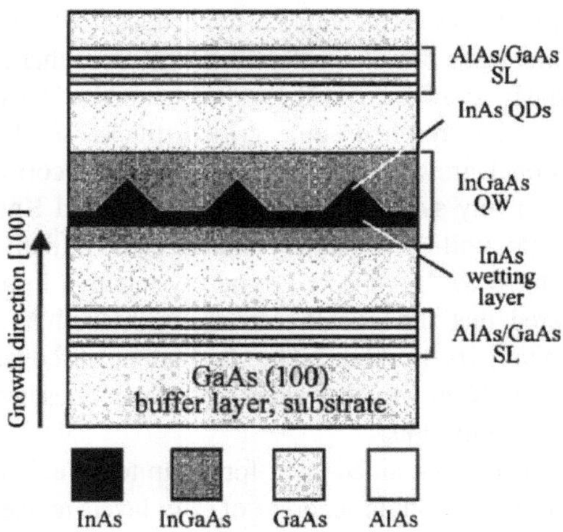

Figure 10.43 Schematic of the structure that contains InAs quantum-dot array in an external InGaAs quantum well on GaAs(100). (Reprinted with permission from [133] © 1999 American Institute of Physics.)

Figure 10.43 shows [133] an example of self-organized InAs quantum dots inserted in InGaAs quantum well grown on a GaAs substrate by molecular-beam epitaxy. Room temperature photoluminescence spectrum shows that QD emission wavelength is controllable from 1.1 to 1.3 μm by varying the composition of $In_xGa_{1-x}As$ quantum-well matrix from $x = 0$ (1.1 μm for GaAs matrix) to $x = 0.3$ (1.3 μm). Figure 10.44a [133] shows the plan-view bright-field TEM image along [100] direction and Fig. 11.44b the cross-section image under dark-field (200) condition for the InAs QD array in a 10-nm $In_{0.12}Ga_{0.88}As$ quantum well. InAs-rich clusters in the region of smaller InAs content are seen in both observation views. The surface density of the islands is estimated as $(3-4)\times 10^{10}$ cm^{-2}. The dots have a typical lateral size of about 20 nm and a height of 5–6 nm. These sizes are larger than those of the InAs islands formed in a GaAs matrix. Because both the QD and QW materials are lattice mismatched to the GaAs substrate, there is a risk of plastic strain relaxation. Nevertheless, no misfit dislocations are revealed by TEM due to the choice of the QW width and the InAs model fraction in the QW [133]. The sample with the QD structure embedded in a vertical optical microcavity shows a Photoluminescence (PL) emission of 1.33 μm [133].

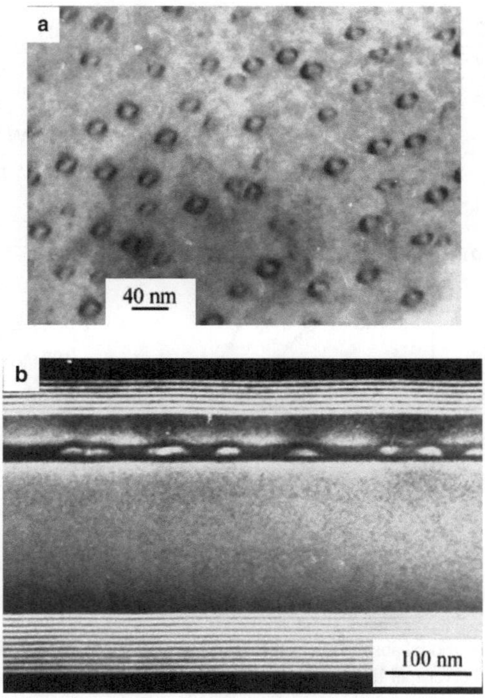

Figure 10.44 (a) Plan-view of the bright-field TEM image along the [100] zone, and (b) the image of the cross section under dark-field (200) condition for the structure containing InAs QD array in a 10-nm $In_{0.12}Ga_{0.88}As$ QW. (Reprinted with permission from [133] © 1999 American Institute of Physics.)

10.5.2 Spontaneous Emission, Gain, and Lasing Spectrum

In Chapter 9, we discussed the absorption and gain of quantum-dot lasers [140–146]. The filling of states of quantum dots is complicated by the inhomogeneous size distribution and the density of dots or the distance between nearby dots [141]. When the inhomogeneous broadening is reduced, clear observation of ground-state and excited-state emission peaks or absorption peaks are observable [141–146]. For example, Fig. 10.45a [141] shows the net model absorption of a quantum-dot sample (dots in a well, or DWELL [139]) showing clear absorption peaks at the ground state and excited state when there is no current injection. In the presence of increasing current, population inversion occurs and a clear gain peak occurs at the ground state or the excited state at a higher current of 200 mA. Analysis of the gain spectrum seems to indicate two quasi-Fermi levels for the ground and excited state. The competition of homogenous broadening and inhomogeneous broadening also leads to

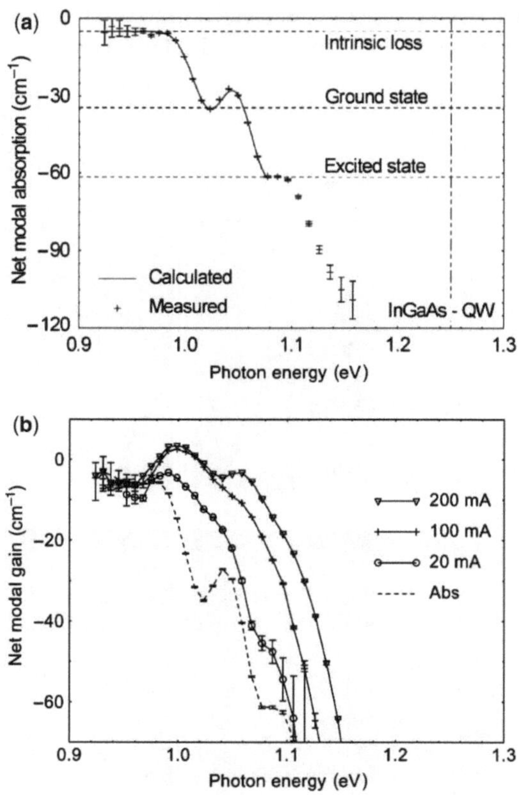

Figure 10.45 (a) The net modal absorption of the three-layer dots-in-well (DWELL) waveguide structure for TE-polarized light. The vertical line indicates the transition energy of the well. The horizontal dashed lines indicate the waveguide loss and the peak absorptions due to the ground and excited states. The solid line is the fitting of the calculated absorption spectrum to the experimental data. (b) The net modal TE gain spectra for pulsed currents per segment from 20 to 200 mA. (Reprinted with permission from [141] © 2004 IEEE.)

interesting lasing spectrum at low and high temperatures [143]. Figure 10.46a–f [142] shows the evolution of the ground state lasing at low current (10 mA) to simultaneous ground and excited state lasing at 20 mA, then excited state lasing at 25 mA. A model taking into account the homogenous and inhomogeneous broadenings explains the experimental data very well. The light output power is shown in Fig. 10.47 [142]. The regions of lasing from only the ground state, both states, and

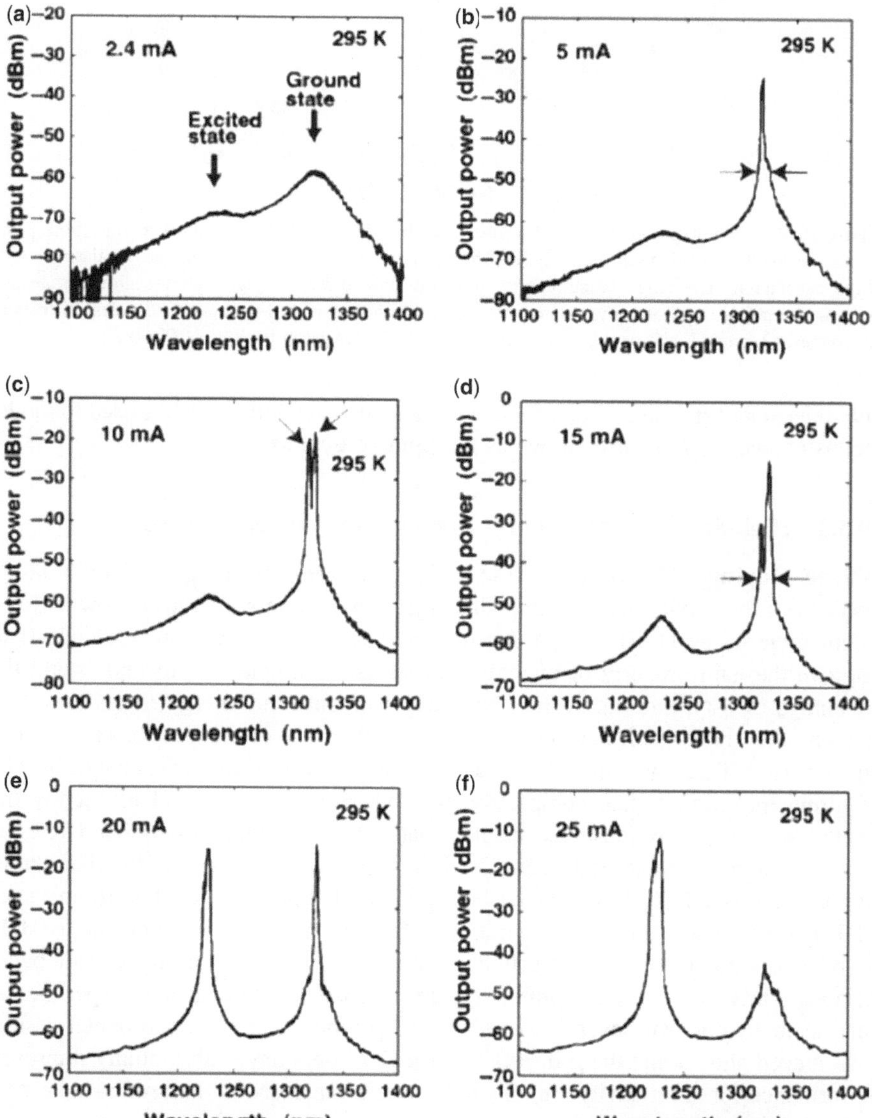

Figure 10.46 Optical emission spectra of the laser at (a) 2.4, (b) 5, (c) 10, (d) 15, (e) 20, and (f) 25 mA. (Reprinted with permission from [142] © 2005 American Institute of Physics.)

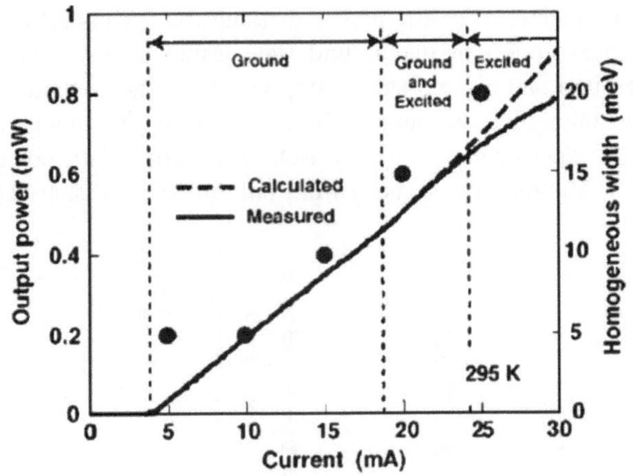

Figure 10.47 Output power and the homogeneous broadening of the ground-state transition (solid circles) versus the injection current at 295K. The solid curve is the measurement, and the dashed curve is the calculation. The range of current for only the ground-state lasing, both the ground-state and excited-state lasings, and only the excited-state lasing is shown, with the perpendicular dashed lines as boundaries. (Reprinted with permission from [142] © 2005 American Institute of Physics.)

only the excited state are marked. The homogenous linewidth, which is used to model the lasing spectrum, is also shown as the right vertical axis.

10.5.3 High-Speed Modulation: *p*-Doping and Tunneling Injection

***p*-Type Doping [147–162]** A few major improvements in quantum-dot lasers result from the *p*-type doping, tunneling injection, and submonolayer growth (for a recent review, see [137]. *p*-type doping was suggested as early as 1988 [127] to improve the gain property of quantum-dot lasers. A simple way to understand the advantage of *p*-type doing is that the heavy-hole subband energy levels are close to each other due to their heavier effective mass than that of the electrons in the conduction band. Therefore, the holes spread out thermally into various subbands instead of filling the ground state completely first, whereas electrons tend to occupy the ground state better because the excited state of the conduction subband (C2) is further away from the ground state (C1). It implies that the number of interband transitions such as C1-HH1 is reduced because not all holes occupy the ground state (HH1) for an undoped quantum dot sample. The holes that spread out to HH2 subband do not recombine effectively with those electrons of the C1 subband. Although this is an oversimplified picture, theoretical models and experimental data seem to confirm several advantages of *p*-type doping. Experimental results have indeed shown that the *p*-doped 1.3-μm QD lasers have substantially improved performance for short cavity lengths, indicating an increase in the ground state gain and exhibit excellent temperature sensitivity [147–149]. Further experimental data [149] on spontaneous and laser emission of *p*-doped and undoped QD heterostructures indeed show that the ground-state spontaneous emission of the carbon-doped sample is larger than that of the excited state when compared with those of an undoped QD sample. The differential gain is increased, but not the modulation bandwidth. It appears that other mechanisms such as the electron transport through the dot wetting layer caused by the nonlasing QDs of the active ensemble may substantially reduce the modulation bandwidth.

High-power and temperature-insensitive quantum-dot lasers with p-type doping have been reported. Continuous-wave high output power (5 W) of GaAs-based

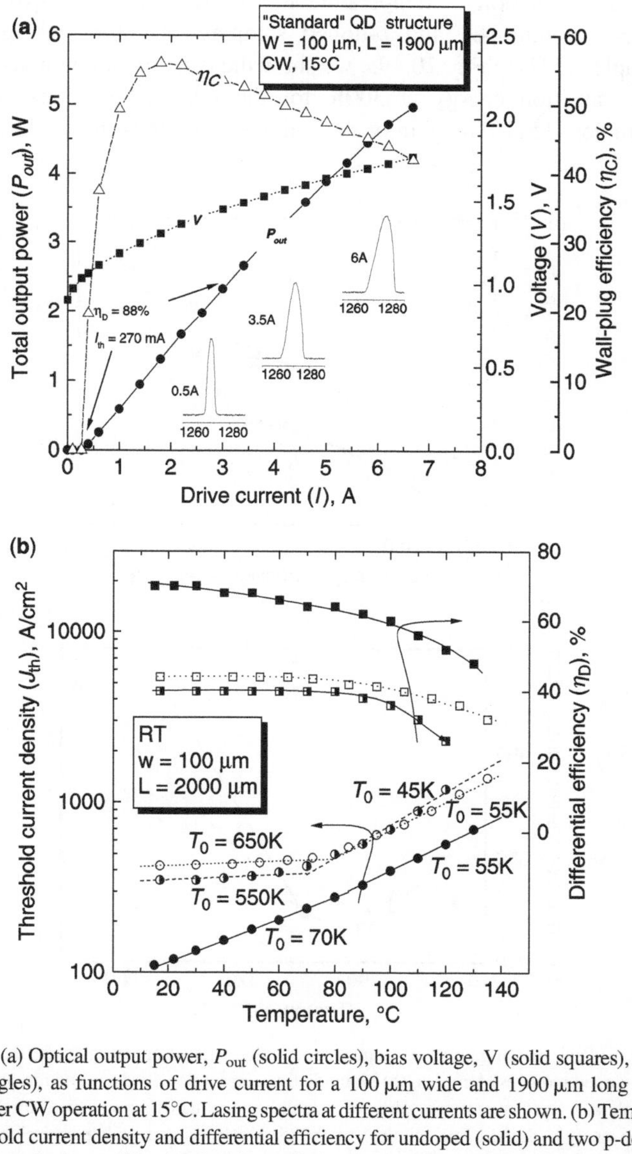

Figure 10.48 (a) Optical output power, P_{out} (solid circles), bias voltage, V (solid squares), and conversion efficiency (triangles), as functions of drive current for a 100 μm wide and 1900 μm long stripe undoped quantum-dot laser CW operation at 15°C. Lasing spectra at different currents are shown. (b) Temperature dependences of threshold current density and differential efficiency for undoped (solid) and two p-doped (half open and open) quantum-dot lasers. (Reprinted with permission from [150] © 2005 IOP Publishing Ltd.)

broad-area (100 μm) QD lasers at 1.3 μm with a wall plug efficiency of 56% have been demonstrated. Figure 10.48a shows the experimental results of the L–I and I–V curves [150]. The wall plug conversion efficiency is also shown. Figure 10.48b shows the threshold current densities of two p-doped QD structures (open and half-open symbols) with that of an undoped QD sample. The threshold of the p-doped sample is higher than that of the undoped sample, yet the characteristic temperature is 650K, showing an almost insensitive threshold behavior.

Experimental analysis has been performed on the gain, radiative and nonradiative components as a function of temperature for 50-μm-wide, 2000-μm-long oxide-striped lasers for undoped and p-doped samples, and for 1000-μm-long and p-doped sample [151]. In Fig. 10.49a, we show the peak modal gain versus transparency minus transition energy at 300K for the undoped (crosses) and p-doped (squares) samples [151]. The gain of the p-doped sample is indeed higher than that

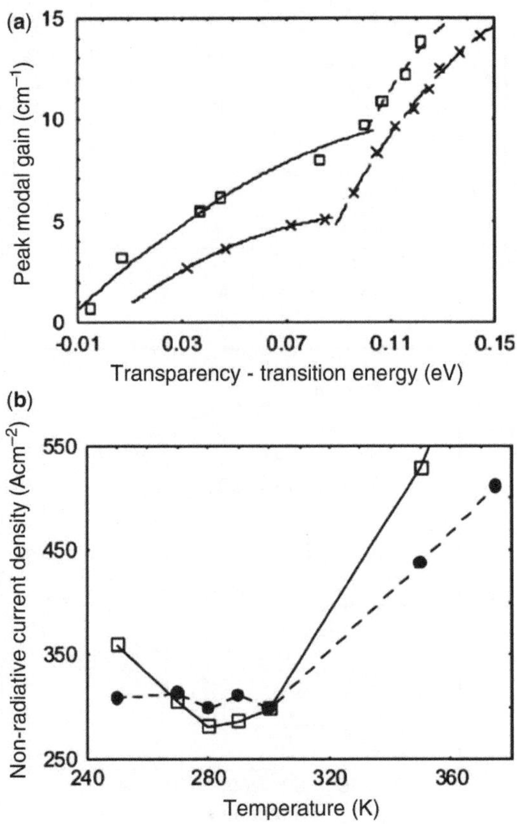

Figure 10.49 (a) Peak modal gain versus transparency minus transition energy at 300K for the undoped (crosses) and p-doped (squares) samples. (b) Nonradiative current density as a function of temperature for the doped sample at a fixed modal gain of 8 cm^{-1} (solid line, squares) and at fixed transparency energy minus transition energy (dashed line, circles) of 0.085 eV. (Reprinted with permission from [151] © 2007 IEEE.)

of the undoped sample. Figure 10.49b shows the nonradiative current density as a function of temperature for the doped sample at a fixed modal gain of 8 cm^{-1} (solid lines, squares) and at fixed transparency energy minus transition energy (dashed lines, circles) of 0.085 eV. It shows that the nonradiative current density as a function of temperature decreases for the *p*-doped sample in a temperature range 240K to 300K.

Figure 10.50a and Fig. 10.50b [152] show a comparison of the optical gain and differential gain of a *p*-doped (solid) and an undoped (dashed) QD laser, indicating clearly that the improvement of the *p*-type doping. The improvement in differential gain also leads to the reduction of the linewidth enhancement factor as indicated by Fig. 10.51 [152]. The horizontal axis is the surface electron density per QD layer [152]. The theory has also been used to fit the measured optical gain spectrum and the linewidth enhancement factor of the *p*-doped quantum-dot laser with a good agreement, as shown in Fig. 9.15 in Chapter 9.

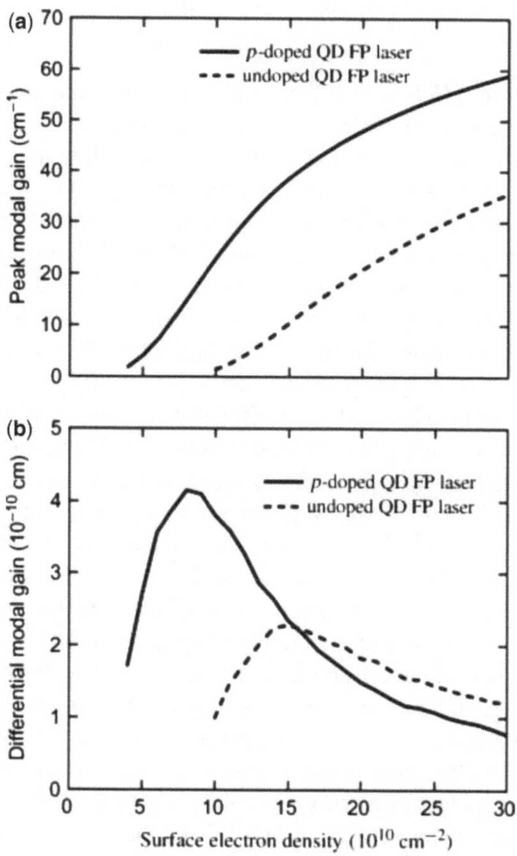

Figure 10.50 The dependence of (a) the peak modal gain and (b) the differential peak modal gain on the injection surface electron density per QD layer for the *p*-doped (solid) and undoped (dashed) QD FP lasers. (Reprinted with permission from [152] © 2006 IEEE.)

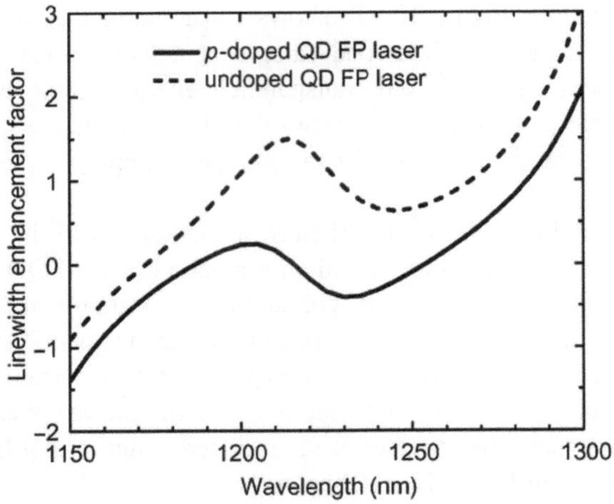

Figure 10.51 Calculated LEF spectra of the *p*-doped (solid) and the undoped (dashed) QD FP laser near the threshold. (Reprinted with permission from [152] © 2006 IEEE.)

Tunneling Injection [154–162] Tunneling injection such as the quantum-well (QW) coupled quantum-dot structures provides advantages such as better carrier collection of carriers by quantum wells and subsequent tunneling injection into quantum dots. The process reduces the threshold current density and avoids the hot carrier effects, which will increase the direct modulation bandwidth. Figure 10.52a shows the energy-diagram schematic of an InP QD-coupled-QW laser [159]. The auxiliary QWs serve as collectors of carriers, which tunnel to QDs for radiative recombination. Tensile strained QWs contribute to strong TM light (with considerable TE component), whereas compressively strained QDs emit TE polarized light. Using a polarization-resolved ASE spectra at very low injection (0.01 mA), Fig. 10.52b, we can probe the QD ground state and the excited state contribution to the TE-polarized emission. The QW tensile strain produces strong TM emission from the wells together with a TE emission from the QD-coupled-QW states. The evolution of the polarization-resolved ASE spectrum (in the logarithmic scale) for the QD-QW laser with tensile-strained QWs at three bias currents is shown in Fig. 10.52c. As the current increases, the TE spectrum shifts to the emission of QD-coupled-QW states. The TM emission spectrum from the QW is clear and increases uniformly at the shorter-wavelength side of the coupled QD-QW TE emission.

Low linewidth enhancement factor has also been measured from a tunneling injection QD laser. In Fig. 10.53a, we show the energy diagram of an InAs QW–QD laser [158]. The auxiliary QWs serve as the collectors of electron and holes that tunnel into QDs for the stimulated recombination. Figure 10.53b shows the measured linewidth enhancement factor spectra of an InAs tunneling-injection QD laser. Pulse measurement at low duty cycle below 2% is needed to avoid the thermal effect. The linewidth enhancement factor is as low as 0.15 near the threshold (23.5 mA).

High-speed performance of a *p*-doped tunneling-injection QD laser has been demonstrated [161]. Figure 10.54a [161] shows the energy-band diagram in the

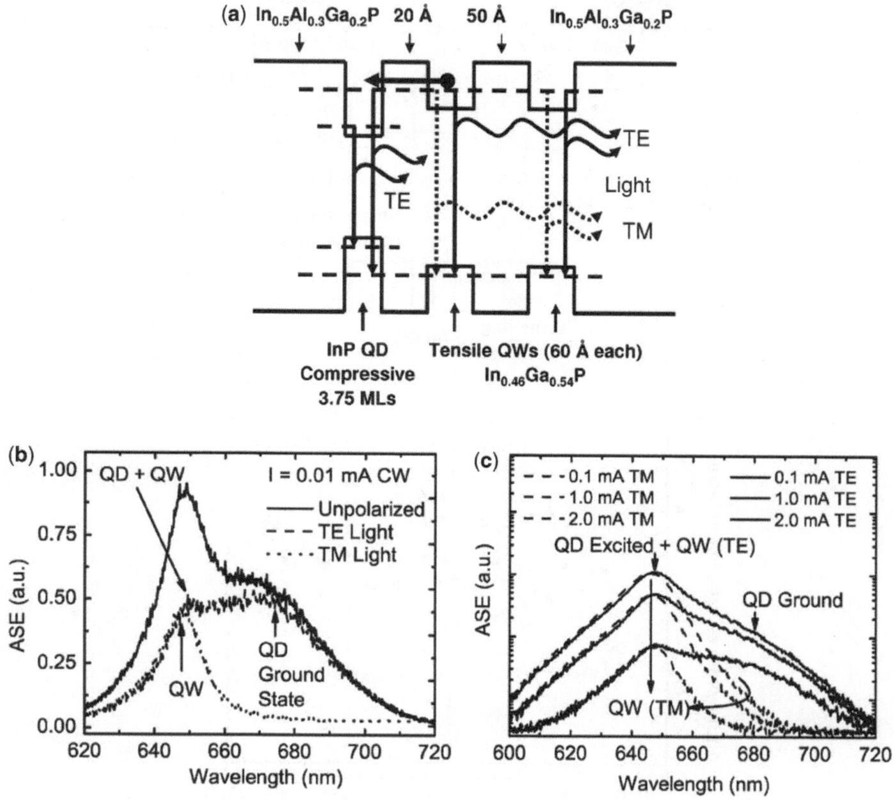

Figure 10.52 (a) Energy diagram schematic of an InP QD-coupled-QW laser. The auxiliary QWs serve as collectors of carriers, which tunnel into QDs for the radiative recombination. Tensile strained QWs contribute to strong TM-polarized light (with considerable TE component), whereas compressively strained QDs emit TE-polarized light. (b) Polarization-resolved ASE spectra at very low injection (0.01 mA) for a QD-QW laser with two tensile strained QWs coupled to a QD layer. The QD ground-state and the excited-state transitions contribute to TE emission. The QW tensile strain produces strong TM emission from the wells together with a TE emission from the QD-coupled-QW states. (c) Evolution of the polarization-resolved ASE spectrum (in the logarithmic scale) for the QD-QW laser with tensile-strained QWs at three bias currents. As the current increases, TE spectrum shifts to QD-coupled-QW-state emission. The TM emission spectrum from the QW is clear and increases uniformly at the shorter wavelength side of the coupled QD-QW TE emission. (Reprinted with permission from [159] © 2005 IEEE.)

active region of 1.3-μm *p*-doped tunneling-injection QD lasers, which illustrates the phonon-assisted tunneling from the injector layer into the first excited state of QDs. The tunneling into the quantum dot state avoids the hot-carrier effect. Figure 10.54b shows the room temperature PL spectrum of the heterostructure, showing distinct peaks from the dot ground and excited states and the injector layer state. The modulation frequency response of the *p*-doped tunneling injection QD laser is shown in Fig. 10.55a [161], with its resonance frequency plotted as a function of the square root of $(I - I_{th})$ in Fig. 10.55b. The QD laser shows a bandwidth of about 25 GHz

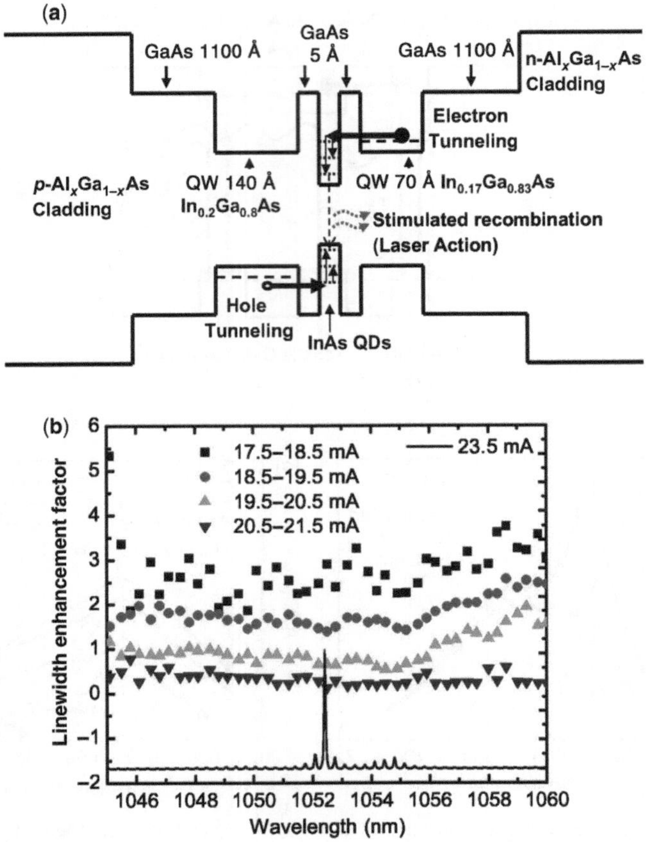

Figure 10.53 (a) Energy diagram of an InAs QW-QD tunneling injection laser. The auxiliary QWs serve as collectors of electron and holes that tunnel into a QD for the stimulated recombination. (b) Linewidth enhancement factor as a function of wavelength at various injection currents obtained for the InAs QW-QD tunneling injection laser. Near threshold, the linewidth enhancement factor goes to almost zero (0.15). (Reprinted with permission from [158] © 2003 American Institute of Physics.)

direct modulation. Carrier dynamics and high-speed modulation of quantum-dot devices remain an intriguing subject of research [163–173].

10.5.4 Quantum-Dot Semiconductor Optical Amplifiers

Quantum-dot semiconductor optical amplifiers (SOAs) [174–182] are of great interest due to the ultrafast (picosecond) gain recovery in the saturation region, which is demonstrated from pump-probe experiments [178, 179]. High bit-rate (>40 Gb/s) and all-optical signal processing functions for multiple wavelengths have been predicted and demonstrated. The expected high performance of QD SOAs is due to the following physical properties, as pointed out in Refs. 176 and 177. First, gain saturation occurs primarily due to the spectral hole burning even under the light pulse

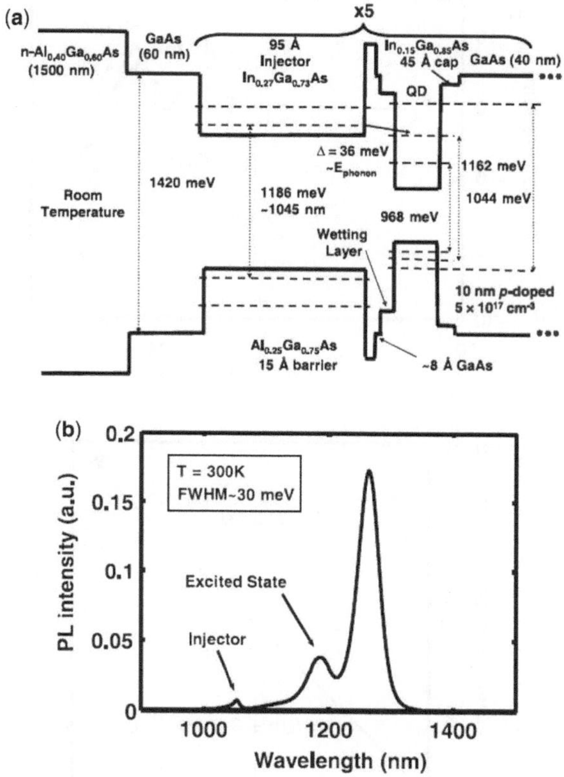

Figure 10.54 (a) The energy band diagram in the active region of 1.3-μm *p*-doped tunneling-injection QD lasers, which illustrates the phonon-assisted tunneling from the injector layer into the first excited state. (b) The room temperature PL spectrum of the heterostructure, which shows distinct peaks from the dot ground and excited states and the injector layer states. (Reprinted with permission from [161] © 2005 IOP Publishing Ltd.)

with a peak power of 0–20 dBm commonly used in optical data transmission. This is due to the low and discrete density of states of QDs. The response time of gain saturation is 100 fs to 1 ps, which is enough for a giga-bit per second (Gb/s) to sub-tera-bit per second (Tb/s) optical transmission system. In addition, the so-called pattern effect is negligible, owing to the compensation of the spectral hole by the carriers relaxing from the excited states, including the states of the wetting layer, or the reservoir states. Second, the spatial isolation of dots prevents the transfer of carriers among dots, leading to a negligible cross-talk between different wavelength channels under gain saturation, when the channels are separated by more than the homogeneous linewidth of the single-dot gain, which is about 10–20 meV at 300K [180]. Third, when the channels are within the linewidth of the homogeneously broadened spectral hole, the interaction between two different wavelength channels via spatially isolated and energetically nonresonant quantum dots through the spectral hole causes the ultrafast cross-gain modulation, which generates switching functions like the wavelength conversion [181]. These features all provide a striking contrast to

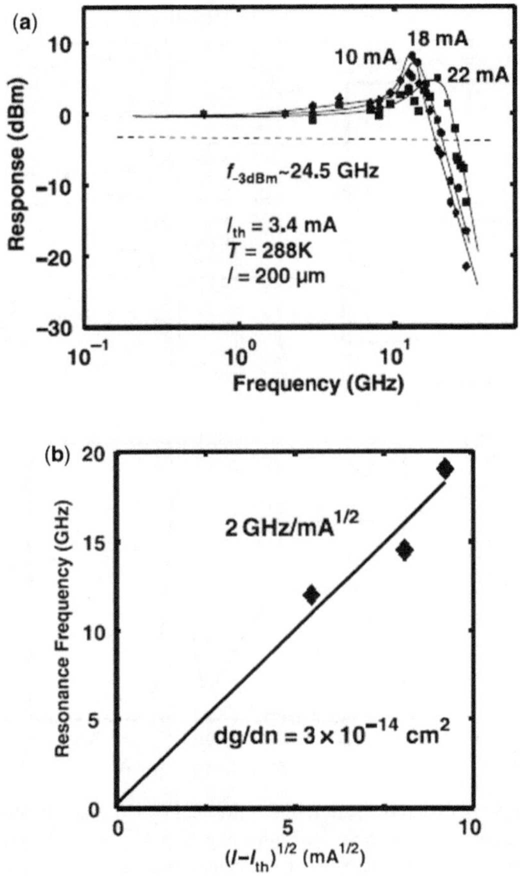

Figure 10.55 (a) Modulation response of a *p*-doped tunneling injection QD laser. (b) Resonance frequency versus the square root of $(I - I_{th})$. (Reprinted with permission from [161] © 2005 IOP Publishing Ltd.)

the bulk or quantum-well SOAs. High performance of QD SOAs for signal processing of 40 Gb/s or even 160 Gb/s have been reported. We refer the readers to comprehensive papers in Refs. 174–177 and 182.

PROBLEMS

10.1 Plot the energy band diagram for a P-$Al_{0.3}Ga_{0.7}As$/i-GaAs/N-$Al_{0.3}Ga_{0.7}As$ double heterojunction under zero bias with the active GaAs layer thickness $d = 1\ \mu m$. Assume that $N_A = 1 \times 10^{18}\ cm^{-3}$ in the P-region and $N_D = 1 \times 10^{18}\ cm^{-3}$ in the N-region. Use the depletion approximation.

10.2 Plot the energy band diagram for a P-InP/i-InGaAsP/N-InP double heterostructure under zero bias with the active layer $d = 1\ \mu m$ using the depletion approximation. Assume that $N_A = 1 \times 10^{18}\ cm^{-3}$ in the P-region and $N_D = 1\times10^{18}\,\text{cm}^{-3}$ in the N-region. The active InGaAsP layer has a band-gap energy of 1.0 eV and is lattice matched to InP. Assume that the band edge discontinuity $\Delta E_c/\Delta E_v$ is 60/40.

10.3 A semiconductor laser operating at 1.55 μm has an internal quantum efficiency $\eta_i = 0.75$, intrinsic absorption coefficient $\alpha_i = 8\,\text{cm}^{-1}$, and the mirror reflectivity $R = 0.3$ on both facets.

(**a**) Plot the inverse external quantum efficiency $1/\eta_e$ versus the cavity length L.

(**b**) Assuming that the threshold current is 3.5 mA at $L = 500\,\mu\text{m}$, plot the optical output power versus the injection current I. Label the slope of the curve.

(**c**) Repeat (**b**) for $I_{th} = 4$ mA and $L = 1000\,\mu\text{m}$.

10.4 If the mirror reflectivities are R_1 and R_2 at the two ends of the semiconductor laser facets, the threshold condition is given by

$$\Gamma g_{th} = \alpha_i + \frac{1}{2L}\ln\frac{1}{R_1 R_2}.$$

(**a**) Find the expressions for the optical output powers, P_1 and P_2, from facets 1 and 2.

(**b**) Plot P_1 versus R_1 for $0.1 \le R_1 \le 1.0$ and P_2 versus R_2 for $0.1 \le R_2 \le 1.0$. You may assume that $\alpha_i = 10\,\text{cm}^{-1}$ and $L = 500\,\mu\text{m}$.

(**c**) Compare P_1 with P_2 if $R_1 = 0.3$ and $R_2 = 0.5$.

10.5 Plot $n(y)$ using (2.4.18) and show that it agrees with the experimental data in Fig. 10.17.

10.6 Describe the mode patterns $|E_y(x, y)|^2$ of the lowest four modes using (10.2.4) for a gain-guided semiconductor laser. Note the Hermite polynomials: $H_0(u) = 1$, $H_1(u) = 2u$, and $H_2(u) = -2 + 4u^2$, $H_3(u) = -12u + 8u^3$.

10.7 (**a**) Explain the physical reasons why $\ln J_{th}$ of a quantum-well laser increases linearly with the inverse cavity length $1/L$ as shown in Fig. 10.28a. What determines the slope of the line?

(**b**) How do you find the cavity length L such that the threshold current I_{th} is minimized? What are the factors determining this minimum threshold current?

(**c**) Find typical numerical values of some quantum-well lasers and replot Fig. 10.28a and Fig. 10.28b.

10.8 Plot the gain spectrum using (10.3.13) for a 100 Å $GaAs/Al_{0.3}Ga_{0.7}As$ quantum-well laser assuming a parabolic band model and an infinite barrier approximation to calculate the subband energies and wave functions. What is the maximum gain coefficient g_{max}?

10.9 (**a**) Calculate the parameters n_c and n_v for a 100 Å GaAs/$Al_{0.3}Ga_{0.7}As$ quantum-well laser at $T = 300K$. Assume single subband occupation.
(**b**) Find the transparency carrier density n_{tr} for part (**a**).

10.10 Discuss how the number of quantum wells can be optimized to minimize the threshold current density at a given threshold gain value using Fig. 10.27.

10.11 Calculate the conduction, heavy-hole and light-hole band-edge energies for $In_{1-x}Ga_xAs/In_{0.52}Al_{0.48}As$ quantum-well structures where $In_{0.52}Al_{0.48}As$ is lattice-matched to the InP substrate, assuming the gallium mole fraction (**a**) $x = 0.37$, (**b**) $x = 0.47$, and (**c**) $x = 0.57$ in the well region. Use the physical parameters in Appendix C (pages 803–804).

10.12 Discuss the advantages of compressively strained and tensile strained quantum-well lasers based on the effective mass, the density of states, and the optical momentum matrix elements.

10.13 Compare the normalized optical momentum matrix element in Fig. 10.35 with the analytical results in Table 9.1 in Section 9.5 using a parabolic band model.

10.14 Discuss the potential advantages of *p*-doping in quantum-dot lasers.

REFERENCES

1. *IEEE J. Quantum Electron.*, special issue on *Semiconductor Lasers*, guest editor, P. D. Dapkus, **QE-23**, June 1987. See the Historical Section, pp. 651–692. It includes the reminiscent accounts on the invention of semiconductor lasers by R. N. Hall, M. I. Nathan, N. Holonyak, Jr., and R. H. Rediker. It also contains the previously unpublished notes of J. von Neumann, written in 1953, which outlined remarkably his ideas on "the photon-disequilibrium-amplification scheme."
2. H. C. Casey, Jr. and M. B. Panish, *Heterostructure Lasers, Part A: Fundamental Principles*, Academic Press, New York, 1978.
3. C. H. Henry, "The origin of quantum wells and the quantum well lasers," Chapter 1 in *Quantum Well Lasers*, P. S. Zory, Jr., Ed. Academic Press, New York, 1993.
4. R. N. Hall, G. E. Fenner, J. D. Kingsley, T. J. Soltys, and R. O. Carlson, "Coherent light emission from GaAs junctions," *Phys. Rev. Lett.* **9**, 366–368 (1962).
5. M. I. Nathan, W. P. Dumke, G. Burns, F. H. Dill, Jr., and G. Lasher, "Stimulated emission of radiation from GaAs p-n junctions," *Appl. Phys. Lett.* **1**, 62–64 (1962).
6. N. Holonyak, Jr., and S. F. Bevacqua, "Coherent (visible) light emission from $Ga(As_{1-x}P_x)$ junctions," *Appl. Phys. Lett.* **1**, 82–83 (1962).
7. T. M. Quist, R. H. Rediker, R. J. Keyes, W. E. Krag, B. Lax, A. L. McWhorter, and H. J. Zeiger, "Semiconductor maser of GaAs," *Appl. Phys. Lett.* **1**, 91–92 (1962).
8. H. Kroemer, "A proposed class of heterojunction injection lasers," *Proc. IEEE* **51**, 1782–1783 (1963).

9. H. Kressel and H. Nelson, "Close confinement gallium arsenide p-n junction lasers with reduced optical loss at room temperature," *RCA Review* **30**, 106–113 (1969).

10. Zh. I. Alferov, V. M. Andreev, E. L. Portnoi, and M. K. Trukan, "AlAs-GaAs heterojunction injection lasers with a low room-temperature threshold," *Sov. Phys. Semicond.* **3**, 1107–1110 (1970).

11. I. Hayashi, M. B. Panish, P. W. Foy, and S. Sumski, "Junction lasers which operate continuously at room temperature," *Appl. Phys. Lett.* **17**, 109–111 (1970).

12. R. Dingle, W. Wiegmann, and C. H. Henry, "Quantum states of confined carriers in very thin $Al_xGa_{1-x}As$-GAs-$Al_xGa_{1-x}As$ heterostructures," *Phys. Rev. Lett.* **33**, 827–830 (1974).

13. E. A. Rezek, H. Shichijo, B. A. Vojak, and N. Holonyak, Jr., "Confined-carrier luminescence of a thin $In_{1-x}GA_xP_{1-z}As_z$ well (x - 0.13, z - 0.29; ~400 Å) in an InP p-n junctions," *Appl. Phys. Lett.* **31**, 534–536 (1977). E. A. Rezek, N. Holonyak, Jr., B. A. Vojak, G. E. Stillman, J. A. Rossi, D. L. Keune, and J. D. Fairing, "LPE $In_{1-x}GA_xP_{1-z}As_zn$ (x - 0.12, z - 0.26) DH laser with multiple thin-layer (<500 Å) active region," *Appl. Phys. Lett.* **31**, 288–290 (1977). E. A. Rezek, N. Holonyak, Jr., B. A. Vojak, and H. Shichijo, "Single and multiple thin layer (L_z<400 Å) $In_{1-x}GA_xP_{1-z}As_z$-InP heterostructure light emitters and lasers ($\lambda - 1.1$ μm, 77K)," *J. Appl. Phys.* **49**, 69–74 (1978).

14. N. Holonyak, Jr., R. M. Kolbas, R. D. Dupuis, and P. D. Dapkus, "Quantum-well heterostructure lasers," *IEEE J. Quantum Electron.* **QE-16**, 170–186 (1980).

15. R. D. Dupuis, P. D. Dapkus, N. Holonyak, Jr., E. A. Rezek, and R. Chin, "Room-temperature laser operation of quantum-well $Ga_{1-x}Al_xAs$-GaAs laser diodes grown by metalorganic chemical vapor deposition," *Appl. Phys. Lett.* **32**, 295–297 (1978).

16. W. T. Tsang, "A graded-index waveguide separate-confinement laser with very low threshold and a narrow Gaussian beam," *Appl. Phys. Lett.* **39**, 134–137 (1981).

17. E. Yablonovitch and E. O. Kane, "Reduction of lasing threshold current density by the lowering of valence band effective mass," *J. Lightwave Technol.* **LT-4**, 504 (1986); Erratum **LT-4**, 961 (1986).

18. A. R. Adams, "Band-structure engineering for low-threshold high-efficiency semiconductor lasers," *Electron. Lett.* **22**, 249–250 (1986).

19. H. K. Choi and C. A. Wang, "InGaAs/AlGaAs strained quantum well diode lasers with extremely low threshold current density and high efficiency," *Appl. Phys. Lett.* **57**, 321–323 (1990).

20. R. L. Williams, M. Dion, F. Chatenoud, and K. Dzurko, "Extremely low threshold current strained InGaAs/AlGaAs lasers by molecular beam epitaxy," *Appl. Phys. Lett.* **58**, 1816–1818 (1991).

21. W. T. Tsang, volume editor, *Lightwave Communications Technology*, in R. K. Willardson and A. C. Beer, Eds., *Semiconductors and Semimetals*, vol. 22, Part A-E, Academic Press, New York, 1985.

22. Y. Arakawa and A. Yariv, "Quantum well lasers—gain, spectra, dynamics," *IEEE J. Quantum Electron.* **QE-22**, 1887–1899 (1986).

23. E. Kapon, "Quantum wire lasers," *Proc. IEEE* **80**, 398–410 (1992).

24. A. Yariv, "Scaling laws and minimum threshold currents for quantum-confined semiconductor lasers," *Appl. Phys. Lett.* **53**, 1033–1035 (1988).

25. W. Wegscheider, L. N. Pfeiffer, M. M. Dignam, A. Pinczuk, K. W. West, S. L. McCall, and R. Hull, "Lasing from excitons in quantum wires," *Phys. Rev. Lett.* **71**, 4071–4074 (1993).

26. K. Iga, S. Ishkawa, S. Ohkouchi, and T. Nishimura, "Room temperature pulsed oscillation of GaAlAs/GaAs surface emitting junction laser," *IEEE J. Quantum Electron.* **QE-21**, 315–320 (1985).

27. S. W. Corzine, R. S. Geels, R. H. Yan, J. W. Scott, L. A. Coldren, and P. L. Gourley, "Efficient, narrow-linewidth distributed-Bragg-reflector surface-emitting laser with periodic gain," *IEEE Photon. Technol. Lett.* **1**, 321–323 (1989).

28. G. A. Evans and J. M. Hammer, Eds., *Surface Emitting Semiconductor Lasers and Arrays*, Academic Press, San Diego, 1993.

29. S. L. McCall, A. F. J. Levi, R. E. Slusher, S. J. Pearton, and R. A. Logan, "Whispering-gallery mode microdisk lasers," *Appl. Phys. Lett.* **60**, 289–291 (1992).

30. A. F. J. Levi, R. E. Slusher, S. L. McCall, T. Tanbun-Ek, D. L. Coblentz, and S. J. Pearton, "Room temperature operation of microdisc lasers with submilliamp threshold current," *Electron. Lett.* **28**, 1010–1011 (1992).

31. Y. Yamamoto and R. E. Slusher, "Optical processes in microcavities," *Physics Today* **46**, 66–73 (1993).

32. M. Yano, H. Imai, and M. Takusagawa, "Analysis of electrical, threshold, and temperature characteristics of InGaAsP/InP double-heterojunction lasers," *IEEE J. Quantum Electron.* **QE-17**, 1954–1963 (1981).

33. G. P. Agrawal and N. K. Dutta, *Long-Wavelength Semiconductor Lasers*, Van Nostrand Reinhold, San Diego, 1986.

34. H. C. Casey, Jr. and M. B. Panish, *Heterostructure Lasers, Part B: Materials and Operating Characteristics*, Academic Press, San Diego, 1978.

35. A. R. Goodwin, J. R. Peters, M. Pion, G. H. B. Thomson, and J. E. A. Whiteway, "Threshold temperature characteristics of double heterostructure $Ga_{1-x}Al_xAs$ lasers," *J. Appl. Phys.* **46**, 3126–3131 (1975).

36. P. J. Anthony, J. R. Pawlik, V. Swaminathan, and W. T. Tsang, "Reduced threshold current temperature dependence in double heterostructure lasers due to separate p-n and heterojunctions," *IEEE J. Quantum Electron.* **QE-19**, 1030–1034 (1983).

37. C. J. Nuese and J. I. Pankove, "Light-emitting diodes-LEDs," Chapter 2 in J. I. Pankov, Ed., *Display Devices*, Springer-Verlag, Berlin, 1980.

38. S. M. Sze and K. Ng, *Physics of Semiconductor Devices*, 2nd ed., Wiley, New York, 2007.

39. J. Wilson and J. F. B. Hawkes, *Optoelectronics: An Introduction*, Prentice-Hall, Englewood Cliffs, NJ, 1983.

40. C. A. Burrus and B. I. Miller, "Small-area, double heterostructure aluminum-gallium arsenide electroluminescent diode sources for optical-fiber transmission lines," *Opt. Commun.* **4**, 307–309 (1971).

41. M. Ettenberg, H. Kressel, and J. P. Wittke, "Very high radiance edge-emitting LED," *IEEE J. Quantum Electron* **QE-12**, 360–364 (1976).

42. J. O'Gorman, S. L. Chuang, and A. F. J. Levi, "Carrier pinning by mode fluctuations in laser diodes," *Appl. Phys. Lett.* **62**, 1454–1456 (1993).

43. S. L. Chuang, J. O'Gorman, and A. F. J. Levi, "Amplified spontaneous emission and carrier pinning in laser diodes," *IEEE J. Quantum Electron.*, special issue on *Semiconductor Lasers* **29**, 1631–1639 (1993).
44. B. Zee, "Broadening mechanism in semiconductor (GaAs) lasers: limitations to single mode power emission," *IEEE J. Quantum Electron.* **QE-14**, 727–736 (1978).
45. B. W. Hakki and T. L. Paoli, "Gain spectra in GaAs double-heterostructure injection lasers," *J. Appl. Phys.* **46**, 1299–1306 (1975).
46. G. Lasher and F. Stern, "Spontaneous and stimulated recombination radiation in semiconductors," *Phys. Rev.* **133**, A553–A563 (1964).
47. C. S. Chang, S. L. Chuang, J. R. Minch, W. W. Fang, Y. K. Chen, and T. Tanbun-Ek, "Amplified spontaneous emission spectroscopy in strained quantum-well lasers," *IEEE J. Sel. Top. Quantum Electron.* **1**, 1106–1107 (1995).
48. T. Keating, S. H. Park, J. Minch, X. Jin, S. L. Chuang, and T. Tanbun-Ek, "Optical gain measurements based on fundamental properties and comparison with many-body theory," *J. Appl. Phys.* **86**, 2945–2952 (1999).
49. H. Yonezu, I. Sakuma, K. Kobayashi, T. Kamejima, M. Ueno, and Y. Nannichi, "A GaAs-$Al_xGa_{1-x}As$ double heterostructure planar stripe laser," *Jpn. J. Appl. Phys.* **12**, 1585–1592 (1973).
50. B. W. Hakki, "Carrier and gain spatial profiles in GaAs stripe geometry lasers," *J. Appl. Phys.* **44**, 5021–5029 (1973).
51. D. D. Cook and F. R. Nash, "Gain-induced guiding and astigmatic output beam of GaAs lasers," *J. Appl. Phys.* **46**, 1660–1672 (1975).
52. N. Chinone, "Nonlinearity in power-output-current characteristics of stripe-geometry injection lasers," *J. Appl. Phys.* **48**, 3237–3243 (1977).
53. W. T. Tsang, "The effects of lateral current spreading, carrier out-diffusion, and optical mode losses on the threshold current density of GaAs-$Al_xGa_{1-x}As$ stripe-geometry DH lasers," *J. Appl. Phys.* **49**, 1031–1044 (1978).
54. T. L. Paoli, "Waveguiding in a stripe-geometry junction laser," *IEEE J. Quantum Electron.* **QE-13**, 662–668 (1977).
55. W. Streifer, D. R. Scifres, and R. D. Burnham, "Analysis of gain-induced waveguiding in stripe-geometry diode lasers," *IEEE J. Quantum Electron.* **QE-14**, 418–427 (1978).
56. W. B. Joyce, "Current-crowded carrier confinement in double-heterostructure lasers," *J. Appl. Phys.* **51**, 2394–2401 (1980).
57. W. Streifer, R. D. Burnham, and D. R. Scifres, "An analytic study of (GaAl)As gain guided lasers at threshold," *IEEE J. Quantum Electron.* **QE-18**, 856–864 (1982).
58. I. P. Kaminow, L. W. Stulz, J. S. Ko, A. G. Dentai, R. E. Nahory, J. C. DeWinter, and R. L. Hartman, "Low-threshold InGaAsP ridge waveguide lasers at 1.3 μm," *IEEE J. Quantum Electron.* **QE-19**, 1312–1319 (1983).
59. H. Hirao, M. S. Tsuji, K. Mizuishi, A. Doi, and M. Nakamura, "Long wavelength InGaAsP/InP lasers for optical fiber communication systems," *J. Opt. Commun.* **1**, 10–14 (1980).
60. H. Kawaguchi and T. Kawakami, "Transverse-mode control in an injection laser by a strip-loaded waveguide," *IEEE J. Quantum Electron.* **QE-13**, 556–560 (1977).
61. M. Asada, A. Kameyama, and Y. Suematsu, "Gain and intervalence band absorption in quantum well lasers," *IEEE J. Quantum Electron.* **20**, 745–753 (1984).

62. K. J. Vahala and C. E. Zah, "Effect of doping on the optical gain and the spontaneous noise enhancement factor in quantum well amplifiers and lasers studied by simple analytical expressions," *Appl. Phys. Lett.* **52**, 1945–1947 (1988).

63. K. Y. Lau, "Ultralow threshold quantum well lasers," Ch. 4, and "Dynamics of quantum well lasers," Chapter 5, in *Quantum Well Lasers*, P. S. Zory, Jr., Ed., Academic Press, San Diego (1993).

64. A. Ghiti, M. Silver, and E. P. O'Reilly, "Low threshold current and high differential gain in ideal tensile- and compressive-strained quantum-well lasers," *J. Appl. Phys.* **71**, 4626–4628 (1992).

65. M. Asada and Y. Suematsu, "The effects of loss and nonradiative recombination on the temperature dependence of threshold current in 1.5–1.6 μm GaInAsP/InP lasers," *IEEE J. Quantum Electron.* **QE-19**, 917–923 (1983).

66. B. Sermage, D. S. Chemla, D. Sivco, and A. Y. Cho, "Comparison of Auger recombination in GaInAs-AlInAs multiple quantum well structure and in bulk GaInAs," *IEEE J. Quantum Electron.* **QE-22**, 774–780 (1986).

67. M. Asada, "Intraband relaxation time in quantum-well lasers," *IEEE J. Quantum Electron.* **25**, 2019–2026 (1989).

68. E. Zielinski, H. Schweizer, S. Hausser, R. Stuber, M. H. Pilkuhn, and G. Weimann, "Systematics of laser operation in GaAs/AlGaAs multiquantum well heterostructures," *IEEE J. Quantum Electron.* **QE-23**, 966–976 (1987).

69. E. Zielinski, F. Keppler, K. Streubel, F. Scholz, R. Sauer, and W. T. Tsang, "Optical gain in strain-free and strained layer $Ga_xIn_{1-x}As/InP$ superlattices," *Superlattices and Microstructures* **5**, 555–559 (1989).

70. P. Blood, S. Colak, and A. I. Kucharska, "Influence of broadening and high-injection effects on GaAs-AlGaAs quantum well lasers," *IEEE J. Quantum Electron.* **24**, 1593–1604 (1988).

71. A. I. Kucharska and D. J. Robbins, "Lifetime broadening in GaAs-AlGaAs quantum well lasers," *IEEE J. Quantum Electron.* **26**, 443–448 (1990).

72. P. W. A. McIlroy, A. Kurobe, and Y. Uematsu, "Analysis and application of theoretical gain curves to the design of multi-quantum-well lasers," *IEEE J. Quantum Electron.* **QE-21**, 1958–1963 (1985).

73. A. Kurobe, H. Furuyama, S. Naritsuka, N. Sugiyama, Y. Kokubun, and M. Nakamura, "Effects of well number, cavity length, and facet reflectivity on the reduction of threshold current of GaAs/AlGsAs multiquantum well lasers," *IEEE J. Quantum Electron.* **QE-24**, 635–640 (1988).

74. M. Rosenzweig, M. Möhrle, H. Düser, and H. Venghaus, "Threshold-current analysis of InGaAs-InGaAsP multiquantum well separate-confinement lasers," *IEEE J. Quantum Electron.* **27**, 1804–1811 (1991).

75. J. S. Osinski, K. M. Dzurko, S. G. Hummel, and P. D. Dapkus, "Optimization of stripe width for low-threshold operation of quantum well laser diodes," *Appl. Phys. Lett.* **56**, 2487–2489 (1990).

76. J. S. Osinski, Y. Zou, P. Grodzinski, A. Mathur, and P. D. Dapkus, "Low-threshold-current-density 1.5 μm lasers using compressively strained InGaAsP quantum wells," *IEEE Photon. Technol. Lett.* **4**, 10–13 (1992).

77. A. Yariv, *Quantum Electronics*, 3rd ed., Wiley, New York, 1989.

78. M. Yamada, K. Tabata, S. Ogita, and M. Yamagishi, "Calculation of lasing gain and threshold current in GaAs-AlGaAs multi-quantum-well lasers," *Trans. IECE Japan* **E68**, 102–108 (1985).

79. G. C. Osbourn, "$In_xGa_{1-x}As$-$In_yGa_{1-y}As$ strained-layer superlattices: a proposal for useful, new electronic materials," *Phys. Rev. B* **27**, 5126–5128 (1983).

80. T. P. Pearsall, Vol. Ed., *Strained Layer Superlattices: Physics*, vol. 32, 1990; and *Strained Layer Superlattices: Materials Science and Technology*, vol. 33, 1991, in R. K. Willardson and A. C. Beer, Eds., *Semiconductor and Semimetals*, Academic Press, New York.

81. E. Yablonovitch and E. O. Kane, "Band structure engineering of semiconductor lasers for optical communications," *J. Lightwave Technol.* **6**, 1292–1299 (1988).

82. P. J. A. Thijs, L. F. Tiemeijer, P. I. Kuindersma, J. J. M. Binsma, and T. Van Dongen, "High-performance 1.5 μm wavelength InGaAs-InGaAsP strained quantum well lasers and amplifiers," *IEEE J. Quantum Electron.* **27**, 1426–1439 (1991).

83. C. E. Zah, R. Bhat, F. J. Favire, Jr., S. G. Menocal, N. C. Andreadakis, K. W. Cheung, D. M. D. Hwang, M. A. Koza, and T. P. Lee, "Low-threshold 1.5 μm compressive-strained multiple- and single-quantum-well lasers," *IEEE J. Quantum Electron.* **27**, 1440–1450 (1991).

84. K. J. Beernink, P. K. York, and J. J. Coleman, "Dependence of threshold current density on quantum well composition for strained-layer InGaAs-GaAs lasers by metalorganic chemical vapor deposition," *Appl. Phys. Lett.* **55**, 2585–2587 (1989).

85. T. Tanbun-Ek, R. A. Logan, H. Temkin, S. N. G. Chu, N. A. Olsson, A. M. Sergent, and K. W. Wecht, "Growth and characterization of continuously graded index separate confinement heterostructure (GRIN-SCH) InGaAs-InP long wavelength strained layer quantum-well lasers by metalorganic vapor phase epitaxy," *IEEE J. Quantum Electron.* **26**, 1323–1327 (1990).

86. C. E. Zah, R. Bhat, K. W. Cheung, N. C. Andreadakis, F. J. Favire, S. G. Menocal, E. Yablonovitch, D. M. Hwang, M. Koza, T. J. Gmitter, and T. P. Lee, "Low-threshold ($\leq$92 A/cm^2) 1.6 μm strained-layer single-quantum-well laser diodes optically pumped by a 0.8 μm laser diode," *Appl. Phys. Lett.* **57**, 1608–1609 (1990).

87. H. Temkin, N. K. Dutta, T. Tanbun-Ek, R. A. Logan, and A. M. Sergent, "InGaAs/InP quantum well lasers with sub-mA threshold current," *Appl. Phys. Lett.* **57**, 1610–1612 (1990).

88. W. T. Tsang, M. C. Wu, T. Tanbun-Ek, R. A. Logan, S. N. G. Chu, and A. M. Sergent, "Low threshold and high power output 1.5 μm InGaAs/InGaAsP separate confinement multiple quantum well laser grown by chemical beam epitaxy," *Appl. Phys. Lett.* **57**, 2065–2067 (1990).

89. A. Kasukawa, R. Bhat, C. E. Zah, M. A. Koza, and T. P. Lee, "Very low threshold current density 1.5 μm GaInAs/AlGaInAs graded-index separate-confinement-heterostructure strained quantum well laser diodes grown by organometallic chemical vapor deposition," *Appl. Phys. Lett.* **59**, 2486–2488 (1991).

90. T. Tanbun-Ek, R. A. Logan, N. A. Olsson, H. Temkin, A. M. Sergent, and K. W. Wecht, "High power output 1.48–1.51 μm continuously graded index separate confinement strained quantum well lasers," *Appl. Phys. Lett.* **57**, 224–226 (1990).

91. M. Joma, H. Horikawa, Y. Matsui, and T. Kamijoh, "High-power 1.48 μm multiple quantum well lasers with strained quaternary wells entirely grown by metalorganic vapor phase epitaxy," *Appl. Phys. Lett.* **58**, 2220–2222 (1991).

92. A. Moser, A. Oosenbrug, E. E. Latta, Th. Forster, and M. Gasser, "High-power operation of strained InGaAs/AlGaAs single quantum well lasers," *Appl. Phys. Lett.* **59**, 2642–2644 (1991).
93. H. K. Choi, C. A. Wang, D. F. Kolesar, R. L. Aggarwal, and J. N. Walpole, "High-power, high-temperature operation of AlInGaAs-AlGaAs strained single-quantum-well diode lasers," *IEEE Photon. Technol. Lett.* **3**, 857–859 (1991).
94. H. Temkin, T. Tanbun-Ek, R. A. Logan, D. A. Cebula, and A. M. Sergent, "High temperature operation of lattice matched and strained InGaAs-InP quantum well lasers," *IEEE Photon. Technol. Lett.* **3**, 100–102 (1991).
95. P. L. Derry, R. J. Fu, C. S. Hong, E. Y. Chan, and L. Figueroa, "Analysis of the high temperature characteristics of InGaAs-AlGaAs strained quantum-well lasers," *IEEE J. Quantum Electron.* **28**, 2698–2705 (1992).
96. H. Temkin, D. Coblentz, R. A. Logan, J. P. van der Ziel, T. Tanbun-Ek, R. D. Yadvish, and A. M. Sergent, "High temperature characteristics of InGaAsP/InP laser structures," *Appl. Phys. Lett.* **62**, 2402–2404 (1993).
97. H. Nobuhara, K. Tanaka, T. Yamamoto, T. Machida, T. Fujii, and K. Wakao, "High-temperature operation of InGaAs/InGaAsP compressive-strained QW lasers with low threshold currents," *IEEE Photon. Technol. Lett.* **5**, 961–962 (1993).
98. D. Ahn and S. L. Chuang, "Optical gain in a strained-layer quantum-well laser," *IEEE J. Quantum Electron.* **24**, 2400–2406 (1988).
99. T. C. Chong and C. G. Fonstad, "Theoretical gain of strained-layer semiconductor lasers in the large strain regime," *IEEE J. Quantum Electron.* **25**, 171–178 (1989).
100. E. P. O'Reilly and A. R. Adams, "Band-structure engineering in strained semiconductor lasers," *IEEE J. Quantum Electron.* **30**, 366–379 (1994).
101. S. W. Corzine, R. H. Yan, and L. A. Coldren, "Theoretical gain in strained InGaAs/AlGaAs quantum wells including valence-band mixing effects," *Appl. Phys. Lett.* **57**, 2835–2837 (1990).
102. N. K. Dutta, J. Lopata, D. L. Sivco, and A. Y. Cho, "Temperature dependence of threshold of strained quantum well lasers," *Appl. Phys. Lett.* **58**, 1125–1127 (1991).
103. I. Suemune, "Theoretical study of differential gain in strained quantum well structures," *IEEE J. Quantum Electron.* **27**, 1149–1159 (1991).
104. E. P. O'Reilly, G. Jones, A. Ghiti, and A. R. Adams, "Improved performance due to suppression of spontaneous emission in tensile-strain semiconductor lasers," *Electron. Lett.* **27**, 1417–1419 (1991).
105. L. F. Lester, S. D. Offsey, B. K. Ridley, W. J. Schaff, B. A. Foreman, and L. F. Eastman, "Comparison of the theoretical and experimental differential gain in strained layer InGaAs/GaAs quantum well lasers," *Appl. Phys. Lett.* **59**, 1162–1164 (1991).
106. J. W. Matthews and A. E. Blakeslee, "Defects in epitaxial multilayers," *J. Crystal Growth* **27**, 118 (1974).
107. R. People and J. C. Bean, "Calculation of critical layer thickness versus lattice mismatch for Ge_xSi_{1-x}/Si strained-layer heterostructures," *Appl. Phys. Lett.* **47**, 322–324 (1985); Erratum, *Appl. Phys. Lett.* **49**, 229 (1986).
108. M. G. A. Bernard and G. Duraffourg, "Laser conditions in semiconductors," *Phys. Stat. Sol.* **1**, 699–703 (1961).

109. S. L. Chuang, "Efficient band-structure calculations of strained quantum wells," *Phys. Rev. B* **43**, 9649–9661 (1991).

110. C. Y. P. Chao and S. L. Chuang, "Spin-orbit-coupling effects on the valence-band structure of strained semiconductor quantum wells," *Phys. Rev. B* **46**, 4110–4122 (1992).

111. T. Tanbun-Ek, N. A. Olsson, R. A. Logan, K. W. Wecht, and A. M. Sergent, "Measurements of the polarization dependent of the gain of strained multiple quantum well InGaAs-InP lasers," *IEEE Photon Technol. Lett.* **3**, 103–105 (1991).

112. N. B. Patel, J. E. Ripper, and P. Brosson, "Behavior of threshold current and polarization of stimulated emission of GaAs injection lasers under uniaxial stress," *IEEE J. Quantum Electron.* **QE-9**, 338–341 (1973).

113. J. M. Liu and Y. C. Chen, "Digital optical signal processing with polarization-bistable semiconductor lasers," *IEEE J. Quantum Electron.* **QE-21**, 298–306 (1985).

114. J. S. Osinski, P. Grodzinski, Y. Zou, and P. D. Dapkus, "Threshold current analysis of compressive strain (0–1.8%) in low-threshold, long-wavelength quantum well lasers," *IEEE J. Quantum Electron.* **29**, 1576–1585 (1993).

115. Y. Zou, J. S. Osinski, P. Grodzinski, P. D. Dapkus, W. Rideout, W. F. Sharfin, and F. D. Crawford, "Effect of Auger recombination and differential gain on the temperature sensitivity of 1.5 μm quantum well lasers," *Appl. Phys. Lett.* **62**, 175–177 (1993).

116. G. Fuchs, C. Schiedel, A. Hangleiter, V. Harle, and F. Scholz, "Auger recombination in strained and unstrained InGaAs/InGaAsP multiple quantum-well lasers," *Appl. Phys. Lett.* **62**, 396–398 (1993).

117. M. C. Wang, K. Kash, C. E. Zah, R. Bhat, and S. L. Chuang, "Measurement of nonradiative Auger and radiative recombination rates in strained-layer quantum-well systems," *Appl. Phys. Lett.* **62**, 166–168 (1993).

118. Y. Zou, J. S. Osinski, P. Grodzinski, P. D. Dapkus, W. C. Rideout, W. F. Sharfin, and F. D. Crawford, "Experimental study of Auger recombination, gain and temperature sensitivity of 1.5 μm compressively strained semiconductor lasers," *Appl. Phys. Lett.* **62**, 175–177 (1993).

119. J. O'Gorman, A. F. J. Levi, T. Tanbun-Ek, D. L. Coblentz, and R. A. Logan, "Temperature dependence of long wavelength semiconductor lasers," *Appl. Phys. Lett.* **60**, 1058–1060 (1992).

120. D. Morris, B. Deveaud, A. Regreny, and P. Auvray, "Electron and hole capture in multiple-quantum-well structures," *Phys. Rev. B* **47**, 6819–6822 (1993).

121. R. Nagarajan, T. Fukushima, M. Ishikawa, J. E. Bowers, R. S. Geels, and L. A. Coldren, "Transport limits in high-speed quantum-well lasers: experiment and theory," *IEEE Photon. Technol. Lett.* **4**, 121–123 (1992).

122. S. Schmitt-Rink, C. Ell, and H. Haug, "Many-body effects in the absorption, gain, and luminescence spectra of semiconductor quantum-well structures," *Phys. Rev. B* **33**, 1183–1189 (1986).

123. M. F. Pereira, Jr., S. W. Koch, and W. W. Chow, "Many-body effects in the gain spectra of strained quantum wells," *Appl. Phys. Lett.* **59**, 2941–2943 (1991).

124. Z. Alferov, "Double heterostructure lasers: early days and future perspectíves," *IEEE J. Sel. Top. Quantum Electron.* **6**, 832–840 (2000).

125. Y. Arakawa and H. Sakaki, "Multidimensional quantum well laser and temperature dependence of its threshold current," *Appl. Phys. Lett.* **40**, 939–941 (1982).

126. M. Asada, Y. Miyamoto, and Y Suamatsu, "Gain and threshold of three-dimensional quantum-box lasers," *IEEE J. Quantum Electron.* **QE-22**, 1915–1921 (1986).

127. T. Takahashi and Y. Arakawa, "Theoretical analysis of gain and dynamic properties of quantum-well box lasers," *Optoelectron. Devices. Technol.* **3**, 155–162 (1988).

128. Ivan N. Stranski and L. Von Krastanow, "Theory of orientation separation of ionic crystals," *Sitzber. Akad. Wiss. Wien, Math.-naturw. Klasse, Abt. IIb* **146**, 797–810, (1938).

129. N. Kirstaedter, N. N. Ledentsov, M. Grundmann, D. Bimberg, V. M. Ustinov, S. S. Rumimov, M. V. Maximov, P. S. Kop'ev, Zh. I. Alferov, U. Richter, P. Werner, U. Gösele, and J. Heydenreich, "Low threshold, large T_0 injection laser emission from (InGa)As quantum dots," *Electron. Lett.* **30**, 1416–1417 (1994).

130. N. Kirstaedter, O. G. Schmidt, N. N. Ledentsov, D. Bimberg, V. M. Ustinov, A. Y. Egorov, A. E. Zhukov, M. V. Maximov, P. S. Kopev, and Z. I. Alferov, "Gain and differential gain of single layer InAs/GaAs quantum dot injection lasers," *Appl. Phys. Lett.* **69**, 1226–1228 (1996).

131. N. N. Ledentsov, V. A. Shchukin, M. Grundmann, N. Kirstaedter, J. Böhrer, O. Schmidt, D. Bimberg, V. M. Ustinov, A. Yu. Egorov, A. E. Zhukov, P. S. Kop'ev, S. V. Zaitsev, N. Yu. Gordeev, Zh. I. Alferov, A. I. Borovkov, A. O. Kosogov, S. S. Ruvimov, P. Werner, U. Gösele, and J. Heydenreich, "Direct formation of vertically coupled quantum dots in Stranski-Krastanow growth," *Phys. Rev. B* **54**, 8743–8750 (1996).

132. D. Bimberg, N. Kirstaedter, N. N. Ledentsov, Zh. I. Alferov, P. S. Kop'ev, and V. M. Ustinov, "InGaAs-GaAs quantum-dot lasers," *IEEE J. Sel. Top. Quantum Electron.* **3**, 196–205 (1997).

133. V. M. Ustinov, N. A. Maleev, A. E. Zhukov, A. R. Kovsh, A. Yu. Egorov, A. V. Lunev, B. V. Volovik, I. L. Krestnikov, Yu. G. Musikhin, N. A. Bert, P. S. Kop'ev, Zh. I. Alferov, N. N. Ledentsov, and D. Bimberg, "InAs/InGaAs quantum dot structures on GaAs substrates emitting at 1.3 μm," *Appl. Phys. Lett.* **74**, 2815–2817 (1999).

134. D. Bimberg, M. Grundmann, and N. N. Ledentsov, *Quantum Dot Heterostructures*, Wiley, New York, 2001.

135. D. Bimberg, M. Grundmann, N. N. Ledentsov, M. H. Mao, Ch. Ribbat, R. Sellin, V. M. Ustinov, A. E. Zhukov, Zh. I. Alferov, and J. A. Lott, "Novel infrared quantum dot Lasers: theory and reality," *Phys. Stat. Sol.* (b) **224**, 787 (2001).

136. V. P. Evtikhiev, A. M. Boiko, I. V. Kudryashov, A. K. Kryganovskii, R. A. Suris, A. N. Titkov, and V. E. Tokranov, "Control of density, size and size uniformity of MBE-grown InAs quantum dots by means of substrate misorientation," *Semicond. Sci. Technol.* **17**, 545–550 (2002).

137. N. N. Ledendsov, "Long-wavelength quantum-dot lasers on GaAs substrate: From media to device concepts," *IEEE J. Sel. Top. Quantum Electron.* **8**, 1015 (2002).

138. N. N. Ledentsov, F. Hopfer, D. Bimberg, "High-speed quantum-dot vertical-cavity surface-emitting lasers," *Proc. IEEE* **95**, 1741–1756 (2007).

139. A. Stintz, G. Liu, H. Li, L. Lester, and K. Malloy, "Low-threshold current density 1.3-μm InAs quantum-dot lasers with the dots-in-a-well (DWELL) structure," *IEEE Photonics Technol. Lett.* **12**, 591–593 (2000).

140. A. Fiore, U. Oesterle, R. P. Stanley, R. Houdre, F. Lelarge, M. Ilegems, P. Borri, W. Langbein, D. Birkedal, J. M. Hvam, M. Cantoni, and F. Bobard, "Structural and electrooptical characteristics of quantum dots emitting at 1.3 μm on gallium arsenide," *IEEE J. Quantum Electron.* **37**, 1050–1058 (2001).

141. S. Osborne, P. Blood, P. Smowton, J. Lutti, Y. C. Xin, A. Stintz, D. Huffaker, and L. F. Lester, "State filling in InAs quantum-dot laser structures," *IEEE J. Quantum Electron.* **40**, 1639–1645 (2004).

142. M. Sugawara, N. Hatori, H. Ebe, M. Ishida, Y. Arakawa, T. Akiyama, K. Otsubo, and Y. Nakata, "Modeling room-temperature lasing spectra of 1.3-μm self-assembled InAs/GaAs quantum-dot lasers: Homogeneous broadening of optical gain under current injection," *J. Appl. Phys.* **97**, 043523 (2005).

143. M. Sugawara, K. Mukai, and Y. Nakata, "Light emission spectra of columnar-shaped self-assembled InGaAs/GaAs quantum-dot lasers: Effect of homogeneous broadening of the optical gain on lasing characteristics," *Appl. Phys. Lett.* **74**, 1561–1563 (1999).

144. A. Capua, L. Rozenfeld, V. Mikhelashvili, G. Eisenstein, M. Kuntz, M. Laemmlin, and D. Bimberg, "Direct correlation between a highly damped modulation response and ultra low relative intensity noise in an InAs/GaAs quantum dot laser," *Opt. Express* **15**, 5388–5393 (2007).

145. A. Sakamoto and M. Sugawara, "Theoretical calculation of lasing spectra of quantum-dot lasers: Effect on homogeneous broadening of optical gain," *IEEE Photon. Technol. Lett.* **12**, 107–109 (2000).

146. A. Markus, J. X. Chen, C. Paranthoen, A. Fiore, C. Platz, and O. Gauthier-Lafaye, "Simultaneous two-state lasing in quantum-dot lasers," *Appl. Phys. Lett.* **82**, 1818–1820 (2003).

147. O. B. Shchekin and D. G. Deppe, "1.3 μm InAs quantum dot laser with $T_0 = 161$K from 0 to 80°C," *Appl. Phys. Lett.* **80**, 3277 (2002).

148. D. G. Deppe, H. Huang, and O. B. Shchekin, "Modulation characteristics of quantum-dot lasers: the influence of p-type doping and the electronic density of states on obtaining high speed," *IEEE J. Quantum Electron.* **38**, 1587–1593 (2002).

149. D. G. Deppe, S. Freisem, H. Huang, and S. Lipson, "Electron transport due to inhomogeneous broadening and its potential impact on modulation speed in p-doped quantum dot lasers," *J. Phys. D Appl. Phys.* **38**, 2119–2125 (2005).

150. S. S. Mikhrin, A. R. Kovsh, I. L. Krestnikov, A. V. Kozhukhov, D. A. Livshits, N. N. Ledentsov, Y. M. Shernyakov, I. I. Novikov, M. V. Maximov, V. M. Ustinov, and Z. I. Alferov, "High power temperature-insensitive 1.3 μm InAs/InGaAs/GaAs quantum dot lasers," *Semicond. Sci. Technol.* **20**, 340–342 (2005).

151. P. M. Smowton, I. C. Sandall, D. J. Mowbray, H. Y. Liu, and Mark Hopkinson, "Temperature-dependent gain and threshold in P-doped quantum dot lasers," *IEEE J. Sel. Top. Quantum Electron.* **13**, 1261–1266 (2007).

152. J. Kim and S. L. Chuang, "Theoretical and experimental study of optical gain, refractive index change, and linewidth enhancement factor of p-doped quantum-dot lasers," *IEEE J. Quantum Electron.* **42**, 942–952 (2006).

153. S. Fathpour, Z. Mi, and P. Bhattacharya, "Small-signal modulation characteristics of p-doped 1.1- and 1.3-μm quantum-dot lasers," *IEEE Photon. Technol. Lett.* **17**, 2250–2252 (2005).

154. T. Chung, G. Walter, and N. Holonyak, Jr., "Coupled strained-Layer InGaAs quantum-well improvement of an InAs quantum dot AlGaAs-GaAs-InAs heterostructure Laser," *Appl. Phys. Lett.* **79**, 4500–4502 (2001).

155. L. V. Asryan and S. Luryi, "Tunneling-injection quantum-dot laser: ultrahigh temperature stability," *IEEE J. Quantum Electron.* **37**, 905–910 (2001).

156. G. Walter, T. Chung, and N. Holonyak, Jr., "High-gain coupled InGaAs quantum well InAs quantum dot AlGaAs–GaAs–InGaAs–InAs heterostructure diode laser operation," *Appl. Phys. Lett.* **80**, 1126 (2002).

157. G. Walter, T. Chung, and N. Holonyak, Jr., "Coupled-stripe quantum-well-assisted AlGaAs–GaAs–InGaAs–InAs quantum-dot laser," *Appl. Phys. Lett.* **80**, 3045 (2002).

158. P. K. Kondratko, S. L. Chuang, G. Walter, T. Chung, and N. Holonyak, "Observations of near-zero linewidth enhancement factor in a quantum-well coupled quantum-dot laser," *Appl. Phys. Lett.* **83**, 4818–4820 (2003).

159. P. K. Kondratko, S. L. Chuang, G. Walter, N. Holonyak, R. D. Heller, X. B. Zhang, and R. D. Dupuis, "Gain narrowing and output behavior of InP-InGaAlP tunneling injection quantum-dot-well laser," *IEEE Photonics Technol. Lett.* **17**, 938–940 (2005).

160. J. Kim, P. K. Kondratko, S. L. Chuang, G. Walter, N. Holonyak, R. D. Heller, X. B. Zhang, and R. D. Dupuis, "Tunneling injection quantum-dot lasers with polarization-dependent photon-mediated carrier redistribution and gain narrowing." *IEEE J. Quantum Electron.* **41**, 1369–1379 (2005).

161. S. Fathpour, Z. Mi, and P. Bhattacharya, "High-speed quantum dot lasers," *J. Phys. D Appl. Phys.* **38**, 2103–2111 (2005).

162. Z. Mi and P. Bhattacharya, "Analysis of the linewidth-enhancement factor of long-wavelength tunnel-injection quantum-dot lasers," *IEEE J. Quantum Electron.* **43**, 363–369 (2007).

163. A. Martinez, A. Lemaître, K. Merghem, L. Ferlazzo, C. Dupuis, A. Ramdane, J.-G. Provost, B. Dagens, O. Le Gouezigou, and O. Gauthier-Lafaye, "Static and dynamic measurements of the alpha-factor of five-quantum-dot-layer single-mode lasers emitting at 1.3 mm on GaAs," *Appl. Phys. Lett.* **86**, 211115 (2005).

164. M. Ishida, M. Sugawara, T. Yamamoto, N. Hatori, H. Ebe, Y. Nakata, and Y. Arakawa, "Theoretical study on high-speed modulation of Fabry-Perot and distributed-feedback quantum-dot lasers: K-factor-limited bandwidth and 10 Gbit/s eye diagrams," *J. Appl. Phys.* **101**, 013108 (2007).

165. P. Borri, W. Langbein, J. M. Hvam, F. Heinrichsdorff, M.-H. Mao, and D. Bimberg, "Ultrafast gain dynamics in InAs/InGaAs quantum dot amplifiers," *IEEE Photon. Technol. Lett.* **12**, 594–596 (2000).

166. A. Markus, J. X. Chen, O. Gauthier-Lafaye, J.-G. Provost, C. Paranthoën, and A. Fiore, "Impact of intraband relaxation on the performance of a quantum-dot laser," *IEEE J. Sel. Top. Quantum Electron.* **9**, 1308–1314 (2003).

167. H. Su, L. Zhang, R. Wang, T. C. Newell, A. L. Gray, and L. F. Lester, "Linewidth study of InAs-InGaAs quantum dot distributed feedback lasers," *IEEE Photon. Technol. Lett.* **16**, 2206–2208 (2004).

168. H. Su, L. Zhang, A. L. Gray, R. Wang, T. C. Newell, K. Malloy, and L. F. Lester, "High external feedback resistance of laterally loss-coupled distributed feedback quantum dot semiconductor lasers," *IEEE Photon. Technol. Lett.* **15**, 1504–1506 (2003).

169. M. Ishida, N. Hatori, T. Akiyama, K. Otsubo, Y. Nakata, H. Ebe, M. Sugawara, and Y. Arakawa, "Photon lifetime dependence of modulation efficiency and K factor in 1.3 μm self-assembled InAs/GaAs quantum-dot lasers: Impact of capture time and maximum modal gain on modulation bandwidth," *Appl. Phys. Lett.* **85**, 4145–4147 (2004).

170. S. Schneider, P. Borri, W. Langbein, U. Woggon, R. L. Sellin, D. Ouyang, and D. Bimberg, "Linewidth enhancement factor in InGaAs quantum-dot lasers," *IEEE J. Quantum Electron.* **40**, 1423–1429 (2004).

171. A. Martinez, A. Lemaitre, K. Merghem, L. Ferlazzo, D. Dupuis, A. Ramdane, J.-G. Provost, B. Dagens, O. L. Gouezigou, and O. Gauthier-Lafaye, "Static and dynamic measurements of the α-factor of five quantum-dot-layer single-mode lasers emitting at 1.3 μm on GaAs," *Appl. Phys. Lett.* **86**, 211115 (2005).

172. J. J. Berry, Martin J. Stevens, R. P. Mirin, and K. L. Silverman, "High-resolution spectral hole burning in InGaAs-GaAs quantum dots," *Appl. Phys. Lett.* **88**, 061114 (2006).

173. A. Capua, L. Rozenfeld, V. Mikhelashvili, G. Eisenstein, M. Kuntz, M. Laemmlin, and D. Bimberg, "Direct correlation between a highly damped modulation response and ultra low relative intensity noise in an InAs/GaAs quantum dot laser," *Opt. Express* **15**, 5388–5393 (2007).

174. T. Akiyama, H. Kuwatsuka, T. Simoyama, Y. Nakata, K. Mukai, M. Sugawara, O. Wada, and H. Ishikawa, "Nonlinear gain dynamics in quantum-dot optical amplifiers and its application to optical communication devices," *IEEE J. Quantum Electron.* **37**, 1059–1065 (2001).

175. T. Akiyama, H. Kuwatsuka, N. Hatori, Y. Nakata, H. Ebe, and M. Sugawara, "Pattern-effect-free semiconductor optical amplifier achieved by using quantum dots," *Electron. Lett.* **38**, 1139–1140 (2002).

176. M. Sugawara, T. Akiyama, N. Hatori, Y. Nakata, H. Ebe, and H. Ishikawa, "Quantum-dot semiconductor optical amplifiers for high-bit-rate signal processing up to 160 Gbs^{-1} and a new scheme of 3R regenerators," *Meas. Sci. Technol.* **13**, 1683–1691 (2002).

177. M. Sugawara, N. Hatori, M. Ishida, H. Ebe, Y. Arakawa, T. Akiyama, K. Otsubo, T. Yamamoto, and Y. Nakata, "Recent progress in self-assembled quantum-dot optical devices for optical telecommunication: temperature-insensitive 10 Gbs^{-1} directly modulated lasers and 40 Gbs^{-1} signal-regenerative amplifiers," *J. Phys. D. Appl. Phys.* **38**, 2126–2134 (2005).

178. S. Dommers, V. V. Temnov, U. Woggon, J. Gomis, J. Martinez-Pastor, M. Laemmlin, and D. Bimberg, "Complete ground state gain recovery after ultrashort double pulses in quantum dot based semiconductor optical amplifier," *Appl. Phys. Lett.* **90**, 033508 (2007).

179. T. Vallaitis, C. Koos, R. Bonk, W. Freude, M. Laemmlin, C. Meuer, D. Bimberg, and J. Leuthold, "Slow and fast dynamics of gain and phase in a quantum dot semiconductor optical amplifier," *Opt. Express* **16**, 170–178 (2008).

180. R. Alizon, D. Hadass, V. Mikhelashvili, G. Eisenstein, R. Schwertberger, A. Somers, J. P. Reithmaier, A. Forchel, M. Calligaro, S. Bansropun, and M. Krakowski, "Multiple wavelength amplification in a wide band high power 1550-nm quantum dash optical amplifier," *Electron. Lett.* **40**, 760–761 (2004).

181. T. Akiyama, N. Hatori, Y. Nakata, H. Ebe, and M. Sugawara, "Pattern-effect-free amplification and cross-gain modulation achieved by using ultrafast gain nonlinearity in quantum-dot semiconductor optical amplifiers," *Phys. Stat. Sol. (b)* **238**, 301–304 (2003).

182. T. Akiyama, M. Sugawara, and Y. Arakawa, "Quantum-dot semiconductor optical amplifiers," *Proc. IEEE* **95**, 1757–1766 (2007).

11

Advanced Semiconductor Lasers

11.1 DISTRIBUTED FEEDBACK LASERS

In Fabry–Perot type semiconductor lasers, there are many longitudinal modes within the gain spectrum, which is generally broad. Although there are some side-mode suppressions in continuous wave (cw) operation of the semiconductor lasers above threshold, the powers of the side modes increase rapidly when the laser diode is pulsed at high bit rates. In optical communication systems with common chromatic dispersion in optical fibers, these undesirable side-mode power partitions limit the bandwidth of the information transmission rate. Therefore, mode selectivity provided by periodic structures such as distributed feedback lasers or distributed Bragg reflectors are attractive for applications for single longitudinal mode operation of diode lasers. In Section 8.5, we have discussed distributed feedback (DFB) structures and studied the electric field and the reflection coefficient. The reflection coefficient depends on the coupling coefficient K and the detuning δ of the propagation constant β from the Bragg wave number $\ell\pi/\Lambda$, where ℓ refers to the order of the grating mode used in the DFB process. We can see that the periodic structures provide an optical frequency selection property.

The basic concepts of the DFB lasers were presented by Kogelnik and Shank [1, 2] in 1971, and they demonstrated the idea by using dyed gelatin on a glass substrate. Optical pumping of semiconductor DFB laser was reported by Nakamura et al. [3] in 1973. Electrical injection of a GaAs/AlGaAs single-heterojunction DFB laser was achieved by Scifres et al. [4] in 1974; and the cw operation at room temperature by Casey et al. [5] and Nakamura et al. [6] in 1975. The DFB laser is the most useful telecommunication source as it has been widely used in commercial long-distance optical communication systems near 1.55 μm wavelength.

11.1.1 Fundamental Formulation for DFB Lasers

A schematic diagram of a distributed feedback (DFB) semiconductor laser [7–12] is illustrated in Fig. 11.1, where a grating above the active region provides the optical distributed feedback coupling. Using the coupled-mode theory in

---

Physics of Photonic Devices, Second Edition. By Shun Lien Chuang
Copyright © 2009 John Wiley & Sons, Inc.

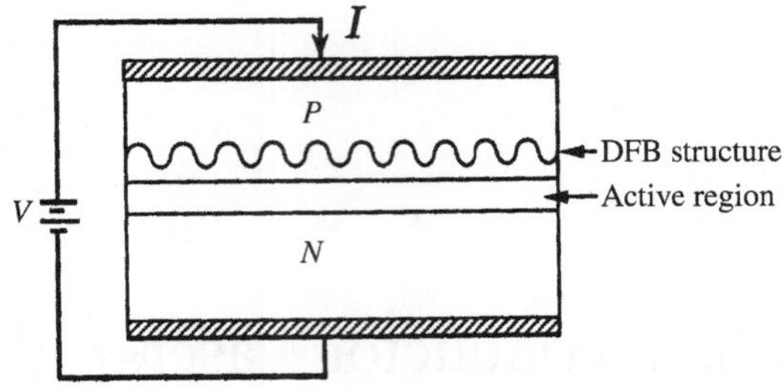

Figure 11.1 A schematic diagram for a distributed feedback (DFB) semiconductor laser, where a periodic grating structure near the active region provides the optical distributed feedback process.

Section 8.5, we summarize the results for the optical electric field in a distributed feedback structure

$$E_y(x) = A(z)e^{i\beta_B z}U(x) + B(z)e^{-i\beta_B z}U(x) \tag{11.1.1}$$

where $U(x) = E_y^{(\mathrm{TE}_0)}(x)$ is the mode pattern of the TE_0 mode. $\beta_B = \ell\pi/\Lambda$ is the Bragg wave vector for the ℓth order grating. The amplitudes $A(z)$ and $B(z)$ of the forward and backward propagating waves satisfy the coupled-mode equation (8.5.17)

$$\frac{d}{dz}\begin{bmatrix} A(z) \\ B(z) \end{bmatrix} = i\begin{bmatrix} \Delta\beta & K_{ab} \\ K_{ba} & -\Delta\beta \end{bmatrix}\begin{bmatrix} A(z) \\ B(z) \end{bmatrix}. \tag{11.1.2}$$

Index Gratings If the index grating inside the waveguide is

$$\Delta n(z) = \Delta n \cos\left(\frac{2\pi}{\Lambda}z + \varphi\right) \tag{11.1.3}$$

we obtain the coupling coefficients (see Section 8.5)

$$K_{ab} = \Gamma k_0 \Delta n_{+1} = K e^{i\varphi} \tag{11.1.4a}$$

$$K_{ba} = -\Gamma k_0 \Delta n_{-1} = -K e^{-i\varphi} \tag{11.1.4b}$$

where $K = \Gamma k_0\ \Delta n/2$ is the magnitude (real), and $K_{ba} = -K_{ab}^*$. The phase of the grating depends on the choice of the period with respect to the coordinates and the facet of the DFB laser. It can be chosen as zero if the period is set appropriately. For lossless index grating, stop band exists when $|\Delta\beta| < |K|$, because

$$q = \sqrt{(\Delta\beta)^2 + K_{ab}K_{ba}} = \sqrt{(\Delta\beta)^2 - |K|^2}, \tag{11.1.5}$$

and the wave propagation becomes highly attenuated within the stop band. For a DFB laser with gain, we use the detuning parameter $\delta = \beta_0 - \beta_B$, where β_0 is the propagation constant of the guided fundamental (TE_0) mode

$$\Delta\beta = \delta - i\frac{g_n}{2} \tag{11.1.6}$$

where

$$g_n = \Gamma g - \alpha \tag{11.1.7a}$$

is the *net modal gain* of the guided mode, Γ is the optical confinement factor, and α is the loss coefficient in the guide α_i and in the substrate α_s

$$\alpha = \Gamma\alpha_i + (1 - \Gamma)\alpha_s. \tag{11.1.7b}$$

Matrix Formulation for DFB Structures The general solution to the coupled-mode equations is expressed as the sum of the two eigenmodes

$$\begin{bmatrix} A(z) \\ B(z) \end{bmatrix} = \begin{bmatrix} A^+ \\ B^+ \end{bmatrix} e^{iqz} + \begin{bmatrix} A^- \\ B^- \end{bmatrix} e^{-iqz} \tag{11.1.8}$$

where

$$\frac{B^+}{A^+} = \frac{\Delta\beta - q}{-K_{ab}} = \frac{K_{ba}}{\Delta\beta + q} \quad \text{for the } e^{+iqz} \text{ eigenmode} \tag{11.1.9a}$$

and

$$\frac{B^-}{A^-} = \frac{\Delta\beta + q}{-K_{ab}} = \frac{K_{ba}}{\Delta\beta - q} \quad \text{for the } e^{-iqz} \text{ eigenmode.} \tag{11.1.9b}$$

We can define the DFB "reflection" coefficient for the "+" eigenmode

$$r_p(q) = \frac{B^+}{A^+} = \frac{K_{ba}}{\Delta\beta + q} \tag{11.1.10a}$$

and the "−" eigenmode

$$r_m(q) = \frac{A^-}{B^-} = \frac{-K_{ab}}{\Delta\beta + q}. \tag{11.1.10b}$$

We can write the general solution (11.1.8) as [8, 11]

$$\begin{bmatrix} A(z) \\ B(z) \end{bmatrix} = A^+ \begin{bmatrix} 1 \\ r_p \end{bmatrix} e^{iqz} + B^- \begin{bmatrix} r_m \\ 1 \end{bmatrix} e^{-iqz}. \tag{11.1.11}$$

By setting $z = 0$, we find

$$\begin{bmatrix} A(0) \\ B(0) \end{bmatrix} = \begin{bmatrix} 1 & r_m \\ r_p & 1 \end{bmatrix} \begin{bmatrix} A^+ \\ B^- \end{bmatrix}. \tag{11.1.12}$$

The general solution can be written as

$$\begin{bmatrix} A(L) \\ B(L) \end{bmatrix} = \begin{bmatrix} F_{11} & F_{12} \\ F_{21} & F_{22} \end{bmatrix} \begin{bmatrix} A(0) \\ B(0) \end{bmatrix} \tag{11.1.13a}$$

where

$$F_{11} = \frac{1}{1 - r_p r_m}\left(e^{iqL} - r_p r_m\, e^{-iqL}\right), \quad F_{12} = \frac{-r_m}{1 - r_p r_m}\left(e^{iqL} - e^{-iqL}\right)$$

$$F_{21} = \frac{r_p}{1 - r_p r_m}\left(e^{iqL} - e^{-iqL}\right), \qquad F_{22} = \frac{1}{1 - r_p r_m}\left(-r_p r_m\, e^{iqL} + e^{-iqL}\right). \tag{11.1.13b}$$

We note the property that the determinant of the **F** matrix is unity.

$$F_{11}F_{22} - F_{12}F_{21} = 1. \tag{11.1.14}$$

11.1.2 Reflection and Transmission Through a DFB Structure

Assume that the DFB structure is sandwiched between two uniform waveguides. There are no sharp transitions at $z = 0$ and $z = L$, Fig. 11.2a. If there is no incident wave at $z = L$, that is, $B(L) = 0$, we can obtain the reflection coefficient

$$\begin{aligned} \Gamma(0) &= \frac{B(0)}{A(0)} = -\frac{F_{21}}{F_{22}} \\ &= \frac{-r_p\left(e^{iqL} - e^{-iqL}\right)}{-r_p r_m\, e^{iqL} + e^{-iqL}}. \end{aligned} \tag{11.1.15}$$

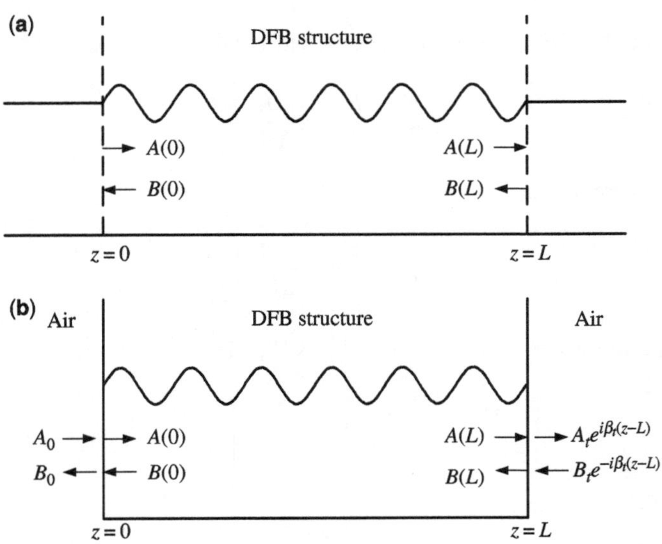

Figure 11.2 (a) A DFB structure sandwiched between two waveguides on each side. (b) A DFB structure with two reflecting facets at two ends.

The transmission coefficient is

$$T = \left|\frac{A(L)}{A(0)}\right|^2 = \left|\frac{1}{F_{22}}\right|^2$$
$$= \left|\frac{1 - r_\mathrm{p} r_\mathrm{m}}{e^{-iqL} - r_\mathrm{p} r_\mathrm{m}\, e^{iqL}}\right|^2. \qquad (11.1.16)$$

We note that, for index grating,

$$qL = \sqrt{\left(\delta L - i\frac{g_\mathrm{n} L}{2}\right)^2 - (KL)^2}. \qquad (11.1.17)$$

Figure 11.3 shows the transmission coefficient (11.1.16) for a uniform grating as a function of detuning δL for various gain coefficients, $G = g_\mathrm{n}L = 0$, 0.2, and 0.4, assuming $KL = 2$. We assume that the uniform grating has zero reflection at two ends ($R_1 = R_2 = 0$). The transmission spectrum is symmetrical with two peaks on both sides of the zero detuning, indicating two degenerate (longitudinal) modes of the DFB structure toward lasing condition. The lasing condition occurs when the denominator of the transmission coefficient vanishes

$$1 - r_\mathrm{p} r_\mathrm{m}\, e^{i2qL} = 0. \qquad (11.1.18)$$

The above lasing condition will be explored later for index- and gain-coupled gratings.

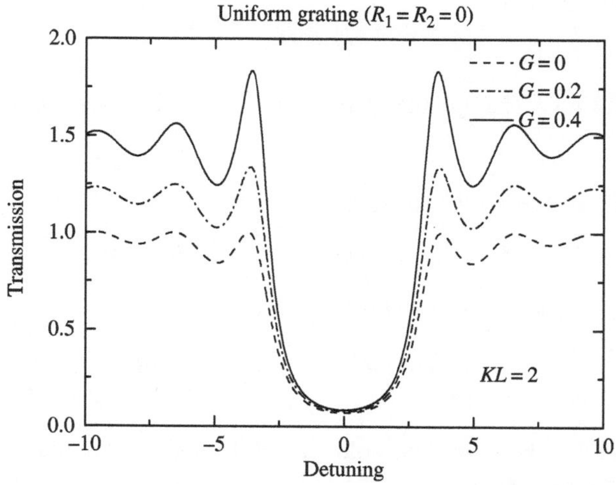

Figure 11.3 The transmission spectrum as a function of detuning δL using Eq. (11.1.16) for a uniform grating with zero reflections at the two ends ($R_1 = R_2 = 0$) for various gain-length products, $G = g_\mathrm{n}L = 0$, 0.2, and 0.4, assuming $KL = 2$.

11.1.3 DFB Structure with Two End Facets

The reflection and transmission of a DFB laser structure with two end facets, Fig. 11.2b, can be formulated using a propagation matrix approach similar to that in Section 5.7. Noting that the electric field for the forward and backward propagation components are $A(z)e^{i\beta_B z}$ and $B(z)e^{-i\beta_B z}$, where $\beta_B = \pi/\Lambda$ is the Bragg vector parameter, we have

$$\begin{aligned}\begin{bmatrix} A_t \\ B_t \end{bmatrix} &= \frac{1}{t_{21}}\begin{bmatrix} 1 & r_{21} \\ r_{21} & 1 \end{bmatrix}\begin{bmatrix} A(L)e^{i\beta_B L} \\ B(L)e^{-i\beta_B L} \end{bmatrix} \\ &= \frac{1}{t_{21}}\begin{bmatrix} 1 & r_{21} \\ r_{21} & 1 \end{bmatrix}\begin{bmatrix} e^{i\beta_B L} & 0 \\ 0 & e^{-i\beta_B L} \end{bmatrix}\begin{bmatrix} F_{11} & F_{12} \\ F_{21} & F_{22} \end{bmatrix}\begin{bmatrix} A(0) \\ B(0) \end{bmatrix} \\ &= \frac{1}{\sqrt{1-|r_1|^2}\sqrt{1-|r_2|^2}}\begin{bmatrix} 1 & -r_2 \\ -r_2 & 1 \end{bmatrix}\begin{bmatrix} e^{i\beta_B L} & 0 \\ 0 & e^{-i\beta_B L} \end{bmatrix}\begin{bmatrix} F_{11} & F_{12} \\ F_{21} & F_{22} \end{bmatrix}\begin{bmatrix} 1 & r_1 \\ r_1 & 1 \end{bmatrix}\begin{bmatrix} A_0 \\ B_0 \end{bmatrix} \\ &= \begin{bmatrix} m_{11} & m_{12} \\ m_{21} & m_{22} \end{bmatrix}\begin{bmatrix} A_0 \\ B_0 \end{bmatrix}. \end{aligned} \tag{11.1.19}$$

The transmission coefficient is

$$T = \left|\frac{A_t}{A_0}\right|^2 = \left|\frac{m_{11}m_{22} - m_{12}m_{21}}{m_{22}}\right|^2 \tag{11.1.20a}$$

where

$$m_{22} = \frac{1}{t_1 t_2}\left[-r_2 e^{i\beta_B L}(r_1 F_{11} + F_{12}) + e^{-i\beta_B L}(r_1 F_{21} + F_{22})\right]. \tag{11.1.20b}$$

Lasing condition occurs at

$$m_{22} = 0. \tag{11.1.21}$$

Lasing Condition The boundary conditions at two sharp ends of the DFB structure are

1. At $z = 0$, $A(0) = r_1 B(0)$, therefore

$$(1 - r_1 r_p)A^+ + (r_m - r_1)B^- = 0. \tag{11.1.22a}$$

2. At $z = L$, $B(L)e^{-i\beta_B L} = r_2 A(L)e^{i\beta_B L}$, thus

$$(\tilde{r}_2 - r_p)\, e^{i2qL}A^+ - (1 - \tilde{r}_2 r_m)B^- = 0 \tag{11.1.22b}$$

where $\tilde{r}_2 = r_2 e^{i2\beta_B L}$. For nontrivial solutions of A^+ and B^-, we must have the determinantal equation satisfied

$$(1 - r_1 r_p)(1 - \tilde{r}_2 r_m) + (\tilde{r}_2 - r_p)(r_m - r_1)e^{i2qL} = 0 \tag{11.1.23}$$

or

$$\left(\frac{r_1 - r_{\rm m}(q)}{1 - r_1 r_{\rm p}(q)}\right)\left(\frac{\tilde{r}_2 - r_{\rm p}(q)}{1 - \tilde{r}_2 r_{\rm m}(q)}\right)e^{i2qL} = 1. \tag{11.1.24}$$

It can be shown that the above Eq. (11.1.23) is identical to (11.1.21) in that the denominator m_{22} of the transmission coefficient in (11.1.20a) vanishes.

For lossless index grating,

$$q \cong \sqrt{\left(\delta - i\frac{g_{\rm n}}{2}\right)^2 - |K|^2} \tag{11.1.25}$$

where $K \equiv K_{ab}$ has been used, and the effect of gain is introduced through the term $g_{\rm n}$ in the detuning $\delta\beta$. The eigenequation (11.1.24) should be solved numerically, in general, for the threshold gain g and the lasing mode spectrum (or detuning), $\delta \simeq \beta_0 - \beta_{\rm B}$.

Case 1: If there is no distributed feedback structure, $K = 0$, $r_{\rm m} = r_{\rm p} = 0$, we recover the eigenequation for the Fabry–Perot lasers,

$$r_1 r_2 e^{i2\beta L} = 1 \tag{11.1.26}$$

where $q \to \delta - ig_{\rm n}/2$ and $q + \beta_{\rm B} = \beta_{\rm TE_0} - ig_{\rm n}/2$.

Case 2: If there are no sharp boundaries at $z = 0$ and $z = L$ such that $r_1 = r_2 = 0$, we have simply

$$r_{\rm m}(q) r_{\rm p}(q) e^{i2qL} = 1 \tag{11.1.27}$$

which is identical to the condition (11.1.18) in that the denominator of the transmission (11.1.16) vanishes. Therefore, $r_{\rm p}(q)$ and $r_{\rm m}(q)$ simply act as reflection coefficients for the forward and backward propagating waves.

Index Grating Equation (11.1.27) is simplified to

$$\frac{|K|^2}{[q + \Delta\beta]^2} e^{i2qL} = 1. \tag{11.1.28}$$

Noting that $q^2 = (\Delta\beta)^2 - |K|^2$, we have

$$q = i\Delta\beta \tan qL. \tag{11.1.29}$$

The above result is the same as the condition that the denominator of the transmission coefficient of the DFB structure (8.5.49) vanishes because $q = iS$. To analyze the solutions for the oscillation condition, we write

$$\Delta\beta L = -iqL \cot(qL) \tag{11.1.30}$$

which can be written as

$$\left(\delta L - i\frac{g_n L}{2}\right) = -i\sqrt{\left(\delta L - i\frac{g_n L}{2}\right)^2 - (KL)^2} \cot \sqrt{\left(\delta L - i\frac{g_n L}{2}\right)^2 - (KL)^2}. \tag{11.1.31}$$

The above equation has to be solved for (δL, $g_n L$) simultaneously for each given KL.

In the high gain (weak coupling) limit, $g_n L \gg KL$, we have $q \simeq \Delta\beta$; therefore, (11.1.28) becomes

$$K^2 e^{i2\Delta\beta L} = (2\Delta\beta)^2. \tag{11.1.32}$$

We compare the magnitude and the phase of the above equation. We obtain [7, 11]

$$K^2 e^{g_n L} = g_n^2 + 4\delta^2 \tag{11.1.33a}$$

and

$$\delta L = \left(m - \frac{1}{2}\right)\pi + \tan^{-1}\left(\frac{2\delta}{g_n}\right). \tag{11.1.33b}$$

Once the threshold gain g_n is obtained from (11.1.33a) for a given K and δ, the threshold current density can be estimated from the model for the gain–current (g–J) relation for a given semiconductor laser structure such as an active double-heterostructure or a quantum-well structure.

Figure 11.4a and Fig. 11.4b show the transmission spectra (11.1.20) as a function of detuning for various cases: (a) symmetrical reflectivities: $R_1 = R_2 = 0.3$, and the gain–length product: $G = g_n L = 0$, 0.2, and 0.4. (b) Asymmetrical coatings: $R_1 = 0.3$, $R_2 = 0.4$ (dashed), and $R_1 = 0.1$, $R_2 = 0.9$ (solid), while $G = 0.4$ is fixed. We can see the side mode suppression ratio increases as R_2 is much larger than R_1. In these plots, we have assumed that $\beta_B L = 2m\pi$ for a symmetrical DFB structure with an exact integral number of periods. We can see that case (a) with a symmetric reflection geometry ($R_1 = R_2$) produces two lasing modes, whereas case (b) $R_1 \neq R_2$ selects only one lasing mode near one side of the stop band. For index coupled DFB lasers, two lasing modes tend to exist on two sides of the stop band, unless there is an asymmetry due to different coatings ($R_1 \neq R_2$) or phase variations at the two end facets. In Fig. 11.4c, we show a typical amplified spontaneous emission spectrum measured from a DFB laser at various bias current below or near threshold. The emission has a peak near one side of the stop band and increases with increasing current.

In Fig. 11.5a and Fig. 11.5b [12] we show a comparison of the measured (dashed) amplified spontaneous emission spectra of two DFB lasers with theoretical results

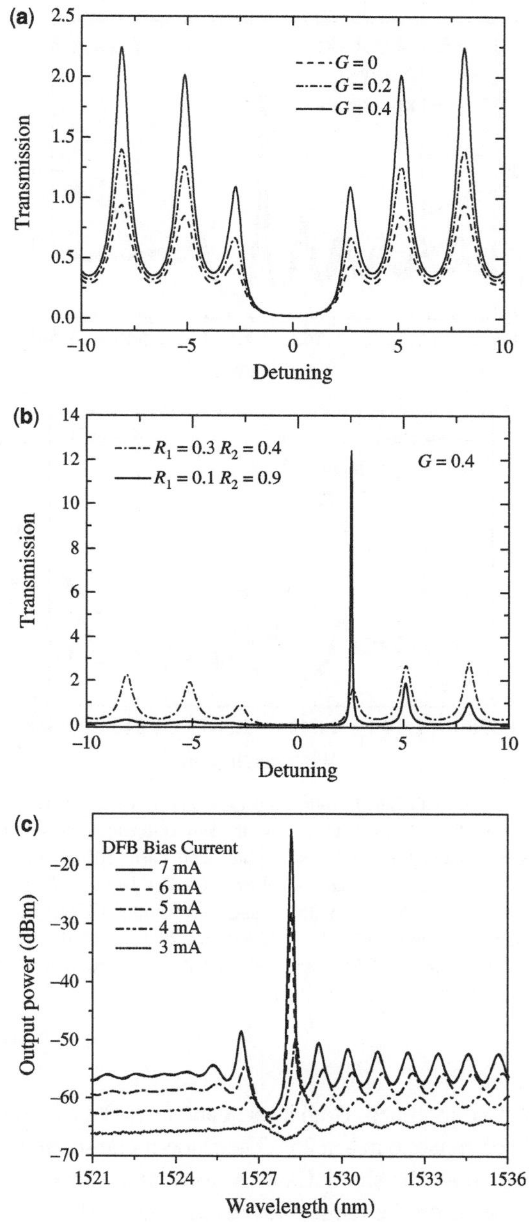

Figure 11.4 (a) The transmission spectrum as a function of detuning δL using Eq. (11.1.20) for a uniform grating with symmetrical end facet reflectivities ($R_1 = R_2 = 0.3$) for various values of gain-length product, $G = g_n L = 0$, 0.2, and 0.4, assuming $KL = 2$. (b) Same as (a) except for asymmetrical coatings: $R_1 = 0.3$, $R_2 = 0.4$ (dashed), and $R_1 = 0.1$, $R_2 = 0.9$ (solid), while the value $G = 0.4$ is fixed. (c) The experimental measured amplified spontaneous emission (ASE) spectra from 5 mA to near threshold in logarithmic scale (dBm).

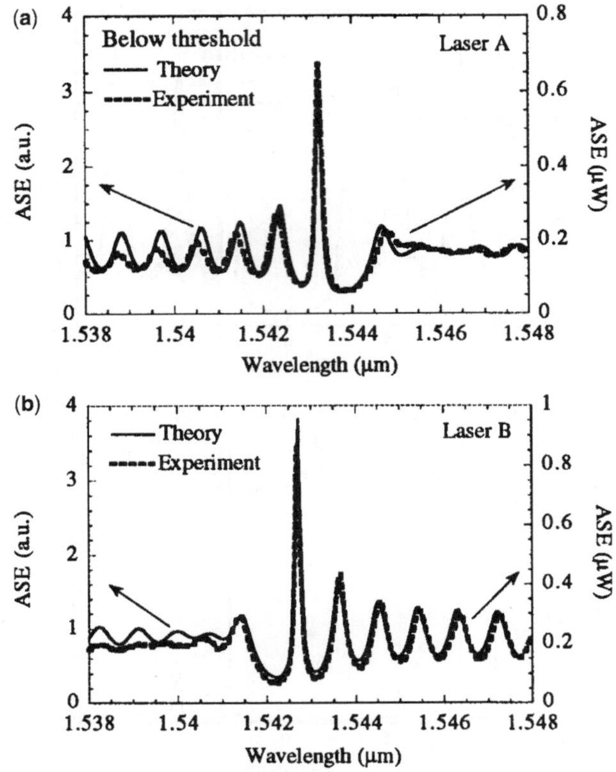

Figure 11.5 The theoretically calculated (solid) and measured (dashed) ASE spectrum for two lasers. Both laser A and laser B are biased at 4.8 mA below (the same) threshold current of 6 mA at 25°. Laser A lases on the shorter wavelength side of the stop band, and Laser B lases on the longer wavelength sides. The parameters used in the model are as follows. (a) For laser A: $R_1 = 0.9$, $\phi_1 = 210°$, $R_2 = 0.01$, $\phi_2 = 350°$, the refractive index = 3.6935, Bragg wavelength = 1.5439 μm, $KL = 1.5$, and the cavity length $L = 360$ μm. (b) For laser B: $R_1 = 0.9$, $\phi_1 = 320°$, $R_2 = 0.01$, $\phi_2 = 180°$, the refractive index = 3.700, Bragg wavelength = 1.5425 μm, $KL = 1.2$, and the cavity length $L = 360$ μm. (Reprinted with permission from [12] © 1997 IEEE.)

(solid) using the matrix approach. Both laser A and laser B are biased at 4.8 mA below (the same) threshold current of 6 mA at 25°. Their side-mode suppression ratios (SMSR) are larger than 50 dB above threshold. Laser A lases on the shorter wavelength side of the stop band and laser B on the longer wavelength sides. The parameters for laser A in Fig. 11.5a are $R_1 = 0.9$, its phase $\phi_1 = 210°$, $R_2 = 0.01$, $\phi_2 = 350°$, the refractive index = 3.6935, Bragg wavelength = 1.5439 μm, $KL = 1.5$, and the cavity length $L = 360$ μm. Note we have defined: $r_1 = R_1^{1/2} e^{i\phi_1}$, $r_2 = R_2^{1/2} e^{i\phi_2}$. For laser B, the parameters are $R_1 = 0.9$, $\phi_1 = 320°$, $R_2 = 0.01$, $\phi_2 = 180°$, the refractive index = 3.700, Bragg wavelength = 1.5425 μm, $KL = 1.2$, and the cavity length $L = 360$ μm. The effects of spatial hole burning on the lasing spectra such as the wavelength shift are also reported in Ref. 12.

11.1.4 Gain Coupled DFB Laser

The gain coefficient in an active DFB medium contains two parts,

$$g(z) = g_{\mathrm{n}} + \Delta g(z), \tag{11.1.34}$$

where g_{n} enters the propagation constant of the guided mode as the net modal gain

$$\Delta\beta = \delta - i\frac{g_{\mathrm{n}}}{2}, \quad g_{\mathrm{n}} = \Gamma g - \alpha_{\mathrm{i}} \tag{11.1.35}$$

where Γ is the optical confinement factor, g is the material gain coefficient, and α_{i} is the intrinsic absorption loss. The gain grating is described by [8, 13, 14]

$$\begin{aligned}\Delta g(z) &= \Delta g \cos\left(\frac{2\pi}{\Lambda}z + \varphi_{\mathrm{g}}\right)\\ &= \left(\frac{\Delta g}{2}e^{i\varphi_{\mathrm{g}}}\right) e^{i\frac{2\pi}{\Lambda}z} + \left(\frac{\Delta g}{2}e^{-i\varphi_{\mathrm{g}}}\right) e^{-i\frac{2\pi}{\Lambda}z}.\end{aligned} \tag{11.1.36}$$

Because Δg enters the change in refractive index through the imaginary part of the complex refractive index as

$$\begin{aligned}\Delta n(z) &= i\left(\frac{-\Delta g(z)}{2k_0}\right)\\ &= \Delta n_{+1}\, e^{i\frac{2\pi}{\Lambda}z} + \Delta n_{-1}\, e^{-i\frac{2\pi}{\Lambda}z},\end{aligned} \tag{11.1.37}$$

we find

$$K_{ab} = \Gamma k_0 \Delta n_{+1} = -iK_{\mathrm{g}}e^{i\varphi_{\mathrm{g}}}, \tag{11.1.38a}$$

$$K_{ba} = -\Gamma k_0 \Delta n_{-1} = +iK_{\mathrm{g}}e^{-i\varphi_{\mathrm{g}}}\left(=K_{ab}^*\right) \tag{11.1.38b}$$

where $K_{\mathrm{g}} = \Gamma\Delta g/4$ is real. We note immediately that

$$\begin{aligned}q &= \sqrt{(\Delta\beta)^2 + K_{ab}K_{ba}}\\ &= \sqrt{(\Delta\beta)^2 + |K_{ab}|^2}\end{aligned} \tag{11.1.39}$$

which does not have a stop band anymore because $K_{ba} = +K_{ab}^*$.

For a gain grating with anti-reflection (AR) coating at two ends, Eq. (11.1.27) simplifies to

$$-\frac{|K_{\mathrm{g}}|^2}{[q+\Delta\beta]^2} e^{i2qL} = 1. \tag{11.1.40a}$$

In the high-gain limit, we simplify the magnitude and phase conditions as

$$\left|K_{\mathrm{g}}^2\right| e^{g_{\mathrm{n}}L} = g_{\mathrm{n}}^2 + 4\delta^2 \tag{11.1.40b}$$

and

$$\delta L = m\pi + \tan^{-1}\left(\frac{2\delta}{g_{\mathrm{n}}}\right). \tag{11.1.40c}$$

The above Eq. (11.1.40c) shows that a lasing mode exists at zero detuning for $m = 0$ (i.e., the lasing occurs at the middle of the stop band for gain-coupled DFB lasers).

In reality, it is difficult to make purely gain-coupled DFB lasers without index grating. When both index and gain gratings exist, the coupling coefficients, K_{ab} and K_{ba}, can be obtained by superposition from (11.1.4) and (11.1.38)

$$K_{ab} = K - iK_{\mathrm{g}}e^{i\varphi_{\mathrm{g}}}, \tag{11.1.41a}$$

$$K_{ba} = -K + iK_{\mathrm{g}}e^{-i\varphi_{\mathrm{g}}}. \tag{11.1.41b}$$

Here, for simplicity, we have taken the phase of the index grating as zero. Therefore, the phase of the gain grating should be the phase difference between the gain and index grating. Complex coupled or partially gain-coupled DFB lasers at 1.3-μm and 1.55-μm wavelengths have been demonstrated using strained-layer multiple quantum-well (MQW) structures [15–19]. The 1.55-μm InGaAsP-InP gain-coupled DFB lasers use a strain-layer MQW active grating for a mixed index and gain coupling. Very good performance such as high single-mode yield (>90%) and high side mode suppression ratio (>55 dB) at 1.55 μm or high power (50 mW) and high-speed modulation (>18 GHz) at 1.3 μm have been reported as well as theoretical analysis [15–19].

11.1.5 Phase-Shifted DFB Laser

Phase-shifted DFB lasers have been proposed and realized for single-mode operation right at the Bragg wavelength [20–22]. In Fig. 11.6 we plot the schematics of a $\lambda/4$-shifted DFB laser structure. The insertion of a phase shift section of $\Lambda/2 = \lambda/(4n_e)$ in the center of the grating causes an extra phase shift of $\pi/2$ for each wave passing through or a round-trip of π. The whole $\lambda/4$-shifted DFB structure

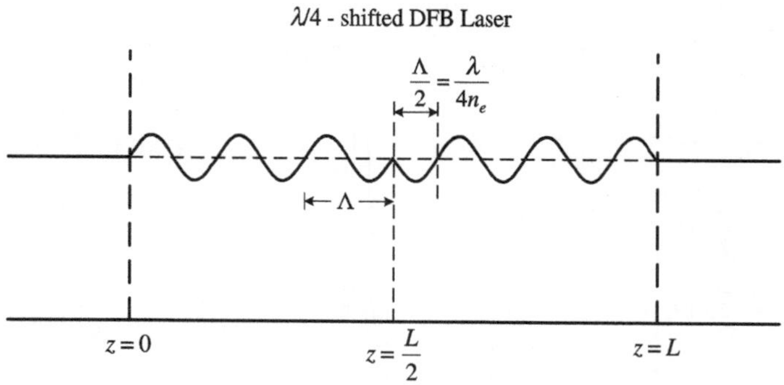

Figure 11.6 Schematic diagram of a quarter-wavelength phase-shifted distributed-feedback semiconductor laser.

can be considered as the cascade of two index gratings with the second grating shifted by π from the first grating.

Section 1: $\varphi = 0$ in $K = |K|\, e^{i\varphi}$. r_m $(\varphi = 0)$ and $r_p(\varphi = 0)$ should be used in the first matrix, $F_{ij}^{(1)}(\varphi = 0)$.

Section 2: $\varphi = \pi$ in $K = |K|\, e^{i\varphi}$. r_m $(\varphi = \pi)$ and $r_p(\varphi = \pi)$ should be used in the second matrix, $F_{ij}^{(2)}(\varphi = \pi)$.

Theoretically, we multiply two **F** matrices in the cascade structure

$$\begin{bmatrix} A(L) \\ B(L) \end{bmatrix} = \begin{bmatrix} F_{11}^{(2)} & F_{12}^{(2)} \\ F_{21}^{(2)} & F_{22}^{(2)} \end{bmatrix} \begin{bmatrix} A(L/2) \\ B(L/2) \end{bmatrix}$$
$$= \begin{bmatrix} F_{11}^{(2)} & F_{12}^{(2)} \\ F_{21}^{(2)} & F_{22}^{(2)} \end{bmatrix} \begin{bmatrix} F_{11}^{(1)} & F_{12}^{(1)} \\ F_{21}^{(1)} & F_{22}^{(1)} \end{bmatrix} \begin{bmatrix} A(0) \\ B(0) \end{bmatrix}$$
$$= \begin{bmatrix} F_{11} & F_{12} \\ F_{21} & F_{22} \end{bmatrix} \begin{bmatrix} A(0) \\ B(0) \end{bmatrix} \quad (11.1.42)$$

$$T = \left|\frac{A(L)}{A(0)}\right|^2 = \left|\frac{1}{F_{22}}\right|^2 \quad (11.1.43a)$$

where

$$F_{22} = F_{21}^{(2)}(\varphi = \pi) F_{12}^{(1)}(\varphi = 0) + F_{22}^{(2)}(\varphi = \pi) F_{22}^{(1)}(\varphi = 0). \quad (11.1.43b)$$

Figure 11.7 plots the transmission coefficient (11.1.43) as a function of detuning δL for various values of gains, $G = g_n L = 0$, 0.2, and 0.5, assuming $KL = 2$. The transmission has a peak at zero detuning ($\delta L = 0$) or center of the stop band.

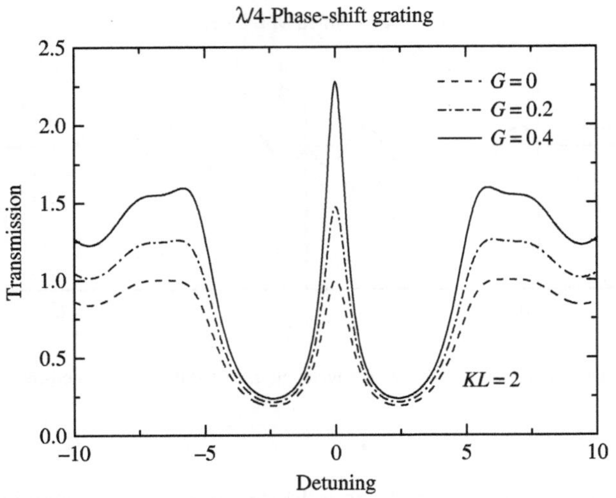

Figure 11.7 Transmission spectra of a quarter-wavelength phase-shifted DFB laser for three gain-length products: $G = g_n L = 0$, 0.2, and 0.4, and KL is fixed at 2.0.

11.1.6 Distributed Bragg Reflector (DBR) Semiconductor Laser

Note that the reflection coefficient for a DBR structure with a finite length L is obtained from (11.1.12)

$$\Gamma(0) = \frac{B(0)}{A(0)} = \frac{r_p A^+ + B^-}{A^+ + r_m B^-}. \tag{11.1.44}$$

Noting that $B(L) = 0$, we have $B^- = -r_p A^+ e^{i2qL}$, or

$$\Gamma(0) = \frac{r_p(1 - e^{i2qL})}{1 - r_m r_p e^{i2qL}} \tag{11.1.45a}$$

$$= \frac{-K^* \sinh SL}{\Delta\beta \sinh SL + iS \cosh SL} \tag{11.1.45b}$$

for $q = iS$ in a passive lossless structure. The last expression (11.1.45b) is the same as that derived in Eq. (8.5.49). We can see that, in general, $\Gamma(0)$ is a complex quantity, $\Gamma(0) = |\Gamma(0)|e^{i\phi}$. In Fig. 11.8, we show a distributed Bragg reflector (DBR) semiconductor laser [23, 24]. The frequency selectivity is provided by the reflection coefficients r_1 and r_2 at both ends, and the active region is the same as a Fabry–Perot edge-emitting laser. As a matter of fact, a DBR laser is just a Fabry–Perot laser with a mirror reflectivity varying with the wavelength. The lasing threshold occurs at the wavelength for which the overall reflectivity is a maximum. The analysis is very similar to that of a Fabry–Perot laser except that r_1 and r_2 are complex numbers and are wavelength dependent.

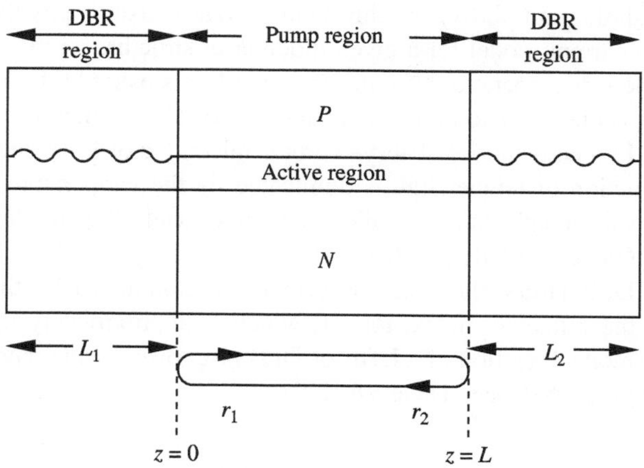

Figure 11.8 Schematic diagram of a distributed Bragg reflector (DBR) semiconductor laser.

The lasing condition occurs at

$$r_1 r_2 e^{i2\beta L} = 1 \tag{11.1.46}$$

where r_1 is the reflection coefficient of the electric field looking into the left DBR reflector (with a length L_1) at $z = 0$, and r_2 is that looking into the right DBR reflector at $z = L$. The complex propagation constant is

$$\beta = \beta_0 - i\frac{\Gamma g - \alpha}{2}. \tag{11.1.47}$$

Note that

$$r_1 = R_1^{1/2} e^{i\phi_1} \qquad r_2 = R_2^{1/2} e^{i\phi_2}. \tag{11.1.48}$$

We obtain

$$\Gamma g_{\text{th}} = \alpha + \frac{1}{2L} \ln\left(\frac{1}{R_1 R_2}\right) \tag{11.1.49}$$

$$2\beta_0 L + \phi_1 + \phi_2 = 2m\pi. \tag{11.1.50}$$

The reflection coefficient for a DBR region with a length L_1 has been obtained in (11.1.45). It should be noted that for an abrupt junction between the active region and the DBR region, the reflection coefficient r_1 (and r_2) has to be multiplied by an effective coupling coefficient C_0 (with $|C_0| \leq 1$) to account for the imperfect power transmission across the junction (i.e., $r_1 \to C_0 r_1$), and the threshold condition (11.1.49) becomes [23]

$$\Gamma g_{\text{th}} = \alpha + \frac{1}{2L} \ln \frac{1}{R_1 R_2 C_0^4}. \tag{11.1.51}$$

From the threshold gain above, the threshold current density J_{th} is then obtainable from the gain–current model for a given diode laser structure.

Stable single-mode operation of semiconductor lasers using DFB and DBR structures has been achieved. The temperature stability of these structures has also been demonstrated. In a conventional Fabry–Perot semiconductor laser, the lasing wavelength as a function of temperature is dominated by the temperature coefficient of the peak gain wavelength $d\lambda_p/dT$, which is approximately 0.5 nm/degree for long-wavelength InGaAsP/InP double-heterostructure lasers [23]. On the other hand, for DBR and DFB lasers, the lasing wavelength is dominated by the temperature coefficient of the refractive index, dn/dT, which is approximately 0.1 nm/degree. A stable fixed-mode operation of a DBR or DFB laser over a wide range of temperature ($>$100 degrees) has been achieved [23].

11.2 VERTICAL CAVITY SURFACE-EMITTING LASERS

In previous chapters, we have discussed semiconductor lasers with emission from the two end facets of the horizontal cavity, so-called edge emitters. Surface-emitting semiconductor lasers [25–27] are attractive for many advantages such as vertical emission from the top or bottom of substrate, ease of coupling to optical fibers due to circular geometry, direct wafer scale probing, and ultralow threshold operation due to the small cavity volume. The fabrication process also eliminates the need of wafer lapping, device cleaving and dicing, facet coating, and wire bonding, which are required for edge emitters. Because of their short cavity lengths, the surface-emitting lasers have inherent single-frequency operation. In addition, the fabrication of two-dimensional laser arrays is possible. Therefore, vertical cavity surface-emitting lasers (VCSELs) have potential applications for optical interconnects, optical communication, and optical processing. VCSELs [28–34] have been developed with performances comparable with those of conventional Fabry–Perot edge-emitting diode lasers. A few schematic diagrams are shown in Fig. 11.9a–c. The basic idea is the light bounces upward and downward within an optical cavity where active layers such as quantum wells are placed near the peak of the longitudinal standing wave pattern. The light output is orthogonal to the plane of the wafers. Front and back surface-emitting lasers are illustrated. This change in the direction of laser emission drastically changes the design, fabrication scalability, and array configurability of semiconductor lasers.

Historically, although the concept of surface emission from a semiconductor laser dates back to 1965 when Melngailis [35] reported on a vertical cavity structure followed later by various reports of different groups on the grating surface emission [36, 37] or 45°-corner turning mirrors etched into the device [38], major work on VCSELs was proposed [25, 39] by Iga in 1977, and the first pulsed lasing operation of a InGaAsP/InP surface-emitting laser at 77K was demonstrated [28] by his group in 1979. A patent, which contains major features of current VCSEL structures, was filed in 1979 by Burnham et al. [40] and awarded in 1982. Room temperature

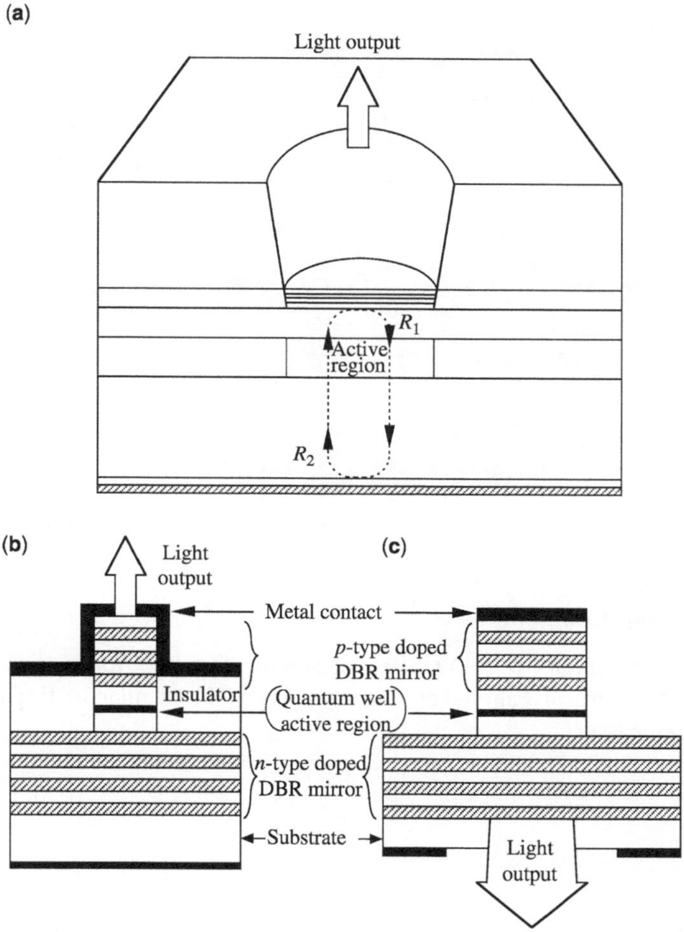

Figure 11.9 (a) Schematic diagram of a surface emitting laser [28]. (b) Front and (c) back surface-emitting lasers using etching through the active region. (Reprinted with permission from [30] © 1991 IEEE.)

pulsed operation [29] and cw operation [41] of a GaAs/AlGaAs surface-emitting laser was realized in 1983 and 1988, respectively. With the advance in crystal growth technologies, low threshold, room temperature cw operation of VCSELs was demonstrated by Jewell et al. at Bell Lab in 1989 [30, 42], which triggered rapid progress in the following years. After the pioneering work of oxidation process of Holonyak's group at the University of Illinois in 1990 [43, 44], various processes using selectively oxidized VCSELs [45–50] were developed with significantly improved laser performance due to the carrier and optical confinements provided by the oxide aperture. Figure 11.10 shows an example of a double-aperture oxidized VCSEL. More recently, many improvements with significant progress on the performances of the VCSELs have been demonstrated.

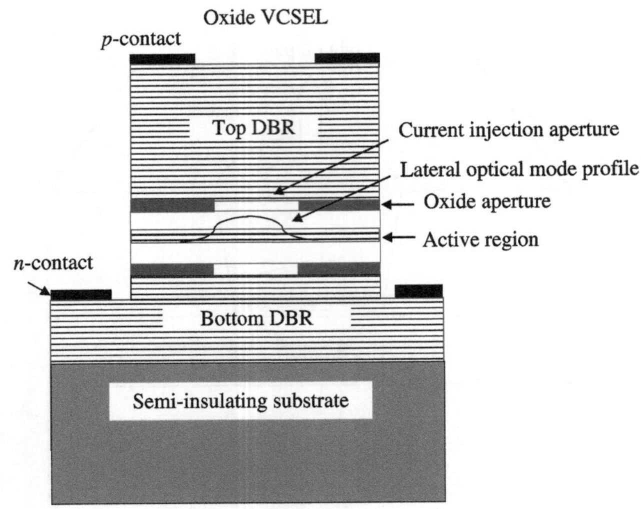

Figure 11.10 An oxide VCSEL with top and bottom oxide apertures near the active region.

11.2.1 Lasing Condition

The lasing condition of a VCSEL is simply that the round-trip resonance condition of a vertical Fabry–Perot cavity, Fig. 11.11, similar to that of a DBR semiconductor laser

$$r_1 r_2 e^{i2\beta L} = 1 \tag{11.2.1a}$$

with a complex propagation constant

$$\beta = \beta_c + i(\alpha - \Gamma g)/2 \tag{11.2.1b}$$

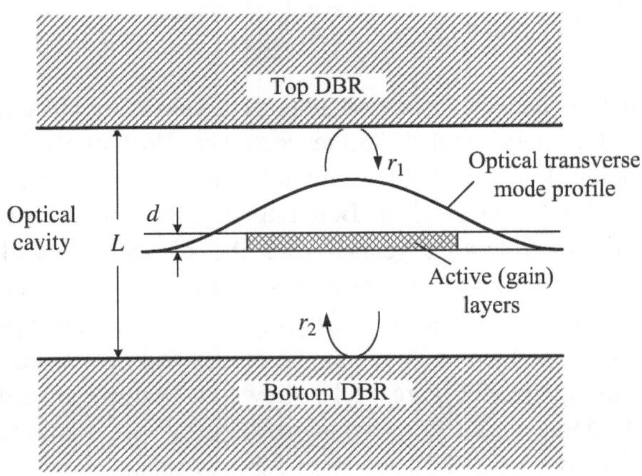

Figure 11.11 Schematic diagram for the round-trip propagation condition for the lasing oscillation of a mode in a VCSEL cavity.

in the optical cavity, and r_1 and r_2 are the complex reflection coefficients of the electric field for the DBR mirrors at the top and bottom of the vertical cavity, $r_1 = |r_1|e^{i\varphi_1}$, $r_2 = |r_2|e^{i\varphi_2}$. Recall from (5.9.49) that the reflection coefficient of the field of a DBR with $(2N + 1)$ quarter-wavelength layers is given by

$$r = \frac{\left(\frac{n_2}{n_1}\right)^{2N} - \left(\frac{n_1^2}{n_0 n_t}\right)}{\left(\frac{n_2}{n_1}\right)^{2N} + \left(\frac{n_1^2}{n_0 n_t}\right)} \tag{11.2.2}$$

which approaches $+1$ when N approaches infinity if $n_1 < n_2$; and -1 if $n_1 > n_2$; where n_1 is the refractive index of the first layer and n_2 is that of the second layer. It implies that the mirror transmission loss can be reduced to almost zero, therefore, the threshold gain can be reduced when we use enough pairs N in the DBR layers.

For lasing oscillation condition, we have the phase condition and the threshold condition.

Phase Condition

$$\varphi_1 + \varphi_2 + 2\beta_c L = 2m\pi. \tag{11.2.3}$$

The reflection coefficient looking into a DBR can have a phase of 0 (or π) depending on the order of quarter-wavelength layers in low-high (or high-low) refractive index.

Case 1: If $\varphi_1 = \varphi_2 = 0 (n_1 < n_2)$, $L = m\dfrac{\lambda}{2n}$. The standing wave pattern will have a peak at the first interface of the top and bottom DBR as $\varphi_1 = \varphi_2 = 0$ (constructive interference between the incident and reflected waves). The shortest cavity length is $\lambda/2n$ or half wavelength, which will have a zero crossing at the center of the cavity, Fig. 11.12a. The next cavity size is λ/n or one wavelength, which will have a maximum at the center of the cavity, Fig. 11.12b.

Case 2: If $\varphi_1 = \varphi_2 = \pi$ $(n_1 > n_2)$, $L = (m-1)\dfrac{\lambda}{2n}$. The standing wave pattern will have a zero at the first interface of the top and bottom DBR as $\varphi_1 = \varphi_2 = \pi$ (destructive interference between the incident and reflected waves). The shortest cavity length is $\lambda/2n$ or half wavelength, which will have a maximum at the center of the cavity, Fig. 11.13a. The next cavity size is λ/n or one wavelength, which will have a zero at the center of the cavity, Fig. 11.13b. Figure 11.14 shows the standing wave pattern of a VCSEL structure with one-wavelength cavity. The optical intensity profile has a peak at the center of the cavity and has nulls or peaks across the DBR quarter-wavelength layers. The intensity becomes very small near the top exit surface due to the high reflectivity of the top DBR.

Threshold Condition The threshold gain is given by

$$\Gamma g = \alpha + \frac{1}{2L}\ln\left(\frac{1}{R_1 R_2}\right) \tag{11.2.4}$$

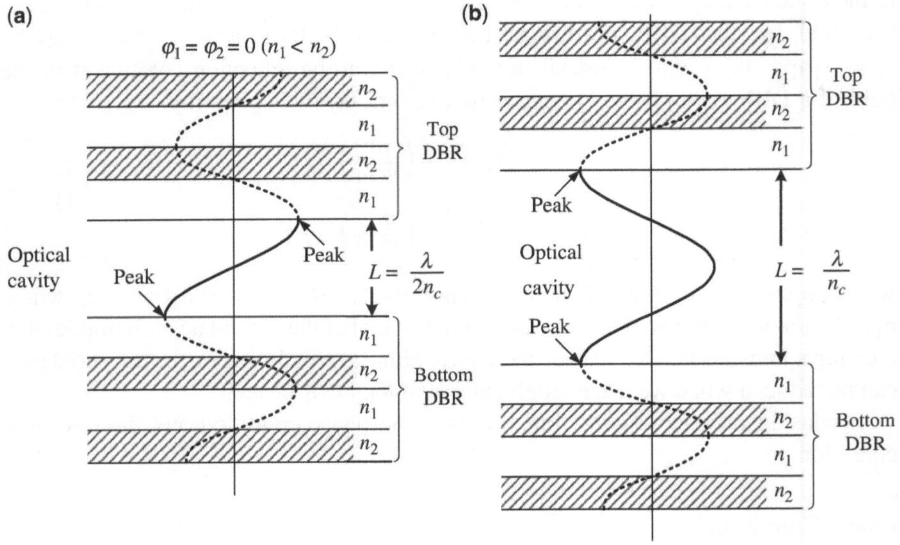

Figure 11.12 The longitudinal standing wave pattern of two optical cavity lengths: (a) half-wavelength (null at the center of the cavity) and (b) one wavelength (peak at the center). Both assume $n_1 < n_2$ (therefore, $\varphi_1 = \varphi_2 = 0$).

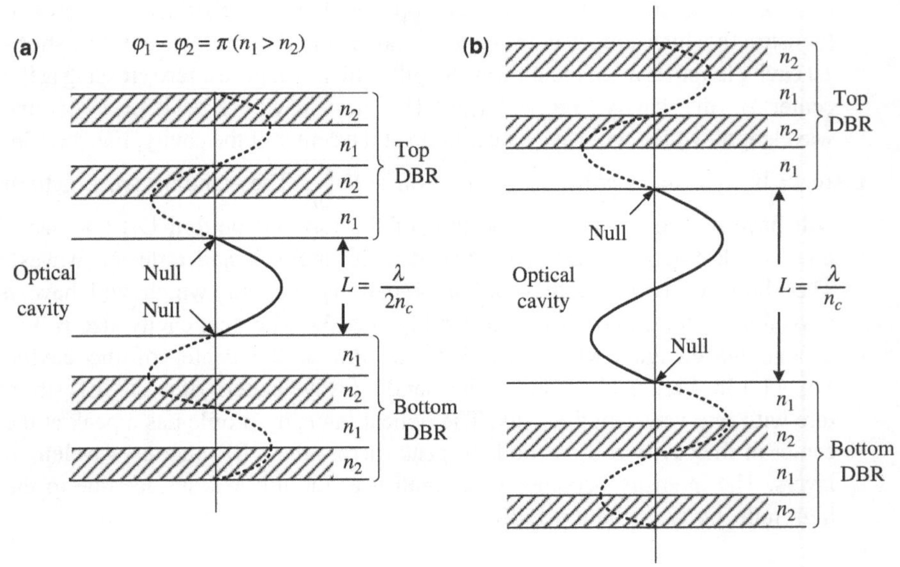

Figure 11.13 The longitudinal standing wave pattern of two optical cavity lengths: (a) half-wavelength (peak at the center of the cavity) and (b) one wavelength (null at the center of the cavity). Both assume $n_1 > n_2$ (therefore, $\varphi_1 = \varphi_2 = \pi$).

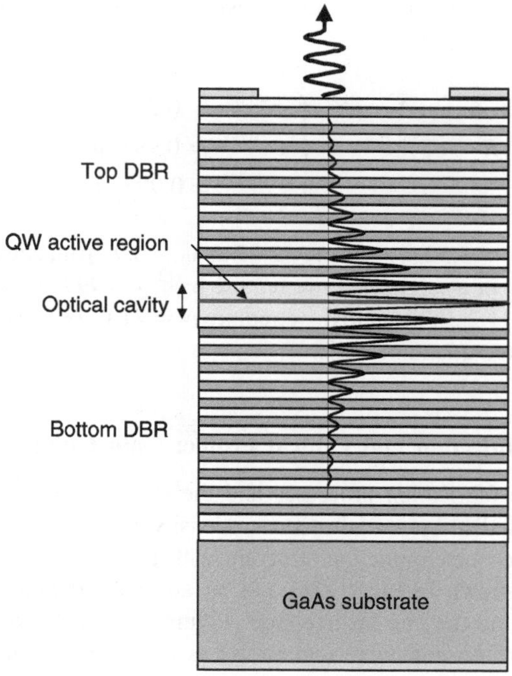

Figure 11.14 The longitudinal standing wave pattern of a VCSEL cavity including the top and bottom DBR regions.

using $R_1 = |r_1|^2$ and $R_2 = |r_2|^2$. Here the optical confinement factor Γ accounts for both the longitudinal and transverse confinements

$$\Gamma = \gamma \frac{d}{L} \Gamma_t \tag{11.2.5}$$

where d = active layer thickness, L = cavity length, $\gamma = 2$ if the thin active layer is placed at the maximum of the standing wave, and $\gamma = 1$ for a thick active layer. The factor d/L accounts for the longitudinal optical confinement, and Γ_t is the transverse optical confinement factor, which accounts for the overlap between the transverse optical intensity profile with the gain profile, Fig. 11.11. The symbol α accounts for the optical absorptions α_i within an effective length (L_{eff}) of the top and bottom DBRs, plus any diffraction loss α_d due to the mode mismatches when the laser mode propagates to the reflecting mirrors. This diffraction loss depends on the size of the diameter of the active region and the locations of the mirrors. We then write αL as

$$\alpha L = 2\alpha_i L_{eff} + \alpha_d L_{cav} (= \text{Loss parameter}). \tag{11.2.6}$$

We can estimate the number of DBR layers needed using a simple threshold analysis. We assume, $R_1 = R_2$, $\Gamma_t = 1$, $\gamma = 2$, and ignore the absorption loss. We have $2dg_{th} = \ln(1/R)$. The active layer thickness $d = n_w L_z$, where n_w is the number of quantum wells and L_z is the thickness of one well. Take $g_{th} = 500\ \text{cm}^{-1}$ and $L_z = 100\ \text{Å}$. We have

$$R = e^{-2n_w L_z g_{th}} = \begin{cases} e^{-0.001} = 0.999, & n_w = 1 \\ e^{-0.002} = 0.998, & n_w = 2 \\ e^{-0.003} = 0.997, & n_w = 3. \end{cases} \tag{11.2.7}$$

Consider $n_1(\text{GaAs}) = 3.52$, $n_2(\text{AlAs}) = 2.95$ at $\lambda_0 = 0.98\ \mu\text{m}$, and assume $n_0 = n_t = n_1(\text{GaAs}) = 3.52$. For a single well, we require $|r_N|^2 = 0.999$, which leads to $N = 23$ periods based on (11.2.2). Similarly, we require $N = 21$ if we use two quantum wells and $N = 20$ for three quantum wells.

11.2.2 Carrier Injection and Optical Profile: Gain Guidance

Depending on the fabrication processes and the structures of the surface-emitting lasers, the current distribution, the carrier density profile, and the optical mode pattern vary. Buried heterojunction structure with the active region surrounded by insulator regions shown in Fig. 11.9b and an air-post structure (Fig. 11.9c) have been used to confine the current injection. Figure 11.15a shows an ion-implanted VCSEL, where the ion-implanted regions act as insulators and the injected current

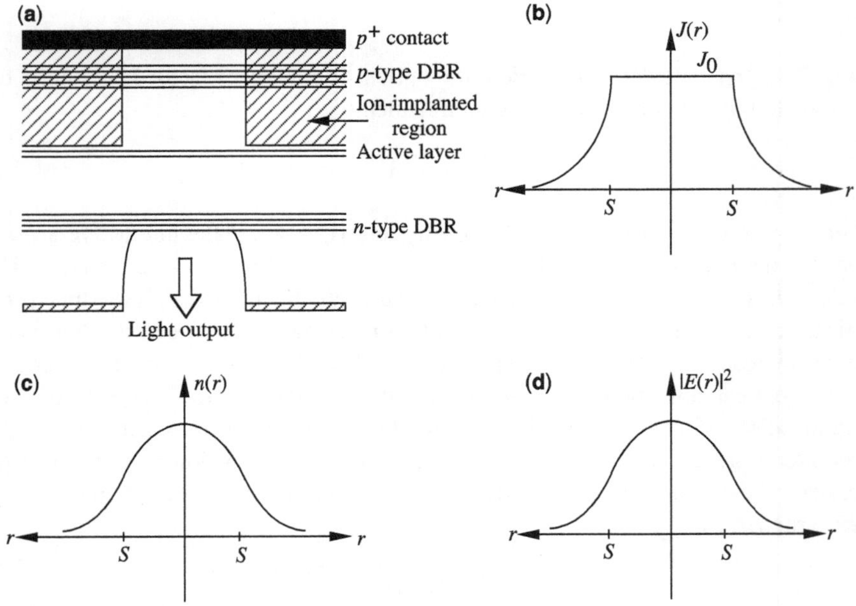

Figure 11.15 (a) Schematic diagram of a gain-guided surface-emitting laser with ion-plantation regions, (b) current density, (c) carrier density, and (d) fundamental optical mode profile. (Reprinted with permission from [51] © 1990 American Institute of Physics.)

into the active layer has a distribution as shown in Fig. 11.15b. The guidance of the surface-emitting laser in the transverse direction is the same as that for the gain-guided stripe geometry diode laser, except that here the injection profile has a circular cross section instead of a long stripe geometry, and the light propagation is along the vertical direction instead of the parallel direction to the active layer. The carrier density profile, Fig. 11.15c, can be solved from the diffusion equation given the current density profile. The fundamental optical mode pattern is shown in Fig. 11.15d once the gain profile is obtained, for example,

$$g(r) \simeq g'(n(r) - n_{\mathrm{tr}}) \tag{11.2.8a}$$

for bulk lasers, where g' is the differential gain and n_{tr} is the transparency carrier density. For quantum-well lasers, one may use

$$g(r) \simeq g_0[\ln(n(r)/n_0) + 1] \tag{11.2.8b}$$

where g_0 and n_0 are constant parameters. Approximate analytic formulas can be found in Ref. 51.

For the surface-emitting laser such as in Fig. 11.9b and Fig. 11.9c, the device has been etched all the way through the active region to provide lateral carrier

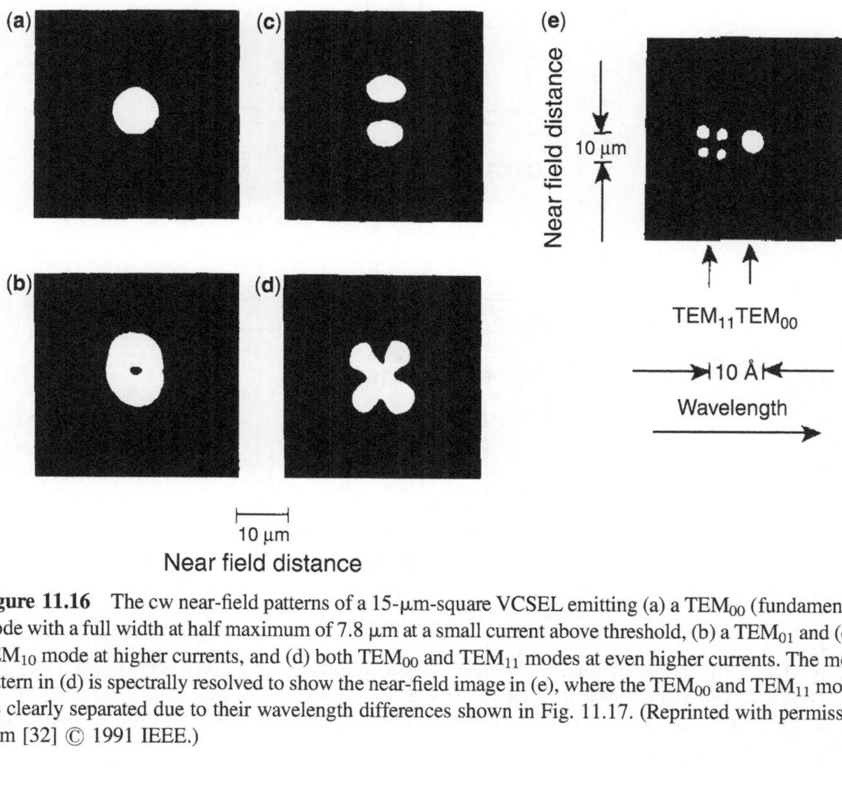

Figure 11.16 The cw near-field patterns of a 15-μm-square VCSEL emitting (a) a TEM$_{00}$ (fundamental) mode with a full width at half maximum of 7.8 μm at a small current above threshold, (b) a TEM$_{01}$ and (c) a TEM$_{10}$ mode at higher currents, and (d) both TEM$_{00}$ and TEM$_{11}$ modes at even higher currents. The mode pattern in (d) is spectrally resolved to show the near-field image in (e), where the TEM$_{00}$ and TEM$_{11}$ modes are clearly separated due to their wavelength differences shown in Fig. 11.17. (Reprinted with permission from [32] © 1991 IEEE.)

confinement. The top structure has a cylindrical semiconductor region with air as the outside region. The interface scattering and diffraction losses also provide a certain degree of mode selectivity, and the fundamental (EH_{11}, HE_{11}, or TEM_{00}) mode of a rectangular or cylindrical waveguide is considered to be very important.

In the following, we show the experimental results from Ref. 32. In Fig. 11.16, we show the near-field patterns of a 15-μm-square VCSEL with a structure (using a shallow implant range of about 1 μm with a proton dosage of $3 \times 10^{15}/\text{cm}^2$ at 100 keV) at various current injection levels above threshold. At a small injection current above threshold condition, the fundamental TEM_{00} mode (or called EH_{00} mode in Chapter 7) lases with a full width at half maximum of 7.8 μm as shown in Fig. 11.16a and its lasing wavelength is 9500.85 Å as shown in the emission spectrum in Fig. 11.17a. As the current increases, the lasing mode changes to a TEM_{01}

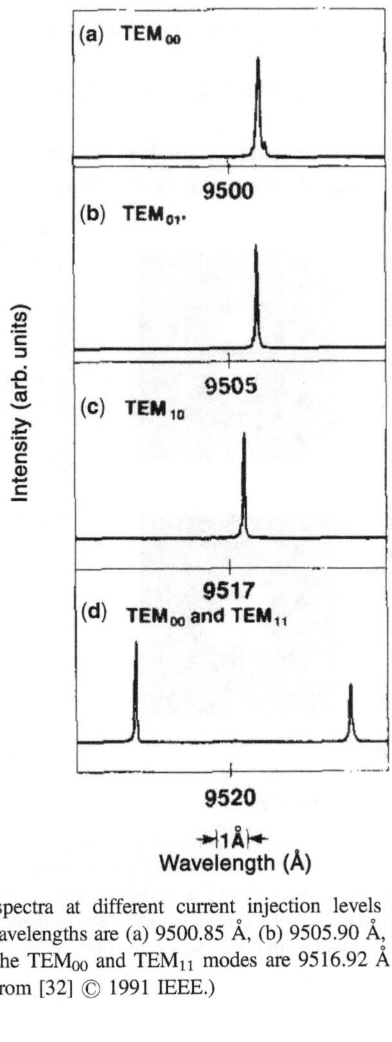

Figure 11.17 The lasing spectra at different current injection levels corresponding with those in Fig. 11.16a–d. The lasing wavelengths are (a) 9500.85 Å, (b) 9505.90 Å, and (c) 9517.40 Å. In (d), the two lasing wavelengths for the TEM_{00} and TEM_{11} modes are 9516.92 Å and 9523.80 Å, respectively. (Reprinted with permission from [32] © 1991 IEEE.)

mode shown in Fig. 11.16b with a lasing wavelength 9505.90 Å shown in Fig. 11.17b, and then shifts to a TEM_{10} lasing mode at 9517.40 Å shown in Fig. 11.16c and in Fig. 11.17c for over a small range of current. At an even higher current level, both the TEM_{00} and TEM_{11} modes lase with the total near-field pattern shown in Fig. 11.16d, and their emission wavelengths are 9516.92 Å and 9523.80 Å, respectively, as shown in Fig. 11.17d. The mode pattern in Fig. 11.16d can be spectrally resolved due to the wavelength difference of the two modes and is shown in Fig. 11.16e, where the TEM_{00} and TEM_{11} modes are clearly separated in the near-field image.

The red shift of the lasing wavelength is caused by the junction heating as the injection current increases. The narrow linewidths of the lasing spectra also indicate that the laser emits a single transverse mode as the wavelength separations between two adjacent transverse modes are around 1 to 2 Å for the gain guided VCSELs.

11.2.3 Oxide-VCSELs: Index Confinement

Wet oxidation of AlGaAs [43] has been employed in the fabrication of edge-emitting lasers [44] and VCSELs [45–50]. Lateral confinement by oxide-apertures forms apertures for current flow into the active region of the optical cavity. In addition, the lower refractive index of the oxide regions provides lateral index confinement of the optical mode. Oxide-confined VCSELs have exhibited high power conversion efficiency and low threshold current density and voltage [45–50]. Figure 11.10 shows a double-aperture oxide-confined VCSEL, which has several advantages [50]. The structure is monolithic, and the low resistance designs using parabolic heterointerface grading and C-doping for the *p*-type DBR can be exploited for current conduction into the active region. The current apertures near the optical cavity eliminate sidewall nonradiative recombination in etched air-post VCSELs and minimize lateral current spreading to outside the laser cavity. Furthermore, the lower refractive index of the oxide layers provides index guidance as well as efficient current flow and heat extraction. Therefore, the enhanced carrier and optical confinements produce reduced threshold current and voltage.

11.2.4 Temperature Dependence and Junction Heating

It should be noted that the junction heating is important in VCSELs. A simplified model to account for the temperature rise due to the injected current from a circular disk above an infinite substrate is [33]

$$\Delta T_{\mathrm{jct}} = \frac{P_{IV} - P_{\mathrm{opt}}}{4\sigma S} \tag{11.2.9}$$

where P_{IV} is the total input electric power (the current–voltage product), P_{opt} is the optical light output, σ is the thermal conductivity of the semiconductor (=0.45 W/cm °C for GaAs), and S is the disk radius.

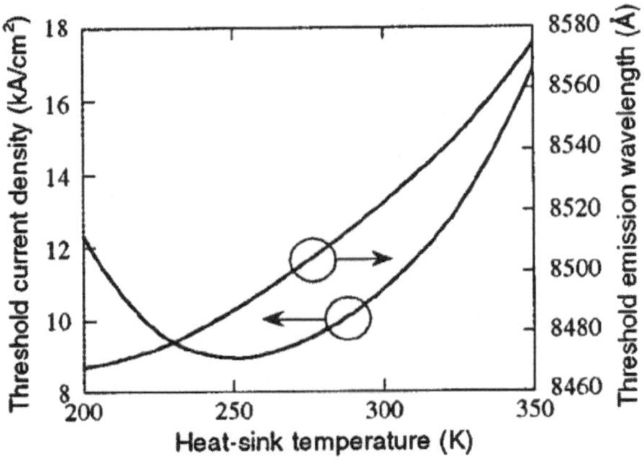

Figure 11.18 Theoretically calculated threshold current density (left axis) and emission wavelength (right axis) as a function of temperature from $T = 200$K to 350K. (Reprinted with permission from [52] © 1997 American Institute of Physics.)

Figure 11.18 shows the dependence of the threshold current density and the emission wavelength as a function of the heat-sink temperature, based on a theoretical model [52]. The results agree with the experimental observations [53, 54]. For GaAs, the gain peak wavelength increases linearly at about 2.7 Å/K; whereas for a GaAs/AlGaAs VCSEL studied here, the mode wavelength increases at only 0.6 Å/K, and the threshold current decreases first from 200K to 250K, then increases at higher temperatures beyond 250K. Figure 11.19 shows a simple explanation based on the interaction between the gain and the cavity resonance mode (wavelength). For temperature smaller than a minimum threshold temperature, the gain peak wavelength is shorter than the cavity resonance wavelength, and the gain is low at the resonance wavelength. As temperature increases toward the minimum threshold current temperature, the gain peak wavelength moves close to the cavity resonance wavelength. Thus the threshold current is reduced because the gain at cavity resonance increases. As temperature increases further, the gain peak wavelength shifts toward the longer wavelength side and the gain at the resonance decreases. Therefore, the threshold current increases with temperature.

11.2.5 Optical Output and Differential Quantum Efficiency

The optical output power is [43]

$$P_{\text{out}} = \frac{\hbar\omega}{q}\eta[I - I_{\text{th}}(I, T_{\text{jct}})] \tag{11.2.10}$$

$$\eta = \eta_{\text{i}}\eta_{\text{opt}} \tag{11.2.11}$$

where η depends on two factors: (1) the injection efficiency η_{i} accounting for the fraction of injected carriers contributing to the emission process (some of the carriers

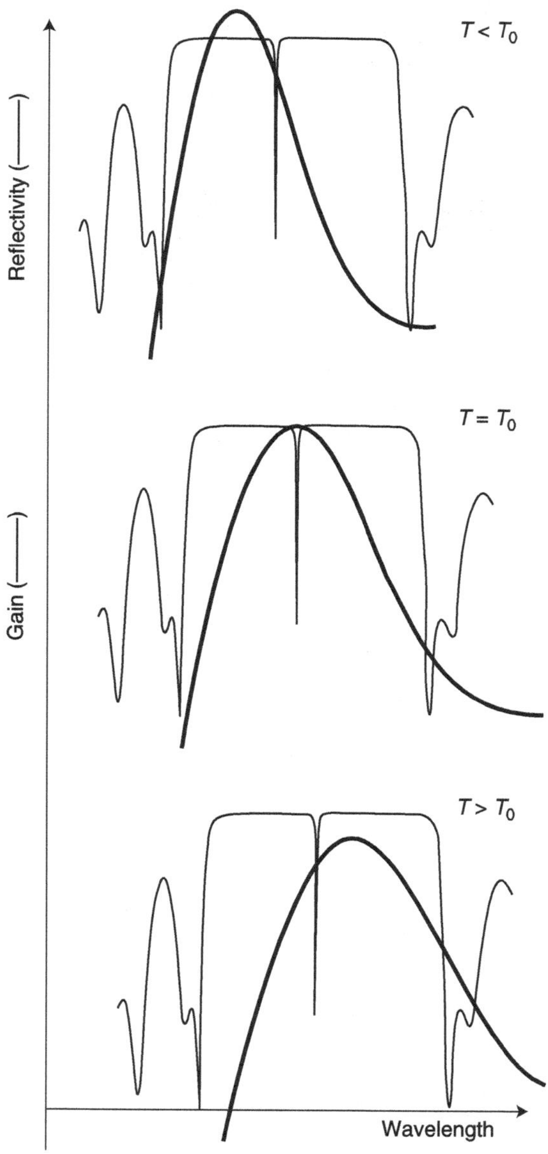

Figure 11.19 Schematic of the interaction between the gain spectrum (dark curves) and the reflectivity spectrum (light) as temperature changes. The dip in the reflectivity spectrum indicates the Bragg wavelength, which increases at 0.6 Å/K, while the gain peak wavelength increases at a higher rate 2.7 Å/K. At a certain temperature (250K in this example), the cavity or emission wavelength aligns with the gain peak, which results in the minimum threshold current density in Fig. 11.18. (Reprinted with permission from [52] © 1997 American Institute of Physics.)

can recombine in the undoped confinement regions where the carriers do not interact with the optical field) and (2) the optical efficiency η_{opt} accounting for the fraction of generated photons that are transmitted out of the cavity

$$A = 2\alpha_i L_{\text{eff}} + \alpha_d L \qquad T_r = \ln\left(\frac{1}{R_1 R_2}\right) \tag{11.2.12}$$

$$\eta_{\text{opt}} = \frac{T_r}{A + T_r}. \tag{11.2.13}$$

Here the absorbance A accounts for the absorption loss in the top and bottom DBR regions with an effective penetration length L_{eff} plus the diffraction loss ($\alpha_d L$) in the DBRs or oxide apertures. L is the cavity length of the active layer. T_r accounts for the top and bottom mirror transmission loss. The factor η_{opt} is similar to that of the external quantum efficiency used in the Fabry–Perot lasers in Section 10.1. Note that the threshold current depends on the injection current as well as on the junction temperature T_{jct}. Models for the laser output taking into account the temperature heating have also been developed in Refs. 55–58.

The power conversion efficiency is given by

$$\eta_P = \frac{P_{\text{out}}}{VI} \tag{11.2.14}$$

where V is the bias voltage and I the bias current. The differential quantum efficiency is then current-dependent

$$\eta_{\text{ext}}(I) = \frac{q}{\hbar\omega}\frac{dP_{\text{out}}}{dI} = \eta\left(1 - \frac{dI_{\text{th}}}{dI}\right). \tag{11.2.15}$$

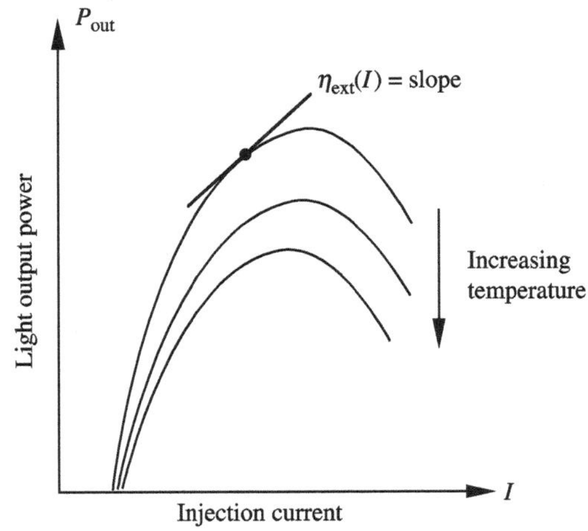

Figure 11.20 A plot of the light output of a surface-emitting laser versus the injection current for different temperatures. The slope at a particular current level I is the differential quantum efficiency, which can be negative at a high current level or a high temperature.

We see that $\eta_{ext}(I)$ can be negative if $\dfrac{dI_{th}}{dI} > 1$. The light output P_{out} versus the injection current I will have a negative slope in this case, as shown in Fig. 11.20. Experimental data showing the negative external quantum efficiency can be found in Refs. 33, 34, and 55. For an update on progress on VCSEL research, we refer to Refs. 59 and 60.

11.3 MICROCAVITY AND PHOTONIC CRYSTAL LASERS

What is the smallest laser you can make?

Ultrasmall semiconductor lasers have attracted research interest because they have many advantages:

1. Small lasers require low power consumption.
2. Suitability for potential large-scale integration.
3. Possible thresholdless laser operation due to efficient coupling of spontaneous emission into a single lasing mode.

Conceptually, the smallest optical cavity is a half wavelength in each direction, so we have the smallest cavity with a volume $(\lambda/2n)^3$ for a material with a refractive index n. The idea of spontaneous emission control using a cavity was proposed by Purcell in 1946 [61]. A microcavity with a well-defined optical mode or a high-quality (Q) cavity allows for the control of the spontaneous emission coupling into the cavity mode. Early experiments using metal cavities to investigate cavity-enhanced single atom spontaneous emission or inhibited spontaneous emission were reported in the 1970s [62–64]. Proposals using photonic crystal structures for controlling the spontaneous emission were published in 1987 [65–67], and some experimental confirmations were reported [68, 69]. With the successful demonstration of VCSELs [28–30, 70], spontaneous emission control in VCSEL cavities has been demonstrated [71, 72]. Microdisk lasers with submilliampere threshold currents operating at room temperature were demonstrated [73–76] in 1992. These lasers operate in the whispering gallery modes. Research on photonic crystals continued to grow in the 1990s [77]. Optical pumping [78–82] as well as electrical injection [83, 84] of photonic crystal lasers were subsequently realized in 1999 and 2004, respectively. Microcavity structures and lasers are extremely interesting for the study of exciton–photon interaction or cavity quantum electrodynamics [85, 86]. By placing quantum dots in a microcavity, single photon sources have also been under intensive research for applications related to quantum information processing and secure optical communications. Single photon sources are expected to emit one photon at a time, instead of multiple photons in a Poisson distribution. The quality of such a source is evaluated based on its high efficiency, small multiphoton probability, and quantum indistiguishability [87–90].

The major idea is to confine the photons into an ultrasmall and high-Q cavity, where the optical gain materials are placed. Engineering the Q factor and the mode volume V is useful for making a low-threshold nanolaser or for the study of cavity quantum electrodynamics. The challenges are [83, 84, 91]:

- How to realize a high-Q cavity with minimal losses due to absorption, surface roughness, and modal leakage?
- How to provide current injection into the gain region of the microcavity?
- How to reduce the thermal effects due to the large current density into an ultrasmall area or volume?

11.3.1 Structures of Microcavity Lasers

(a) **Microdisk Lasers,** Fig. 11.21a [73–76, 91, 92]: The cavity is a dielectric disk where the gain region such as a quantum well structure is placed in the center layer of the microdisk region. The optical modes are the so-called whispering gallery modes traveling around the perimeter of the disk, and the mode confinement is based on total internal reflection from the microdisk walls. The active region is supported by upper and lower pedestals. By current injection

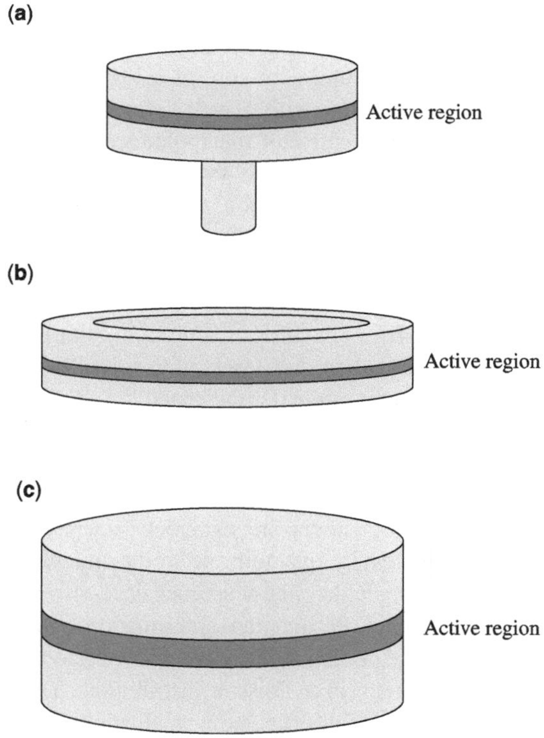

Figure 11.21 Schematics of (a) a microdisk laser, (b) a ring laser, and (c) a mesa-type laser, where the shaded layer denotes the active region for electron–hole recombination.

from the contact, the carriers diffuse to the edge of the disk, where the stimulated emission or optical gain occurs because the whispery gallery modes travel around the perimeter of the disk. Similar structures to the microdisk lasers are the ring laser, Fig. 11.21b, and the mesa-type of lasers, Fig. 11.21c. The modes are essentially guided around the perimeter or the ring waveguide. Figure 11.22 shows an example of a whispering gallery mode of a microdisk laser [74].

(b) Photonic Crystal Lasers [78–84, 91, 93–95]: Figure 11.23a shows a schematic diagram of the free-standing slab structure for current injection, and Fig. 11.23b shows the layered parameters of the designed wafer. The photonic crystal is constructed by a periodic arrangement of refractive index variation, such as air holes in semiconductors, which form a photonic band gap. In semiconductor photonic crystals, the electronic transition spectrum in the photonic band gap is inhibited, and the spontaneous emission rate is reduced. Photonic defects such as missing air holes in an otherwise periodic arrangement of air holes in a semiconductor layer produce resonant optical modes with fields localized near the defects, and the resonant frequencies locate within the photonic band gap. One-dimensional and 2D photonic crystals have been fabricated, whereas a 3D photonic crystal is more difficult to implement. Usually, a spontaneous coupling factor β is defined as the coupling efficiency of the spontaneous emission energy into the resonant mode. Two-dimensional photonic crystals or the photonic band structures are designed by solving the

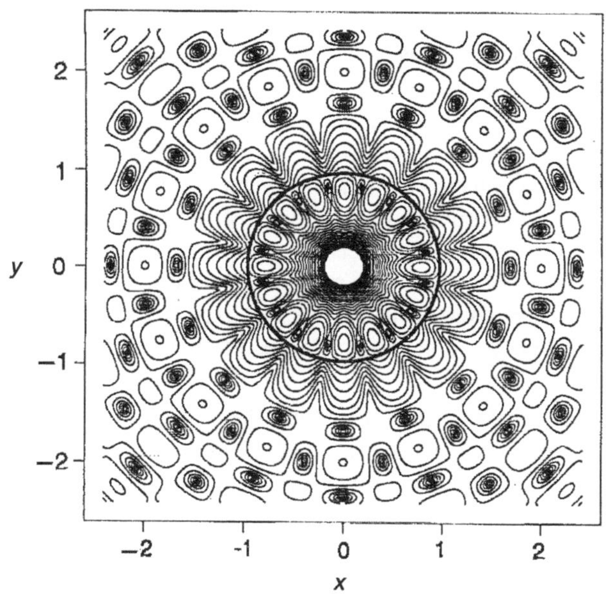

Figure 11.22 A whispering gallery mode of a microdisk laser. (Reprinted with permission from [74] © 1992 American Institute of Physics.)

(a)

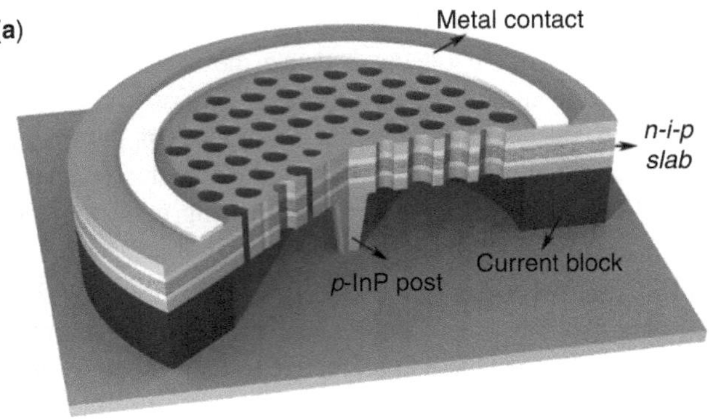

(b)

Material	Thickness	Doping (cm^{-3})	Remark
n-InP	0.1 μm	~ 1.1×10^{18}	Capping layer, removed
n-1.24Q InGaAsP	600 Å	~ 2.7×10^{19}	Delta doping
i-1.24Q InGaAsP	200 Å	Undoped	
1.30Q (−0.5%) InGaAsP	115 Å	Barrier	6 pairs, λc = 1.55 μm
1.65Q (+0.8%) InGaAsP	70 Å	Well	
1.30Q (−0.5%) InGaAsP	115 Å	Barrier	
i-1.24Q InGaAsP	200 Å	Undoped	
p-1.24Q InGaAsP	600 Å	~ 3×10^{18}	
p-InP	1.0 μm	~ 1.5×10^{18}	Sacrificial layer (post)
p-InGaAs	0.1 μm	~ 5×10^{18}	Etch-stop layer
p-Buffer InP	0.3 μm	~ 1×10^{18}	
p-InP substrate			

Figure 11.23 (a) Schematic diagram of a photonic crystal laser, which consists of a free-standing slab with a single-defect photonic crystal pattern with a small post near the center. (b) The designed layered parameters of the sample wafer. (Reprinted with permission from [84] © 2005 IEEE.)

Maxwell wave equation for the eigenvalues and corresponding eigenvectors by taking into account the periodic refractive index pattern. Optical modes of a photonic crystal with defects can also be solved using the finite-difference time-domain method (FDTD) method or the finite-element method (FEM), which are discrete modes if they are localized when the resonant frequencies fall into the photonic band gap. More details will be discussed later in this subsection.

(c) **Micropillar or Micropost Cavities,** Fig. 11.24 [87–90, 96]: The vertical cavity is essentially the same as a VCSEL structure, and the lateral dimension is a cylinder of a radius a or diameter $D = 2a$. Current injection is through the top and bottom DBRs. The transverse optical modes are similar to those of the optical fiber modes, HE_{mn} or EH_{mn} modes, of the cylindrical geometry. An extremely high-Q cavity with a very small mode volume of the order $(\lambda/n)^3$ can be realized. For example, when the diameter D of a micropillar cavity consisting of a top and bottom GaAs/AlAs DBR mirror is decreased from 2 μm to 0.5 μm, the mode volume of the fundamental mode HE_{11} drops

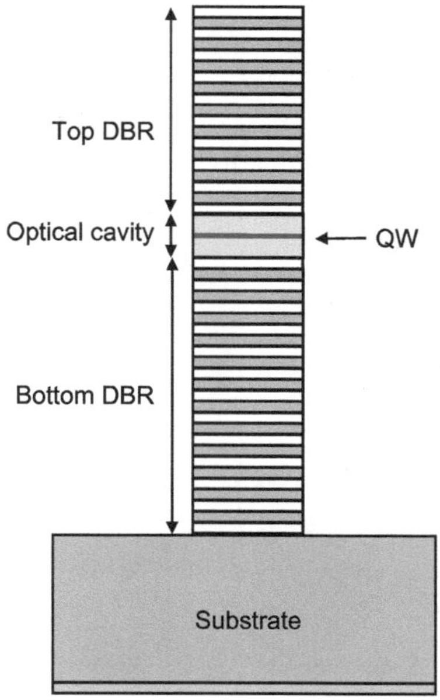

Figure 11.24 A micropillar or micropost optical cavity with the top and bottom distributed-Bragg reflectors (DBRs). The structure is essentially the same as that of a VCSEL, except for its diameter *D*.

from $19.2(\lambda/n)^3$ to $2(\lambda/n)^3$ while the Q factor drops by only a factor of 2 from 11,500 to 5000, where n is the refractive index in the cavity [87, 90].

(d) Nanolasers [97]: Figure 11.25 shows a nanolaser that has essentially a small microcylinder coated with gold metal on the top and side walls. The laser light emission is from the bottom substrate. The metal serves as confinement of the optical waveguide, and the active region is a layer of bulk semiconductor (a disk) in the middle of the optical circular waveguide. This laser has an ultrasmall threshold of 6 μA at a temperature of 77K.

11.3.2 Purcell Factor and Spontaneous Emission Factor

The Purcell factor predicts the enhancement of spontaneous emission rate into a resonant cavity mode compared with that of an unbound medium. Consider a single localized radiation dipole, which is weakly coupled to the field. It is placed in (i) a uniform lossless medium with a volume V (free space modes), Fig. 11.26a, or (ii) in a cavity with a volume V_c with a single mode, Fig. 11.26b. The ratio of the spontaneous emission rate of a single cavity mode to that of the free space modes is called the Purcell factor. We show the basic derivation of the Purcell factor as follows.

If the emission linewidth of the dipole radiation is much narrower than that of the cavity resonance modes, we can consider that the dipole sees a continuum of modes and calculate the spontaneous emission rate based on Fermi's golden rule. The

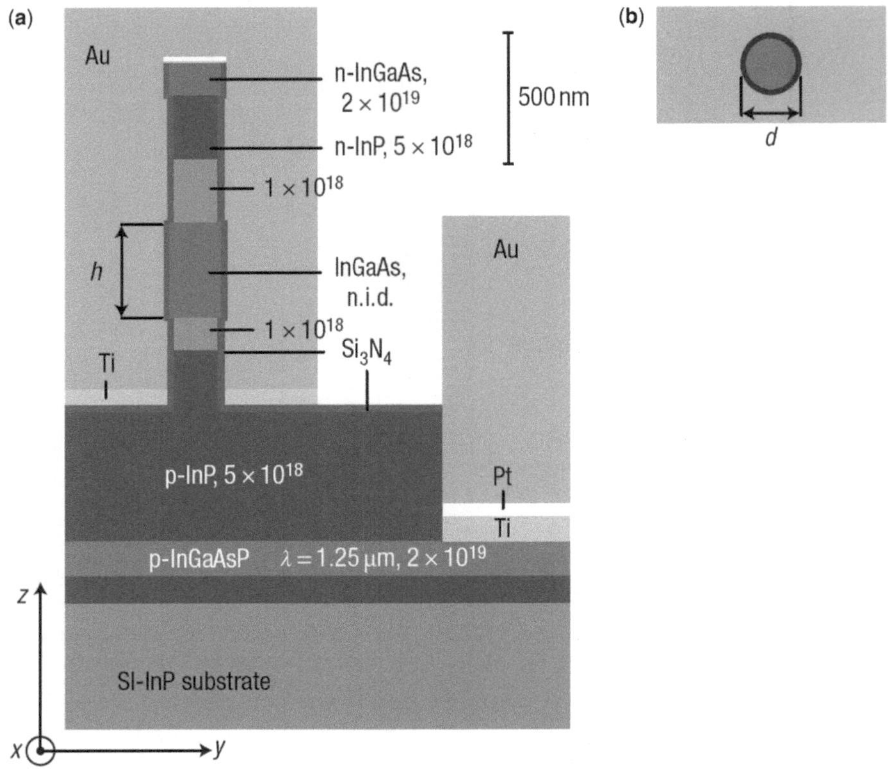

Figure 11.25 (a) A nanolaser structure and (b) its cross section seen from the top. The laser active cavity has an insulator layer protection followed by gold coating surrounding the side wall of the cylindrical geometry. The laser light is emitted toward the bottom and exits from the substrate. (Adapted by permission from [97] © 2007 Macmillan Publishers Ltd: Nature Photonics.)

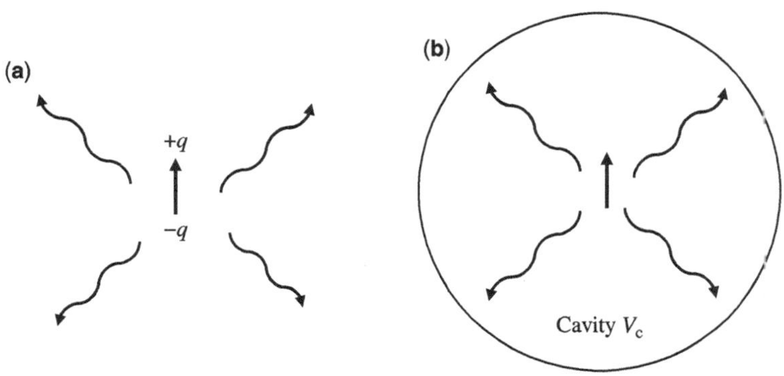

Figure 11.26 (a) A radiating dipole (emitter) such as that due to an electron–hole pair radiative recombination in free space or an unbounded medium with a refractive index n. (b) A radiating dipole radiating in a microcavity with a cavity volume V_c, where the majority of the optical intensity is confined.

transition rate from an initial state $|i\rangle$ to a final state $|f\rangle$ due to an optical dipole μ at position $\mathbf{r}_e$ in the volume V is given by

$$\frac{1}{\tau} = \frac{2\pi}{\hbar}\rho(\hbar\omega)|\langle f|\boldsymbol{\mu}\cdot\mathbf{E}(\mathbf{r}_e)|i\rangle|^2 \tag{11.3.1}$$

where the photon density of states $\rho(\hbar\omega)$ (1/eV) is the number of modes per energy interval (eV) at the emitter frequency ω in a volume V. We have

$$\rho_{\text{free}}(\hbar\omega) = \frac{\omega^2 n^3}{\pi^2 \hbar\, c^3} V \tag{11.3.2}$$

for the "free-space" mode density. The optical electric field is

$$\begin{aligned} \mathbf{E}(\mathbf{r}, t) &= \mathbf{E}(\mathbf{r}, \omega)\, e^{-i\omega t} + \mathbf{E}^*(\mathbf{r}, \omega)\, e^{i\omega t} \\ \mathbf{E}(\mathbf{r}, \omega) &= iE_{\max}\mathbf{f}(\mathbf{r}) \end{aligned} \tag{11.3.3}$$

where the amplitude $E_{\max}$ is the maximum of the electric field and $|\mathbf{f}(\mathbf{r}_{\max})| = 1$ at the location of the field maximum. $E_{\max}$ and $\mathbf{f}(\mathbf{r})$ satisfy the normalization condition, which requires that the time-averaged storage (electric and magnetic) energy for a single photon of energy $\hbar\omega$ equals

$$2\varepsilon_0 \int_V n^2(\mathbf{r})|\mathbf{E}(\mathbf{r}, \omega)|^2 d^3\mathbf{r} = \hbar\omega. \tag{11.3.4}$$

We note that the time-averaged electrical energy density, $\varepsilon_0 n^2(\mathbf{r})|\mathbf{E}(\mathbf{r}, \omega)|^2$, is equal to that of the magnetic energy. For a plane wave in a volume V, we have

$$E_{\max} = \sqrt{\frac{\hbar\omega}{2\varepsilon_0 n^2 V}}, \qquad \mathbf{f}(\mathbf{r}) = \hat{e}\, e^{i\mathbf{k}\cdot\mathbf{r}}. \tag{11.3.5}$$

For a microcavity, the optical modes $\mathbf{E}(\mathbf{r})$ are not plane waves anymore, and an effective volume V_{eff} is defined. First, we define $n_{\max} = n(\mathbf{r}_{\max})$, where $\mathbf{r}_{\max}$ is the location of the electrical field at its maximum

$$V_{\text{eff}} = \frac{\int_V n^2(\mathbf{r})|\mathbf{E}(\mathbf{r})|^2 d^3\mathbf{r}}{n_{\max}^2 E_{\max}^2} = \frac{\int_V n^2(\mathbf{r})|\mathbf{f}(\mathbf{r})|^2 d^3\mathbf{r}}{n_{\max}^2}. \tag{11.3.6}$$

Note that the definition of V_{eff} is only for convenience, and the position $\mathbf{r}_{\max}$ of the maximum electric field (not the refractive index) is chosen to define $E_{\max}$ and $n_{\max}$

$$E_{\max} = \sqrt{\frac{\hbar\omega}{2\varepsilon_0 n_{\max}^2 V_{\text{eff}}}}. \tag{11.3.7}$$

For a microcavity, the spontaneous emission rate is changed in at least three ways.

1. The spectral density of modes;
2. The amplitude of the vacuum field;
3. The orientation of the field with respect to the dipole direction.

Consider a microcavity with a single mode:

ω_c = Resonance cavity frequency,
$\Delta\omega$ = The full width at half maximum (FWHM or full linewidth) of the optical mode spectral density.

The quality factor Q is defined as

$$Q = \frac{\omega_c \text{ (Energy stored in the cavity)}}{\text{Power dissipation per cycle}} \simeq \frac{\omega_c}{\Delta\omega}. \tag{11.3.8}$$

The mode density seen by the dipole emitter is a Lorentzian function, Fig. 11.27,

$$\rho_{cav}(\hbar\omega) = \frac{(\hbar\Delta\omega)/(2\pi)}{(\hbar\omega - \hbar\omega_c)^2 + (\hbar\Delta\omega/2)^2} \tag{11.3.9a}$$

which satisfies the normalization condition

$$\int_{-\infty}^{+\infty} \rho_{cav}(\hbar\omega)\, d(\hbar\omega) = 1. \tag{11.3.9b}$$

Note at the peak of the cavity mode frequency

$$\rho_{cav}(\hbar\omega_c) = \frac{2}{\pi\hbar\Delta\omega} = \frac{2Q}{\pi\hbar\omega_c}. \tag{11.3.10}$$

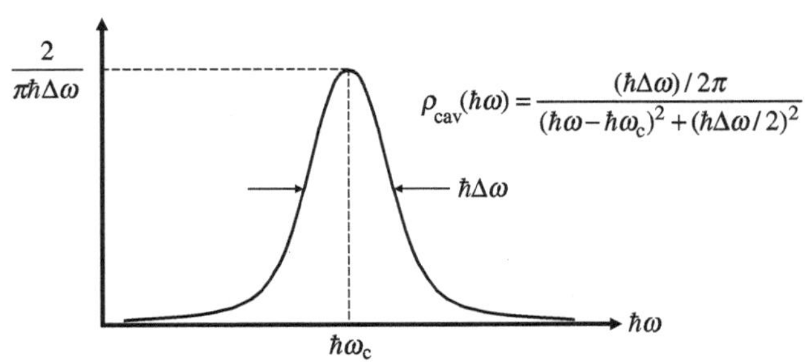

Figure 11.27 A plot of the photon density spectrum or cavity mode spectrum as a function of the photon energy $\hbar\omega$ near the cavity resonance $\hbar\omega_c$. The full width at half maximum is $\hbar\Delta\omega$.

The spontaneous emission rate in a "free-space" of volume V is, after taking into account a factor of $1/3$ for the dipole strength for averaging over x, y, and z directions,

$$\begin{aligned}\frac{1}{\tau_{\text{free}}} &= \frac{2\pi}{\hbar}\rho_{\text{free}}(\hbar\omega)\frac{|\boldsymbol{\mu}|^2}{3}|\mathbf{E}|^2 \\ &= \frac{2\pi}{\hbar}\rho_{\text{free}}(\hbar\omega)\frac{|\boldsymbol{\mu}|^2}{3}\frac{\hbar\omega}{2\varepsilon_0 n^2 V} \\ &= \frac{2\pi}{\hbar}\frac{\omega^2 n^3}{\pi^2\hbar c^3}\frac{|\boldsymbol{\mu}|^2}{3}\frac{\hbar\omega}{2\varepsilon_0 n^2}.\end{aligned} \tag{11.3.11}$$

The spontaneous emission rate in a microcavity is

$$\begin{aligned}\frac{1}{\tau_{\text{cav}}} &= \frac{2\pi}{\hbar}\rho_{\text{cav}}(\hbar\omega)\left\langle|\boldsymbol{\mu}\cdot\mathbf{E}(\mathbf{r}_e)|^2\right\rangle \\ &= \frac{2\pi}{\hbar}\rho_{\text{cav}}(\hbar\omega)\eta^2\frac{|\boldsymbol{\mu}|^2|\mathbf{E}(\mathbf{r}_e)|^2}{E_{\max}^2}E_{\max}^2\end{aligned} \tag{11.3.12}$$

where the factor

$$\eta^2 = \frac{\left\langle|\boldsymbol{\mu}\cdot\mathbf{E}(\mathbf{r}_e)|^2\right\rangle}{|\boldsymbol{\mu}|^2|\mathbf{E}(\mathbf{r}_e)|^2} = \cos^2\theta \tag{11.3.13}$$

accounts for the relative angle between the optical dipole vector and the modal electrical field direction. The ratio of the microcavity spontaneous emission rate to that of the "free-space" is thus

$$\begin{aligned}\frac{\tau_{\text{free}}}{\tau_{\text{cav}}} &= \frac{3\pi^2\hbar c^3}{\omega^2 n^3}\rho_{\text{cav}}(\hbar\omega)\frac{|\mathbf{E}(\mathbf{r}_e)|^2}{E_{\max}^2}\frac{n^2}{V_{\text{eff}}n_{\max}^2}\eta^2 \\ &\simeq \frac{3Q}{4\pi^2 V_{\text{eff}}}\left(\frac{\lambda}{n}\right)^3\frac{n^2}{n_{\max}^2}\frac{(\hbar\Delta\omega/2)^2}{(\hbar\omega-\hbar\omega_c)^2+(\hbar\Delta\omega/2)^2}\frac{|\mathbf{E}(\mathbf{r}_e)|^2}{E_{\max}^2}\eta^2\end{aligned} \tag{11.3.14}$$

where we have taken the expression near the cavity resonance $Q = \omega_c/\Delta\omega \simeq \omega/\Delta\omega$. If the emitting dipole satisfies the following conditions,

1. The dipole is placed at the maximum electrical intensity position,
2. The dipole direction is aligned with the field direction, and
3. The dipole emits photons at the resonance of the cavity, $\hbar\omega = \hbar\omega_c$,

we have the famous Purcell factor (assume $n_{\max} = n$ for simplicity)

$$F_p = \frac{3Q}{4\pi^2 V_{\text{eff}}}\left(\frac{\lambda}{n}\right)^3 \tag{11.3.15}$$

where the wavelength $\lambda = \lambda_c$ at the cavity resonance wavelength, $\omega_c/c = 2\pi/\lambda_c$. For a microcavity with a volume V_c, we can set $V_{eff} = V_c$ for convenience, when the modal confinement in the cavity is close to unity.

Increasing the Q/V_{eff} ratio can potentially reduce the laser threshold if other non-radiative losses remain constant because the spontaneous emission coupling into the lasing mode is increased. The spontaneous emission coupling factor is defined as:

$$\beta = \frac{\text{Sponatneous emission power coupled into a single mode}}{\text{Sponatneous emission power coupled into all modes}}. \tag{11.3.16}$$

It is approximated as [98]

$$\beta = \frac{F_p}{F_p + 1}. \tag{11.3.17}$$

Increasing the Purcell factor F_p will increase the spontaneous emission coupling into the lasing (cavity) mode. The fraction of the spontaneous emission into all other modes $(1 - \beta)$ is one of the fundamental losses in a laser cavity. By reducing $(1 - \beta)$ or increasing β, we can reduce the laser threshold.

Figure 11.28 shows the measured photoluminescence lifetime due to emission from quantum dot when it is placed on-resonance or off-resonance with the fundamental mode of a micropillar (VCSEL-type) cavity (inset of Fig. 11.28) [86]. When the quantum dot or quantum box (QB) emission is in resonance with the cavity fundamental mode, the photoluminescence (PL) lifetime is as short as 0.25 ns, which is five times shorter than the lifetime when a reference quantum dot is placed in a GaAs matrix (1.3 ns). On the other hand, the off-resonance QD emission

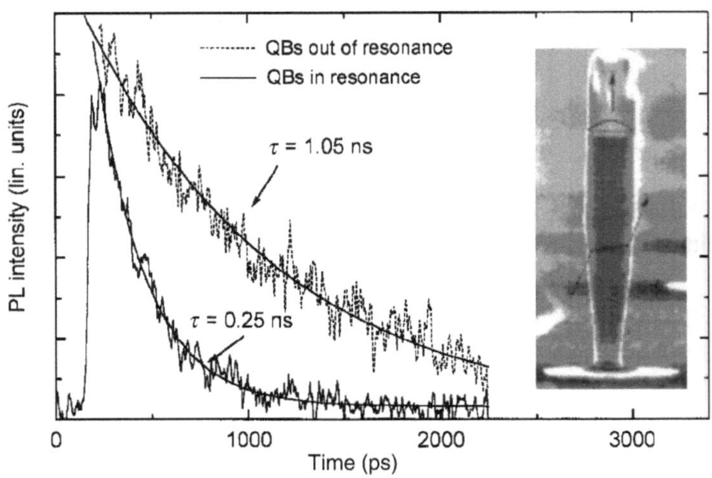

Figure 11.28 The measured photoluminescence (PL) lifetime due to emission from a quantum dot when it is on-resonance (0.25 ns) or off-resonance (1.05 ns) with the fundamental mode of the micropillar cavity. The insert shows a micropillar laser cavity. (Reprinted with permission from [86] © 1999 IEEE.)

PL lifetime (1.05 ns) due to the quantum dot emission coupling into the leaky mode is close to the reference quantum dot's lifetime (1.3 ns).

11.3.3 Two-dimensional Photonic Crystal Lasers

Figure 11.29 shows the band structure of a free-standing slab of a square-lattice photonic crystal for both TE-like and TM-like polarizations. The radius of the air hole is 0.35a and the slab thickness is 0.40a, where a is the lattice constant. The two dotted horizontal lines represent the resonant frequency positions of the whispering gallery mode (WGM) and the dipole mode when a single defect (i.e., one missing air hole) is present. The electrical field intensity patterns of the WGM and dipole mode are shown in Fig. 11.30a and Fig. 11.30b, respectively. The C_{4V} symmetry of the square lattice allows the creation of two nondegenerate modes and one doubly degenerate mode, among which a quadrupole-shaped nondegenerate mode and a dipole-shaped degenerate mode have been observed experimentally and confirmed theoretically by FDTD simulations. The nondegenerate mode in Fig. 11.30a has the characteristics of the WGM with the azimuthal mode number of 2 (i.e., exp($im\varphi$), $m = 2$). The resonant mode in Fig. 11.30b is a dipole mode polarized along the x direction. (The other degenerate dipole mode is polarized along the y direction.)

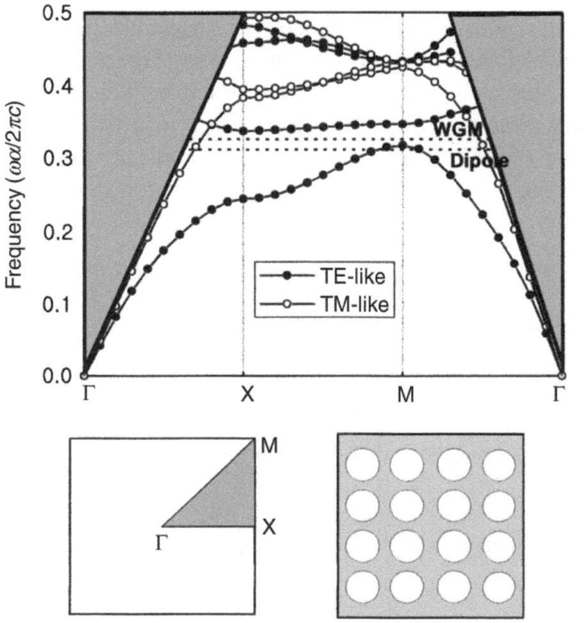

Figure 11.29 The band structure of a free-standing square lattice photonic crystal for both TE-like and TM-like polarizations. The two dotted horizontal lines represent the resonant frequency values of two optical modes, the whispering gallery mode (WGM) and the dipole-like mode, when a single defect (one missing hole) is present in an otherwise periodic square lattice in a two-dimensional slab structure. (Reprinted with permission from [93] © 2003 American Institute of Physics.)

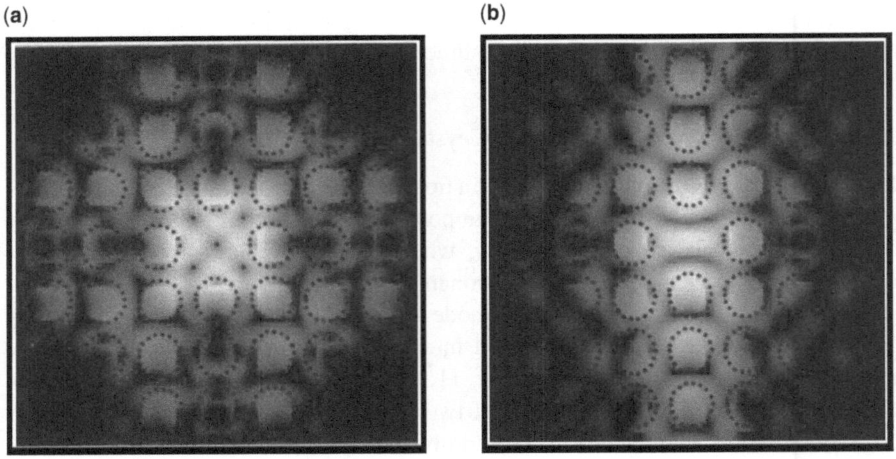

Figure 11.30 The optical mode patterns of (a) the whispering gallery mode (WGM) and (b) the dipole-like mode of a single defect square lattice photonic crystal laser in Fig. 11.23. (Reprinted with permission from [93] © 2003 American Institute of Physics.)

Figure 11.23a shows a schematic diagram of the free-standing slab structure with a small central post for current injection. It is designed such that the selected optical mode has a central intensity node and Q is therefore not degraded. On the other hand, those modes with central antinodes are discarded because they encounter unavoidable optical losses through the central post. Figure 11.31 shows the SEM picture of the fabricated photonic crystal structure with a single defect (one missing air hole) surrounded by chirped air holes in a two-dimensional photonic crystal slab. The radius of the air holes varies monotonically from the center: air hole radius = 0.28a (region I), 0.35a (region II), 0.385a (region III), 0.4a (region IV), and

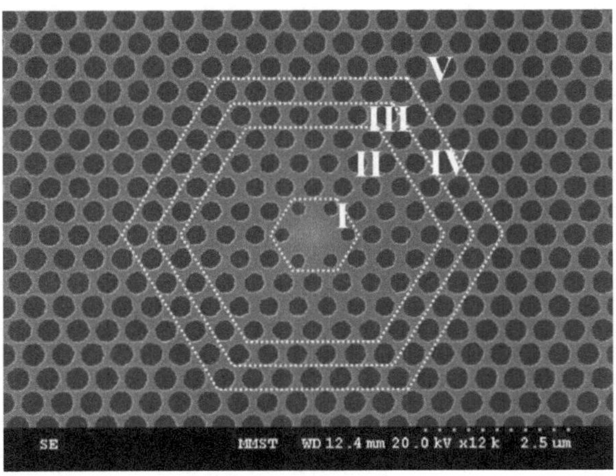

Figure 11.31 A SEM picture of a single-defect chirped air-holes photonic-crystal laser structure. (Reprinted with permission from [84] © 2005 IEEE.)

0.41a (region V), where $a = 510$ nm. A FDTD simulation for the fabricated structure in Fig. 11.31 allows for the estimation of the quality (Q) factor, the modal volume (V), and the Purcell factor (F_p) as a function of the post size (normalized to p_0), as shown in Fig. 11.32a–c. To compare the measured Q with the theoretical value, the size of the diamond-like post (p_0) is estimated from the SEM image, $p_0 = 0.64a \times 0.51a$ in the diagonal directions. In Fig. 11.32a, the Q factor (=3480) at the post size = 1.0 p_0 agrees well with the experimental value. In addition, Q decreases rapidly when the post size becomes larger than p_0. On the other hand, the electrical and thermal resistances increase with reduced post size. Thus, an optimization of the post size is necessary for a stable laser operation with better optical, electrical, and thermal properties.

11.3.4 Rate Equations and *L–I* Curve

The rate equations [99, 100] are commonly used to model the light output power versus the injection current (*L–I* curve) of a microcavity laser. Here we present a simple consistent model for the two rate equations, which have the same form for both edge emitters and microcavity lasers. The subtle differences will be explained below

$$\frac{dn}{dt} = \eta_i \frac{J}{qd} - \frac{n}{\tau(n)} - R_{st}(n)S \tag{11.3.18}$$

$$\frac{dS}{dt} = \Gamma R_{st}(n)S - \frac{S}{\tau_p} + \Gamma \beta R_{sp}(n) \tag{11.3.19}$$

where

n = electron density (cm^{-3})
S = photon density (cm^{-3})
η_i = injection quantum efficiency
J = current density (A/cm^2)
q = unit charge (Coulomb)
d = thickness of the active region (cm)
$\tau(n)$ = carrier lifetime (s)
τ_p = photon lifetime (s)
$R_{st}(n)$ = stimulated emission rate (s^{-1})
$R_{sp}(n)$ = spontaneous emission rate ($s^{-1}cm^{-3}$)
Γ = optical confinement factor
β = spontaneous emission coupling factor.

The lifetime $\tau(n)$ is defined as

$$\frac{n}{\tau(n)} = An + R_{sp}(n) + Cn^3. \tag{11.3.20}$$

The spontaneous emission rate into all modes, $R_{sp}(n)$, can be calculated rigorously if we know the cavity modes and the density of photon modes for a cavity.

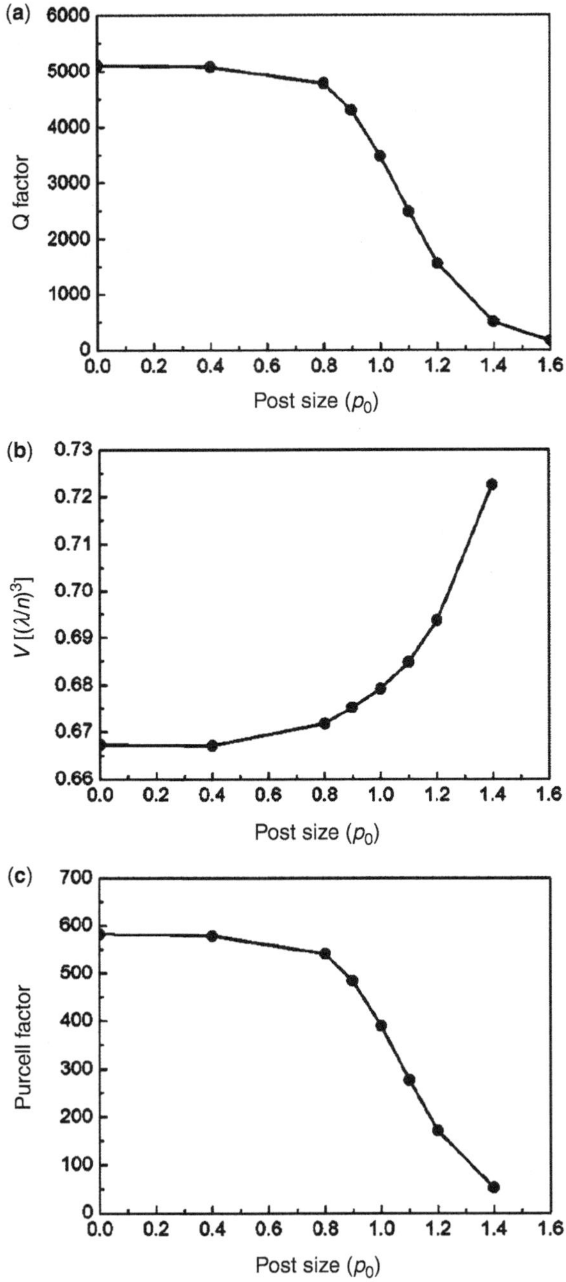

Figure 11.32 Theoretical results based on the FDTD simulation of (a) the quality (Q) factor, (b) the modal volume (V), and (c) the Purcell factor (F_p) as a function of the post size (normalized to p_0) for the fabricated photonic crystal laser structure in Fig. 11.31. (Reprinted with permission from [84] © 2005 IEEE.)

For simplicity, it is usually parameterized as

$$R_{sp}(n) = \gamma B n^2 \tag{11.3.21}$$

which accounts for electron–hole pair radiative recombination. Here a (dimensionless) field enhancement factor γ is added to account for the case when the gain region is placed at the peak of the standing wave of the microcavity (such as $\gamma = 2$ when the gain is right at the peak of the VCSEL cavity). B is the radiative recombination coefficient ($\text{cm}^3\,\text{s}^{-1}$). The spontaneous and stimulated emission rate can be calculated more rigorously from the expressions in Section 11.3.2 by integrating the spontaneous and stimulated emission rates over all the emitting dipoles over the active region and taking into account the occupation probability of the initial and final states. For spontaneous emission rate per unit active volume (V_a) coupling into the cavity modes $\mathbf{E}$, we use the photon density of the states of the cavity:

$$\beta R_{sp}(n) = \frac{2\pi}{\hbar}\frac{2}{V_a}\sum_{c,v}\int_0^\infty d(\hbar\omega)\left|\langle c|\boldsymbol{\mu}\cdot\mathbf{E}(\mathbf{r}_e)|v\rangle\right|^2\delta(E_c - E_v - \hbar\omega)f_c(1-f_v)\rho_{cav}(\hbar\omega) \tag{11.3.22}$$

where 2 accounts for the two spins. For coupling into the continuum modes of the cavity, the photon density of modes of the continuum $\rho_{cont}(\hbar\omega)$ with corresponding electric field mode has to be used, which is usually an unknown or difficult to find except for the case of the free space modes:

$$(1-\beta)R_{sp}(n) = \frac{2\pi}{\hbar}\frac{2}{V_a}\sum_{c,v}\int_0^\infty d(\hbar\omega)\left|\langle c|\boldsymbol{\mu}\cdot\mathbf{E}_{cont}(\mathbf{r}_e)|v\rangle\right|^2\delta(E_c - E_v - \hbar\omega)f_c(1-f_v)\rho_{cont}(\hbar\omega). \tag{11.3.23}$$

The stimulated emission rate per unit active volume ($\text{s}^{-1}\,\text{cm}^{-3}$) is

$$R_{st}(n)S = \frac{2\pi}{\hbar}\frac{2}{V_a}\sum_{c,v}\left|\langle c|\boldsymbol{\mu}\cdot\mathbf{E}(\mathbf{r}_e)|v\rangle\right|^2\delta(E_c - E_v - \hbar\omega)(f_c - f_v)SV_{eff}. \tag{11.3.24}$$

where the effective volume factor comes from the rate equations and the balance of units. For a traveling wave mode, such as that in a Fabry–Perot cavity with a uniform gain medium, $R_{st}(n)S$ can be simplified if we define the optical energy confinement factor and the effective volume as

$$\Gamma = \frac{2\varepsilon_0\int_{V_a} n_a^2|\mathbf{E}(\mathbf{r})|^2 d^3\mathbf{r}}{2\varepsilon_0\int_V n^2(\mathbf{r})|\mathbf{E}(\mathbf{r})|^2 d^3\mathbf{r}} = \frac{V_a}{V_{eff}} \tag{11.3.25}$$

Following a similar procedure as in Section 9.3.1 using the Bloch functions in the active region (V_a), the $\boldsymbol{k}$-selection rule, and (9.1.27), we obtain

$$R_{st}(n)S = v_g g(n)S \tag{11.3.26}$$

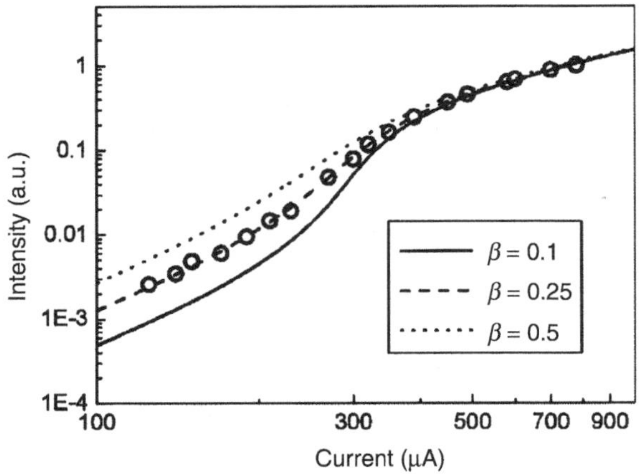

Figure 11.33 A comparison of the theoretically calculated light output vs. the injection current for three values of the spontaneous emission coupling factor $\beta = 0.1$, 0.25, and 0.5. It can be seen that the 0.25 curve fits the experimental data very well. (Reprinted with permission from [84] © 2005 IEEE.)

where v_g is the group velocity and $g(n)$ is the optical gain coefficient (1/cm) of the optical mode at the photon wavelength. The rate equations are similar to those in Chapter 10 except for the field enhancement factor for a microcavity, and the stimulated emission rate can be calculated more rigorously for a cavity mode. The gain for quantum-well active layers can be approximated by

$$g(n) = g_0 \ln\left(\frac{n}{n_{\text{tr}}}\right) \tag{11.3.27}$$

where n_{tr} = transparent carrier density and g_0 is the gain constant at the carrier density n_0 such that $n_0/n_{\text{tr}} = e^1$ and $g(n_0) = g_0$.

In Fig. 11.33, a comparison of the stimulated light output power versus the injection current (L–I curve) with the measured data is shown [84] for the spontaneous coupling factor $\beta = 0.1$, 0.25, and 0.5. It can be shown that $\beta = 0.25$ fits the experimental data very well. The data are for a monopole mode in a photonic crystal laser with chirped air holes near the defect, Fig. 11.31.

11.4 QUANTUM-CASCADE LASERS

Quantum-cascade lasers (QCLs) were realized by a team of researchers at Bell Labs in 1994 [101]. The theory for light amplification using a superlattice with a bias voltage was proposed back in 1971 [102]. The key idea is that the coherent light generation by intersubband transitions of only electrons (unipolar processes), instead of electrons and holes (bipolar processes), in the energy ladder of a biased superlattice, or called photon-assisted tunneling, can be realized. For a historical review, please see the review papers by Capasso and others [103–107]. It is interesting to see that a few researchers at Bell Labs have also demonstrated quantum-well infrared photodetectors (QWIPs) (see Chapters 9 and 15), in which the intersubband absorption occurs via the bound-to-continuum transitions in GaInAs/AlInAs quantum-well systems.

The first demonstration of QC lasers used a three-quantum-well active region [105]. Figure 11.34 shows various active regions proposed by different groups [108] for intersubband optical transitions: they include a single quantum well, two quantum wells, three quantum wells, photon-assisted tunneling structures, doped superlattice, and chirped superlattice active regions. In a quantum-cascade structure, multiple stages are required. Each stage or period usually consists of an injector region that is doped, an active region where the radiative transition occurs, and a carrier transport region (such as a miniband) into the next stage. The basic ideas are the selection of carrier (electron) injected into the excited state of the quantum-well or biased superlattice, the photon stimulated emission in the active region, and the depopulation of the ground-state carriers through tunneling into the next stage of the quantum-cascade structure. Therefore, an injected electron, after producing a photon emission at one stage, is recycled in the next or the following stages, resulting in multiple photon emissions.

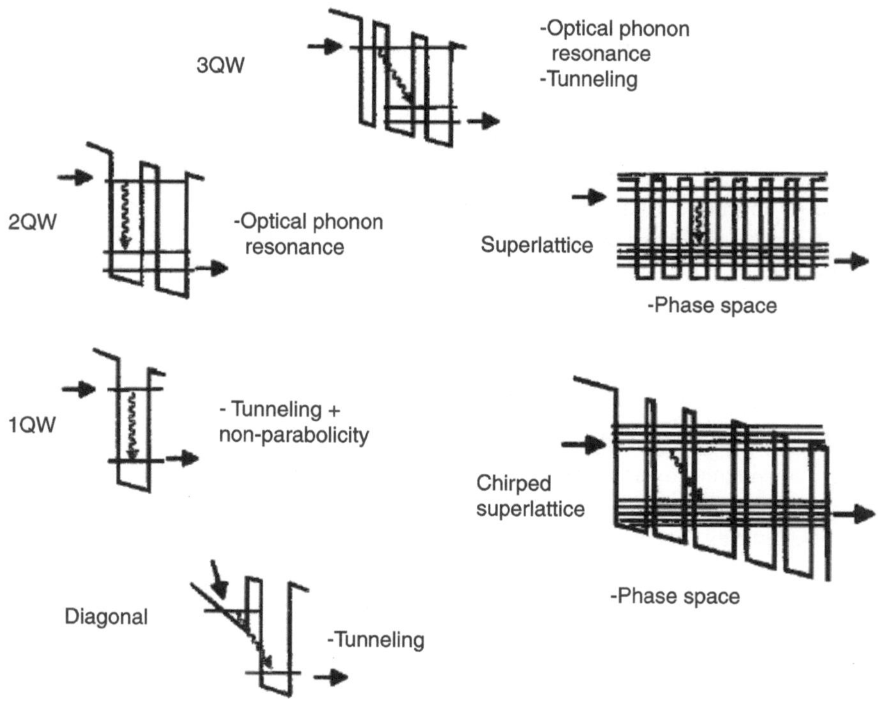

Figure 11.34 Various active regions for intersubband lasers proposed in the literature. They are based on (clockwise) diagonal photon-assisted tunneling structures, one quantum well, two quantum wells, three quantum wells, photon-assisted tunneling, doped superlattices, and chirped superlattices. (Reprinted with permission from [108] © 2002 IEEE.)

The first QC laser was based on the material system GaInAs/AlInAs ($\Delta E_c \sim 520$ meV) grown on InP with a lasing wavelength of 4.2 μm [101]. The first cw operation at room temperature was reported in 2002 [107] with an output power of 10–20 mW at 298 K at a wavelength of 9.1 μm and showed a maximum cw operation temperature of 312K. High power (>100 mW) and high temperature (>358K) were reported in 2007 [106] using the same material system. Other material systems such as GaAs/AlGaAs ($\Delta E_c \sim 290$ meV) on GaAs substrate with a longer lasing wavelength λ at 9 μm [109], InAs/AlSb (a large $\Delta E_c \sim 2.1$ eV for Γ-valley) grown on n-InAs substrate with a low-temperature wavelength at 10 μm [110] or at 2.95–2.97 μm from low to high temperatures [111], or GaInAs/AlAsSb (a large $\Delta E_c \sim 1.6$ eV for Γ-valley) lattice-matched to InP substrate with a room-temperature short lasing wavelength (3.7–3.9 μm) [112, 113] were realized. Distributed-feedback lasers with a tunability of 0.3–1.0 nm/K heat-sink temperature or 20–40 nm/A cw current were reported [114]. Microfluidic tuning of DFB QC lasers were also realized [115]. Novel lasers such as microdisk [116] and photonic-crystal [117] QC lasers have also been demonstrated. Although most of the QC lasers were grown by MBE, there has been much advancement in metal–organic vapor phase epitaxy (MOVPE) growth of quantum-cascade lasers [118–120] with comparable performance of similar structures. MOVPE is a well-established platform for high-volume production of low-cost and reliable semiconductor lasers. The technology allows for the deposition of uniform and thick layers at high rates and offers excellent stability over long growth runs and short machine down-time. Therefore, MOVPE is suitable for the industrial production of QC lasers [118].

In this section, we focus on type I QC lasers for mid-IR, far-IR, and THz ranges. We discuss the fundamental designs for the active region in Section 11.4.1, rate equations and optical gain in Section 11.4.2, waveguide designs in Section 11.4.3, and QC laser performance in Section 11.4.4. We then describe THz QC lasers in Section 11.4.5. The last section (11.4.6) addresses the type II QC lasers, which are based on interband transitions [121].

11.4.1 Type I Mid-IR Quantum-Cascade Laser Structures

There are a few designs for the active region of quantum-cascade lasers. The major design considerations are

- The energy levels and corresponding wave functions.
- How the carriers are injected (into the excited state) in the presence of bias.
- The selection of states for radiative transition and their intersubband oscillator strength.
- The depopulation of carriers of the lower energy state of the radiative transition.
- The extraction or tunneling of the carriers from the ground state into the injection/relaxation layer of the next stage.

Below, we focus on three major designs of the active region of QC lasers.

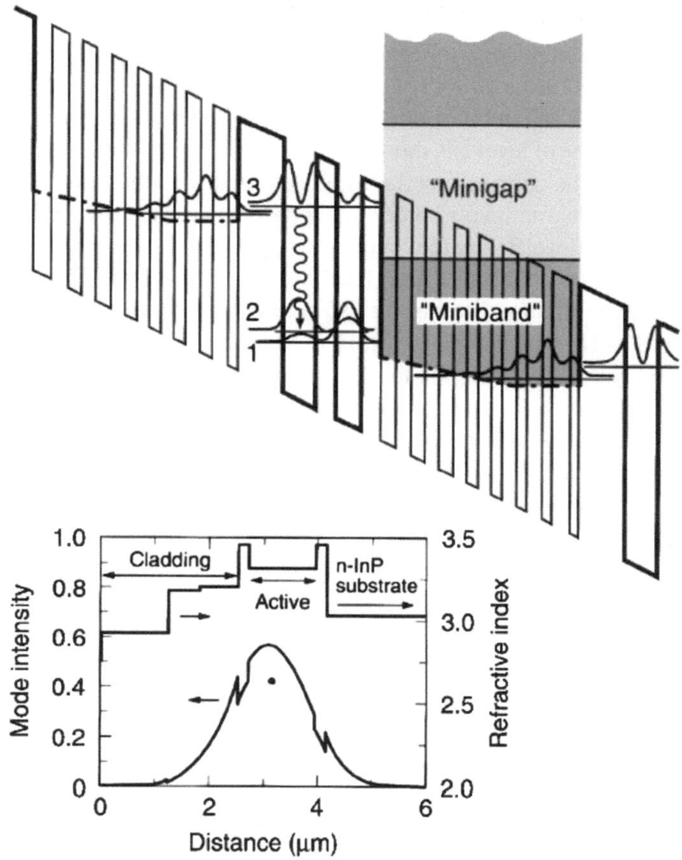

Figure 11.35 Schematic diagram for the conduction-band edge and wave functions of two periods of a quantum-cascade laser. The design uses a vertical transition in a coupled quantum well active region in the presence of a bias field of 70 kV/cm. The radiative transition occurs between level 3 and level 2; and levels 2 and 1 are separated by one optical phonon energy. Level 3 falls into the minigap of the (right) chirped superlattice, and level 1 falls in the miniband for subsequent carrier relaxation or injection into the excited state (level 3) of the next stage. The layer sequence of one period of the ($Al_{0.48}In_{0.52}As/Ga_{0.47}In_{0.53}As$) alternating lasers in nanometers (nm) from the (thick) injection barrier is (6.8/4.8)(2.8/3.9)(2.7/2.2)(2.2/2.1)(2.1/2.0) (2.0/1.8)(1.8/1.7)(2.0/1.6)(2.2/1.6)(2.4/1.4). The underlined layers are doped n-type to $3 \times 10^{17}\,\text{cm}^{-3}$. The bottom inset shows the refractive index profile and the corresponding optical mode profile. (Reprinted with permission from [122] © 1995 American Institute of Physics.)

(a) Bound-to-Bound Transition [122–125]: Figure 11.35 (top) shows a design where the active region contains a coupled quantum-well structure with three major bound states, levels 1, 2, and 3, followed by a graded-gap superlattice injection/relaxation region. The radiative transition occurs between level 3 and level 2, which are discrete bound states, and they have a strong overlap integral, so-called vertical transition in real space. (This is an improvement over the "diagonal" transition in the first version of QC lasers using three quantum wells, where the transition between the higher state $n = 3$ wave function centered in the first well closest to the injection barrier, and the $n = 2$ state wave function is in the middle well; therefore, the overlap integral or the oscillator strength is relatively weaker.) At the threshold bias field (70 kV/cm), the ground states of the two 4.8- and 3.9-nm quantum wells are anti-crossed (or coupled) with an energy splitting of an optical phonon energy. Therefore, the carrier lifetime of level 2 is short (0.6 ps) due to the resonant phonon depopulation [122]. The chirped superlattice is designed such that level 3 of the active region is in its minigap in order to suppress the escape of electrons from the excited state (level 3) into the continuum states, while allowing for the extraction of electrons from the lower state (level 1 of the coupled wells) into the miniband for carrier relaxation. Most of the QC layers are undoped with the exception of layers 12–18, which are doped to $3 \times 10^{17}\ \text{cm}^{-3}$. The injection region or the doped layers (12–18) fall in the chirped SL region.

By inserting an extra quantum well, a double resonance phonon depopulation scheme has also been realized [107, 108, 126] in a four-level system, where the radiative transition occurs between level 4 and level 3, while level 3 and 2 as well as level 2 and 1 are both separated by one phonon energy at the biased threshold field.

(b) **Bound-to-Continuum Transition** [108, 127, 128], Fig. 11.36: In such a design, the radiative transition occurs between an isolated bound state and a continuum of miniband, which provides an efficient carrier extraction mechanism. This approach provides another solution to maintain a high population inversion and low-threshold current density. As shown in Fig. 11.36a [127], the active region spans the whole period and consists of a chirped superlattice showing a tilted lower miniband, the width of which is a maximum in the center and decreases on both sides close to the injection barriers. The upper state is created in the first minigap by a small well adjacent to the injection barrier, and its wave function has a maximum close to the injection barrier and decreases smoothly in the active region. This upper state is well separated from the higher states of the superlattice. The computed oscillator strengths for various transitions from the upper state into the miniband states are shown as the lowest curve in Fig. 11.36b, which gives the general features of the luminescence spectrum at 80K and 300K.

(c) **Interminibands Transition** [129–131]: Figure 11.37a shows a design of interminiband quantum-cascade lasers. The radiative transition occurs between two minibands in the chirped superlattices. The bottom few states of the miniband 1 are coupled to the miniband 2 of the next stage in the presence of a bias voltage near threshold. Compared with the previous designs, which require resonant tunneling into a discrete excited state, superlattices offer the advantage of wide energy minibands, which can carry large current density without tuning into misaligned levels when the applied voltage is increased. In earlier designs, the superlattice (SL) active regions are doped in order to minimize electric field penetration, which results in higher optical losses and reduced population inversion at high temperature. An improved design by placing dopants in the injection/relaxation regions outside the active region has demonstrated high-power performance at room temperature [131, 132].

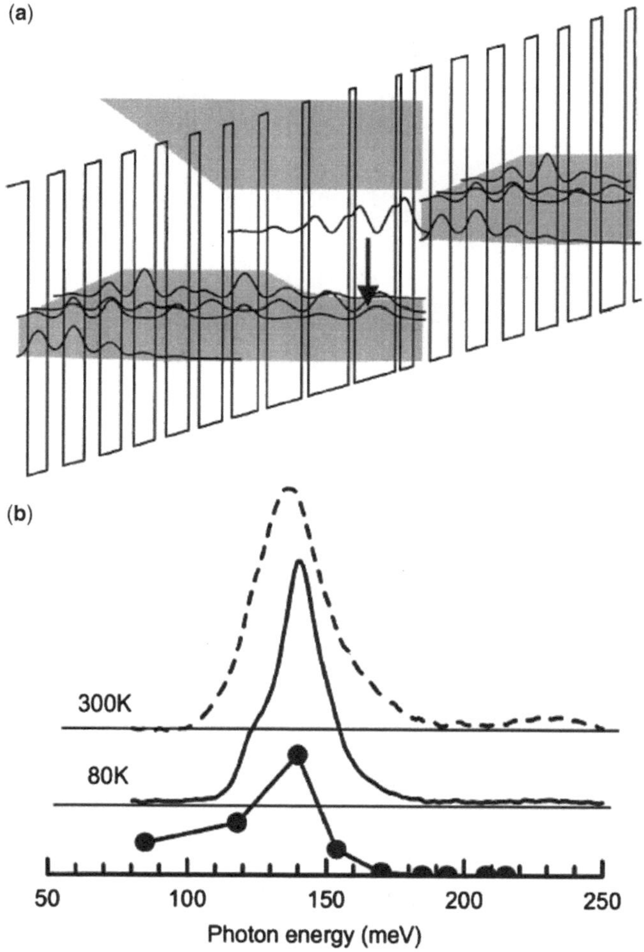

Figure 11.36 (a) Schematic diagram for the conduction band of (more than) one stage of a bound-to-continuum QC laser structure at a bias field of 35 kV/cm. The squared moduli of the wave functions are plotted at the corresponding energy levels. The sequence for the alternating layers $\mathbf{Al_{0.48}In_{0.52}As}$/$Ga_{0.47}In_{0.53}As$ (boldface is for barriers) starting from the injection barrier is **4.0**/2.0/**0.7**/6.0/**0.9**/5.9/**1.0**/5.2/**1.3**/3.8/**1.4**/3.5/**1.5**/3.3/**1.6**/3.1/**1.9**/3.1/**2.3**/3.0/**2.4**/2.9. The underlined regions are doped n-type with Si to 3×10^{17} cm^{-3}. (b) Electroluminescence spectrum from the active region at 300K and 80K, with the bottom curve indicating the calculated oscillator strength of the various transitions from the upper (bound) state. (Reprinted with permission from [127] © 2001 American Institute of Physics.)

11.4.2 Rate Equations, Optical Gain, and Threshold Current

We start with the rate equations for a three-level system of the active region, which has an injector state from the left and has some mechanism to depopulate the lower level. In a QCL, the depopulation of the lower level is based on the transport into the injector of the next stage. Figure 11.38 [127] shows a model for the key parameters used in the rate equation of a three-level model (as in a bound-to-bound transition) of a quantum-cascade laser. We assume an injection efficiency

(a)

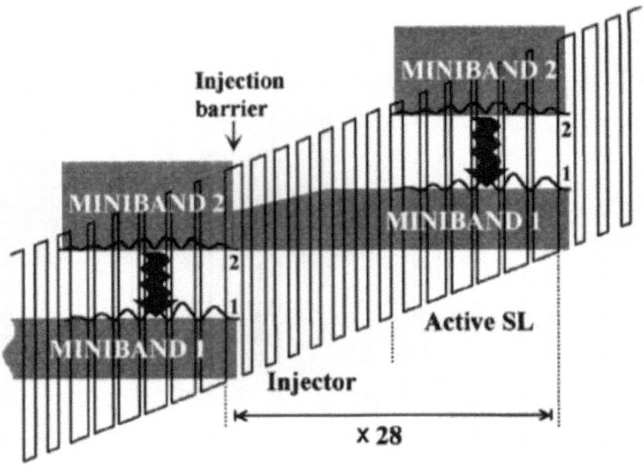

(b) **Complete structure**

GaInAs	$n = 1 \times 10^{20}$ cm^{-3}	100 Å
GaInAs	7×10^{18} cm^{-3}	5000 Å
Graded	2×10^{17} cm^{-3}	300 Å
AlInAs	2×10^{17} cm^{-3}	10000 Å
AlInAs	1×10^{17} cm^{-3}	12000 Å
Graded	1×10^{17} cm^{-3}	300 Å
GaInAs	5×10^{16} cm^{-3}	3300 Å
Injector	2×10^{17} cm^{-3}	302 Å ×28
Active SL region		375 Å ×28
Graded	2×10^{17} cm^{-3}	302 Å
GaInAs	5×10^{16} cm^{-3}	5000 Å
Graded	1×10^{17} cm^{-3}	250 Å

Injector

GaInAs		23 Å
AlInAs		25 Å
GaInAs	2×10^{17} cm^{-3}	23 Å
AlInAs	2×10^{17} cm^{-3}	25 Å
GaInAs	2×10^{17} cm^{-3}	22 Å
AlInAs	2×10^{17} cm^{-3}	26 Å
GaInAs	2×10^{17} cm^{-3}	20 Å
AlInAs	2×10^{17} cm^{-3}	26 Å
GaInAs		19 Å
AlInAs		27 Å
GaInAs		19 Å
AlInAs		29 Å
GaInAs		18 Å
AlInAs		35 Å

Active SL

GaInAs	51 Å
AlInAs	11 Å
GaInAs	48 Å
AlInAs	11 Å
GaInAs	44 Å
AlInAs	11 Å
GaInAs	41 Å
AlInAs	12 Å
GaInAs	38 Å
AlInAs	13 Å
GaInAs	35 Å
AlInAs	25 Å

Figure 11.37 (a) Conduction-band profile showing two stages of a interminiband quantum-cascade laser at a bias of 45 kV/cm. The shaded regions indicate the energy and spatial extension of the manifold of miniband-like states. The radiative transition occurs between the edges of the minibands 2 and 1, and the wave function moduli are also plotted at the corresponding energy levels. (b) The details of the materials compositions, layer thicknesses, and doping concentrations of the complete structure. (Reprinted with permission from [129] © 1998 American Institute of Physics.)

into level i, η_i, and the lasing transition occurs between levels 3 and 2. The rate equations are given as

$$
\begin{aligned}
\frac{dn_3}{dt} &= \eta_3 \frac{J}{qd} - \frac{n_3}{\tau_3} - v_g g(n_3 - n_2)S \\
\frac{dn_2}{dt} &= \eta_2 \frac{J}{qd} + \frac{n_3}{\tau_{32}} - \frac{n_2}{\tau_2} + v_g g(n_3 - n_2)S \\
\frac{dn_1}{dt} &= \eta_1 \frac{J}{qd} + \frac{n_3}{\tau_{31}} + \frac{n_2}{\tau_{21}} - \frac{n_1}{\tau_1} \\
\frac{dS}{dt} &= \Gamma v_g g(n_3 - n_2)S - \frac{S}{\tau_p} + \beta R_{sp}(n_3, n_2)
\end{aligned}
\tag{11.4.1}
$$

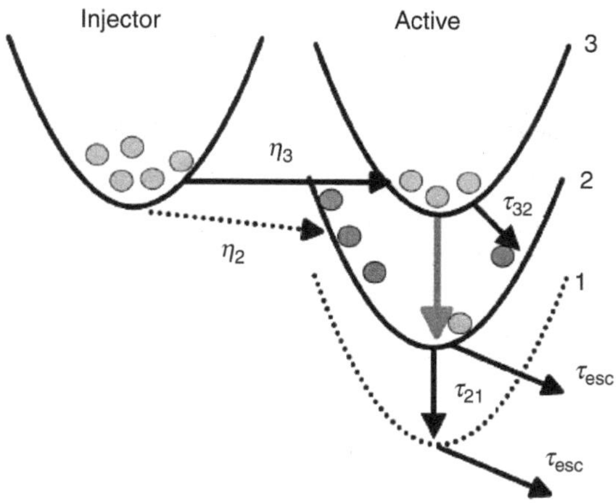

Figure 11.38 Schematic diagram to illustrate the parameters used in the rate equations. (Reprinted with permission from [127] © 2001 American Institute of Physics.)

where n_i are the electron concentrations ($1/\text{cm}^3$) in level i, J is the current density (A/cm^2), d is the thickness (cm) of the active region, and $1/\tau_i$ is the total scattering rate for level i ($1/\text{s}$). Here g is the gain coefficient and is a function of the population inversion, $n_3 - n_2$. S is the photon density, and τ_p is the photon lifetime. R_{sp} is the spontaneous emission rate, and β is the spontaneous emission coupling coefficient. Below threshold, the photon density S is negligible. In steady state, we can set $d/dt = 0$, which gives

$$\begin{aligned} n_3 &= \eta_3 \frac{J\tau_3}{qd} \\ n_2 &= \eta_2 \frac{J\tau_2}{qd} + \frac{n_3\tau_2}{\tau_{32}} \\ n_1 &= \eta_1 \frac{J\tau_1}{qd} + \frac{n_3\tau_1}{\tau_{31}} + \frac{n_2\tau_1}{\tau_{21}}. \end{aligned} \tag{11.4.2}$$

The gain spectrum for intersubband transition between levels 3 and 2 is based on (9.7.5)

$$g(\hbar\omega) = \frac{\omega}{n_r c\varepsilon_0} \frac{|\mu_{32}|^2 \gamma}{(E_3 - E_2 - \hbar\omega)^2 + \gamma^2}(n_3 - n_2) \tag{11.4.3}$$

where n_r is the refractive index, μ_{32} is the intersubband dipole moment, and γ is the half width at half maximum (HWHM) or the linewidth. The peak gain for intersubband transitions is given as

$$g_{\text{peak}} = \frac{\omega}{n_r c\varepsilon_0} \frac{|\mu_{32}|^2}{\gamma}(n_3 - n_2). \tag{11.4.4}$$

If we insert (11.4.2) into (11.4.4), we obtain

$$g_{\text{peak}} = \frac{\omega}{n_r c \varepsilon_0} \frac{|\mu_{32}|^2}{\gamma} \frac{J}{qd} \left[\eta_3 \tau_3 \left(1 - \frac{\tau_2}{\tau_{32}} \right) - \eta_2 \tau_2 \right]. \tag{11.4.5}$$

Equation (11.4.5) indicates that, in order to achieve population inversion (gain), we want $\eta_3\ \tau_3$ to be large, and $\eta_2 \tau_2$ and τ_2/τ_{32} to be small. At threshold, the peak gain must equal the threshold gain

$$g_{\text{th}} = \frac{\alpha_i + \alpha_m}{\Gamma} \tag{11.4.6}$$

where α_{i} is the intrinsic loss, α_{m} is the mirror loss, and Γ is the optical confinement factor. Equations (11.4.5) and (11.4.6) give the threshold current density

$$J_{\text{th}} = \frac{\alpha_{\text{i}} + \alpha_{\text{m}}}{\Gamma} \frac{qdn_r c \varepsilon_0}{\omega} \frac{\gamma}{|\mu_{32}|^2} \frac{1}{\eta_3 \tau_3 (1 - \tau_2/\tau_{32}) - \eta_2 \tau_2}. \tag{11.4.7}$$

Figure 11.39a [122] shows the measured net modal gain as a function of wavenumber of a QC laser with a 1.3-mm cavity length at various bias currents from 40 to 200 mA below threshold (230 mA) at a temperature of 15K. The gain spectrum appears Lorentzian or Gaussian-like [132, 133]. Figure 11.39b shows the peak net modal gain as a function of the injection current density J. The linear relation is quite clear. Thermal effects and carrier heating effects have been investigated experimentally [134] and theoretically [135]. A comprehensive model for the rate equations of QC lasers and the laser linewidth has been recently developed and compared well with experimental data [136].

11.4.3 Optical Waveguides and Modes of Quantum-Cascade Lasers

Because the intersubband transition is dominated by the z-polarization of the quantum-well or superlattice structures, where z is the growth axis of the epilayers, the optical waveguide modes are TM polarized. Because metals and doped semiconductors behave like plasma in mid-infrared or terahertz frequencies, surface plasma or plasmonics is used as part of the waveguide mechanism. The solution of Maxwell's equations for the TM polarization is usually based on finite-difference time-domain method (FDTD), finite-element method (FEM), or a simple one-dimensional finite-difference frequency domain method or propagation matrix method.

Three types of optical waveguides are used for quantum-cascade lasers:

1. A conventional dielectric waveguide with a high-index core (consisting of the active region) sandwiched between lower-index cladding layers. The optical mode has a relatively symmetrical profile, Fig. 11.40a.
2. A single surface-plasmon waveguide consisting of one metal surface next to the core of the dielectric waveguide, Fig. 11.40b.

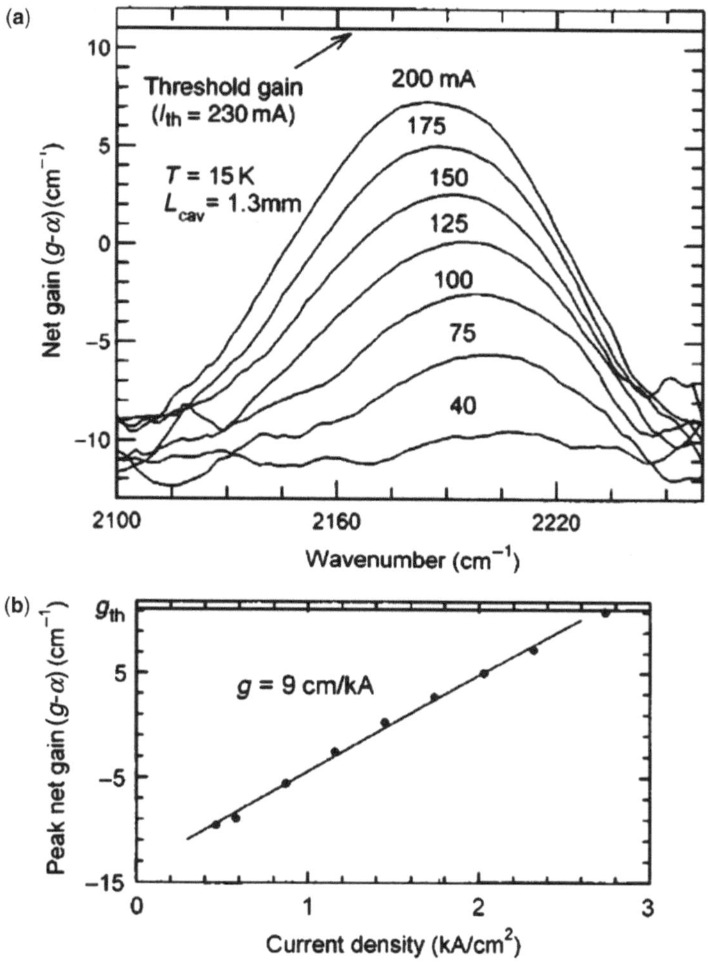

Figure 11.39 (a) Net modal gain of a Fabry–Perot quantum-cascade laser of Fig. 11.35 measured by the Hakki–Paoli method. (b) The peak modal gain of (a) is plotted as a function of the injection current density. (Reprinted with permission from [122] © 1995 American Institute of Physics.)

3. A double-surface plasmon (metal–semiconductor–metal) waveguide or a double metal claddings. The optical confinement is close to unity, and the waveguide structure is essentially the same as a microwave metal waveguide with the optical mode decays exponentially into the metal, Fig. 11.40c.

The metal cladding or surface plasmon waveguides (see Chapter 7) are especially useful for wavelength longer than 15 μm, including the terahertz range. The reason is that the metal cladding allows for the realization of a large optical confinement factor without growing the relatively thick epitaxial layers. Research at Bell Labs on QC lasers has pushed the growth of epitaxial layer thickness close to the upper practical limits of modern molecular-beam epitaxy (about 10 μm) using a

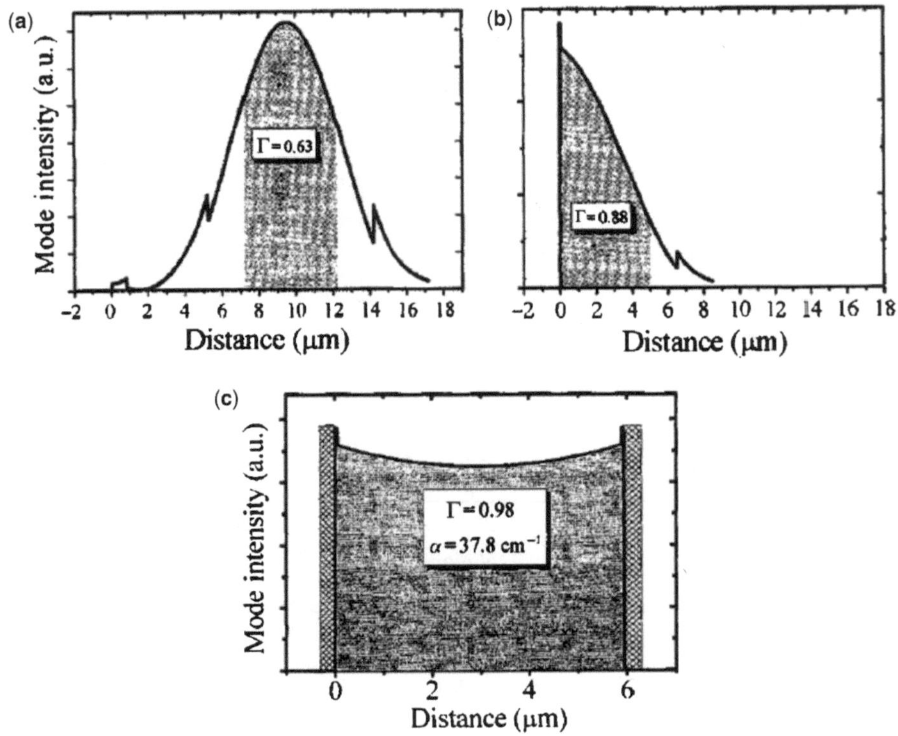

Figure 11.40 Normalized optical intensity profile of the fundamental mode for (a) a conventional dielectric waveguide (the optical confinement factor $\Gamma = 0.63$, and the waveguide loss $\alpha = 41\ \text{cm}^{-1}$) and (b) a surface-plasmon waveguide ($\Gamma = 0.88$ and $\alpha = 41\ \text{cm}^{-1}$) designed for a quantum-cascade laser at 24-μm wavelength. The shaded areas indicate the multiple stages of active and injection regions. (c) The optical intensity profile for a double-surface plasmon (two metal claddings) waveguide ($\Gamma = 0.98$ and $\alpha = 37.8\ \text{cm}^{-1}$) designed for a wavelength of 21 μm. (Reprinted with permission from [103] © 2002 IEEE.)

conventional dielectric waveguide. Using a single metal on top of an active region with 35 quantum-cascade periods, which supports surface plasmon modes, only 4-μm-thick epitaxial material was grown to realize a QC laser at $\lambda = 17$ μm, as opposed to the 9-μm-thick dielectric waveguide used for the same wavelength. The much improved confinement factor reduces the threshold substantially and increases the optical output power.

The calculated optical losses for both waveguides in Fig. 11.40a and 11.40b for a QC laser at $\lambda = 24$ μm are $41\ \text{cm}^{-1}$. The optical confinement factor is 0.88 using only 6.5 μm of epitaxial material, as opposed to the prohibitively thick 14 μm for the conventional waveguide with an optical confinement factor of 0.63. The double-plasmon waveguide in Fig. 11.40c is designed for a wavelength of 21 μm. The optical confinement factor is 0.98 with a waveguide loss of $37.8\ \text{cm}^{-1}$.

For long-wavelength QC lasers using a double-plasmon waveguide with a lasing wavelength of 76 μm from a vertical-transition active region, waveguide loss has been measured and the loss is $42 \pm 20\ \text{cm}^{-1}$ [137].

The near-field scanning optical microscope (NSOM) imaging of QC lasers with wavelengths of 5.3 μm and 7.0 μm has also been reported [138]. The measured modal patterns and theoretical results seem to agree well. NSOM provides a nice optical characterization tool for the optical modes and identifies the features near the boundaries of the active regions and cladding regions.

11.4.4 Performance of Quantum-Cascade Lasers

Figure 11.41 shows the experimental data of the voltage ($I-V$) and the light output power ($L-I$) as a function of the injection current for a QC laser using the bound-to-continuum active region shown in Fig. 11.36. The QC laser was grown by MBE using InGaAs/AlGaAs lattice-matched to InP substrate. It consists of 35 periods of active layers (Fig. 11.36) embedded in an optical waveguide formed on one side by the InP substrate and the other side by an InP top cladding grown by MOCVD. The agreement between the calculated oscillator strength and the measured electroluminescence spectrum (Fig. 11.36) is very good. The QC lasers were mounted junction-side up onto a copper submount. The cavity length is 3 mm long and the width is 28 μm. The lasing wavelength is 9.1 μm. The peak optical output power is measured with a pulsed current at a duty cycle of 1.5% at various temperatures from −30°C to 150°C. The inset shows the threshold current as a function of the temperature, showing a characteristic temperature T_0 of 190K. The slope efficiency is $dP/dI = 200\,\text{mW/A}$ at $T = 30°\text{C}$ and a maximum power of 700 mW. Peak power above 1 W at −30°C was also reported.

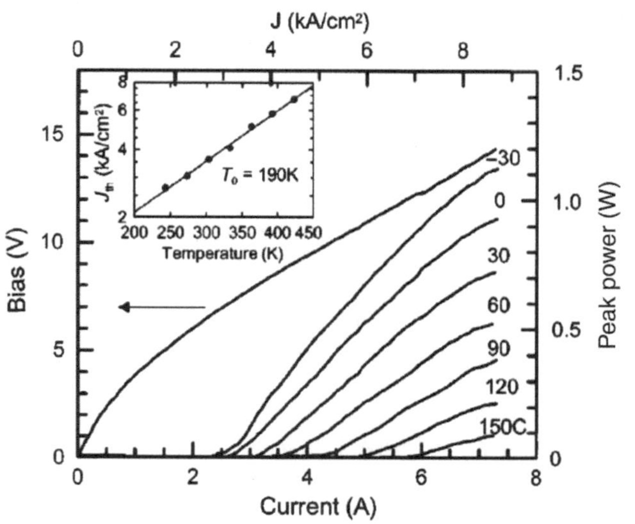

Figure 11.41 Peak optical output power versus injection current of a bound-to-continuum QC laser at various temperatures. The measurement was performed with a pulsed mode at 1.5% duty cycle. The inset shows the temperature dependence of the threshold current with a characteristic temperature of 190K. (Reprinted with permission from [127] © 2001 American Institute of Physics.)

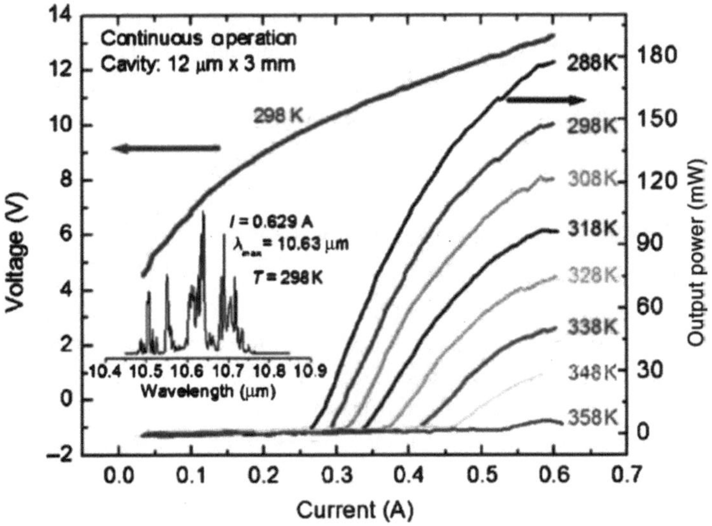

Figure 11.42 The L–I curves of a cw operational QC laser with a 12-μm-wide and 3-mm-long cavity. The V–I curve is also shown for $T = 298\text{K}$ along with the emission spectrum at 0.629 A injection current. The peak wavelength is 10.6 μm. The material system is $Al_{0.48}In_{0.52}As/Ga_{0.47}In_{0.53}As$ grown by gas-source MBE with InP as the upper cladding and cap layer. (Reprinted with permission from [106] © 2007 American Institute of Physics.)

Figure 11.42 shows the L–I curves of a high-power (>100 mW) and high-temperature (358K) QC laser near a wavelength of 10.6 μm using the $Al_{0.48}In_{0.52}As/Ga_{0.47}In_{0.53}As$ material system [106]. The laser device is 12 μm wide and 3 mm long. The V–I curve is shown for $T = 298\text{K}$ along with the emission spectrum at 0.629 A injection current. At room temperature (298K), the cw output power is 148 mW with a threshold current density of 0.81 kA/cm^2 and a slope efficiency of 0.717 W/A. The threshold voltage is 10.23 V, and the differential resistance above threshold is 10 Ω. The maximum output power ranges from 176 mW at 288K to 6 mW at 358K.

Shorter wavelength (3.7∼3.9 μm) GaInAs/AlAsSb QC lasers [112] using a 3 QW vertical transition active region show a peak power of 31 mW at 300K with a T_0 of 170K for temperature range of 220K to 310K. Shorter wavelength below 3 μm using InAs/AlSb based on the bound-to-continuum design shows a threshold of 3 kA/cm^2 at 84K and operates up to room temperature [111].

11.4.5 Terahertz Quantum-Cascade Lasers

Terahertz frequencies (1 − 10 THz, photon energy 4 − 40 meV, or wavelength 300 − 30 μm) cover the gap between microwave (GHz) and optical (10^{15} Hz) frequencies, and they are among the underdeveloped electromagnetic spectra. They have potential applications in areas such as astrophyics, biomedical imaging, biochemical agent detection, remote atmospheric sensing and monitoring, noninvasive inspection of semiconductor wafers and circuits, as well as high-speed and free-space communications [139]. Conventional semiconductor devices for GHz and optical devices are relatively more well-developed than are those of the terahertz devices. The reason is that semiconductor electronic devices such as transistors or millimeter-wave devices are limited to below 1 THz due to the transport and parasitic RC time constants. Semiconductor photonic devices, such as laser diodes based on the quantum electronics or the particle nature of photons, are typically limited by the band gap (larger than 10 THz or 40 meV). Therefore, 1 to 10 THz frequency range presents the opportunities for the monocarrier type of photonic devices such as quantum-cascade lasers based on the intersubband transition. Although mid-IR quantum-cascade lasers were realized in 1994 [101], it took several years to realize the 4.4 THz QC lasers in 2002 [140]. The major challenges are as follows [139]. First, because the photon energy (of the order 10–20 meV) in the THz range is smaller or comparable with phonon energy, the selective depopulation mechanism based on longitudinal optical (LO)-phonon scattering is difficult to implement. Second, a low-loss optical waveguide is difficult to make in the THz frequencies. The first THz laser was realized in 2002 using a chirped superlattice [140]. The optical confinement was based on a double-surface plasmon waveguide grown on a semi-insulating (SI) substrate. A THz QC laser based on the bound-to-continuum intersubband transition was soon realized with higher powers and operating temperatures [141]. An idea using resonant LO-phonon scattering to selectively depopulate the lower radiative level while maintaining a long upper level lifetime was realized [142]. Using the double-plasmon waveguide, significant improvements in the performance have been made. Much progress has also been achieved by a few groups [143–156] in order to achieve higher power and operation temperature or to lower the frequency below 2 THz. For examples, low frequency such as 1.39 THz [152], 1.6 − 1.8 THz [154], and 1.9 THz [153] have been reported. High power (145 mW at 5K) and lasing up to 160K have been demonstrated [156].

Figure 11.43a shows the MQW active design from Ref. 149. Figure 11.43b and Fig. 11.43c show the optical (THz) waveguide modal profiles (solid) and the real part of the dielectric constants for the semi-insulating surface plasmon (SISP) waveguide and the double-metal waveguide. The structure was grown by MBE on a semi-insulating substrate with 178 cascade modules. The cladding and contact layers were chosen such that the device could be processed into either a SISP waveguide or a double-sided metal waveguide. First, an $Al_{0.5}Ga_{0.5}As$ etch-stop layer of 0.3 μm was grown then follow by an 0.8-μm GaAs contact layer doped at $n^+ = 2 \times 10^{18}$ cm^{-3}.

The device was fabricated into a metal–semiconductor–metal waveguide using In-Au metallic wafer bonding as described in Fig. 11.44 [147]. The ridges were dry etched using the top Ti/Au (200/4000 Å) metallization as a self-aligned etch mask.

The SISP waveguide is formed by sandwiching the 178 QC periods between a 60-nm upper contact layer (doped $n^+ = 5 \times 10^{18}$ cm^{-3}) and a bottom 800-nm contact layer ($n = 2 \times 10^{18}$ cm^{-3}) on a semi-insulating GaAs substrate [142, 148]. A top non-alloyed ohmic contact was made by evaporating Ti/Au layers on low-temperature-grown GaAs. Wet etching was used to define a 150-μm-wide ridge,

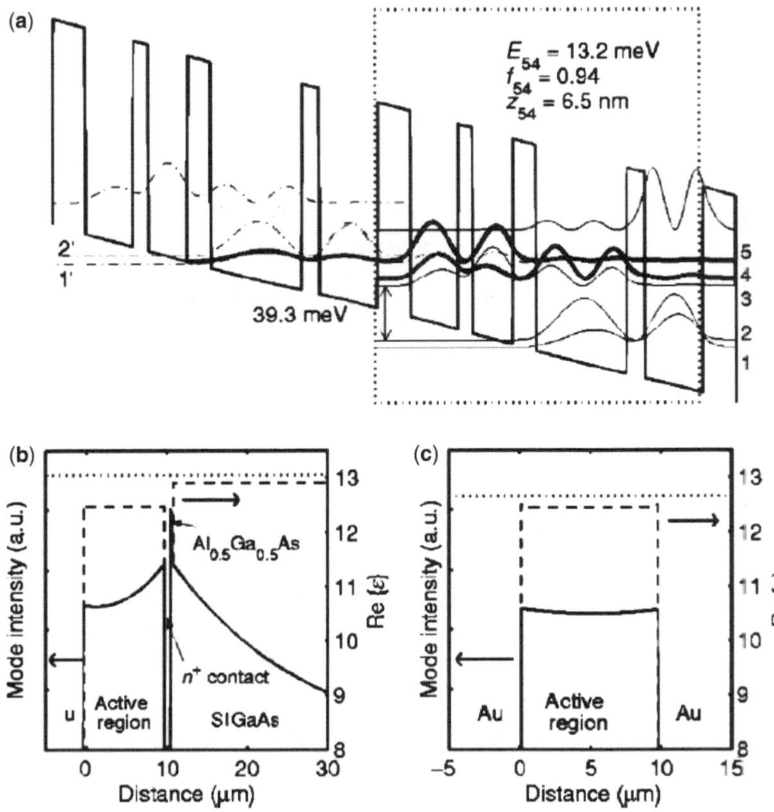

Figure 11.43 (a) Conduction-band profile of a THz quantum-cascade laser structure calculated using a self-consistent Schrödinger–Poisson solver with a 72% band edge discontinuity rule. The dotted box indicates the four well module: (from left to right in Angstrom, beginning with the barrier): 54/78/24/65/38/149/30/95. The underlined 149 Å well is doped with $n = 1.9 \times 10^{16}$ cm^{-1}. The optical mode profiles (solid) and the real part of the complex dielectric constants (dashed) are plotted for (b) SI-surface plasmon waveguide and (c) a double-metal-claddings waveguide. (Reprinted with permission from [148] © 2003 American Institute of Physics.)

and a Ni/Ge/Au alloyed contact was made to the lower contact layer at the bottom of the ridge. Because the plasma frequency for the lower contact is above the frequency of interest, a waveguide is formed between the top metallic contact and the surface plasmon mode associated with the quasi-metallic lower layer. A high-reflectivity (HR) coating (Al_2O_3/Ti/Au) was evaporated on the cleaved rear facet [148].

At 3.8 THz, a waveguide loss of $\alpha_w = 14.2$ cm^{-1} and an optical confinement factor of $\Gamma = 0.98$ were obtained for the metal–semiconductor–metal waveguide. On the other hand, $\alpha_w = 8.2$ cm^{-1} and $\Gamma = 0.324$ were obtained for the SISP waveguide. Drude model (see Chapter 5) with a relaxation time of 0.1, 0.5, and 0.05 ps was used for the heavily doped semiconductor, lightly doped semiconductor, and gold, respectively.

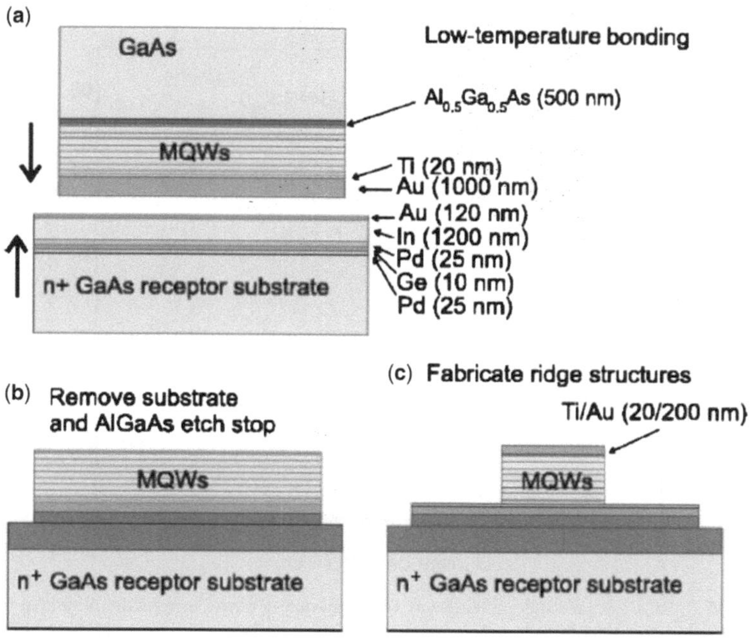

Figure 11.44 Schematic diagram for the fabrication of the metal–semiconductor–metal waveguide or double-metal waveguide for quantum-cascade lasers. (Reprinted with permission from [147] © 2003 American Institute of Physics.)

The better performance of the metal–semiconductor–metal waveguide THz laser is shown in Fig. 11.45 [147]. The laser is a 150-μm-wide, 2.74-mm-long Fabry–Perot ridge lasing up to the heat-sink temperature $T_{max} = 137$K when biased with 200-ns pulses repeated at 1 kHz. The threshold current density is 625 A/cm^2 at 5K. The voltage versus current ($I-V$) curve is also shown for a smaller device (100 μm wide and 1.45 mm long). In these devices, lasing occurs up to a peak current density of 1400 A/cm^2, whereupon the injector subband $n = 1'$ becomes misaligned with the upper radiative state $n = 5$, and negative differential resistance was observed. The shoulders in the $V-I$ curves at 500 A/cm^2 correspond to the parasitic current channel. Electrons are not injected into the upper state until the device is biased beyond this current level, and lasing begins at a current density slightly higher than this parasitic channel. It indicates that the threshold current density is still limited by the parasitic channel rather than by the intrinsic gain or loss. Increasing the intrainjector barrier thickness seemed to suppress the parasitic channel and reduce threshold current density.

High-power THz lasers using a SISP waveguide at 4.4 THz show a 248 mW peak power in pulsed mode, and 138 mW cw power at a heat sink temperature of 10K [151]. High-power and high-temperature THz QC lasers based on lens-coupled metal–metal waveguides have also been reported. The QC laser at 4.1 THz produced 145 mW of peak pulsed power at 5K with a wall plug power efficiency of 0.7%, lasing

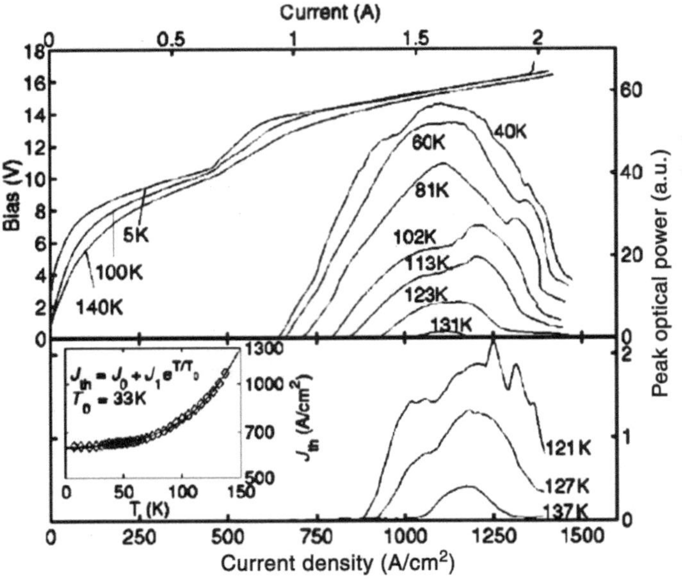

Figure 11.45 (Top) The measured voltage and THz light output power using a double-metal waveguide of Fig. 11.43c as a function of the injection current density using 200-ns pulses at 1 kHz repetition rate. The L–I curve is from a QC laser of 60-μm-wide and 2.48-mm-long ridge. The V–I curve is obtained from a different device: 100 μm wide and 1.45 mm long. (Bottom) The lower panel shows the L–I curve using a 150-μm-wide and 2.74-mm-long ridge, along with its threshold current density versus temperature. The current density axis only applies to the 60-μm-wide device. (Reprinted with permission from [147] © 2003 American Institute of Physics.)

up to a maximum operation temperature of 160K, as shown in Fig. 11.46 [156]. The same laser produced 26 mW of peak power using a Winston cone instead of a lens, lasing up to 165K. The large increase in output power for the case of lens is attributed to an increase of the collection efficiency.

Theoretical work includes the active-layer design, Monte Carlo simulation to account for various carrier scattering mechanisms in order to estimate the carrier density in each state, followed by the gain and threshold analyses [157–159]. The design of the active region allows for the determination of the energy levels, the corresponding wavefunctions, and oscillator strengths. The determination of the carrier populations on the energy levels are challenging because various scattering mechanisms such as electron–phonon, electron–impurity, and electron–electron scatterings play important roles for each injection current or bias voltage. Therefore, Monte Carlo simulations provide a solution for estimating the carrier population for a given bias at a given lattice temperature.

Experimentally, DFB lasers in the THz range have been developed [160]. Periodic structures have been employed in the waveguide of THz lasers [161] to control the lasing properties such as the DFB effect without the back-facet reflection, to achieve vertical emission of the radiation with second-order gratings, and to increase the facet reflectivity by fabricating passive distributed Bragg reflectors. Real-time THz imaging using QC lasers has also been demonstrated [162, 163].

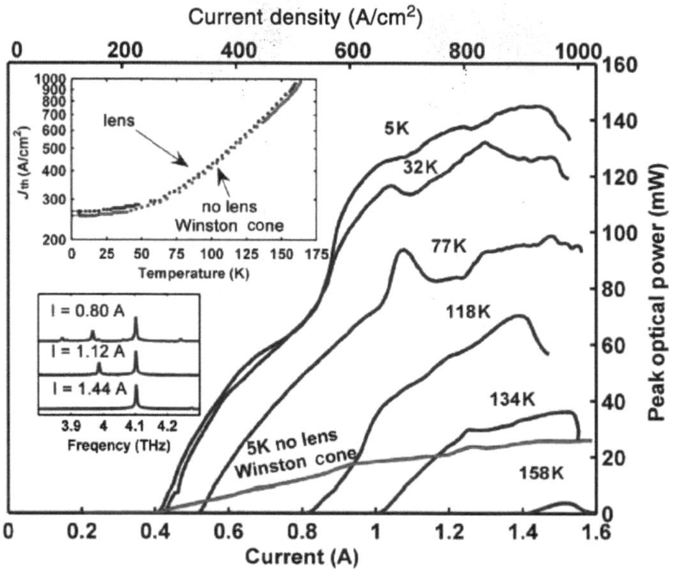

Figure 11.46 Light output power of a 4.1 THz double-metal waveguide QC laser at various temperatures from 5K to 158K with lens-coupled and Winston cone-coupled designs. The insets show the threshold current density versus temperature (top) and typical lasing spectra with lens (bottom). (Reprinted with permission from [156] © 2007 Optical Society of America.)

11.4.6 Type II Quantum-Cascade Lasers

Type II quantum-cascade lasers provide an alternative solution for mid-infrared lasers because the idea of cascading provides carrier recycling for multiple photon emissions. The multiple stages also allow for the increase of confinement factor. The concept of the interband cascade lasers were proposed in 1994 [164] and experimentally demonstrated in 1997 [165]. Theoretical simulations were also published [166, 167]. The first laser exhibited a lasing wavelength of 3.8 μm at 80K and operated up to 170K. For a review, please see Ref. 121.

Figure 11.47 shows a schematic diagram [168] to illustrate the physical principles. The injected electrons from the left tunnel through the first barrier into the conduction band of the InAs quantum well, where the electrons recombine with holes, which populate mostly in the nearby GaInSb layer due to the type II band alignment. The electron subband energy falls into the band gap of the GaInSb layer to avoid the tunneling out of the electrons into the following GaInSb and GaSb layers and force the electrons to recombine with the holes in the GaInSb layer. The overlap between the electrons and holes in the type II alignment depends on the leakage of the electron wave function from the InAs layer into the GaInSb layer and the hole wave function from the GaInSb layer into the InAs layer. The electron–hole recombination for stimulated photon emission requires electron transitions into the valence band, followed by tunneling through the AlSb barrier into the next GaSb quantum-well hole subband, then subsequent tunneling through the AlSb barrier back into the conduction band of the next injection/relaxation region consisting of InAs/AlInSb multilayered

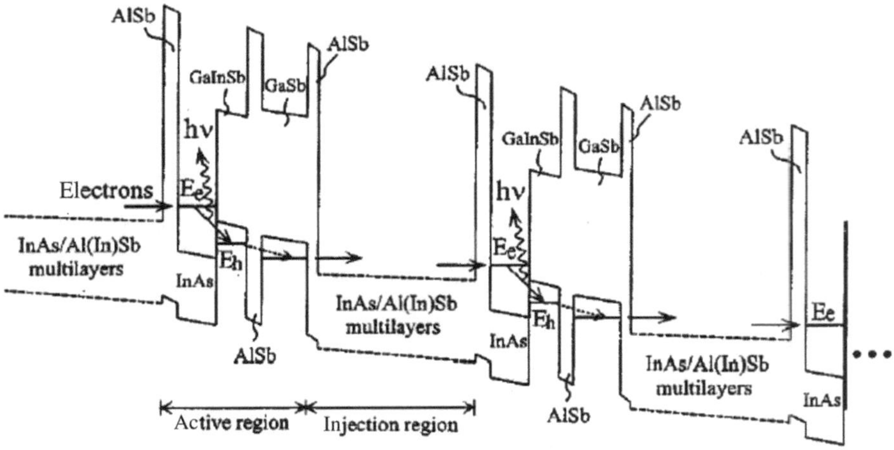

Figure 11.47 Schematic of a type II interband cascade laser structure. (Reprinted with permission from [168] © 2007 IEEE.)

structure. Therefore, "electron recycling" allows for an increase of the injection quantum efficiency to produce multiple photons depending on the number of stages. The difference between type I and type II quantum-cascade lasers are that type II cascade lasers require electrons and holes, therefore, interband radiative process is involved. The curvature of the energy versus momentum relations for the electrons and holes are opposite: upward for electrons and downward for holes. The band structures, wave functions, and optical gain model can be found in Ref. 169.

The selective operating characteristics of interband cascade lasers are briefly summarized. Laser (pulsed) operation with wavelength at 3.6 to 4.3 μm in a temperature range of 78K to 330K has been reported [170, 171]. For 150-μm-wide stripes, the threshold current densities are as low as $4.8\ \mathrm{A/cm^2}$ at 78K (cw) and $1.15\ \mathrm{kA/cm^2}$ at room temperature (pulsed). At 78K, the cw wall-plug efficiency for an 11-μm-wide ridge with 0.5-mm-long cavity and coated facets is 27%, while the other with a 3-mm-long cavity emits a maximum cw power of 200 mW. Devices have a maximum cw operating temperature of 269K at $\lambda = 4.05$ μm [171]. Continuous wave operation of DFB type II interband-cascade lasers has also been demonstrated [168] with a side-mode suppression ratio >20 dB at temperatures up to 261K near 3.3 μm. The electric power consumption (less than 1.1 W) for these DFB lasers is relatively low over the entire operating range compared with that for the mid-IR DFB type I QC lasers, which allows for cw operation with simple one-stage TE coolers at ambient temperature.

11.5 GaN-BASED BLUE–GREEN LASERS AND LEDs

In Chapter 1, we presented a historical development of semiconductor laser diodes and LEDs since Holonyak's first red GaAsP devices in 1962 [172, 173]. In the visible range, blue semiconductor lasers and LEDs [174] were absent until the 1990s, as shown in Fig. 1.3 in Chapter 1. The major breakthroughs relied on the development of GaN-based technologies by Akasaki and Nakamura [175–180], although II–VI blue–green lasers have also been demonstrated [181–183] with a record of 100 hours cw operation at a wavelength of 530 nm by Sony [184]. II–VI lasers and LEDs suffer from defects and recombination-enhanced defect motion [185], and therefore, reliability becomes an important issue of research. For a comparison of GaN with II–VI blue–green lasers, see Refs. 186 and 187. GaN forms in both cubic (zinc-blende) and hexagonal (wurtzite) crystal structures. However, we will discuss the wurtzite band structure, which is used in semiconductor lasers. Figure 11.48 shows a schematic diagram of a wurtzite crystal [188, 189]. It is a hexagonal structure and is similar to that of a zinc-blende crystal grown along the (111) direction, except that the atomic layers has a 60° rotation on the ABAB sequence versus ABAC sequence of a zinc-blende crystal [190].

Figure 11.49 shows a plot of the band-gap energy of nitride-based materials versus the lattice constant for wurtzite nitride-based materials. The band gap and lattice parameters are taken from Refs. 191–194, noting that the band gap of InN has been found to be close to 0.7 eV [195–197] in 2002 instead of 1.9 eV (see Appendix D on p. 807). We see that the AlGaInN is a wide band-gap III–V alloy, which covers a broad spectrum useful for lasers and photodetectors, but it is difficult to find substrates with closely lattice-matched lattice constants. The alloy has been successfully prepared on sapphire or SiC substrates by two-stage epitaxial growth of a buffer layer followed by GaN. The quality of the GaN grown on sapphire was improved by incorporating an AlN buffer layer or GaN buffer layer grown at a low temperature (450°C to 600°C). Although *P*-type GaN has been difficult to achieve, the work by Akasaki and Nakamura has made it possible to achieve a *p-n* junction in a GaN material system. With these developments, LEDs were commercialized in 1992. Despite the dislocation problem associated with lattice mismatch, the LEDs had a long lifetime. However, it was rather difficult to fabricate reliable laser diodes because of the high threshold current density of 5–6 $\mathrm{kA/cm^2}$, high operating voltage, and high dislocation density around 10^8 to $10^{10}\ \mathrm{cm^{-2}}$. Continuous

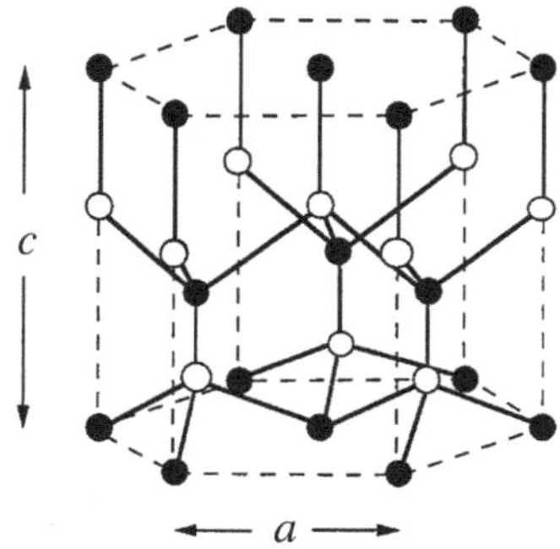

Figure 11.48 A wurtzite crystal consists of two interpenetrating hexagonal closely packed sublattices displaced by 5/8 of the *c*-axis along the (0001) direction. The dashed lines show the boundary of a unit cell. (Reprinted with permission from [188] © 1996 American Physics Society.)

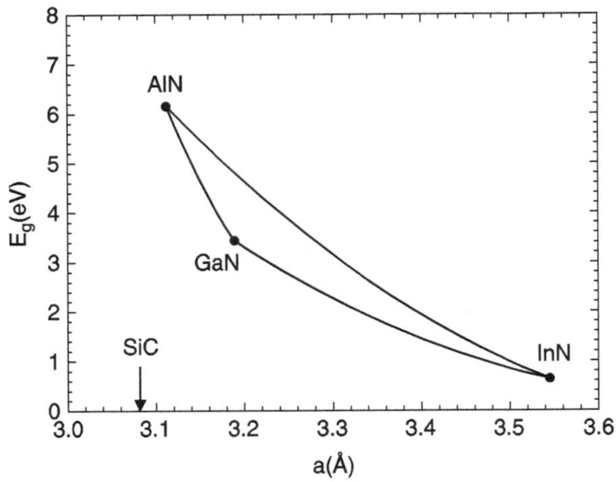

Figure 11.49 A plot of the band-gap energy versus the lattice constant for selective nitride-based materials. The triangular areas represent AlInGaN quarternary compounds. The parameters are from Appendix D on page 807. SiC substrate has a lattice constant of 3.0806 Å (labeled) while sapphire has a much larger lattice mismatch (not shown).

wave room-temperature AlGaInN lasers were first demonstrated in 1996 at Nichia. It was reported that ultrathin QWs less than 5 nm helped to decrease threshold current densities, and a Mg-doped GaN–AlGaN superlattice cladding layer reduced the series resistance. Progress has been rapid [198–212], and GaN-based lasers were developed as a light source for optical disk recording, laser printing, and display applications. As reviewed in Ref. 201, the advent of the epitaxial lateral overgrowth (ELOG) technique and improvements in the quality of GaN substrates have subsequently led to a remarkable reduction in dislocation density to around 10^5 to 10^6 cm^{-2}. The reduction of dislocation densities has led to greatly improved reliability of GaN lasers and led to the mean time to failure of more than 10,000 hours. Impressive performance such as high power with long lifetime (e.g., >500 mW at 445-nm wavelength at room temperature by Nichia; 750 mW AlGaInN blue LDs cw operation at 440–450 nm at 35°C over 10,000 hours by Sony) have also been reported (see Refs. 202–210 for performances of recent blue laser diodes). Continuous wave operation of GaN/InGaN microdisk lasers at 300K [211] and electrical injection of GaN-based VCSELs at 77K [212] have also been demonstrated.

11.5.1 Band Structure of Wurtzite Crystals

The wurtzite structure differs from the zinc-blende structure in several aspects [188–197, 213–241]. The anisotropic hexagonal structure leads to energy splitting in the valence band. First there is a crystal field splitting, which leads to two valence band edges

$$\begin{aligned} \langle X|H_0|X\rangle &= \langle Y|H_0|Y\rangle = E_v + \Delta_1 \\ \langle Z|H_0|Z\rangle &= E_v \end{aligned} \tag{11.5.1}$$

where E_v is the reference energy in the valence band, and Δ_1 is called the crystal-field splitting energy, which is caused by the anisotropic nature of the hexagonal crystal or the z-dependent wave function along the c-axis, while the x–y plane is isotropic. Second, the spin–orbit interaction H_{so} leads to two spin–orbit coupling energies:

$$\begin{aligned} \langle X|H_{so,z}|Y\rangle &= -i\Delta_2 \\ \langle Y|H_{so,x}|Z\rangle &= \langle Z|H_{so,y}|X\rangle = -i\Delta_3. \end{aligned} \tag{11.5.2}$$

We usually take:

$$\Delta_1 = \Delta_{cr} \qquad \Delta_2 = \Delta_3 = \Delta_{so} \tag{11.5.3}$$

where we have assumed $\Delta_2 = \Delta_3$, which is part of the cubic approximation (see Appendix 11A). The band-edge energies can be described under the following consideration. Without the spin–orbit interaction, $\Delta_2 = \Delta_3 = 0$, we have the top two degenerate bands, plus the lower band as the reference level E_v

$$\begin{aligned} E_1 &= E_2 = E_v + \Delta_1 = \Delta_1 \\ E_3 &= E_v = 0 \end{aligned} \tag{11.5.4}$$

where we have set the reference energy E_v to 0. We plot the band edge energies in Fig. 11.50a for a positive Δ_1. When we include the spin–orbit interaction, the E_1 and E_2 bands split, Fig. 11.50b, and we have

$$\begin{aligned} E_1 &= \Delta_1 + \Delta_2 \\ E_2 &= \frac{\Delta_1 - \Delta_2}{2} + \sqrt{\left(\frac{\Delta_1 - \Delta_2}{2}\right)^2 + 2\Delta_3^2} \\ E_3 &= \frac{\Delta_1 - \Delta_2}{2} - \sqrt{\left(\frac{\Delta_1 - \Delta_2}{2}\right)^2 + 2\Delta_3^2}. \end{aligned} \tag{11.5.5}$$

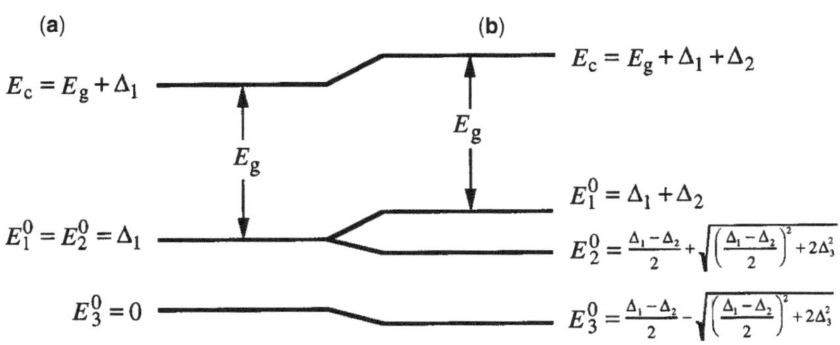

Figure 11.50 The band-edge energies for heavy-hole (HH), light-hole (LH), and crystal-field split-off hole (CH) bands: (a) without spin–orbit interaction ($\Delta_2 = \Delta_3 = 0$) and (b) with spin–orbit interaction ($\Delta_2 = \Delta_3 \neq 0$) for a GaN wurtzite semiconductor crystal. The corresponding band-edge energies are also listed. (Reprinted with permission from [188] © 1996 American Physics Society.)

The conduction-band edge is given by adding the band gap energy to E_1

$$E_c = E_g + \Delta_1 + \Delta_2. \tag{11.5.6}$$

The three bands are labeled according to their zone center wave functions. E_1 is labeled as heavy hole (HH), E_2 is light hole (LH), and E_3 is called the crystal-field split-off hole (CH) band. The HH and LH splitting is caused by the spin–orbit interaction.

The energy splitting, $E_1 - E_2$ and $E_1 - E_3$, are measured from the differences between the interband optical transition energies, which then determine the values of Δ_{cr} and Δ_{so}. A reported experimental value of 22 meV for Δ_{cr} of GaN is commonly used.

The band structures for GaN and AlN are shown in Fig. 11.51 [218] for the valence-band energies as a function of k_x $(=k_y)$ and k_z. We can see that the top

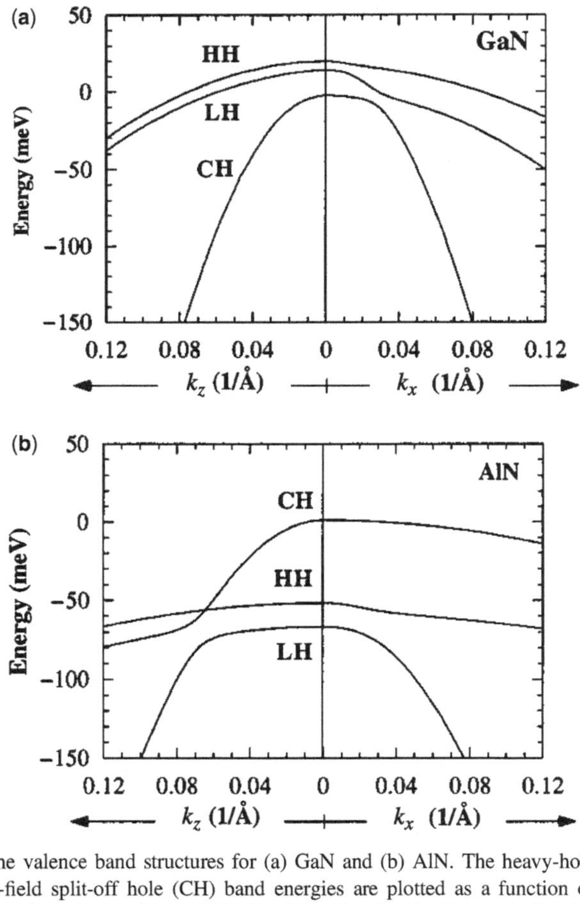

Figure 11.51 The valence band structures for (a) GaN and (b) AlN. The heavy-hole (HH), light-hole (LH), and crystal-field split-off hole (CH) band energies are plotted as a function of the wave vector along the transverse k_x and crystal growth k_z direction. (Reprinted with permission from [218] © 1996 American Institute of Physics.)

valence band for GaN is a heavy-hole (HH) band, whereas for AlN it is the crystal-field split-off hole (CH) band. The reason is that AlN has a negative Δ_1, therefore, its top valence band is E_2, and the valence bands from top to bottom are CH, HH, and LH, respectively. The conduction-band energy is $E_c = E_g + E_2$ measured from the reference energy E_v of AlN. To obtain the band structure near the zone center, we find the eigenvalues from the Hamitonian matrix (Appendix 11A), which can be obtained analytically [218].

Figure 11.52 shows the interband transition energies of wurtzite and cubic (zinc-blende) GaN as a function of temperature [241]. The net band-gap is determined by $E_c - E_1$ and it can be compared with the lowest band-edge transition energy (A line) of the photoluminescence (PL) or phororeflectance (PR) spectra, with an exciton correction energy of about 28 meV ($E_c - E_1 - 28$ meV). Similarly, $E_c - E_2$ and $E_c - E_3$ with a similar exciton energy correction can be used to compare with the transition energies of the B and C lines in the PL or PR spectrum. The exciton energy from a cubic GaN is also shown for comparison.

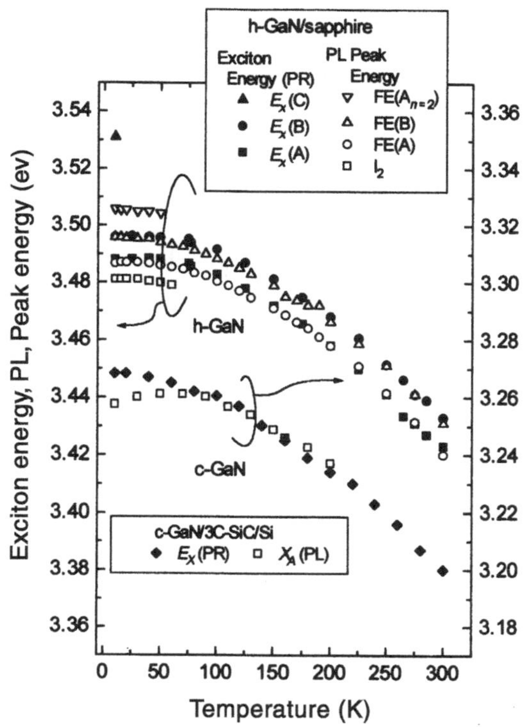

Figure 11.52 The photoluminescence (PL) peak energies and excitonic resonance energies obtained from the analysis of PL and corresponding photoreflectance (PR) spectra of hexagonal-GaN (left axis) and cubic-GaN (right axis) as a function of temperature. Exciton energies are shown by solid symbols and PL peak energies by open symbols. (Reprinted with permission from [241] © 1997 Japanese Journal Applied Physics.)

11.5.2 Strained Wurtzite Crystals

The general relation or Hooke's law between the stress $\boldsymbol{\tau}$ and strain $\boldsymbol{\varepsilon}$ tensors for the hexagonal crystal is [214]

$$\begin{bmatrix} \tau_{xx} \\ \tau_{yy} \\ \tau_{zz} \\ \tau_{yz} \\ \tau_{zx} \\ \tau_{xy} \end{bmatrix} = \begin{bmatrix} C_{11} & C_{12} & C_{13} & 0 & 0 & 0 \\ C_{12} & C_{11} & C_{13} & 0 & 0 & 0 \\ C_{13} & C_{13} & C_{33} & 0 & 0 & 0 \\ 0 & 0 & 0 & C_{44} & 0 & 0 \\ 0 & 0 & 0 & 0 & C_{44} & 0 \\ 0 & 0 & 0 & 0 & 0 & \frac{C_{11}-C_{12}}{2} \end{bmatrix} \begin{bmatrix} \varepsilon_{xx} \\ \varepsilon_{yy} \\ \varepsilon_{zz} \\ 2\varepsilon_{yz} \\ 2\varepsilon_{zx} \\ 2\varepsilon_{xy} \end{bmatrix} \tag{11.5.7}$$

where C_{ij}'s are the elastic stiffness constants. Let us consider a few cases for which the strain tensor is diagonal, that is,

$$\varepsilon_{xy} = \varepsilon_{yz} = \varepsilon_{zx} = 0. \tag{11.5.8}$$

Case (a): A strained-layer wurtzite crystal pseudomorphically grown along the (0001) direction (c-axis) of a sapphire substrate. The diagonal elements are

$$\varepsilon_{xx} = \varepsilon_{yy} = \frac{a_0 - a}{a} \qquad \varepsilon_{zz} = -\frac{2C_{13}}{C_{33}}\varepsilon_{xx} \tag{11.5.9}$$

where a_0 and a are the lattice constants of the substrate and the well-layer material, respectively. The last relation in (11.5.9) is obtained from the fact that the stress along the z direction is zero on the surface with a normal along the z direction: $\tau_{zz} = 0$.

Case (b): A wurtzite crystal layer under an external biaxial in-plane stress $\tau_{xx} = \tau_{yy} = T$ and $\tau_{zz} = 0$; we have [236]

$$\varepsilon_{xx} = \varepsilon_{yy} = \frac{C_{33}}{(C_{11} + C_{12})C_{33} - 2C_{13}^2}T \qquad \varepsilon_{zz} = \frac{-2C_{13}}{C_{33}}\varepsilon_{xx}. \tag{11.5.10}$$

Case (c): A wurtzite crystal layer under an external uniaxial stress $\tau_{zz} = T$ along the [0001] direction and $\tau_{xx} = \tau_{yy} = 0$; we have

$$\begin{aligned} \varepsilon_{xx} = \varepsilon_{yy} &= \frac{-C_{13}}{(C_{11} + C_{12})C_{33} - 2C_{13}^2}T \\ \varepsilon_{zz} &= \frac{C_{11} + C_{12}}{(C_{11} + C_{12})C_{33} - 2C_{13}^2}T. \end{aligned} \tag{11.5.11}$$

The band-edge energies can be obtained easily from the three-by-three Hamiltonian at $k = 0$:

$$
\begin{aligned}
E_1 &= \Delta_1 + \Delta_2 + \lambda_\varepsilon + \theta_\varepsilon \\
E_2 &= \frac{\Delta_1 - \Delta_2 + \theta_\varepsilon}{2} + \lambda_\varepsilon + \sqrt{\left(\frac{\Delta_1 - \Delta_2 + \theta_\varepsilon}{2}\right)^2 + 2\Delta_3^2} \\
E_3 &= \frac{\Delta_1 - \Delta_2 + \theta_\varepsilon}{2} + \lambda_\varepsilon - \sqrt{\left(\frac{\Delta_1 - \Delta_2 + \theta_\varepsilon}{2}\right)^2 + 2\Delta_3^2}
\end{aligned}
\tag{11.5.12}
$$

where the strain induced shifts are (Appendix 11A)

$$
\lambda_\varepsilon = D_1\varepsilon_{zz} + D_2(\varepsilon_{xx} + \varepsilon_{yy}), \qquad \theta_\varepsilon = D_3\varepsilon_{zz} + D_4(\varepsilon_{xx} + \varepsilon_{yy}). \tag{11.5.13}
$$

The conduction-band edge has a hydrostatic energy shift, $P_{c\varepsilon}$,

$$
\begin{aligned}
E_c &= E_g + \Delta_1 + \Delta_2 + P_{c\varepsilon} \\
P_{c\varepsilon} &= a_{cz}\varepsilon_{zz} + a_{ct}(\varepsilon_{xx} + \varepsilon_{yy}).
\end{aligned}
\tag{11.5.14}
$$

For example, the interband transition energy is

$$
\begin{aligned}
E_c - E_1 &= E_g + P_{c\varepsilon} - \lambda_\varepsilon - \theta_\varepsilon \\
&= (a_{cz} - D_1)\varepsilon_{zz} + (a_{ct} - D_2)(\varepsilon_{xx} + \varepsilon_{yy}) - D_3\varepsilon_{zz} - D_4(\varepsilon_{xx} + \varepsilon_{yy}).
\end{aligned}
\tag{11.5.15}
$$

Figure 11.53a and Fig. 11.53b show the plots of the band edge energy of the conduction band and HH, LH, and CH band as a function of strain. Figure 11.54 shows the A-, B-, and C-line exciton transition energies as a function of the c-axis strain ($\varepsilon_{zz} = (c - c_0)/c_0$) measured and compared with the theoretical formula using $E_c - E_1$, $E_c - E_2$, and $E_c - E_3$, minus the exciton correction energy of about 28 meV. Because the experimental data provide the total band-gap shift as a function of an externally applied pressure, only the total value for the interband deformation potential, a, is reported. Theoretically, the hydrostatic deformation potentials for interband $a = \partial E_g/\partial(\ln V) = a_c - a_v$, a_c (conduction band), a_v (valence band), and the shear deformation potential, b, have been used for zinc-blende structures. For wurtzite structure, the parameters D_1 and D_2 play a similar role to the hydrostatic deformation potential, a_v, and D_3 and D_4 play a similar role to the shear deformation potential, b, of zinc-blende crystals. Because wurtzite is anisotropic, we have

$$
a_1 = a_{cz} - D_1 \qquad a_2 = a_{ct} - D_2 \tag{11.5.16}
$$

for the longitudinal ε_{zz} and transverse $\varepsilon_{xx} + \varepsilon_{yy}$ interband deformation potentials. The theoretical results (lines) agree well with the experimental data (symbols) [240].

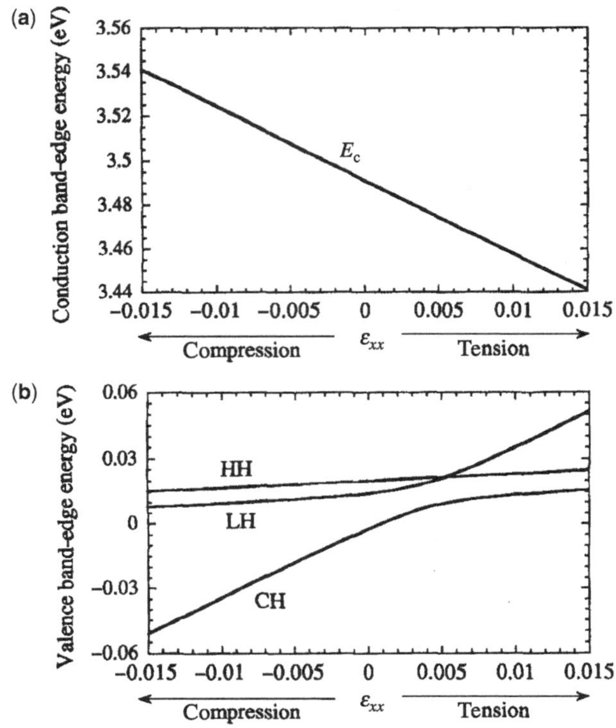

Figure 11.53 The band edges of (a) the conduction band and (b) the HH, LH, and CH bands of a strained GaN crystal as a function of the in-plane strain ε_{xx}. The reference energy is set at E_v^0 of an unstrained GaN crystal. (Reprinted with permission from [189] © 1996 IEEE.)

For a GaN quantum well sandwiched between two thick (relaxed) $Al_xGa_{1-x}N$ layers, the strain of the GaN well is compressive in nature because the lattice constant of the GaN layer before deformation is larger than the lattice constant a_0 of the $Al_xGa_{1-x}N$ layers. Therefore, ε_{xx} is negative, and the band edge shifts will be positive for the conduction band and negative for the top valence bands.

Valence-Band Dispersions of Strained Wurtzite Crystals Figure 11.55 shows the valence-band dispersions of the HH, LH, and CH bands as a function of the transverse k_x and longitudinal $k_z = c(0001)$ axis wave vector for (a) biaxial compressively strained ($\varepsilon_{zz} = 0.2\%$) GaN, (b) unstrained, (c) a small biaxial tensile strain ($\varepsilon_{zz} = -0.09\%$), and (d) a large tensile strain ($\varepsilon_{zz} = -0.2\%$). The difference in symmetry between the wurtzite and zinc-blende structures causes some fundamental differences in their electronic and optical properties. The degeneracy between the heavy-hole and light-hole bands at the zone center (Γ point) for the zinc-blende structure is broken for the wurtzite structure. For a zinc-blende layer grown along the (001)

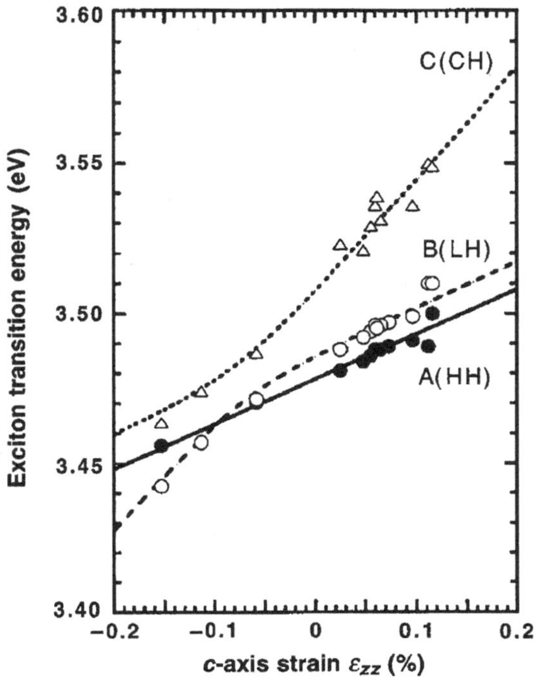

Figure 11.54 Transition energies of the A-, B-, and C-line excitons as a function of the strain along the *c*-axis. The symbols are experimental data from Ref. 240, and the lines are theoretical results. (Reprinted with permission from [216] © 1998 American Physics Society.)

direction with a compressive strain, the deformation potentials lift the HH band up and reduce its in-plane effective mass. However, for a wurtzite layer grown along the *c*-axis, the compressive strain shifts both the HH and LH bands by almost the same amount, and the in-plane effective masses remain almost the same as those in the unstrained case.

11.5.3 Wurtzite Quantum Wells and Polarization Fields [242–259]

Polarity of the Ga-face and N-face For binary AB compound such as the GaN wurtzite structure, the sequence of the atomic layers of the constituents A and B is reversed along the [0001] and $[000\bar{1}]$ directions. The corresponding (0001) and $(000\bar{1})$ faces are the A-face and B-face, respectively. For GaN epitaxial layers and GaN-based heterostructures grown along a direction normal to the {0001} basal plane, the atoms are arranged in bilayers. These bilayers consist of two closely spaced hexagonal layers, one formed by cations and the other by anions, leading to polar faces. Thus, for GaN, a basal surface should be either Ga- or N-faced. By Ga-face, we mean Ga on the top position of the {0001} bilayer, corresponding to the [0001] polarity (which is also defined

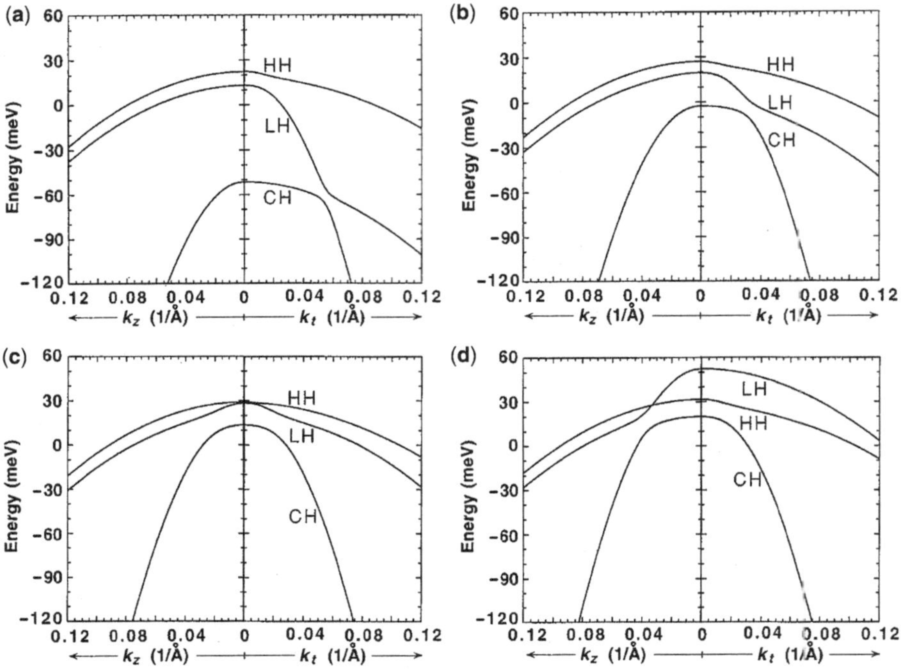

Figure 11.55 Valence-band dispersions of the HH, LH, and CH bands as a function of the transverse k_x and longitudinal $k_z = c(0001)$ axis wave vector for (a) biaxial compressively strained ($\varepsilon_{zz} = 0.2\%$) GaN, (b) unstrained, (c) a small biaxial tensile strain ($\varepsilon_{zz} = -0.09\%$), and (d) a large tensile strain ($\varepsilon_{zz} = -0.2\%$). (Reprinted with permission from [216] ©1998 American Physics Society.)

as the $+z$ direction with a vector pointing from a Ga atom to the nearest-neighbor N atom), as shown in Fig. 11.56a [256]. For an N-face on the top of the bilayer, Fig. 11.56b, the axis is $[000\bar{1}]$ or the $-z$ direction. Research by various groups indicates that PIMBE grown GaN films are of N-face, whereas MOCVD grown films are of Ga-face materials [256].

Wurtzite crystals have an induced piezoelectric field. For GaN, two polarization fields exist.

(a) The first one is the piezoelectric (PZ) polarization, which is related to the strain components by the PZ coefficients:

$$\begin{bmatrix} P_x^{\mathrm{pz}} \\ P_y^{\mathrm{pz}} \\ P_z^{\mathrm{pz}} \end{bmatrix} = \begin{bmatrix} 0 & 0 & 0 & 0 & e_{15} & 0 \\ 0 & 0 & 0 & e_{15} & 0 & 0 \\ e_{31} & e_{31} & e_{33} & 0 & 0 & 0 \end{bmatrix} \begin{bmatrix} \varepsilon_{xx} \\ \varepsilon_{yy} \\ \varepsilon_{zz} \\ \varepsilon_{yz} \\ \varepsilon_{zx} \\ \varepsilon_{xy} \end{bmatrix}. \tag{11.5.17}$$

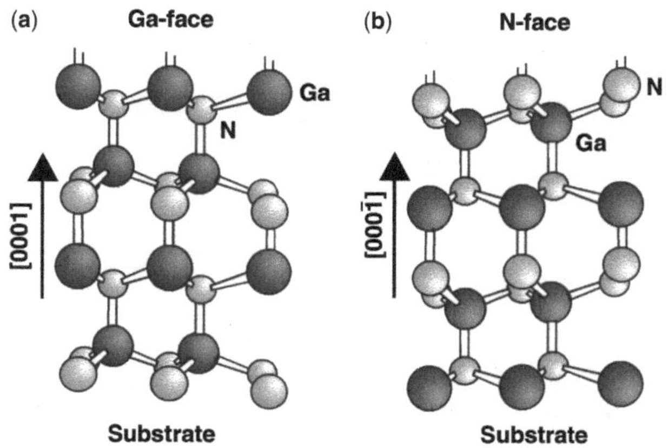

Figure 11.56 (a) Ga-faced growth of GaN wurtzite crystal structure on substrate along the [0001] (=z) direction and (b) N-faced top layer growth along the [000$\bar{1}$] or $-z$ direction. (Reprinted with permission from [256] © 1999 American Institute of Physics.)

Using the strain components in (11.5.8) and (11.5.9), we find the induced piezoelectric field polarization:

$$\begin{aligned} P_x^{\mathrm{pz}} &= P_y^{\mathrm{pz}} = 0 \\ P_z^{\mathrm{pz}} &= e_{31}(\varepsilon_{xx} + \varepsilon_{yy}) + e_{33}\varepsilon_{zz} \\ &= 2\left(e_{31} - \frac{C_{13}}{C_{33}} e_{33}\right)\varepsilon_{xx}. \end{aligned} \tag{11.5.18}$$

Because $[e_{31} - e_{33}\ (C_{13}/C_{33})] < 0$ for the whole range of $Al_xGa_{1-x}N$ compositions, the PZ polarization is negative for tensile strain and positive for compressive strain.

(b) The second is the spontaneous polarization P_z^{sp} along the growth z-axis. The spontaneous polarization for GaN and AlN has been found to be negative, meaning that for the Ga-face or Al-face, the spontaneous polarization is pointing toward the substrate or the negative z direction, Fig. 11.57a–c [256]. As a result, the alignment of the PZ polarization and the spontaneous polarization is parallel in the case of tensile strain (Fig. 11.57b) and antiparallel in the case of compressive strained top layers (Fig. 11.57c). If the polarity flips a sign from the Ga-face to N-face, both the SP and PZ polarizations flip a sign, Fig. 11.57d–f. Surface charges may exist in the interface, and the surface charge density is determined by

$$\rho_{\mathrm{s}} = P(\text{bottom}) - P(\text{top}). \tag{11.5.19}$$

If there is no additional surface charge, the induced electric field is determined by the boundary condition at the surface:

$$D_z = \varepsilon F_z + P_z^{\mathrm{pz}} + P_z^{\mathrm{sp}} = 0. \tag{11.5.20}$$

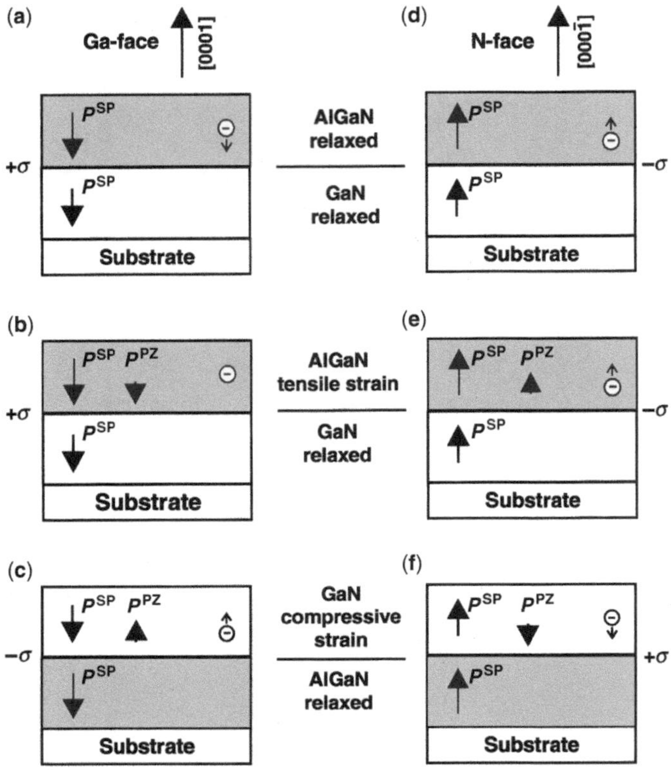

Figure 11.57 Spontaneous (SP) and piezoelectric (PZ) polarization and the induced surface charge density in Ga-face and N-face strained and relaxed GaN/AlGaN heterostructures. (a)(b)(c) are for Ga-faced and (d)(e)(f) are for N-faced surfaces. (a) and (d) are relaxed AlGaN layer on top of relaxed GaN. (b) and (e) are tensile strain AlGaN on relaxed GaN. (c) and (f) are compressive strain GaN on relaxed AlGaN. (Reprinted with permission from [256] © 1999 American Institute of Physics.)

We have a total induced piezoelectric and spontaneous polarization field:

$$F_z = -\frac{\left(P_z^{\mathrm{pz}} + P_z^{\mathrm{sp}}\right)}{\varepsilon}. \tag{11.5.21}$$

In the following, we discuss a quantum well structure such as GaN/AlGaN or InGaN/GaN quantum wells. We drop the subscript "z" with the understanding that all the fields are along the z-axis and use subscript "w" for well and "b" for barrier fields

$$P_{\mathrm{w}} = P_{\mathrm{w}}^{\mathrm{pz}} + P_{\mathrm{w}}^{\mathrm{sp}} \qquad P_{\mathrm{b}} = P_{\mathrm{b}}^{\mathrm{pz}} + P_{\mathrm{b}}^{\mathrm{sp}}. \tag{11.5.22}$$

The continuity of the displacement vector (D_z) normal to the surface without free carrier surface charge gives

$$D_z = \varepsilon_{\mathrm{w}} F_{\mathrm{w}} + P_{\mathrm{w}} = \varepsilon_{\mathrm{b}} F_{\mathrm{b}} + P_{\mathrm{b}}. \tag{11.5.23}$$

Two simple cases are considered.

A Single Quantum Well with Infinite Barriers The fields due to interface charges cancel each other in the barriers, therefore,

$$F_w = \frac{P_b - P_w}{\varepsilon_w}. \tag{11.5.24}$$

A Periodic Superlattice The periodic boundary condition implies the net voltage drop over one period is zero, which means the line integral of the electric field over one period vanishes

$$F_w L_w + F_b L_b = 0. \tag{11.5.25}$$

Solving for (11.5.23) and (11.5.25), we find the electric field in the well:

$$F_w = \frac{L_b}{\varepsilon_b L_w + \varepsilon_w L_b}(P_b - P_w) \tag{11.5.26a}$$

$$F_b = \frac{-L_w}{\varepsilon_b L_w + \varepsilon_w L_b}(P_b - P_w). \tag{11.5.26b}$$

The fields in the well and barrier are opposite in direction. Figure 11.58 shows an example of an InGaN/GaN quantum well grown with a Ga-faced top layer. The spontaneous polarization fields in the $In_xGa_{1-x}N$ region is taken as [228, 259] (see Appendix D on page 807 for more interpolation formulas for nitride compounds)

$$P^{sp}(x) = -0.042x - 0.034(1-x) + 0.037x(1-x)(\mathrm{C/m^2}). \tag{11.5.27}$$

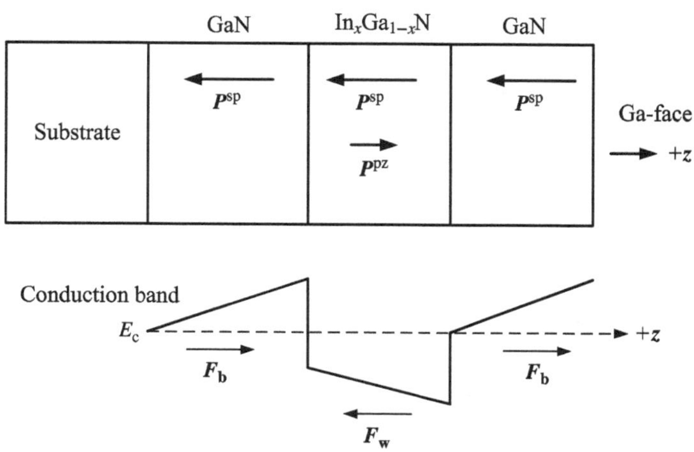

Figure 11.58 An InGaN/GaN quantum well grown with Ga-faced top layer and the conduction-band profile for a superlattice model.

The piezoelectric polarization (C/m^2) of a biaxial-compressively strained $In_xGa_{1-x}N$ is

$$P^{pz}(x) = 0.148x - 0.0424x(1-x)(C/m^2). \tag{11.5.28}$$

The conduction band profile is shown schematically as the bottom curve in Fig. 11.58.

11.5.4 Optical Gain of Strained Wurtzite Quantum-Well Lasers

To model the band structure of InGaN/GaN and GaN/AlGaN quantum wells, we need to solve for the quantized subband energy levels and their corresponding wave functions in the conduction and valence bands, as shown in Fig. 11.59. For a given carrier density due to current injection, we then solve the Schrödinger and Poisson equations self-consistently to account for the carrier screening effects. The optical gain spectrum is then calculated. In Fig. 11.59a, we show a self-consistent potential profile (solid) with corresponding C1, HH1, and HH2 subband energy levels and wave functions at a carrier density of $2 \times 10^{19}/cm^3$ of a 5-nm GaN/$Al_{0.3}Ga_{0.7}N$ quantum well laser [223]. The flat band profile (dashed lines) is also shown. At a higher carrier density, Fig. 11.59b, we see that the screened potential

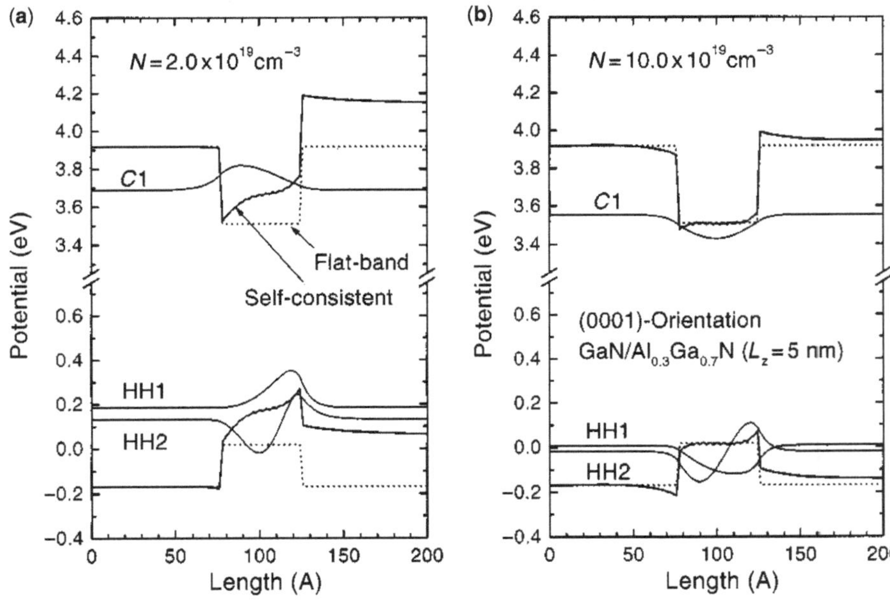

Figure 11.59 (a) A self-consistent potential profile (solid) with corresponding C1, HH1, and HH2 subband energy levels and wave functions at a carrier density of 2×10^{19} cm^{-3} of a 5-nm GaN/$Al_{0.3}Ga_{0.7}N$ quantum well laser. The flat band profile (dashed lines) is also shown. (b) At a higher carrier density, the screened potential profile (solid) approaches that of the flat band condition (dashed) due to the carrier screening effect. (Reprinted with permission from [223] © 1998 American Institute of Physics.)

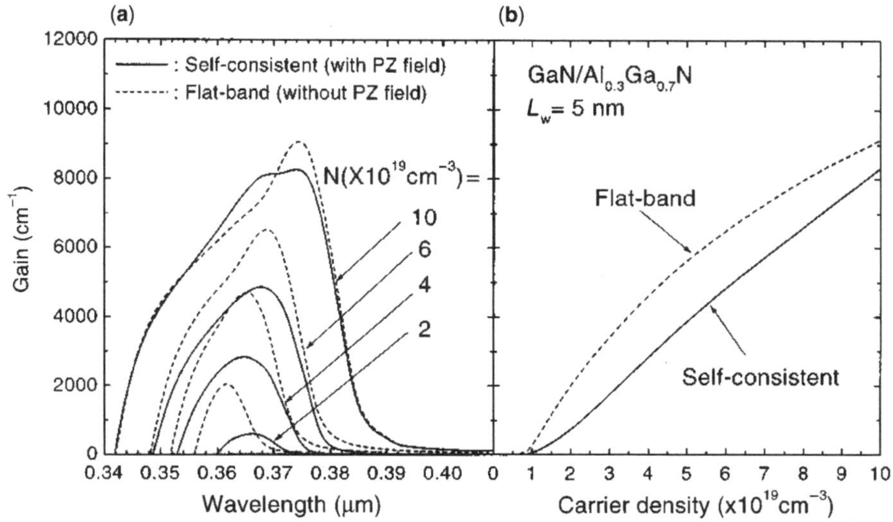

Figure 11.60 (a) Optical gain spectrum with (solid) and without (dashed) PZ polarization field screening. The PZ field reduces the optical gain by a significant factor, especially when the screening due to carriers is not large enough. (b) Peak gain as a function of the carrier density for the PZ polarization with self-consistent solution and the flat band case (no PZ field). (Reprinted with permission from [223] © 1998 American Institute of Physics.)

profile (solid curves) approaches that of the flat band condition (dashed) due to the carrier screening effects. The optical gain spectrum with (solid) and without (dashed) PZ polarization field screening is plotted in Fig. 11.60a [223]. The PZ field reduces the optical gain by a significant factor, especially when the screening due to carriers is not large enough. We plot the peak gain as a function of the carrier density in Fig. 11.60b for the PZ polarization with a self-consistent solution and the flat band case (no PZ field).

11.5.5 Blue–Green Lasers and LEDs [175–180, 198–210]

In Fig. 11.61, we show the blue, green, and yellow laser emission spectra of single quantum-well structures from the early data of Nakamura in 1997 [177]. Figure 11.62a shows an InGaN MQW laser diode structure and its typical L–I and I–V curves (Fig. 11.62b) in 1997 [177]. At that time, the threshold current and voltage were high. Over the past decade, impressive performances of blue or violet GaN-based lasers and LEDs have been reported. The driving market is because GaN-based 400-nm high-power lasers have been developed for application as a light source for optical disk recording. A new disk format using these lasers was recently proposed [203], which was called "Blu-ray Disc" and supported by 11 companies, including SONY, Matsushita, and Philips. Both the Blu-ray Disc system and the DVD recording system require lasers with low relative intensity noise (RIN; which is mainly caused by spontaneous emission noise and will be discussed in

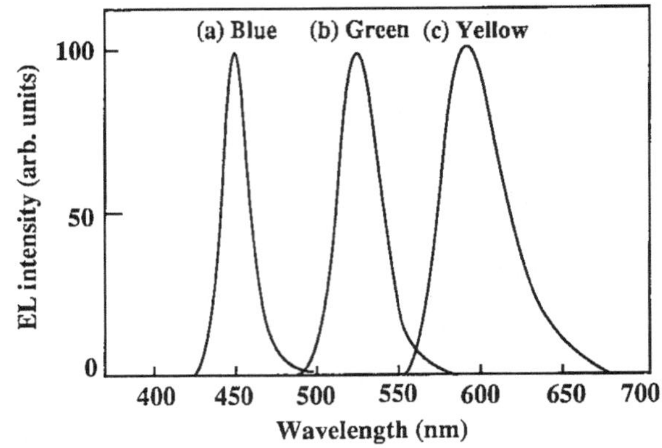

Figure 11.61 Electroluminescence (EL) of blue, green, and yellow single quantum-well LEDs at a forward bias of 20 mA. (Reprinted with permission from [177] © 1997 IEEE.)

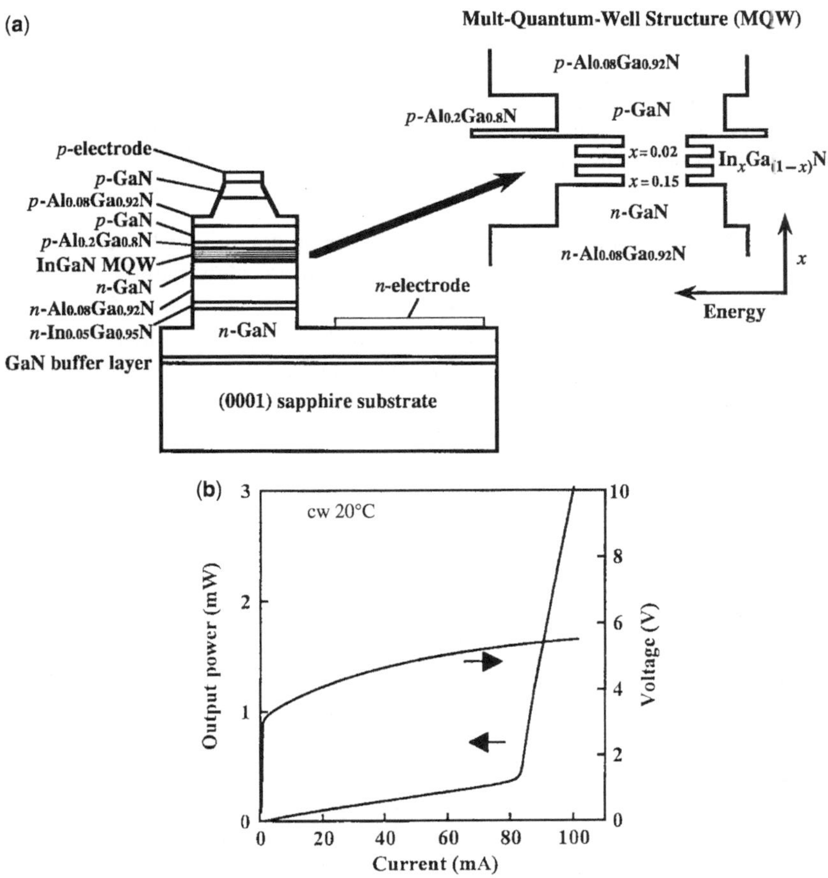

Figure 11.62 (a) A schematic diagram for an InGaN MQW laser diode and the band structure. (b) The L–I and I–V curve of the laser diode. (Reprinted with permission from [177] © 1997 IEEE.)

Chapter 12) of below 125 dB/Hz, a beam profile with low aspect ratio, and reliability of greater than 5000 hours. In addition, lasers for use in optical disks capable of capacities greater than 50 GB are required to operate at an output power exceeding 100 mW in order to write information to a dual-layer disk [203]. Therefore, laser diodes with low operating current, a stable beam profile, and high reliability are necessary to satisfy these requirements. Figure 11.63a and Fig. 11.63b show a schematic diagram of an InGaN-based blue laser diode with the energy band diagram of the active layer and the corresponding optical intensity profile [203]. The high barrier in the conduction band on the right side of the quantum wells serves as an electron blocking layer to avoid electron overflow outside of the active region at high bias current. The free carrier absorption in the Mg-doped superlattice plays an important

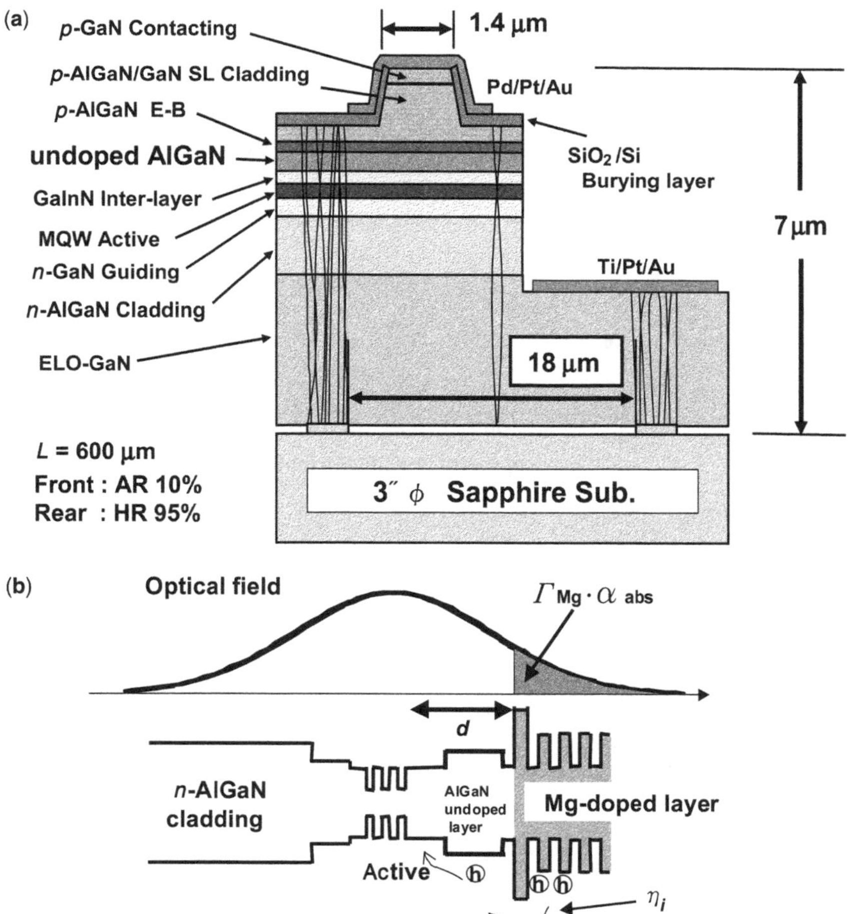

Figure 11.63 (a) A schematic diagram showing an InGaN-based blue laser diode. (b) The energy band diagram of the active layer with corresponding optical intensity profile. The high barrier in the conduction band on the right side of the quantum wells serves as an electron blocking layer to avoid electron overflow outside of the active region at high bias current. (Reprinted with permission from [203] © 2003 IEEE.)

role. Therefore, the optical confinement factor in that region is one of the important design parameters for the laser. In Fig. 11.64a, we show a comparison of the threshold current density $\ln(J_{th})$ of blue–violet with red (690 nm) lasers with various cavity lengths versus the threshold gain. Figure 11.64b shows the extracted logarithmic peak gain per well (1/cm) versus the current density per well (A/cm^2) for blue–violet and red lasers [203] as shown in (10.3.24) of Chapter 10

$$g = g_0\left[\ln\left(\frac{J}{J_0}\right) + 1\right] \tag{11.5.29}$$

where the values $(g_0, J_0) = (1614, 1381)$ for GaN-based blue laser diodes; $(881, 299)$ for GaInP red (690 nm) laser diodes, and (1500, 105) for GaAs laser diodes.

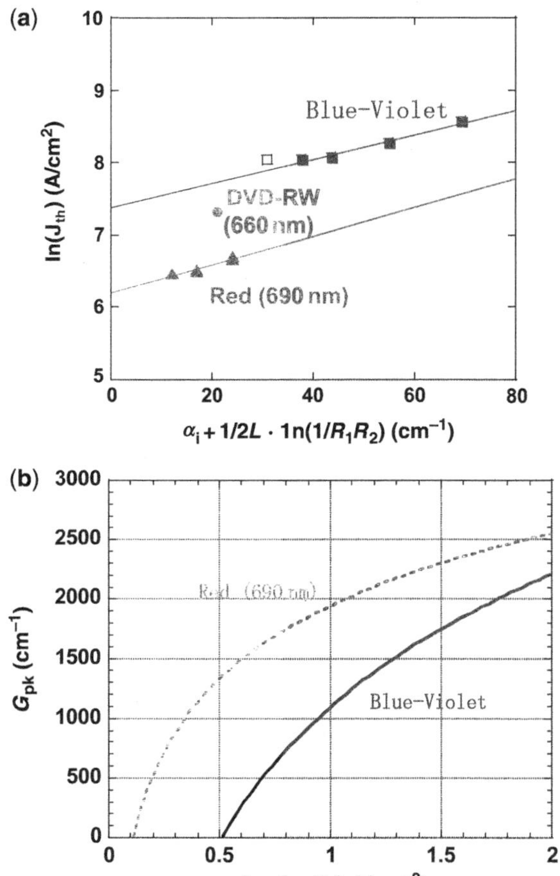

Figure 11.64 (a) Comparison of the threshold current density $\ln(J_{th})$ of blue–violet with red (690 nm) lasers with various cavity lengths versus the threshold gain. (b) The extracted logarithmic peak gain versus the current density for blue–violet and red lasers. (Reprinted with permission from [203] © 2003 IEEE.)

Figure 11.65a and Fig. 11.65b show the L–I curves of two laser diodes with different thicknesses d between the MQW active region and the AlGaN current blocking layer, indicating a reduction of threshold current, improved slope efficiency (from 1.1 to 1.4 W/A), and characteristic temperature (T_0 from 146K to 235K). Figure 11.65c shows the measured I–V curves at 20°C to 80°C for the improved device. Figure 11.66 shows the measured relative intensity noise (RIN) as a function of light output power (a) without high-frequency modulation and (b) with a 400-MHz signal modulation [203].

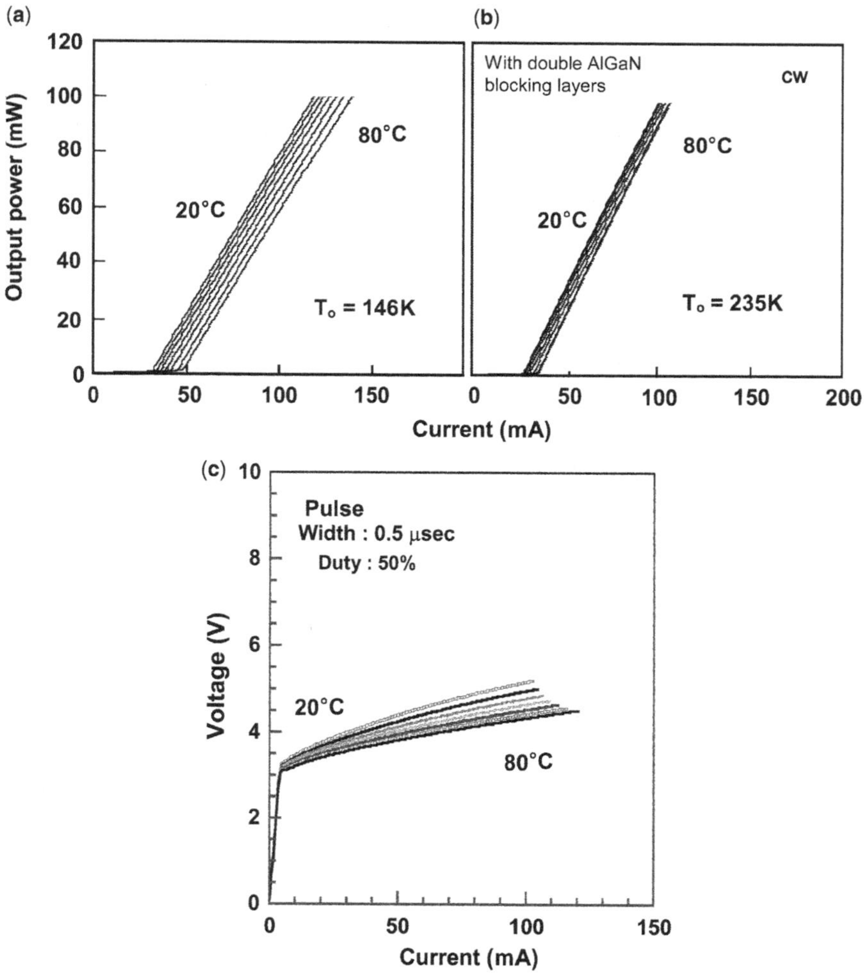

Figure 11.65 (a)(b) The L–I curves of two laser diodes with different thicknesses d between the MQW active region and the current blocking AlGaN layer, indicating a reduction of threshold current, improved slope efficiency (from 1.1 to 1.4 W/A), and characteristic temperature (T_0 from 146K to 235K) from (a) to (b). (c) The measured I–V curves at 20°C to 80°C. (Reprinted with permission from [203] © 2003 IEEE.)

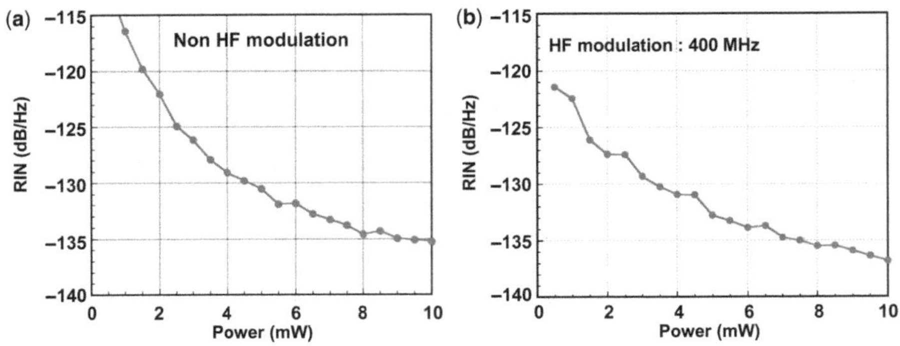

Figure 11.66 Measured relative intensity noise (RIN) as a function of light output power (a) without high-frequency modulation and (b) with a 400 MHz signal modulation. (Reprinted with permission from [203] © 2003 IEEE.)

Figure 11.67a shows a more recent laser diode grown on a (0001) *n*-GaN substrate [209]. The conduction band profile is shown in Fig. 11.67b. The device has a width of 12 μm and a cavity length of 2000 μm. The laser shows impressive performance at 440-nm wavelength with a light output power of 1000 mW at 1700 mA at 25°C. Aging tests at 750 mW cw operation at 35°C show a lifetime of more than 10,000 hours has also been realized.

11.5.6 Nonpolar (m-Plane and a-Plane) and Semipolar (r-plane) Devices

In order to avoid the polarization fields, growth of nitride-based epilayers along nonpolar directions have been intensively researched in the past decade [260–280].

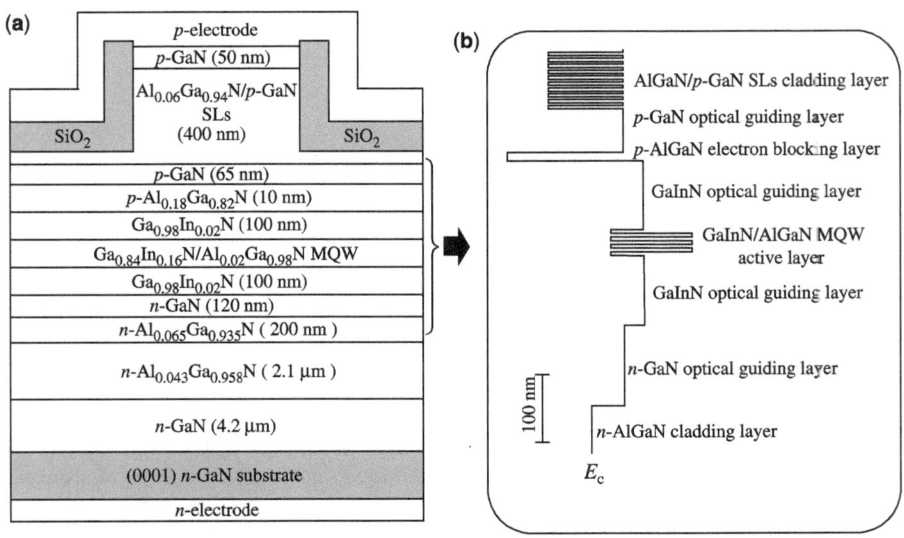

Figure 11.67 (a) Schematics of a laser diode structure grown on (0001) *n*-GaN substrate and (b) the conduction band profile. (Reproduced with permission from [209] © 2007 Wiley-VCH Verlag GmbH & Co. KGaA.)

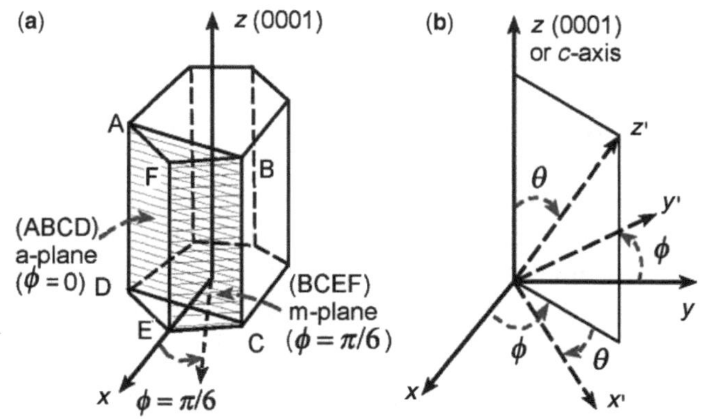

Figure 11.68 (a) A diagram for the a-plane and m-plane of a hexagonal structure. The c-plane is the hexagonal facet perpendicular to the *c*-axis or *z*(0001) axis. (b) A configuration for the Euler angles for rotation of the crystal axis. (Reprinted with permission from [227] © 2007 IEEE.)

Nonpolar planes include the a-plane $(11\bar{2}0)$ and m-plane $((10\bar{1}0)$, which show an absence of the polarization field and reduce the energy band profile to the flat band condition (except for the bias field). Semipolar planes such as $(10\bar{1}\bar{1})$, $(10\bar{1}\bar{3})$, $(1\bar{1}02)$, and $(11\bar{2}2)$ have been investigated [274–277, 280]. Figure 11.68a shows the crystal orientation and the definition of the m-plane and a-plane [227]. Figure 11.68b shows the definition of the Euler angles for the coordinate rotation for the different growth z'-axis. The c-plane is perpendicular to the (0001) axis. Crystal growth along semipolar planes such as the r-plane has also been reported. In Fig. 11.69a and Fig. 11.69b, we show the energy band profile based on a self-consistent Schrödinger–Poisson model

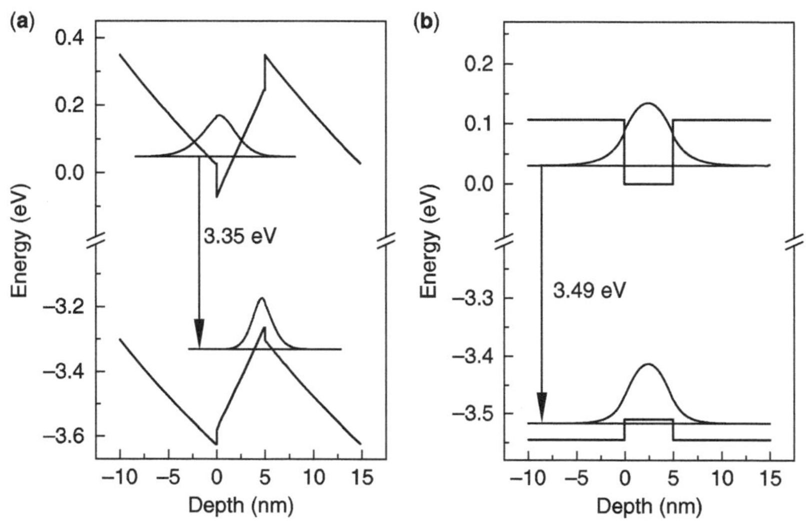

Figure 11.69 (a) The energy band profile based on a self-consistent Schrödinger–Poisson model for a $GaN/Al_{0.1}Ga_{0.9}N$ quantum well structure grown on (a) c-plane and (b) m-plane. The m-plane sample has a zero polarization field; therefore, it shows a flat band energy profile. (Adapted by permission from [261] © 2000 Macmillan Publishers Ltd: *Nature*.)

for a $GaN/Al_{0.1}Ga_{0.9}N$ quantum well structure grown on c-plane and m-plane [261]. The m-plane sample has a zero polarization field, therefore, it exhibits a flat band energy profile. Figure 11.70 shows the measured effects of the PZ field on the emission characteristics of GaN/AlGaN multiple quantum wells [261]. The CL spectra at 5K show that the c-plane sample has a red shift due to the field effect (peak at 3.36 eV) compared with the peak energy 3.48 eV of the m-plane sample. The inset shows the prolonged carrier lifetime of the c-plane sample due to the separation of the electron and hole wave functions, as indicated in Fig. 11.69.

To understand the improvement of the optical gain for samples grown on nonpolar planes, we show in Fig. 11.71a the optical gain spectra on the quantum-well plane (TE polarization along the x' and y' direction) for the a-plane (solid) and m-plane (dashed) samples [227]. The c-plane sample (dotted curve) is isotropic on the quantum well plane. The structure is an $In_{0.15}Ga_{0.85}N/GaN$ QW with a well width of 30 Å. The carrier density is $2 \times 10^{13}/cm^2$. The increase of the optical gain for the m-plane or a-plane due to the absence of spontaneous and piezoelectric polarization fields is clear. We plot in Fig. 11.71b the corresponding normalized optical momentum matrix elements as a function of the in-plane wave vector for x' and y' polarizations, which shows the anisotropic nature on the $x'-y'$ plane. Unlike the c-plane (or called $x - y$ plane) on which the optical polarization is isotropic, the m-plane and a-plane show the extra anisotropic polarization dependence.

The first (pulsed) operation of m-plane nitride laser diodes was realized on low extended defect bulk m-plane GaN substrates in 2007 [270]. The laser diodes had a

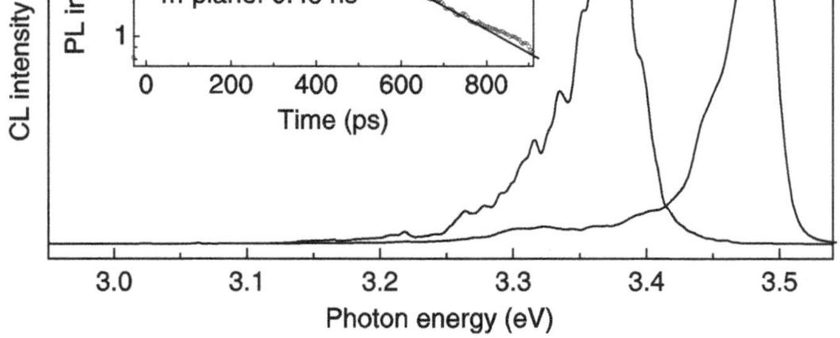

Figure 11.70 The effects of the PZ field on the emission characteristics of GaN/AlGaN multiple quantum wells. The catholuminescence (CL) spectra at 5 K show that the c-plane sample has a red shift due to the field effect (peak at 3.36 eV) compared with the peak energy 3.48 eV of the m-plane sample. The inset shows the prolonged carrier lifetime of the c-plane sample due to the separation of the electron and hole wave functions, as indicated in Fig. 11.69. (Adapted by permission from [261] © 2000 Macmillan Publishers Ltd: *Nature*.)

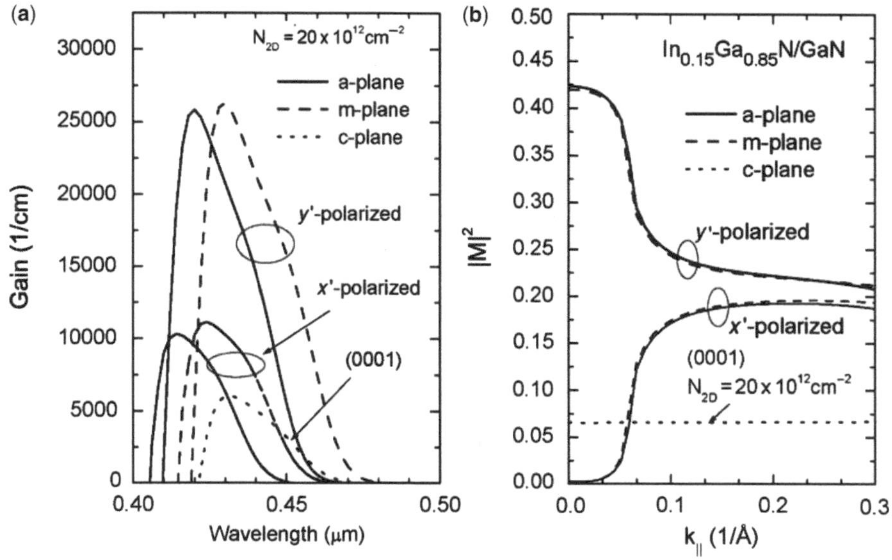

Figure 11.71 (a) The optical gain spectra on the quantum-well plane (TE polarization along the x' and y' direction) for the a-plane (solid) and m-plane (dashed) samples. The c-plane sample (dotted curve) is isotropic on the quantum well plane. (b) The corresponding normalized optical momentum matrix elements as a function of the in-plane wave vector for x' and y' polarizations. The structure is an $In_{0.15}Ga_{0.85}N/GaN$ QW with a well width of 30 Å. The surface carrier density is 2×10^{13} cm^{-2}. (Reprinted with permission from [227] © 2007 IEEE.)

threshold current as low as $7.5\,kA/cm^2$. Continuous wave operation of nonpolar m-plane growth InGaN/GaN multi-quantum-well structure laser diodes has also been demonstrated. Because thin wells (<4 nm) are used for c-plane growth to reduce the polarization fields, nonpolar plane quantum well growth allows for thick quantum wells without Al-containing waveguide cladding layers. For example, in 2007 an AlGaN-cladding-free m-plane InGaN/GaN laser diode was realized with a cw operation of more than 15 hours [271, 273]. The laser diodes structure consists of a thick Si-doped *n*-GaN cladding layer, a five-period InGaN/GaN MQW structure using 8-nm InGaN QWs and 8-nm GaN barriers with a 10-nm Mg-doped *p*-AlGaN electron blocking layer, a 750-nm Mg-doped *p*-GaN cladding layer, and a 20-nm highly Mg-doped p^{++}-GaN contact layer. The devices show a peak output power of more than 25 mW with a threshold current density of $6.8\,kA/cm^2$ and voltage of 5.6 V. Intensive research continues in order to demonstrate superior performance than that of the c-plane growth MQW lasers and LEDs.

11.6 COUPLED LASER ARRAYS

Coherent high-power semiconductor laser arrays [281–294] have been of considerable interest recently. Shown in Fig. 11.72a and Fig. 11.72b are two examples of gain-guided and buried-ridge semiconductor laser arrays [282, 287]. The concept of

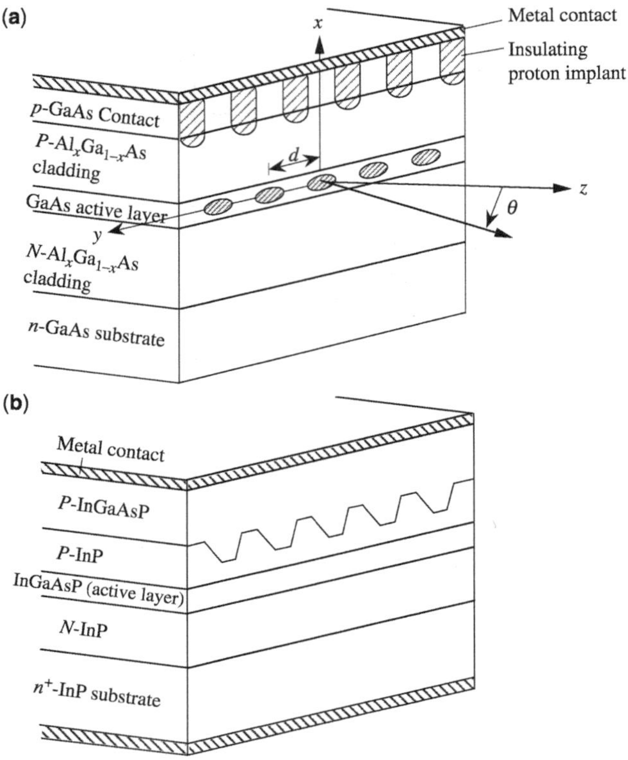

Figure 11.72 (a) A schematic diagram for a coupled gain-guided semiconductor laser array [281]. (b) Semiconductor buried-ridge laser array [286].

coupled laser arrays is very similar to that of the antenna arrays. As the amount of power of a single-element semiconductor laser is limited because of its small size in the radiation aperture (facet), a coupled array is expected to deliver a higher power with a narrower radiation pattern (i.e., a better directivity) if the elements are properly excited. The challenging research issues in laser arrays are how to excite the desired (fundamental) mode without wasting optical powers to different sidelobes.

In principle, for a given semiconductor laser array, the optical modes should be calculated [291] by solving Maxwell's equations for a given gain and index profile, which is determined by the injection and coupling conditions. Here we present a supermode analysis for its simplicity and ease of understanding the operation of laser arrays. The supermode analysis for a coupled laser array follows closely the coupled-mode theory [283]. The optical electric field can be written as a linear combination of the field of each element,

$$\mathbf{E}(x, y, z) = \sum_{n=1}^{N} a_n(z)\, \mathbf{E}^{(n)}(x, y) \tag{11.6.1}$$

where $\mathbf{E}^{(n)}(x, y)$ refers to the optical electric field in the nth element before the coupling effect is taken into account. The coupling effect is included in the coupling coefficient K assuming only nearby coupling.

11.6.1 Solution of the Coupled-Mode Equations

The coupled-mode equations for N identical equally spaced waveguides, assuming only adjacent waveguide coupling with a coupling coefficient K, can be written as

$$\frac{d}{dz}\mathbf{A}(z) = i\,\mathbf{MA}(z) \tag{11.6.2}$$

where

$$\mathbf{A}(z) = \begin{bmatrix} a_1(z) \\ a_2(z) \\ \vdots \\ a_N(z) \end{bmatrix}, \quad \mathbf{M} = \begin{bmatrix} \beta & K & & \mathbf{0} & \\ K & \beta & K & & \\ & \ddots & \ddots & \ddots & \\ & & K & \beta & K \\ & \mathbf{0} & & K & \beta \end{bmatrix} \tag{11.6.3}$$

and β is the propagation constant of the individual waveguide mode. The general solutions for the normalized eigenvectors $\mathbf{A}^{(\ell)}(z)$ and their corresponding eigenvalues β_ℓ for an arbitrary N are

$$\mathbf{A}^{(\ell)}(z) = \begin{bmatrix} a_1^{(\ell)} \\ a_2^{(\ell)} \\ \vdots \\ a_N^{(\ell)} \end{bmatrix} e^{i\beta_\ell z} \qquad \ell = 1, 2, \ldots, N \tag{11.6.4}$$

where

$$\beta_\ell = \beta + 2K\cos\left(\frac{\ell\pi}{N+1}\right) \tag{11.6.5}$$

$$a_n^{(\ell)} = \sqrt{\frac{2}{N+1}}\sin\left(\frac{n\ell\pi}{N+1}\right) \quad n = 1, 2, \ldots, N. \tag{11.6.6}$$

The order of β_ℓ is chosen such that $\beta_1 > \beta_2 > \beta_3 \cdots > \beta_N$.

To prove this, we start with the general form (11.6.4) and substitute it into the differential Eq. (11.6.2)

$$\begin{bmatrix} (\beta-\beta_\ell)/K & 1 & & & \mathbf{0} \\ 1 & (\beta-\beta_\ell)/K & 1 & & \\ & 1 & (\beta-\beta_\ell)/K & & \\ & & \ddots & & 1 \\ \mathbf{0} & & & 1 & (\beta-\beta_\ell)/K \end{bmatrix} \begin{bmatrix} a_1 \\ a_2 \\ a_3 \\ \vdots \\ a_N \end{bmatrix} = \begin{bmatrix} 0 \\ 0 \\ 0 \\ \vdots \\ 0 \end{bmatrix}. \tag{11.6.7}$$

We use the property of the $N \times N$ determinant, $D_N(2\cos\theta)$, which is a polynomial of order N with an argument $2\cos\theta$

$$D_N(2\cos\theta) = \det \begin{bmatrix} 2\cos\theta & 1 & & \mathbf{0} & \\ 1 & 2\cos\theta & 1 & & \\ & \ddots & \ddots & \ddots & \\ & & 1 & 2\cos\theta & 1 \\ & \mathbf{0} & & 1 & 2\cos\theta \end{bmatrix} = \frac{\sin[(N+1)\theta]}{\sin\theta} \tag{11.6.8}$$

which obeys the recursion relation

$$D_N(2x) = 2xD_{N-1}(2x) - D_{N-2}(2x) \tag{11.6.9}$$

where $x = \cos\theta$. This recursion relation can be proved easily using the definition of the $N \times N$ determinant and expanding it in terms of the two elements in the first row (or column). The above recursion relation (11.6.9) is satisfied by the Nth order Chebyshev polynomial [295] $U_n(\cos\theta)$, or

$$D_N(2\cos\theta) = \frac{\sin(N+1)\theta}{\sin\theta}. \tag{11.6.10}$$

We define

$$2\cos\theta = \frac{\beta-\beta_\ell}{K}, \quad \text{or} \quad \beta_\ell = \beta - 2K\cos\theta. \tag{11.6.11}$$

The eigenvalue equation can have nontrivial solutions only if the determinant is zero

$$\frac{\sin(N+1)\theta}{\sin\theta} = 0. \tag{11.6.12}$$

Therefore,

$$(N+1)\theta = m\pi, \quad m = 1, 2, 3, \ldots, N. \tag{11.6.13}$$

We can choose $m = N + 1 - \ell$, $\ell = 1, 2, \ldots, N$, such that

$$\begin{aligned} \theta_\ell &= \frac{\ell}{N+1}\pi, \qquad \theta = (\pi - \theta_\ell) \\ \beta_\ell &= \beta + 2K \cos\theta_\ell, \quad \ell = 1, 2, 3, \ldots, N \end{aligned} \tag{11.6.14}$$

and β_1, β_2, $\beta_3, \ldots,$ and β_N are in descending order for $K > 0$.

The eigenvectors can be obtained by substituting β_ℓ into (11.6.7)

$$\begin{cases} -2a_1 \cos\theta_\ell + a_2 = 0 \\ a_{n-1} - 2a_n \cos\theta_\ell + a_{n+1} = 0 \quad n = 2, 3, \ldots, N-1 \\ a_{N-1} - 2a_N \cos\theta_\ell = 0. \end{cases} \tag{11.6.15}$$

The above difference equation can be solved assuming the solution a_n of the form, $a_n = r^n$. Therefore

$$\frac{1}{r} - 2\cos\theta_\ell + r = 0 \tag{11.6.16}$$

or $r = \exp(\pm i\theta_\ell)$. The general solution for a_n is therefore $\exp(in\theta_\ell)$ or $\exp(-in\theta_\ell)$, or their linear combinations

$$a_n = A\cos n\theta_\ell + B\sin n\theta_\ell. \tag{11.6.17}$$

Because $a_0 = 0$, we find $A = 0$. Therefore, $a_n = B \sin n\theta_\ell$. The amplitude B can be found from the normalization condition, $\sum_{n=1}^{N} |a_n|^2 = 1$. Because

$$\begin{aligned} \sum_{n=1}^{N} \sin^2 n\theta_\ell &= \sum_{n=1}^{N} \frac{1 - \cos 2n\theta_\ell}{2} \\ &= \frac{N}{2} - \frac{1}{2}\mathrm{Re}\left[\sum_{n=1}^{N} e^{i2n\theta_\ell}\right] \\ &= \frac{N+1}{2} \end{aligned} \tag{11.6.18}$$

where $\theta_\ell = \ell\pi/(N+1)$ has been used, and the normalization constant $B = \sqrt{2}/\sqrt{N+1}$. We conclude that the N eigenvectors with corresponding eigenvalues β_ℓ are described by (11.6.4)–(11.6.6).

11.6.2 Far-Field Radiation Pattern

To calculate the far-field radiation pattern, recall from Chapter 6 that the optical electric field under far-field approximation is given by the Fourier transform of the aperture field,

$$\mathbf{E}(\mathbf{r}) = \frac{-ike^{ikr}}{4\pi r}\hat{r} \times \iint dx'dy'\, e^{-i\mathbf{k}\cdot\mathbf{r}'}\mathbf{M}_s(\mathbf{r}'), \tag{11.6.19}$$

where the equivalent magnetic surface current density is related to the aperture field by

$$\mathbf{M}_s = -2\hat{n} \times \mathbf{E}_A(\mathbf{r}') = -2\hat{z} \times \mathbf{E}_A(\mathbf{r}'). \tag{11.6.20}$$

If $\mathbf{E}_A(\mathbf{r}') = \hat{y}\, E_A\, (x', y')$ for TE modes in the waveguides, we have

$$\begin{aligned} \mathbf{M}_s &= 2\hat{x}\, E_A \\ \hat{r} \times \hat{x} &= \sin\theta \sin\phi(-\hat{z}) + \cos\theta \hat{y}. \end{aligned} \tag{11.6.21}$$

Considering the observation point in the $y-z$ plane, ($\phi = \pi/2$), we have

$$\mathbf{E}(\theta) \propto (-\sin\theta\hat{z} + \cos\theta\hat{y}) \iint dxdy\, e^{-i\mathbf{k}\cdot\mathbf{r}} E_A(\mathbf{r}). \tag{11.6.22}$$

The aperture field is given by (11.6.1), noting that $E^{(n)}(\mathbf{r}) = E^{(1)}(\mathbf{r} - n\mathbf{d})$, where $E^{(1)}$ is the optical aperture field of the first element

$$E_A(\mathbf{r}) = \sum_n a_n^{(\ell)} E^{(1)}(\mathbf{r} - n\mathbf{d}). \tag{11.6.23}$$

Therefore, the far-field radiation pattern is

$$|\mathbf{E}(\theta)|^2 = \underbrace{\left|\sum_{n=1}^{N} a_n^{(\ell)} e^{-i\mathbf{k}\cdot n\mathbf{d}}\right|^2}_{\text{Arraypattern}} \underbrace{\left|\iint d\mathbf{r}\, e^{-i\mathbf{k}\cdot\mathbf{r}} E^{(1)}(\mathbf{r})\right|^2}_{\text{Unitpattern}}. \tag{11.6.24}$$

Using $\mathbf{k}\cdot\mathbf{d} = kd \sin\theta$, the array pattern of the ℓth supermode, $A_\ell(\theta)$, can be obtained from

$$\begin{aligned} A_\ell(\theta) &= \left|\sum_{n=1}^{N} a_n^{(\ell)} e^{-inkd\sin\theta}\right|^2 \\ &= \frac{1}{2(N+1)}\left|\sum_{n=1}^{N} e^{in(\theta_\ell - kd\sin\theta)} - \sum_{n=1}^{N} e^{-in(\theta_\ell + kd\sin\theta)}\right|^2 \\ &= \frac{1}{2(N+1)}\left|(-1)^\ell \frac{\sin\frac{N}{2}(\theta_\ell - kd\sin\theta)}{\sin\frac{1}{2}(\theta_\ell - kd\sin\theta)} - \frac{\sin\frac{N}{2}(\theta_\ell + kd\sin\theta)}{\sin\frac{1}{2}(\theta_\ell + kd\sin\theta)}\right|^2. \end{aligned} \tag{11.6.25}$$

In Fig. 11.73a, the magnitudes of the 10 eigenvectors $A_\ell(\theta)$, for $\ell = 1, 2, \ldots, 10$ of a 10-element array are shown (from Ref. 283). Each mode is labeled by ℓ. For each eigenmode ℓ, the magnitudes $a_n^{(\ell)}$, $n = 1, 2, \ldots, 10$ are represented by vertical bars.

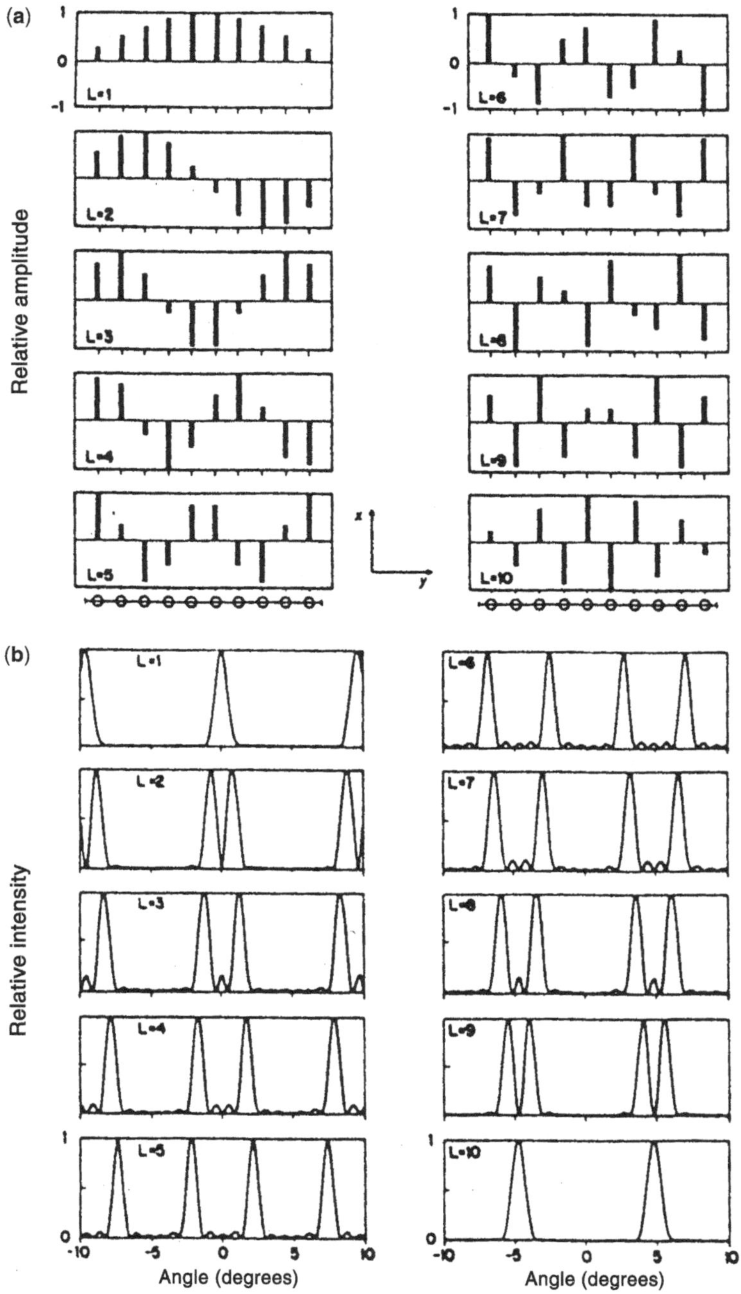

Figure 11.73 (a) The magnitudes of the 10 eigenvectors $A^{(\ell)}(z)$ for $\ell = 1, 2, 3, \ldots, 10$ of a 10-element array. (b) The corresponding far-field patterns of the 10 array modes. (Reprinted with permission from [282] © 1984 American Institute of Physics.)

The corresponding far-field patterns are shown in Fig. 11.73b. We can see that for the fundamental ($\ell = 1$) and the highest ($\ell = 10$) order modes, the amplitudes are not uniform throughout the elements.

To improve the uniformity of the near-field profile, the two outermost waveguides are chosen [285] to have a smaller spacing to increase the coupling coefficient by a $\sqrt{2}$ factor, with all the other guides equally spaced. The lowest and the highest supermodes have a uniform near-field intensity envelope, and the injected charges will be used more efficiently. These supermodes should be relatively stable with increased pumping above threshold. Another design is to use a Y-junction waveguide [286] structure such that all waveguide modes will lase in phase with equal amplitudes.

Using closely spaced antiguided waveguides [288, 289], a new class of phase-locked diode laser arrays has been developed. Fundamental array-mode operation in a diffraction-limit-beam pattern is obtained up to 200 mW. The out-of-phase mode is also shown to lase with 110 mW per uncoated facet. The threshold currents of these antiguided arrays have been shown to be in the 270 to 320 mA range with an external quantum efficiency near 30% to 35% and cw operation up to 200 mW. As pointed out before, the real optical modal profiles in the laser arrays using gain-guided [289] or antiguided [291–293] structures have to be calculated properly to compare with experimental data. The supermode analysis is presented in this section only to provide some basic understanding of the laser array characteristics.

APPENDIX 11A: HAMILTONIAN FOR STRAINED WURTZITE CRYSTALS

We choose the six valence-band wave functions, $\{u_1, u_2, \ldots, u_6\}$ as [188, 189]

$$\begin{aligned} u_1 &= \frac{-1}{\sqrt{2}}|(X+iY)\uparrow\rangle \quad u_2 = \frac{1}{\sqrt{2}}|(X-iY)\uparrow\rangle \quad u_3 = |Z\uparrow\rangle \\ u_4 &= \frac{1}{\sqrt{2}}|(X-iY)\downarrow\rangle \quad u_5 = \frac{-1}{\sqrt{2}}|(X+iY)\downarrow\rangle \quad u_6 = |Z\downarrow\rangle, \end{aligned} \tag{11A.1}$$

which are the states of interest (class A), and all other states of no interest are called class B. Again we use Löwdin's method (see Appendix 3A) and obtain the 6×6 Hamiltonian matrix for the valence bands. The full Hamiltonian can be written as [188, 189]

$$\mathbf{H} = \begin{bmatrix} F & -K^* & -H^* & 0 & 0 & 0 \\ -K & G & H & 0 & 0 & \Delta \\ -H & H^* & \lambda & 0 & \Delta & 0 \\ 0 & 0 & 0 & F & -K & H \\ 0 & 0 & \Delta & -K^* & G & -H^* \\ 0 & \Delta & 0 & H^* & -H & \lambda \end{bmatrix} \begin{matrix} |u_1\rangle \\ |u_2\rangle \\ |u_3\rangle \\ |u_4\rangle \\ |u_5\rangle \\ |u_6\rangle \end{matrix} \tag{11A.2}$$

where the matrix elements are defined as

$$
\begin{aligned}
F &= \Delta_1 + \Delta_2 + \lambda + \theta & \lambda &= \lambda_k + \lambda_\varepsilon \\
G &= \Delta_1 - \Delta_2 + \lambda + \theta & \theta &= \theta_k + \theta_\varepsilon \\
\lambda_k &= \frac{\hbar^2}{2m_0}\left[A_1 k_z^2 + A_2\left(k_x^2 + k_y^2\right)\right] & \lambda_\varepsilon &= D_1\varepsilon_{zz} + D_2(\varepsilon_{xx} + \varepsilon_{yy}) \\
\theta_k &= \frac{\hbar^2}{2m_0}\left[A_3 k_z^2 + A_4\left(k_x^2 + k_y^2\right)\right] & \theta_\varepsilon &= D_3\varepsilon_{zz} + D_4(\varepsilon_{xx} + \varepsilon_{yy}) \\
K &= \frac{\hbar^2}{2m_0}A_5(k_x + ik_y)^2 + D_5\varepsilon_+ & \varepsilon_\pm &= \varepsilon_{xx} \pm 2i\varepsilon_{xy} - \varepsilon_{yy} \\
H &= \frac{\hbar^2}{2m_0}A_6 k_z(k_x + ik_y) + D_6\varepsilon_{z+} & \varepsilon_{z\pm} &= \varepsilon_{zx} \pm i\varepsilon_{zy}.
\end{aligned}
\tag{11A.3}
$$

The strain effects are included by considering the wurtzite crystal symmetry and a straightforward addition of corresponding terms: $k_\alpha k_\beta \rightarrow \varepsilon_{\alpha\beta}$ with the deformation potentials, $D_1, D_2, \ldots, D_6$, at the corresponding positions of $A_1, A_2, \ldots, A_6$, which play the same roles as the Luttinger–Kohn parameters for zinc-blende crystals. A simplification based on the cubic approximation [190] is to make use of the similarity between the wurtzite structure and the cubic crystal. If we map the $c(0001)$ axis of WZ crystal to the (111) direction (called z'-axis) of ZB crystal and set the x' and y' axes along the $(11\bar{2})$ and $(\bar{1}10)$ directions, respectively, in the new coordinate system, we can derive the relations

$$
\begin{aligned}
&A_1 - A_2 = -A_3 = 2A_4 \quad A_3 + 4A_5 = \sqrt{2}A_6 \quad \Delta_2 = \Delta_3 \\
&D_1 - D_2 = -D_3 = 2D_4 \quad D_3 + 4D_5 = \sqrt{2}D_6.
\end{aligned}
\tag{11A4}
$$

Therefore, only five band-structure parameters such as A_1, A_2, A_5, Δ_1, and Δ_2 and three deformation potentials are necessary for the calculation of the valence-band structures. It has also been found that a seventh coefficient A_7 for the linear k terms vanishes; therefore, we discard it at the beginning and keep only the quadratic terms of k in addition to the band-edge energy terms.

The band-edge energies of the conduction, heavy-hole (HH), light-hole (LH), and crystal-field split-off hole (CH) bands and their corresponding (double fold degenerate wavefunctions) at the zone center are given by

$$
\begin{array}{llll}
\mathrm{C} & E_c & u_c = |iS\uparrow\rangle & |iS\downarrow\rangle \\
\mathrm{HH} & E_1 & u_{hh} = u_1 & u_4 \\
\mathrm{LH} & E_2 & u_{lh} = pu_2 + qu_6 & qu_3 + pu_5 \\
\mathrm{CH} & E_3 & u_{ch} = qu_2 - pu_6 & -pu_3 + qu_5
\end{array}
\tag{11A.5}
$$

where we define p and q to include the strain effects

$$
p = \frac{E_2 - \lambda_\varepsilon}{\sqrt{(E_2 - \lambda_\varepsilon)^2 + 2\Delta_3^2}} \qquad q = \frac{\sqrt{2}\Delta_3}{\sqrt{(E_2 - \lambda_\varepsilon)^2 + 2\Delta_3^2}}.
\tag{11A.6}
$$

It is straightforward to show that

$$p^2 = \frac{E_2 - \lambda_\varepsilon}{E_2 - E_3} \qquad q^2 = \frac{E_3 - \lambda_\varepsilon}{E_3 - E_2}. \tag{11A.7}$$

Note that $p^2 + q^2 = 1$.

As shown in Ref. 218, the 6×6 Hamiltonian matrix (11A.2) can be block-diagonalized to two 3×3 matrices by a change of bases. It is then much easier to find the eigenvalues and eigenvectors of a 3×3 matrix because the eigenvalue equation will have a polynomial of order 3 in the eigenenergy, which has two degenerate analytical solutions. The results are then shown in Fig. 11.51 using the band structure parameters listed in Appendix D at the end of the book.

Model for the Quantum-Well Heterojunction Consider a wurtzite GaN/$Al_xGa_{1-x}N$ quantum-well structure, Fig. 11A.1 [188, 189]. We have to find (1) the reference band-edge energies E_v^0 in the well and the barrier, and the difference between the two values and (2) the conduction band-edge energies E_c^0 in the well and the barrier. An earlier experimental report shows a value of 0.8 ± 0.3 eV for the valence-band edge discontinuity between GaN and AlN wurtzite semiconductors. Therefore, a band-edge discontinuity ratio of 33% for the valence band and 67% for the conduction band has been proposed. The top valence band of GaN is the E_1^0 (HH)

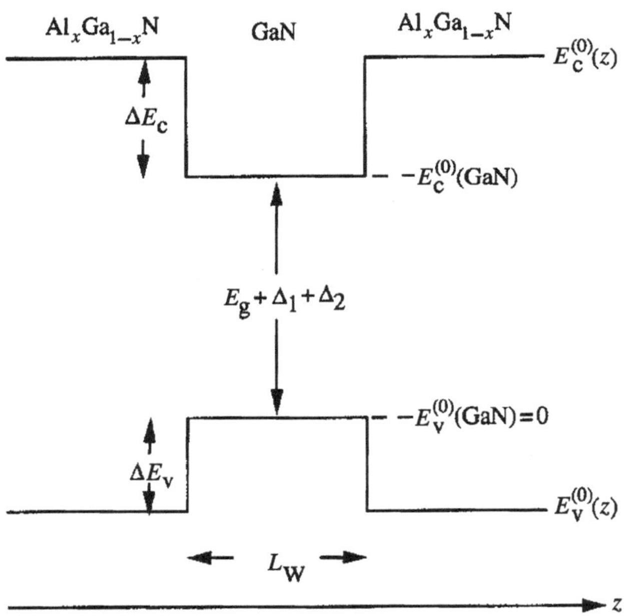

Figure 11A.1 Band-edge energies of a wurtzite GaN/AlN heterojunction. Δ_1 is positive for GaN and negative for AlN. The band-edge discontinuities ΔE_c^0 and ΔE_v^0 are labeled for a GaN/AlN heterojunction. (Reprinted with permission from [189] © 1996 IEEE.)

band and the top valence band of AlN is the E_2^0 (CH) band. Therefore, the difference between the two band edges is $E_1^0(\text{GaN}) - E_2^0(\text{AlN}) = 0.8 \pm 0.3$ eV. Because the error (0.3 eV) is much larger than Δ_1, Δ_2, and Δ_3 of both GaN and AlN, we can simply assume that $E_v^0(\text{GaN}) - E_v^0(\text{AlN}) \simeq 0.8$ eV and use the interpolation formula for the location of E_v^0 for the $Al_xGa_{1-x}N$ materials

$$E_v^0(\text{Al}_x\text{Ga}_{1-x}\text{N}) = xE_v^0(\text{AlN}) + (1-x)E_v^0(\text{GaN}) + b_v x(1-x) \tag{11A.8}$$

where b_v accounts for the contribution of the bowing effects of the band-gap discontinuity to the valence band. Because the band gap of the $Al_xGa_{1-x}N$ has been found from experimental data,

$$E_g(\text{Al}_x\text{Ga}_{1-x}\text{N}) = xE_g(\text{AlN}) + (1-x)E_g(\text{GaN}) - bx(1-x) \tag{11A.9}$$

and a value of 0.98 eV has been reported for the bowing parameter b ($=b_c + b_v$), we assume that $b_v = 0.33b$ and use a positive sign in front of b_v because the band gap is related to $E_c - E_v$. The conduction band edge is then:

$$E_c^0(\text{Al}_x\text{Ga}_{1-x}\text{N}) = xE_c^0(\text{AlN}) + (1-x)E_c^0(\text{GaN}) - b_c x(1-x). \tag{11A.10}$$

The other parameters, Δ_i, $i = 1, 2, 3$, are linearly interpolated from their GaN and AlN values. The valence-band quantum-well potential is

$$E_v^0(z) = \begin{cases} E_v^0(\text{GaN}) & \text{well} \\ E_v^0(\text{Al}_x\text{Ga}_{1-x}\text{N}) & \text{barriers} \end{cases} \tag{11A.11}$$

and a similar assignment for the conduction-band edge profile. Note that the difference between E_c^0 and E_v^0 is $E_g + \Delta_1 + \Delta_2$. The contributions due to strain and the difference in the band-structure parameters and Δ_i's are taken into account explicitly in the remaining Hamiltonians. Again, for convenience, the reference can be set at the valence-band edge of the unstrained well value, that is, subtract all energies by $E_v^0(\text{GaN})$, which is equivalent to set $E_v^0(\text{GaN}) = 0$ in the above equations, and find the energies of all other quantities.

APPENDIX 11B: BAND-EDGE OPTICAL TRANSITION MATRIX ELEMENTS

The polarization selection rules for the conduction band to the light-hole band are different from those of the zinc-blende crystals. For the zinc-blende structure, the dominant C-LH transitions are TM polarized (along the growth axis). However, for the wurtzite structure, TE polarization is favored for the C-LH transitions. Two Kane's parameters, P_1 and P_2, are related to the interband momentum matrix elements by

$$\left\langle iS\left|\frac{\hbar}{i}\frac{\partial}{\partial z}\right|Z\right\rangle = \frac{m_0}{\hbar}P_1 \tag{11B.1}$$

$$\left\langle iS\left|\frac{\hbar}{i}\frac{\partial}{\partial x}\right|X\right\rangle = \left\langle iS\left|\frac{\hbar}{i}\frac{\partial}{\partial y}\right|Y\right\rangle = \frac{m_0}{\hbar}P_2. \tag{11B.2}$$

We obtain modified Kane's parameters for wurtzite crystals,

$$P_1^2 = \frac{\hbar^2}{2m_0}\left(\frac{m_0}{m_e^z}-1\right)\frac{(E_g+\Delta_1+\Delta_2)(E_g+2\Delta_2)-2\Delta_3^2}{(E_g+2\Delta_2)} \tag{11B.3}$$

$$P_2^2 = \frac{\hbar^2}{2m_0}\left(\frac{m_0}{m_e^t}-1\right)\frac{E_g\left[(E_g+\Delta_1+\Delta_2)(E_g+2\Delta_2)-2\Delta_3^2\right]}{(E_g+\Delta_1+\Delta_2)(E_g+\Delta_2)-\Delta_3^2} \tag{11B.4}$$

which are related to the optical momentum matrix elements

$$E_{pz} = \frac{2m_0}{\hbar^2}P_1^2 \qquad E_{px} = \frac{2m_0}{\hbar^2}P_2^2. \tag{11B.5}$$

Using the band-edge wavefunctions with p and q given by (11A.6), we obtain the momentum matrix elements at the band edge ($k_t = 0$) including the strain effects in Table 11B.1 for the optical transition from the conduction band edge to the three valence subband edges (C-HH, C-LH, and C-CH transitions labeled as E_1, E_2, and E_3, respectively), when the optical polarization is perpendicular to the c-axis (TE polarization) and parallel to the c-axis (TM polarization). We also calculate the energy parameters for the interband optical transition oscillator strengths and obtain $E_{px} = 15.7$ eV and $E_{pz} = 13.9$ eV. Notice the sum rules for each polarization in

Table 11B.1 Polarization-Dependent Interband Optical-Matrix Elements

Valence Energy	TM ($\hat{\mathbf{e}} \parallel c$-axis)	TE ($\hat{\mathbf{e}} \perp c$-axis)
E_1 (HH band)	0	$\frac{m_0}{4}E_{px}$
E_2 (LH band)	$q^2\left(\frac{m_0}{2}E_{pz}\right)$	$p^2\left(\frac{m_0}{4}E_{px}\right)$
E_3 (CH band)	$p^2\left(\frac{m_0}{2}E_{pz}\right)$	$q^2\left(\frac{m_0}{4}E_{px}\right)$
Sum	$\frac{m_0}{2}E_{pz}$	$\frac{m_0}{2}E_{px}$

The energy parameters E_{px} and E_{pz} for the interband transition elements are related to Kane's parameters, P_1 and P_2, in Eq. (11B.5). p and q satisfy $p^2 + q^2 = 1$. $p^2 = \frac{E_2-\lambda_\varepsilon}{E_2-E_3}$, $q^2 = \frac{E_3-\lambda_\varepsilon}{E_3-E_2}$.

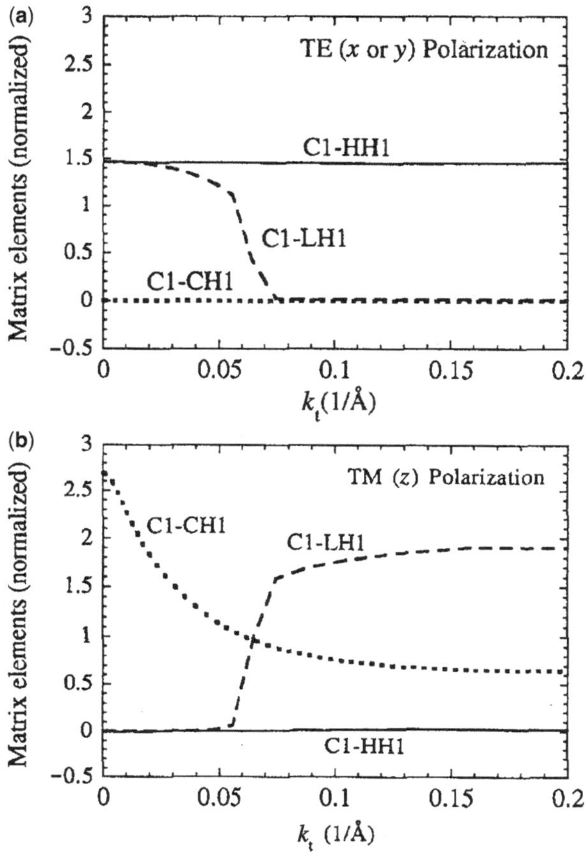

Figure 11B.1 The normalized momentum matrix elements for the first conduction subband to the first HH, LH, and CH subband transitions for (a) TE polarization on the quantum-well (x–y) plane and (b) TM polarization along the c-axis or z direction for a 26 Å GaN/$Al_{0.3}Ga_{0.7}N$ quantum well. (Reprinted with permission from [189] © 1996 IEEE.)

Table 11B.1. Figure 11B.1 shows the normalized momentum matrix elements as a function of the in-plane wavevector for the first conduction subband to the first HH, LH, and CH subband transitions for (a) TE polarization on the quantum-well ($x-y$) plane and (b) TM polarization along the c-axis or z direction for a 26 Å GaN/$Al_{0.3}Ga_{0.7}N$ quantum well [189].

PROBLEMS

11.1 Show that the lasing condition (11.1.23) is identical to (11.1.21), which corresponds with the zero of the denominator of the transmission coefficient of the DFB structure.

11.2 Derive the oscillation condition (11.1.31) for a DFB laser. Discuss the solution for the detuning δL and the gain $g_n L$.

11.3 Discuss the coupling coefficients, K_{ab} and K_{ba} in (11.1.41), for a complex grating with both index- and gain-grating couplings. Do they satisfy a simple relation as that in the lossless condition? Why or why not?

11.4 Plot the transmission spectrum (11.1.43a) of a phase-shifted DFB laser structure as a function of detuning δL for $G = g_n L = 0$, 0.5, and 1.0 assuming $KL = 1.5$.

11.5 Derive the lasing conditions (11.1.46) to (11.1.50) for a DBR semiconductor laser in Fig. 11.8.

11.6 Compare the threshold condition for a surface-emitting laser with that for a DBR semiconductor laser. What are their differences and similarities?

11.7 Explain why the optical output power versus injection current (L–I) curve of a surface-emitting laser has a negative differential quantum efficiency at a high injection level.

11.8 Compare the current density, the carrier concentration, and the optical modal profiles of a gain-guided surface emitting laser (Fig. 11.15) with those of a gain-guided stripe-geometry semiconductor laser in Section 10.2 (Fig. 10.15).

11.9 Derive the Purcell factor (11.3.15) and summarize the assumptions used.

11.10 Prove (11.3.26) using (11.3.24) and the definition of the gain coefficient.

11.11 Derive the threshold current density (11.4.7) for a quantum-cascade laser.

11.12 Compare the physical principles and emission polarizations of type-I QC lasers with type-II QC lasers.

11.13 Discuss the difference between the spontaneous polarization and the piezoelectric polarization in a GaN film.

11.14 Derive (11.5.26).

11.15 Using the coupled-mode equations in Section 11.6 assuming only nearby-element coupling, write the explicit eigenvalues and eigenvectors for **(a)** three coupled waveguides and **(b)** four coupled waveguides. Plot the modal profiles for the supermodes.

11.16 Derive the array pattern for the ℓth supermode, $A_\ell(\theta)$, in (11.6.25).

REFERENCES

1. H. Kogelnik and C. V. Shank, "Stimulated emission in a periodic structure," *Appl. Phys. Lett.* **18**, 152–154 (1971).

2. H. Kogelnik and C. V. Shank, "Coupled-wave theory of distributed feedback lasers," *J. Appl. Phys.* **43**, 2327–2335 (1972).
3. M. Nakamura, A. Yariv, H. W. Yen, S. Somekh, and H. L. Garvin, "Optically pumped GaAs surface laser with corrugation feedback," *Appl. Phys. Lett.* **22**, 515–516 (1973).
4. D. R. Scifres, R. D. Burnham, and W. Streifer, "Distributed-feedback single heterojunction GaAs diode laser," *Appl. Phys. Lett.* **25**, 203–205 (1974).
5. H. C. Casey, Jr., S. Somekh, and M. Ilegems, "Room-temperature operation of low-threshold separate-confinement heterostructure injection laser with distributed feedback," *Appl. Phys. Lett.* **27**, 142–144 (1975).
6. M. Nakamura, A. Aiki, J. Umeda, and A. Yariv, "cw operation of distributed-feedback GaAs-GaAlAs diode lasers at temperatures up to 300 K," *Appl. Phys. Lett.* **27**, 403–405 (1975).
7. G. P. Agrawal and N. K. Dutta, *Long-Wavelength Semiconductor Lasers*, Van Nostrand Reinhold, San Diego, 1986.
8. N. Chinone and M. Okai, "Distributed feedback semiconductor lasers," in *Semiconductor Lasers—Past, Present, and Future*, G. P. Agrawal, Ed., Chapter 2, AIP Press, New York, 1995.
9. G. Morthier and P. Vankwikelberge, *Handbook of Distributed Feedback Laser Diodes*, Artech House, Boston, MA, 1997.
10. H. Ghafouri-Shiraz and B. S. K. Lo, *Distributed Feedback Laser Diodes- Principles and Physical Modelling*, Wiley, New York, 1996.
11. A. Yariv, *Quantum Electronics*, 3rd ed., Wiley, New York, 1989.
12. W. Fang, A. Hsu, S. L. Chuang, T. Tanbun-Ek, and A. K. Sergent, "Measurement and modeling of distributed-feedback lasers with spatial hole burning," *IEEE J. Sel. Top. Quantum Electron.* **3**, 547–554 (1997).
13. Y. Luo, Y. Nakano, K. Tada, T. Inoue, H. Hosomatsu, and H. Iwaoka, "Fabrication and characteristics of gain-coupled distributed feedback lasers with a corrugated active layer," *IEEE J. Quantum Electron.* **27**, 1724–1731 (1991).
14. K. David, J. Buus, and R. Baets, "Basic analysis of AR-coated partly gain-coupled DFB lasers: the standing wave effect," *IEEE J. Quantum Electron.* **28**, 427–433 (1992).
15. G. P. Li, T. Makino, R. Moore, and N. Puetz, "1.55 μm index/gain coupled DFB lasers with strained layer multiquantum-well active grating," *Electron. Lett.* **28**, 1726–1727 (1992).
16. G. P. Li, T. Makino, R. Moore, N. Puetz, K. W. Leong, and H. Lu, "Partially gain-coupled 1.55 μm strained-layer multiquantum-well DFB lasers," *IEEE J. Quantum Electron.* **29**, 1736–1742 (1993).
17. H. Lu, T. Makino, and G. P. Li, "Dynamic properties of partly gain-coupled 1.55-μm DFB lasers," *IEEE J. Quantum Electron.* **31**, 1443–1450 (1995).
18. H. Lu, C. Blaauw, B. Benyon, G. P. Li, and T. Makino, "High-power and high-speed performance of 1.3-μm strained MQW gain-coupled DFB lasers," *IEEE J. Sel. Top. Quantum Electron.* **1**, 375–381 (1995).
19. T. Makino, "Threshold condition of DFB semiconductor lasers by the local-normal-mode transfer-matrix method: correspondence to the coupled-wave method," *IEEE J. Lightwave Technol.* **12**, 2092–2099 (1994).

20. H. Haus and C. V. Shank, "Antisymmetric taper of distributed feedback lasers," *IEEE J. Quantum Electron.* **QE-12**, 532–539 (1976).
21. H. A. Haus, *Waves and Fields in Optoelectronics*, Prentice Hall, Englewood Cliffs, NJ, 1984.
22. K. Sekartedjo, N. Eda, K. Furuya, Y. Suematsu, F. Koyama, and T. Tanbun-Ek, "1.5 μm phase-shifted DFB lasers for single-mode operation," *Electron. Lett.* **20**, 80–81 (1984).
23. Y. Suematsu, K. Kishino, S. Arai, and F. Koyama, "Dynamic single-mode semiconductor lasers with a distributed reflector," in *Semiconductors and Semimetals*, vol. 22, R. K. Willardson and A. C. Beer, Eds., Chapter 4, Academic Press, San Diego, 1985.
24. S. Wang, "Design considerations of the DBR injection laser and the waveguiding structure for integrated optics," *IEEE J. Quantum Electron.* **QE-13**, 176–186 (1977).
25. K. Iga, "Surface-emitting laser- its birth and generation of new optoelectronics fields," *IEEE J. Sel. Top. Quantum Electron.* **6**, 1201–1215 (2000).
26. C. Wilmsen, H. Temkin, and L. A. Coldren, *Vertical-Cavity Surface-Emitting Lasers: Design, Fabrication, Characterization, and Applications*, Cambridge University Press, Cambridge, UK, 1999.
27. G. A. Evans and J. M. Hammer, Eds., *Surface Emitting Semiconductor Lasers and Arrays*, Academic Press, San Diego, 1993.
28. H. Soda, K. Iga, C. Kitahara, and Y. Suematsu, "GaInAsP/InP surface emitting injection lasers," *Jpn. J. Appl. Phys.* **18**, 2329–2330 (1979).
29. K. Iga, S. Ishkawa, S. Ohkouchi, and T. Nishimura, "Room temperature pulsed oscillation of GaAlAs/GaAs surface emitting junction laser," *IEEE J. Quantum Electron.* **QE-21**, 315–320 (1985).
30. J. L. Jewell, J. P. Harbison, A. Scherer, Y. H. Lee, and L. T. Florez, "Vertical-cavity surface-emitting lasers: design, growth, fabrication, characterization," *IEEE J. Quantum Electron.* **27**, 1332–1346 (1991).
31. R. S. Geels, S. W. Corzine, and L. A. Coldren, "InGaAs vertical-cavity surface-emitting lasers," *IEEE J. Quantum Electron.* **27**, 1359–1367 (1991).
32. C. J. Chang-Hasnain, J. P. Harbison, G. Hasnain, A. C. Von Lehmen, L. T. Florez, and N. G. Stoffel, "Dynamic, polarization, and transverse mode characteristics of vertical cavity surface emitting lasers," *IEEE J. Quantum Electron.* **27**, 1402–1409 (1991).
33. J. W. Scott, R. S. Geels, S. W. Corzine, and L. A. Coldren, "Modeling temperature effects and spatial hole burning to optimize vertical-cavity surface-emitting laser performance," *IEEE J. Quantum Electron.* **29**, 1295–1308 (1993).
34. K. Tai, K. F. Huang, C. C. Wu, and J. D. Wynn, "Continuous wave visible InGaP/InGaAlP quantum well surface emitting laser diodes," *Electron. Lett.* **29**, 1314–1316 (1993).
35. I. Melngailis, "Longitudinal injection-plasma laser of InSb," *Appl. Phys. Lett.* **6**, 59–60 (1965).
36. R. D. Burnham, D. R. Scifres, and W. Streifer, "Single-heterostructure distributed-feedback GaAs-diode lasers," *IEEE J. Quantum Electron.* **QE-11**, 439–449 (1975).
37. Zh. I. Alferov, V. M. Andreyev, S. A. Gurevich, R. F. Kazarinov, V. R. Larionov, M. N. Mizerov, and E. L. Portnoy, "Semiconductor lasers with the light output

through the diffraction grating on the surface of the waveguide layer," *IEEE J. Quantum Electron.* **QE-11**, 449–451 (1975).

38. A. J. Springthorpe, "A novel double-heterostructure p-n junction laser," *Appl. Phys. Lett.* **31**, 524–525 (1977).

39. K. Iga and F. Koyama, "Vertical-cavity surface emitting lasers and arrays," in *Surface Emitting Semiconductor Lasers and Arrays*, G. A. Evans and J. M. Hammer, Eds., Chapter 3, Academic Press, San Diego, 1993.

40. R. D. Burnham, D. R. Scifres, and W. Streifer, "Transverse light emitting electroluminescent devices," United State Patent No. 4309670, Jan. 1982 (Filed Sept. 13, 1979).

41. F. Koyama, S. Kinoshita, and K. Iga, "Room temperature continuous wave lasing characteristics of GaAs vertical cavity surface-emitting laser," *Appl. Phys. Lett.* **55**, 221–222 (1989).

42. J. L. Jewell, A. Scherer, S. L. McCall, Y. H. Lee, S. Walker, J. P. Harbison, and L. T. Florez, "Low-threshold electrically pumped vertical-cavity surface-emitting microlasers," *Electron. Lett.* **25**, 1123–1124 (1989).

43. J. M. Dallesasse, N. Holonyak, Jr., A. R. Sugg, T. A. Richard, and N. El-Zein, "Hydrolyzation oxidation of $Al_xGa_{1-x}As$-AlAs-GaAs quantum well heterostructures and superlattices," *Appl. Phys. Lett.* **57**, 2844–2846 (1990).

44. J. M. Dallesasse and N. Holonyak, Jr., "Native-oxide stripe-geometry $Al_xGa_{1-x}As$-GaAs quantum well heterostructure lasers," *Appl. Phys. Lett.* **58**, 394–396 (1991).

45. D. L. Huffaker, D. G. Deppe, K. Kumar, and T. J. Rogers, "Native-oxide defined ring contact for low threshold vertical-cavity lasers," *Appl. Phys. Lett.* **65**, 97–99 (1994).

46. K. D. Choquette, R. P. Schneider, Jr., K. L. Lear, and K. M. Geib, "Low threshold voltage vertical-cavity lasers fabricated by selective oxidation," *Electron. Lett.* **30**, 2043–2044 (1994).

47. M. H. MacDougal, P. D. Dapkus, V. Pudikov, H. Zhao, ad G. M. Yang, "Ultralow threshold current vertical-cavity surface emitting lasers with AlAs-oxide-GaAs distributed Bragg reflectors," *IEEE Photon. Technol. Lett.* **7**, 229–231 (1995).

48. Y. Hayashi, T. Mukaihara, N. Hatori, N. Ohnoki, A. Matsutani, F. Koyama, and K. Iga, "Record low threshold index-guided InGaAs/AlGaAs vertical-cavity surface emitting laser with a native oxide confinement structure," *Electron. Lett.* **31**, 560–561 (1995).

49. K. L. Lear, K. D. Choquette, R. P. Schneider, Jr., S. P. Kilcoyne, and K. M. Geib, "Selectively oxidized vertical-cavity surface emitting lasers with 50% power conversion efficiency," *Electron. Lett.* **31**, 208–209 (1995).

50. K. D. Choquette, K. L. Lear, R. P. Schneider, Jr., K. M. Geib, J. J. Figiel, and R. Hull, "Fabrication and performance of selectively oxidized vertical-cavity lasers," *IEEE Photon. Technol. Lett.* **7**, 1237–1239 (1995).

51. N. K. Dutta, "Analysis of current spreading, carrier diffusion, and transverse mode guiding in surface emitting lasers," *J. Appl. Phys.* **68**, 1961–1963 (1990).

52. J. F. Seurin and S. L. Chuang, "Discrete Bessel transform and beam propagation method for modeling of vertical-cavity surface-emitting lasers," *J. Appl. Phys.* **82**, 2007–2016 (1997).

53. B. Tell, K. F. Brown-Goebeler, R. E. Leibenguth, F. M. Baez, and Y. H. Lee, "Temperature dependence of GaAs-AlGaAs vertical cavity surface emitting lasers," *Appl. Phys. Lett.* **60**, 683–685 (1992).

54. P. Zhou, B. Lu, J. Cheng, K. J. Malloy, S. Z. Sun, S. D. Hersee, and J. C. Zolper, "Vertical-cavity surface-emitting lasers with thermally stable electrical characteristics," *J. Appl. Phys.* **77**, 2264–2267 (1995).
55. G. Hasnain, K. Tai, L. Yang, Y. H. Wang, R. J. Fischer, J. D. Wynn, B. Weir, N. K. Dutta, and A. Y. Cho, "Performance of gain-guided surface emitting lasers with semiconductor distributed Bragg reflectors," *IEEE J. Quantum Electron.* **27**, 1377–1385 (1991).
56. R. Michalzik and K. J. Ebeling, "Modeling and design of proton-implanted ultralow-threshold vertical-cavity laser diodes," *IEEE J. Quantum Electron.* **29**, 1963–1974 (1993).
57. W. Nakwaski and M. Osiński, "Thermal analysis of GaAs-AlGaAs etched-well surface-emitting double-heterostructure lasers with dielectric mirrors," *IEEE J. Quantum Electron.* **29**, 1981–1995 (1993).
58. J. W. Scott, D. B. Young, B. J. Thibeault, M. G. Peters, and L. A. Coldren, "Design of index-guided vertical-cavity lasers for low temperature-sensitivity, sub-milliamp thresholds, and single-mode operation," *IEEE J. Sel Top. Quantum Electron.* **1**, 638–647 (1995).
59. N. N. Ledentsov, F. Hopfer, and D. Bimberg, "High-speed quantum-dot vertical-cavity surface-emitting lasers," *Proc. IEEE* **95**, 1741–1756 (2007).
60. F. Koyama, "Recent advances in surface-emitting lasers," Chapter 4 in *Optical Fiber Telecommunications* V, a:Components, I. P. Kaminow, Kaminow, T. Li, and A. E. Willner, eds., Elsevier Science & Technology, 2008.
61. E. M. Purcell, "Spontaneous emission probabilities at radio frequencies," *Phys. Rev.* **69**, 681 (1946).
62. K. H. Drexhage, "Interaction of light with monomolecular dye layers," in *Progress in Optics*, E. Wolf, Ed., vol. 12, 165–232, North-Holland, New York, 1974.
63. P. Goy, J. M. Raimond, M. Gross, and S. Haroche, "Observation of cavity-enhanced single-atom spontaneous emission," *Phys. Rev. Lett.* **50**, 1903–1909 (1983).
64. R. G. Hulet, E. S. Hilfer, and D. Kleppner, "Inhibited spontaneous emission by Rydberg atom," *Phys. Rev. Lett.* **55**, 2137–2140 (1985).
65. E. Yablonovitch, "Inhibited spontaneous emission in solid-state physics and electronics," *Phys. Rev. Lett.* **58**, 2059–2062 (1987).
66. S. John, "Strong localization of photons in certain disordered dielectric superlattices," *Phys. Rev. Lett.* **58**, 2486–2489 (1987).
67. E. Yablonovitch, T. J. Gmitter, R. D. Meade, A. M. Rappe, K. D. Brommer, and J. D. Joannopoulos, "Donor and acceptor modes in photonic band structure," *Phys. Rev. Lett.* **67**, 3380–3383 (1991).
68. F. De Martini, G. Innocenti, G. R. Jacobovitz, and P. Mataloni, "Anomalous spontaneous emission time in a microscopic optical cavity," *Phys. Rev. Lett.* **59**, 2955–2958 (1987).
69. F. De Martini and G. R. Jacobovitz, "Anomalous spontaneous-stimulated-decay phase transition and zero-threshold laser action in a microscopic cavity," *Phys. Rev. Lett.* **60**, 1711–1714 (1988).
70. K. Iga, F. Koyama, and S. Kinoshita, "Surface emitting semiconductor lasers," *IEEE J. Quantum Electron.* **24**, 1845–1855 (1988).
71. Y. Yamamoto, S. Machida, K. Igeta, and Y. Horikoshi, "Enhanced and inhibited spontaneous emission of free excitons in GaAs quantum wells in a micro-cavity," *Proc. 6th Rochester Conf. Quantum Optics.*, eds. J. H. Eberly, L. Mandel, and E. Wolf, Plenum Press, 1989.

72. Y. Yamamoto, S. Machida, K. Igeta, and G. Björk, "Controlled spontaneous emission in microcavity semiconductor lasers," in *Coherence, Amplification and Quantum Effects in Semiconductor Lasers*, Y. Yamamoto, Ed. Wiley, New York, 1991.

73. A. F. J. Levi, R. E. Slusher, S. L. McCall, T. Tanbun-Ek, D. L. Coblentz, and S. J. Pearton, "Room temperature operation of microdisk lasers with submilliamp threshold current," *Electron. Lett.* **28**, 1010–1012 (1992).

74. S. L. McCall, A. F. J. Levi, R. E. Slusher, S. J. Pearton, and R. A. Logan, "Whispering gallery mode microdisk lasers," *Appl. Phys. Lett.* **60**, 289–291 (1992).

75. A. F. J. Levi, R. E. Slusher, S. L. McCall, S. J. Pearton, and W. S. Hobson, "Room temperature lasing in $In_{0.51}Ga_{0.49}P/In_{0.2}Ga_{0.8}As$ microcylinder laser diodes," *Appl. Phys. Lett.* **62**, 2021–2023 (1993).

76. A. F. J. Levi, S. L. McCall, S. J. Pearton, and R. A. Logan, "Room temperature operation of submicrometer radius disk laser," *Electron. Lett.* **29**, 1666–1667 (1993).

77. J. D. Joannopoulos, R. D. Meade, and J. N. Winn, *Photonic Crystals*. Princeton Universiy Press, Princeton, NJ, 1995.

78. O. Painter, R. K. Lee, A. Scherer, A. Yariv, J. D. O'Breien, P. D. Dapkus, and I. Kim, "Two-dimensional photonic bandgap defect laser," *Science* **284**, 1819–1821 (1999).

79. O. Painter, J. Vuckovic, and A. Scherer, "Defect modes of a two-dimensional photonic crystal in an optically thin dielectric slab," *J. Opt. Soc. Am. B* **16**, 275–285 (1999).

80. O. Painter, A. Husain, A. Scherer, J. D. O'Brien, I. Kim, and P. D. Dapkus, "Room temperature photonic crystal defect lasers at near-infrared wavelengths in InGaAsP," *J. Lightwave Technol.* **17**, 2082–2088 (1999).

81. J. K. Hwang, H. Y. Ryu, D. S. Song, I. Y. Han, H. K. Park, Y. H. Lee, and D. H. "Room temperature triangular-lattice two-dimensional photonic band gap lasers operating at 1.54 μm," *Appl. Phys. Lett.* **76**, 2982–2984 (2000).

82. P. T. Lee, J. R. Cao, S. J. Choi, Z. J. Wei, J. D. O'Brein, and P. D. Dapkus, "Operation of photonic crystal membrane lasers above room temperature," *Appl. Phys. Lett.* **81**, 3311–3313 (2002).

83. H. G. Park, S. H. Kim, S. H. Kwon, Y. G. Ju, J. K. Yang, J. H. Baek, S. B. Kim, and Y. H. Lee, "Electrically driven single-cell photonic crystal laser," *Science* **305**, 1444–1447 (2004).

84. H. G. Park, S. H. Kim, M. K. Seo, Y. G. Ju, S. B. Kim, and Y. H. Lee, "Characteristics of electrically driven two-dimensional photonic crystal lasers," *IEEE J. Quantum Electron.* **41**, 1131–1141 (2005).

85. C. Weisbuch, M. Nishioka, A. Ishikawa, and Y. Arakawa, "Observation of the coupled exciton-photon mode splitting in a semiconductor quantum microcavity," *Phys. Rev. Lett.* **69**, 3314–3317 (1992).

86. J.-M. Gerard and B. Gayral, "Strong Purcell effect for InAs quantum boxes in three-dimensional solid-state microcavities," *J. Lightwave Technol.* **17**, 2089–2095 (1999).

87. J. Vuckovic, C. Santori, D. Fattal, M. Pelton, G. S. Solomon, and Y. Yammamoto, "Cavity-enhanced single photons from a quantum dot," in *Optical Microcavities*, K. Vahala, Ed., Chapter 4, World Scientific, Singapore, 2004.

88. J. M. Gerard, B. Sermage, B. Gayral, B. Legrand, E. Costard, and V. Thierry-Mieg, "Enhanced spontaneous emission by quantum boxes in a monolithic optical microcavity," *Phys. Rev. Lett.* **81**, 1110–1113 (1998).

89. G. S. Solomon, M. Pelton, and Y. Yamamoto, "Single-mode spontaneous emission from a single quantum dot in a three-dimensional microcavity," *Phys. Rev. Lett.* **86**, 3903 (2001).
90. M. Pelton, J. Vuckovic, G. S. Solomon, A. Scherer, and Y. Yamamoto, "Three-dimensionally confined modes in micropost microcavities: quality factors and Purcell factors," *IEEE J. Quantum Electron.* **38**, 170–177 (2002).
91. T. Baba, "Photonic crystals and microdisk cavities based on GaInAsP-InP system," *IEEE J. Sel. Top. Quantum Electron.* **3**, 808–830 (1997).
92. T. Baba and D. Sano, "Low-threshold lasing and Purcell effects in microdisk lasers at room temperature," *IEEE J. Sel. Top. Quantum Electron.* **9**, 1340–1346 (2003).
93. H. Y. Ryu, S. H. Kim, H. G. Park, and Y. H. Lee, "Characteristics of single defect laser modes in a two-dimensional square lattice photonic crystal slab," *J. Appl. Phys.* **93**, 831–837 (2003).
94. K. Nozaki and T. Baba, "Carrier and photon analyses of photonic microlasers by two-dimensional rate equations," *IEEE J. Sel. Areas Commun.* **23**, 1413–1417, (2005).
95. J. O'Brien, W. Kuang, J. R. Cao, M. H. Shih, P. T. Lee, M. Bagheri, A. Mock, and W. K. Marshall, "Photonic crystal micropcavity lasers," *J. Phys. D Appl. Phys.* **40**, 2671–2682 (2007).
96. A. Scherer, J. L. Jewell, and J. P. Harbison, "Lasing in submicron wide vertical cavity microlasers," *Opt. Photon. News* **2**, 9 (1991).
97. M. T. Hill, Y. S. Oei, B. Smalbrugge, Y. Zhu, T. de Vries, P. J. Van Veldhoven, F. W. M. Van Otten, T. J. Eijkemans, J. P. Turkiewicz, H. De Waardt, E. J. Geluk, S. H. Kwon, Y. H. Lee, R. Notzel, and M. K. Smit, "Lasing in metallic-coated nanocavities," *Nature Photon.* **1**, 589–594 (2007).
98. J. M. Gerard and B. Gayral, "InAs quantum dots: artificial atoms for solid-state cavity-quantum electrodynamics," *Physica* **E 9**, 131–139 (2001).
99. H. Yokoyama and S. D. Brorson, "Rate equation analysis of microcavity lasers," *J. Appl. Phys.* **66**, 4801–4805 (1989).
100. G. Björk and Y. Yamamoto, "Analysis of semiconductor microcavity lasers using rate equations," *IEEE J. Quantum Electron.* **27**, 2386–2396 (1991).
101. J. Faist, F. Capasso, D. L. Sivco, C. Sirtori, A. Hutchinson, and A. Y. Cho, "Quantum cascade laser," *Science*, **264**, 553–556, (1994).
102. R. Kazarinov and R. A. Suris, "Amplification of electromagnetic waves in a semiconductor superlattice," *Sov. Phys. Semicond.* **5**, 707–709 (1971).
103. F. Capasso, R. Paiella, R. Martini, R. Colombelli, C. Gmachl, T. L. Myers, M. S. Taubman, R. M. Williams, C. G. Bethea, K. Unterrainer, H. Y. Hwang, D. L. Sivco, A. Y. Cho, A. M. Sergent, H. C. Liu, and E. A. Whittaker, "Quantum cascade lasers: Ultrahigh-speed operation, optical wireless communication, narrow linewidth, and far-infrared emission," *IEEE J. Quantum Electron.* **38**, 511–532 (2002).
104. F. Capasso, C. Gmachl, D. L. Sivco, and A. Y. Cho, "Quantum cascade lasers," *Phys. Today*, **55**, 34–40 (2002).
105. F. Capasso, C. Gmachl, R. Paiella, A. Tredicucci, A. L. Hutchinson, D. L. Sivco, J. N. Baillargeon, A. Y. Cho, and H. C. Liu, "New frontiers in quantum cascade lasers and applications," *IEEE J. Select. Topics Quantum Electron.* **6**, 931–947 (2000).

106. S. Silvken, A. Evans, W. Zhang, and M. Razeghi, "High-power, continuous-operation intersubband laser for wavelengths greater than 10 μm," *Appl. Phys. Lett.* **90**, 151115 (2007).

107. M. Beck, D. Hofstetter, T. Aellen, J. Faist, U. Oesterle, M. Ilegems, E. Gini, and H. Melchior, "Continuous wave operation of a mid-infrared semiconductor laser at room temperature," *Science.* **295**, 301–305 (2002).

108. J. Faist, D. Hofstetter, M. Beck, T. Aellen, M. Rochat, and S. Blaser, "Bound-to-continuum and two-phonon resonance quantum-cascade lasers for high duty cycle, high-temperature operation," *IEEE J. Quantum Electron.* **38**, 533–546 (2002).

109. C. Sirtori, H. Page, C. Becker, and V. Ortiz, "GaAs-AlGaAs quantum cascade lasers: Physics, technology, and prospects," *IEEE J. Quantum Electron.* **38**, 547–558 (2002).

110. K. Ohtani and H. Ohno, "InAs/AlSb quantum cascade lasers operating at 10 μm," *Appl. Phys. Lett.* **82**, 1003–1005 (2003).

111. J. Devenson, R. Teissier, O. Cathabard, and A. N. Baranov, "InAs/AlSb quantum cascade lasers emitting below 3 μm," *Appl. Phys. Lett.* **90**, 111118 (2007).

112. Q. Yang, C. Manz, W. Bronner, K. Schäuble, Ch. Mann, K. Schwarz, K. Köhler, and J. Wagner, "Bound-to-continuum GaInAs-AlAsSb quantum cascade lasers with reduced electric injection power density," *IEEE Photon. Technol. Lett.* **18**, 1356–1358 (2006).

113. Q. Yang, C. Manz, W. Bronner, K. Köhler, and J. Wagner, "Room-temperature short-wavelength ($\lambda \sim 3.7$–3.9 μm) GaInAs/AlAsSb quantum-cascade lasers," *Appl. Phys. Lett.* **88**, 121127 (2006).

114. C. Gmachl, A. Straub, R. Colombelli, F. Capasso, D. L. Sivco, A. M. Sergent, and A. Y. Cho, "Single-mode, tunable distributed-feedback and multiple-wavelength quantum cascade lasers," *IEEE J. Quantum Electron.* **38**, 569–581 (2002).

115. L. Diehl, B. G. Lee, P. Behroozi, M. Lončar, M. A. Belkin, F. Capasso, T. Aellen, D. Hofstetter, M. Beck, and J. Faist, "Microfluidic tuning of distributed feedback quantum cascade lasers," *Opt. Express*, **14**, 11660–11667 (2006).

116. C. Gmachl, J. Faist, F. Capasso, C. Sirtori, D. L. Sivco, and A. Y. Cho, "Long-wavelength (9.5–11.5 μm) microdisk quantum-cascade lasers," *IEEE J. Quantum Electron.* **33**, 1567–1573 (1997).

117. M. Lončar, B. G. Lee, L. Diehl, M. Belkin, F. Capasso, M. Giovannini, J. Faist, and E. Gini, "Design and fabrication of photonic crystal quantum cascade lasers for optofluidics," *Opt. Express*, **15**, 4499–4514 (2007).

118. L. Diehl, D. Bour, S. Corzine, J. Zhu, G. Höfler, M. Lončar, M. Troccoli, and F. Capasso, "High-temperature continuous wave operation of strain-balanced quantum cascade lasers grown by metal organic vapor-phase epitaxy," *Appl. Phys. Lett.* **89**, 081101-1–081101-3 (2006).

119. Z. Liu, D. Wasserman, S. S. Howard, A. J. Hoffman, C. F. Gmachl, X. Wang, T. Tanbun-Ek, L. Cheng, and F.-S. Choa, "Room-temperature continuous-wave quantum cascade lasers grown by MOCVD without lateral regrowth," *IEEE Photon. Technol. Lett.* **18**, 1347–1349 (2006).

120. J. S. Roberts, R. P. Green, L. R. Wilson, E. A. Zibik, D. G. Revin, J. W. Cockburn, and R. J. Airey, "Quantum cascade lasers grown by metalorganic vapor phase epitaxy," *Appl. Phys. Lett.* **82**, 4221–4223 (2003).

121. R. Q. Yang, J. L. Bradshaw, J. D. Bruno, J. T. Pham, and D. E. Wortman, "Mid-infrared type-II interband cascade lasers," *IEEE J. Quantum Electron.* **38**, 559–568 (2002).

122. J. Faist, F. Capasso, C. Sirtori, D. L. Sivco, A. L. Hutchinson, and A. Y. Cho, "Continuous wave operation of a vertical transition quantum cascade laser about T = 80K," *Appl. Phys. Lett.* **67**, 3057–3059 (1995).

123. C. Sirtori, J. Faist, F. Capasso, D. L. Sivco, A. L. Hutchinson, S. N. G. Chu, and A. Y. Cho, "Continuous wave operation of midinfrared (7.4–8.6 μm) quantum cascade lasers up to 110K temperature," *Appl. Phys. Lett.* **68**, 1745–1747 (1996).

124. J. Faist, F. Capasso, D. L. Sivco, A. L. Hutchinson, S.-N. G. Chu, and A. Y. Cho, "Short wavelength ($\lambda \sim 3.4$ μm) quantum cascade laser based on strained compensated InGaAs/AlInAs," *Appl. Phys. Lett.* **72**, 680–682 (1998).

125. C. Gmachl, A. Tredicucci, F. Capasso, A. L. Hutchinson, D. L. Sivco, J. N. Baillargeon, and A. Y. Cho, "High-power $\lambda \approx 8$ μm quantum cascade lasers with near optimum performance," *Appl. Phys. Lett.* **72**, 3130–3132 (1998).

126. T. Aellen, M. Beck, N. Hoyler, M. Giovannini, J. Faist, and E. Gini, "Doping in quantum cascade lasers. I. InAlAs-InGaAs/InP midinfrared devices," *J. Appl. Phys.* **100**, 043101 (2006).

127. J. Faist, M. Beck, T. Aellen, and E. Gini, "Quantum-cascade lasers based on a bound-to-continuum transition," *Appl. Phys. Lett.* **78**, 147–149 (2001).

128. R. Maulini, A. Mohan, M. Giovannini, J. Faist, and E. Gini, "External cavity quantum-cascade laser tunable from 8.2 to 10.4 μm using a gain element with heterogeneous cascade," *Appl. Phys. Lett.* **88**, 201113 (2006).

129. A. Tredicucci, F. Capasso, C. Gmachl, D. L. Sivco, A. L. Hutchinson, and A. Y. Cho, "High performance interminiband quantum cascade lasers with graded superlattices," *Appl. Phys. Lett.* **73**, 2101–2103 (1998).

130. A. Tredicucci, C. Gmachl, F. Capasso, D. L. Sivco, A. L. Hutchinson, and A. Y. Cho, "Long wavelength superlattice quantum cascade lasers at $\lambda \cong 17$ μm," *Appl. Phys. Lett.* **74**, 638–640 (1999).

131. A. Tredicucci, F. Capasso, C. Gmachl, D. L. Sivco, A. L. Hutchinson, A. Y. Cho, J. Faist, and G. Scamarcio, "High-power inter-miniband lasing in intrinsic superlattices," *Appl. Phys. Lett.* **72**, 2388–2390 (1998).

132. M. Lerttamrab, S. L. Chuang, C. Gmachl, D. L. Sivco, F. Capasso, and A. Y. Cho, "Linewidth enhancement factor of a type-I quantum-cascade laser," *J. Appl. Phys.* **94**, 5426–5428 (2003).

133. J. Kim, M. Lerttamrab, S. L. Chuang, C. Gmachl, D. L. Sivco, F. Capasso, and A. Y. Cho, "Theoretical and experimental study of optical gain and linewidth enhancement factor of type-I quantum-cascade lasers," *IEEE J. Quantum Electron.* **40**, 1663–1674 (2004).

134. M. S. Vitiello, G. Scamarcio, V. Spagnolo, T. Losco, R. P. Green, A. Tredicucci, H. E. Beere, and D. A. Ritchie, "Electron-lattice coupling in bound-to-continuum THz quantum-cascade lasers," *Appl. Phys. Lett.* **88**, 241109 (2006).

135. V. B. Gorfinkel, S. Luryi, and B. Gelmont, "Theory of gain spectra for quantum cascade lasers and temperature dependence of their characteristics at low and moderate carrier concentrations," *IEEE J. Quantum Electron.* **32**, 1995–2003 (1996).

136. M. Yamanishi, T. Edamura, K. Fujita, N. Akikusa, and H. Kan, "Theory of the intrinsic linewidth of quantum cascade lasers: hidden reason for the narrow linewidth and line-broadening by thermal photons," *IEEE J. Quantum Electron.* **44**, 12–29 (2008).

137. M. Rochat, M. Beck, J. Faist, and U. Oesterle, "Measurement of far-infrared waveguide loss using a multisection single-pass technique," *Appl. Phys. Lett.* **78**, 1967–1969 (2001).

138. N. Yu, L. Diehl, E. Cubukcu, C. Pflügl, D. Bour, S. Corzine, J. Zhu, G. Höfler, K. B. Crozier, and F. Capasso, "Near-field imaging of quantum cascade laser transverse modes," *Opt. Express*, **15**, 13227–13235 (2007).

139. B. S. Williams, S. Kumar, and Q. Hu, "Operation of terahertz quantum-cascade lasers at 164K in pulsed mode and at 117K in continuous-wave mode," *Opt. Express*, **13**, 3331–3339 (2005).

140. R. Köhler, A. Tredicucci, F. Beltram, H. E. Beere, E. H. Linfield, A. G. Davies, D. A. Ritchie, R. C. Iotti, and F. Rossi, "Terahertz semiconductor-heterostructure laser," *Nature*, **417**, 156–159 (2002).

141. M. Rochat, L. Ajili, H. Willenberg, J. Faist, H. Beere, G. Davies, E. Linfield, and D. Ritchie, "Low-threshold terahertz quantum-cascade lasers," *Appl. Phys. Lett.* **81**, 1381–1383 (2002).

142. B. S. Williams, H. Callebaut, S. Kumar, Q. Hu, and J. L. Reno, "3.4-THz quantum cascade laser based on longitudinal-optical-phonon scattering for depopulation," *Appl. Phys. Lett.* **82**, 1015–1017 (2003).

143. R. Köhler, A. Tredicucci, F. Beltram, H. E. Beere, E. H. Linfield, A. G. Davies, D. A. Ritchie, S. S. Dhillon, and C. Sirtori, "High-performance continuous-wave operation of superlattice terahertz quantum-cascade lasers," *Appl. Phys. Lett.* **82**, 1518–1520 (2003).

144. S. Barbieri, J. Alton, S. S. Dhillon, H. E. Beere, M. Evans, E. H. Linfield, A. G. Davies, D. A. Ritchie, R. Köhler, A. Tredicucci, and F. Beltram, "Continuous-wave operation of tehahertz quantum-cascade lasers," *IEEE J. Quantum Electron.* **39**, 586–591 (2003).

145. G. Scalari, L. Ajili, J. Faist, H. Beere, E. Linfield, D. Ritchie, and G. Davies, "Far-infrared ($\lambda \sim 87$ μm) bound-to-continuum quantum-cascade lasers operating up to 90K," *Appl. Phys. Lett.* **82**, 3165–3167 (2003).

146. H. Callebaut, S. Kumar, B. S. Williams, Q. Hu, and J. L. Reno, "Analysis of transport properties of tetrahertz quantum cascade lasers," *Appl. Phys. Lett.* **83**, 207–209 (2003).

147. B. S. Williams, S. Kumar, H. Callebaut, Q. Hu, and J. L. Reno, "Terahertz quantum-cascade laser at $\lambda \approx 100$ μm using metal waveguide for mode confinement," *Appl. Phys. Lett.* **83**, 2124–2126 (2003).

148. B. S. Williams, S. Kumar, H. Callebaut, Q. Hu, and J. L. Reno, "Terahertz quantum-cascade laser operating up to 137K," *Appl. Phys. Lett.* **83**, 5142–5144 (2003).

149. G. Scalari, N. Hoyler, M. Giovannini, and J. Faist, "Terahertz bound-to-continuum quantum-cascade lasers based on optical-phonon scattering extraction," *Appl. Phys. Lett.* **86**, 181101 (2005).

150. C. Worrall, J. Alton, M. Houghton, S. Barbieri, H. E. Beere, D. Ritchie, and C. Sirtori, "Continuous wave operation of a superlattice quantum cascade laser emitting at 2 THz," *Opt. Express*, **14**, 171–181 (2006).

151. B. S. Willimas, S. Kumar, Q. Hu, and J. L. Reno, "High-power terahertz quantum-cascade lasers," *Electron. Lett.* **42**, 89–91 (2006).

152. G. Scalari, C. Walther, J. Faist, H. Beere, and D. Ritchie, "Electrically switchable, two-color quantum cascade laser emitting at 1.39 and 2.3 THz," *Appl. Phys. Lett.* **88**, 141102 (2006).

153. S. Kumar, B. S. Williams, Q. Hu, and J. L. Reno, "1.9 THz quantum-cascade lasers with one-well injector," *Appl. Phys. Lett.* **88**, 121123 (2006).

154. C. Walther, G. Scalari, J. Faist, H. Beere, and D. Ritchie, "Low frequency terahertz quantum cascade laser operating from 1.6 to 1.8 THz," *Appl. Phys. Lett.* **89**, 231121 (2006).

155. Y. Chassagneux, J. Palomo, R. Colombelli, S. Barbieri, S. Dhillon, C. Sirtori, H. Beere, J. Alton, and D. Ritchie, "Low threshold THz QC lasers with thin core regions," *Electron. Lett.* **43**, 41–42 (2007).

156. A. W. M. Lee, Q. Qin, S. Kumar, B. S. Williams, Q. Hu, and J. L. Reno, "High-power and high-temperature THz quantum-cascade lasers based on lens-coupled metal-metal waveguides," *Opt. Lett.* **32**, 2840–2842 (2007).

157. R. Köhler, R. C. Iotti, A. Tredicucci, and F. Rossi, "Design and simulation of terahertz quantum cascade lasers," *Appl. Phys. Lett.* **79**, 3920–3922 (2001).

158. S.-C. Lee and A. Wacker, "Theoretical analysis of spectral gain in a terahertz quantum-cascade laser: Prospects for gain at 1 THz," *Appl. Phys. Lett.* **83**, 2506–2508 (2003).

159. C. Jirauschek, G. Scarpa, P. Lugli, M. S. Vitiello, and G. Scamarcio, "Comparative analysis of resonant phonon THz quantum cascade lasers," *J. Appl. Phys.* **101**, 086109 (2007).

160. S. Kumar, B. S. Williams, Q. Qin, A. W. M. Lee, Q. Hu, and J. L. Reno, "Surface-emitting distributed feedback terahertz quantum-cascade lasers in metal-metal waveguides," *Opt. Express*, **15**, 113–128 (2007).

161. O. Demichel, L. Mahler, T. Losco, C. Mauro, R. Green, J. Xu, A. Tredicucci, F. Beltram, H. E. Beere, D. A. Ritchie, and V. Tamošiūnas, "Surface plasmon photonic structures in terahertz quantum cascade lasers," *Opt. Express*, **14**, 5335–5345 (2006).

162. A. W. M. Lee, Q. Qin, S. Kumar, B. S. Williams, Q. Hu, and J. L. Reno, "Real-time terahertz imaging over a standoff distance (>25 meters)," *Appl. Phys. Lett.* **89**, 141125 (2006).

163. A. W. M. Lee, B. S. Williams, S. Kumar, Q. Hu, and J. L. Reno, "Real-time imaging using a 4.2-THz quantum cascade laser and a 320×240 microbolometer focal-plane array," *IEEE Photon. Technol. Lett.* **18**, 1415–1417 (2006).

164. R. Q. Yang, "Infrared laser based on intersubband transitions in quantum wells," *Superlattices Microstruc.* **77**, 77–83 (1995).

165. C.-H. Lin, R. Q. Yang, D. Zhang, S. J. Murry, S. S. Pei, A. A. Allerma, and S. R. Kurtz, "Type-II interband quantum cascade laser at 3.8 μm," *Electron. Lett.* **33**, 598–599 (1997).

166. J. R. Meyer, I. Vurgaftman, R. Q. Yang, and L. R. Ram-Mohan, "Type-II and type-I interband cascade lasers," *Electron. Lett.* **32**, 45–46 (1996).

167. I. Vurgaftman, J. R. Meyer, and L. R. Ram-Mohan, "High-power/low-threshold type-II interband cascade mid-IR laser-design and modeling," *IEEE Photon. Technol. Lett.* **9**, 170–172 (1997).

168. R. Q. Yang, C. J. Hill, K. Mansour, Y. Qiu, A. Soibel, R. E. Muller, and P. M. Echternach, "Distributed feedback mid-IR interband cascade lasers at thermoelectric cooler temperatures," *IEEE J. Select. Topics Quantum Electron.* **13**, 1074–1078 (2007).

169. G. Liu and S. L. Chuang, "Modeling of Sb-based type-II quantum-cascade lasers," *Phys. Rev. B*, **65**, 165220 (2002).

170. C. L. Canedy, W. W. Bewley, M. Kim, C. S. Kim, J. A. Nolde, D. C. Larrabee, J. R. Lindle, I. Vurgaftman, and J. R. Meyer, "High-temperature interband cascade lasers emitting at $\lambda = 3.6-4.3$ μm," *Appl. Phys. Lett.* **90**, 181120 (2007).

171. W. W. Bewley, C. L. Canedy, M. Kim, C. S. Kim, J. A. Nolde, J. R. Lindle, I. Vurgaftman, and J. R. Meyer, "Interband cascade laser operating to 269K at $\lambda = 4.05\ \mu m$," *Electron. Lett.* **43**, 39–40 (2007).

172. N. Holonyak Jr., "Is the light emitting diode (LED) an ultimate lamp?" *Am. J. Phys.* **68**, 864–866 (2000).

173. N. Holonyak Jr., "From transistors to lasers and light-emitting diodes," *MRS Bull.* **30**, 509–517 (2005).

174. E. F. Schubert, *Light-Emitting Diodes*, Cambridge University Press, Cambridge, UK, 2003.

175. I. Akasaki and H. Amano, "Widegap column-III nitride semiconductors for UV/blue light emitting devices," *J. Electrochem. Soc.* **141**, 2266–2271 (1994).

176. S. Nakamura, M. Senoh, S. Nagahama, N. Iwasa, T. Yamada, T. Mukai, Y. Sugimoto, and H. Kiyoku, "Room-temperature continuous-wave operation of InGaN multi-quantum-well structure laser diodes," *Appl. Phys. Lett.* **69**, 4056–4058 (1996).

177. S. Nakamura, "GaN-based blue/green semiconductor laser," *IEEE J. Select. Topics Quantum Electron.* **3**, 435–442 (1997).

178. S. Nakamura, "RT-CW operation of InGaN multi-quantum-well structure laser diodes," *Mater. Sci. Eng. B.* **50**, 277–284 (1997).

179. S. Nakamura, "Materials issues for InGaN-based lasers," *J. Elect. Mater.* **27**, 160–165 (1998).

180. S. Nakamura and G. Fasol, *The Blue Laser Diode*, Springer, Berlin, 1997.

181. M. Hasse, J. Qiu, J. DePuydt, and H. Cheng, "Blue-green laser diodes," *Appl. Phys. Lett.* **59**, 1272–1274, (1991).

182. H. Jeon, J. Ding, W. Patterson, A. V. Nurmikko, W. Xie, D. C. Grillo, M. Kobayashi, and R. L. Gunshor, "Blue-green injection laser diodes in (Zn,Cd)Se/ZnSe quantum wells," *Appl. Phys. Lett.* **59**, 3619–3621 (1991).

183. N. Nakamura, S. Itoh, K. Nakano, H. Okuyama, M. Ozawa, A. Ishibashi, M. Ikeda, and Y. Mori, "Room temperature continuous operation of blue-green laser diodes," *Electron. Lett.* **29**, 1488–1489 (1993).

184. S. Taniguchi, T. Hino, S. Itoh, K. Nakano, N. Nakayama, A. Ishibashi, and M. Ikeda, "100 h II–VI blue-green laser diode," *Electron. Lett.* **32**, 552 (1996).

185. S. L. Chuang, M. Ukita, S. Kijima, S. Taniguchi, and A. Ishibashi, "Universal curves for optical power degradation of II–VI light-emitting diodes," *Appl. Phys. Lett.* **69**, 1588–1590 (1996).

186. T. Matsuoka, "InGaAlN and II–VI systems for blue-green light emitting devices," *Adv. Mater.* **8**, 469–479 (1996).

187. A. Nurmikko and R. L. Gunshor, "Blue and green semiconductor lasers: a status report," *Semicond. Sci. Technol.* **12**, 1337–1347 (1997).

188. S. L. Chuang, and C. S. Chang, "k.p Method for strained wurtzite semiconductors," *Phys. Rev B* **54**, 2491–2504 (1996).

189. S. L. Chuang, "Optical gain of strained wurtzite GaN quantum-well lasers," *IEEE J. Quantum Electron.* **32**, 1791–1800 (1996).

190. S. H. Park and S. L. Chuang, "Comparison of zinc-blende and wurtzite GaN semiconductors with spontaneous polarization and piezoelectric field effects," *J. Appl. Phys.* **87**, 353–364 (2000).

191. I. Vurgaftman, J. R. Meyer, and L. R. Ram-Mohan, "Band parameters for III–V compound semiconductors and their alloys," *J. Appl. Phys.* **89**, 5815–5875 (2001).

192. I. Vurgaftman, and J. R. Meyer, "Band parameters for nitrogen-containing semiconductors," *J. Appl. Phys.* **94**, 3675–3696 (2003).

193. K. H. Hellwege, *Landolt-Börnstein Numerical Data and Functional Relationships in Science and Technology*, New Series, Group III, vol. 17a, Springer Verlag, Berlin, 1982; Physics of Group IV Elements and III–V Compounds, vol. 22a, Springer Verlag, Berlin, 1986.

194. J. H. Edgar, Ed., *Properties of Group III Nitrides*, INSPEC, IEE, London, 1994.

195. J. Wu, W. Walukiewicz, K. M. Yu, J. W. Ager III, E. E. Haller, H. Lu, W. J. Schaff, Y. Saito, Y. Nanishi, "Unusual properties of the fundamental band gap of InN," *Appl. Phys. Lett.* **80**, 3967–3969 (2002).

196. J. Wu, W. Walukiewicz, K. M. Yu, J. W. Ager III, E. E. Haller, H. Lu, and W. J. Schaff,, "Small bandgap bowing in $In_{1-x}Ga_xN$," *Appl. Phys. Lett.* **80**, 3967–3969 (2002).

197. J. Wu, W. Walukiewicz, W. Shan, K. M. Yu, J. W. Ager III, S. X. Li, E. E. Haller, H. Lu, and W. J. Schaff, "Temperature dependence of the fundamental band gap of InN," *J. Appl. Phys.* **94**, 4457–4460 (2003).

198. D. P. Bour, M. Kneissl, L. T. Romano, M. McCluskey, C. G. Van de Walle, B. S. Krusor, R. Donaldson, J. Walker, C. Dunnrowicz, and N. M. Johnson, "Characteristics of InGaN/AlGaN multiple quantum well laser diodes," *IEEE J. Select. Topics Quantum Electron.* **4**, 498–504 (1998).

199. S. F. Chichibu, A. C. Abare, M. P. Mack, M. S. Minsky, T. Deguchi, D. Cohen, P. Kozodoy, S. B. Fleischer, S. Keller, J. S. Speck, J. E. Bowers, E. Hu, U. K. Mishra, L. A. Coldren, S. P. DenBaars, K. Wada, T. Sota, and S. Nakamura, "Optical properties of InGaN quantum wells," *Mater. Sci. Eng. B* **59**, 298–306 (1999).

200. S. Nakamura, M. Senoh, S. Nagahama, N. Iwasa, T. Matsushita, and T. Mukai, "Blue InGaN-based laser diodes with an emission wavelength of 450 nm," *Appl. Phys. Lett.* **76**, 22–24 (2000).

201. N. M. Johnson, A. V. Nurmikko, and S. P. DenBaars, "Blue diode laser," *Physics Today.* **33**, 31 (2000).

202. T. Asano, T. Tojyo, T. Mizuno, M. Takeya, S. Ikeda, K. Shibuya, T. Hino, S. Uchida, and M. Ikeda, "100-mW kink-free blue-violet laser diodes with low aspect ratio," *IEEE J. Quantum Electron.* **39**, 135–140 (2003).

203. S. Uchida, M. Takeya, S. Ikeda, T. Mizuno, T. Fujimoto, O. Matsumoto, S. Goto, T. Tojyo, and M. Ikeda, "Recent progress in high-power blue-violet lasers," *IEEE J. Sel. Topics Quantum Electron.* **9**, 1252–1259 (2003).

204. S. Nagahama, Y. Sugimoto, T. Kozaki, and T. Mukai, "Recent progress of AlInGaN laser diodes," *Proc. SPIE*, **5738**, 57–62 (2005).

205. M. Asif Khan, M. Shatalov, H. P. Maruska, H. M. Wang, and. E. Kuokstis, "III-nitride UV devices," *Jpn. J. Appl. Phys.* **44**, 7191–7206 (2005).

206. T. Kozaki, H. Matsumura, Y. Sugimoto, S. Nagahama, and T. Mukai, "High-power and wide wavelength range GaN-baser laser diodes," *Proc. SPIE.* **6133**, 613306-1 (2006).

207. H. Y. Ryu, K. H. Ha, S. N. Lee, T. Jang, H. K. Kim, J. H. Chae, K. S. Kim, K. K. Choi, J. K. Son, H. S. Paek, Y. J. Sung, T. Sakong, O. H. Nam, and Y. J. Park, "Highly stable temperature characteristics of InGaN blue laser diodes," *Appl. Phys. Lett.* **89**, 031122 (2006).

208. S. N. Lee, S. Y. Cho, H. Y. Ryu, J. K. Son, H. S. Paek, T. Sakong, T. Jang, K. K. Choi, K. H. Ha, M. H. Yang, O. H. Nam, Y. Park, and E. Yoon, "High-power GaN-based blue-violet laser diodes with AlGaN/GaN multiquantum barriers," *Appl. Phys. Lett.* **88**, 111101 (2006).

209. M. Ohta, Y. Ohizumi, Y. Hoshina, T. Tanaka, Y. Yabuki, K. Funato, S. Tomiya, S. Goto, and M. Ikeda, "High-power pure blue laser diodes," *Phys. Stat. Sol.* (a) **206**, 2068–2072 (2007).

210. H. Y. Ryu, K. H. Haleem, S. N. Lee, T. Jang, J. K. Son, H. S. Paek, Y. J. Sung, H. K. Kim, K. S. Kim, O. H. Nam, Y. J. Park, and J. I. Shim, "High-performance blue InGaN laser diodes with single-quantum-well active layers," *IEEE Photon. Technol. Lett.* **19**, 1717–1719 (2007).

211. A. C. Tamboli, E. D. Haberer, R. Sharma, K. H. Lee, S. Nakamura, and E. L. Hu, "Room-temperature continuous-wave lasing in GaN/InGaN microdisks," *Nat. Photonics* **1**, 61–64 (2007).

212. T. C. Lu, C. C. Kao, H. C. Kuo, G. S. Huang, S. C. Wang, "CW lasing of current injection blue GaN-based vertical cavity surface emitting laser," *Appl. Phys. Lett.* **92**, 141102 (2008).

213. J. F. Nye, *Physical Properties of Crystals: Their Representation by Tensors and Matrices*, Clarendon Press, Oxford, 1985.

214. A. E. H. Love, *A Treatise on the Mathematical Theory of Elasticity*, pp. 151–160, Dover, New York, 1944.

215. M. Suzuki, T. Uenoyama, and A. Yanase, "First-principles calculations of effective-mass parameters of AlN and GaN," *Phys. Rev. B* **52**, 8132 (1995).

216. M. Kumagai, S. L. Chuang, and H. Ando, "Analytical solutions of the block-diagonalized Hamiltonian for strained wurtzite semiconductors," *Phys. Rev. B* **57**, 15303–15314 (1998).

217. M. Suzuki and T. Uenoyama, "Reduction of threshold current density of wurtzite GaN/AlGaN quantum well lasers by uniaxial strain in (0001) plane," *Jpn. J. Appl. Phys.* **35**, L953–L955 (1996).

218. S. L. Chuang, and C. S. Chang, "Effective-mass Hamiltonian for strained wurtzite GaN and analytical solutions," *Appl. Phys. Lett.* **68**, 1657–1659 (1996).

219. A. Niwa, T. Ohtoshi and T. Kuroda, "Tight-binding analysis of the optical matrix element in wurtzite- and zincblende-GaN quantum wells," *Jpn. J. Appl. Phys.* **35**, L599–L601 (1996).

220. F. Jain and W. Huang, "Modeling of optical gain in InGaN-AlGaN and $In_xGa_{1-x}N$-$In_yGa_{1-y}N$ quantum-well lasers," *IEEE J. Quantum Electron.* **32**, 859–864 (1996).

221. S. L. Chuang, and C. S. Chang, "A band-structure model of strained quantum-well wurtzite semiconductors," *Semicond. Sci. Technol.* **12**, 252–263 (1997).

222. Y. C. Yeo, T. C. Chong, M. F. Li, and W. J. Fan, "Analysis of optical gain and threshold current density of wurtzite InGaN/GaN/AlGaN quantum well lasers," *J. Appl. Phys.* **84**, 1813–1819 (1998).

223. S. H. Park and S. L. Chuang, "Piezoelectric fields on electrical and optical properties of wurtzite GaN/AlGaN quantum-well lasers," *Appl. Phys. Lett.* **72**, 3103–3105 (1998).

224. S. H. Park and S. L. Chuang, "Crystal-orientation effects on the piezoelectric field and electronic properties of strained wurtzite semiconductors," *Phys. Rev. B* **59**, 4725–4737 (1999).

225. J. Y. Chang and Y. K. Kuo, "Simulation of blue InGaN quantum-well lasers," *J. Appl. Phys.* **93**, 4992–4998 (2003).
226. S. J. Huang, and S. T. Yen, "Improvement in threshold of InGaN/GaN quantum-well lasers by *p*-type modulation doping," *J. Appl. Phys.* **102**, 113103 (2007).
227. S. H. Park, D. Ahn, and S. L. Chuang, "Electronic and optical properties of a- and m-plane wurtzite InGaN-GaN quantum wells," *IEEE J. Quantum Electron.* **43**, 1175–1182 (2007).
228. R. A. Arif, Y. K. Ee, and N. Tansu, "Polarization engineering via staggered InGaN quantum wells for radiative efficiency enhancement of light emitting diodes," *Appl. Phys. Lett.* **91**, 091110 (2007).
229. R. Dingle, D. D. Sell, S. E. Stokowski, and M. Ilegems, "Absorption, reflectance, and luminescence of GaN epitaxial layers," *Phys. Rev. B* **4**, 1211–1218 (1971).
230. F. P. Kesamanly, "Gallium nitride: band structure, properties, and potential applications (review)," *Sov. Phys. Semicond.* **8**, 147–156 (1974).
231. N. E. Christensen and I. Gorczyca, "Optical and structural properties of III–V nitrides under pressure," *Phys. Rev. B* **50**, 4397–4415 (1994).
232. C. Wetzel, T. Takeuchi, S. Yamaguchi, H. Katoh, H. Amano, and I. Akasaki, "Optical band gap in $Ga_{1-x}In_xN$ ($0 < x < 0.2$) on GaN by photoreflection spectroscopy," *Appl. Phys. Lett.* **73**, 1994–1996 (1998).
233. S. J. Hwang, W. Shan, R. J. Hauenstein, J. J. Song, M.-E. Lin, S. Strite, B. N. Sverdlov, and H. Morkoç, "Photoluminescence of zinc-blende GaN under hydrostatic pressure," *Appl. Phys. Lett.* **64**, 2928–2930 (1994).
234. W. Shan, T. J. Schmidt, X. H. Yang, S. J. Hwang, J. J. Song, and B. Goldenberg, "Temperature dependence of interband transitions in GaN grown by metalorganic chemical vapor deposition," *Appl. Phys. Lett.* **66**, 985–987 (1995).
235. W. Shan, T. J. Schmidt, R. J. Hauenstein, J. J. Song, and B. Goldenberg, "Pressure-dependent photoluminescence study of wurtzite GaN," *Appl. Phys. Lett.* **66**, 3492–3494 (1995).
236. B. Gil, O. Briot, and R.-L. Aulombard, "Valence-band physics and the optical properties of GaN epilayers grown onto sapphire with wurtzite symmetry," *Phys. Rev. B* **52**, R17028–R17031 (1995).
237. S. Chichibu, T. Azuhata, T. Sota, and S. Nakamura, "Excitonic emissions from hexagonal GaN epitaxial layers," *J. Appl. Phys.* **79**, 2784–2786 (1996).
238. K. Kim, W. R. L. Lambrecht, and B. Segall, "Elastic constants and related properties of tetrahedrally bonded BN, AlN, GaN, and InN," *Phys. Rev. B* **53**, 16310–16326 (1996).
239. K. Kim, W. R. L. Lambrecht, B. Segall, and M. Schilfgaarde, "Effective masses and valence-band splittings in GaN and AlN," *Phys. Rev. B* **56**, 7363–7375 (1997).
240. A. Shikanai, T. Azuhata, T. Sota, S. Chichibu, A. Kuramata, K. Horino, and S. Nakamura, "Biaxial strain dependence of exciton resonance energies in wurtzite GaN," *J. Appl. Phys.* **81**, 417–424 (1997).
241. S. Chichbu, H. Okumura, S. Nakamura, G. Feuillet, T. Azuhata, T. Sota, and S. Yoshida, "Exciton spectra of cubic and hexagonal GaN epitaxial films," *Jpn. J. Appl. Phys.* **36**, 1976–1983 (1997).

242. D. L. Smith and C. Mailhiot, "Piezoelectric effects in strained-layer superlattices," *J. Appl. Phys.* **63**, 2717–2719 (1988).

243. A. Bykhovski, B. Gelmont, and M. Shur, "Strain and charge distribution in GaN-AlN-GaN semiconductor-insulator-semiconductor structure for arbitrary growth orientation," *Appl. Phys. Lett.* **63**, 2243–2245 (1993).

244. A. Bykhovski, B. Gelmont, and M. Shur, "The influence of the strain-induced electric field on the charge distribution in GaN-AlN-GaN structure," *J. Appl. Phys.* **74**, 6734–6739 (1993).

245. A. Bykhovski, B. Gelmont, M. Shur, and A. Khan, "Current-voltage characteristics of strained piezoelectric structures," *J. Appl. Phys.* **77**, 1616–1620 (1995).

246. A. Bykhovski, B. L. Gelmont, and M. S. Shur, "Elastic strain relaxation and piezoeffect in GaN-AlN, GaN-AlGaN and GaN-InGaN superlattices," *J. Appl. Phys.* **81**, 6332–6338 (1997).

247. T. Takeuchi, S. Sota, M. Katsuragawa, M. Komori, H. Takeuchi, H. Amano, and I. Akasaki, "Quantum-confined Stark effect due to piezoelectric fields in GaInN strained quantum wells," *Jpn. J. Appl. Phys.* **36** (Part 2), L382–L385 (1997).

248. T. Takeuchi, H. Takeuchi, S. Sota, H. Sakai, H. Amano, and I. Akasaki, "Optical properties of strained AlGaN and GaInN on GaN," *Jpn. J. Appl. Phys.*, **36** (Part 2), L177–L179 (1997).

249. F. Bernardini and V. Fiorentini, "Polarization-based calculation of the dielectric tensor of polar crystals," *Phys. Rev. Lett.* **79**, 3958–3961 (1997).

250. F. Bernardini, V. Fiorentini, and D. Venderbilt, "Spontaneous polarization and piezoelectric constants of III–V nitrides," *Phys. Rev. B* **56**, R10024–R10027 (1997).

251. T. Takeuchi, C. Wetzel, S. Yamaguchi, H. Sakai, H. Amano, I. Akasaki, Y. Kaneko, S. Nakagawa, Y. Yamaoka, and N. Yamada, "Determination of piezoelectric fields in strained GaInN quantum wells using the quantum-confined Stark effect," *Appl. Phys. Lett.* **73**, 1691–1693 (1998).

252. S. F. Chichibu, A. C. Abare, M. S. Minsky, S. Keller, S. B. Fleichser, J. E. Bowers, E. Hu, U. K. Mishra, L. A. Coldren, S. P. DenBarrs, and T. Sota, "Effective band gap inhomogeneity and piezoelectric field in InGaN/GaN multiquantum well structures," *Appl. Phys. Lett.* **73**, 2006–2008 (1998).

253. H. S. Kim, J. Y. Lin, H. X. Jiang, W. W. Chow, A. Botchkarev, and H. Morkoç, "Piezoelectric effects on the optical properties of GaN/$Al_xGa_{1-x}N$ multiple quantum wells," *Appl. Phys. Lett.* **73**, 3426–3428 (1998).

254. R. Langer, J. Simon, V. Ortiz, N. T. Pelekanos, A. Barski, R. Andrè, and M. Godlewski, "Giant electric fields in unstrained GaN single quantum wells," *Appl. Phys. Lett.* **74**, 3827–3829 (1999).

255. F. Bernardini and V. Fiorentini, "Macroscopic polarization and band offsets at nitride heterojunctions," *Phys. Rev. B* **57**, R9427–R9430 (1998).

256. O. Ambacher, J. Smart, J. R. Shealy, N. G. Weimann, K. Chu, M. Murphy, W. J. Schaff, L. F. Eastman, R. Dimitrov, L. Wittmer, M. Stutzmann, W. Rieger, and J. Hilsenbeck, "Two-dimensional electron gases induced by spontaneous and piezoelectric polarization charges in N- and Ga-face AlGaNGaN heterostructures," *J. Appl. Phys.* **85**, 3222–3233 (1999).

257. O. Ambacher, J. Smart, J. R. Shealy, N. G. Weimann, K. Chu, M. Murphy, W. J. Schaff, L. F. Eastman, R. Dimitrov, L. Wittmer, M. Stutzmann, W. Rieger, and J. Hilsenbeck,

"Two-dimensional electron gases by spontaneous and piezoelectric polarization charges in N- and Ga-face AlGaN/GaN heterostructures," *J. Appl. Phys.* **85**, 3222–3233 (1999).

258. S. H. Park and S. L. Chuang, "Spontaneous polarization effects in wurtzite GaN/AlGaN quantum wells and comparison with experiment," *Appl. Phys. Lett.* **76**, 1981–1983 (2000).

259. O. Ambacher, J. Majewski, C. Miskys, A. Link, M. Hermann, M. Eickhoff, M. Stutzmann, F. Bernardini, V. Fiorentini, V. Tilak, B. Schaff and L. F. Eastman "Pyroelectric properties of Al (In) GaN/GaN hetero and quantum well structure," *J. Phys. Condens. Matter* **14**, 3399–3434 (2002).

260. S. Tripathy, R. K. Soni, H. Asahi, K. Iwata, R. Kuroiwa, K. Asami, and S. Gonda, "Optical properties of GaN layers grown on C-, A-, R-, and M-plane sapphire substrates by gas source molecular beam epitaxy," *J. Appl. Phys.* **85**, 8386–8399 (1999).

261. P. Waltereit, O. Brandt, A. Trampert, H. T. Grahn, J. Menniger, M. Ramsteiner, M. Reiche, and K. H. Ploog, "Nitride semiconductors free of electrostatic fields for efficient white-light-emitting diodes," *Nature* **406**, 865–868 (2000).

262. P. Waltereit, O. Brandt, M. Ramsteiner, and K. H. Ploog, "M-plane GaN(1 $\bar{1}$ 00): A way to evade electrical polarization in nitrides," *Phys. Stat. Sol.* (a) **180**, 133 (2000).

263. S. Ghosh, P. Waltereit, O. Brandt, H. T. Grahn, and K. H. Ploog, "Electronic band structure of wurtzite GaN under biaxial strain in the M plane investigated with photoreflectance spectroscopy," *Phys. Rev. B* **65**, 075202 (2002).

264. S. Ghosh, O. Brandt, H. T. Grahn, and K. H. Ploog, "Strained M-plane GaN for the realization of polarization-sensitive photodetectors," *Appl. Phys. Lett.* **81**, 3380–3382 (2002).

265. Y. J. Sun, O. Brandt, U. Jahn, T. Y. Liu, A. Trampert, S. Cronenberg, S. Dhar, and K. H. Ploog, "Impact of nucleation conditions on the structural and optical properties of *M*-plane GaN(1 $\bar{1}$ 00) grown on γ-$LiAlO_2$," *J. Appl. Phys.* **92**, 5714–5719 (2002).

266. N. F. Gardner, J. C. Kim, J. J. Wierer, Y. C. Shen, and M. R. Krames, "Polarization anisotropy in the electroluminescence of m-plane InGaN–GaN multiple-quantum-well light-emitting diodes," *Appl. Phys. Lett.* **86**, 111101 (2005).

267. B. A. Haskell, T. J. Baker, M. B. McLaurin, F. Wu, P. T. Fini, S. P. DenBaars, J. S. Speck, and S. Nakamura, "Defect reduction in (1 $\bar{1}$ 00) m-plane gallium nitride via lateral epitaxial overgrowth by hydride vapor phase epitaxy," *Appl. Phys. Lett.* **86**, 111917 (2005).

268. K. Okamoto, H. Ohta, S. F. Chichibu, J. Ichihara, and H. Takasu, "Continuous-wave operation of m-plane InGaN multiple quantum well laser diodes," *Jpn. J. Appl. Phys.* **46**, L187–L189 (2007).

269. A. Chakraborty, B. A. Haskell, S. Keller, J. S. Speck, S. P. DenBaars, S. Nakamura, and U. K. Mishra, "Demonstration of nonpolar m-plane InGaN/GaN light-emitting diodes on free-standing m-plane GaN substrates," *Jpn. J. Appl. Phys.* **44**, L173–L175 (2005).

270. M. C. Schmidt, K. C. Kim, R. M. Farrell, D. F. Feezell, D. A. Cohen, M. Saito, K. Fujito, J. S. Speck, S. P. DenBaars, and S. Nakamura, "Demonstration of nonpolar m-plane InGaN/GaN laser diodes," *Jpn. J. Appl. Phys.* **46**, L190–L191 (2007).

271. R. M. Farrell, D. F. Feezell, M. C. Schmidt, D. A. Haeger, K. M. Kelchner, K. Iso, H. Yamada, M. Saito, K. Fujito, D. A. Cohen, J. S. Speck, S. P. DenBaars, and

S. Nakamura, "Continuous-wave operation of AlGaN-cladding-free nonpolar m-plane InGaN/GaN laser diodes," *Jpn. J. Appl. Phys.* **46**, L761–L763 (2007).

272. M. C. Schmidt, K. C. Kim, H. Sato, N. Fellows, H. Masui, S. Nakamura, S. P. DenBaars, and J. S. Speck, "High power and high external efficiency m-plane InGaN light emitting diodes," *Jpn. J. Appl. Phys.* **46**, L126–L128 (2007).

273. D. F. Feezell, M. C. Schmidt, R. M. Farrell, K. C. Kim, M. Saito, K. Fujito, D. A. Cohen, J. S. Speck, S. P. DenBaars, and S. Nakamura, "AlGaN-cladding-free nonpolar InGaN/GaN laser diodes," *Jpn. J. Appl. Phys.* **46**, L284–L286 (2007).

274. A. Tyagi, H. Zhong, R. B. Chung, D. F. Feezell, M. Saito, K. Fujito, J. S. Speck, S. P. DenBaars, S. Nakamura, "Semipolar ($10\bar{1}\bar{1}$) InGaN/GaN laser diodes on bulk GaN substrates," *Jpn. J. Appl. Phys.* **46**, L444–L445 (2007).

275. F. Wu, M. D. Craven, S. H. Lim, and J. S. Speck, "Polarity determination of a-plane GaN on r-plane sapphire and its effects on lateral overgrowth and heteroepitaxy," *J. Appl. Phys.* **94**, 942–947 (2003).

276. A. Chitnis, C. Chen, V. Adivarahan, M. Shatalov, E. Kuokstis, V. Mandavilli, J. Yang, and M. A. Khan, "Visible light-emitting diodes using a-plane GaN–InGaN multiple quantum wells over r-plane sapphire," *Appl. Phys. Lett.* **84**, 3663–3665 (2004).

277. M. D. Craven, S. H. Lim, F. Wu, J. S. Speck, and S. P. DenBaars, "Structural characterization of nonpolar ($11\bar{2}0$) a-plane GaN thin films grown on ($1\ \bar{1}\ 02$) r-plane sapphire," *Appl. Phys. Lett.* **81**, 469–471 (2002).

278. M. D. Craven, P. Waltereit, F. Wu, J. S. Speck, and S. P. DenBaars, "Characterization of a-plane GaN/(Al,Ga)N multiple quantum wells grown via metalorganic chemical vapor Deposition," *Jpn. J. Appl. Phys.* **Part 2 42**, L235–L238 (2003).

279. M. D. Craven, P. Waltereit, J. S. Speck, and S. P. DenBaars, "Well-width dependence of photoluminescence emission from a-plane GaN/AlGaN multiple quantum wells," *Appl. Phys. Lett.* **84**, 496–498 (2004).

280. H. Sato, A. Tyagi, H. Zhong, N. Fellows, R. B. Chung, M. Saito, K. Fujito, J. S. Speck, S. P. Denbaars, S. Nakamura, "High power and high efficiency green light emitting diode on free-standing semipolar ($11\bar{2}2$) bulk GaN substrate," *Phys. Stat. Sol. (RRL).* **1**, 162–164 (2007).

281. D. R. Scifres, W. Streifer, and R. D. Burnham, "High-power coupled-multiple-stripe phase-locked injection laser," *Appl. Phys. Lett.* **34**, 259–261 (1979).

282. D. R. Scifres, R. D. Burnham, C. Lindström, W. Streifer, and T. L. Paoli, "Phase-locked (GaAl)As laser emitting 1.5 W cw per mirror," *Appl. Phys. Lett.* **42**, 645–647 (1983).

283. J. K. Butler, D. E. Ackley, and D. Botez, "Coupled-mode analysis of phase-locked injection laser arrays," *Appl. Phys. Lett.* **44**, 293–295, 1984; and Erratum, **44**, 935 (1984).

284. S. Wang, J. Z. Wilcox, M. Jansen, and J. J. Yang, "In-phase locking in diffraction-coupled phased-array diode lasers," *Appl. Phys. Lett.* **48**, 1770–1772 (1986).

285. W. Streifer, M. Orsiński, D. R. Scifres, D. F. Welch and P. S. Cross, "Phased-array lasers with a uniform, stable supermode," *Appl. Phys. Lett.* **49**, 1496–1498 (1986).

286. W. Streifer, P. S. Cross, D. F. Welch, and D. R. Scifres, "Analysis of a Y-junction semiconductor laser array," *Appl. Phys. Lett.* **49**, 58–60 (1986).

287. E. Kapon, Z. Rav-Noy, S. Margalit, and A. Yariv, "Phase-locked arrays of buried-ridge InP/InGaAsP diode lasers," *J. Lightwave Technol.* **LT-4**, 919–925 (1986).

288. G. R. Hadley, J. P. Hohimer, and A. Owyoung, "High-order ($\nu > 10$) eigenmodes in ten-stripe gain-guided diode laser arrays," *Appl. Phys. Lett.* **49**, 684–686 (1986).

289. W. K. Marshall and J. Katz, "Direct analysis of gain-guided phase-locked semiconductor laser arrays," *IEEE J. Quantum Electron.* **QE-22**, 827–832 (1986).

290. I. J. Mawst, D. Botez, T. J. Roth, G. Peterson, and J. J. Yang, "Diffraction-coupled, phase-locked arrays of antiguided, quantum-well lasers grown by metalorganic chemical vapour deposition," *Electron. Lett.* **24**, 958–959 (1988).

291. D. Botez, L. Mawst, P. Hayashida, G. Peterson, and T. J. Roth, "High-power, diffraction-limited-beam operation from phase-locked diode-laser arrays of closely spaced "leaky" waveguides (antiguides)," *Appl. Phys. Lett.* **53**, 464–466 (1988).

292. G. R. Hadley, "Index-guided arrays with a large index step," *Opt. Lett.* **14**, 308–310 (1989).

293. D. Botez, L. J. Mawst, G. L. Peterson, and T. J. Roth, "Phase-locked arrays of antiguides: modal content and discrimination," *IEEE J. Quantum Electron.* **26**, 482–495 (1990).

294. M. Sakamoto, J. G. Endriz, and D. R. Scifres, "120 W CW output power from a monolithic AlGaAs (800 nm) laser diode array mounted on a diamond heat sink," *Electron. Lett.* **28**, 197–199 (1992).

295. M. Abramowitz and I. A. Stegun, Eds., *Handbook of Mathematical Functions*, Dover, New York, 1972.

Part IV
Modulation of Light

12

Direct Modulation of Semiconductor Lasers

For a semiconductor laser, the output power of light intensity P_{out} increases linearly with the injection current above threshold as discussed in Section 10.1, Eq. (10.1.21),

$$P_{out} = \eta_i \frac{\hbar\omega}{q} \frac{\alpha_m}{\alpha_i + \alpha_m}(I - I_{th}). \tag{12.1.1}$$

Therefore, for an injection current with a dc component and a small signal ac modulation, Fig. 12.1,

$$I = I_0 + i(t), \tag{12.1.2}$$

we expect the optical output power to have corresponding dc and ac components

$$P_{out}(t) = P_0 + p(t). \tag{12.1.3}$$

In this chapter, we study the direct current modulation of diode lasers and some intrinsic effects such as relaxation oscillations, modulation speed, and the laser linewidth theory. Our goal is to understand how the laser light output characteristics vary as we increase the modulation frequency of the injection current.

12.1 RATE EQUATIONS AND LINEAR GAIN ANALYSIS

Assume only one mode is lasing; the rate equations for the carrier density $n(t)$ $(1/\text{cm}^3)$ and the photon density $S(t)$ $(1/\text{cm}^3)$ can be written as [1–4]

$$\frac{dn(t)}{dt} = \eta_i \frac{J(t)}{qd} - \frac{n(t)}{\tau} - v_g g(n) S(t) \tag{12.1.4}$$

$$\frac{dS(t)}{dt} = \Gamma v_g g(n) S(t) - \frac{S(t)}{\tau_p} + \beta R_{sp}(n) \tag{12.1.5}$$

Physics of Photonic Devices, Second Edition. By Shun Lien Chuang
Copyright © 2009 John Wiley & Sons, Inc.

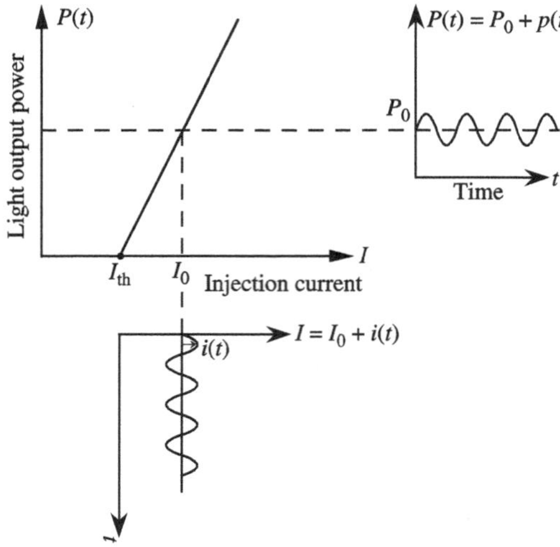

Figure 12.1 For an injection current $I = I_0 + i(t)$, the optical output power is $P(t) = P_0 + p(t)$, where (I_0, P_0) is the bias point for the direct current modulation of the semiconductor laser.

where

$n(t)$ = electron density (cm^{-3})
η_i = injection quantum efficiency
q = unit charge (1.6×10^{-19} Coulomb)
τ = carrier lifetime (s)
$\mathrm{v}_g = c/n_g$ = the group velocity of light (cm/s)
Γ = optical confinement factor
$R_{sp}(n)$ = spontaneous emission rate per unit volume ($\mathrm{cm}^{-3}\,\mathrm{s}^{-1}$)
$S(t)$ = photon density (cm^{-3})
J = injection current density ($\mathrm{A/cm}^2$)
d = thickness of the active region (cm)
τ_p = photon lifetime (s)
$g(n)$ = the gain coefficient (cm^{-1})
β = the spontaneous emission factor.

In the first rate equation for the carrier density, the first term $\eta_i J/qd$ is the injected number of carriers into the active layer per unit volume per second. The recombination rate

$$R(n) = \frac{n}{\tau(n)} = An + R_{sp}(n) + Cn^3 \tag{12.1.6}$$

accounts for the carrier loss due to the radiative recombination from the spontaneous emission of photons

$$R_{sp}(n) = Bn^2 \tag{12.1.7}$$

and nonradiative recombinations at defects (An) and the Auger processes (Cn^3). The time constant τ generally depends on n but is assumed to be a constant for simplicity. The term $\mathrm{v}_g g(n)S$ is the carrier loss due to stimulated emissions.

In the second rate equation for photon density S, the first term $\Gamma v_g g(n)S(t)$ is the increasing rate of the number of photons per unit volume due to the stimulated emission. The second term $(-S/\tau_p)$ is the decreasing rate of the photon density due to absorptions and transmission through the mirrors of the laser cavity. The photon lifetime in the cavity τ_p is determined physically by the rate due to absorption and mirror transmission losses:

$$\frac{1}{\tau_p} = v_g\left(\alpha + \frac{1}{L}\ln\frac{1}{R}\right). \tag{12.1.8}$$

In other words, photons disappear from the cavity via absorption processes or transmission out of the two end facets. The last term βR_{sp} is the fraction of spontaneous emission entering the lasing mode, which is generally very small. The β factor in (12.1.5) includes the effect of optical confinement factor Γ. The spontaneous emission factor β can be calculated with a simple analytical expression for index-guided laser modes with a plane phase front [5]. For a gain-guided laser with a cylindrical constant phase front, the spontaneous emission is shown to be enhanced by another factor [6, 7].

Note that the optical confinement factor Γ appears only in (12.1.2). This is because the optical mode extends beyond the active region d by a factor d/Γ. The two rate equations ensure that the total photon and carrier numbers (not densities) are balanced taking into account difference in the occupied volumes by the carriers and photons (d vs. d/Γ).

12.1.1 Linear Gain Theory

If we assume the photon density S is not high so that the nonlinear gain saturation effect can be ignored, we have a simplified gain model

$$g(n) = g(n_0) + g'(n(t) - n_0) \tag{12.1.9}$$

where $g_0 = g(n_0)$, and $g' = \partial g/\partial n$ is the differential gain at $n = n_0$. Assume

$$\begin{aligned} J(t) &= J_0 + j(t) \\ n(t) &= n_0 + \Delta n(t) \\ S(t) &= S_0 + s(t) \end{aligned} \tag{12.1.10}$$

where $j(t)$, $\Delta n(t)$, and $s(t)$ are small signals compared with their corresponding dc values, J_0, n_0, and S_0, respectively.

dc Solutions The photon density S_0 and the carrier density n_0 at steady state are

$$S_0 = \frac{\beta R_{sp}}{\dfrac{1}{\tau_p} - \Gamma v_g g_0} \tag{12.1.11a}$$

$$n_0 = \tau\left(\eta_i \frac{J_0}{qd} - v_g g_0 S_0\right). \tag{12.1.11b}$$

Note that for negligible βR_{sp}, the inverse photon lifetime is

$$\frac{1}{\tau_p} \simeq \Gamma v_g g_0 \tag{12.1.12}$$

where $g_0 = g_{th}$ is pinned at threshold gain value above threshold.

Small-Signal ac Analysis The small-signal equations are

$$\frac{d}{dt}\Delta n(t) = \eta_i \frac{j(t)}{qd} - \frac{\Delta n(t)}{\tau} - v_g[g' S_0 \Delta n(t) + g_0 s(t)] \tag{12.1.13}$$

$$\frac{d}{dt}s(t) = \Gamma v_g[g' S_0 \Delta n(t) + g_0 s(t)] - \frac{s(t)}{\tau_p}. \tag{12.1.14}$$

For a sinusoidal excitation,

$$j(t) = \mathrm{Re}[j(\omega)e^{-i\omega t}] \tag{12.1.15}$$

the steady-state solution can be obtained by substituting the phasor notations in the frequency domain

$$\begin{aligned} \Delta n(t) &= \mathrm{Re}[\Delta n(\omega)e^{-i\omega t}] \\ s(t) &= \mathrm{Re}[s(\omega)e^{-i\omega t}] \end{aligned} \tag{12.1.16}$$

into (12.1.13) and (12.1.14). We find the complex amplitude $\Delta n(\omega)$ for the carrier density taking into account (12.1.12)

$$\Delta n(\omega) \simeq \frac{-i\omega}{\Gamma v_g g' S_0} s(\omega) \tag{12.1.17}$$

and the complex magnitude $s(\omega)$ for the photon density

$$\frac{s(\omega)}{j(\omega)} \simeq \frac{\eta_i \Gamma v_g g' S_0/(qd)}{\omega_r^2 - \omega^2 - i\gamma\omega}. \tag{12.1.18}$$

Here we define ω_r as:

$$\omega_r^2 \simeq \frac{v_g g' S_0}{\tau_p} \tag{12.1.19}$$

and a damping factor

$$\gamma \simeq \frac{1}{\tau} + v_g g' S_0. \tag{12.1.20}$$

The relaxation frequency f_r is

$$f_r \simeq \frac{1}{2\pi}\sqrt{\frac{v_g g' S_0}{\tau_p}} \tag{12.1.21}$$

which is proportional to $\sqrt{S_0}$ or the square root of the optical output power as it is linearly proportional to the photon density S_0.

12.1.2 Small-Signal Modulation Response

The frequency response function is

$$\frac{s(\omega)}{j(\omega)} = |M(\omega)|e^{i\phi(\omega)} \tag{12.1.22a}$$

$$M(\omega) = \frac{(\eta_i \Gamma \tau_p/qd)\omega_r^2}{[(\omega_r^2 - \omega^2)^2 + \gamma^2\omega^2]^{1/2}} \tag{12.1.22b}$$

$$\phi(\omega) = \tan^{-1}\left(\frac{\gamma\omega}{\omega_r^2 - \omega^2}\right) \tag{12.1.22c}$$

or in terms of the normalized response to the dc value

$$\left|\frac{M(\omega)}{M(0)}\right| = \frac{\omega_r^2}{[(\omega_r^2 - \omega^2)^2 + \gamma^2\omega^2]^{1/2}} \tag{12.1.23}$$

which has a flat response at low frequencies (low pass) and peaks near $\omega = \omega_r$, then rolls off as the frequency increases further. The exact peak position is to find the minimum of the denominator in the magnitude of the frequency response (12.1.22b). The resultant relaxation frequency peak will occur at a frequency close to that in (12.1.19). The 3-dB frequency f_{3dB} ($= \omega_{3dB}/2\pi$) or bandwidth occurs when the magnitude of the modulation response is reduced by a factor of $\sqrt{2}$. In other words, the denominator of (12.1.23) is $\sqrt{2}\omega_r^2$, or

$$(\omega_{3\,dB}^2 - \omega_r^2)^2 + \omega_{3dB}^2\left(\frac{1}{\tau} + \tau_p\omega_r^2\right)^2 = 2\omega_r^4 \tag{12.1.24}$$

for which the only unknown ω_{3dB} can be solved.

Example The frequency response for a distributed feedback semiconductor laser [8] is shown in Fig. 12.2 for different optical output powers at 5, 10, 15, and 20 mW. The peak response occurs at the relaxation oscillation frequency f_r and decreases to -3 dB at f_{3dB} compared with the normalized response at low frequencies (0 dB). The bandwidth increases as the optical output power P_0 increases with a $\sqrt{P_0}$ dependence because P_0 is proportional to the photon density S_0, as shown in Fig. 12.3. The 3-dB frequency f_{3dB} versus the square root of the output power $\sqrt{P_0}$ are shown

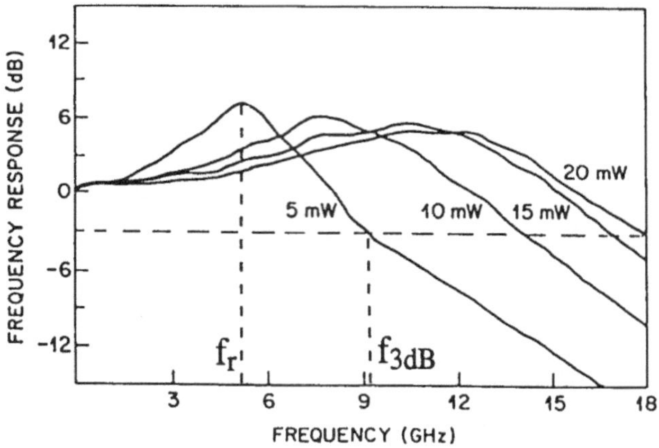

Figure 12.2 Small-signal frequency response of a distributed feedback laser at different light output powers. The submount temperature was 20°C. The peak response determines the relaxation frequency f_r, and the -3-dB response gives the 3-dB frequency f_{3dB}. (Reprinted with permission from [8] © 1989 American Institute of Physics.)

with the relaxation frequency f_r. Because the output power P_0 is proportional to $I - I_{th}$, the plot f_r versus $\sqrt{I - I_{th}}$ also shows a linear relationship. Therefore, f_r can also be plotted versus $\sqrt{I - I_{th}}$ for semiconductor lasers and a linear relationship is shown [9]. Improvements of the modulation bandwidth using quantum wells and quantum wires or strained quantum wells have been discussed [4, 10, 11].

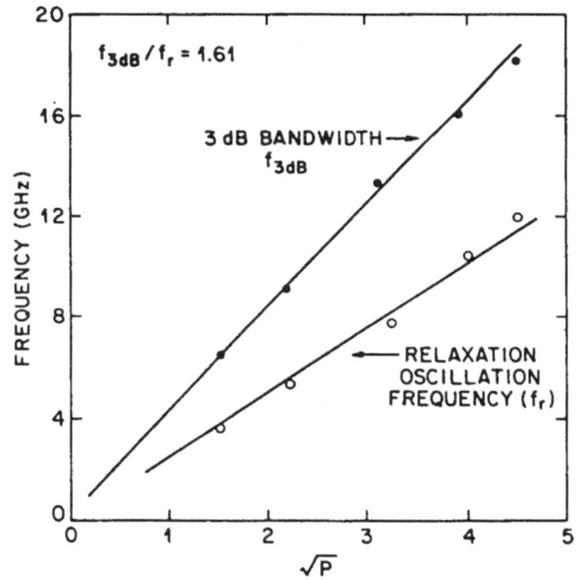

Figure 12.3 The relaxation oscillation frequency and the 3-dB bandwidth of the distributed feedback laser in Fig. 12.2 are plotted as a function of square root of light output power P. (Reprinted with permission from [8] © 1989 American Institute of Physics.)

12.2 HIGH-SPEED MODULATION RESPONSE WITH NONLINEAR GAIN SATURATION

12.2.1 Nonlinear Gain Saturation

The nonlinear gain model $g(n, S)$ at high photon density S can be taken from [12, 13]

$$g(n, S) = \frac{g(n_0) + g'(n(t) - n_0)}{1 + \varepsilon S(t)} \tag{12.2.1}$$

where $g_0 = g(n_0)$ and $g' = [\partial g/\partial n]_{n=n_0}$ is the differential gain at n_0. The factor $1 + \varepsilon S$ accounts for nonlinear gain saturation, which is important when the photon density is large. The factor ε is called the gain suppression coefficient.

dc Solution The steady-state solution at $I = I_0$ is obtained from $d/dt = 0$ in the rate equations

$$\eta_i \frac{J_0}{qd} = \frac{n_0}{\tau} + \frac{v_g g_0 S_0}{1 + \varepsilon S_0} \tag{12.2.2}$$

$$\Gamma \frac{v_g g_0 S_0}{1 + \varepsilon S_0} = \frac{S_0}{\tau_p} - \beta R_{sp}(n_0). \tag{12.2.3}$$

If $\varepsilon \neq 0$, the general solution for S_0 can be found analytically from (12.2.3) (see Problem 12.5). Then n_0 is obtained from (12.2.2).

Small Signal ac Analysis Using the linearized expression by substituting (12.1.10) into (12.2.1), we obtain

$$g(n, S) = \frac{g(n_0)}{1 + \varepsilon S_0} + \frac{g'}{1 + \varepsilon S_0} \Delta n(t) - \frac{g_0}{(1 + \varepsilon S_0)^2} \varepsilon s(t). \tag{12.2.4}$$

The small signal ac responses $\Delta n(t)$ and $s(t)$ satisfy the following equations

$$\frac{d}{dt}\begin{bmatrix} \Delta n(t) \\ s(t) \end{bmatrix} + \begin{bmatrix} A & D \\ -C & B \end{bmatrix}\begin{bmatrix} \Delta n(t) \\ s(t) \end{bmatrix} = \begin{bmatrix} \eta_i j(\omega)/qd \\ 0 \end{bmatrix} \tag{12.2.5}$$

where A, B, C, and D are defined as

$$\begin{aligned} A &= \frac{1}{\tau} + \frac{v_g g' S_0}{1 + \varepsilon S_0} & B &= \frac{1}{\tau_p} - \frac{\Gamma v_g g_0}{(1 + \varepsilon S_0)^2} \\ C &= \frac{\Gamma v_g g' S_0}{1 + \varepsilon S_0} & D &= \frac{v_g g_0}{(1 + \varepsilon S_0)^2}. \end{aligned} \tag{12.2.6}$$

12.2.2 Sinusoidal Steady-State Solution of the Small-Signal Equations

The small-signal response in the sinusoidal steady-state can be found from (12.2.5) and (12.2.6) by solving for $s(\omega)$ in terms of $j(\omega)$. We replace d/dt by $-i\omega$ and obtain two algebraic equations of the form

$$\begin{bmatrix} -i\omega + A & D \\ -C & -i\omega + B \end{bmatrix} \begin{bmatrix} \Delta n(\omega) \\ s(\omega) \end{bmatrix} = \begin{bmatrix} \eta_i j(\omega)/qd \\ 0 \end{bmatrix}. \tag{12.2.7}$$

Therefore, solving for the small-signal photon density function, we find

$$s(\omega) = \frac{C\eta_i j(\omega)/(qd)}{(-i\omega + A)(-i\omega + B) + CD}. \tag{12.2.8}$$

Define the relaxation frequency as

$$\begin{aligned} \omega_r^2 &= AB + CD \\ &= \frac{v_g g' S_0}{\tau_p(1+\varepsilon S_0)} + \frac{1}{\tau}\left[\frac{1}{\tau_p} - \frac{\Gamma v_g g_0}{(1+\varepsilon S_0)^2}\right] && (12.2.9a) \\ &\simeq \frac{v_g g' S_0}{\tau_p(1+\varepsilon S_0)} && (12.2.9b) \end{aligned}$$

where the second term in (12.2.9a) is negligible compared with the leading term because

$$B = \frac{1}{\tau_p} - \frac{\Gamma v_g g_0}{(1+\varepsilon S_0)^2} = \frac{1}{(1+\varepsilon S_0)}\left[\frac{\varepsilon S_0}{\tau_p} + \beta \frac{R_{sp}}{S_0}\right]. \tag{12.2.10}$$

We define the damping factor

$$\begin{aligned} \gamma &= A + B \\ &= \frac{1}{\tau} + \frac{v_g g' S_0}{(1+\varepsilon S_0)}\left(1 + \frac{\varepsilon}{v_g g' \tau_p}\right) + \beta \frac{R_{sp}}{S_0(1+\varepsilon S_0)} && (12.2.11a) \\ &\simeq \frac{1}{\tau} + K f_r^2 && (12.2.11b) \end{aligned}$$

where we have ignored the last term in (12.2.11a) and a K factor (ns) is defined,

$$K = 4\pi^2\left(\tau_p + \frac{\varepsilon}{v_g g'}\right), \tag{12.2.12}$$

and $\omega_r = 2\pi f_r$ has been used. The relaxation frequency based on the nonlinear gain model has an extra factor $(1 + \varepsilon S_0)$ in the denominator of the expression:

$$\omega_r \simeq \sqrt{\frac{v_g g' S_0}{\tau_p(1+\varepsilon S_0)}}. \tag{12.2.13}$$

We obtain the frequency response function for semiconductor lasers

$$M(\omega) = \frac{s(\omega)}{j(\omega)} = \frac{\omega_r^2(\eta_i \Gamma \tau_p)/(qd)}{\omega_r^2 - \omega^2 - i\gamma\omega} = |M(\omega)|e^{i\phi(\omega)} \tag{12.2.14}$$

$$\left|\frac{M(\omega)}{M(0)}\right| = \frac{\omega_r^2}{[(\omega_r^2 - \omega^2)^2 + \gamma^2\omega^2]^{1/2}} \tag{12.2.15a}$$

$$\phi(\omega) = \tan^{-1}\frac{\gamma\omega}{\omega_r^2 - \omega^2}. \tag{12.2.15b}$$

The 3-dB cutoff frequency occurs when (12.1.24) is satisfied using ω_r^2 in (12.2.9b) and γ in (12.2.11b). The maximum possible bandwidth occurs when the following condition is satisfied,

$$2\omega_r^2 = \gamma^2 \simeq K^2 f_r^4, \tag{12.2.16}$$

such that the frequency response function is a monotonic decreasing function, (12.2.15) $\propto \omega_r^2/(\omega^4 + \omega_r^4)^{1/2}$. We then have the maximum relaxation frequency

$$f_{r,\max} = \frac{2\pi\sqrt{2}}{K}. \tag{12.2.17}$$

By fitting the frequency response function (12.2.15) to the experimental data, the damping factor γ can be found together with the relaxation frequency f_r. Equation (12.2.11b) shows that a linear relation holds with a slope K if the damping factor γ is plotted versus f_r^2 and the intercept with the vertical axis gives the inverse carrier lifetime $1/\tau$.

Example Figure 12.4 [14] shows the experimental results for two strained quantum-well lasers with (1) a slope $K = 0.22$ ns for a tensile strain laser with four quantum wells, and the maximum bandwidth is $f_{r,\max} = 2\pi\sqrt{2}/K = 40$ GHz, and (2) a slope with $K = 0.58$ ns for a compressive strain laser with four quantum wells, and $f_{r,\max} = 15$ GHz. Both lines intercept the vertical axis at $1/\tau \simeq 5$ GHz, or the carrier lifetime τ at threshold is 0.2 ns for both samples. The difference of the slopes of the two laser structures can be explained from the difference in the differential gain of over a factor of 2 because $f_r^2/P_0 = 3.6\ \text{GHz}^2/\text{mW}$ for the compressive strain laser and $f_r^2/P_0 = 7.7\ \text{GHz}^2/\text{mW}$ for the tensile strain sample.

Because the damping factor depends on the K factor, relaxation frequency, and the inverse carrier lifetime, the experimental data provide very good guidance for the design of high-speed semiconductor lasers. The K factor can also be used to determine the nonlinear gain suppression coefficient ε. Theoretical models and experimental data on strained and unstrained quantum-well lasers have been presented with interesting results [15–51]. The measured nonlinear gain suppression coefficient

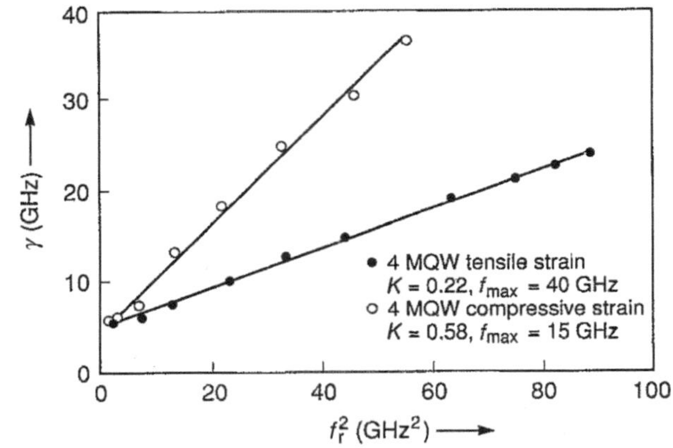

Figure 12.4 Experimental results showing the linear relation between the damping factor γ and f_r^2 for two quantum-well lasers: $\gamma = Kf_r^2 + 1/\tau$. The slope gives the K factor (in ns) and the intercept with the vertical (γ) axis gives the inverse carrier lifetime, $1/\tau$, at threshold. (Reprinted with permission from [14] © 1991 American Institute of Physics.)

ε ranges from 2×10^{-7} to 13×10^{-7} cm^3 for InGaAs or InGaAsP materials in the quantum wells [15, 17, 18]. Various physical mechanisms such as the well-barrier hole burning effects [20], carrier heating and spectral hole burning [21–24], carrier transport [25–33], and carrier capture by and escape from quantum wells [35–39] and circuit parasitic [40] have been investigated and are shown to affect by varying degrees the high-speed modulation of semiconductor quantum-well lasers. The major sources of carrier heating in semiconductor lasers are injection heating, stimulated recombination heating, and free carrier heating. These changes in the carrier temperature will be reflected in the changes in gain. An increase of a few degrees will result in a decrease in gain of several percent, which appears in the rate equations as a nonlinear gain suppression coefficient. Spectral hole burning theory also predicts an increase in the nonlinear gain suppression coefficient with increasing carrier temperature. We will investigate the carrier transport, capture, and escape in quantum-well lasers in the next section.

12.3 TRANSPORT EFFECTS ON QUANTUM-WELL LASERS: ELECTRICAL VERSUS OPTICAL MODULATION

In the previous section, we discuss the modulation response of a semiconductor laser using two coupled rate equations governing the carrier and photon densities. The carrier diffusion, capture, and escape times are usually used to characterize the carrier transport processes [25–33], which give a parasitic-like roll-off and are indistinguishable from parasitic effects. Parasitic effects come from the bias circuit and the shunting of modulation current around the active layer, which will cause a low-frequency roll-off of modulation response. In quantum well (QW) lasers, the carrier transport effect is an important limit for the modulation bandwidth of multiple quantum-well (MQW) lasers. In general, when the number of wells increases, the modulation bandwidth of the device initially increases, but it is ultimately saturated by the carrier transport effect. This saturation effect was observed in gain-coupled InGaAsP distributed feedback (DFB) lasers with more than eight QWs [41].

In this section, the high-speed modulation response of a QW DFB laser operating at 1.55 μm is investigated for two types of modulation [32, 33]: electrical [25–30] and optical [42–45]. Electrical modulation is performed by applying a microwave signal directly on the injection current of the test laser. Optical modulation is performed by injecting light from a second pump laser operating at 1.3 μm that itself is being directly modulated electrically, so that the injected light is absorbed in the QWs of the 1.55-μm test laser. Both sets of responses are compared with theory in order to extract the values of the fundamental parameters that determine as well as limit the modulation bandwidth. Optical injection at this wavelength eliminates circuit parasitic and carrier transport effects, and the intrinsic laser response can be directly measured [42–45].

A schematic of high-speed modulation by electrical injection and optical injection is shown in Fig. 12.5 [32]. We model the carrier transport (diffusion, capture, and emission) in such a laser structure and study its effects on the small-signal modulation with electrical injection and optical injection. Our objective is to calculate the

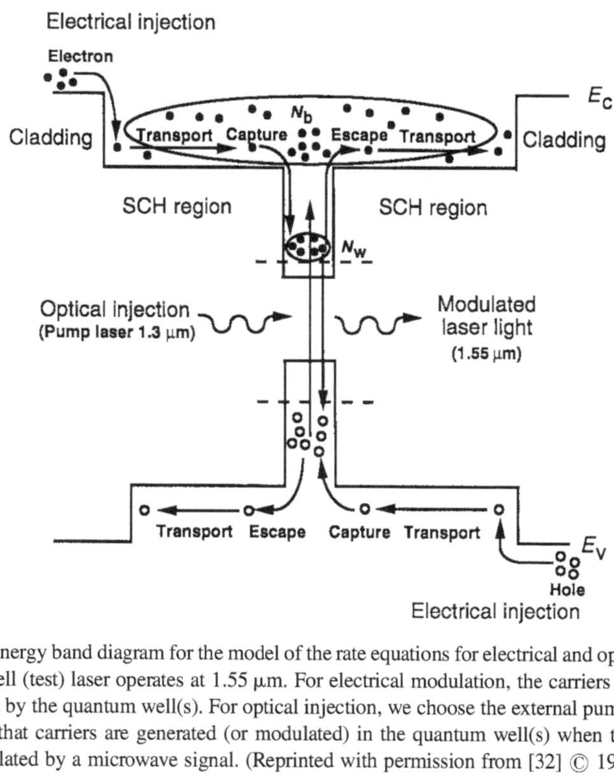

Figure 12.5 Energy band diagram for the model of the rate equations for electrical and optical modulation. The quantum-well (test) laser operates at 1.55 μm. For electrical modulation, the carriers have to transport and get captured by the quantum well(s). For optical injection, we choose the external pump laser operating at 1.3 μm such that carriers are generated (or modulated) in the quantum well(s) when the pump laser is externally modulated by a microwave signal. (Reprinted with permission from [32] © 1999 IEEE.)

small-signal frequency response of the test laser for optical modulation from the pump laser and use the same rate equations with a different source term to derive the electrical modulation response of the test laser.

Electrical Injection For electrical microwave modulation, the electrons are injected from the outer edge of the left separate-confinement-heterostructure (SCH) region and the holes from the outer edge of the right SCH region. The injected carriers diffuse through the SCH region and are captured into the QWs before recombining via the stimulated emission process. The transport effects can be modeled by taking into account the carrier density in the SCH region, the carrier density in the well region, and the photon density separately. Coupling of the carrier density in the barrier states above the QWs to the carrier density in the QWs is modeled by two terms representing carrier capture and escape into or from the wells, respectively. In this case, three rate equations are needed. The source term enters through the injection current in electrical modulation. The model considers carrier injection from the outer edges of the SCH region, diffusion across the SCH region, and the subsequent capture and emission of carriers by the QW.

Optical Injection For optical injection using an external pump laser, the pump photon density acts as the source term. Because the optical energy of the pump laser determines whether the photons are absorbed in the well or in the barriers, we choose the pump wavelength to be longer than the band-gap wavelength of the barriers and shorter than the band-edge wavelength of the wells so that absorption occurs only in the wells. Compared with electrical modulation, optical modulation with an optical energy in the absorption range of the QW directly produces photon-generated carriers inside the test laser's active region via the injection of a modulated laser beam through one of the test laser's mirror facets. Therefore, the majority of carriers transporting through the SCH region is not required for lasing action, although the coupling between SCH and QW states still exists for optical modulation. In this way, optical modulation removes the severe low-frequency roll-off due to the transport and parasitic effects and helps to clarify the intrinsic response.

The rate equations for both electrical and optical injections are written as

$$\frac{dN_{\mathrm{b}}(t)}{dt} = \eta_{\mathrm{i}}\frac{I(t)}{qV_{\mathrm{b}}} - \frac{N_{\mathrm{b}}(t)}{\tau_{\mathrm{b}}} - \frac{N_{\mathrm{b}}(t)}{\tau_{\mathrm{bw}}} + \frac{N_{\mathrm{w}}(t)V_{\mathrm{w}}}{\tau_{\mathrm{wb}}V_{\mathrm{b}}} \tag{12.3.1}$$

$$\frac{dN_{\mathrm{w}}(t)}{dt} = \frac{N_{\mathrm{b}}(t)V_{\mathrm{b}}}{\tau_{\mathrm{bw}}V_{\mathrm{w}}} - \frac{N_{\mathrm{w}}(t)}{\tau_{\mathrm{w}}} - \frac{N_{\mathrm{w}}(t)}{\tau_{\mathrm{wb}}} - \mathrm{v}_g\frac{g(N_{\mathrm{w}})}{1+\varepsilon S}S + \mathrm{v}_{\mathrm{g}}\alpha_{\mathrm{p}}S_{\mathrm{p}} \tag{12.3.2}$$

$$\frac{dS(t)}{dt} = \Gamma\mathrm{v}_g\frac{g(N_{\mathrm{w}})}{1+\varepsilon S}S - \frac{S(t)}{\tau_{\mathrm{p}}} \tag{12.3.3}$$

where q is the electron unit charge, $g(N_{\mathrm{w}})$ is the optical gain at the carrier concentration N_{w} in the bound states of the wells, N_{b} is the carrier density in the barrier (continuum) states including the SCH and active layers, S is the photon density of the test laser, ε is the nonlinear gain suppression coefficient, Γ is the optical confinement factor, τ_p is the photon lifetime, η_i is the injection efficiency, $I(t)$ is the injection electrical current, V_b is the volume of the SCH and active region, V_w is the volume of the well region, τ_b is the carrier recombination lifetime in the barrier region, τ_w is the carrier recombination lifetime in the well, τ_{bw} is the effective carrier diffusion across the SCH region and capture time by the wells, and τ_{wb} is the thermionic emission and carrier diffusion time from the well to the barrier states. The last term in (12.3.2) is a source term that represents the photon generation due to optical pumping at the wavelength (1.3 μm) where the optically injected photons are absorbed in the well regions. S_p is the pump photon density, and α_p is the absorption at the pump wavelength. This model is often referred to as the reservoir model [28] and is equivalent to models that incorporate additional effects such as diffusive transport [44]. The spontaneous emission term has been ignored in (12.3.3) for above-threshold operation. Assuming a small-signal optical injection due to the external pump laser

$$S_p(t) = S_{p0} + \mathrm{Re}[s_p(\omega)e^{-i\omega t}] \tag{12.3.4}$$

where the test laser is biased at a dc current above threshold. The responses for the photon density and the carrier densities in the barrier and well are

$$S(t) = S_0 + \mathrm{Re}[s(\omega)e^{-i\omega t}] \tag{12.3.5}$$

$$N_b(t) = N_{b0} + \mathrm{Re}[n_b(\omega)e^{-i\omega t}] \tag{12.3.6}$$

$$N_w(t) = N_{w0} + \mathrm{Re}[n_w(\omega)e^{-i\omega t}]. \tag{12.3.7}$$

The linear portion of the optical gain is written as:

$$g(N_w) = g(N_{w0}) + g'(N_w(t) - N_{w0}) \tag{12.3.8}$$

where g' is the differential gain of the test laser, and N_{w0} is steady-state carrier density in the QW. The nonlinear gain suppression effect has been included explicitly in the rate equations. Substituting (12.3.4)–(12.3.8) into (12.3.1)–(12.3.3), the steady-state quantities can be obtained by setting the time-derivative terms to zero. Then, the small-signal equations for the time-dependent terms are obtained similar to those in (12.2.5). The set of small-signal equations can be solved to give a relationship between the test laser signal and the pump signal. The optical modulation response, $M(\omega) = s(\omega)/s_p(\omega)$, can be written in a normalized form

$$\frac{M(\omega)}{M(0)} = \frac{\omega_r^2}{\omega_r^2 - \omega^2 - i\omega\gamma} \quad \text{(Optical modulation)} \tag{12.3.9}$$

where the major parameters are

$$\omega_r^2 = \frac{(\mathrm{v}_g g'/\chi)S_0}{\tau_p(1+\varepsilon S_0)}\left(1 + \frac{\varepsilon}{\mathrm{v}_g g' \tau_w}\right) \tag{12.3.10}$$

$$\gamma = \frac{1}{\chi\tau_w} + \frac{(v_g g'/\chi)S_0}{1+\varepsilon S_0}\left(1 + \frac{\varepsilon\chi}{v_g g'\tau_p}\right) \tag{12.3.11}$$

$$\chi = 1 + \frac{\tau_{bw}}{\tau_{wb}}. \tag{12.3.12}$$

For a small-signal electrical injection,

$$I(t) = I_0 + \mathrm{Re}[i(\omega)e^{-i\omega t}] \tag{12.3.13}$$

with no external pump, $S_p(t) = 0$. The resulting modulation response is given by [9, 21] $M(\omega) = s(\omega)/i(\omega)$, which can be written in the form

$$\frac{M(\omega)}{M(0)} = \frac{\omega_r^2}{(1 - i\omega\tau_{bw})(\omega_r^2 - \omega^2 - i\omega\gamma)} \quad \text{(Electrical modulation)} \tag{12.3.14}$$

where τ_{bw} is a constant that contributes to low-frequency roll-off. The roll-off in the modulation response derived above is equivalent to the roll-off caused by a simple electrical parasitic, such as a capacitance in parallel to the device. However, in real devices, the effective capacitance may be bias-dependent because the device capacitance is related to the storage of charge in the forward-biased junction. If the transport effect dominates, the roll-off frequency of the transport effect is $1/(2\pi\tau_{bw})$. In this case, we can measure the carrier diffusion and capture time of the QW. Because the stimulated emission occurs in the QWs, the carrier lifetime in the SCH or barrier region is usually very long, so we set $\tau_{bw}/\tau_b \ll 1$ in deriving (12.3.14).

The modulation responses for the optical and electrical injection modulations are (12.3.9) and (12.3.14), respectively. The comparison of the two modulations is clear. The relaxation frequency ω_r and damping factor γ for both cases are the same. The transport factor $\chi = 1 + \tau_{bw}/\tau_{wb}$ has also been introduced to the equations. The only difference between the two responses is that optical injection has no low frequency roll-off, which corresponds to transport effects. There would also be a difference when the number of wells is varied. Thus, optical modulation gives an intrinsic modulation response of a laser because the injected pump light is chosen such that the pump light is absorbed only in the well region. For optical injection, the carriers are injected directly into the QW region, which removes the requirement of carrier transport from the SCH region to the QW region before lasing. In other words, optical injection removes the low-frequency roll-off caused by carriers transporting through SCH region. However, the SCH and QW regions are still coupled together. There are always carriers being captured by and escaping from the QWs. The rate of capture and escape will influence the modulation response intrinsically. Therefore, because of the SCH structure, carrier diffusion capture–escape still affects the optical modulation response, which comes in through the χ factor. Thus, the effective differential gain is reduced by a factor of χ for the same photon density. This reduction is present even in the absence of low-frequency roll-off in the optical injection, which results in the reduction of the resonance frequency. The effective carrier recombination lifetime in the well is also increased by a factor of χ, and the nonlinear gain suppression coefficient remains unchanged. Another important factor in high-speed modulation is the K factor, which is the slope of the damping factor versus relaxation frequency squared

$$\gamma \simeq \frac{1}{\tau} + Kf_r^2 \tag{12.3.15}$$

$$K = 4\pi^2\left(\tau_p + \frac{\varepsilon}{v_g g' \chi}\right). \tag{12.3.16}$$

The K factor is usually used to determine the maximum possible modulation bandwidth $f_{max} = 2\pi\sqrt{2}/K$. A small K factor means a large laser bandwidth. Even if we use optical injection modulation, the carrier capture and escape processes still increase the K factor and limit the maximum modulation. The above parameters characterize the intrinsic modulation response of semiconductor lasers.

The high-speed modulation response of a buried heterostructure MQW DFB laser was measured. The active region consists of 7 InGaAsP (−1.6% compressive strain) quantum wells with a well width of 70 Å and a photoluminescence (PL) wavelength of 1.5564 μm. The barriers are 100 Å InGaAsP lattice-matched to InP with a PL wavelength peak at 1.255 μm. Therefore, the pump light at 1.3 μm will be absorbed in the QWs. More details are described in Ref. 32. A schematic diagram of the modulation response experiments is shown in Fig. 12.6. The laser is held at constant temperature and measurements were made from 20°C to 40°C at a 5°C interval. In the electrical modulation experiments, shown as path A in Fig. 12.6, the HP 8510 network analyzer provides a small microwave signal (0 dBm) at frequencies swept from 45 MHz to 10 GHz, which is coupled to the test laser electrodes through a bias-T and a high-speed probe. The laser light is coupled to a fiber using a lensed

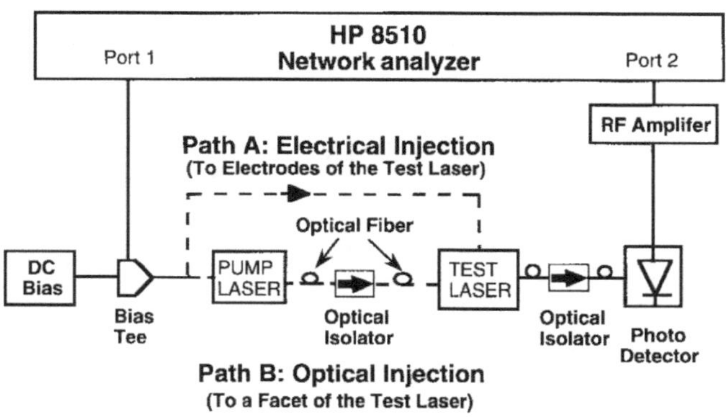

Figure 12.6 Schematic diagram of electrical and optical modulation experiment. The test laser is a quantum-well laser operating at 1.55 μm wavelength. Path A is for electrical injection with a microwave modulation. Path B is for optical injection with a microwave modulation using a 1.3-μm wavelength pump laser. (Reprinted with permission from [32] © 1999 IEEE.)

fiber-optical interface and travels through an optical isolator before being measured by a high-speed (29 GHz) photodetector. The small signal response is increased by an 18-dB-gain RF amplifier before entering the network analyzer, which measures the magnitude of the modulation response. The data are averaged to reduce noise. The optical response measurements shown as path B in Fig. 12.6 are similar to electrical modulation experiments, except that a 1.3-μm pump laser with a wide bandwidth is modulated electrically, and its modulated optical output is injected into the test DFB laser facet. The modulation of the pump laser is recorded at a fixed bias and stored for the calibration of the optical response of the 1.55-μm test laser. The pump light is completely absorbed in the test laser, so the light coupled out of the test laser does not require additional filtering of the pump light.

The electrical and optical modulation responses are first compared at a fixed temperature of 20°C. Figure 12.7 shows the least-square fits of the measured modulation responses using the theory derived above for electrical and optical injections. The fits are used to extract the differential gain and nonlinear gain suppression factor. By comparing the two responses, the roll-off frequency can be obtained more accurately. Both modulation responses show clearly that the relaxation peaks increase in frequency as the dc bias is increased. The modulation responses due to optical injection exhibit a slower roll-off at high frequencies, which indicates the absence of low-frequency pole on the optical response. Also, the peak of the optical pump response is generally higher than that of the electrical, and the differences between the two responses increase with increasing current. For 35-mA current injection, electrical modulation has a clear roll-off. The roll-off frequency is 3.52 GHz at this temperature (20°C). The experimental data [32] also show that the roll-off frequency is relatively independent of the current bias in the temperature range 20°C to 40°C.

In Fig. 12.8a, the relaxation frequency squared is plotted versus optical power. The slope of the linear fit will be used later to extract the differential gain. The damping factor is calculated using extracted parameters from the modulation responses and is plotted in Fig. 12.8b. The slope of the linear fit at large relaxation frequency is the K

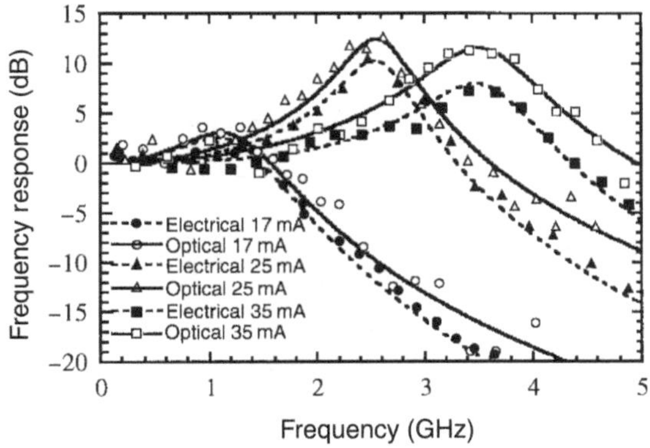

Figure 12.7 Comparison of the modulation responses of the electrical and optical injection. The solid (optical modulation) and dashed (electrical modulation) curves are theoretical results and the symbols are experimental data. (Reprinted with permission from [32] © 1999 IEEE.)

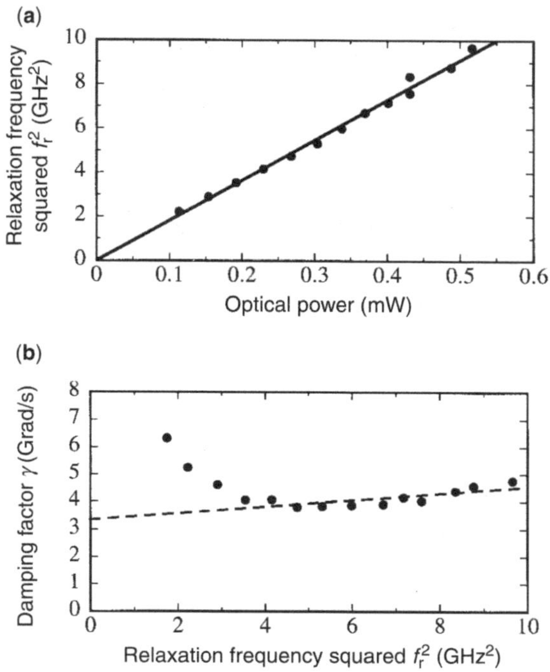

Figure 12.8 (a) The relaxation frequency squared versus optical output power of the test laser at 20°C. The solid line is the theoretical result, and the symbols are experimental data extracted from the modulation responses. (b) The damping factor versus the relaxation frequency squared. The dashed line is the theoretical curve (ignoring the spontaneous emission coupling), and its slope determines K. (Reprinted with permission from [32] © 1999 IEEE.)

factor. The deviation at the low-frequency end between the data and the linear dashed line is due to the negligence of the spontaneous emission term. If we include the spontaneous emission factor, the damping factor should have an additional term that can be neglected at high photon density (corresponding to a large relaxation frequency). The complete set of extracted parameters at 20°C is listed in Ref. 32. The differential gain and nonlinear gain suppression coefficient are within the values typically found in the literature. More experimental data on high-speed modulation of strained and unstrained quantum well lasers using InGaAsP and InAlGaAs materials and their dependence on temperature are reported in Ref. 46. When the external pump laser wavelength is tuned in the gain region of the test laser, the optical modulation becomes a cross-gain modulation [47], and enhancement of the modulation bandwidth has also been observed and analyzed theoretically. In addition, injection locking [52, 53] occurs when the detuning wavelength is very close to the test laser wavelength. A further increase of the modulation bandwidth by a significant factor has been realized [54–57] under injection locking condition. For example, a bandwidth of 66 GHz of a vertical cavity surface-emitting laser (VCSEL) at 1.55-μm wavelength using a cascaded injection locking approach has been realized [57].

A direct modulation response at 40 Gb/s bandwidth using a 1.3-μm DFB ridge-waveguide laser has been reported [48]. The device has a short cavity length of 100 μm using a MQW InAlGaAs active layer, which has a high differential gain. The laser operates at 60°C under a 40 Gb/s direct modulation.

12.4 SEMICONDUCTOR LASER SPECTRAL LINEWIDTH AND THE LINEWIDTH ENHANCEMENT FACTOR

The spectral properties of semiconductor lasers have been investigated since the early 1980s. Experiments by Fleming and Mooradian [58] showed that the laser spectral linewidth has a Lorentzian shape and the linewidth is inversely proportional to the optical output power. However, the magnitude of the linewidth was much larger than they had expected based on the conventional formula from the Townes–Schawlow theory [59]. A model proposed by Henry [60, 61], which has the same form as that derived by Haken in 1963 [62, 63], explained the phenomena by noting that the semiconductor laser is similar to a detuned oscillator, and there is a spectrum linewidth enhancement by a factor $(1+\alpha_e^2)$ due to the coupling between the amplitude and phase fluctuations of the optical field. A more formal derivation was given by Vahala and Yariv [64]. In this section, we present the model of Henry. The linewidth of a semiconductor laser with an output power P_{out} and a mirror transmission loss α_m at a photon energy $\hbar\omega$ is

$$\Delta\nu = \frac{\hbar\omega v_g \alpha_m R_{sp}}{8\pi P_{out}}(1+\alpha_e^2) \tag{12.4.1}$$

where v_g is the group velocity, R_{sp} is the spontaneous emission rate (1/s), and the linewidth enhancement factor α_e is defined as:

$$\alpha_e = \frac{\partial n'/\partial n}{\partial n''/\partial n} = -\frac{4\pi}{\lambda}\frac{\partial n'/\partial n}{\partial g/\partial n}. \tag{12.4.2}$$

Here n' and n'' are the real and imaginary parts of the refractive index due to the carrier injection into the active region, and n is the carrier density. $\partial g/\partial n$ is the differential gain of the semiconductor laser at the lasing wavelength λ.

12.4.1 Basic Equations for the Optical Intensity and Phase in the Presence of Spontaneous Emission

Consider an optical field $E(z, t)$ given by

$$E(z,t) = E(t)e^{ikz-i\omega t} \tag{12.4.3a}$$

$$E(t) = \sqrt{I(t)}e^{i\phi(t)} \tag{12.4.3b}$$

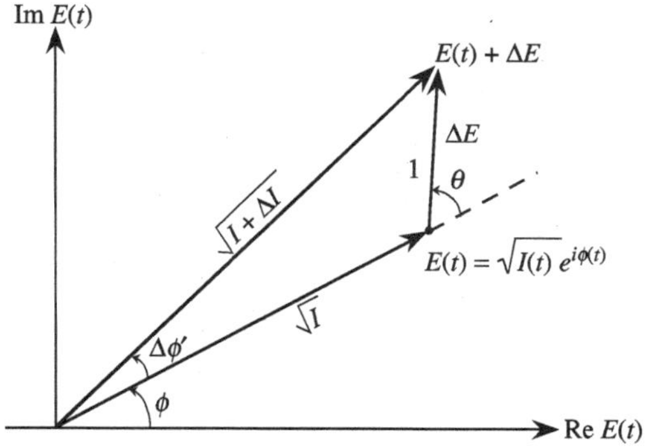

Figure 12.9 A plot on the complex optical electric field domain [60] $E(t) = \sqrt{I(t)}e^{i\phi(t)}$ showing its magnitude $\sqrt{I(t)}$ and phase ϕ can be changed by a spontaneous emission of a photon (magnitude is one because the intensity I has been normalized to represent the photon number in the cavity) with a phase change $\Delta\phi'$.

where $I(t)$ represents the intensity and $\phi(t)$ the phase of laser field. We assume that the optical intensity $I(t) = E(t)E^*(t)$ has been normalized such that it represents the average number of photons in the laser cavity. The time-dependent magnitude $E(t)$ is a complex phasor assuming its time variation is much slower than the optical frequency ω. The phasor $E(t)$ is plotted in the complex plane as a vector with a magnitude $\sqrt{I(t)}$ and a phase $\phi(t)$ as shown in Fig. 12.9. The basic assumption is that a random spontaneous emission alters $E(t)$ by ΔE, which adds a unit magnitude (one photon) and a phase θ, which is random

$$\Delta E = e^{i(\phi+\theta)}. \tag{12.4.4}$$

There are two contributions to the phase change $\Delta\phi$,

$$\Delta\phi = \Delta\phi' + \Delta\phi'' \tag{12.4.5}$$

where $\Delta\phi'$ is due to the out-of-phase component of ΔE and $\Delta\phi''$ is due to the intensity change, which is coupled to the phase change. To obtain the first contribution $\Delta\phi'$, we note from Fig. 12.9, that $\sqrt{I}\Delta\phi' \simeq \sin\theta$ or

$$\Delta\phi' = \frac{\sin\theta}{\sqrt{I}}. \tag{12.4.6}$$

To obtain the second contribution $\Delta\phi''$ due to the intensity change, we start from the wave equation

$$\frac{\partial^2}{\partial z^2}E(z, t) = \frac{1}{c^2}\frac{\partial^2}{\partial t^2}\varepsilon_r E(z, t) \tag{12.4.7}$$

where c is speed of light in free space and ε_r is the relative permittivity $\varepsilon_r = \varepsilon/\varepsilon_0$ of the semiconductor. We obtain

$$\frac{2i\omega}{c^2}\varepsilon_r\frac{\partial E(t)}{\partial t} = -\left(\frac{\omega^2}{c^2}\varepsilon_r - k^2\right)E(t) \tag{12.4.8}$$

neglecting the second-order derivative $\partial^2 E(t)/\partial t^2$ because $E(t)$ is a slowly varying function. We can also express ε_r in terms of the complex refractive index

$$\sqrt{\varepsilon_r} = n' + in'' \text{ (or } n + i\kappa) \tag{12.4.9}$$

and the real and imaginary parts of the propagation constants give

$$k = \frac{\omega}{c}n' \qquad \frac{1}{2}(g - \alpha) = -\frac{\omega}{c}n'' \tag{12.4.10}$$

where g is the gain coefficient and α is the absorption coefficient of the optical intensity. At threshold, the gain is balanced by the absorption, $g = \alpha$, $n'' = 0$, and ε_r is real. A change in carrier density n will cause n' and n'' to deviate from the threshold values

$$\varepsilon_r = (n' + \Delta n' + i\Delta n'')^2 \simeq n'^2 + 2in'\Delta n''(1 - i\alpha_e) \tag{12.4.11}$$

where we have defined a linewidth enhancement factor α_e as the ratio of the change in the real part of the refractive index to the change in the imaginary part

$$\alpha_e = \frac{\Delta n'}{\Delta n''}. \tag{12.4.12}$$

Therefore

$$\begin{aligned}\frac{\partial E}{\partial t} &= -\frac{\omega\Delta n''}{n'}(1 - i\alpha_e)E(t)\\ &= \frac{g-\alpha}{2}\mathrm{v_g}(1 - i\alpha_e)E(t)\end{aligned} \tag{12.4.13}$$

where the group velocity $\mathrm{v_g} = c/n'$ has been used, and we ignore the dispersion effect for simplicity. If we include the effect of the medium dispersion, (12.4.8) has to be modified [60], and the result for (12.4.13) still holds with the group velocity given by $\mathrm{v_g} = c/(n' + \omega\partial n'/\partial\omega)$, which is derived using $k = \omega n'/c$ and $\mathrm{v_g} = 1/(\partial k/\partial\omega)$. Substituting $E(t) = \sqrt{I(t)}e^{i\phi(t)}$ into (12.4.13) and separating the real and imaginary parts, we find

$$\frac{1}{2}\frac{dI}{dt} = \frac{g-\alpha}{2}\mathrm{v_g}I(t) \tag{12.4.14a}$$

$$\frac{d\phi}{dt} = -\frac{g-\alpha}{2}\mathrm{v_g}\alpha_e. \tag{12.4.14b}$$

Therefore, we obtain

$$\frac{d\phi}{dt} = -\frac{\alpha_e}{2I}\frac{dI}{dt}. \tag{12.4.15}$$

Initially at $t = 0$, $I(0) = I + \Delta I$, and at $t = \infty$, the relaxation oscillations die out, $I(\infty) = I$, we obtain

$$\Delta\phi'' = \frac{\alpha_e}{2I}\Delta I = \frac{\alpha_e}{2I}(1 + 2\sqrt{I}\cos\theta) \tag{12.4.16}$$

which can be derived from the three sides of the triangle in Fig. 12.9. The total phase change is then

$$\Delta\phi = \Delta\phi' + \Delta\phi'' = \frac{\alpha_e}{2I}\Delta I + \frac{1}{\sqrt{I}}(\sin\theta + \alpha_e\cos\theta). \tag{12.4.17}$$

The ensemble average of the spontaneous emission events at a time duration t is contributed from the constant term $(\alpha_e/2I)$ multiplied by the total number of the events, $R_{sp}t$,

$$\langle\Delta\phi\rangle = \frac{\alpha_e}{2I}R_{sp}t \tag{12.4.18}$$

as $\langle\sin\theta\rangle = \langle\cos\theta\rangle = 0$. Here R_{sp} is the spontaneous emission rate (1/s). Equation (12.4.18) gives an angular frequency shift

$$\Delta\omega = \frac{d}{dt}\langle\Delta\phi\rangle = \frac{\alpha_e}{2I}R_{sp}. \tag{12.4.19}$$

The total phase fluctuation for $R_{sp}t$ spontaneous events gives the variance

$$\begin{aligned}\langle\Delta\phi^2\rangle &= \left\langle\frac{1}{I}(\sin\theta + \alpha_e\cos\theta)^2\right\rangle \\ &= \frac{(1+\alpha_e^2)}{2I}R_{sp}|t|.\end{aligned} \tag{12.4.20}$$

We put an absolute value for $|t|$ because $\langle\Delta\phi^2\rangle$ is a positive quantity. Therefore, we found the mean $\langle\Delta\phi\rangle$ and its variance $\langle(\Delta\phi)^2\rangle$ or $\langle\Delta\phi^2\rangle$ as above.

12.4.2 Power Spectrum and Semiconductor Laser Spectral Linewidth

The power spectrum of the laser is the Fourier transform of the correlation function

$$\begin{aligned} W(\omega) &= \int_{-\infty}^{\infty} dt e^{i\omega t} \langle E^*(t)E(0)\rangle \\ &= \int_{-\infty}^{\infty} dt e^{i\omega t} \langle [I(t)I(0)]^{1/2} e^{-i\Delta\phi(t)}\rangle \\ &\simeq I(0) \int_{-\infty}^{\infty} dt e^{i\omega t} \langle e^{-i\Delta\phi(t)}\rangle \end{aligned} \tag{12.4.21}$$

where the small intensity fluctuation is neglected and the amplitude function for the field $E(t) = [I(t)]^{1/2}\ e^{i\phi(t)}$ has been used, which does not include the central frequency of the laser, that is, ω in (12.4.2), and

$$\Delta\phi(t) = \phi(t) - \phi(0).$$

Because the spontaneous emission events are random, the phase ϕ should have a Gaussian probability distribution function, $P(\Delta\phi) =$ a Gaussian function. The ensemble average for a Gaussian distribution is [61, 65]

$$\left\langle e^{-i\Delta\phi(t)}\right\rangle = \int_{-\infty}^{\infty} d(\Delta\phi) P(\Delta\phi) e^{-i\Delta\phi} = e^{-\frac{1}{2}\langle\Delta\phi^2\rangle}. \tag{12.4.22}$$

Using the result for the variance $\langle\Delta\phi^2\rangle$ in (12.4.20), we can define a coherence time as

$$\frac{1}{t_c} = \frac{R_{sp}}{4I}(1 + \alpha_e^2) \tag{12.4.23}$$

such that $\langle\Delta\phi^2\rangle/2 = |t|/t_c$. The power spectrum (12.4.21) gives a Lorentzian function

$$W(\omega) \simeq I(0)\frac{2t_c}{1 + \omega^2 t_c^2} \tag{12.4.24}$$

with a full width at half maximum (FWHM) of

$$\Delta\omega = \frac{2}{t_c} = \frac{R_{sp}}{2I}(1 + \alpha_e^2) \tag{12.4.25}$$

or

$$\Delta\nu = \frac{R_{sp}}{4\pi I}(1 + \alpha_e^2). \tag{12.4.26}$$

The number of photons I in the laser cavity (the photon density multiplied by the volume) is related to the optical output power by

$$P_{\text{out}} = \frac{1}{2}(I\hbar\omega)\text{v}_{\text{g}}\alpha_{\text{m}} \tag{12.4.27}$$

where $I\hbar\omega$ is the total photon energy and $\text{v}_{\text{g}}\alpha_{\text{m}}$ is the escaping rate of the photon out of the cavity with a length L where the mirror transmission loss is

$$\alpha_{\text{m}} = \frac{1}{L}\ln\left(\frac{1}{R}\right) \tag{12.4.28}$$

and R is the mirror reflectivity at both ends. We find

$$\Delta\nu = (\Delta\nu)_{\text{ST}}(1 + \alpha_{\text{e}}^2) \tag{12.4.29a}$$

$$(\Delta\nu)_{\text{ST}} = \frac{\hbar\omega\text{v}_{\text{g}}\alpha_{\text{m}}R_{\text{sp}}}{8\pi P_{\text{out}}} \tag{12.4.29b}$$

where $(\Delta\nu)_{\text{ST}}$ is the Schawlow–Townes limit of the laser linewidth. The expression (12.4.29a) is commonly used to explain the spectral linewidth of semiconductor lasers. It indicates a reduction of linewidth with increasing optical power and a broadening factor determined by $(1 + \alpha_{\text{e}}^2)$. The spontaneous emission rate (1/s) is related to the gain coefficient g (cm^{-1}) by a dimensionless spontaneous emission factor n_{sp}

$$R_{\text{sp}} = \text{v}_{\text{g}}g n_{\text{sp}} \tag{12.4.30}$$

where

$$n_{\text{sp}} = \frac{1}{1 - e^{(\hbar\omega - \Delta F)/k_{\text{B}}T}} \tag{12.4.31}$$

and $\Delta F = F_n - F_p$ is the separation of the quasi-Fermi levels between the electron and the hole. Here the photon number and the spontaneous emission rate are defined using the effective volume of the optical mode.

12.4.3 Linewidth Enhancement Factor in Semiconductor Lasers

Experimental data [66] for the laser linewidth, which depends on the optical output power with an inverse law, are shown in Fig. 12.10. The surprisingly large linewidth measured in this set of data was explained by Henry [60] using the correction factor $(1 + \alpha_{\text{e}}^2)$, where $\alpha_{\text{e}} \simeq 5$. More measurements have been done for various semiconductor lasers including index-guided double-heterostructure, unstrained and strained quantum-well lasers, with reduced linewidth enhancement factor [67, 68].

In a semiconductor laser, the injected carrier-induced refractive index change is associated with the change in the gain. The linewidth enhancement factor α_{e} in (12.4.2) can be directly expressed in terms of the ratio of the differential change of

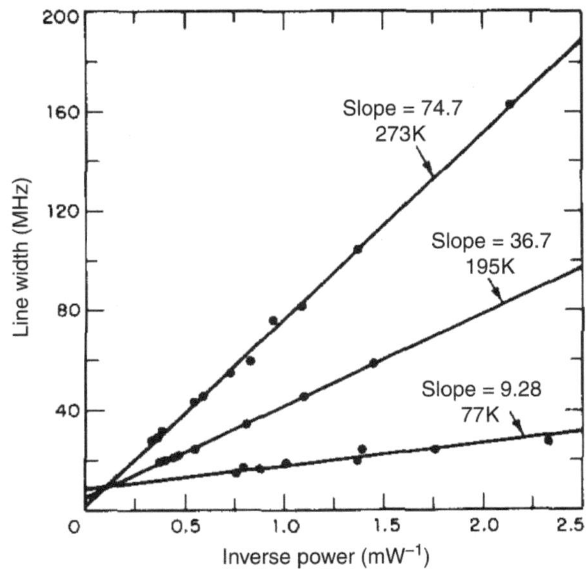

Figure 12.10 Semiconductor laser linewidth versus inverse power at three temperatures exhibiting the linear behavior. The magnitude of the large linewidth was explained [60] using the correction factor $1+\alpha_e^2$ with $\alpha_e \simeq 5$ at room temperature. (Reprinted with permission from [66] © 1982 American Institute of Physics.)

the refractive index per injected carrier and the differential gain. The linewidth enhancement factor varies from about 1.5 to 10 and is dependent on the lasing wavelength. Experimental data of α_e for two strained quantum-well lasers are shown in Fig. 12.11. The high-speed modulation characteristics of these lasers have been described in Fig. 12.4.

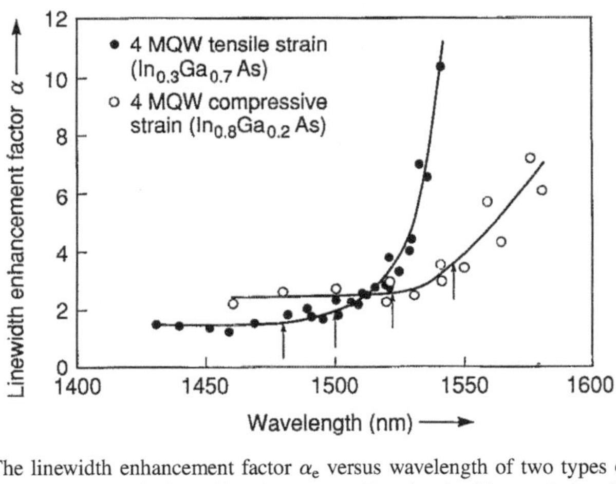

Figure 12.11 The linewidth enhancement factor α_e versus wavelength of two types of strained MQW lasers. The lasing wavelength is indicated by the arrows. (Reprinted with permission from [14] © 1991 American Institute of Physics.)

The dependence of α_e on the loss or the Fermi levels of semiconductor lasers has also been shown to be important [69]. For strained quantum wells, a theoretical analysis [70] shows that α_e can be reduced to about 1.1 using tensile strains with TM polarization of semiconductor quantum-well lasers. Similar behavior to the spectral broadening in semiconductor lasers also occurs in an intensity modulator using the electroabsorption effects [71–73]. This is because a phase modulation due to the change in the refractive index is associated with the change in the absorption coefficient. More recent measurements on strained quantum-well lasers [74] and quantum-dot lasers [75–78] have been reported. Quantum-dot lasers have extremely low chirp (close to zero). An approach to measure the linewidth enhancement factor based on the injection method has been demonstrated [79]. Measurements of the linewidth enhancement factors of type I and type II quantum-cascade lasers in 4 μm or 8 μm indicate low chirp as well [80–82]. The type I quantum-cascade lasers have a symmetrical (Lorentzian or Gaussian) lineshape near the gain peak, therefore, the refractive index change with increasing current near the peak gain wavelength is small (almost zero) due to the Kramers–Kronig relations. Thus the linewidth enhancement factor is small. Care should be taken such as using a pulsed measurement at low duty cycle to avoid the refractive index change caused by thermal effects [81].

12.5 RELATIVE INTENSITY NOISE SPECTRUM

In this section, we discuss the theory for the relative intensity noise (RIN) spectrum. The laser output light contains random fluctuations in intensity, phase, and frequency even when the laser is biased at a constant current without any current fluctuation. There are two major noise mechanisms. The dominant noise is due to spontaneous emission and the second is the electron–hole recombination or shot noise. Because spontaneous emission occurs at an extremely rapid rate, the intensity and phase of the laser exhibit fluctuations over a timescale as short as 100 ps. Light intensity fluctuation affects the signal to noise ratio (SNR), and the phase fluctuations lead to spectral linewidth broadening, as we have discussed in Section 12.4. The relative intensity noise is defined as [83, 84]

$$\mathrm{RIN} = \frac{\langle \delta P(t)^2 \rangle}{P_0^2} \tag{12.5.1}$$

where P_0 is the mean intensity of the laser output power and $\langle \delta P(t)^2 \rangle$ is mean squared power fluctuation. The rate equation (12.2.5) derived for the small-signal modulation can be applied by considering the noise sources as small fluctuations of the carrier density and photon density, $F_n(t)$ and $F_s(t)$, respectively.

$$\frac{d}{dt}\begin{bmatrix} \delta n(t) \\ \delta s(t) \end{bmatrix} + \begin{bmatrix} A & D \\ -C & B \end{bmatrix}\begin{bmatrix} \delta n(t) \\ \delta s(t) \end{bmatrix} = \begin{bmatrix} F_n(t) \\ F_s(t) \end{bmatrix}. \tag{12.5.2}$$

The solution in the Fourier frequency domain is written as:

$$\begin{bmatrix} -i\omega + A & D \\ -C & -i\omega + B \end{bmatrix} \begin{bmatrix} \delta n(\omega) \\ \delta s(\omega) \end{bmatrix} = \begin{bmatrix} F_n(\omega) \\ F_s(\omega) \end{bmatrix}. \tag{12.5.3}$$

We can define the normalized frequency response as

$$H(\omega) = \frac{M(\omega)}{M(0)} = \frac{\omega_r^2}{\Delta} \tag{12.5.4}$$

where we have used the determinant

$$\Delta = \begin{vmatrix} -i\omega + A & D \\ -C & -i\omega + B \end{vmatrix} = \omega_r^2 - \omega^2 - i\gamma\omega. \tag{12.5.5}$$

The solutions for the carrier density and photon density fluctuations are

$$\delta n(\omega) = \frac{H(\omega)}{\omega_r^2}[(-i\omega + B)F_n(\omega) - DF_s(\omega)] \tag{12.5.6}$$

$$\delta s(\omega) = \frac{H(\omega)}{\omega_r^2}[(-i\omega + A)F_s(\omega) + CF_n(\omega)]. \tag{12.5.7}$$

The source fluctuations are random and their expectation values are related by

$$\left\langle F_i(\omega)F_j(\omega)\right\rangle = \frac{1}{2\pi}\int_{-\infty}^{\infty}\left\langle F_i(\omega)F_j^*(\omega')\right\rangle d\omega'. \tag{12.5.8}$$

Using the Fourier transform of the power fluctuation

$$\langle \delta P(t)\rangle = \frac{1}{2\pi}\int_{-\infty}^{\infty} \delta P(\omega)e^{-i\omega t}d\omega \tag{12.5.9}$$

and the spectral density function

$$\langle \delta P(\omega)\delta P^*(\omega')\rangle = S_{\delta p}(\omega)2\pi\delta(\omega - \omega') \tag{12.5.10}$$

we obtain the spectral density function [83]

$$\begin{aligned} \left\langle \delta P(t)^2\right\rangle &= \frac{1}{2\pi}\int_{-\infty}^{\infty} S_{\delta p}(\omega')|B(\omega')|^2 d\omega' \\ &\simeq S_{\delta p}(\omega)\frac{1}{2\pi}\int_{-\infty}^{\infty}|B(\omega')|^2 d\omega' \\ &\simeq S_{\delta p}(\omega)2\Delta f \end{aligned} \tag{12.5.11}$$

where a narrow band pass filter $B(\omega') \simeq 1$ near $\omega' = \omega$ with a bandwidth of Δf is used. There is a factor of 2 because the spectral density function is an even function

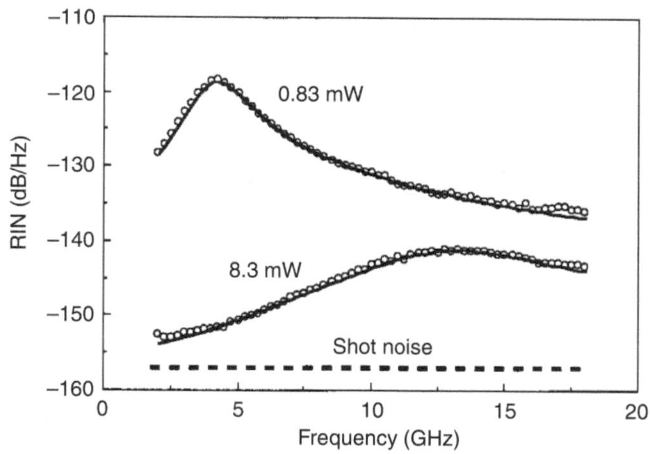

Figure 12.12 The relative intensity noise (RIN) spectrum as two bias currents above threshold. The symbols are experimental data and the solid curves are the theoretical fits. The shot noise background is also shown as the dashed line. (Reprinted with permission from [86] © 1992 IEEE.)

of the frequency. The RIN spectrum is thus given by

$$\frac{\mathrm{RIN}}{\Delta f} = \frac{2S_{\delta p}(\omega)}{P_0^2}. \tag{12.5.12}$$

Further derivations for the spectral density using (12.5.6) and (12.5.7) lead to [83, 85]

$$S_{\delta p}(\omega) = \hbar\omega P_0\left[(a_1 + a_2\omega^2)\frac{|H(\omega)|^2}{\omega_{\mathrm{r}}^4} + 1\right] \tag{12.5.13}$$

$$\frac{\mathrm{RIN}}{\Delta f} = \frac{2\hbar\omega}{P_0}\left[(a_1 + a_2\omega^2)\frac{|H(\omega)|^2}{\omega_{\mathrm{r}}^4} + 1\right]. \tag{12.5.14}$$

Typically, the last term (one) in the square bracket of (12.5.14) is much smaller than the leading two terms, and is thus negligible [85]. The expression requires essentially four parameters: a_1, a_2, ω_{r}, and γ, which can be extracted by fitting the experimental RIN spectrum at various bias current above threshold. Figure 12.12 shows an example of the RIN spectrum at two bias currents (symbols: experimental data; solid curves: theory) [86]. The extracted relaxation frequency from the peak of the RIN spectrum shows a linear dependence on the square root of the optical power output. The damping factor γ also shows a linear dependence on the squared relation frequency with the slope determined by the K factor. The RIN characteristics of a semiconductor laser under the injection locking condition shows a reduction in the RIN noise floor as well as an increase of the relaxation frequency controllable by the injection laser power and detuning [87].

PROBLEMS

12.1 (**a**) Calculate the photon lifetime for a semiconductor laser assuming the following parameters: $\alpha_i = 10\,\text{cm}^{-1}$, refractive index = 3.2, and the cavity length $L = 200\,\mu\text{m}$. (**b**) If we increase the intrinsic absorption coefficient to $20\,\text{cm}^{-1}$, what is the photon lifetime? (**c**) What if we increase the cavity length to 400 μm while keeping $\alpha_i = 10\,\text{cm}^{-1}$?

12.2 Find (**a**) the dc solutions for S_0 and n_0 and (**b**) the ac solutions for $s(t)$ and $n(t)$ in the linear gain theory.

12.3 Discuss how one may improve the relaxation frequency f_r of a semiconductor laser.

12.4 Derive an expression for the 3-dB angular frequency $\omega_{3\text{dB}}$ for the small-signal frequency response of a semiconductor laser using the linear gain theory.

12.5 Find the dc solutions for S_0 and n_0 in the nonlinear gain theory for the frequency response of a semiconductor laser.

12.6 Derive the small-signal ac solutions for $s(t)$ and $n(t)$ in the nonlinear gain theory.

12.7 Discuss the effect of decreasing the damping factor γ.

12.8 Plot the frequency response curve when the condition $2\omega_r^2 = \gamma^2$ is satisfied.

12.9 Derive the linear relation between the damping factor γ and the relaxation frequency squared f_r^2 in (12.2.11b). How would the parameters, K, and the inverse carrier lifetime affect the bandwidth of the frequency response?

12.10 Discuss the high-speed optical modulation vs. electrical modulation in terms of the physics of carrier transport, theoretical modulation response, and the experimental setup.

12.11 Derive (12.4.13) and (12.4.14) taking into account the material dispersion.

12.12 Derive (12.4.22) for a Gaussian distribution function $P(\Delta\phi)$.

12.13 Describe the factors determining the spectral linewidth of a semiconductor laser. How may one decrease the semiconductor laser linewidth in terms of the device parameters such as laser length, the laser threshold, and the power? How can one reduce the linewidth enhancement factor?

REFERENCES

1. T. Ikegami and Y. Suematsu, "Resonance-like characteristics of the direct modulation of a junction laser," *Proc. IEEE* **55**, 122–123 (1967).
2. A. Yariv, *Quantum Electronics*, 3rd ed., Wiley, New York, 1989.

3. K. Lau and A. Yariv, "High-frequency current modulation of semiconductor lasers," in W. T. Tsang, volume editor, *Lightwave Communications Technology*, vol. 22, Part B, Chapter 2, in R. K. Willardson and A. C. Beer, Eds., *Semiconductors and Semimetals*, Academic Press, New York, 1985.
4. K. Y. Lau, "Ultralow threshold quantum well lasers," Chapter 4, and "Dynamics of quantum well lasers," Chapter 5, in P. S. Zory, Jr., Ed., *Quantum Well Lasers*, Academic Press, San Diego, 1993.
5. T. P. Lee, C. A. Burrus, J. A. Copeland, A. G. Dentai, and D. Marcuse, "Short-cavity InGaAsP injection lasers: Dependence of mode spectra and single-longitudinal-mode power on cavity length," *IEEE J. Quantum Electron.* **QE-18**, 1101–1112 (1982).
6. K. Petermann, "Calculated spontaneous emission factor for double-heterostructure injection lasers with gain-induced waveguiding," *IEEE J. Quantum Electron.* **QE-15**, 566–570 (1979).
7. W. Streifer, D. R. Scifres, and R. D. Burnham, "Spontaneous emission factor of narrow-strip gain-guided diode lasers," *Electron. Lett.* **17**, 933 (1981).
8. N. K. Dutta, S. J. Wang, A. B. Piccirilli, R. F. Karlicek, Jr., R. L. Brown, M. Washington, U. K. Chakrabarti, and A. Gnauck, "Wide-bandwidth and high-power InGaAsP distributed feedback lasers," *J. Appl. Phys.* **66**, 4640–4644 (1989).
9. K. Uomi, M. Aoki, T. Tsuchiya, and A. Takai, "Dependence of high-speed properties on the number of quantum wells in 1.55 μm InGaAs-InGaAsP MQW $\lambda/4$-shifted DFB lasers," *IEEE J. Quantum Electron.* **29**, 355–360 (1993).
10. Y. Arakawa, K. Vahala, A. Yariv and K. Lau, "Enhanced modulation bandwidth of GaAlAs double heterostructure lasers in high magnetic fields: Dynamic response with quantum wire effects," *Appl. Phys. Lett.* **47**, 1142–1144 (1985).
11. Y. Arakawa and A. Yariv, "Quantum well lasers- gain, spectra, dynamics," *IEEE J. Quantum Electron.* **QE-22**, 1887–1899 (1986).
12. R. Olshansky, P. Hill, V. Lanzisera, and W. Powazinik, "Frequency response of 1.3 μm InGaAsP high speed semiconductor lasers," *IEEE J. Quantum Electron.* **QE-23**, 1410–1418 (1987).
13. R. Olshansky, P. Hill, V. Lanzisera, and W. Powazinik, "Universal relationship between resonant frequency and damping rate of 1.3 μm InGaAsP semiconductor lasers," *Appl. Phys. Lett.* **50**, 653–655 (1987).
14. L. F. Tiemeijer, P. J. A. Thijs, P. J. de Waard, J. J. M. Binsma, and T. V. Dongen, "Dependence of polarization, gain, linewidth enhancement factor, and K factor on the sign of the strain of InGaAs/InP strained-layer multiquantum well lasers," *Appl. Phys. Lett.* **58**, 2738–2740 (1991).
15. T. Fukushima, J. E. Bowers, R. A. Logan, T. Tanbun-Ek, and H. Temkin, "Effect of strain on the resonant frequency and damping factor in InGaAs/InP multiple quantum well lasers," *Appl. Phys. Lett.* **58**, 1244–1246 (1991).
16. S. D. Offsey, W. J. Schaff, L. F. Lester, L. F. Eastman, and S. K. McKernan, "Strained-layer InGaAs-GaAs-AlGaAs lasers grown by molecular beam epitaxy for high-speed modulation," *IEEE J. Quantum Electron.* **27**, 1455–1462 (1991).
17. H. Yasaka, K. Takahata, N. Yamamoto and M. Naganuma, "Gain saturation coefficients of strained-layer multiple quantum-well distributed feedback lasers," *IEEE Photon. Technol. Lett.* **3**, 879–882 (1991).

18. J. Zhou, N. Park, J. W. Dawson, K. Vahala, M. A. Newkirk, U. Koren, and B. I. Miller, "Highly nondegenerate four-wave mixing and gain nonlinearity in a strained multiple-quantum-well optical amplifier," *Appl. Phys. Lett.* **62**, 2301–2303 (1993).
19. S. R. Chinn, "Measurement of nonlinear gain suppression and four-wave mixing in quantum well lasers," *Appl. Phys. Lett.* **59**, 1673–1675 (1991).
20. W. Rideout, W. F. Sharfin, E. S. Koteles, M. O. Vassell, and B. Elman, "Well-barrier hole burning in quantum well lasers," *IEEE Photon. Technol. Lett.* **3**, 784–786 (1991).
21. A. Uskov, J. Mork, and J. Mark, "Theory of short-pulse gain saturation in semiconductor laser amplifiers," *IEEE Photon. Technol. Lett.* **4**, 443–446 (1992).
22. M. Grupen and K. Hess, "Severe gain suppression due to dynamic carrier heating in quantum well lasers," *Appl. Phys. Lett.* **70**, 808–810 (1997).
23. M. P. Kesler and E. P. Ippen, "Subpicosecond gain dynamics in GaAlAs laser diodes," *Appl. Phys. Lett.* **51**, 1765–1767 (1987).
24. Y. Matsui, H. Murai, S. Arahira, Y. Ogawa, and A. Suzuki, "Enhanced modulation bandwidth for strain-compensated InGaAlAs–InGaAsP MQW lasers," *IEEE J. Quantum Electron.* **34**, 1970–1978 (1998).
25. R. Nagarajan, T. Fukushima, S. W. Corzine, and J. E. Bowers, "Effects of carrier transport on high-speed quantum well lasers," *Appl. Phys. Lett.* **59**, 1835–1837 (1991).
26. R. Nagarajan, T. Fukushima, M. Ishikawa, J. E. Bowers, R. S. Geels, and L. A. Coldren, "Transport limits in high-speed quantum-well lasers: experiment and theory," *Photon. Technol. Lett.* **4**, 121–123 (1992).
27. R. Nagarajan, M. Ishikawa, and J. E. Bowers, "Effect of carrier transport on relative intensity noise and critique of K factor predictions of modulation response," *Electron. Lett.* **28**, 845–848 (1992).
28. R. Nagarajan, M. Ishikawa, T. Fukushima, R. Geels, and J. Bowers, "High speed quantum-well lasers and carrier transport effects," *IEEE J. Quantum Electron.* **28**, 1990–2008 (1992).
29. M. Ishikawa, R. Nagarajan, T. Fukushima, J. Wasserbauer, and J. Bowers, "Long wavelength high-speed semiconductor laser with carrier transport effects," *IEEE J. Quantum Electron.* **28**, 2230–2241 (1992).
30. R. Nagarajan and J. E. Bowers, "Effects of carrier transport on injection efficiency and wavelength chirping in quantum-well lasers," *IEEE J. Quantum Electron.* **29**, 1601–1608 (1993).
31. A. P. Wright, B. Garrett, G. H. B. Thompson, and J. E. A. Whiteaway, "Influence of carrier transport on wavelength chirp of InGaAs/InGaAsP MQW lasers," *Electron. Lett.* **28**, 1911–1913 (1992).
32. T. Keating, X. Jin, S. L. Chuang, and K. Hess, "Temperature dependence of electrical and optical modulation responses of quantum-well lasers," *IEEE J. Quantum Electron.* **35**, 1526–1534 (1999).
33. X. Jin and S. L. Chuang, "Microwave modulation of a quantum-well laser with and without external optical injection," *IEEE Photon. Technol. Lett.* **13**, 648–650 (2001).
34. E. Meland, R. Holmstrom, J. Schlafer, R. B. Lauer, and W. Powazinik, "Extremely high-frequency (24 GHz) InGaAsP diode lasers with excellent modulation efficiency," *Electron. Lett.* **26**, 1827–1829 (1990).
35. P. W. M. Blom, J. E. M. Haverkort, P. J. van Hall, and J. H. Wolter, "Carrier-carrier scattering induced capture in quantum-well lasers," *Appl. Phys. Lett.* **62**, 1490–1492 (1993).

36. D. Morris, B. Deveaud, A. Regreny, and P. Auvray, "Electron and hole capture in multiple-quantum-well structures," *Phys. Rev. B* **47**, 6819–6822 (1993).

37. S. C. Kan, D. Vassilovski, T. C. Wu, and K. Y. Lau, "On effects of carrier diffusion and quantum capture in high speed modulation of quantum well lasers," *Appl. Phys. Lett.* **61**, 752–754 (1992).

38. S. C. Kan, D. Vassilovski, T. C. Wu, and K. Y. Lau, "Quantum capture limited modulation bandwidth of quantum well, wire, and dot lasers," *Appl. Phys. Lett.* **62**, 2307–2309 (1993).

39. D. Klotzkin, K. Kamath, and P. Bhattacharya, "Quantum capture times at room temperature in high-speed InGaAs–GaAs self-organized quantum-dot lasers," *IEEE Photon. Technol. Lett.* **9**, 1301–1303 (1997).

40. R. Olshansky, P. Hill, V. Lanzisera, and W. Powazink, "Frequency response of 1.3 μm InGaAsP high speed semiconductor lasers," *IEEE J. Quantum Electron.* **QE-23**, 1410–1418 (1987).

41. H. Lu, C. Blaauw, B. Benyon, G. P. Li, and T. Makino, "High-power and high-speed performance of 1.3-μm strained MQW gain-coupled DFB lasers," *IEEE J. Sel. Top. Quantum Electron.* **1**, 375–381 (1995).

42. C. B. Su, J. Eom, C. H. Lange, C. B. Kim, R. B. Lauer, W. C. Rideout, and J. S. LaCourse, "Characterization of the dynamics of semiconductor lasers using optical modulation," *IEEE J. Quantum Electron.* **28**, 118–127 (1992).

43. K. J. Vahala and M. A. Newkirk, "Parasitic-free modulation of semiconductor lasers," *IEEE J. Quantum Electron.* **25**, 1393–1398 (1989).

44. N. Tessler, M. Margalit, R. Ben Michael, M. Orenstein, and G. Eisenstein, "Modulation of quantum well lasers by short optical excitation: energy and spatial dependent effects," *IEEE Photon. Technol. Lett.* **7**, 23–25 (1995).

45. D. Vassilovski, T. C. Wu, S. Kan, K. Y. Lau, and C. E. Zah, "Unambiguous determination of quantum capture, carrier diffusion, and intrinsic effect in quantum-well laser dynamics using wavelength-selective optical modulation," *IEEE Photon. Technol. Lett.* **7**, 706–708 (1995).

46. G. Liu and S. L. Chuang, "High-speed modulation of long-wavelength $In_{1-x}Ga_xAs_yP_{1-y}$ and $In_{1-x-y}Ga_xAl_yAs$ strained quantum-well lasers," *IEEE J. Quantum Electron.* **37**, 1283–1291 (2001).

47. X. Jin, T. Keating, and S. L. Chuang, "Small signal cross-gain modulation in semiconductors," *IEEE J. Quantum Electron.* **36**, 1485–1492 (2000).

48. K. Nakahara, T. Tsuchiya, T. Kitatani, K. Shinoda, T. Taniguchi, T. Kikawa, M. Aoki, and M. Mukaikubo, "40-Gb/s direct modulation with high extinction ratio operation of 1.3-μm InGaAlAs multiquantum well ridge waveguide distributed feedback lasers," *Photon. Technol. Lett.* **19**, 1436–1438 (2007).

49. C. E. Zah, R. Bhat, B. N. Pathak, F. Favire, W. Lin, M. C. Wang, N. C. Andreadakis, D. M. Hwang, M. A. Koza, T. P. Lee, Z. Wang, D. Darby, D. Flanders, and J. J. Hsieh, "High-performance uncooled 1.3-μm AlGaInAs/InP strained-layer quantum-well laser for subscriber loop applications," *IEEE J. Quantum Electron.* **30**, 511–523 (1994).

50. S. Kakimoto and H. Watanabe, "Threshold current, differential gain, and relaxation resonance frequency of 1.55-μm bulk and MQW DFB lasers diodes," *IEEE J. Quantum Electron.* **34**, 1231–1239 (1998).

51. S. J. Sweeney, A. F. Philips, A. R. Adams, E. P. O'Reilly, and P. J. A. Thijs, "The effect of temperature dependent processes on the performance of 1.5-μm compressively strained InGaAs(P) MQW semiconductor diode lasers," *IEEE Photon. Technol. Lett.* **10**, 1076–1078 (1998).
52. R. Lang, "Injection locking properties of a semiconductor laser," *IEEE J. Quantum Electron.* **18**, 976–983 (1982).
53. T. B. Simpson and J. M. Liu, "Enhanced modulation bandwidth in injection-locked semiconductor lasers," *IEEE Photon. Technol. Lett.* **9**, 1322–1324 (1997).
54. X. J. Meng, T. Chau, and M. C. Wu, "Experimental demonstration of modulation bandwidth enhancement in distributed-feedback lasers with external light injection," *Electron. Lett.* **34**, 2031–2032 (1998).
55. X. Jin and S. L Chuang, "Bandwidth enhancement of Fabry-Perot quantum-well lasers by injection-locking," *Solid State Electron.* **50**, 1141–1149 (2006).
56. L. Chrostowski, X. Zhao, and C. J. Chang-Hasnain, "Microwave performance of optically injection-locked VCSELs," *IEEE Trans. Microwave Theory Techniq.* **54**, 788 (2006).
57. X. Zhao, D. Parekh, E. K. Lau, H. K. Sung, M. C. Wu, W. Hofmann, M. C. Amann, and C. J. Chang-Hasnain, "Novel cascaded injection-locked 1.55-μm VCSELs with 66 GHz modulation bandwidth," *Optics Express* **15**, 14810–14816 (2007).
58. M. W. Fleming and A. Mooradian, "Fundamental line broadening of single-mode (GaAl)As diode lasers," *Appl. Phys. Lett.* **38**, 511 (1981).
59. A. L. Schawlow and C. H. Townes, "Infrared and optical masers," *Phys. Rev.* **112**, 1940–1949 (1958).
60. C. H. Henry, "Theory of the linewidth of semiconductor lasers," *IEEE J. Quantum Electron.* **QE-18**, 259–264 (1982).
61. C. H. Henry, "Spectral properties of semiconductor lasers," Chapter 3, in W. T. Tsang, Vol. Ed., *Lightwave Communications Technology*, vol. 22 Part B, in R. K. Willardson and A. C. Beer, Eds., *Semiconductors and Semimetals*, Academic Press, New York, 1985.
62. E. Haken and H. Haken, "Zur theorie des halbleiter lasers," *Z. Phys.* **176**, 421–428 (1963).
63. H. Haug and H. Haken, "Theory of noise in semiconductor laser emission," *Z. Phys.* **204**, 262–275 (1967).
64. K. Vahala and A. Yariv, "Semiclassical theory of semiconductor lasers- Part I," *IEEE J. Quantum Electron.* **QE-19**, 1096–1101 (1983).
65. M. Lax, "Classical noise. V. Noise in self-sustained oscillators," *Phys. Rev.* **160**, 290–307 (1967).
66. D. Welford and A. Mooradian, "Output power and temperature dependence of the linewidth of single-frequency cw (GaAl)As diode lasers," *Appl. Phys. Lett.* **40**, 865–867 (1982).
67. Y. Asai, J. Ohya, and M. Ogura, "Spectral linewidth and resonant frequency characteristics of InGaAsP/InP multiquantum well lasers," *IEEE J. Quantum Electron.* **25**, 662–667 (1989).
68. N. K. Dutta, H. Temkin, T. Tanbun-Ek, and R. Logan, "Linewidth enhancement factor for InGaAs/InP strained quantum well lasers," *Appl. Phys. Lett.* **57**, 1390–1391 (1990).
69. Y. Arakawa and A. Yariv, "Fermi energy dependence of linewidth enhancement factor of GaAlAs buried heterostructure lasers," *Appl. Phys. Lett.* **47**, 905–907 (1985).
70. Y. Huang, S. Arai, and K. Komori, "Theoretical linewidth enhancement factor of $Ga_{1-x}In_xAs$/GaInAsP/InP strained-quantum-well structures," *IEEE Photon. Technol. Lett.* **5**, 142–145 (1993).

71. F. Koyama and K. Iga, "Frequency chirping of external modulation and its reduction," *Electron. Lett.* **21**, 1065–1066 (1985).

72. Y. Noda, M. Suzuki, Y. Kushiro, and S. Akiba, "High-speed electroabsorption modulator with strip-loaded GaInAsP planar waveguide," *J. Lightwave Technol.* **LT-4**, 1445–1453 (1986).

73. T. H. Wood, "Multiple quantum well (MQW) waveguide modulators," *J. Lightwave Technol.* **6**, 743–757 (1988).

74. C. S. Chang, S. L. Chuang, J. Minch, Y. K. Chen, and T. Tanbun-Ek, "Amplified spontaneous emission spectroscopy in strained quantum-well lasers," *IEEE J. Sel. Topics Quantum Electron.* **1**, 1100–1107 (1995).

75. P. Kondratko, S. L. Chuang, G. Walter, T. Chung, and N. Holonyak, Jr., "Observation of near-zero linewidth enhancement factor in a quantum-well coupled quantum-dot laser," *Appl. Phys. Lett.* **83**, 4818–4820 (2003).

76. J. Kim, H. Su, S. Minin, and S. L. Chuang, "Comparison of linewidth enhancement factor between p-doped and undoped quantum-dot lasers," *IEEE Photon. Technol. Lett.* **18**, 1022–1024 (2006).

77. J. Kim and S. L. Chuang, "Theoretical and experimental study of optical gain, refractive index change, and linewidth enhancement factor of p-doped quantum-dot lasers," *IEEE J. Quantum Electron.* **42**, 942–952 (2006).

78. H. Su, L. Zhang, R. Wang, T. C. Newell, A. L. Gray, and L. F. Lester, "Linewidth study of InAs-InGaAs quantum dot distributed-feedback lasers," *Photon. Technol. Lett.* **16**, 2206–2208 (2004).

79. G. Liu, X. Jin, and S. L. Chuang, "Novel techniques for measurement of linewidth enhancement factors of strained QW lasers using injection locking," *IEEE Photon. Tech. Lett.* **13**, 430–432 (2001).

80. M. Lerttamrab, S. L. Chuang, C. Gmachl, D. L. Sivco, F. Capasso, and A. Y. Cho, "Linewidth enhancement factor of a type-I quantum-cascade laser," *J. Appl. Phys.* **94**, 5426–5428 (2003).

81. J. Kim, M. Lerttamrab, S. L. Chuang, C. Gmachl, D. L. Sivco, F. Capasso, and A. Y. Cho, "Theoretical and experimental study of linewidth enhancement factor of type-I quantum cascade lasers," *IEEE J. Quantum Electron.* **40**, 1663–1674 (2004).

82. M. Lerttamrab, S. L. Chuang, R. Q. Yang, and C. J. Hill, "Linewidth enhancement factor of a type-II interband-cascade laser," *J. Appl. Phys.* **96**, 3568–3570 (2004).

83. L. A. Coldren and S. Corzine, *Diode Lasers and Photonic Integrated Circuits*, Wiley, New York, 1995.

84. G. P. Agrawal, *Lightwave Technology—Components and Devices*, Wiley, New York, 2004.

85. Y. Yamamoto and N. Imoto, "Internal and external field fluctuations of a laser oscillator—Part I: quantum mechanical Langevin treatment," *IEEE J. Quantum Electron.* **22**, 2032–2042 (1986).

86. M. C. Tatham, I. F. Lealman, C. P. Seltzer, L. D. Westbrook, and D. M. Cooper, "Resonance frequency, damping, and differential gain in 1.5 μm multiple quantum-well lasers," *IEEE J. Quantum Electron.* **28**, 408–414 (1992).

87. X. Jin and S. L. Chuang, "Relative intensive noise characteristics of injection-locked semiconductor lasers," *Appl. Phys. Lett.* **77**, 1250–1252 (2000).

13

Electrooptic and Acoustooptic Modulators

In this chapter, we discuss electrooptic effects and modulators. The bulk electrooptic effects are discussed first, and their applications as amplitude and phase modulators are presented. These devices using waveguide structures are then shown. The basic idea is that the optical refractive index of electrooptic materials, such as $LiNbO_3$, KH_2PO_4, or GaAs, and ZnS semiconductors, can be changed by an applied electric field. Therefore, an incident optical field propagating through the crystal with a proper polarization experiences efficient electrooptic effects. The transmitted field changes in either phase or polarization, which can be used in the designs of phase modulators as well as amplitude modulators. We then discuss scattering of light by sound and present a coupled-mode analysis of acoustooptic modulators.

13.1 ELECTROOPTIC EFFECTS AND AMPLITUDE MODULATORS

To understand the electrooptic effects [1, 2], we consider a crystal described by the constitutive relation associating the displacement vector **D** to the electric field **E** by a permittivity tensor $\boldsymbol{\varepsilon}$,

$$\mathbf{D} = \boldsymbol{\varepsilon} \cdot \mathbf{E} \qquad \boldsymbol{\varepsilon} = \begin{bmatrix} \varepsilon_{xx} & \varepsilon_{xy} & \varepsilon_{xz} \\ \varepsilon_{yx} & \varepsilon_{yy} & \varepsilon_{yz} \\ \varepsilon_{zx} & \varepsilon_{zy} & \varepsilon_{zz} \end{bmatrix} \tag{13.1.1}$$

and the permeability is μ_0. Let us define the inverse of the permittivity tensor $\boldsymbol{\varepsilon}$ as **K**.

$$\mathbf{K} = \boldsymbol{\varepsilon}^{-1}. \tag{13.1.2}$$

The index ellipsoid of the crystal is described by

$$\varepsilon_0 \sum_{i,j=1}^{3} K_{ij} x_i x_j = 1 \tag{13.1.3}$$

where $x_1 = x$, $x_2 = y$, and $x_3 = z$ for convenience. For most crystals, the permittivity matrix $\boldsymbol{\varepsilon}$ is symmetric due to the symmetry property of the structure. Therefore, $\boldsymbol{\varepsilon}$ can

Physics of Photonic Devices, Second Edition. By Shun Lien Chuang
Copyright © 2009 John Wiley & Sons, Inc.

be diagonalized as

$$\boldsymbol{\varepsilon} = \begin{bmatrix} \varepsilon_x & 0 & 0 \\ 0 & \varepsilon_y & 0 \\ 0 & 0 & \varepsilon_z \end{bmatrix}. \tag{13.1.4}$$

The coordinate system in which $\boldsymbol{\varepsilon}$ is diagonalized is called the principal system. In this system, $\varepsilon_0 \mathbf{K}$ is a diagonal matrix with the diagonal elements equal to the reciprocals of the squared refractive indices of the three characteristic polarizations along the principal axes

$$\varepsilon_0 \mathbf{K} = \varepsilon_0 \begin{bmatrix} 1/\varepsilon_x & 0 & 0 \\ 0 & 1/\varepsilon_y & 0 \\ 0 & 0 & 1/\varepsilon_z \end{bmatrix} = \begin{bmatrix} 1/n_x^2 & 0 & 0 \\ 0 & 1/n_y^2 & 0 \\ 0 & 0 & 1/n_z^2 \end{bmatrix} \tag{13.1.5}$$

and the index ellipsoid (13.1.3) is

$$\varepsilon_0 \left(\frac{x_1^2}{\varepsilon_x} + \frac{x_2^2}{\varepsilon_y} + \frac{x_3^2}{\varepsilon_z} \right) = \frac{x^2}{n_x^2} + \frac{y^2}{n_y^2} + \frac{z^2}{n_z^2} = 1 \tag{13.1.6}$$

where $n_i = \sqrt{\varepsilon_i/\varepsilon_0}$, $i = x$, y, and z.

13.1.1 Electrooptic Effects

In a linear electrooptic material, the index ellipsoid is changed in the presence of an applied electric field $\mathbf{F}$, and K_{ij} becomes $K_{ij} + \Delta K_{ij}$, where the change ΔK_{ij} is linearly proportional to the electric field

$$\varepsilon_0 \Delta K_{ij} = \sum_{k=1}^{3} r_{ijk} F_k. \tag{13.1.7}$$

The linear electrooptic effect is also called the Pockels effect, after Friedrich Pockels [3] (1865–1913), who described it in 1893. These r_{ijk} coefficients are also called the Pockels coefficients or electrooptic coefficients.

Equation (13.1.7) can also be generalized to include the quadratic electrooptic effects, which are usually smaller than the linear effects,

$$\varepsilon_0 \Delta K_{ij} = \sum_{k=1}^{3} r_{ijk} F_k + \sum_{k,l=1}^{3} S_{ijkl} F_k F_l. \tag{13.1.8}$$

However, for materials with centrosymmetry, the index ellipsoid function must be an even function of the applied electric field because it must remain invariant upon the sign reversal of the electric field. Therefore, r_{ijk} vanishes, and the quadratic electrooptic effects dominate. The quadratic electrooptic effect is also called the Kerr effect [3], after John Kerr (1824–1907), who discovered the effect in 1875.

In this chapter, we focus on linear electrooptical effects. From the symmetry property of the crystal, the following matrix correspondence is defined, so-called the engineering notation:

$$\begin{bmatrix} 11 & 12 & 13 \\ & 22 & 23 \\ & & 33 \end{bmatrix} \longleftrightarrow \begin{bmatrix} 1 & 6 & 5 \\ & 2 & 4 \\ & & 3 \end{bmatrix}, \tag{13.1.9}$$

that is, we have $(ij) \leftrightarrow I = 1, 2, \ldots, 6$, and $r_{ijk} \equiv r_{Ik}$. For example, the matrix elements $(23) = (32) = 4$, $(13) = (31) = 5$, and $(12) = (21) = 6$. Note that r_{Ik} is a 6×3 matrix, and Eq. (13.1.7) can be rewritten as

$$\varepsilon_0 \begin{bmatrix} \Delta K_1 \\ \Delta K_2 \\ \Delta K_3 \\ \Delta K_4 \\ \Delta K_5 \\ \Delta K_6 \end{bmatrix} = \begin{bmatrix} r_{11} & r_{12} & r_{13} \\ r_{21} & r_{22} & r_{23} \\ r_{31} & r_{32} & r_{33} \\ r_{41} & r_{42} & r_{43} \\ r_{51} & r_{52} & r_{53} \\ r_{61} & r_{62} & r_{63} \end{bmatrix} \begin{bmatrix} F_x \\ F_y \\ F_z \end{bmatrix}. \tag{13.1.10}$$

We interchange $F_x = F_1$, $F_y = F_2$, and $F_z = F_3$ for convenience. Depending on the symmetry of the crystal, many of the matrix elements r_{Ik} may vanish. We usually refer to some databook [4] or reference tables [5–8] for the crystal symmetry and nonvanishing r_{Ik} values. A few important electrooptic materials are shown in Table 13.1 for illustration purposes.

Example The potassium dihydrogen phosphate (KDP or KH_2PO_4) crystal is uniaxial in the absence of an applied field

$$\varepsilon_0 \sum_{i,j=1}^{3} K_{ij} x_i x_j = \frac{x^2}{n_o^2} + \frac{y^2}{n_o^2} + \frac{z^2}{n_e^2} = 1 \tag{13.1.11}$$

where $n_x = n_y = n_o$, and $n_z = n_e$. With an applied field, we have $r_{Ik} = 0$ except for r_{63} and $r_{41} = r_{52}$. Note that $x_1 = x$, $x_2 = y$, $x_3 = z$, and use the mapping table in (13.1.9). Equation (13.1.10) gives

$$\varepsilon_0 \begin{bmatrix} \Delta K_1 \\ \Delta K_2 \\ \Delta K_3 \\ \Delta K_4 \\ \Delta K_5 \\ \Delta K_6 \end{bmatrix} = \begin{bmatrix} 0 & 0 & 0 \\ 0 & 0 & 0 \\ 0 & 0 & 0 \\ r_{41} & 0 & 0 \\ 0 & r_{52} & 0 \\ 0 & 0 & r_{63} \end{bmatrix} \begin{bmatrix} F_x \\ F_y \\ F_z \end{bmatrix} = \begin{bmatrix} 0 \\ 0 \\ 0 \\ r_{41}F_x \\ r_{52}F_y \\ r_{63}F_z \end{bmatrix}. \tag{13.1.12}$$

TABLE 13.1 A Few Electrooptic Materials with Their Parameters [1, 4, 6, 9]

Point-Group Symmetry	Material	Refractive Index n_o	Refractive Index n_e	Wavelength, λ_0 (μm)	Nonzero Electrooptic Coefficients (10^{-12} m/V)
3m	$LiNbO_3$	2.297	2.208	0.633	$r_{13} = r_{23} = 8.6$, $r_{33} = 30.8$ $r_{42} = r_{51} = 28$, $r_{22} = 3.4$ $r_{12} = r_{61} = -r_{22}$
32	Quartz (SiO_2)	1.544	1.553	0.589	$r_{41} = -r_{52} = 0.2$ $r_{62} = r_{21} = -r_{11} = 0.93$
$\bar{4}2m$	KH_2PO_4 (KDP)	1.5115	1.4698	0.546	$r_{41} = r_{52} = 8.77$, $r_{63} = 10.3$
		1.5074	1.4669	0.633	$r_{41} = r_{52} = 8$, $r_{63} = 11$
$\bar{4}2m$	$NH_4H_2PO_4$ (ADP)	1.5266	1.4808	0.546	$r_{41} = r_{52} = 23.76$, $r_{63} = 8.56$
		1.5220	1.4773	0.633	$r_{41} = r_{52} = 23.41$, $r_{63} = 7.828$
$\bar{4}2m$	KD_2PO_4 (KD*P)	1.5079	1.4683	0.546	$r_{41} = r_{52} = 8.8$, $r_{63} = 26.8$
$\bar{4}2m$	GaAs	3.60	$= n_o$	0.9	$r_{41} = r_{52} = r_{63} = 1.1$
		3.42	$= n_o$	1.0	$r_{41} = r_{52} = r_{63} = 1.5$
		3.34	$= n_o$	10.6	$r_{41} = r_{52} = r_{63} = 1.6$
$\bar{4}3m$	InP	3.29	$= n_o$	1.06	$r_{41} = r_{52} = r_{63} = 1.45$
		3.20	$= n_o$	1.35	$r_{41} = r_{52} = r_{63} = 1.3$
$\bar{4}3m$	ZnSe	2.60	$= n_o$	0.633	$r_{41} = r_{52} = r_{63} = 2.0$
$\bar{4}3m$	β-ZnS	2.36	$= n_o$	0.6	$r_{41} = r_{52} = r_{63} = 2.1$

Therefore, $\varepsilon_0 \Delta K_4 = r_{41} F_x$, $\varepsilon_0 \Delta K_5 = r_{52} F_y$, and $\varepsilon_0 \Delta K_6 = r_{63} F_z$

$$\varepsilon_0 \sum_{i,j=1}^{3} \Delta K_{ij} x_i x_j = 2r_{41} F_x yz + 2r_{52} F_y xz + 2r_{63} F_z xy \tag{13.1.13}$$

where a factor of 2 accounts for the symmetry of the matrix elements. Therefore, the new index ellipsoid

$$\varepsilon_0 \sum_{i,j=1}^{3} (K_{ij} + \Delta K_{ij}) x_i x_j = 1 \tag{13.1.14}$$

becomes

$$\frac{x^2}{n_o^2} + \frac{y^2}{n_o^2} + \frac{z^2}{n_e^2} + 2r_{41} F_x yz + 2r_{52} F_y xz + 2r_{63} F_z xy = 1. \tag{13.1.15}$$

In general, the above index ellipsoid may not have the principal axes along the x, y, or z directions any more, as will be shown in the following examples.

13.1.2 Longitudinal Amplitude Modulator

Consider the setup as in Fig. 13.1 with the applied electric field along the propagation direction (z) of light $\mathbf{F} = \hat{z}F_z$. The index ellipsoid is

$$\frac{x^2}{n_o^2} + \frac{y^2}{n_o^2} + \frac{z^2}{n_e^2} + 2r_{63}F_z xy = 1. \qquad (13.1.16)$$

The above equation shows that the principal axes along the x and y directions are rotated because of the cross term, $2r_{63}F_z xy$. We have to find the new principal optical axes such that the index ellipsoid can be described by a diagonal $(K + \Delta K)_{ij}$ matrix. A coordinate rotation of 45° on the x–y plane gives

$$\begin{aligned} x &= \frac{1}{\sqrt{2}}(x' + y') \\ y &= \frac{1}{\sqrt{2}}(-x' + y'). \end{aligned} \qquad (13.1.17)$$

Substituting (13.1.17) into (13.1.16), we find

$$x'^2\left(\frac{1}{n_o^2} - r_{63}F_z\right) + y'^2\left(\frac{1}{n_o^2} + r_{63}F_z\right) + \frac{z^2}{n_e^2} = 1. \qquad (13.1.18)$$

We may rewrite the index ellipsoid as

$$\frac{x'^2}{n_{x'}^2} + \frac{y'^2}{n_{y'}^2} + \frac{z^2}{n_e^2} = 1 \qquad (13.1.19)$$

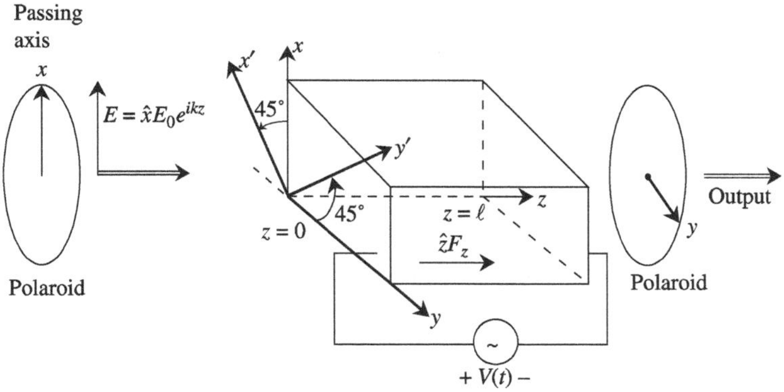

Figure 13.1 A longitudinal amplitude modulator in which an applied electric field is biased along the direction of optical wave propagation.

where

$$n_{x'} = \frac{n_o}{(1 - n_o^2 r_{63} F_z)^{1/2}} \simeq n_o + \frac{\Delta n}{2}, \quad n_{y'} \simeq n_o - \frac{\Delta n}{2}$$

$$\Delta n = n_{x'} - n_{y'} = n_o^3 r_{63} F_z. \tag{13.1.20}$$

We note that in the new coordinate system $(x'-y'-z)$, the matrices $(K + \Delta K)_{ij}$ and ε_{ij} are diagonalized

$$\varepsilon_0(\mathbf{K} + \Delta\mathbf{K}) = \begin{bmatrix} 1/n_{x'}^2 & 0 & 0 \\ 0 & 1/n_{y'}^2 & 0 \\ 0 & 0 & 1/n_e^2 \end{bmatrix}, \quad \boldsymbol{\varepsilon} = \begin{bmatrix} \varepsilon_{x'} & 0 & 0 \\ 0 & \varepsilon_{y'} & 0 \\ 0 & 0 & \varepsilon_z \end{bmatrix} \tag{13.1.21}$$

where $\varepsilon_{x'} = n_{x'}^2 \varepsilon_0$, $\varepsilon_{y'} = n_{y'}^2 \varepsilon_0$ and $\varepsilon_z = n_e^2 \varepsilon_0$. The incident optical field, after passing through the polarizer, can be expressed as

$$\mathbf{E} = \hat{x} E_0 e^{ikz} = \left(\frac{\hat{x}' + \hat{y}'}{\sqrt{2}}\right) E_0 e^{ikz} \tag{13.1.22}$$

where $k = \omega/c = 2\pi/\lambda_0$, and λ_0 is the wavelength in free space. Upon hitting the surface at $z = 0$, the wave is decomposed into two orthonormal polarizations along the $\hat{x}'$ and $\hat{y}'$ directions; each satisfies all of the Maxwell's equations independently because each is a characteristic polarization of the crystal. The propagation constants of the $\hat{x}'$- and $\hat{y}'$-polarized components are $kn_{x'}$ and $kn_{y'}$, respectively. Neglecting the reflections at the surfaces $z = 0$ and $z = \ell$, the optical field inside the crystal can be written as

$$\mathbf{E} = \left(\hat{x}' e^{ikn_{x'}z} + \hat{y}' e^{ikn_{y'}z}\right) \frac{E_0}{\sqrt{2}}. \tag{13.1.23}$$

The transmitted field passing through the second polaroid at $z = \ell$ is the y component of $\mathbf{E}$ in (13.1.23), or $(\hat{y} \cdot \mathbf{E})$ using

$$\hat{x}' = \frac{\hat{x} - \hat{y}}{\sqrt{2}} \qquad \hat{y}' = \frac{\hat{x} + \hat{y}}{\sqrt{2}} \tag{13.1.24}$$

$$\begin{aligned} E_y &= \left(-e^{ikn_{x'}\ell} + e^{ikn_{y'}\ell}\right) \frac{E_0}{2} \\ &= (-i \sin\phi) e^{ikn_o\ell} E_0 \end{aligned} \tag{13.1.25}$$

where the angle is defined as

$$\phi = \frac{k\ell}{2}(n_{x'} - n_{y'}) = \frac{k\ell}{2}\Delta n = \frac{k\ell}{2} n_o^3 r_{63} F_z. \tag{13.1.26}$$

The transmitted power intensity divided by the incident power intensity is proportional to

$$T = \frac{P_t}{P_i} = \frac{|E_y|^2}{|E_0|^2} = \sin^2 \phi(t). \tag{13.1.27}$$

Noting that $F_z(t)\ell = V(t)$ is the applied voltage, we define V_π, the voltage yielding a phase difference of π between the two characteristic polarizations, that is, $k\ell(n_{x'} - n_{y'}) = \pi$,

$$V_\pi = \frac{\pi}{kn_o^3 r_{63}} = \frac{\lambda_0}{2n_o^3 r_{63}} \tag{13.1.28}$$

and obtain

$$\phi(t) = \frac{\pi}{2}\frac{V(t)}{V_\pi}. \tag{13.1.29}$$

By varying $V(t)$, the output light intensity is modulated. To obtain a linear response, $V(t)$ has to be biased at a dc voltage $= V_\pi/2$ (so $\phi = \pi/4$) where the transmission factor is 50% as shown in Fig. 13.2, that is,

$$V(t) = \frac{V_\pi}{2} + v_0 \sin \omega_m t \tag{13.1.30}$$

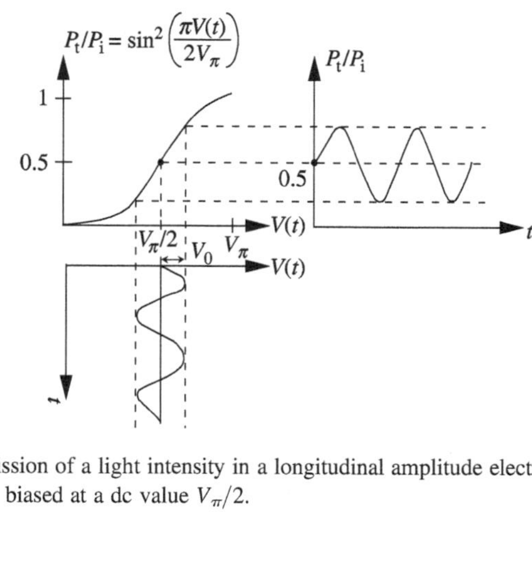

Figure 13.2 Transmission of a light intensity in a longitudinal amplitude electrooptical modulator. The applied voltage $V(t)$ is biased at a dc value $V_\pi/2$.

and the transmission factor is

$$\begin{aligned} T &= \sin^2\left(\frac{\pi}{4}+\frac{\pi v_0}{2V_\pi}\sin\omega_{\mathrm{m}}t\right) \\ &= \frac{1}{2}\left[1+\sin\left(\frac{\pi v_0}{V_\pi}\sin\omega_{\mathrm{m}}t\right)\right] \\ &\simeq \frac{1}{2}\left[1+\frac{\pi v_0}{V_\pi}\sin\omega_{\mathrm{m}}t\right]. \end{aligned} \tag{13.1.31}$$

We see that for a small input signal $\pi v_0 \ll V_\pi$, and the bias at the 50% point, $V_\pi/2$, a linear response can be achieved. However, because V_π is typically very large, a better way is to add a quarter-wave-plate between the electrooptic crystal and the output polaroid, with the two principal axes of the plate along the x' and y' directions such that an extra phase difference of $\pi/2$ is introduced between the x' and y' components of (13.1.23)

$$\mathbf{E} = \left(\hat{x}' e^{i\pi/2} e^{ikn_{x'}\ell} + \hat{y}' e^{ikn_{y'}\ell}\right)\frac{E_0}{\sqrt{2}} \tag{13.1.32}$$

$$\frac{P_{\mathrm{t}}}{P_{\mathrm{i}}} = \frac{|E_y|^2}{|E_0|^2} = \sin^2\left(\frac{\pi}{4}+\frac{k\Delta n\ell}{2}\right) = \sin^2\left(\frac{\pi}{4}+\frac{\pi V(t)}{2V_\pi}\right). \tag{13.1.33}$$

In this case, the modulation voltage is $V(t) = v_0 \sin\omega_{\mathrm{m}}t$ instead of (13.1.30). No dc bias is necessary.

13.1.3 Transverse Amplitude Modulator

A transverse amplitude modulator is shown in Fig. 13.3 in which the applied field is biased in a direction perpendicular to the propagation direction of light. The incident

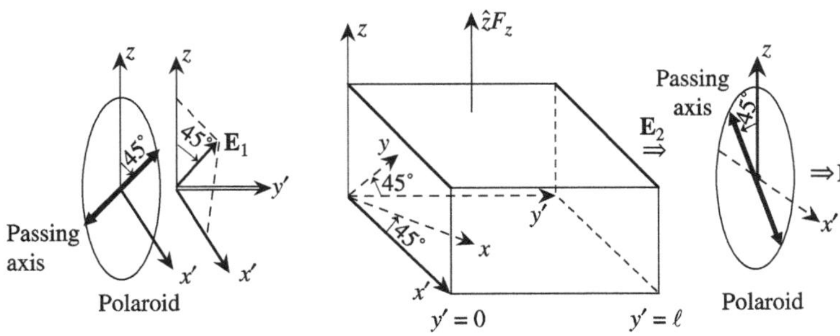

Figure 13.3 A transverse amplitude modulator. The bias field $\hat{z}F_z$ is perpendicular to the direction of light wave propagation.

optical electric field after passing through the polaroid is

$$\mathbf{E}_1 = \left(\frac{\hat{x}' + \hat{z}}{\sqrt{2}}\right) E_0 e^{iky'} \tag{13.1.34}$$

propagating in free space. Note that the direction of propagation has been chosen to be along the y' direction of the electrooptic crystal, which makes an angle of 45° with the principal x and y axes of the unbiased crystal. After passing through the crystal, the x' and z components gain different phases at $y' = \ell$

$$\mathbf{E}_2 = \left(\hat{x}' e^{ikn_{x'}\ell} + \hat{z} e^{ikn_e\ell}\right) \frac{E_0}{\sqrt{2}}. \tag{13.1.35}$$

Here, we still use $LiNbO_3$ crystal in our analysis. The transmitted field through the second polaroid with the passing axis given by $(-\hat{x}' + \hat{z})/\sqrt{2}$ is

$$E_t = \mathbf{E}_2 \cdot \left(\frac{-\hat{x}' + \hat{z}}{\sqrt{2}}\right) = \left(-e^{ikn_{x'}\ell} + e^{ikn_e\ell}\right) \frac{E_0}{2}. \tag{13.1.36}$$

Define

$$\phi = \frac{k\ell}{2}(n_{x'} - n_e) = \frac{k\ell}{2}(n_o - n_e) + \frac{k\ell}{4}\Delta n \tag{13.1.37}$$

where $\Delta n = n_o^3 r_{63} F_z$. We obtain the transmission factor

$$\frac{P_t}{P_i} = \frac{|E_t|^2}{|E_0|^2} = \sin^2\phi. \tag{13.1.38}$$

Here the time-dependent field $F_z(t)$ modulates the phase $\phi(t)$, which determines the output light intensity as a function of time. Because $F_z(t) = V(t)/d$, where d is the thickness of the crystal, we can define a half-wave voltage V_π as the voltage required to introduce an extra phase shift of π

$$V_\pi = \frac{2\pi}{kn_o^3 r_{63}}\left(\frac{d}{\ell}\right) = \frac{\lambda_0}{r_{63} n_o^3}\left(\frac{d}{\ell}\right). \tag{13.1.39}$$

The factor d/ℓ can be chosen to be small; therefore, the half-wave voltage V_π is reduced compared with that for a longitudinal amplitude modulator. We write

$$\phi = \frac{k\ell}{2}(n_o - n_e) + \frac{\pi}{2}\frac{V(t)}{V_\pi}. \tag{13.1.40}$$

A linear response is obtainable if we choose

$$\frac{k\ell}{2}(n_o - n_e) = \frac{\pi}{4} \tag{13.1.41}$$

and the input signal $V(t) = v_0 \sin \omega_m t$ is small, $v_0 \ll V_\pi$. The transmission factor is the same as (13.1.33) and is similar to Fig. 13.2.

13.2 PHASE MODULATORS

13.2.1 Optical Phase Modulation

Consider an incident optical field propagating along the z direction and passing through the polaroid as shown in Fig. 13.4.

$$\mathbf{E} = \hat{x}' E_0 e^{ikz}. \tag{13.2.1}$$

Because the polaroid has been aligned such that the passing axis is along one of the characteristic polarizations of the electrooptic crystal, $\hat{x}'$, the transmitted field at $z = \ell$ is simply

$$\mathbf{E}_t = \hat{x}' E_0 e^{ikn_{x'}\ell} = \hat{x}' E_0 e^{ikn_o\ell} e^{ik\Delta n\ell/2}. \tag{13.2.2}$$

Therefore, if we modulate the applied electric field as

$$F_z(t) = F_0 \sin \omega_m t \tag{13.2.3}$$

we find

$$\mathbf{E}_t = \hat{x}' E_0 e^{ikn_o\ell} e^{i\delta \sin \omega_m t} \tag{13.2.4}$$

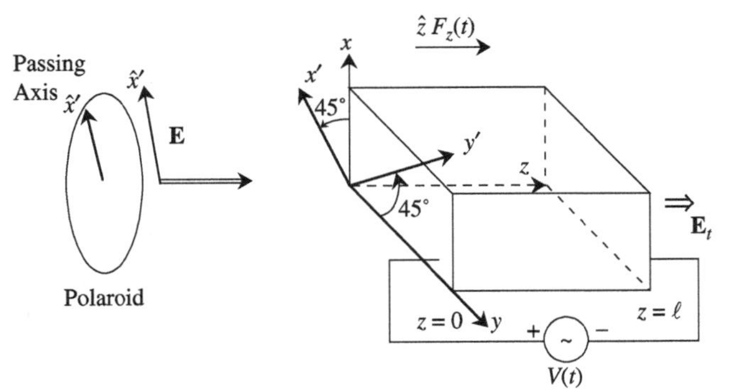

Figure 13.4 A longitudinal phase modulator with an applied electric field $\hat{z}F_z(t)$ along the propagation direction of optical field **E**.

where

$$\delta = \frac{kn_o^3 r_{63} F_0 \ell}{2}. \tag{13.2.5}$$

The electric field in the time domain is

$$\begin{aligned}\mathbf{E}_t(\ell, t) &= \text{Re}[\mathbf{E}_t e^{-i\omega t}] \\ &= \hat{x}' E_0 \cos(kn_o\ell + \delta \sin \omega_m t - \omega t).\end{aligned} \tag{13.2.6}$$

The phase of the output optical field is modulated by the factor $\delta \sin \omega_m t$. If we use the mathematical identity [10]

$$e^{i\delta \cos \phi} = \sum_{m=-\infty}^{\infty} J_m(\delta) e^{im\phi} e^{im\pi/2} \tag{13.2.7}$$

or, equivalently,

$$e^{i\delta \sin \phi} = \sum_{m=-\infty}^{\infty} J_m(\delta) e^{im\phi}. \tag{13.2.8}$$

Equation (13.2.6) can also be written as

$$\begin{aligned}\mathbf{E}_t(\ell, t) &= \text{Re}[\hat{x}' E_0 e^{i(kn_o\ell - \omega t)} e^{i\delta \sin \omega_m t}] \\ &= \text{Re}[\hat{x}' E_0 e^{i(kn_o\ell - \omega t)} \sum_{m=-\infty}^{\infty} J_m(\delta) e^{im\omega_m t}] \\ &= \hat{x}' E_0 \sum_{m=-\infty}^{\infty} J_m(\delta) \cos[kn_o\ell - (\omega - m\omega_m)t].\end{aligned} \tag{13.2.9}$$

The output optical field contains, in addition to the fundamental frequency ω with an amplitude $J_0(\delta)E_0$, various sidebands with frequencies, $\omega \pm \omega_m$, $\omega \pm 2\omega_m, \ldots,$ and so forth, with corresponding amplitudes $\mp J_1(\delta)E_0$, $J_2(\delta)E_0, \ldots$. Note that $J_{-m}(\delta) = (-1)^m J_m(\delta)$. If the magnitude $\delta \simeq 2.4048$, the root of the zeroth-order Bessel function, $J_0(\delta) = 0$, all of the power in the fundamental frequency ω is transferred to the nonzero-order harmonics.

Example $LiNbO_3$ has a 3m point-group symmetry and its r_{Ik} matrix has nonzero elements [1, 11]: $r_{12} = -r_{22}$, $r_{23} = r_{13}$, r_{33}, $r_{42} = r_{51}$, $r_{61} = -r_{22}$ from Table 13.1 and their r-values at a wavelength $\lambda_0 = 0.633$ μm are

$$r_{13} = 8.6 \times 10^{-12}\,\text{m/V}, \qquad r_{33} = 30.8 \times 10^{-12}\,\text{m/V}$$

$$r_{22} = 3.4 \times 10^{-12}\,\text{m/V}, \qquad r_{51} = 28.0 \times 10^{-12}\,\text{m/V}.$$

The refractive indices of $LiNbO_3$ have a uniaxial form

$$n_x = n_y = n_o = 2.297 \quad \text{and} \quad n_z = n_e = 2.208.$$

For an applied electric field $\mathbf{F} = \hat{x}F_x + \hat{y}F_y + \hat{z}F_z$, the new index ellipsoid is derived as follows:

$$\varepsilon_0 \begin{bmatrix} \Delta K_1 \\ \Delta K_2 \\ \Delta K_3 \\ \Delta K_4 \\ \Delta K_5 \\ \Delta K_6 \end{bmatrix} = \begin{bmatrix} 0 & r_{12} & r_{13} \\ 0 & r_{22} & r_{23} \\ 0 & 0 & r_{33} \\ 0 & r_{42} & 0 \\ r_{51} & 0 & 0 \\ r_{61} & 0 & 0 \end{bmatrix} \begin{bmatrix} F_x \\ F_y \\ F_z \end{bmatrix} = \begin{bmatrix} r_{12}F_y + r_{13}F_z \\ r_{22}F_y + r_{23}F_z \\ r_{33}F_z \\ r_{42}F_y \\ r_{51}F_x \\ r_{61}F_x \end{bmatrix} \tag{13.2.10}$$

$$x^2\left(\frac{1}{n_o^2} + r_{12}F_y + r_{13}F_z\right) + y^2\left(\frac{1}{n_o^2} + r_{22}F_y + r_{23}F_z\right) + z^2\left(\frac{1}{n_e^2} + r_{33}F_z\right)$$
$$+ 2r_{42}F_y yz + 2r_{51}F_x xz + 2r_{61}F_x xy = 1. \tag{13.2.11}$$

Because r_{33} is the largest coefficient, an applied electric field along the z direction will be most efficient for the electrooptic control. Therefore, for $F_x = F_y = 0$, the index ellipsoid is given by

$$x^2\left(\frac{1}{n_o^2} + r_{13}F_z\right) + y^2\left(\frac{1}{n_o^2} + r_{23}F_z\right) + z^2\left(\frac{1}{n_e^2} + r_{33}F_z\right) = 1 \tag{13.2.12}$$

which is of the form

$$\frac{x^2}{n_x^2} + \frac{y^2}{n_y^2} + \frac{z^2}{n_z^2} = 1 \tag{13.2.13}$$

where the new refractive indices are

$$n_x = n_y \simeq n_o - \frac{1}{2}n_o^3 r_{13}F_z \tag{13.2.14a}$$

$$n_z \simeq n_e - \frac{1}{2}n_e^3 r_{33}F_z. \tag{13.2.14b}$$

13.2.2 X-cut $LiNbO_3$ Crystal

For an X-cut $LiNbO_3$ crystal, as shown in Fig. 13.5a, two electrodes are placed symmetrically on both sides of the waveguides such that the bias field $\mathbf{F} = \hat{z}F_z$ is along the z direction, and the index ellipsoid is described by (13.2.13). An incident optical electric field with TE polarization will transmit as

$$\mathbf{E} = \hat{z}E_0 e^{ikn_z y}\big|_{y=\ell} = \hat{z}E_0 e^{ikn_e \ell} e^{-ikn_e^3 r_{33}F_z \ell/2}. \tag{13.2.15a}$$

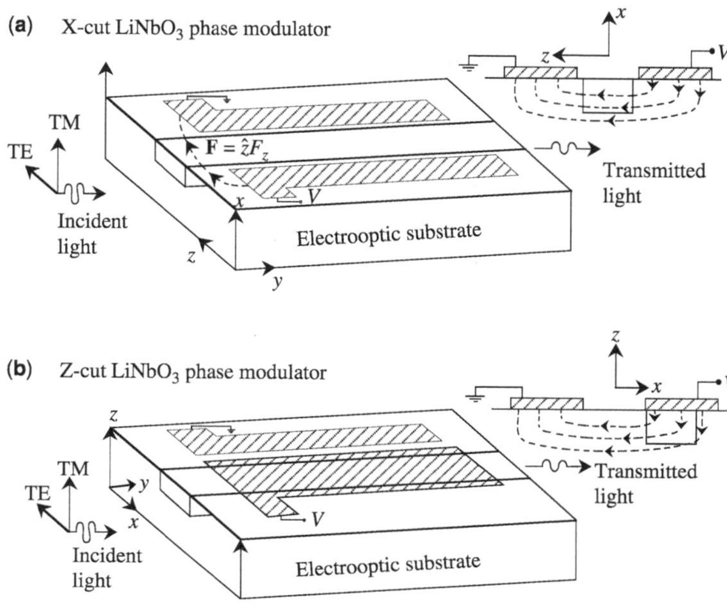

Figure 13.5 Electrooptic phase modulator using (a) an X-cut $LiNbO_3$ substrate, where the electrodes are placed symmetrically on both sides of the waveguide, such that the bias field is along the z direction; and (b) a Z-cut $LiNbO_3$ substrate, where one electrode is placed directly above the waveguide such that the bias field in the waveguide is along the positive (or negative) z direction for the most efficient phase modulation, as r_{33} is the largest electrooptic coefficient.

Similarly, for TM polarization,

$$\mathbf{E} = \hat{x}E_0 e^{ikn_x y}\big|_{y=\ell} = \hat{x}E_0 e^{ikn_o\ell} e^{-ikn_o^3 r_{13} F_z \ell/2}. \qquad (13.2.15b)$$

Therefore, TE polarization should be used for most efficient phase modulation as $r_{33} > r_{13}$.

13.2.3 Z-cut $LiNbO_3$ Crystal

For a Z-cut $LiNbO_3$ crystal as shown in Fig. 13.5b, the electrodes are placed such that the waveguide is below one of the two electrodes where the field is perpendicular to the Z-cut surface and $\mathbf{F} = \hat{z}F3$. The optical transmission field will be

$$\mathbf{E} = \hat{x}E_0 e^{ikn_o\ell} e^{-ikn_o^3 r_{13} F_z \ell/2} \quad \text{(TE polarization)} \qquad (13.2.16a)$$

and

$$\mathbf{E} = \hat{z}E_0 e^{ikn_e\ell} e^{-ikn_e^3 r_{33} F_z \ell/2} \quad \text{(TM polarization).} \qquad (13.2.16b)$$

In this case, TM polarization is preferred for most efficient phase modulation. For more discussions on the polarization and electrode designs, including the TE and TM polarization conversions, see Refs. 11–16.

13.3 ELECTROOPTIC EFFECTS IN WAVEGUIDE DEVICES

In Chapter 8, we discussed optical directional couplers using parallel waveguides. Here we discuss briefly the applications of electrooptic effects in waveguide structures. We have discussed KDP and $LiNbO_3$ as examples for electrooptic amplitude and phase modulators. Some of these materials such as $LiNbO_3$ as electrooptic crystals have been used in many commercial devices. Here, we will discuss GaAs as an example for electrooptic waveguide devices. For integrated optoelectronics, semiconductor materials, especially III–V compounds, are attractive because many active and passive components, such as semiconductor lasers, photodetectors, and field effect transistors, are made of these compound semiconductors. However, considerable research work is still necessary for integrating passive and active devices with desired operation characteristics.

Example GaAs at 10.6 μm wavelength. $n_o = 3.34$, $r_{41} = r_{52} = r_{63} = 1.6 \times 10^{-12}$ m/V. All other r components are zero. For a biased field $\boldsymbol{F} = \hat{x}F_x + \hat{y}F_y + \hat{z}F_z$, the new index ellipsoid is, following (13.1.12)–(13.1.15),

$$\frac{x^2}{n_o^2} + \frac{y^2}{n_o^2} + \frac{z^2}{n_o^2} + 2r_{63}F_x yz + 2r_{63}F_y xz + 2r_{63}F_z xy = 1. \tag{13.3.1}$$

If we choose $F_x = F_y = 0$, $\mathbf{F} = \hat{z}F_z$, we then have

$$\frac{x^2}{n_o^2} + \frac{y^2}{n_o^2} + 2r_{63}F_z xy + \frac{z^2}{n_o^2} = 1. \tag{13.3.2}$$

This is similar to that of the KDP materials, except that the crystal in the absence of field is isotropic, or $n_e = n_o$. Again a rotation of 45° in the x–y plane gives $x = (x' + y')/\sqrt{2}$, $y = (-x' + y')/\sqrt{2}$,

$$\frac{x'^2}{n_{x'}^2} + \frac{y'^2}{n_{y'}^2} + \frac{z'^2}{n_o^2} = 1 \tag{13.3.3}$$

where

$$n_{x'} \simeq n_o + \frac{\Delta n}{2}, \quad n_{y'} \simeq n_o - \frac{\Delta n}{2}, \quad \Delta n = n_o^3 r_{63} F_z. \tag{13.3.4}$$

The previous analysis for *longitudinal* amplitude modulators for $LiNbO_3$ is applicable to GaAs materials.

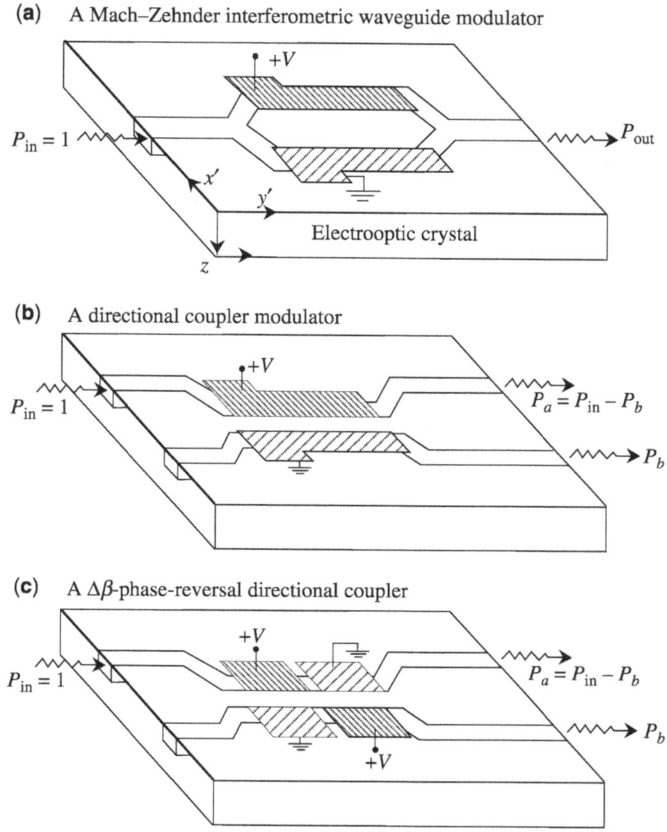

Figure 13.6 Waveguide electrooptic devices with electrode designs. (a) A Mach–Zehnder interferometric modulator, (b) a directional coupler modulator, and (c) a $\Delta\beta$-phase-reversal directional coupler.

In the following examples, we consider (a) a Mach–Zehnder interferometric waveguide modulator, (b) a directional coupler modulator, and (c) a $\Delta\beta$-phase-reversal directional coupler, as shown in Fig. 13.6 [11–17]. The input power P_{in} is taken as 1 in each case. The electrodes are designed such that the applied electric field is along the z-direction $\mathbf{F} = \pm\hat{z}F_z$ and the bias fields in two waveguides are opposite in signs. Therefore, the difference between the two propagation factors is approximately

$$\Delta\beta \simeq k\left(n_{\text{o}} + \frac{\Delta n}{2}\right) - k\left(n_{\text{o}} - \frac{\Delta n}{2}\right) = kn_{\text{o}}^3 r_{63}\frac{V(t)}{d} \tag{13.3.5}$$

where an effective width d is defined for the electric field ($F_z \simeq V(t)/d$). The important point is that $\Delta\beta \propto V(t)$. We will study the transmission characteristics of these devices as a function of the detuning factor $\Delta\beta\ell$.

13.3.1 Mach–Zehnder Interferometric Waveguide Modulator

As shown in Fig. 13.6a, a single waveguide is branched into two arms for a distance ℓ and combined again into one arm as the output waveguide. The waveguide dimension can be chosen to guide the fundamental mode only. With a proper choice of the polarization of the incident wave and the electrode design, the transmitted intensity is

$$P_{\text{out}} = \frac{1}{4}\left|e^{i\beta_a \ell} + e^{i\beta_b \ell}\right|^2 = \cos^2\left(\frac{\Delta\beta\ell}{2}\right) \tag{13.3.6}$$

where the output intensity is normalized such that the peak transmission factor is 1 for perfect power transmission. One way to understand this transmission behavior is that if the guided modes are in phase at the exit of the Y junction, Fig. 13.7, they add up constructively and transmit with the maximum power. If they are out of phase by 180°, they will cancel each other. Another way to look at this is, if they add up to a first-order mode, it will leak out over a very short distance because the waveguide is designed to guide the fundamental mode only, resulting in a destructive output. A plot of the output power P_{out} versus the mismatch $\Delta\beta\ell$ is shown in Fig. 13.8. The interferometric behavior is clearly seen.

Broadband Ti-diffused $LiNbO_3$ electrooptic modulators with a traveling-wave electrode using X-cut and Z-cut crystal facets have been fabricated with

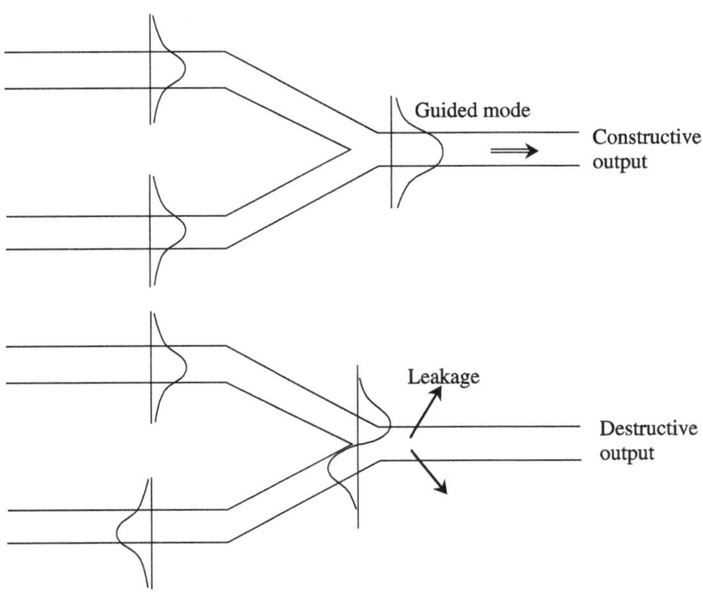

Figure 13.7 An illustration of the Y-junction constructive and destructive outputs in a Mach–Zehnder interferometric waveguide modulator.

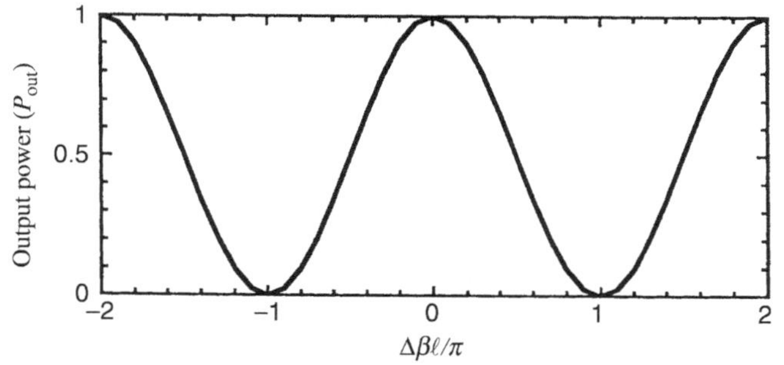

Figure 13.8 The output power from a Mach–Zehnder interferometric waveguide modulator as a function of the mismatch factor $\Delta\beta\ell$ (normalized by π).

impressive performance for telecommunications at 10 Gb/s or 40 Gb/s [18–23]. For an overview, we refer to Ref. 21. $LiNbO_3$ has been an important material of choice for the fabrication of electrooptic modulators due to its combination of high electrooptic coefficients and high optical transparency in the near-infrared wavelengths used for telecommunications [21]. It has a high Curie temperature (1100°C to 1180°C) making it practical for fabrication of low-loss optical waveguides through interdiffusion of metals such as titanium. $LiNbO_3$ is also thermally, chemically, and mechanically stable and is compatible with conventional integrated-circuit processing technology. $LiNbO_3$ waveguides have been traditionally fabricated by interdiffusion of Ti at temperatures near 1000°C. An alternative method is through the annealed proton exchange (APE) process, which is a low-temperature process (120°C to 250°C) whereby Li ions from the $LiNbO_3$ wafer are exchanged with protons from an acid bath. Proton exchange can only be applied to X- and Z-cut wafers; the acid chemically etches Y-cut wafers. Exchanged layers exhibit an increase of the extraordinary refractive index; the ordinary index is left virtually undisturbed or even reduced. As a result, the waveguides are polarization dependent. The APE process is followed by a high-temperature annealing, which stabilizes the index. RF electrodes using gold are fabricated either directly on the surface of the $LiNbO_3$ wafer or on an optically transparent buffer layer to reduce optical loss due to metal loading and provide a means for optical/RF velocity matching.

Figure 13.9a shows a schematic diagram of a Mach–Zehnder optical modulator using coplanar waveguide (CPW) electrodes [19]. Two types of traveling-wave electrodes are commonly used: the coplanar waveguide electrodes, Fig. 13.9b, or asymmetric coplanar strip-line (A-CPS) electrodes, Fig. 13.9c [19]. The bandwidth is limited by the velocity mismatch between microwave and optic wave in the devices, which is caused by the high dielectric constant of $LiNbO_3$ itself. Various modulator structures have been used to reduce the velocity mismatch or the drive

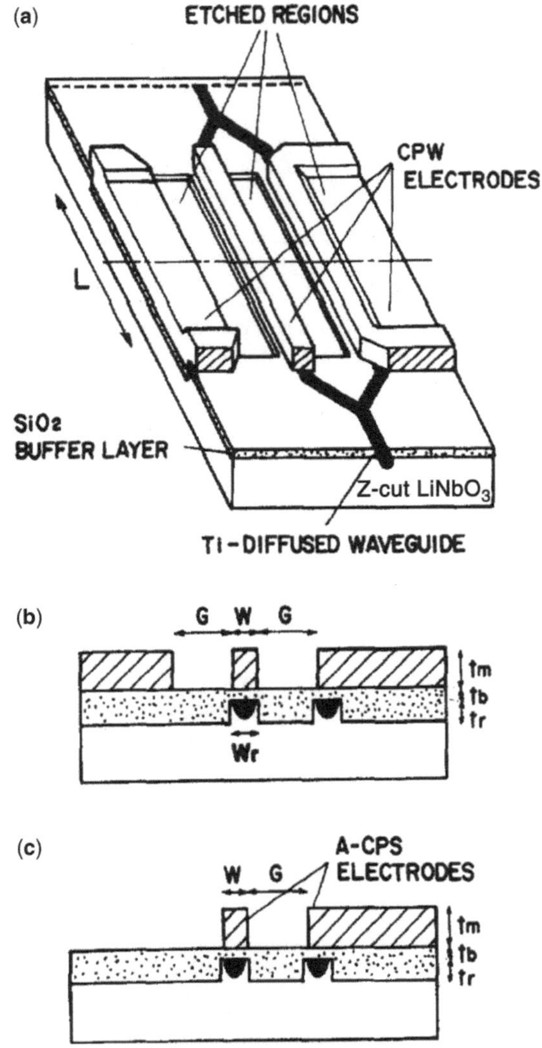

Figure 13.9 (a) Schematic diagram for a Mach–Zehnder interferometric optical waveguide modulator with coplanar waveguide (CPW) electrodes. (b) Cross-sectional view of the CPW electrodes, and (c) cross-sectional view of asymmetric planar strip-line (A-CPS) electrodes for ridge-type optical waveguide modulators. (Reprinted with permission from [19] © 1995 IEEE.)

voltage: structures with a shielding plane on the CPW electrodes, structures using a thicker A-CPS or CPW electrode, ridge-type modulators [18–20] with an etched Z-cut $LiNbO_3$ substrate between the center and ground electrodes or with a thicker SiO_2 buffer layer and CPW electrodes, and structures using X-cut crystal with two-step back slots without a buffer layer [22, 23].

13.3.2 Directional Coupler Modulator

For an incident optical beam into waveguide a in a directional coupler modulator, the output power is

$$P_b = |b(\ell)|^2 = \frac{|K|^2}{\Psi^2}\sin^2(\Psi\ell) \tag{13.3.7a}$$

$$\Psi = \left[\left(\frac{\Delta\beta}{2}\right)^2 + K^2\right]^{1/2} \tag{13.3.7b}$$

and

$$P_a = P_{\text{in}} - P_b = 1 - |b(\ell)|^2 \tag{13.3.8}$$

where the input power is assumed to be 1. Because $\Delta\beta \simeq \beta_a - \beta_b = kn_o^3 r_{63} V/d$, we plot the output power P_a versus $\Delta\beta\ell$. Suppose we design the modulator with a length ℓ such that $P_a = 0$, and $P_b = 1$, at $\Delta\beta = 0$ (i.e., $K\ell = \pi/2$). In order to switch to $P_a = 1$, and $P_b = 0$, we require at least $\Delta\beta\ell = \sqrt{3}\pi$, assuming the field-induced change in the refractive index affects the coupling coefficient negligibly. (Otherwise, we can calculate the field-dependent K and still use the expressions for P_a and P_b in (13.3.7) and (13.3.8) to find the output powers.) To switch from a cross state to a parallel state, the applied voltage has to be large enough such that $\Delta\beta\ell = \sqrt{3}\pi$ is satisfied. A plot of P_a versus $\Delta\beta\ell$ for $K\ell = \pi/2$ is shown as the thick solid curve in Fig. 13.10. We also plot P_a versus $\Delta\beta\ell$ for $K\ell = \pi$, and $K\ell = 3\pi/2$. We see that complete switching from the $\otimes$ (cross) state to the $\ominus$ (parallel) state is possible (for $K\ell = \pi/2$ or $3\pi/2$). For $K\ell = \pi$, where we start with the parallel state at $\Delta\beta\ell = 0$, it is impossible to switch to the $\otimes$ state simply by changing $\Delta\beta\ell$ alone. This fact can also be checked with the switching diagram in Fig. 8.11.

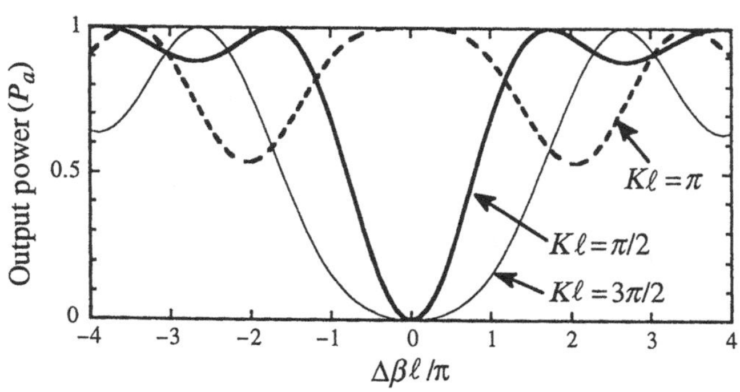

Figure 13.10 The output power from waveguide a as a function of $\Delta\beta L$ for $K\ell = \pi/2$, π, and $3\pi/2$ for a directional coupler modulator.

13.3.3 Δβ-Phase-Reversal Directional Coupler

The $\Delta\beta$-phase-reversal directional coupler [11–17] is shown in Fig. 13.6c, and its analysis can be found in Section 8.6. The switching diagram is shown in Fig. 8.13. Suppose we start with the parallel state at $K\ell = \pi$, $\Delta\ell = 0$, where $\Delta = (\beta_a - \beta_b)/2$. The output power is

$$P_a = |a(\ell)|^2 = \left| \cos^2\left(\frac{\Psi\ell}{2}\right) + \frac{|\Delta^2 - K|^2}{\Psi^2} \sin^2\left(\frac{\Psi\ell}{2}\right) \right|^2. \tag{13.3.9}$$

We plot P_a versus $\Delta\beta\ell$ as the solid curve in Fig. 13.11 for $K\ell = \pi$ and also for $K\ell = \pi/2$ (dashed curve) for comparison.

13.4 SCATTERING OF LIGHT BY SOUND: RAMAN–NATH AND BRAGG DIFFRACTIONS

The refractive index of a medium can be modulated by a mechanical strain produced by an acoustic wave; this is called the acoustooptic effect. A sound wave creates a sinusoidal perturbation of the density, or strain or pressure of the material. The induced change in refractive index can be described as

$$\Delta n(\mathbf{r}, t) = \Delta n \cos(\mathbf{k}_s \cdot \mathbf{r} - \omega_s t) \tag{13.4.1}$$

with ω_s = the angular frequency, $\mathbf{k}_s$ = the wave vector, $k_s = 2\pi/\lambda_s$, λ_s = wavelength, and $v_s = \omega_s/k_s$ is the velocity of sound in the medium.

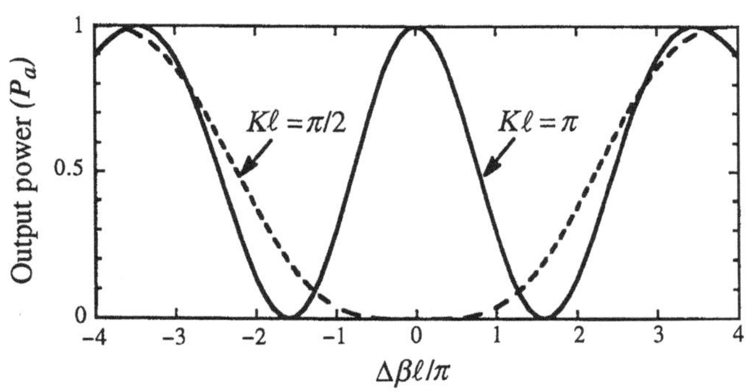

Figure 13.11 The output power P_a of a $\Delta\beta$-phase-reversal directional coupler as a function of $\Delta\beta\ell$ for $K\ell = \pi/2$ and π.

13.4.1 Raman–Nath Diffraction

Assume that the length of the interaction between the light and the acoustic wave ℓ is small [9]

$$\ell \ll \frac{kn}{k_s^2} \tag{13.4.2}$$

where $k = 2\pi/\lambda_0$, λ_0 is the optical wavelength in free space, and n is the refractive index. This is called the Raman–Nath regime of diffraction (Fig. 13.12). In this case, the thin region in which the acoustic wave propagates acts like a phase grating, and the diffracted lights can go to many different directions determined by the generalized Snell's law for a grating

$$kn \sin \theta_m = m\frac{2\pi}{\lambda_s}, \qquad m = \text{an integer}$$

$$\theta_m = \sin^{-1}\left(m\frac{\lambda_0}{n\lambda_s}\right). \tag{13.4.3}$$

A simple analysis of the diffraction efficiency for this case is to consider a y-polarized light at $x = \ell$

$$E_y(x = \ell, z, t) = E_0 e^{i[kn(z,t)\ell - \omega t]} \tag{13.4.4}$$

where

$$n(z, t) = n + \Delta n \cos(k_s z - \omega_s t). \tag{13.4.5}$$

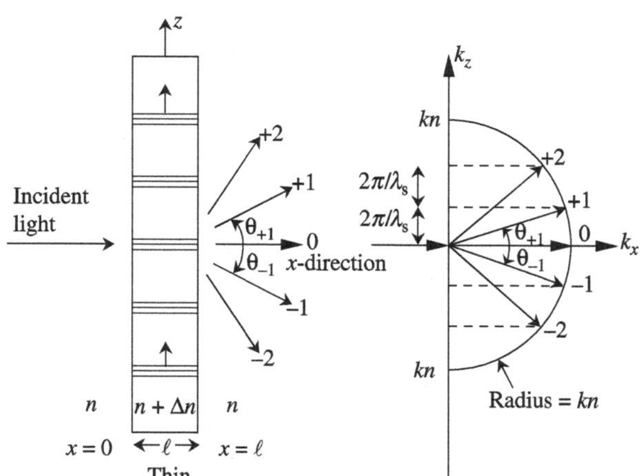

Figure 13.12 Raman–Nath diffraction. The interaction length ℓ is short and the thin region acts like an optical phase grating with the period equal to the acoustic wavelength λ_s.

We then write the field at $x = \ell$ using the mathematical identity [10] in (13.2.7)

$$e^{i\delta\cos\phi} = \sum_{m=-\infty}^{\infty} i^m J_m(\delta) e^{im\phi} \tag{13.4.6}$$

and set

$$\delta = k\Delta n\ell \tag{13.4.7}$$

$$\phi = k_s z - \omega_s t. \tag{13.4.8}$$

The electric field at $x = \ell$ then becomes

$$E_y(x = \ell, z, t) = E_0 e^{ikn\ell} \sum_{m=-\infty}^{\infty} i^m J_m(k\Delta n\ell) e^{imk_s z} e^{-i(\omega + m\omega_s)t}. \tag{13.4.9}$$

Because for $x \geq \ell$, the electric field has to satisfy the wave equation in the medium described by the refractive index n, we should have the solution of the form

$$E_y(x \geq \ell, z, t) = \sum_{m=-\infty}^{\infty} E_m e^{ik_{xm}(x-\ell) + ik_{zm}z} e^{-i\omega_m t} \tag{13.4.10}$$

where

$$\omega_m = \omega + m\omega_s \qquad k_{zm} = mk_s$$

$$k_{xm} = \sqrt{\left(\frac{\omega_m}{c}n\right)^2 - k_{zm}^2} \tag{13.4.11}$$

and

$$E_m = E_0 e^{ikn\ell} i^m J_m(k\Delta n\ell). \tag{13.4.12}$$

We note that $\omega_s \ll \omega$; therefore, $\omega_m \simeq \omega$, and

$$k_{xm} = \sqrt{\left(\frac{\omega}{c}n\right)^2 - (mk_s)^2}. \tag{13.4.13}$$

The diffraction angle of the mth order is therefore

$$k_{zm} = kn\sin\theta_m \simeq mk_s \tag{13.4.14}$$

The diffraction angles are given by (13.4.3).

13.4.2 Bragg Diffraction

When the interaction length λ between the optical and acoustic waves is long compared with kn/k_s^2, we have the Bragg diffraction. In this case, the incident $\mathbf{k}_i$ vector has to come from a particular direction satisfying the Bragg condition

$$2kn\sin\theta = k_s \tag{13.4.15}$$

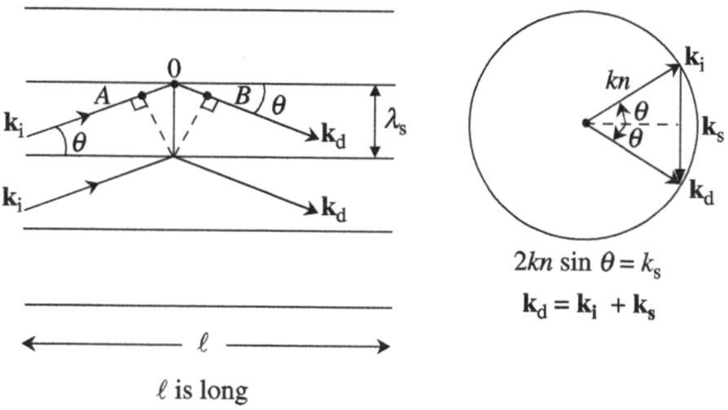

Figure 13.13 Bragg diffraction. When the interaction length ℓ is long, only one particular angle of incidence with one diffracted beam satisfying the Bragg condition $2kn \sin \theta = k_s$ will be observed. The path $AO + OB = \lambda/n$ for constructive interference. Therefore, $2\lambda_s \sin \theta = \lambda/n$.

where θ is the angle of incidence, which is also the angle of diffraction. There is only one diffracted beam determined by the above Bragg condition (Fig. 13.13). The analysis of the Bragg diffraction will be presented in Section 13.5.

13.5 COUPLED-MODE ANALYSIS FOR BRAGG ACOUSTOOPTIC WAVE COUPLERS

The analysis for Bragg diffraction can be based on the coupled-mode theory [1, 2]. We start with Maxwell's equations

$$\nabla \times \mathbf{E} = -\frac{\partial}{\partial t}\mu \mathbf{H} \qquad \nabla \times \mathbf{H} = \frac{\partial}{\partial t}\mathbf{D} \tag{13.5.1}$$

where the displacement vector is

$$\mathbf{D} = \varepsilon_0 n^2(\mathbf{r}, t)\mathbf{E} \tag{13.5.2}$$

and the refractive index variation is

$$\begin{aligned} n(\mathbf{r}, t) &= n + \Delta n(\mathbf{r}, t) \\ \Delta n(\mathbf{r}, t) &= \Delta n \cos(\mathbf{k}_s \cdot \mathbf{r} - \omega_s t). \end{aligned} \tag{13.5.3}$$

Here the background refractive index n and the amplitude of variation Δn are independent of the position and t.

Consider a TE polarized wave $\boldsymbol{E} = \hat{y}\, E_y(x, z, t)$ and assume that both the acoustic wave and the optical wave propagate in the x–z plane ($\mathbf{r} = x\hat{x} + z\hat{z}$). This solution satisfies the Gauss law because

$$\nabla \cdot \mathbf{D} = \frac{\partial}{\partial y}[\varepsilon_0 n^2(x, z, t)\, E_y(x, z, t)] = 0. \tag{13.5.4}$$

The wave equation is derived from (13.5.1) and (13.5.2)

$$\nabla \times \nabla \times \mathbf{E} = -\mu_0 \varepsilon_0 \frac{\partial^2}{\partial t^2}[n^2 + 2n\,\Delta n(\mathbf{r}, t)]\mathbf{E} \tag{13.5.5}$$

$$\left(\nabla^2 - \frac{n^2}{c^2}\frac{\partial^2}{\partial t^2}\right)\mathbf{E} = \frac{2n}{c^2}\frac{\partial^2}{\partial t^2}[\Delta n(\mathbf{r}, t)\mathbf{E}]. \tag{13.5.6}$$

We assume the incident electric field to be

$$\mathbf{E}_\mathrm{i}(r, t) = \hat{y}\frac{1}{2}E_\mathrm{i}(\mathbf{r})e^{i(\mathbf{k}_\mathrm{i}\cdot\mathbf{r} - \omega_\mathrm{i} t)} + \text{c.c.} \tag{13.5.7}$$

and the diffracted electric field

$$\mathbf{E}_\mathrm{d}(\mathbf{r}, t) = \hat{y}\frac{1}{2}E_\mathrm{d}(\mathbf{r})e^{i(\mathbf{k}_\mathrm{d}\cdot\mathbf{r} - \omega_\mathrm{d} t)} + \text{c.c.} \tag{13.5.8}$$

The variation of the refractive index can be put in the form

$$\Delta n(\mathbf{r}, t) = \frac{\Delta n}{2}e^{i(\mathbf{k}_\mathrm{s}\cdot\mathbf{r} - \omega_\mathrm{s} t)} + \frac{\Delta n}{2}e^{-i(\mathbf{k}_\mathrm{s}\cdot\mathbf{r} - \omega_\mathrm{s} t)}. \tag{13.5.9}$$

Then

$$\begin{aligned}\nabla^2\mathbf{E}_\mathrm{i} &= \hat{y}\frac{1}{2}\left[-k_\mathrm{i}^2 E_\mathrm{i} + 2i\mathbf{k}_\mathrm{i}\cdot\nabla E_\mathrm{i} + \nabla^2 E_\mathrm{i}\right]e^{i\left(\mathbf{k}_\mathrm{i}\cdot\mathbf{r} - \omega_\mathrm{i} t\right)} + \text{c.c.}\\ &\simeq \hat{y}\frac{1}{2}\left(-k_\mathrm{i}^2 E_\mathrm{i} + 2ik_\mathrm{i}\frac{\partial E_\mathrm{i}}{\partial r_\mathrm{i}}\right)e^{i\left(\mathbf{k}_\mathrm{i}\cdot\mathbf{r} - \omega_\mathrm{i} t\right)} + \text{c.c.}\end{aligned} \tag{13.5.10}$$

where the second derivative of E_i has been ignored because we assume the amplitude $E_\mathrm{i}(r)$ is slowly varying compared with the $\exp(i\mathbf{k}_\mathrm{i}\cdot\mathbf{r})$ dependence and $\mathbf{r}_\mathrm{i}$ is now along the direction of $\mathbf{k}_\mathrm{i}$. A similar expression holds for $\nabla^2\mathbf{E}_\mathrm{d}$. The term containing the product of $\Delta n(\mathbf{r}, t)\mathbf{E}$ will give rise to four terms

$$\Delta n(\mathbf{r}, t)E_\mathrm{i} = \frac{\Delta n}{4}E_\mathrm{i}(r)\left[e^{i(\mathbf{k}_\mathrm{i}+\mathbf{k}_\mathrm{s})\cdot\mathbf{r} - i(\omega_\mathrm{i}+\omega_\mathrm{s})t} + e^{i(\mathbf{k}_\mathrm{i}-\mathbf{k}_\mathrm{s})\cdot\mathbf{r} - i(\omega_\mathrm{i}-\omega_\mathrm{s})t}\right] + \text{c.c.} \tag{13.5.11}$$

and a similar expression holds for $\Delta n(\mathbf{r}, t)\mathbf{E}_\mathrm{d}$. Noting that the total electric field $\mathbf{E} = \mathbf{E}_\mathrm{i} + \mathbf{E}_\mathrm{d}$, we compare the terms of the same spatial and time variations and find

$$\mathbf{k}_\mathrm{d} = \mathbf{k}_\mathrm{i} + \mathbf{k}_\mathrm{s} \qquad \omega_\mathrm{d} = \omega_\mathrm{i} + \omega_\mathrm{s} \tag{13.5.12}$$

or

$$\mathbf{k}_\mathrm{d} = \mathbf{k}_\mathrm{i} - \mathbf{k}_\mathrm{s} \qquad \omega_\mathrm{d} = \omega_\mathrm{i} - \omega_\mathrm{s}. \tag{13.5.13}$$

These results are illustrated in Fig. 13.14a and Fig. 13.14b. Equation (13.5.12) shows the conservations of momentum and energy for a photon with initial wave vector $\mathbf{k}_\mathrm{i}$ absorbing a phonon with a wave vector $\mathbf{k}_\mathrm{s}$ resulting in a final photon state with momentum $\hbar\mathbf{k}_\mathrm{d} = \hbar\mathbf{k}_\mathrm{i} + \hbar\mathbf{k}_\mathrm{s}$ and energy $\hbar\omega_\mathrm{d} = \hbar\omega_\mathrm{i} + \hbar\omega_\mathrm{s}$.

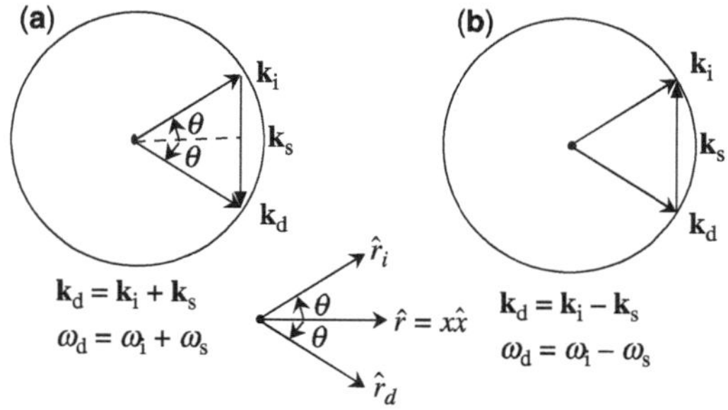

Figure 13.14 The diagrams for the diffraction of light by sound: (a) $\mathbf{k}_d = \mathbf{k}_i + \mathbf{k}_s$, $\omega_d = \omega_i + \omega_s$, and (b) $\mathbf{k}_d = \mathbf{k}_i - \mathbf{k}_s$, $\omega_d = \omega_i - \omega_s$.

Similarly, (13.5.13) corresponds to the emission of a phonon from the incident photon. Here $\hbar$ is the Planck constant. Also noting that $k_i = \frac{\omega_i}{c}n$, and $k_d = \frac{\omega_d}{c}n$, we find from (13.5.6) and (13.5.10)

$$i\mathbf{k}_i \cdot \nabla E_i = ik_i \frac{\partial E_i}{\partial r_i} = -\frac{\omega_i^2 n}{2c^2} \Delta n E_d(r).$$

Because $\mathbf{r}_i$ is along the direction of $\mathbf{k}_i$, and $\mathbf{r}_d$ is along the direction of $\mathbf{k}_d$, we take $\mathbf{r}$ along the x direction, and

$$r_i \cos\theta = x, \qquad r_d \cos\theta = x. \tag{13.5.14}$$

We obtain

$$\frac{dE_i}{dx} = iK_{id}E_d \qquad K_{id} = \frac{\omega_i \Delta n}{2c\cos\theta} \tag{13.5.15a}$$

$$\frac{dE_d}{dx} = iK_{di}E_i \qquad K_{di} = \frac{\omega_d \Delta n}{2c\cos\theta}. \tag{13.5.15b}$$

Because $\omega_s \ll \omega_i$, ω_d we have $\omega_d \simeq \omega_i \equiv \omega$ and $K_{di} \simeq K_{id} \equiv K$

$$K = \frac{\omega \Delta n}{2c\cos\theta}. \tag{13.5.16}$$

The solutions for the coupled-mode equation given the initial conditions $E_i(0)$ and $E_d(0)$ are

$$\begin{aligned} E_i(x) &= E_i(0)\cos Kx + iE_d(0)\sin Kx \\ E_d(x) &= E_d(0)\cos Kx + iE_i(0)\sin Kx. \end{aligned} \tag{13.5.17}$$

If initially, $E_d(0) = 0$, the field amplitudes are

$$E_i(x) = E_i(0)\cos Kx \qquad E_d(x) = iE_i(0)\sin Kx. \tag{13.5.18}$$

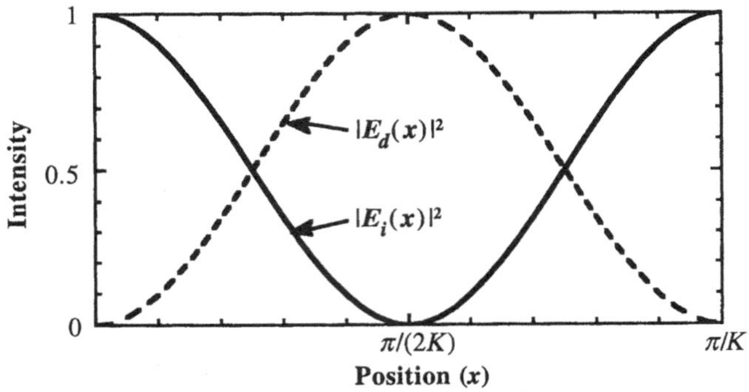

Figure 13.15 The coupling of energies between the incident and diffracted optical waves in an acoustooptic medium in which a sound wave propagates.

The energy $|E_i(0)|^2$ is coupled to $|E_d(x)|^2$ and backward during the interaction as the optical waves propagate along the x direction, as shown in Fig. 13.15. We can write that the diffraction efficiency at a length ℓ is

$$\eta = \frac{|E_d(\ell)|^2}{|E_i(0)|^2} = \sin^2 K\ell. \tag{13.5.19}$$

PROBLEMS

13.1 Calculate and compare the voltage parameter $V_\pi = \lambda_0/(2n_0^3 r_{63})$ for the materials and wavelengths with the nonzero r_{63} coefficients in Table 13.1.

13.2 Show from Maxwell's equations that for a permittivity tensor in the principal axis system,

$$\boldsymbol{\varepsilon} = \begin{bmatrix} \varepsilon_x & 0 & 0 \\ 0 & \varepsilon_y & 0 \\ 0 & 0 & \varepsilon_z \end{bmatrix}$$

(**a**) a plane wave polarized along the principal axis $\hat{x}$ and propagating along the z direction, $\mathbf{E} = \hat{x}E_0e^{i\beta z}$, the propagation constant is $\beta = \omega\sqrt{\mu\varepsilon_x}$;

(**b**) a plane wave of the form $\mathbf{E} = \hat{y}E_0e^{i\beta z}$ will have a propagation constant $\beta = \omega\sqrt{\mu\varepsilon_y}$.

13.3 For a longitudinal amplitude modulator as shown in Fig. 13.1,

(**a**) if the bias voltage is $V(t) = (0.5 + 0.1 \sin \omega_m t)V_\pi$, plot the output light intensity as a function of time.

(**b**) Repeat part (**a**) if $V(t) = 0.5V_\pi \sin \omega_m t$.

13.4 Modify (sketch) the design of optical setup in Fig. 13.1 by adding a quarter-wave plate such that the transfer function (13.1.33) can be realized with a linear response and the bias voltage $V(t)$ will not require a dc bias voltage.

13.5 For the transverse modulator shown in Fig. 13.3, plot the transmission factor P_t/P_i versus time assuming that

$$\frac{k\ell}{2}(n_o - n_e) = \frac{\pi}{4} \quad \text{and} \quad V(t) = 0.1V_\pi \sin \omega_m t.$$

13.6 A quarter-wave plate is added immediately after the first polaroid in the transverse amplitude modulator in Fig. 13.3, and the electrooptical material is GaAs ($n_e = n_o = 3.42$) assuming the wavelength λ_0 is 1.0 μm. The electric field $\mathbf{E}_1$ is circularly polarized

$$\mathbf{E}_1 = (\hat{x}' + \hat{z}e^{i\pi/2})\frac{E_0}{\sqrt{2}}e^{iky'}$$

before impinging on the GaAs crystal.

(a) Find the electric field $\mathbf{E}_2$ at $y' = \ell$ in Fig. 13.3.

(b) Find the transmitted field $\mathbf{E}_t$ after passing the exit polaroid.

(c) Obtain the transmission factor P_t/P_i and plot it versus time for $V(t) = (V_\pi/4) \sin \omega_m t$.

13.7 Consider a transverse electrooptic modulator as shown in Fig. 13.16. The incident electric field is randomly polarized, and only half of its power passes through the polaroid. The crystal is a KDP with an ac electric field applied in the z direction, and the refractive index ellipsoid is described by n_o on the $x - y$ plane and n_e along the z axis before the ac field is applied.

(a) Find the expressions for the optical electric fields $\mathbf{E}_1$ and $\mathbf{E}_2$.

(b) Find the expressions for the electric fields $\mathbf{E}_3$ and $\mathbf{E}_4$ after reflecting from the perfect mirror.

(c) Assume that the applied ac electric field across the modulator in this problem is $F_z(t) = F_{z0} \cos \omega t$. Find the ratio of the output optical intensity to the incident optical intensity as a function of time. Use a graphical

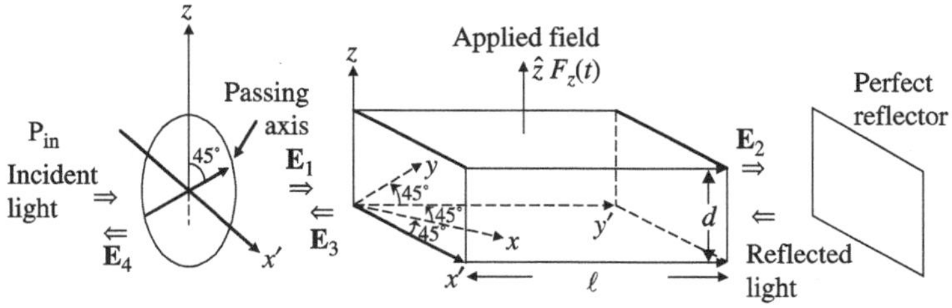

Figure 13.16 Diagram for Problem 13.7.

approach to illustrate your solution assuming that

$$\frac{k\ell}{2}(n_o - n_e) = \frac{\pi}{8} \qquad \frac{kn_o^3}{2} r_{63} F_{z0} \ell = \frac{\pi}{8}.$$

(d) If we have a dc applied field, $F_z = E_0$, find the value E_0 such that the incident light P_{in} is completely absorbed by the system.

13.8 **(a)** For a GaAs transverse modulator, derive the index ellipsoid for

$$\mathbf{F} = F_x \hat{x} + F_y \hat{y} + F_z \hat{z}.$$

(b) If **F** is along the (111) direction, that is,

$$\mathbf{F} = \frac{1}{\sqrt{3}}(\hat{x} + \hat{y} + \hat{z}) F_0,$$

design a transverse modulator and calculate the voltage parameter V_π.

13.9 Discuss the design of a phase modulator using GaAs compared with that for $LiNbO_3$ used in the text.

13.10 For a GaAs phase modulator, compare the longitudinal configuration in Fig. 13.4 versus a possible transverse configuration such that the direction of the applied electric field **F** is perpendicular to the direction of the optical wave propagation.

13.11 Derive (13.3.5) and (13.3.6), and explain how a Mach–Zehnder interferometer can be designed to function as a modulator. How do you achieve the best extinction ratio (or on-off ratio)?

13.12 **(a)** Check the output power P_a in Fig. 13.10 using (13.3.8) for $K\ell = \pi/2$.
(b) Plot P_a versus $(\Delta\beta\ell/\pi)$ for $K\ell = 2\pi$.

13.13 Plot the output power P_a versus $\Delta\beta\ell/\pi$ for a $\Delta\beta$-phase-reversal coupler using (13.3.9) for $K\ell = 3\pi/2$.

13.14 Derive (13.4.9)–(13.4.12).

13.15 Derive the coupled-mode equations in (13.5.15a) and (13.5.15b).

REFERENCES

1. A. Yariv, *Optical Electronics*, 3rd ed., Holt-Rinehart and Winston, New York, 1985.
2. H. A. Haus, *Waves and Fields in Optoelectronics*, Prentice-Hall, Englewood Cliffs, NJ, 1984.
3. B. E. A. Saleh and M. C. Teich, *Fundamentals of Photonics*, Wiley, New York, 1991.

4. S. Adachi, *Physical Properties of III–V Semiconductor Compounds*, Wiley, New York, 1992.

5. K. H. Hellwege, Ed., *Landolt–Börnstein Numerical Data and Functional Relationships in Science and Technology*, New Series, Group III **17a**, Springer-Verlag, Berlin, 1982; Groups III–V **22a**, Springer-Verlag, Berlin, 1986.

6. K. Tada and N. Suzuki, "Linear electrooptical properties of InP," *Jpn. J. Appl. Phys.* **19**, 2295–2296 (1980); and N. Suzuki and K. Tada, "Electrooptic properties and Raman scattering in InP," *Jpn. J. Appl. Phys.* **23**, 291–295 (1984).

7. S. Adachi and K. Oe, "Linear electro-optic effects in zincblende-type semiconductors: key properties of InGaAsP relevant to device design," *J. Appl. Phys.* **56**, 74–80 (1984); and "Quadratic electrooptic (Kerr) effects in zincblende-type semiconductors: Key properties of InGaAsP relevant to device design," *J. Appl. Phys.* **56**, 1499–1504 (1984).

8. S. Adachi, *Properties of Indium Phosphide*, INSPEC, The Institute of Electrical Engineers, London and New York, 1991.

9. A. K. Ghatak and K. Thyagarajan, *Optical Electronics*, Cambridge University Press, Cambridge, UK, 1989.

10. M. Abramowitz and I. A. Stegun, Eds., *Handbook of Mathematical Functions with Formulas, Graphs, and Mathematical Tables*, Chapter 9, Dover, New York, 1972.

11. S. Thaniyavarn, "Optical modulation: electrooptical Devices," Chapter 4 in K. Chang, Ed., *Handbook of Microwave and Optical Components*, vol. 4 of *Fiber and Electro-Optical Components*, Wiley, New York, 1991.

12. H. Nishihara, M. Haruna, and T. Suhara, *Optical Integrated Circuits*, McGraw-Hill, New York, 1989.

13. T. Tamir, Ed., *Guided-Wave Optoelectronics*, 2nd ed., Springer-Verlag, Berlin, 1990.

14. R. C. Alferness, "Guided-wave devices for optical communication," *IEEE J. Quantum Electron.* **QE-17**, 946–959 (1981).

15. O. G. Ramer, "Integrated optic electrooptic modulator electrode analysis," *IEEE J. Quantum Electron.* **QE-18**, 386–392 (1982).

16. D. Marcuse, "Optimal electrode design for integrated optics modulators," *IEEE J. Quantum Electron.* **QE-18**, 393–398 (1982).

17. H. Kogelnik and R. V. Schmidt, "Switched directional couplers with alternating $\Delta\beta$," *IEEE J. Quantum Electron.* **QE-12**, 396–401 (1976).

18. K. Noguchi, O. Mitomi, H. Miyazawa, and S. Seki, "A broadband Ti:$LiNbO_3$ optical modulator with a ridge structure," *J. Lightwave Technol.* **13**, 1164–1168 (1995).

19. O. Mitomi, K. Noguchi, and H. Miyazawa, "Design of ultra-broadband $LiNbO_3$ optical modulators with ridge structure," *IEEE Trans. Microwave The. Techn.* **41**, 2203–2207 (1995).

20. O. Mitomi, K. Noguchi, and H. Miyazawa, "Broadband and low driving-voltage $LiNbO_3$ optical modulators," *IEE Proc. Optoelectron.* **145**, 360–364 (1998).

21. E. L. Wooten, K. M. Kissa, A. Yi-Yan, E. J. Murphy, D. A. Lafaw, P. F. Hallemeier, D. Maack, D. V. Attanasio, D. J. Fritz, G. J. McBrien, and D. E. Bossi, "A review of lithium niobate modulators for fiber-opticcommunications systems," *IEEE J. Sel. Top. Quantum Electron.* **6**, 69–82 (2000).

22. J. Kondo, A. Kondo, K. Aoki, S. Takatsuji, O. Mitomi, M. Imaeda, Y. Kozuka, and M. Minakata, "High-speed and low-driving-voltage X-cut $LiNbO_3$ optical modulator with two step backside slot," *Electron. Lett.* **38**, 472–473 (2002).
23. J. Kondo, A. Kondo, K. Aoki, M. Imaeda, T. Mori, Y. Mizuno, S. Takatsuji, Y. Kozuka, O. Mitomi, and M. Minakata, "40-Gb/s X-cut $LiNbO_3$ optical modulator with two-step back-slot structure," *J. Lightwave Technol.* **20**, 2110–2114 (2002).

14

Electroabsorption Modulators

Electroabsorption (EA) effects near the semiconductor band edges have been an interesting research subject for many years. These include the interband photon-assisted tunneling or Franz–Keldysh effects [1–3] and the exciton absorption effects [4–9]. With the recent development of research in semiconductor quantum-well structures, optical absorptions in quantum wells have been shown to exhibit a drastic change by an applied electric field [10–13]. Whereas previous excitonic electroabsorptions in bulk semiconductors were mostly observed at low temperatures, sharp excitonic absorption spectra in quantum wells have been observed at room temperature. This so-called quantum-confined Stark effect (QCSE) [11, 12] shows a significant amount of change of the absorption coefficient with an applied voltage bias because of the enhanced exciton binding energy in a quasi-two-dimensional structure using quantum wells. The quantum-well barriers confine both the electrons and holes within the wells; therefore, the exciton binding energy is increased and the exciton is more difficult to ionize. The analytic solutions for pure two-dimensional (2D) and three-dimensional (3D) hydrogen models in Chapter 3 show that the exciton binding energy of the $1s$ ground state is four times larger in the 2D case than that in the 3D case [14]. The sharp excitonic absorption spectrum with a small scattering linewidth shows the possibility of a big change of the absorption coefficient by an applied voltage bias. The change in the absorption coefficient can be as large as $104\,\text{cm}^{-1}$ in $\text{GaAs}/\text{Al}_x\text{Ga}_{1-x}\text{As}$ quantum wells [10–13]. Interesting quantum-well electroabsorption modulators at room temperature have been undergoing intensive research recently.

In this chapter, we will discuss the theory for electroabsorptions with and without excitonic effects. We first present the effective mass theory for a two-particle system: an electron–hole pair in Section 14.1. The general formulation for the optical absorption due to an electron–hole pair is presented. We show that a change of variables from the electron and hole position coordinates $\mathbf{r}_e$ and $\mathbf{r}_h$ to their difference coordinates $\mathbf{r} = \mathbf{r}_e - \mathbf{r}_h$ and their center-of-mass coordinates $\mathbf{R}$ leads to possible analytical solutions [4, 5, 8, 9] when the interaction potential is due to (1) an electric field only, which leads to electroabsorption effects in which a light is incident, or (2) the Coulomb interaction between the electron and the hole, which gives the

Physics of Photonic Devices, Second Edition. By Shun Lien Chuang
Copyright © 2009 John Wiley & Sons, Inc.

excitonic absorption when a light is incident, or (3) both an electric field bias and the exciton effects.

Case 1 is the Franz–Keldysh effect and will be discussed in Section 14.2. The exciton effect, case 2, will then be presented in Section 14.3. Both have analytical solutions for direct band-gap semiconductors near the absorption edge. Case 3 in a quantum well, which will be presented in Section 14.4, is the quantum-confined Stark effect. The general solutions will be obtained using two methods: one is based on a numerical solution of the Schrödinger equation in the momentum space for an electron–hole pair confined in a quantum well with an applied electric field [15]. The other is based on a variational method [11, 12], which is commonly used in the literature because of its relative simplicity and accuracy especially for the bound state energy of the 1s excitons. Device applications including quantum-well electroabsorption modulators [16, 17] and integrated EA modulator-lasers (EMLs) will be presented in Sections 14.5 and 14.6, respectively, followed by self-electrooptic effect devices (SEEDs) [18–20] in Section 14.7.

14.1 GENERAL FORMULATION FOR OPTICAL ABSORPTION DUE TO AN ELECTRON–HOLE PAIR

In Chapter 9, we derive the general formula for absorption coefficient in SI units,

$$\alpha(\hbar\omega) = C_0 \frac{2}{V} \sum_{i,f} \left| \langle f | e^{i\mathbf{k}_{op}\cdot\mathbf{r}} \hat{e} \cdot \mathbf{p} | i \rangle \right|^2 \delta(E_f - E_i - \hbar\omega)[f(E_i) - f(E_f)] \tag{14.1.1a}$$

$$C_0 = \frac{\pi e^2}{n_r c \varepsilon_0 m_0^2 \omega} \tag{14.1.1b}$$

where $\hat{e}$ is the polarization of the optical electric field, and $\mathbf{p}$ is the momentum operator. The absorption coefficient depends on the initial state $|i\rangle$ with corresponding energy E_i and the final state $|f\rangle$ with energy E_f. The summation over the initial and final states taking into account the Fermi occupation factor $f(E)$ of these states gives the overall absorption spectrum. We also note that the delta function accounts for the energy conservation, and the matrix element in (14.1.1a) takes into account the momentum conservation automatically, as has been discussed in Chapter 9, where no interaction between the electrons and holes is considered.The general theory for the solution of a two-particle system such as an electron–hole pair is derived in Appendix 14A. The key idea is that, when the interaction Hamiltonian depends on the distance between the two particles (i.e., the vector difference between the two interacting particles), the two-particle wavefunction can be expressed in terms of the product of two parts: a center-of-mass coordinate part, which behaves like a plane wave, and a difference-coordinate part, which depends on the interaction type, such as an electric field or a Coulomb potential.

14.1.1 Absorption Formula

The absorption spectrum depends on the wavefunction in the difference coordinate of the two-particle system at the origin due to the k-selection rule and the integration over the density of states. The absorption coefficient (14.1.1) can be simplified as (see Appendix 14A)

$$\alpha(\hbar\omega) = A_0 2 \sum_n |\phi_n(0)|^2 \delta(E_n + E_g - \hbar\omega) \tag{14.1.2a}$$

$$A_0 = C_0 |\hat{e} \cdot \mathbf{p}_{cv}|^2 = \frac{\pi e^2 |\hat{e} \cdot \mathbf{p}_{cv}|^2}{n_r c\, \varepsilon_0\, m_0^2 \omega} \tag{14.1.2b}$$

where n corresponds to the discrete and continuum states of $\phi(r)$ satisfying the effective mass equation in the difference coordinate system, (14A.17),

$$\left[-\frac{\hbar^2}{2m_r^*}\nabla^2 + V(\mathbf{r})\right]\phi(\mathbf{r}) = E\phi(\mathbf{r}). \tag{14.1.3}$$

Three potentials will be considered:

Free Electrons and Holes

$$V(\mathbf{r}) = 0. \tag{14.1.4}$$

An Applied Electric Field

$$V(\mathbf{r}) = e\mathbf{F} \cdot \mathbf{r}. \tag{14.1.5}$$

Coulomb Interaction (Exciton Effect)

$$V(\mathbf{r}) = -\frac{e^2}{4\pi\varepsilon_s r}. \tag{14.1.6}$$

The equation for a Coulomb potential is the Schrödinger equation for a hydrogen atom and its solutions for both bound and continuum states have been presented in Chapter 3.

14.1.2 Physical Interpretation of $2|\phi_n(0)|^2$: Density of States

Consider the case of a free electron and a free hole without Coulomb interaction, that is, $V(\mathbf{r}) = 0$ in (14.1.3). We use an energy E measured from the band gap E_g. In the discrete picture, we have the plane wave solution (free carrier)

$$\phi_n(\mathbf{r}) = \frac{e^{ik_x x + ik_y y + ik_z z}}{\sqrt{V}} \tag{14.1.7}$$

with a corresponding energy spectrum

$$E_n = \frac{\hbar^2 k^2}{2m_\mathrm{r}^*} \tag{14.1.8}$$

where the quantum numbers $n \equiv (n_x, n_y, n_z)$, $k_x = n_x \frac{2\pi}{L}$, $k_y = n_y \frac{2\pi}{L}$, $k_z = n_z \frac{2\pi}{L}$, and the normalization rule

$$\int \phi_n^*(\mathbf{r}) \phi_n(\mathbf{r}) d^3\mathbf{r} = \delta_{nn'} \tag{14.1.9}$$

has been used. We check $\phi_n(0) = 1/\sqrt{V}$ and

$$\begin{aligned} 2\sum_n |\phi_n(0)|^2 \delta(E_n - E) &= 2\int \frac{d^3\mathbf{k}}{(2\pi)^3/V} \frac{1}{V} \delta\left(\frac{\hbar^2 k^2}{2m_\mathrm{r}^*} - E\right) \\ &= \frac{1}{2\pi^2} \left(\frac{2m_\mathrm{r}^*}{\hbar^2}\right)^{3/2} \sqrt{E} \end{aligned} \tag{14.1.10}$$

which is the three-dimensional reduced density of states, where the sum over the discrete quantum number has been replaced by integration over the continuum k-space as in (14.1.10).

14.1.3 Optical Absorption Spectrum for Interband Free Electron–Hole Transitions

The optical absorption is given by integrating (14.1.2a) over the energy

$$\alpha(\hbar\omega) = A_0 \rho_\mathrm{r}^{3\mathrm{D}}(\hbar\omega - E_\mathrm{g}) \tag{14.1.11a}$$

$$\rho_\mathrm{r}^{3\mathrm{D}}(\hbar\omega - E_\mathrm{g}) = \frac{1}{2\pi^2} \left(\frac{2m_\mathrm{r}^*}{\hbar^2}\right)^{3/2} \sqrt{\hbar\omega - E_\mathrm{g}} \tag{14.1.11b}$$

which gives the absorption coefficient due to a free electron and hole. The momentum matrix element of a bulk semiconductor is

$$|\hat{e} \cdot \mathbf{p}_{\mathrm{cv}}|^2 = M_\mathrm{b}^2 = \frac{m_0}{6} E_\mathrm{p} \tag{14.1.12}$$

where the energy parameter E_p (in electron volt) for the matrix element is tabulated in the Appendix C in page 803.

14.2 FRANZ–KELDYSH EFFECT: PHOTON-ASSISTED TUNNELING

The Franz–Keldysh effect [1–3, 21–26] refers to optical absorption near the semiconductor band gap. In the presence of a uniform electric field, the energy band profile tilts along the direction of the electric field, Fig. 14.1a. A photon with energy slightly below the band gap can be absorbed because of the tunneling effect. The wave function for the conduction-band electron can penetrate through the band gap; so does the hole wave function. Therefore, an absorption tail below the band gap exists. The absorption spectrum above the band edge shows an oscillatory behavior as a function of the photon energy due to the wave nature of the electron and hole wave functions in the presence of the biased electric field (Fig. 14.1b).

Let us consider the case of a uniform applied electric field, $V(\mathbf{r}) = e\mathbf{F} \cdot \mathbf{r}$. The Schrödinger equation for the wave function $\phi(\mathbf{r})$ in the difference coordinate system (14A.17) is

$$\left[-\frac{\hbar^2}{2m_r^*}\nabla^2 + e\mathbf{F}\cdot\mathbf{r}\right]\phi(\mathbf{r}) = E\phi(\mathbf{r}). \tag{14.2.1}$$

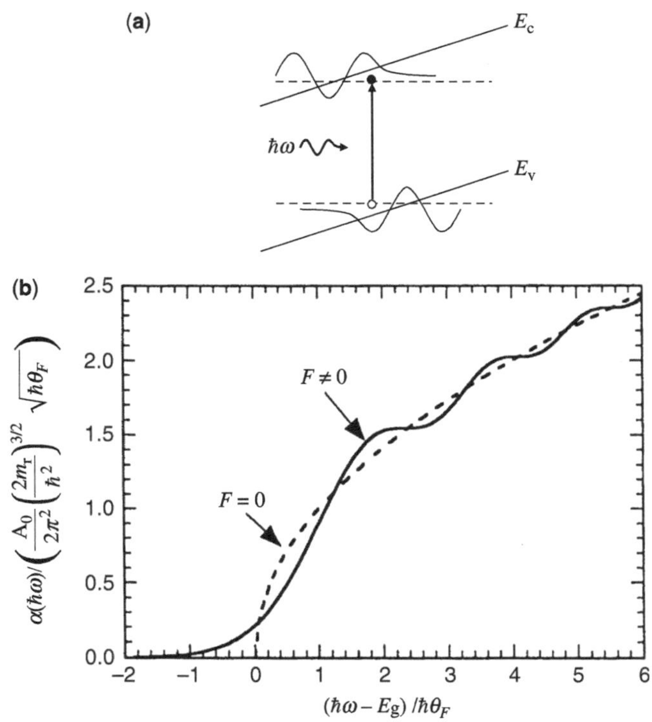

Figure 14.1 (a) Franz–Keldysh effect or photon-assisted absorption in a bulk semiconductor with a uniform electric field bias. (b) Absorption spectrum for a finite field $F \neq 0$ (solid curve). The dashed line is the free electron and hole absorption spectrum without an applied electric field ($F = 0$).

Assume that the applied field is in the z direction, $\mathbf{F} = \hat{z}F$. Because the potential energy eFz is independent of x and y, the wave function is plane-wave like in the x–y plane, and it can be written in the form

$$\phi(\mathbf{r}) = \frac{e^{ik_x x + ik_y y}}{\sqrt{A}} \phi(z) \tag{14.2.2}$$

where the z-dependent wave function $\phi(z)$ satisfies

$$\left[\frac{-\hbar^2}{2m_r^*}\frac{d^2}{dz^2} + eFz\right]\phi(z) = E_z\phi(z). \tag{14.2.3}$$

The total energy E is related to E_z for the z-dependent wave function

$$E = \frac{\hbar^2}{2m_r^*}\left(k_x^2 + k_y^2\right) + E_z. \tag{14.2.4}$$

14.2.1 Solution of the Schrödinger Equation for a Uniform Electric Field

The solution of the Schrödinger Eq. (14.2.3) with a uniform field can be obtained by a change of variable

$$Z = \left(\frac{2m_r^* eF}{\hbar^2}\right)^{1/3}\left(z - \frac{E_z}{eF}\right). \tag{14.2.5}$$

Therefore,

$$\frac{d^2\phi(Z)}{dZ^2} - Z\phi(Z) = 0. \tag{14.2.6}$$

The Airy functions [27] Ai(Z) or Bi(Z) are the solutions. Because the wave function has to decay as z approaches $+\infty$ (because of the potential $+eFz$), the Airy function Ai(Z) has to be chosen. The energy spectrum is continuous because the potential is not bounded as $z \to -\infty$. Therefore, the (real) wave function satisfying the normalization condition

$$\int_{-\infty}^{\infty} dz\,\phi_{E_{z1}}(z)\phi_{E_{z2}}(z) = \delta(E_{z1} - E_{z2}) \tag{14.2.7}$$

for a continuum spectrum is

$$\phi_{E_z}(z) = \left(\frac{2m_r^*}{\hbar^2}\right)^{1/3}\frac{1}{(eF)^{1/6}}\mathrm{Ai}\left[\left(\frac{2m_r^* eF}{\hbar^2}\right)^{1/3}\left(z - \frac{E_z}{eF}\right)\right]. \tag{14.2.8}$$

To prove that $\phi_{E_z}(z)$ satisfies the normalization condition, we use the integral representation [27] of the Airy function

$$\mathrm{Ai}(t) = \int_{-\infty}^{\infty} \frac{dk}{2\pi} e^{i\left(kt + \frac{k^3}{3}\right)}. \tag{14.2.9}$$

Therefore,

$$\begin{aligned}
&\int_{-\infty}^{\infty} \mathrm{Ai}(t-\alpha_1)\,\mathrm{Ai}(t-\alpha_2)dt \\
&= \int_{-\infty}^{\infty} \frac{dk}{2\pi} \int_{-\infty}^{\infty} \frac{dk'}{2\pi} \int_{-\infty}^{\infty} dt\, e^{i(t-\alpha_1)k + ik^3/3} e^{i(t-\alpha_2)k' + ik'^3/3} \\
&= \int_{-\infty}^{\infty} \frac{dk}{2\pi} e^{-i(\alpha_1-\alpha_2)k} \\
&= \delta(\alpha_1 - \alpha_2)
\end{aligned} \tag{14.2.10}$$

where the identity

$$\int_{-\infty}^{\infty} dt\, e^{i(k+k')t} = 2\pi\delta(k+k') \tag{14.2.11}$$

has been used. Using

$$t = \left(\frac{2m_r^* eF}{\hbar^2}\right)^{1/3} z, \qquad \alpha = \left(\frac{2m_r^* eF}{\hbar^2}\right)^{1/3} \frac{E_z}{eF}$$

and

$$\delta(\alpha_1 - \alpha_2) = \left(\frac{2m_r^*}{\hbar^2 e^2 F^2}\right)^{-1/3} \delta(E_{z1} - E_{z2})$$

in (14.2.10), we obtain the normalization condition (14.2.7).

14.2.2 Summation of the Density of States and Absorption Spectrum

Because the quantum number is determined by (k_x, k_y, E_z) as described above in the wave function (14.2.2) and the corresponding energy spectrum (14.2.4), the sum over all the states n for the absorption spectrum in (14.1.2a) has to be replaced by the sum over all the quantum numbers

$$\sum_n \to \sum_{k_x} \sum_{k_y} \int dE_z$$

where the sum over the energy E_z is an integral because E_z is a continuous spectrum and a delta function normalized rule (14.2.7) has been adopted. Therefore,

$$\alpha(\hbar\omega) = A_0 2 \sum_{k_x k_y} \int dE_z |\phi(\mathbf{r}=0)|^2 \delta\left[\frac{\hbar^2}{2m_r^*}\left(k_x^2 + k_y^2\right) + E_z + E_g - \hbar\omega\right] \quad (14.2.12)$$

where A_0 is given by (14.1.2b). Because

$$\frac{2}{A}\sum_{k_x k_y} = 2\int \frac{d^2\,\mathbf{k}_t}{(2\pi)^2} = \frac{m_r^*}{\pi\hbar^2}\int dE_t \quad (14.2.13)$$

where $E_t = \hbar^2\left(k_x^2 + k_y^2\right)/2m_r^* = \hbar^2 k_t^2/2m_r^*$, we carry out the integration over E_t with the delta function and obtain the expression for the absorption coefficient

$$\alpha(\hbar\omega) = A_0 \frac{m_r^*}{\pi\hbar^2} \int_{-\infty}^{\hbar\omega - E_g} dE_z |\phi_{E_z}(z=0)|^2$$

$$= A_0 \frac{m_r^*}{\pi\hbar^2} \int_{-\infty}^{\hbar\omega - E_g} dE_z \left(\frac{2m_r^*}{\hbar^2}\right)^{2/3} \frac{1}{(eF)^{1/3}} \mathrm{Ai}^2\left[\left(\frac{2m_r^*}{\hbar^2 e^2 F^2}\right)^{1/3}(-E_z)\right]. \quad (14.2.14)$$

Let

$$\hbar\theta_F = \left(\frac{\hbar^2 e^2 F^2}{2m_r^*}\right)^{1/3} \qquad \tau = -E_z/(\hbar\theta_F) \qquad \eta = (E_g - \hbar\omega)/\hbar\theta_F. \quad (14.2.15)$$

We find the absorption coefficient

$$\alpha(\hbar\omega) = \frac{A_0}{2\pi}\left(\frac{2m_r^*}{\hbar^2}\right)^{3/2} \sqrt{\hbar\theta_F} \int_{\eta}^{\infty} d\tau \mathrm{Ai}^2(\tau)$$

$$= \frac{A_0}{2\pi}\left(\frac{2m_r^*}{\hbar^2}\right)^{3/2} \sqrt{\hbar\theta_F}[-\eta \mathrm{Ai}^2(\eta) + \mathrm{Ai}'^2(\eta)] \quad (14.2.16)$$

where $\mathrm{Ai}'(\eta)$ is the divative of $\mathrm{Ai}(\eta)$ with respect to η.

It is interesting to show that in the limit when $F \to 0$, we have

$$\lim_{F \to 0} \left[\pi\sqrt{\hbar\theta_F} \int_{\eta}^{\infty} \mathrm{Ai}^2(\tau) d\tau \right] = \sqrt{\hbar\omega - E_g}$$

for $\hbar\omega > E_g$ as expected, because

$$\alpha(\hbar\omega) \underset{F \to 0}{\longrightarrow} \alpha_0(\hbar\omega) = A_0 \frac{1}{2\pi^2} \left(\frac{2m_r^*}{\hbar^2} \right)^{3/2} \sqrt{\hbar\omega - E_g}. \quad (14.2.17)$$

Notice that the prefactor A_0 depends on the bulk momentum-matrix element $|\hat{e} \cdot \mathbf{p}_{cv}|^2 = M_b^2$, which can be determined experimentally [24] by fitting the measured absorption spectrum with the above theoretical results with $F = 0$ and $F \neq 0$. The Franz–Keldysh absorption spectrum (14.2.16) is plotted in Fig. 14.1b (solid curve) schematically and compared with the zero-field spectrum (dashed curve) using (14.2.17). It shows the Franz–Keldysh oscillation phenomena in the absorption spectrum above the band gap and the exponentially decaying behavior below the band gap.

14.3 EXCITON EFFECT

When we consider the Coulomb interaction between the electron and the hole

$$V(\mathbf{r}) = \frac{-e^2}{4\pi\varepsilon_s r}, \quad (14.3.1)$$

the wave function $\phi(\mathbf{r})$ satisfies the Schrödinger equation for the hydrogen atom

$$\left[-\frac{\hbar^2}{2m_r^*} \nabla^2 + V(\mathbf{r}) \right] \phi(\mathbf{r}) = E\phi(\mathbf{r}). \quad (14.3.2)$$

The solutions $\phi(\mathbf{r})$ for both the three-dimensional (3D) and the two-dimensional (2D) cases have been studied in Section 3.4 or in the first edition of this book [4–15].

14.3.1 Three-dimensional Exciton

We use the general formula (14.1.2) for the absorption coefficient, where the summation over n includes both the bound and continuum states of the solution to (14.3.2) with the Coulomb potential. The wave functions $\phi_n(\mathbf{r})$ should be normalized properly for both bound and continuum states as discussed in Sections 14.1 and 14.2, Appendix 14A, and Section 3.4 [4, 5]. For bound state contributions, we have the oscillator strength

$$|\phi_n(0)|^2 = \frac{1}{\pi a_0^3 n^3}, \qquad a_0 = \frac{4\pi\varepsilon_s\hbar^2}{m_r^* e^2} = \text{the exciton Bohr radius} \tag{14.3.3a}$$

and the exciton binding energy

$$E_n = -\frac{R_y}{n^2}, \qquad R_y = \frac{m_r^* e^4}{2\hbar^2(4\pi\varepsilon_s)^2} = \text{the exciton Rydberg energy.} \tag{14.3.3b}$$

Therefore,

$$\begin{aligned}\alpha_B(\hbar\omega) &= A_0 \sum_{n=1}^{\infty}\left[\frac{2}{\pi a_0^3 n^3}\right]\delta\left(\hbar\omega - E_g + \frac{R_y}{n^2}\right)\\ &= A_0 \sum_{n=1}^{\infty}\frac{1}{R_y \pi a_0^3 n^3}\delta\left(\varepsilon + \frac{1}{n^2}\right)\end{aligned} \tag{14.3.4}$$

where

$$\varepsilon = (\hbar\omega - E_g)/R_y \tag{14.3.5}$$

is a normalized energy measured from the band gap E_g.

For the continuum-state contributions, we obtain

$$\begin{aligned}\alpha_C(\hbar\omega) &= A_0 \int_0^{\infty} dE \frac{e^{\pi/\sqrt{E/R_y}}}{2\pi R_y a_0^3 \sinh\left(\dfrac{\pi}{\sqrt{E/R_y}}\right)} \delta(\hbar\omega - E_g - E)\\ &= \frac{A_0}{2\pi^2 R_y a_0^3}\sqrt{\varepsilon}\frac{(\pi/\sqrt{\varepsilon})e^{\pi/\sqrt{\varepsilon}}}{\sinh\left(\frac{\pi}{\sqrt{\varepsilon}}\right)}\end{aligned} \tag{14.3.6}$$

where the first bracket is $A_0\rho_r^{3D}(E = \hbar\omega - E_g)$, and the second bracket is called the Sommerfeld enhancement factor for the 3D case [4–9, 28]

$$\begin{aligned}S_{3D}(\varepsilon) &= \frac{(\pi/\sqrt{\varepsilon})e^{\pi/\sqrt{\varepsilon}}}{\sinh(\pi/\sqrt{\varepsilon})}\\ &= \frac{2\pi/\sqrt{\varepsilon}}{1 - e^{-2\pi/\sqrt{\varepsilon}}}.\end{aligned} \tag{14.3.7}$$

As $\varepsilon \to \infty$, we find

$$\alpha_C(\hbar\omega) \longrightarrow \frac{A_0}{2\pi^2 R_y a_0^3}\sqrt{\varepsilon}\, e^{\pi/\sqrt{\varepsilon}} \longrightarrow \frac{A_0}{2\pi^2 R_y a_0^3}\left(\sqrt{\varepsilon}+\pi\right) \tag{14.3.8}$$

which approaches the 3D joint density-of-states in an interband transition without the exciton effects, $\sqrt{(\hbar\omega - E_g)/R_y}$, plus a constant π. As $\varepsilon \to 0$,

$$\alpha_C(\hbar\omega) \longrightarrow \frac{A_0}{2\pi^2 R_y a_0^3}(2\pi) \tag{14.3.9}$$

which gives a finite value in contrast to the vanishing result of the interband absorption at $\hbar\omega = E_g$.

The total absorption due to both the bound and continuum states is given by

$$\alpha(\hbar\omega) = \frac{A_0}{2\pi^2 R_y a_0^3}\left[4\pi\sum_{n=1}^{\infty}\frac{1}{n^3}\delta\left(\frac{\hbar\omega - E_g}{R_y}+\frac{1}{n^2}\right) + S_{3D}\left(\frac{\hbar\omega - E_g}{R_y}\right)\sqrt{\frac{\hbar\omega - E_g}{R_y}}\right]. \tag{14.3.10}$$

If we include the finite linewidth due to scatterings by replacing the delta function by a Lorentzian function, $\delta(x) = (\gamma/\pi)/(x^2+\gamma^2)$, where γ is the half-linewidth normalized by Rydberg if x is a normalized energy, we find

$$\alpha(\hbar\omega) = \frac{A_0}{2\pi^2 R_y a_0^3}\left\{4\sum_{n=1}^{\infty}\frac{\gamma/n^3}{\left(\varepsilon+\frac{1}{n^2}\right)^2+\gamma^2} + \int_0^{\infty}\frac{d\varepsilon'}{\pi}\frac{\gamma S_{3D}(\varepsilon')\sqrt{\varepsilon'}}{(\varepsilon-\varepsilon')^2+\gamma^2}\right\}. \tag{14.3.11}$$

14.3.2 Two-dimensional Exciton

The absorption spectrum for a two-dimensional structure with exciton effects [14, 29] can also be obtained using (14.1.2). For bound state contributions, we have

$$|\phi_n(0)|^2 = \frac{1}{\pi a_0^2\left(n-\frac{1}{2}\right)^3} \qquad E_n = -\frac{R_y}{\left(n-\frac{1}{2}\right)^2} \tag{14.3.12}$$

and obtain

$$\alpha_B(\hbar\omega) = A_0\sum_{n=1}^{\infty}\frac{2}{R_y a_0^2\pi\left(n-\frac{1}{2}\right)^3}\delta\left(\varepsilon+\frac{1}{\left(n-\frac{1}{2}\right)^2}\right) \tag{14.3.13}$$

where $\varepsilon = (\hbar\omega - E_g)/R_y$ again. The continuum-state contributions give

$$\alpha_C(\hbar\omega) = A_0 \int_0^\infty dE \frac{S_{2D}(E)}{2\pi R_y a_0^2} \delta(E + E_g - \hbar\omega)$$
$$= A_0 \frac{m_r^*}{\pi\hbar^2} S_{2D}(\varepsilon) \tag{14.3.14}$$

where the two-dimensional Sommerfeld enhancement factor is

$$S_{2D}(\varepsilon) = \frac{2}{1 + e^{-2\pi/\sqrt{\varepsilon}}}. \tag{14.3.15}$$

The total absorption is the sum of $\alpha_B(\hbar\omega)$ and $\alpha_C(\hbar\omega)$

$$\alpha(\hbar\omega) = \frac{A_0}{2\pi R_y a_0^2} \left[4 \sum_{n=1}^{\infty} \frac{1}{\left(n - \frac{1}{2}\right)^3} \delta\left(\varepsilon + \frac{1}{\left(n - \frac{1}{2}\right)^2} \right) + S_{2D}(\varepsilon) \right]. \tag{14.3.16}$$

Notice that without the Sommerfeld enhancement factor, $S_{2D}(\varepsilon)$ is set to 1 and $\alpha_f(\hbar\omega) = A_0/(2\pi R_y a_0^2)$. If we include the finite linewidth effect, we have

$$\alpha(\hbar\omega) = \frac{A_0}{2\pi R_y a_0^2} \left[4 \sum_{n=1}^{\infty} \frac{1}{\left(n - \frac{1}{2}\right)^3} \frac{\gamma}{\left[\varepsilon + \frac{1}{\left(n - \frac{1}{2}\right)^2} \right]^2 + \gamma^2} + \int_0^\infty \frac{d\varepsilon'}{\pi} \frac{\gamma S_{2D}(\varepsilon')}{(\varepsilon' - \varepsilon)^2 + \gamma^2} \right]. \tag{14.3.17}$$

The above results for both 2D and 3D excitons are summarized in Table 14.1. The absorption spectra for a finite linewidth and zero linewidth are plotted in Fig. 14.2a–d for comparison.

14.3.3 Experimental Results for 3D and Quasi-2D Excitons

Experimentally, the linewidth γ is always finite and increases with temperature. For example, in Fig. 14.3a, we show the absorption spectra [30] of a bulk (3D) GaAs at four different temperatures, $T = 21$K, 90K, 186K, and 294K. We can see that the

Table 14.1 Absorption Coefficients due to Exciton Bound and Continuum States

$$\varepsilon = (\hbar\omega - E_g)/R_y, \quad A_0 = \frac{\pi e^2 |\hat{e} \cdot \mathbf{p}_{cv}|^2}{n_r\, c\, \varepsilon_0\, \omega m_0^2}, \quad a_0 = \frac{\hbar^2}{m_r^*}\left(\frac{4\pi\varepsilon_s}{e^2}\right), \quad R_y = \frac{m_r^* e^4}{2\hbar^2(4\pi\varepsilon_s)^2}$$

Bound States	Continuum States
Two-dimensional exciton:	
Zero linewidth	
$A_0 \sum_{n=1}^{\infty} \frac{2}{\pi a_0^2 (n-\frac{1}{2})^3} \frac{1}{R_y} \delta\left(\varepsilon + \frac{1}{(n-\frac{1}{2})^2}\right)$	$\frac{A_0}{2\pi R_y a_0^2} S_{2D}(\varepsilon)$ $S_{2D}(\varepsilon) = \frac{2}{1+\exp(-2\pi/\sqrt{\varepsilon})}$
Finite linewidth	
$A_0 \sum_{n=1}^{\infty} \frac{2}{\pi a_0^2 (n-\frac{1}{2})^3} \frac{1}{R_y} \frac{\gamma/\pi}{\left[\varepsilon + \frac{1}{\left(n-\frac{1}{2}\right)^2}\right]^2 + \gamma^2}$	$\frac{A_0}{2\pi R_y a_0^2} \int_0^{\infty} \frac{d\varepsilon'}{\pi} \frac{\gamma S_{2D}(\varepsilon')}{(\varepsilon'-\varepsilon)^2+\gamma^2}$
Three-dimensional exciton:	
Zero linewidth	
$A_0 \sum_{n=1}^{\infty} \left(\frac{2}{\pi a_0^3 n^3}\right) \frac{1}{R_y} \delta\left(\varepsilon + \frac{1}{n^2}\right)$	$\frac{A_0}{2\pi^2 R_y a_0^3} \sqrt{\varepsilon}\, S_{3D}(\varepsilon)$ $S_{3D}(\varepsilon) = \frac{2\pi/\sqrt{\varepsilon}}{1-e^{-2\pi/\sqrt{\varepsilon}}}$
Finite linewidth	
$A_0 \sum_{n=1}^{\infty} \left(\frac{2}{\pi a_0^3 n^3}\right) \frac{1}{R_y} \frac{\gamma/\pi}{(\varepsilon+\frac{1}{n^2})^2+\gamma^2}$	$\frac{A_0}{2\pi^2 R_y a_0^3} \int_0^{\infty} \frac{d\varepsilon'}{\pi} \frac{\gamma\sqrt{\varepsilon'}\, S_{3D}(\varepsilon')}{(\varepsilon'-\varepsilon)^2+\gamma^2}$

exciton linewidth is broadened with an increasing temperature. The absorption edge has a red shift because the GaAs band gap decreases with increasing temperature [31]

$$E_g(T) = E_g(0) - \frac{aT^2}{T+b} \text{ (eV)} \tag{14.3.18}$$

where $E_g(0) = 1.519$ eV, $a = 5.405 \times 10^{-4}$ eV/K and $b = 204$K. For comparison, we show the exciton absorption spectra [32] of an $In_{0.53}Ga_{0.47}As/In_{0.52}Al_{0.48}As$ (lattice-matched to InP substrate) quantum-well structure at different temperatures in Fig. 14.3b. The quantum-well structure has a quasi-two-dimensional character because the electron and hole wave functions are confined in the z direction with a

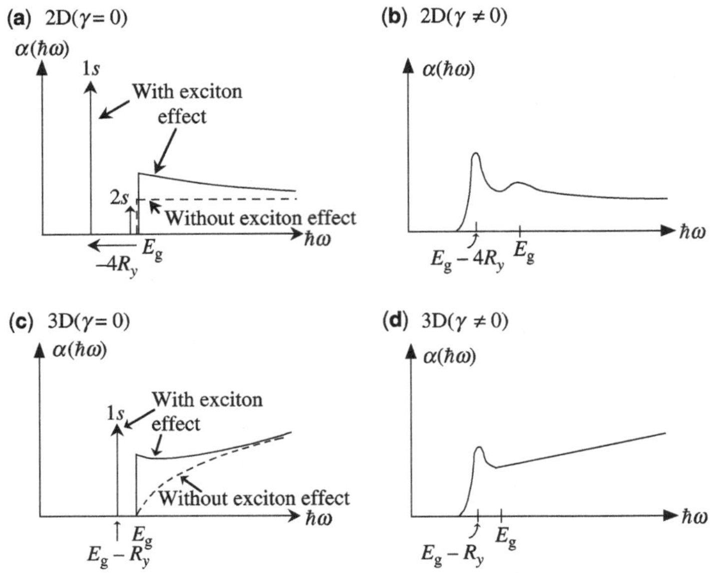

Figure 14.2 Absorption spectra for a two-dimensional (2D) exciton with (a) a zero linewidth and (b) a finite linewidth, and a three-dimensional (3D) exciton with (c) a zero linewidth, and (d) a finite linewidth.

finite well width instead of being restricted to the $x-y$ plane as in the "pure" 2D case. It is expected that the binding energy of the $1s$ exciton in the quasi-2D structure will be between the 3D value ($=R_y$) and 2D value ($=4R_y$). Furthermore, we also see the splitting of the heavy-hole (HH) exciton and light-hole (LH) exciton in a quasi-2D structure, whereas we do not observe the HH and LH exciton splittings in a bulk GaAs sample because of the degeneracy of the HH and LH bands at the zone center of the valence-band structure. The energies and absorption spectra of the HH and LH excitons will be further investigated in Section 14.4.

The insert of Fig. 14.3b shows the measured (squares) half width at half maximum (HWHM) of the first heavy-hole absorption line as a function of temperature. The solid line is a fit to the expression

$$\gamma(=\text{HWHM}) = \Gamma_0 + \frac{\Gamma_{\text{ph}}}{\exp\left(\dfrac{\hbar\omega_{\text{LO}}}{k_B T}\right) - 1} \tag{14.3.19}$$

where $\Gamma_0 = 2.3$ meV accounts for the inhomogeneous broadenings such as scatterings by interface roughness and alloy fluctuations, and the second term represents the homogeneous broadening due to InGaAs longitudinal optical (LO) phonon scatterings with $\hbar\omega_{\text{LO}} = 35$ meV and $\Gamma_{\text{ph}} = 15.3$ meV.

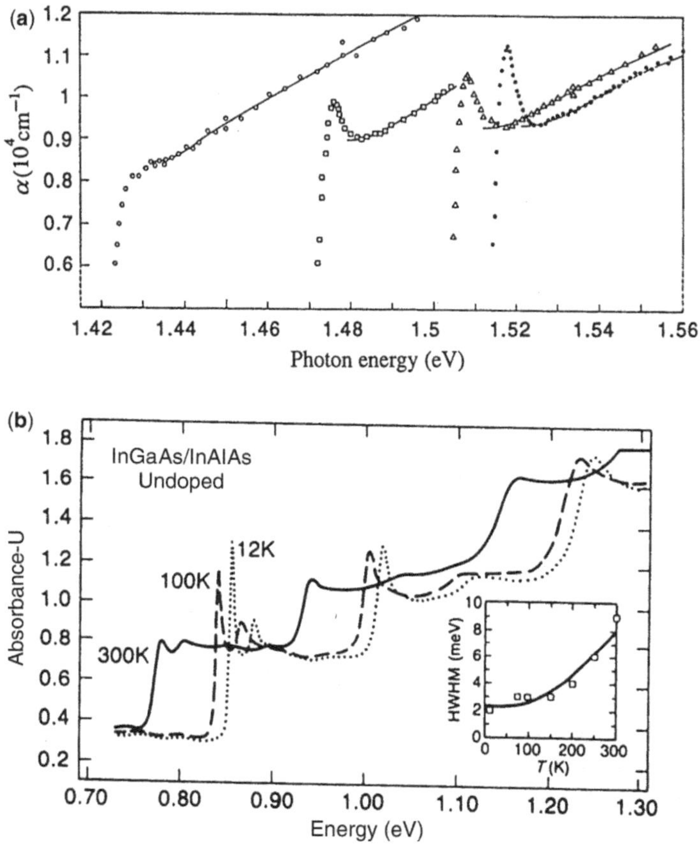

Figure 14.3 (a) Band-edge absorption spectra of a bulk GaAs sample at $T = 294$K (circles), 186K (squares), 90K (triangles), and 21K (dots). (Reprinted with permission from [30] © 1962 American Physical Society.) (b) Absorption spectra of an $In_{0.53}Ga_{0.47}As/In_{0.52}Al_{0.48}As$ quantum-well sample at $T = 300$K (solid curve), 100K (dashed curve), and 12K (dotted curve). The insert shows the half width at half maximum (HWHM) of the first absorption peak as a function of temperature. (Squares are measured data, curve is calculated.) (Reprinted with permission from [32] © 1988 IEEE.)

14.4 QUANTUM CONFINED STARK EFFECT (QCSE)

In this section, we consider the exciton absorption in a quantum-well structure in the presence of a uniform applied electric field [11–13, 15]. The effective mass equation, similar to (14A.4), can be written as [12]

$$\left[H_e - H_h + E_g - \frac{e^2}{4\pi\varepsilon_s|\mathbf{r}_e - \mathbf{r}_h|}\right]\Phi(\mathbf{r}_e, \mathbf{r}_h) = E\Phi(\mathbf{r}_e, \mathbf{r}_h) \tag{14.4.1}$$

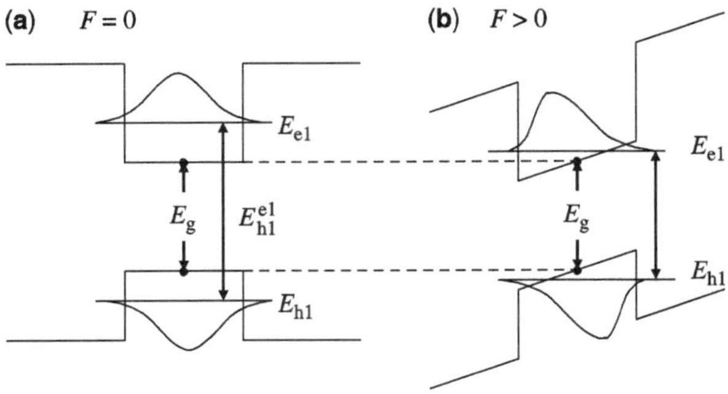

Figure 14.4 Quantum-well energy subbands and wave functions (a) in the absence of an applied electric field and (b) in the presence of an applied electric field.

where

$$H_e = -\frac{\hbar^2}{2m_e^*}\nabla_e^2 + V_e(\mathbf{r}_e) \tag{14.4.2a}$$

$$-H_h = -\frac{\hbar^2}{2m_h^*}\nabla_h^2 + V_h(\mathbf{r}_h) \tag{14.4.2b}$$

The electron potential $V_e(\mathbf{r}_e)$ or the hole potential $V_h(\mathbf{r}_h)$ may also include the effect of the electric field $|e|\mathbf{F}\cdot\mathbf{r}_e$ or $-|e|\mathbf{F}\cdot\mathbf{r}_h$ in addition to the quantum confinement potentials. The interaction between the electron and the hole is due to the Coulomb potential.

Let us assume that the quantum well is grown along the z axis and the uniform electric field is also applied along the z direction, the electron and hole potential can be written as (Fig. 14.4)

$$V_e(\mathbf{r}_e) = V_e(z_e) \tag{14.4.3a}$$

and

$$V_h(\mathbf{r}_h) = V_h(z_h). \tag{14.4.3b}$$

For solution of the electron–hole effective-mass equation with excitonic effects, please see Appendix 14B for details [15, 33–35].

14.4.1 Optical Absorption Spectrum with Exciton Effects in Quantum Wells

Without the Coulomb interaction, the electron subbands are determined by the Hamiltonian H_e, and we denote the electron subbands by $n =$ C1, C2, C3, and so forth. Similarly, the hole subbands are determined by H_h and the subband levels are $m =$ HH1, HH2, and so forth, for heavy holes and LH1, LH2, and so forth, for light holes. The solution to the eigenvalue Eq. (14B.16) is a set of exciton binding energies and corresponding eigenfunctions for 1s, 2s, 3s as well as continuum states. We denote the quantum number for the exciton state as x, which includes both bound and continuum states. The absorption coefficient for a quantum-well structure can be obtained by substituting the matrix element $\langle f|\hat{e}\cdot\mathbf{p}|i\rangle$ into (14.1.1)

$$\alpha(\hbar\omega) = C_0 \frac{2}{V}\sum_x \left| \sqrt{A}\sum_{nm} \phi_{nm}^x(\rho=0)\hat{e}\cdot\mathbf{p}_{\rm cv}\, I_{nm}\right|^2 \delta(E_x - \hbar\omega)$$
$$C_0 \equiv \frac{\pi e^2}{n_{\rm r} c\varepsilon_0 \omega m_0^2} \tag{14.4.4}$$

where the exciton transition energy is

$$E_x = E_{hm}^{en} + E_{\rm ex} \tag{14.4.5a}$$

and the band edge transition energy is

$$E_{hm}^{en} = E_{\rm g} + E_{en} - E_{hm}. \tag{14.4.5b}$$

The exciton binding energy $E_{\rm ex}$ is a discretized set of $1s$, $2s$, $3s$, ... states and continuum-state energies. Assuming that there is no mixing between different subbands, consider only the pair $n = C1$ and $m =$ HH1, for example. We may drop the summation over nm and treat each pair of n and m independently. This assumption is valid only if the subband energy difference is much larger than the exciton binding energy. The absorption coefficient becomes

$$\alpha(\hbar\omega) = C_0 \frac{2}{L}\sum_x \left|\phi_{nm}^x(\rho=0)\right|^2 |\hat{e}\cdot\mathbf{p}_{\rm cv}|^2 |I_{nm}|^2 \delta(E_x - \hbar\omega). \tag{14.4.6}$$

Exciton Discrete (1*s*) State Contribution For the $1s$ state, we use the variational approach and find

$$\alpha(\hbar\omega) = C_0 \frac{2}{L}\left(\frac{2}{\pi}\frac{1}{\lambda^2}\right)|\hat{e}\cdot\mathbf{p}_{\rm cv}|^2 |I_{nm}|^2 \frac{\gamma/\pi}{(E_x - \hbar\omega)^2 + \gamma^2} \tag{14.4.7}$$

where λ is the variational parameter of the trial function, see (14B.23) in Appendix 14B, and a finite linewidth (2γ) has been assumed and the delta function replaced by a Lorentzian function, and $x = 1s$ state. The matrix element $|\hat{e}\cdot\mathbf{p}_{\rm cv}|^2$ is obtained [15, 36] from (9.5.14)–(9.5.15) in Section 9.5.

For TE Polarization ($\hat{e} = \hat{x}$ or $\hat{y}$)

$$|\hat{e}\cdot\mathbf{p}_{\rm cv}|^2 = M_{\rm b}^2 \begin{cases} \frac{3}{2} & \text{Heavy-hole exciton} \\ \frac{1}{2} & \text{Light-hole exciton.} \end{cases} \tag{14.4.8}$$

For TM Polarization ($\hat{e} = \hat{z}$)

$$|\hat{e} \cdot \mathbf{p}_{cv}|^2 = M_b^2 \begin{cases} 0 & \text{Heavy-hole exciton} \\ 2 & \text{Light-hole exciton} \end{cases} \tag{14.4.9}$$

where M_b^2 is the bulk matrix element discussed in Chapter 9

$$M_b^2 \left(\cong \frac{m_0^2 E_g (E_g + \Delta)}{6\, m_e^* (E_g + 2\Delta/3)} \right) = \frac{m_0}{6} E_p. \tag{14.4.10}$$

Exciton Continuum-State Contributions For continuum-state contributions, the wave function $|\phi^x\ (\rho = 0)|^2$ gives a Sommerfeld enhancement factor [13–15]. Noting that the sum over the states x becomes the sum over a continuum distribution of states k_x and k_y,

$$2\sum_x = 2\sum_{k_x}\sum_{k_y} = A\frac{m_r^*}{\pi\hbar^2}\int dE_t \tag{14.4.11}$$

and $E_{ex} \rightarrow \hbar^2 k_t^2/(2m_r^*) = E_t$, we obtain the absorption coefficient due to the continuum states

$$\alpha_C(\hbar\omega) = C_0 \frac{m_r^*}{\pi\hbar^2 L} M_b^2 |I_{nm}|^2 \int_0^\infty dE_t M(E_t) |\phi^x(0)|^2 \frac{\gamma/\pi}{(E_x - \hbar\omega)^2 + \gamma^2} \tag{14.4.12}$$

where $E_x = E_{hm}^{en} + E_t$ and $|\phi^x(0)|^2$ is usually approximated by the Sommerfeld enhancement factor for a 2D exciton

$$|\phi^x(0)|^2 \simeq \frac{s_0}{1 + e^{-2\pi/(ka_0)}} \tag{14.4.13}$$

where $1 \leq s_0 \leq 2$. For a pure 2D exciton, $s_0 = 2$, and $ka_0 = \sqrt{E_t/R_y}$. The matrix-element $M(E_t)$ is defined [15, 37, 38] as $|\hat{e} \cdot \mathbf{p}_{cv}|^2 = M(E_t) M_b^2$, which has been derived in Section 9.5 and is tabulated in Table 9.1

TE polarization
$$M(E_t) = \begin{cases} \frac{3}{4}(1 + \cos^2\theta_{nm}) & \text{Heavy-hole exciton} \\ \frac{1}{4}(5 - 3\cos^2\theta_{nm}) & \text{Light-hole exciton} \end{cases} \tag{14.4.14}$$

TM polarization
$$M(E_t) = \begin{cases} 0 & \text{Heavy-hole exciton} \\ \frac{1}{2}(1 + 3\cos^2\theta_{nm}) & \text{Light-hole exciton} \end{cases} \tag{14.4.15}$$

where $\cos^2\theta_{nm} \simeq (E_{en} + |E_{hm}|)/(E_{en} + |E_{hm}| + E_t)$. The heavy-hole exciton contribution to the TM polarization is taken to be zero instead of $(3/2)\sin^2\theta_{nm}$ because a rigorous valence-band mixing model shows that the heavy-hole exciton has a negligible contribution to the TM case [39–43]. Note that $k_0 = 1/a_0$ and

$$\frac{m_r^*}{\pi\hbar^2} = \frac{k_0^2}{2\pi R_y}. \tag{14.4.16}$$

Total Absorption Spectrum The complete absorption spectrum can be written as the sum of the bound-state and continuum-state contributions [15]

$$\alpha(\omega) = C_0\frac{2}{L}M_b^2|I_{nm}|^2\frac{k_0^2}{2\pi R_y}\left[4\sum_{\substack{x=\text{discrete}\\ \text{states}}} M(0)|a_0\phi^x(0)|^2\frac{R_y\gamma}{(E_x-\hbar\omega)^2+\gamma^2}\right.$$
$$\left.+\int_0^\infty\frac{dE_t}{\pi}M(E_t)|\phi^x(0)|^2\frac{\gamma}{(E_{hm}^{en}-E_t-\hbar\omega)^2+\gamma^2}\right]. \tag{14.4.17}$$

Let us look at the band-edge transition energy E_{hm}^{en}

$$E_{hm}^{en} = E_g + E_{en} - E_{hm} \tag{14.4.18}$$

for the electron subband n and the hole subband m in the presence of an applied electric field. Note that the subband energies are measured from the band edges at the center of the quantum well (positive for electrons and negative for holes).

14.4.2 Perturbation Method

A simple second-order perturbation theory shows [44] that (see the example in Section 3.5 and its references)

$$\begin{aligned} E_n &= E_n^{(0)} + C_n m^* e^2F^2L_{\text{eff}}^4/\hbar^2 \\ C_n &= \frac{32}{\pi^6}\sum_{m\neq n}\frac{|1-(-1)^{n-m}|^2m^2n^2}{(n^2-m^2)^5} \\ &= \frac{n^2\pi^2-15}{24n^4\pi^4} \end{aligned} \tag{14.4.19}$$

in an infinite quantum-well model assuming an effective well width L_{eff}. Therefore, the band edge transition energy for the first conduction and the first heavy-hole subband is determined by

$$E_{\text{h1}}^{\text{el}}(F) = E_{\text{h1}}^{\text{el}}(F=0) + C_1\left(m_{\text{e}}^* + m_{\text{h}}^*\right)e^2F^2L_{\text{eff}}^4/\hbar^2 \tag{14.4.20}$$

where $C_1 = -2.19 \times 10^{-3}$. As an example, for a GaAs/Al$_x$Ga$_{1-x}$As quantum well with an effective well width of 100 Å, the above change in the transition energy is $E_{\text{h1}}^{\text{el}}(F) - E_{\text{h1}}^{\text{el}}(F=0) = -11.8\,\text{meV}$ at $F = 100\,\text{kV/cm}$ assuming $m_{\text{e}}^* = 0.0665\,m_0$ and $m_{\text{hh}}^* = 0.34\,m_0$. The exciton binding energy E_{ex} also depends on the electric field strength. In general, the band-edge shift due to the electric field is quite clear and appears stronger than the binding energy change with the electric field. Variational methods instead of the above perturbation method for the band edge energies have also been used [44–47].

14.4.3 Exciton Absorption Spectrum and Comparison with Experimental Data

Theoretical absorption spectrum for a single pair of transition from the first conduction subband to the first heavy-hole subband is shown in Fig. 14.5, where the discrete $1s$ bound state $E_x = E_{1s}$ contributes as a Lorentzian spectrum, and the step-like density of states enhanced by the Sommerfeld factor contributes as the high energy tail.

With an applied electric field, the quantum confined Stark effects can be measured from the shift of the peak absorption coefficient as a function of the applied electric field. Figure 14.6 shows the polarization-dependent optical absorption spectra for (a) TE and (b) TM polarizations [17, 48] with estimated electric fields [13]. Theoretical results using the parabolic band model [15] presented in this section and using a valence-band mixing model [43] have been used to successfully match these

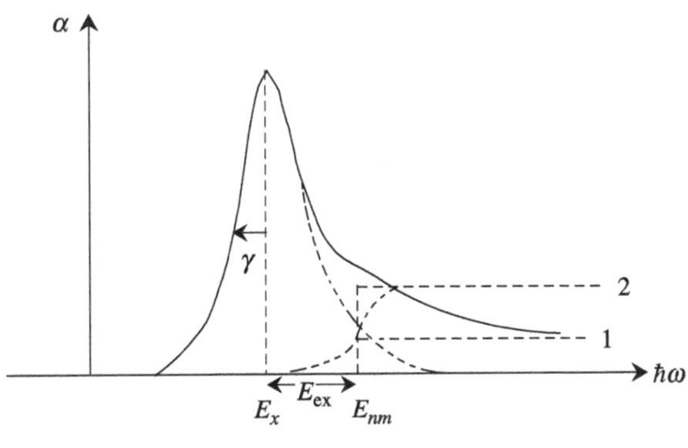

Figure 14.5 Exciton absorption spectrum of a quantum-well structure in the absence of an applied electric field. The contributions due to the discrete and continuum states are shown separately as dashed lines.

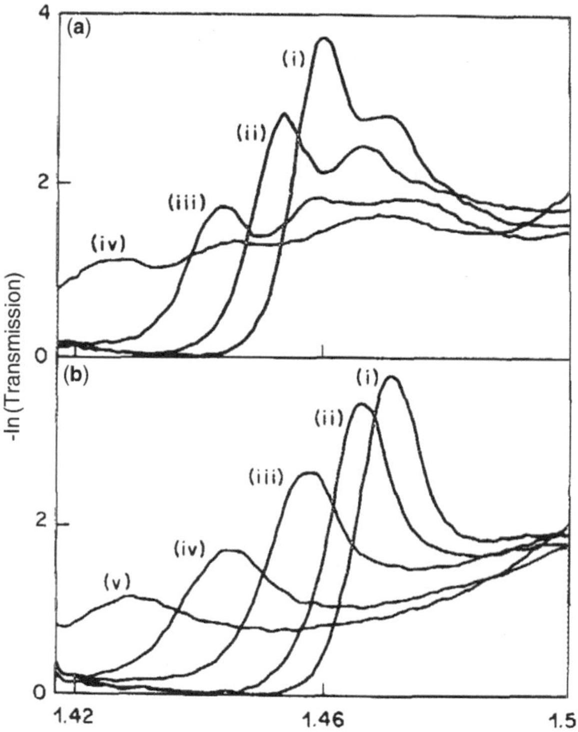

Figure 14.6 Experimental absorption spectra of a GaAs/$Al_{0.3}$ $Ga_{0.7}As$ quantum-well waveguide modulator as a function of field: (a) TE polarization for (i) 0 kV/cm, (ii) 60 kV/cm, (iii) 100 kV/cm, (iv) 150 kV/cm. (b) TM polarization for (i) 0 kV/cm, (ii) 60 kV/cm, (iii) 110 kV/cm, (iv) 150 kV/cm, and (v) 200 kV/cm. (Reprinted with permission from [48] © 1985 American Institute of Physics.)

experimental data, as shown in Fig. 14.7. To understand these data, we make the following observations:

1. The exciton absorption peak energy depends on the band-edge transition energy $E_{\rm h1}^{\rm e1}(F)$, which shifts quadratically as a function of the field, minus the amount of the $1s$ state exciton binding energy E_{1s}, which is about 8 meV for the heavy hole exciton. The transition energies of the heavy-hole and light-hole exciton peaks versus the applied electric field are shown in Fig. 14.8. The binding energy for a bulk (3D) GaAs R_y is about 4.2 meV and is $4R_{\rm y} \simeq 16.8$ meV for a pure 2D exciton. A quasi-two-dimensional quantum-well structure gives a binding energy corresponding to an *effective dimension* between 2D and 3D. Therefore, the binding energy of the quantum well is somewhere between 2D and 3D.

   As a matter of fact, because the hydrogen atom equation in an α-dimensional space has analytical solutions for an integer α such as 2 and 3, the idea is to

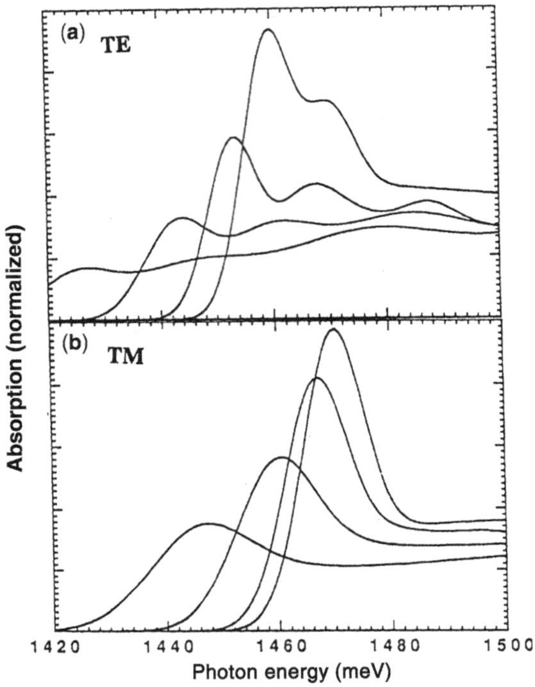

Figure 14.7 Theoretically calculated absorption spectra for the quantum-well waveguide modulator with experimental data shown in Fig. 14.6 (a) TE polarization and (b) TM polarization. (Reprinted with permission from [43] © 1993 American Physics Society.)

extend the general result for a given α using the concept of analytical continuation. Using this approach, the oscillator strength for the bound and continuum states can be obtained using the analytical formulas once the effective dimension is determined. The effective dimension can be extracted by comparing the binding energy calculated variationally as discussed in this section with the analytical formula for binding energy [49].

2. The TE oscillator strength for the heavy-hole exciton is approximately three times that of the light-hole exciton. However, because the light-hole exciton transition energy is already in the continuum states of the heavy-hole transition, the spectrum would not show a 3 : 1 ratio.
3. The TM polarization spectra show that it is the light-hole exciton transition that is dominant for this polarization as a result of the optical momentum matrix-selection rule.
4. At a fixed optical energy $\hbar\omega$ of the incident light, the absorption coefficient can change drastically especially when $\hbar\omega$ is near an exciton peak absorption energy. This enhanced change of the absorption by an applied voltage will be further discussed in the next section on interband electroabsorption modulators.

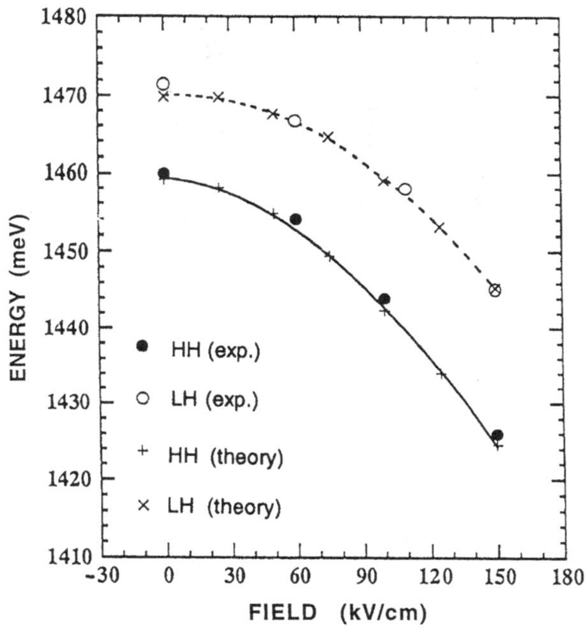

Figure 14.8 Comparison of experimental and theoretically calculated HH and LH exciton energies versus the electric field obtained from the peak energies in Fig. 14.6 and Fig. 14.7. (Reprinted with permission from [43] © 1993 American Physics Society.)

14.5 ELECTROABSORPTION MODULATOR

The electroabsorption modulators can be designed [10, 50–56] using a waveguide configuration and a transverse transmission (or reflection) configuration, as shown in Figs. 14.9a and b, respectively. Suppose the operating optical energy $\hbar\omega_0$ is chosen near the exciton peak when a voltage V is applied, Fig. 14.9c. The transmission coefficient is proportional to

$$T(V) = e^{-\alpha(V)L}. \tag{14.5.1}$$

For the waveguide modulator, $\alpha(V)$ is the absorption coefficient of the waveguide region multiplied by the optical confinement factor Γ, and L is the total length of the guide for the waveguide modulator. For the transverse transmission modulator, $\alpha(V)$ is the average absorption coefficient of the multiple-quantum-well region, and L is the total thickness of the multiple quantum well (MQW) region. The couplings or reflections at the facets are ignored here for convenience. The on/off ratio (or contrast ratio) $R_{\text{on/off}}$ is defined as (Fig. 14.9d)

$$R_{\text{on/off}} = \frac{P_{\text{out}}(V_{\text{on}} = 0)}{P_{\text{out}}(V_{\text{off}} = V)} = \frac{T(0)}{T(V)} \tag{14.5.2a}$$

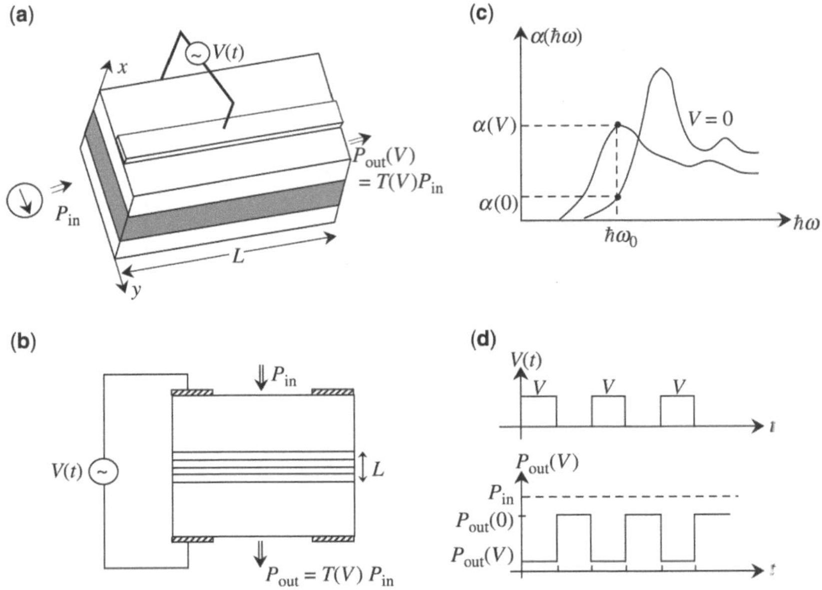

Figure 14.9 (a) A waveguide electroabsorption modulator. (b) A transverse transmission electroabsorption modulator. (c) The absorption coefficient $\alpha(\hbar\omega)$ of a quantum-well modulator at two different bias voltages, V and 0, for example. (d) With a bias voltage $V(t)$ across the modulator, the transmitted optical power (solid square wave) is modulated. Dashed line is the input power P_{in}.

or in decibels

$$\begin{aligned} R_{on/off}(\text{dB}) &= 10\log_{10}\frac{T(V_{on}=0)}{T(V_{off}=V)} \\ &= 4.343[\alpha(V)-\alpha(0)]L. \end{aligned} \tag{14.5.2b}$$

Therefore, in principle, the magnitude of the extinction ratio or the on/off ratio can be made as large as possible by increasing the cavity length L. However, we note that the insertion loss L_{in} is defined as

$$L_{in} = \frac{P_{in} - P_{out}(0)}{P_{in}} = 1 - T(0) = 1 - e^{-\alpha(0)L} \tag{14.5.3}$$

at the transmission (on) state. Because $\alpha(0)$ is always finite, a large cavity length L will decrease the transmissivity $T(0)$ exponentially. We may achieve an infinite on/off ratio using an infinitely long cavity but obtain no light transmission even for the on-state. If $T(0) \rightarrow 0$, the insertion loss approaches 100%. Naturally, this is not desirable and an optimum design, which maximizes the extinction ratio and minimizes the insertion loss, is necessary. Note that V_{on} is set to zero here only for illustration purposes. For high-speed switching, it is usually not chosen to be zero.

Another useful figure of merit in the design of the electroabsorption modulator is the change in absorption coefficient per unit applied voltage.

$$\frac{\Delta\alpha}{\Delta V} = \frac{\alpha(V_{\text{off}}) - \alpha(V_{\text{on}})}{\Delta V} \tag{14.5.4}$$

where the change in voltage $\Delta V = |V_{\text{on}} - V_{\text{off}}|$. This occurs because the on/off ratio per unit applied voltage is

$$\begin{aligned}\frac{R_{\text{on/off}}}{\Delta V} &= 4.343\frac{(\alpha(V) - \alpha(0))L}{\Delta V} \\ &= 4.343\frac{\Delta\alpha}{\Delta F}\end{aligned} \tag{14.5.5}$$

where the electric field across the multiple-quantum-well region is approximately given by $F = V/L$, if the built-in voltage V_{bi} is ignored. (Otherwise $\Delta F = (V - V_{\text{bi}})/L$ has to be used.) For more discussions on the figure of merits such as $\Delta\alpha/(\Delta F)^2$, $C(\Delta V)^2$, and the optimization of the contrast ratio, drive voltage, bandwidth, and total insertion loss, see Ref. 56. An important limiting factor in the high-speed response is the time constant for the holes be to be swept out of the quantum wells. A detailed analysis with experimental results on the carrier swept out and the barrier design have been reported [57].

EA modulators operating at 40 Gb/s with low driving voltage have been realized using ideas including strained InGaAs/InAlAs MQWs. Optimization of the design and demonstration of 40 Gb/s modulators and their EA modulation characteristics have been reported in Refs. 58–63. QCSE using Ge/SiGe quantum wells on silicon substrate has also been demonstrated with electroabsorption effects as large as III–V [64]. The Ge layers are compressively strained whereas the SiGe layers are tensile strained. The design is such the compressive and tensile strains are balanced in each quantum well pair so no strain energy is accumulated.

14.6 INTEGRATED ELECTROABSORPTION MODULATOR-LASER (EML)

The integrated electroabsorption modulator-laser (EML) is probably the first practical optoelectronic integrated circuit (OEIC) component, which was produced in large volume as the dominant source for dense-wavelength division multiplexing (DWDM) communications systems. The basic idea and the simplest design is to integrate a DFB laser with an EA modulator, creating an EML, which is used as a laser transmitter. The DFB laser is cw biased above threshold, and the high-speed modulated data such as 2.5 Gb/s, 10 Gb/s, or 40 Gb/s electrical signals are applied to the EA modulator at a reverse biased voltage. The laser can be a DFB laser [65–68], a DBR laser [69], a $\lambda/4$-shifted DFB laser [70], or a sample-grating tunable DBR laser [71]. For DWDM applications, tunable lasers play an important role. For example, a tunable sample-grating DBR laser with an integrated EA modulator covering 51 channels at 100 GHz spacing over a 40-nm tuning range has been reported [71]. A fully functional 2.5 Gb/s EA modulator integrated with tunable DBR laser transmitter over 20 channels on a 50 GHz grid meeting all long-haul transmission requirements for stability, chirp, power, and linewidth has been demonstrated as well [72]. In recent years, integration of 40 channels of InP-transmitter photonic integrated circuit at 40 Gb/s each channel with an aggregate data rate of 1.6 Tbit/s has been realized [73–76]. Figure 14.10 shows the scaling of InP-based transmitter chips used in telecommunications network from LED through EML to recent 40 × 40 Gb/s photonic integrated circuits. Figure 14.11a shows a schematic diagram of the 40-channel transmitter photonic integrated-circuit architecture with 240 device elements on a single chip. Figure 14.11b shows the measured output optical spectrum of the 40 wavelengths. Each channel consists of a tunable DFB laser with a back facet power monitor, an EA modulator, a power flattening element (PFE), and front power monitor. The PFE is used to level the output power from the individual channels. The 40 channels are then combined using the arrayed waveguide grating (AWG) multiplexer.

In this section, we discuss the essential ideas and the physical principles of the EML devices. The major reason for using the external EA modulator is that the chirp of an EA modulator is much smaller than that of a directly modulated laser. In an EA modulator, the linewidth enhancement factor or the chirp is related to the change of the real (n') and imaginary (n'') parts of the refractive index due to a reverse biased voltage (V):

$$\alpha_e = \frac{\partial n'/\partial V}{\partial n''/\partial V} = \frac{4\pi}{\lambda}\frac{\partial n'/\partial V}{\partial \alpha/\partial V}. \tag{14.6.1}$$

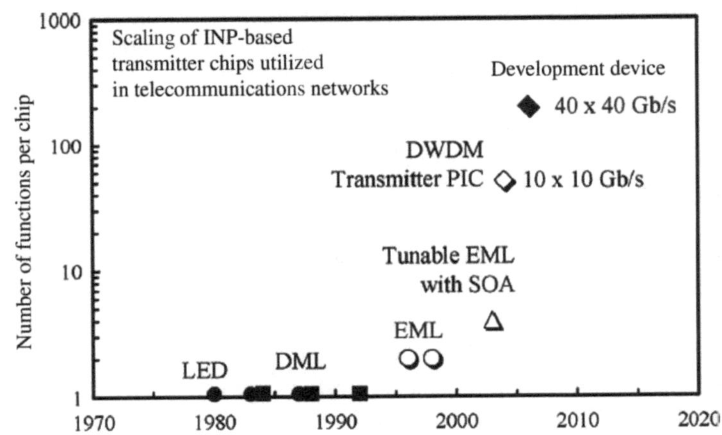

Figure 14.10 A plot of the scaling of photonic integrated transmitter chips in terms of the number of functions per chip. (Reprinted with permission from [76] © 2007 IEEE.)

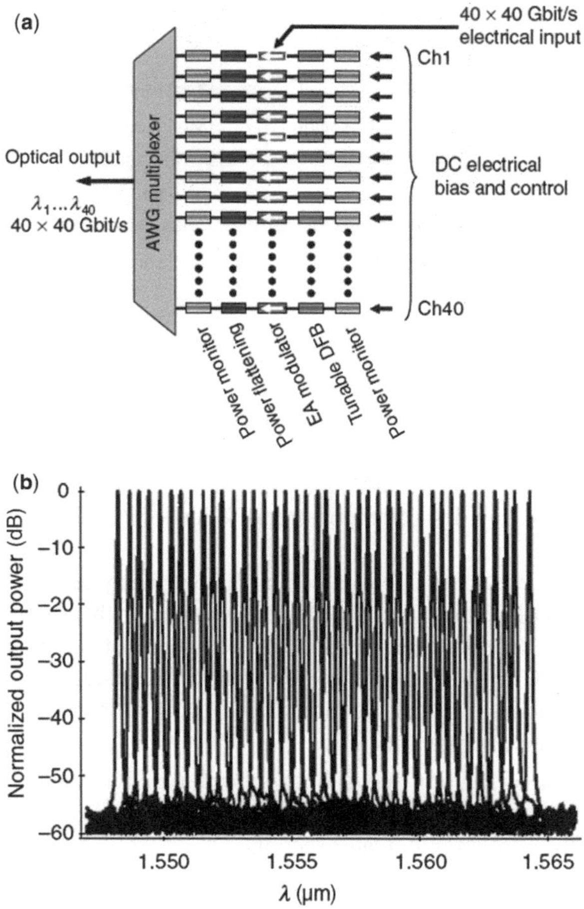

Figure 14.11 (a) An architecture for a 40-channel transmitter photonic integrated circuit. (b) The measured normalized optical output spectrum of the 40 wavelengths. (Reprinted with permission from [74] © 2006 Institute of Engineering and Technology: Electron. Lett.)

Because the modulator is reverse-biased, the carrier density is quite small except for carrier generation due to absorption of the optical signal. In the case of a directly modulated laser, the device is under forward bias and the carrier density is very high. The chirp parameter of a forward biased semiconductor laser is typically 2 to 5 for strained quantum-well lasers and even higher for bulk semiconductor diode lasers. This chirp causes significant broadening of the laser linewidth, resulting in a significant dispersion for long haul optical communication. Direct laser modulation results in multiple-angstrom wavelength chirp; therefore, external modulation using an EA modulator or $LiNbO_3$ electrooptical modulator is preferred. The α_e parameter of an EA modulator is much smaller than unity (0.2–0.8), which leads to the wavelength chirp smaller than 0.2 Å [77, 78]. An integrated EA modulator with a DFB laser is especially attractive due to the ability to perform monolithic integration

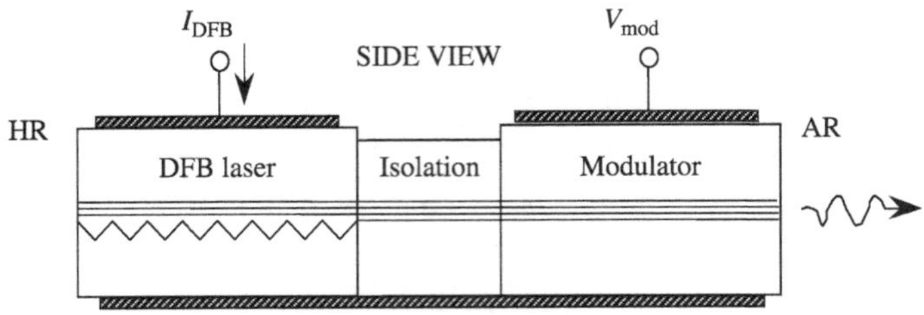

Figure 14.12 A schematic diagram of an integrated electroabsorption modulator-laser (EML).

and mass production. This EML was the first practical photonic integrated circuit component with a large-volume production.

Figure 14.12 shows a schematic of an integrated EML [79]. The DFB laser is forward biased at a cw operation current far above threshold, whereas the EA modulator is reverse biased with an additional high-speed data signal (such as 10 Gb/s data) applied. A few key design criteria are as follows [66, 80–82]:

(a) The DFB laser section operates with a photon energy below the band edge of the modulator section with a detuning wavelength (between the laser wavelength and the modulator band-edge wavelength) of the order varying from 40 nm to 80 nm for optimized performance. Considerations of extinction ratio, insertion loss when the modulator is in the ON state, and chirp of the modulator are important factors.

(b) The DFB laser section has a high reflection (HR) coating; the modulator section has an antireflection (AR) coating; and the laser section is isolated electrically from the modulator section.

(c) The optical transmission from the laser to the modulator waveguide should be maximized. The cross-talk between the laser and modulator due to reflection from the exit facet of the modulator should be minimized.

A few integration techniques [83] have been used for the integration of a DFB (or DBR) laser and an EA modulator: the selective-area growth (SAG) [66, 67, 84], the butt-joint technology [68, 85], the stacked-layer technology [86–88], the quantum-well interdiffusion [89], and the identical-layer design [83, 90, 91]. Below we discuss some of these techniques.

1. **Selective-Area Growth (SAG):** The key idea is to use MOCVD selective-area growth for the multiple quantum well growth in the active regions with a single growth step over a masked substrate. This method is based on the variation of the band gap of MQW materials in the same plane on a masked substrate. Source material that does not nucleate on the dielectric masked areas is deposited nearby, inducing a local increase in growth rate. Epitaxial band-gap energy shifts are hence mainly controlled by the geometry of the lithographically defined dielectric pattern. This method is most popular due to its processing

simplicity as a planar technology and its high performance due to the precise control of the wavelength shift and the smooth taper-like optical transition between two waveguides. A potential drawback is the difficulty of independent optimization of each device in terms of a precise strain or composition adjustment.

Figure 14.13a shows a structure of an integrated EA modulator with a DFB laser based on the selective-area growth technique [66]. It consists of a separate-confinement-heterojunction (SCH)-MQW structure using InGaAs/InGaAsP

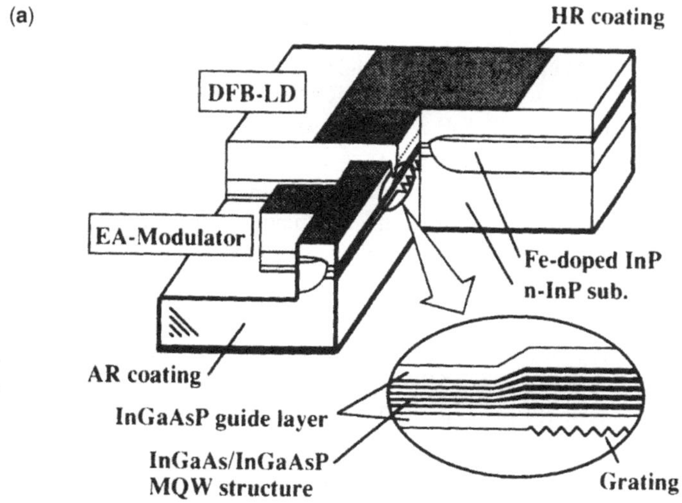

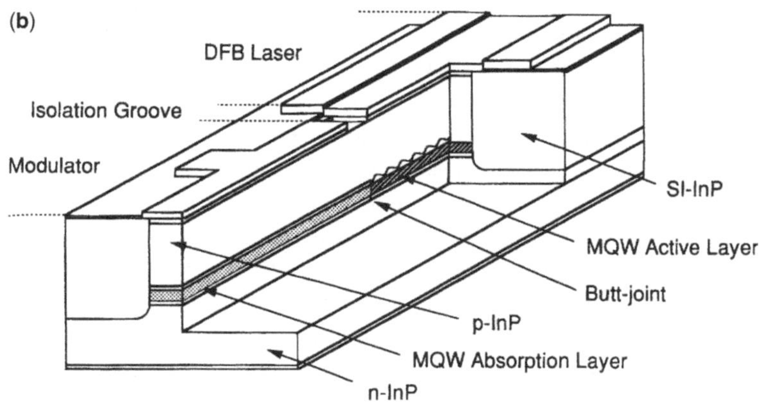

Figure 14.13 (a) A structure of an integrated EA modulator with a DFB laser based on the selective-area growth technique. The materials are InGaAs quantum wells with InGaAsP barriers with slightly different compositions and well widths in the laser and modulator sections. (Reprinted with permission from [66] © 1993 IEEE.) (b) An integrated EA modulator–DFB laser based on a butt-joint technology. The modulator section may have different number of quantum wells from that of the laser section in this approach. (Reprinted with permission from [68] © 1997 IEEE.)

with slightly different thicknesses and materials compositions. The band edges of the modulator and laser sections are characterized by photoluminescence (PL) and they are 1.49 and 1.565 μm, respectively, which are controlled by the SAG mask dimensions. The fabrication process [66] based on the SAG is summarized in Figs. 14.14a–d.

(a) Wafer preparation: Initially, a first-order grating was partially etched in the laser section on a (100) *n*-InP substrate. Then an 0.15-μm-thick *n*-InGaAsP ($\lambda_g = 1.15$ μm) lower optical waveguide layer and a 10-nm-thick *n*-InP cap layer were grown on the partially corrugated substrate. A SAG mask made of a 0.2-μm-thick SiO_2 mask is formed, which consists of a pair of 15-μm-wide SiO_2 strips with a 10-μm gap between the strips only in the laser section, and no mask for the modulator section.

(b) Selective-area growth of the MQW layers by MOCVD on the patterned wafer: On the patterned wafer, successive MOCVD-SAG growth was performed to grow an InGaAs/InGaAsP MQW structure (with five wells), an InGaAsP upper optical waveguide layer ($\lambda_g = 1.15$ μm), a *p*-InP cladding layer, and an InGaAsP cap layer. The growth conditions for the wells were adjusted to produce tensile strain (0.3%) in the modulator section in the unmasked area. This produces a wider well to enhance the QCSE and for a better crystal quality in the laser section. The SiO_2 mask is designed such that the PL wavelengths are 1.49 μm and 1.565 μm for the modulator and laser section, respectively. The transition region between the laser and modulator section was found to be less than 50 μm, corresponding roughly with the length of the vapor phase diffusion in the MOCVD growth conditions.

(c) Buried-heterostructure (BH) processing or growth: After SAG, the wafer was processed into a planar BH structure with a 3.5-μm-thick semi-insulating (SI) layer doped by $1.5 \times 10^{17}\,\text{cm}^{-3}$ Fe. To reduce the parasitic capacitance further, an 0.3-μm-thick *n*-InP hole blocking layer was introduced between the SI and *p*-InP cladding layers.

(d) Electrical isolation etching and electrode formation: The *p*-InP layer is etched partially for electrical isolation between the laser and modulator. The resultant isolation resistance is typically 10 kΩ. The modulator is AR coated ($R = 0.2\%$) and the laser section is HR coated ($R = 90\%$).

2. **Butt-Coupling Technology:** This method involves a multiple epitaxy growth of distinct active layers for the laser and modulator and allows for independent design of the active layers of the two devices. It consists of the growing the laser active layers on the whole wafer, followed by selective etching and subsequent regrowth of the modulator layers. A regrowth of the *p*-InP confining layer is finally performed. Figure 14.13b is based on a butt-joint technology [68], which has the advantage of allowing for optimization of the modulator and laser structures independently. The numbers of the quantum wells in the laser and modulator sections can be chosen differently, for example. The fabrication starts with the growth of six pairs of InGaAsP MQWs on an *n*-type (100) InP substrate as the active layer of the DFB laser by MOVPE. The

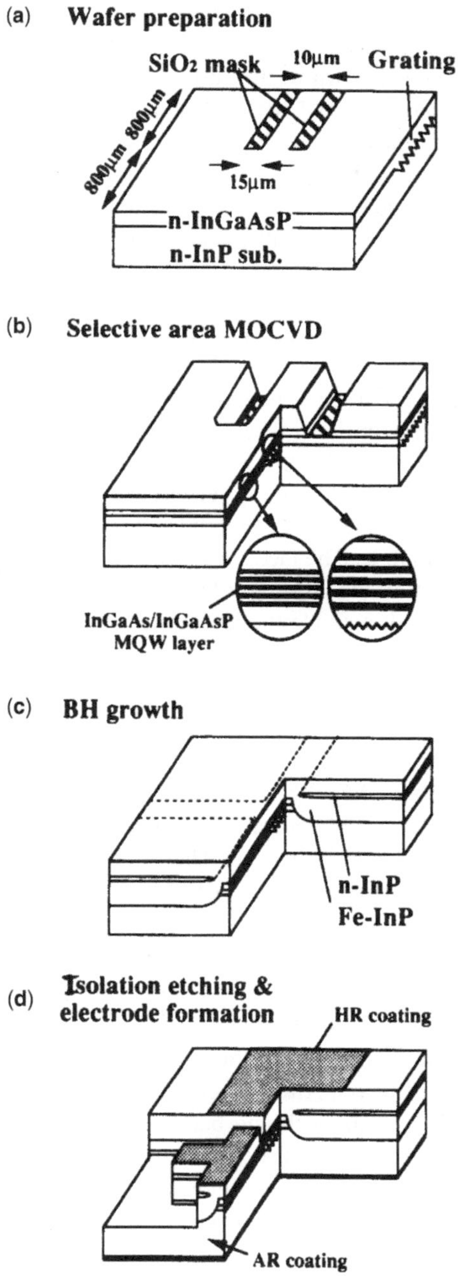

Figure 14.14 The fabrication process based on the selective-area growth. (a) Wafer preparation using an 0.2-μm-thick SiO_2 mask, which consists of a pair of 15-μm-wide SiO_2 strips with 10-μm-gap between the strips formed only in the laser section, and no mask for the modulator section. (b) Selective-area growth of the MQW layers by MOCVD on the patterned wafer. (c) Buried-heterostructure (BH) processing or growth. (d) Electrical isolation etching and electrode formation. The modulator is AR coated ($R = 0.2\%$) and the laser section is HR coated ($R = 90\%$). (Reprinted with permission from [66] © 1993 IEEE.)

wells were 6.7-nm-thick InGaAsP with compressive strain and the barriers were 15.1-nm-thick InGaAsP (1.25-μm PL wavelength). The MQW active layer was then etched down to the substrate except in the DFB laser section. Next, an additional InGaAsP MQW structure, with a 9.7-nm-thick InGaAsP well and a 5-nm-thick InGaAsP barrier (in 8 or 14 pairs), was selectively grown as the absorption layer of the modulator. The well was compressively strained by 0.5%, and the barrier was tensile strained to compensate for the total strain of the MQW. A butt-joint configuration is thus formed between the modulator and the DFB laser section. After a corrugation grating is formed in the DFB laser section, a *p*-type InP cladding layer and *p*-type contact layer were successively grown. The epitaxial layers including the MQW active and the MQW absorption layers were formed in a 2-μm-wide high-mesa ridge waveguide structure by a dry etching technique. The modulator and laser were buried in Fe-doped InP to reduce electric capacitance and to form the surface plane. An isolation groove was formed between the laser and modulator, followed by electrode formation. The devices were cleaved into a chip, in which the modulator is from 90 to 250 μm long, and the laser is 450 μm long. The facets were coated with an AR film for the modulator and a HR film for the DFB laser facet. Finally, the EML is packaged into a compact module with a single-mode fiber (SMF) pigtail for measurement.

3. **Double Stack Approach:** In this method, two MQW active layers are stacked on top of each other in one growth step, separated by a thin InP space layer. Selective wet chemical etching allows the localized removal of the top (laser) active layer, followed by a regrowth of *p*-type cladding and contact layers after waveguide processing.
4. **Identical Active Layer Approach:** This method is simple because it uses the same active layers for both devices. The design requires a positive detuning (i.e., longer wavelength) of the DFB laser Bragg wavelength relative to that of the exciton peak wavelength of the modulator absorption spectrum. The amount of detuning such as 40 nm for example is achievable owing to the broad gain spectrum of the laser as well as the sharp excitonic absorption spectrum of the modulator with the current MOVPE growth.
5. **Quantum-Well Interdiffusion:** This method uses quantum well intermixing by selective interdiffusion. After the growth of a single MQW stack, selective interdiffusion between the well and barrier material is induced, resulting in a blue shift of the absorption edge, which depends on the optical functions to be integrated. The research issues include the reproducibility, compatibility with a regrowth step, and ability to achieve a specific modulator structure for optimized performance.

Figure 14.15 shows an example of the laser emission spectrum of another EML, where the DFB laser is biased at a constant current and the EA modulator is biased at 0, −1, and −2 volts. We can see the reduction of the output power

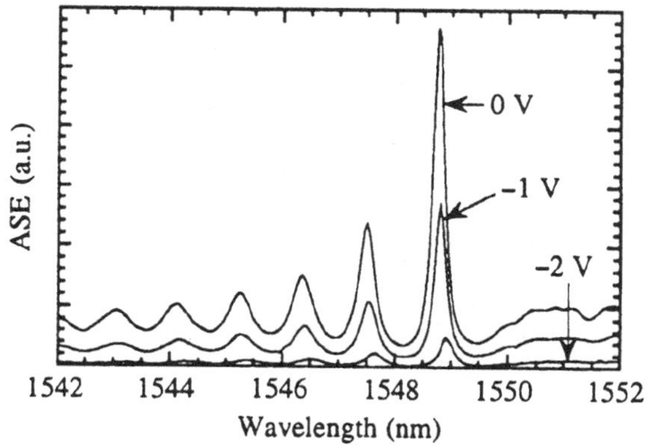

Figure 14.15 The laser emission spectrum of an EML, where the DFB laser is biased at a constant current and the electroabsorption modulator is biased at 0, -1, and -2 volts. (Reprinted with permission from [75] © 1997 SPIE.)

with increasing reverse bias voltage due to the increase of the absorption via QCSE. For a theoretical model of an EML, the propagation matrix approach as we described in Section 11.1 for DFB lasers has been applied [79, 92, 93]. The effects on the dynamic chirp have also been investigated in Refs. 77, 78, 81, 82. The light output power of the lasing mode from the EML of Fig. 14.13a is shown in Fig. 14.16a for various reverse biased voltages of the EA modulator. The reduction of the light output power with increasing reverse bias is again clear. Figure 14.16b shows the transmission or output power of the EML at three bias currents (20, 50, and 80 mA) of the DFB laser as a function of the EA modulator bias voltage. The modulation bandwidth or frequency response of the EA modulator at three bias currents of the DFB laser is shown in Fig. 14.16c. The design gives a 3-dB bandwidth of about 14 GHz. High bandwidth of 40 Gb/s or above has been realized in the past years. The bandwidth is limited mostly by the RC time constant of the modulator. One important effect is the hole pileup in the valence band of the quantum wells of the EA modulator section when the electron–hole pairs are generated due to the absorption of the DFB laser light by the modulator quantum wells. If the escape rate of the holes is too slow when the modulator is under high-speed modulation, the response of the modulator will be damped because the absorption of the modulator is reduced due to Pauli blocking caused by the hole pileup. Therefore, the design of the quantum wells to reduce the band edge discontinuity of the valence (heavy hole) band can be employed to reduce hole pileup. Different materials systems such as InGaAs/InAlAs [94], InAlGaAs/InAlGaAs [95] and InGaAs/InGaAsP [66], InGaAsP/InGaAsP [68] quantum well/barrier have been investigated and compared for EML applications near 1.55-μm telecommunication applications. Wide-bandwidth (larger than 50 GHz) using a traveling wave electrode EA modulator integrated DFB laser to overcome the RC-induced bandwidth

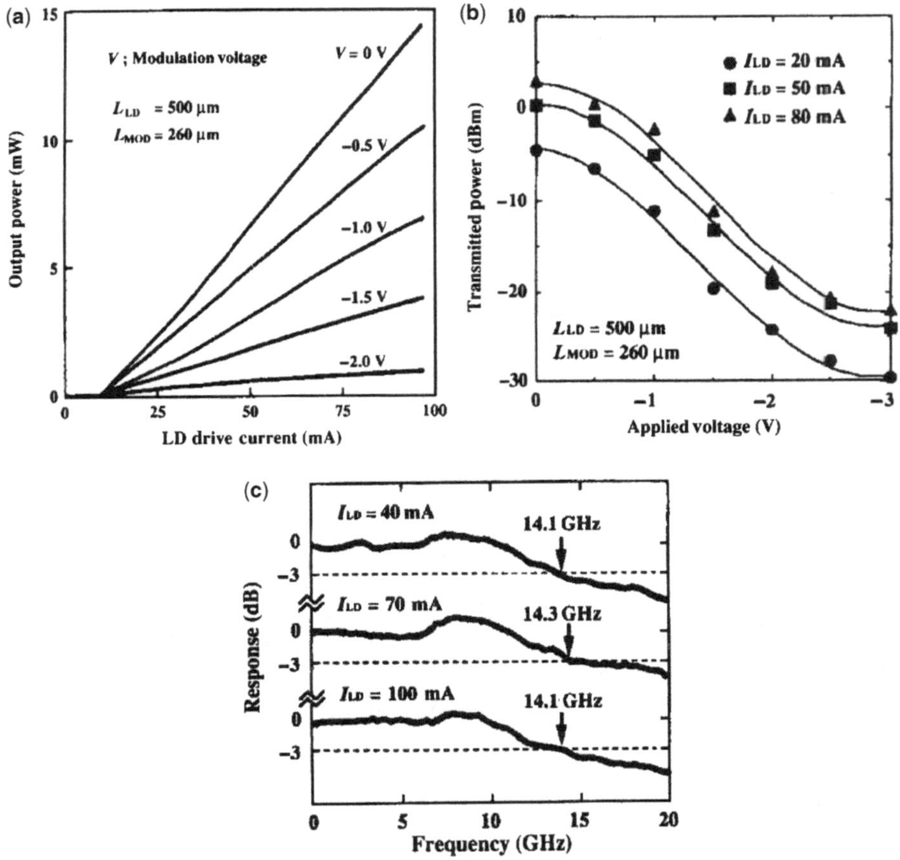

Figure 14.16 (a) The light output power of an EML (Fig. 14.13a) versus the bias current of the laser (L–I curve) under various reverse biased voltages of the EA modulator. (b) The transmission or output power of EML at three bias currents of the DFB laser as a function of the EA modulator bias voltage. (c) The modulation bandwidth or frequency response of the EA modulator at three bias currents of the DFB laser. (Reprinted with permission from [66] © 1993 IEEE.)

limitation has also been demonstrated [96]. Optimal design without the traveling wave electrode to achieve 40 Gb/s bandwidth is also possible [97].

14.7 SELF-ELECTROOPTIC EFFECT DEVICES (SEEDs)

Using the quantum confined Stark effects, interesting optical switch devices such as the self-electrooptic effect devices (SEEDs) [18–20, 50–56] have been demonstrated. These devices show optical bistability and consist of a multiquantum-well *p-i-n* diode structure with different possible loads, such as a resistor (R-SEED), a constant current source, or another *p-i-n* multiple-quantum-well diode (Symmetric or

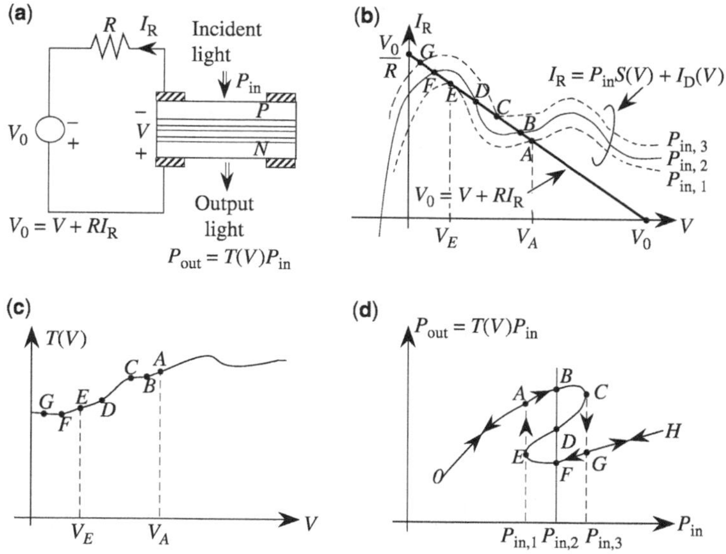

Figure 14.17 (a) A circuit diagram for a self-electrooptic effect device with a resistor load (R-SEED) in the presence of an incident laser light at the input P_{in}. (b) A graphical solution for the photocurrent response $I_R = S(V)\,P_{in} + I_D(V)$ and the load line $V_0 = V + RI_R$. (c) The transmission $T(V)$ of the laser light passing through the MQW *p-i-n* diode is plotted as a function of the voltage drop V across the diode. (d) The switch diagram for the optical output power versus the optical input power with the arrows showing the path of switching.

S-SEED). In this section, we discuss basic physical principles of a R-SEED. As shown in Fig. 14.17a, the circuit equation is given by Kirchoff's voltage law

$$V_0 = V + RI_R \tag{14.7.1}$$

where V is the voltage drop across the diode. The reverse bias voltage V and the reversed current I_R are defined as shown in Fig. 14.17a. For an incident laser light with a power P_{in}, the current I_R is the sum of the photocurrent $S(V)\,P_{in}$ and the dark current $I_D\,(V)$

$$I_R = P_{in}S(V) + I_D(V) \tag{14.7.2}$$

where $S(V)$ is the responsivity of the diode and is defined as

$$S(V) = q\frac{\eta_{int}A}{\hbar\omega} \tag{14.7.3}$$

where

$$A \simeq (1 - R_p)(1 - e^{-\alpha L} + R_n e^{-\alpha L}) \tag{14.7.4}$$

is the absorbance, that is, the fraction of absorbed power for a unit incident power, assuming a single path absorption. The absorbance can be derived noting that the reflectivity at the front surface R_p, plus the transmission after passing through the substrate $(1 - R_p)(1 - R_n)e^{-\alpha L}$, plus the absorbance A, must equal 1. At steady state, the voltage drop across the multiple-quantum-well diode V and the current I_R is determined by the simultaneous solutions of the above two equations. A graphical illustration of these two equations is plotted in Fig. 14.17b, where the intersection points stand for the possible solutions for I_R and V.

As can be seen from Fig. 14.17b, when the optical input power P_{in} is increased, the photocurrent is increased, in general, and the number of intersection points varies. For an optical power equal to $P_{in,1}$, we have two intersection points A and E with corresponding voltages V_A and V_E. Therefore, there are two possible output states: $P_{out} = T(V_A)\,P_{in,1}$, which is higher in output power and $P_{out} = T(V_E)\,P_{in,1}$, which is lower in output power because of the transmissivity $T(V_A) > T(V_E)$, Fig. 14.17c. Once we increase the optical input power to $P_{in,2}$, we find that there are three intersection points, B, D, and F, with corresponding voltages V_B, V_D, and V_F. Therefore, the optical output powers through the diode are given by three possible values, $P_{out} = T(V_B)\,P_{in,2} > T(V_D)\,P_{in,2} > T(V_F)\,P_{in,2}$. Increasing the optical input power to $P_{in,3}$, we have only two intersection points C and G with corresponding voltages V_C and V_G. Therefore, we obtain the output powers $P_{out} = T(V_C)\,P_{in,3} > T(V_G)\,P_{in,3}$.

The switching curve and the directions are shown in Fig. 14.17d. At a small input power, $P_{in} < P_{in,1}$, the optical transmission power $P_{out} = T(V)\,P_{in}$ is monotonically increasing with the input power P_{in}, as the photocurrent is small for a small incident optical power. Therefore, most of the reverse bias voltage is across the diode and V is close to V_0 ($V_A < V < V_0$). In this range of voltage, the transmission coefficient is rather flat or monotonic. As the input optical power exceeds $P_{in,1}$, $P_{in,2}$ and $P_{in,3}$, photocurrent will appear in the circuit, and the voltage drop across the diode will drop from V_C to V_G, where the transmission will drop from a high value $T(V_C)$ to a low value $T(V_G)$. After $P_{in} > P_{in,3}$, the output power starts to increase again because $P_{out} = T(V)\,P_{in}$ increases as P_{in} increases. This switch sequence is therefore $A \rightarrow B \rightarrow C \rightarrow G \rightarrow H$.

In the reverse direction, if we switch down the optical input power from a large value of P_{in} at H, it will go through $H \rightarrow G \rightarrow F \rightarrow E \rightarrow A \rightarrow 0$, because at point E the voltage has to switch from V_E to V_A as we decrease the input power $P_{in} \leq P_{in,1}$. Therefore, the optical output power will switch from a low state E to a high state A due to the large transmission coefficient $T(V_A) > T(V_E)$.

Symmetric self-electrooptic effect devices (S-SEEDs) [53] using another *p-i-n* multiple-quantum-well diode as the load, and field-effect transistor self-electrooptic effect devices (F-SEEDs) [20] have also been demonstrated to show interesting physics and applications. For example, the S-SEED can act as a differential logic gate capable of NOR, OR, NAND, and AND functions. These devices made by maximizing the ratio of the absorption coefficients in the high and low states while minimizing the change in electric field can give nearly optimum performance [53]. From the I–V curve of a SEED, Fig. 14.17a and Fig. 14.17b, we see that negative differential conductivity exists. A SEED oscillator [19, 52] can also be designed by a series connection of a SEED and an LC resonator circuit and the SEED is optically pumped to produce a negative electric conductance in the photocurrent response. For example, oscillators with oscillation frequencies from 8.5 to 110 MHz have been demonstrated [52]. For the 8.5 MHz oscillator, frequency tuning by changing the bias voltage of the SEED has a tuning rate 16.7 kHz/V. This frequency tuning is caused by the change in the capacitance in the depletion layer of the SEED with the voltage change. The capacitance can also be changed optically by changing the optical power coupled to the SEED.

APPENDIX 14A: TWO-PARTICLE WAVE FUNCTION AND THE EFFECTIVE MASS EQUATION

To describe an electron–hole pair state, the two-particle wave function $\Psi(\mathbf{r}_e, \mathbf{r}_h)$ for an electron at position $\mathbf{r}_e$ and a hole at position $\mathbf{r}_h$ can be expressed as a linear combination of the direct product of the single (uncorrelated) electron and hole Bloch functions, $\Psi_{c,\mathbf{k}_e}(\mathbf{r}_e)$ and $\Psi_{v,-\mathbf{k}_h}(\mathbf{r}_h)$, respectively.

$$\Psi(\mathbf{r}_e, \mathbf{r}_h) = \sum_{\mathbf{k}_e}\sum_{\mathbf{k}_h} A(\mathbf{k}_e, \mathbf{k}_h)\Psi_{c,\mathbf{k}_e}(\mathbf{r}_e)\Psi^*_{v,-\mathbf{k}_h}(\mathbf{r}_h) \tag{14A.1}$$

where $A(\mathbf{k}_e, \mathbf{k}_h)$ represents the amplitude function. Note that the Bloch functions $\Psi_{c,\mathbf{k}_e}(\mathbf{r}_e)$ and $\Psi_{v,-\mathbf{k}_h}(\mathbf{r}_h)$ contain both the slowly varying plane-wavelike envelope and fast-varying Bloch periodic functions. In the effective mass approximation for electron and hole pairs, an envelope function $\Phi(\mathbf{r}_e, \mathbf{r}_h)$ is defined as the inverse Fourier transform of the amplitude function $A(\mathbf{k}_e, \mathbf{k}_h)$

$$\Phi(\mathbf{r}_e, \mathbf{r}_h) = \sum_{\mathbf{k}_e}\sum_{\mathbf{k}_h} A(\mathbf{k}_e, \mathbf{k}_h)\frac{e^{i\mathbf{k}_e\cdot\mathbf{r}_e}}{\sqrt{V}}\frac{e^{i\mathbf{k}_h\cdot\mathbf{r}_h}}{\sqrt{V}} \tag{14A.2}$$

which is the plane-wave expansion of the two-particle wave function. The Fourier transform of the wave function $\Phi(\mathbf{r}_e, \mathbf{r}_h)$ is

$$A(\mathbf{k}_e, \mathbf{k}_h) = \int d^3\mathbf{r}_e \int d^3\mathbf{r}_h \Phi(\mathbf{r}_e, \mathbf{r}_h)\frac{e^{-i\mathbf{k}_e\cdot\mathbf{r}_e}}{\sqrt{V}}\frac{e^{-i\mathbf{k}_h\cdot\mathbf{r}_h}}{\sqrt{V}}. \tag{14A.3}$$

The major difference between Ψ and Φ in (14A.1) and (14A.2) is the basis functions used in their expansions. In the envelope function Φ for the electron–hole pair states, the fast-varying Bloch periodic parts $u_c(\mathbf{r})$ and $u_v(\mathbf{r})$, which behave like $|iS\rangle$ and $|\frac{3}{2}, \pm\frac{3}{2}\rangle$ or $|\frac{3}{2}, \pm\frac{1}{2}\rangle$, have been dropped from the basis functions and only the plane-wave parts are kept. The envelope wave function $\Phi(\mathbf{r}_e, \mathbf{r}_h)$ satisfies the effective mass equation

$$[E_g + E_c(-i\nabla_e) - E_v(-i\nabla_h) + V(\mathbf{r}_e, \mathbf{r}_h)]\Phi(\mathbf{r}_e, \mathbf{r}_h) = E\Phi(\mathbf{r}_e, \mathbf{r}_h) \tag{14A.4}$$

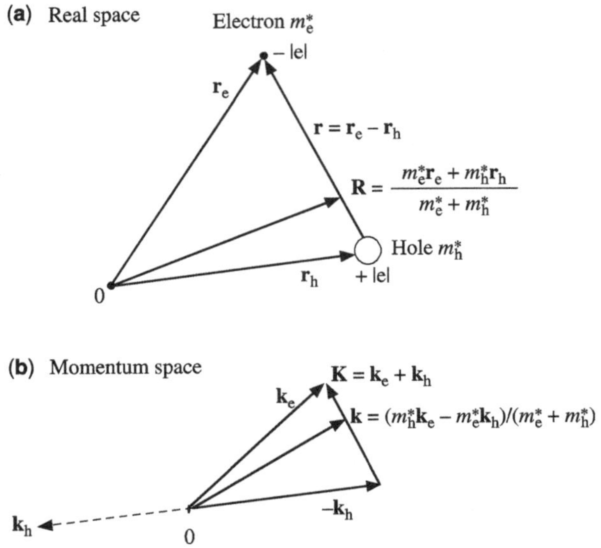

Figure 14A.1 (a) An illustration for the electron position vector $\mathbf{r}_e$, the hole position vector $\mathbf{r}_h$, the difference coordinate vector $\mathbf{r} = \mathbf{r}_e - \mathbf{r}_h$, and the center-of-mass coordinate vector $\mathbf{R} = (m_e^*\mathbf{r}_e + m_h^*\mathbf{r}_h)/(m_e^* + m_h^*)$. (b) The relations between the wave vectors, $\mathbf{k}_e$, $\mathbf{k}_h$, $\mathbf{k}$, and $\mathbf{K}$, in the Fourier transform space.

where we have replaced $\mathbf{k}_e$ in the dispersion relation $E_c \equiv E_c(\mathbf{k}_e)$ by the differential operator $-i\nabla_e$ for the $\mathbf{r}_e$ variables, and $\mathbf{k}_h$ in $E_v \equiv E_v(\mathbf{k}_h)$ by $-i\nabla_h$ for the $\mathbf{r}_h$ variables. Using the parabolic model, we have $E_c(\mathbf{k}_e) = \hbar^2 k_e^2/2m_e^*$ and $E_v(\mathbf{k}_h) = -\hbar^2 k_h^2/2m_h^*$.

The interaction potential $V(\mathbf{r}_e, \mathbf{r}_h)$ may be of the form

(1)
$$V(\mathbf{r}_e, \mathbf{r}_h) = e\mathbf{F} \cdot (\mathbf{r}_e - \mathbf{r}_h). \tag{14A.5}$$

It is the potential energy of a free electron and a free hole in the presence of a uniform electric field $\mathbf{F}$. This will lead to the Franz–Keldysh effect [1–3] for the optical absorption, as discussed in Section 14.2. The interaction potential can also be of the form

(2)
$$V(\mathbf{r}_e, \mathbf{r}_h) = -\frac{e^2}{4\pi\varepsilon_s|\mathbf{r}_e - \mathbf{r}_h|}. \tag{14A.6}$$

It is the Coulomb interaction between an electron at $\mathbf{r}_e$ and a hole at $\mathbf{r}_h$, where ε_s is the permittivity of the semiconductor. This potential leads to the exciton effect [4–9] in the optical absorption, which is discussed in Section 14.3.

Solution of the Two-Particle Effective-Mass Equation In general, for $V(\mathbf{r}_e, \mathbf{r}_h) = V(\mathbf{r}_e - \mathbf{r}_h)$, which depends only on the difference between the electron and hole position vectors, we may change the variables into the difference coordinate and the center-of-mass coordinate system, $\mathbf{r}$ and $\mathbf{R}$, respectively, as shown in Fig. 14A.1a

$$\mathbf{r} = \mathbf{r}_e - \mathbf{r}_h \qquad \mathbf{R} = (m_e^* \, \mathbf{r}_e + m_h^* \, \mathbf{r}_h)/M \tag{14A.7}$$

where $M = m_e^* + m_h^*$. The corresponding Fourier transform variables of $\mathbf{r}$ and $\mathbf{R}$ in the momentum space are, Fig. 14A.1b,

$$\mathbf{k} = (m_h^* \, \mathbf{k}_e - m_e^* \, \mathbf{k}_h)/M \quad \text{and} \quad \mathbf{K} = \mathbf{k}_e + \mathbf{k}_h \tag{14A.8}$$

respectively, which can also be checked using

$$\mathbf{k} \cdot \mathbf{r} + \mathbf{K} \cdot \mathbf{R} = \mathbf{k}_e \cdot \mathbf{r}_e + \mathbf{k}_h \cdot \mathbf{r}_h. \tag{14A.9}$$

We can also express the above relations as

$$\mathbf{r}_e = R + \frac{m_h^*}{M}\mathbf{r} \qquad \mathbf{r}_h = \mathbf{R} - \frac{m_e^*}{M}\mathbf{r} \tag{14A.10}$$

$$\mathbf{k}_e = \frac{m_e^*}{M}\mathbf{K} + \mathbf{k} \qquad \mathbf{k}_h = \frac{m_h^*}{M}\mathbf{K} - \mathbf{k} \tag{14A.11}$$

From the corresponding differential operators

$$\begin{aligned} \mathbf{k}_e &= -i\nabla_e \quad & \mathbf{k}_h &= -i\nabla_h \\ \mathbf{k} &= -i\nabla_{\mathbf{r}} \quad & \mathbf{K} &= -i\nabla_{\mathbf{R}} \end{aligned} \tag{14A.12}$$

we obtain

$$\frac{\hbar^2 k_e^2}{2m_e^*} + \frac{\hbar^2 k_h^2}{2m_h^*} = \frac{\hbar^2 K^2}{2M} + \frac{\hbar^2 k^2}{2m_r^*} \tag{14A.13a}$$

$$-\frac{\hbar^2}{2m_e^*}\nabla_e^2 - \frac{\hbar^2}{2m_h^*}\nabla_h^2 = -\frac{\hbar^2}{2M}\nabla_{\mathbf{R}}^2 - \frac{\hbar^2}{2m_r^*}\nabla_r^2. \tag{14A.13b}$$

Here m_r^* is the reduced effective mass, defined by

$$\frac{1}{m_r^*} = \frac{1}{m_e^*} + \frac{1}{m_h^*}. \tag{14A.14}$$

Therefore, the effective mass equation (14A.4) becomes

$$\left[-\frac{\hbar^2}{2M}\nabla_R^2 - \frac{\hbar^2}{2m_r^*}\nabla_r^2 + V(\mathbf{r}) - (E - E_g)\right]\Phi(\mathbf{R}, \mathbf{r}) = 0. \tag{14A.15}$$

The solution to the above equation can then be obtained using the method of the separation of variables, noting that the **R** dependence is a simple free particle wave function,

$$\Phi(\mathbf{R}, \mathbf{r}) = \frac{e^{i\mathbf{K}\cdot\mathbf{R}}}{\sqrt{V}}\phi(\mathbf{r}) \tag{14A.16}$$

where $\phi(\mathbf{r})$ satisfies

$$\left[-\frac{\hbar^2}{2m_r^*}\nabla_r^2 + V(\mathbf{r}) - E'\right]\phi(\mathbf{r}) = 0 \tag{14A.17}$$

and the energy E' is

$$E' = E - E_g - \frac{\hbar^2 K^2}{2M}. \tag{14A.18}$$

Define the Fourier transform pair as

$$\phi(\mathbf{r}) = \sum_{\mathbf{k}} a(\mathbf{k})\frac{e^{i\mathbf{k}\cdot\mathbf{r}}}{\sqrt{V}} \quad a(\mathbf{k}) = \int d^3\mathbf{r}\,\phi(\mathbf{r})\frac{e^{-i\mathbf{k}\cdot\mathbf{r}}}{\sqrt{V}}. \tag{14A.19}$$

We find

$$\Phi(\mathbf{r}_e, \mathbf{r}_h) = \Phi(\mathbf{R}, \mathbf{r}) = \frac{e^{i\mathbf{K}\cdot\mathbf{R}}}{\sqrt{V}}\sum_{\mathbf{k}} a(\mathbf{k})\frac{e^{i\mathbf{k}\cdot(\mathbf{r}_e-\mathbf{r}_h)}}{\sqrt{V}}. \tag{14A.20}$$

Optical Matrix Element of the Two-Particle Transition Picture The optical matrix element between the ground state (all electrons are in the valence band) and the final state (the electron–hole pair state) is described by [4, 5, 13]

$$\begin{aligned}\langle f|e^{i\mathbf{k}_{op}\cdot\mathbf{r}}\hat{e}\cdot\mathbf{p}|i\rangle &= \sum_{\mathbf{k}_e}\sum_{\mathbf{k}_h} A^*(\mathbf{k}_e, \mathbf{k}_h)\langle c, \mathbf{k}_e|e^{i\mathbf{k}_{op}\cdot\mathbf{r}}\hat{e}\cdot\mathbf{p}|\mathrm{v}, -\mathbf{k}_h\rangle \\ &= \sum_{\mathbf{k}_e}\sum_{\mathbf{k}_h} A^*(\mathbf{k}_e, \mathbf{k}_h)\hat{e}\cdot\mathbf{p}_{cv}(\mathbf{k}_e)\delta_{\mathbf{k}_e+\mathbf{k}_h,\mathbf{k}_{op}} \\ &= \sum_{\mathbf{k}} A^*(\mathbf{k}, -\mathbf{k})\hat{e}\cdot\mathbf{p}_{cv}(\mathbf{k})\end{aligned} \tag{14A.21}$$

where the long wavelength (or dipole) approximation $\mathbf{k}_{op} \simeq 0$ has been used. Therefore, the **k** selection rule

$$\mathbf{k}_e + \mathbf{k}_h = \mathbf{k}_{op} \simeq \mathbf{0} \tag{14A.22}$$

has been adopted. Comparing (14A.20) with the definition in (14A.2), and using $\mathbf{K} = \mathbf{k}_e + \mathbf{k}_h \simeq 0$, we find that the matrix element is

$$\langle f | e^{i\mathbf{k}_{op}\cdot\mathbf{r}} \hat{e}\cdot\mathbf{p} | i \rangle \simeq \hat{e}\cdot\mathbf{p}_{cv} \sum_{\mathbf{k}} A^*(\mathbf{k}, -\mathbf{k}) = \hat{e}\cdot\mathbf{p}_{cv} \sum_{\mathbf{k}} a^*(\mathbf{k})$$

$$= \hat{e}\cdot\mathbf{p}_{cv}\sqrt{V}\phi^*(0) \tag{14A.23}$$

where we have assumed that $\hat{e}\cdot\mathbf{p}_{cv}$ is independent of $\mathbf{k}$. The absorption spectrum is given by substituting the matrix element (14A.23) into (14.1.1), and we obtain (14.1.2) in the text.

APPENDIX 14B: SOLUTION OF THE ELECTRON–HOLE EFFECTIVE-MASS EQUATION WITH EXCITONIC EFFECTS

Using the transformations for the difference coordinate and the center-of-mass coordinate systems for the x and y components,

$$\boldsymbol{\rho} = \boldsymbol{\rho}_e - \boldsymbol{\rho}_h \quad \mathbf{R}_t = \left(m_e^* \boldsymbol{\rho}_e + m_h^* \boldsymbol{\rho}_h\right)/M \tag{14B.1}$$

where $M = m_e^* + m_h^*$, and $\boldsymbol{\rho} = x\hat{x} + y\hat{y}$, we obtain after following similar procedures as in (14A.7) to (14A.15)

$$\left[-\frac{\hbar^2}{2M}\nabla_{R_t}^2 - \frac{\hbar^2}{2m_r^*}\nabla_\rho^2 - \frac{\hbar^2}{2m_e^*}\frac{\partial^2}{\partial z_e^2} - \frac{\hbar^2}{2m_h^*}\frac{\partial^2}{\partial z_h^2} + V_e(z_e)\right.$$
$$\left. + V_h(z_h) - \frac{e^2}{4\pi\varepsilon_s|\mathbf{r}_e - \mathbf{r}_h|} - (E - E_g)\right]\Phi(\mathbf{r}_e, \mathbf{r}_h) = 0. \tag{14B.2}$$

Because the dependence on $\mathbf{R}_t$ comes from only the leading kinetic energy term, the solution can be written as [33]

$$\Phi(\mathbf{r}_e, \mathbf{r}_h) = \frac{e^{i\mathbf{K}_t\cdot\mathbf{R}_t}}{\sqrt{A}} F(\boldsymbol{\rho}, z_e, z_h) \tag{14B.3}$$

where the exciton envelop function $F(\boldsymbol{\rho}, z_e, z_h)$ satisfies

$$\left[-\frac{\hbar^2}{2m_r^*}\nabla_\rho^2 + H_e(z_e) - H_h(z_h) - \frac{e^2}{4\pi\varepsilon_s|\mathbf{r}_e - \mathbf{r}_h|}\right.$$
$$\left. -\left(E - E_g - \frac{\hbar^2 K_t^2}{2M}\right)\right] F(\boldsymbol{\rho}, z_e, z_h) = 0. \tag{14B.4}$$

The center of mass of the electron–hole pair is thus moving freely with a kinetic energy $\hbar^2 K_t^2/(2M)$. Here, identical relations to (14A.7) to (14A.11) for the parallel components of the real-space and momentum-space vectors exist, for example,

$$\mathbf{K}_t = \mathbf{k}_{\mathrm{et}} + \mathbf{k}_{\mathrm{ht}} \qquad \mathbf{k}_t = \left(m_{\mathrm{h}}^* \,\mathbf{k}_{\mathrm{et}} - m_{\mathrm{e}}^* \,\mathbf{k}_{\mathrm{ht}}\right)/M. \tag{14B.5}$$

Note that

$$H(z_{\mathrm{e}}) = -\frac{\hbar^2}{2m_{\mathrm{e}}^*}\frac{d^2}{dz_{\mathrm{e}}^2} + V_{\mathrm{e}}(z_{\mathrm{e}}) \qquad -H(z_{\mathrm{h}}) = -\frac{\hbar^2}{2m_{\mathrm{h}}^*}\frac{d^2}{dz_{\mathrm{h}}^2} + V_{\mathrm{h}}(z_{\mathrm{h}}) \tag{14B.6}$$

which are simply one-dimensional Schrödinger equations for a particle in-a-box model. Let us consider

$$H(z_{\mathrm{e}})f_n(z_{\mathrm{e}}) = E_{en} f_n(z_{\mathrm{e}}) \qquad H(z_{\mathrm{h}})g_m(z_{\mathrm{h}}) = E_{hm} g_m(z_{\mathrm{h}}) \tag{14B.7}$$

where $f_n(z_{\mathrm{e}})$ and $g_m(z_{\mathrm{h}})$ are the free-electron and the free-hole wave functions in the absence of any interactions. With the Coulomb interaction term, the solution $F(\boldsymbol{\rho}, z_{\mathrm{e}}, z_{\mathrm{h}})$ is more complicated. However, using the completeness properties of the solutions $\{f_n(z_{\mathrm{e}})\}$ and $\{g_m(z_{\mathrm{h}})\}$, we can expand the exciton envelop function as [12, 15, 19]

$$F(\boldsymbol{\rho},\ z_{\mathrm{e}},\ z_{\mathrm{h}}) = \sum_n \sum_m \phi_{nm}(\boldsymbol{\rho})\, f_n(z_{\mathrm{e}})\, g_m(z_{\mathrm{h}}). \tag{14B.8}$$

The Fourier transform pairs for the $\boldsymbol{\rho}$ dependence can be written as

$$\begin{aligned} \phi_{nm}(\boldsymbol{\rho}) &= \sum_{\mathbf{k}_t} G_{nm}(\mathbf{k}_t)\frac{e^{i\mathbf{k}_t \cdot \boldsymbol{\rho}}}{\sqrt{A}} \\ G_{nm}(\mathbf{k}_t) &= \int d^2\boldsymbol{\rho}\, \phi_{nm}(\boldsymbol{\rho})\frac{e^{-i\mathbf{k}_t \cdot \boldsymbol{\rho}}}{\sqrt{A}}. \end{aligned} \tag{14B.9}$$

The envelope function $F(\boldsymbol{\rho}, z_{\mathrm{e}}, z_{\mathrm{h}})$ becomes

$$F(\boldsymbol{\rho},\ z_{\mathrm{e}},\ z_{\mathrm{h}}) = \sum_{nm} \sum_{\mathbf{k}_t} G_{nm}(\mathbf{k}_t)\frac{e^{i\mathbf{k}_t \cdot \boldsymbol{\rho}}}{\sqrt{A}} f_n(z_{\mathrm{e}})\, g_m(z_{\mathrm{h}}). \tag{14B.10}$$

The complete envelope function is obtained from (14B.3), noting that all possible $\mathbf{K}_t$ can be included

$$\begin{aligned} \Phi(\mathbf{r}_{\mathrm{e}},\ \mathbf{r}_{\mathrm{h}}) &= \sum_{\mathbf{K}_t}\sum_{\mathbf{k}_t}\sum_{nm} G_{nm}(\mathbf{k}_t)\frac{e^{i\mathbf{K}_t \cdot \mathbf{R}_t}}{\sqrt{A}}\frac{e^{i\mathbf{k}_t \cdot \boldsymbol{\rho}}}{\sqrt{A}} f_n(z_{\mathrm{e}})\, g_m(z_h) \\ &= \sum_{\mathbf{k}_{\mathrm{et}}}\sum_{\mathbf{k}_{\mathrm{ht}}}\sum_{nm} G_{nm}(\mathbf{k}_t)\frac{e^{i\mathbf{k}_{\mathrm{et}} \cdot \boldsymbol{\rho}_{\mathrm{e}}}}{\sqrt{A}}\frac{e^{i\mathbf{k}_{\mathrm{ht}} \cdot \boldsymbol{\rho}_h}}{\sqrt{A}} f_n(z_e)\, g_m(z_{\mathrm{h}}). \end{aligned} \tag{14B.11}$$

Comparing (14B.10) and (14B.11) with (14A.1) and (14A.2), we find that the original electron–hole pair state can be written as [8, 9, 15, 34]

$$\Psi(\mathbf{r}_e, \mathbf{r}_h) = \sum_{\mathbf{k}_{et} n} \sum_{\mathbf{k}_{ht} m} G_{nm}(\mathbf{k}_t)\, \psi_{n\mathbf{k}_{et}}(\mathbf{r}_e)\, \psi^*_{m-\mathbf{k}_{ht}}(\mathbf{r}_h) \tag{14B.12a}$$

$$\psi_{n\mathbf{k}_{et}}(\mathbf{r}_e) = \frac{e^{i\mathbf{k}_{et}\cdot\boldsymbol{\rho}_e}}{\sqrt{A}} f_n(z_e)\, u_c(\mathbf{r}_e) \tag{14B.12b}$$

$$\psi_{m-\mathbf{k}_{ht}}(r_h) = \frac{e^{-i\mathbf{k}_{ht}\cdot\boldsymbol{\rho}_h}}{\sqrt{A}} g_m(z_h)\, u_v(\mathbf{r}_h). \tag{14B.12c}$$

Optical Matrix Element for Excitonic Transitions in a Quantum Well The optical matrix element is obtained using $|i\rangle = |\text{ground state}\rangle$, $|f\rangle = |\Psi(\mathbf{r}_e, \mathbf{r}_h)\rangle$, and

$$\begin{aligned}
\langle f|\hat{e}\cdot\mathbf{p}|i\rangle &= \sum_{nm}\sum_{\mathbf{k}_{et}}\sum_{\mathbf{k}_{ht}} G^*_{nm}(\mathbf{k}_t)\langle \psi_{n\mathbf{k}_{et}}(\mathbf{r})|\hat{e}\cdot\mathbf{p}|\psi_{m-\mathbf{k}_{ht}}(\mathbf{r})\rangle \\
&= \sum_{nm}\sum_{\mathbf{K}_t}\sum_{\mathbf{k}_t} G^*_{nm}(\mathbf{k}_t)\hat{e}\cdot\mathbf{p}_{cv}\, I_{nm}\, \delta_{\mathbf{k}_{et}+\mathbf{k}_{ht},0} \\
&= \sum_{nm}\sum_{\mathbf{k}_t} G^*_{nm}(\mathbf{k}_t)\hat{e}\cdot\mathbf{p}_{cv}\, I_{nm} \\
&= \sum_{nm}\left[\sqrt{A}\,\phi^*_{nm}(\boldsymbol{\rho}=0)\right]\hat{e}\cdot\mathbf{p}_{cv}\, I_{nm}
\end{aligned} \tag{14B.13}$$

where $I_{nm} = \int_{-\infty}^{\infty} f_n^*(z) g_m(z) dz$, and $\mathbf{p}_{cv} = \langle u_c(\mathbf{r})|\mathbf{p}|u_v(\mathbf{r})\rangle$. The above matrix element can be expressed alternatively in terms of $F(\boldsymbol{\rho}, z_e, z_h)$

$$\langle f|\hat{e}\cdot\mathbf{p}|i\rangle = \int_{-\infty}^{\infty} dz F^*(\boldsymbol{\rho}=0, z, z)\sqrt{A}\,\hat{e}\cdot\mathbf{p}_{cv}. \tag{14B.14}$$

Noting that $\mathbf{K}_t = \mathbf{k}_{et} + \mathbf{k}_{ht} = 0$ from (14B.13), we substitute the expression (14B.8) for $F(\boldsymbol{\rho}, z_e, z_h)$ into (14B.4), and obtain

$$\begin{aligned}
&\sum_{n'm'}\left[-\frac{\hbar^2}{2m_r^*}\nabla_\rho^2 + E_{en'} - E_{hm'} - \frac{e^2}{4\pi\varepsilon_s|\mathbf{r}_e - \mathbf{r}_h|} - (E - E_g)\right] \\
&\quad\times f_{n'}(z_e) g_{m'}(z_h)\phi_{n'm'}(\boldsymbol{\rho}) = 0.
\end{aligned} \tag{14B.15}$$

Multiplying the above equation by $f_n^*(z_e)$ and $g_m^*(z_h)$ and integrating over z_e and z_h, we obtain

$$\left[-\frac{\hbar^2}{2m_r^*}\nabla_\rho^2 - V_{nm}(\rho)\right]\phi_{nm}(\boldsymbol{\rho}) = \mathrm{E}_{ex}\,\phi_{nm}(\boldsymbol{\rho}) \tag{14B.16}$$

where

$$V_{nm}(\rho) \equiv \sum_{n'm'} \int dz_e f_n^*(z_e) f_{n'}(z_e) \int dz_h g_m^*(z_h) g_{m'}(z_h) \frac{e^2}{4\pi\varepsilon_s \left[\rho^2 + |z_e - z_h|^2\right]^{1/2}}$$

$$\simeq \int dz_e |f_n(z_e)|^2 \int dz_h |g_m(z_h)|^2 \frac{e^2}{4\pi\varepsilon_s \left[\rho^2 + |z_e - z_h|^2\right]^{1/2}} \tag{14B.17}$$

$$E_{ex} = E - (E_g + E_{en} - E_{hm}). \tag{14B.18}$$

Above, we have ignored coupling among different subbands so that only $n' = n$ and $m' = m$ are taken into account in (14B.17). The wave function $\phi_{nm}(\boldsymbol{\rho})$ satisfies the normalization condition in a two-dimensional space

$$\int d^2\boldsymbol{\rho} |\phi_{nm}(\boldsymbol{\rho})|^2 = \int_0^\infty 2\pi\rho d\rho |\phi_{nm}(\boldsymbol{\rho})|^2 = 1. \tag{14B.19}$$

This was derived using

$$\int d^2\mathbf{r}_e \int d^3\mathbf{r}_h |\Phi(\mathbf{r}_e, \mathbf{r}_h)|^2 = 1 \tag{14B.20}$$

and $d^3\mathbf{r}_e\, d^3\mathbf{r}_h = d^2\mathbf{R}_t\, d^2\boldsymbol{\rho}\, dz_e\, dz_h$; therefore,

$$\int d^2\boldsymbol{\rho} \int dz_e \int dz_h |F(\boldsymbol{\rho}, z_e, z_h)|^2 = 1. \tag{14B.21}$$

Variational Method for Exciton Problem Two different methods are commonly used to find the solution for the exciton Eq. (14B.16). The most common approach is a variational method, which is very useful to find the bound state solution. Noting that the $1s$ state solution of $\phi(\boldsymbol{\rho})$ behaves like $e^{-\rho/a_0}$ or $e^{-\rho/(2a_0)}$, the following variational form is assumed in the variational approach [12]

$$E_{ex}(\lambda) = \frac{\left\langle \phi \left| -\frac{\hbar^2}{2m_r}\nabla_\rho^2 - V(\rho) \right| \phi \right\rangle}{\langle \phi | \phi \rangle} \tag{14B.22}$$

where

$$\phi(\boldsymbol{\rho}) = \sqrt{\frac{2}{\pi}} \frac{1}{\lambda} e^{-\rho/\lambda} \tag{14B.23}$$

which satisfies the normalization condition $\langle\phi|\phi\rangle = 1$ in (14B.19). We find

$$E_{\mathrm{ex}}(\lambda) = \frac{\hbar^2}{2m_{\mathrm{r}}\lambda^2} - \frac{e^2}{4\pi\varepsilon_{\mathrm{s}}}\frac{4}{\lambda^2}\int_{-\infty}^{\infty} dz_{\mathrm{e}}|f_n(z_{\mathrm{e}})|^2 \times \int_{-\infty}^{\infty} dz_{\mathrm{h}}|g_m(z_{\mathrm{h}})|^2 \int_{0}^{\infty} \rho d\rho \frac{e^{-2\rho/\lambda}}{\left[\rho^2 + (z_{\mathrm{e}} - z_{\mathrm{h}})^2\right]^{1/2}}. \tag{14B.24}$$

It is convenient to write in terms of the normalized parameters

$$\beta = \frac{a_0}{\lambda} \qquad a_0 = \frac{4\pi\varepsilon_{\mathrm{s}}\hbar^2}{m_{\mathrm{r}}^* e^2} \tag{14B.25}$$

and the normalized exciton binding energy

$$\varepsilon_{\mathrm{ex}} = E_{\mathrm{ex}}/R_y \qquad R_y = \frac{m_{\mathrm{r}}^* e^4}{2\hbar^2(4\pi\varepsilon_{\mathrm{s}})^2} \tag{14B.26}$$

$$\varepsilon_{\mathrm{ex}} = \beta^2 - 4\beta \int_{-\infty}^{\infty} dz_{\mathrm{e}}|f_n(z_{\mathrm{e}})|^2 \int_{-\infty}^{\infty} dz_{\mathrm{h}}|g_m(z_{\mathrm{h}})|^2 G(2\beta|z_{\mathrm{e}} - z_{\mathrm{h}}|/a_0) \tag{14B.27}$$

where the function $G(x)$ is defined as an integral

$$G(x) = \int_{0}^{\infty} dt \frac{te^{-t}}{[t^2 + x^2]^{1/2}}, \tag{14B.28}$$

which is a smooth monotonic function and can be approximated [55] analytically. It has the properties that $G(0) = 1$ and $G(\infty) = 0$. Typically, for a quantum-well problem, $|z_{\mathrm{e}} - z_{\mathrm{h}}|$ is finite and the argument in $G(x)$ is over a finite range. The pure two-dimensional limit can be obtained by ignoring the z-dependence and using $G(0) = 1$,

$$\varepsilon_{\mathrm{ex}} = \beta^2 - 4\beta. \tag{14B.29}$$

Thus we find the minimum at $\beta = 2$ and $\varepsilon_{\mathrm{ex}} = -4$ as expected for a pure 2D exciton binding energy.

Momentum-Space Solution for the Exciton Problem Alternatively, the real space exciton differential Eq. (14B.16) can also be written in terms of the momentum space integral equation [15, 35]. A direct numerical solution of the above equation has been discussed in Ref. 15.

PROBLEMS

14.1 Following the derivations in Eq. (14.1.2), show that the physical interpretation of $2|\phi_n(0)|^2$ gives the reduced density of states for (**a**) a one-dimensional and (**b**) a two-dimensional case.

14.2 Derive Eq. (14A.13).

14.3 Plot the Franz–Keldysh absorption spectrum for GaAs with an applied field $F = 100\,\mathrm{kV/cm}$ at $T = 300\mathrm{K}$.

14.4 Compare the oscillator strengths and the exciton binding energies of the bound states of the 2D and 3D excitons.

14.5 Compare the absorption spectra of the continuum-state contributions of the 2D and 3D excitons.

14.6 (**a**) Show that for interband absorption in a pure two-dimensional structure without exciton effects, the absorption spectrum taking into account the finite linewidth broadening is given by

$$\alpha(\hbar\omega) = \frac{A_0}{2\pi R_y a_0^2} \int_0^\infty \frac{d\varepsilon'}{\pi} \frac{\gamma}{(\varepsilon' - \varepsilon)^2 + \gamma^2}$$

where

$$\varepsilon = (\hbar\omega - E_g)/R_y.$$

(**b**) Carry out the integration analytically and plot the absorption spectrum $\alpha(\hbar\omega)$.

14.7 Derive Eq. (14B.14).

14.8 Show that $G(x)$ defined in Eq. (14B.28) gives $G(0) = 1$ and $G(\infty) = 0$.

14.9 Use the perturbation result (14.4.20) for the band-edge transition energy $E_{\mathrm{h1}}^{\mathrm{e1}}(F)$ to compare with the data shown in Fig. 14.6. Estimate the exciton binding energies for the heavy-hole and light-hole excitons separately. Discuss the accuracy of this simple method.

14.10 Compare the advantages and disadvantages of the waveguide modulator versus the transverse transmission modulator.

14.11 Discuss the physics and operation principles of a SEED.

REFERENCES

1. W. Franz, "Einfluss eines elektrichen Feldes auf eine optische Absorptionskante," *Z. Naturforsch.* **13a**, 484–489 (1958).

2. L. V. Keldysh, "The effect of a strong electric field on the optical properties of insulating crystals," *Soviet Phys. JETP* **34**, 788–790 (1958).
3. K. Tharmalingam, "Optical absorption in the presence of a uniform field," *Phys. Rev.* **130**, 2204–2206 (1963).
4. R. J. Elliot, "Intensity of optical absorption by excitons," *Phys. Rev.* **108**, 1384–1389 (1957).
5. R. J. Elliot, Theory of Excitons: I, in *Polarons and Excitons, Scottish Universities' Summer School*, C. G. Kuper and G. D. Whitfield, Eds., Plenum Press, New York, pp. 269–293, 1962.
6. R. S. Knox, "Theory of Excitons," in *Solid State Physics*, Suppl. 5, Academic Press, New York, 1963.
7. J. D. Dow and D. Redfield, "Electroabsorption in semiconductors: the excitonic absorption edge," *Phys. Rev. B* **1**, 3358–3371 (1970).
8. E. J. Johnson, "Absorption near the fundamental edge," Chapter 6 in R. K. Williardson and A. C. Beer, Eds., *Semiconductors and Semimetals*, vol. 3, Academic Press, New York, pp. 153–258, 1967.
9. J. O. Dimmock, "Introduction to the theory of exciton states in semiconductors," Chapter 7 in R. K. Williardson and A. C. Beer, Eds., *Semiconductors and Semimetals*, vol. 3, Academic Press, New York, pp. 259–319, 1967.
10. T. H. Wood, C. A. Burrus, D. A. B. Miller, D. S. Chemla, T. C. Damen, A. C. Gossard, and W. Wiegmann, "High-speed optical modulation with GaAs/GaAlAs quantum wells in a p -i-n diode structure," *Appl. Phys. Lett.* **44**, 16–18 (1984).
11. D. A. B. Miller, D. S. Chemla, T. C. Damen, A. C. Gossard, W. Wiegmann, T. H. Wood, and C. A. Burrus, "Band-edge electroabsorption in quantum well structures: the quantum-confined Stark effect," *Phys. Rev. Lett.* **53**, 2173–2176 (1984).
12. D. A. B. Miller, D. S. Chemla, T. C. Damen, A. C. Gossard, W. Wiegmann, T. H. Wood, and C. A. Burrus, "Electric field dependence of optical absorption near the band gap of quantum well structures," *Phys. Rev. B* **32**, 1043–1060 (1985).
13. S. Schmitt-Rink, D. S. Chemla, and D. A. B. Miller, "Linear and nonlinear optical properties of semiconductor quantum wells," *Adv. Phys.* **38**, 89–188 (1989).
14. M. Shinada and S. Sugano, "Interband optical transitions in extremely anisotropic semiconductors. I. Bound and unbound exciton absorption," *J. Phys. Soc. Jpn.* **21**, 1936–1946 (1966).
15. S. L. Chuang, S. Schmitt-Rink, D. A. B. Miller, and D. S. Chemla, "Exciton Green's-function approach to optical absorption in a quantum well with an applied electric field," *Phys. Rev. B* **43**, 1500–1509 (1991).
16. T. H. Wood, "Multiple quantum well (MQW) waveguide modulators," *J. Lightwave Technol.* **6**, 743–757 (1988).
17. D. A. B. Miller, J. S. Weiner, and D. S. Chemla, "Electric-field dependence of linear optical properties in quantum well structures: waveguide electroabsorption and sum rules," *IEEE J. Quantum Electron.* **QE-22**, 1816–1830 (1986).
18. D. A. B. Miller, D. S. Chemla, T. C. Damen, A. C. Gossard, W. Wiegmann, T. H. Wood, and C. A. Burrus, "Novel hybrid optically bistable switch: The quantum well self-electro-optic effect device," *Appl. Phys. Lett.* **45**, 13–15 (1984).

19. D. A. B. Miller, D. S. Chemla, T. C. Damen, T. H. Wood, C. A. Burrus, A. C. Gossard, and W. Wiegmann, "The quantum well self-electrooptic effect device: Optoelectronic bistability and oscillation, and self-linearized modulation," *IEEE J. Quantum Electron.* **QE-21**, 1462–1476 (1985).

20. D. A. B. Miller, M. D. Feuer, T. Y. Chang, S. C. Shunk, J. E. Henry, D. J. Burrows, and D. S. Chemla, "Field-effect transistor self-electrooptic effect device: integrated photodiode, quantum well modulator and transistor," *IEEE Photon. Technol. Lett.* **1**, 62–64 (1989).

21. D. E. Aspnes, "Electric-field effects on the dielectric constant of solids," *Phys. Rev.* **153**, 972–982 (1967).

22. D. E. Aspnes and N. Bottka, "Electric-field effects on the dielectric function of semiconductors and insulators," in R. K. Williardson and A. C. Beer, Eds., *Semiconductors and Semimetals*, vol. 9: *Modulation Techniques*, Academic Press, New York, pp. 457–543, 1972.

23. M. Cardona, *Modulation Spectroscopy*, in *Solid State Physics*, Suppl. 11, Academic, New York, pp. 165–275, 1969.

24. B. R. Bennett and R. A. Soref, "Electrorefraction and electroabsorption in InP, GaAs, GaSb, InAs, and InSb," *IEEE J. Quantum Electron.* **QE-23**, 2159–2166 (1987).

25. D. A. B. Miller, D. S. Chemla, and S. Schmitt-Rink, "Relation between electroabsorption in bulk semiconductors and in quantum wells: the quantum-confined Franz-Keldysh effect," *Phys. Rev. B* **33**, 6976–6982 (1986).

26. H. Shen and F. H. Pollak, "Generalized Franz-Keldysh theory of electromodulation," *Phys. Rev. B* **42**, 7097–7102 (1990).

27. M. Abramowitz and I. A. Stegun, *Handbook of Mathematical Functions*, Chapter 10, Dover, New York, 1972.

28. F. Bassani and G. Pastori Parravicini, *Electronic States and Optical Transitions in Solids*, Pergamon Press, Oxford, 1975.

29. C. Y. P. Chao and S. L. Chuang, "Analytical and numerical solutions for a two-dimensional exciton in momentum space," *Phys. Rev. B* **43**, 6530–6543 (1991).

30. M. D. Sturge, "Optical absorption of gallium arsenide between 0.6 and 2.75 eV," *Phys. Rev.* **127**, 768–773 (1962).

31. J. S. Blakemore, "Semiconducting and other major properties of gallium arsenide," *J. Appl. Phys.* **53**, R123–R181 (1982).

32. G. Livescu, D. A. B. Miller, D. S. Chemla, M. Ramaswamy, T. Y. Chang, N. Sauer, A. C. Gossard, and J. H. English, "Free carrier and many-body effects in absorption spectra of modulation-doped quantum wells," *IEEE J. Quantum Electron.* **24**, 1677–1689 (1988).

33. G. Bastard and J. A. Brum, "Electronic states in semiconductor heterostructures," *IEEE J. Quantum Electron.* **QE-22**, 1625–1644 (1986).

34. G. D. Sanders and Y. C. Chang, "Theory of photoabsorption in modulation-doped semiconductor quantum wells," *Phys. Rev. B* **35**, 1300–1315 (1987).

35. R. Zimmermann, "On the dynamic Stark effect of excitons, the low field limit," *Phys. Stat. Sol. (b)* **146**, 545–554 (1988).

36. Y. Kan, H. Nagai, M. Yamanishi, and I. Suemune, "Field effects on the refractive index and absorption coefficient in AlGaAs quantum well structures and their feasibility for electrooptic device applications," *IEEE J. Quantum Electron.* **QE-23**, 2167–2180 (1987).

37. M. Yamanishi and I. Suemune, "Comment on polarization dependent momentum matrix elements in quantum well lasers," *Jpn. J. Appl. Phys.* **23**, L35–L36 (1984).

38. M. Asada, A. Kameyama, and Y. Suematsu, "Gain and intervalence band absorption in quantum-well lasers," *IEEE J. Quantum Electron.* **QE-20**, 745–753 (1984).

39. B. Zhu and K. Huang, "Effect of valence-band hybridization on the exciton spectra in GaAs-$Ga_{1-x}Al_xAs$ quantum wells," *Phys. Rev. B* **36**, 8102–8108 (1987).

40. G. E. W. Bauer and T. Ando, "Exciton mixing in quantum wells," *Phys. Rev. B* **38**, 6015–6030 (1988).

41. H. Chu and Y. C. Chang, "Theory of line shapes of exciton resonances in semiconductor superlattices," *Phys. Rev. B* **39**, 10861–10871 (1989).

42. L. C. Andreani and A. Pasquarello, "Accurate theory of excitons in GaAs-$Ga_{1-x}Al_xAs$ quantum wells," *Phys. Rev. B* **42**, 8928–8938 (1990).

43. C. Y. P. Chao and S. L. Chuang, "Momentum-space solution of exciton excited states and heavy-hole-light-hole mixing in quantum wells," *Phys. Rev. B* **48**, 8210–8221 (1993).

44. M. Matsuura and T. Kamizato, "Subbands and excitons in a quantum well in an electric field," *Phys. Rev. B* **33**, 8385–8389 (1986).

45. G. Bastard, E. E. Mendez, L. L. Chang, and L. Esaki, "Variational calculations on a quantum well in an electric field," *Phys. Rev. B* **28**, 3241–3245 (1983).

46. S. Nojima, "Electric field dependence of the exciton binding energy in GaAs/$Al_xGa_{1-x}As$ quantum wells," *Phys. Rev. B* **37**, 9087–9088 (1988).

47. D. Ahn and S. L. Chuang, "Variational calculations of subbands in a quantum well with uniform electric field: Gram-Schmidt orthogonalization approach," *Appl. Phys. Lett.* **49**, 1450–1452 (1986).

48. J. S. Weiner, D. A. B. Miller, D. S. Chemla, T. C. Damen, C. A. Burrus, T. H. Wood, A. C. Gossard, and W. Wiegmann, "Strong polarization-sensitive electroabsorption in GaAs/AlGaAs quantum well waveguides," *Appl. Phys. Lett.* **47**, 1148–1150 (1985).

49. P. Lefebvre, P. Christol, and H. Mathieu, "Unified formulation of excitonic absorption spectra of semiconductor quantum wells, superlattices and quantum wires," *Phys. Rev. B* **48**, 17308–17315 (1993).

50. H. S. Cho and P. R. Prucnal, "Effect of parameter variations on the performance of GaAs/AlGaAs multiple-quantum-well electroabsorption modulators," *IEEE J. Quantum Electron.* **25**, 1682–1690 (1989).

51. G. Lengyel, K. W. Jelley, and R. W. H. Engelmann, "A semi-empirical model for electroabsorption in GaAs/AlGaAs multiple quantum well modulator structures," *IEEE J. Quantum Electron.* **26**, 296–304 (1990).

52. C. R. Giles, T. H. Wood, and C. A. Burrus, "Quantum-well SEED optical oscillators," *IEEE J. Quantum Electron.* **26**, 512–518 (1990).

53. A. L. Lentine, D. A. B. Miller, L. M. F. Chirovsky, and L. A. D'Asaro, "Optimization of absorption in symmetric self-electrooptic effect devices: a system perspective," *IEEE J. Quantum Electron.* **27**, 2431–2439 (1991).

54. P. J. Mares and S. L. Chuang, "Comparison between theory and experiment for InGaAs/InP self-electrooptic effect devices," *Appl. Phys. Lett.* **61**, 1924–1926 (1992).

55. P. J. Mares and S. L. Chuang, "Modeling of self-electrooptic-effect devices," *J. Appl. Phys.* **74**, 1388–1397 (1993).

56. M. K. Chin and W. S. C. Chang, "Theoretical design optimization of multiple-quantum-well electroabsorption waveguide modulators," *IEEE J. Quantum Electron.* **29**, 2476–2488 (1993).

57. A. M. Fox, D. A. B. Miller, G. Livescu, J. E. Cunningham, and W. Y. Jan, "Quantum well carrier sweep out: relation to electroabsorption and exciton saturation," *IEEE J. Quantum Electron.* **27**, 2281–2295 (1991).

58. T. Ido, H. Sano, D. J. Moss, S. Tanaka, and A. Takai, "Strained InGaAs/InAlAs MQW electroabsorption modulators with large bandwidth and low driving voltage," *IEEE Photon. Technol. Lett.* **6**, 1207–1209 (1994).

59. T. Ido, S. Tanaka, M. Suzuki, M. Koizumi, H. Sano, and H. Inoue, "Ultra-high-speed multiple-quantum-well electro-absorption optical modulators with integrated waveguides," *IEEE J. Lightwave Technol.* **14**, 2026–2034 (1996).

60. T. Ido, H. Sano, S. Tanaka, D. J. Moss, and H. Inoue, "Performance of strained InGaAs/InAlAs multiple-quantum-well electroabsorption modulators," *IEEE J. Lightwave Technol.* **14**, 2324–2331 (1996).

61. D. G. Moodie, M. J. Harlow, M. J. Guy, S. D. Perrin, C. W. Ford, and M. J. Robertson, "Discrete electroabsorption modulators with enhanced modulation depth," *IEEE J. Lightwave Technol.* **14**, 2035–2043 (1996).

62. B. Mason et al., "40-Gb/s tandem electroabsorption modulator," *IEEE Photon. Technol. Lett.* **14**, 27–29 (2002).

63. H. Fukano, T. Yamanaka, and M. Tamura, "Design and fabrication of low-driving-voltage electroabsorption modulators operating at 40 Gb/s," *IEEE J. Lightwave Technol.* **25**, 1961–1969 (2007).

64. Y. H. Kuo, "Quantum-confined Stark effects in Ge/SiGe quantum wells on Si for optical modulators," *IEEE J. Sel. Top. Quantum Electron.* **12**, 1503–1513 (2006).

65. Y. Kawamura, K. Wakita, Y. Itaya, Y. Yoshikuni, and H. Asahi, "Monolithic integration of InGaAs/InP DFB lasers and InGaAs/InAlAs MQW optical modulators," *Electron. Lett.* **22**, 242–243 (1986).

66. M. Aoki, M. Suzuki, H. Sano, T. Kawano, T. Ido, T. Taniwatari, K. Uomi, and A. Takai, "InGaAs/InGaAsP MQW electroabsorption-modulator integrated with a DFB laser fabricated by band-gap energy control selective area MOCVD," *IEEE J. Quantum Electron.* **29**, 2088–2096 (1993).

67. M. Aoki, M. Takahashi, M. Suzuki, H. Sano, K. Uomi, T. Kawano, and A. Takai, "High-extinction-ratio MQW electroabsorption-modulator integrated DFB laser fabricated by in-plane bandgap energy control technique," *IEEE Photon. Technol. Lett.* **4**, 580–582 (1992).

68. H. Takeuchi, K. Tsuzuki, K. Sato, M. Yamamoto, Y. Itaya, A. Sano, M. Yoneyama, and T. Otsuji, "Very high speed light source module up to 40 Gb/s containing an MQW electroabsorption-modulator integrated with a DFB laser," *IEEE J. Sel. Top. Quantum Electron.* **3**, 336–343 (1997).

69. T. Tanbun-Ek, Y. K. Chen, J. A. Grenko, E. K. Byrne, J. E. Johnson, R. A. Logan, A. Tate, A. M. Sergent, K. W. Wecht, P. F. Sciortino, Jr., and S. N. G. Chu, "Integrated DFB-DBR laser modulator grown by selective area metalorganic vapor phase epitaxy growth technique," *J. Crystal Growth* **145**, 902–906 (1994).

70. M. Suzuki, H. Tanaka, H. Taga, S. Yamamoto, and Y. Matsushima, "$\lambda/4$-shifted DFB laser/electroabsorption modulator integrated light source for multigigabit transmission," *IEEE Trans. Lightwave Technol.* **10**, 90–95 (1992).

71. B. Mason, G. A. Fish, S. P. DenBaars, and L. A. Coldren, "Widely tunable sampled grating DBR laser with integrated electroabsorption modulator," *IEEE Photon. Technol. Lett.* **11**, 638–640 (1999).

72. J. E. Johnson et al., "Fully stabilized electroabsorption-modulated tunable DBR lasertransmitter for long-haul optical communications," *J. Sel. Top. Quantum Electron.* **7**, 168–177 (2001).

73. R. Nagarajan et al., "400 Gbit/s (10 channel × 40 Gbit/s) DWDM photonic integrated circuits," *Electron. Lett.* **41**, 347–349 (2005).

74. R. Nagarajan, M. Kato, J. Pleumeekers, P. Evans, D. Lambert, A. Chen, V. Dominic, A. Mathur, P. Chavarkar, M. Missey, A. Dentai, S. Hurtt, J. Back, R. Muthiah, S. Murthy, R. Salvatore, S. Grubb, C. Joyner, J. Rossi, R. Schneider, M. Ziari, F. Kish, and D. Welch, "Single-chip 40-channel InP transmitter photonic integrated circuit capable of aggregate data rate of 1.6 Tbit/s," *Electron. Lett.* **42**, 771–773 (2006).

75. D. F. Welch, F. A. Kish, R. Nagarajan, C. H. Joyner, R. P. Schneider, Jr., V. G. Dominic, M. L. Mitchell, S. G. Grubb, T.-K. Chiang, D. D. Perkins, and A. C. Nilsson, "The realization of large-scale photonic integrated circuits and the associated impact on fiber-optic communication systems," *J. Lightwave Technol.* **24**, 4674–4683 (2006).

76. D. F. Welch et al., "Large-scale InP photonic integrated circuits: enabling efficient scaling of optical transport networks," *IEEE J. Sel. Top. Quantum Electron.* **13**, 22–31 (2007).

77. L. M. Zhang and J. E. Carroll, "Semiconductor 1.55 um laser source with Gigabit/second integrated electroabsorptive modulator," *IEEE J. Quantum Electron.* **30**, 2573–2577 (1994).

78. Y. Kim, H. Lee, J. Lee, J. Han, T. W. Oh, and J. Jeong, "Chirp characteristics of 10-Gb/s electroabsorption modulator integrated DFB lasers," *IEEE J. Quantum Electron.* **36**, 900–908 (2000).

79. W. Fang, S. L. Chuang, T. Tanbun-Ek, and Y. K. Chen, "Modeling and experiment of 1.55 μm integrated electrabsorption modulator with distributed-feedback lasers," *Proc. SPIE*, vol. **3006**, Photonics West, Optoelectronics'97 Symposium, 1997.

80. M. Suzuki, H. Tanaka, S. Akiba, and Y. Kushiro, "Electrical and optical interactions between integrated InGaAsP/InPDFB lasers and electroabsorption modulators," *J. Lightwave Technol.* **6**, 779–785 (1988).

81. M. Aoki, S. Takashima, Y. Fujiwara, and S. Aoki, "New transmission simulation of EA-modulator integrated DFB-lasers considering the facet reflection-induced chirp," *IEEE Photon. Technol. Lett.* **9**, 380–382 (1997).

82. R. A. Salvatore, R.T. Sahara, M. A. Bock, and I. Libenzon, "Electroabsorption modulated laser for long transmission spans," *IEEE J. Quantum Electron.* **38**, 464–476 (2002).

83. A. Ramdane, F. Devaux, N. Souli, D. Delprat, and A. Ougazzaden, "Monolithic integration of multiple-quantum-well lasers and modulators for high-speed transmission," *IEEE J. Sel. Top. Quantum Electron.* **2**, 326–335 (1996).

84. D. Delprat, A. Ramdane, L. Silvestre, A. Ougazzaden, F. Delorme, and S. Slempkes, "20-Gb/s integrated DBR laser-EA modulator by selective area growth for 1.55-/spl mu/m WDM applications," *J. Lightwave Technol.* **9**, 898–900 (1997).

85. H. Soda, K. Nakai, and H. Ishikawa, "Frequency response of an optical intensity modulated monolithically integrated with a DFB laser," *Proc. 14th Eur. Conf. Optical Communication (ECOC'88)*, pp. 227–230, 1988.

86. K. Wakita, K. Sato, I. Kotaka, M. Yamamoto, and T. Kataoka, "20 Gbit/s 1.55 um strained InGaAsP MQW modulator integrated DFB laser module," *Electron. Lett.* **4**, 302–303 (1994).

87. B. Stegmueller, E. Baur, and M. Kicherer, "15-GHz modulation performance of integrated DFB laser diode EA modulator with identical multiple-quantum-well double-stack active layer," *IEEE Photon. Technol. Lett.* **14**, 1647–1649 (2002).

88. B. Stegmueller and C. Hanke, "Integrated 1.3um DFB laser electroabsorption modulator based on identical MQW double-stack active layer with 25-GHz modulation performance," *IEEE Photon. Technol. Lett.* **15**, 1029–1031 (2003).

89. A. Ramdane, P. Krauz, E.V.K. Rao, A. Hamoudi, A. Ougazzaden, D. Robein, A. Gloukhian, and M. Carre, "Monolithic integration of InGaAsP-InP strained-layer distributed feedback laser and external modulator by selective quantum-well interdiffusion," *IEEE Photon. Technol. Lett.* **7**, 1016–1018 (1995).

90. A. Ramdane, A. Ougazzaden, F. Devaux, F. Delorme, M. Schneider, and J. Landreau, "Very simple approach for high performance DFB laser-electroabsorption modulator monolithic integration," *Electron. Lett.* **30**, 1980–1981 (1994).

91. M. Le Pallec, C. Kazmierski, E. Vergnol, S. Perrin, J. G. Provost, P. Doussiere, G. Glastre, D. Carpentier, and S. Fabre, "New integrated buried laser-ridge modulator with identical active layer," *IEEE Photon. Technol. Lett.* **15**, 362–364 (2003).

92. W. Fang, A. Hsu, S. L. Chuang, T. Tanbun-Ek, and A. M. Sergent, "Measurement and modeling of distributed-feedback lasers with spatial hole burning," *IEEE J. Sel. Top. Quantum Electron.* **3**, 547–554 (1997).

93. A. Hsu, S. L. Chuang, W. Fang, L. Adams, G. Nykolak, and T. Tanbun-Ek, "A wavelength-tunable curved waveguide DFB laser with integrated modulator," *IEEE J. Quantum Electron.* **35**, 961–969 (1999).

94. T. Ido, H. Sano, S. Tanaka, D. J. Moss, and H. Inoue, "Performance of strained InGaAs/InAlAs multiple-quantum-well electroabsorption modulators," *J. Lightwave Technol.* **14**, 2324–2331 (1996).

95. K. Nakahara, T. Tsuchiya, T. Kitatani, K. Shinoda, T. Taniguchi, T. Kikawa, M. Aoki, and M. Mukaikubo, "40-Gb/s direct modulation with high extinction ratio operation of 1.3-um InGaAlAs multiquantum well ridge waveguide distributed feedback lasers," *IEEE Photon. Technol. Lett.* **19**, 1436–1438 (2007).

96. A. Akage, K. Kawano, S. Oku, R. Iga, H. Okamoto, Y. Miyamoto, and H. Takeuchi, "Wide bandwidth of over 50 GHz traveling wave electrode electroabsorption modulator integrated DFB laser," *Electron. Lett.* **37**, 299–300 (2001).

97. H. Kawanishi, Y. Yamauchi, N. Mineo, Y. Shibuya, H. Mural, K. Yamada, and H. Wada, "EAM-integrated DFB laser modules with more than 40-GHz bandwidth," *IEEE Photon. Technol. Lett.* **13**, 954–956 (2001).

Part V

Detection of Light and Solar Cells

15

Photodetectors and Solar Cells

Photodetectors play important roles in optical communication systems. The major physical mechanism of photodetectors is the absorption of photons, which changes the electric properties of the electronic system such as the generation of a photocurrent in a photoconductor or a photovoltage in a photovoltaic detector. The performance of a photodetector depends on the optical absorption process, the carrier transport, and the interaction with the circuit system. Similar physical processes exist in solar cells made of semiconductor *p-n* junctions, which play an important role in conversion of solar radiation into electrical energy.

For intrinsic optical absorptions, such as interband processes in a direct semiconductor, the general theory for the absorption spectrum has been presented in Chapter 9. The interband absorption creates electron–hole pairs. The carrier transport of these electrons and holes after generation depends on the design of the photodetectors. In this chapter, we study photoconductors, photodiodes using *p-n* junctions and *p-i-n* structures, avalanche photodiodes, and intersubband quantum-well photodetectors. Our focus is on the understanding of the physical processes of the carrier generation and transport. We derive the relation between the optical power and the photocurrents. Some important noises in these photodetectors are also discussed. More extensive treatment of photodetectors can be found in Refs. 1–5. Finally, we discuss photovoltaic solar cells based on *p-n* junctions, including quantum-well solar cells and multijunction solar cells.

15.1 PHOTOCONDUCTORS

15.1.1 Photoconductivity

Consider a uniform *p*-type semiconductor with a uniform optical illumination. The total electron and hole carrier concentrations, n and p, deviate from their thermal equilibrium values, n_0 and p_0, due to the optical excitation

$$n = n_0 + \delta n \qquad p = p_0 + \delta p \tag{15.1.1}$$

Physics of Photonic Devices, Second Edition. By Shun Lien Chuang
Copyright © 2009 John Wiley & Sons, Inc.

where the excess electron and hole concentrations are equal $\delta n = \delta p$ because each absorbed photon breaks a bond and creates an electron–hole pair. Because $p_0 \gg n_0$ in an extrinsic p-type semiconductor, the net recombination rate (2.3.6) can be reduced to

$$R_n = \frac{\delta n}{\tau_n} \tag{15.1.2}$$

where the low level injection condition (i.e., δn, $\delta p \ll p_0$) has been assumed. Thus the excess electron concentration satisfies the rate equation

$$\frac{\partial}{\partial t}\delta n = G_0(t) - \frac{\delta n}{\tau_n} \tag{15.1.3}$$

where G_0 is the net optical generation rate. For a uniform semiconductor with a dc voltage bias V, as shown in Fig. 15.1, the electron and hole current densities are given only by the drift components because there is no diffusion current due to the absence of spatial dependence ($\partial/\partial x = 0$). The total current density is

$$J = J_n + J_p = q(\mu_n n + \mu_p p)E = \sigma E \tag{15.1.4}$$

where the conductivity is

$$\sigma = q(\mu_n n + \mu_p p). \tag{15.1.5}$$

The photoconductivity $\Delta\sigma$ is defined as the difference between the conductivity when there is an optical injection and the dark conductivity $\sigma_0 = q(\mu_n n_0 + \mu_p p_0)$

$$\Delta\sigma = q(\mu_n \delta n + \mu_p \delta p) = q(\mu_n + \mu_p)\delta n. \tag{15.1.6}$$

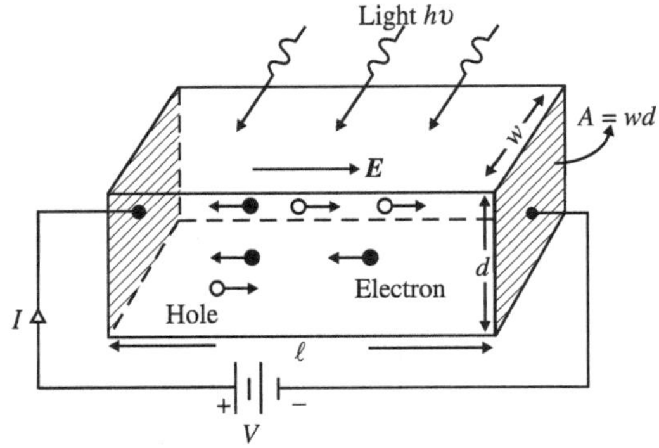

Figure 15.1 A simple photoconductor with an external bias voltage V. The optically generated electrons and holes transport in opposite directions in the presence of the bias field.

15.1.2 Photocurrent Responses in the Time Domain

Case 1. A Constant Light Intensity If G_0 is independent of time, $\partial/\partial t = 0$ at steady state, we obtain

$$\delta n = G_0 \tau_n. \tag{15.1.7}$$

The total current I is the current density J multiplied by the cross-sectional area $A = wd$. The photocurrent ΔI is defined as the difference between the total current in the presence of optical excitation and the dark current $I_0 = \sigma_0 AV/\ell$

$$\Delta I = \Delta\sigma A \frac{V}{\ell} = q(\mu_n + \mu_p)\delta n A \frac{V}{\ell} \simeq q\mu_n G_0 \tau_n A \frac{V}{\ell} \tag{15.1.8}$$

where the electric field $E = V/\ell$, with ℓ being the length of the sample and $\mu_n \gg \mu_p$, have been used. The above expression can also be expressed in terms of the transit time of the electrons

$$\tau_t = \tau_{t,n} = \frac{\ell}{v_n} = \frac{\ell}{\mu_n E} = \frac{\ell^2}{\mu_n V} \tag{15.1.9}$$

such that

$$\Delta I = q(G_0 \ell A)\frac{\tau_n}{\tau_t}. \tag{15.1.10}$$

The above expression has a good physical interpretation. The term $G_0\ell A$ is the total number of electron–hole pairs created per second in the sample with a volume ℓA. The ratio τ_n/τ_t gives the photoconductive gain, which is determined by how fast the electrons can transit across the electrodes and contribute to the photocurrent in the circuit before they can recombine with holes. If we include both electron and hole contributions:

$$\frac{1}{\tau_t} = \frac{1}{\tau_{t,n}} + \frac{1}{\tau_{t,p}} = (\mu_n + \mu_p)\frac{E}{\ell}. \tag{15.1.11}$$

As shown in Fig. 15.1, when an electron–hole pair is created, the photocurrent will be small if the electron and the hole immediately recombine before they can be collected by the electrodes (i.e., if the recombination lifetime τ_n is much shorter than the transit time). On the other hand, when the transit time τ_t is short, a significant amount of photogenerated electrons will be able to reach the left electrode before they recombine with holes in the semiconductor. To preserve charge neutrality in the semiconductor, the left electrode will provide the same number of holes at the same rate as the electrons reaching that electrode per second, resulting in the photoconductor current measured by the external circuit. For example, $\tau_n/\tau_t = 10$ is equivalent to having 10 round-trips taken by the electron before it disappears through the recombination in the photoconductor.

The optical generation rate is equal to the number of injected photons per second, or photon flux ($P_{opt}/h\nu$) per unit volume (ℓwd) multiplied by the quantum efficiency η

$$G_0 = \eta \frac{P_{opt}/h\nu}{\ell wd} \tag{15.1.12}$$

where w is the width, d is the depth of the sample ($wd = A$), P_{opt} is the optical power (watt) of the injected light, and η is the quantum efficiency or the fraction of photons creating electron–hole pairs. If the surface reflections and the finite thickness of the detector are considered, the quantum efficiency η is just the intrinsic quantum efficiency η_i, i.e., the number of electron-hole pairs generated by each absorbed photon, multiplied by the absorbance

$$\eta = \eta_i(1 - R)(1 - e^{-\alpha d}) \tag{15.1.13}$$

where R is the optical reflectivity between the air and semiconductor and α is the absorption coefficient of the optical intensity. The injected primary photocurrent I_{ph} is defined as

$$I_{ph} = \eta q \frac{P_{opt}}{h\nu}. \tag{15.1.14}$$

The photocurrent is

$$\Delta I = \eta q \frac{P_{opt}}{h\nu}\frac{\tau_n}{\tau_t} = I_{ph}\frac{\tau_n}{\tau_t}. \tag{15.1.15}$$

Again the photoconductive gain is given by

$$\frac{\Delta I}{I_{ph}} = \frac{\tau_n}{\tau_t}. \tag{15.1.16}$$

If there are more electrons traveling across the electrodes before recombining with the holes, there will be more photocurrent appearing in the external circuit. The current responsivity R_λ (A/W) is the photocurrent response per unit optical incident power

$$R_\lambda = \frac{\Delta I}{P_{opt}} = \eta \frac{q}{h\nu}\frac{\tau_n}{\tau_t} \tag{15.1.17}$$

and depends on the operation wavelength λ.

Case 2. Transient Response If the constant light illumination is switched off at $t = 0$, that is, $G_0(t) = G_0$, $t \leq 0$, and 0 for $t > 0$, the response for $t > 0$ satisfies

$$\frac{\partial}{\partial t}\delta n(t) = -\frac{\delta n(t)}{\tau_n}. \tag{15.1.18}$$

The solution for the excess carrier density is

$$\begin{aligned}\delta n(t) &= \delta n(0)e^{-t/\tau_n} \\ &= G_0\tau_n e^{-t/\tau_n}\end{aligned} \tag{15.1.19a}$$

and the photocurrent response obtained from (15.1.8) is

$$\Delta I(t) = q\mu_n G_0\tau_n\left(\frac{AV}{\ell}\right)e^{-t/\tau_n}. \tag{15.1.19b}$$

The result is shown in Fig. 15.2a. The decay time constant is the minority carrier recombination lifetime τ_n.

Impulse Response If the light injection is an impulse function,

$$G_0(t) = g_0\delta(t), \tag{15.1.20}$$

then,

$$\frac{\partial}{\partial t}\delta n(t) = g_0\delta(t) - \frac{\delta n(t)}{\tau_n}. \tag{15.1.21}$$

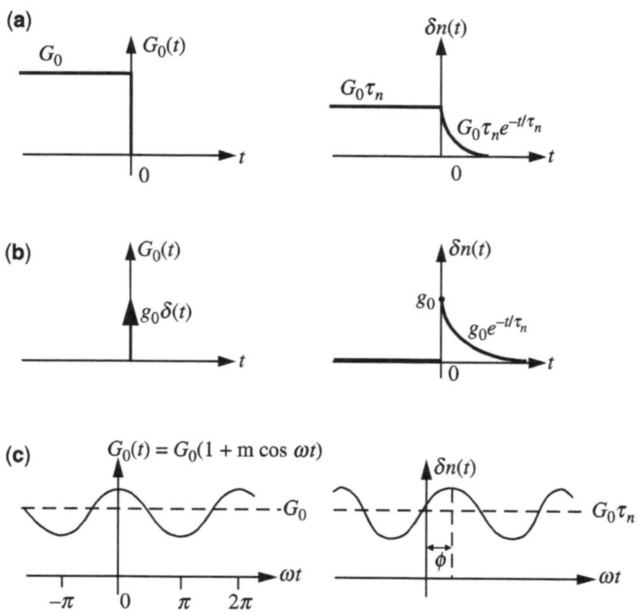

Figure 15.2 The generation rate $G_0(t)$ and the photogenerated excess electron concentration $\delta n(t)$ for (a) switch-off, (b) impulse response, and (c) sinusoidal steady-state response of a photoconductor.

Integrating the above equation from $t = 0_-$ to 0_+ gives

$$\delta n(0_+) - \delta n(0_-) = g_0.$$

Assuming that initially $\delta n(0_-) = 0$ (i.e., the semiconductor is at thermal equilibrium before $t = 0$), we obtain

$$\delta n(t) = g_0 e^{-t/\tau_n} \quad \text{for} \quad t \geq 0. \tag{15.1.22}$$

This result is plotted in Fig. 15.2b.

Case 3. Sinusoidal Steady-State Response If the optical intensity is modulated by a sinusoidal microwave signal such that

$$G_0(t) = G_0 \cos \omega t = \text{Re}[G_0 e^{-i\omega t}] \tag{15.1.23}$$

where Re means the real part of the quantity, the sinusoidal steady-state response can be found by the phasor notation

$$\delta n(t) = \text{Re}[\delta n(\omega) e^{-i\omega t}]. \tag{15.1.24}$$

We have

$$\frac{\partial}{\partial t}\delta n(t) = G_0 \cos \omega t - \frac{\delta n(t)}{\tau_n}. \tag{15.1.25}$$

The solution for $\delta n(\omega)$ using (15.1.23) – (15.1.25) is

$$\delta n(\omega) = \frac{G_0 \tau_n}{1 - i\omega\tau_n}. \tag{15.1.26}$$

Thus

$$\begin{aligned} \delta n(t) &= \text{Re}\left[\frac{G_0 \tau_n}{1 - i\omega\tau_n} e^{-i\omega t}\right] \\ &= \frac{G_0 \tau_n}{\sqrt{1 + \omega^2 \tau_n^2}} \cos(\omega t - \phi) \end{aligned} \tag{15.1.27}$$

where $\phi = \tan^{-1}(\omega\tau_n)$ is the phase delay in the ac response.

On the other hand, if the optical intensity or the optical generation rate consists of a dc and an ac component, Fig. 15.2c,

$$P(t) = P_{\text{opt}}(1 + m \cos \omega t) \tag{15.1.28}$$

$$G_0(t) = G_0(1 + m \cos \omega t) \tag{15.1.29}$$

where m is the modulation index, the photocurrent response using (15.1.15) gives

$$I(t) = I_p \left[1 + \frac{m}{\sqrt{1 + \omega^2 \tau_n^2}} \cos(\omega t - \phi) \right] \tag{15.1.30a}$$

$$I_p = q\eta \frac{P_{opt}}{h\nu} \frac{\tau_n}{\tau_t}. \tag{15.1.30b}$$

In other words, for a root-mean-square (rms) optical power, we replace the cosine function in (15.1.28) by $1/\sqrt{2}$,

$$p_{rms} = \frac{m\, P_{opt}}{\sqrt{2}} \tag{15.1.31}$$

the rms photocurrent signal is

$$i_p = q\eta \frac{p_{rms}}{h\nu} \left(\frac{\tau_n}{\tau_t} \right) \frac{1}{\sqrt{1 + \omega^2 \tau_n^2}}. \tag{15.1.32}$$

15.1.3 Noises in Photoconductors

In this subsection, we present some fundamental concepts including the spectral density function for noises in photodetectors [6]. Important noises such as generation–recombination or shot and thermal noise are discussed.

Spectral Density Function S(f) The signal such as the photocurrent in the time domain $i(t)$ of the photoconductor is related to its Fourier transform $i(f)$ in the frequency domain by

$$i(t) = \int_{-\infty}^{\infty} i(f) e^{-i2\pi ft} df \tag{15.1.33a}$$

$$i(f) = \int_{-\infty}^{\infty} i(t) e^{i2\pi ft} dt \tag{15.1.33b}$$

The average power P over a time duration T is

$$P = \frac{1}{T} \int_{-T/2}^{T/2} i^2(t) dt = \frac{1}{T} \int_{-\infty}^{\infty} |i(f)|^2 df$$

$$= \frac{2}{T} \int_0^{\infty} |i(f)|^2 df = \int_0^{\infty} S(f) df \tag{15.1.34}$$

where we have used

$$\lim_{T\to\infty}\int_{-T/2}^{T/2} e^{i2\pi(f-f')t}dt = \delta(f-f') \tag{15.1.35}$$

and defined the spectral density function $S(f)$ as

$$S(f) = \frac{2}{T}|i(f)|^2. \tag{15.1.36}$$

Shot Noise For a sequence of events such as collection of charges within a time interval T, say

$$i(t) = \sum_{i=1}^{N_t} h(t-t_i), \quad 0 \le t \le T \tag{15.1.37}$$

where N_t = total number of events within time T, we obtain the Fourier transform $i(f)$

$$i(f) = \sum_{i=1}^{N_t} h(f)e^{i2\pi f t_i}. \tag{15.1.38}$$

The ensemble average of $|i(f)|^2$ is

$$\begin{aligned}\left\langle |i(f)|^2 \right\rangle &= \left\langle |h(f)|^2 \left\{ N_t + \sum_{i\neq j}^{N_t}\sum_{j}^{N_t} \exp\left[i2\pi f(t_j - t_i)\right] \right\} \right\rangle \\ &= |h(f)|^2 \langle N_t \rangle = \langle N \rangle T |h(f)|^2\end{aligned} \tag{15.1.39}$$

where $\langle N \rangle = \langle N_t \rangle / T$ is the average number of events occurring per second. The random distributions of t_i and t_j give the cancellation in the second term in (15.1.39) when $i \neq j$. The spectral density function is

$$S(f) = \frac{2}{T}\left\langle |i(f)|^2 \right\rangle = 2\langle N \rangle |h(f)|^2. \tag{15.1.40}$$

Now we consider an injected electron between two capacitor plates with a separation ℓ. The current is

$$h(t) = \frac{q}{\ell}v(t) \qquad 0 \le t \le \tau_t \tag{15.1.41}$$

where $v(t)$ is the instantaneous velocity of the electron and τ_t is the transit time. The average current $\langle I \rangle$ is related to the average number of electron injections per second $\langle N \rangle$ by

$$\langle I \rangle = q\langle N \rangle. \tag{15.1.42}$$

The Fourier transform of $h(t)$ reduces to

$$h(f) = \int_0^{\tau_t} \frac{q}{\ell} v(t) e^{i2\pi ft} dt \simeq \int \frac{q}{\ell} \frac{dx(t)}{dt} dt = q \tag{15.1.43}$$

assuming the frequency is low enough such that $f\tau_t \ll 1$. The spectral density function is

$$S(f) = 2\langle N\rangle |h(f)|^2 = 2\langle N\rangle q^2 = 2q\langle I\rangle. \tag{15.1.44}$$

The power of this shot noise within a frequency interval between f and $f + \Delta f$ associated with the current is denoted by

$$\left\langle i_S^2(f) \right\rangle \equiv S(f)\Delta f = 2q\langle I\rangle \Delta f. \tag{15.1.45}$$

Generation–Recombination Noise In a photoconductor, the photogenerated carriers have a finite lifetime τ, which is a random variable with a mean value $\langle\tau\rangle$ denoted as τ_n for the electrons. The photocurrent due to *one* injected electron is

$$h(t) = \begin{cases} qv/\ell & 0 \le t \le \tau \\ 0 & \text{otherwise.} \end{cases} \tag{15.1.46}$$

Here we use v as the mean drift velocity and $\tau_t = \ell/v$ as the mean transit time. The Fourier transform of $h(t)$ is

$$h(f) = \int_0^{\tau} e^{i2\pi ft} \frac{q}{\tau_t} dt = \frac{q}{\tau_t} \left(\frac{e^{i2\pi f\tau} - 1}{i2\pi f} \right)$$

$$|h(f)|^2 = \left(\frac{q}{\tau_t}\right)^2 \frac{1}{(2\pi f)^2} \left(2 - e^{-i2\pi f\tau} - e^{i2\pi f\tau}\right). \tag{15.1.47}$$

If we assume that the probability function of the random variable τ obeys Poisson statistics

$$p(\tau) = \frac{1}{\tau_n} e^{-\tau/\tau_n} \tag{15.1.48}$$

with an average $\langle\tau\rangle = \int_0^{\infty} \tau p(\tau)\, d\tau = \tau_n$, we find

$$\left\langle |h(f)|^2 \right\rangle = \int_0^{\infty} |h(f)|^2 p(\tau) d\tau = \frac{2q^2(\tau_n/\tau_t)^2}{1 + (2\pi f\tau_n)^2}. \tag{15.1.49}$$

Because the average photocurrent with the photoconductive gain τ_n/τ_t due to $\langle N\rangle$ injected (primary) electrons per second is

$$I_p \equiv \langle I\rangle = \langle N\rangle q \frac{\tau_n}{\tau_t} \tag{15.1.50}$$

the spectral density function due to $\langle N \rangle$ injected electrons per second is obtained from (15.1.40), (15.1.49), and (15.1.50)

$$S(f) = 2\langle N \rangle \left\langle |h(f)|^2 \right\rangle = \frac{4qI_{\mathrm{p}}(\tau_n/\tau_t)}{1 + (2\pi f \tau_n)^2}. \tag{15.1.51}$$

The generation–recombination noise current is, therefore,

$$\left\langle i_{\mathrm{GR}}^2 \right\rangle \equiv S(f)\Delta f = \frac{4qI_{\mathrm{p}}(\tau_n/\tau_t)\Delta f}{1 + (2\pi f \tau_n)^2}. \tag{15.1.52}$$

Thermal Noise (also Johnson Noise or Nyquist Noise) In a photodetector, the random thermal motion of charge carriers contribute to a thermal noise current [6–8]. In other words, the thermal noise of a resistor R results in a random electric current $i(t)$ that is characterized by a power spectral density at temperature T

$$S(f) = \frac{4}{R} \frac{hf}{e^{(hf/k_{\mathrm{B}}T)} - 1}. \tag{15.1.53}$$

This thermal noise current adds to the photocurrent signal $i_{\mathrm{p}}(t)$ and affects the clarity of the detected signals. At low frequency $hf \ll k_{\mathrm{B}}T$, we have

$$S(f) \simeq \frac{4\,k_{\mathrm{B}}T}{R}. \tag{15.1.54}$$

The thermal noise is described by

$$\left\langle i_{\mathrm{T}}^2 \right\rangle = \int_0^{\Delta f} S(f)dt \simeq \frac{4k_{\mathrm{B}}T}{R}\Delta f \tag{15.1.55}$$

where the bandwidth Δf of the circuit is assumed to be much smaller than $k_{\mathrm{B}}T/h$.

Signal-to-Noise Ratio The power signal-to-noise ratio is therefore [8]

$$(S/N)_{\mathrm{power}} = \frac{i_{\mathrm{p}}^2}{\left\langle i_{\mathrm{T}}^2 \right\rangle + \left\langle i_{\mathrm{GR}}^2 \right\rangle} = \frac{\eta m^2 (P_{\mathrm{opt}}/h\nu)}{8\Delta f \left[1 + \frac{k_{\mathrm{B}}T}{qRI_{\mathrm{p}}} \frac{\tau_t}{\tau_n} \left(1 + \omega^2 \tau_n^2 \right) \right]}. \tag{15.1.56}$$

Noise-Equivalent Power (NEP) The noise-equivalent power is defined as the power that corresponds to the incident rms optical power ($p_{\mathrm{rms}} = m\ P_{\mathrm{opt}}/\sqrt{2}$) required such that the signal-to-noise ratio is one in a bandwidth of 1 Hz.

Detectivity (D^*) The detectivity is defined as

$$D^* = \frac{\sqrt{A\Delta f}}{\text{NEP}} \text{ cm (Hz)}^{1/2}/\text{W} \tag{15.1.57}$$

where A is the detector cross-sectional area in cm^2. More discussions on photoconductive detectors such as $Hg_xCd_{1-x}Te$ can be found in Ref. 9.

15.1.4 *n-i-p-i* Superlattice Photoconductor

Recently, modulation doping in semiconductors has been introduced for novel device applications. An interesting example is a GaAs doping superlattice used as a photoconductor. For an extensive review of the compositional and doping superlattices, see

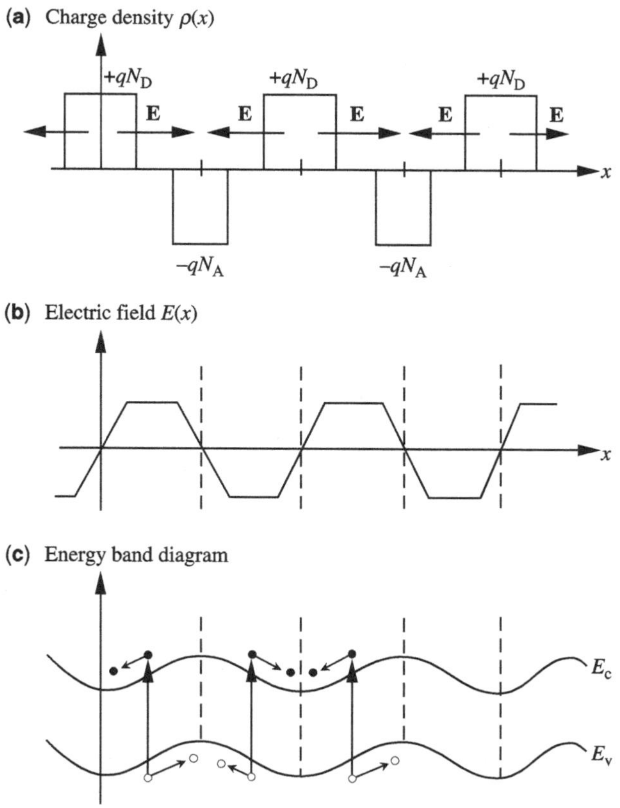

Figure 15.3 (a) The charge density profile due to ionized donors and acceptors in an *n-i-p-i* doping superlattice using the depletion approximation. (b) The electric field profile due to the charge distribution in (a). Notice that the electric field is alternating between positive and negative directions. (c) The energy band diagram of the *n-i-p-i* superlattice. The photogenerated electron–hole pairs are separated in real space because of the band profiles.

Refs. 10 and 11. Here we consider a GaAs doped periodically n-type and p-type separated by intrinsic regions as shown in Fig. 15.3a. The electric field profile $E(x)$ can be obtained noting that $dE(x)/dx = \rho(x)/\varepsilon$, where $\rho(x) = +qN_D$ in the n-doped regions, $-qN_A$ in the p-doped regions, and zero in the intrinsic regions. Therefore $E(x)$ is either a linear profile with a positive slope qN_D/ε in n-regions or a negative slope $-qN_A/\varepsilon$ in p-regions or a zero slope in the intrinsic regions, Fig. 15.3b. The potential profiles for the conduction band edge

$$E_C(x) = -q\varphi(x) = q \int_{-\infty}^{x} E(x')dx' \tag{15.1.58}$$

is the integral of the electric field profile and is shown in Fig. 15.3c.

For an incident light with energy above the band gap, the photogenerated carriers fall to the band edge and drift or diffuse to the valleys of each band as shown in Fig. 15.3c. The electrons are separated from the holes in real space, resulting in a long recombination lifetime τ_n. This enhancement of lifetime has been found to be many orders of magnitude larger than the bulk carrier lifetime. Because the photocurrent response is proportional to $G_0\tau_n$, an extremely large responsivity using the n-i-p-i superlattice as a photoconductor can be designed.

15.2 *p-n* JUNCTION PHOTODIODES

Consider a p-n junction photodiode [12–15] as shown in Fig. 15.4. The charge distribution $\rho(x)$, the electric field $E(x)$, and the potential energy profile under the depletion approximation have been discussed in Chapter 2. Here we investigate the photocurrent response if the diode is illuminated by a uniform light intensity, described by a generation rate $G(x, t)$, which is the number of electron–hole pairs created per unit time per unit volume.

15.2.1 *I–V* Curve and Photocurrent

Let us focus first on the n-side of the diode. The charge continuity equation is given by (2.4.2)

$$\frac{\partial p_n}{\partial t} = G(x, t) - \frac{\delta p_n}{\tau_p} - \frac{1}{q}\frac{\partial}{\partial x}J_p(x) \tag{15.2.1}$$

where $p_n = p_{n0} + \delta p_n$ is the total hole concentration in the n-region, p_{n0} is the hole concentration in the absence of any electric or optical injection, and δp_n is the excess hole concentration due to the external injections. The minority (hole) current density in the quasi-neutral region ($x \geq x_n$) is dominated by the diffusion component [12–14], as discussed in Chapter 2

$$J_p(x) \simeq -qD_p\frac{\partial p_n}{\partial x}. \tag{15.2.2}$$

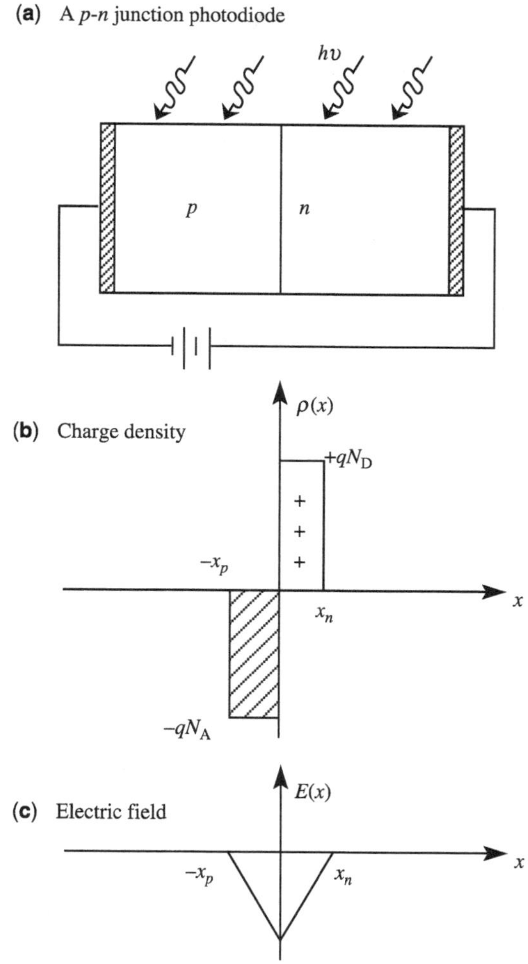

Figure 15.4 (a) A *p-n* junction diode under the illumination of a uniform light. (b) The charge distribution $\rho(x)$ under depletion approximation. (c) The electric field $E(x)$ obtained from Gauss's law.

Because p_{n0} is independent of x and t, we have at steady state if $G(x, t) = G_0$ is independent of x and t,

$$D_p \frac{\partial^2}{\partial x^2} \delta p_n - \frac{\delta p_n}{\tau_p} = -G_0. \tag{15.2.3}$$

The above equation can be solved by summing the homogeneous and particular solutions

$$\delta p_n(x) = c_1 e^{-(x-x_n)/L_p} + c_2 e^{(x-x_n)/L_p} + G_0 \tau_p \tag{15.2.4}$$

where $L_p = \sqrt{D_p \tau_p}$ is the diffusion length for holes. The particular solution is due to the optical generation. If the n-region is very long, we can set $c_2 = 0$; otherwise,

$\delta p_n(x \rightarrow \infty) \rightarrow +\infty$, which is unphysical. We expect as $x \rightarrow +\infty$, $\delta p_n(x)$ will approach $G_0\tau_p$ = total photogenerated holes. At $x = x_n$, the hole concentration is pinned by the voltage bias V with the exponential dependence

$$p_n(x = x_n) = p_{n_0} e^{qV/k_B T} \tag{15.2.5}$$

if the Boltzmann statistics are assumed. Therefore, we obtain

$$\delta p_n(x_n) = p_{n_0}(e^{qV/k_B T} - 1) \tag{15.2.6}$$

and

$$\delta p_n(x) = \left[p_{n_0}(e^{qV/k_B T} - 1) - G_0\tau_p\right] e^{-(x-x_n)/L_p} + G_0\tau_p. \tag{15.2.7}$$

The hole current density $J_p(x)$ is

$$\begin{aligned} J_p(x) &\simeq -qD_p \frac{\partial}{\partial x} \delta p_n(x) \\ &= q\frac{D_p}{L_p}\left[p_{n0}(e^{qV/k_B T} - 1) - G_0\tau_p\right] e^{-(x-x_n)/L_p}. \end{aligned} \tag{15.2.8}$$

We obtain $J_p(x)$ at the boundary of the depletion region x_n as

$$J_p(x_n) = q\frac{D_p}{L_p} p_{n0}(e^{qV/k_B T} - 1) - qG_0L_p \tag{15.2.9}$$

where the first term is due only to the voltage bias, and the last term is due to optical generation. We see that only the portion of the photogenerated holes within a diffusion length L_p away from the depletion boundary can diffuse (and survive) to the depletion region and be swept across the depletion region by the electric field and collected as the photocurrent by the external circuits. This means that the majority carriers on the p-side have to supply this current immediately. A dual approach for the electron current density at $x = -x_p$ gives

$$J_n(-x_p) = q\frac{D_n}{L_n} n_{p0}(e^{qV/k_B T} - 1) - qG_0L_n. \tag{15.2.10}$$

The total current I is the sum of $J_p(x_n)$ and $J_n(-x_p)$ multiplied by the cross-sectional area of the diode A

$$\begin{aligned} I &= A[J_p(x_n) + J_n(-x_p)] \\ &= I_0(e^{qV/k_B T} - 1) - I_{\text{ph}} \end{aligned} \tag{15.2.11}$$

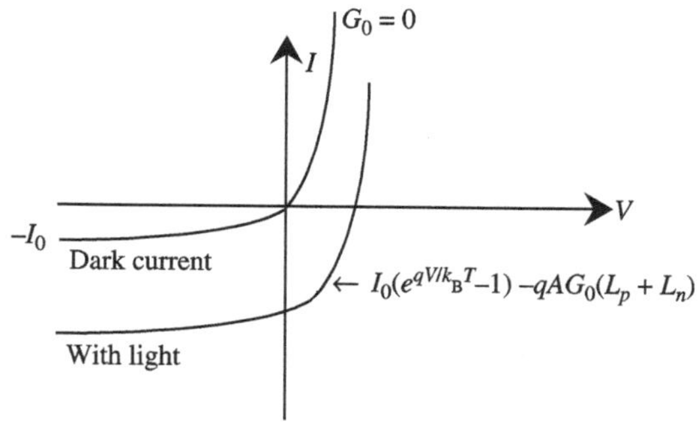

Figure 15.5 The I–V curves of a photodiode with and without illumination of light.

where

$$I_0 = qA\left[\frac{D_p}{L_p}p_{n0} + \frac{D_n}{L_n}n_{p0}\right] \tag{15.2.12a}$$

$$I_{\text{ph}} = qAG_0(L_p + L_n). \tag{15.2.12b}$$

I_0 is the diode reverse current. I_{ph} is the photocurrent of the diode (in the negative direction of the polarity), which is the generation of the electron–hole pairs in the volume defined by $A(L_p + L_n)$. The I–V curves of a photodiode with and without the illumination of light are plotted in Fig. 15.5. When $G_0 = 0$, the diode reverse current $-I_0$ is the dark current, which is usually very small compared with the photocurrent I_{ph} under a reverse bias condition. Therefore, the photocurrent is proportional to the generation rate G_0, which is proportional to the incident optical power P_{opt}.

The problems with the *p-n* junction photodiodes are

1. Optical absorption within the diffusion lengths L_p and L_n is very small (i.e., over narrow regions of L_p and L_n near the depletion region). Because L_p and L_n are very small, the contributions of the photocurrents are not effective.
2. The diffusion process is slow, which results in a slow photoresponse if the optical intensity varies with time.
3. The junction capacitance C_j is simply $\varepsilon A/x_{\text{w}}$, where x_{w} is the total depletion width derived in Eq. (2.5.25) simplified for a homojunction diode with $\varepsilon_{\text{p}} = \varepsilon_{\text{N}} = \varepsilon$

$$C_j = A\left[\frac{q\varepsilon}{2(V_0 - V)}\frac{N_{\text{D}}N_{\text{A}}}{N_{\text{D}} + N_{\text{A}}}\right]^{1/2} \tag{15.2.13}$$

which can slow down the response by the RC_j time delay. For example, if $A = (1\,\text{mm})^2$, $\varepsilon = 11.7\varepsilon_0$, $N_{\text{D}} = 10^{15}\,\text{cm}^{-3} \ll N_{\text{A}}$ for a p^+-n photodiode, and $-V = 10\,V \gg V_0$, we obtain $C_j \simeq 30\,\text{pF}$ and the 3-dB cutoff frequency $f_{\text{c}} = 1/(2\pi RC_j) = 100\,\text{MHz}$ for $R = 50\,\Omega$.

15.2.2 R_0A Product

A useful figure of merit for the *p-n* junction photodiodes is the R_0A product [15]. Because the photodiode is operated at zero-bias voltage in many direct detection applications, the differential resistance at zero-bias voltage R_0 multiplied by the junction area A is commonly used

$$(R_0A)^{-1} = \frac{1}{A}\frac{dI}{dV}\bigg|_{V=0} = \frac{dJ}{dV}\bigg|_{V=0}. \tag{15.2.14}$$

Thus far, we have derived the dark current I_0 contributed by the diffusion processes in this section. Using (15.2.11) and (15.2.12), we obtain

$$\begin{aligned}(R_0A)^{-1} &= \frac{q^2}{k_BT}\left(\frac{D_p}{L_p}p_{n0} + \frac{D_n}{L_n}n_{p0}\right)\\ &= \frac{q^2}{k_BT\,n_i^2}\left(\frac{D_pN_D}{L_p} + \frac{D_nN_A}{L_n}\right)\\ &= \frac{q}{n_i^2}\sqrt{\frac{q}{k_BT}}\left(N_D\sqrt{\frac{\mu_p}{\tau_p}} + N_A\sqrt{\frac{\mu_n}{\tau_n}}\right)\end{aligned} \tag{15.2.15}$$

where we have used the relations $p_{n0} = N_D/n_i^2$, $n_{p0} = N_A/n_i^2$, and the Einstein relations $D_p/\mu_p = D_n/\mu_n = k_BT/q$. The first term in (15.2.15) is the contribution to $1/(R_0A)$ from the diffusion current on the *n*-side of the photodiode, and the second term is from the diffusion current on the *p*-side. There can also be other contributions to R_0A products such as the generation–recombination current in the space-charge region, the surface leakage current, and the interband tunneling current, which depend on the material properties, device geometry, and surface conditions.

15.2.3 Current Responsivity and Detectivity

For a root-mean-square photon flux density $\Phi(\lambda)$ (= the number of incident photons per second per unit area) of monochromatic radiation at a wavelength λ,

$$\Phi(\lambda) = \frac{P_\lambda}{h\nu A} \tag{15.2.16}$$

where A is the photodetector illumination area, and P_λ is the root-mean-square input optical power at a wavelength λ, $P_\lambda = p_{\text{rms}} = mP_{\text{opt}}/\sqrt{2}$, we can write the rms photocurrent

$$i_p = q\eta\Phi(\lambda)A = q\eta\frac{P_\lambda}{h\nu} \tag{15.2.17}$$

and η is the quantum efficiency of the photodiode including the effects of the intrinsic quantum efficiency, the reflection, and the absorption depth (15.1.13). The current responsivity is therefore

$$R_\lambda = \frac{i_\mathrm{p}}{P_\lambda} = \eta \frac{q}{h\nu}. \tag{15.2.18}$$

The signal-to-noise ratio S/N (current) is

$$\frac{S}{N} = \frac{i_\mathrm{p}}{\sqrt{\langle i_\mathrm{n}^2 \rangle}} = \frac{R_\lambda P_\lambda}{\sqrt{\langle i_\mathrm{n}^2 \rangle}}. \tag{15.2.19}$$

The detectivity for the above S/N is defined as

$$D_\lambda^* = \frac{R_\lambda \sqrt{A \Delta f}}{\sqrt{\langle i_\mathrm{n}^2 \rangle}} \text{ cm (Hz)}^{1/2}/\text{W}. \tag{15.2.20}$$

For a photodiode at thermal equilibrium (i.e., no externally applied voltage and no illumination of light), the thermal noise depends on the zero bias resistance R_0 using (15.1.56)

$$\langle i_\mathrm{n}^2 \rangle = \frac{4\, k_\mathrm{B} T}{R_0} \Delta f. \tag{15.2.21}$$

When not in thermal equilibrium, the $I-V$ curve is

$$I(V) = I_0 (e^{qV/k_\mathrm{B}T} - 1) - I_\mathrm{ph} \tag{15.2.22a}$$

$$I_\mathrm{ph} = q \eta \Phi_\mathrm{B} A \tag{15.2.22b}$$

where Φ_B is the photon flux density due to the background radiation. The mean-squared shot noise current has contributions from three additive terms [15]: (1) a forward current that depends on voltage, $I_0 \exp(qV/k_\mathrm{B}T)$, (2) a reverse diode saturation current, and (3) the background radiation–induced photocurrent. Because these shot noise currents fluctuate independently, the total mean-squared shot noise current is

$$\langle i_\mathrm{n}^2 \rangle = 2q \left[I_0 e^{qV/k_\mathrm{B}T} \Delta f + I_0 \Delta f + I_\mathrm{ph} \Delta f \right]. \tag{15.2.23}$$

At an operation voltage $V = 0$, $R_0^{-1} = (dI/dV)_{V=0} = qI_0/k_\mathrm{B}T$ and (15.2.23) can be written as

$$\langle i_\mathrm{n}^2 \rangle = \left(\frac{4\, k_\mathrm{B} T}{R_0} + 2q^2 \eta \Phi_\mathrm{B} A \right) \Delta f. \tag{15.2.24}$$

The detectivity at zero bias voltage is then obtained from (15.2.20)

$$D_\lambda^* = \left(\frac{q\eta}{h\nu}\right)\frac{1}{\left[\left(\dfrac{4\,k_B T}{R_0 A}\right) + 2q^2\eta\Phi_B\right]^{1/2}}. \tag{15.2.25}$$

For a thermally limited case (i.e., when the thermal noise is dominant over the background radiation induced signal and other noises), we have

$$\left(D_\lambda^*\right)_T = \frac{q\eta}{h\nu}\sqrt{\frac{R_0\,A}{4\,k_B T}} \tag{15.2.26}$$

which relates the R_0A product to the thermally limited detectivity. If the photodiode is background radiation limited, which means that the background radiation–induced photocurrent is dominant, we have

$$\left(D_\lambda^*\right)_{\mathrm{BLIP}} = \frac{1}{h\nu}\sqrt{\frac{\eta}{2\Phi_B}} \tag{15.2.27}$$

which is the detectivity of the background limited infrared photodetector (BLIP).

15.3 *p-i-n* PHOTODIODES

To enhance the responsivity of the photodiode, an intrinsic region used as the major absorption layer is added, Fig. 15.6a. For a light injected from the p^+-side with an optical power intensity I_{opt} (W/cm^2), the generation rate of electron–hole pairs is [8]

$$G(x) = \eta_i(1 - R)\,\Phi(\lambda)\alpha e^{-\alpha x} \tag{15.3.1a}$$

$$\Phi(\lambda) = \frac{P_{\mathrm{opt}}/A}{h\nu} = \frac{I_{\mathrm{opt}}}{h\nu} \tag{15.3.1b}$$

where $\Phi(\lambda)$ is the photon flux density (i.e., the number of photons injected per unit area per second) and I_{opt} is the incident optical power P_{opt} divided by the area A. Here η_i is the intrinsic quantum efficiency for the probability of creating an electron–hole pair for each incident photon, and $\alpha(\lambda)$ is the absorption coefficient (1/cm) of the semiconductors. Note that the total injected number of photoelectrons per unit area per second is

$$S_0 = \int_0^\infty G(x)dx = \eta_i(1 - R)\Phi(\lambda). \tag{15.3.2}$$

The energy band profiles for the *p-i-n* diode can be obtained graphically based on the well-known depletion approximation as shown in Fig. 15.6b–e for the charge density $\rho(x)$, the electric field $E(x)$, the potential $\phi(x)$, and the band diagram.

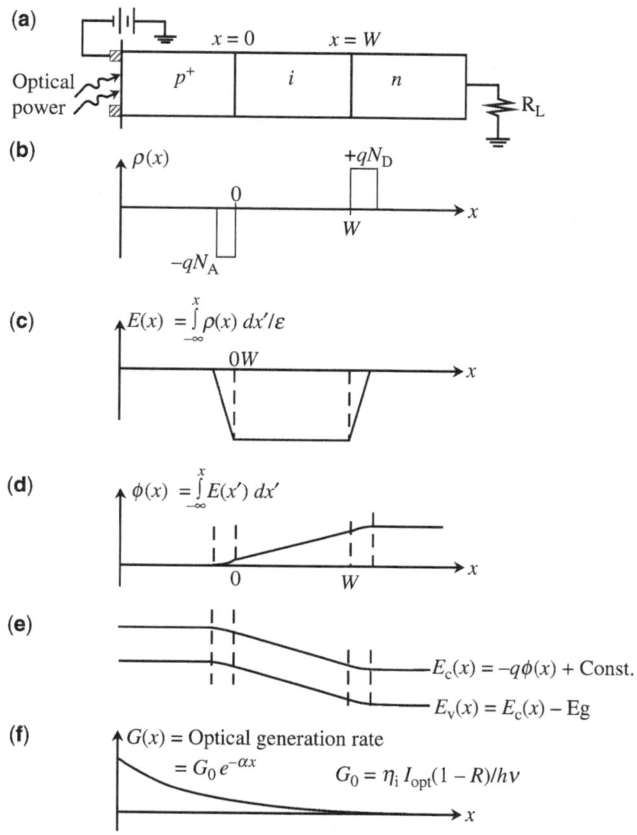

Figure 15.6 (a) A p-i-n photodiode under optical illumination from the p^+-side, (b) the charge density $\rho(x)$ under depletion approximation, (c) the static electric field profile $E(x)$, (d) the electrostatic potential $\varphi(x)$, (e) the conduction and valence band edge profiles, and (f) the optical generation rate $G(x)$.

At steady state, the total photocurrent density consists of both a drift and a diffusion component

$$J = J_{dr} + J_{diff}. \tag{15.3.3}$$

Considering the p^+-region to be of negligible thickness, we look at the contribution in the intrinsic region $0 < x < W$

$$J_{dr} = -q \int_0^W G(x)dx = -qS_0(1 - e^{-\alpha W}) = -q\eta\Phi(\lambda) \tag{15.3.4a}$$

$$\eta = \eta_i(1 - R)(1 - e^{-\alpha W}) \tag{15.3.4b}$$

where the minus sign of the drift current density accounts for the fact that the drift current flows in the $-x$ direction, and η is the quantum efficiency including the effects of surface reflectivity R and finite thickness of the absorption layer W.

The above expression also shows that an increase of $W \gg 1/\alpha$ enhances the photocurrent because of the increasing amount of absorption.

For $x > W$, the analysis is similar to that of the *p-n* junction diode; the hole (minority) current density on the *n*-side is due only to diffusion

$$J_{\text{diff}} \simeq -q\, D_p \frac{\partial}{\partial x} P_n(x) \tag{15.3.5}$$

$$0 = \frac{\partial P_n}{\partial t} = G(x) - \frac{\delta P_n}{\tau_p} - \frac{1}{q}\frac{\partial}{\partial x} J_p(x). \tag{15.3.6}$$

Therefore, we solve

$$\frac{\partial^2}{\partial x^2}\delta P_n - \frac{1}{L_p^2}\delta P_n = -\frac{1}{D_p}G(x). \tag{15.3.7}$$

The solution for $\delta P_n(x)$ consists of the homogeneous solution and the particular solution

$$\delta P_n(x) = Ae^{-(x-W)/L_p} + Be^{(x-W)/L_p} + Ce^{-\alpha x}. \tag{15.3.8}$$

We set $B = 0$ because $\delta P_n(x \to \infty)$ should be finite assuming that the *n*-region is much thicker than the hole diffusion length. Otherwise, an additional boundary condition at the finite end of the *n*-region has to be added for a finite thickness, which will be considered in Section 15.6. The particular solution is obtained by substituting $Ce^{-\alpha x}$ into (15.3.7)

$$C = \frac{S_0}{D_p}\frac{\alpha L_p^2}{1 - \alpha^2 L_p^2}. \tag{15.3.9}$$

The coefficient A is then determined by the boundary condition

$$\delta P_n(W) = P_{n0}\left(e^{qV/k_B T} - 1\right) \simeq -P_{n0} \tag{15.3.10}$$

which is pinned by the reverse bias voltage, and we find

$$A = -P_{n0} - Ce^{-\alpha W}. \tag{15.3.11}$$

Thus, the solution for the excess hole concentration in the *n*-side is

$$\delta P_n(x) = -\left(P_{n0} + Ce^{-\alpha W}\right) e^{-(x-W)/L_p} + Ce^{-\alpha x}. \tag{15.3.12}$$

The hole current density on the n-side is

$$\begin{aligned} J_{\text{diff}} &= -qD_p \frac{d}{dx} P_n(x)\big|_{x=W} \\ &= qD_p\alpha\left(1 - \frac{1}{\alpha L_p}\right) Ce^{-\alpha W} - q\frac{D_p}{L_p} P_{n0} \\ &= -q\left[\frac{S_0 \alpha L_p}{1 + \alpha L_p} e^{-\alpha W} + P_{n0}\frac{D_p}{L_p}\right]. \end{aligned} \tag{15.3.13}$$

The total current density is

$$\begin{aligned} J &= J_{\text{dr}} + J_{\text{diff}}(\text{at } x = W) \\ &= -q\, S_0\left[1 - \frac{e^{-\alpha W}}{1 + \alpha L_p}\right] - q\, P_{n0}\frac{D_p}{L_p}. \end{aligned} \tag{15.3.14}$$

The quantum efficiency is

$$\eta = \frac{J/q}{\Phi(\lambda)} = \eta_{\text{i}}(1 - R)\left(1 - \frac{e^{-\alpha W}}{1 + \alpha L_p}\right) \tag{15.3.15}$$

neglecting the contribution of the last term (dark current) in (15.3.14). Note that if $W \to \infty$, the current density is dominated by

$$J = -qS_0 = -q\eta_{\text{i}}(1 - R)\Phi(\lambda) \tag{15.3.16}$$

as expected. Long-wavelength *p-i-n* photodiodes for high-speed receiver applications are of great interest for optical communication systems. For high-sensitivity optical receivers, photodiodes with a small junction area around a few tens of micrometers in diameter and a low doping depletion region of a few micrometers are desired [16]. At low doping, the center absorption region can be depleted at a small bias voltage and reduce the tunneling leakage current. Most of the long-wavelength photodetectors use $In_{0.47}Ga_{0.53}As$ grown on an InP substrate as the absorption region. The leakage current in InGaAs *p-i-n* diodes is dominated by the interband tunneling at high reverse-bias voltages and generation–recombination processes at low voltages [17, 18]. Hybrid *p-i-n*/FET receivers have been assembled for high-speed photoreceivers, which offer better sensitivity than other *p-i-n* photodiode receivers [16, 19]. Ultra-broadband *p-i-n* photodetectors have also been shown to have a great potential for high-speed applications [20] with an impulse response in the picosecond scale.

15.4 AVALANCHE PHOTODIODES

To enhance the photocurrent response, some built-in multiplication processes may be used such that more photocurrents can be extracted in the external circuits at a given optical illumination. Ideally, we would have a single-carrier-type photomultiplier; the carrier concentration will grow if impact ionization occurs in a region with a large electric field. In semiconductors, both electrons and holes can impact ionize more electron–hole pairs as shown in Fig. 15.7a. A schematic diagram for an avalanche diode is shown in Fig. 15.7b. A feedback process occurs because electrons and holes travel in opposite directions. Let us define

$$\begin{aligned}\alpha_n &= \text{the electron ionization coefficient (1/cm)}\\ &= \text{the number of electron–hole pairs}\\ &\quad\ \text{generated per unit distance per incident electron.}\\ \beta_p &= \text{the hole ionization coefficient (1/cm)}\\ &= \text{the number of electron–hole pairs}\\ &\quad\ \text{generated per unit distance per incident hole.}\end{aligned}$$

In general $\alpha_n \neq \beta_p$ for most semiconductors. They are functions of the applied electric field E with an exponential dependence:

$$\alpha_n(E) = \alpha_0 e^{-C_n/E} \quad \text{and} \quad \beta_p(E) = \beta_0 e^{-C_p/E}$$

where the constants α_0, β_0, C_n, and C_p depend on the materials. For a general overview of the fundamentals of avalanche photodiodes, see Refs. 21–23.

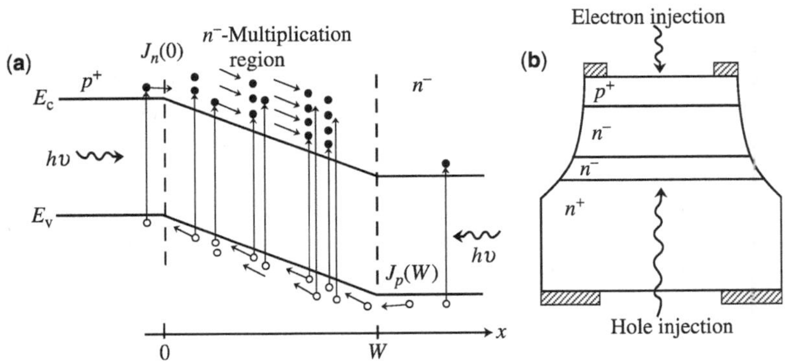

Figure 15.7 (a) The energy band diagram for an avalanche photodiode with the electron and hole ionization coefficients α_n and β_p. The electron and hole injections are given by $J_n(0)$ and $J_p(W)$. (b) A schematic diagram for an avalanche photodiode.

15.4.1 Ideal Avalanche Photodiode: Single-Carrier-Type Capable of Ionizing Collisions

We now consider the simplified case in which only electrons cause impact ionization [22]. Suppose we have an incident current density $J_n(x)$ at a plane located at position x, Fig. 15.8a. Over an incremental distance Δx, the total generated electron–hole pairs is $\alpha_n \Delta x$ multiplied by $J_n(x)$ because $\alpha_n \Delta x$ is the number of ionized electrons per incident electron per second in a distance Δx. Therefore, the current density $J_n(x + \Delta x)$ is the sum of the incident (or primary) current density and the ionized (or secondary) current density

$$J_n(x + \Delta x) = J_n(x) + \alpha_n \Delta x J_n(x) \tag{15.4.1}$$

or

$$\frac{d}{dx} J_n(x) = \alpha_n J_n(x). \tag{15.4.2}$$

Its solution is

$$J_n(x) = J_n(0) e^{\int_0^x \alpha_n(x')dx'}. \tag{15.4.3}$$

In general, $\alpha_n \equiv \alpha_n(x)$ is not a constant because the electric field $E(x)$ may not be uniform. The multiplication factor M_n for the electrons is defined as

$$M_n = \frac{J_n(W)}{J_n(0)} = e^{\int_0^W \alpha_n(x')dx'}. \tag{15.4.4}$$

For a uniform α_n, we have

$$J_n(x) = J_n(0) e^{\alpha_n x} \qquad M_n = e^{\alpha_n W}. \tag{15.4.5}$$

Here the multiplication factor is finite because W is a finite width.

A similar procedure for holes propagating in the $-x$ direction as shown in Fig. 15.8b leads to

$$-\frac{d}{dx} J_p(x) = \beta_p J_p(x) \tag{15.4.6}$$

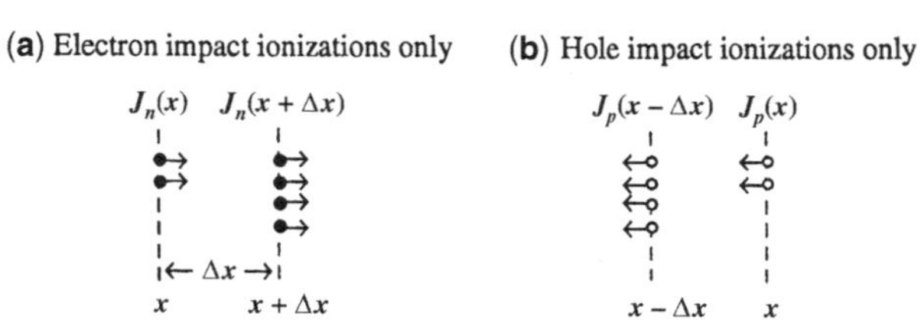

Figure 15.8 Schematic diagrams for only (a) electron and (b) hole impact ionizations.

and its solution is

$$J_p(x) = J_p(W)e^{\int_x^W \beta_p(x')dx'}. \tag{15.4.7}$$

The hole multiplication ratio is defined as

$$M_p = \frac{J_p(0)}{J_p(W)} = e^{\int_0^W \beta_p(x')dx'}. \tag{15.4.8}$$

For the special case in which β_p is independent of the position x, we have

$$J_p(x) = J_p(W)e^{\beta_p(W-x)} \qquad M_p = e^{\beta_p W}. \tag{15.4.9}$$

15.4.2 Both Electron and Hole Capable of Impact Ionization

Let us write the complete coupled equations when both electrons and holes cause impact ionizations in the presence of optical generation

$$\frac{d}{dx}J_n(x) = \alpha_n(x)J_n(x) + \beta_p(x)J_p(x) + qG(x) \tag{15.4.10a}$$

$$-\frac{d}{dx}J_p(x) = \alpha_n(x)J_n(x) + \beta_p(x)J_p(x) + qG(x). \tag{15.4.10b}$$

Note that each impact ionization process creates an electron–hole pair, as does the optical generation rate $G(x)$ per unit volume. The above two equations lead to

$$\frac{d}{dx}\left(J_n(x) + J_p(x)\right) = 0 \tag{15.4.11}$$

or the total current density

$$J = J_n(x) + J_p(x) = \text{constant},$$

which is independent of the position, as it should be as this is a one-dimensional problem. Any current passing through a surface at x has to be the same at steady state. The two coupled first-order differential equations can be solved using one of the two variables, for example, $J_n(x)$,

$$\frac{d}{dx}J_n(x) - [\alpha_n(x) - \beta_p(x)]J_n(x) = \beta_p J + qG(x) \tag{15.4.12}$$

where J is independent of x and is determined later by the boundary conditions. Noting that the first-order differential equation

$$\frac{d}{dx}y(x) + p(x)\,y(x) = Q(x) \tag{15.4.13}$$

has a solution of the form

$$y(x) = \frac{\int_{x_0}^{x} dx' Q(x') e^{\int_{x_0}^{x'} p(x'')dx''} + y(x_0)}{\exp\left(\int\limits_{x_0}^{x} p(x')dx'\right)}, \tag{15.4.14}$$

we obtain by setting all initial conditions at $x_0 = 0$:

$$J_n(x) = e^{\phi(x)}\left[J\int_0^x \beta_p(x')e^{-\phi(x')}dx' + q\int_0^x G(x')e^{-\phi(x')}dx' + J_n(0)\right] \tag{15.4.15}$$

where

$$\phi(x) = \int_0^x [\alpha_n(x') - \beta_p(x')]dx'. \tag{15.4.16}$$

We then match the boundary conditions at $x = W$. Suppose $J_n(0)$ and $J_p(W)$ are given; we find $J_n(x = W)$ from (15.4.15) and use

$$J_n(W) = J - J_p(W) \tag{15.4.17}$$

$$\varphi(W) - \varphi(x') = \int_{x'}^{W} [\alpha_n(x'') - \beta_p(x'')]dx''. \tag{15.4.18}$$

J is found as

$$J = \left(\frac{1}{1-B}\right)\left[J_p(W) + J_n(0)e^{\phi(W)} + q\int_0^W G(x')e^{\int_{x'}^{W}[\alpha_n(x'')-\beta_p(x'')]dx''}dx'\right] \tag{15.4.19}$$

$$B = \int_0^W dx' \beta_p(x')e^{\int_{x'}^{W}[\alpha_n(x'')-\beta_p(x'')]dx''}. \tag{15.4.20}$$

Then, we have $J_n(x)$ from (15.4.15). The hole current density is

$$J_p(x) = J - J_n(x). \tag{15.4.21}$$

Alternatively, we can start with (15.4.10b) using J_p as the variable and find

$$J_p(x) = \frac{\int_x^W dx'[\alpha_n(x')J + qG(x')]\, e^{\int_{x'}^{W}[\alpha_n(x'')-\beta_p(x'')]dx''} + J_p(W)}{\exp\left(\int_x^W [\alpha_n(x') - \beta_p(x')]dx'\right)} \tag{15.4.22}$$

and obtain another expression for J using the boundary condition at $x = 0$, $J_p(0) = J - J_n(0)$

$$J = \left(\frac{1}{1-A}\right)\left[J_n(0) + J_p(W)e^{-\phi(W)} + q\int_0^W G(x')e^{-\int_0^{x'}[\alpha_n(x'')-\beta_p(x'')]dx''}dx'\right] \quad (15.4.23)$$

$$A = \int_0^W dx'\alpha_n(x')e^{-\int_0^{x'}[\alpha_n(x'')-\beta_p(x'')]dx''}. \quad (15.4.24)$$

Let us consider three special cases with only electron injection at $x = 0$, or hole injection at $x = W$, or optical injection $G(x)$ at a position x.

Case 1. Only Electron Injection at $x = 0$ ($J_p(W) = 0$ and $G(x) = 0$): The electron multiplication ratio is obtained from (15.4.19)

$$M_n = \frac{J}{J_n(0)} = \frac{e^{\varphi(W)}}{1-B}. \quad (15.4.25)$$

An alternative form, if we start with $J_p(x)$ from the beginning and find J in (15.4.23),

$$M_n = \frac{J}{J_n(0)} = \frac{1}{1-A}. \quad (15.4.26)$$

Case 2. Only Hole Injection at $x = W$ ($J_n(0) = 0$ and $G(x) = 0$): The hole multiplication ratio is, using (15.4.23),

$$M_p = \frac{J}{J_p(W)} = \frac{e^{-\varphi(W)}}{1-A} \quad (15.4.27)$$

which is also a dual form of (15.4.25). A dual form of (15.4.26) is

$$M_p = \frac{J}{J_p(W)} = \frac{1}{1-B}. \quad (15.4.28)$$

A simple case is, if we consider $\alpha_n = \beta_p = \text{constant}$, and $G(x) = 0$, we find

$$J_n(x) = J\beta_p x + J_n(0). \quad (15.4.29)$$

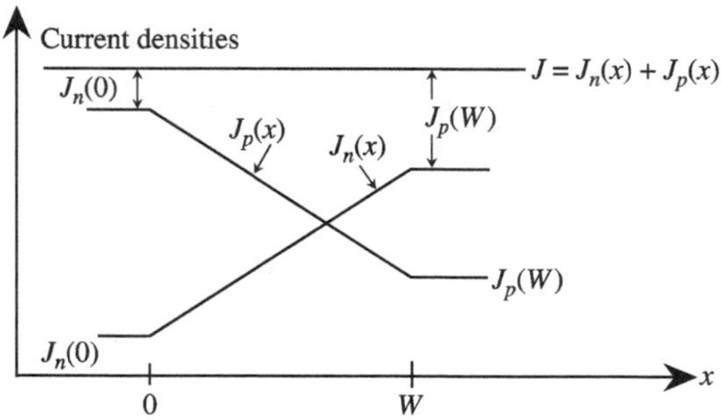

Figure 15.9 The electron and hole current densities, $J_n(x)$ and $J_p(x)$, and the total current density J as a function of position x for a special case $\alpha_n = \beta_p =$ constant, and the only injections are determined by the values $J_n(0)$ and $J_p(W)$.

Using $J_n(W) = J - J_p(W)$, we find

$$J = \frac{J_n(0) + J_p(W)}{1 - \beta_p W}. \tag{15.4.30}$$

The results of $J_n(x)$ and $J_p(x)$ are plotted in Fig. 15.9.

Case 3. Only Optical Injection $G(x') = G_0\,\delta(x' - x)$, $(J_n(0) = 0, J_p(W) = 0)$: We find that the total current density from (15.4.19) and the multiplication rate depends on the position of the optical injection x

$$M(x) = \frac{J}{qG_0} = \left(\frac{1}{1-B}\right) e^{\int_x^W [\alpha_n(x') - \beta_p(x')]dx'}. \tag{15.4.31}$$

We can also check two special optical injection positions $x = 0$ and $x = W$. We find

$$M(x = 0) = M_n \tag{15.4.32}$$

and

$$M(x = W) = M_p \tag{15.4.33}$$

which are the same expressions as in (15.4.26) and (15.4.28), as expected.

By controlling the electron injection in the p-region or the hole injection in the n-region, the multiplication factors $M_n(V)$ and $M_p(V)$ as a function of the reverse biased voltage V can be determined [24–31]. They usually increase very slowly at a low reverse bias and show an exponentially increasing behavior above a certain large bias voltage, as shown in Fig. 15.10. These reverse bias voltages can be of the order 30 or 40 V. Once M_n and M_p are measured, α_n and β_p are found from (15.4.25) to (15.4.28). If both α_n and β_p are independent of the position x, the integrations can be carried out analytically. We can express α_n and β_p in terms

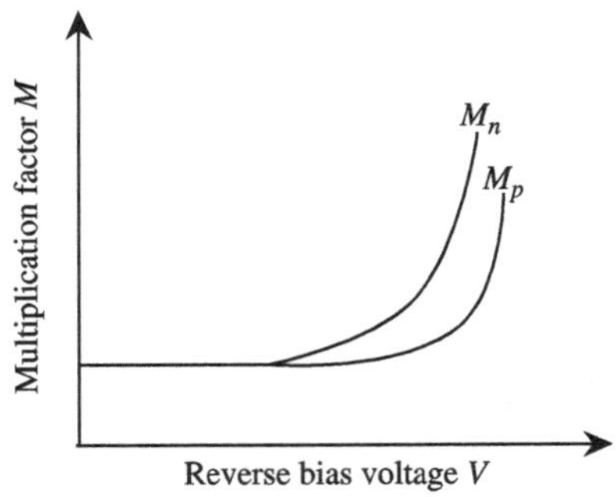

Figure 15.10 Multiplication factors for the electron and hole injections, M_n and M_p, respectively, are plotted as a function of the reverse bias voltage.

of M_n and M_p

$$\alpha_n = \left(\frac{M_n - 1}{M_n - M_p}\right)\frac{1}{W}\ln\left(\frac{M_n}{M_p}\right) \tag{15.4.34}$$

$$\beta_p = \left(\frac{M_p - 1}{M_n - M_p}\right)\frac{1}{W}\ln\left(\frac{M_n}{M_p}\right). \tag{15.4.35}$$

From the above discussions, we see that in order to determine the impact ionization coefficients from the multiplication of photocurrent, proper experimental conditions [24–26] have to be met such as: (1) $M_n(V)$ and $M_p(V)$ for pure electron injection and pure hole injection, respectively, have to be measured in the same diode, but not in complementary p^+-n and n^+-p devices; (2) the primary injected photocurrent (without multiplication) must be determined accurately as a function of bias voltage; (3) the electric field should be slowly varying in space and uniform in the active region. The dependence of the electric field on the position and bias voltage must be accurately known. It is not easy to meet all of these conditions without approximations. Experimental data for Si [27], GaAs [24, 25, 28], InP [26, 29, 30], InGaAs, and InGaAsP have been reported. We list some results below:

GaAs [25]:

$$\alpha_n = 1.899 \times 10^5 e^{-(5.75\times10^5/E)^{1.82}} (1/\text{cm})$$

$$\beta_p = 2.215 \times 10^5 e^{-(6.57\times10^5/E)^{1.75}} (1/\text{cm})$$

InP [26]:

1. $240\,\text{kV/cm} < E < 380\,\text{kV/cm}$, $N = 1.2 \times 10^{15}\,\text{cm}^{-3}$

$$\alpha_n = 1.12 \times 10^7 e^{-3.11\times10^6/E} (1/\text{cm})$$

$$\beta_p = 4.76 \times 10^6 e^{-2.55\times10^6/E} (1/\text{cm})$$

2. $360\,\text{kV/cm} < E < 560\,\text{kV/cm}$, $N = 3.0 \times 10^{16}\,\text{cm}^{-3}$

$$\alpha_n = 2.93 \times 10^6 e^{-2.64\times10^6/E}(1/\text{cm})$$

$$\beta_p = 1.62 \times 10^6 e^{-2.11\times10^6/E}(1/\text{cm})$$

3. $530\,\text{kV/cm} < E < 770\,\text{kV/cm}$, $N = 1.2 \times 10^{17}\,\text{cm}^{-3}$

$$\alpha_n = 2.32 \times 10^5 e^{-7.16\times10^{11}/E^2}(1/\text{cm})$$

$$\beta_p = 2.48 \times 10^5 e^{-6.23\times10^{11}/E^2}(1/\text{cm})$$

InP [30]:

$$\alpha_n = 5.55 \times 10^6 e^{(-3.10\times10^6/E)}(1/\text{cm})$$

$$\beta_p = 1.98 \times 10^6 e^{(-2.29\times10^6/E)}(1/\text{cm})$$

InGaAsP ($E_g = 0.92$ eV) [30]:

$$\alpha_n = 3.37 \times 10^6 e^{(-2.29\times10^6/E)}(1/\text{cm})$$

$$\beta_p = 2.94 \times 10^6 e^{(-2.40\times10^6/E)}(1/\text{cm})$$

$In_{0.53}Ga_{0.47}As$ [30]:

$$\alpha_n = 2.27 \times 10^6 e^{(-1.13\times10^6/E)}(1/\text{cm})$$

$$\beta_p = 3.95 \times 10^6 e^{(-1.45\times10^6/E)}(1/\text{cm})$$

The electric field E is in V/cm in the above expressions. For strained $In_{0.2}Ga_{0.8}As$ and $In_{0.15}Ga_{0.63}Al_{0.22}As$ channels embedded in $Al_{0.3}Ga_{0.7}As$ material, the α_n and β_p values have been found to be higher in $In_{0.2}Ga_{0.8}As$ channels and lower in the $In_{0.15}Ga_{0.63}Al_{0.22}As$ channels compared with the unstrained GaAs channels using hole injection in a lateral *p-i-n* diode configuration [31].

For an rms optical power (15.1.31)

$$P_{\text{rms}} = \frac{m\,P_{\text{opt}}}{\sqrt{2}} \tag{15.4.36}$$

where m is the microwave modulation depth in (15.1.30) and P_{opt} is the optical input power. The multiplied rms photocurrent response is

$$i_p = q\eta\frac{P_{\text{rms}}}{h\nu}M \tag{15.4.37}$$

where $M = M_n$ for the electron multiplication and $M = M_p$ for the hole multiplication.

Excess Noise Because the multiplication processes are random, an excess-noise factor can be defined for the multiplication factor M, which is treated as a random variable,

$$F(M) = \langle M^2 \rangle / \langle M \rangle^2 \tag{15.4.38}$$

where $\langle \, \rangle$ means ensemble average. It has been shown that these excess-noise factors [32, 33] can be written in terms of the ratio of the impact ionization coefficients α_n/β_p

$$F_n = \frac{\beta_p}{\alpha_n}\langle M_n \rangle + \left(1 - \frac{\beta_p}{\alpha_n}\right)\left(2 - \frac{1}{\langle M_n \rangle}\right) \quad \text{electron injection} \tag{15.4.39}$$

$$F_p = \frac{\alpha_n}{\beta_p}\langle M_p \rangle + \left(1 - \frac{\alpha_n}{\beta_p}\right)\left(2 - \frac{1}{\langle M_p \rangle}\right) \quad \text{hole injection.} \tag{15.4.40}$$

Note that if no avalanche multiplications exist, $\langle M_n \rangle$, $\langle M_p \rangle$, F_n, and F_p are all equal to unity. In Fig. 15.11, we plot F_n versus $\langle M_n \rangle$ for various $\beta_n/\alpha_n = k$ ratios. We see the increase of the excess-noise factor F_n with increasing β_p/α_n for electron injection. The physical reason for this is that, in the case of electron injection from the p^+-region in Fig. 15.7, the secondary electron–hole pairs also cause impact ionizations. The holes propagate in the opposite direction to that of the electrons. Therefore, if β_p/α_n is increased, the back-propagating holes will create more impact ionization currents. The measured amplified current at the end electrodes will have more fluctuating signals because these secondary or higher-order impact

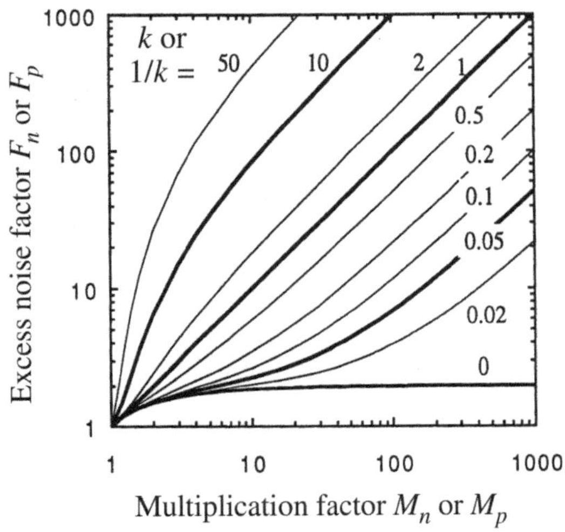

Figure 15.11 The excess-noise factor F versus the multiplication factor $\langle M \rangle$ for different values of the ratio of the electron and hole ionization coefficients. For electron injection $k = \beta_p/\alpha_n$ and for hole injection $\alpha_n/\beta_p = 1/k$ should be used in the ratio.

ionization processes will contain more random characteristics. For hole injection, $1/k$ should be used in the same diagram. The multiplication or gain noise is given by

$$\langle i_{\mathrm{M}}^2\rangle = 2q\langle M\rangle I_{\mathrm{p}}\, F(M) I_{\mathrm{p}}. \tag{15.4.41}$$

The mean-squared shot-noise current after multiplication is generalized from the shot noise in (15.1.45) by multiplying the factor $\langle M^2\rangle$

$$\begin{aligned}\langle i_{\mathrm{S}}^2\rangle &= 2q(I_{\mathrm{P}} + I_{\mathrm{B}} + I_{\mathrm{D}})\langle M^2\rangle B \\ &= 2q(I_{\mathrm{P}} + I_{\mathrm{B}} + I_{\mathrm{D}})\langle M\rangle^2 FB\end{aligned} \tag{15.4.42}$$

where I_{P} is the average steady-state photocurrent (15.1.30b), I_{B} is the background current, and I_{D} is the dark current [34–36], and $B = \Delta f$ is the bandwidth. The thermal noise is

$$\langle i_{\mathrm{T}}^2\rangle = \frac{4\, k_{\mathrm{B}} T}{R_{\mathrm{eq}}} B \tag{15.4.43}$$

where $1/R_{\mathrm{eq}} = 1/R_{\mathrm{j}} + 1/R_{\mathrm{L}} + 1/R_{\mathrm{i}}$ accounts for the junction resistance R_{j}, the external load resistance R_{L}, and the input resistance R_{i} of the following amplifier of the photodiode.

For the modulation depth $m = 1$, the signal-to-noise ratio (power) for the avalanche diode is [8, 21]

$$\frac{S}{N} = \frac{i_{\mathrm{P}}^2}{\langle i_{\mathrm{S}}^2\rangle + \langle i_{\mathrm{T}}^2\rangle} = \frac{\frac{1}{2}\left(q\eta \frac{P_{\mathrm{opt}}}{\mathrm{h}\nu}\right)^2 \langle M\rangle^2}{2q(I_{\mathrm{P}} + I_{\mathrm{B}} + I_D)\langle M\rangle^2 FB + 4\, k_{\mathrm{B}} T\, B/R_{\mathrm{eq}}}. \tag{15.4.44}$$

Because $F(M) \geq 1$ and is a monotonically increasing function of the average multiplication ratio $\langle M\rangle$, the above signal-to-noise ratio can be optimized at a particular value of $\langle M\rangle$.

High-speed detections using avalanche photodiodes and their time dependence or frequency response have been investigated [37–40]. For $M_0 > \alpha_n/\beta_p$, the frequency-dependent multiplication factor is

$$M(\omega) = \frac{M_0}{\sqrt{1 + (M_0 \omega \tau_t)^2}} \tag{15.4.45}$$

where τ_t is an effective transit time through the avalanche region. The effects of the avalanche build-up time have also been reported [39].

15.4.3 Separate Absorption and Multiplication Avalanche Photodiode (SAM APD) [29, 30, 41, 42]

Various contributions to the dark current of an APD have been investigated, which include the generation–recombination via midgap traps in the depletion region, tunneling of the carriers across the band gap, and a surface leakage current across the *p-n* junction in InGaAs APD, for example. When the reverse bias is above a certain value before the breakdown voltage V_B, it has been found that the tunneling current is dominant, unless the doping density N_D in the absorption region can be reduced below a certain value in which the tunneling current can be reduced to be smaller than the generation–recombination current. A separate absorption and multiplication structure has been proposed [41] to reduce the tunneling current. The geometry is shown in Fig. 15.12a, where a low-field InGaAs region is used as

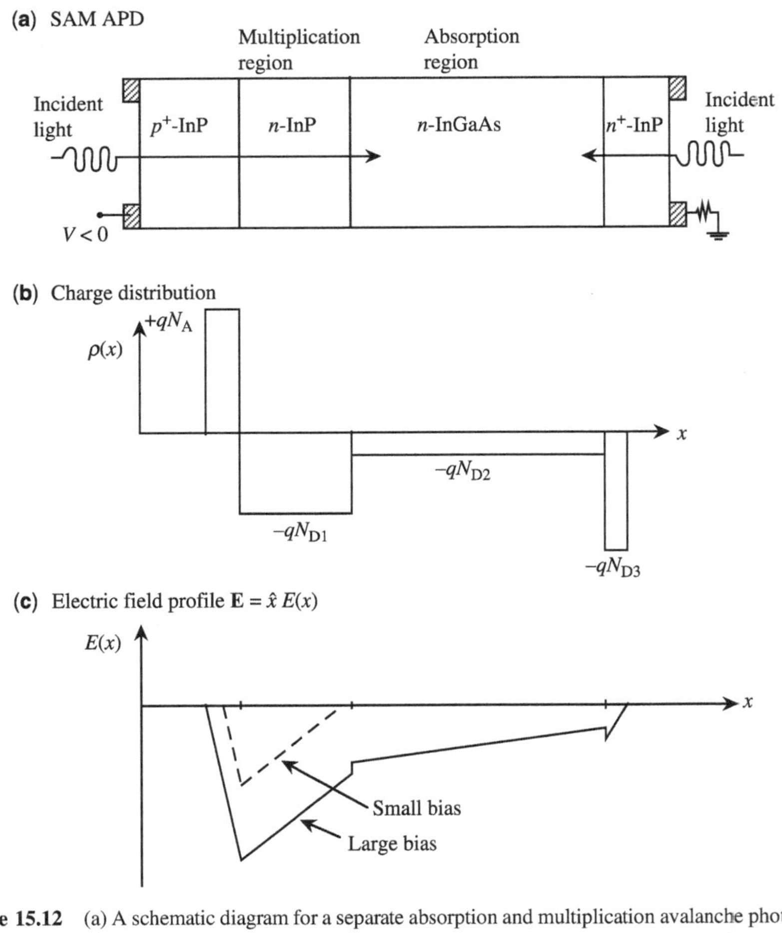

Figure 15.12 (a) A schematic diagram for a separate absorption and multiplication avalanche photodiode (SAM APD), where the absorption occurs at the narrow band gap InGaAs region and the photogenerated carriers are swept into the InP multiplication region where the electric field is larger. (b) Charge density profile under a large reverse bias. (c) The electric field profile under a large (solid lines) or small (dashed lines) reverse bias.

the absorption region and the photogenerated carriers are swept into the high-field InP binary region where avalanche multiplications occur [16]. Because the InP layer has a larger band gap than that of the InGaAs absorption region, the tunneling current can be reduced. The electric field profile $\mathbf{E} = \hat{x}E(x)$ can be obtained using the charge density profile $\rho(x)$ based on the depletion approximation at a large reverse bias voltage because $\partial \varepsilon E(x)/\partial x = \rho(x)$, Fig. 15.12b. Because ε is slightly different in InGaAs and InP layers, a slight discontinuity in $E(x)$ occurs at the InGaAs/InP interface, Fig.15.12c.

15.4.4 Multiple-Quantum-Well APD

Heterostructure avalanche photodiodes have been fabricated for high-speed low dark current operations [41–43] since the late 1970s. Research on quantum-well photodiodes was started in the early 1980s. A plot of the excess factor F shows that the excess noise factor is minimized if $\beta_p/\alpha_n \ll 1$ using electron injection or $\alpha_n/\beta_p \ll 1$ for hole injection. In those limits, an ideal single-carrier-type multiplication process will dominate, and the excess noise caused by the feedback process of the impact ionization caused by the secondary electrons or holes in the opposite direction can be minimized. Ideas using multiple-quantum-well (MQW) structures for APD applications have been proposed and explored both theoretically and experimentally [44–52]. For example, consider GaAs/$Al_xGa_{1-x}As$ MQW structures as the impact ionization region for electron injection (Fig. 15.13). Because $\Delta E_c/\Delta E_v \simeq 2:1$ ($\Delta E_c \simeq 0.67\Delta E_g$, $\Delta E_v \simeq 0.33\Delta E_g$), the electrons coming from the left barrier region will gain a larger kinetic energy ΔE_c when entering the barrier region than that of the holes traveling in the opposite direction. Therefore, the electron impact ionization coefficient α_n will be enhanced compared with β_p. We expect $\alpha_n/\beta_p \gg 1$. The excess noise factor is expected to be minimized. Measurements of the effective ionization coefficients α_n and β_p show an enhancement of α_n/β_p from 2 in a bulk GaAs to about 8 in a GaAs/$Al_xGa_{1-x}As$ multiple-quantum-well structure [45, 46] and $M_n = 10$ at an electric field $E = 250\,\text{kV/cm}$. It has also been reported [47] that this ratio α_n/β_p varies (not monotonically) with

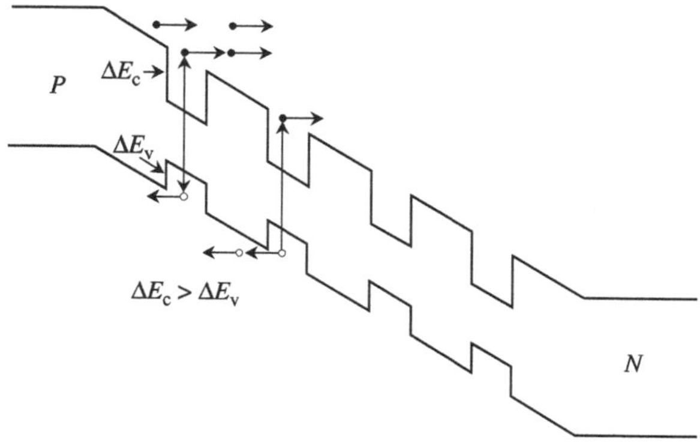

Figure 15.13 A multiple-quantum-well avalanche photodiode using GaAs/$Al_xGa_{1-x}As$ with the property that the ratio of the impact ionization coefficients α_n/β_p is much larger than 1 as ΔE_c is larger than ΔE_v.

the aluminum mole fraction x. At higher values of x (≥ 0.45) above the onset of indirect electron transitions, the noise is increased.

Intersubband Avalanche Photomultiplier Avalanche photomultipliers using an intersubband type (bound-to-continuum state transition) in an n-type doped or p-type doped multiple-quantum-well structure have also been proposed and investigated both theoretically [50, 51] and experimentally [52–54]. The idea is to introduce electrons in the quantum-well regions by doping the wells n-type. Incident photogenerated carriers will "impact ionize" those carriers confined in the wells and kick them out of the wells contributing to the avalanche multiplication current. Because it is a single-carrier type of photomultiplication, the excess noise is expected to be minimized. Experimental results on this intersubband avalanche multiplication have been reported [52–54].

15.5 INTERSUBBAND QUANTUM-WELL PHOTODETECTORS

In Section 9.7, we discussed intersubband absorption in a quantum-well structure. In order to provide carriers for the intersubband transitions, donors for n-type electronic transitions have to be introduced in the quantum wells (or barriers) to provide free electrons that will be confined in the well regions at steady state without a bias voltage. When an incident infrared radiation illuminates the QW detector, electrons may absorb the photon energy and jump to a higher energy subband and be collected by the electrodes with an applied voltage. Theory and experiments on intersubband absorption and quantum-well infrared photodetectors (QWIPs) [55–87] have been investigated for long wavelength applications, which may be competitive with HgCdTe detectors. The advantages include the mature GaAs growth and processing technologies for high uniformity and reproducibility. For an extensive review of the subject, see Refs. 86 and 87 and the references therein.

15.5.1 Intersubband Optical Absorption

For n-type multiple-quantum-well photodetectors, the optical matrix selection rule shows that the optical polarization must have a component along the growth (z) axis (i.e., TM polarized as discussed in Section 9.7). For TE polarized light, the absorption is expected to be small. However, for p-type doped quantum-well photodiodes, the valence-band mixing effects due to the heavy-hole and light-hole states show that the x- and y-polarized light can have as large an absorption coefficient as the z-polarized light [87–90]. Therefore, a normal incidence geometry is possible for p-type QWIP.

Here we discuss mostly n-type QWIPs because of their potential for 3- to 5-μm and 8- to 12-μm photodetector applications. In Fig. 15.14, we show two examples using (a) a 45°-coupled QWIP and (b) a two-dimensional grating-coupled QWIP. These designs provide the necessary polarization selection rule such that the infrared radiation has a component along the growth direction of the multiple-quantum-well

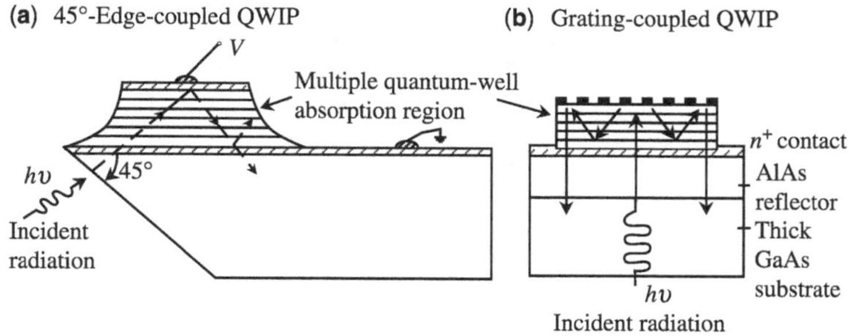

Figure 15.14 Schematic diagrams of (a) a 45°-edge-coupled quantum-well infrared photodetector (QWIP) and (b) a two-dimensional grating-coupled QWIP. The growth axis of the quantum well layers is defined as the z axis. Light polarized along the x–y plane is not absorbed due to the dipole selection rule of the intersubband transition.

absorption region. The theory in Section 9.7 shows that the absorption spectrum is a Lorentzian function

$$\alpha(\hbar\omega) = \alpha_0 \frac{\gamma/\pi}{(E_{21} - \hbar\omega)^2 + \gamma^2} \tag{15.5.1}$$

where the intersubband energy $E_{21} = E_2 - E_1$ is the subband spacing in a simple single-particle model as presented in Section 9.6. If the Coulomb interactions and screening effects are included, E_{21} will have a slight shift due to the many-body effects [91–94]. The measured absorption spectrum [87] for a bound-to-bound transition is shown in Fig. 15.15 as a Lorentzian shape expected from the theory. Note that the absorption α_0 is dependent on $|\langle\varphi_2|z|\varphi_1\rangle|^2$; therefore, a rotation of the polarization as a function of the polarization angle ϕ measured from the TM polarized wave incident at a fixed angle of incidence set at the Brewster angle $\theta_B = 73°$ shows the polarization dependence [95, 96] in Fig. 15.16 $\propto \cos^2\phi$, as expected from the theory. At $\phi = 90°$, the incident light is polarized along the TE direction (i.e., the polarization of the incident radiation is perpendicular to the plane of incidence; therefore, it is parallel to the quantum-well plane of the QWIP), and no absorption occurs for this polarization.

15.5.2 Photoconductive Gain in Quantum Wells

For a bound-to-bound state transition as shown in Fig. 15.17a, the photoexcited electrons in the E_2 level have to get out of the well either by tunneling or thermionic emission (or by other phonon or impurity scattering processes). It is therefore desirable to design the E_2 level to be close to the barrier energy such that the absorption oscillator strength (i.e., the intersubband dipole moment) and the escape probability can be optimized.

For a bound-to-continuum state transition, Fig. 15.17b, the electrons have a greater probability to transport into the barrier region and be collected by the electrode and

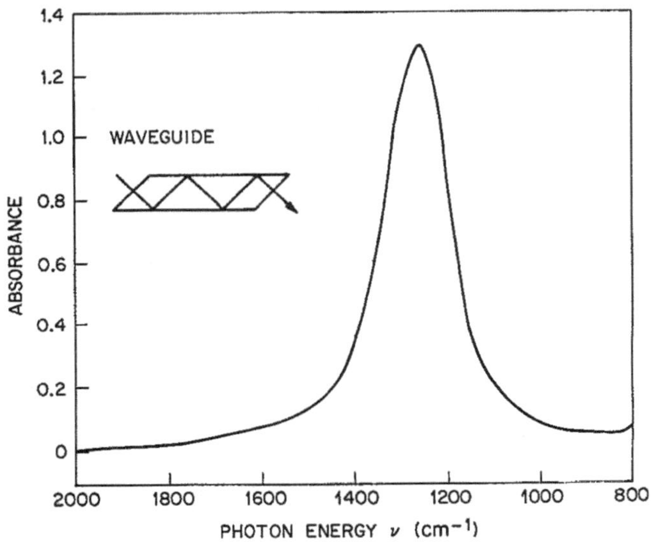

Figure 15.15 Measured QWIP absorption spectrum for a multipass waveguide geometry. (Reprinted with permission from [87] © 1993 American Institute of Physics.)

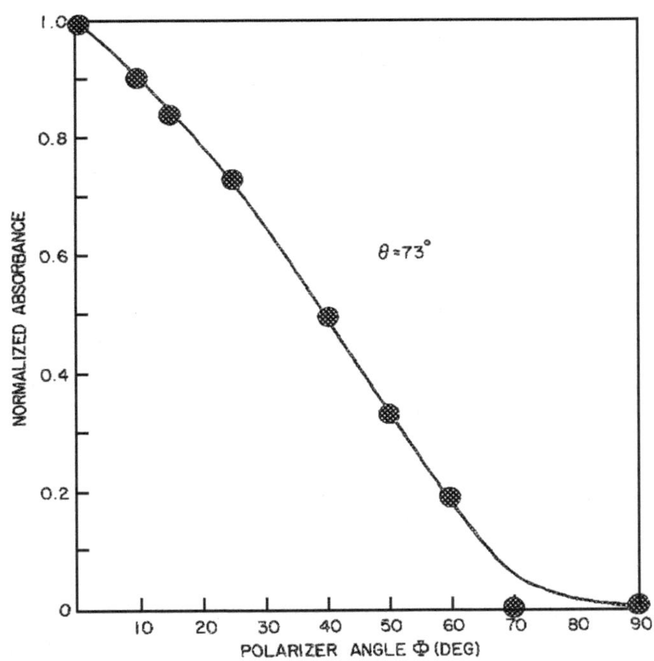

Figure 15.16 Experimental results for the polarization selection rule showing the peak absorption versus the polarization angle ϕ where $\phi = 0°$ is TM polarization and $\phi = 90°$ is TE polarization, all at an angle of incidence $\theta_B = 73°$, the Brewster angle. (Reprinted with permission from [95] © 1987 American Institute of Physics.)

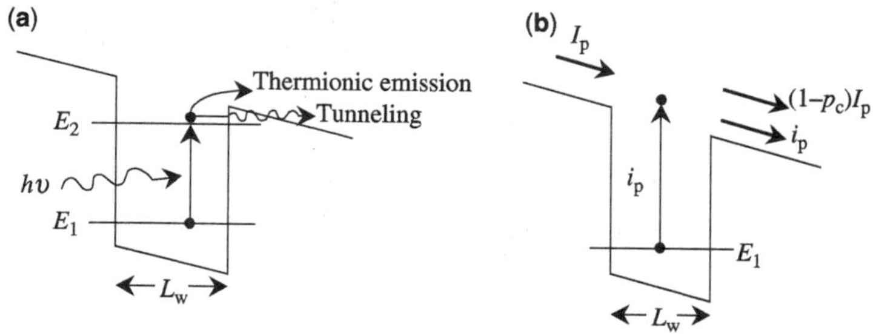

Figure 15.17 (a) Bound-to-bound state transition and (b) bound-to-continuum state transition in a biased quantum-well infrared photodetector (QWIP). The well width L_w in (b) is designed to be small enough such that only one bound state exists in the quantum well and the second level E_2 is pushed into the continuum.

contribute to the photocurrent, although the intersubband dipole matrix element between the ground state wave function and the highly oscillatory continuum-state wave function may be smaller. For a simplified analysis [87, 97, 98], we look at Fig. 15.17b, where p_c is treated as an effective capture probability for an incident current I_p for both the case (a) bound-to-bound and case (b) bound-to-continuum transitions. We obtain $p_c I_p$ as the fraction of incident current captured by the well and $(1 - p_c)I_p$ as the remaining current transmitted to the next period. The incident infrared radiation creates a photocurrent i_p

$$i_p = q\eta_w \Phi A \tag{15.5.2}$$

where Φ is the photon flux density (number of photons per second per unit area), A is the area of the photon illumination, and η_w is the net quantum efficiency of a single well including the effect of the escape probability from the well.

From current continuity, we have

$$I_p = (1 - p_c)I_p + i_p. \tag{15.5.3}$$

Therefore,

$$i_p = p_c I_p. \tag{15.5.4}$$

The total net photocurrent is

$$I_p = q\eta\Phi A g \tag{15.5.5}$$

where g is defined to be the overall photoconductive gain and η is the overall quantum efficiency of the MQW photodetector consisting of N_w quantum wells. We have $\eta \simeq N_w \eta_w$ if $\eta \ll 1$ as η is proportional to the absorbance of the structure.

From (15.5.2) to (15.5.5), we find

$$g = \frac{1}{p_c}\frac{\eta_w}{\eta} \simeq \frac{1}{p_c N_w} \tag{15.5.6}$$

which gives the value for gain. The capture probability p_c is found to decrease almost exponentially with the applied voltage [87].

15.5.3 Dark Current

A simple model for the bias-dependent dark current I_D [87, 99, 100] is to take the "effective" number of electrons $n^*(V)$, which tunnel out of the well or are thermally excited out of the well into the continuum states, multiplied by the average transport velocity $v(V)$, the cross-sectional area of the detector A, and the electron charge q

$$I_D(V) = qn^*(V)v(V)A \tag{15.5.7}$$

where

$$n^*(V) = \frac{m_e^*}{\pi\hbar^2 L_p}\int_{E_1}^{\infty} f(E)T(E, V)dE \tag{15.5.8}$$

$$f(E) = \frac{1}{1 + e^{(E-E_F)/k_B T}} \tag{15.5.9}$$

where E_F is the Fermi level measured from the conduction band edge (same as the first subband level E_1) and L_p is the length of a period. $T(E, V)$ is the tunneling probability through the triangular barrier with a bias voltage V. The velocity is

$$v = \frac{\mu F}{\sqrt{1 + \left(\frac{\mu F}{v_s}\right)^2}} \tag{15.5.10}$$

where μ is the electron mobility, F is the average field determined by the bias voltage V and the overall MQW width, and v_s is the saturation drift velocity.

The above simplified model has been used to explain the dark current of experimental data with a very good agreement, as shown in Fig. 15.18 for a 10.7-μm QWIP [99]. An even more simplified model [100, 101] assumes

$$T(E) = \begin{cases} 1, & E > E_B \\ 0, & E < E_B \end{cases} \tag{15.5.11}$$

where E_B is the barrier height on the right-hand side of the quantum well. We obtain

$$n^* = \frac{m_e^* k_B T}{\pi\hbar^2 L_p} e^{-(E_C - E_F)/k_B T} \tag{15.5.12}$$

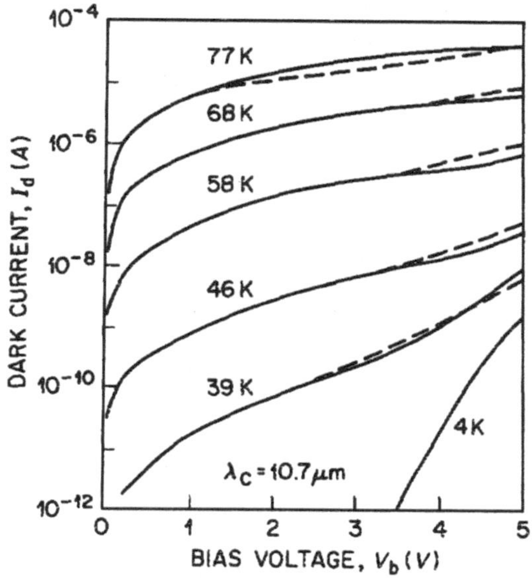

Figure 15.18 Dark currents from measured (solid curves) and calculated (dashed) data as a function of the bias voltage at various temperatures for a QWIP with a cutoff wavelength 10.7 μm. (Reprinted with permission from [99] © 1990 American Institute of Physics.)

where $E_C = E_B - E_1$ is the spectral cut-off energy. The dark current becomes

$$I_D = qvA\left(\frac{m_e^* k_B T}{\pi \hbar^2 L_p}\right) e^{-(E_C - E_F)/k_B T} \qquad (15.5.13)$$

and the Fermi level E_F is determined from

$$N_D = \left(\frac{m_e^* k_B T}{\pi \hbar^2 L_w}\right) \ln\left[1 + e^{(E_F - E_1)/k_B T}\right]. \qquad (15.5.14)$$

Because $I_D/T \propto e^{-(E_C - E_F)/k_B T}$, plotting $\ln(I_D/T)$ versus $(E_C - E_F)$ should give a slope of $-1/k_B T$, and the result of $E_C - E_F$ can also be compared with the optically measured spectral cutoff energy E_C. This simple model has been reported to agree with experimental observations [87, 99, 102].

15.6 SOLAR CELLS

Solar energy is one of the most important sources of power supply. The major idea is to develop clean-energy resources as an alternative to fossil fuels and stop air pollution resulting from the mass consumption of fossil fuels and maintain the ecological cycles of the biosystems on the earth [103]. Historically, the Industrial Revolution was initiated by the invention of the steam engine by James Watt in 1765. From an observation of the relation of the industrial development in the following centuries and the form of energy resources, from solid (coal) to liquid (oil) to gases (LNG, LPG), the important driving forces are the mass consumption of energy due to mass production technology, transportation, and storage. The resultant environmental load becomes a significant issue. The pollutant emission factors for electricity generation on the unit of carbon equivalent gram per kWh are 322.8 for coal, 258.5 for oil, and 178.0 for LNG, respectively [103]. As is well-known, electrical energy is the most convenient form of energy in modern civilization. As a result, direct conversion of solar radiation to electricity by photovoltaic solar cells plays an important role in the development of clean and renewable energy technologies for civilization in the 21st century.

In this section, we investigate the fundamentals of solar cells with a focus on the basic physics of *p-n* junctions and recent research on III–V compound semiconductors related to photonics. The reader is referred to books such as Refs. 103–107 for a more detailed treatment on the many materials, technologies, and system designs.

15.6.1 Solar Radiation and Basic Device Parameters

Solar radiation follows the Planck radiation spectrum from the surface of the sun with a source temperature of about 5800K plus modification due to various factors such as absorption by the atmosphere. Figure 15.19a and Fig. 15.19b show the solar radiation spectrum for various conditions [108, 109]. The illumination is given in AMX (air mass X) where

$$X = \frac{1}{\cos\theta} \tag{15.6.1}$$

and θ is the angle between the zenith and the position of the sun under terrestrial conditions. Air mass zero (AM0) is the solar irradiation above the earth's atmosphere. The accepted standard for AM0 is $135.3\,\mathrm{mW/cm^2}$ or $1.353\,\mathrm{kW/m^2}$, which is known as the solar constant. Air mass one (AM1) is when the sun is directly overhead and about 70% of the sunlight incident on the earth's atmosphere reaches the earth's surface on a clear day [108] (another 7% reaches the ground after scattering). The rest is absorbed or backscattered. AM2 is when the sun is at 60° from the zenith. AM1.5 is when the sun is at 48.19° from the zenith.

Ideal Conversion Efficiency The current versus the bias voltage of a *p-n* junction in the presence of light illumination is the same as what we have derived for a photodiode,

$$I = I_0(e^{qV/\gamma k_B T} - 1) - I_{\mathrm{ph}}. \tag{15.6.2}$$

Figure 15.20 shows the current–voltage curve with and without the optical illumination.

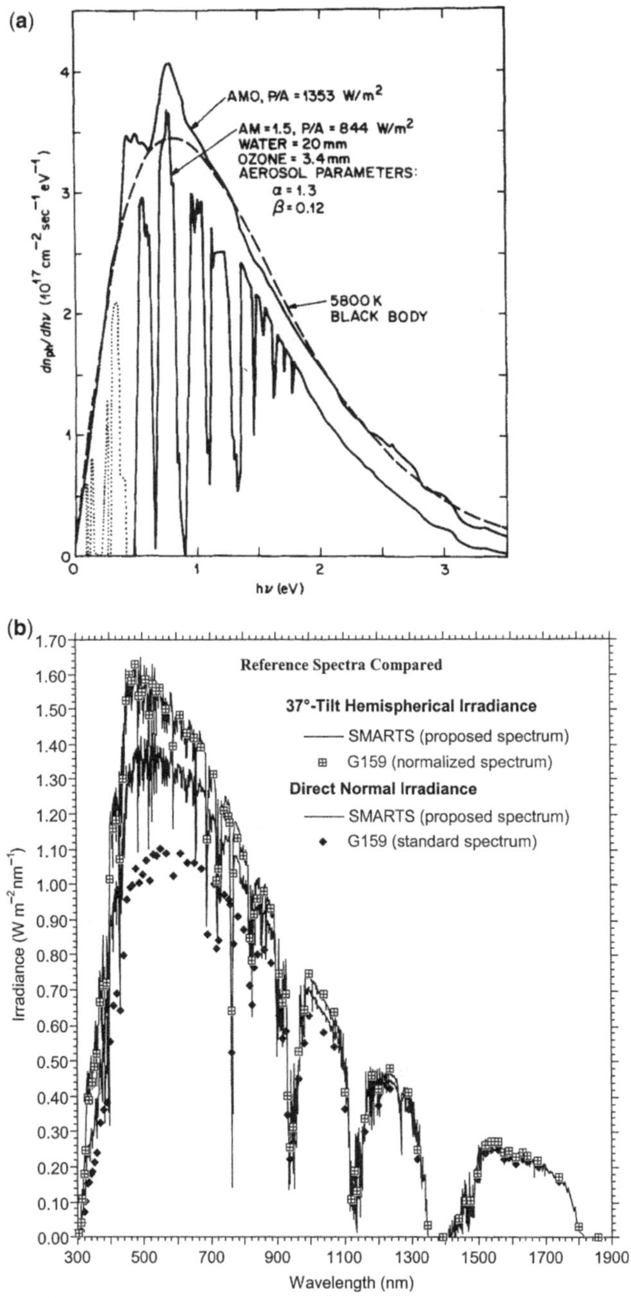

Figure 15.19 (a) The solar radiation spectrum for various conditions. Air mass zero (AM0) is the solar irradiation above the earth's atmosphere ($= 135.3\ \mathrm{mW/cm^2}$, also known as solar constant). It follows close to the black body radiation of a 5800K source from the sun's surface. (Reprinted with permission from [109] © 1980 American Institute of Physics.) (b) The solar radiation spectrum as a function of wavelength. (Reprinted with permission from [108] © 2002 Elsevier: Pergamon.)

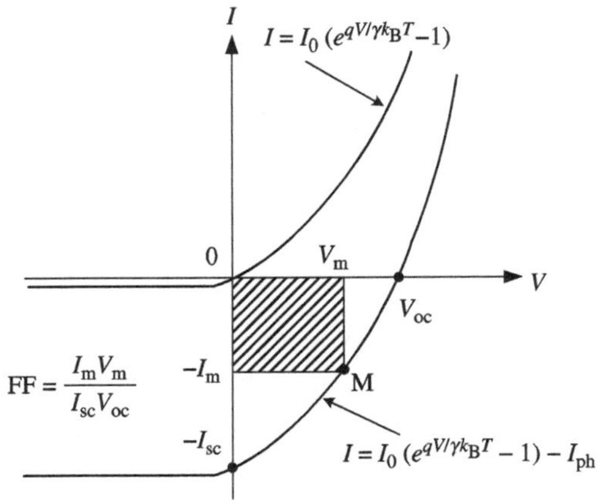

Figure 15.20 The current–voltage curves of a *p-n* junction solar cell with and without optical illumination.

Open circuit voltage is

$$V_{\text{oc}} = \frac{\gamma k_{\text{B}} T}{q} \ln\left(\frac{I_{\text{ph}}}{I_0} + 1\right). \tag{15.6.3}$$

Short-circuit current $I = -I_{\text{sc}}$

$$I_{\text{sc}} = I_{\text{ph}}. \tag{15.6.4}$$

The maximum power output occurs at an operation point M on the I–V curve such that $I_{\text{m}} V_{\text{m}}$ is a maximum, which is delivered to a properly matched load resistance R. The operation point M can be obtained by finding the maximum value of the output electrical power:

$$P = IV = [I_0(e^{qV/\gamma k_{\text{B}} T} - 1) - I_{\text{ph}}]V \tag{15.6.5}$$

which is simply obtained by setting the derivative of the above power with respect to the voltage to zero (see Problem 15.12). The fill factor (FF) or the curve factor (CF) is defined as the ratio

$$\text{FF} = \frac{V_{\text{m}} I_{\text{m}}}{V_{\text{oc}} I_{\text{sc}}}. \tag{15.6.6}$$

The ideal conversion efficiency [107, 110] is

$$\eta = \frac{V_{\text{m}} I_{\text{m}}}{P_{\text{opt}}} = \frac{V_{\text{oc}} I_{\text{sc}}}{P_{\text{opt}}} \text{FF} \tag{15.6.7}$$

where P_{opt} is the incident optical power (W) on the solar cell surface.

15.6.2 *p-n* Junction Solar Cells and Spectral Response [103–107]

Figure 15.21a shows the schematics of an *n*-on-*p* solar cell, where the solar light illumination is from the top surface with an antireflection coating to improve photon collection, and the electrode fingers are for current collection. The theory of the spectral response of the photocurrent is discussed below. Figure 15.21b shows a one-dimensional model for the analysis of the *p-n* junction: W_n is the thickness of the *n*-region; W_p is that of the *p*-region; the total depletion width is $x_w = x_p + x_n$. The figure also indicates one diffusion length of the minority carrier (L_p and L_n) on each side of the depletion region. The boundaries at the two edges of the depletion region are $x_j = W_n - x_n$ and $x_j + x_w = W_n + x_p$. Figure 15.21c shows the generation rate of electron–hole pairs as a function of the optical penetration distance into the semiconductor *p-n* junction. The generation rate of electron–hole pairs is

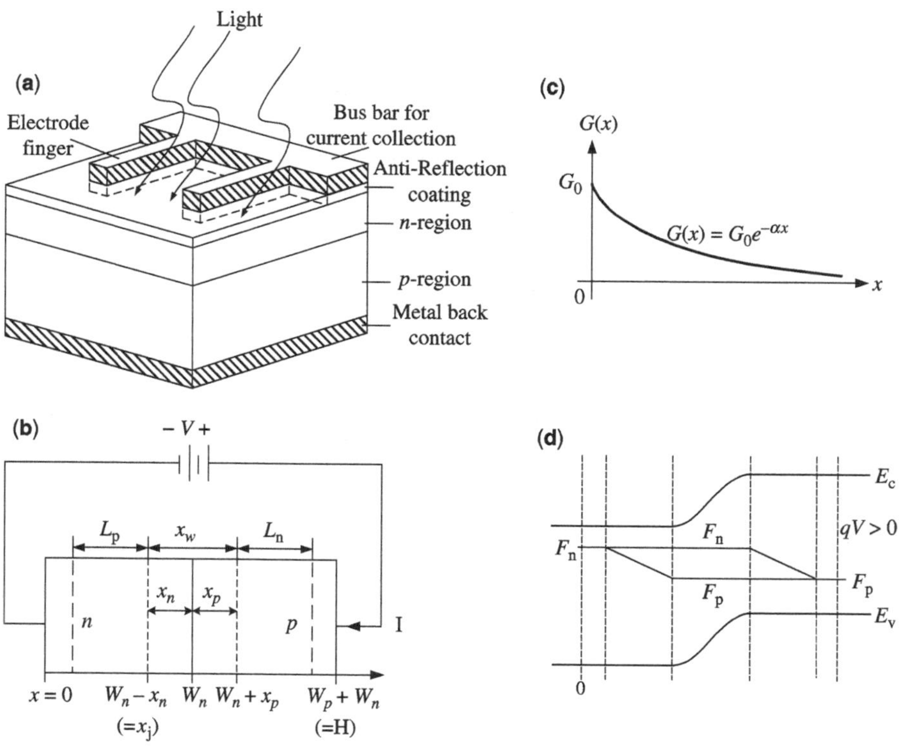

Figure 15.21 (a) Schematics of an *n*-on-*p* solar cell. (b) A one-dimensional model for the analysis of the *p-n* junction, the depletion width $x_w = x_p + x_n$, and one diffusion length of the minority carrier on each side of the depletion region. W_n is the thickness of the *n*-region and W_p is that of the *p*-region. We define the boundary at each side of the depletion region using $x_j = W_n - x_n$ and $x_j + x_w = W_n + x_p$. (c) The generation rate of electron–hole pairs, $G(x)$, as a function of the optical penetration distance into the *p-n* junction. (d) The energy band diagram and the quasi-Fermi levels for a small forward bias voltage.

proportional to the optical intensity profile in the device

$$
\begin{aligned}
G(x, \lambda) &= G_0\, e^{-\alpha(\lambda)x} \\
G_0 &= \eta_{\mathrm{i}}[1 - R(\lambda)]\Phi(\lambda)\alpha(\lambda)
\end{aligned}
\tag{15.6.8}
$$

where α is the absorption spectrum, $\Phi(\lambda) = I_{\mathrm{opt}}/\hbar\omega$ is the optical flux density for an incident optical power intensity I_{opt} ($\mathrm{W/cm^2}$), and η_i is the intrinsic quantum efficiency to account for the average number (100% maximum) of electron–hole pairs generated per incident photon. Figure 15.21d shows the energy band diagram and the quasi-Fermi levels for a small forward bias voltage (defined as positive for the p-electrode). To obtain the I–V curve of a p-n junction solar cell, the key steps are essentially the same as those for photodiodes.

1. Find the minority carrier density in the quasi-neutral region of the n and p region in the presence of optical illumination with an incident photon flux (optical intensity) at a given wavelength λ for the given generation rate $G(x, \lambda)$. The major equations are the diffusion equations in the presence of carrier generation. The carrier densities or current densities have to satisfy the required boundary conditions.
2. Calculate the minority current density at the edge of the depletion region, that is, $J_p(\lambda)$ at $x = x_{\mathrm{j}}$ ($= W_{\mathrm{n}} - x_{\mathrm{n}}$) on the n-side, and $J_n(\lambda)$ at $x = x_{\mathrm{j}} + x_{\mathrm{w}}$ the p-side, where x_{w} is the depletion width.
3. Calculate the contribution due to the drift current density due to optical generation in the depletion region $J_{\mathrm{dr}}(\lambda)$

$$
\begin{aligned}
J_{\mathrm{dr}}(\lambda) &= q\int_{x_{\mathrm{j}}}^{x_{\mathrm{j}}+x_{\mathrm{w}}} G_0 e^{-\alpha x}\,dx \\
&= q\eta_{\mathrm{i}}[1 - R(\lambda)\,]\Phi(\lambda)e^{-\alpha x_{\mathrm{j}}}(1 - e^{-\alpha x_{\mathrm{w}}}).
\end{aligned}
\tag{15.6.9}
$$

4. The total current density for an incident photon flux at a given wavelength is

$$
J(\lambda) = J_p(\lambda) + J_n(\lambda) + J_{\mathrm{dr}}(\lambda). \tag{15.6.10}
$$

5. The spectral response is defined as

$$
S_{\mathrm{R}}(\lambda) = \frac{J(\lambda)}{q[1 - R(\lambda)\,]\Phi(\lambda)}. \tag{15.6.11}
$$

6. The total photocurrent is obtained by integrating the product of the transmitted photon flux into the solar cell and the spectral response to the maximum

wavelength λ_m

$$J_{ph} = q \int_0^{\lambda_m} [1 - R(\lambda)]\Phi(\lambda)S_R(\lambda)\, d\lambda. \tag{15.6.12}$$

Below, we carry out the procedures for step 1 and 2 first for the n-side, then for the p-side by a similar procedure. In the quasi-neutral region of the n-side, $0 < x < x_j$ $(= W_n - x_n)$

$$\frac{\partial^2}{\partial x^2}\delta p_n - \frac{1}{L_p^2}\delta p_n = -\frac{1}{D_p}G(x, \lambda). \tag{15.6.13}$$

The solution for the excess minority (hole) concentration is

$$\delta p_n(x) = A\cosh\left(\frac{x - x_j}{L_p}\right) + B\sinh\left(\frac{x - x_j}{L_p}\right) + Ce^{-\alpha x} \tag{15.6.14a}$$

$$C = \frac{G_0\tau_p}{(1 - \alpha^2 L_p^2)} \tag{15.6.14b}$$

where C comes from the particular solution due to the inhomogeneous source term. The two constants, A and B, are determined by the two boundary conditions.

1. At $x = x_j$, the excess carrier density vanishes, that is,

$$\delta p_n(x_j) = p_{n0}(e^{qV/k_BT} - 1)(= 0, \text{ at } V = 0). \tag{15.6.15}$$

The above boundary condition determines

$$A = -Ce^{-\alpha x_j}. \tag{15.6.16}$$

2. At $x = 0$, the hole current density (flowing *into* the surface) satisfies the surface recombination current density, $-J_p(0) = qS_p\delta p_n(0)$, which gives

$$D_p\frac{d}{dx}\delta p_n\bigg|_{x=0} = S_p\delta p_n(0). \tag{15.6.17}$$

The above boundary condition then determines B. The final current density $J_p(\lambda)$ is given by the hole current density at the edge of the depletion region, $J_p(x = x_j)$, which is

$$J_p(\lambda) = -qD_p\frac{d}{dx}\delta p_n\bigg|_{x=x_j}. \tag{15.6.18}$$

Similarly, in the quasi-neutral region of the p-side, $x_\mathrm{j} + x_\mathrm{w} < x < W_\mathrm{p} + W_\mathrm{n} = H$

$$\frac{\partial^2}{\partial x^2}\delta n_p - \frac{1}{L_n^2}\delta n_p = -\frac{1}{D_n}G(x, \lambda). \tag{15.6.19}$$

The solution for the excess minority (electron) concentration is

$$\delta n_p(x) = A' \cosh\left(\frac{x - x_\mathrm{j} - x_\mathrm{w}}{L_n}\right) + B' \sinh\left(\frac{x - x_\mathrm{j} - x_\mathrm{w}}{L_n}\right) + C' e^{-\alpha x} \tag{15.6.20a}$$

$$C' = \frac{G_0 \tau_n}{(1 - \alpha^2 L_n^2)}. \tag{15.6.20b}$$

The two constants, A' and B', are determined by the two boundary conditions.

1. At $x = x_\mathrm{j} + x_\mathrm{w}$, the excess carrier density is given by the condition

$$\delta n_p(x_\mathrm{j} + x_\mathrm{w}) = n_{p0}(e^{qV/k_\mathrm{B}T} - 1)(=0 \text{ at } V = 0). \tag{15.6.21}$$

   We obtain

$$A' = -C' e^{-\alpha(x_\mathrm{j} + x_\mathrm{w})}. \tag{15.6.22}$$

2. At $x = W_\mathrm{p} + W_\mathrm{n} = H$, the electron current density flowing into the surface satisfies the surface recombination current density: $J_n(H) = (-q)S_n \delta n_p(H)$, which leads to

$$D_n \frac{d}{dx}\delta n_p\bigg|_{x=H} = -S_n \delta n_p(H). \tag{15.6.23}$$

The above boundary condition on the current density determines B'. The final electron current density $J_n(\lambda)$ is given by its value at the edge of the depletion region, $J_n(x = x_\mathrm{j} + x_\mathrm{w})$, which is

$$J_n(\lambda) = qD_n \frac{d}{dx}\delta n_p\bigg|_{x=x_\mathrm{j}+x_\mathrm{w}}. \tag{15.6.24}$$

Following the above procedures, the total photocurrent in the presence of the broad solar light radiation is to integrate the spectral response with the transmitted photon flux in (15.6.12) for a given bias voltage V. Therefore, the general I–V curve of a p-n junction solar cell can be obtained.

Back-Surface Field Solar Cells By adding a heavily doped p^+-layer between the collecting metal contact and the p-region, Fig. 15.22a, a back surface field [104, 107] exists and improves the short-circuit current and the open-circuit voltage.

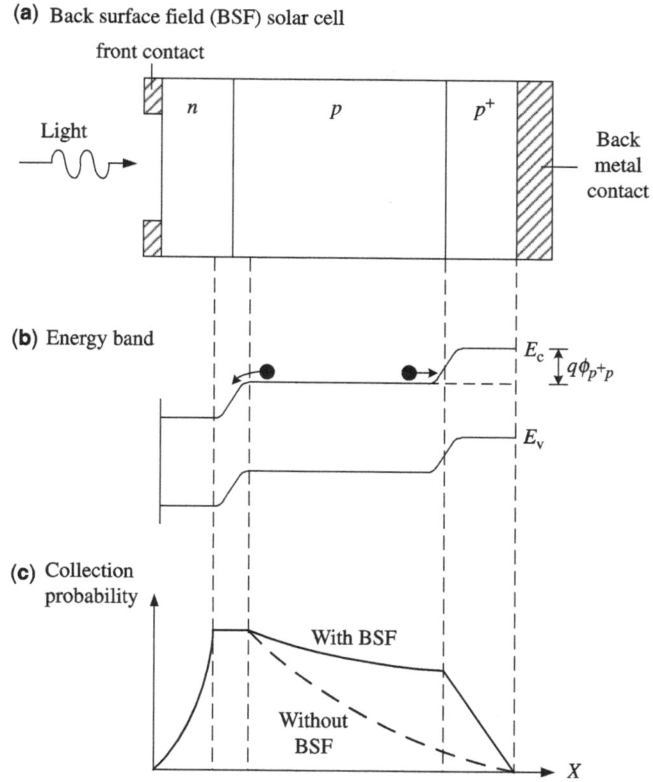

Figure 15.22 (a) A back surface field solar cell. (b) The energy band diagram. (c) A plot of the collection efficiency of each region of a solar cell with and without a back surface field.

Figure 15.22b shows the energy band diagram and Fig. 15.22c shows a simple plot of the collection efficiency of each region. The analysis is similar to the above and that of Section 15.2 based on the depletion approximation. The BSF cell helps confine the electrons more in the p-region due to the extra p^+-barrier and is similar to a cell of a larger width with a small surface electron recombination velocity. The photocurrent is therefore enhanced. The open circuit voltage is increased due to the extra barrier caused by p^+-doping, the increased short-circuit current, and the reduced (electron) surface recombination current.

***I–V* Curves and Equivalent Circuit** Figures 15.23a and 15.23b show the equivalent circuit of a solar cell and the I–V curve in the presence of shunt (R_{sh}) and series (R_s) resistances. An ideal solar cell assumes that the shunt resistance is infinity and the series resistance is zero, and the conversion efficiency has a larger maximum value than otherwise. When either the shunt resistance is not infinity or the series resistance is not zero, the I–V curve is shifted toward the origin and

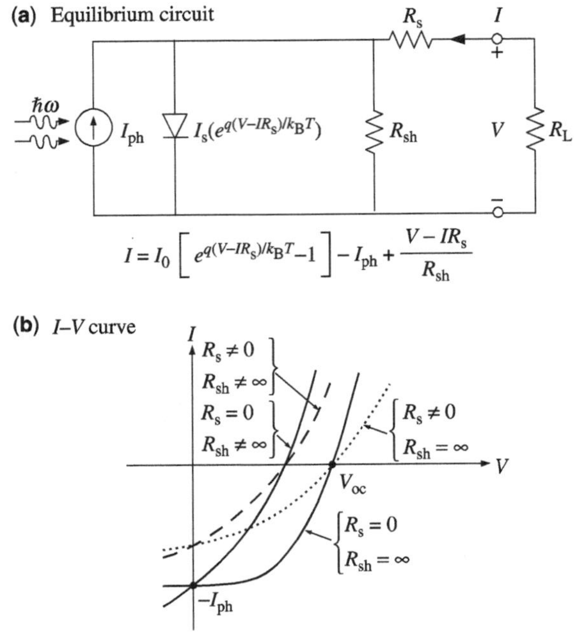

Figure 15.23 (a) The equivalent circuit of a solar cell. (b) The $I-V$ curve in the presence of shunt (R_{sh}) and series (R_s) resistances. In an ideal solar cell, the shunt resistance is infinity and the series resistance is zero.

conversion efficiency is reduced, as shown in Fig. 15.23b

$$I = I_0[e^{q(V-IR_s)/\gamma k_B T} - 1] - I_{ph} + \frac{V - IR_s}{R_{sh}}. \tag{15.6.25}$$

15.6.3 Quantum-Well and Multijunction Solar Cells

Quantum-Well Solar Cells By placing multiple quantum wells [111–121] in the intrinsic region of a *p-i-n* structure, it is possible to realize solar cells with improved quantum efficiency, Fig. 15.24. It is important to maintain the built-in field across

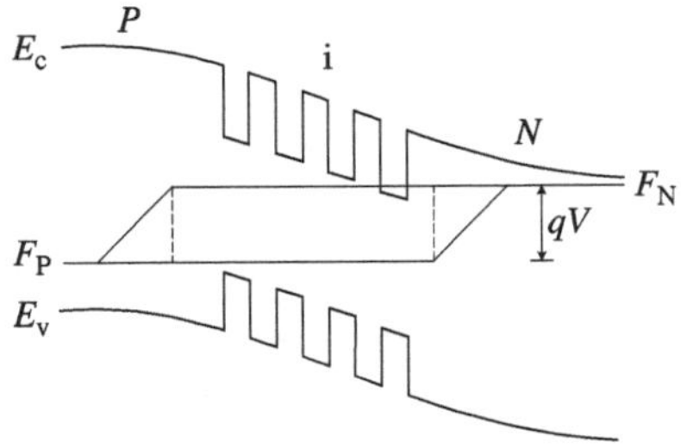

Figure 15.24 A multiple quantum-well (MQW) solar cell in a *P-I-N* structure. The intrinsic region contains many quantum wells for enhanced absorption.

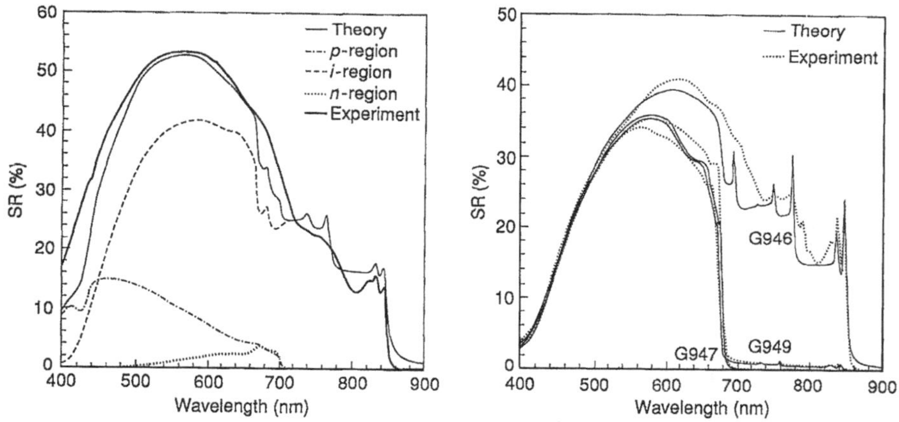

Figure 15.25 (a) The spectral response of a *p-i-n* quantum-well solar cell with contributions from the *p*, *i*, and *n* regions separately and comparison with experimental data. (b) The comparison of three samples with difference only in the intrinsic region: all three samples have the same total thickness in the intrinsic layer except that sample G947 has a bulk uniform AlGaAs; sample G949 contains a single quantum well replacing an equivalent thickness of AlGaAs; and sample G946 has 30 quantum wells. (Reprinted with permission from [111] © 1993 American Institute of Physics.)

the MQW region such that the photogenerated carriers in the quantum wells can escape to the barriers and get collected by the *p-n* junction at room temperature. There is some voltage loss. Figure 15.25a shows the theoretical spectral response of a *P-i-N* quantum well solar cell with contributions from the *P*, *I*, and *N* regions respectively and compares that with experimental data [111]. Figure 15.25b shows the comparison of three samples with differences only in the intrinsic region: all three samples have the same total thickness in the intrinsic layer except that the reference sample (G947) has a bulk AlGaAs (no quantum wells); sample G949 contains a single quantum well replacing an equivalent thickness of AlGaAs; and sample G946 has 30 quantum wells. It is interesting to see the increase in the absorption spectrum in the quantum well region due to 1 and 30 quantum wells below the band edge of the reference sample (with only the barrier AlGaAs material without any quantum well). Theoretical analyses on the improvement in spectral absorption, open circuit voltage, short-circuit current, and conversion efficiency have been investigated and debated intensively since the early 1990s. A critical review on the issue of the enhancement of photovoltaic conversion efficiency has been reported [121].

Multijunction Solar Cells In recent years, terrestrial concentrator multijunction III–V solar cells have shown remarkable performance in conversion efficiency [122–128]. Conversion efficiencies beyond 40% has been realized using metamorphic GaInP/GaInAs/Ge multijunction solar cells [127, 128]. As a result, there is a resurgence of research in multijunction cells and commercial interest in concentrator III–V photovoltaics. New materials [129] incorporating nitride (III-N-V) have potential applications in next-generation multijunction solar cells because these materials

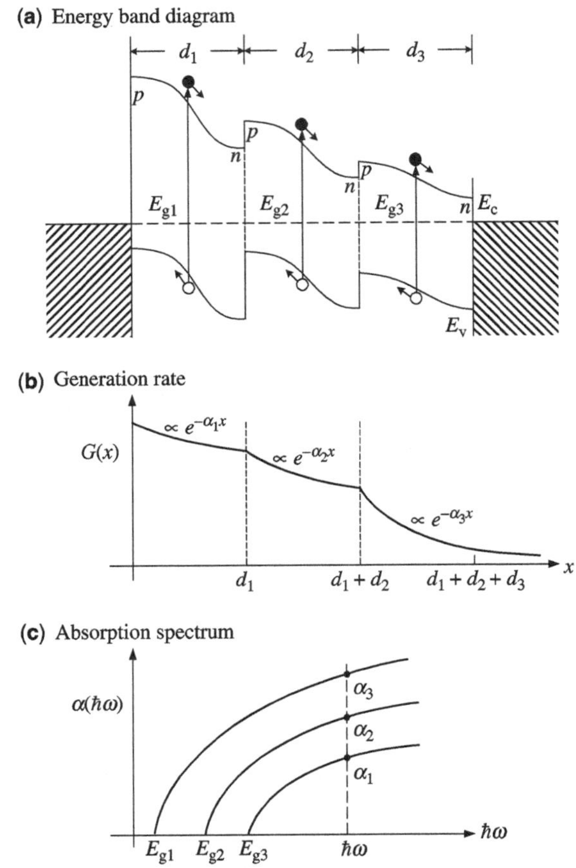

Figure 15.26 (a) The schematics for the electron–hole generation rate as a function of optical depth into the multijunction solar cell from the top wide band gap into the bottom narrow band gap region. (b) The generation rate of electron–hole pairs is proportional to the optical intensity, which decays as a function of position x. (c) The absorption spectra of the three band gaps.

can be lattice-matched to substrates such as Si, Ge, and GaAs, with band gaps that are complementary to those of other III–V compound semiconductors. The absorption spectra of some of these materials have been shown in Chapter 2, Fig. 2.4. Poor minority carrier transport in III-N-V materials is a critical research issue.

Figure 15.26a shows the schematic for the electron–hole generation rate as a function of optical depth into the multijunction solar cell from the top wide band gap into the bottom narrow band gap region. Figure 15.26b shows the generation rate per unit volume of the electron–hole pairs, which is proportional to the optical intensity that decays into the solar cell depth. The absorption spectra of all three band gaps are illustrated in Fig. 15.26c. The absorption of the solar radiation spectrum by different band gap layers allows for the conversion of more photons into electron and hole pairs. Carrier transport and collection become important

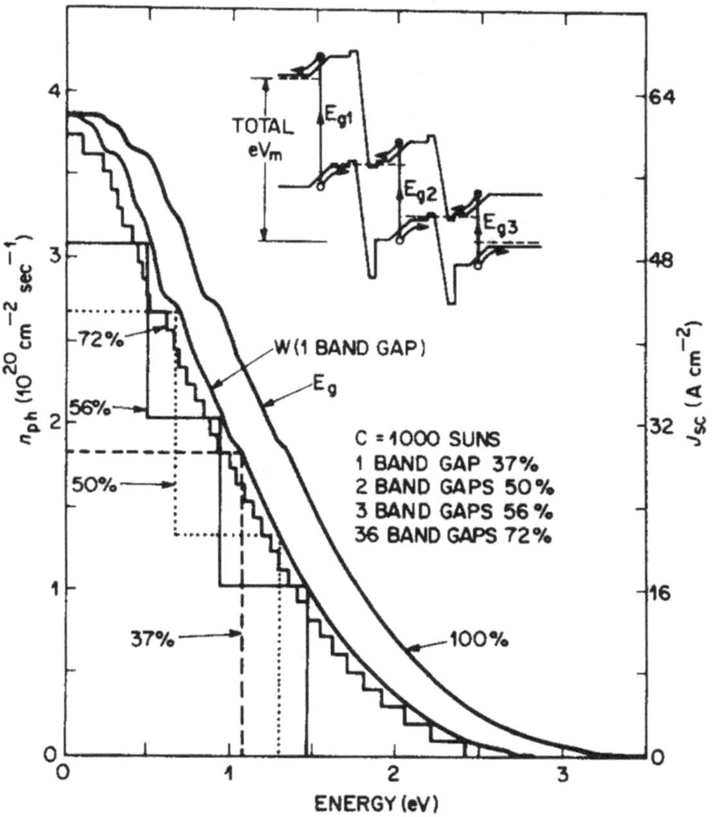

Figure 15.27 A graphical method to illustrate the improvement of conversion efficiency with increasing number of junctions. The insert shows the energy band profile of a three-junction solar cell. (Reprinted with permission from [109] © 1980 American Institute of Physics.)

issues. Thus the tunneling junction between different cells plays a significant role in the device design. Figure 15.27 shows a graphical method [109] to extract the efficiency of multijunction solar cells and indicates improvement as the number of junctions increases.

Figure 15.28a and Fig. 15.28b [128] show a three-junction solar cell using (a) lattice-matched (LM) and (b) metamorphic (MM) GaInP/GaInAs/Ge three-junction solar cells, respectively. These solar cells show high conversion efficiencies (from 30% to 40.7%). Figure 15.28c shows the efficiency, open-circuit voltage (V_{oc}), and filling factor (FF) of high performance MM and LM three-junction solar cells as a function of incident intensity [128]. Figure 15.29a shows the semiconductor layered structure of a four-junction (Al)GaInP/AlGa(In)As/Ga(In)As/Ge terrestrial concentration solar cell. Figure 15.29b shows the quantum efficiency as a function of photon energy [128]. The contribution of each cell, as determined by its band gap, is clearly seen.

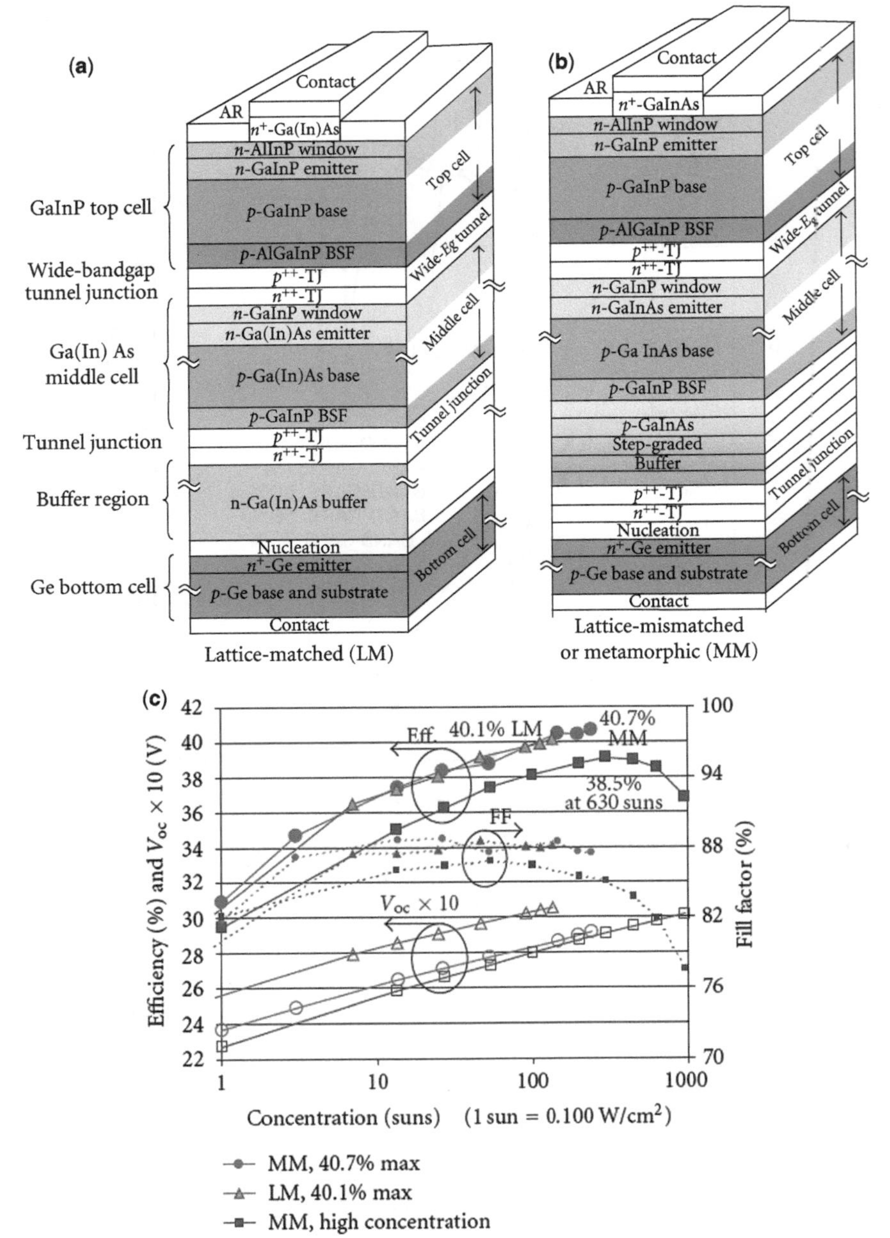

Figure 15.28 A three-junction solar cell using (a) lattice-matched (LM) and (b) metamorphic (MM) GaInP/GaInAs/Ge three-junction solar cells, respectively. (c) The efficiency, V_{oc}, and FF of high-performance MM and LM three-junction solar cells as a function of incident intensity. (Reprinted with permission from [128] © 2007 Hindawi Publishing Corporation.)

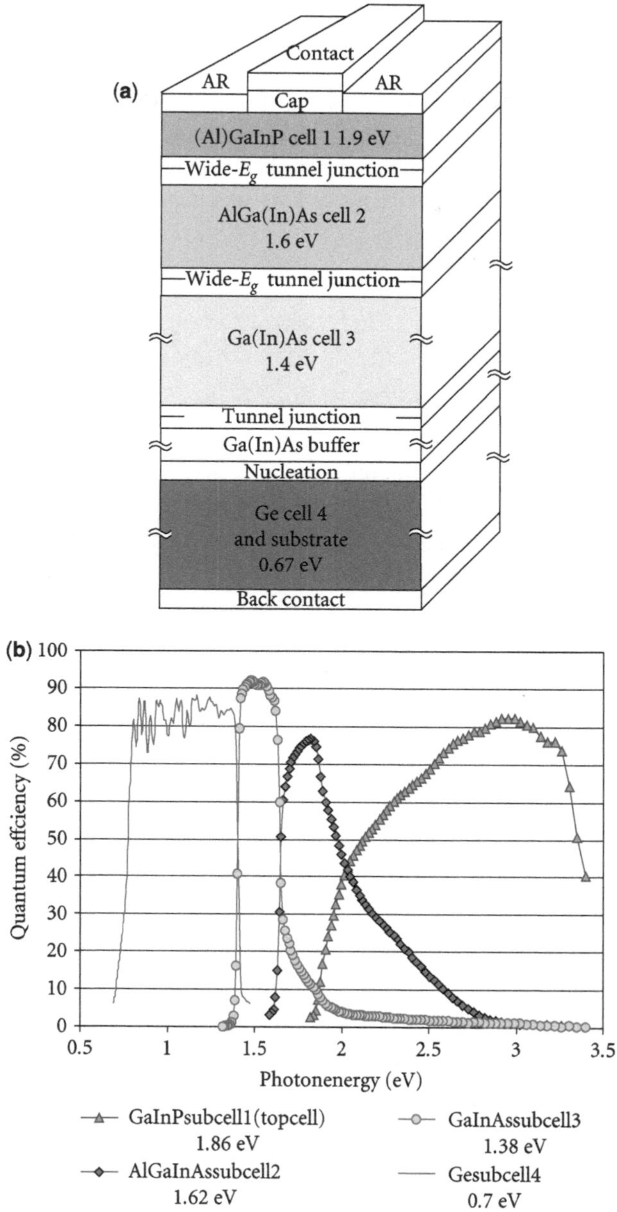

Figure 15.29 (a) The semiconductor layered structure of a four-junction (Al)GaInP/AlGa(In)As/Ga(In)As/Ge terrestrial concentration solar cell. (b) The quantum efficiency of a four-junction solar cell as a function of photon energy. (Reprinted with permission from [128] © 2007 Hindawi Publishing Corporation.)

Quantum-Dot Solar Cells More recently, quantum dots such as colloidal PbSe and PbS quantum dots [130, 131] have also been used for solar cell applications with impressive performance. Very efficient multiple exciton generation (MEG) for single photon absorption in these dots has been reported. An impact ionization process is used to explain the intrinsic quantum yield of 300% or, equivalently, three electron–hole pairs are generated per absorbed incident photon. The research appears to be attractive for novel applications of quantum dots to solar cells with improved conversion efficiency.

PROBLEMS

15.1 Consider a photoconductor (Fig. 15.1) that is an extrinsic semiconductor bar with a thickness $d = 0.1$ mm, a width $w = 1$ mm, a length $\ell = 4$ mm, and an acceptor doping concentration 10^{15}cm^{-3}. Assume that the electron mobility $\mu_n = 3000\ \text{cm}^2\ \text{V}^{-1}\ \text{s}^{-1}$ ($\gg \mu_p$) and the applied voltage $V = 4$ V.

(a) If the photoconductor is illuminated by a uniform steady light such that the optical generation rate of electrons is G_n, find an expression for the photocurrent $I_{\text{ph}} = I - I_0$, where I_0 is the dark current and I is the current when there is illumination of light. Find a numerical value for I_{ph} if $G_n = 10^{16}\ \text{cm}^{-3}\ \text{s}^{-1}$ and $\tau_n = 10^{-3}$ s.

(b) If the photoconductor is illuminated by a uniform light with a sinusoidal time variation, that is, $G_n(t) = g \cos \omega t$, show that the photocurrent is given by the form

$$i(t) = \frac{I_{\text{p}}}{\sqrt{1 + \omega^2 \tau_n^2}} \cos(\omega t - \varphi).$$

What are I_{p} and ϕ in terms of g and τ_n? What determines the 3-dB cutoff frequency in the frequency response of the photocurrent?

(c) If the light has a dc (steady) and an ac component as may be used in optical communication, $G_n(t) = G_0(1 + m \cos \omega t)$, find an expression for the photocurrent using the results from parts (a) and (b).

15.2 Explain the physical meaning of the photoconductive gain τ_n/τ_t and why it can be much larger than 1.

15.3 Derive Eqs. (15.1.30a) and (15.1.30b).

15.4 Derive the junction capacitance C_{j} for a heterojunction using the depletion approximation in Section 2.5 for a *p-N* junction.

15.5 Derive the R_0A product in Eq. (15.2.15).

15.6 Replot Fig. 15.6a–e for a p^+-n^--n^+ photodiode, where a superscript "+" means heavy doping concentration and a superscript "–" means light doping concentration.

15.7 Derive Eqs. (15.4.23) and (15.4.25).

15.8 An avalanche photodiode with the electron and hole ionization coefficients α_n and β_p is assumed to have a uniform field in the impact ionization region such that α_n and β_p are *independent of the position x*. The electron and hole injections are given by $J_n(0)$ and $J_p(W)$, and the generation rate due to optical injection is $G(x)$.

(a) Write the two equations for the electron and hole current densities and solve for *the hole current density* as a function of x in terms of the injection conditions $J_n(0)$ and $J_p(W)$. Find the total current density J.

(b) We assume that the electric field in the avalanche region is uniform such that α_n and β_p are independent of the position x.

(i) If $G(x) = 0$ for all x, and $J_n(0) = 0$, find the multiplication factor for holes, M_p, defined by $M_p = \frac{J}{J_p(W)}$.

(ii) On the other hand, if $G(x) = 0$ for all x, $J_p(W) = 0$, find the multiplication factor for electrons by $M_n = \frac{J}{J_n(0)}$.

(c) Using the results in (b) for M_n and M_p, show that α_n and β_p can be derived as expressed in (15.4.34) and (15.4.35) once M_n and M_p are measured.

15.9 Discuss the physics for the excess noise factor $F(M) = \langle M^2 \rangle / \langle M \rangle^2$. How can this excess noise be minimized?

15.10 Derive Eqs. (15.5.12) and (15.5.13).

15.11 Discuss the polarization selection rule for an n-type doped quantum-well infrared detector using intersubband transitions. Why are the configurations such as a 45°-edge-coupled structure or a grating-coupled structure used in the designs of these intersubband photodetectors?

15.12 Using the I–V curve of a solar cell, find the optimal power conversion point M.

15.13 (a) Find the minority current densities at the edge of the depletion region, $J_p(x = x_j)$ in the presence of optical generation and zero voltage bias ($V = 0$) following the procedures in (15.6.18). (b) Repeat part (a) for $J_n(x = x_j + W)$ using (15.6.24). (c) Find the total current (15.6.10) using parts (a), (b), and (15.6.9).

15.14 The dark current of a p-n junction solar cell in the absence of optical illumination, $G = 0$, can be obtained by following the same steps 1 and 2 in Section 15.6. The same boundary conditions (15.6.15), (15.6.17), (15.6.21), and (15.6.23) can be used, except that V is not necessarily zero. Find the dark current using the above procedure.

15.15 Find the open-circuit voltage and short-circuit current for the four cases in Fig. 15.23b.

REFERENCES

1. R. K. Willardson and A. C. Beer, Eds., *Semiconductors and Semimetals*, Vol. 5, *Infrared Detectors*, Academic Press, New York, 1970.
2. R. K. Willardson and A. C. Beer, Eds., *Semiconductors and Semimetals*, Vol. 12, *Infrared Detectors II*, Academic Press, New York, 1977.
3. A. Rogalski, Ed., *Selected Papers on Semiconductor Infrared Detectors*, SPIE Milestone Series, Vol. MS66, SPIE Optical Engineering Press, Bellingham, WA, 1992.
4. W. T. Tsang, Volume Editor, *Part D, Photodetectors*, in Vol. 22, *Lightwave Communications Technology*, R. K. Willardson and A. C. Beer, Eds., *Semiconductors and Semimetals*, Academic Press, New York, 1985.
5. J. D. Vincent, *Fundamentals of Infrared Detector Operation and Testing*, Wiley, New York, 1990.
6. A. Yariv and P. Yeh, *Photonics—Optical Electronics in Modern Communications*, 6th ed., Oxford University Press, New York, 2007.
7. B. E. A. Saleh and M. C. Teich, *Fundamentals of Photonics*, 2nd ed., Wiley, New York, 2007.
8. S. M. Sze and K. K. Ng, *Physics of Semiconductor Devices*, Wiley, New York, 2007.
9. R. K. Willardson and A. C. Beer, Eds., *Semiconductors and Semimetals*, Vol. 18, *Mercury Cadmium Telluride*, Academic Press, New York, 1981.
10. G. H. Döhler, "Doping superlattices (n-i-p-i crystals)," *IEEE J. Quantum Electron.* **QE-22**, 1682–1695 (1986).
11. R. A. Street, G. H. Döhler, J. N. Miller, and P. P. Ruden, "Luminescence of n-i-p-i heterostructures," *Phys. Rev. B* **33**, 7043–7046 (1986).
12. P. E. Gray, D. DeWitt, A. R. Boothroyd, and J. F. Gibbons, *Physical Electronics and Circuit Models of Transistors*, Vol. 2 in *Semiconductor Electronics Education Committee Books*, Wiley, New York, 1964.
13. G. W. Neudeck, *The PN Junction Diode*, Vol. 2 in *Modular Series on Solid State Devices*, G. W. Neudeck and R. F. Pierret, Eds., Addison-Wesley, Reading, MA, 1983.
14. B. G. Streetman and S. Banerjee, *Solid State Electronic Devices*, 6th ed., Prentice Hall, Upper Saddle River, NJ, 2005.
15. M. B. Reine, A. K. Sood, and T. J. Tredwell, "Photovoltaic infrared detectors," Vol. 18, *Mercury Cadmium Telluride*, in R. K. Willardson and A. C. Beer, Eds., *Semiconductors and Semimetals*, Academic Press, New York, 1981.
16. M. Brain and T. P. Lee, "Optical receivers for lightwave communication systems," *J. Lightwave Technol.* **LT-3**, 1281–1300 (1985).
17. S. R. Forrest, R. F. Leheny, R. E. Nahory, and M. A. Pollack, "$In_{0.53}Ga_{0.47}As$ photodiodes with dark current limited by generation-recombination and tunneling," *Appl. Phys. Lett.* **37**, 322–325 (1980).
18. S. R. Forrest, "Performance of $In_xGa_{1-x}As_yP_{1-y}$ photodiodes with dark current limited by diffusion, generation-recombination and tunneling," *IEEE J. Quantum Electron.* **QE-717**, 217–226 (1981).
19. M. C. Brain, "Comparison of available detectors for digital optical fiber systems for the 1.2–1.55 μm wavelength range," *IEEE J. Quantum Electron.* **QE-18**, 219–224 (1982).

20. J. E. Bowers and C. A. Burrus, Jr., "Ultrawide-band long-wavelength p-i-n photodetectors," *J. Lightwave Technol.* **LT-5**, 1339–1350 (1987).

21. G. E. Stillman and C. M. Wolfe, "Avalanche photodiodes," pp. 291–393, *Infrared Detectors II*, Vol. 12 in R. K. Willardson and A. C. Beer, Eds., *Semiconductors and Semimetals*, Academic Press, New York, 1977.

22. P. P. Webb, R. J. McIntyre, and J. Conradi, "Properties of avalanche photodiodes," *RCA Review* **35**, 234–278 (1974).

23. T. P. Lee and T. Li, "Photodetectors," pp. 593–626 in S. E. Miller and A. G. Chynoweth, Eds., *Optical Fiber Telecommunications*, Academic, New York, 1979.

24. G. E. Stillman, "Impact ionization coefficients in InGaAs," pp. 76–83 in P. Bhattacharya, Ed., *Properties of Lattice-Matched and Strained Indium Gallium Arsenide*, INSPEC, the Institute of Electrical Engineers, London, UK, 1993.

25. G. E. Bulman, V. M. Robbins, K. F. Brennan, K. Hess, and G. E. Stillman, "Experimental determination of impact ionization coefficients in (100) GaAs," *IEEE Electron Device Lett.* **4**, 181–185 (1983).

26. L. W. Cook, G. E. Bulman, and G. E. Stillman, "Electron and hole impact ionization coefficients in InP determined by photomultiplication measurements," *Appl. Phys. Lett.* **40**, 589–591 (1982).

27. M. H. Woods, W. C. Johnson, and M. A. Lampert, "Use of a Schottky barrier to measure impact ionization coefficients in semiconductors," *Solid State Electron.* **16**, 381–394 (1973).

28. H. Ando and H. Kanbe, "Ionization coefficient measurement in GaAs by using multiplication noise characteristics," *Solid State Electron.* **24**, 629–634 (1981).

29. F. Osaka, T. Mikawa, and T. Kaneda, "Impact ionization coefficients of electrons and holes in (100)-oriented $Ga_{1-x}In_xAs_yP_{1-y}$," *IEEE J. Quantum Electron.* **QE-21**, 1326–1338 (1985).

30. F. Osaka and T. Mikawa, "Excess noise design of InP/GaInAsP/GaInAs avalanche photodiodes," *IEEE J. Quantum Electron.* **QE-22**, 471–478 (1986).

31. Y. C. Chen and P. K. Bhattacharya, "Impact ionization coefficients for electrons and holes in strained $In_{0.2}Ga_{0.8}As$ and $In_{0.15}Ga_{0.63}Al_{0.22}As$ channels embedded in $Al_{0.3}Ga_{0.7}As$," *J. Appl. Phys.* **73**, 465–467 (1993).

32. R. J. McIntyre, "Multiplication noise in uniform avalanche diodes," *IEEE Trans. Electron. Devices* **ED-13**, 164–168 (1966).

33. R. J. McIntyre, "The distribution of gains in uniformly multiplying avalanche photodiodes," *IEEE Trans. Electron. Devices* **ED-19**, 703–712 (1972).

34. H. Ando, H. Kanbe, M. Ito, and T. Kaneda, "Tunneling current in InGaAs and optimum design for InGaAs/InP avalanche photodiode," *Jpn. J. Appl. Phys.* **19**, L277–L280 (1980).

35. S. R. Forrest, O. K. Kim, and R. G. Smith, "Analysis of the dark current and photoresponse of $In_{0.53}Ga_{0.47}As$/InP avalanche photodiodes," *Solid State Electron.* **26**, 951–968 (1983).

36. F. Capasso, A. Y. Cho, and P. W. Foy, "Low-dark-current low-voltage 1.3–1.6 μm avalanche photodiode with high-low electric field profile and separate absorption and multiplication regions by molecular beam epitaxy," *Electron. Lett.* **20**, 635–637 (1984).

37. C. A. Lee, R. L. Batdorf, W. Wiegmann, and G. Kaminsky, "Time dependence of avalanche processes in silicon," *J. Appl. Phys.* **38**, 2787–2796 (1967).

38. R. B. Emmons, "Avalanche-photodiode frequency response," *J. Appl. Phys.* **38**, 3705–3714 (1967).

39. H. Ando and H. Kanbe, "Effect of avalanche build-up time on avalanche photodiode sensitivity," *IEEE J. Quantum Electron.* **QE-21**, 251–255 (1985).

40. G. Kahraman, B. E. A. Saleh, W. L. Sargeant, and M. C. Teich, "Time and frequency response of avalanche photodiodes with arbitrary structure," *IEEE Trans. Electron. Devices* **39**, 553–560 (1992).

41. K. Nishida, K. Taguchi, and Y. Matsumoto, "InGaAsP heterostructure avalanche photodiodes with high avalanche gain," *Appl. Phys. Lett.* **35**, 251–252 (1979).

42. N. Susa, H. Nakagome, O. Mikami, H. Ando, and H. Kanbe, "New InGaAs/InP avalanche photodiode structure for the 1–1.6 μm wavelength region," *IEEE J. Quantum Electron.* **QE-16**, 864–870 (1980).

43. G. E. Stillman, V. M. Robbins, and N. Tabatabaie, "III-V compound semiconductor devices: optical detectors," *IEEE Trans. Electron. Devices* **ED-31**, 1643–1655 (1984).

44. R. Chin, N. Holonyak, Jr., G. E. Stillman, J. Y. Tang, and K. Hess, "Impact ionization in multilayered heterojunction structures," *Electron. Lett.* **16**, 467–469 (1980).

45. F. Capasso, W. T. Tsang, A. L. Hutchinson, and G. F. Williams, "Enhancement of electron impact ionization in a superlattice: a new avalanche photodiode with a large ionization rate ratio," *Appl. Phys. Lett.* **40**, 38–40 (1982).

46. F. Capasso, W. T. Tsang, and G. F. Williams, "Staircase solid-state photomultipliers and avalanche photodiodes with enhanced ionization rates ratio," *IEEE Trans. Electron. Devices* **ED-30**, 381–390 (1983).

47. T. Kagawa, H. Iwamura, and O. Mikami, "Dependence of the GaAs/AlGaAs superlattice ionization rate on Al content," *Appl. Phys. Lett.* **54**, 33–35 (1989).

48. T. Tanoue and H. Sakaki, "A new method to control impact ionization rate ratio by spatial separation of avalanching carriers in multilayered heterostructures," *Appl. Phys. Lett.* **41**, 67–70 (1982).

49. K. Brennan, "Theory of electron and hole impact ionization in quantum well and staircase superlattice avalanche photodiode structures," *IEEE Trans. Electron. Devices* **ED-32**, 2197–2205 (1985).

50. S. L. Chuang and K. Hess, "Impact ionization across the conduction band edge discontinuity of quantum-well heterostructures," *J. Appl. Phys.* **59**, 2885–2894 (1986).

51. S. L. Chuang and K. Hess, "Tunneling-assisted impact ionization for a superlattice," *J. Appl. Phys.* **61**, 1510–1515 (1987).

52. F. Capasso, J. Allam, A. Y. Cho, K. Mohammed, R. J. Malik, A. L. Hutchinson, and D. Sivco, "New avalanche multiplication phenomenon in quantum well superlattices: Evidence of impact ionization across the band-edge discontinuity," *Appl. Phys. Lett.* **48**, 1294–1296 (1986).

53. M. Toivonen, M. Jalonen, A. Salokatve, and M. Pessa, "Unipolar avalanche multiplication phenomenon in multiquantum well structures," *Appl. Phys. Lett.* **62**, 1664–1666 (1993).

54. B. F. Levine, K. K. Choi, C. G. Bethea, J. Walker, and R. J. Malik, "Quantum well avalanche multiplication initiated by 10 μm intersubband absorption and photoexcited tunneling," *Appl. Phys. Lett.* **51**, 934–936 (1987).

55. A. Kamgar, P. Kneschaurek, G. Dorda, and J. F. Koch, "Resonance spectroscopy of electronic levels in a surface accumulation layer," *Phys. Rev. Lett.* **32**, 1251–1254 (1974).

56. S. J. Allen, Jr., D. C. Tsui, and B. Vinter, "On the absorption of infrared radiation by electrons in semiconductor inversion layers," *Solid State Commun.* **20**, 425–428 (1976).

57. L. L. Chang, L. Esaki, and G. A. Sai-Halasz, "Infrared optical devices of layered structure," *IBM Tech. Disclosure Bull.* **20**, 2019–2020 (1977).

58. L. Esaki and H. Sakaki, "New photoconductor," *IBM Tech. Disc. Bull.* **20**, 2456 (1977).

59. T. Ando, A. B. Fowler, and F. Stern, "Electronic properties of two-dimensional systems," *Rev. Mod. Phys.* **54**, 437–672 (1982).

60. L. C. Chiu, J. S. Smith, S. Margalit, A. Yariv, and A. Y. Cho, "Application of internal photoemission from quantum-well and heterojunction superlattices to infrared photodetectors," *Infrared Phys.* **23**, 93–97 (1983).

61. D. D. Coon and R. P. G. Karunasiri, "New mode of IR detection using quantum wells," *Appl. Phys. Lett.* **45**, 649–651 (1984).

62. L. C. West and S. J. Eglash, "First observation of an extremely large-dipole infrared transition within the conduction band of a GaAs quantum well," *Appl. Phys. Lett.* **46**, 1156–1158 (1985).

63. A. Harwit and J. S. Harris, Jr., "Observation of Stark shifts in quantum well intersubband transitions," *Appl. Phys. Lett.* **50**, 685–687 (1987).

64. D. Ahn and S. L. Chuang, "Calculation of linear and nonlinear intersubband optical absorption in a quantum well model with an applied electric field," *IEEE J. Quantum Electron.* **QE-23**, 2196–2204 (1987).

65. E. J. Roan and S. L. Chuang, "Linear and nonlinear intersubband electroabsorptions in a modulation-doped quantum well," *J. Appl. Phys.* **69**, 3249–3260 (1991).

66. K. Bajema, R. Merlin, F. Y. Juang, S. C. Hong, J. Singh, and P. K. Bhattacharya, "Stark effect in GaAs-$Al_xGa_{1-x}As$ quantum wells: Light scattering by intersubband transitions," *Phys. Rev. B* **36**, 1300–1302 (1987).

67. K. K. Choi, B. F. Levine, C. G. Bethea, J. Walker, and R. J. Malik, "Multiple quantum well 10 μm GaAs/$Al_xGa_{1-x}As$ infrared detector with improved responsivity," *Appl. Phys. Lett.* **50**, 1814–1816 (1987).

68. M. Nakayama, H. Kuwahara, H. Kato, and K. Kubota, "Intersubband transitions in GaAs-$Al_xGa_{1-x}As$ modulation-doped superlattices," *Appl. Phys. Lett.* **51**, 1741–1743 (1987).

69. A. D. Wieck, J. C. Maan, U. Merkt, J. P. Kotthaus, K. Ploog, and G. Weimann, "Intersubband energies in GaAs-$Ga_{1-x}Al_xAs$ heterojunctions," *Phys. Rev. B* **35**, 4145–4148 (1987).

70. M. Zachau, "Electronic eigenstates and intersubband transitions in doped heterostructure superlattices," *Semicond. Sci. Technol.* **3**, 879–885 (1988).

71. Z. Ikonic, V. Milanovic, and D. Tjapkin, "Bound-free intraband absorption in GaAs—$Al_xGa_{1-x}As$ semiconductor quantum wells," *Appl. Phys. Lett.* **54**, 247–249 (1989).

72. X. Zhou, P. K. Bhattacharya, G. Hugo, S. C. Hong, and E. Gulari, "Intersubband absorption in strained InGaAs/AlGaAs ($0 < x < 0.15$) multiquantum wells," *Appl. Phys. Lett.* **54**, 855–856 (1989).

73. J. D. Ralston, M. Ramsteiner, B. Dischler, M. Maier, G. Brandt, P. Koidl, and D. J. As, "Intersubband transitions in partially interdiffused GaAs/AlGaAs multiple quantum-well structures," *Appl. Phys. Lett.* **70**, 2195–2199 (1991).

74. F. H. Julien, P. Vagos, J. M. Lourtioz, and D. D. Yang, "Novel all-optical 10 μm waveguide modulator based on intersubband absorption in GaAs/AlGaAs quantum wells," *Appl. Phys. Lett.* **59**, 2645–2647 (1991).

75. E. Rosencher, E. Martinet, F. Luc, Ph. Bois, and E. Bockenhoff, "Discrepancies between photocurrent and absorption spectroscopies in intersubband photoionization from GaAs/AlGaAs multiquantum wells," *Appl. Phys. Lett.* **59**, 3255–3257 (1991).

76. K. K. Choi, L. Fotiadis, M. Taysing-Lara, and W. Chang, "Infrared absorption and photoconductive gain of quantum well infrared photodetectors," *Appl. Phys. Lett.* **60**, 592–594 (1992).

77. L. S. Yu, Y. H. Wang, and S. S. Li, "Low dark current step-bound-to-miniband transition InGaAs/GaAs/AlGaAs multiquantum-well infrared detector," *Appl. Phys. Lett.* **60**, 992–994 (1992).

78. W. Chen and T. G. Andersson, "Intersubband transitions for differently shaped quantum wells under an applied electric field," *Appl. Phys. Lett.* **60**, 1591–1593 (1992).

79. H. C. Liu, "Noise gain and operating temperature of quantum well infrared photodetectors," *Appl. Phys. Lett.* **61**, 2703–2705 (1992).

80. H. C. Liu, "Dependence of absorption spectrum and responsivity on the upper state position in quantum well intersubband photodetectors," *J. Appl. Phys.* **73**, 3062–3067 (1993).

81. J. P. Peng, Y. M. Mu, and X. C. Shen, "Performance of the multiquantum well infrared photodetector," *J. Appl. Phys.* **74**, 1421–1425 (1993).

82. M. Helm, W. Hilber, T. Fromherz, F. M. Peeters, K. Alavi, and R. N. Pathak, "Infrared absorption in superlattices: a probe of the miniband dispersion and the structure of the impurity band," *Phys. Rev. B* **48**, 1601–1606 (1993).

83. R. P. G. Karunasiri, J. S. Park, and K. L. Wang, "$Si_{1-x}Ge_x/Si$ multiple quantum well infrared detector," *Appl. Phys. Lett.* **59**, 2588–2590 (1991).

84. E. Rosencher, B. Vinter, and B. Levine, Eds., *Intersubband Transitions in Quantum Wells*, Sept. 9–14, Cargese, France, Plenum, New York, 1992.

85. H. C. Liu, B. F. Levine, and J. Y. Andersson, Eds., *Quantum Well Intersubband Transition Physics and Devices*, Plenum, New York, 1994.

86. M. O. Manasreh, Ed., *Semiconductor Quantum Wells and Superlattices for Long-Wavelength Infrared Detectors*, Artech House, Norwood, MA, 1993.

87. B. F. Levine, "Quantum-well infrared photodetectors," *J. Appl. Phys.* **74**, R1–R81 (1993).

88. Y. C. Chang and R. B. James, "Saturation of intersubband transitions in p-type semiconductor quantum wells," *Phys. Rev. B* **39**, 12672–12681 (1989).

89. B. F. Levine, S. D. Gunapala, J. M. Kuo, S. S. Pei, and S. Hui, "Normal incidence hole intersubband absorption long wavelength $GaAs/Al_xGa_{1-x}As$ quantum well infrared photodetectors," *Appl. Phys. Lett.* **59**, 1864–1866 (1991).

90. P. Man and D. S. Pan, "Analysis of normal-incident absorption in p-type quantum-well infrared photodetectors," *Appl. Phys. Lett.* **61**, 2799–2801 (1992).

91. K. M. S. V. Bandara, D. D. Coon, Byungsung O, Y. F. Lin, and M. H. Francombe, "Exchange interactions in quantum well subbands," *Appl. Phys. Lett.* **53**, 1931–1933 (1988).

92. M. O. Manasreh, F. Szmulowicz, T. Vaughan, K. R. Evans, C. E. Stutz, and D. W. Fischer, "Origin of the blueshift in the intersubband infrared absorption in GaAs/$Al_{0.3}Ga_{0.7}As$ multiple quantum wells," *Phys. Rev. B* **43**, 9996–9999 (1991).

93. J. W. Choe, Byungsung O, K. M. S. V. Bandara, and D. D. Coon, "Exchange interaction effects in quantum well infrared detectors and absorbers," *Appl. Phys. Lett.* **56**, 1679–1681 (1990).

94. S. L. Chuang, M. S. C. Luo, S. Schmitt-Rink, and A. Pinczuk, "Many-body effects on intersubband transitions in semiconductor quantum-well structures," *Phys. Rev. B* **46**, 1897–1900 (1992).

95. B. F. Levine, R. J. Malik, J. Walker, K. K. Choi, C. G. Bethea, D. A. Kleinman, and J. M. Vandenberg, "Strong 8.2 μm infrared intersubband absorption in doped GaAs/AlAs quantum well waveguides," *Appl. Phys. Lett.* **50**, 273–275 (1987).

96. J. Y. Andersson and G. Landgren, "Intersubband transitions in single AlGaAs/GaAs quantum wells studied by Fourier transform infrared spectroscopy," *J. Appl. Phys.* **64**, 4123–4127 (1988).

97. H. C. Liu, "Photoconductive gain mechanism of quantum-well intersubband infrared detectors," *Appl. Phys. Lett.* **60**, 1507–1509 (1992).

98. B. F. Levine, A. Zussman, S. D. Gunapala, M. T. Asom, J. M. Kuo, and W. S. Hobson, "Photoexcited escape probability, optical gain, and noise in quantum well infrared photodetectors," *J. Appl. Phys.* **72**, 4429–4443 (1992).

99. B. F. Levine, C. G. Bethea, G. Hasnain, V. O. Shen, E. Pelve, R. R. Abbott, and S. J. Hsieh, "High sensitivity low dark current 10 μm GaAs quantum well infrared photodetectors," *Appl. Phys. Lett.* **56**, 851–853 (1990).

100. M. A. Kinch and A. Yariv, "Performance limitations of GaAs/AlGaAs infrared superlattices," *Appl. Phys. Lett.* **55**, 2093–2095 (1989).

101. S. D. Gunapala, B. F. Levine, L. Pfeiffer, and K. West, "Dependence of the performance of GaAs/AlGaAs quantum well infrared photodetectors on doping and bias," *Appl. Phys. Lett.* **69**, 6517–6520 (1991).

102. B. F. Levine, A. Zussman, J. M. Kuo, and J. de Jong, "19 μm cutoff long-wavelength GaAs/$Al_xGa_{1-x}As$ quantum-well infrared photodetectors," *Appl. Phys. Lett.* **71**, 5130–5135 (1992).

103. Y. Hamakawa, Ed., *Thin-Film Solar Cells—Next Generation Photovoltaics and Its Applications*, Chapter 1, Springer, Berlin, 2004.

104. C. Hu and R. M. White, *Solar Cells: From Basic to Advanced Systems*, McGraw Hill, 1983.

105. S. J. Fonash, *Solar Cell Device Physics*, Academic Press, New York, 1981.

106. M. A. Green, *Solar Cells—Operating Principles, Technology, and System Applications*, Prentice-Hall, Englewood Cliffs, NJ, 1982.

107. S. M. Sze and K. K. Ng, *Physics of Semiconductor Devices*, Wiley, New York, 2007.

108. C. A. Gueymard, D. Myers, and K. Emery, "Proposed reference irradiance spectra for solar energy systems testing," *Solar Energy* **73**, 443–467 (2003).
109. C. H. Henry, "Limiting efficiencies of ideal single and multiple energy gap terrestrial solar cells," *J. Appl. Phys.* **51**, 4494–4500 (1980).
110. E. Bucher, "Solar cell materials and their basic parameters," *Appl. Phys.* **17**, 1–25 (1978).
111. M. Paxma, J. Nelson, B. Braun, J. Connolly, K. W. J. Barnham, C. T. Foxon, and J. S. Roberts, "Modeling the spectral response of the quantum well solar cell," *J. Appl. Phys.* **74**, 614–621 (1993).
112. G. L. Araújo and A. Martí, "Electroluminescence coupling in multiple quantum well diodes and solar cells," *Appl. Phys. Lett.* **66**, 894–895 (1995).
113. N. G. Anderson, "Ideal theory of quantum well solar cells," *J. Appl. Phys.* **78**, 1850–1861 (1995).
114. N. G. Anderson and S. J. Wojtczuk, "Open-circuit voltage characteristics of InP-based quantum well solar cells," *J. Appl. Phys.* **79**, 1973–1978 (1996).
115. K. Barnham, J. Connolly, P. Griffin, G. Haarpaintner, J. Nelson, E. Tsui, A. Zachariou, J. Osborne, C. Button, G. Hill, M. Hopkinson, M. Pate, J. Roberts, and T. Foxon, "Voltage enhancement in quantum well solar cells," *J. Appl. Phys.* **80**, 1201–1206 (1996).
116. J. Nelson, J. Barnes, N. Ekins-Daukes, B. Kluftinger, E. Tsui, K. Barnham, C. T. Foxon, T. Cheng, and J. S. Roberts, "Observation of suppressed radiative recombination in single quantum well *p-i-n* photodiodes," *J. Appl. Phys.* **82**, 6240–6246 (1997).
117. B. Kluftinger, K. Barnham, J. Nelson, T. Fozon, and T. Cheng, "Temperature-dependent study of the quasi-Fermi level separation in double quantum well *P-I-N* structures," *Microelectron. Eng.* **51-52**, 265–274 (2000).
118. K. W. J. Barnham, I. Ballard, J. P. Connolly, N. J. Ekins-Daukes, B. G. Kluftinger, J. Nelson, and C. Rohr, "Quantum well solar cells," *Physica E* **14**, 27–36 (2002).
119. A. C. Varonides, "Thermionic escape of net photogenerated carriers and current densities from illuminated lightly doped single quantum wells," *Physica E* **14**, 142–149 (2002).
120. J. Nelson, I. Ballard, K. Barnham, J. P. Connolly, J. S. Roberts, and M. Pate, "Effect of quantum well location on the single quantum well p-i-n photodiode dark currents," *J. Appl. Phys.* **86**, 5898–5905 (1999).
121. N. G. Anderson, "On quantum well solar cell efficiencies," *Physica E* **14**, 126–131 (2002).
122. J. M. Olson, S. R. Kurtz, A. E. Kibbler, and P. Faine, "A 27.3% efficient $Ga_{0.5}In_{0.5}P$/GaAs tandem solar cell," *Appl. Phys. Lett.* **56**, 623–625 (1990).
123. S. R. Kurtz, A. A. Allerman, E. D. Jones, J. M. Gee, and J. J. Banas, "InGaAsN solar cells with 1.0 eV band gap, lattice matched to GaAs," *Appl. Phys. Lett.* **74**, 729–731 (1999).
124. M. Yamaguchi, "III-V compound multi-junction solar cells: present and future,". *Sol. Energy Mater. Sol. Cells* **75**, 261–269 (2003).
125. J. M. Román, "State-of-the-art of III-V solar cell fabrication technologies, device designs and applications," *Advanced Photovoltaic Cell Design*, **EN548**, 1–8 (2004).
126. M. A. Green, K. Emery, D. L. King, S. Igari, and W. Warta, "Solar cell efficiency tables (version 24)," *Prog. Photovolt. Res. Appl.* **12** 365–372 (2004).

127. R. R. King, D. C. Law, K. M. Edmondson, C. M. Fetzer, G. S. Kinsey, H. Yoon, R. A. Sherif, and N. H. Karam, "40% efficient metamorphic GaInP/GaInAs/Ge multijunction solar cells," *Appl. Phys. Lett.* **90**, 183516 (2007).

128. R. R. King, D. C. Law, K. M. Edmondson, C. M. Fetzer, G. S. Kinsey, H. Yoon, D. D. Krut, J. H. Ermer, R. A. Sherif, and N. H. Karam, "Advances in high-efficiency III-V multijunction solar cells," *Advances in OptoElectronics*, Vol. 2007, Article ID 29523, 8 pages. Doi:10.1155/2007/79523.

129. J. F. Geisz and D. J. Friedman, "III-N-V semiconductors for solar photovoltaic applications," *Semicond. Sci. Technol.* **17**, 769–777 (2002).

130. R. J. Ellingson, M. C. Beard, J. C. Johnson, P. Yu, O. I. Micic, A. J. Nozik, A. Shabaev, and A. L. Efros, "High efficient multiple exciton generation in colloidal PbSe and PbS quantum dots," *Nano Lett.* **5**, 865–871 (2005).

131. R. D. Schaller, V. M. Agranovich, and V. I. Klimov, "High efficiency carrier multiplication through direct photogeneration of multi-excitons via virtual single-exciton states," *Nature Phys.* **1**, 189–194 (2005).

Appendix A

Semiconductor Heterojunction Band Lineups in the Model–Solid Theory

Semiconductor heterojunctions and superlattices have been under intensive investigation both theoretically and experimentally in the past three decades. There are tremendous potential device applications using heterojunctions. In this appendix, we discuss a simplified model to determine the energy band lineups of semiconductor heterojunctions based on the model–solid theory [1–4]. The goal is to develop a reliable model to predict band offsets for a wide variety of heterojunctions without the need for difficult calculations such as in the local-density-functional theory or *ab initio* pseudopotential method. The relation of the model–solid theory to the fully self-consistent first-principles calculations can be found in Refs. 2 and 3.

The major idea is to set up an absolute reference energy level. All calculated energies can then be put on an absolute energy scale, allowing us to derive band lineups. In the model–solid theory, an average energy over the three uppermost valence bands (the heavy-hole, the light-hole, and the spin–orbit split-off bands) $E_{v,av}$ is obtained from theory and it is referred to as the absolute energy level. The values of $E_{v,av}$ for different semiconductors are usually tabulated [1] (Table C.2 in Appendix C) so that no calculations for these values are necessary. These results have to be compared with those of the first-principle calculations whenever possible to justify the model. An estimate of the maximum possible error is about 0.1 eV. It should be noted that band offsets should be checked with experimental data such as those in Refs. 5–19. The model–solid theory provides a simple method for estimating the band offsets for materials, especially ternary compounds with varying compositions for which experimental data may not be always available.

UNSTRAINED SEMICONDUCTORS

If the materials A and B have the same lattice constants, we may have an ideal heterojunction and there is no strain in the semiconductors. For this case, the

Physics of Photonic Devices, Second Edition. By Shun Lien Chuang
Copyright © 2009 John Wiley & Sons, Inc.

heavy-hole and light-hole band edges (E_{HH} and E_{LH}) are degenerate at the zone center, and their energy position is denoted as E_v

$$E_v = E_{v,av} + \frac{\Delta}{3} \tag{A.1}$$

where Δ is the spin–orbit splitting energy, and the spin–orbit split-off band edge energy E_{SO} is

$$E_{SO} = E_v - \Delta = E_{v,av} - \frac{2\Delta}{3}. \tag{A.2}$$

The conduction-band edge is obtained by adding the band-gap energy E_g to E_v

$$E_c = E_v + E_g. \tag{A.3}$$

Note that in the model–solid theory, the spin–orbit splitting energy Δ and the band-gap energy E_g are taken from experimental results. The only input provided by the model–solid theory is the tabulated $E_{v,\,av}$ value. This $E_{v,\,av}$ value is essentially the same as the p-state energy E_p in Fig. 4.3a.

With the above results, the band lineups between materials A and B are shown in Fig. A.1. We have

$$\Delta E_g = E_g^A - E_g^B$$

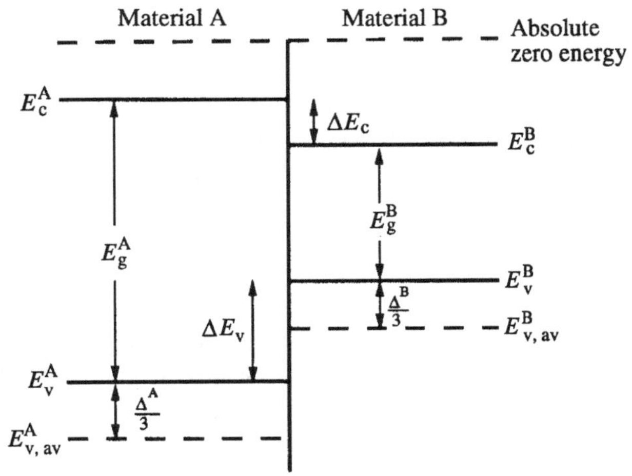

Figure A.1 Band lineups in the model–solid theory. $E_{v,av}$ in each material region is obtained from the model–solid theory and is tabulated in Appendix C. The band-gap energy E_g and the spin–orbit splitting Δ of each material are taken from experimental results.

and the band-edge discontinuities are

$$\Delta E_c = E_c^A - E_c^B, \qquad \Delta E_v = E_v^B - E_v^A \tag{A.4}$$

$$\Delta E_c + \Delta E_v = \Delta E_g. \tag{A.5}$$

The partition ratios of the band-edge discontinuities, $Q_c = \Delta E_c/\Delta E_g$ and $Q_v = \Delta E_v/\Delta E_g$, are obtained from this theory and can also be compared with experimental data.

STRAINED SEMICONDUCTORS

If a material A with a lattice constant a is grown on a substrate with a lattice constant a_0 along the z direction, we have

$$\varepsilon_{xx} = \varepsilon_{yy} = \frac{a_0 - a}{a} \tag{A.6a}$$

and

$$\varepsilon_{zz} = -2\frac{C_{12}}{C_{11}}\varepsilon_{xx}. \tag{A.6b}$$

The band-edge shifts are

$$\Delta E_{v,av} = a_v(\varepsilon_{xx} + \varepsilon_{yy} + \varepsilon_{zz}) \equiv -P_\varepsilon \tag{A.7a}$$

$$\Delta E_c = a_c(\varepsilon_{xx} + \varepsilon_{yy} + \varepsilon_{zz}) \equiv P_c. \tag{A.7b}$$

The position of the average energy of the valence bands $E_{v,av}$ under strain is shifted from its unstrained position $E_{v,av}^0$ in (A.1) by $-P_\varepsilon$.

$$E_{v,av} = E_{v,av}^0 - P_\varepsilon. \tag{A.8}$$

We thus have the center of the valence-band edge energy

$$E_v = E_{v,av}^0 + \frac{\Delta}{3} = E_v^0 - P_\varepsilon. \tag{A.9}$$

The heavy-hole, light-hole, and spin–orbit split-off band edges are

$$E_{HH} = E_v^0 - P_\varepsilon - Q_\varepsilon \tag{A.10}$$

$$E_{LH} = E_v^0 - P_\varepsilon - \frac{\Delta}{2} + \frac{Q_\varepsilon}{2} + \frac{1}{2}\left[\Delta^2 + 2\Delta Q_\varepsilon + 9Q_\varepsilon^2\right]^{1/2} \tag{A.11}$$

$$E_{SO} = E_v^0 - P_\varepsilon - \frac{\Delta}{2} + \frac{Q_\varepsilon}{2} - \frac{1}{2}\left[\Delta^2 + 2\Delta Q_\varepsilon + 9Q_\varepsilon^2\right]^{1/2}. \tag{A.12}$$

The conduction band edge is shifted by P_c given by (A.7b)

$$E_c = E_v^0 + E_g(x) + P_c. \tag{A.13}$$

Note that in the limit of a large spin–orbit split-off energy $\Delta \gg |Q_\varepsilon|$, we can ignore the coupling of the spin–orbit split-off band and

$$E_{\mathrm{LH}} \approx E_{\mathrm{v}}^{0} - P_\varepsilon + Q_\varepsilon \tag{A.14a}$$

$$E_{\mathrm{SO}} \approx E_{\mathrm{v}}^{0} - P_\varepsilon - \Delta. \tag{A.14b}$$

For a ternary alloy such as $\mathrm{A}_x\mathrm{B}_{1-x}\mathrm{C}$ with a lattice constant $a(x)$

$$a(x) = xa(\mathrm{AC}) + (1-x)a(\mathrm{BC}) \tag{A.15}$$

which is a linear interpolation of the lattice constants, $a(\mathrm{AC})$ and $a(\mathrm{BC})$, of the binary compound semiconductors, we use the following formula to calculate an energy level $E(= E_{\mathrm{v,av}}^{0}$ for example):

$$E(\mathrm{A}_x\mathrm{B}_{1-x}\mathrm{C}) = xE(\mathrm{AC}) + (1-x)E(\mathrm{BC}) + 3x(1-x)[-a_{\mathrm{v}}(\mathrm{AC}) + a_{\mathrm{v}}(\mathrm{BC})]\frac{\Delta a}{a_0} \tag{A.16}$$

where the last term accounts for a *strain* contribution to the ternary alloy, and $\Delta a = a(\mathrm{AC}) - a(\mathrm{BC})$ is the difference between the lattice constants of two compounds AC and BC. Once $E_{\mathrm{v,av}}^{0}$ is determined, the band-edge energies for the strained ternary compound can be calculated following (A.6)–(A.13).

Many theoretical parameters for the electronic and optical properties such as those listed in Table C.2 in Appendix C can be found in the data books compiled by various groups such as Refs. 20–23, review papers such as Refs. 24–27, and research papers such as Refs. 28–37.

Example. GaAs/AlAs Heterojunction GaAs and AlAs have almost the same lattice constants. Therefore, the heterojunction has a negligible strain. We see from Table C.2 in Appendix C,

$$E_{\mathrm{v,av}}(\mathrm{GaAs}) = -6.92\,\mathrm{eV},\ \Delta(\mathrm{GaAs}) = 0.34\,\mathrm{eV},\ E_{\mathrm{g}}(\mathrm{GaAs}) = 1.52\,\mathrm{eV}$$

$$E_{\mathrm{v,av}}(\mathrm{AlAs}) = -7.49\,\mathrm{eV},\ \Delta(\mathrm{AlAs}) = 0.28\,\mathrm{eV},\ E_{\mathrm{g}}^{\mathrm{dir}}(\mathrm{AlAs}) = 3.13\,\mathrm{eV}.$$

Therefore,

$$E_{\mathrm{v}}(\mathrm{GaAs}) = -6.92 + \frac{0.34}{3} = -6.81\,\mathrm{eV}$$

$$E_{\mathrm{v}}(\mathrm{AlAs}) = -7.49 + \frac{0.28}{3} = -7.40\,\mathrm{eV}$$

$$\Delta E_{\mathrm{v}} = -6.81 + 7.40 = 0.59\,\mathrm{eV}.$$

Also, the band-gap discontinuity is $\Delta E_{\mathrm{g}} = 1.61$ eV, and the valence-band discontinuity ratio is $\Delta E_{\mathrm{v}}/\Delta E_{\mathrm{g}} = 0.37$.

Example. $In_{0.53}Ga_{0.47}As/InP$ Heterojunction

$$\begin{aligned} a(\text{GaAs}) &= 5.6533\ \text{Å},\ a(\text{InAs}) = 6.0584\ \text{Å},\ a(\text{InP}) = 5.8688\ \text{Å} \\ E_{v,av}(\text{In}_{1-x}\text{Ga}_x\text{As}) &= xE_{v,av}(\text{GaAs}) + (1-x)E_{v,av}(\text{InAs}) \\ &\quad + 3x(1-x)[-a_v(\text{GaAs}) + a_v(\text{InAs})]\frac{\Delta a}{a} \\ \Delta a &= 5.6533 - 6.0584 = -0.4051\ \text{Å} \\ a(\text{In}_{1-x}\text{Ga}_x\text{As}) &= xa(\text{GaAs}) + (1-x)a(\text{InAs}) \\ \Delta(\text{In}_{1-x}\text{Ga}_x\text{As}) &\approx x\Delta(\text{GaAs}) + (1-x)\Delta(\text{InAs}). \end{aligned}$$

For $x = 0.47$, $In_{0.53}Ga_{0.47}As$ is lattice matched to InP. Therefore, we do not have the strain terms ($P_\varepsilon = 0$, $Q_\varepsilon = 0$). We obtain

$$E_{v,av}(\text{In}_{0.53}\text{Ga}_{0.47}\text{As}) = -6.779\ \text{eV}, \quad \Delta = 0.361\ \text{eV}.$$

Using $E_{v,av}(\text{InP}) = -7.04$ eV and $E_v(\text{InP}) = -7.003$ eV, we find $\Delta E_v = 0.344$ eV. From room temperature data for the band gap, $E_g(\text{In}_{0.53}\text{Ga}_{0.47}\text{As}) = 0.73$ eV, $E_g(\text{InP}) = 1.35$ eV, and $\Delta E_g = 0.62$ eV, we obtain the ratio $\Delta E_v/\Delta E_g = 0.55 = 55\%$.

Some Experimental Reports on Band-Edge Discontinuities There has been a considerable amount of experimental data on band offsets, mostly on unstrained systems. For strained semiconductors, the band offsets are complicated by the deformation potentials, which also shift the conduction- and valence-band edges. Therefore, fewer data are available for strained heterojunctions.

1. $GaAs/Al_xGa_{1-x}As$ system:

$$\begin{aligned} E_g(\text{GaAs}) &= 1.424\ \text{eV (300K)} \\ E_g(\text{Al}_x\text{Ga}_{1-x}\text{As}) &= 1.424 + 1.247x\,\text{eV (300K)} \\ \Delta E_g(x) &= 1.247x\,\text{eV} \\ \Delta E_c &= 0.67\ \Delta E_g, \qquad \Delta E_v = 0.33\ \Delta E_g \\ (\Delta E_c &= 0.69\ \Delta E_g,\ \Delta E_v \cong 0.31\ \Delta E_g;\ \text{Ref. 11}). \end{aligned}$$

2. $In_{0.53}Ga_{0.47}As/InP$ (~0K) [18]:

$$\begin{aligned} E_g(\text{InP}) &= 1.423\ \text{eV} \\ E_g(\text{In}_{0.53}\text{Ga}_{0.47}\text{As}) &= 0.811\ \text{eV} \\ \Delta E_g &= 0.612\ \text{eV} \\ \Delta E_c &= 0.252\ \text{eV} = 0.41\ \Delta E_g,\ \Delta E_v = 0.360\ \text{eV} = 0.59\ \Delta E_g. \end{aligned}$$

3. $In_{0.52}Al_{0.48}As/InP$ (~0K) [18]:

$$E_g(\text{InP}) = 1.423 \text{ eV}$$
$$E_g(\text{In}_{0.52}\text{Al}_{0.48}\text{As}) = 1.511 \text{ eV}$$
$$\Delta E_g = 0.088 \text{ eV}$$
$$\Delta E_c = 0.252 \text{ eV} = 2.86\, \Delta E_g,\ \Delta E_v = -0.164 \text{ eV (type II)}.$$

The above results for $In_{0.53}Ga_{0.47}As/InP/In_{0.52}Al_{0.48}As$ band offsets and their transitivity relations can be illustrated [18] in Fig. A.2. The transitivity relations give $\Delta E_c = 0.504$ eV $= 0.72\, \Delta E_g$ and $\Delta E_v = 0.196$ eV $= 0.28\, \Delta E_g$ for an $In_{0.53}Ga_{0.47}As/In_{0.52}Al_{0.48}As$ heterojunction.

4. $In_{1-x}Ga_xAs_yP_{1-y}/InP$ lattice-matched system [22]: For $In_{1-x}Ga_xAs_yP_{1-y}$ quaternary semiconductor lattice matched to InP substrate,

$$x = \frac{0.1896y}{0.4176 - 0.0125y}$$
$$E_g(\text{In}_{1-x}\text{Ga}_x\text{As}_y\text{P}_{1-y}) = 1.35 - 0.775y + 0.149y^2 \text{ eV}$$
$$\Delta E_g(y) = 0.775y - 0.149y^2 \text{ eV}$$
$$\Delta E_v(y) = 0.502y - 0.152y^2 \text{ eV}$$
$$\Delta E_c(y) = \Delta E_g(y) - \Delta E_v(y) = 0.273y + 0.003y^2 \text{ eV}$$

where $\Delta E_v(y)$ was determined experimentally.

Some reports on strained $In_xGa_{1-x}As/InP$, $In_xGa_{1-x}As/In_{0.52}Al_{0.48}As$, InGaAs/InGaAsP, and $In_xGa_{1-x}As/GaAs$ can be found in Refs. 7–9, 14, 15, and 19.

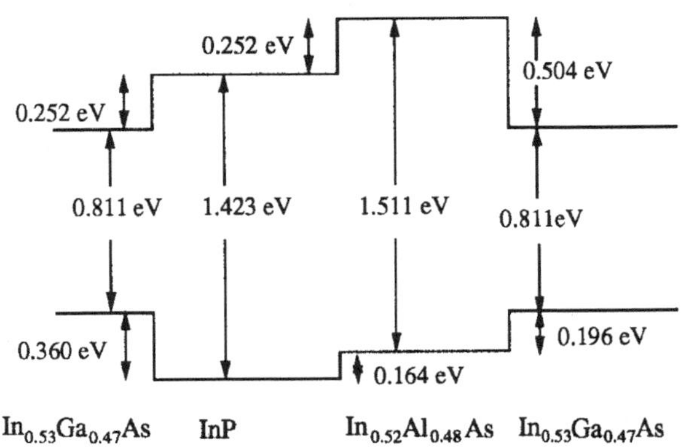

Figure A.2 Band offsets and transivity of $In_{0.53}Ga_{0.47}As/InP$ ($\Delta E_c = 0.41\, \Delta E_g$, $\Delta E_v = 0.59\, \Delta E_g$) and $In_{0.52}Al_{0.48}As/InP$ ($\Delta E_c = 0.252$ eV $= 2.86\, \Delta E_g > \Delta E_g = 0.088$ eV) at low temperatures (0K) [18].

REFERENCES

1. C. G. Van de Walle, "Band lineups and deformation potentials in the model-solid theory," *Phys. Rev. B* **39**, 1871–1883 (1989).
2. C. G. Van de Walle and R. M. Martin, "Theoretical calculations of semiconductor heterojuntion discontinuities," *J. Vac. Sci. Technol. B* **4**, 1055–1059 (1986).
3. C. G. Van de Walle and R. M. Martin, "Theoretical study of band offsets at semiconductor interfaces," *Phys. Rev. B* **35**, 8154–8165 (1987).
4. C. G. Van de Walle, K. Shahzad, and D. J. Olego, "Strained-layer interfaces between II–VI compound semiconductors," *J. Vac. Sci. Technol. B* **6**, 1350–1353 (1988).
5. R. People, K. W. Wecht, K. Alavi, and A. Y. Cho, "Measurement of the conduction-band discontinuity of molecular beam epitaxial grown $In_{0.52}Al_{0.48}As$ /$In_{0.53}Ga_{0.47}As$, N-n heterojunction by C-V profiling," *Appl. Phys. Lett.* **43**, 118–120 (1983).
6. R. C. Miller, A. C. Gossard, D. A. Kleinman, and O. Munteanu, "Parabolic quantum wells with the GaAs-$Al_xGa_{1-x}As$ system," *Phys. Rev. B* **29**, 3740–3743 (1984).
7. R. People, "Effects of coherency strain on the band gap of pseudomorphic $In_xGa_{1-x}As$ on (001) InP," *Appl. Phys. Lett.* **50**, 1604–1606 (1987).
8. R. People, "Band alignments for pseudomorphic InP/$In_xGa_{1-x}As$ heterostructures for growth on (001) InP," *J. Appl. Phys.* **62**, 2551–2553 (1987).
9. G. Ji, D. Huang, U. K. Reddy, T. S. Henderson, R. Houdré, and H. Morkoç, "Optical investigation of highly strained InGaAs-GaAs multiple quantum wells," *J. Appl. Phys.* **62**, 3366–3373 (1987).
10. C. D. Lee and S. R. Forrest, "Effects of lattice mismatch on $In_xGa_{1-x}As$/InP heterojunctions," *Appl. Phys. Lett.* **57**, 469–471 (1990).
11. L. Hrivnák, "Determination of Γ electron and light hole effective masses in $Al_xGa_{1-x}As$ on the basis of energy gaps, band-gap offsets, and energy levels in $Al_xGa_{1-x}As$/GaAs quantum wells," *Appl. Phys. Lett.* **56**, 2425–2427 (1990).
12. B. R. Nag and S. Mukhopadhyay, "Band offset in InP/$Ga_{0.47}In_{0.53}As$ heterostructures," *Appl. Phys. Lett.* **58**, 1056–1058 (1991).
13. M. S. Hybertsen, "Band offset transitivity at the InGaAs/InAlAs/InP(001) heterointerfaces," *Appl. Phys. Lett.* **58**, 1759–1761 (1991).
14. B. Jogai, "Valence-band offset in strained GaAs-$In_xGa_{1-x}As$ superlattices," *Appl. Phys. Lett.* **59**, 1329–1331 (1991).
15. J. H. Huang, T. Y. Chang, and B. Lalevic, "Measurement of the conduction-band discontinuity in pseudomorphic $In_xGa_{1-x}As$/$In_{0.52}Al_{0.48}As$ heterostructures," *Appl. Phys. Lett.* **60**, 733–735 (1992).
16. S. Tiwari and D. J. Frank, "Empirical fit to band discontinuities and barrier heights in III–V alloy systems," *Appl. Phys. Lett.* **60**, 630–632 (1992).
17. R. F. Kopf, M. H. Herman, M. L. Schnoes, A. P. Perley, G. Livescu, and M. Ohring, "Band offset determination in analog graded parabolic and triangular quantum wells of GaAs/AlGaAs and GaInAs/AlInAs," *J. Appl. Phys.* **71**, 5004–5011 (1992).
18. J. Bohrer, A. Krost, T. Wolf, and D. Bimberg, "Band offsets and transitivity of $In_xGa_{1-x}As$/$In_{1-y}Al_yAs$/InP heterostructures," *Phys. Rev. B.* **47**, 6439–6443 (1993).

19. M. Nido, K. Naniwae, T. Terakado, and A. Suzuki, "Band-gap discontinuity control for InGaAs/InGaAsP multiquantum-well structures by tensile-strained barriers," *Appl. Phys. Lett.* **62**, 2716–2718 (1993).

20. K. H. Hellwege, Ed., *Landolt-Börnstein Numerical Data and Functional Relationships in Science and Technology*, New Series, Group III **17a**, Springer-Verlag, Berlin, 1982; Groups III–V **22a**, Springer-Verlag, Berlin, 1986.

21. For a brief version of the data book in Ref. 20, see: O. Madelung, Ed., *Semiconductors, Group IV Elements and III–V Compounds*, in R. Poerschke, Ed., *Data in Science and Technology*, Springer-Verlag, Berlin, 1991.

22. S. Adachi, *Physical Properties of III–V Semiconductor Compounds*, Wiley, New York, 1992.

23. S. Adachi, *Properties of Indium Phosphide*, INSPEC, The Institute of Electrical Engineers, London and New York, 1991.

24. J. S. Blakemore, "Semiconducting and other major properties of gallium arsenide," *J. Appl. Phys.* **53**, R123–R181 (1982).

25. S. Adachi, "Material parameters of $In_xGa_{1-x}As_yP_{1-y}$ and related binaries," *J. Appl. Phys.* **53**, 8775–8792 (1982).

26. S. Adachi and K. Oe, "Internal strain and photoelastic effects in $Ga_{1-x}Al_xAs$/GaAs and $In_xGa_{1-x}As_yP_{1-y}$ /InP crystals," *J. Appl. Phys.* **54**, 6620–6627 (1983).

27. S. Adachi, "GaAs, AlAs, and $Ga_{1-x}Al_xAs$: material parameters for use in research and device applications," *J. Appl. Phys.* **58**, R1–R28 (1985).

28. P. Lawaetz, "Valence-band parameters in cubic semiconductors," *Phys. Rev. B.* **4**, 3460–3467 (1971).

29. R. E. Nahory, M. A. Pollack, and W. D. Johnston, Jr., "Band gap versus composition and demonstration of Vegard's law for $In_{1-x}Ga_xAs_yP_{1-y}$ lattice matched to InP," *Appl. Phys. Lett.* **33**, 659–661 (1978).

30. K. Alavi and R. L. Aggarwal, "Interband magnetoabsorption of $In_{0.53}GA_{0.47}As$," *Phys. Rev. B* **21**, 1311–1315 (1980).

31. A. Raymond, J. L. Robert, and C. Bernard, "The electron effective mass in heavily doped GaAs," *J. Phys. C Solid State Phys.* **12**, 2289–2293 (1979).

32. L. G. Shantharama, A. R. Adams, C. N. Ahmad and R. J. Nicholas, "The $\mathbf{k}\cdot\mathbf{p}$ interaction in InP and GaAs from the band-gap dependence of the effective mass," *J. Phys. C Solid State Phys.* **17**, 4429–4442 (1984).

33. W. Stolz, J. C. Maan, M. Altarelli, L. Tapfer, and K. Ploog, "Absorption spectroscopy on $Ga_{0.47}In_{0.53}As/Al_{0.48}In_{0.52}As$ multi-quantum-well hetero-structures. I. Excitonic transitions," *Phys. Rev. B* **36**, 4301–4309 (1987).

34. W. Stolz, J. C. Maan, M. Altarelli, L. Tapfer, and K. Ploog, "Absorption spectroscopy on $Ga_{0.47}In_{0.53}As/Al_{0.48}In_{0.52}As$ multi-quantum-well hetero-structures. II. Subband structure," *Phys. Rev. B* **36**, 4310–4315 (1987).

35. L. W. Molenkamp, R. Eppenga, G. W. 't Hooft, P. Dawson, C. T. Foxon, and K. J. Moore, "Determination of valence-band effective-mass anisotropy in GaAs quantum wells by optical spectroscopy," *Phys. Rev. B* **38**, 4314–4317 (1988).

36. D. Gershoni and H. Temkin, "Optical properties of III–V strained-layer quantum wells," *J. Luminescence* **44**, 381–398 (1989).

37. R. Sauer, S. Nilsson, P. Roentgen, W. Heuberger, V. Graf, A. Hangleiter, and R. Spycher, "Optical study of extended-molecular-layer flat islands in lattice-matched $In_{0.53}Ga_{0.47}As$/InP and $In_{0.53}Ga_{0.47}As/In_{1-x}Ga_xAs_yP_{1-y}$ quantum wells grown by low-pressure metal-organic vapor-phase epitaxy with different interruption cycles," *Phys. Rev. B* **46**, 9525–9537 (1992).

Appendix B

Optical Constants of GaAs and InP

Table B.1 Optical Constants of GaAs

Optical Energy, $\hbar\omega$ (eV)	Wavelength, λ (μm)	Refractive Index, n	Extinction Coeff., κ	Absorption Coeff., $\alpha = 4\pi\kappa/\lambda$ (10^4 cm^{-1})	References*
0.5	2.4797	3.3240			*a*, *b*
0.6	2.0664	3.3378			
0.7	1.7712	3.3543			
0.8	1.5498	3.3737			
0.9	1.3776	3.3965			
1.0	1.2399	3.4232			
1.1	1.1271	3.4546			
1.2	1.0332	3.4920			
1.3	0.9537	3.5388			
1.35	0.9184	3.5690			
1.4	0.8856	3.6140	0.0017	0.0240	*a*, *c*
1.42	0.8731		0.0271	0.3900	
1.425	0.8701		0.0554	0.8001	
1.43	0.8670		0.0572	0.8290	
1.435	0.8640		0.0557	0.8101	
1.44	0.8610		0.0568	0.8290	
1.45	0.8551		0.0612	0.8994	
1.47	0.8434		0.0664	0.9893	
1.5	0.8266	3.666	0.080	1.216	*a*, *d*
1.6	0.7749	3.700	0.091	1.476	
1.7	0.7293	3.742	0.112	1.930	
1.8	0.6888	3.785	0.151	2.755	
1.9	0.6526	3.826	0.179	3.447	
2.0	0.6199	3.878	0.211	4.277	
2.1	0.5904	3.940	0.240	5.108	
2.2	0.5636	4.013	0.276	6.154	
2.3	0.5391	4.100	0.320	7.460	

(*Continued*)

Physics of Photonic Devices, Second Edition. By Shun Lien Chuang
Copyright © 2009 John Wiley & Sons, Inc.

Table B.1 *Continued*

Optical Energy, $\hbar\omega$ (eV)	Wavelength, λ (μm)	Refractive Index, n	Extinction Coeff., κ	Absorption Coeff., $\alpha = 4\pi\kappa/\lambda$ (10^4 cm^{-1})	References*
2.4	0.5166	4.205	0.371	9.025	
2.5	0.4959	4.333	0.441	11.174	
2.6	0.4769	4.492	0.539	14.204	

*References are indicated on the first row for items in rows that follow.

[a]E. D. Palik, "Gallium arsenide (GaAs)," pp. 429–443 in E. D. Palik, Ed., *Handbook of Optical Constants of Solids*, Academic, New York, 1985.

[b]A. N. Pikhtin and A. D. Yas'kov, "Dispersion of the refractive index of semiconductors with diamond and zinc–blende structures," *Sov. Phys. Semicond.* **12**, 622 (1978).

[c]H. C. Casey, D. D. Sell, and K. W. Wecht, "Concentration dependence of the absorption coefficient for n– and p–type GaAs between 1.3 and 1.6 eV," *J. Appl. Phys.* **46**, 250 (1975).

[d]D. E. Aspnes and A. A. Studna, "Dielectric functions and optical parameters of Si, Ge, GaP, GaAs, GaSb, InP, InAs, and InSb from 1.5 to 6.0 eV," *Phys. Rev. B* **27**, 985 (1983).

The refractive index n, the extinction coefficient κ, and the absorption coefficient of GaAs as a function of photon energy are plotted in Fig. B.1.

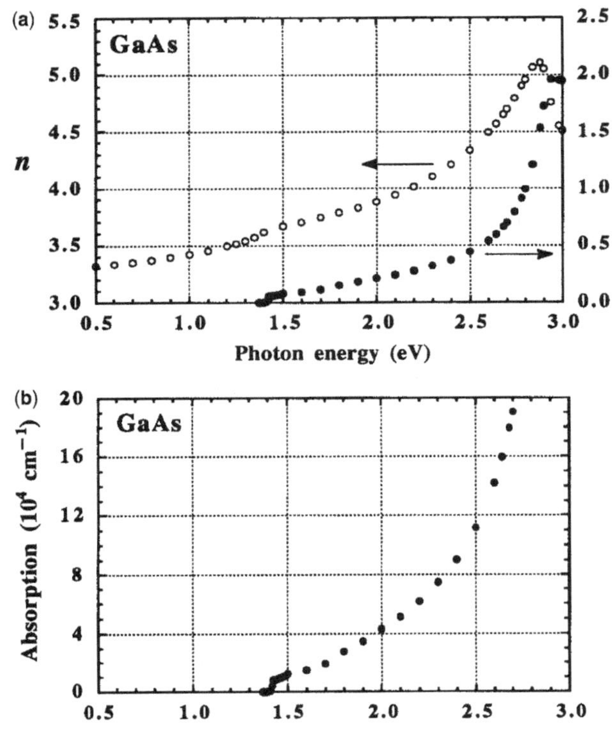

Figure B.1 (a) Real and imaginary parts, n and κ, of the complex refractive index, and (b) the absorption coefficient of GaAs versus photon energy.

Table B.2 Optical Constants of InP

Optical Energy, $\hbar\omega$ (eV)	Wavelength, λ (μm)	Refractive Index, n	Extinction Coeff., κ	Absorption Coeff., $\alpha = 4\pi\kappa/\lambda$ ($10^4 cm^{-1}$)	References*
0.6	2.0664	3.129			*a, b*
0.7	1.7712	3.146			
0.8	1.5498	3.167			
0.9	1.3776	3.191			
1.0	1.2399	3.22			
1.1	1.1271	3.254			
1.2	1.0332	3.297			
1.25	0.9919	3.324			
1.272	0.975	3.346	0.0000113	0.0001456	*a, c, d*
1.301	0.953	3.362	0.000281	0.003705	*a, b, d*
1.326	0.935	3.385	0.0059	0.07930	*a, c, d*
1.333	0.930	3.390	0.0109	0.1473	
1.340	0.925	3.396	0.0355	0.4822	
1.345	0.9218	3.399	0.0571	0.7791	
1.5	0.8266	3.456	0.203	3.086	*a, e*
1.6	0.7749	3.467	0.218	3.535	
1.7	0.7293	3.476	0.242	4.170	
1.8	0.6888	3.492	0.270	4.926	
1.9	0.6526	3.517	0.293	5.739	
2.0	0.6199	3.549	0.317	6.426	
2.1	0.5904	3.585	0.347	7.386	
2.2	0.5636	3.629	0.380	8.473	
2.3	0.5391	3.682	0.416	9.697	
2.4	0.5166	3.745	0.457	11.117	
2.5	0.4959	3.818	0.511	12.948	

*References are indicated on the first row for items in rows that follow.

[a]O. J. Glembocki and H. Piller, "Indium Phosphide (InP)," pp. 429–443 in E. D. Palik, Ed., *Handbook of Optical Constants of Solids*, Academic Press, New York, 1985.

[b]A. N. Pikhtin and A. D. Yas'kov, "Dispersion of the refractive index of semiconductors with diamond and zinc–blende structures," *Sov. Phys. Semicond.* **12**, 622 (1978).

[c]G. D. Pettit and W. J. Turner, "Optical Constant," *J. Appl. Phys.* **36**, 2081 (1965).

[d]B. O. Seraphin and H. E. Bennett, "Refractive index of InP," pp. 499–543 in R. K. Willardson and A. C. Beer, Eds., *Semiconductors and Semimetals*, vol. 3, Academic Press, New York, 1967.

[e]D. E. Aspnes and A. A. Studna, "Dielectric functions and optical parameters of Si, Ge, GaP, GaAs, GaSb, InP, InAs, and InSb from 1.5 to 6.0 eV," *Phys. Rev. B* **27**, 985 (1983).

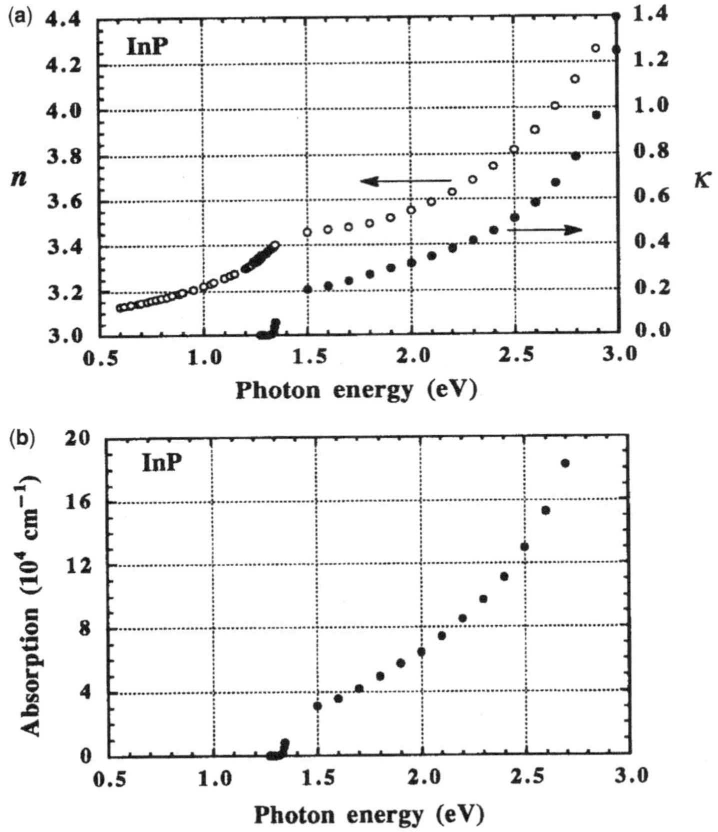

Figure B.2 (a) Real and imaginary parts, n and κ, of the complex refractive index, and (b) the absorption coefficient of InP versus photon energy.

The refractive index n, the extinction coefficient κ, and the absorption coefficient of InP as a function of photon energy are plotted in Fig. B.2.

Appendix C

Electronic Properties of Si, Ge, and a Few Binary, Ternary, and Quaternary Compounds

Physics of Photonic Devices, Second Edition. By Shun Lien Chuang
Copyright © 2009 John Wiley & Sons, Inc.

Table C.1 Important Physical Properties[a,b]

Physical Properties	Si	Ge	GaAs	InAs	InP	GaP
Band gap, E_g (eV)	(Indirect)	(Indirect)	(Direct)	(Direct)	(Direct)	(Indirect)
0K	1.170	0.744	1.519	0.42	1.424	2.350
300K	1.124	0.664	1.424	0.354	1.344	2.272
Dielectric constant, ε_s (F/m)	$11.9\varepsilon_0$	$16.2\varepsilon_0$	$13.1\varepsilon_0$	$15.15\varepsilon_0$	$12.56\varepsilon_0$	$11.11\varepsilon_0$
Effective mass (m^*/m_0)						
Electrons						
(Longitudinal m_ℓ^*/m_0)	0.9163	1.59	0.0665	0.023	0.077	0.254
(Transverse m_t^*/m_0)	0.1905	0.0823				4.8
Holes						
(Heavy, m_{hh}^*/m_0)	0.537	0.284	0.5	0.4	0.6	0.67
(Light, $m_{\ell h}^*/m_0$)	0.153	0.043	0.087	0.026	0.12	0.17
Intrinsic carrier concentration						
n_i (cm^{-3}) at 300K	1.02×10^{10}	2.33×10^{13}	2.1×10^6	1.3×10^{15}	1.2×10^8	
Lattice constant (Å)	5.43102	5.6579	5.6533	6.0584	5.8688	5.4505
Minority carrier lifetime(s)						
at 300K	2.5×10^{-3}	10^{-3}	$\sim 10^{-8}$			
Mobility ($cm^2\,V^{-1}\,s^{-1}$) at 300K						
Electron	1450	3900	9200	2×10^4 to 3.3×10^4	5370	160
Hole	370	1800	400	100 – 450	150	135

[a] O. Madelung, Ed., *Semiconductors, Group IV Elements and III–V Compounds*, in R. Poerschke, Ed., *Data in Science and Technology*, Springer, Berlin, 1991.

[b] S. Sze, *Physics of Semiconductor Devices*, Wiley, New York, 1982.

Table C.2 Important Band Structure Parameters for GaAs, AlAs, InAs, InP, and GaP[a–f]

	Materials				
	GaAs	AlAs	InAs	InP	GaP
Parameters					
a_0 (Å)	5.6533	5.6600	6.0584	5.8688	5.4505
E_g (eV)					
0K	1.519	3.13 2.229*	0.42	1.424	2.90 2.35*
300K	1.424	3.03 2.168[b]*	0.354	1.344	2.78 2.27[c]*
Δ (eV)	0.34	0.28	0.38	0.11	0.08
$E_{v,av}$ (eV)	−6.92	−7.49	−6.67	−7.04	−7.40
Optical matrix Parameter E_p (eV)	25.7 (25.0)[f]	21.1	22.2	20.7 (16.7)[f]	22.2
Deformation Potentials (eV)					
a_c (eV)	−7.17	−5.64	−5.08	−5.04	−7.14
a_v (eV)	1.16	2.47	1.00	1.27	1.70
$a = a_c - a_v$ (eV)	−8.33	−8.11	−6.08	−6.31	−8.83
b (eV)	−1.7	−1.5	−1.8	−1.7	−1.8
d (eV)	−4.55	−3.4	−3.6	−5.6	−4.5
C_{11} (10^{11}dyne/cm^2)	11.879	12.5	8.329	10.11	14.05
C_{12} (10^{11}dyne/cm^2)	5.376	5.34	4.526	5.61	6.203
C_{44} (10^{11}dyne/cm^2)	5.94	5.42	3.96	4.56	7.033
Effective Masses					
m_e^*/m_0	0.067	0.15	0.023	0.077	0.25
m_{hh}^*/m_0	0.50	0.79	0.40	0.60	0.67
$m_{\ell h}^*/m_0$	0.087	0.15	0.026	0.12	0.17
$m_{hh,z}^*/m_0 = 1/(\gamma_1 - 2\gamma_2)$	0.333	0.478	0.263	0.606	0.326
$m_{\ell h,z}^*/m_0 = 1/(\gamma_1 + 2\gamma_2)$	0.094	0.208	0.027	0.121	0.199
γ_1	6.8 (6.85)	3.45	20.4	4.95	4.05
γ_2	1.9 (2.1)	0.68	8.3	1.65	0.49
γ_3	2.73 (2.9)	1.29	9.1	2.35	1.25

*Indirect band gap, $E_g(X)$ value.

[a]C. G. Van de Walle, "Band lineups and deformation potentials in the model–solid theory," *Phys. Rev. B* **39**, 1871–1883 (1989).

[b]P. Lawaetz, "Valence–band parameters in cubic semiconductors," *Phys. Rev. B* **4**, 3460–3467 (1971).

[c]S. Adachi, "GaAs, AlAs, and $Ga_{1-x}Al_xAs$: material parameters for use in research and device applications," *J. Appl. Phys.* **58**, R1–R28 (1985).

[d]O. Madelung, Ed., *Semiconductors, Group IV Elements and III–V Compounds*, in R. Poerschke, Ed. in Chief, *Data in Science and Technology*, Springer, Berlin, 1991.

[e]K. H. Hellwege, Ed., *Landolt–Börnstein Numerical Data and Functional Relationships in Science and Technology*, New Series, Group III **17a**, Springer, Berlin, 1982; Groups III–V **22a**, Springer, Berlin, 1986.

[f]L. G. Shantharama, A. R. Adams, C. N. Ahmad, and R. J. Nicholas, "The **k** · **p** interaction in InP and GaAs from the band–gap dependence of the effective mass," *J. Phys. C Solid State Phys.* **17**, 4429–4442 (1984).

Table C.3 Important Band Structure Parameters for $Al_xGa_{1-x}As$, $In_{1-x}Ga_xAs$, $Al_xIn_{1-x}As$, $Ga_xIn_{1-x}As_yP_{1-y}$ and $In_{1-x-y}Al_xGa_y$ As Compounds[a-f]

General Interpolation Formula for Ternary Compound Parameters $P(III_xIII_{1-x}V)$ $P(A_xB_{1-x}C) = xP(AC) + (1-x)P(BC)$	Ref.
$Al_xGa_{1-x}As$	
$E_g(\Gamma) = 1.424 + 1.247x$ (eV) at 300K for $x < 0.4$	*c*
$1.519 + 1.447x - 0.15x^2$ (eV) at 0K for $x < 0.4$	*b*
$m_e^*/m_0 = 0.067 + 0.083x$	*a*
$m_{hh}^*/m_0 = 0.50 + 0.29x$ (denisty of states mass)	interpol.
$m_{\ell h}^*/m_0 = 0.087 + 0.063x$	*f*
$m_{so}^*/m_0 = 0.15 + 0.09x$	*a*
$\gamma_i(x) = x\gamma_i(AlAs) + (1-x)\gamma_i(GaAs)$ (for calculating transport masses)	interpol.
$In_{1-x}Ga_xAs$	
$E_g(\Gamma) = 0.36 + 0.505x + 0.555x^2$ (eV) at 300K	*c*
$0.324 + 0.7x + 0.4x^2$ (eV) at 300K	*a*
$0.422 + 0.7x + 0.4x^2$ (eV) at 2K	
$m_e^*/m_0 = 0.025(1-x) + 0.071x - 0.0163x(1-x)$	
or	
$1/m_e^*(x) = x/m_e^*(GaAs) + (1-x)/m_e^*(InAs)$	
$In_{0.53}Ga_{0.47}As$	
$E_g(\Gamma) = 0.813$ (eV) at 2K	
0.75 (eV) at 300K	*a*
$m_e^*/m_0 = 0.041$	
$m_{hh}^*/m_0 = 0.465$ //[001]	
0.56 //[110]	
$m_{\ell h}^*/m_0 = 0.0503$	
$Al_xIn_{1-x}As$	
$E_g(\Gamma) = 0.36 + 2.35x + 0.24x^2$ (eV) at 300K	*a*
$0.357 + 2.29x$ (eV) at 300K for $0.44 < x < 0.54$	
$0.447 + 2.22x$ (eV) at 4K for $0.44 < x < 0.54$	
$Al_{0.48}In_{0.52}As$	
$E_g(\Gamma) = 1.508$ (eV) at 4K	
1.450 (eV) at 300K	*a*
$m_e^*/m_0 = 0.075$	*d*
$m_{hh}^*/m_0 = 0.41$	
$m_{\ell h}^*/m_0 = 0.096$	

(*Continued*)

Table C.3 *Continued*

General Interpolation Formula for Quaternary Compound Parameters $P(III_xIII_{1-x}V_y V_{1-y})$:
$P(A_xB_{1-x}C_yD_{1-y}) = xy\, P(AC) + (1-x)(1-y)\, P(BD) + (1-x)y\, P(BC) + x(1-y)\, P(AD)$ — Ref.

$Ga_xIn_{1-x}As_yP_{1-y}$

$$E_g(x, y) = 1.35 + 0.668x - 1.068y + 0.758x^2 + 0.078y^2 - 0.069xy - 0.322x^2y + 0.03xy^2 \text{ (eV)} \quad \text{at 300K}$$ — *a, g*

$$m_e^*(x, y)/m_0 = 0.08 - 0.116y + 0.026x - 0.059xy + (0.064 - 0.02y)x^2 + (0.06 + 0.032x)y^2$$

$$a(x, y) = 5.8688 - 0.4176x + 0.1896y + 0.0125xy \text{ (Å)}$$

Lattice-matched to InP:

$$x = 0.1894y/(0.4184 - 0.013y)$$ — *a, g*

$$E_g(y) = 1.35 - 0.775y + 0.149y^2 \text{ (eV)} \quad \text{at 298K}$$

$$1.425 - 0.7668y + 0.149y^2 \text{ (eV)} \quad \text{at 4.2K}$$

$$m_e^*/m_0 = 0.080 - 0.039y$$ — *e*

$$m_{hh}^*/m_0 = 0.46$$

$$m_{\ell h}^*/m_0 = 0.12 - 0.099y + 0.030y^2$$

$$m_{so}^*/m_0 = 0.21 - 0.01y - 0.05y^2$$

General Interpolation Formula for Quaternary Compound Parameters $P(III_{1-x-y}III_xIII_y\ V)$
$P(A_{1-x-y}B_xC_yD) = (1 - x - y)\, P(AD) + x\, P(BD) + y\, P(CD) + x(1 - y)\, P(AD)$ — Ref.

$In_{1-x-y}Al_xGa_yAs$

$$E_g(x, y) = 0.36 + 2.093x + 0.629y + 0.577x^2 + 0.436y^2 + 1.013xy - 2.0xy(1 - x - y) \text{ (eV)} \quad \text{at 300K}$$ — *a, g*

Latticed-matched to InP:

$(In_{0.52}Al_{0.48})_z(In_{0.53}Ga_{0.47})_{1-z}As$

$$x = 0.48z \quad 0.983x + y = 0.468$$

$$E_g(z) = 0.76 + 0.49z + 0.20z^2 \text{ (eV)} \quad \text{at 300K}$$

$$m_e^*/m_0 = 0.0427 + 0.0328z$$

[a] K. H. Hellwege, Ed., *Landolt–Börnstein Numerical Data and Functional Relationships in Science and Technology*, New Series, Group III **17a**, Springer, Berlin, 1982; Groups III–V **22a**, Springer-Verlag, Berlin, 1986.

[b] M. El Allai, C. B. Sorensen, E. Veje, and P. Tidemand–Petersson, "Experimental determination of the GaAs and $Ga_{1-x}Al_xAs$ band–gap energy dependence on temperature and aluminum mole fraction in the direct band–gap region," *Phys. Rev. B* **48**, 4398–4404 (1993).

[c] S. Adachi, "Material parameters of $In_xGa_{1-x}As_yP_{1-y}$ and related binaries," *J. Appl. Phys.* **53**, 8775–8792 (1982).

[d] W. Stolz, J. C. Maan, M. Altarelli, L. Tapfer, and K. Ploog, "Absorption spectroscopy on $Ga_{0.47}In_{0.53}As/Al_{0.48}In_{0.52}As$ multi–quantum–well hetero–structures. II. Subband structure," *Phys. Rev. B* **36**, 4310–4315 (1987).

[e] S. Adachi, *Physical Properties of III–V Semiconductor Compounds*, Wiley, New York, 1992.

[f] H. C. Casey, Jr., and M. B. Panish, *Heterostructure Lasers Part A: Fundamental Principles*, Academic Press, Orlando, 1978.

[g] O. Madelung, *Semiconductors–Group IV Elements and III-V Compounds*, in *Data in Science and Technology*, R. Poerschke, Editor in Chief, Springer-Verlag, Berlin, 1991.

Appendix D

Parameters for InN, GaN, AlN, and Their Ternary Compounds

Table D.1 Recommended Band Structure Parameters for Wurtzite Nitride Binaries[a]

Parameters	GaN	AlN	InN
a (Å) at $T = 300$K	3.189	3.112	3.545
c (Å) at $T = 300$K	5.185	4.982	5.703
E_g (eV) at $T = 0$K	3.510	6.25	0.69[b]
at $T = 300$K	3.44	6.16	0.64
α (meV/K)	0.909	1.799	0.41[b]
β (K)	830	1462	454[b]
Δ_{cr} (eV)	0.010	−0.169	0.040
Δ_{so} (eV)	0.017	0.019	0.005
$m_e^{\parallel}/m_0$	0.20	0.32	0.07
$m_e^{\perp}/m_0$	0.20	0.30	0.07
A_1	−7.21	−3.86	−8.21
A_2	−0.44	−0.25	−0.68
A_3	6.68	3.58	7.57
A_4	−3.46	−1.32	−5.23
A_5	−3.40	−1.47	−5.11
A_6	−4.90	−1.64	−5.96
A_7 (eV Å)	0.0937	0	0
a_1 (eV)	−4.9	−3.4	−3.5
a_2 (eV)	−11.3	−11.8	−3.5
D_1 (eV)	−3.7	−17.1	−3.7
D_2 (eV)	4.5	7.9	4.5
D_3 (eV)	8.2	8.8	8.2
D_4 (eV)	−4.1	−3.9	−4.1
D_5 (eV)	−4.0	−3.4	−4.0
D_6 (eV)	−5.5	−3.4	−5.5
C_{11} (GPa)	390	396	223
C_{12} (GPa)	145	137	115
C_{13} (GPa)	106	108	92

(*Continued*)

Physics of Photonic Devices, Second Edition. By Shun Lien Chuang
Copyright © 2009 John Wiley & Sons, Inc.

Table D.1 *Continued*

Parameters	GaN	AlN	InN
C_{33} (GPa)	398	373	224
C_{44} (GPa)	105	116	48
d_{13} (pm/V)	−1.6	−2.1	−3.5
d_{33} (pm/V)	3.1	5.4	7.6
d_{15} (pm/V)	3.1	3.6	5.5
P_{sp} (C/m^2)	−0.034	−0.090	−0.042

Note: $E_g(T) = E_g(0) - \frac{\alpha T^2}{T+\beta}$

Source: This table (except for InN bandgap parameters, as noted from Ref. b) is reproduced with permission from

[a]I. Vurgaftman and J. R. Meyer, "Band parameters for nitrogen-containing semiconductors," *J. Appl. Phys.* **94**, 3675–3696 (2003). (© 2003 American Institute of Physics.)

[b]J. Wu, W. Walukiewicz, W. Shan, K. M. Yu, J. W. Ager III, S. X. Li, E. E. Haller, H. Lu, and W. J. Schaff, "Temperature dependence of the fundamental band gap of InP," *J. Appl. Phys.* **94**, 4457–4460 (2003).

Table D.2 Band Gap of $In_xGa_{1-x}N$, $Al_xGa_{1-x}N$, and $Al_xIn_{1-x}N$

General Interpolation Formula for Ternary Compound Parameters P:

$$P(A_xB_{1-x}C) = xP(AC) + (1-x)P(BC)$$

Except Band-Gap Energy:

$$P(A_xB_{1-x}C) = xP(AC) + (1-x)P(BC) - bx(1-x)$$

$In_xGa_{1-x}N$

$$E_g(x) = xE_g(\text{InN}) + (1-x)E_g(\text{GaN}) - 1.4x(1-x)\,(\text{eV})$$

$$P^{SP}(x) = [-0.042x - 0.034(1-x) + 0.037x(1-x)]\,\text{C/m}^2$$

$$P^{Pz}_{In_xGa_{1-x}N/InN}(x) = [-0.113(1-x) - 0.0276x(1-x)]\,\text{C/m}^2$$

$$P^{Pz}_{In_xGa_{1-x}N/GaN}(x) = [0.148x - 0.0424x(1-x)]\,\text{C/m}^2$$

$$P^{Pz}_{In_xGa_{1-x}N/AlN}(x) = [0.182x - 0.026(1-x) - 0.0456x(1-x)]\,\text{C/m}^2$$

$Al_xGa_{1-x}N$

$$E_g(x) = xE_g(\text{AlN}) + (1-x)E_g(\text{GaN}) - 0.7x(1-x)\,(\text{eV})$$

$$P^{SP}(x) = [-0.090x - 0.034(1-x) + 0.021x(1-x)]\,\text{C/m}^2$$

$$P^{Pz}_{Al_xGa_{1-x}N/InN} = [-0.28x - 0.113(1-x) + 0.42x(1-x)]\,\text{C/m}^2$$

$$P^{Pz}_{Al_xGa_{1-x}N/GaN} = [-0.0525x + 0.0282x(1-x)]\,\text{C/m}^2$$

$$P^{Pz}_{Al_xGa_{1-x}N/AlN} = [0.026(1-x) - 0.0248x(1-x)]\,\text{C/m}^2$$

(*Continued*)

Table D.2 *Continued*

$Al_xIn_{1-x}N$

$$E_g(x) = xE_g(\text{AlN}) + (1-x)E_g(\text{InN}) - 2.5x(1-x)\,(\text{eV})$$

$$\text{P}^{\text{SP}}(x) = [-0.090x - 0.042(1-x) + 0.070x(1-x)]\,\text{C/m}^2$$

$$\text{P}^{\text{Pz}}_{\text{Al}_x\text{In}_{1-x}\text{N/InN}}(x) = [-0.28x + 0.104x(1-x)]\,\text{C/m}^2$$

$$\text{P}^{\text{Pz}}_{\text{Al}_x\text{In}_{1-x}\text{N/GaN}}(x) = [-0.0525x + 0.148(1-x) + 0.0938x(1-x)]\,\text{C/m}^2$$

$$\text{P}^{\text{Pz}}_{\text{Al}_x\text{In}_{1-x}\text{N/AlN}} = [0.182(1-x) + 0.092x(1-x)]\,\text{C/m}^2$$

Sources: I. Vurgaftman and J. R. Meyer, "Band parameters for nitrogen-containing semiconductors," *J. Appl. Phys.* **94**, 3675–3696 (2003). O. Ambacher, J. Majewski, C. Miskys, A. Link, M. Hermann, M. Eickhoff, M. Stutzmann, F. Bernardini, V. Fiorentini, V. Tilak, B. Schaff, and L. F. Eastman, "Pyroelectric properties of Al(In)GaN/GaN hetero- and quantum well structures," *J. Phys. Condens. Matter* **14**, 3399–3434 (2002).

Index

Physics of Photonic Devices, Second Edition. By Shun Lien Chuang
Copyright © 2009 John Wiley & Sons, Inc.

Physical Constants	Symbol	Numerical Values
Speed of light in free space	c	$= 2.9979 \times 10^8$ m/s
Permittivity in free space	ε_0	$= 8.8542 \times 10^{-12}$ F/m $\left(\simeq \frac{1}{36\pi} \times 10^{-9}\ \text{F/m}\right)$
Permeability in free space	μ_0	$= 4\pi \times 10^{-7}$ H/m
Boltzmann constant	k_B	$= 1.3807 \times 10^{-23}$ J/K
Elementary charge	q or e	$= 1.60219 \times 10^{-19}$ C
Free electron mass	m_0	$= 9.1095 \times 10^{-31}$ kg
Planck constant	h	$= 6.6262 \times 10^{-34}$ Js
Reduced Planck constant	$\hbar = \frac{h}{2\pi}$	$= 1.05459 \times 10^{-34}$ Js; $= 6.5822 \times 10^{-16}$ eVs
Angstrom unit	1 Å	$= 10^{-10}$ m $= 10^{-8}$ cm $= 10^{-4}$ μm
Bohr radius	$a_0 = \frac{4\pi\varepsilon_0\hbar^2}{e^2 m_0}$	$= 0.529177$ Å
Ryberg energy	$R_y = \frac{m_0 e^4}{2(4\pi\varepsilon_0)^2\hbar^2}$	$= \frac{\hbar^2}{2m_0}\left(\frac{1}{a_0}\right)^2 = 13.6058$ eV
Energy unit (electron-volt)	1 eV	$= 1.60219 \times 10^{-19}$ J
Thermal energy at 300 K	$k_B T$	$= 25.853$ meV

Useful Formulas and Physical Quantities	
Photon energy	$\hbar\omega = h\frac{c}{\lambda} = \frac{1.2398}{\lambda}$ eV, where λ is in (μm)
Rydberg of an exciton	$R_y = \frac{m_r e^4}{2(4\pi\varepsilon_s)^2\hbar^2} = \frac{(m_r/m_0)}{(\varepsilon_s/\varepsilon_0)^2} \times 13.6058$ eV
Bohr radius of an exciton	$a_0 = \frac{4\pi\varepsilon_s\hbar^2}{e^2 m_r} = \frac{\varepsilon_s/\varepsilon_0}{m_r/m_0} \times 0.529177$ Å
Quantized subband energy in a quantum well (infinite barrier model)	$E = \frac{\hbar^2}{2m^*}\left(\frac{n\pi}{L}\right)^2 = \frac{n^2}{(m^*/m_0)}\frac{1}{L^2} \times 37.6033$ eV (L is in Å)
Conduction band density parameter	$N_c = 2\left(\frac{m_e^* k_B T}{2\pi\hbar^2}\right)^{3/2} = 2.51 \times 10^{19}\left(\frac{m_e^*}{m_0}\frac{T}{300}\right)^{3/2}$ cm^{-3}

(1) Maxwell's Equations

$$\nabla \times \mathbf{E} = -\frac{\partial}{\partial t}\mathbf{B}$$

$$\nabla \times \mathbf{H} = \mathbf{J} + \frac{\partial \mathbf{D}}{\partial t}$$

$$\nabla \cdot \mathbf{D} = \rho$$

$$\nabla \cdot \mathbf{B} = 0$$

Continuity Equation

$$\nabla \cdot \mathbf{J} + \frac{\partial}{\partial t}\rho = 0$$

(2) Semiconductor Electronics Equations

$$\nabla \cdot (\varepsilon \nabla \phi) = -q(p - n + N_D^+ - N_A^-)$$

$$\frac{\partial n}{\partial t} = G_n - R_n + \frac{1}{q}\nabla \cdot \mathbf{J}_n$$

$$\frac{\partial p}{\partial t} = G_p - R_p - \frac{1}{q}\nabla \cdot \mathbf{J}_p$$

$$\mathbf{J}_n = q\mu_n n\mathbf{E} + qD_n \nabla n$$

$$\mathbf{J}_p = q\mu_p p\mathbf{E} - qD_p \nabla p$$

$$\mathbf{E} = -\nabla \phi$$

(3) Density-of-States, $\rho_e(E)$

3D $$\frac{1}{2\pi^2}\left(\frac{2m_e^*}{\hbar^2}\right)^{3/2}\sqrt{E}$$

2D $$\frac{m_e^*}{\pi \hbar^2 L_z}\sum_n H(E - E_n)$$

1D $$\frac{1}{\pi L_x L_y}\sqrt{\frac{2m_e^*}{\hbar^2}}\sum_{n_x, n_y}\frac{1}{\sqrt{E - E_{n_x} - E_{n_y}}}$$

(4) Fermi Distribution

Electron in conduction band

$$f_c(E) = \frac{1}{1 + e^{(E - F_c)/k_B T}}$$

Electron occupation probability in valence band

$$f_v(E) = \frac{1}{1 + e^{(E - F_v)/k_B T}}$$

Hole in valence band

$$f_h(E) = 1 - f_v(E) = \frac{1}{1 + e^{(F_v - E)/k_B T}}$$